Molecular
Approaches to
Malaria

Molecular Approaches to
Malaria

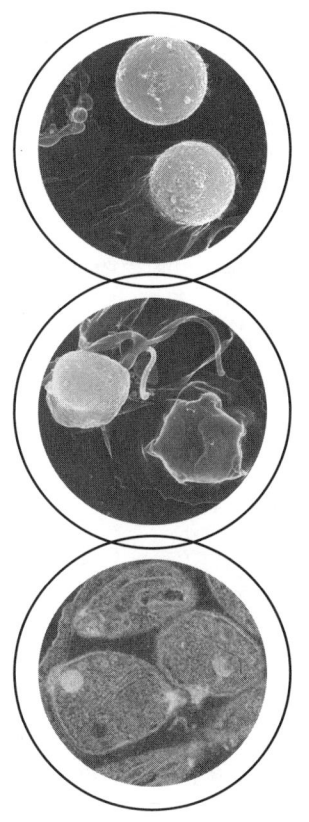

Edited by
Irwin W. Sherman

ASM PRESS

Washington, D.C.

Copyright © 2005 ASM Press
American Society for Microbiology
1752 N St., N.W.
Washington, DC 20036-2904

Library of Congress Cataloging-in-Publication Data

Molecular approaches to malaria/edited by Irwin W. Sherman.
   p. ; cm.
  Includes bibliographical references and index.
  ISBN 13: 978-1-55581–330–7 (alk. paper)
  ISBN 10: 1-55581–330–5
  1. Malaria—Molecular aspects
  [DNLM: 1. Malaria—genetics. 2. Malaria—therapy. 3. Molecular Biology.
  4. Plasmodium—genetics. WC 750 M7178 2005] I. Sherman, Irwin W.

QR201.M3M647 2005
616.9′ 362—dc22

                                                                         2005010990

10 9 8 7 6 5 4 3 2 1

All Rights Reserved
Printed in the United States of America

Address editorial correspondence to ASM Press, 1752 N St., N.W., Washington, DC
20036–2904, USA

Send orders to ASM Press, P.O. Box 605, Herndon, VA 20172, USA
Phone: 800-546-2416; 703-661-1593
Fax: 703-661-1501
E-mail: books@asmusa.org
Online: www.asmpress.org

DEDICATION

This book is dedicated to the memory of William Trager (1910-2005). Professor Trager had an unflagging curiosity about organisms "living together." He was the examplar of a bench scientist perhaps best known for his outstanding achievement of the in vitro cultivation of Plasmodium falciparum. *Modest and humane in dealing with colleagues, young and old, he remained a constant source of inspiration to those whose lives he touched.*

CONTENTS

Contributors xi
Preface xvii

I. INTRODUCTION   1

1. The Life of *Plasmodium*: an Overview
Irwin W. Sherman
3

2. PlasmoDB: the *Plasmodium* Genome Resource
Patricia L. Whetzel, Shailesh V. Date, Kobby Essien, Martin J. Fraunholz, Bindu Gajria, Gregory R. Grant, John Iodice, Jessica C. Kissinger, Philip T. Labo, Arthur J. Milgram, Christian J. Stoeckert, Jr., and David S. Roos
12

3. Making a Home for *Plasmodium* Post-Genomics: Ultrastructural Organization of the Blood Stages
Lawrence H. Bannister, Gabriele Margos, and John M. Hopkins
24

4. Genetic Manipulation of *Plasmodium falciparum*
Alan F. Cowman and Brendan S. Crabb
50

5. The Transcriptome of the Malaria Parasite *Plasmodium falciparum*
Karine Le Roch and Elizabeth Winzeler
68

6. The *Plasmodium* Proteome
Jeffrey R. Johnson and John R. Yates III
85

7. **Evolutionary History and Population Genetics of Human Malaria Parasites**
Martine Zilversmit and Daniel L. Hartl
95

II. INVASION AND GAMETE FORMATION   111

8. **A Mechanistic Approach to Merozoite Invasion of Red Blood Cells: Merozoite Biogenesis, Rupture, and Invasion of Erythrocytes**
Mary R. Galinski, Anton R. Dluzewski, and John W. Barnwell
113

9. **The Sporozoite**
R. E. Sinden and K. Matuschewski
169

10. **Gametocytes and Gametes**
Pietro Alano and Oliver Billker
191

III. GROWTH AND METABOLISM   221

11. **Molecular Approaches to Malaria: Glycolysis in Asexual-Stage Parasites**
Charles J. Woodrow and Sanjeev Krishna
223

12. **The Mitochondrion**
Akhil B. Vaidya
234

13. **Trafficking and the Tubulovesicular Membrane Network**
Kasturi Haldar, Narla Mohandas, Souvik Bhattacharjee, Travis Harrison, N. Luisa Hiller, Konstantinos Liolios, Sean Murphy, Pamela Tamez, and Christiaan van Ooij
253

14. **The Apicoplast**
Stuart A. Ralph
272

15. **Protein Kinases Regulating *Plasmodium* Proliferation and Development**
Christian Doerig
290

16. **Proteases and Hemoglobin Degradation**
Philip J. Rosenthal
311

17. ***Plasmodium* Lipids: Metabolism and Function**
Henri J. Vial and Choukri Ben Mamoun
327

18. *Plasmodium* Ribosomes and Opportunities
for Drug Intervention
Indu Sharma and Thomas F. McCutchan
353

19. Oxidative Stress and Antioxidant Defense
in Malarial Parasites
Katja Becker, Sasa Koncarevic, and Nicholas H. Hunt
365

20. New Permeation Pathways
Serge L. Thomas and Stéphane Egée
384

IV. IMMUNE EVASION 397

21. Molecular Aspects of Antigenic Variation
in *Plasmodium falciparum*
Paul Horrocks, Susan A. Kyes, Peter C. Bull, and Kirk W. Deitsch
399

22. Rosetting
J. Alexandra Rowe
416

V. PROTECTION 427

23. Mechanisms of Antimalarial Drug Action and Resistance
Anne-Catrin Uhlemann, Yongyuth Yuthavong, and David A. Fidock
429

24. Host Genetic Factors in Resistance and Susceptibility
to Malaria
Dominic P. Kwiatkowski and Gaia Luoni
462

25. Progress in Development of a Vaccine To Aid Malaria Control
Vasee S. Moorthy and Filip Dubovsky
480

VI. VECTOR 499

26. The *Anopheles gambiae* Genome
Frank H. Collins and Catherine A. Hill
501

27. The Transcriptome of Human Malaria Vectors
Osvaldo Marinotti and Anthony A. James
516

Index 531

CONTRIBUTORS

Pietro Alano
Dipartimento di Malattie Infettive, Parassitarie e Immunomediate, Istituto Superiore di Sanità, Viale Regina Elena 299, 00161 Rome, Italy

Lawrence H. Bannister
Centre for Neuroscience Research, Guy's, King's and St. Thomas's School of Biomedical Science, Hodgkin Building, Guy's Hospital Campus, London SE1 1UL, United Kingdom

John W. Barnwell
Malaria Branch, Division of Parasitic Disease, Centers for Disease Control and Prevention, Mailstop F-36, Bldg. 109, Room 1121, 4770 Buford Highway, NE, Atlanta, GA 30341

Katja Becker
Interdisciplinary Research Center, Heinrich-Buff-Ring 26-32, Justus-Liebig University, D-35392 Giessen, Germany

Choukri Ben Mamoun
Center for Microbial Pathogenesis, E7041, Department of Genetics and Development Biology, University of Connecticut Health Center, 263 Farmington Ave., Farmington, CT 06030-3710

Souvik Bhattacharjee
Departments of Pathology and Microbiology-Immunology, Northwestern University, 303 E. Chicago Ave., Chicago, IL 60611

Oliver Billker
Department of Biological Sciences, Imperial College London, London SW7 2AZ, United Kingdom

Peter C. Bull
Weatherall Institute of Molecular Medicine, University of Oxford, John Radcliffe Hospital, Oxford OX3 9DS, United Kingdom

Frank H. Collins
Center for Tropical Disease Research and Training, Galvin Life Sciences Building, University of Notre Dame, Notre Dame, IN 46556-0369

Alan F. Cowman
The Walter and Eliza Hall Institute of Medical Research, 1G Royal Parade,
Parkville 3050, Melbourne, Australia

Brendan S. Crabb
The Walter and Eliza Hall Institute of Medical Research, 1G Royal Parade,
Parkville 3050, Melbourne, Australia

Shailesh V. Date
Genomics Institute, Center for Bioinformatics, and Department of Biology,
University of Pennsylvania, Philadelphia, PA 19104

Kirk W. Deitsch
Department of Microbiology and Immunology, Weill Medical College of Cornell
University, 1300 York Ave., W-704, Box 62, New York, NY 10021

Anton R. Dluzewski
Department of Immunobiology, Guy's, King's and St. Thomas's School of Medicine, 2nd
Floor, New Guy's House, King's College London, Guy's Hospital, London Bridge,
London SE1 9RT, United Kingdom

Christian Doerig
INSERM U 609, Wellcome Centre for Molecular Parasitology, University of Glasgow,
56 Dumbarton Rd., Glasgow, G11 6NU Scotland, United Kingdom

Filip Dubovsky
Malaria Vaccine Initiative, PATH, 7500 Old Georgetown Rd., 12th Floor,
Bethesda, MD 20814

Stéphane Egée
CNRS, FRE 2775, Station Biologique, Place G. Teissier, BP 74, 29682 Roscoff, France

Kobby Essien
Genomics Institute, Center for Bioinformatics, and Department of Biology,
University of Pennsylvania, Philadelphia, PA 19104

David A. Fidock
Department of Microbiology and Immunology, Albert Einstein College of Medicine,
Forchheimer 403, 1300 Morris Park Ave., The Bronx, NY 10461

Martin J. Fraunholz
Genomics Institute, Center for Bioinformatics, and Department of Biology,
University of Pennsylvania, Philadelphia, PA 19104

Bindu Gajria
Genomics Institute, Center for Bioinformatics, and Department of Biology,
University of Pennsylvania, Philadelphia, PA 19104

Mary R. Galinski
Emory Vaccine Center at Yerkes, Division of Infectious Diseases, Department of Medicine,
Emory University, 954 Gatewood Rd., Atlanta, GA 30329

Gregory R. Grant
Genomics Institute, Center for Bioinformatics and Department of Biology,
University of Pennsylvania, Philadelphia, PA 19104

Kasturi Haldar
Departments of Pathology and Microbiology-Immunology, Northwestern University,
303 E. Chicago Ave., Chicago, IL 60611

Travis Harrison
Departments of Pathology and Microbiology-Immunology, Northwestern University,
303 E. Chicago Ave., Chicago, IL 60611

Daniel L. Hartl
Department of Organismic and Evolutionary Biology, Harvard University,
Cambridge, MA 02138

Catherine A. Hill
Department of Entomology, 901 West State St., Purdue University,
West Lafayette, IN 47907-2089

N. Luisa Hiller
Departments of Pathology and Microbiology-Immunology, Northwestern University,
303 E. Chicago Ave., Chicago, IL 60611

John M. Hopkins
Centre for Neuroscience Research, Guy's, King's and St. Thomas's School of Biomedical
Science, Hodgkin Building, Guy's Hospital Campus, London SE1 1UL, United Kingdom

Paul Horrocks
Weatherall Institute of Molecular Medicine, University of Oxford, John Radcliffe Hospital,
Oxford OX3 9DS, United Kingdom

Nicholas H. Hunt
Institute for Biomedical Research, Department of Pathology (D06), University of Sydney,
New South Wales 2006, Australia

John Iodice
Genomics Institute, Center for Bioinformatics, and Department of Biology,
University of Pennsylvania, Philadelphia, PA 19104

Anthony A. James
Departments of Microbiology & Molecular Genetics and Molecular Biology &
Biochemistry, University of California, Irvine, 3205 McGaugh
Hall, Irvine, CA 92697-3900

Jeffrey R. Johnson
Department of Cell Biology, The Scripps Research Institute, La Jolla, CA 92037

Jessica C. Kissinger
Center for Tropical and Emerging Global Diseases and Department of Genetics,
University of Georgia, Athens, GA 30602

Sasa Koncarevic
Interdisciplinary Research Center, Heinrich-Buff-Ring 26-32, Justus-Liebig University,
D-35392 Giessen, Germany

Sanjeev Krishna
Department of Cellular and Molecular Medicine, St. George's Hospital Medical School,
Cranmer Terrace, London SW17 0RE, United Kingdom

Dominic P. Kwiatkowski
Wellcome Trust Centre for Human Genetics and University Department of Paediatrics,
University of Oxford, Oxford, United Kingdom

Susan A. Kyes
Weatherall Institute of Molecular Medicine, University of Oxford, John Radcliffe Hospital,
Oxford OX3 9DS, United Kingdom

Philip T. Labo
Genomics Institute, Center for Bioinformatics, and Department of Biology,
University of Pennsylvania, Philadelphia, PA 19104

Karine Le Roch
Genomics Institute of the Novartis Research Foundation, San Diego, CA 92121

Konstantinos Liolios
Departments of Pathology and Microbiology-Immunology, Northwestern University,
303 E. Chicago Ave., Chicago, IL 60611

Gaia Luoni
Wellcome Trust Centre for Human Genetics and University Department of Paediatrics,
University of Oxford, Oxford, United Kingdom

Gabriele Margos
Department of Immunology, Guy's, King's and St. Thomas's School of Medicine, New
Guy's House, Guy's Hospital Campus, London SE1 9RT, United Kingdom

Osvaldo Marinotti
Department of Molecular Biology & Biochemistry, University of California, Irvine,
3205 McGaugh Hall, Irvine, CA 92697-3900

K. Matuschewski
Department of Parasitology, Heidelberg University School of Medicine, Im Neuenheimer
Feld 324, 69120 Heidelberg, Germany

Thomas F. McCutchan
Molecular Biology Section, Laboratory of Malaria Vector Research, National Institute of
Allergy and Infectious Diseases, National Institutes of Health, Bethesda, MD 20892

Arthur J. Milgram
Genomics Institute, Center for Bioinformatics, and Department of Biology,
University of Pennsylvania, Philadelphia, PA 19104

Narla Mohandas
New York Blood Center, 310 East 67th Street, New York, NY 10021

Vasee S. Moorthy
Malaria Vaccine Initiative, PATH, 7500 Old Georgetown Rd., 12th Floor,
Bethesda, MD 20814

Sean Murphy
Departments of Pathology and Microbiology-Immunology, Northwestern University,
303 E. Chicago Ave., Chicago, IL 60611

Stuart A. Ralph
Biology of Host Parasite Interactions Unit, Institut Pasteur, Batiment Nicolle, 25 rue du
Docteur Roux, 75724 Paris Cedex 15, France

David S. Roos
Genomics Institute, Center for Bioinformatics, and Department of Biology, University of Pennsylvania, Philadelphia, PA 19104

Philip J. Rosenthal
Department of Medicine, San Francisco General Hospital, University of California, San Francisco, CA 94143

J. Alexandra Rowe
Institute of Immunology and Infection Research, University of Edinburgh, West Mains Rd., Edinburgh EH9 3JT, United Kingdom

Indu Sharma
Molecular Biology Section, Laboratory of Malaria Vector Research, National Institute of Allergy and Infectious Diseases, National Institutes of Health, Bethesda, MD 20892

Irwin W. Sherman
Department of Biology, University of California, Riverside, CA 92521

R. E. Sinden
Department of Biological Sciences, Imperial College London, Sir Alexander Fleming Building, Imperial College Rd., South Kensington, London SW7 2AZ, United Kingdom

Christian J. Stoeckert, Jr.
Genomics Institute, Center for Bioinformatics, and Department of Biology, University of Pennsylvania, Philadelphia, PA 19104

Pamela Tamez
Departments of Pathology and Microbiology-Immunology, Northwestern University, 303 E. Chicago Ave., Chicago, IL 60611

Serge L. Thomas
CNRS, FRE 2775, Station Biologique, Place G. Teissier, BP 74, 29682 Roscoff, France

Anne-Catrin Uhlemann
Department of Infectious Diseases, St. George's Hospital Medical School, Cranmer Terrace, London SW17 0RE, United Kingdom

Akhil B. Vaidya
Center for Molecular Parasitology, Department of Microbiology and Immunology, Drexel University College of Medicine, Philadelphia, PA 19129

Christiaan van Ooij
Departments of Pathology and Microbiology-Immunology, Northwestern University, 303 E. Chicago Ave., Chicago, IL 60611

Henri J. Vial
Dynamique Moléculaire des Interactions Membranaires, UMR 5539 CNRS/Université Montpellier II, case 107, Place Eugène Bataillon, F-34095 Montpellier Cedex 5, France

Patricia L. Whetzel
Genomics Institute, Center for Bioinformatics, and Department of Biology, University of Pennsylvania, Philadelphia, PA 19104

Elizabeth Winzeler
Department of Cell Biology, ICND202, The Scripps Research Institute, 10550 North Torrey Pines Rd., La Jolla, CA 92037

Charles J. Woodrow
Department of Cellular and Molecular Medicine, St. George's Hospital Medical School, Cranmer Terrace, London SW17 0RE, United Kingdom

John R. Yates III
Department of Cell Biology, The Scripps Research Institute, La Jolla, CA 92037

Yongyuth Yuthavong
National Center for Genetic Engineering and Biotechnology (BIOTEC), National Science and Technology Development Agency, Thailand Science Park, 111 Paholyothin Rd., Pathumthani 12120, Thailand

Martine Zilversmit
Department of Organismic and Evolutionary Biology, Harvard University, Cambridge, MA 02138

PREFACE

In 2002, thanks to the prodigious efforts of an international consortium, the genetic blueprint of the human malaria parasite, *Plasmodium falciparum*, and its vector, *Anopheles gambiae*, became available to the entire scientific community. With the sequencing of these genomes (and that of *Homo sapiens*) came the power to understand this host-parasite relationship in ways few of us could have imagined less than a decade ago. These achievements, on their own, could be marked by a book on molecular aspects of malaria, but when I was approached to be the editor for such a volume I was faced with two questions. What would be the book's specific objectives? What topics should it contain?

My aim as editor was to provide readers with more than a fleeting glimpse of the genomes and the abundant research that they have spawned. This book, *Molecular Approaches to Malaria*, has been designed to serve as a convenient and accessible resource for molecular biologists (as well as biochemists, cell biologists, chemists, pathologists, parasitologists, entomologists, and immunologists) who want to redirect their research toward malaria and to inform and expand the research horizons of malariologists already working in the field. It is my fond hope that *Molecular Approaches to Malaria* will provide grist for the mills that will lead to the development of potent and specific antimalarials and insecticides and the preparation of sensitive and inexpensive diagnostics, as well as assist in the production of protective and practical vaccines. And it is expected that the molecular approaches described in these pages will be translated into interventions to block mosquito transmission of malaria, prevent parasite invasion, and allow a deeper understanding of what it takes for a human to become immune.

To produce this book, we harnessed the talents of many creative authors (many of whom march to very different drummers). We also had the very difficult task of choosing the specific topics to be included. Through the entire process—from concept to selection to completion—I have been gratified by the enthusiasm of the many contributors to this book and have profited immensely from their thoughtful suggestions. *Molecular Approaches to Malaria* is neither a compendium nor an encyclopedia, but it is an overview of many rapidly advancing fields of inquiry with a single focus: malaria. As such, this volume is expected to be a prologue for the conquest of one of the world's deadliest diseases.

IRWIN W. SHERMAN

INTRODUCTION

I

THE LIFE OF *PLASMODIUM:* AN OVERVIEW

Irwin W. Sherman

I

FROM BAD AIR TO MOSQUITOES

Historical records, some >3,000 years old, attest to the antiquity of the disease malaria (Sherman, 1998a; Harrison, 1978; Bruce-Chwat, 1988). Although there are many descriptions of people with enlarged spleens, periodic fevers, headaches, chills, and weakness, the discovery of the causative agent of the disease did not take place until 20 October 1880, when Alphonse Laveran (1845-1922), a military physician stationed in Bone, Algeria, examined a drop of blood from a soldier suffering from an intermittent fever. Using a light microscope, Laveran noticed some crescent-shaped bodies among the red blood cells that were almost entirely transparent, save for some pigment inclusions. On 6 November 1880, while examining a drop of blood from a feverish artilleryman, he saw several transparent mobile filaments emerging from a clear spherical body. He recognized that these bodies were alive, and that he was looking at an animal parasite, not a bacterium or a fungus. Subsequently, he examined blood samples from 192 malaria patients: in 148 of these, he found the telltale crescents. Where there were no crescents, there were no symptoms of malaria. He named the parasite *Oscillaria malariae* and communicated his findings to the Societé Medicale des Hopitaux on 24 December 1880. The drawings in his paper (Fig. 1) provided convincing evidence that, without use of stains or a microscope fitted with an oil immersion lens, Laveran had seen the development of the parasite.

At first, Laveran's announcement was received with skepticism. Indeed, in 1882 when he visited Rome and showed his slides to the Italian parasitologists, they scoffed at him and told him that the spherical bodies were nothing more than degenerating red blood cells. Initially, the Italians examined only preparations that had been heat fixed and stained with methylene blue, so they did not see any movement of the parasite that had caused Laveran to give it the name *Oscillaria*. However, 2 years later when, like Laveran, they began to examine fresh preparations of blood, they were able to observe the ameboid movements of the parasite within the red blood cell, as well as emerging whiplike filaments from the clear spherical bodies within the cell.

In 1886, using thin smears of fresh blood, Camillo Golgi (1843-1926) discovered that the

Irwin S. Sherman, Department of Biology, University of California, Riverside, CA 92521.

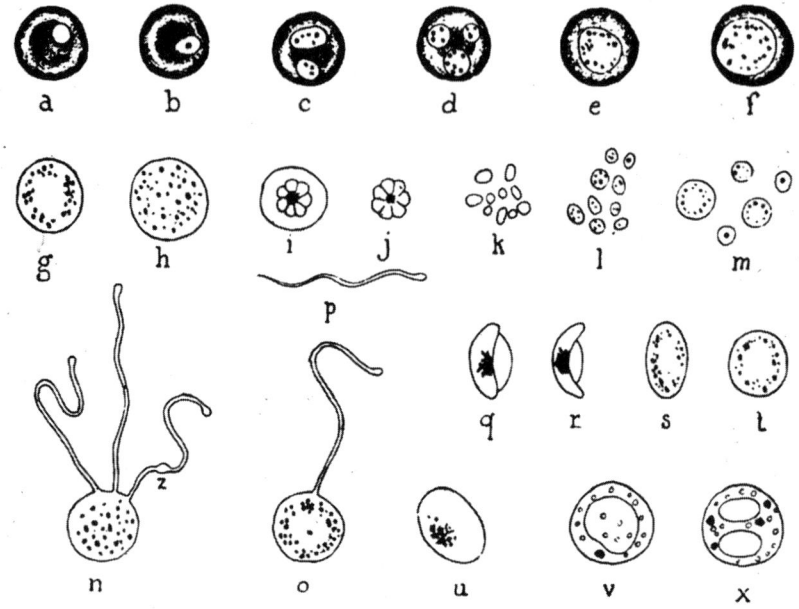

FIGURE 1 Laveran's drawings of *P. falciparum* in the blood.

parasite reproduced asexually by multiple fission and showed that the fever coincided with erythrocyte lysis and parasite release. In 1891, Dimitri Romanovsky prepared heat-fixed thin blood films and used a combination of methylene blue and eosin to stain the nucleus and cytoplasm of the parasite (Color Plate 1). The significance of Laveran's observation of the release of motile filaments went unappreciated until 1896 to 1897, when William MacCallum and Eugene L. Opie, students at Johns Hopkins University, found that the blood of sparrows and crows infected with *Haemoproteus* (a bird parasite closely related to malaria) contained two kinds of sex cells (crescent-shaped gametocytes) and that filament extrusion (called exflagellation) reflected the release of microgametes from the male gametocyte. They also correctly interpreted their observations: gametocytes occur in the blood; when ingested by a mosquito, the gametes are released in the mosquito stomach, where fertilization takes place, producing a wormlike zygote, the ookinete.

Neither Laveran nor MacCallum solved the problem of malaria transmission. It was Ronald Ross, a surgeon-major in the Indian Medical Service, who showed how "bad air" could cause malaria. After many years of frustrating efforts,

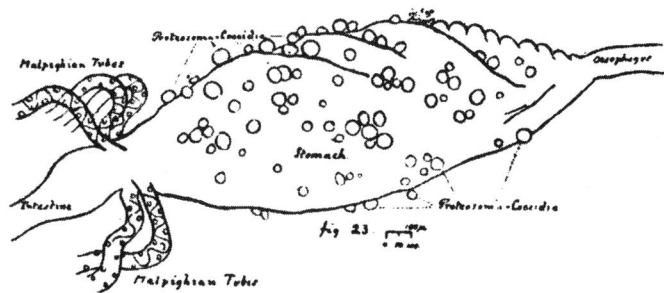

a. Mosquito stomach with oocysts

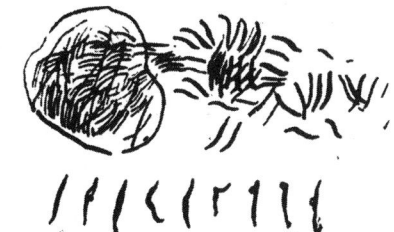

b. Oocyst bursting to release sporozoites

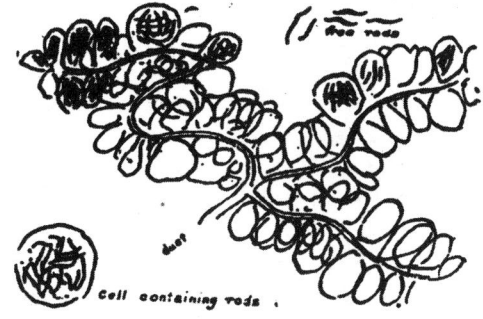

c. Mosquito salivary gland with sporozoites

FIGURE 2 Ross's drawing of the development of malaria parasites in the mosquito. (a) Mosquito stomach with oocysts; (b) oocyst bursting to release sporozoites; (c) mosquito salivary gland with sporozoites.

Ross decided to see whether mosquitoes could become infected with malaria. He wrote (Ross, 1923): "I fed them on . . . a patient who had crescents in his blood. . . . At about 1 p.m., I determined to sacrifice the last mosquito. . . . The dissection was excellent and I went carefully through the tissues, now so familiar to me, searching every micron with the same passion and care as one would have in searching some vast ruined palace for a little hidden treasure. . . . I had scarcely commenced the search again when I saw a clear and almost perfectly circular outline before me of about 12 microns in diameter. The outline was too sharp, the cell too small to be an ordinary stomach cell of a mosquito. I looked a little further. Here was another, and another exactly similar cell. In each of these, there was a cluster of small granules, black as jet, and exactly like the black pigment granules of the. . . crescents" (Fig. 2). Ross had shown that 4 or 5 days after a mosquito fed on infected blood, it had wart-like oocysts on its stomach. He reported his findings to the *British Medical Journal* in a paper entitled "On some peculiar pigmented cells found in two mosquitoes fed on malarial blood." It appeared in print on 18 December 1897.

Before he could complete this work on human malaria, Ross was posted to Calcutta, India. However, there were not a large number of malaria cases in the Calcutta hospitals, so he turned to something that his mentor Patrick Manson had suggested earlier: the study of mosquitoes and malaria, as seen in birds. Before long, Ross found that crows and pigeons had malaria parasites in their blood; he also found malaria-pigmented cells in the stomachs of mosquitoes that had been fed on infected birds. The parasites grew to their maximum size within about 6 days. Then, he noticed that some of the oocysts seemed to have stripes or ridges in them; this happened on the seventh or eighth day after the mosquito had been fed on infected blood. On 4 July 1898, Ross got something of value. Near the mosquito's head, he found a large, branching salivary gland. He wondered whether this gland would pour parasites into the blood of a healthy creature. During 21 July and 22 July of that year, he took some uninfected sparrows, allowed malaria-carrying mosquitoes to bite them, and was able to show that the healthy sparrows became infected in a few days. This was the proof that malaria was not conveyed by bad air. Ross communicated his findings on 28 July 1898.

Ross wanted the glory of discovery for himself and for England, but he was not alone in his quest to discover the cause of malaria. The German government had dispatched a team of scientists under the leadership of Robert Koch to work in the Campagna region near Rome, Italy, an area where malaria was notoriously endemic. They isolated a bacillus from the air and the mud of the marshes and, rejecting the claims of Laveran, named the causative agent for malaria *Bacillus malariae*. However, when the bacillus could not be grown in the laboratory Koch discarded it as the cause of the disease; undeterred, he continued to look further. Koch then visited the laboratory of Professor Giovanni Battista Grassi at the University of Rome, told him of his failure with the "germ" of malaria, and mentioned Ross's communication. At that moment, Grassi had what is today called an "ah-ha moment."

Where Ross was patient, perseverant, and willing to carry out a seemingly endless series of trial-and-error experiments, Grassi was methodical and analytical. He was also able to distinguish the different kinds of mosquitoes. Grassi observed, "There was not a single place where there is malaria—where there aren't mosquitoes too, and either malaria is carried by one particular blood sucking mosquito out of the forty different kinds of mosquitoes in Italy—or it isn't carried by mosquitoes at all" (De Kruif, 1926). He recognized there were still two tasks left: to identify the mosquito that transmitted malaria to humans and to demonstrate the mosquito cycle for this form of malaria. Working with Amico Bignami, Giovanni Bastianelli, Angelo Celli, and Antonio Dionisi, he went into the highly malarious Roman Campagna and the area surrounding it, collecting mosquitoes and at the same time recording information on the incidence of malaria among the inhabitants. It soon became apparent that most of the mosquitoes could be eliminated as carriers of the disease because they occurred where there was no malaria. But there was an exception.

Where there were *zanzarone*, as the Italians called the large, brown-spotted, winged mosquitoes, there was always malaria. Grassi recognized that the *zanzarone* were *Anopheles* and he wrote: "It is the anopheles mosquito that carries malaria. . ." (De Kruif, 1926). Grassi and his team were able to infect clean anopheles mosquitoes by having them feed on patients with crescents in their blood, and he was able to trace the development of the parasite from the mosquito stomach to the salivary glands. The life cycle in humans was, as Ross had correctly surmised, similar to that of the bird malaria with which he had worked. Their work showed that the association of the disease with swampy, marshy areas of the world is due to the fact that these areas are ideal breeding sites for mosquitoes.

With this work, Grassi demolished Koch's theory and was able to prove that "It is not the mosquito's children, but only the mosquito who herself bites a malaria sufferer—it is only that mosquito who can give malaria to healthy people" (De Kruif, 1926). Grassi wanted his beloved country of Italy to receive recognition for his work but it did not. Instead it provoked a bitter and very nasty disagreement with Ross. Ross claimed that it was only after Grassi had read his work on the transmission of malaria in birds that he recognized that malaria in humans was only found in those areas where *Anopheles* was. Ross (1923) wrote in his memoirs, "They . . . had this paper of mine before them when they wrote their note. Their statement was . . . a deliberate and intentional lie, told in order to discredit my work and so to obtain priority. . . . Many of the items . . . are directly pirated from my . . . results . . . stolen straight from me . . ." Ross did not complete the proof of mosquito transmission in malaria in humans; the Italians did that. But he flagged the dapple-winged mosquito as the vector. Grassi's contribution was to recognize that *Anopheles* was the vector. In effect, Ross was the explorer at the helm of the ship, and the Italians rode the decks and helped make a landing. Ross, not Grassi, received the Nobel Prize in 1902; Laveran received the Nobel Prize in 1907.

Although malaria can be induced in a host by the introduction of parasites (called sporozoites) through the bite of an infectious female mosquito, the parasites do not immediately appear in the blood. This was surprising in view of the fact that in 1903 Fritz Schaudinn claimed to have seen sporozoites directly invade erythrocytes. Schaudinn was a dominant figure in parasitology, so few doubted his word; but no one could repeat his observation, and over time it was called into question. In 1948, H. E. Shortt, P. C. C. Garnham, and their colleagues in England inoculated rhesus monkeys with a syringe full of sporozoites that they had obtained from the salivary glands of mosquitoes infected with *Plasmodium cynomolgi* (a parasite similar to the benign tertian malaria in humans caused by *Plasmodium vivax*). In a week, parasites, at what are called preerythrocytic stages, were found in the livers of the monkeys. Later, they demonstrated similar stages in biopsy material taken from the livers of human volunteers who had been infected by the bite of mosquitoes carrying *P. vivax*. After infected mosquitoes had fed on other volunteers, this stage was found at the same site in patients with malignant tertian malaria caused by *Plasmodium falciparum*. It was now clear that when an infected female anopheline mosquito feeds, it injects sporozoites that go first to the liver, where they live and multiply for several weeks.

All human malarial agents *(P. falciparum, P. vivax, Plasmodium ovale,* and *Plasmodium malariae)* are transmitted through the bite of an infected female anopheline mosquito when she injects sporozoites from her salivary glands during blood feeding (Fig. 3). The number of sporozoites inoculated is usually <25. These travel via the bloodstream to the liver where they enter liver cells. The entire process takes <1 h. Within the liver cell, the parasite multiplies asexually to produce ≥10,000 infective offspring. These do not return to their spawning ground, the liver, but instead invade erythrocytes. The asexual reproduction of parasites in red blood cells (called schizogony or merogony) and the release of infectious offspring (merozoites) when the red blood cell lyses are responsible for the pathogenesis of this disease. Merozoites released from erythrocytes can invade other red blood cells and continue the cycle of 10-fold parasite multiplication, with extensive red blood cell destruction. In some cases, the merozoites enter red blood cells but do not divide. Instead, they differentiate into male or female gametocytes (the crescents of Laveran). When ingested by the female mosquito, the male gametocyte divides into eight flagellated microgametes, which escape from the enclosing red blood cell (exflagellation). These swim to the macrogamete, one microgamete fertilizes the macrogamete, and the resultant motile zygote (the ookinete) moves between or through the cells of the stomach wall. This encysted zygote, resembling a wart on the outside of the mosquito stomach, is an oocyst. Through asexual multiplication, threadlike sporozoites are produced in the oocyst, which bursts and releases sporozoites into the body cavity of the mosquito. These sporozoites quickly find their way to the salivary glands. When the female mosquito feeds again, the transmission cycle is complete.

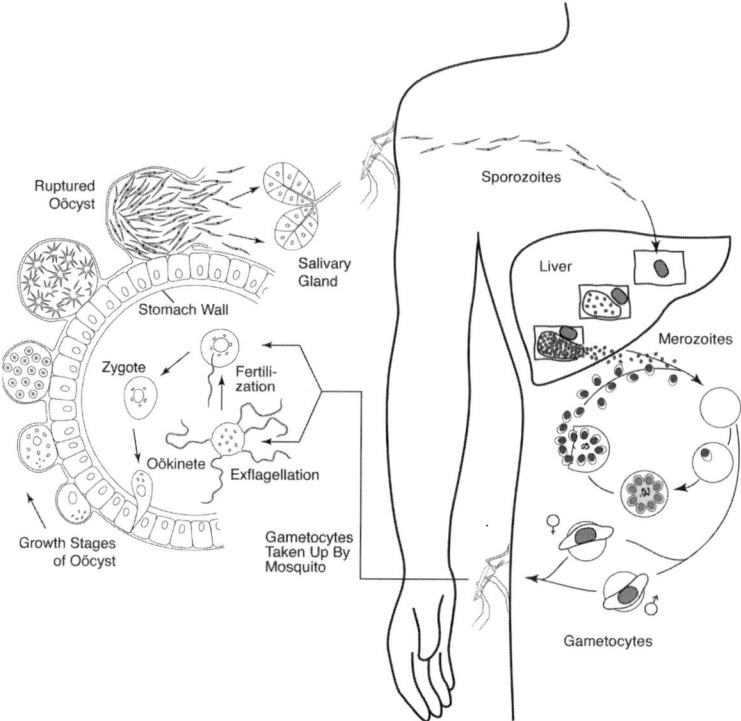

FIGURE 3 The life cycle of *Plasmodium falciparum*.

FROM BIOCHEMISTRY TO GENOMICS

Twenty-five years ago, I wrote, "The microscopic size and inaccessibility of the liver forms of malaria have limited our knowledge about the biochemistry of these stages; similarly, the minuteness of the mosquito stages has also prevented biochemists from studying these forms. Consequently, most of the information on the malaria-host relationship is derived from studies of parasites growing within or removed from erythrocytes since such stages are both accessible and abundant. Therefore, of necessity, our knowledge of the biochemistry of *Plasmodium* is limited almost exclusively to the interactions of malaria parasites with one host cell type, the erythrocyte" (Sherman, 1979).

The earliest studies of malaria biochemistry (mostly enzymatic and metabolic) were carried out with animal models such as the monkey malarial agent *Plasmodium knowlesi* or the bird malaria agents *Plasmodium lophurae* and *Plasmodium gallinaceum*, since these could provide larger amounts of starting material. Later, the rodent malarial agents from Africa, especially *Plasmodium berghei* and *Plasmodium chabaudi*, served as surrogates for the human malarial agent *P. falciparum* (Scheibel, 1988; Scheibel and Sherman, 1988; Sherman, 1998; Vial and Ancelin, 1998; Rosenthal and Meshnick, 1998; Vaidya, 1998). Biochemical work on organelles, carbohydrates, lipids, nucleotides, amino acids, hemoglobin digestion, transport, and cofactors in these species has been updated periodically over the past 50 years. Before 1976, almost no information was available for the most virulent of the human malaria agents, *P. falciparum*, simply because this parasite could not be grown in vivo in the amounts needed for conventional biochemical and metabolic analyses and it could not be continuously cultured in vitro.

Today, however, it is possible to grow all the erythrocytic developmental stages of *P. falciparum* in the laboratory. The "taming" of *P. falciparum* to a life in the laboratory took place in the Rockefeller University laboratory of William Trager. In the 1970s, Trager began a systematic evaluation of the suitability of commercially available media for growing the parasites; the medium RPMI 1640 was found to be superior to all others. He also found that an atmosphere of 7% carbon dioxide and 5% oxygen with the balance nitrogen was better than 7% carbon dioxide and 93% air and that HEPES buffer was effective in maintaining the pH of the serum-containing medium. In 1975, a postdoctoral fellow, James B. Jensen, joined Trager's laboratory, and together they set up stationary cultures in Petri dishes and held these in a candle jar. (In the candle jar, a white candle is placed in a desiccator equipped with a stopcock, the candle is lit, and the desiccator cover is replaced. When the candle flame goes out, the stopcock is closed. The atmosphere in the candle jar is 1 to 3% carbon dioxide and 4 to 15% oxygen.) Maintained at 37°C, the parasites grow with a 48-h asexual cycle, as they do in the human body (Trager and Jensen, 1976).

The Trager-Jensen method has been used to answer many questions, including how sickle cell hemoglobin protects humans with sickle cell trait (Friedman, 1978; Friedman et al., 1979). When AS red blood cells, taken from a heterozygous individual with sickle cell trait, are infected with *P. falciparum* in vitro and the O_2 levels in the candle jar are lowered to 3% or less, potassium and water leak out of the host cell and parasite growth is inhibited. When SS cells from a homozygous individual with sickle cell anemia are infected and the oxygen levels are lowered in the candle jar, the sickle hemoglobin forms solid rods and these rigid "spears" kill the parasite. When red blood cells from a normal individual (AA) are infected in vitro, the parasite grows and divides normally at 3% oxygen. In vivo, when a malaria parasite gets into an AS cell and that infected cell sequesters in the capillaries of the deep tissues where the oxygen concentration is low, the host red blood cell sickles, the potassium level drops, and the parasite dies. In addition, some infected AS cells may sickle in the circulation, and these are eliminated by the filtering action of the spleen. (Uninfected AS or AA red blood cells do not sickle under these conditions.) The phagocytes in organs like the spleen and liver recognize the sickled infected red blood cell and remove it. Consequently, when an individual with sickle cell trait is infected with falciparum malaria, there is selective removal of malaria-infected red blood cells. The protection—innate immunity—is especially important in the young (those <10 years of age), since they have not had sufficient time to build up immunity as a result of infection; in these individuals, such infections are also milder.

In vitro-grown *P. falciparum* also provided material for the first definitive ultrastructural description of the parasite (Langreth et al., 1978) and now supplies sufficient quantities for detailed cytochemical, biochemical, and genetic analyses. Falciparum genes, the proteins they encode, and the metabolic pathways in which they participate have been placed in a convenient and very useful website (http://sites.huji.ac.il/malaria/). The site was originally constructed to provide information on metabolic pathways, but it has been expanded to include cell-cell interactions (cytoadherence, rosetting, and invasion). There is also an artistic rendition of the ultrastructure of the developmental stages of *P. falciparum* in the erythrocyte.

In 1996, an international consortium was established to sequence the entire 26-Mb genome of *P. falciparum* and the 278-Mb genome of *Anopheles gambiae*. Six years later, an almost complete annotated sequence was published for the parasite and its vector (Hall and Gardner, 2004; Holt et al., 2002). The annotation predicted 5,409 falciparum genes with 60% lacking a known function or homolog in any other organism. PlasmoDB (the *Plasmodium* genome resource) is the official database of the malaria genome project. It provides access to finished sequences for *P. falciparum* (strain 3D7) and can be accessed at http://www.plasmodb.org. The annotation for *A. gambiae* had an estimated 14,000 genes.

In 2003, Le Roch et al. and Bozdech et al. described gene expression using microarray profiles (transcriptomes) for the entire life cycle of *P. falciparum*. By performing statistically supported cluster analyses, it was possible to group together annotated genes with similar expression profiles and of known function with many hypothetical genes of unknown function. Further analyses will enable a better understanding of the parasite's response to environmental cues such as temperature, pH, and redox potential, as well as to stressors (antimalarial drugs and oxidative stress). The transcriptomes of *A. gambiae* could be grouped into those involved in host seeking and blood feeding, transmission, cuticle and peritophic membrane components, digestion of the blood meal, lipids, insecticide resistance, and nuclear regulation. More than 200 genes have been putatively implicated in innate immunity (Osta et al., 2004).

Investigations at the molecular level of enzyme structure, gene sequences, chromosomal arrangements, and transcriptional control will permit an uncovering of the adhesive molecules that mediate cell-cell interactions, determine the mechanisms of protein trafficking, and identify putative drug targets and vaccine candidates. Molecular approaches (genomics, bioinformatics, and proteomics), as well as high-throughput screening methods, DNA microarrays, comparative expression profiles, knockouts, transient and stable transfections, and mutants, will undoubtedly provide answers to the following key questions:

1. What are the unique biochemical targets in *P. falciparum* that can serve as a basis for the rational design of antimalarials?
2. What are the molecular characteristics of adhesins that could be the basis for developing receptor blockers for merozoite and sporozoite invasion?
3. Is there an in vitro test that will correlate with protective immunity?
4. What are the molecular mechanisms of protective immunity in humans?
5. What are the disease-producing molecules of the plasmodium parasite?
6. What molecular and immunologic means can be used to block mosquito transmission?
7. Can agents be developed that can reverse parasite resistance to antimalarials?
8. What antigens are critical to protective immunity?
9. What are the malaria resistance genes in the human host?
10. What molecular mechanisms are involved in parasite egress from the red blood cell and hepatocyte?

Although there is still much to do, our best hope is that by employing molecular approaches the path to effectively controlling malaria will be both shorter and quicker.

REFERENCES

Bozdech, Z., M. Llinas, B. Pulliam, E. Wong, J. Zhu, and J. DeRisi. 2003. The transcriptome of the intraerythrocytic developmental cycle of *Plasmodium falciparum*. *PLoS Biol.* **1**:1–15.

Bruce-Chwatt, L. 1988. History of malaria from prehistory to eradication, p. 1–59. *In* W. Wernsdorfer and I. McGregor (ed.), *Malaria: Principles and Practice of Malariology*. Churchill Livingstone, Edinburgh, United Kingdom.

De Kruif, P. 1926. *Microbe Hunters*. Harcourt Brace, New York, N.Y.

Friedman, M. 1978. Erythrocyte mechanism of sickle cell resistance to malaria. *Proc. Natl. Acad. Sci. USA* **75**:1994–1997.

Friedman, M., E. F. Roth, R. L. Nagel, and W. Trager. 1979. *Plasmodium falciparum*: physiological interaction with the human sickle cell. *Exp. Parasitol.* **47**:73–80.

Hall, N., and M. Gardner. 2004. The genome of *Plasmodium falciparum*. *In* A. P. Waters and C. Janse (ed.), *Malaria Parasites: Genomes and Molecular Biology*. Caister Academic, Wynmondham, England.

Harrison, G. 1978. *Mosquitoes, Malaria and Man: A History of the Hostilities since 1880*. Dutton, New York, N.Y.

**Holt, R. A., G. M. Subramanian, A. Halpern, G. G. Sutton, R. Charlab, D. R. Nusskern, P. Wincker, A. G. Clark, J. M. Ribeiro, R. Wides, S. L. Salzberg, B. Loftus, M. Yandell, W. H. Majoros, D. B. Rusch, Z. Lai, C. L. Kraft, J. F. Abril, V. Anthouard, P. Arensburger, P. W. Atkinson, H. Baden, V. de Berardinis, D. Baldwin, V. Benes, J. Biedler, C. Blass, R. Bolanos, D. Boscus, M. Barnstead, S. Cai, A. Center, K. Chaturverdi, G. K. Christophides, M. A. Chrystal, M. Clamp, A. Cravchik, V. Curwen, A.

Dana, A. Delcher, I. Dew, C. A. Evans, M. Flanigan, A. Grundschober-Freimoser, L. Friedli, Z. Gu, P. Guan, R. Guigo, M. E. Hillenmeyer, S. L. Hladun, J. R. Hogan, Y. S. Hong, J. Hoover, O. Jaillon, Z. Ke, C. Kodira, E. Kokoza, A. Koutsos, I. Letunic, A. Levitsky, Y. Liang, J. J. Lin, N. F. Lobo, J. R. Lopez, J. A. Malek, T. C. McIntosh, S. Meister, J. Miller, C. Mobarry, E. Mongin, S. D. Murphy, D. A. O'Brochta, C. Pfannkoch, R. Qi, M. A. Regier, K. Remington, H. Shao, M. V. Sharakhova, C. D. Sitter, J. Shetty, T. J. Smith, R. Strong, J. Sun, D. Thomasova, L. Q. Ton, P. Topalis, Z. Tu, M. F. Unger, B. Walenz, A. Wang, J. Wang, M. Wang, X. Wang, K. J. Woodford, J. R. Wortman, M. Wu, A. Yao, E. M. Zdobnov, H. Zhang, Q. Zhao, S. Zhao, S. C. Zhu, I. Zhimulev, M. Coluzzi, A. della Torre, C. W. Roth, C. Louis, F. Kalush, R. J. Mural, E. W. Myers, M. D. Adams, H. O. Smith, S. Broder, M. J. Gardner, C. M. Fraser, E. Birney, P. Bork, P. T. Brey, J. C. Venter, J. Weissenbach, F. C. Kafatos, F. H. Collins, and S. L. Hoffman. 2002. The genome sequence of the malaria mosquito *Anopheles gambiae*. *Science* **298**:129–149.

Langreth, S. G., J. B. Jensen, R. T. Reese, and W. Trager. 1978. Fine structure of human malaria in vitro. *J. Protozool.* **25**:443–452.

Le Roch, K., Y. Zhou, P. Blair, M. Grainger, J.K. Moch, J.D. Haynes. P. De la Vega, A.A. Holder. S. Batalov, D. Carucci, and E. A. Winzeler. 2003. Discovery of gene function by expression profiling of the malaria parasite life cycle. *Science* **301**:1503–1508.

Osta, M. A., G. K. Christophides, and F. C. Kafatos. 2004. Effects of mosquito genes on *Plasmodium* development. *Science* **303**:2030-2032.

Rosenthal, P., and S. R. Meshnick. 1998. Hemoglobin processing and the metabolism of amino acids, heme and iron, p. 145–158. *In* I. W. Sherman (ed.), *Malaria: Parasite Biology, Pathogenesis, and Protection*. ASM Press, Washington, D.C.

Ross, R. 1923. *Memoirs: with a Full Account of the Great Malaria Problems and Its Solution*. Murray, London, United Kingdom.

Scheibel, L. 1988. Plasmodial metabolism and related organellar function during various stages of the life cycle: carbohydrates, p. 171–217. *In* W. Wernsdorfer and I. McGregor (ed.), *Malaria: Principles and Practice of Malariology*. Churchill Livingstone, Edinburgh, United Kingdom.

Scheibel, L. W., and I. W. Sherman. 1988. Plasmodial metabolism and related organellar function during various stages of the life cycle: proteins, lipids, nucleic acids and vitamins, p. 219–252. *In* W. Wernsdorfer and I. McGregor (ed.), *Malaria: Principles and Practice of Malariology*. Churchill Livingstone, Edinburgh, United Kingdom.

Sherman, I. 1979. Biochemistry of *Plasmodium* (malarial parasites). *Microbiol. Rev.* **43**:453–495.

Sherman, I. 1998a. A brief history of malaria and discovery of the parasite's life cycle, p. 3–10. *In* I. W. Sherman (ed.), *Malaria: Parasite Biology, Pathogenesis, and Protection*. ASM Press, Washington, D.C.

Sherman, I. 1998b. Carbohydrate metabolism of asexual stages, p. 135–144. *In* I. W. Sherman (ed.), *Malaria: Parasite Biology, Pathogenesis, and Protection*. ASM Press, Washington, D.C.

Trager, W., and J. Jensen. 1976. Human malaria parasites in continuous culture. *Science* **193**:673–675.

Vaidya, A. 1998. Mitochondrial physiology as a target for atovaquone and other antimalarials, p. 355–368. *In* I. W. Sherman (ed.), *Malaria: Parasite Biology, Pathogenesis, and Protection*. ASM Press, Washington, D.C.

Vial, H., and M. L. Ancelin. 1998. Malarial lipids, p. 159–176. *In* I. W. Sherman (ed.), *Malaria: Parasite Biology, Pathogenesis, and Protection*. ASM Press, Washington, D.C.

PlasmoDB: THE *PLASMODIUM* GENOME RESOURCE

Patricia L. Whetzel, Shailesh V. Date, Kobby Essien, Martin J. Fraunholz, Bindu Gajria, Gregory R. Grant, John Iodice, Jessica C. Kissinger, Philip T. Labo, Arthur J. Milgram, Christian J. Stoeckert, Jr., and David S. Roos

2

The *Plasmodium* Genome Database (http://PlasmoDB.org) was introduced in 2000, in response to emerging needs of the malaria research community for access to genomic-scale datasets. In its earliest manifestations, prior to completion of the *Plasmodium falciparum* genome sequence and curated annotation, PlasmoDB focused on automated analysis of available sequence data (Plasmodium Database Collaborative, 2001; Bahl et al., 2002; Milgram et al., 2003), enabling researchers to identify draft sequences for specific genes of interest, even in the absence of the surrounding genomic context. With the release of a reference sequence for *P. falciparum* strain 3D7 (Gardner et al., 2002), the focus of PlasmoDB shifted to providing access to completed sequence and curated annotations, compiling all available information associated with individual genes (predicted gene and protein sequence and structure, annotations, etc.) and enabling genome-wide studies, such as the identification of all predicted genes with a particular sequence motif or chromosomal location (Kissinger et al., 2002; Bahl et al., 2003).

The availability of an effectively complete *P. falciparum* genome sequence has stimulated a wide range of functional genomics research (cf. Bozdech et al., 2003; Coppel et al., 2004; Doolan et al., 2003; Florens et al., 2002; Florens et al., 2004; Hiller et al., 2004; Lasonder et al., 2002; Le Roch et al., 2003; Le Roch et al., 2004; Marti et al., 2004; Roos et al., 2002; Sam-Yellowe et al., 2004), and PlasmoDB has endeavored to keep pace with these studies, providing access to the underlying datasets, and allowing a variety of integrative queries, e.g., finding all genes for which both transcript and proteomics data suggest expression in gametocyte stage parasites (Kissinger and Roos, 2004).

In parallel with the rapid expansion of genomic-scale functional genomics datasets for *P. falciparum*, sequencing efforts have been dedicated to several other *Plasmodium* species, including *P. berghei*, *P. chabaudi*, *P. gallinaceum*, *P. knowlesi*, *P. reichenowi*, *P. vivax*, and *P. yoelii* (Berry et al., 2004; Carlton et al., 2003; Carlton et al., 2005; Hall et al., 2005). Accordingly, pipelines for

---

Patricia L. Whetzel, Shailesh V. Date, Kobby Essien, Martin J. Fraunholz, Bindu Gajria, Gregory R. Grant, John Iodice, Philip T. Labo, Arthur J. Milgram, Christian J. Stoeckert, Jr., and David S. Roos, Genomics Institute, Center for Bioinformatics, and Department of Biology, University of Pennsylvania, Philadelphia, PA 19104. *Jessica C. Kissinger*, Center for Tropical and Emerging Global Diseases and Department of Genetics, University of Georgia, Athens, GA 30602.

the automated analysis of unfinished sequences continue to be important, and PlasmoDB has grown to include genes from additional *Plasmodium* species; functional genomics data for these organisms are also beginning to emerge (Hall et al., 2005; Kaiser et al., 2004). PlasmoDB also provides a variety of tools for cross-species analysis, including the identification of ortholog and paralog groups (Li et al., 2003).

To facilitate sophisticated queries of *Plasmodium* data, all of this information is stored in a relational database, utilizing the Genomics Unified Schema (GUS) (Davidson et al., 2001; http://www.gusdb.org). The design of GUS is based on the DNA→RNA→ protein central dogma of molecular biology, using this principle to organize, store, and integrate sequence and functional genomics data so that researchers can access the entirety of information available for individual DNA sequences, genes, and proteins (or groups of these entities). PlasmoDB presents genomic data via Gene Pages and also incorporates a variety of Analysis Tools and data mining queries, which take advantage of the underlying database structure (for a brief introduction to the architecture underlying PlasmoDB, see Kissinger et al., 2002, and Kissinger and Roos, 2004). Data of interest can be saved using the History feature and downloaded using the ReportMaker tool. PlasmoDB will continue to integrate new datasets as they emerge, developing tools of interest to the malaria researcher and facilitating the discovery of new diagnostics, drugs, and vaccines.

GENE PAGES

Gene Pages provide access to the complete collection of information for each gene in the PlasmoDB database. Because of the diversity of data available, a table of contents is provided at the top of each Gene Page, and multiple alternative views of the data may be considered: a Summary view, an Annotation view, a Protein view, an Expression view, and a Sequence view (Color Plate 2). The Summary view presents the most commonly requested information about the gene in question, including curated annotations, Basic Local Alignment Tool (BLAST) results (Altschul et al., 1997), Ortholog/Paralog predictions (Li et al., 2003), a graphic representation of the genomic context and predicted protein features, and highlights of functional genomics data such as microarray and proteomics results. More detailed information about genes is obtained using the specialized views (Annotation, Protein, Expression, and Sequence) and can be accessed with a single click from the Summary page.

The Annotation view displays all known information related to a gene's annotation, including official annotations from curators at the genome sequencing centers (Berry et al., 2004), unofficial comments from PlasmoDB users, and a variety of automated analyses. This view also provides the history of the gene identifier. As sequencing efforts proceed, provisional names are often used to identify genes before the assignment of permanent identifiers once the genome assemblies are completed and annotated. For example, in the name PFF0480w, PFF designates a gene on the *P. falciparum* chromosome 6 (PFA = chromosome 1, PFF = chromosome 6, etc.), 0480 reflects gene position reading from left to right on the chromosome, and *w* versus *c* reflects the coding strand. Prior to the assignment of this stable identifier based on the complete genome sequence, however, PFF0480w was known as MAL6P1.100. The Annotation view also displays assignments based on structured ontologies, including enzyme commission numbers, and gene ontology (GO) terms describing biological process, cellular component, and function (see http://www.geneontology.org). Additional information available under the Annotation view includes RefSeq assignments, a graphic view of BLAST matches to the nonredundant GenBank-EMBL-DDBJ sequence databases for all species, and predicted orthologs and paralogs (Li et al., 2003). Links are also provided to external resources, including the GeneDB database at the Sanger Institute (http://www.genedb.org), the Apicomplexan Expressed Sequence Tag (EST) Database (Li et al., 2004), the Malaria Parasite Metabolic Pathway database at Hebrew University (http://sites.huji.ac.il/malaria/), and the MR4 reagent repository at the American

Type Culture Collection (http://www.malaria.mr4.org).

Protein views link to any structures available in the Protein Data Bank and models for *Plasmodium* orthologs of proteins in other species for which structural information is available (http://bioinfo.icgeb.res.in/codes/model.html). This view also lists and provides a graphic representation of predicted protein secondary structure, predicted Pfam and ProSite domains, putative signal peptide transmembrane domains, and other targeting predictions (Bender et al., 2003; Foth et al., 2003; Hiller et al., 2004; Marti et al., 2004), such as those mediating targeting to the apicoplast, a promising drug target (Roos et al., 2002; Ralph et al., 2004). In addition, the Protein view shows BLASTP results comparing the protein against the motif databases ProDom and CDD (Servant et al., 2002). Protein expression data are provided as a map illustrating the location of peptide fragments identified through proteomics studies of different developmental stages (Florens et al., 2002) or subcellular fractions (Florens et al., 2004).

Expression views display information from both RNA microarrays (on a variety of platforms) and proteomics data. Three distinct probe sets—clone arrays (Mamoun et al., 2001), spotted oligonucleotide arrays (Bozdech et al., 2003), and photolithographic (Affymetrix) oligonucleotide arrays (Le Roch et al., 2003)—have been mapped to the *P. falciparum* genome. Several array studies have been loaded into PlasmoDB and are available for analysis, including both glass slide and photolithographic arrays interrogating whole-genome expression throughout the intraerythrocytic cycle and entering into gametocytogenesis. Many additional experiments are currently under way in various laboratories, and numerous microarray data sets are expected over the coming years. The expression views also display a graphic representation of expression at various life cycle stages, as well as oligonucleotides that map to a given gene, which are displayed in the microarray gene page view.

Sequence views provide a graphic representation of predicted genes, the percent AT plot, and predictions of low complexity sequence. For each gene, the sequence view also presents exon predictions (based on a variety of algorithms, along with curated annotation), information about single nucleotide polymorphisms (limited single nucleotide polymorphism data are available at present, but this class of data is likely to become increasingly prominent in the coming years), and predicted mRNA and protein sequences. As with all results found in other gene views, more detailed information, along with analysis of the entire genome, can be gleaned by querying the data.

ANALYSIS TOOLS

PlasmoDB provides the user with a variety of analysis tools for examining and extracting information from the genome and predicted proteome, using BLAST, electronic PCRs (Schuler, 1997), defined motif searches, and tools for the analysis of microarray and proteomics data.

BLAST tools (Altschul et al., 1997) available at PlasmoDB allow users to search against a variety of nucleotide and amino acid databases of *P. falciparum* and additional *Plasmodium* species. The results of these searches are stored in the History, from which they may be downloaded or combined with other queries for further analysis (see below). An oligonucleotide search tool uses the WU-BLAST algorithm, with parameters optimized for query sequences as short as 15 to 20 bp. An additional tool provides a graphic view of sequence similarity as a function of taxonomic divergence, i.e., the similarity between a given gene and genes in other organisms of varying taxonomic relatedness. In the near future, PlasmoDB is also expected to incorporate additional phylogenetic profile analyses (Pellegrini et al., 1999; Date and Marcotte, 2003).

The electronic PCR tool permits specified sequence-tagged site markers to be identified within finished and draft genome sequences; other tools allow the user to search for oligonucleotide locations based on the name of the sequence probe. Given a user-defined span (or spans) of genomic DNA, sequence(s) may be downloaded using the multiple sequence retrieval (SRT) tool or a list of features may be compiled for that region. PlasmoDB also includes tools to search for motifs, including organellar targeting signals or motifs defined by the user. For example, the investigator may wish to identify any predicted protein with multiple instances where two cysteines are 11 amino acids apart.

Tools designed to analyze functional genomics data include XCluster for microarray analysis and Lutefisk and EMOWSE for proteomics analysis. XCluster (http://genetics.stanford.edu/~sherlock/cluster.html) generates clusters from gene expression data based on either self-organizing map or hierarchical clustering. Lutefisk (Johnson and Taylor, 2002) is a proteomics tool designed to analyze mass spectrum data, with considerable flexibility provided in the form of user-specified parameters for mass tolerance, spectral processing, etc. The results of this analysis can then be used to search a selected protein database for peptide sequence matches. The EMOWSE tool (http://www.hgmp.mrc.ac.uk/Software/EMBOSS/) is used to identify proteins based on their peptide mass fingerprint. By pasting in the masses of the tryptic peptides from the mass spectrum and indicating the enzyme used for protein digestion, this analysis identifies proteins matching the spectrum in order of decreasing probability.

DATA MINING AND QUERIES

Gene Pages provide an encyclopedic view of the *Plasmodium* genome, where it is possible to look up all information about a specified gene. But this is not the only method that PlasmoDB employs to display information about *Plasmodium* genes. Indeed, the rationale behind storing data in a relational database (rather than in an encyclopedia) is the potential to formulate queries that interrogate the entire corpus of available information. Numerous queries have been developed by the PlasmoDB team, based largely on inquiries from malaria researchers seeking to mine genomic-scale data sets for lists of genes that match specified parameters. For example, some users might want to search for all genes that are annotated as transcription factors. Although this information is provided under the Annotation view for each individual gene, looking through each individual Gene Page for this information would be quite a tedious task!

Data Mining Queries are accessible either as a series of pull-down menus on the PlasmoDB home page (Fig. 1), or via the Queries link in the blue tool bar at the top of each page. On the Query page, options are organized under a variety of categories based on the datum type and the kind of question to be asked. For example, in seeking putative transcription factors, users may query PlasmoDB to search the annotation text of *Plasmodium* genes, including GO terms and any uncurated comments provided by PlasmoDB users (if so desired). Users may also search for all genes that meet a particular set of (user-defined) criteria for chromosomal location, gene structure, gene features (e.g., signal peptides or transmembrane domains), the availability of crystal structures or structural models, presence in specified metabolic pathways, etc. Tools are also available for searching the results of microarray and proteomics data and for identifying genes with a specific species distribution of putative orthologs.

In addition to conducting simple queries designed to identify, for example, all putative transcription factors, or all genes with five or more exons, or all genes that are transcribed in gametocytes, more complex questions can be formulated by combining queries, as illustrated in Fig. 1 and 2. In this example, a user interested in identifying potential vaccine targets (Tongren et al., 2004) might wish to begin by searching for proteins located on the surface of the cell, as specified by the presence of a secretory signal sequence, a transmembrane domain, or both. A further criterion for inclusion might include abundant transcripts in schizont-stage parasites, on the assumption that these may be the most likely to be expressed as antigen in free merozoites. One might also be interested in genes that are conserved in other *Plasmodium* species but absent from the human and mouse genome. Each of these queries is shown in Fig. 1: this user has selected "Signal Peptide" from the Sequence Feature query menu, bringing up a page that may be configured based on the database and chromosome to be searched and the algorithm

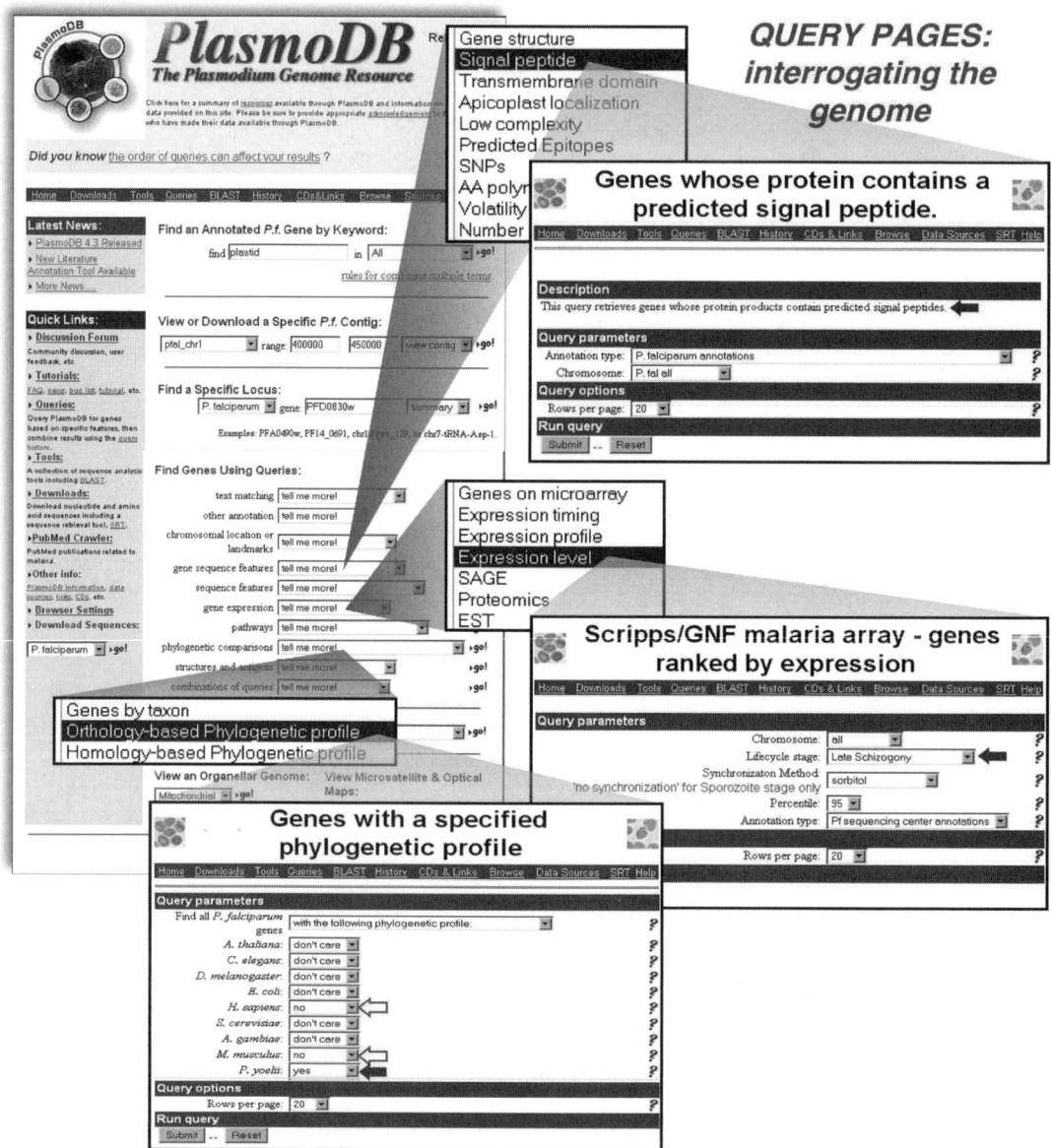

FIGURE 1 Gene Queries using PlasmoDB. A wide variety of dynamic queries may be formulated to interrogate the PlasmoDB database, using pull-down menus available on the home page (also accessible via the Queries button on the blue tool bar, via the help and tutorial pages, and from links at relevant locations throughout the site). Three queries are shown, focused on characteristics that might be of interest to researchers seeking to mine the database for candidate vaccine antigens: (i) a search for genes predicted to encode a secretory signal sequence, (ii) a search for genes that are abundantly transcribed in late schizonts, and (iii) a search for genes that are conserved in *P. yoelii* but not in the human or mouse genomes.

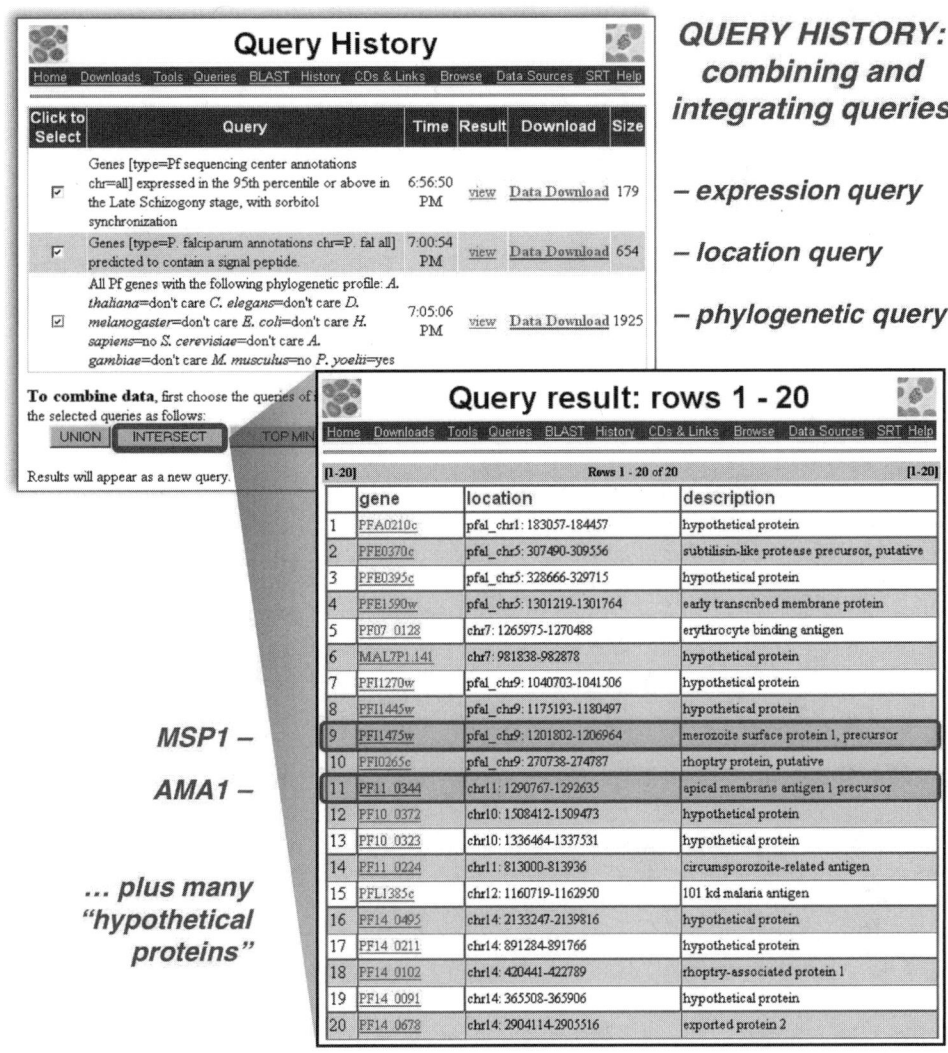

FIGURE 2 Integrating gene queries by using the History function. The History page, accessible via the blue tool bar at the top of each page, provides a list of all queries conducted during the current session. Using the boxes at left to specify individual queries of continuing interest permits these datasets to be combined (union), subtracted, or intersected. Intersecting the three queries defined in Fig. 1 yields 20 *P. falciparum* genes that satisfy the specified criteria with respect to phylogenetic distribution, expression, and predicted subcellular location. This list includes merozoite surface protein 1, apical membrane antigen 1, and numerous hypothetical proteins that may warrant further investigation as candidate vaccine antigens.

of interest. Executing this query retrieves a long list of candidate secretory proteins (not shown, but see below).

In addition to returning the results from individual queries, all queries in each PlasmoDB session are entered into the History page, accessible from the blue tool bar. As shown in Fig. 2, a simple signal sequence query identifies 654 hits, the expression query noted above yields 179 hits, and the phylogenetic query produces

1,925 hits, when formulated as above. Taking the intersection of these queries (union and subtraction operations are also permitted) yields a total of 20 hits, including the leading vaccine antigens merozoite surface protein 1 and apical membrane antigen 1 and a variety of other genes that might be of interest to explore as candidate antigens.

UNDERLYING ARCHITECTURE

While flat-file documents (such as encyclopedias, static Web pages, and spreadsheets) are easy to generate and share with others, they do not easily model complex relationships between entities such as genes, proteins, pathways, expression, and annotation. As a result, manipulation of data stored in flat files is usually limited to basic search functions, such as "sort" and "find." Relational databases are useful for storing large amounts of complex data because they can model the variety of relationships that exist within the data. Each table in the database is designed to store a specific type of data, and relationships between tables are defined. For example, unique names are associated with individual genes (a one-to-one relationship), but proteins may be found in multiple pathways (a one-to-many relationship). Genes and GO terms are defined by many-to-many relationships: one gene may be annotated by many GO terms, and one GO term may be used to annotate many genes. Storing data in a relational database enables the integration of complex biological queries, such as those highlighted above.

The PlasmoDB database is managed using a relational database management system, implementing the Genomics United Schema (GUS) to store a variety of data types, including sequence and functional genomics data (Davidson et al., 2001). GUS is composed of five name spaces (DoTS, shared resource [SRes], RAD, TESS, and Core), each of which is modeled to store a specific type of data (http://www.gusdb.org; Kissinger et al., 2002). The DoTS namespace is designed to store sequence information, such as genomic sequence and EST data, while the RAD namespace is designed to store microarray data. The SRes namespace stores controlled vocabularies, which are used for annotating the data. The TESS namespace stores data related to transcription factor binding sites, and the Core namespace stores information about data provenance. The design of the DoTS namespace derives from the biological concept of the central dogma. For example, separate tables store nucleotide sequences, RNA sequences, translated amino acid sequences, and the type and location of various gene features and annotations. Thus, data are grouped and presented in a way that approximates the way biologists think about this information. Since the schema is based on the central dogma, each step of the process from genomic DNA to protein is tracked, and the steps of this process are annotated.

Data organization in the GUS schema is highly defined, or strongly typed, in several ways. Concepts are modeled so that specific attributes are represented by individual columns in the relevant schema tables and filled with the appropriate data. Tables and attributes are named such that they specify the type of data to be stored. For example, the DoTS namespace table "TranslatedAASequence" stores translated protein sequence data (as its name implies) and also contains specific columns for the protein sequence itself and the sequence length. Keys are used to explicitly state the relationships between individual tables, such as those defining intron-exon structure, the resulting mRNA, and the predicted protein sequence. Such strong typing of the schema provides a measure of self-documentation and allows researchers who use this system to share data with some degree of confidence that they know what they are getting.

WHAT'S NEW IN PlasmoDB

Responding to the expanding scope and range of genomic-scale data sets available for malaria parasites, PlasmoDB has grown to incorporate not only new data, but also a wide range of new data types. In addition, a variety of tools, queries, and other resources and features have been implemented to facilitate effective mining of this information.

New and updated sequence data are now available for *P. vivax* and *P. yoelii*, in addition to the sequence for *P. falciparum*. The Find a Specific Locus query on the home page provides access to specific genes for these species. Other queries can also be configured to interrogate various *Plasmodium* species whenever the relevant data are available. All Gene Pages for all *Plasmodium* species are structured as described above (see Color Plate 2), providing a variety of protein and sequence features. Newly added analyses include a display of protein secondary structure predictions, based on PsiPred (a neural network-based algorithm using position-specific scoring matrices) (McGuffin et al., 2000). Secondary structure predictions are represented on both the Summary and Protein views of the Gene Page.

Functional genomics data available at PlasmoDB include genome-wide expression profiles for *P. falciparum* at various time points throughout the intraerythrocytic cycle, as well as (more limited) data for gametocytes and sporozoites (LeRoch et al., 2003; Bozdech et al., 2003). Data for different strains (HB3 versus 3D7), synchronization protocols (sorbitol versus temperature synchronization), and platforms (spotted oligonucleotide [glass slide] versus photolithographic [Affymetrix] arrays) have been transformed to facilitate direct comparison between these results (Color Plate 2). While these datasets differ, chiefly in the high degree of time resolution available in the Bozdech dataset, versus the large number of specific probes available in the LeRoch dataset, the high degree of concordance between data from these two groups lends high confidence to the results obtained. Protein expression data are also available in PlasmoDB, covering many developmental stages (Florens et al., 2002; Florens et al., 2004; Lasonder et al., 2002).

Several queries enable the user to identify genes that exhibit a desired expression profile, based on the profile of individual genes or user-specified profiles. For example, one researcher might wish to search for those genes whose expression in the glass slide data set is closest to that of PFD0830w (DHFR-TS) (Color Plate 2) or for genes whose expression matches a more complicated profile as drawn by the user on the Expression Profile query page. Alternatively, another researcher might search for genes whose expression is maximal in late schizonts in the LeRoch data set, or during the 40- to 48-h interval in the Bozdech dataset. A third researcher might seek genes for which experimental evidence supports expression of both RNA and protein in gametocytes.

A phylogenetic query allows PlasmoDB users to retrieve genes for which orthologs are conserved in (or absent from) species of specific interest. For example, an evolutionary biologist might wish to identify genes conserved in eukaryotes but absent from prokaryotes. A drug developer might seek genes found in both *P. falciparum* and *P. yoelii* but absent from the human (or murine) host (Fig. 1). The volatility or mutability of a gene can also be queried, identifying cases where the protein sequence may have been more or less subject to positive selection (e.g., essential housekeeping genes) or negative selection (e.g., immunodominant genes) (Plotkin et al., 2004).

The results obtained from most types of queries conducted during a single session are automatically stored on the History page, accessible via the blue tool bar. Recent enhancements allow users to add individual genes to the History page by clicking "Add this gene to your History" on the Gene Page view (Color Plate 2) and to delete query results no longer of interest. Applying Boolean set operators (union, intersection, and subtraction) to these results enables the user to build more complex queries, as illustrated in Fig. 2. Results of interest (protein, mRNA, DNA sequence, chromosomal location, predicted transmembrane domains, GO annotations, etc.) may then be extracted from the database using the ReportMaker tool. For example, the user might wish to download 1 kb of genomic sequence upstream of each gene in a defined set. Sequences can also be retrieved using the multiple sequence retrieval tool (SRT).

Recently added user support features include an expanded tutorial section, user comments, additional annotations and links, and a PlasmoDB Support System. Annotations provided by curators at the genome sequencing centers (who are responsible for the official sequence records) are loaded into the database before each new release of PlasmoDB. In addition, the User Comments feature allows users to enter further comments at any time during a release cycle; to ensure proper attribution, registration is required before a user comment is entered. These comments are visible to all users of PlasmoDB via the Summary and Annotation gene page views and may be queried using text search tools. User comments are also passed along to curators at the genome sequencing centers for review and possible inclusion in the official genome annotation.

Additional information may be obtained from links to external malaria resources. Gene Pages contain links to the GeneDB database maintained by the Sanger Institute's Pathogen Sequencing Unit, including data for *P. falciparum*, *P. berghei*, *P. chabaudi*, *P. knowlesi*, and many other organisms. A prototype implementation of natural language processing algorithms from the Mining the Bibliome project (R. MacDonald and F. Pereira, unpublished data) provides links to published literature, ranked according to potential relevance. As this is a prototype tool, users are invited to rank the actual utility of these articles, so this feedback may be employed to train future versions of the automatic article annotation software. The MR4 Malaria Resource and Reagent Repository (http://www.malaria.mr4.org) provides a resource of reagents for malaria including parasites, mosquito vectors, antibodies, antigens, gene libraries, molecular probes, and constructs. MR4 reagents corresponding to a specific gene(s) in PlasmoDB are listed under the Annotation View of individual Gene Pages and may be searched with the MR4 Reagent query.

The Support System, accessible at the bottom of every PlasmoDB Web page, provides a forum for users to report bugs, post messages, ask questions, and view comments posted by others. Questions may be posted publicly (anonymously, if desired), enabling others who enter the forum to search for commonly asked questions and learn more about PlasmoDB. The system also enables users to communicate with the PlasmoDB development team privately with questions, comments, and bug reports (describing the error encountered while using PlasmoDB and the URL of the broken page). Each inquiry is entered into a tracking system and assigned to the appropriate developer at PlasmoDB so that answers may be provided directly to the individual user or publicly in cases of general interest. All users are encouraged to take advantage of this feature to ask questions, report bugs, and make suggestions for improvements to the PlasmoDB database.

FUTURE PLANS

As more *Plasmodium* species are sequenced and additional functional genomics data are generated in the malaria field, this information will be incorporated into PlasmoDB. New genomic and EST data will be added for several *Plasmodium* species, including *P. vivax*, and this information will be available for queries similar to those now focused on *P. falciparum*. The database will also be configured to permit organism-specific views, so that the data may be approached from a *P. vivax*- or murine malaria-centric viewpoint. With support from the Bioinformatics Resource Center initiative of the U.S. NIH (http://www.niaid.nih.gov/dmid/genomes/brc/default.htm), PlasmoDB will also be incorporated into an integrated Apicomplexan parasite database, including information on the related parasites *Toxoplasma* and *Cryptosporidium*.

New functional genomics datasets anticipated in the near future include (i) numerous expression profiling studies, including analysis of additional life cycle stages, mutant parasite lines, drug-treated samples, and studies of rodent malaria species; (ii) additional proteomics studies, focusing on various life cycle stages in several *Plasmodium* species; (iii) protein-protein interaction datasets, including the results from yeast two-hybrid studies, immunoprecipitation experiments, and computational predictions; (iv) additional protein structural data and reagents available from structural genomics projects; (v) population diversity data sets, including whole-genome microsatellite studies; (vi) genome-scale analyses of subcellular localization (both experimentally determined and computationally predicted); and (vii) genome-scale analysis of predicted and experimentally determined indices of immunogenicity.

This integrated architecture will support queries based on cross-genome comparisons. For example, when asking for all genes expressed in a particular developmental stage, a researcher will be able to integrate across multiple *Plasmodium* species. The ability to correlate the results from multiple experimental approaches will also be supported, providing (for example) a more detailed comparison of whole-genome microarray data sets performed on different array platforms. PlasmoDB will also support comparisons of microarray and protein expression studies (LeRoch et al., 2004).

PlasmoDB will continue to provide analysis methods aimed at predicting protein function, including the identification of orthologs and paralogs of proteins that have been annotated in other species (Li et al., 2003), phylogenetic profiling (Date and Marcotte, 2003), and ontology-based pattern identification (Zhou et al., 2005). In addition, these data will be incorporated into a framework for studying protein and metabolic pathway networks, enhancing our understanding of parasite biology and the identification of diagnostics, drugs, and vaccines.

CONCLUSION

PlasmoDB is designed as a resource for sequence and functional genomics data related to *Plasmodium* species. The underlying architecture exploits a relational database management system that implements the Genomics Unified Schema (GUS), enabling complex queries intended to elucidate the biology of malaria. Information on the genomes of various *Plasmodium* parasites may be gleaned by browsing the genome in Sequence View mode, and information on specific genes may be obtained from the various Gene Pages. Powerful analysis tools and user-defined queries allow researchers to formulate and obtain answers to their own questions.

The PlasmoDB database receives many thousands of hits daily, >6 million in 2004, from >40,000 unique users, in >100 countries worldwide. But this resource will only remain valuable to the extent that it keeps pace with the questions that researchers wish to ask. All of those using PlasmoDB are therefore encouraged to communicate their comments, questions, and concerns using the Support System noted above.

As this resource is entirely dependent on the underlying data, we—as database developers and members of the malaria research community—wish to express our gratitude to the many researchers in genome sequencing centers, academic laboratories, government institutes, pharmaceutical firms, and clinical centers who have made these data available. We hope that all users of the database will take advantage of the User Comments option to enter new information. Scientists engaged in the production of genomic-scale datasets are encouraged to contact the PlasmoDB team as early as possible during the design of such experiments, to ensure that the results of their research can be accommodated, displayed, and queried as soon as they are ready for release.

ACKNOWLEDGMENTS

The utility of PlasmoDB depends on the many researchers worldwide who generate the relevant data and consent to make these data available to the community, often well in advance of formal publication. Particular thanks go to those engaged in the production of genomic-scale datasets, including researchers at the genome centers responsible for generating, assembling, annotating, and refining the primary sequence data. Researchers developing new datasets are encouraged to contact PlasmoDB—as early as possible—via various mechanisms available on the website.

We are also grateful for financial support from the Burroughs Wellcome Fund and the National Institutes of Health.

REFERENCES

Altschul, S. F., T. J. Madden, A. A. Schaffer, J. Zhang, Z. Zhang, W. Miller, and D. J. Lipman. 1997. Gapped BLAST and PSI-BLAST: a new generation of protein database search programs. *Nucleic Acids Res.* 25:3389–3402.

Bahl, A., B. P. Brunk, R. L. Coppel, J. Crabtree, S. J. Diskin, M. J. Fraunholz, G. Grant, D. Gupta, R. L. Huestis, J. C. Kissinger, P. Labo, L. Li, S. K. McWeeney, A. J. Milgram, D. S. Roos, J. Schug, and C. J. Stoeckert, Jr. 2002. PlasmoDB: the *Plasmodium* genome resource. An integrated database providing tools for accessing and analyzing mapping, expression and sequence data (both finished and unfinished). *Nucleic Acids Res.* **30:**87–90.

Bahl, A., B. P. Brunk, J. Crabtree, M. J. Fraunholz, B. Gajria, H. Ginsburg, G. R. Grant, D. Gupta, J. C. Kissinger, P. Labo, L. Li, M. D. Mailman, A. J. Milgram, D. S. Pearson, D. S. Roos, J. Schug, C. J. Stoeckert, Jr., and P. Whetzel. 2003. PlasmoDB: the *Plasmodium* genome resource. Tools for integrating experimental and computational data. *Nucleic Acids Res.* **31:**212–215.

Bender, A, G. G. van Dooren, S. A. Ralph, G. I. McFadden, and G. Schneider. 2003. Properties and prediction of mitochondrial transit peptides from *Plasmodium falciparum*. *Mol. Biochem. Parasitol.* **132:**59–66.

Berry, A. E., M. F. Gardner, G. J. Caspers, D. S. Roos, and M. Berriman. 2004. Curation of the *Plasmodium falciparum* genome. *Trends Parasitol.* **20:**548–552.

Bozdech, Z., M. Llinas, B. L. Pulliam, E. D. Wong, J. Zhu, and J. L. DeRisi. 2003. The transcriptome of the intraerythrocytic developmental cycle of *Plasmodium falciparum*. *PLoS Biol.* **1:**85–100.

Carlton, J. 2003. *Plasmodium vivax* genome sequencing project. *Trends Parasitol.* **19:**227–231.

Carlton, J., J. Silva, and N. Hall. 2005. The genome of model malaria parasites, and comparative genomics. *Curr. Issues Mol. Biol.* **7:**23–37.

Coppel, R. L., D. S. Roos, and Z. Bozdech. 2004. The genomics of malaria infection. *Trends Parasitol.* **20:**553–557.

Date, S. V., and E. M. Marcotte. 2003. Discovery of uncharacterized cellular systems by genome-wide analysis of functional linkages. *Nat. Biotechnol.* **21:**1055–1062.

Davidson, S., J. Crabtree, B. P. Brunk, J. Schug, V. Tannen, G. C. Overton, and C. J. Stoeckert. 2001. K2/Klesli and GUS: experiments in integrated access to genomic data sources. *IBM Syst. J.* **40:**512–531.

Doolan, D. L., S. Southwood, D. A. Freilich, J. Sidney, N. L. Graber, L. Shatney, L. Bebris, L. Florens, C. Dobano, A. A. Witney, E. Appella, S. L. Hoffman, J. R. Yates III, D. J. Carucci, and A. Sette. 2003. Identification of *Plasmodium falciparum* antigens by antigenic analysis of genomic and proteomic data. *Proc. Natl. Acad. Sci. USA* **100:**9952–9957.

Florens, L., M. P. Washburn, J. D. Raine, R. M. Anthony, M. Grainger, J. D. Haynes, J. K. Moch, N. Muster, J. B. Sacci, D. L. Tabb, A. A. Witney, D. Wolters, Y. Wu, M. J. Gardner, A. A. Holder, R. E. Sinden, J. R. Yates, and D. J. Carucci. 2002. A proteomic view of the *Plasmodium falciparum* life cycle. *Nature* **419:**520–526.

Florens, L., X. Liu, Y. Wang, S. Yang, O. Schwartz, M. Peglar, D. J. Carucci, J. R. Yates, and Y. Wu. 2004. Proteomics approach reveals novel proteins on the surface of malaria-infected erythrocytes. *Mol. Biochem. Parasitol.* **135:**1–11.

Foth, B. J., S. A. Ralph, C. J. Tonkin, N. Struck, M. J. Fraunholz, D. S. Roos, A. F. Cowman, and G. I. McFadden. 2003. Dissecting apicoplast targeting in the malaria parasite *Plasmodium falciparum*. *Science* **299:**705–708.

Gardner, M. J., N. Hall, E. Fung, O. White, M. Berriman, R. Hyman, J. Carlton, A. Pain, K. E. Nelson, S. Bowman, I. T. Paulsen, K. James, J. A. Eisen, K. Rutherford, S. Salzberg, A. Craig, V. Nene, S. Shallom, B. Suh, J. Peterson, S. Angiuoli, M. Pertea, J. Allen, J. Selengut, D. Haft, A. Vaidya, A. Fairlamb, D. S. Roos, G. I. McFadden, L. M. Cummings, C. Mungall, A. A. Kanapin, J. C. Venter, D. J. Carucci, S. L. Hoffman, C. Newbold, R. W. Davis, C. M. Fraser, and B. Barrell. 2002. The genome sequence of the human malaria parasite *Plasmodium falciparum*. *Nature* **419:**498–511.

Hall, N., M. Karras, J. D. Raine, J. M. Carlton, T. W. Kooij, M. Berriman, L. Florens, C. S. Janssen, A. Pain, G. K. Christophides, K. James, K. Rutherford, B. Harris, D. Harris, C. Churcher, M. A. Quail, D. Ormond, J. Doggett, H. E. Trueman, J. Mendoza, S. L. Bidwell, M. A. Rajandream, D. J. Carucci, J. R. Yates, F. C. Kafatos, C. J. Janse, B. Barrell, C. M. Turner, A. P. Waters, and R. E. Sinden. 2005. A comprehensive survey of the *Plasmodium* life cycle by genomic, transcriptomic, and proteomic analyses. *Science* **307:**82–86.

Hiller, N. L., S. Bhattacharjee, C. van Ooij, K. Liolios, T. Harrison, C. L. Estrano, and K. Haldar. 2004. A host-targeting signal in virulence proteins reveals a 'secretome' in malarial infection. *Science* **306:**1934–1937.

Johnson, R. S., and J. A. Taylor. 2002. Searching sequence databases via de novo peptide sequencing by tandem mass spectrometry. *Mol. Biotechnol.* **22:**301–315.

Kaiser, K., K. Matuschewski, N. Camargo, J. Ross, and S. H. Kappe. 2004. Transcriptome profiling identifies *Plasmodium* genes encoding pre-erythrocytic stage-specific proteins. *Mol. Microbiol.* **51:**1221–1232.

Kissinger, J. C., B. P. Brunk, J. Crabtree, M. J. Fraunholz, B. Gajria, A. J. Milgram, D. S. Pearson, J. Schug, A. Bahl, S. J. Diskin, H. Ginsburg, G. R. Grant, D. Gupta, P. Labo, L. Li, M. D. Mailman, S. K. McWeeney, P. Whetzel, C. J. Stoeckert, Jr., and D. S. Roos. 2002. The *Plasmodium* genome database: designing and mining a eukaryotic genomics resource. *Nature* **419**:490–492.

Kissinger, J. C., and D. S. Roos. 2004. Getting the most out of bioinformatics resources. *In* A. P. Waters and C. J. Janse (ed.), *Malaria Parasites*. Horizon, Norfolk, United Kingdom.

Lasonder, E., Y. Ishihama, J. S. Andersen, A. M. Vermunt, A. Pain, R. W. Sauerwein, W. M. Eling, N, Hall, A. P. Waters, H. G. Stunnenberg, and M. Man. 2002. Analysis of the *Plasmodium falciparum* proteome by high-accuracy mass spectrometry. *Nature* **419**:537–542.

Le Roch, K. G., Y. Zhou, P. L. Blair, M. Grainger, J. K. Moch, J. D. Haynes, P. De la Vega, A. A. Holder, S. Batalov, D. J. Carucci, and E. A. Winzeler. 2003. Discovery of gene function by expression profiling of the malaria parasite life cycle. *Science* **301**:1503–1508.

Le Roch, K.G., J. R. Johnson, L. Florens, Y. Zhou, A. Santrosyan, M. Grainger, S. F. Yan, K. C. Williamson, A. A. Holder, D. J. Carucci, J. R. Yates, and E. A. Winzeler. 2004. Global analysis of transcript and protein levels across the Plasmodium falciparum life cycle. *Genome Res.* **11**:2308–2318.

Li, L., C. J. Stoeckert, Jr., and D. S. Roos. 2003. OrthoMCL: identification of ortholog groups for eukaryotic genomes. *Genome Res.* **13**:2178–2190.

Li, L., J. Crabtree, S. Fischer, D. Pinney, C. J. Stoeckert, Jr., L. D. Sibley, and D. S. Roos. 2004. ApiESTDB: analyzing clustered EST data of the apicomplexan parasites. *Nucleic Acids Res.* **32**:326–328.

Mamoun, C. B., I. Y. Gluzman, C. Hott, S. K. MacMillan, A. S. Amarkone, D. L. Anderson, J. M. Carlton, J. B. Dame, D. Chakrabarti, R. K. Martin, B. H. Brownstein, and D. E. Goldberg. 2001. Co-ordinated programme of gene expression during asexual intraerythrocytic development of the human malaria parasite *Plasmodium falciparum* revealed by microarray analysis. *Mol. Microbiol.* **39**:26–36.

Marti, M., R. T. Good, M. Rug, E. Knuepfer, and A. F. Cowman. 2004. A unique export signal targets virulence and remodeling proteins from the malaria parasite to the host erythrocyte. *Science* **306**:1930–1933.

McGuffin, L. J., K. Bryson, and D.T. Jones. 2000. The PsiPred protein structure prediction server. *Bioinformatics* **16**:404–405.

Milgram, A. J., J. C. Kissinger, B. Gajria, D. S. Pearson, A. Bahl, P. Labo, and D. S. Roos. 2003. *Plasmodium falciparum* GenePlot: Internet-independent access to the malaria parasite genome. *Nature* **422**:CD–ROM. Distributed with the 6 March issue.

Pellegrini, M., E. M. Marcotte, M. J. Thompson, D. Eisenberg, and T. O. Yeates. 1999. Assigning protein functions by comparative genome analysis: protein phylogenetic profiles. *Proc. Natl. Acad. Sci. USA* **96**:4285–4288.

Plasmodium Genome Database Collaborative. 2001. PlasmoDB: an integrative database of the *Plasmodium falciparum* genome. Tools for accessing and analyzing finished and unfinished sequence data. *Nucleic Acids Res.* **29**:66–69.

Plotkin, J. B., J. Dushoff, and H. B. Fraser. 2004. Detecting selection using a single genome sequence of *M. tuberculosis* and *P. falciparum*. *Nature* **428**:942–945.

Ralph, S. A., G. G. van Dooren, R. F. Waller, M. J. Crawford, M. J. Fraunholz, B. J. Foth, C. J. Tonkin, D. S. Roos, and G. I. McFadden. 2004. Metabolic pathway maps and functions of the *Plasmodium falciparum* apicoplast. *Nat. Rev. Microbiol.* **2**:203–216.

Roos, D. S., M. J. Crawford, R. G. K. Donald, M. Fraunholz, O. S. Harb, C.Y. He, J. C. Kissinger, M. K. Shaw, and B. Striepen. 2002. Mining the *Plasmodium* genome database to define organellar function: what does the apicoplast do? *Phil. Trans. R. Soc. Lond. B Biol. Sci.* **357**:35–46.

Sam-Yellowe, T.Y., L. Florens T. Wang, J. D. Raine, D. J. Carucci, R. Sinden, and J. R. Yates. 2004. Proteome analysis of rhoptry-enriched fractions isolated from *Plasmodium* merozoites. *J. Proteome Res.* **3**:995–1001.

Schuler, G. D. 1997. Sequence mapping by electronic PCR. *Genome Res.* **7**:541–550.

Servant, F., C. Bru, S. Carrere, E. Courcelle, J. Gouzy, D. Peyruc, and D. Kahn. 2002. ProDom: automated clustering of homologous domains. *Brief. Bioinformatics* **3**:246–251.

Tongren, J. E., F. Zavala, D. S. Roos, and E. M. Riley. 2004. Malaria vaccines: if at first you don't succeed.... *Trends Parasitol.* **20**:604–610.

Zhou, Y., J. A. Young, A. Santrosyan, K. Chen, F. S. Yan, and E. A. Winzeler. 2005. *In silico* gene function prediction using ontology-based pattern identification. *Bioinformatics* **21**:1237–1245.

MAKING A HOME FOR *PLASMODIUM* POST-GENOMICS: ULTRASTRUCTURAL ORGANIZATION OF THE BLOOD STAGES

Lawrence H. Bannister, Gabriele Margos, and John M. Hopkins

3

Clearly, research into the molecular biology of malaria parasites has flourished during the last two decades. Made possible by the completion of the *Plasmodium falciparum* genomic sequence, the development of powerful new techniques for the analysis of gene expression is generating new molecular data in considerable quantities. This revolution has also created a dilemma: how best to use this stream of information to gain deeper insights into the parasite's functional makeup and how to find new ways of combatting malaria. Two fruitful approaches to this problem are currently the localization of newly discovered molecules by immunolabeling or fluorescent tagging and the analysis of mutant phenotypes. Both require a good understanding of the parasite's cellular organization in its different stages, which in turn depends on detailed morphological analysis. Since the discovery of the malaria parasite, microscopy has played a major part in defining the biology of its life cycle (Garnham, 1966). The application of electron microscopy (EM) to malaria parasites from the 1950s onwards settled many often heated controversies about the nature of the malaria parasite; by about 1980, electron microscopists had described the main ultrastructural features of this genus (see, e.g., Langreth et al., 1978; Aikawa and Seed, 1980; Sinden, 1982a; Bannister et al., 2001). Since then, the growing interest in matters molecular have drawn attention away from EM as a tool for exploring malaria biology, and it has been left to a few enthusiasts to follow the ultrastructural trail, most fruitfully in collaboration with multidisciplinary research teams. Paradoxically, the growth of molecular biology has created new roles for EM, especially the localization of novel parasite molecules by immunocytochemistry and the analysis of mutant phenotypes.

With the development of confocal scanning and related fluorescence techniques, light microscopy has taken a new lease on life, particularly in the exploration of trafficking and other dynamic cellular processes. However, because of

Lawrence H. Bannister, Wolfson Centre, Guy's, King's and St. Thomas's School of Biomedical and Health Sciences, Hodgkin Building, Guy's Hospital Campus, London SE1 1UL, United Kingdom. *Gabriele Margos,* Department of Immunobiology, Guy's, King's and St. Thomas's School of Medicine, New Guy's House, Guy's Hospital Campus, London SE1 9RT, United Kingdom. *John M. Hopkins,* Wolfson Centre, Guy's, King's and St. Thomas's School of Biomedical and Health Sciences, Hodgkin Building, Guy's Hospital Campus, London SE1 1UL, United Kingdom.

Molecular Approaches to Malaria, Edited by Irwin W. Sherman
© 2005 ASM Press, Washington, D.C.

the resolution limitations of photon-dependent methods, EM remains the technique of choice for the localization of molecules by immuno-EM (IEM) and for studies of detailed cellular organization. New techniques of specimen preparation, e.g., by cryofixation, new breeds of electron microscopes, and advanced computer-assisted image analysis are now becoming available for malaria research, with much potential for structural and molecular exploration. It is perhaps timely to assess how much we know about parasite structure and where we need to go from here.

In this chapter, we briefly outline the cellular architecture of the blood stages of *P. falciparum*, for which there is a considerable body of molecular data. Space limitations forbid a comprehensive survey of the other major species of *Plasmodium*, although it is clear from the literature (see, e.g., Aikawa and Seed, 1980) that many structural features are shared across this genus. Similarly, we do not attempt to review fluorescence imaging or IEM localization data, and we restrict ourselves to morphological analysis, chiefly by EM, to build a picture of the internal organization of the different blood stages. We hope that this approach will help to provide a structural background for the molecular chapters concerning *Plasmodium* blood stages in this book and that it may stimulate further activity in this important, although somewhat neglected, area of research.

THE MAIN PHASES OF ERYTHROCYTIC LIFE

The asexual forms, which dominate the relationship between the human host and parasite in terms of time spent and pathological impact, are traditionally classified by their detailed light microscopic features as seen in Giemsa-stained blood films. These can be related to some extent to the special activities of the different parasite stages—invasion of red blood cells (RBC) by merozoites, feeding and growth by rings and trophozoites, and multiplicative formation of new merozoites in schizonts. Much more information can be gained from EM, which can give many detailed clues about the functional states of cells and correlate more accurately with complex patterns of gene expression currently becoming available (see, e.g., Florens et al., 2002).

THE MEROZOITE

General Morphology

Merozoites have a remarkable simplicity of design, with a minimal size and organelle content, yet after invasion they are able to launch rapidly the next trophic intracellular phase within a few minutes (as seen most clearly in *Plasmodium knowlesi* (Dvorak et al., 1975). In *P. falciparum*, they are very small—about 1.3 μm long and 0.9 μm wide (Fig. 1)—although somewhat larger in *P. knowlesi*, among other species (Mitchell and Bannister, 1988). They are ellipsoidal with a short flat-ended apical prominence at one pole (Fig. 2 and 3; Color Plate 3A). An irregularly hemispherical nucleus is located in the basal half of the cell, while in the region of the apical prominence a group of secretory vesicles (apical organelles) is gathered, namely, rhoptries, micronemes, and dense granules.

The Nucleus

This rather irregular and variably shaped structure contains a homogeneous matrix of fine chromatin strands, with no evidence of a nucleolus. It is surrounded by a typical nuclear envelope punctuated by nuclear pores (Fig. 2 and 6; Color Plate 3A). Close to the nucleus, and probably continuous with the nuclear envelope, a complex mass of folded membranes is also usually visible. The lack of any clearly condensed chromatin (heterochromatin) in this or any other blood stage is mysterious and suggests the presence of different mechanisms for the control of gene expression than are found in classical eukaryotic cells.

Rhoptries

Two of these organelles are present, typically, club-shaped membrane-bounded vesicles about 400 nm long by 200 nm at their widest diameter (Fig. 2 to 5; see Fig. 43) (Bannister et al., 2001). Mature rhoptries contain many different

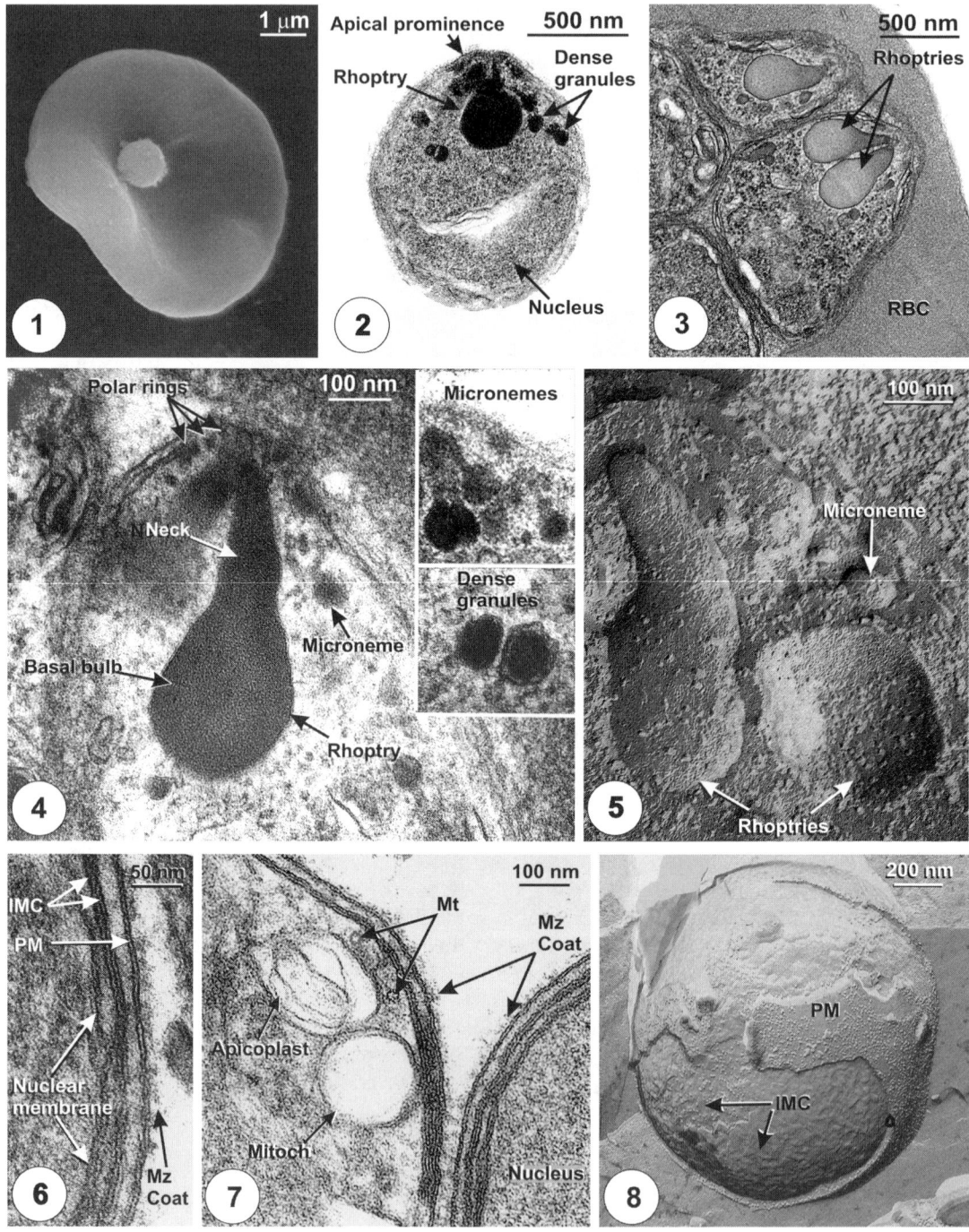

proteins (Etzion et al., 1991; Sam-Yellowe et al., 2004; Topolska et al., 2004) within a single limiting membrane, some located in the basal bulb (e.g., RAP-1 or Clag9) (Bushell et al., 1988; Ling et al., 2004) and others located in the narrower neck (or duct) (Roger et al., 1988), which has a distinctive development and contents (Bannister et al., 2000). The rhoptry membrane has many intramembrane particles (IMPs) when viewed by freeze fracture (Fig. 5) (Bannister et al., 2000), indicating assemblies of transmembrane proteins, especially those functioning as channels. The IMPs of its two fracture faces have different sizes and frequencies, indicating a complex molecular mixture, yet to be defined chemically and functionally.

Micronemes

These are much smaller and more numerous (up to 40 per merozoite) than the rhoptries and converge around the tips of rhoptry ducts near their junction with the apical membrane of the parasite (Fig. 2, 4, and 5; see Fig. 43). They vary in shape and size with species. In *P. falciparum*, they have bulbous bases about 60 nm wide and long narrow necks up to 100 nm long (Bannister et al., 2003). In some other species, e.g., *Plasmodium berghei*, they do not have bulbous ends and have banana-like shapes (Ladda, 1969). Micronemal secretory activity has yet to be analyzed in detail, although it is known from the IEM (of *P. knowlesi*) that some of their multiple proteins are discharged at the apical prominence prior to invasion, e.g., apical membrane antigen 1 (Thomas et al., 1990; see also Healer et al., 2002, and Bannister et al., 2003), which in *P. knowlesi* appears to be essential for tight apical attachment to RBCs prior to entry (Mitchell et al., 2004).

Dense Granules

These are dense, rounded vesicles, larger (about 90 nm) than micronemes and scattered through the apical part of the merozoite, although not usually in the apical prominence (Fig. 2 and 4; see Fig. 43). They are usually spheroidal in shape, but can also be elliptical or rectangular and often have wrinkled membranes. Their contents are finely granular, and there is usually a narrow, less-dense perimeter beneath their bounding membrane (Fig. 4). In sections, they can be (and have been) mistaken for micronemes, although the larger size of the dense granules is a helpful, although not infallible, distinguishing feature. The contents of these structures, which include invasion-related molecules, are exocytosed through the pellicle into the parasitophorous vacuole late in RBC invasion (Bannister et al., 1975; Torii et al., 1989; Bannister and Dluzewski, 1990; Culvenor et al., 1991), causing parasitophorous vacuole membrane (PVM) expansion and other modifications of the RBC.

Merozoite Surface

The merozoite surface is covered with a coat, about 15 nm thick, of projecting tufts of fine filaments (Fig. 6 and 7; Color Plate 3A), as in other species of *Plasmodium* (Bannister and Mitchell, 1986). Beneath this is the pellicle, consisting of three membranes in sequence, the plasma membrane and the two membranes of the inner membrane complex (IMC). The IMC

FIGURES 1 TO 8 Merozoite ultrastructure of *P. falciparum*. Figure 1 is a scanning EM of a merozoite attached to an RBC, demonstrating the relative sizes of the two cells. Figure 2 is a transmission EM of a longitudinal section through a free merozoite, showing a number of organelles. In Fig. 3, typical mature rhoptries are present within merozoites shortly to be released from a schizont. Figure 4 and insets within it show the apical region of a merozoite and the three types of apical organelles (magnifications in all images are the same). In Fig. 5, a merozoite has been freeze fractured to show numerous intramembranous particles in the two rhoptry membrane faces, representing intramembranous protein domains. In Fig. 6 and 7, the merozoite pellicle and adjacent structures are shown at a higher magnification; in Fig. 6, the section includes the nuclear envelope as well as the two membranes of the IMC, the merozoite plasma membrane. In Fig. 7, the close relation of the apicoplast to the subpellicular microtubules and mitochondrion is seen in transverse section. Figure 8 shows a freeze-fracture preparation of a merozoite, revealing the different membrane faces and intramembranous particle distributions of the pellicle including the plasma membrane and IMC. Mitoch, mitochondrion; Mt, microtubules; Mz, merozoite; PM, plasma membrane.

is really a flat cisterna underlying the whole surface except at the apical prominence, where its anterior edge is attached to the outer surface of the third polar ring (Fig. 4; Color Plate 3A). The plasma membrane is connected to the IMC by numerous fine (4-nm-thick) filaments. Freeze fracture demonstrates the trilaminar structure of the merozoite pellicle (Fig. 8) and, in *P. knowlesi*, has shown the plasma membrane and IMC to have distinctive populations of IMPs (McLaren et al., 1979; Aikawa et al., 1981), although with no evidence of the linear arrays associated with the cytoskeleton seen, for example, in the *Toxoplasma* pellicle (Morrissette et al., 1997).

The merozoite surface is chemically complex, with an increasing number of proteins assigned to it (see the review by Topolska et al., 2004), including merozoite surface protein 1, which has been localized by the IEM to the merozoite coat (Heidrich et al., 1986). There may also be differences of composition within the coat, since the apical region has a more positive surface charge than elsewhere (Akaki et al., 2002), although this could be due to secretion from the apical organelles on to the apical end. The merozoite coat is removed during RBC invasion as it enters the invasion pit (*P. knowlesi*) (Aikawa et al., 1978).

Subpellicular Microtubules
Also connected to the third polar ring are two or sometimes three subpellicular microtubules (Fig. 7; Color Plate 3A) which extend (in *P. falciparum*) parallel to each other along one side of the merozoite for about two-thirds its length, attached to the underside of the IMC by a series of short cross-bridges (Bannister and Mitchell, 1995) (collectively termed the f-mast, for falciparum merozoite assemblage of subpellicular microtubules). In other species such as *P. knowlesi*, the microtubules diverge posteriorly (Mitchell and Bannister, 1988). This small number of subpellicular microtubules contrasts with the much larger array found in the bigger merozoites of the avian parasite *Plasmodium fallax* (Aikawa, 1967), reflecting perhaps a greater need for mechanical stabilization.

The subpellicular microtubules are also important in merozoite development, where they are trackways for the translocation of micronemes to the apical prominence (see the discussion of schizont structure below). A dynein(s) and a kinesin(s) which could mediate this movement are present at the apical end of the mature merozoite, where the microtubule minus ends (and γ-tubulin) (Fowler et al., 2001) are located at the microtubule organizing center (Russell and Burns, 1984).

Mitochondrion
This is a single, elongated mitchondrion that in *P. falciparum* lies longitudinally just beneath the pellicle (Fig. 7; Color Plate 3A). It has a smooth outline, but the internal of its two membranes bears a few irregular tubular cristae. In some species, such as *P. knowlesi*, it is often folded into a spiral shape and positioned more deeply away from surface attachments (Mitchell and Bannister, 1988).

Apicoplast
This is similar in shape to the mitochondrion (Fig. 7; Color Plate 3A), though wider and shorter and with a rather irregular outline. The apicoplast often has a more complex interior (Hopkins et al., 1999). It is closely associated with the mitochondrion in all species of *Plasmodium* studied, and in *P. falciparum* it lies alongside and parallel to it (Color Plate 3A). In this species, it is also attached along its length to the overlying subpellicular microtubule band (Fig. 7), although this relationship is not present in *P. berghei* or *P. knowlesi* (Hopkins et al., 1999). The apicoplast can be distinguished from the mitochondrion by its more numerous bounding membranes. The precise number of these membranes has been a matter of some vigorous debate (Waller and McFadden, 2005). *Toxoplasma gondii* is reported to have four (Kohler et al., 1997), while *Plasmodium* appears to have three (Fig. 7; see Fig. 20) (Hopkins et al., 1999). Sections of *Plasmodium* (*P. falciparum*, *P. knowlesi*, and *P. berghei*) apicoplasts also show complex internal lamellae and vesicular structures, some clearly due to the interior folding of the apicoplast wall

(Hopkins et al., 1999), indicating a large surface area and perhaps the formation of export (or import) vesicles. We have suggested that the difficulty in determining the wall membrane number is due to the very wrinkled nature of apicoplast membranes, which generates complex images when the (relatively thick) sections are viewed by EM. However, the disparity between *Toxoplasma* and *Plasmodium* apicoplasts and phylogenetic ideas about them remain to be definitively agreed upon.

Other Merozoite Structures

A single cytostomal ring is located on the merozoite's side just beneath the plasma membrane in a small gap in the IMC (Color Plate 3A), but it is endocytically inactive in this stage. The cytoplasm contains numerous free ribosomes, often quite closely packed in readiness for the renewed burst of protein synthesis in the ring stage. Little or no rough endoplasmic reticulum (RER) is present in mature merozoites and must be generated afresh after invasion. In addition, various other small clear vesicles are present of unknown significance, although they are potential sites of novel antigen location. Some of them have dense eccentrically placed granules within them and probably represent acidocalciosomes that have been visualized in whole-merozoite preparations (Ruiz et al., 2004). A Golgi complex appears to be completely absent from the mature merozoite.

THE RING-STAGE PARASITE

General Morphology

The ring stage has been described in a number of species, including rodent and avian forms (see Aikawa, 1971, for a general review of the earlier literature; Slomianny et al., 1985) and *P. falciparum* (Langreth et al., 1978; Bannister et al., 2004). Ring-stage parasites are defined as small (in *P. falciparum*, they are 2 to 3.7 μm in diameter) discoidal forms, usually with a biconcave shape resembling in miniature the RBC they have invaded (Fig. 9 to 11; Color Plate 3B). The precise form of the disk varies, in some instances flat (Fig. 10 and 11), but more commonly cup shaped (Fig. 9) and sometimes almost curved into a hollow sphere. In all these forms, the thicker periphery of the disk contains most of the organelles, including the nucleus and the majority of the ribosomes, while the thinner center contains more abundant smooth membranes but fewer ribosomes. The shape of the disk and the predominantly peripheral position of its ribosomal (and therefore RNA) content are reflected in the basophilic ring-like appearance in Giemsa-stained preparations. Occasionally, young parasites take a completely different form, with long lobar extensions reaching into the RBC (the dendritic form). Whether this represents some pathological change in the parasite is not certain; it may be an indication of the ability to change shape in an actively motile manner.

Nucleus, Ribosomes, and RER

In Giemsa preparations (Color Plate 3B), the nucleus takes various forms, often an elongated body curved along the parasites' perimeter, but sometimes more rounded (see Fig. 50). Sometimes, there appear to be two nuclei, although the two lobes are seen by EM to be still connected with each other by a narrow band of nuclear material. Ultrastructurally, its interior is uniformly pale staining and lacks a nucleolus (Fig. 10 and 11). Membranous vesicles and lamellar whorls are often present within dilatations of the nuclear envelope (Bannister et al., 2004).

After invasion, free ribosomes and RER begin to proliferate, especially in the thick rim of the ring (Fig. 10, 11, and 14; Color Plate 3B), with evidence of accelerating cytosolic and membrane-associated protein synthesis. As in other parasite stages, the nuclear envelope is central to the trafficking apparatus, being continuous with the RER and itself bearing ribosomes.

Vacuoles and Vesicles

These are striking features of the cytoplasm and include a number of classes distinguishable by their size and content (Fig. 11 to 14; Color Plate 3B). The most obvious are food vacuoles, from 150 to 300 nm across, endocytosed through the single cytostome, close to which most of

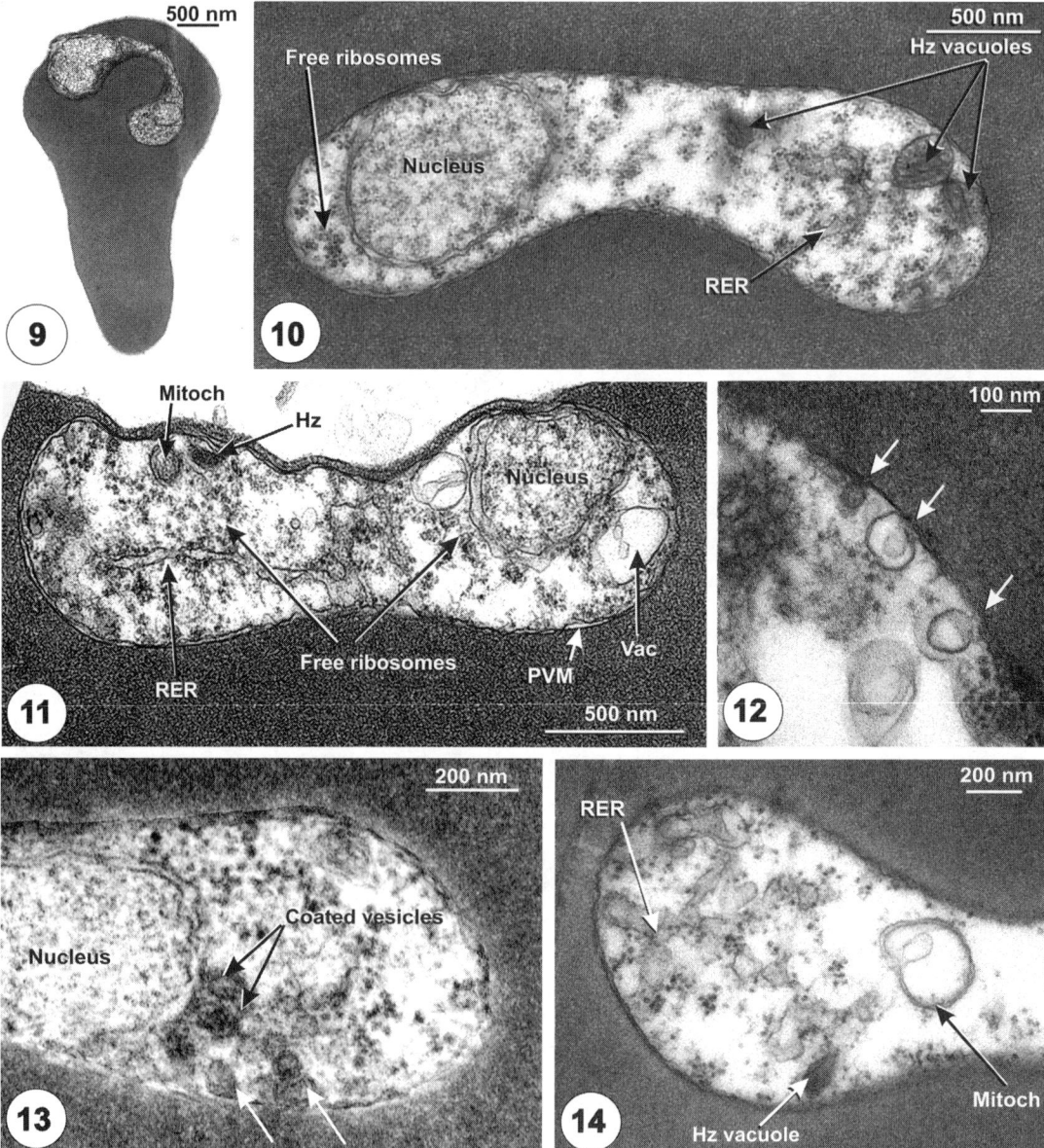

FIGURES 9 TO 14 Ring-stage ultrastructure. Figure 9 shows an early ring and RBC at low magnification, the parasite curved into a cup-like form in this example. In Fig. 10, a larger ring shows the development of ribosomes and RER and the accumulation of hemozoin-containing vacuoles. In Fig. 11, a ring stage situated adjacent to the RBC membrane contains more highly developed endoplasmic reticulum, a mitochondrion, and a large vacuole. Figure 12 details part of the ring surface at high magnification, showing three double-membraned vesicles (arrows). In Fig. 13, the zone of coated vesicle budding from the nuclear envelope is seen on the left and a series of double-membrane vesicles lies close to the parasite's surface (white arrows). In Fig. 14, detail of a maturing ring, with a complex network of RER, a mitochondrion in transverse section, and a dense hemozoin-containing vacuole are shown. Hz, hemozoin; Mitoch, mitochondrion; vac, large vacuole.

them are located (Fig. 10, 11, and 14; Color Plate 3). They contain material identical to RBC cytosol, ranging to crystals of hemozoin, representing stages in hemoglobin uptake and degradation. Early cytostomal vesicles (food vacuoles) are lined by two membranes, the inner membrane derived from the PVM and the outer membrane derived from the parasite plasma membrane. As processing of their contents proceeds, the inner membrane breaks down and small crystals of hemozoin appear within single membrane vacuoles. Hemozoin vacuoles remain separate from each other until the end of the ring stage, when all hemozoin is thereafter concentrated in a single pigment vacuole.

Other vacuoles include large irregular sacs 300 to 500 nm across (Fig. 11) often containing membranous or other cellular material but of unknown function. Double membrane vesicles (60 to 200 nm across) (Fig. 12) are present near the parasite plasma membrane, with which they are sometimes seen to fuse. They are especially interesting, since by their lack of dense contents they appear unrelated to hemoglobin uptake as suggested for *Plasmodium chabaudi* (Slomianny et al., 1985) and may instead be mechanisms for membrane export into the PVM (Bannister et al., 2004; Cooke et al., 2004). Such a device could provide membrane for the growth of the PVM and the further export of PVM-derived vesicles into the RBC (Cooke et al., 2004), as seen in the formation of clefts and knobs in later stages (see below). Small (80- to 150-nm) single-membrane clear vesicles (Fig. 13) are probably a heterogeneous population of vesicles engaged in various trafficking activities, including endocytosis, export, and movements within the cytoplasm.

Golgi Complex

There is molecular evidence for vesicular trafficking via a Golgi system in the ring stage (Adisa et al., 2002); by EM, a single, small, Golgi-like complex is visible, consisting of a cluster of coated vesicles budding from the nuclear envelope, close to a group of larger, irregular, uncoated vesicles (Fig. 13) or, in more mature rings, a tubular cisterna. In more advanced rings, the coated vesicle budding zone is located on a long extension of the outer nuclear membrane, so that the Golgi-like complex is more distant from the main nuclear mass (Color Plate 3B), although still connected to it (Bannister et al., 2004).

Mitochondrion and Apicoplast

The mitochondrion is a single tubular structure of variable diameter, increasing in length as the ring enlarges (Fig. 14). It contains a few tubular cristae of variable shape. The apicoplast is connected to it at one end or along its whole length, and it is also in contact with the nuclear envelope. Both the mitochondrion and the apicoplast are longer than in the merozoite (Hopkins et al., 1999).

The Parasitophorous Vacuole Membrane (PVM)

The PVM ensheathes the parasite throughout its erythrocytic phase, but it undergoes some major changes in its structure, composition, and relation to the parasite during the asexual cycle. It is formed initially during invasion as an invagination of the RBC surface, which separates from the host cell surface to enclose the parasite in a bubble-like membrane but goes on to expand considerably as the parasite first flattens into an early ring and thereafter increases in volume. In freeze-fracture preparations of merozoite (*P. knowlesi*) invasion into RBC, the early PVM is smooth and almost devoid of IMPs and therefore much of its protein content, but afterwards it regains a distinctive population of IMPs and antigens. The PVM around the maturing parasite (in *P. berghei*) has a very irregular surface in freeze-fracture preparations and comes very close to or touches the parasite's plasma membrane (Wunderlich et al., 1982) as well containing newly imported parasite antigens. It is clearly a very unusual structure both structurally and functionally (Lingelbach and Joiner, 1998).

THE TROPHOZOITE

General Morphology

The chief characteristics of this stage in Giemsa preparations of *P. falciparum* are its rounded shape, its possession of a single pigment vacuole, and increased overall basophilia (Fig. 15 to 26; Color Plate 3C). Within the RBC, Maurer's clefts consisting of basophilic dots and dashes are also visible (see Fig. 51). In EM sections, the parasite is seen to have lost its biconcave form and is rounded up into an ellipsoidal or spheroidal shape (Fig. 15). The RBC it inhabits has developed surface irregularities known prosaically as knobs (Fig. 15, 16, 23, and 26). The perimeter of the parasite is often deeply invaginated to form finger-like or bulbous intrusions of RBC cytosol into its interior, sometimes with only a thin lining of parasite cytoplasm, which gives the superficial appearance of a cleft (Fig. 17; Color Plate 3C) (Slomianny, 1990; Elford et al., 2004; see also below).

Nucleus and Ribosomes

The nucleus has a rounded, although irregularly ellipsoidal, shape and still has a pale staining interior with no hint of a nucleolus or heterochromatin (Fig. 15 and 17). The protein synthetic apparatus includes numerous unattached ribosomes, which increase in number to high density as the trophozoite develops (Color Plate 3C). Toward the end of this stage, the amount of RER increases markedly (Fig. 18) and free ribosomes become fewer, presumably reflecting preparation for membrane-dependent processes in schizogony. As in other stages, there is continuity between the RER and the nuclear envelope, which also bears ribosomes on its outer membrane.

Golgi Complex

The identity of this set of structures is indicated by its IEM labeling for the *cis* Golgi marker ERD2 (Van Wye et al., 1996). It is larger and considerably more complex than in rings, although basically constructed in the same way. In sections, it appears as a prominent mass of coated and uncoated vesicles, together with membrane tubules lying close to the nuclear envelope from which the coated vesicles bud (Fig. 19; Color Plate 3C). The whole apparatus is very unlike the stacked cisternal complexes classically described for most eukaryote cells, including *Toxoplasma*, and its structure suggests some fundamental molecular specialization worthy of closer investigation.

The Mitochondrion and Apicoplast

The mitochondrion and apicoplast (Fig. 20) are much longer than in the ring stage, though still attached to each other, usually at one end, and intermittently in membrane contact, a possible route for the exchange of metabolites. Internally, the mitochondrion is the same as found in rings. The apicoplast is connected at one end to the nuclear envelope and comes into contact with various other organelles elsewhere in the cell, including the food vacuole (Hopkins et al., 1999). Its internal structure is more complex, with the wall membranes folded internally into complex whorls and clusters of membrane vesicles accumulating in the space between the external and middle membranes of the apicoplast.

Cytostomes

In the trophozoite, these are two or three in number and appear to be maximally active, with the formation of many food vacuoles containing hemoglobin near them. When first endocytosed, each food vacuole is surrounded by two membranes (the inner from the PVM and the outer from the parasite's plasma membrane) (Fig. 21; Color Plate 3C). They then move to the pigment vacuole (Yayon et al., 1984), their inner membrane first breaking down, and then the vesicle interiorized (in a manner not yet clear) within the pigment vacuole for further destruction (Fig. 15; Color Plate 3C). Finally, small crystals of hemozoin appear within the pigment vacuole and grow larger, where the accompanying membrane debris may play a part in their formation and growth (Hempelmann et al., 2004). In *P. falciparum*, hemozoin crystals have square cross sections and sharply angular ends (Fig. 22), although the detailed form varies between species (Noland et al., 2003). One or

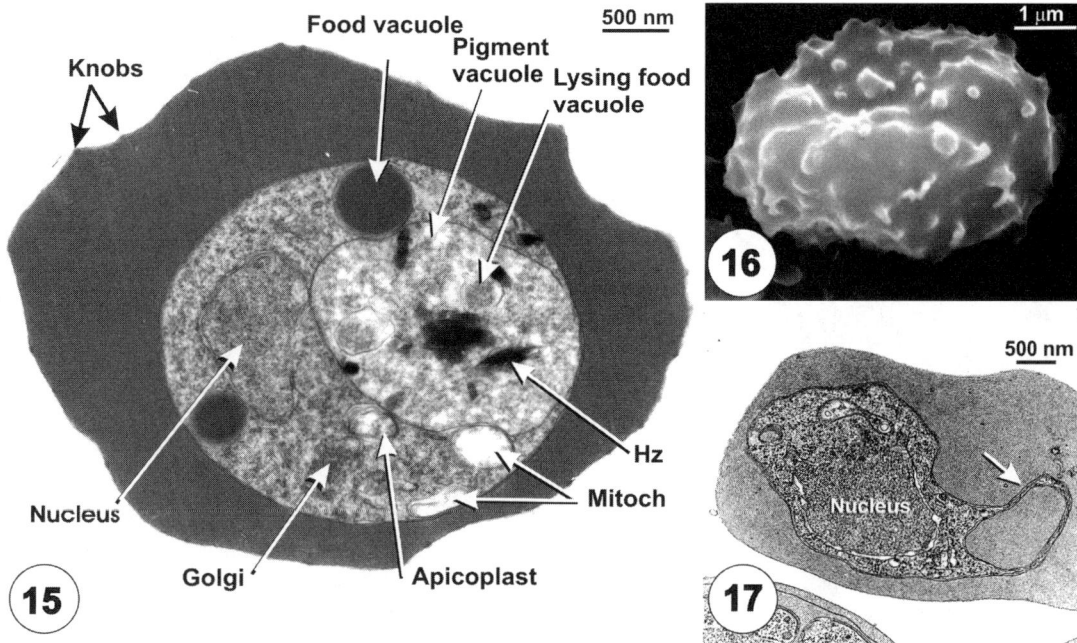

FIGURES 15 TO 17 Trophozoite stage ultrastructure. Figure 15 shows the arrangement of organelles in a mature trophozoite. The parasite contains a large pigment vacuole with lysing food vacuoles and hemozoin within it and other structures as detailed in Fig. 18 to 26. Figure 16 is a scanning EM of a trophozoite-infected RBC, showing the characteristic irregular surface, including small knob-like protruberances. In Fig. 17, a more mature trophozoite is shown, with a thin layer of cytoplasm superficially resembling a circular cleft delimiting a large mass of partially enclosed RBC cytosol (white arrow). Hz, hemozoin; Mitoch, mitochondrion.

more lipid vacuoles are associated with the pigment vacuole towards the end of the trophozoite stage (Jackson et al., 2004; Vielemeyer et al., 2004) related perhaps to hemozoin formation or to lipid recovery from the pigment vacuole. The single pigment vacuole as a repository for all food vacuoles is typical of *P. falciparum*, although not of many other species of *Plasmodium* such as *P. berghei*, where multiple, small pigment-containing vacuoles persist through the asexual period (Slomianny, 1990); however, even in *P. falciparum*, some isolated hemozoin vacuoles are often found, and there may be variations in the details of this process due to different parasite lines or culture conditions.

Structures Exported by the Parasite into the RBC

As the parasite continues to grow and differentiate (Fig. 23 to 26; Color Plate 3C), it exports membranes and other structures into the surrounding RBC (for recent reviews, see Przyborski et al., 2003, and Cooke et al., 2004). In *P. falciparum*, these include membranes—the clefts of Maurer, circular clefts, small vesicles, and dense protein-containing projections from the surface of the RBC, termed knobs. Similar structures exist in other species, such as Schüffner's dots (cleft-like membranes) in *Plasmodium vivax* and caveolae analogous to knobs in *P. knowlesi* and *Plasmodium cynomolgi* (Atkinson and Aikawa, 1990), although these are invaginations of the membrane rather than protrusions.

The term tubulovesicular network (TVN) is used to denote the whole assembly of exported membranous structures in the RBC by some authors (Haldar et al., 2001), while others (Elford et al., 1995) use this name for only one component, the circular clefts (see below).

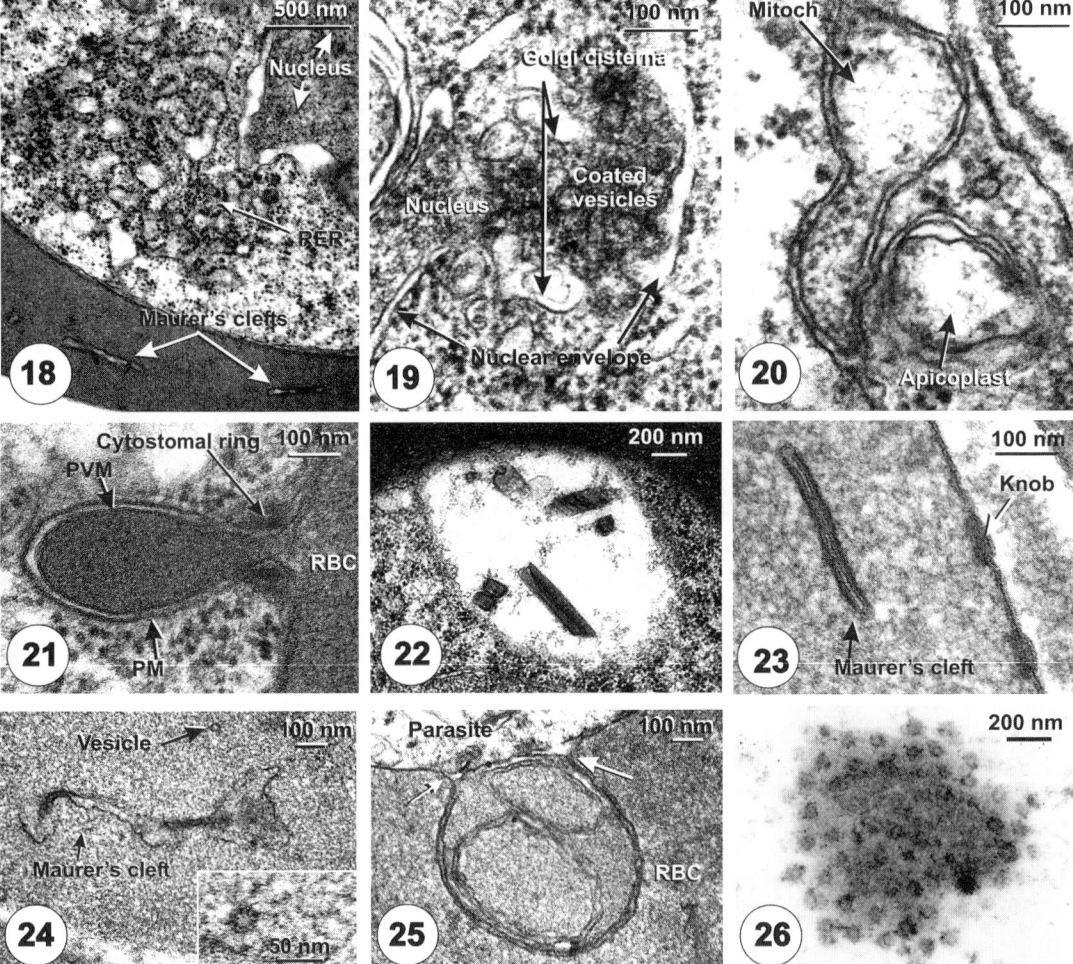

FIGURES 18 TO 26 Details of trophozoite ultrastructure. In Fig. 18, a branched complex series of RER cisternae are seen to be continuous with the nuclear envelope. Two Maurer's clefts are also visible. Figure 19 shows the Golgi complex of a mature trophozoite, where an extension of the nuclear envelope is the site of multiple coated vesicle budding. A tubular structure, corresponding to a Golgi cisterna, is closely associated with this mass of vesicles. In Fig. 20, the close association between the two-membrane mitochondrion and three-membrane apicoplast is depicted. Figure 21 shows a cytostomal vacuole in the process of formation through the cytostomal ring at the trophozoite's surface. The cytosol of the RBC, with the PVM and parasite's plasma membrane, is seen to be invaginated into the vacuole. In Fig. 22, the pigment vacuole of a maturing trophozoite contains angular crystals of hemozoin. Figures 23 to 26 depict structures exported from the parasite into the RBC. In Fig. 23, a typical Maurer's cleft with an external dense coating in transverse section is present in the RBC cytosol; two surface knobs are shown. In Fig. 24, a Maurer's cleft is visible in oblique section, revealing an irregular plate-like form. The insert shows a small exported vesicle with a spiky surface, present in the RBC cytosol. Figure 25 shows a circular cleft, continuous with the PVM (arrows). This example is more complex internally than usual, because of the presence of smaller circular cleft membranes within it. Figure 26 depicts a tangential section through the surface of an infected RBC, showing a high density of knobs associated with its membrane, a configuration typical of a multiply infected RBC. Mitoch, mitochondrion; PM, parasite plasma membrane.

As mentioned above, there are also cleft-like loops of membrane which are in fact thin extensions of the parasite within a PVM sheath (Fig. 17) (Elford et al., 1995) and so consist of four membranes (the two inner ones are the parasite plasma membrane, and the outer ones are the PVM); these are not therefore exported structures per se. These various structures will now be considered briefly.

(i) MAURER'S CLEFTS

These vary in form depending on the strain of parasite (Fig. 23 and 24; Color Plate 3C). At their simplest, they are flat, plate-like cisternae or tubules, mostly lying at a short distance and parallel to the RBC plasma membrane. They are coated patchily on their external (RBC-contacting) surfaces by a thin layer of dense material, with occasionally more extensive dense masses attached, a feature likely to be responsible for their basophilic reaction to Giemsa staining (making them visible by light microscopy). In lysed, infected RBC, Maurer's clefts maintain their relationship to the RBC surface, to which they are attached by fine filamentous links (Kriek et al., 2003). In some parasite strains, Maurer's clefts are more complex, forming what appear to be stacks of cisternae resembling Golgi complexes. However, serial sectioning indicates that each stack really consists of an interconnected series of folded cisternae or tubules (Wickert et al., 2003). It is not yet settled how these structures are exported into the RBC cytoplasm or whether Maurer's clefts are permanently confluent with the PVM (Wickert et al., 2003), although light microscopic fluorescence studies and IEM indicate that they are antigenically distinct (see the review by Cooke et al., 2004). It is possible that they are formed in late ring to early trophozoite stages by budding from the PVM and then become independent entities as the asexual cycle proceeds, subsequently receiving their distinctive antigenic markers.

(ii) SMALL VESICLES

Within the infected RBC cytoplasm, there are numerous small (15- to 25-nm) vesicles, often associated with Maurer's clefts and the RBC plasma membrane (Fig. 24; Color Plate 3C). Treatment of RBC with G-protein-stimulating chemicals (e.g., aluminium fluoride) favoring vesicular export from the RBC increases the numbers (and visualization) of these (Taraschi, 1999), and they are likely to be concerned with the export of molecules such as *P. falciparum* erythrocyte membrane protein 1 (PfEMP-1) to the RBC surface (and to Maurer's clefts). The vesicles bear short spikes, which radiate out from their external surfaces. The precise nature and destinations of these structures are not clear, although it is likely that they act as delivery systems from the PVM to the Maurer's clefts and RBC surface (for reviews, see Kriek et al., 2003; Wickert et al., 2003; and Cooke et al., 2004).

(iii) CIRCULAR CLEFTS (TVN)

In this chapter, we follow the terminology of Atkinson and Aikawa (1990) to distinguish the circular clefts from the (short) clefts of Maurer, avoiding the name TVN because of its alternative usage to describe the complete system of exported membranes (Fig. 25; Color Plate 3C). Circular clefts are spirals and loops of double membrane, which do not bear dense material externally or leaflets of parasite cytoplasm internally. In lysed infected RBC, they often retain unaffected hemoglobin in their embrace, indicating that they can actually form separate sealed compartments within the RBC cytosol. They can often be traced to continuity with the PVM, and it appears that they are, in effect, looped extensions of that membrane, as also shown by their antigenic similarity (e.g., with Exp-1 and Exp-2, unlike Maurer's clefts, where these are absent) (Johnson et al., 1994).

(iv) KNOBS

These are dense masses of material associated with the RBC membrane skeleton and surface membrane, which they bend to produce an angular prominence (Fig. 23 and 26; Color Plate 3C). Knobs are first seen in trophozoites, increasing over time to a sustained maximum near the beginning of the schizont period and in greater numbers in multiple infections, as

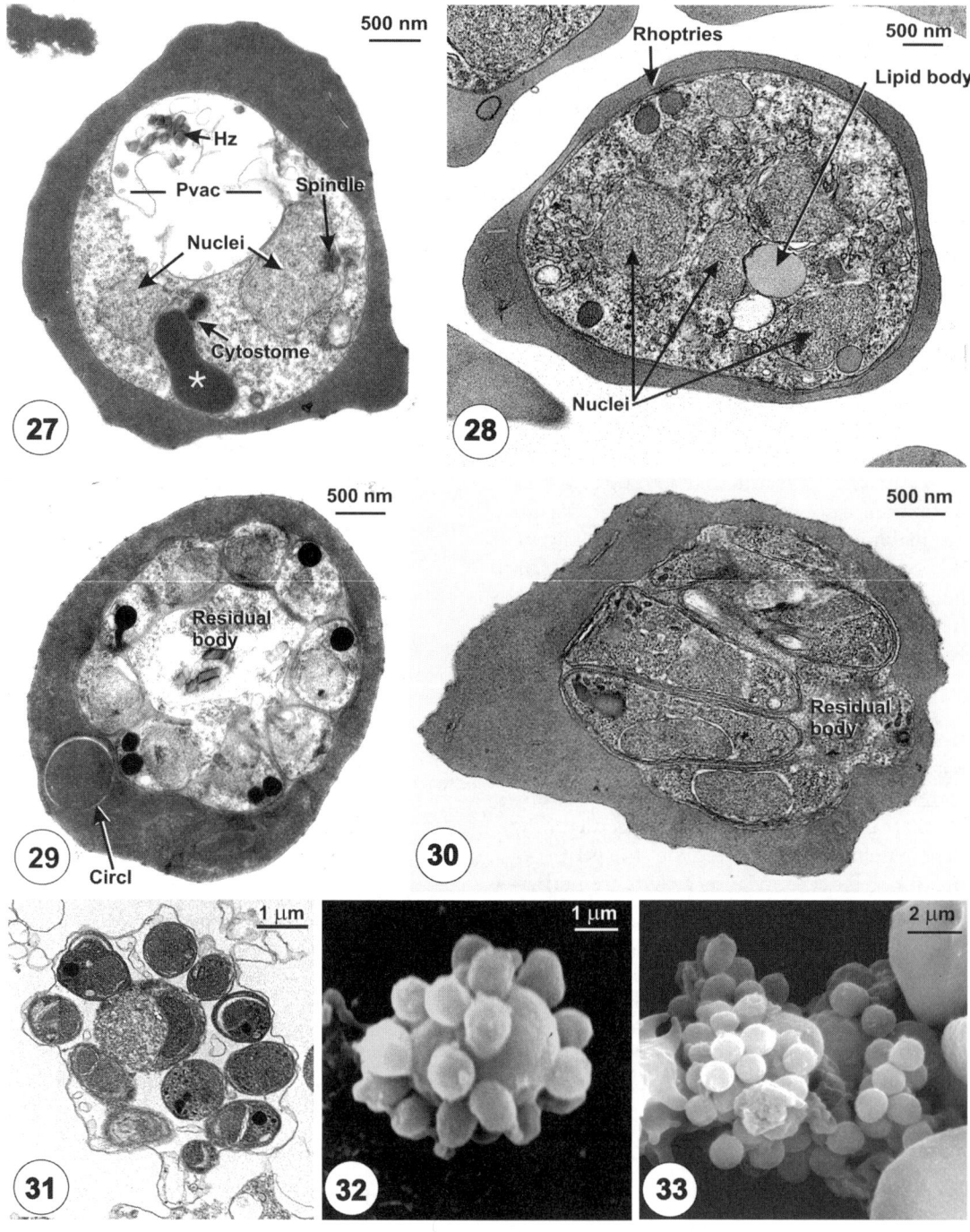

determined by a combination of atomic force microscopy and immunofluorescence (Nagao et al., 2000). The knobs have a complex substructure (Aikawa et al., 1996; Garcia et al., 1997), which may be related to a positive surface charge and consequent tendency to bind to electronegative surfaces such as those of other cells. If the RBC membrane is removed experimentally with detergent, the knobs remain attached to the membrane skeleton (Chishti et al., 1992), and they are clearly mechanically stable structures able to transmit adhesive forces through the whole RBC surface.

THE SCHIZONT

General Morphology

The schizont period is first detectable ultrastructurally by the appearance of a spindle pole body (SPB) in the nuclear envelope and the beginnings of mitotic spindle formation within it, heralding the onset of nuclear division and merozoite formation (Fig. 27 to 37; Color Plate 3D and E). During this phase, the nucleus, mitochondrion, apicoplast, and Golgi complex multiply, while the parasite synthesizes and assembles merozoite components in readiness for the next cycle of RBC infection (Florent et al., 2004). The schizont gradually moves away from trophic activities, including cytostomal feeding, to the maturation and finally release of merozoites; during these processes, the parasite becomes structurally more complex and developmentally dynamic. The nucleus undergoes approximately four rounds of mitotic divisions to create about 16 nuclei; the number may vary from 8 to 32 in a single schizont, often with intermediate numbers indicating a rather asynchronous process with relaxed control of final numbers (Margos et al., 2004). When the full number is reached, synthesis and placement of organelles continue as the merozoites bud off from the parent cytoplasm (the residual body containing the pigment vacuole), but they are fully mature only just before release. At the same time, the mitochondrion and apicoplast divide by forming forked branches, which eventually separate into multiple organelles (Color Plate 3D) (Divo et al. 1985; Waller and McFadden, 2005). The ultrastructural changes which have been analyzed recently in *P. falciparum* (Margos et al., 2004) are summarized briefly here for that species.

Stages of Schizont Development

These can be conveniently divided into 7 main steps. Schizont stages (SzS) 1 to 4 represent the typical four rounds of mitotic nuclear division to generate 16 nuclei, and SzS 5 to 7 define the subsequent maturation of merozoites, including the formation of merozoite buds (Szs 5), merozoite separation from the residual body (SzS 6),

---

FIGURES 27 TO 33 Ultrastructure of schizont development. Figure 27 shows an early, two-nucleus schizont, one of the nuclei with a part of a mitotic spindle. Also visible are a cytostomal vacuole formed at the end of a deep intrusion of RBC cytoplasm into the parasite's surface (asterisk) and a large pigmented vacuole containing a small cluster of hemozoin crystals. The surface of the infected RBC shows numerous knobs. Figure 28 shows a more advanced schizont at approximately the eight-nucleus stage with spheroidal rhoptries in pairs around the periphery indicating the positions of future merozoite apices. Also present is a prominent lipid body. Note the absence of knobs from the surface of the RBC, as expected from this knobless strain (C10) of parasite, contrasting with the IT04 strain depicted in Fig. 27. Figure 29 shows a schizont immediately after the end of nuclear division, with the beginning of merozoite budding from the central mass (residual body) containing the hemozoin. Note that the rhoptries are now elongated club-like forms approaching their mature state. Figure 30 shows merozoites, here elongated and connected only by narrow stalks to the residual body, but the PVM and RBC hemoglobin are still intact. Figure 31 shows a schizont with the merozoites now separate from the residual body, surrounded by only a single membrane after the hemoglobin has been released. In Fig. 32 and 33, scanning EMs of schizonts from which the membranes of the RBC and PV have been mechanically removed during specimen preparation are shown. Figure 32 shows an early budding stage, with merozoite apices protruding from the residual body; Fig. 33 depicts a rather later stage with two more mature clusters of merozoites, probably representing a double RBC infection. Circl, circular cleft; Hz, hemozoin crystals; Pvac, pigment vacuole.

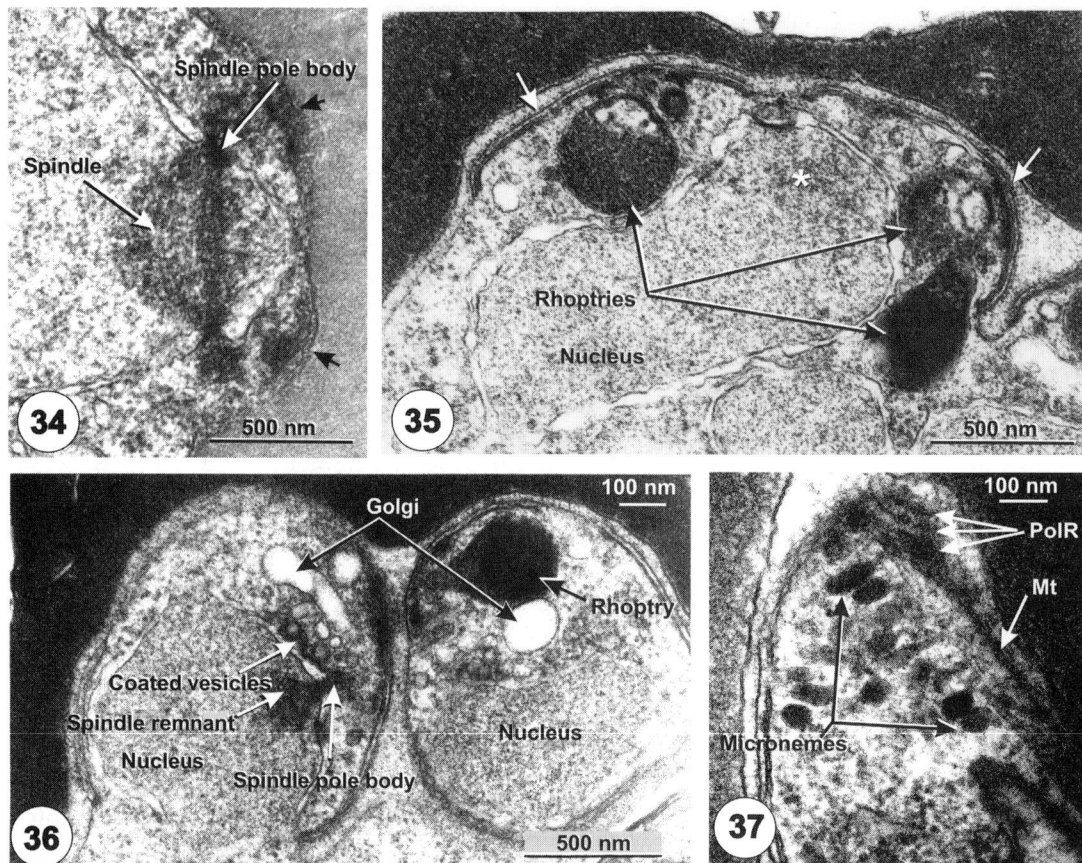

FIGURES 34 TO 37 Details of merozoite development. In Fig. 34, the mitotic spindle and associated SPBs and early rhoptry centers (black arrows) are situated in an early (four-nucleus) schizont. In Fig. 35, at a late final divisional state (8 nuclei, progressing to 16) merozoite apices with pairs of rhoptries are developing at either end of a dividing nucleus (the spindle is not visible in this specimen). In each pair, one rhoptry is typically more mature than the other. Pellicles are assembled around the merozoite apices, giving their membranes a dense multilayered appearance (white arrows). Figure 36 shows two merozoites at a late schizont stage, although they are still attached to the residual body. Coated vesicles are budding from the nuclear envelope close to the SPBs, and a Golgi cisterna lies more apically near each, enlarged on the right to form a rounded vesicle close to a developing rhoptry. In Fig. 37, the apical migration of micronemes along a subpellicular microtubule is shown in a longitudinal section of a late merozoite. The section also passes through the edge of the apical prominence and its polar rings. Mt, subpellicular microtubule; PolR, polar rings. Figures 35 and 36 are reproduced from Bannister et al., 2000, by kind permission of Cambridge University Press. Figure 37 is reproduced from Bannister et al., 2003, by kind permission of the *Journal of Cell Science*.

the loss of RBC cytosol, and often loss of one of the two membranes enclosing the merozoites, prior to release (SzS 7).

SPBs and Nuclear Division

Mitosis in *Plasmodium* schizonts (Fig. 34) occurs entirely endomitotically (Vickerman and Cox, 1967), i.e., within intact nuclear membranes, the mitotic spindles being intranuclear (as in *Saccharomyces* species and many other unicellular organisms). SPBs are formed at the nuclear surface within nuclear pores and then replicate side by side into pairs. One SPB then migrates around the nuclear envelope to the opposite side of the nucleus, and spindle microtubules sprout from the opposite SPBs inwards toward each other to form a complete spindle. The chromosomes attach to spindle microtubules by their kinetochores at the spindle equator, and then the chromatids separate and migrate to opposite poles. The nucleus pinches across its middle to create two nuclei. The SPBs again replicate and the process is repeated until the full nuclear number is reached. Unlike mitosis in most eukaryotes, the chromosomes remain as highly extended thin strands which cannot be distinguished from the rest of the nuclear contents by standard EM, although their kinetochores are visible, and their number has been counted by serial sectioning to determine the number (14) of *P. falciparum* chromosomes (Prensier and Slomianny, 1986).

SPBs and Merozoite Assembly

The positions of SPBs clearly determine spindle orientation within the schizont, important for the assembly of merozoites. Centers of merozoite assembly (Fig. 28, 29, 35; Color Plate 3D) are formed close to SPBs, marked by the budding of a cluster of coated vesicles from the outer nuclear membrane and a simple Golgi complex with a single cisterna (Bannister et al., 2000). These centers form around the periphery of the parasite, so the SPBs (and spindles) have to be positioned correctly. How this is achieved is unknown, but the process indicates a considerable level of spatial coordination within the developing schizont. If spindle formation is disrupted, e.g., by Taxol (paclitaxel) treatment, merozoite development can proceed but becomes highly abnormal in both merozoite number and final structure (Taraschi et al., 1998).

Merozoite Maturation

The formation of merozoite organelles is visible from about the second nuclear division (SzS 2), when the first signs of rhoptry development are seen close to the ends of mitotic spindles (Fig. 34). Rhoptries are at first small (80-nm) spheroidal structures with granular contents (Jaikaria et al., 1993), but they expand into larger spheres (Fig. 35) and then elongate at one pole (beginning about SzS 3) towards the schizont surface (Fig. 35 and 36) to create the characteristic club-like shape of mature merozoites (Bannister et al., 2000). Micronemes and dense granules are formed late in schizont development, with micronemes beginning at about SzS 5 and dense granules at SzS 6. All the apical organelles (rhoptries, micronemes, and dense granules) are formed from the Golgi cisterna: rhoptries by fusion of small Golgi-derived vesicles and micronemes and dense granules by direct budding of individual organelles from its cisterna.

Merozoite Assembly

The final assembly of merozoites begins towards the end of the last mitosis (Fig. 29, 35 to 37; Color Plate 3D) when surface clefts, lined by the three-membrane pellicle, start to deepen around the foci of apical organelle formation. Nuclei now complete their division and move into each bud, accompanied by the mitochondrion and apicoplast (both of which multiplied during earlier schizont stages) (Color Plate 3D) along the merozoite's subpellicular microtubules, which are now completely assembled along one side of the bud. The rhoptry maturation completes (Fig. 36) and micronemes formed near the nucleus at the Golgi cisterna are transported to their apical destination along the band of subpellicular microtubules (Fig. 37; Color Plate 3D) (Bannister et al., 2003). The dense granules move out from the Golgi complex into the apical half of the merozoite without microtubular involvement. An inactive cytostome also appears at one side of the merozoite, and the polar rings and apical prominence (Fig. 37) are fully formed. The merozoite by now is nearly complete in shape and is attached to the residual body only by a narrow pedicle, with a ring of dense material beneath its membrane indicating a cytoskeletal constriction annulus presumably responsible for pinching the merozoite off from the residual body, an event occurring as soon as all the organelles are in place. The residual body consists of a mass of hemozoin crystals enclosed in the pigment vacuole membrane, surrounded by parasite plasma membrane (lacking a pellicular structure) bearing a filamentous coat. The

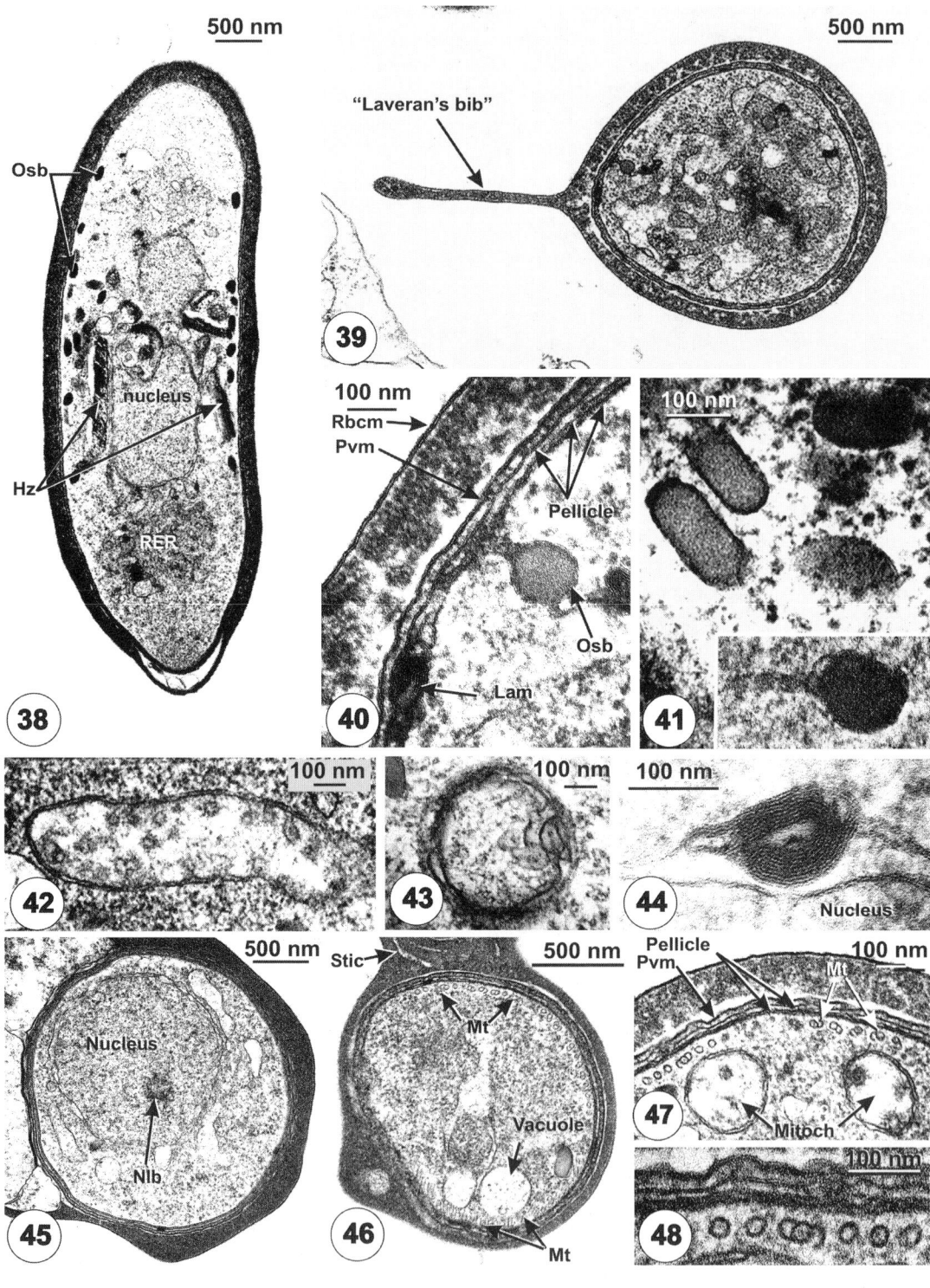

merozoite is not, however, completely mature internally until release from the RBC, as the Golgi body and some ER are still visible (Bannister et al., 2000).

Merozoite Exit

Toward the end of these developmental processes, the schizont consists of a membranous sac consisting of two membranes, an inner PVM and an outer RBC membrane enclosing the merozoites and residual body (Fig. 30; Color Plate 3E). Exactly what happens to these membranes and the mechanism of merozoite release under normal circumstances have yet to be definitively settled. Experimental treatment with the cysteine protease inhibitor E64 allows the lysis of the RBC membrane but inhibits PVM breakdown (Salmon et al., 2001), whereas serine protease inhibitors allow the PVM but not the RBC to lyse (Wickham et al., 2003).

Direct observation of living cells suggests that hemoglobin is present in the infected RBC until merozoite release, indicating that the PVM breaks down first (Winograd et al., 1999), as also found in fluorescently labeled RBCs (Wickham et al., 2003). However, in sections from normal cultures, it is possible to find late schizonts where the PVM but not the RBC membrane has lysed, side by side with those with only a PVM surrounding them, or alternatively, with both membranes but no hemoglobin.

GAMETOCYTES

These are complex dynamic stages, which mature progressively over a period of many days and undergo striking changes of structure as they do so (see Chapter 10). Although switching from asexual to sexual forms appears to occur at the ring stage or even earlier, gametocytes are only clearly recognizable microscopically much later, and the distinction between macrogametocytes (female) and microgametocytes (male) is apparent only near the time of their final maturation. The details of structure and developmental timing differ considerably with species (see Chapter 10). In *P. falciparum*, the process takes 7 days or more (Ponnudurai et al., 1986). The chief ultrastructural features of gametocytes are illustrated in Fig. 39 to 48 and in Color Plate 3 (see also Chapter 10).

When mature (Fig. 38 and 39; Color Plate 3F), *P. falciparum* gametocytes of both genders are similar in length (about 9 μm), curving in a crescentic, sausage-like shape along the edge of their RBC host, unlike the more-rounded gametocytes of some other species (Aikawa et al., 1969). In Giemsa-stained blood films (Color Plate 3E), male and female forms differ from

---

FIGURES 38 TO 48 Gametocyte ultrastructure. Figure 38 shows a stage V macrogametocyte in longitudinal section. Hemozoin crystals (Hz) are present around the nucleus, and osmiophilic bodies (Osb) situated around the parasite's periphery. Clusters of RER are present at either end. Figure 39 depicts a transverse section through an RBC containing a mature gametocyte; RER is present in the parasite's cytoplasm. On the left, the RBC is flattened into a flap-like extension (Laveran's bib) to one side; around the parasite, the hemoglobin is beginning to disappear. Figure 40 details the surface region of a mature gametocyte and adjacent RBC. Visible is the PVM closely adhering to the three-membraned pellicle of the gametocyte. An osmiophilic body (Osb) with a narrow stalk is in contact with the pellicle, which also contains a multilamellar membranous structure (Lam). Note that a hemoglobin-free zone is present immediately outside the PVM. In Fig. 41, a higher magnification of a group of osmiophilic bodies, one of them with a duct-like extension, is shown. Figure 42 shows a mitochondrion in longitudinal section with tubular cristae. Figure 43 shows the transverse section of an apicoplast. In Fig. 44, a multilamellar membranous inclusion in the nuclear envelope is shown. Figure 45 shows the transverse section of a macrogametocyte nucleus containing a nucleolus-like body. In Fig. 46, the transverse section of a stage IV gametocyte is shown with a band of subpellicular microtubules (Mt) and vacuolar structures. Gametocyte-specific membranous clefts are also visible in the RBC. In Fig. 47, higher magnification through the surface region of a stage IV gametocyte shows the pellicle with a band of subpellicular microtubules (Mt), some of them doublets or triplets, attached to the pellicle. Two transversely sectioned mitochondrial profiles are also present. The PVM encloses the gametocyte. In Fig. 48, a higher magnification of part of Fig. 47 shows details of the subpellicular microtubules and their pellicle attachments. Mitoch, mitochondrion; Nlb, nucleolus-like body; Stic, sexual-stage tubular intraerythrocytic compartment.

each other—macrogametocytes are narrower and more basophilic (blue staining) and have a central, small, oval nucleus surrounded and often obscured by hemozoin particles. Male microgametocytes are broader and less basophilic (pink staining) and contain a larger, often lobated nucleus; their hemozoin particles are scattered throughout the cytoplasm. These differences reflect more complex ultrastructural distinctions related to their respective destinies and sexual roles (see, e.g., Sinden, 1982a). Microgametocyte structure is directed towards the eventual production of eight microgametes after transmission to the mosquito, entailing three very rapid rounds of DNA replication and mitosis and the generation of the appropriate apparatus (including a flagellum) for each microgamete (Sinden et al., 1976; see also Aikawa et al., 1984, for *Plasmodium gallinaceum*). The microgametocyte cytoplasm has free polyribosome clusters, but there is relatively little RER, suggesting a dominance of cytosolic protein synthesis (e.g., in the generation of cytoskeletal proteins). In contrast, the macrogametocyte cytoplasm contains the metabolic equipment needed to launch the ookinete stage after fertilization, being rich in both free polyribosomes and RER (hence the basophilia after Giemsa staining), as well as numerous free polyribosomes and various membranous organelles. There are four times as many ribosomes as in the microgametocyte (Shaw et al., 1996), a fact reflecting the later transformation into a zygote and, later, an ookinete.

Other reported cytoplasmic components include mitochondria (Fig. 42), reported to number up to six per cell (Krungkrai et al., 2000), an apicoplast(s) (Fig. 43), a Golgi complex(es), an active cytostome(s), etc. In both gametocyte forms, the cytosol is also pervaded by finely granular material (of unknown composition) which adds to the density of the cytoplasm. Smooth membraned vacuoles of various shapes are present (Fig. 47), often dilated to form vacuoles loosely filled with granular material, and smooth membranes wound into complex whorls are also commonly present (Color Plate 3F). The nucleus is typically pale and evenly granular in appearance, except in macrogametocytes where a central dense mass resembling a nucleolus is present (Fig. 45). In situ hybridization has failed to find ribosomal RNA in this structure, so it is unlikely to be a classical nucleolus; at present, the nature of the gametocyte remains obscure (Shaw et al., 1996).

Specific Gametocyte Organelles

Both types of gametocyte contain specific organelles distinguishing them clearly from the asexual trophic stages. In later gametocyte stages, these include a three-membrane pellicle and a set of dense cytoplasmic secretory vesicles termed osmiophilic bodies (i.e., densely stainable with osmium tetroxide).

The Gametocyte Pellicle

The pellicle has a structure similar to that described for merozoites—a superficial plasma membrane attached by fine filaments to the deeper IMC (Fig. 40 and 47). Freeze-fracture preparations show that the IMC consists of a number of flattened compartments which abut each other at seam-like junctions (Meszoely et al.,1987). External to the gametocyte, the PVM lies close to the parasite's surface but at a uniform distance of about 20 nm (Fig. 40, 47, and 48; Color Plate 3F), with a coat about 20 nm thick, consisting of regular projections connecting the plasma membrane and the PVM (Fig. 41; Color Plate 3F). This arrangement shows the gametocyte has a structured coat, which binds the PVM to itself tightly throughout the later stages of maturation until its final exit from the RBC host. The significance of this combination of membranes is not known, but it perhaps helps to maintain cell shape in the turbulent peripheral circulation. Occasionally, masses of concentric myelin-like lamellae are present in the pellicle (Fig. 40) and may be related to osmiophilic body discharge.

The three-membraned pellicle present in other stages of the parasite (merozoites, ookinetes, and sporozoites) forms part of the mechanism for gliding motility, a function unlikely here. It could be needed to anchor the microtubular cytoskeleton when it is assembled, perhaps for translocating organelles such as the osmiophilic bodies (see below) to their correct positions for secretion, as seen in micronemal transport in merozoites, already mentioned, but this requires further analysis.

In both male and female forms, there is active cytostomal feeding on the RBC cytosol as in asexual stages, but the resulting hemozoin is contained in numerous small vesicles (Fig. 38; Color Plate 3F) rather than the single common pigment vacuole seen in asexual *P. falciparum* trophozoites (Aikawa et al., 1969; el Shoura and al Amari, 1993).

Osmiophilic Bodies

These are membrane-enclosed secretory vesicles, elliptical or spheroidal in shape, and 100 to 300 nm in diameter with finely granular densely staining contents (Fig. 38, 40, and 41). They originate around the Golgi apparatus towards the end of gametocytogenesis (from stage III onwards; see below) and move peripherally to lie close to the pellicle, sometimes with long neck-like protrusions reminiscent of merozoite micronemes, although much larger. Secretion of their contents into the parasitophorous vacuole (PV) is likely to be the trigger for the lysis of the membranes surrounding the gametocyte within the mosquito gut as it transforms into gametes (Scherf et al., 1992; Alano et al., 1995).

Structures Exported into the RBC

Another important feature of the gametocyte stages is the absence of knobs and Maurer's clefts from the RBC they infect, reflecting the non-sequestering nature of this stage; free circulation of the gametocyte in cutaneous blood vessels is essential for their ingestion by the mosquito. However, other membranous structures are present in the RBC cytosol, including cleft-like membrane-enclosed spaces (Fig. 46 and 54), sometimes enwrapping the gametocyte in concentric layers. Their membranes are antigenically distinct from the PVM and from Maurer's clefts of asexual stages and have been termed the sexual-stage tubular intraerythrocytic compartment in recognition of their unique status (Eksi et al., 2002). Other membranes form dense multilamellar masses within either the PV or the adjacent RBC cytosol, evidence of considerable level of membrane export during this stage and perhaps a structural clue to the mechanism of eventual lysis of the membranes surrounding the gametes within the mosquito gut.

Stages of Gametocyte Formation

The above description applies to mature gametocytes, but to achieve this state there are many striking developmental changes (for *P. falciparum*, see Sinden, 1982b; Kass et al., 1971; Ruangjirachuporn et al., 1992). In terms of structure, they are divisible according to the light microscopic terminology of Hawking et al. (1971) into five phases, as follows.

In stage 1, early differentiation occurs, with stage-specific gene expression beginning at the early ring or even the merozoite stage. Gametocytes are detectable microscopically only when they reach the size of trophozoites, when they contain multiple small hemozoin vacuoles scattered through the cytoplasm instead of the single pigment vacuole typical of asexual *P. falciparum*. There is also an absence of knobs or true Maurer's clefts in the gametocyte-infected RBC.

In stage II, gametocytes begin to develop an IMC beneath the plasma membrane to create a three-membrane pellicle, and their distinctive gametocyte nature is now clearly recognizable. At the same time, microtubules begin to polymerize, starting at an SPB located in a nuclear pore. Some microtubules are intranuclear, and others cytoplasmic, including a contingent which attaches to the underside of the pellicle by short cross-bridges. There, they begin to form a sheath of parallel longitudinally oriented, evenly spaced microtubules which extend to elongate the gametocyte to a straight spindle-like form pointed at both ends (the oat-grain configuration). Within the cytoplasm, large numbers of ribosomes and some endoplasmic reticulum are generated to give it a patchily dense appearance.

At stage III, the subpellicular microtubules continue to grow in number and length in both micro- and macrogametocytes. The RBC becomes modified by the parasite, which exports membranous whorls and other cleft-like material into the RBC cytoplasm. At this stage, the difference between micro- and macrogametocytes is accentuated by the proliferation of mitochondria, RER, and ribosome profiles in the cytoplasm, particularly in macrogametocytes. The mitochondria develop prominent vesicular cristae (Fig. 42), indicating a change in metabolism from acristate mitochondria of asexual forms (Krungkrai et al., 2000).

At stage IV, both micro- and macrogametocytes are elongate, straight, fusiform cells 8 to 9 μm long (Color Plate 3F). Their longitudinal growth, due to microtubule lengthening, stretches the surrounding RBC further, and the microtubules now form a parallel array (Fig. 46 to 48) beneath much of the parasite's perimeter (Sinden, 1982a; Kaidoh et al., 1995). Individual microtubules are mostly single, but doublets and triplets also occur, suggesting a complexity of tubulin assembly (Fig. 48). Dense transverse bands cross the microtubule array at intervals, suggesting a means of mechanically stabilizing this cytoskeletal array (Kaidoh et al., 1993). At this stage, the osmiophilic bodies approach the pellicle. The subpellicular microtubules are clearly important for gametocyte maturation. Treatment with antimicrotubule agents (Sinden et al., 1985) or α-tubulin II gene mutation (of males) (Guinet et al., 1996; Ponnudurai et al., 1986) blocks gametocyte maturation. The microtubules may be important in the trafficking of organelles within the gametocyte, as in merozoites (see above), but this needs further investigation.

Within the RBC host, there are also various changes. Lamellar membranous inclusions become visible in the PV space, while outside the PVM the RBC cytosol begins to disappear, at first clearing a lucent zone close to the PVM and then creating irregular gaps towards the RBC surface (Fig. 40, 46, and 47), although shrinkage during preparation for EM may also contribute to this appearance. These events are accompanied by an alteration in RBC shape, the gametocyte now occupying a tubular bulge along one side of the RBC, while the rest of the host cell flattens into a thin flap-like extension (Laveran's bib) (Fig. 39).

In stage V, the final period of maturation, there are further developments. The subpellicular microtubules depolymerize, and the parasite, no longer cytoskeletally stiffened, bends into a crescentic shape. In macrogametocytes, osmiophilic bodies increase in number and density around the parasite's periphery, becoming positioned with their ducts inserted into the pellicle (Fig. 40), as though in readiness for the discharge of contents. In the nuclear membrane, the SPB enlarges, anticipating spindle formation at gametogenesis. Within the RBC, major changes also occur, its cytosol eventually disappearing completely, although the RBC membrane and PVM remain intact until they enter the mosquito gut. The stage V gametocyte is the end point of the erythrocytic phase of the parasite's life.

CONCLUSIONS AND FUTURE PROSPECTS

This brief description of the morphological changes occurring within the parasite during the asexual and sexual erythrocytic periods indicates the great complexity and continual modulation of cellular form typical of *P. falciparum*, a statement which can be repeated for the other species of malaria parasite and other stages not considered here. Although the major structural features of the parasite and its interactions with RBCs are now broadly established, the level of structural information available must be considerably improved if morphology is to really serve the needs of genome-based research. For this, we need the new tools and techniques of light microscopy and EM, including, for example, improved imaging of living parasites, better preparation methods for general ultrastructure and localization of gene expression, more quantitative analysis (currently, there is hardly any), and three-dimensional visualization to determine spatial organization. Immunolocalization is still a chancy business, and much better methods for antigen preservation and molecular labeling are needed, for both light microscopy and EM. We have passed the era of early ultrastructural discovery and need to engage in thorough, rigorous, morphology-based analysis of the parasite as an important part of the assault on malaria.

This is a big challenge, but there are hopeful signs. We have new instruments, for example, cryofixation for the preservation of structure and antigenicity and EM tomography. Techniques of image analysis are becoming increasingly powerful and accessible, and information technology is increasingly providing a firm infrastructure for global research awareness and action. These are needed to bring structure-based investigations to a level high enough to make biological sense of the genome-triggered explosion of molecular data. There are few young electron microscopists available to attack such a demanding challenge, and few funded opportunities to do so. It is hoped that the research community will recognize the importance of this approach and find ways of promoting it. There is no substitute for creative, careful observation, rigorous analysis, and imaginative understanding or for the great but very worthwhile effort undoubtedly needed for the final push towards mastery of malaria.

ACKNOWLEDGMENTS

We thank our colleagues for the provision of parasites from which the electron micrographs were prepared, including Anton Dluzewski, Ian Williams, Sanjeev Krishna, Iain Wilson, and Robert Sinden. We are grateful to Tony Brain for the preparation of freeze-fracture material shown in Fig. 5 and 8 and to the Electron Microscope Unit at Guy's Hospital for general assistance throughout this study. We thank Dan Osborn and Simon Hughes for photographic assistance and the use of the Zeiss Axiophot microscope and Axiocam camera for the photomicrography of Giemsa-stained material shown in Color Plate 3.

We gladly acknowledge the continued support, encouragement, and provision of facilities at the Malaria Research Laboratory by Graham Mitchell.

The study was funded by the Wellcome Trust (grant no. 069515).

REFERENCES

Adisa, A., M. Rug, M. Foley, and L. Tilley. 2002. Characterisation of a delta-COP homologue in the malaria parasite, *Plasmodium falciparum*. *Mol. Biochem. Parasitol.* **123:**11-21.

Aikawa, M. 1967. Ultrastructure of the pellicular complex of *Plasmodium fallax*. *J. Cell Biol.* **35:**103-113.

Aikawa, M. 1971. *Plasmodium*: the fine structure of malarial parasites. *Exp. Parasitol.* **30:**284-320.

Aikawa, M., and T. M. Seed. 1980. Morphology of plasmodia, p. 285-344. *In* J. P. Kreier (ed.), *Malaria*, vol. 1. Academic Press, New York, N.Y.

Aikawa, M., C. G. Huff, and H. Sprinz. 1969. Comparative fine structure study of the gametocytes of avian, reptilian, and mammalian malarial parasites. *J. Ultrastruct. Res.* **26:**316-331.

Aikawa, M., L. H. Miller, J. Johnson, and J. Rabbege. 1978. Erythrocyte entry by malarial parasites. A moving junction between erythrocyte and parasite. *J. Cell Biol.* **77:**72-82.

Aikawa, M., L. H. Miller, J. R. Rabbege, and N. Epstein. 1981. Freeze-fracture study on the erythrocyte membrane during malarial parasite invasion. *J. Cell Biol.* **91:**55-62.

Aikawa, M., J. R. Rabbege, I. Udeinya, and L. H. Miller. 1983. Electron microscopy of knobs in *Plasmodium falciparum*-infected erythrocytes. *J. Parasitol.* **69:**435-437.

Aikawa, M., R. Carter, Y. Ito, and M. M. Nijhout. 1984. New observations on gametogenesis, fertilization, and zygote transformation in *Plasmodium gallinaceum*. *J. Protozool.* **31:**403-413.

Aikawa, M., K. Kamanura, S. Shiraishi, Y. Matsumoto, H. Arwati, M. Torii, Y. Ito, T. Takeuchi, and B. Tandler. 1996. Membrane knobs of unfixed *Plasmodium falciparum* infected erythrocytes: new findings as revealed by atomic force microscopy and surface potential spectroscopy. *Exp. Parasitol.* **84:**339-343.

Akaki, M., E. Nagayasu, Y. Nakano, and M. Aikawa. 2002. Surface charge of *Plasmodium falciparum* merozoites as revealed by atomic force microscopy with surface potential spectroscopy. *Parasitol. Res.* **88:**16-20.

Alano, P., D. Read, M. Bruce, M. Aikawa, T. Kaido, T. Tegoshi, S. Bhatti, D. K. Smith, C. Luo, S. Hansra, R. Carter, and J. F. Elliott. 1995. COS cell expression cloning of Pfg377, a *Plasmodium falciparum* gametocyte antigen associated with osmiophilic bodies. *Mol. Biochem. Parasitol.* **74:**143-156.

Atkinson, C. T., and M. Aikawa. 1990. Ultrastructure of malaria-infected erythrocytes. *Blood Cells* **16:**351-368.

Bannister, L. H., and G. H. Mitchell. 1986. Lipidic vacuoles in *Plasmodium knowlesi* erythrocytic schizonts. *J. Protozool.* **33:**271-275.

Bannister, L. H., and A. R. Dluzewski. 1990. The ultrastructure of red cell invasion in malaria infections: a review. *Blood Cells* **16:**257-292.

Bannister, L. H., and G. H. Mitchell. 1995. The role of the cytoskeleton in *Plasmodium falciparum* merozoite biology: an electron-microscopic view. *Ann. Trop. Med. Parasitol.* **89:**105-111.

Bannister, L. H., G. A. Butcher, E. D. Dennis, and G. H. Mitchell. 1975. Structure and invasive behaviour of *Plasmodium knowlesi* merozoites in vitro. *Parasitology* **71**:483–491.

Bannister, L. H., J. M. Hopkins, R. E. Fowler, S. Krishna, and G. H. Mitchell. 2000. Ultrastructure of rhoptry development in *Plasmodium falciparum* erythrocytic merozoites. *Parasitology* **121**:273–287.

Bannister, L. H., J. M. Hopkins, R. E. Fowler, S. Krishna, and G. H. Mitchell. 2001. A brief illustrated guide to the ultrastructure of *Plasmodium falciparum* asexual blood stages. *Parasitol. Today* **16**:427–433.

Bannister, L. H., J. M. Hopkins, A. R. Dluzewski, G. Margos, I. T. Williams, M. J. Blackman, C. H. Kocken, A. W. Thomas, and G. H. Mitchell. 2003. *Plasmodium falciparum* apical membrane antigen 1 (PfAMA-1) is translocated within micronemes along subpellicular microtubules during merozoite development. *J. Cell Sci.* **116**:3825–3834.

Bannister, L. H., J. M. Hopkins, G. Margos, A. R. Dluzewski, and G. H. Mitchell. 2004. Three-dimensional ultrastructure of the ring stage of *Plasmodium falciparum*: evidence for export pathways. *Microsc. Microanal.* **10**:551–562.

Bushell, G. R., L. T. Ingram, C. A. Fardoulys, and J. A. Cooper. 1988. An antigenic complex in the rhoptries of *Plasmodium falciparum*. *Mol. Biochem. Parasitol.* **28**:105–112.

Chishti, A. H., K. I. Andrabi, L. H. Derick, J. Palek, and S. C. Liu. 1992. Isolation of skeleton-associated knobs from human red blood cells infected with malaria parasite *Plasmodium falciparum*. *Mol. Biochem. Parasitol.* **52**:283–287.

Cooke, B. M., K. M. Lingelbach, L. H. Bannister, and L. Tilley. 2004. Protein trafficking in *Plasmodium falciparum*-infected red blood cells. *Trends Parasitol.* **20**:581–589.

Culvenor, J. G., K. P. Day, and R. F. Anders. 1991. *Plasmodium falciparum* ring-infected erythrocyte surface antigen is released from merozoite dense granules after erythrocyte invasion. *Infect. Immun.* **59**:1183–1187.

Divo, A. A., T. G. Geary, J. B. Jensen, and H. Ginsburg. 1985. The mitochondrion of *Plasmodium falciparum* visualized by rhodamine 123 fluorescence. *J. Protozool.* **32**:442–446.

Eksi, S., A. Stump, S. L. Fanning, M. I. Shenouda, H. Fujioka, and K. C. Williamson. 2002. Targeting and sequestration of truncated Pfs230 in an intraerythrocytic compartment during *Plasmodium falciparum* gametocytogenesis. *Mol. Microbiol.* **44**:1507–1516.

el Shoura, S. M., and O. M. al Amari. 1993. Falciparum malaria in naturally infected human patients. I. Ultrastructural differences between malaria pigments in intraerythrocytic asexual and sexual forms. *J. Morphol.* **215**:201–206.

Elford, B. C., G. M. Cowan, and D. J. P. Ferguson. 2004. Parasite-regulated membrane transport processes and metabolic control in malaria-infected erythrocytes. *Biochem. J.* **308**:361–374.

Etzion, Z., M. C. Murray, and M. E. Perkins. 1991. Isolation and characterization of rhoptries of *Plasmodium falciparum*. *Mol. Biochem. Parasitol.* **47**:51–61.

Florens, L., M. P. Washburn, J. D. Raine, R. M. Anthony, M. Grainger, J. D. Haynes, J. K. Moch, N. Muster, J. B. Sacci, D. L. Tabb, A. A. Witney, D. Wolters, Y. Wu, M. J. Gardner, A. A. Holder, R. E. Sinden, J. R. Yates, and D. J. Carucci. 2002. A proteomic view of the *Plasmodium falciparum* life cycle. *Nature* **419**:520–526.

Florent, I., S. Charneau, and P. Grellier. 2004. *Plasmodium falciparum* genes differentially expressed during merozoite morphogenesis. *Mol. Biochem. Parasitol.* **135**:143–148.

Fowler, R. E., A. M. Smith, J. Whitehorn, I. T. Williams, L. H. Bannister, and G. H. Mitchell. 2001. Microtubule associated motor proteins of *Plasmodium falciparum* merozoites. *Mol. Biochem. Parasitol.* **117**:187–200.

Garcia, C. R., M. Takeuschi, K. Yoshioka, H. Miyamoto. 1997. Imaging *Plasmodium falciparum*-infected ghost and parasite by atomic force microscopy. *J. Struct. Biol.* **119**:92–98.

Garnham, P. C. C. 1966. *Malaria Parasites and Other Haemosporidia*. Blackwell, Oxford, United Kingdom.

Guinet, F., J. A. Dvorak, H. Fujioka, D. B. Keister, O. Muratova, D. C. Kaslow, M. Aikawa, A. B. Vaidya, and T. E. Wellems. 1996. A developmental defect in *Plasmodium falciparum* male gametogenesis. *J. Cell Biol.* **135**:269–278.

Haldar, K., B. U. Samuel, N. Mohandas, T. Harrison, and N. L. Hiller. 2001. Transport mechanisms in *Plasmodium*-infected erythrocytes: lipid rafts and a tubovesicular network. *Int. J. Parasitol.* **31**:1393–1401.

Healer, J., S. Crawford, S. Ralph, G. McFadden, and A. F. Cowman. 2002. Independent translocation of two micronemal proteins in developing *Plasmodium falciparum* merozoites. *Infect. Immun.* **70**:5751–5758.

Heidrich, H. G., M. Matzner, A. Miettinen-Baumann, and W. Strych. 1986. Immunoelectron microscopy shows that the 80,000-dalton antigen of *Plasmodium falciparum* merozoites is localized in the surface coat. *Z. Parasitenkd.* **72**:681–683.

Hempelmann, E., C. Motta, R. Hughes, S. A. Ward, and P. Bray. 2004. *Plasmodium falciparum*: sacrificing membrane to grow crystals? *Trends Parasitol.* **19**:23–26.

Hopkins, J. M., R. E. Fowler, S. Krishna, I. Wilson, G. H. Mitchell, and L. H. Bannister. 1999. The plastid in *Plasmodium falciparum* asexual blood stages: a three-dimensional ultrastructural analysis. *Protista.* **150**:283–295.

Jackson, K. E., N. Klonis, D. J. P. Ferguson, A. Adisa, C. Dogovski, and L. Tilley. 2004. Food vacuole-associated lipid bodies and heterogeneous lipid environments in the malaria parasite, *Plasmodium falciparum. Mol. Microbiol.* **54**:109–122.

Jaikaria, N. S., C. Rozario, R. G. Ridley, and M. E. Perkins. 1993. Biogenesis of rhoptry organelles in *Plasmodium falciparum. Mol. Biochem. Parasitol.* **57**:269–279.

Johnson, D., K. Gunther, I. Ansorge, J. Benting, A. Kent, L. Bannister, R. Ridley, and K. Lingelbach. 1994. Characterization of membrane proteins exported from *Plasmodium falciparum* into the host erythrocyte. *Parasitology* **109**:1–9.

Kaidoh, T., J. Nath, H. Fujioka, V. Okoye, and M. Aikawa. 1995. Effect and localization of trifluralin in *Plasmodium falciparum* gametocytes: an electron microscopic study. *J. Eukaryot. Microbiol.* **42**:61–64.

Kaidoh, T., J. Nath, V. Okoye, and M. Aikawa. 1993. Novel structure in the pellicular complex of *Plasmodium falciparum* gametocytes. *J. Eukaryot. Microbiol.* **40**:269–271.

Kass, L., D. Willerson, Jr., K. H. Rieckmann, P. E. Carson, and R. P. Becker. 1971. *Plasmodium falciparum* gametocytes. Electron microscopic observations on material obtained by a new method. *Am. J. Trop. Med. Hyg.* **20**:187–194.

Kohler, S., C. F. Delwiche, P. D. Denny, L. G. Tilney, P. Webster, R. J. M. Wilson, J. D. Palmer, and D. S. Roos. 1997. A plastid of probable green algal origin in apicomplexan parasites. *Science* **275**:1485–1487.

Kriek, N., L. Tilley, P. Horrocks, R. Pinches, B. C. Elford, D. J. P. Ferguson, K. Lingelbach, and C. Newbold. 2003. Characterization of the pathway for transport of the cytoadherence-mediating protein, PfEMP1, to the host cell surface in malaria parasite-infected erythrocytes. *Mol. Microbiol.* **50**:1215–1227.

Krungkrai, J., P. Prapunwattana, and S. R. Krungkrai. 2000. Ultrastructure and function of mitochondria in gametocytic stage of *Plasmodium falciparum. Parasite* **7**:19–26.

Ladda, R. L. 1969. New insights into the fine structure of rodent malarial parasites. *Mil. Med.* **134**:825–865.

Langreth, S. G., J. B. Jensen, R. T. Reese, and W. Trager. 1978. Fine structure of human malaria in vitro. *J. Protozool.* **25**:443–452.

Ling, I. T., L. Florens, A. R. Dluzewski, O. Kaneko, M. Grainger, B. Y. S. Yim Lim, T. Tsuboi, J. M. Hopkins, J. R. Johnson, M. Torii, L. H. Bannister, I. I. I. Yates, Jr., A. A. Holder, and D. Mattei. 2004. The *Plasmodium falciparum* clag9 gene encodes a rhoptry protein that is transferred to the host erythrocyte upon invasion. *Mol. Microbiol.* **52**:107–118.

Lingelbach, K., and K. A. Joiner. 1998. The parasitophorous vacuole membrane surrounding *Plasmodium* and *Toxoplasma*: an unusual compartment in infected cells. *J. Cell Sci.* **111**:1467–1475.

Margos, G., L. H. Bannister, A. R. Dluzewski, J. M. Hopkins, I. T. Williams, and G. H. Mitchell. 2004. Correlation of structural development and differential expression of invasion-related molecules in schizonts of *Plasmodium falciparum. Parasitology* **129**:273–287.

McLaren, D. J., L. H. Bannister, P. I. Trigg, and G. A. Butcher. 1979. Freeze fracture studies on the interaction between the malaria parasite and the host erythrocyte in *Plasmodium knowlesi* infections. *Parasitology* **79**:125–139.

Meszoely, C. A., E. F. Erbe, R. L. Steere, J. Trosper, and R. L. Beaudoin. 1987. *Plasmodium falciparum*: freeze-fracture of the gametocyte pellicular complex. *Exp. Parasitol.* **64**:300–309.

Mitchell, G. H., and L. H. Bannister. 1988. Malaria parasite invasion: interactions with the red cell membrane. *Crit. Rev. Oncol. Hematol.* **8**:225–310.

Mitchell, G. H., A. W. Thomas, G. Margos, A. R. Dluzewski, and L. H. Bannister. 2004. Apical membrane antigen 1, a major malaria vaccine candidate, mediates the close attachment of invasive merozoites to host red blood cells. *Infect. Immun.* **72**:154–158.

Morrissette, N. S., J. M. Murray, and D. S. Roos. 1997. Subpellicular microtubules associate with an intramembranous particle lattice in the protozoan parasite *Toxoplasma gondii. J. Cell Sci.* **110**:35–42.

Nagao, E., O. Kaneko, and J. A. Dvorak. 2000. *Plasmodium falciparum*-infected erythrocytes: qualitative and quantitative analyses of parasite-induced knobs by atomic force microscopy. *J. Struct. Biol.* **130**:34–44.

Noe, A. R., D. J. Fishkind, and J. H. Adams. 2000. Spatial and temporal dynamics of the secretory pathway during differentiation of the *Plasmodium yoelii* schizont. *Mol. Biochem. Parasitol.* **108**:169–185.

Noland, G. S., N. Briones, and D. J. Sullivan, Jr. 2003. The shape and size of hemozoin crystals distinguishes diverse *Plasmodium* species. *Mol. Biochem. Parasitol.* **130**:91–99.

Ponnudurai, T., A. H. Lensen, J. F. Meis, and J. H. Meuwissen. 1986. Synchronization of *Plasmodium falciparum* gametocytes using an automated suspension culture system. *Parasitology* **93**:263–274.

Prensier, G., and C. Slomianny. 1986. The karyotype of *Plasmodium falciparum* determined by ultrastructural serial sectioning and 3D reconstruction. *J. Parasitol.* **72**:731–736.

Przyborski, J. M., H. Wickert, G. Krohne, and M. Lanzer. 2003. Maurer's clefts—a novel secretory organelle? *Mol. Biochem. Parasitol.* **132:**17–26.

Roger, N., J. F. Dubremetz, P. Delplace, B. Fortier, G. Tronchin, and A. Vernes. 1988. Characterization of a 225 kilodalton rhoptry protein of *Plasmodium falciparum*. *Mol. Biochem. Parasitol.* **27:**135–141.

Ruangjirachuporn, W., B. A. Afzelius, H. Helmby, A. V. S. Hill, B. M. Greenwood, J. Carlson, K. Berzins, P. Perlmann, and M. Wahlgren. 1992. Ultrastructural analysis of fresh *Plasmodium falciparum*-infected erythrocytes and their cytoadherence to human leukocytes. *Am. J. Trop. Med. Hyg.* **46:**511–519.

Ruiz, F. A., S. Luo, S. N. Moreno, and R. Docampo. 2004. Polyphosphate content and fine structure of acidocalcisomes of *Plasmodium falciparum*. *Microsc. Microanal.* **10:**563–567.

Russell, D. G., and R. G. Burns. 1984. The polar ring of coccidian protozoans: a unique microtubule-organizing centre. *J. Cell Sci.* **65:**193–207.

Salmon, B. L., A. Oksman, and D. E. Goldberg. 2001. Malaria parasite exit from the host erythrocyte: a two-step process requiring extraerythrocytic proteolysis. *Proc. Nat. Acad. Sci. USA* **98:**271–276.

Sam-Yellowe, T. Y., L. Florens, T. Wang, J. D. Raine, D. J. Carucci, R. Sinden, and I. I. I. Yates, Jr. 2004. Proteome analysis of rhoptry-enriched fractions isolated from plasmodium merozoites. *J. Proteome Res.* **3:**995–1001.

Scherf, A., R. Carter, C. Petersen, P. Alano, R. Nelson, M. Aikawa, D. Mattei, D. S. Pereira, and J. Leech. 1992. Gene inactivation of Pf11–1 of *Plasmodium falciparum* by chromosome breakage and healing: identification of a gametocyte-specific protein with a potential role in gametogenesis. *EMBO J.* **11:**2293–2301.

Shaw, M. K., J. Thompson, and R. E. Sinden. 1996. Localization of ribosomal RNA and Pbs21-mRNA in the sexual stages of *Plasmodium berghei* using electron microscope in situ hybridization. *Eur. J. Cell Biol.* **71:**270–276.

Sinden, R. E. 1982a. Gametocytogenesis of *Plasmodium falciparum* in vitro: an electron microscopic study. *Parasitology.* **84:**1–11.

Sinden, R. E. 1982b. Gametocytogenesis of *Plasmodium falciparum* in vitro: ultrastructural observations on the lethal action of chloroquine. *Ann. Trop. Med. Parasitol.* **76:**15–23.

Sinden, R. E., E. U. Canning, and B. J. Spain. 1976. Gametogenesis and fertilization in *Plasmodium yoelii nigeriensis*: a transmission electron microscope study. *Proc. R. Soc. Lond. B Biol. Sci.* **193:**55–76.

Sinden, R. E., R. H. Hartley, and N. J. King. 1985. Gametogenesis in *Plasmodium*; the inhibitory effects of anticytoskeletal agents. *Int. J. Parasitol.* **15:**211–217.

Slomianny, C. 1990. Three-dimensional reconstruction of the feeding process of the malaria parasite. *Blood Cells* **16:**669–378.

Slomianny, C., G. Prensier, and P. Charet. 1985. Ingestion of erythrocytic stroma by *Plasmodium chabaudi* trophozoites: ultrastructural study by serial sectioning and 3-dimensional reconstruction. *Parasitology* **90:**579–588.

Taraschi, T. F. 1999. Macromolecular transport in malaria-infected erythrocytes. *Novartis Found. Symp.* **226:**114–120.

Taraschi, T. F., D. Trelka, T. Schneider, and I. Matthews. 1998. *Plasmodium falciparum*: characterization of organelle migration during merozoite morphogenesis in asexual malaria infections. *Exp. Parasitol.* **88:**184–193.

Thomas, A. W., L. H. Bannister, and A. P. Waters. 1990. Sixty-six kilodalton-related antigens of *Plasmodium knowlesi* are merozoite surface antigens associated with the apical prominence. *Parasite Immunol.* **12:**105–113.

Topolska, A. E., L. Wang, C. G. Black, and R. L. Coppel. 2004. Merozoite cell biology, p. 200–215. *In* A. P. Waters and C. J. Janse (ed.), *Malaria Parasites: Genomes and Molecular Biology*. Caister Academic Press, New York, N.Y.

Torii, M., J. H. Adams, L. H. Miller, and M. Aikawa. 1989. Release of merozoite dense granules during erythrocyte invasion by *Plasmodium knowlesi*. *Infect. Immun.* **57:**3230–3233.

Van Wye, J., N. Ghori, P. Webster, R. R. Mitschler, H. G. Elmendorf, and K. Haldar. 1996. Identification and localization of rab6, separation of rab6 from ERD2 and implications for an 'unstacked' Golgi, in *Plasmodium falciparum*. *Mol. Biochem. Parasitol.* **83:**107–120.

Vickerman, K., and F. E. G. Cox. 1967. Merozoite formation in the erythrocytic stages of the malaria parasite *Plasmodium vinckei*. *Trans. R. Soc. Trop. Med. Hyg.* **61:**303–312.

Vielemeyer, O., M. T. McIntosh, K. A. Joiner, and I. Coppens. 2004. Neutral lipid synthesis and storage in the intraerythrocytic stages of *Plasmodium falciparum*. *Mol. Biochem. Parasitol.* **135:**197–209.

Waller, R. F., and G. I. McFadden. 2005. The apicoplast: a review of the derived plastid of apicomplexan parasites. *Curr. Issues Mol. Biol.* **7:**57–80.

Wickert, H., F. Wissing, K. T. Andrews, A. Stich, G. Krohne, and M. Lanzer. 2003. Evidence for trafficking of PfEMP1 to the surface of *P. falciparum*-infected erythrocytes via a complex membrane network. *Eur. J. Cell Biol.* **82:**271–284.

Wickham, M. E., J. G. Culvenor, and A. F. Cowman. 2003. Selective inhibition of a two-step egress of malaria parasites from the host erythrocyte. *J. Biol. Chem.* **278:**37658–37663.

Winograd, E., C. A. Clavijo, L. Y. Bustamante, and M. Jaramillo. 1999. Release of merozoites from *Plasmodium falciparum*-infected erythrocytes could be mediated by a non-explosive event. *Parasitol. Res.* **85:**621–624.

Wunderlich, F., H. Stubig, and E. Konigk. 1982. Development of *Plasmodium chabaudi* in mouse red blood cells: structural properties of the host and parasite membranes. *J. Protozool.* **29:**60–66.

Yayon, A., R. Timberg, S. Friedman, and H. Ginsburg. 1984. Effects of chloroquine on the feeding mechanism of the intraerythrocytic human malarial parasite *Plasmodium falciparum*. *J. Protozool.* **31:**367–372.

GENETIC MANIPULATION OF *PLASMODIUM FALCIPARUM*

Alan F. Cowman and Brendan S. Crabb

4

Methods to genetically manipulate the genome of the human malaria parasite *Plasmodium falciparum* have only been developed in the last decade. Although the technical options remain somewhat limited relative to other eukaryotic systems, much has already been achieved with this technology (Table 1). In this chapter, we review historical and technical issues, achievements to date, and future possibilities with respect to transformation of *P. falciparum* blood stages. With the complete sequence of the 23-Mb *P. falciparum* genome now available (Gardner et al., 2002), the need (and possibilities) for functional genomic studies in this organism has never been greater.

BACKGROUND AND HISTORY

Unlike most other *Plasmodium* species, *P. falciparum* blood stages can be cultured in the laboratory (Trager and Jensen, 1976), a feature that is key to the ability to transform this organism. Although this in vitro culture system has been used widely for many years, the parasite remained intransigent to genetic manipulation for the better part of 20 years. It was not until 1995 that Wu and colleagues (1995) first described an electroporation method to introduce plasmid DNA into erythrocytic stage parasites via a transient transfection procedure. This was a crucial advance; soon after, stable transfection systems were described that used this same strategy to introduce DNA parasites (Crabb and Cowman, 1996; Wu et al., 1996).

Despite the relatively low efficiency of the *P. falciparum* transfection system, it was sufficient for both gene targeting and transgene expression approaches to be developed and utilized to analyze gene function in this organism (Wu et al., 1995; Crabb and Cowman, 1996; Wu et al., 1996; Crabb et al., 1997a; Fidock and Wellems, 1997; Duraisingh et al., 2002). Numerous publications describing *P. falciparum* gene knockouts have been described, beginning with one targeting disruption of the gene encoding the cytoadherence-associated knob-associated histidine-rich protein (Crabb et al., 1997a). Gene targeting for the purposes of allelic replacement has also been extensively utilized, especially to study *P. falciparum* drug resistance (beginning with Triglia et al., 1998),

---

Alan F. Cowman and Brendan S. Crabb, The Walter and Eliza Hall Institute of Medical Research, 1G Royal Parade, Parkville 3050, Melbourne, Australia.

Molecular Approaches to Malaria, Edited by Irwin W. Sherman
© 2005 ASM Press, Washington, D.C.

TABLE 1 Biological processes investigated by *P. falciparum* transfection

Process	Sequence[a]	Approach[a]	Reference(s)
Cytoadherence- antigenic variation	KAHRP	KO, GFP	Crabb et al., 1997; Wickham et al., 2001
	PfEMP-3	KO	Waterkeyn et al., 2000
	var/Rep20	KO, RA, GS	Deitsch et al., 1999; Deitsch et al., 2001; Horrocks et al., 2002; Andrews et al., 2003; O'Donnell et al., 2002
Merozoite-RBC invasion			
Surface	MSP-1	AR	O'Donnell et al., 2000; Drew et al., 2004
	MSP-2	Trans.	Wickham et al., 2003
	MSP-3	KO	Mills et al., 2002
	MSP-5	KO	P. R. Sanders and B. S. Crabb, unpublished data
Rhoptry	RAP complex	KO	Baldi et al., 2000; Baldi et al., 2002
	RhopH1 (Clag9)	KO	Trenholme et al., 2000
	Rh2a/Rh2b	KO	Duraisingh et al., 2003
Microneme	AMA-1	Trans.	Triglia et al., 2000; Healer et al., 2004
	EBA-175	KO, AR	Kaneko et al., 2000; Reed et al., 2000; Duraisingh et al., 2003; Gilberger et al., 2003
	EBA-140	KO	Maier et al., 2003
	EBA-181	KO	Gilberger et al., 2003
Dense granules	RESA	GFP	Rug et al., 2004
Enzymes	Falcipain 1	KO	Eksi et al., 2004; Sijwali et al., 2004
	Falcipain 2	KO	Sijwali and Rosenthal, 2004
	Plasmepsins	KO	Liu et al., 2004; Omara-Opyene et al., 2004
	SERAs	KO	Miller et al., 2002
	CDPK1	GFP	Moskes et al., 2004
	Thioredoxin red	KO	Krnajski et al., 2002
Sexual stages	Pfg27	KO	Lobo et al., 1999
	Pfs48/45	KO	van Dijk et al., 2001
	Chitinase	KO	Tsai et al., 2001
	PfCTRP	KO	Templeton et al., 2000
	PfCCp2 and -3	KO	Pradel et al., 2004
	Pfs230	KO	Eksi et al., 2002
	Falcipain 1	KO	Eksi et al., 2004
	Pfs16	KO	Kongkasuriyachai et al., 2004
Protein export	Numerous	GFP	Burghaus and Lingelbach, 2001; Wickham et al., 2001; Akompong et al., 2002; Cheresh et al., 2002; Adisa et al., 2003; Lopez-Estrano et al., 2003; Cooke et al., 2004; Klemba et al., 2004; Marti et al., 2004; Hiller et al., 2004
Plastid/mitochondrion	Numerous	GFP	Waller et al., 2000; Foth et al., 2003; Sato et al., 2003; Sato et al, 2004; Sato and Wilson, 2004; Tonkin et al., 2004
Immunity	MSP-1	AR	O'Donnell et al., 2001; Singh et al., 2003; Corran et al., 2004; John et al., 2004
	AMA-1	Trans.	Healer et al., 2004
Drug resistance	DHFR-TS	Trans.	Wu et al., 1996; Fidock and Wellems, 1997
	PPPK-DHPS	AR	Triglia et al., 1998; Wang et al., 2004
	Pfmdr-1	AR	Reed et al., 2000
	Pfcrt	AR	Fidock et al., 2000; Sidhu et al., 2002; Waller et al., 2003
	Pfcg1, Pfcg2	AR	Fidock et al., 2000

[a]KO, gene knockout (or, more accurately, targeted disruption); AR, allelic replacement; GFP, GFP targeting approaches; GS, genetic screen; Trans., transgene expression; RA, reporter assays; RBC, red blood cell.

and antigen function and immunity (beginning with O'Donnell et al., 2000). An important adaptation of transgene expression systems in this organism was the utilization of green fluorescent protein (GFP) to visualize protein localization and follow trafficking in live blood stage parasites (beginning with Waller et al., 2000). Although still considered somewhat in its infancy and to a degree limited to blood stages, these gene targeting and transgenic approaches have been crucial to many studies that have yielded insight into a wide range of parasite and associated biological processes, including antigenic variation and cytoadherence, erythrocyte invasion, organelle (e.g., apicoplast and food vacuole) function, gametocytogenesis, insect stage biology, protein export and/or trafficking, drug resistance, and blood stage immunity. Table 1 summarizes the different proteins and processes that have been investigated with *P. falciparum* transfection technology.

Transfection of the rodent malaria parasite *Plasmodium berghei* emerged at about the same time as it did in *P. falciparum*; in fact, stable transformation of *Plasmodium* parasites was described first for this organism (van Dijk et al., 1995; van Dijk et al., 1996). In many ways, the development and utilization of transfection in *P. berghei* have paralleled (and are similar to) that in *P. falciparum*, but there are some significant differences. First, *P. berghei* transfection appears to be more efficient, allowing transfection of linearized DNA and direct integration into the genome. This efficiency probably relates to the fact that it is the extracellular merozoite that is transformed, a form likely to be much more amenable to DNA uptake than the intraerythrocytic forms transfected in *P. falciparum* (discussed further below). Second, *P. berghei* blood stage parasites cannot be maintained continuously in culture; therefore, selection, maintenance, and cloning of parasites are performed in vivo using a susceptible rodent. Although different, the *P. berghei* transfection system has complemented that in *P. falciparum*, especially as it allows blood stage phenotypes to be analyzed in vivo. Also, the dissection of generic aspects of *Plasmodium* sexual stage biology can be addressed in much more detail using the *P. berghei* system than is readily possible with *P. falciparum*. In this chapter, we will generally confine our comments to *P. falciparum* transformation systems, although on occasion, advances and lessons learned from the *P. berghei* system will be discussed.

GENERAL METHODOLOGY AND TECHNICAL FEATURES

Efficiency

Unlike many other systems, including rodent malaria *P. berghei*, *P. falciparum* parasites can, up to now, only be transformed with circular plasmid DNA. Stable transformation with linear DNA has not been achieved, although such DNA does enter parasites, as evidenced by the expression of reporter genes transiently transfected into *P. falciparum* (our unpublished data). The reasons for the requirement to transfect circular DNA to obtain stable transformation have not been established, although they most likely relate to the relatively low efficiency of transfection in this organism, recently calculated as $\sim 10^{-6}$ (O'Donnell et al., 2002). The reasons for the low efficiency have not been definitively determined; however, they may be related to the requirement of performing transfection at the intracellular ring stage of the blood stage cycle. Here, plasmid DNA must pass through four lipid bilayers (specifically the erythrocyte, parasitophorous vacuole, parasite, and [ultimately] nuclear membranes) before plasmid-encoded genes can be transcribed. It is possible that transit through all these membranes does not even occur at the time of transfection. Indeed, it has been demonstrated that nontransformed parasites that infect erythrocytes preloaded with DNA (under the same electroporation conditions used normally for a *P. falciparum* transfection) will take up the DNA and ultimately be transformed at a rate that is similar, if not slightly improved, to that of normal transfection (Deitsch et al., 2001). Hence, it is probable that such spontaneous but inefficient uptake of DNA from the erythrocyte cytosol is the normal occurrence in transfection of *P. falciparum* ring stages. As mentioned above, the improved efficiency achieved in rodent parasite *P. berghei* relates at least in part to the ability to harvest and transform large numbers of viable extracellular merozoites. To date, it has not proved possible to prepare sufficient viable *P. falciparum* merozoites to use this strategy.

Episomal Replication

Given the inability to transfect linear DNA and achieve fast integration into the *P. falciparum* genome, a key to the success of *P. falciparum* transfection is the somewhat mysterious ability of transfected plasmids to replicate episomally in parasites. In the presence of a selection agent, plasmids are maintained in parasites in an episomally replicating form. Although replication appears to be relatively efficient, these forms do not partition equally between daughter merozoites; hence, episomally replicating plasmids are rapidly lost from the population when drug pressure is removed (O'Donnell et al., 2001; O'Donnell et al., 2002). Concatamerization and the mechanism of replication in *P. falciparum* (Kadekoppala et al., 2001; O'Donnell et al., 2001) and in *P. berghei* (Williamson et al., 2002) have been examined to some degree. In both cases, transfected plasmids appear to replicate as circular, double-stranded concatamers that are interconnected via DNA that is at least partially single-stranded. These forms resemble and have been termed "handcuffs" (Williamson et al., 2002). The characteristics of replicative episomal forms implicate both rolling circle- and recombination-dependent replication mechanisms. Interestingly, transformed plasmids do not appear to possess a single origin of replication; rather, bacterial plasmid replication in *Plasmodium* parasites can originate from multiple sites and can occur independently of parasite sequences. Importantly, stably replicating episomes appear to assemble into phased nucleosomes in a manner that allows appropriate temporal activity of promoters encoded on these plasmids (Horrocks et al., 2002).

Positive Selectable Markers for Transfection in *P. falciparum*

The two most commonly used positive selectable markers in *P. falciparum* are a mutated *Toxoplasma gondii dhfr-ts* gene, which confers resistance to pyrimethamine (Donald and Roos, 1993), and the human *dhfr* gene, which confers resistance to the experimental antimalarial drug WR99210, a selection agent that remains effective when used with *P. falciparum* lines that are naturally resistant to pyrimethamine (Fidock and Wellems, 1997). Three other positive selectable markers have been successfully used to derive drug-resistant parasite populations. These markers, blasticidin S deaminase (Ben Mamoun et al., 1999), neomycin phosphotransferase II (Ben Mamoun et al., 1999; Wang et al., 2002), and puromycin-N-acetyltransferase (de Koning-Ward et al., 2001), confer resistance to blasticidin S, geneticin (G418), and puromycin, respectively.

FUNCTIONAL ANALYSIS OF *P. FALCIPARUM* BY TRANSFECTION

Analysis of Promoter Elements in *P. falciparum* by Transient Transfection

Transfection of *P. falciparum* blood stages, using transient transfection with the reporter genes chloramphenicol acetyl transferase (Wu et al., 1995; Lucas and Holder, 2004) and luciferase (Deitsch et al., 2001; Militello and Wirth, 2003), has been an important tool in the identification of functional characterization of a number of transcriptional control elements from this organism. A range of 5' untranslated regions have been demonstrated to contain promoters active in erythrocytic stages, including those driving expression of genes encoding 86-kDa heat shock protein (Wu et al., 1995; Militello et al., 2004), histidine-rich protein 3 (Wu et al., 1995), calmodulin (Crabb and Cowman, 1996), dihydrofolate reductase (Crabb and Cowman, 1996), proliferating cell nuclear antigen (Horrocks et al., 1996), glycophorin-binding protein 130 (Horrocks and Lanzer, 1999), apical membrane antigen 1 (AMA-1) (Triglia et al., 2000), CDP-diacylglycerol synthase (Osta et al., 2002), merozoite surface protein 2 (Wickham et al., 2003), and *P. falciparum* erythrocyte membrane protein 1 (PfEMP1; *var*) (Deitsch et al., 1999; Voss et al., 2000; Deitsch et al., 2001). These promoters appear to operate across *Plasmodium* species and are obviously centrally important elements in all *P. falciparum* transfection vectors.

A consistent theme emerges from the analysis of these *P. falciparum* promoters of a bipartite structure comprising a core basal promoter, regulated upstream by *cis*-acting sequences such as enhancers that control expression levels or other elements controlling timing of expression in the blood stage cycle (Triglia et al., 2000; Horrocks et al., 2002; Wickham et al., 2003). Although this is a typical structure for eukaryotic promoters, very few binding elements resembling those in other organisms are found in *P. falciparum* 5′ untranslated regions. It is possible that *Plasmodium* regulatory sequences encode a unique set of *cis*-acting elements (Osta et al., 2002; Militello et al., 2004).

Stable Transgene Expression in *P. falciparum*

Plasmid transfection vectors designed to express transgenes have been used extensively in *P. falciparum*, and this has been useful in a wide variety of studies (Crabb and Cowman, 1996; Crabb et al., 1997; Cowman et al., 2000; Cowman et al., 2002). Stable expression of a protein encoded by an introduced transgene has provided the means to analyze promoter elements involved in temporal expression of proteins (Wickham et al., 2003). Additionally, use of fluorescent chimeric proteins has allowed direct visualization of trafficking to subcellular and extracellular locations in *P. falciparum*-infected erythrocytes (Kadekoppala et al., 2000; Waller et al., 2000; Wickham et al., 2001). It has also been possible to address functional questions with respect to the role of parasite proteins and that of polymorphisms in the evasion of host immune responses (Triglia et al., 2000; Healer et al., 2004).

Analysis of Promoter Function and Temporal Expression of Proteins

Identification of the sequence elements within a promoter is an important biological question in all organisms, as it provides an understanding of the regulation of gene expression required for the development of the cell. This is of interest for parasitic protozoa such as *P. falciparum*, as they have complex life cycles involving different levels of gene expression required for their growth and development. This is particularly evident in the asexual life cycle in which a cascade of gene activation occurs after merozoite invasion that allows the expression of proteins responsible for remodelling the host cell and growth and development of daughter merozoites for a new round of invasion (Bozdech et al., 2003; Le Roch et al., 2003). Most promoter analysis has been performed using transient transfection; this has identified elements within promoters, but it is limited in analyzing sequences important for specific timing of gene expression. This was overcome for the merozoite surface protein 2 gene (*MSP-2* gene) by using a system of stable episomal expression of a transgene in *P. falciparum* (Wickham et al., 2003). The *MSP-2* gene is dimorphic, and antibodies are available that allow specific detection of each allelic form of the protein. This was utilized to express the FC27 allele in the 3D7 strain of parasite, so that expression of the transgenic MSP-2 protein could be detected. This provided a system to dissect the important regions of the *MSP-2* promoter with respect to function and temporal timing of expression.

Analysis of Protein Function and Polymorphisms in Immune Evasion by Transgene Expression

The ability to disrupt genes in *P. falciparum* is important in terms of analyzing the function of specific proteins; however, when a gene is essential this has not been possible (Cowman et al., 2000). The development of a system for conditional expression of genes is an important step in solving this problem but has until recently not been available (see below). Consequently, alternative approaches have been used, including allelic replacement to insert specific sequences or mutations within the endogenous gene (Triglia et al., 1998). When this has not been possible, trans-species complementation has been a useful technique in some limited examples (Triglia et al., 2000).

AMA-1 is expressed in merozoites of *P. falciparum*, and current evidence suggests this protein plays a role in parasite reorientation during invasion of the host erythrocyte (Mitchell et al., 2004). AMA-1 is an important vaccine candidate; however, it shows a high degree of polymorphism most likely driven by the selective pressure of the host immune system (Escalante et al., 2001; Polley and Conway, 2001; Cortes et al., 2003; Healer et al., 2004). Understanding the role of specific polymorphisms in evasion of the immune system is important for the design of a vaccine. AMA-1 plays an essential role in merozoite invasion as attempts to disrupt the corresponding gene have been unsuccessful, and additionally, antibodies to this protein are potent inhibitors of merozoite invasion. The role of AMA-1 in merozoite invasion has been analyzed using trans-species complementation by transfecting a *Plasmodium chabaudi AMA-1* (*PcAMA-1*) transgene gene into *P. falciparum* (Triglia et al., 2000). AMA-1 is highly conserved throughout *Plasmodium* spp. and is also present in other members of the Apicomplexa such as *T. gondii* (Hehl et al., 2000), and it was thought likely that it would be possible to directly complement the function of this protein using an ortholog from other *Plasmodium* spp. (Triglia et al., 2000). This has proved not to be possible for both *P. falciparum* and *P. berghei*, where plasmids providing expression of an AMA-1 protein from a different species were utilized in attempts to disrupt and complement the function of the endogenous gene (Kocken et al., 1998; Triglia et al., 2000). To circumvent this problem, the function of the *P. falciparum* AMA-1 (PfAMA-1) protein has been inhibited directly with antibodies specific to this protein to demonstrate that PcAMA-1 is capable of providing approximately 40% functional complementation for merozoite invasion (Triglia et al., 2000). This has provided a means to analyze the role of specific domains of AMA-1 by constructing transgenes that express chimeric forms of *P. chabaudi* and *P. falciparum* AMA-1 (Healer et al., 2004).

Additionally, it has been demonstrated that expression of PcAMA-1 in *P. falciparum* allows this parasite to invade mouse erythrocytes more efficiently, consistent with an important role in the invasion process (Triglia et al., 2000).

An understanding of the role and importance of polymorphisms within specific *P. falciparum* proteins such as AMA-1 is important for the development of a vaccine, and transgene expression can provide specific strategies to probe this important question. It has been demonstrated that antibodies to specific alleles of PfAMA-1 can inhibit merozoite invasion of the homologous strain, but other strains are less susceptible to inhibition (Crewther et al., 1996; Hodder et al., 2001). This has suggested that the polymorphisms within PfAMA-1 are responsible for resistance of some parasite strains to heterologous PfAMA-1 antibody inhibition of merozoite invasion. Transgene expression of specific alleles of PfAMA-1 in different *P. falciparum* strains has shown directly that the polymorphisms in the protein are responsible for the evasion of immune responses that inhibit parasite invasion (Healer et al., 2004). The use of transgenic expression on the background of an endogenous gene copy has proved to be very useful and provides a means to more finely dissect the role of genetic diversity and susceptibility to protective antibodies in *P. falciparum*. These findings have important implications for the development of an AMA-1-based malaria vaccine, and this strategy will be useful for future analyses of antigen diversity in evasion of immune responses.

Expression of Fluorescent Transgenes for Analysis of Trafficking

A major advance in understanding the molecular dynamics of cellular systems has derived from elegant studies in which the gene encoding GFP is appended to a gene encoding a protein of interest and transfected into a living cell (Lippincott-Schwartz et al., 2000; Klonis et al., 2002). In *P. falciparum*, this has provided the means to investigate the protein sequences required for trafficking to specific locations within the parasite and also out to the host erythrocyte (Kadekoppala et al., 2000; Waller et al., 2000; Wickham et al., 2001). The tagging of specific proteins with fluorescent proteins has also enabled the use of fluorescence recovery after photobleaching (Axelrod, 1977) to measure the physical state of the chimeric proteins in different locations (Wickham et al., 2001). This has provided additional information on the connectivity of specific compartments within the parasite-infected erythrocyte.

The identification of sequences required for trafficking not only has provided information on the mechanisms involved but also has allowed the development of predictive tools for all of the proteins being targeted to specific locations (Foth et al., 2003; Marti et al., 2004; Hiller et al., 2004). The sequences required for trafficking of proteins to the apicoplast have been analyzed using GFP-tagged proteins to demonstrate that targeting is a two-step process involving a bipartite N-terminal sequence consisting of a signal sequence for entry to the endoplasmic reticulum and a plant-like transit peptide for subsequent entry into the apicoplast (Waller et al., 2000; Cheresh et al., 2002; Foth et al., 2003). This has provided important information to predict most of the proteins in *P. falciparum* targeted to this organelle (Ralph et al., 2004). Similar approaches have identified the sequences required for trafficking of proteins through the parasitophorous vacuole membrane into the host erythrocyte, including the virulence proteins *P. falciparum* erythrocyte membrane protein 1 (PfEMP1), RIFIN, and STEVOR (Marti et al., 2004; Hiller et al., 2004). This motif has been termed the *Plasmodium* export element and appears to be conserved throughout *Plasmodium* spp. This information has enabled the prediction of many of the exported proteins in *P. falciparum* and will allow the identification of a core set of proteins in *Plasmodium* spp. required for host cell remodelling and parasite survival (Marti et al., 2004; Hiller et al., 2004).

The use of GFP-tagged proteins in *P. falciparum* for analyzing trafficking is becoming more widely used; consequently, there are a number of vectors available (Waller et al., 2000; Wickham et al., 2001; Sato et al., 2003; Sato and Wilson, 2004), including those that use the Multisite Gateway system to provide a more adaptable set of vectors and easier construction (Marti et al., 2004; Tonkin et al., 2004). These have been used for analysis of trafficking to the apicoplast (Tonkin et al., 2004) and mitochondrion and across the parasitophorous vacuole membrane to the host erythrocyte (Marti et al., 2004). The tagging of specific proteins targeted to organelles or different compartments has provided a visual means of following the development of these organelles and compartments (Waller et al., 2000; Wickham et al., 2001). This has given important insights into the development of the apicoplast (Waller et al., 2000), mitochondrion (Sato et al., 2003; Sato and Wilson, 2004), food vacuole (Cheresh et al., 2002; Klemba et al., 2004), and components outside of the parasite in the infected erythrocyte such as Maurer's clefts (Wickham et al., 2001) through the growth and division of these structures in the asexual life cycle.

Gene Targeting for Disruption

The ability to target specific gene sequences in *P. falciparum* has allowed gene disruption (Crabb et al., 1997), as well as introduction of specific sequences or mutations by allelic replacement (Triglia et al., 1998). Insertion of sequences into the *P. falciparum* genome by transfection occurs almost exclusively by homologous recombination; this has been important as it allows easier strategies for gene disruption and allelic replacement. Although insertion by illegitimate recombination appears to be possible, it is rare and only appears to be recovered in transfection experiments when the homologous recombination event is lethal or deleterious.

The first gene disruption in *P. falciparum* was of the gene encoding knob-associated histidine-rich protein (the *KAHRP* gene), and this proved that the corresponding protein was required for knob formation on the parasite-infected erythrocyte surface (Crabb et al., 1997a). It also demonstrated that knobs were required for cytoadherence via PfEMP1 under physiologically relevant flow conditions that mimic those found in the microvasculature of the human host.

These data suggested that the biological role of knobs is adhesion of flowing parasitized cells in the hemodynamic environment of the microcirculation and, once adhered, strengthening the interaction so that parasitized erythrocytes can withstand detachment. The strategy designed for disrupting the *KAHRP* gene in this study was for double-crossover recombination; however, only single-recombination events were recovered. This suggested either that double recombination was not possible or, more likely, that the event was very rare and, because of the low transfection efficiency, single recombination events dominated.

Once the ability to disrupt genes in *P. falciparum* was demonstrated, it provided the vectors and strategies to disrupt many other genes (Table 1). This has included genes encoding proteins involved in a broad range of functions including merozoite invasion (Cowman et al., 2000; Cowman et al., 2002), sporozoite development, gametocyte formation, and host cell remodelling. There have been a relatively small number of genes disrupted for which corresponding proteins are expressed in the sexual or mosquito stages of the *P. falciparum* life cycle; however, as their function is generally not required for asexual development the gene disruptions are more straightforward to obtain and the parasites can then be tested for their ability to develop into gametocytes and subsequent forms after transmission in the mosquito. Using this approach the genes encoding chitinase (Tsai et al., 2001), Pfs230 (Eksi et al., 2002), Pfs16 (Kongkasuriyachai et al., 2004), Pfg27 (Lobo et al., 1999), Pfs48/45 (van Dijk et al., 2001), PfCTRP (Trottein et al., 1995; Templeton et al., 2000), and PfCCp2 and PfCCp3 (Pradel et al., 2004) were disrupted by single-crossover recombination. The construction of these parasites that grow normally through the asexual stage allows testing of the encoded protein in development of gametocytes and transmission in the mosquito. For example, using this approach the PfCCp2 and -3 proteins have been found to be capable of forming oocyst sporozoites but were blocked in transition to the salivary gland (Pradel et al., 2004). These proteins have multiple adhesive domains and are highly conserved with orthologs in other apicomplexan parasites, including *P. berghei*. In this rodent malaria species, the gene equivalent to PfCCp3 has been identified and is called PbSR; disruption of these genes leads to blockage of sporozoite development (Claudianos et al., 2002). These results have suggested that the PfCCp family of proteins plays an essential role in mosquito transmission.

The construction of *P. falciparum* parasite lines that lack expression of specific merozoite proteins by gene disruption has provided important insights into their function (Cowman et al., 2000; Cowman et al., 2002). Initially, it was believed that disruption of genes involved in merozoite invasion would be lethal, as this is an obligatory step in the parasite life cycle. Interestingly, a large number of genes that express proteins in merozoite stages have been disrupted, and many disruptions have resulted in *P. falciparum* lines in which either there is no measurable phenotype or the loss of function is not deleterious to the survival of the parasite. This is most likely due to the redundant nature of the merozoite invasion process where many proteins are expressed that serve overlapping functions. An example of this is disruption of the erythrocyte-binding antigen 175 (*EBA175*) gene in the W2mef parasite line that results in a significant switch in receptor usage on the host cell for invasion from sialic acid dependent to independent (Reed et al., 2000; Duraisingh et al., 2003). This most likely provides a mechanism to evade host immune responses and host receptor polymorphisms that block or decrease invasion. The exceptions to the redundancy of merozoite protein function so far include the *MSP-1* and *AMA-1* genes that encode proteins that appear to be essential for the invasion process (Cowman et al., 2000; O'Donnell et al., 2000; Triglia et al., 2000). Interestingly, although it is possible to disrupt the *MSP-3* gene, a gene expressing a protein located in the parasitophorous vacuole and merozoite surface, these parasites show a decreased ability to invade erythrocytes. This suggested either that this protein was important for the invasion process or that aberrant expression of the resulting truncated protein caused a dominant negative effect, reducing the efficiency of the invasion process (Mills et al., 2002).

Despite the ability to disrupt many genes expressing merozoite proteins, a weakness of the initial gene disruption strategies was that only single-recombination events could be obtained, which required a long process of selection to obtain parasites with integrated plasmids rather than those that maintain them as episomes (Crabb et al., 1997a). It also had the disadvantage of obtaining potential reversion events by looping out of the plasmid in the absence of drug selection. Additionally, the chances of obtaining a truncated protein were increased despite integration of a plasmid into the gene with potential dominant negative mutations as have been demonstrated for truncation of PfEMP3 (Waterkeyn et al., 2000). Parasites with integrated copies of the plasmid were selected over those with episomes by sequential periods of growth on and off the drug used for selection. Episomal plasmids are segregated randomly to the daughter merozoites, resulting in some cells obtaining many copies and other cells obtaining none. In contrast, plasmids that have integrated into the expected site by homologous recombination will be segregated normally to all daughter merozoites, resulting in a distinct advantage in maintenance over episomal plasmids. Removal of selective pressure by growth without drug results in the rapid loss of episomal plasmids; after a period of time, reintroduction of drug selection will favor the parasite population with integrated copies of the plasmid. This strategy has enabled many gene disruption events to be recovered and has also provided a strategy for allelic replacement (see below).

Although gene disruption by single-crossover recombination was very useful, the disadvantages made it important to develop vector systems that favored double recombination events. This would have several major advantages, including irreversible deletion of a portion of the targeted gene, making it impossible to obtain parasites with reversion of the integration event. Also, double-crossover recombination would result in deletion of the plasmid backbone and direct insertion of the selection cassette into the gene. This would provide a much cleaner system of insertion and subsequent gene knockouts using different drug selection genes. The absence of the plasmid backbone in the first gene knockout parasites would enable easier targeting of a second gene by homologous recombination, whereas if it was present it was highly likely that the favored event would be integration into the plasmid backbone present within the genome of these parasites.

The selection of gene disruption events in *P. falciparum* using a double-crossover recombination strategy was obtained by introducing a negative selectable marker into the transfection plasmid (Duraisingh et al., 2002). Both thymidine kinase (TK) and cytidine deaminase (CD) were tested for their ability to provide a negative selection when expressed in transfected *P. falciparum*. These genes are not present in *P. falciparum*; in the presence of the appropriate drug, the enzyme activity of TK or CD synthesizes a toxic compound that inhibits normal metabolic processes such as DNA replication. Therefore, when these genes are incorporated into a transfection vector the enzymes are expressed normally and are not detrimental to parasite growth and development. However, in the presence of ganciclovir for TK and 5-fluorouracil for CD, only parasites can survive where the corresponding gene has been deleted. The positive drug-selectable marker is placed between two homologous target regions that act as templates for double-crossover recombination. Therefore, the only parasites that can survive in the presence of both positive and negative drug selection are those in which the episomal plasmid has undergone integration via a double-recombination event, deleting either TK or CD and incorporating the positive drug-selectable marker. Experiments testing both TK and CD showed that both could provide powerful negative selection in the presence of the appropriate drug; however, TK was much more successful, as vectors using CD resulted in the growth of mutant parasites resistant to the effect of 5-fluorouracil (Duraisingh et al., 2002). The vector using TK has provided a transfection system for gene knockouts that is very powerful and allowed disruption of genes that were not previously possible using single-crossover recombination strategies (Duraisingh et al., 2003).

While the *TK* gene was successful in allowing construction of parasites with specific gene disruptions, it had some shortcomings. It was found that in some cases the negative selection provided by TK and ganciclovir was not potent enough, and parasites with a single plasmid inserted via single recombination could survive in high concentrations of drug (Duraisingh et al., 2003; Maier et al., 2003). This is an important issue because of the strategy of transfection and the low efficiency. Transfected parasites are initially obtained by selecting with drugs such as WR99210 that provide a positive selection by virtue of the *hDHFR* gene (Fidock and Wellems, 1997) between the two target sequences for homologous recombination. Parasites initially maintain the plasmid as episomes; these are usually concatemers of at least three copies (O'Donnell et al., 2001b). In some cases, integration of these can occur via single-crossover recombination inserting multiple copies of the plasmid. When ganciclovir is added, there is a large amount of TK enzyme present in both of these subpopulations because of the presence of multiple copies of the gene, resulting in their death. The only parasites that should survive are those that have deleted the *TK* gene via homologous double-crossover recombination into the target gene. However, in some cases there will be some parasites that have integrated a single copy of the plasmid, and it appears that they do not express enough TK enzyme to cause death in the presence of ganciclovir. This resulted in examples of integration events that have occurred via single-crossover recombination and retention of the *TK* gene even in the presence of ganciclovir. This suggests that a more potent negative selection gene is required. Recently, the yeast *CD* gene was tested and proved to be successful; it may be better than the current negative selection plasmids using TK (A. G. Maier, A. P. Waters, and A. F. Cowman, unpublished data).

Despite the shortcomings of the TK-negative selection vector, it has proved to be very successful, as it has allowed disruption of genes previously not possible using single-crossover recombination strategies. For example, the *P. falciparum* reticulocyte-binding homolog gene family (designated Rh, also known as reticulocyte-binding-like and normalocyte-binding proteins) is involved in merozoite invasion, and attempts to disrupt the genes by single recombination transfection were not successful (Duraisingh et al., 2002; Duraisingh et al., 2003; Triglia et al., 2005). However, by using the pHTk vector that incorporated a negative selection approach with ganciclovir, four members of this gene family were disrupted. These results have provided important new information on the role of this protein family in merozoite invasion.

Gene Targeting for Allelic Replacement

While gene disruption is an important technique to address the function of specific proteins, it is not always possible because the proteins are essential for the viability of the parasite (Cowman et al., 2000; Cowman et al., 2002). An example of this is MSP-1, a protein that appears to be essential, as the gene cannot be disrupted (O'Donnell et al., 2000). This limits reverse genetic strategies to probe the function of this protein, but allelic replacement via single-crossover recombination has provided not only important information with respect to this protein but also a system to address the role of antibodies in immunity that was not previously possible (O'Donnell et al., 2001a). MSP-1 is located on the merozoite surface and undergoes a complex series of processing events, resulting in a 19-kDa glycosylphosphatidyl inositol domain remaining on the surface in the ring stage after the completion of invasion (Holder and Freeman, 1984). The rest of the processed peptides of the 195-kDa protein are shed during the invasion process. MSP-1 is an important vaccine candidate, and the MSP-1_{19} domain contains two epidermal growth factor-like domains that appear to be key to the function of the protein. Antibodies to this domain are potent inhibitors of merozoite invasion, and much effort has been directed to understanding the role of this domain and antibodies directed against it in immunity to *P. falciparum* infection.

To probe the function of the MSP-1$_{19}$ domain, allelic replacement strategies were used to insert the *P. chabaudi* MSP-1$_{19}$ domain into the *P. falciparum MSP-1* gene (O'Donnell et al., 2000; O'Donnell et al., 2001a; Drew et al., 2004). The *P. chabaudi* MSP-1$_{19}$ domain shows approximately 30% homology with the same domain from *P. falciparum*; however, it was able to replace and complement its function. This domain in *P. falciparum* appears to be highly conserved, as it is relatively invariant when compared with many different isolates, with the exception of three amino acid positions that show some variation in the population. This led to the view that the sequence of MSP-1$_{19}$ domain may be constrained, perhaps by a crucial function. The replacement of the MSP-1$_{19}$ domain of *P. falciparum* with the equivalent region of the homologous protein from *P. chabaudi* (and more recently with an even more divergent domain from *P. berghei* MSP-8) (Drew et al., 2004) clearly showed that this domain could vary significantly. The *P. falciparum* line expressing a chimeric MSP-1 protein with a *P. chabaudi* MSP-1$_{19}$ region provided an important tool to analyze the immune response of individuals using inhibition of invasion as a surrogate marker of protection (O'Donnell et al., 2001a). This showed that antibodies to the MSP-1$_{19}$ domain are a major component of the inhibitory response in *P. falciparum*-immune humans (O'Donnell et al., 2001a; Corran et al., 2004; John et al., 2004) and *P. chabaudi*- or *P. berghei*-immune mice (O'Donnell et al., 2001a; de Koning-Ward et al., 2003). This provided a system in which antibody immune responses could be compared using inhibition of merozoite invasion in two isogenic lines that differ by only one gene. This is a powerful approach that can be utilized for many other *P. falciparum* antigens.

Allelic replacement has been extremely useful for analyzing the role of mutations in drug resistance, particularly as the genes involved are usually essential and the altered amino acids can be inserted without disrupting the sequence and function of the target proteins (Cowman, 2001). This strategy was first used for the gene encoding dihydropteroate synthase (*dhps*) to show that specific polymorphisms were responsible for resistance to the sulfone and sulfonamide group of antimalarial drugs (Triglia et al., 1998). As the mutations involved are towards the 3' end of the gene, it was possible to utilize replacement of the 3' end of the gene. With this approach, a series of mutant *dhps* alleles that mirror *P. falciparum* variants found in field isolates were found to confer different levels of sulfadoxine resistance (Triglia et al., 1997). This analysis showed that alteration of Ala-437 to Gly (A437G) conferred a fivefold increase in sulfadoxine resistance on the parasite; the addition of further mutations increased the level of resistance 24-fold above that seen for the transfectant expressing the wild-type *dhps* allele (Triglia et al., 1998). This indicated that resistance to high levels of sulfadoxine in *P. falciparum* has arisen by an accumulation of mutations and that Gly-437 is a key residue, consistent with its occurrence in most DHPS enzymes from resistant isolates. These studies provided proof that the mechanism of resistance to sulfadoxine in *P. falciparum* involves mutations in the *dhps* gene and determined the relative contribution of these mutations to this phenotype.

Allelic replacement of polymorphisms identified in the *pfmdr1* gene has also been used to show that they influence chloroquine resistance and can modulate resistance or sensitivity to quinine, mefloquine, and halofantrine (Reed et al., 2000). Polymorphisms in *pfmdr1*, the gene encoding the P-glycoprotein homologue 1 (Pgh1) protein of *P. falciparum*, had previously been linked to chloroquine resistance. Pgh1 had also been implicated in resistance to mefloquine and halofantrine. However, conclusive evidence of a direct causal association between *pfmdr1* and resistance to these antimalarials remained elusive; a genetic cross suggested that Pgh1 played no role in resistance to chloroquine and mefloquine (Wellems et al., 1990). Insertion of the polymorphisms into the 3' end of the *pfmdr1* gene provided direct proof that mutations in Pgh1 can modulate resistance to mefloquine, quinine, and halofantrine (Reed et al., 2000). The same mutations influence parasite resistance towards chloroquine in a strain-specific manner, as well as the level of sensitivity to the structurally unrelated compound, artemisinin.

While allelic replacement of polymorphisms into the *pfmdr1* gene showed that they were associated with mediating higher levels of chloroquine resistance, it also proved that this gene was not sufficient for resistance to this important antimalarial (Reed et al., 2000). A genetic cross between a chloroquine-sensitive and -resistant *P. falciparum* line was used to map (Wellems et al., 1990) and eventually isolate the *pfcrt* (*P. falciparum* chloroquine resistance transporter) gene that was central to this drug resistance mechanism. A complex series of mutations were identified in *pfcrt*, although the key amino acid change present in all chloroquine-resistant isolates was a threonine mutation at position 76 (Fidock et al., 2000). Direct proof that this mutation and others in the same gene in *P. falciparum* were required for chloroquine resistance was obtained by a 3′ allelic replacement strategy that inserted these mutations into a chloroquine-sensitive strain to show that they conferred resistance to this drug (Sidhu et al., 2002).

CONDITIONAL MUTAGENESIS IN *PLASMODIUM* PARASITES

Since the development of transfection for *P. berghei* and *P. falciparum*, the need for an inducible expression/conditional knockout system was recognized as a crucial next step to allow the functional analysis of many blood stage genes. The current systems are not amenable to the expression of toxic transgenes (such as dominant-negative transgenes) or the deletion or modification of genes that are essential or important to maintenance of the erythrocytic cycle. Very recently two advances, one with *P. berghei* (Carvalho et al., 2004) and the other with *P. falciparum* (Meissner et al., 2005), have been described that demonstrate significant progress toward this goal.

Carvalho and colleagues (2004) described the successful employment of the *Flp/FRT* site-specific recombination system that allows the deletion of sequences of interest in the insect stages. This deletion occurs between sequences flanked by FRT recombination sites and only in parasites that also express the Flp recombinase in the sexual stages. To generate parasites that contain both the target locus and the recombinase, a cross-fertilization approach in mosquitoes was used. Parasites in which the target sequence has been deleted express GFP in the blood stages, allowing these to be distinguished from other progeny. Theoretically, this system should allow deleterious mutations, including knockouts, to be effected in essential blood stage genes. In practical terms, this technology is at present probably only useful in the rodent malaria where all life cycle stages can be reasonably maintained. However, future development of this and related recombination systems may allow circumstances where the recombinase is expressed in an inducible manner such that both recombinase and the FRT target gene can be contained in one parasite. Such a system would be transferable to *P. falciparum* where it is practical in most laboratories to only maintain the blood stages.

The approach toward development of a conditional mutagenesis system in *P. falciparum* was somewhat different (Meissner et al., 2005). Here, transactivation domains isolated in the related apicomplexan parasite *T. gondii* were utilized to establish a tetracycline-regulated expression system in *P. falciparum* blood stages. In this system, silent minimal *P. falciparum* promoters, which contain Tet operators fused upstream, become active in the presence of a tetracycline repressor–*T. gondii* transactivator fusion. Expression of the transgene under the control of the inducible minimal promoter is switched off by addition of the nontoxic tetracycline analogue anhydrotetracycline. Moreover, the stage specificity of expression of the inducible transgene is dependent on the timing of expression of the transactivator. This inducible system should have immediate practical value to express transgenes of interest strongly and in a stage-specific manner, whether or not these have a deleterious effect on blood stage growth. There is the prospect that the system could be adapted to generate conditional knockout lines as performed with *T. gondii* (Meissner et al., 2002).

FUTURE OUTLOOK

The ability to manipulate the genome of *P. falciparum* by reverse genetics has provided an important array of tools to address functional questions with respect to the biological processes in this parasite. The powerful approaches that can be utilized have already been demonstrated; however, much progress is still needed to improve their efficiency and utility, as well as to develop new genetic tools. In particular, there is a great need for a system to construct conditional mutants of *P. falciparum* using inducible promoters such as those described above. There have been great strides made in developing this system for *P. falciparum* (Meissner et al., 2005), and its wide availability has the potential to further advance reverse genetic approaches to analyze this parasite, enabling specific genes to be deactivated and the resulting phenotypes measured. The efficiency of transfection in *P. falciparum* is still very low, making it a difficult method; improvement would broaden the number of laboratories that can successfully use this methodology but would also allow new approaches such as library complementation to identify specific genes involved in phenotypes. While the use of reverse genetic technologies for *P. falciparum* is a very useful approach, it is important to remember that it is a tool and by itself limited; however, when combined with the full array of bioinformatic, proteomic, microarray, cell biological, and biochemical techniques, it becomes very powerful. This technology has already provided important biological insights into this parasitic protozoan, and the continued development of the tools will provide us with even more powerful approaches to understand this infectious agent and its interaction with the mosquito vector and its human host.

REFERENCES

Adisa, A., M. Rug, N. Klonis, M. Foley, A. F. Cowman, and L. Tilley. 2003. The signal sequence of exported protein-1 directs the green fluorescent protein to the parasitophorous vacuole of transfected malaria parasites. *J. Biol. Chem.* **278:**6532–6542.

Akompong, T., M. Kadekoppala, T. Harrison, A. Oksman, D. E. Goldberg, H. Fujioka, B. U. Samuel, D. Sullivan, and K. Haldar. 2002. trans expression of a *Plasmodium falciparum* histidine-rich protein II (HRPII) reveals sorting of soluble proteins in the periphery of the host erythrocyte and disrupts transport to the malarial food vacuole. *J. Biol. Chem.* **277:**28923-28933.

Andrews, K. T., L. A. Pirrit, J. M. Przyborski, C. P. Sanchez, Y. Sterkers, S. Ricken, H. Wickert, C. Lepolard, M. Avril, A. Scherf, J. Gysin, and M. Lanzer. 2003. Recovery of adhesion to chondroitin-4-sulphate in *Plasmodium falciparum* varCSA disruption mutants by antigenically similar PfEMP1 variants. *Mol. Microbiol.* **49:**655-669.

Axelrod, D. 1977. Cell surface heating during fluorescence photobleaching recovery experiments. *Biophys. J.* **18:**129-131.

Baldi, D. L., K. T. Andrews, R. F. Waller, D. S. Roos, R. F. Howard, B. S. Crabb, and A. F. Cowman. 2000. RAP1 controls rhoptry targeting of RAP2 in the malaria parasite *Plasmodium falciparum*. *EMBO J.* **19:**2435-2443.

Baldi, D. L., R. Good, M. T. Duraisingh, B. S. Crabb, and A. F. Cowman. 2002. Identification and disruption of the gene encoding the third member of the low-molecular-mass rhoptry complex in *Plasmodium falciparum*. *Infect. Immun.* **70:** 5236–5245.

Ben Mamoun, C., I. Y. Gluzman, S. Goyard, S. M. Beverley, and D. E. Goldberg. 1999. A set of independent selectable markers for transfection of the human malaria parasite *Plasmodium falciparum*. *Proc. Natl. Acad. Sci. USA* **96:**8716–8720.

Bozdech, Z., M. Llinas, B. L. Pulliam, E. D. Wong, J. Zhu, and J. L. DeRisi. 2003. The transcriptome of the intraerythrocytic developmental cycle of *Plasmodium falciparum*. *PLoS Biol.* **1:**E5.

Burghaus, P. A., and K. Lingelbach. 2001. Luciferase, when fused to an N-terminal signal peptide, is secreted from transfected *Plasmodium falciparum* and transported to the cytosol of infected erythrocytes. *J. Biol. Chem.* **276:**26838–26845.

Carvalho, T. G., S. Thiberge, H. Sakamoto, and R. Menard. 2004. Conditional mutagenesis using site-specific recombination in *Plasmodium berghei*. *Proc. Natl. Acad. Sci. USA* **101:**14931–14936.

Cheresh, P., T. Harrison, H. Fujioka, and K. Haldar. 2002. Targeting the malarial plastid via the parasitophorous vacuole. *J. Biol. Chem.* **277:**16265-16277.

Claudianos, C., J. T. Dessens, H. E. Trueman, M. Arai, J. Mendoza, G. A. Butcher, T. Crompton, and R. E. Sinden. 2002. A malaria scavenger receptor-like protein essential for parasite development. *Mol. Microbiol.* **45:**1473–1484.

Cooke, B. M., K. Lingelbach, L. H. Bannister, and L. Tilley. 2004. Protein trafficking in *Plasmodium falciparum*-infected red blood cells. *Trends Parasitol.* **20:**581–589.

Corran, P. H., R. A. O'Donnell, J. Todd, C. Uthaipibull, A. A. Holder, B. S. Crabb, and E. M. Riley. 2004. The fine specificity, but not the invasion inhibitory activity, of 19-kilodalton merozoite surface protein 1-specific antibodies is associated with resistance to malarial parasitemia in a cross-sectional survey in The Gambia. *Infect. Immun.* **72:**6185–6189.

Cortes, A., M. Mellombo, I. Mueller, A. Benet, J. C. Reeder, and R. F. Anders. 2003. Geographical structure of diversity and differences between symptomatic and asymptomatic infections for *Plasmodium falciparum* vaccine candidate AMA1. *Infect. Immun.* **71:**1416–1426.

Cowman, A. F. 2001. Functional analysis of drug resistance in *Plasmodium falciparum* in the post-genomic era. *Int. J. Parasitol.* **31:**871–878.

Cowman, A. F., D. L. Baldi, M. Duraisingh, J. Healer, K. E. Mills, R. A. O'Donnell, J. Thompson, T. Triglia, M. E. Wickham, and B. S. Crabb. 2002. Functional analysis of *Plasmodium falciparum* merozoite antigens: implications for erythrocyte invasion and vaccine development. *Philos. Trans. R. Soc. Lond. B Biol. Sci.* **357:**25–33.

Cowman, A. F., D. L. Baldi, J. Healer, K. E. Mills, R. A. O'Donnell, M. B. Reed, T. Triglia, M. E. Wickham, and B. S. Crabb. 2000. Functional analysis of proteins involved in *Plasmodium falciparum* merozoite invasion of red blood cells. *FEBS Lett.* **476:**84–88.

Crabb, B. S., and A. F. Cowman. 1996. Characterization of promoters and stable transfection by homologous and nonhomologous recombination in *Plasmodium falciparum*. *Proc. Natl. Acad. Sci. USA* **93:**7289–7294.

Crabb, B. S., B. M. Cooke, J. C. Reeder, R. F. Waller, S. R. Caruana, K. M. Davern, M. E. Wickham, G. V. Brown, R. L. Coppel, and A. F. Cowman. 1997a. Targeted gene disruption shows that knobs enable malaria-infected red cells to cytoadhere under physiological shear stress. *Cell* **89:**287–296.

Crabb, B. S., T. Triglia, J. G. Waterkeyn, and A. F. Cowman. 1997b. Stable transgene expression in *Plasmodium falciparum*. *Mol. Biochem. Parasitol.* **90:**131–144.

Crewther, P. E., M. L. S. M. Matthew, R. H. Flegg, and R. F. Anders. 1996. Protective immune responses to apical membrane antigen 1 of *Plasmodium chabaudi* involve recognition of strain-specific epitopes. *Infect. Immun.* **64:**3310–3317.

Deitsch, K., C. Driskill, and T. Wellems. 2001a. Transformation of malaria parasites by the spontaneous uptake and expression of DNA from human erythrocytes. *Nucleic Acids Res.* **29:**850–853.

Deitsch, K. W., M. S. Calderwood, and T. E. Wellems. 2001b. Malaria. Cooperative silencing elements in *var* genes. *Nature* **412:**875–876.

Deitsch, K. W., A. del Pinal, and T. E. Wellems. 1999. Intra-cluster recombination and *var* transcription switches in the antigenic variation of *Plasmodium falciparum*. *Mol. Biochem. Parasitol.* **101:**107–116.

de Koning-Ward, T. F., R. A. O'Donnell, D. R. Drew, R. Thomson, T. P. Speed, and B. S. Crabb. 2003. A new rodent model to assess blood-stage immunity to the *Plasmodium falciparum* antigen MSP-1$_{19}$ reveals a protective role for invasion inhibitory antibodies. *J. Exp. Med.* **198:**869–875.

de Koning-Ward, T. F., A. P. Waters, and B. S. Crabb. 2001. Puromycin-*N*-acetyltransferase as a selectable marker for use in *Plasmodium falciparum*. *Mol. Biochem. Parasitol.* **117:**155–160.

Donald, R. G. K., and D. S. Roos. 1993. Stable molecular transformation of *Toxoplasma gondii*: a selectable dihydrofolate reductase-thymidylate synthase marker based on drug-resistance mutations in malaria. *Proc. Natl. Acad. Sci. USA* **90:**11703–11707.

Drew, D. R., R. A. O'Donnell, B. J. Smith, and B. S. Crabb. 2004. A common cross-species function for the double EGF-like modules of the highly divergent *Plasmodium* surface proteins MSP-1 and MSP-8. *J. Biol. Chem.* **279:**20147–20153.

Duraisingh, M., A. Maier, T. Triglia, and A. F. Cowman. 2003a. Erythrocyte-binding antigen 175 mediates invasion in Plasmodium falciparum utilizing sialic acid-dependent and -independent pathways. *Proc. Natl. Acad. Sci. USA* **100:**4796–4801.

Duraisingh, M. T., T. Triglia, and A. F. Cowman. 2002. Negative selection of *Plasmodium falciparum* reveals targeted gene deletion by double crossover recombination. *Int. J. Parasitol.* **32:**81–89.

Duraisingh, M. T., T. Triglia, S. A. Ralph, J. C. Rayner, J. W. Barnwell, G. I. McFadden, and A. F. Cowman. 2003b. Phenotypic variation of *Plasmodium falciparum* merozoite proteins directs receptor targeting for invasion of human erythrocytes. *EMBO J.* **22:**1047–1057.

Eksi, S., B. Czesny, D. C. Greenbaum, M. Bogyo, and K. C. Williamson. 2004. Targeted disruption of *Plasmodium falciparum* cysteine protease, falcipain 1, reduces oocyst production, not erythrocytic stage growth. *Mol. Microbiol.* **53:**243–250.

Eksi, S., A. Stump, S. L. Fanning, M. I. Shenouda, H. Fujioka, and K. C. Williamson. 2002. Targeting and sequestration of truncated Pfs230 in an intraerythrocytic compartment during *Plasmodium falciparum* gametocytogenesis. *Mol. Microbiol.* **44:**1507–1516.

Escalante, A. A., H. M. Grebert, S. C. Chaiyaroj, M. Magris, S. Biswas, B. L. Nahlen, and A. A.

Lal. 2001. Polymorphism in the gene encoding the apical membrane antigen-1 (AMA-1) of *Plasmodium falciparum* X. Asembo Bay Cohort Project. *Mol. Biochem. Parasitol.* **113:**279–287.

Fidock, D. A., T. Nomura, R. A. Cooper, X. Su, A. K. Talley, and T. E. Wellems. 2000a. Allelic modifications of the *cg2* and *cg1* genes do not alter the chloroquine response of drug-resistant *Plasmodium falciparum*. *Mol. Biochem. Parasitol.* **110:**1–10.

Fidock, D. A., T. Nomura, A. K. Talley, R. A. Cooper, S. M. Dzekunov, M. T. Ferdig, L. M. Ursos, A. bir Singh Sidhu, B. Naude, K. W. Deitsch, X. Su, J. C. Wootton, P. D. Roepe, and T. E. Wellems. 2000b. Mutations in the *P. falciparum* digestive vacuole transmembrane protein PfCRT and evidence for their role in chloroquine resistance. *Mol. Cell* **6:**861–871.

Fidock, D. A., and T. E. Wellems. 1997. Transformation with human dihydrofolate reductase renders malaria parasites insensitive to WR99210 but does not affect the intrinsic activity of proguanil. *Proc. Natl. Acad. Sci. USA* **94:**10931–10936.

Foth, B. J., S. A. Ralph, C. J. Tonkin, N. S. Struck, M. Fraunholz, D. S. Roos, A. F. Cowman, and G. I. McFadden. 2003. Dissecting apicoplast targeting in the malaria parasite *Plasmodium falciparum*. *Science* **299:**705–708.

Gardner, M. J., N. Hall, E. Fung, O. White, M. Berriman, R. W. Hyman, J. M. Carlton, A. Pain, K. E. Nelson, S. Bowman, I. T. Paulsen, K. James, J. A. Eisen, K. Rutherford, S. L. Salzberg, A. Craig, S. Kyes, M. S. Chan, V. Nene, S. J. Shallom, B. Suh, J. Peterson, S. Angiuoli, M. Pertea, J. Allen, J. Selengut, D. Haft, M. W. Mather, A. B. Vaidya, D. M. Martin, A. H. Fairlamb, M. J. Fraunholz, D. S. Roos, S. A. Ralph, G. I. McFadden, L. M. Cummings, G. M. Subramanian, C. Mungall, J. C. Venter, D. J. Carucci, S. L. Hoffman, C. Newbold, R. W. Davis, C. M. Fraser, and B. Barrell. 2002. Genome sequence of the human malaria parasite *Plasmodium falciparum*. *Nature* **419:**498–511.

Gilberger, T., J. Thompson, T. Triglia, R. Good, M. Duraisingh, and A. Cowman. 2003a. A novel erythrocyte binding antigen-175 paralogue from *Plasmodium falciparum* defines a new trypsin-resistant receptor on human erythrocytes. *J. Biol. Chem.* **278:**14480–14486.

Gilberger, T. W., J. K. Thompson, M. B. Reed, R. T. Good, and A. F. Cowman. 2003b. The cytoplasmic domain of the *Plasmodium falciparum* ligand EBA-175 is essential for invasion but not protein trafficking. *J. Cell Biol.* **162:**317–327.

Healer, J., V. Murphy, R. Masciantonio, A. N. Hodder, A. W. Gemmill, R. Anders, A. F. Cowman, and A. H. Batchelor. 2004. Allelic polymorphisms in apical membrane antigen-1 are responsible for evasion of antibody-mediated inhibition in *Plasmodium falciparum*. *Mol. Microbiol.* **52:**159–168.

Hehl, A. B., C. Lekutis, M. E. Grigg, P. J. Bradley, J. F. Dubremetz, E. Ortega-Barria, and J. C. Boothroyd. 2000. *Toxoplasma gondii* homologue of *Plasmodium* apical membrane antigen 1 is involved in invasion of host cells. *Infect. Immun.* **68:**7078–7086.

Hiller, N. L., S. Bhattacharjee, C. van Ooij, K. Liolios, T. Harrison, C. Lopez-Estrano, and K. Haldor. 2004. A host-targeting signal in virulence proteins reveals a secretome in malarial infection. *Science* **306:**1934–1937.

Hodder, A. N., P. E. Crewther, and R. F. Anders. 2001. Specificity of the protective antibody response to apical membrane antigen 1. *Infect. Immun.* **69:**3286–3294.

Holder, A. A., and R. R. Freeman. 1984. The three major antigens on the surface of *Plasmodium falciparum* merozoites are derived from a single high molecular weight precursor. *J. Exp. Med.* **160:**624–629.

Horrocks, P., M. R. Jackson, S. Cheesman, J. H. White, and B. J. Kilbey. 1996. Stage specific expression of proliferating cell nuclear antigen and DNA polymerase δ from *Plasmodium falciparum*. *Mol. Biochem. Parasitol.* **79:**177–182.

Horrocks, P., and M. Lanzer. 1999. Mutational analysis identifies a five base pair cis-acting sequence essential for GBP130 promoter activity in *Plasmodium falciparum*. *Mol. Biochem. Parasitol.* **99:**77–87.

Horrocks, P., R. Pinches, N. Kriek, and C. Newbold. 2002a. Stage-specific promoter activity from stably maintained episomes in *Plasmodium falciparum*. *Int. J. Parasitol.* **32:**1203–1206.

Horrocks, P., R. Pinches, S. Kyes, N. Kriek, S. Lee, Z. Christodoulou, and C. I. Newbold. 2002b. Effect of *var* gene disruption on switching in *Plasmodium falciparum*. *Mol. Microbiol.* **45:**1131–1141.

John, C. C., R. A. O'Donnell, P. O. Sumba, A. M. Moormann, T. F. de Koning-Ward, C. L. King, J. W. Kazura, and B. S. Crabb. 2004. Evidence that invasion-inhibitory antibodies specific for MSP-$1_{(19)}$ can play a protective role against blood-stage *Plasmodium falciparum* infection in individuals in a malaria endemic area of Africa. *J. Immunol.* **173:**666–672.

Kadekoppala, M., P. Cheresh, D. Catron, D. D. Ji, K. Deitsch, T. E. Wellems, H. S. Seifert, and K. Haldar. 2001. Rapid recombination among transfected plasmids, chimeric episome formation and trans gene expression in *Plasmodium falciparum*. *Mol. Biochem. Parasitol.* **112:**211–218.

Kadekoppala, M., K. Kline, T. Akompong, and K. Haldar. 2000. Stable expression of a new chimeric fluorescent reporter in the human malaria parasite *Plasmodium falciparum*. *Infect. Immun.* **68:**2328–2332.

Kaneko, O., D. A. Fidock, O. M. Schwartz, and L. H. Miller. 2000. Disruption of the C-terminal region of EBA-175 in the Dd2/Nm clone of *Plas-*

modium falciparum does not affect erythrocyte invasion. *Mol. Biochem. Parasitol.* **110:**135–146.

Klemba, M., W. Beatty, I. Gluzman, and D. E. Goldberg. 2004. Trafficking of plasmepsin II to the food vacuole of the malaria parasite *Plasmodium falciparum. J. Cell Biol.* **164:**47–56.

Klonis, N., M. Rug, M. Wickham, I. Harper, A. F. Cowman, and L. Tilley. 2002. Fluorescence photobleaching analysis for the study of cellular dynamics. *Eur. J. Biophys.* **31:**36–51.

Kocken, C. H. M., A. M. van der Wel, M. A. Dubbeld, D. L. Narum, F. M. van de Rijke, G.-J. van Gemert, X. van der Linde, L. Bannister, C. Janse, A. P. Waters, and A. W. Thomas. 1998. Precise timing of expression of a *Plasmodium falciparum*-derived transgene in *Plasmodium berghei* is a critical determinant of subsequent subcellular localization. *J. Biol. Chem.* **273:**15119–15124.

Kongkasuriyachai, D., H. Fujioka, and N. Kumar. 2004. Functional analysis of *Plasmodium falciparum* parasitophorous vacuole membrane protein (Pfs16) during gametocytogenesis and gametogenesis by targeted gene disruption. *Mol. Biochem. Parasitol.* **133:**275–285.

Krnajski, Z., T. W. Gilberger, R. D. Walter, A. F. Cowman, and S. Muller. 2002. Thioredoxin reductase is essential for the survival of Plasmodium falciparum erythrocytic stages. *J. Biol. Chem.* **277:**25970–25975.

Le Roch, K. G., Y. Zhou, P. L. Blair, M. Grainger, J. K. Moch, J. D. Haynes, P. De La Vega, A. A. Holder, S. Batalov, D. J. Carucci, and E. A. Winzeler. 2003. Discovery of gene function by expression profiling of the malaria parasite life cycle. *Science* **301:**1503–1508.

Lippincott-Schwartz, J., T. H. Roberts, and K. Hirschberg. 2000. Secretory protein trafficking and organelle dynamics in living cells. *Annu. Rev. Cell Dev. Biol.* **16:**557–589.

Liu, J., I. Y. Gluzman, M. E. Drew, and D. E. Goldberg. 2005. The role of *Plasmodium falciparum* food vacuole plasmepsins. *J. Biol. Chem.* **280:**1432–1437.

Lobo, C. A., H. Fujioka, M. Aikawa, and N. Kumar. 1999. Disruption of the Pfg27 locus by homologous recombination leads to loss of the sexual phenotype in *Plasmodium falciparum. Mol. Cell* **3:**793–798.

Lopez-Estrano, C., S. Bhattacharjee, T. Harrison, and K. Haldar. 2003. Cooperative domains define a unique host cell-targeting signal in *Plasmodium falciparum*-infected erythrocytes. *Proc. Natl. Acad. Sci. USA* **100:**12402–12407.

Lucas, S. J., and A. A. Holder. 2004. An improved chloramphenicol acetyltransferase assay for *Plasmodium falciparum* transfection. *Mol. Biochem. Parasitol.* **136:**287–296.

Maier, A. G., M. T. Duraisingh, J. C. Reeder, S. S. Patel, J. W. Kazura, P. A. Zimmerman, and A. F. Cowman. 2003. *Plasmodium falciparum* erythrocyte invasion through glycophorin C and selection for Gerbich negativity in human populations. *Nat. Med.* **9:**87–92.

Marti, M., R. T. Good, M. Rug, E. Knuepfer, and A. F. Cowman. 2004. Targeting malaria virulence and remodeling proteins to the host erythrocyte. *Science* **306:**1930–1933.

Meissner, M., D. Schluter, and D. Soldati. 2002. Role of *Toxoplasma gondii* myosin A in powering parasite gliding and host cell invasion. *Science* **298:**837–840.

Meissner, M., E. Krejany, P. R. Gilson, T. F. De koning-Ward, D. Soldati, and B. S. Crabb. 2005. Tetracycline analogue-regulated transgene expression in *Plasmodium falciparum* blood stages using *Toxoplasma gondii* transactivators. *Proc. Natl. Acad. Sci. USA* **102:**2980–2985

Militello, K. T., M. Dodge, L. Bethke, and D. F. Wirth. 2004. Identification of regulatory elements in the *Plasmodium falciparum* genome. *Mol. Biochem. Parasitol.* **134:**75–88.

Militello, K. T., and D. F. Wirth. 2003. A new reporter gene for transient transfection of *Plasmodium falciparum. Parasitol. Res.* **89:**154–157.

Miller, S. K., R. T. Good, D. R. Drew, M. Delorenzi, P. R. Sanders, A. N. Hodder, T. P. Speed, A. F. Cowman, T. F. de Koning-Ward, and B. S. Crabb. 2002. A subset of *Plasmodium falciparum* SERA genes are expressed and appear to play an important role in the erythrocytic cycle. *J. Biol. Chem.* **277:**47524–47532.

Mills, K. E., J. A. Pearce, B. S. Crabb, and A. F. Cowman. 2002. Truncation of merozoite surface protein 3 disrupts its trafficking and that of acidic-basic repeat protein to the surface of *Plasmodium falciparum* merozoites. *Mol. Microbiol.* **43:**1401–1411.

Mitchell, G. H., A. W. Thomas, G. Margos, A. R. Dluzewski, and L. H. Bannister. 2004. Apical membrane antigen 1, a major malaria vaccine candidate, mediates the close attachment of invasive merozoites to host red blood cells. *Infect. Immun.* **72:**154–158.

Moskes, C., P. A. Burghaus, B. Wernli, U. Sauder, M. Durrenberger, and B. Kappes. 2004. Export of Plasmodium falciparum calcium-dependent protein kinase 1 to the parasitophorous vacuole is dependent on three N-terminal membrane anchor motifs. *Mol. Microbio.l* **54:**676–691.

O'Donnell, R. A., L. H. Freitas-Junior, P. R. Preiser, D. H. Williamson, M. Duraisingh, T. F. McElwain, A. Scherf, A. F. Cowman, and B. S. Crabb. 2002. A genetic screen for improved plasmid segregation reveals a role for Rep20 in the interaction of Plasmodium falciparum chromosomes. *EMBO J.* **21:**1231–1239.

O'Donnell, R. A., T. F. de Koning-Ward, R. A. Burt, M. Bockarie, J. C. Reeder, A. F. Cowman, and B. S. Crabb. 2001a. Antibodies against merozoite surface protein (MSP)-1 19 are a major component of the invasion-inhibitory response in individuals immune to malaria. *J. Exp. Med.* **193:**1403–1412.

O'Donnell, R. A., P. R. Preiser, D. H. Williamson, P. W. Moore, A. F. Cowman, and B. S. Crabb. 2001b. An alteration in concatameric structure is associated with efficient segregation of plasmids in transfected *Plasmodium falciparum* parasites. *Nucleic Acids Res.* **29:**716–724.

O'Donnell, R. A., A. Saul, A. F. Cowman, and B. S. Crabb. 2000. Functional conservation of the malaria vaccine antigen MSP-1_{19} across distantly related *Plasmodium* species. *Nat. Med.* **6:**91–95.

Omara-Opyene, A. L., P. A. Moura, C. R. Sulsona, J. A. Bonilla, C. A. Yowell, H. Fujioka, D. A. Fidock, and J. B. Dame. 2004. Genetic disruption of the *Plasmodium falciparum* digestive vacuole plasmepsins demonstrates their functional redundancy. *J. Biol. Chem.* **279:**54088–54096.

Osta, M., L. Gannoun-Zaki, S. Bonnefoy, C. Roy, and H. J. Vial. 2002. A 24 bp *cis*-acting element essential for the transcriptional activity of *Plasmodium falciparum* CDP-diacylglycerol synthase gene promoter. *Mol. Biochem. Parasitol.* **121:**87–98.

Polley, S. D., and D. J. Conway. 2001. Strong diversifying selection on domains of the *Plasmodium falciparum* apical membrane antigen 1 gene. *Genetics* **158:**1505–1512.

Pradel, G., K. Hayton, L. Aravind, L. M. Iyer, M. S. Abrahamsen, A. Bonawitz, C. Mejia, and T. J. Templeton. 2004. A multidomain adhesion protein family expressed in *Plasmodium falciparum* is essential for transmission to the mosquito. *J. Exp. Med.* **199:**1533–1544.

Ralph, S. A., G. G. Van Dooren, R. F. Waller, M. J. Crawford, M. J. Fraunholz, B. J. Foth, C. J. Tonkin, D. S. Roos, and G. I. McFadden. 2004. Tropical infectious diseases: metabolic maps and functions of the *Plasmodium falciparum* apicoplast. *Nat. Rev. Microbiol.* **2:**203–216.

Reed, M. B., S. R. Caruana, A. H. Batchelor, J. K. Thompson, B. S. Crabb, and A. F. Cowman. 2000a. Targeted disruption of an erythrocyte binding antigen in *Plasmodium falciparum* is associated with a switch toward a sialic acid independent pathway of invasion. *Proc. Natl. Acad. Sci. USA* **97:**7509–7514.

Reed, M. B., K. J. Saliba, S. R. Caruana, K. Kirk, and A. F. Cowman. 2000b. Pgh1 modulates sensitivity and resistance to multiple antimalarials in *Plasmodium falciparum*. *Nature* **403:**906–909.

Rug, M., M. E. Wickham, M. Foley, A. F. Cowman, and L. Tilley. 2004. Correct promoter control is needed for trafficking of the ring-infected erythrocyte surface antigen to the host cytosol in transfected malaria parasites. *Infect. Immun.* **72:**6095–6105.

Sato, S., B. Clough, L. Coates, and R. J. Wilson. 2004. Enzymes for heme biosynthesis are found in both the mitochondrion and plastid of the malaria parasite *Plasmodium falciparum*. *Protist* **155:**117–125.

Sato, S., K. Rangachari, and R. J. Wilson. 2003. Targeting GFP to the malarial mitochondrion. *Mol. Biochem. Parasitol.* **130:**155–158.

Sato, S., and R. J. Wilson. 2004. The use of DsRED in single- and dual-color fluorescence labeling of mitochondrial and plastid organelles in *Plasmodium falciparum*. *Mol. Biochem. Parasitol.* **134:**175–179.

Sidhu, A. B., D. Verdier-Pinard, and D. A. Fidock. 2002. Chloroquine resistance in *Plasmodium falciparum* malaria parasites conferred by pfcrt mutations. *Science* **298:**210–213.

Sijwali, P. S., K. Kato, K. B. Seydel, J. Gut, J. Lehman, M. Klemba, D. E. Goldberg, L. H. Miller, and P. J. Rosenthal. 2004. *Plasmodium falciparum* cysteine protease falcipain-1 is not essential in erythrocytic stage malaria parasites. *Proc. Natl. Acad. Sci. USA* **101:**8721–8726.

Sijwali, P. S., and P. J. Rosenthal. 2004. Gene disruption confirms a critical role for the cysteine protease falcipain-2 in hemoglobin hydrolysis by *Plasmodium falciparum*. *Proc. Natl. Acad. Sci. USA* **101:**4384–4389.

Singh, S., M. C. Kennedy, C. A. Long, A. J. Saul, L. H. Miller, and A. W. Stowers. 2003. Biochemical and immunological characterization of bacterially expressed and refolded *Plasmodium falciparum* 42-kilodalton C-terminal merozoite surface protein 1. *Infect. Immun.* **71:**6766–6774.

Templeton, T. J., D. C. Kaslow, and D. A. Fidock. 2000. Developmental arrest of the human malaria parasite *Plasmodium falciparum* within the mosquito midgut via *CTRP* gene disruption. *Mol. Microbiol.* **36:**1–9.

Tonkin, C. J., G. G. van Dooren, T. P. Spurck, N. S. Struck, R. T. Good, E. Handman, A. F. Cowman, and G. I. McFadden. 2004. Localization of organellar proteins in *Plasmodium falciparum* using a novel set of transfection vectors and a new immunofluorescence fixation method. *Mol. Biochem. Parasitol.* **137:**13–21.

Trager, W., and J. B. Jensen. 1976. Human malaria parasites in continuous culture. *Science* **193:**673–675.

Trenholme, K. R., D. L. Gardiner, D. C. Holt, E. A. Thomas, A. F. Cowman, and D. J. Kemp. 2000. clag9: a cytoadherence gene in *Plasmodium falciparum* essential for binding of parasitized erythrocytes to CD36. *Proc. Natl. Acad. Sci. USA* **97:** 4029–4033.

Triglia, T., M. Duraisingh, R. Good, and A. F. Cowman. 2005. Reticulocyte-binding protein homologue 1 is required for sialic acid-dependent invasion into human erythrocytes by *Plasmodium falciparum*. *Mol. Microbiol.* **55:**162–174.

Triglia, T., J. Healer, S. R. Caruana, A. N. Hodder, R. F. Anders, B. S. Crabb, and A. F. Cowman. 2000. Apical membrane antigen 1 plays a central role in erythrocyte invasion by *Plasmodium* species. *Mol. Microbiol.* **38:**706–718.

Triglia, T., J. G. T. Menting, C. Wilson, and A. F. Cowman. 1997. Mutations of dihydropteroate synthase are responsible for sulfone and sulfonamide resistance in *Plasmodium falciparum*. *Proc. Natl. Acad. Sci. USA* **94:**13944–13949.

Triglia, T., P. Wang, P. F. G. Sims, J. E. Hyde, and A. F. Cowman. 1998. Allelic exchange at the endogenous genomic locus in *Plasmodium falciparum* proves the role of dihydropteroate synthase in sulfadoxine-resistant malaria. *EMBO J.* **17:**3807–3815.

Trottein, F., T. Triglia, and A. F. Cowman. 1995. Molecular cloning of a gene from *Plasmodium falciparum* that codes for a protein sharing motifs found in adhesive molecules from mammals and plasmodia. *Mol. Biochem. Parasitol.* **74:**129–141.

Tsai, Y. L., R. E. Hayward, R. C. Langer, D. A. Fidock, and J. M. Vinetz. 2001. Disruption of *Plasmodium falciparum* chitinase markedly impairs parasite invasion of mosquito midgut. *Infect. Immun.* **69:**4048–4054.

van Dijk, M. R., C. J. Janse, J. Thompson, A. P. Waters, J. A. Braks, H. J. Dodemont, H. G. Stunnenberg, G. J. van Gemert, R. W. Sauerwein, and W. Eling. 2001. A central role for P48/45 in malaria parasite male gamete fertility. *Cell* **104:**153–164.

van Dijk, M. R., C. J. Janse, and A. P. Waters. 1996. Expression of a *Plasmodium* gene introduced into subtelomeric regions of *Plasmodium berghei* chromosomes. *Science* **271:**662–664.

van Dijk, M. R., A. P. Waters, and C. J. Janse. 1995. Stable transfection of malaria parasite blood stages. *Science* **268:**1358–1362.

Voss, T. S., J. K. Thompson, J. Waterkeyn, I. Felger, N. Weiss, A. F. Cowman, and H. P. Beck. 2000. Genomic distribution and functional characterisation of two distinct and conserved *Plasmodium falciparum var* gene 5' flanking sequences. *Mol. Biochem. Parasitol.* **107:**103–115.

Waller, K. L., R. A. Muhle, L. M. Ursos, P. Horrocks, D. Verdier-Pinard, A. B. Sidhu, H. Fujioka, P. D. Roepe, and D. A. Fidock. 2003. Chloroquine resistance modulated in vitro by expression levels of the *Plasmodium falciparum* chloroquine resistance transporter. *J. Biol. Chem.* **278:**33593–33601.

Waller, R. F., M. B. Reed, A. F. Cowman, and G. I. McFadden. 2000. Protein trafficking to the plastid of *Plasmodium falciparum* is via the secretory pathway. *EMBO J.* **19:**1794–1802.

Wang, P., Q. Wang, T. V. Aspinall, P. F. Sims, and J. E. Hyde. 2004. Transfection studies to explore essential folate metabolism and antifolate drug synergy in the human malaria parasite *Plasmodium falciparum*. *Mol. Microbiol.* **51:**1425–1438.

Wang, P., Q. Wang, P. F. Sims, and J. E. Hyde. 2002. Rapid positive selection of stable integrants following transfection of *Plasmodium falciparum*. *Mol. Biochem. Parasitol.* **123:**1–10.

Waterkeyn, J. F., M. E. Wickham, K. Davern, B. M. Cooke, J. C. Reeder, J. G. Culvenor, R. F. Waller, and A. F. Cowman. 2000. Targeted mutagenesis of *Plasmodium falciparum* erythrocyte membrane protein 3 (PfEMP3) disrupts cytoadherence of malaria-infected red blood cells. *EMBO J.* **19:**2813–2823.

Wellems, T. E., L. J. Panton, I. Y. Gluzman, R. V. do Rosario, R. W. Gwadz, J. A. Walker, and D. J. Krogstad. 1990. Chloroquine resistance not linked to *mdr*-like genes in a *Plasmodium falciparum* cross. *Nature* **345:**253–255.

Wickham, M. E., M. Rug, S. A. Ralph, N. Klonis, G. I. McFadden, L. Tilley, and A. F. Cowman. 2001. Trafficking and assembly of the cytoadherence complex in *Plasmodium falciparum*-infected human erythrocytes. *EMBO J.* **20:**5636–5649.

Wickham, M. E., J. K. Thompson, and A. F. Cowman. 2003. Characterisation of the merozoite surface protein-2 promoter using stable and transient transfection in *Plasmodium falciparum*. *Mol. Biochem. Parasitol.* **129:**147–156.

Williamson, D. H., P. R. Preiser, P. W. Moore, S. McCready, M. Strath, and R. J. Wilson. 2002. The plastid DNA of the malaria parasite *Plasmodium falciparum* is replicated by two mechanisms. *Mol. Microbiol.* **45:**533–542.

Wu, Y., L. A. Kirkman, and T. E. Wellems. 1996. Transformation of *Plasmodium falciparum* malaria parasites by homologous integration of plasmids that confer resistance to pyrimethamine. *Proc. Natl. Acad. Sci. USA* **93:**1130–1134.

Wu, Y., C. D. Sifri, H.-H. Lei, X.-S. Su, and T. E. Wellems. 1995. Transfection of *Plasmodium falciparum* within human red blood cells. *Proc. Natl. Acad. Sci. USA* **92:**973–977.

THE TRANSCRIPTOME OF THE MALARIA PARASITE *PLASMODIUM FALCIPARUM*

Karine Le Roch and Elizabeth Winzeler

5

Studies of gene expression in the malaria parasite have advanced significantly in recent years as the result of reporter gene constructs in transfection experiments, the release of the genome sequence, computational genome analysis, and the development of high-throughput techniques such as serial analysis of gene expression (SAGE), microarrays, and multidimensional protein identification technology (MudPIT). In this chapter, we explain how such data, when combined, can give a comprehensive overview of the transcriptome of the malaria parasite throughout its life cycle. These data are expected to lead to the identification of key *Plasmodium*-specific regulatory targets, a fundamental step for the development of new antimalarials.

Malaria parasites, like so many other vector-borne pathogens, have evolved a complex life cycle consisting of the mosquito stage, the liver stage, and the erythrocytic stage (see chapter 1 for a complete description of the life cycle). During the life cycle, the parasites invade cells, undergo morphological changes, and multiply within several distinct cell types. The ability of the malaria parasite to complete such a life cycle in a vertebrate and an invertebrate implies a high degree of transcriptional and/or posttranscriptional regulations to ensure that the necessary proteins are present at the appropriate time and in adequate amounts for precise developmental progression.

Gene expression throughout the cell cycle is usually regulated by complex interactions between DNA regulatory motifs and transcription factors that bind to these motifs, as well as regulation by distinct chromatin structures. Until recently, gene expression and regulation in malaria parasites were poorly understood, a result of the complexity of the life cycle, the fact that the organism is haploid during most of its development, and difficulties in the use of traditional forward and reverse genetic tools. Indeed, genetic tools, such as transfection and genetic modification, have only recently been developed and are still technically challenging and time consuming for use with the malaria parasite. Inducible promoters and methods for rapidly generating mutants, which have been used in other models, are not yet available for

---

Karine Le Roch, Genomics Institute of the Novartis Research Foundation, San Diego, CA 92121. *Elizabeth Winzeler,* Department of Cell Biology ICND202, The Scripps Research Institute, 10550 North Torrey Pines Rd., La Jolla, CA 92037.

Molecular Approaches to Malaria, Edited by Irwin W. Sherman
© 2005 ASM Press, Washington, D.C.

Plasmodium. RNA interference experiments, proven to be valuable in characterizing transcriptional regulation in several parasites (Kuwabara and Coulson, 2000; McRobert and McConkey, 2002; Robinson and Beverley, 2003), are not conclusive for *Plasmodium*.

In an international effort to increase the understanding of the malaria parasite and to accelerate the discovery of drugs and protective vaccines, the entire 22.8-Mb genome of the most lethal malaria species, *Plasmodium falciparum* (consisting of 14 chromosomes, a linear mitochondrial genome, and a circular plastid-like genome), was sequenced and published in October 2002. The *P. falciparum* genome is twice the size of the yeast (*Schizosaccharomyces pombe*) genome, with the richest (80.6%) A+T content to date (90% in the introns and intergenic regions). There are at least 5,409 predicted open reading frames (ORFs), but over 60% lack sequence similarity to genes from any other sequenced organism (Gardner et al., 2002). Detection of over 2,400 proteins by mass spectrometry showed that a large number of these hypothetical ORFs are transcribed and validated gene prediction algorithms (Florens et al., 2002). Thus, almost two-thirds of the plasmodial proteins appear to be unique to this organism. This is a reflection of a either greater evolutionary distance from other organisms or the reduction in sequence similarity, due to the A+T richness of the genome (Gardner et al., 2002). Particular categories of genes appeared to be overrepresented in the *P. falciparum* genome, i.e., those involved in immune evasion and host-parasite interactions. Other categories were underrepresented, i.e., those associated with cell cycle, cell organization and biogenesis, enzymes, transporters, and transcription factors (Gardner et al., 2002). The presence of underrepresented gene families does not necessarily mean that fewer genes are involved in these processes than in other organisms but highlights a lack of biological knowledge. Although defining putative roles for these unannotated ORFs in the absence of homologs in other organisms remains challenging, discovery of their roles and identification of *Plasmodium*-specific key regulatory elements will be fundamental to understanding how these parasites regulate their complex and varied developmental progression.

Understanding a complex eukaryote such as *P. falciparum* at the biological level requires more than just knowledge of genes and genome; it must incorporate a full appreciation of the complex parasite-host interactions that occur throughout the life cycle. In the past few years, high-throughput functional techniques (transcriptomic and/or proteomic techniques) have emerged as significant tools for rapidly and cost-effectively elucidating gene functions. SAGE or DNA microarray technologies represent powerful tools for generating quantitative gene expression information. Variations in gene expression may reflect important aspects of biological function. Up- or down-regulation of a gene or set of genes in a particular stage of the life cycle may provide information about a gene's function, its involvement in a metabolic pathway, or interaction with the host. Systematic characterization of expression patterns may also provide information for interpreting biological significance and lead to an understanding of how developmental events are controlled.

Although gene expression information may aid in the identification of transcription factors or DNA regulatory motifs within the *Plasmodium* genome, the high A+T content of the genome (80.6%) and its lack of similarity to known eukaryotic genomes (Gardner et al., 2002) may make this somewhat more difficult. Until recently, the most complete but not always highly comprehensive data regarding gene expression and regulation for *Plasmodium* were of the *var* gene family, located at the subtelomeric regions of all 14 chromosomes and involved in antigenic variation (see chapter 21). Despite the importance of the *var* gene family in pathogenicity and immune evasion, the mechanisms by which *var* gene expression is controlled may not be entirely relevant to other genes. Indeed, transcription of the *var* genes is likely to be influenced by immune pressure from the host and epigenetic regulatory mechanisms. Together with the *rifin* and *stevor* gene families, *var* genes are certainly unique because of their high copy number and their subtelomeric location.

The development of high-throughput techniques (SAGE, microarrays, and proteomic tools) together with computational analysis has facilitated the global analysis of gene expression in *P. falciparum*. Indeed, these techniques have demonstrated that steady-state RNA levels change throughout the parasite life cycle, indicating control of gene expression at the level of mRNA transcription and/or stability. In this chapter, we explain how, when combined, such data can give a comprehensive overview of the transcriptome of the malaria parasite throughout its life cycle. These data will almost certainly lead to the identification of key *Plasmodium*-specific regulatory targets that could be fundamental for the development of new antimalarials.

UNDERSTANDING TRANSCRIPTION WITHIN THE *var* GENE FAMILY

Due to the role of the *var* gene family in antigenic variation and consequently its link to virulence and pathogenicity, *var* transcripts and transcriptional regulation have been extensively analyzed. The *var* gene family consists of approximately 40 to 60 genes found either as a single copy in the subtelomeric regions or as a cluster of tandem repeats in the central region of the chromosome. All *var* genes seem to have an identical structure with two exons separated by an intron containing the splicing consensus sites; the first exon corresponds to the polymorphic extracellular portion and the second corresponds to the highly conserved transmembrane anchor (Gardner et al., 2002; Su et al., 1995).

Parasite isolates possess a high degree of diversity within their *var* genes because of duplications, deletions, or recombination events that provide the parasite with a huge repertoire of antigenic determinants. Most of the *var* genes are transcribed monocistronically at the ring stage (Scherf et al., 1998). However, only one gene appears to be fully transcribed and translated at the trophozoite stage, and this determines the antigenic phenotype of the infected red blood cell (Gardner et al., 1996; Kyes et al., 2000). *var* gene switching, regulated at the transcription level, occurs at a frequency estimated at 2 to 18% per generation (Gatton et al., 2003). It is interesting that some silent *var* transcripts in the ring stage do not seem to have a consensus start codon. An additional interesting feature is that a few *var* sterile transcripts also start within the intron and are able to be read through an internal stop codon (Bischoff et al., 2000). This finding is noteworthy because it suggests that *Plasmodium* pseudogenes may be, in fact, fully functional. The role of truncated and untranslated *var* transcripts in the erythrocytic stages remains unclear, but such transcripts have also been detected for other genes, indicating that this phenomenon is not specific to the *var* gene family (Chen et al., 1998).

Although the structure of the different *var* genes seems to be identical, the sequence of the upstream region differs and is dependent on the gene's location, i.e., subtelomeric or within the internal chromosome cluster (Voss et al., 2000). The transcription start site was identified to be 1 kb upstream of the methionine start codon, indicating a relatively long 5' untranslated region. The exact significance of these two promoter sequences, subtelomeric or internal chromosome cluster, with a 30-bp conserved sequence remains unknown. Also, microarray analysis (see below; Le Roch et al., 2003) using the 3D7 parasite strain found that subtelomeric *var* genes appear to be relatively transcriptionally silent at the trophozoite stage, whereas several of the internal *var* genes showed a dramatic change in transcript level across the life cycle, suggesting that allelic exclusion could be occurring at the transcriptional level between these loci (Le Roch et al., 2003).

To understand a possible mechanism of allelic exclusion, a 5' regulatory region of a *var* gene was cloned upstream of a luciferase reporter gene and transfected into the malaria parasite. It was observed that the *var* promoter on the episomal construct appeared to be always transcriptionally active (Deitsch et al., 1999), suggesting that silent *var* promoters become active when removed from the chromosomal context. This experiment also showed that transcription factors are certainly present and that epigenetic regulation involving chromatin structure may occur in the malaria parasite. To further analyze transcriptional silencing in *var* genes, the possibility of a regulatory element downstream of the *var* promoters was considered. Initially, the single conserved intron found in all *var* genes was inserted downstream of a *var* promoter-luciferase transcription unit in a plasmid construct. The intron was placed downstream of a stop codon and polyadenylation signal of the transcript. Transfecting *P. falciparum* with this plasmid resulted in complete *var* promoter silencing (Deitsch et al., 2001). These findings led to the conclusion that this was a cooperative effect requiring both the intron and an additional element existing somewhere in the 5′ regulatory region of the *var* gene. Also, this silencing effect required transition of the transfected parasite through the S phase of the cell cycle, suggesting that expression is regulated by chromatin structure.

In addition to the *var* gene family, two other large multicopy gene families have been described in the *P. falciparum* genome: the *rifin* and *stevor* gene families (Cheng et al., 1998; Fernandez et al., 1999; Kyes et al., 1999). Both have interesting and complex expression patterns and seem to be overexpressed in samples from patients, leading to the proposal that host factors play an additional role in the transcriptional regulation of those genes involved in immune evasion (Fernandez et al., 1999; Daily et al., 2005).

These extensive studies of the *var* gene family have improved our understanding of gene expression and transcriptional regulation in the malaria parasite. Experiments with these gene families led to the conclusion that transcriptional regulation is tightly regulated and involves several factors. As antigenic variation is intimately linked to the pathogenicity of the parasite, determining the molecular mechanisms involved in the transcriptional controls of the cell cycle or life cycle progression will unquestionably lead to a general appreciation of transcriptional regulation, as well as an increased understanding of the parasite's biology. Although tools such as microarrays and high-throughput proteomic methods could shed light on how *var* gene expression is regulated, the extreme diversity of these genes will make the task a challenging one.

SAGE AND ANTISENSE DETECTION

Before the complete release of the *P. falciparum* genome sequence, several genomic efforts were accomplished to investigate the transcriptome of the malaria parasite. The most complete and technically advanced project was realized by the application of the SAGE technique (Munasinghe et al., 2001; Patankar et al., 2001). SAGE was one of the first techniques used to collect genome-wide expression data. In SAGE, short sequence tags are isolated from the 3′ ends of transcripts, concatenated, and then sequenced before being mapped back to the nucleotide sequence (Velculescu et al., 1995). SAGE detected 4,866 transcripts in an asynchronous culture of the stages of erythrocytic *P. falciparum*. This work demonstrated that a large proportion of the *P. falciparum* genes were expressed during the erythrocytic stages of the life cycle and gave an estimate of those genes that were more highly expressed than others. Interestingly, this analysis showed that 17% of the tags corresponded to the antisense strand of annotated genes (Gunasekera et al., 2004; Patankar et al., 2001). Detection of such antisense transcripts has been confirmed. Indeed, although stable *MSP2* antisense transcripts have been found in all asexual stages (Kyes et al., 2002), it has been suggested that they represent read-through from the adjacent adenylosuccinate lyase gene, which is situated tail to tail with *MSP2*. The identification of widespread antisense transcription has led to speculation that these transcripts play a role in gene expression regulation. Nevertheless, if the antisense transcripts do have a role in regulation of specific transcripts, a few of them could possibly be the result of undetected ORFs and/or overlaps between adjacent genes (Kyes et al., 2002). Indeed, when microarray data containing both sense and antisense probes were used (see the description of the Affymetrix technique described in greater detail below), antisense transcripts in the *P. falciparum* genome were detected slightly above background for 10% of the transcripts from a mixed population of erythrocytic stages. Hybridization of cRNA, cDNA, and direct labeling of mRNA (Rosenow et al., 2001) concluded that a high level of antisense transcripts detected using standard protocols (cDNA or cRNA hybridizations) was likely to be an artifact introduced by the reverse retrotranscriptase and subsequent amplification steps (Le Roch, unpublished results). The analysis of specific antisense transcripts throughout the life cycle of the malaria parasite will require careful experimental protocols.

EXPRESSION ANALYSIS BEFORE COMPLETION OF THE GENOME SEQUENCE

The first large-scale expression study of *P. falciparum* was accomplished with a DNA microarray platform constructed using a nuclease-generated genomic library (Hayward et al., 2000). This array, containing 3,648 random sequences, identified stage-specific transcripts between the asexual trophozoite and the sexual gametocyte stages. One year later, a 944-element expressed sequence tag-based DNA microarray produced observations of the first temporal change in gene expression throughout the erythrocytic cycle. These first experimental procedures, using microarray techniques as well as the construction of a cDNA library to explore the transcriptome of the malaria sporozoite stage (Kappe et al., 2001), showed changes and specificities within the transcriptome of the malaria parasite throughout the life cycle. The major limitations for these experiments are the incomplete sizes of the cDNA libraries or expressed sequence tags and the small number of different stages subjected to analysis.

MICROARRAY TECHNOLOGIES AND FULL-GENOME COVERAGE APPROACH

With the completion of the *P. falciparum* genome, two transcriptional analyses covering the entire genome were published (Bozdech et al., 2003a; Le Roch et al., 2003). These two alternative formats for oligonucleotide-based microarrays are now commonly utilized. Both techniques allow effective and complete genome design. Despite differences in the technologies, short oligonucleotide (Le Roch et al., 2002; Le Roch et al., 2003) versus long-oligomer microarray (Bozdech et al., 2003a; Bozdech et al., 2003b), both studies showed comparable expression patterns for the erythrocytic stages and emphasized that microarray technologies are both reliable and powerful (Labo et al., 2004). When well designed and analyzed, biochips can be a critical tool for answering significant biological questions.

Le Roch et al. (2003) used a customized high-density oligonucleotide array (constructed using in situ synthesis by photochemistry and mask-based photolithography; Affymetrix, Santa Clara, Calif.) (Color Plate 4). This array consisted of a variable number of probes matching the predicted ORFs (from 1 to >200 probes, depending of the length of the genes) and probes in the intergenic and antisense coding regions. All probes were considered a perfect match; control probes for a novel background calculation algorithm were also included. All together, the array contained single-stranded 25-mer probes covering the 23-Mb genome with an average density of one probe per 75 bases for double-stranded samples and 150 bases for single-stranded samples.

About the same time that we produced the custom-made malaria array, Bozdech et al. constructed a gene-specific microarray of the *P. falciparum* genome sequence, using the publicly available resources from the Malaria Genome Consortium. For each of the predicted ORFs, they designed 70-mer oligonucleotide array elements for the entire genome (Bozdech et al., 2003b). Their DNA microarray used 7,462 70-mer oligonucleotides, representing 4,488 of the 5,409 annotated ORFs.

EXPERIMENTAL DESIGN AND EXPRESSION LEVEL

For the high-density oligonucleotide array experiment, total RNA was extracted from different stages of the malaria parasite life cycle including the sporozoite stage; seven time points spanning the intraerythrocytic cell cycle (using two independent synchronization methods to obtain replicates and reveal genes that were under true cell cycle control), and mature gametocytes. To analyze the expression level within the array, the match-only integral distribution algorithm was used (Zhou and Abagyan, 2002; Zhou and Abagyan, 2003) to give an absolute expression level for each gene and to establish that 88% of the predicted genes were expressed in at least one stage of the life cycle. Expression levels throughout this life cycle varied by 5 orders of magnitude. Fifty-one percent of the genes were considered constitutively expressed. We speculate these are important for the maintenance of the parasite through its life cycle. The remaining 49% of the expressed genes were found to be life cycle regulated.

The 70-mer oligonucleotide array study was confined to a high-resolution analysis of the erythrocytic cell cycle (samples were collected hourly for a 48-h period). Because experimental variability is high for glass slide arrays, expression values were normalized throughout the intraerythrocytic developmental cycle (IDC) of *P. falciparum* to a common pool control in a standard two-color competitive hybridization (Eisen and Brown, 1999). The relative abundance of individual mRNAs varied continuously throughout the IDC with a single maximum and a single minimum. In contrast to the short oligonucleotide array, Bozdech et al. (2003b) found only 20% of the genes with a relative constant expression profile, but discrepancies may have been due to differences in the statistical criteria used.

CLUSTER ANALYSIS OF THE TRANSCRIPTOME VERSUS FAST FOURIER TRANSFORM (FFT) ANALYSIS

To obtain a comprehensive view of genome-wide expression, various methods of cluster analysis using statistical algorithms have been used to organize and group genes according to similarities in expression patterns. For the high-density oligonucleotide array, genes whose expression was regulated were first grouped on the basis of expression time through the life cycle. Groups were assigned using a robust k-mean algorithm (Color Plate 5A). The cluster number, $k = 15$, was arbitrarily chosen as a reasonable estimate for the biological conditions analyzed. Despite the fact that most of the genes in these clusters were hypothetical proteins (48 to 88%), a sufficient number had already been described experimentally, had homology with other species, and/or were suggested to have specific cellular roles. By comparing gene ontology (GO) rosters with gene cluster rosters, it was observed that the overlap was much greater, i.e., by several orders of magnitude, than could be expected by chance (Le Roch et al., 2003). This observation suggested that a gene's expression profile provides insight into that gene's cellular role.

Despite the usefulness of hierarchical clustering for comparing sets of expression data, Bozdech et al. (2003b) used a different approach in the analysis of expression patterns. They applied the simple Fourier analysis technique to calculate the apparent phase and the frequency of expression for each gene during the IDC (Color Plate 5B). To create a phaseogram of the IDC transcriptome of *P. falciparum*, a score for each expression profile was calculated based upon the tightness in the periodicity and the amplitude of the peak. The IDC phaseogram showed a cascade of expression from the ring to the schizont stage (Bozdech et al., 2003a). Using this FFT method of analysis, they demonstrated a programmed cascade of cellular processes that ensured the completion of the *P. falciparum* IDC; their work showed that functionally related genes usually have common expression profiles.

Despite the use of dissimilar technologies, different *P. falciparum* strains (3D7 versus HB3), various methods of synchronization (sorbitol versus thermocycling incubation), and various sample time points throughout the erythrocytic cycle, the expression profiles were almost identical. In addition, both studies came to the same conclusions: the malaria parasite has a tight cascade of gene expression, and genes with correlated temporal expression patterns often share similar functional roles. To further elucidate the biological relevance of the data sets and to explore the program of gene regulation as well as for the discovery of novel biochemical pathways, additional experiments and computational tools will be needed.

BIOLOGICAL RELEVANCE OF EXPRESSION PROFILING

The advantages of using gene expression profiling for the malaria parasite are enormous. In contrast to the analysis of human tissues, malaria parasites can be synchronized and thus represent a homogenous population. In addition, it appears to be difficult to perturb the cascade of gene expression throughout the erythrocytic cell cycle by applying stress or drugs (see below). These phenomena may explain why the malaria parasite represents an excellent model for understanding of gene expression with microarray technology and why the results are exceptionally reproducible from platform to platform. Indeed, microarray technology together with bioinformatics and data processing can correlate a change in gene expression with a strong biological relevance. Although cluster analysis together with the principle of guilt by association (Waters, 2003) constitutes a significant advance for the better understanding of many hypothetical proteins in *Plasmodium*, there is still a need to maximize the vast amount of gene expression data sets to fully exploit the potential of high-throughput genomic approaches and to narrow our understanding of parasite gene expression. To make the most of high-throughput techniques, Zhou et al. (2004) applied a novel data-mining algorithm, ontology-based pattern identification (OPI), to the malaria life cycle data. OPI is a partially supervised clustering method that systematically discovers the expression patterns that best represent functionally related genes based on the principle of guilt by association. OPI uses the Gene Ontology (GO) consortium to partially organize clusters. Using the malaria life cycle expression data sets with the *Plasmodium* gene annotation and some a priori knowledge of the functional classification of a subset of genes, OPI creates a dynamical cluster maximized for its biological information content. With this technique, rather than the k-mean algorithm, genes are grouped into multiple functional categories according to their expression profiles and their association with multiple biological functions—a model closely related to biological reality. Similar methods to analyze microarray data have been published recently with a similar idea: challenging in silico biological interpretations of microarray experiments with the hierarchical nature of GO terms (Breitling et al., 2004a and 2004b; Lee et al., 2004; Toronen, 2004).

METABOLIC PATHWAY ANALYSIS

While it may be obvious that genes involved in multiprotein complexes or with a similar function should be coexpressed, it might be unexpected to find that genes involved in a single metabolic pathway would cluster together. Indeed, genes involved in a single metabolic pathway may be posttranscriptionally regulated or specifically activated at the protein level only when needed. To investigate possible coexpression of genes involved in single metabolic pathways, Young et al. (submitted for publication) used the OPI for this analysis. They found, for example, that the cluster for carbohydrate metabolism included 7 of the 10 enzymes involved in glycolysis, with the exception of the hexokinase aldolase (first step) and pyruvate kinase (last step). The absence of aldolase is not surprising because this enzyme is known to have a role in the invasion pathway as well (Sibley, 2004). Overall, this result demonstrated that genes associated with similar biochemical pathways are generally expressed together. One surprising observation was that, in addition to genes involved in glycolysis, this cluster possessed several ribosomal proteins that represent a significant overlap between protein synthesis and carbohydrate metabolism. Although the two processes could possibly be under similar transcriptional regulation, this result may highlight the fact that the numbers of biological conditions examined thus far were insufficient to distinguish all molecular processes. Although OPI attempts to separate functionally unrelated coexpressed genes by clustering according to biological knowledge, single hypothetical genes may be found in multiple clusters that contain overlapping genes. This observation is somewhat troublesome when attempting to predict functions of coclustering genes, but such data are extremely valuable for generating hypotheses about gene networks and multifunctional genes. The new computational tool of OPI will generate a more complete knowledge of the parasite's metabolic pathways or "metabolome."

TRANSCRIPTIONAL CONTROL IN THE MALARIA PARASITE

Mechanisms controlling transcriptional or posttranscriptional activations are fundamental in eukaryotic cells. Gene expression in parasitic protozoa has shown itself to be unique when it comes to mechanisms of regulation. For instance, *Leishmania* and other members of the Trypanosomatidae have polycistronic transcription with maturation of their mRNA by transplicing events (Agabian, 1990). Nuclear run analyses of *Leishmania* or *Trypanosoma* species have shown that the steady-state level of mRNA is essentially regulated by posttranscriptional regulation and/or mRNA stability (Beetham et al., 1997; Burchmore and Landfear, 1998; Martinez-Calvillo et al., 2003; Soto et al., 2003; Beetham et al., 2003). While polycistronic transcription is common in bacteria and the Archaea, it was assumed for many years that this phenomenon did not occur in eukaryotic cells. However, recent work has shown that polycistronic transcription exists for 15% of the genes in the nematode *Caenorhabditis elegans* (Blumenthal et al., 2002). Whereas trypanosomatide species show a transcriptional regulation closely related to bacterial or archaeal cells, transcriptional regulation of *P. falciparum* genes appears more closely related to eukaryotic cells. Indeed, transcription in *P. falciparum* has been shown to be generally monocistronic and developmentally regulated (Horrocks et al., 1998; Horrocks et al., 1996; Lanzer et al., 1993; Scherf et al. 1998). This has been confirmed by the latest microarray data sets, which show that there is a good correlation between the time when a gene is expressed and when its product is required by the cell (Bozdech et al., 2003a; Le Roch et al., 2003). If transcriptional regulation is an important control point for gene expression in the malaria parasite, how is *Plasmodium* able to control it?

Even with the publication of the *Plasmodium* genome, the identification of transcription factors has been challenging (Gardner et al., 2002). A recent study using multiple sequence alignments of transcriptional regulator domains employing the profile-hidden Markov model was able to identify 156 protein hits (Coulson et al., 2004). Although the identification of 156 proteins represents a third of the transcriptional control elements expected for the genome size of *Plasmodium* (Coulson et al., 2004), this together with microarray data confirmed that transcriptional control by specific transcription factors occurs within the malaria parasite. Indeed, electromobility shift assays performed with *Plasmodium* nuclear extracts and the promoter's regions have identified several specific interactions between DNA elements and proteins (Dechering et al., 1999; Horrocks and Lanzer, 1999).

Despite the progress in bioinformatics analysis, recognition of promoter regions remains difficult in the *Plasmodium* genome due to the high A+T content (>90% in the noncoding region). Under such conditions, the relevance of the eukaryotic consensus TATA box remains weak, even if an overrepresentation of homopolymeric tracts (dA-dT) has been detected (Horrocks et al., 1998; Porter, 2001). Nevertheless, simple genome sequence comparisons, transient transfection experiments, and electromobility shift assays have been useful in detecting large upstream DNA fragments that are necessary to drive the expression of several *Plasmodium* genes (Crabb and Cowman, 1996; Crabb et al., 1997; Goonewardene et al., 1993; Horrocks and Kilbey, 1996; Wu et al., 1995). A recent study using a new bioinformatics strategy identified regulatory elements upstream of *P. falciparum* ORFs within the *hsp* gene family (Militello et al., 2004). This palindromic G-box element was also found to be conserved among *Plasmodium* spp., suggesting its importance in the biology of these parasites. Together, the data support the conclusion that *Plasmodium* genes, as in most eukaryotic cells, consist of a bipartite structure with an upstream element that drives transcriptional regulation in a sequence-dependent manner.

CHROMATIN STRUCTURE AND GENE REGULATION

The nucleosomal organization of *P. falciparum* is typical of eukaryotic cells with a nucleosome phasing of 155 ± 5 bp (Cary et al., 1994; Lanzer et al., 1994). Malaria gene sequences, with functions that have been shown to alter chromatin structure (specifically histone-related genes) have been identified, and these show a high degree of homology with other eukaryotes (Bennett et al., 1995; Creedon et al., 1992). The role of chromatin in gene regulation in the malaria parasite has been highlighted by the identification of components of chromatin regulatory pathways such as histone acetyl transferases and deacetylases, which are able to disrupt nucleosomes in an ATP-dependent manner and facilitate access to transcription factors (Ji and Arnot, 1997; Struhl, 1996). The importance of histone regulation in the malaria parasite has been confirmed by the use of agents such as apicidin, an inhibitor of histone deacetylases which also shows antiparasitic activity (Darkin-Rattray et al., 1996). The presence of adjacent cotranscriptional genes (Le Roch et al., 2003), as well as chromosomal clusters of coexpressed proteins (Florens et al., 2004), supports chromatin structure as a potential mechanism for gene control in *Plasmodium*.

Analysis of the expression profiles of the histone-related genes across the life cycle shows a high degree of stage specificity with maximum detection at the schizont stage (Le Roch et al., 2003). It is interesting that detection of histone H2A (PFC0920w), H2B (PF07_0054), H3 (MAL6P1.106), and H4 (PF11_0061) proteins is maximal in the ring and merozoite stages (Florens et al., 2002; Le Roch et al., 2004; Le Roch et al., 2003) with a slight delay between the appearance of RNA transcripts and protein detection. The decrease in histone abundance observed during transition from the ring stage to the trophozoite stage is correlated with increased DNA synthesis during this developmental period (Inselburg and Banyal, 1984). In addition to the role of acetyl transferases and deacetylases in the regulation of histone tails, histone downregulation could be associated with the increase in overall transcription in trophozoites, whereas histones in the ring stage could function in a general regulatory manner by repressing overall transcription in the highly packed nucleosomes.

POSTTRANSCRIPTIONAL CONTROL IN THE MALARIA PARASITE

Transcriptional control has been shown to be predominant in the *Plasmodium* genome, but posttranscriptional controls have also been identified. Indeed, several putative RNA-binding proteins have been identified in the *P. falciparum* genome, including two members of the Puf family of RNA-binding proteins known to regulate translation and RNA stability (Cui et al., 2002) and proteins with sequence similarity to UBA2, a promoter-independent mRNA stabilizing protein originally identified in *Arabidopsis thaliana* (Lambermon et al., 2002). In a study using multiple sequence alignments of protein domains obtained from transcriptional regulators and using the profile-hidden Markov model, Coulson et al. (2004) were able to identify only a third of the transcriptional control elements expected for the size of the *Plasmodium* genome. Proteins modulating mRNA decay and translation rates were found to be the most abundant, suggesting that posttranscriptional regulation may impart advantages for *Plasmodium*, such as a rapid adaptation to the hosts. Indeed, posttranscriptional control at the level of mRNA translation is not unlikely in *P. falciparum*, considering the presence of structurally distinct ribosome populations during different stages of development and differentiation (Waters et al., 1989; Waters et al., 1995). In addition, posttranscriptional regulation has been described for genes involved in sexual differentiation (Dechering et al., 1997; Vervenne et al., 1994), mitochondrial RNA processing (Rehkopf et al., 2000), and the stability of RNAs encoding surface antigens (Lanzer et al., 1993; Levitt et al.,

1993). Analysis of mRNA transcripts using microarray data and of protein abundance levels using the MudPIT technology for seven different stages of the parasite life cycle show a common discrepancy. Indeed, for 45% of the genes, a delay was observed between mRNA and protein accumulation within a family of functionally related genes, highlighting the importance of posttranscriptional control within the malaria parasite (Le Roch et al., 2004).

The identification of a consensus sequence conferring RNA stability in the untranslated regions of *Plasmodium* transcripts remains difficult because of the high A+T content in the *Plasmodium* genome. And identification of distinguishing functional motifs from random intergenic sequence is near impossible. However, a negative regulatory element-binding sequence recognized by the PfPuf family was identified in several genes where there was a poor correlation between mRNA and protein expression profiles. It has also been suggested that differential polyadenylation of transcripts in *P. falciparum* could confer some regulatory effects, but it remains to be determined whether putative mRNA stabilization sequences such as the negative regulatory element sequence modulate stabilization by recruiting polyadenylation factors.

CAN A STRESS INDUCE TRANSCRIPTIONAL CHANGE?

An interesting way to understand transcriptional regulation in an organism is by the analysis of transcriptional changes under different kinds of stress. When a microorganism or a tissue is treated with small molecules that inhibit basic cellular processes, genes in the inhibited pathway may be transcriptionally up- or down-regulated (Evans and Guy, 2004; Gunther et al., 2003; Hatzixanthis et al., 2003; Reinoso-Martin et al., 2003; Schuller et al., 2004). The effects of stress provide the means for evaluating the global transcriptional response to a drug treatment and may be a useful tool for identifying the cellular processes affected by the drug, as well as possible positive or negative feedback regulation. Recent studies have been undertaken to investigate environmental perturbations and their effects on transcription in *P. falciparum*. Temperature studies which have a significance in malaria host-parasite interaction, i.e., fever, suggest that the rate of transcription of A- and S-type rRNA varies as a function of temperature (Fang and McCutchan, 2002). Experiments addressing transcriptional changes caused by alterations in glucose concentration identified changes in rRNA levels and over 550 other genes (Fang and McCutchan, 2002). But in analyzing these data, care must be exercised. Transcriptional changes throughout the life cycle of the malaria parasite can vary by 5 orders of magnitude (Le Roch et al., 2003), and an arrest in the cell cycle induced by stress of the erythrocytic stages can induce nonspecific changes. In the last few years, several laboratories have used microarray technology to identify transcriptional changes during drug treatments. Ganesan et al. (2003) were the first to analyze the effect of the lethal antifolate WR99210, specific for the *P. falciparum* dihydrofolate reductase-thymidylate synthase, by the long oligonucleotide microarray. RNA synthesis for the de novo pyrimidine biosynthesis and folate biosynthesis pathway showed only subtle changes (<25%) in dying cells. Additional work presented at the Molecular Parasitology Meeting, Woods Hole, Mass., in 2004 (Painter et al., 2004) drew the same conclusion: RNA synthesis is not affected by drug treatment, regardless of the drug used. An extensive study using the antimalarial choline analog T4, a compound in preclinical trials which targets the inhibition of phosphatidylcholine biosynthesis (Le Roch, unpublished), has shown a significant induction (2- to 69-fold-changes) in stress-related genes and genes involved in sexual differentiation after more than >30 h of incubation with synchronized parasites. No significant changes were observed for the enzymes involved in the lipid biosynthesis pathway, but an arrest of the genes involved in the cell cycle progression was detected. This illustrates that the parasites can respond to a chemical stress; but instead of inducing a specific response in the metabolic pathway that is theoretically involved in the drug's action, the drug may only stimulate sexual development. The induction of sexual development is not surprising when one considers that gametocytogenesis can be induced by stress. Gamete formation may be a mechanism whereby the parasite is able to escape the death of the vertebrate host through a more rapid transmission by the *Anopheles* mosquito.

Despite the absence of significant transcriptional changes under drug pressure, these recent analyses do demonstrate the existence of a tight and specific transcriptional regulation across the *P. falciparum* life cycle. The malaria parasite, as an obligate intracellular parasite, has evolved in a buffered intracellular environment where evolutionary forces may have induced the loss of genes involved in transcriptional feedback responses. Therefore, this organism may be an excellent model for the analysis of a tight, programmed, and inflexible regulation of the cell cycle progression. This may also explain why *Plasmodium* has a greater susceptibility to specific new antimetabolites.

THE USE OF THE TRANSCRIPTOME TO IDENTIFY NEW DRUG TARGETS

The genome sequence, expression profiles, and functional characterization of hypothetical genes using rational databases and computational queries have allowed an increased understanding of the biology of the malaria parasite and have provided a glimpse of unique biochemical targets in *P. falciparum*. The increased understanding of the parasite transcriptome can serve as the basis for the rational design of novel antimalarials.

Plasmodium contains a unique organelle, an apicoplast, a plastid structure homologous to chloroplasts acquired by the process of endosymbiosis (Foth and McFadden, 2003; Foth et al., 2003). The apicoplast has been implicated in various metabolic functions, including synthesis of lipids, heme, and isoprenoids (Ralph et al., 2004). Inhibitors of plastid-associated proteins have been shown to kill the parasite and to demonstrate that the apicoplast is an essential organelle. Computational analysis predicted that 550 nuclear proteins targeted to the plastid may potentially be excellent drug targets. A large percentage of these plastid proteins have unknown functions; however, analysis of the life cycle expression data using computational algorithms such as the OPI will allow a functional characterization of a subset, thereby narrowing the search for the best drug candidates.

Proteases have also been shown to play an important role in the metabolism of the erythrocytic cell cycle (Rosenthal, 2002; Rosenthal et al., 2002), such as degrading the host cell hemoglobin in the food vacuole to produce amino acids essential for protein synthesis. Inhibition of growth by protease inhibitors validates their importance for parasite development (Rosenthal et al., 2002). Ninety-two proteases have been identified in the *P. falciparum* genome (Wu et al., 2003). Variation in their expression profiles shows that they are involved in cellular processes in addition to hemoglobin degradation. Association with OPI clusters will allow the identification of protease functions throughout the life cycle and may suggest protease inhibitors with efficacy for all or many parasite stages.

Molecular mechanisms regulating cell proliferation and development in the malaria parasite are still largely unknown. Cell cycle controls and signal transduction pathways responsible for the developmental stage transitions have been difficult to analyze thus far. Identification of putative homologs for a number of eukaryotic cell cycle regulators such as cyclines, cycline-dependent kinases, and components involved in transduction pathways (e.g., the Map kinase pathways) has been limited by the fact that the plasmodial sequence homologies, due to their high A+T content, are usually weak. The importance of the cell cycle and life cycle progression in the malaria parasite and the fact that inhibition by kinase inhibitors kills the parasite validate the drug potential of these genes (Doerig, 2004). Gene expression profiling and cluster analysis will help to elucidate these complex pathways and identify specific metabolic targets.

Lipid metabolism is also clearly of interest for drug design, since the intraerythrocytic growth of *P. falciparum* is associated with a dramatic increase in total membrane content; this is a result of parasite enzymatic activities. Parasite membranes are associated with essential structures and specific processes (cell invasion, nutrient acquisition, trafficking, modulation of the host membranes, or immune evasion against the host immune system). In addition, *Plasmodium* has unique phospholipid metabolic pathways (Vial et al., 2003). The identification of fatty acid biosynthesis, as well as the isoprenoid biosynthesis (in the apicoplast), has confirmed the uniqueness of lipid metabolism in the malaria parasite. In addition, drugs that target the phosphatidylcholine biosynthesis pathway inhibit parasite growth and validate this metabolic pathway as a potential drug target (Wengelnik et al., 2002). Complete elucidation of lipid metabolism in *Plasmodium* will highlight possible new antimalarial candidates.

The possibilities for discovering drugs against the malaria parasite are vast. We have not discussed the potential of inhibitors targeting the parasite's transport mechanisms, the possible inhibitors of the invasion process, or the ubiquitin regulation system, a largely underinvestigated pathway but one likely to have a central role in the cell cycle progression. Expression profile analyses and transcriptional regulation have already elucidated the function of several genes within specific metabolic pathways; but with new identification of the function of the significant number of hypothetical proteins, insights will be on the horizon. There is no doubt that the complete functional characterization of a specific pathway or those enzymes crucial to parasite survival (and not that of the host) will contribute to the rational design of new chemotherapeutic agents.

THE TRANSCRIPTOME AND VACCINE DEVELOPMENT

Over the last 20 years, extensive research has focused on vaccine development, but so far the outlook for vaccines is less optimistic than for drug discovery. An effective malaria vaccine must induce a protective immune response equivalent to or better than that provided by natural immunity. Indeed, when an adult with acquired natural immunity returns to the area where the disease is endemic after a few months away, protective immunity is usually lost. For this reason, an effective malaria vaccine requires new methods of maximizing the longevity of the protective immune response. The winning vaccine will possess multiantigenic determinants with multistage expression. The most promising antigens under evaluation for use in vaccine development against the erythrocytic stage belong to the invasive stages (e.g., merozoite surface proteins, erythrocyte-binding antigens, and rhoptry proteins). Interestingly, all of these potential targets have similar expression profiles across the cell cycle and are extensively expressed at the late schizont stage. Identification of hypothetical genes coexpressed with cell invasion genes may have great potential as vaccine candidates. Approximately 100 to 200 hypothetical proteins have been identified with such profiles (Bozdech et al., 2003a; Le Roch et al., 2003).

Detection of single-feature polymorphism within the *Plasmodium* genome will provide new insights for the identification of possible vaccine candidates. Indeed, large-scale identification of single-feature polymorphisms has been successfully tested for four isolates of *P. falciparum*. For chromosome 2, variations were mostly concentrated within genes under genetic selection pressure (Volkman et al., 2002). This study concluded that a whole genome analysis of many malaria strains together with expression profiles data will certainly identify new and efficient vaccine candidates under genetic selection pressure across all stages of the malaria life cycle. This work is currently under progress and promises to provide some exciting results (C. Kidgell et al., submitted for publication).

CONCLUSIONS

Functional genomic analyses using high-throughput studies of the whole genome have generated huge data sets. Complete genome sequences for *Plasmodium* species, genome annotations, steady-state levels using SAGE, microarray technologies, and cluster analysis of coexpressed genes contribute to the understanding of patterns of gene expression and development of the malaria parasite. The challenge today is the complete understanding of mechanisms that drive the regulation of gene expression throughout the parasite life cycle. A complete understanding of the parasite biology requires the integration of additional data sets such as the proteome analysis of the parasite life cycle (Florens et al., 2002; Lasonder et al., 2002). A full description of the proteomic data is presented in chapter 6. The combination of both data sets, transcriptome and proteome, has been shown to offer an additional level of information (Le Roch et al., 2004). Analysis of protein-protein interactions using yeast double hybrid data sets (D. LaCount et al., manuscript in preparation) is expected to assist the construction of the *Plasmodium* interactome and to yield fundamental biological information. Today most of these data are stored, organized, analyzed, and widely accessible on-line at the Plasmodium Genome Database, PlasmoDB (http://plasmodb.org) and provide researchers with the most comprehensive collection of related data sets. These data have changed the traditional gene-by-gene approaches and are expected to bring success to the discovery of the antiparasitic agents.

REFERENCES

Agabian, N. 1990. Trans splicing of nuclear premRNAs. *Cell* **61:**1157-1160.

Beetham, J. K., J. E. Donelsin, and R. R. Dahlin. 2003. Surface glycoprotein PSA (GP46) expression during short- and long-term culture of *Leishmania chagasi*. *Mol. Biochem. Parasitol.* **131:**109–117.

Beetham, J. K., K. S. Myung, J. J. McCoy, M. E. Wilson, and J. E. Donelson. 1997. Glycoprotein 46 mRNA abundance is post-transcriptionally regulated during development of Leishmania chagasi promastigotes to an infectious form. *J. Biol. Chem.* **272:**17360–17366.

Bennett, B. J., J. Thompson, and R. L. Coppel. 1995. Identification of Plasmodium falciparum histone 2B and histone 3 genes. *Mol. Biochem. Parasitol.* **70:**231–233.

Bischoff, E., M. Guillotte, O. Mercereau-Puijalon, and S. Bonnefoy. 2000. A member of the Plasmodium falciparum Pf60 multigene family codes for a nuclear protein expressed by readthrough of an internal stop codon. *Mol. Microbiol.* **35:**1005–1016.

Blumenthal, T., D. Evans, C. D. Link, A. Guffanti, D. Lawson, J. Thierry-Mieg, D. Thierry-Mieg, W. L. Chiu, K. Duke, M. Kiraly, and S. K. Kim. 2002. A global analysis of Caenorhabditis elegans operons. *Nature* **417:**851–854.

Bozdech, Z., M. Llinas, B. L. Pulliam, E. D. Wong, J. Zhu, and J. L. DeRisi. 2003a. The transcriptome of the intraerythrocytic developmental cycle of Plasmodium falciparum. *PLoS Biol* **1:**E5.

Bozdech, Z., J. Zhu, M. P. Joachimiak, F. E. Cohen, B. Pulliam, and J. L. DeRisi. 2003b. Expression profiling of the schizont and trophozoite stages of Plasmodium falciparum with a long-oligonucleotide microarray. *Genome Biol.* **4:**R9.

Breitling, R., A. Amtmann, and P. Herzyk. 2004a. Graph-based Iterative Group Analysis enhances microarray interpretation. *BMC Bioinformatics* **5:**100.

Breitling, R., A. Amtmann, and P. Herzyk. 2004b. Iterative Group Analysis (iGA): a simple tool to enhance sensitivity and facilitate interpretation of microarray experiments. *BMC Bioinformatics* **5:**34.

Burchmore, R. J., and S. M. Landfear. 1998. Differential regulation of multiple glucose transporter genes in Leishmania mexicana. *J. Biol. Chem.* **273:**29118–29126.

Cary, C., D. Lamont, J. P. Dalton, and C. Doerig. 1994. Plasmodium falciparum chromatin:nucleosomal organisation and histone-like proteins. *Parasitol. Res.* **80:**255–258.

Chen, Q., V. Fernandez, A. Sundstrom, M. Schlichtherle, S. Datta, P. Hagblom, and M. Wahlgren. 1998. Developmental selection of *var* gene expression in Plasmodium falciparum. *Nature* **394:**392–395.

Cheng, Q., N. Cloonan, K. Fischer, J. Thompson, G. Waine, M. Lanzer, and A. Saul. 1998. stevor and rif are Plasmodium falciparum multicopy gene families which potentially encode variant antigens. *Mol. Biochem. Parasitol.* **97:**161–176.

Coulson, R. M., N. Hall, and C. A. Ouzounis. 2004. Comparative genomics of transcriptional control in the human malaria parasite Plasmodium falciparum. *Genome Res.* **14:**1548–1554.

Crabb, B. S., and A. F. Cowman. 1996. Characterization of promoters and stable transfection by homologous and nonhomologous recombination in Plasmodium falciparum. *Proc. Natl. Acad. Sci. USA* **93:**7289–7294.

Crabb, B. S., T. Triglia, J. G. Waterkeyn, and A. F. Cowman. 1997. Stable transgene expression in Plasmodium falciparum. *Mol. Biochem. Parasitol.* **90:**131–144.

Creedon, K. A., D. C. Kaslow, P. K. Rathod, and T. E. Wellems. 1992. Identification of a Plasmodium falciparum histone 2A gene. *Mol. Biochem. Parasitol.* **54:**113–115.

Cui, L., Q. Fan, and J. Li. 2002. The malaria parasite Plasmodium falciparum encodes members of the Puf RNA-binding protein family with conserved RNA binding activity. *Nucleic Acids Res.* **30:**4607–4617.

Daily, J. P., K. G. Le Roch, O. Sarr, D. Ndiaye, A. Lukens, Y. Zhou, O. Ndir, S. Mboup, A. Sultan,

E. A. Winzeler, and D. F. Wirth. 2005. In vivo transcriptome of Plasmodium falciparum reveals overexpression of transcripts that encode surface proteins. *J. Infect. Dis.* **191:**1196–1203.

Darkin-Rattray, S. J., A. M. Gurnett, R. W. Myers, P. M. Dulski, T. M. Crumley, J. J. Allocco, C. Cannova, P. T. Meinke, S. L. Colletti, M. A. Bednarek, S. B. Singh, M. A. Goetz, A. W. Dombrowski, J. D. Polishook, and D. M. Schmatz. 1996. Apicidin: a novel antiprotozoal agent that inhibits parasite histone deacetylase. *Proc. Natl. Acad. Sci. USA* **93:**13143–13147.

Dechering, K. J., A. M. Kaan, W. Mbacham, D. F. Wirth, W. Eling, R. N. Konings, and H. G. Stunnenberg. 1999. Isolation and functional characterization of two distinct sexual-stage-specific promoters of the human malaria parasite Plasmodium falciparum. *Mol. Cell. Biol.* **19:**967–978.

Dechering, K. J., J. Thompson, H. J. Dodemont, W. Eling, and R. N. Konings. 1997. Developmentally regulated expression of pfs16, a marker for sexual differentiation of the human malaria parasite Plasmodium falciparum. *Mol. Biochem. Parasitol.* **89:**235–244.

Deitsch, K. W., M. S. Calderwood, and T. E. Wellems. 2001. Malaria. Cooperative silencing elements in *var* genes. *Nature* **412:**875–876.

Deitsch, K. W., A. del Pinal, and T. E. Wellems. 1999. Intra-cluster recombination and var transcription switches in the antigenic variation of Plasmodium falciparum. *Mol. Biochem. Parasitol.* **101:**107–116.

Doerig, C. 2004. Protein kinases as targets for antiparasitic chemotherapy. *Biochim. Biophys. Acta* **1697:**155–168.

Eisen, M. B., and P. O. Brown. 1999. DNA arrays for analysis of gene expression. *Methods Enzymol.* **303:**179–205.

Evans, W. E., and R. K. Guy. 2004. Gene expression as a drug discovery tool. *Nat. Genet.* **36:**214–215.

Fang, J., and T. F. McCutchan. 2002. Thermoregulation in a parasite's life cycle. *Nature* **418:**742.

Fernandez, V., M. Hommel, Q. Chen, P. Hagblom, and M. Wahlgren. 1999. Small, clonally variant antigens expressed on the surface of the Plasmodium falciparum-infected erythrocyte are encoded by the rif gene family and are the target of human immune responses. *J. Exp. Med.* **190:**1393–1404.

Florens, L., X. Liu, Y. Wang, S. Yang, O. Schwartz, M. Peglar, D. J. Carucci, J. R. Yates III, and Y. Wub. 2004. Proteomics approach reveals novel proteins on the surface of malaria-infected erythrocytes. *Mol. Biochem. Parasitol.* **135:**1–11.

Florens, L., M. P. Washburn, J. D. Raine, R. M. Anthony, M. Grainger, J. D. Haynes, J. K. Moch, N. Muster, J. B. Sacci, D. L. Tabb, A. A. Witney, D. Wolters, Y. Wu, M. J. Gardner, A. A. Holder, R. E. Sinden, J. R. Yates, and D. J. Carucci. 2002. A proteomic view of the Plasmodium falciparum life cycle. *Nature* **419:**520–526.

Foth, B. J., and G. I. McFadden. 2003. The apicoplast: a plastid in Plasmodium falciparum and other Apicomplexan parasites. *Int. Rev. Cytol.* **224:**57–110.

Foth, B. J., S. A. Ralph, C. J. Tonkin, N. S. Struck, M. Fraunholz, D. S. Roos, A. F. Cowman, and G. I. McFadden. 2003. Dissecting apicoplast targeting in the malaria parasite Plasmodium falciparum. *Science* **299:**705–708.

Ganesan, K., L. Jiang, J. White, and P. K. Rathod. 2003. Rigidity of the plasmodium transcriptome revealed by a lethal antifolate. Presented at the Molecular Parasitology Meeting, Woods Hole, Mass.

Gardner, J. P., R. A. Pinches, D. J. Roberts, and C. I. Newbold. 1996. Variant antigens and endothelial receptor adhesion in Plasmodium falciparum. *Proc. Natl. Acad. Sci. USA* **93:**3503–3508.

Gardner, M. J., N. Hall, E. Fung, O. White, M. Berriman, R. W. Hyman, J. M. Carlton, A. Pain, K. E. Nelson, S. Bowman, I. T. Paulsen, K. James, J. A. Eisen, K. Rutherford, S. L. Salzberg, A. Craig, S. Kyes, M. S. Chan, V. Nene, S. J. Shallom, B. Suh, J. Peterson, S. Angiuoli, M. Pertea, J. Allen, J. Selengut, D. Haft, M. W. Mather, A. B. Vaidya, D. M. Martin, A. H. Fairlamb, M. J. Fraunholz, D. S. Roos, S. A. Ralph, G. I. McFadden, L. M. Cummings, G. M. Subramanian, C. Mungall, J. C. Venter, D. J. Carucci, S. L. Hoffman, C. Newbold, R. W. Davis, C. M. Fraser, and B. Barrell. 2002. Genome sequence of the human malaria parasite Plasmodium falciparum. *Nature* **419:**498–511.

Gatton, M. L., J. M. Peters, E. V. Fowler, and Q. Cheng. 2003. Switching rates of Plasmodium falciparum *var* genes: faster than we thought? *Trends Parasitol.* **19:**202–208.

Goonewardene, R., J. Daily, D. Kaslow, T. J. Sullivan, P. Duffy, R. Carter, K. Mendis, and D. Wirth. 1993. Transfection of the malaria parasite and expression of firefly luciferase. *Proc. Natl. Acad. Sci. USA* **90:**5234–5236.

Gunasekera, A. M., S. Patankar, J. Schug, G. Eisen, J. Kissinger, D. Roos, and D. F. Wirth. 2004. Widespread distribution of antisense transcripts in the Plasmodium falciparum genome. *Mol. Biochem. Parasitol.* **136:**35–42.

Gunther, E. C., D. J. Stone, R. W. Gerwien, P. Bento, and M. P. Heyes. 2003. Prediction of clinical drug efficacy by classification of drug-induced genomic expression profiles in vitro. *Proc. Natl. Acad. Sci. USA* **100:**9608–9613.

Hatzixanthis, K., M. Mollapour, I. Seymour, B. E. Bauer, G. Krapf, C. Schuller, K. Kuchler, and P. W. Piper. 2003. Moderately lipophilic carboxylate compounds are the selective inducers of the

Saccharomyces cerevisiae Pdr12p ATP-binding cassette transporter. *Yeast* **20:**575–585.

Hayward, R. E., J. L. Derisi, S. Alfadhli, D. C. Kaslow, P. O. Brown, and P. K. Rathod. 2000. Shotgun DNA microarrays and stage-specific gene expression in Plasmodium falciparum malaria. *Mol. Microbiol.* **35:**6–14.

Horrocks, P., K. Dechering, and M. Lanzer. 1998. Control of gene expression in Plasmodium falciparum. *Mol. Biochem. Parasitol.* **95:**171–181.

Horrocks, P., M. Jackson, S. Cheesman, J. H. White, and B. J. Kilbey. 1996. Stage specific expression of proliferating cell nuclear antigen and DNA polymerase delta from Plasmodium falciparum. *Mol. Biochem. Parasitol.* **79:**177–182.

Horrocks, P., and B. J. Kilbey. 1996. Physical and functional mapping of the transcriptional start sites of Plasmodium falciparum proliferating cell nuclear antigen. *Mol. Biochem. Parasitol.* **82:**207–215.

Horrocks, P., and M. Lanzer. 1999. Differences in nucleosome organization over episomally located plasmids coincide with aberrant promoter activity in P. falciparum. *Parasitol. Int.* **48:**55–61.

Inselburg, J., and H. S. Banyal. 1984. Synthesis of DNA during the asexual cycle of Plasmodium falciparum in culture. *Mol. Biochem. Parasitol.* **10:**79–87.

Ji, D. D., and D. E. Arnot. 1997. A Plasmodium falciparum homologue of the ATPase subunit of a multi-protein complex involved in chromatin remodelling for transcription. *Mol. Biochem. Parasitol.* **88:**151–162.

Kappe, S. H., M. J. Gardner, S. M. Brown, J. Ross, K. Matuschewski, J. M. Ribeiro, J. H. Adams, J. Quackenbush, J. Cho, D. J. Carucci, S. L. Hoffman, and V. Nussenzweig. 2001. Exploring the transcriptome of the malaria sporozoite stage. *Proc. Natl. Acad. Sci. USA* **98:**9895–9900.

Kidgell, C., J. Borevitz, D. Plouffe, S. Volkman, J. Johnson, K. Le Roch, D. Wirth, Y. Zhou, S. Batalov, and E. Winzeler. Genome-wide analysis of Plasmodium falciparum strains reveals novel polymorphic proteins under selection and identifies genetic heterogeneity within the parasite genome. Submitted for publication.

Kuwabara, P. E., and A. Coulson. 2000. RNAi—prospects for a general technique for determining gene function. *Parasitol. Today* **16:**347–349.

Kyes, S., Z. Christodoulou, R. Pinches, and C. Newbold. 2002. Stage-specific merozoite surface protein 2 antisense transcripts in Plasmodium falciparum. *Mol. Biochem. Parasitol.* **123:**79–83.

Kyes, S., R. Pinches, and C. Newbold. 2000. A simple RNA analysis method shows var and rif multigene family expression patterns in Plasmodium falciparum. *Mol. Biochem. Parasitol.* **105:**311–315.

Kyes, S. A., J. A. Rowe, N. Kriek, and C. I. Newbold. 1999. Rifins: a second family of clonally variant proteins expressed on the surface of red cells infected with Plasmodium falciparum. *Proc. Natl. Acad. Sci. USA* **96:**9333–9338.

Labo, P., G. Grant, and D. Ross. 2004. Integrating functional genomics data sets: *Plasmodium falciparum* gene expression. Presented at the Molecular Parasitology Meeting, Woods Hole, Mass.

Lambermon, M. H., Y. Fu, D. A. Wieczorek Kirk, M. Dupasquier, W. Filipowicz, and Z. J. Lorkovic. 2002. UBA1 and UBA2, two proteins that interact with UBP1, a multifunctional effector of pre-mRNA maturation in plants. *Mol. Cell. Biol.* **22:**4346–4357.

Lanzer, M., S. P. Wertheimer, D. de Bruin, and J. V. Ravetch. 1993. Plasmodium: control of gene expression in malaria parasites. *Exp. Parasitol.* **77:**121–128.

Lanzer, M., S. P. Wertheimer, D. de Bruin, and J. V. Ravetch. 1994. Chromatin structure determines the sites of chromosome breakages in Plasmodium falciparum. *Nucleic Acids Res.* **22:**3099–3103.

Lasonder, E., Y. Ishihama, J. S. Andersen, A. M. Vermunt, A. Pain, R. W. Sauerwein, W. M. Eling, N. Hall, A. P. Waters, H. G. Stunnenberg, and M. Mann. 2002. Analysis of the Plasmodium falciparum proteome by high-accuracy mass spectrometry. *Nature* **419:**537–542.

Le Roch, K., J. Johnson, Z. Zhou, K. Henson, Y. Yates, H. Vial, and E. Winzeler. Functional genomics of the antimalarial choline analogue T4 on Plasmodium falciparum reveal potential therapeutic targets. Unpublished data.

Le Roch, K. G., J. R. Johnson, L. Florens, Y. Zhou, A. Santrosyan, M. Grainger, S. F. Yan, K. C. Williamson, A. A. Holder, D. J. Carucci, J. R. Yates III, and E. A. Winzeler. 2004. Global analysis of transcript and protein levels across the Plasmodium falciparum life cycle. *Genome Res.* **14:**2308–2318.

Le Roch, K. G., Y. Zhou, S. Batalov, and E. A. Winzeler. 2002. Monitoring the chromosome 2 intraerythrocytic transcriptome of Plasmodium falciparum using oligonucleotide arrays. *Am. J. Trop. Med. Hyg.* **67:**233–243.

Le Roch, K. G., Y. Zhou, P. L. Blair, M. Grainger, J. K. Moch, J. D. Haynes, P. De La Vega, A. A. Holder, S. Batalov, D. J. Carucci, and E. A. Winzeler. 2003. Discovery of gene function by expression profiling of the malaria parasite life cycle. *Science* **301:**1503–1508.

Lee, S. G., J. U. Hur, and Y. S. Kim. 2004. A graph-theoretic modeling on GO space for biological interpretation of gene clusters. *Bioinformatics* **20:**381–388.

Levitt, A., F. O. Dimayuga, and V. R. Ruvolo. 1993. Analysis of malarial transcripts using cDNA-directed polymerase chain reaction. *J. Parasitol.* **79:** 653–662.

Martinez-Calvillo, S., S. Yan, D. Nguyen, M. Fox, K. Stuart, and P. J. Myler. 2003. Transcription of Leishmania major Friedlin chromosome 1 initiates in both directions within a single region. *Mol. Cell* **11:**1291–1299.

McRobert, L., and G. A. McConkey. 2002. RNA interference (RNAi) inhibits growth of Plasmodium falciparum. *Mol. Biochem. Parasitol.* **119:**273–278.

Militello, K. T., M. Dodge, L. Bethke, and D. F. Wirth. 2004. Identification of regulatory elements in the Plasmodium falciparum genome. *Mol. Biochem. Parasitol.* **134:**75–88.

Munasinghe, A., S. Patankar, B. P. Cook, S. L. Madden, R. K. Martin, D. E. Kyle, A. Shoaibi, L. M. Cummings, and D. F. Wirth. 2001. Serial analysis of gene expression (SAGE) in Plasmodium falciparum: application of the technique to A-T rich genomes. *Mol. Biochem. Parasitol.* **113:**23–34.

Painter, H. J., J. A. Morrisey, J. Stumhofer, and A. B. Vaidya. 2004. Early response to antimalaria drug treatment examined by whole genome transcription profiling. Presented at the Molecular Parasitology Meeting, Woods Hole, Mass.

Patankar, S., A. Munasinghe, A. Shoaibi, L. M. Cummings, and D. F. Wirth. 2001. Serial analysis of gene expression in Plasmodium falciparum reveals the global expression profile of erythrocytic stages and the presence of anti-sense transcripts in the malarial parasite. *Mol. Biol. Cell* **12:**3114–3125.

Porter, M. E. 2001. The DNA polymerase delta promoter from Plasmodium falciparum contains an unusually long 5′ untranslated region and intrinsic DNA curvature. *Mol. Biochem. Parasitol.* **114:**249–255.

Ralph, S. A., B. J. Foth, N. Hall, and G. I. McFadden. 2004. Evolutionary pressures on apicoplast transit peptides. *Mol. Biol. Evol.* **21:**2183–2194.

Rehkopf, D. H., D. E. Gillespie, M. I. Harrell, and J. E. Feagin. 2000. Transcriptional mapping and RNA processing of the Plasmodium falciparum mitochondrial mRNAs. *Mol. Biochem. Parasitol.* **105:**91–103.

Reinoso-Martin, C., C. Schuller, M. Schuetzer-Muehlbauer, and K. Kuchler. 2003. The yeast protein kinase C cell integrity pathway mediates tolerance to the antifungal drug caspofungin through activation of Slt2p mitogen-activated protein kinase signaling. *Eukaryot. Cell* **2:**1200–1210.

Robinson, K. A., and S. M. Beverley. 2003. Improvements in transfection efficiency and tests of RNA interference (RNAi) approaches in the protozoan parasite Leishmania. *Mol. Biochem. Parasitol.* **128:**217–228.

Rosenow, C., R. M. Saxena, M. Durst, and T. R. Gingeras. 2001. Prokaryotic RNA preparation methods useful for high density array analysis: comparison of two approaches. *Nucleic Acids Res.* **29:**E112.

Rosenthal, P. J. 2002. Hydrolysis of erythrocyte proteins by proteases of malaria parasites. *Curr. Opin. Hematol.* **9:**140–145.

Rosenthal, P. J., P. S. Sijwali, A. Singh, and B. R. Shenai. 2002. Cysteine proteases of malaria parasites: targets for chemotherapy. *Curr. Pharm. Des.* **8:**1659–1672.

Scherf, A., R. Hernandez-Rivas, P. Buffet, E. Bottius, C. Benatar, B. Pouvelle, J. Gysin, and M. Lanzer. 1998. Antigenic variation in malaria: in situ switching, relaxed and mutually exclusive transcription of *var* genes during intra-erythrocytic development in Plasmodium falciparum. *EMBO J.* **17:**5418–5426.

Schuller, C., Y. M. Mamnun, M. Mollapour, G. Krapf, M. Schuster, B. E. Bauer, P. W. Piper, and K. Kuchler. 2004. Global phenotypic analysis and transcriptional profiling defines the weak acid stress response regulon in Saccharomyces cerevisiae. *Mol. Biol. Cell* **15:**706–720.

Sibley, L. D. 2004. Intracellular parasite invasion strategies. *Science* **304:**248–253.

Soto, M., L. Quijada, R. Larreta, S. Iborra, C. Alonso, and J. M. Requena. 2003. Leishmania infantum possesses a complex family of histone H2A genes: structural characterization and analysis of expression. *Parasitology* **127:**95–105.

Struhl, K. 1996. Chromatin structure and RNA polymerase II connection: implications for transcription. *Cell* **84:**179–182.

Su, X. Z., V. M. Heatwole, S. P. Wertheimer, F. Guinet, J. A. Herrfeldt, D. S. Peterson, J. A. Ravetch, and T. E. Wellems. 1995. The large diverse gene family var encodes proteins involved in cytoadherence and antigenic variation of Plasmodium falciparum-infected erythrocytes. *Cell* **82:**89–100.

Toronen, P. 2004. Selection of informative clusters from hierarchical cluster tree with gene classes. *BMC Bioinformatics* **5:**32.

Velculescu, V. E., L. Zhang, B. Vogelstein, and K. W. Kinzler. 1995. Serial analysis of gene expression. *Science* **270:**484–487.

Vervenne, R. A., R. W. Dirks, J. Ramesar, A. P. Waters, and C. J. Janse. 1994. Differential expression in blood stages of the gene coding for the 21-kilodalton surface protein of ookinetes of Plasmodium berghei as detected by RNA in situ hybridisation. *Mol. Biochem. Parasitol.* **68:**259–266.

Vial, H. J., P. Eldin, A. G. Tielens, and J. J. van Hellemond. 2003. Phospholipids in parasitic protozoa. *Mol. Biochem. Parasitol.* **126:**143–154.

Volkman, S. K., D. L. Hartl, D. F. Wirth, K. M. Nielsen, M. Choi, S. Batalov, Y. Zhou, D. Plouffe, K. G. Le Roch, R. Abagyan, and E. A. Winzeler. 2002. Excess polymorphisms in genes for membrane proteins in Plasmodium falciparum. *Science* **298:**216–218.

Voss, T. S., J. K. Thompson, J. Waterkeyn, I. Felger, N. Weiss, A. F. Cowman, and H. P. Beck. 2000. Genomic distribution and functional characterisation of two distinct and conserved Plasmodium falciparum *var* gene 5′ flanking sequences. *Mol. Biochem. Parasitol.* **107:**103–115.

Waters, A. P. 2003. Parasitology. Guilty until proven otherwise. *Science* **301:**1487–1488.

Waters, A. P., C. Syin, and T. F. McCutchan. 1989. Developmental regulation of stage-specific ribosome populations in Plasmodium. *Nature* **342:**438–440.

Waters, A. P., W. White, and T. F. McCutchan. 1995. The structure of the large subunit rRNA expressed in blood stages of Plasmodium falciparum. *Mol. Biochem. Parasitol.* **72:**227–237.

Wengelnik, K., V. Vidal, M. L. Ancelin, A. M. Cathiard, J. L. Morgat, C. H. Kocken, M. Calas, S. Herrera, A. W. Thomas, and H. J. Vial. 2002. A class of potent antimalarials and their specific accumulation in infected erythrocytes. *Science* **295:**1311–1314.

Wu, Y., C. D. Sifri, H. H. Lei, X. Z. Su, and T. E. Wellems. 1995. Transfection of Plasmodium falciparum within human red blood cells. *Proc. Natl. Acad. Sci. USA* **92:**973–977.

Wu, Y., X. Wang, X. Liu, and Y. Wang. 2003. Datamining approaches reveal hidden families of proteases in the genome of malaria parasite. *Genome Res.* **13:**601–616.

Young, J., Q. Fivelman, P. Blair, K. Chen, K. Le Roch, F. Yan, D. Baker, D. Carucci, Y. Zhou, and E. Winzeler. Elucidation of Plasmodium falciparum gene function using ontology-based pattern identification. Submitted for publication.

Zhou, Y., and R. Abagyan. 2002. Match-only intesity oligonucleotide array analysis. *BMC Bioinformatics* **3:**3.

Zhou, Y., and R. Abagyan. 2003. Algorithms for high-density oligonucleotide array. *Curr. Opin. Drug Discov. Devel.* **6:**339–345.

Zhou, Y., J. A. Young, A. Santrosyan, K. Chen, F. S. Yan, and E. A. Winzeler. 2004. In silico gene function prediction using ontology-based pattern identification. *Bioinformatics* **21:**1237–1245.

THE *PLASMODIUM* PROTEOME

Jeffrey R. Johnson and John R. Yates III

6

An organism's proteome (the protein complement to the genome) provides a multivariate description of the molecular factors contributing to a particular condition or biologically relevant state. In the search for an effective malaria vaccine, the major proteins contributing to a particular stage make attractive targets for intervention strategies, and their elucidation on a high-throughput scale is now feasible by modern proteome-wide approaches. However, the utility of proteome-wide approaches in malaria research is not limited to cataloging proteins represented by different *Plasmodium* stages. Proteome-wide studies of malaria have also lent a great deal of insight into protein-protein and protein-drug interactions, subcellular localization, functional characterization of unknown genes, and mechanisms of posttranscriptional regulation through comparisons with mRNA expression profiles.

FUNDAMENTALS OF PROTEOMICS

Mass spectrometry is the workhorse of high-throughput proteomics. The advent of ionization techniques that allowed for large macromolecules to be accurately measured within a mass spectrometer, namely matrix-assisted laser desorption ionization (MALDI) (Karas and Hillencamp, 1988) and electrospray ionization (ESI) (Fenn et al., 1989), quickly gave way to methods for analyzing proteins and nucleic acids and eventually to automated platforms for identifying the proteins comprising a biological sample.

A typical proteomics experiment can be divided into three general areas: (i) the front end, which includes sample preparation, proteolytic digestion, and separations; (ii) mass analysis, which involves measuring masses of peptides and peptide fragments in a mass analyzer; and (iii) the back end, which involves the bioinformatic interpretation of mass spectra to assign protein identifications. Each area consists of diverse strategies with strengths and weaknesses for different applications, and there are near-infinite combinations of strategies to combine front-end, mass analysis, and back-end strategies to create one experiment. The most common strategies are described in this section.

The starting material for a proteomics experiment defines what the goal of the experiment should be and can be composed of anything

---

Jeffrey R. Johnson and John R. Yates III, Department of Cell Biology, The Scripps Research Institute, La Jolla, CA 92037.

Molecular Approaches to Malaria, Edited by Irwin W. Sherman
© 2005 ASM Press, Washington, D.C.

from an isolated protein or protein complex, to subcellular fractions purified for specific organelles, to crude cellular lysates. For complex protein mixtures, most front-end strategies follow one of two approaches. The conventional method involves the separation of proteins by one- or two-dimensional gel electrophoresis, followed by excision of protein spots, proteolytic digestion, and mass analysis of peptides for identification. The second approach, often referred to as shotgun proteomics, involves the proteolytic digestion of protein mixtures in solution en masse, followed by liquid chromatographic separations of the peptides either offline or directly coupled with mass analysis (liquid chromatography-tandem mass spectrometry [LC-MS/MS]), where peptides elute directly into a mass spectrometer through an ESI source. While gel-based approaches are still popular in proteomics, they have significant limitations, including an inherently limited range of pH and molecular weight and a bias towards soluble proteins. Both of these gel- and non-gel-based strategies are considered part of the bottom-up dogma to proteomics, where proteins are digested into peptides and all analyses are performed at the peptide level. While this would seem to increase the complexity of the experiment, peptides are much more readily separated and analyzed in a mass spectrometer than intact proteins, which are much more diverse in size and biochemical properties, making chromatographic separations much more cumbersome than peptide separations. These separations are also much more disparate in terms of the conditions required to obtain informative mass spectra, whereas informative mass spectra of peptides can be readily acquired under fairly standard conditions. However, new methods are arising in the top-down arena of proteomics that omit the proteolytic digestion step and analyze proteins intact, with the advantage that characterization of native, intact proteins inherently implies that posttranslational modifications will be characterized as well. With bottom-up analyses, most proteins are often identified by relatively few peptides that rarely span the entirety of a protein sequence, such that gaps in the sequence that are not covered by peptides may contain information that is lost. Nonetheless, top-down approaches require quite sophisticated (and expensive) analytical tools and have not yet demonstrated the potential of bottom-up proteomics in terms of throughput, sensitivity, and automation; thus, the work outlined in this chapter focuses on bottom-up approaches to study the malaria proteome. The future of front-end strategies will likely involve fusions of top-down and bottom-up approaches allowing the advantages of each technique to complement each other.

Strategies for mass analysis are often dictated by the strengths of the type of mass spectrometer available. High mass accuracy analyzers like time-of-flight (TOF) or Fourier-transform ion cyclotron resonance analyzers can measure the intact mass of peptides with high accuracy, such that proteins can be identified by matching measured peptide masses to a database of theoretical peptide masses and then matching back to the proteins the peptides are derived from. This strategy, referred to as peptide-mass mapping or mass fingerprinting, is only effective for very simple protein mixtures, as the probability of obtaining an unambiguous identification decreases as the size of the database increases. Alternatively, identifications from spectra generated by fragmenting intact peptides (MS/MS spectra) provide further information about the identity of the peptide, as each peptide sequence generates a characteristic MS/MS spectrum. Collision-induced dissociation is the most common method for generating fragmentation spectra and is most commonly performed in ion trap or triple quadrupole instruments. While ion trap and triple quadrupole instruments are very sensitive and robust, they offer low mass accuracy and would not be amenable to peptide mass-mapping strategies. However, the information contained within an MS/MS spectrum combined with an intact peptide mass (albeit a low-accuracy measurement) generally results in a more confident identification than simply measuring the mass of an intact peptide alone. Novel combinations of classical mass analyzers are gaining popularity, such as TOF-TOF or quadrupole-TOF instruments, for their ability to obtain high-quality MS/MS spectra as well as high-accuracy mass measurements.

Considered by many to be the bottleneck of any proteomics experiment, back-end bioinformatics comprises the daunting task of sifting through tremendous amounts of data to make protein identifications. As discussed above, relatively simple protein mixtures (i.e., protein spots excised from a gel) can be deconvolved by peptide mass-mapping strategies. Complex protein mixtures, however, require more sophisticated tools, since the databases of potential proteins present are often quite large, and commonly represent the entire six-frame-translated genome of an organism. With few exceptions, complex mixtures demand the information contained within MS/MS spectra to confidently assign protein identifications, which can be performed using a variety of software tools. Perhaps the most famous of these is Sequest (Eng et al., 1994), an algorithm that generates a theoretical MS/MS spectrum for every peptide sequence in a database and then performs a cross-correlation to judge which sequence best fits with the experimental MS/MS spectrum. The intact peptide mass helps the algorithm to limit the number of peptide sequences needed to compare in the database (the search space) and then ranks the best-fitting sequences by a cross-correlation score. Other software tools that use probabilistic models to find the best-fitting peptide sequence based on the fragment ions detected in an MS/MS spectrum are also popular. In general, it is not straightforward to assign a peptide sequence directly from an MS/MS spectrum (de novo sequencing), but algorithms that generate partial sequences, or sequence tags, from an MS/MS spectrum are useful for their flexibility in matching peptides with posttranslational modifications because they can limit the search space by the sequence tag instead of the intact peptide mass. The actual process of generating a list of protein identifications from peptide identifications is a hotly contested issue in the field. Most published protein lists apply empirically derived criteria sets, such as minimum cross-correlation scores or a minimum number of peptides per protein, to build a list of confidently identified proteins. There is certainly no standard for this process, which undoubtedly leads to inconsistencies in the field, but algorithms are constantly arising for calculating the probability or confidence of protein identifications based on peptide identification parameters.

PROFILING STUDIES IN *PLASMODIUM*

It should be clear from the description of peptide and protein identification strategies outlined above that a sequenced genome is a prerequisite for high-throughput proteomics studies. In October 2002, the genome sequence of *Plasmodium falciparum* was released, paving the way not only for genome-wide studies of mRNA (discussed in chapter 5) but also for protein profiling (Gardner et al., 2002). Accordingly, the stage-specific analysis of the *P. falciparum* proteome was performed concurrently with the assembly of the genome sequence and was published in two studies along with the genome sequence in a special issue of *Nature* dedicated to *P. falciparum* (Florens et al., 2002; Lasonder et al., 2002). Two slightly different LC-MS/MS approaches were used to study the proteomes of various life cycle stages. One used a unique implementation of LC-MS/MS called multidimensional protein identification technology, or MudPIT for short, where two-dimensional chromatographic separations are performed online with mass analysis (Washburn et al., 2001). The other study separated proteins first by one-dimensional sodium dodecyl sulfate-polyacrylamide gel electrophoresis, cut bands into 1-cm pieces, performed in-gel digestions of those proteins, and then analyzed the peptides by one-dimensional LC-MS/MS.

The results of these studies described the stage-specific expression of proteins throughout the life cycle of the parasite, with 2,415 and 1,289 proteins identified in each study, respectively. In particular, there were a number of unexpected proteins identified in the sporozoite stages, mainly belonging to the *P. falciparum* erythrocyte membrane protein 1 (PfEMP1) and *rifin* multigene families of polymorphic surface proteins that had only been previously described as being expressed at the surface of the red blood cell at the erythrocytic stages of the parasite's life cycle. One hypothesis for this unexpected observation is that the sporozoites, which do not replicate until they reach the liver, employ the expression of these multigene families to evade immune detection, whereas the asexual stages of the parasite replicate frequently, allowing for antigenic switching of surface antigens to avoid immune detection. Significantly, a large number of proteins identified in both profiling studies contained predicted transmembrane domains, of which over half were annotated as hypothetical proteins. In terms of vaccine development strategies, the verified expression of these proteins that are likely to be surface-expressed or secreted proteins makes them attractive antigens for follow-up immunization studies.

In one approach to identify novel, immunogenic *P. falciparum* antigens, a subset of proteins detected in the protein profiling datasets was subjected to scanning by an HLA supertype antigen-binding algorithm and tested by in vitro HLA-peptide-binding assays (Doolan et al., 2003). As an alternative to conventional approaches to identify immunogenic antigens, such as expression cloning or mass spectrometric sequencing of major histocompatibility complex-eluted peptides, this approach circumvents a dynamic range problem that arises due to the fact that most antigens detected by conventional approaches are dominant epitopes, whereas lower-abundance antigens often go unidentified. In essence, the analysis begins with expressed antigens (represented by the proteome data) and works backward to predict immunogenicity and then to empirically test those predictions. The proteins chosen for this analysis had a wide range of mRNA expression levels, stage specificity, putative transmembrane domains, and estimated protein abundance. Of those tested, 16 proteins were identified that generated an immune response in immunized volunteers, whereas mock-infected controls generated no such response. These antigenic proteins demonstrated a more complex and multifaceted immune response to infection than was previously assumed, as opposed to a targeted response against a few dominant antigens.

The accumulation of data collected from large-scale profiling experiments has also demonstrated ways that proteome data can aid the annotation of recently sequenced *Plasmodium* genomes. The *P. falciparum* genome showed markedly different properties from any other organism, namely, that it is the most A-T-rich genome sequenced to date, making the bioinformatic prediction of gene products from the genome sequence difficult based heavily on knowledge from other organisms. By verifying the expression of protein products and mapping their locations back to the genome, proteome datasets provide a means to improve gene prediction algorithms with empirical evidence of expressed genomic loci. Within the *P. falciparum* MS/MS datasets were a significant number of peptides that were identified in a six-frame translation of the genome but were not identified in the annotated protein database (Lasonder et al., 2002). Similarly, 83 regions of the *Plasmodium yoelii yoelii* genome that were annotated as noncoding regions had peptides detected, yielding a more-accurate refinement of the genome annotation (Carlton et al., 2002). These "orphan" peptides can result from incorrect intron-exon splice site predictions or from incorrect predictions of coding and noncoding regions of the genome; thus, their identification emphasizes areas for improving and training newer, more-efficient gene prediction algorithms.

SUBCELLULAR PROTEOMICS

While proteome-wide studies of whole-cell lysates lend a great deal of information about an organism from a holistic perspective, it is equally informative to combine cell fractionation methods with high-throughput protein identifications to define the proteome of a particular subcellular location or organelle. The genome sequence of *P. falciparum* has an astonishing 60% of genes annotated as hypothetical, stressing an urgent need for methods to characterize gene functions on a large scale. Whereas mRNA expression profiles are often clustered to help assign function to hypothetical genes, operating on the "guilt-by-association" principle that functionally related genes often share similar patterns of transcriptional regulation (Le Roch et al., 2003), clustering protein expression profiles is generally not as successful. However, *Plasmodium* proteins are often compartmentalized into function-specific organelles; thus, the inference of protein function based on subcellular localization is often a powerful means to initially characterize hypothetical proteins.

As mentioned previously, the identification of surface-expressed and secreted proteins is particularly useful for selecting out of the myriad proteins expressed by the parasite those potentially immunogenic proteins that would make promising vaccine targets. The apical organelles of *Plasmodium* are the secretory organelles of the parasite and are associated with cell invasion and host cell modification. The rhoptries are the largest of these organelles and can be efficiently fractionated from most other cellular components, allowing for the characterization of the rhoptry proteome. Traditional methods of raising antibodies against purified rhoptries to identify novel rhoptry proteins were successful in defining the major protein components of the organelle, but these methods are not comprehensive and are generally biased towards abundant proteins (Sam-Yellowe and Ndengele, 1993; Sam-Yellowe et al., 1995; Sam-Yellowe et al., 2001). To define the full protein repertoire, cell fractionation was combined with MudPIT (described above) to identify the proteins comprising rhoptry-enriched fractions of infective *Plasmodium berghei*, *P. yoelii yoelii*, and *P. chabaudi* merozoites (Sam-Yellowe et al., 2004b). As it is not reasonable to expect complete purity using any cellular fractionation approach, contaminating cytoplasmic components were distinguished from enriched proteins by comparing the rhoptry-enriched proteins detected with the whole-cell proteome analysis of *P. berghei* mixed asexual blood stages (Fig. 1). In doing so, a list of 148 proteins detected in the rhoptry-enriched fractions was reduced to 36 proteins considered specifically enriched by comparison

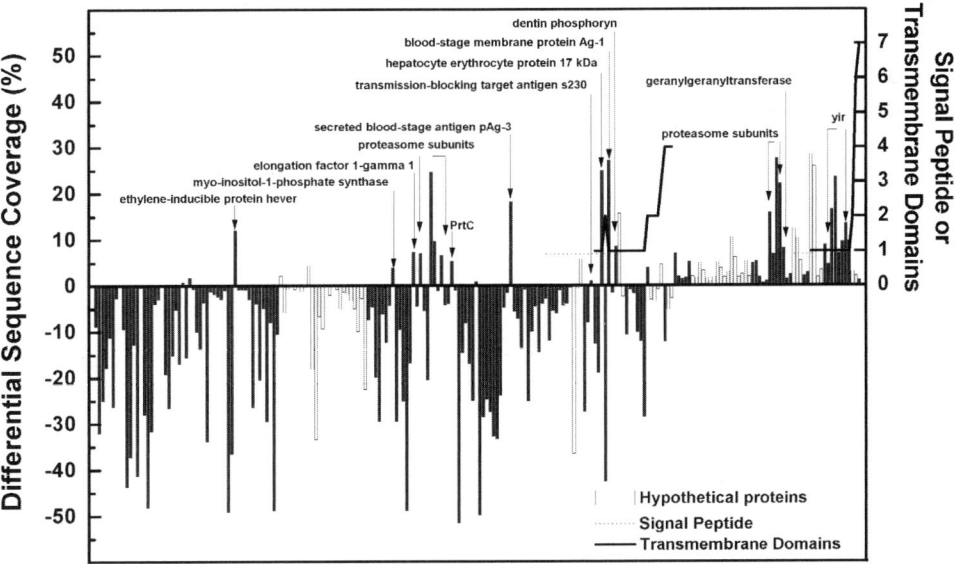

FIGURE 1 Subtractive analysis of rhoptry-enriched proteins. Plotted is the percent difference in sequence coverage between proteome analyses of rhoptry-enriched proteins and whole cell extracts of *P. berghei* asexual blood stages. Transmembrane domains and signal peptides are indicated, as are proteins representative of the rhoptry-enriched fraction.

with the asexual proteome. A large proportion of these proteins demonstrated a C-terminal sequence motif (YXXϕ, where ϕ is a large, hydrophobic residue) that had been shown to localize proteins to the rhoptry organelles of *Toxoplasma* parasites (Joiner and Roos, 2002). In addition, a number of proteins known to be secreted outside the parasite plasma membrane were detected in the analysis, including both proteins detected in culture supernatants and proteins known to be localized to the parasitophorous vacuolar membrane. The hypothetical proteins detected in the rhoptry-enriched fractions will make interesting targets for follow-up studies, in terms of both identifying novel vaccine targets and learning about how the parasite employs proteins to invade and remodel the host cell.

In addition to subcellular fractionation, innovative methods to label and purify parasite proteins expressed at the surface of the infected erythrocyte have been developed and applied to identify novel surface molecules (Florens et al., 2003). Two nonpermeable biotins with labeling specificities for lysine and cysteine residues were used to label intact surface proteins on parasite-infected erythrocytes, subsequently purified with a streptavidin column, eluted, digested, and subjected to MudPIT analysis. The proteins identified were scrutinized for novelty, transmembrane domains, and inter- and intraspecies conservation. As above, the surface-enriched protein mixtures were compared with whole-cell proteome analyses of trophozoites and schizonts to identify those proteins that were unique or enriched in the surface protein sample. A total of 36 uncharacterized parasite-infected erythrocyte surface proteins were detected that met these criteria, with two of these candidate genes selected for further studies. Most known surface proteins, such as the PfEMP1 and rifin proteins, belong to highly polymorphic, multigene families, limiting their applicability as potential vaccine antigens. By this approach, several uncharacterized proteins were detected, most of which were encoded by single-copy genes, including the two candidates. With immunofluorescence microscopy to characterize subcellular localization, the candidate genes were found at the erythrocyte plasma membrane in both knobless and rosetting-negative-type *P. falciparum* strains. Further, one candidate gene appeared to belong to a chromosomal cluster of genes forming a complex implicated in translocating parasite proteins to the parasite-infected erythrocyte surface. Intervention strategies targeting this protein may therefore prevent translocation of other essential proteins to the erythrocyte surface.

Targeted approaches, while less comprehensive, have been successful at identifying novel proteins in more ambiguous subcellular localizations that are not amenable to purification by conventional fractionation methods. The tubovesicular membrane network, for example, is a highly complex network that extends from the parasite's self-constructed parasitophorous vacuole, connecting it to the host cell membrane. Within this network are the loosely defined Maurer's clefts, flattened vesicular structures found beneath the erythrocyte membrane that are implicated in trafficking parasite-secreted proteins to the host cell surface. Compared with other organelles of the parasite, the Maurer's clefts are extremely difficult to separate from the rest of the tubovesicular membrane and the host cell, so alternative strategies to identify proteins here must circumvent the purification process. In this case, a panel of monoclonal antibodies was raised against asexual stage parasites, identifying particular isolates that labeled subcellular localizations of interest (i.e., Maurer's clefts) by immunofluorescence microscopy (Sam-Yellowe et al., 2001). Once identified, the antibodies were employed to immunoprecipitate the protein target(s), which was then analyzed by MudPIT (Sam-Yellowe et al., 2004a). The result was the identification of a novel, multicopy gene family of transmembrane proteins, the PfMC-2TM family, with several similarities to other families of antigenically variant erythrocyte surface proteins, such as the *var*, *rifin*, *stevor*, and *sep/etramp* families. The majority of known *P. falciparum* erythrocyte surface proteins are encoded by multiple genes in the subtelomeric regions of chromosomes, with very high sequence variation in the surface-exposed regions of the proteins. These classes of proteins are highly immunogenic but have limited utility for therapeutics, given the parasite's ability to switch to the expression of a different family member encoding a different extracellular domain. The PfMC-2TM genes share these features, but working out how the different multigene families work together to express proteins at the erythrocyte surface is still unclear. While this method has lower throughput than other organellar proteome approaches, the feasibility of targeting highly specific processes in this manner lends itself to a wide array of applications with *Plasmodium*.

COMPARISONS WITH mRNA ABUNDANCE PROFILES

There are several indications that posttranscriptional regulation plays an important role in regulating the expression of gene products in *Plasmodium*. In a bioinformatic survey of transcription-associated proteins (TAPs) represented in the *P. falciparum* genome, it was discovered that TAPs are much less abundant than other eukaryotic organisms and that TAPs involved in mRNA stability and translation were determined to be the most abundant TAPs detected in the genome (Coulson et al., 2004). Indeed, the posttranscriptional regulation of several genes involved in sexual differentiation (Vervenne et al., 1994; Dechering et al., 1997), mitochondrial RNA processing (Rehkopf et al., 2000), and the stability of mRNA transcripts encoding surface antigens (Lanzer et al., 1993; Levitt et al., 1993) has been demonstrated previously. To gain a broader perspective on the prevalence of posttranscriptional regulation in the parasite, it is informative to make genome-wide comparisons of the parasite transcriptome with its proteome. Both data sets provide profiles of mRNA and protein expression, with abundance correlated to specific points throughout the parasite life cycle; thus, an integrated analysis of both should identify those genes regulated posttranscriptionally.

Comparing the mRNA expression data for nine stages (early ring, late ring, early trophozoite, late trophozoite, early schizont, late schizont, merozoite, gametocyte, and sporozoite) (Le Roch et al., 2003) and protein expression data for seven stages (ring, trophozoite, schizont, merozoite, gametocyte, gamete, and sporozoite) (Florens et al., 2002) revealed prevalent discrepancies between mRNA and protein abundance for many genes (Le Roch et al., 2004). A total of 4,294 transcripts and 2,904 proteins were detected, with 2,584 genes in common between both data sets. A comparison of the median mRNA intensity for genes where the corresponding proteins were or were not detected indicated that transcripts were, on average, 2.5 times more abundant when the protein was detected than transcripts where the protein was not detected for genes in the asexual erythrocytic cycle. This was not the case, however, for genes detected in gametocytes and sporozoites, where median mRNA intensities were not significantly different for genes where protein products were or were not detected; this suggested that discrepancies between mRNA and protein abundance in these stages are more likely a result of posttranscriptional regulatory phenomena rather than the detection limits of the respective technologies.

To compare mRNA and protein datasets, a metric for protein abundance must be established. While mRNA abundance values are inherently quantitative, reflecting the fluorescence intensity of hybridized transcripts detected on the oligonucleotide array and therefore directly indicating transcript abundance, protein abundance derived from proteome analyses is not as clearly quantitative. The physical principles enabling ionization techniques like ESI and MALDI are poorly understood with respect to how efficiently different proteins or peptides are ionized and detected in a mass spectrometer; thus, spectral abundance of a peptide ion does not necessarily reflect the abundance of the peptide in the sample from which it was derived. Strategies employing stable isotope labeling have been applied in the field to circumvent this issue; by simultaneously analyzing identical peptides from different sample preparations that differ only in their isotopic composition, the

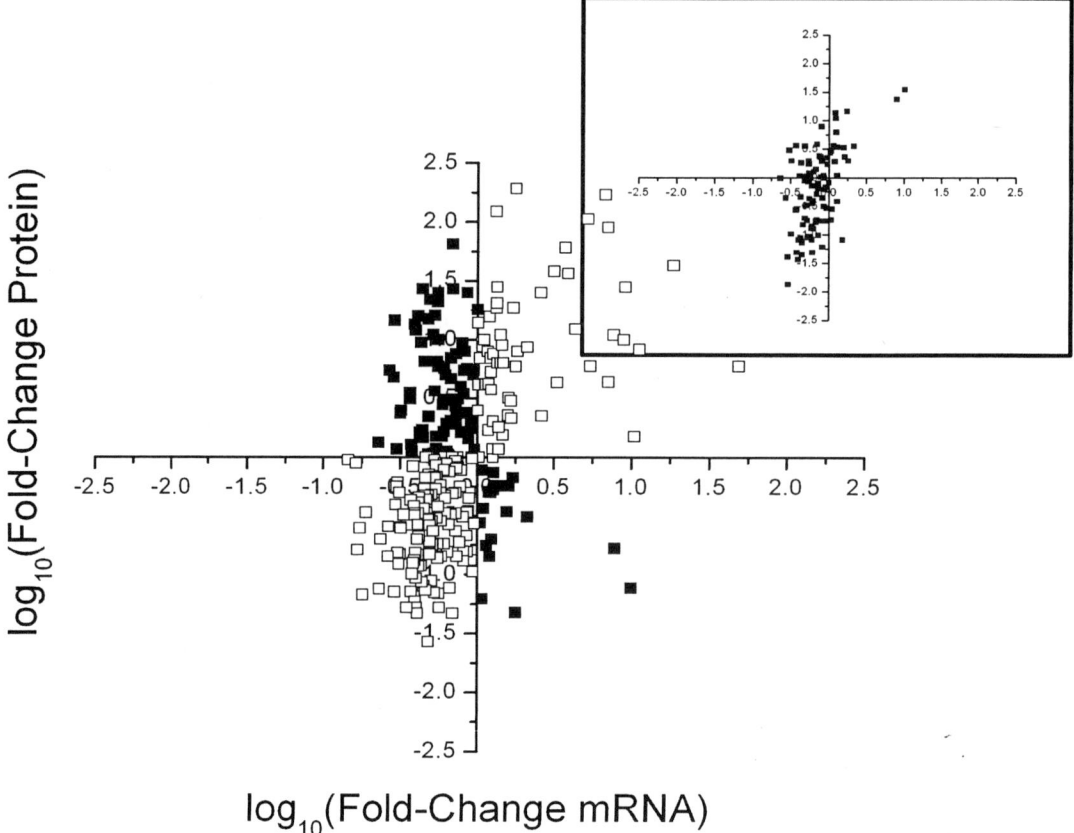

FIGURE 2 Scatterplots of mRNA and protein fold changes during trophozoite-to-schizont transitions. Quadrants are indicated by Roman numerals. Data points falling into quadrants I and III are shown by open squares, and data points falling into quadrants II and IV are shown by closed squares. Data points that fell into quadrants II and IV were replotted (inset) as the mRNA fold change of the transition indicated versus the protein fold change of the following transition (i.e., the inset is a plot of the fold change in mRNA abundance for the trophozoite-to-schizont versus the fold change in protein abundance for the schizont-to-merozoite transition).

mass difference separates peptides from different samples within the same mass spectrum such that the ratio of their signal intensities represents the relative amount of each peptide present in the original samples (Gygi et al., 1999; Ong et al., 2002). These types of experiments are complex and require multiple replicates; in the case of the *P. falciparum* proteome, there is a wealth of data already collected that did not use stable isotope dilution strategies. More commonly, protein abundance is estimated by semiquantitative parameters of protein identification, such as the sequence coverage of a protein (the percentage of protein sequence detected by peptides in an experiment), or as was used for this comparison, the number of MS/MS spectra identified per protein (spectral count). These semiquantitative parameters have, in fact, been shown to provide a reliable estimate of protein abundance when more precise quantitative approaches are unavailable or unfeasible (Florens et al., 2002; Gao et al., 2003; Durr et al., 2004; Liu et al., 2004).

Using a statistical method commonly used to compare unrelated datasets, the Spearman rank correlation, a pairwise comparison of stage-specific transcriptome and proteome datasets was performed. This revealed that, in many cases, the transcriptome of a particular stage correlated better with the proteome of the following stage rather than with the proteome of the same stage, suggesting a delay between the maximal accumulation of an mRNA transcript and its cognate protein. However, the Spearman rank analysis only reflects the correlation of whole datasets, so to investigate the prevalence for this type of effect in terms of individual genes, scatterplots of life cycle transitions in the asexual erythrocytic cycle were constructed (Fig. 2). On average, 55% of data points fell into quadrants of the scatterplot corresponding to complementary changes in mRNA and protein abundance (quadrants I and III). For those data points that fell into quadrant II or IV, replotting the same mRNA fold changes against the protein fold changes for the next life cycle transition fixed the majority of these data points, relocating them into quadrant I or III 74% of the time (Fig. 2, inset).

Most significantly, correlating mRNA and protein expression profiles for individual genes revealed particular genes and families of functionally related genes that appeared to observe similar patterns of mRNA and protein accumulation. For example, almost all genes involved in glycolysis observed low correlations between mRNA and protein expression profiles for the asexual erythrocytic stages, but when a time shift was considered by shifting the mRNA stages forward, the correlation coefficients improved significantly for all but one glycolysis gene. Interestingly, the one gene that did not seem to observe a time shift was the aldolase gene, which is known to have a dual function as part of the cell invasion machinery (Jewett and Sibley, 2003). This gene-by-gene correlation analysis identified those genes that are most likely under posttranscriptional control and will be extremely useful for future studies trying to unravel the molecular mechanisms underlying this type of gene regulation.

REFERENCES

Carlton, J. M., S. V. Angiuoli, B. B. Suh, T. W. Kooij, M. Pertea, J. C. Silva, M. D. Ermolaeva, J. E. Allen, J. D. Selengut, H. L. Koo, J. D. Peterson, M. Pop, D. S. Kosack, M. F. Shumway, S. L. Bidwell, S. J. Shallom, S. E. van Aken, S. B. Riedmuller, T. V. Feldblyum, J. K. Cho, J. Quackenbush, M. Sedegah, A. Shoaibi, L. M. Cummings, L. Florens, J. R. Yates, J. D. Raine, R. E. Sinden, M. A. Harris, D. A. Cunningham, P. R. Preiser, L. W. Bergman, A. B. Vaidya, L. H. van Lin, C. J. Janse, A. P. Waters, H. O. Smith, O. R. White, S. L. Salzberg, J. C. Venter, C. M. Fraser, S. L. Hoffman, M. J. Gardner, and D. J. Carucci. 2002. Genome sequence and comparative analysis of the model rodent malaria parasite Plasmodium yoelii yoelii. *Nature* **419**:512–519.

Coulson, R. M., N. Hall, and C. A. Ouzounis. 2004. Comparative genomics of transcriptional control in the human malaria parasite Plasmodium falciparum. *Genome Res.* **14**:1548–1554.

Dechering, K. J., J. Thompson, H. J. Dodemont, W. Eling, and R. N. Konings. 1997. Developmentally regulated expression of pfs16, a marker for sexual differentiation of the human malaria parasite Plasmodium falciparum. *Mol. Biochem. Parasitol.* **89**:235–244.

Doolan, D. L., S. Southwood, D. A. Freilich, J. Sidney, N. L. Graber, L. Shatney, L. Bebris, L. Florens, C. Dobano, A. A. Witney, E. Appella, S. L. Hoffman, J. R. Yates III, D. J. Carucci, and A. Sette. 2003. Identification of Plasmodium falciparum antigens by antigenic analysis of genomic and proteomic data. *Proc. Natl. Acad. Sci. USA* **100**: 9952–9957.

Durr, E., J. Yu, K. M. Krasinska, L. A. Carver, J. R. Yates, J. E. Testa, P. Oh, and J. E. Schnitzer. 2004. Direct proteomic mapping of the lung microvascular endothelial cell surface in vivo and in cell culture. *Nat. Biotechnol.* **22**:985–992.

Eng, J. K., A. L. McCormack, and J. R. Yates III. 1994. An approach to correlate tandem mass spectral data of peptides with amino acid sequences in a protein database. *J. Am. Soc. Mass Spectrom.* **5**:976–989.

Fenn, J. B., M. Mann, C. K. Meng, S. F. Wong, and W. C. Whitehouse. 1989. Electrospray ionization for the mass spectrometry of large biomolecules. *Science* **246**:64–71.

Florens, L., X. Liu, Y. Wang, S. Yang, O. Schwartz, M. Peglar, D. J. Carucci, J. R. Yates III, and Y. Wu. 2004. Proteomics approach reveals novel antigens on the surface of malaria-infected erythrocytes. *Mol. Biochem. Parasitol.* **135**:1–11.

Florens, L., M. P. Washburn, J. D. Raine, R. M. Anthony, M. Grainger, J. D. Haynes, J. K. Moch, N. Muster, J. B. Sacci, D. L. Tabb, A. A. Witney, D. Wolters, Y. Wu, M. J. Gardner, A. A. Holder, R. E. Sinden, J. R. Yates, and D. J. Carucci. 2002. A proteomic view of the Plasmodium falciparum life cycle. *Nature* **419**:520–526.

Gao, J., G. J. Opiteck, M. S. Friedrichs, A. R. Dongre, and S. A. Hefta. 2003. Changes in the protein expression of yeast as a function of carbon source. *J. Proteome Res.* **2:**643–649.

Gardner, M. J., N. Hall, E. Fung, O. White, M. Berriman, R. W. Hyman, J. M. Carlton, A. Pain, K. E. Nelson, S. Bowman, I. T. Paulsen, K. James, J. A. Eisen, K. Rutherford, S. L. Salzberg, A. Craig, S. Kyes, M. S. Chan, V. Nene, S. J. Shallom, B. Suh, J. Peterson, S. Angiuoli, M. Pertea, J. Allen, J. Selengut, D. Haft, M. W. Mather, A. B. Vaidya, D. M. Martin, A. H. Fairlamb, M. J. Fraunholz, D. S. Roos, S. A. Ralph, G. I. McFadden, L. M. Cummings, G. M. Subramanian, C. Mungall, J. C. Venter, D. J. Carucci, S. L. Hoffman, C. Newbold, R. W. Davis, C. M. Fraser, and B. Barrell. 2002. Genome sequence of the human malaria parasite *Plasmodium falciparum*. *Nature* **419:**498–511.

Gygi, S. P., B. Rist, S. A. Gerber, F. Turecek, M. H. Gelb, and R. Aebersold. 1999. Quantitative analysis of complex protein mixtures using isotope-coded affinity tags. *Nat. Biotechnol.* **17:**994–999.

Jewett, T. J., and L. D. Sibley. 2003. Aldolase forms a bridge between cell surface adhesins and the actin cytoskeleton in apicomplexan parasites. *Mol. Cell* **11:**885–894.

Joiner, K. A., and D. S. Roos. 2002. Secretory traffic in the eukaryotic parasite Toxoplasma gondii: less is more. *J. Cell Biol.* **157:**557–563.

Karas, M., and F. Hillenkamp. 1988. Laser desorption ionization of proteins with molecular mass exceeding 10000 daltons. *Anal. Chem.* **60:**2299–2301.

Lanzer, M., S. P. Wertheimer, D. de Bruin, and J. V. Ravetch. 1993. Plasmodium: control of gene expression in malaria parasites. *Exp. Parasitol.* **77:**121–128.

Lasonder, E., Y. Ishihama, J. S. Andersen, A. M. Vermunt, A. Pain, R. W. Sauerwein, W. M. Eling, N. Hall, A. P. Waters, H. G. Stunnenberg, and M. Mann. 2002. Analysis of the Plasmodium falciparum proteome by high-accuracy mass spectrometry. *Nature* **419:**537–542.

Le Roch, K. G., J. R. Johnson, L. Florens, Y. Zhou, A. Santrosyan, M. Grainger, S. F. Yan, K. C. Williamson, A. A. Holder, D. J. Carucci, J. R. Yates, and E. A. Winzeler. 2004. Global analysis of transcript and protein levels across the Plasmodium falciparum life cycle. *Genome Res.* **14:**2308–2318.

Le Roch, K. G., Y. Zhou, P. L. Blair, M. Grainger, J. K. Moch, J. D. Haynes, P. De La Vega, A. A. Holder, S. Batalov, D. J. Carucci, and E. A. Winzeler. 2003. Discovery of gene function by expression profiling of the malaria parasite life cycle. *Science* **301:**1503–1508.

Levitt, A., F. O. Dimayuga, and V. R. Ruvolo. 1993. Analysis of malarial transcripts using cDNA-directed polymerase chain reaction. *J. Parasitol.* **79:**653–662.

Liu, H., R. G. Sadygov, and J. R. Yates III. 2004. A model for random sampling and estimation of relative protein abundance in shotgun proteomics. *Anal. Chem.* **76:**4193–4201.

Ong, S. E., B. Blagoev, I. Kratchmarova, D. B. Kristensen, H. Steen, A. Pandey, and M. Mann. 2002. Stable isotope labeling by amino acids in cell culture, SILAC, as a simple and accurate approach to expression proteomics. *Mol. Cell. Proteomics* **1:**376–386.

Rehkopf, D. H., D. E. Gillespie, M. I. Harrell, and J. E. Feagin. 2000. Transcriptional mapping and RNA processing of the Plasmodium falciparum mitochondrial mRNAs. *Mol. Biochem. Parasitol.* **105:**91–103.

Sam-Yellowe, T., L. Florens, J. R. Johnson, T. Wang, J. A. Drazba, K. G. Le Roch, Y. Zhou, S. Batalov, D. J. Carucci, E. A. Winzeler, and J. R. Yates. 2004a. A Plasmodium gene family encoding Maurer's cleft membrane proteins: structural properties and expression profiling. *Genome Res.* **14:**1052–1059.

Sam-Yellowe, T. Y., L. Florens, T. Wang, J. D. Raine, D. J. Carucci, R. E. Sinden, and J. R. Yates. 2004b. Proteome analysis of rhoptry-enriched fractions isolated from Plasmodium merozoites. *J. Proteome Res.* **3:**995–1001.

Sam-Yellowe, T. Y., H. Fujioka, M. Aikawa, T. Hall, and J. A. Drazba. 2001. A Plasmodium falciparum protein located in Maurer's clefts underneath knobs and protein localization in association with Rhop-3 and SERA in the intracellular network of infected erythrocytes. *Parasitol Res* **87:**173–185.

Sam-Yellowe, T. Y., H. Fujioka, M. Aikawa, and D. G. Messineo. 1995. Plasmodium falciparum rhoptry proteins of 140/130/110 kd (Rhop-H) are located in an electron lucent compartment in the neck of the rhoptries. *J. Eukaryot. Microbiol.* **42:**224–231.

Sam-Yellowe, T. Y., and M. M. Ndengele. 1993. Monoclonal antibody epitope mapping of Plasmodium falciparum rhoptry proteins. *Exp. Parasitol.* **76:**46–58.

Vervenne, R. A., R. W. Dirks, J. Ramesar, A. P. Waters, and C. J. Janse. 1994. Differential expression in blood stages of the gene coding for the 21-kilodalton surface protein of ookinetes of Plasmodium berghei as detected by RNA in situ hybridisation. *Mol. Biochem. Parasitol.* **68:**259–266.

Washburn, M. P., D. Wolters, and J. R. Yates III. 2001. Large-scale analysis of the yeast proteome by multidimensional protein identification technology. *Nat. Biotechnol.* **19:**242–247.

EVOLUTIONARY HISTORY AND POPULATION GENETICS OF HUMAN MALARIA PARASITES

Martine Zilversmit and Daniel L. Hartl

7

Early work on the systematics of malaria parasites was based on morphological studies and information on parasite life histories (the details and timing of the life cycle); however, most current views about the evolution of malaria parasites are based on comparisons of DNA sequences among the organisms. Over the last few decades, molecular phylogenetics and population genetics have revealed a great deal about the evolutionary relationships and population structure of the four malaria parasites infecting humans, their sister species, and more distantly related malaria parasites. During this period, there have also been controversies concerning these relationships. Some disputes have been resolved (and others long forgotten), but some issues remain contentious even after >20 years of intensive study.

The central theme of this chapter is the molecular evolution of species of the genus *Plasmodium*—the experimental methods, conclusions, and confounding elements peculiar to *Plasmodium*. One section ("Human Malaria Parasites in Broad Evolutionary Context") is a brief summary of studies on the evolutionary history of *Plasmodium* and related organisms; another section ("Ever Since Speciation—Population Dynamics of Human Malarias") is a summary of the molecular population genetics of contemporary world populations of the two most virulent human malaria parasites. We emphasize consensus views where they exist but also point out areas of continuing controversy.

MOLECULAR EVOLUTION AND THE MALARIA PARASITE

Before discussing the relationships deduced from molecular data, it is necessary to cover some of the peculiar aspects of the molecular evolution of malaria parasites. Molecular phylogenetics cannot resolve all ambiguities even in groups of organisms that include such well-studied species as the budding yeast *Saccharomyces cerevisiae* or the vinegar fly *Drosophila melanogaster* (Remsen et al., 2002; Zilversmit et al., 2002; Rokas et al., 2003). The inability to resolve evolutionary relationships typically results from (i) including species that are too distantly related or species that are too closely related, (ii) reliance on DNA or protein sequences that change over evolutionary

Martine Zilversmit and Daniel L. Hartl, Department of Organismic and Evolutionary Biology, Harvard University, Cambridge, MA 02138.

Molecular Approaches to Malaria, Edited by Irwin W. Sherman
© 2005 ASM Press, Washington, D.C.

time either too rapidly or too slowly relative to the time of divergence of the species whose relationships are to be resolved, or (iii) including in the analysis one or more molecules whose sequences have changed as a result of natural selection in a way that contradicts or obscures the evolutionary relationships among the species themselves. These limitations are especially relevant to studies of the genus *Plasmodium*, owing to the high A+T content of some of the species (McCutchan et al., 1984) and many other atypical aspects of their molecular biology, such as their enigmatic mechanisms of gene regulation, strong natural selection, and special genetic mechanisms for immune evasion (Coulson et al., 2004; Rasti et al., 2004). Although special features of the molecular genetics of *Plasmodium* have been problematical in the past and may continue to cause some ambiguity, the past several years have seen great advances in our understanding, owing to improved and large-scale methods for DNA sequence acquisition, as well as the continuing refinement of computational and statistical methods in molecular phylogenetics. Progress in understanding the evolutionary history of *Plasmodium* is also gaining momentum, due to a renewed emphasis on integrated studies of the biology of infectious disease.

HUMAN MALARIA PARASITES IN BROAD EVOLUTIONARY CONTEXT

In addition to humans, *Plasmodium* parasites infect a range of vertebrate hosts, including birds, lizards, rodents, and nonhuman primates. This genus has recently been shown to be a paraphyletic group—a group whose most recent common ancestor does not include all members of the group—and should properly include the *Hepatocystis* group of mammalian parasites (sister to the nonhuman primate *Plasmodium* species) and the *Haemoproteus* parasites of birds and lizards (grouping with the bird and lizard *Plasmodium* species) (Escalante et al., 1998; Perkins and Schall, 2002).

Molecular Evolution and the Genus *Plasmodium*: Critical Review

The field of *Plasmodium* phylogenetics has been particularly dynamic over the past decade. Multiple unexpected relationships have emerged as older paradigms have been challenged. Recent studies have shown that characters previously used to classify and track the evolution of malaria parasites, such as life cycle patterns, virulence level, and morphology (e.g., Russell et al., 1963; Coatney et al., 1971), are either uninformative or in some cases misleading (Escalante et al., 1998; Kissinger et al., 1998; Perkins and Schall, 2002).

The study of the molecular phylogenetics of malaria parasites began in 1991 (Waters et al., 1991), and the field developed significantly in the subsequent time. The conclusions of many earlier studies were affected by difficulties stemming from insufficient taxon sampling, use of distantly related outgroups, and problems with the genes chosen for analysis. The choice of both the ingroup taxa (the organisms of interest) and outgroup taxa (organisms outside the group of interest) for phylogenetic analysis can have a significant influence on the strength of the resulting evolutionary hypothesis. In some cases, if a few ingroup taxa in a study are distantly related to the others or more quickly evolving so that their lineages include more changes, these taxa may end up grouped together in the phylogeny in a single clade (subgroup). This phenomenon is known as long branch attraction (Philippe, 2000). Even if the distantly related or rapidly evolving taxa are not grouped together, they may cause spurious associations in the tree topology. Likewise, although an outgroup must be more distantly related to the ingroup than any pair of members in the ingroup, it is important that the outgroup have an appropriate divergence time from the common ancestor of the ingroup. A divergence time that is too great results in loss of the phylogenetic signal and an unreliable tree topology (Siddall and Barta, 1992).

In addition to problems of taxon sampling, for years the study of malaria systematics was also stymied by inherent problems in the loci chosen for analysis. Small subunit (SSU) rRNA and the circumsporozoite protein (*csp*) gene have been the workhorses of *Plasmodium* phylogenetics (Waters et al., 1991; Waters et al., 1993; Escalate and Ayala, 1994; Escalante et al., 1995; McCutchan et al. 1996; Qari et al., 1996; Escalante et al., 1996; Escalante et al., 1997). However, recent research indicates that neither of these genes may be appropriate for evolutionary studies. SSU rRNA is a standard locus used in high-level molecular systematics, but it was later found that *Plasmodium* species possess several separate gene families of rRNA, each expressed at a different point in the life cycle, which are not evolving in a concerted manner and may still be exchanging genetic information with each other (Corredor and Enea, 1993; Paul et al., 2003). This is an issue because several of the older studies appear to have included a mixture of paralogs (duplicate genes) and orthologs (proper homologs), and only orthologs yield reliable gene trees. Some studies may still be using stage-specific SSU loci that can exchange genetic information with paralogous loci along part of their length. (See Corredor and Enea, 1993, for an excellent discussion of the peculiar features of rRNA in *Plasmodium*.)

Likewise, the *csp* locus commonly used for evolutionary studies has also proven problematic. Because the CSP protein was a popular candidate for vaccine development, data on *csp* sequences were relatively abundant at a time when sufficient data sets for other molecules were difficult to generate. The problem with the use of the *csp* gene in molecular systematics stems from the fact that the gene encodes a surface protein. Under ideal circumstances, a gene chosen for phylogenetic analysis should be subjected to at most weak selection pressure, so that changes in the gene over evolutionary time are selectively neutral or nearly neutral. For genes whose changes through evolutionary time are selectively neutral, the changes reflect the occurrence of neutral mutations; over a long enough time, the pattern of change in the molecules reflects the pattern of speciation among the taxa. Because CSP is a surface protein, it is under strong selective pressure from the vertebrate immune system; therefore, selectively driven, nonneutral changes in the gene may either obscure the phylogenetic signal or lead to incorrect phylogenetic inferences. This problem is compounded further in the case of *csp*, which contains a large repeat region that makes homologous genes in divergent species difficult to align. Though useful in some circumstances, these considerations render *csp* inappropriate for use in molecule systematics of highly divergent species.

At this time, new loci are being developed for molecular systematics that appear to be evolving neutrally and that provide enough variation to be informative, such as the mitochondrial gene for cytochrome *b* (Escalante et al., 1998; Perkins and Schall, 2002) and the gene for the housekeeping enzyme adenylosuccinate lyase (Kedseierki et al., 2002). One caveat is that mitochondrial cytochrome *b* is the target of some antimalarial drugs, and mutations in the gene are known to be associated with resistance (Vaidya et al., 1993). Studies of these and other suitably chosen genes will hopefully lead to *Plasmodium* phylogenies based on the combined data from multiple loci, which can produce much more robust conclusions (Zilversmit et al., 2002; Coulson, 2003). On the other hand, the chance of identifying true neutrally evolving loci in *Plasmodium* species is compromised by the possibility of unrecognized selection pressure from the host immune system or (in the case of human parasites) drugs; hence, an extra measure of caution is called for.

Toward a Better Understanding of the Relationships of Human Malaria Parasites

One seemingly obvious conclusion that was quickly challenged by molecular evolutionary studies was the assumption that human malaria parasites are closely related. The first studies to employ significant taxon sampling made clear that, although primate parasites (with the exception of *Plasmodium falciparum* and *Plasmodium reichenowi*; see below) grouped together, those parasites with a human host were not close to each other within this group (Fig. 1). This pattern indicates that malaria most likely has an ancient association with primates (Escalante et al., 1998; Perkins and Schall, 2002).

Malaria has also been found in African rodents, and many of these *Plasmodium* species have been adapted to laboratory mice and rats and used as models for human malarias. Because of their use as models, it is important to

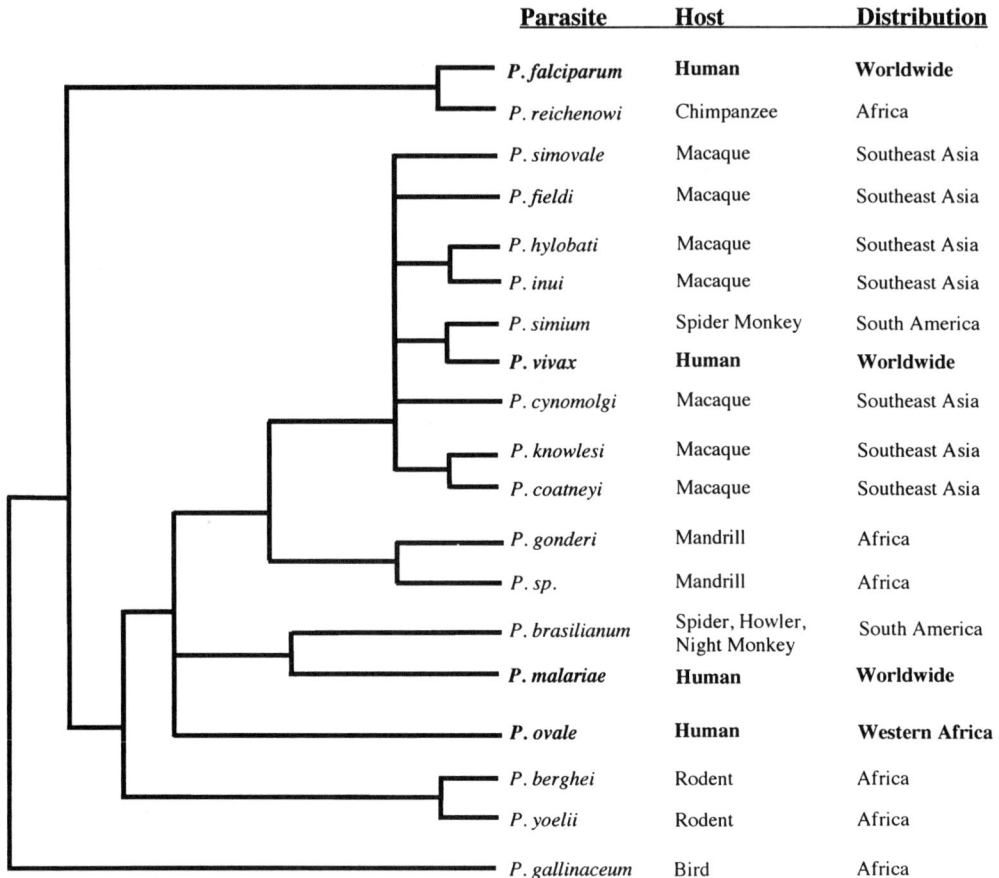

FIGURE 1 Summary phylogenetic tree showing the relationship between malaria parasites of mammals. Note that primate *Plasmodium* species are monophyletic, with the exception of the *P. falciparum*-*P. reichenowi* pairing, which is outside both rodent and primate clades. Rodent and primate parasites form monophyletic groups. Branch lengths are not proportional; only nodes with statistical support of ≥85% are shown (Escalante et al., 1995; Perkins and Schall, 2002; Vargas-Serrato et al., 2003).

note that rodent malarias are relatively closely related to each other but are distinct from the primate malarias (Fig. 1) (Escalante et al., 1998; Perkins and Schall, 2002).

The Emergence of *P. falciparum*

One of the most intriguing issues in malaria evolution is the mysterious origin of *P. falciparum,* and few issues have gone through as many significant revisions in the past 15 years. Older textbooks, and even some recent ones, refer to *P. falciparum* as arising from a recent horizontal transfer from birds, and it is to this recent host shift that the high virulence of *P. falciparum* is attributed (White, 2003). Recent research, however, refutes this hypothesis, because in the tree shown in Fig. 1 *P. falciparum* falls within a clade of mammalian parasites—distinct from and outside the group affecting birds and lizards. On the other hand, the species is not within the primate or rodent malarias. Rather, *P. falciparum* is off in a clade that includes only itself and its close sister species, the chimpanzee parasite *P. reichenowi*.

The avian origin hypothesis emerged originally from some of the early molecular phylogenies which, as discussed earlier, had problems due to taxon sampling and choice of genes (Waters et al., 1991; Waters et al., 1993; Escalante et al., 1995; Escalante et al., 1996; Escalante et al., 1997). First in 1996 (Qari et al.) and then again in 2002 (Perkins and Schall), new evidence came to light refuting an avian origin. Perkins and Schall (2002) published a relatively strong phylogeny based on cytochrome *b* with a large sample size of vertebrate parasites and an outgroup from a sister order. This cytochrome *b* phylogeny placed *P. falciparum* within the clade of mammalian parasites (Fig. 1). In addition to the phylogenetics, this analysis included an explicit test of the avian origin hypothesis using the Shimodaira-Hasegawa test, a phylogenetic tree-based method that calculates the likelihood of alternate trees, and rejected it.

At this time, most of the evidence (Qari et al., 1996; Escalante et al., 1997; Perkins and Schall, 2002) indicates that the only species closely related to *P. falciparum* is *P. reichenowi*; their divergence took place between 5 and 8 million years ago, based on fossil dates of the human-chimpanzee split (Escalante and Ayala, 1994; Escalante et al., 1995). Another parasite, *Plasmodium coatneyi*, has been classified as "falciparum like," due to its lack of a hypnozoite stage, the production of knob proteins on the red blood cell surface, and the presence of cytoadherence and rosetting (Coatney et al., 1971). This parasite, however, has been found to be a sister species to *Plasmodium knowlesi*, an Asian macaque parasite that falls within the standard primate parasite clade (Vargas-Serrato et al., 2003). The falciparum-like characteristics of *P. coatneyi* provide additional support for the idea that parasite strategies evolve too quickly to be informative characters for deducing evolutionary relationships.

The Source of *Plasmodium vivax*: Asian Origin

The origin of the species *P. vivax* has not been as hotly debated as that of *P. falciparum*, but it seems destined to be controversial because recent data reveal some unexpected relationships that have called previous assumptions into question. Specifically, phylogenetic evidence (Fig. 1) indicates that *P. vivax* branched from its relatives between 2 and 3 million years ago in Asia (Escalante et al., 1998). This parasite is found nested within a clade of exclusively Asian parasites (with the exception of its very close sister species, *Plasmodium simium*, from South America) supporting an Asian origin. This grouping, though, may reflect a bias in taxon sampling toward Asian parasites.

The emergence of *P. vivax* in Asia, however, contradicts the common belief that there was an African origin for the parasite. The theory that *P. vivax* emerged in Africa is an inference based on the high frequency with which humans in West Africa show lack of expression of the Duffy antigen on the surface of their red blood cells (Miller et al., 1976). Since the Duffy antigen is a receptor for *P. vivax* red blood cell invasion and lack of it confers resistance from the disease, the Duffy-negative phenotype has been interpreted as circumstantial evidence that *P. vivax* was present in West Africa in ancient times. The key word here is "ancient," because it is conceivable that *P. vivax* originated in Asia, followed by subsequent transfer to an African population. Recently, an Asian form of the Duffy-negative genotype was discovered in an area where *P. vivax* malaria is endemic (Zimmerman et al., 1999). This finding weakens the argument for an African origin based on the Duffy-negative phenotype. Why this phenotype is so widespread in Africa but so restricted in Asia may reflect the time that the relevant mutation occurred. The mutation occurs at only a very low frequency because it eliminates the Duffy antigen only on the surface of red blood cells but does not affect its presence on other cells in which the antigen occurs (Iwamoto et al., 1996).

Additional support for an Asian origin of *P. vivax* comes from studies of a sister species, *P. simium*. This species infects New World monkeys and is so similar to *P. vivax* that it has long been held that the two species are virtually identical, owing to a recent host shift (see discussion below) (Escalante et al., 1995). However, *P. simium* is more like Asian-African *P. vivax* than the *P. vivax* found in the New World (Li et al., 2001), supporting a non-American origin. This result is of additional interest as it indicates the possibility of two separate colonizations of the Americas by *P. vivax*. The understanding of the population history and structure of contemporary *P. vivax* parasites is very incomplete (discussed further below), and additional studies at the population level will be needed to resolve this issue.

Other Primate Malarias: Did Humans Bring Malaria to the New World?

We have already discussed the origin of human malarias by means of horizontal transfer from other vertebrates, and there is also evidence that humans have been the source for other primates. The inference emerges from evidence that humans brought malaria to the New World. Although there are multiple malaria parasites of Old World monkeys and apes, with several parasites specialized for one host, the situation in the New World is the reverse (Coatney et al., 1971). There are only two known forms of malaria that infect a range of New World monkey hosts, namely, *P. simium* and *P. brasilianum* (Coatney et al., 1971). What is astonishing about this pair of species is that each is a very close sister species of a human malaria parasite: *P. simium* is a sister of *P. vivax*, and *P. brasilianum* is a sister of *Plasmodium malariae* (Fig. 1) (Escalante et al., 1995; Escalante et al., 1997; Perkins and Schall, 2002). DNA sequence comparisons indicate that there is statistically no genetic distance between each simian species and its human sister species; hence, the species pairs are extremely closely related. Application of the molecular clock to divergence data based on the *csp* locus dates both speciation events to around 15,000 years ago (Escalante et al., 1995), which is coincident with human colonization of the Americas (Ingman et al., 2000). (Issues concerning the use of the *csp* locus for evolutionary studies have already been discussed; although calculating dates based on more neutrally evolving genes is necessary, the problems are of less concern when the locus is used to study closely related species.) Because *P. vivax* and *P. malariae* are much older than 15,000 years (Escalante et al., 1995), it appears that this time point represents the emergence of the simian species; however, it is difficult to deduce the direction of the host shift.

The Relationship of Human Malarias to Model Systems and Other Parasites

All primate malaria parasites, including those that infect humans, belong to the genus *Plasmodium*. Relative to larger, more inclusive (macroevolutionary) assemblages of organisms, the genus *Plasmodium* is placed within the family Plasmodiidae, which is in the order Haemosporidia, class Aconoidasida, and the phylum Apicomplexa. The Apicomplexa are part of the infrakingdom Alveolata, which includes the Dinoflagellates and Euglenozoa (Fig. 2 and 3) (Cavalier-Smith, 2004).

Similar to problems concerning the systematics of the genus *Plasmodium*, discussed above, the organization and nomenclature of the higher-level categories among eukaryotes are far from settled, in part because of new insights provided by molecular systematics. Although this appears to be an esoteric concern, it has some significant bearing on malaria research based on model systems and work on other parasites. As some of the higher-order classifications of *Plasmodium* are new, they may seem unfamiliar, and some are not universally accepted. In particular, the category infrakingdom (as opposed to kingdom as the ultimate classification) was introduced by Cavalier-Smith (1991; cited in Cavalier-Smith and Chao, 2003) and seems to be increasing in popularity. This classification was proposed in part to circumvent problems in the Linnaean system resulting from the discovery that Protista is a paraphyletic group, making it a "trash bin" category of otherwise not closely affiliated organisms that happen to possess a true nucleus but are not plants, animals, or fungi.

The conclusion of the past 15 years of molecular research (for example, Woese et al., 1990; Pace, 1997) is that the organisms assigned to Protista are far more diverse and divergent among themselves than are other eukaryotic groups. Problems in classification caused by the paraphyly of the Protista have important con-

FIGURE 2 Unrooted tree showing the relationship of infrakingdom Alveolata (*Plasmodium*) to other infrakingdoms (including parasites and genomic models) within the domain Eukarya (Cavalier-Smith, 1991, 2004; Cavalier-Smith and Chao, 2003; Baldauf et al., 2000; Hedges, 2002; Leander et al., 2003). The dotted line indicates the possible sister relationship of Plantae to some protist groups (Bastien et al., 2004). Branch lengths are not proportional; only nodes showing statistical support of ≥85% are included.

sequences for human health as the evolutionary relationship of *Plasmodium* to other protozoan parasites (and other single-celled organisms, such as yeasts) is far more complex and distant than the designation protist implies (Baldauf et al., 2000; Cavalier-Smith, 2004). One implication of the revised views of evolutionary relationships among the eukaryotes is that malaria parasites may have more in common with some plants than they do with *Entamoeba*, and another is that malaria parasites are more closely related to diatoms (Chromista) than they are to *S. cerevisiae*. (The unicellular lifestyle of *S. cerevisiae* is evidently derived, as its closest fungal relatives are

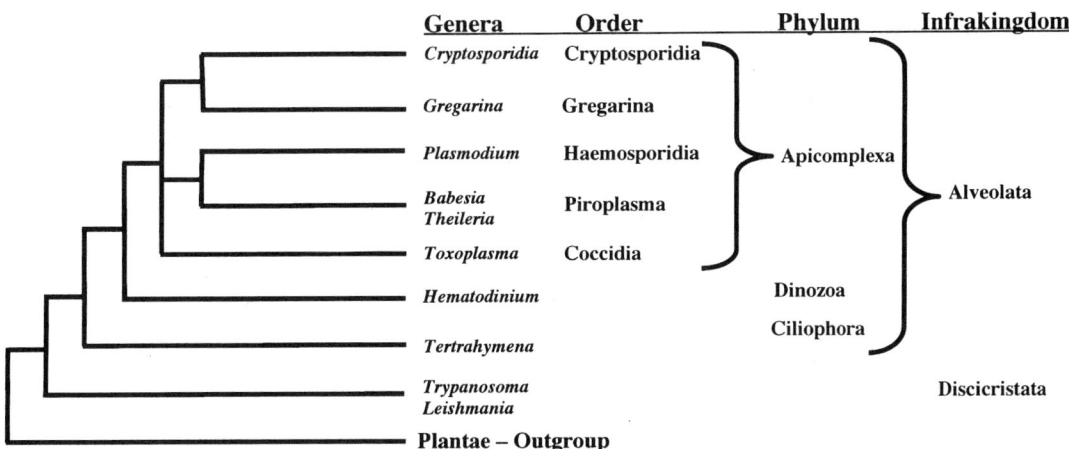

FIGURE 3 Placement of the order Haemosporidia with the phylum Apicomplexa and relation of Apicomplexa within Alveolata. The infrakingdom Discicristata is the sister group to Alveolata and includes several well-studied parasites. Branch lengths are not proportional; only nodes showing statistical support of ≥85% are included (Baldauf et al., 2000).

multicellular.) Most importantly, a proper placement of malaria parasites in relation to the rest of Eukarya facilitates comparisons among genomic DNA sequences and the correct identification of corresponding (orthologous) genes.

The Phylum Apicomplexa and the Origin of Parasitism in the *Plasmodium* Lineage

The phylum Apicomplexa is a monophyletic group: it includes all of the descendants of its most recent common ancestor. The group consists exclusively of unicellular parasites, and it is defined morphologically by the presence of an apical complex used for cell invasion that consists of rhoptries, micronemes, and polar rings. The organisms have no visible means of locomotion (e.g., flagella or cilia). They are thought to have originated 395 to 929 million years ago, based on the SSU rRNA molecular model (Escalante and Ayala, 1994). Although there are no free-living close relatives of *Plasmodium* within Apicomplexa, the sister groups *Colpodella* and the dinoflagellates include both parasitic and free-living forms, which suggests that the common ancestor of the Apicomplexa was itself parasitic (Kuvardina et al., 2002; Leander et al., 2003). As the origin of vertebrates has recently been hypothesized to be later than the origin of the Apicomplexa—estimated at about 564 million years ago by Kumar et al. (1998)—it is likely that Apicomplexa arose as a group of invertebrate parasites that acquired the ability to exploit vertebrate hosts, thereafter evolving with their vertebrate hosts in parallel.

Within Apicomplexa, the most closely related group of organisms to the *Plasmodium*-containing haemosporids are the piroplasms, which include well-known parasites of the genera *Theileria* and *Babesia* (Baldauf et al., 2000; Hedges, 2002; Leander et al., 2003). The other three orders within Apicomplexa are the cryptosporids, the gregarines (a group of invertebrate blood parasites), and the coccidians (which include *Toxoplasma*). Current molecular studies have been unable to resolve the relationships among these groups, and so their relative affinity to *Plasmodium* is equivocal. It is worth noting here that although *Toxoplasma* is often used as a model organism for *Plasmodium*, it is not the closest relative or the best outgroup for phylogenetic studies. It is the superior tools and experimental tractability of *Toxoplasma*, rather than evolutionary proximity, that make *Toxoplasma* an important model for *Plasmodium* research (Fig. 3).

EVER SINCE SPECIATION— POPULATION DYNAMICS OF HUMAN MALARIAS

Some of the most interesting and medically relevant studies of human parasites are studies at the population level. Understanding genetic diversity, as well as the rates and patterns of molecular evolution of parasite genomes, is critical to extending the useful clinical lifetime of vaccines and drugs before they become compromised by vaccine evasion or drug resistance. Understanding population structure also addresses epidemiological issues such as the spread of drug-resistant and highly virulent strains.

Plasmodium ovale and *P. malariae* cause the rarest and least virulent forms of malaria. The population genetics of these parasites has therefore received far less attention than that of *P. falciparum* and *P. vivax* and is not covered here. Population studies of *P. falciparum* and *P. vivax*, however, have already yielded unexpected results, paradoxical conclusions, and questions calling for additional research.

P. falciparum: Population Structure and Genome Diversity

Consistent with its importance in public health, *P. falciparum* has been studied in the most detail and with the greatest rigor at the population level. The availability of the *P. falciparum* genome sequence (Gardner et al., 2002) has allowed for a rapid expansion in the number of studies that are now possible, resulting in an efflorescence of evolutionary work on this parasite. This research has gone far to clarify both the population structure and the diversity of the world population of *P. falciparum*.

A number of critical questions should be addressed when examining the population structure and history of any pathogen: how diverse the world subpopulations are, how distinct they are, whether there is significant exchange of genetic information from one subpopulation to another, and how far back in time the current subpopulations can be traced before merging into a single, common ancestral population.

In principle, populations of *P. falciparum* parasites in different parts of the world could be very similar to each other (i.e., have similar frequencies of alternative alleles of polymorphic genes) or they could be genetically differentiated (indicated by differing allele frequencies among subpopulations). Early studies to estimate the extent of genetic differentiation yielded incongruous results, but further studies have brought some resolution. One study examined population structure by using the F_{ST} statistic (a statistic used to detect population structure by examining the variance in allele frequency at a locus among subpopulations) for *msp1* and *msp2*, encoding merozoite surface proteins (Conway, 1997), and another, similar, study examined the gene *pfs48/45*, encoding a gametocyte surface protein (Drakeley et al., 1996). Although the *msp* loci reflected little population structure (low F_{ST}), the *pfs48/45* study showed strong population subdivision (high F_{ST}). Later work by Anderson et al. (2000) used microsatellite polymorphisms to examine population structure more widely throughout the genome. The microsatellite data revealed that *P. falciparum* displays a spectrum of population structures throughout the world (Fig. 4). Strong linkage disequilibrium (nonrandom associations among alleles), low genetic diversity, and extensive population subdivision were found in areas where transmission is low (for example, South America), whereas the opposite characteristics are found in areas of very high transmission (as is the case in Africa). Intermediate values were found for the third region studied, Papua New Guinea, where transmission rates are between those of Africa and South America.

The differing levels of interpopulation and intrapopulation genetic variation may reflect the amount of inbreeding in areas of low transmission. If a population undergoes inbreeding and thereby avoids mixing genetic material with neighboring populations or gene pools, population genetic theory predicts that there will be a reduction of variation (heterozygosity) within the population as a consequence of random genetic drift. In the same way, inbreeding increases the difference among subpopulations (Hartl and Clark, 1997).

Inbreeding also decreases the effective amount of recombination between genes, but in the case of *P. falciparum* estimates of effective recombination have been inconsistent. In 1997, Rich et al. published a study that examined the pattern of substitution at the *csp* locus. They found that the correlation between polymorphic nucleotide sites along the gene was independent of the distance between the nucleotide sites. Because recombination results in a decrease in the expected correlation with distance, they inferred that natural populations of *P. falciparum* undergo little genetic recombination and therefore that the mode of reproduction of *P. falciparum* is essentially clonal. This conclusion was challenged by Conway et al. (1999), who found that polymorphic nucleotide sites across the sequence of *msp1* (the gene for merozoite surface protein 1) were statistically independent, indicating linkage equilibrium (random association among alleles) and a high rate of recombination. Studies of *csp* and *msp1* and other major antigenic loci are not necessarily representative of other genes, due to the high degree of immune selection, and therefore may show atypical patterns of linkage disequilibrium. At this point, the effective rate of recombination in different subpopulations of *P. falciparum* and in different regions of the *P. falciparum* genome remains unresolved.

Population History of *P. falciparum*

Although *P. falciparum* is an ancient parasite of the human population, the origin of the modern epidemic form as we know it today is another matter. Some studies suggest that all extant *P. falciparum* originate from a single common ancestor coincident with the advent of agriculture <50,000 years ago (Rich et al., 1998; Rich and Ayala, 2000), whereas other studies imply that *P. falciparum* has maintained a very large population

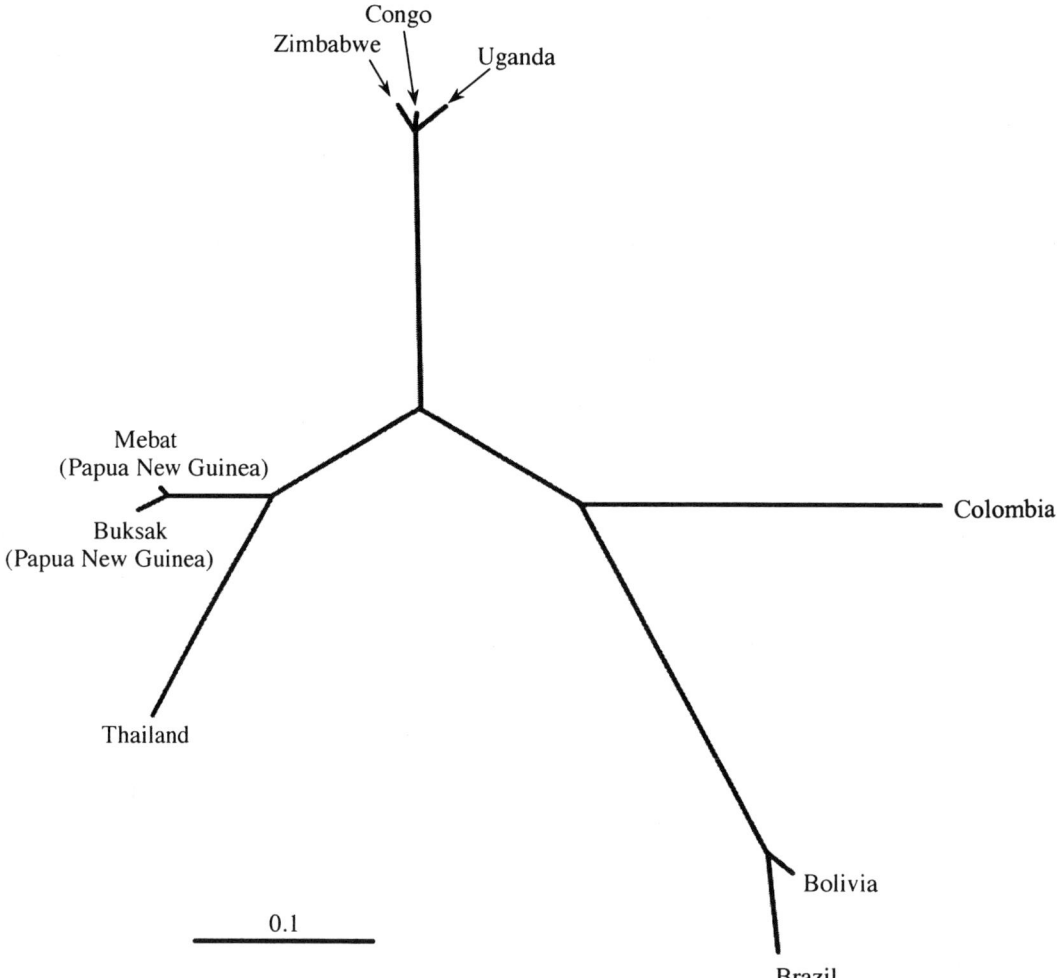

FIGURE 4 An unrooted neighbor-joining tree showing the relationships between nine *P. falciparum* populations. Only nodes with a bootstrap value of >75% are shown. Note that the African isolates (Zimbabwe, Congo, Uganda) from an area of high transmission show a large overall difference from the other isolates, but the relationship of the individual isolates cannot be distinguished with any statistical certainty. However, in South America, the Colombian and Bolivian-Brazilian isolates show a greater distance from each other than even overall distances from isolates from elsewhere in the world. (Figure adapted from Anderson et al., 2000.)

size continuously for at least 300,000 years (Hughes and Verra, 2001), predating the origin of modern humans (Ingman et al., 2000).

The first argument for a recent common ancestor of modern *P. falciparum* came from medical anthropology. A classic paper by Livingstone (1958) presented an extensive geographic mapping of falciparum malaria in Africa, the frequency of the sickle-cell trait (hemoglobin S) associated with malaria resistance, linguistic groupings, and the spread of agriculture. He found overlapping gradients of sickle-cell hemoglobin and falciparum malaria that coincided with the predicted progression of swidden (slash and burn) agriculture about 6,000 years ago. This form of agriculture greatly increases the open, sunlit areas within forests, which promotes the formation and persistence of pools of fresh water that are the breeding grounds of the anopheline mosquitoes that transmit the parasite. Additional support to this hypothesis comes from studies of glucose-6-phosphate dehydrogenase deficiency, another genetic condition associated with malaria resistance. Recent studies of human X chromosomes carrying this mutation are consistent with Livingstone's scenario in suggesting that the original mutant allele arose some 10,000 years ago (Tishkoff et al., 2001; Saunders et al., 2002).

Evidence for a relatively recent common ancestor of today's *P. falciparum* also comes from the unexpectedly low level of nucleotide sequence variation found in the parasite genome. The neutral theory of molecular evolution predicts that genomes evolve by accumulating random mutations, many of which are neutral in their effects on reproductive fitness (Kimura, 1968). Unaffected or only weakly affected by natural selection, these neutral mutations accumulate over time in parts of the genome such as intergenic regions, introns, and third-codon positions in protein-encoding regions (synonymous mutations). However, Rich et al. (1998) reported that *P. falciparum* showed an unexpectedly low level of synonymous polymorphisms in the coding regions of 10 genes examined, which included rapidly evolving antigenic genes, as well as more slowly evolving housekeeping genes. They attributed this finding to the emergence of modern *P. falciparum* from a single common ancestor more recently than 50,000 years ago, an interval over which very little neutral genetic variation would be expected to accumulate. Consistent with this result, Volkman et al. (2001) reported a low level of nucleotide polymorphism in 25 introns of nine genes, mostly on chromosome 2. On the other hand, the level of nucleotide polymorphisms across chromosome 3 suggests an older age of 100,000 to 180,000 years (Mu et al., 2002). One possible explanation for the discrepancy is that the genome of *P. falciparum* has a mosaic structure, with some regions sharing a common ancestor more recently than others. This would be consistent with a model of population history in which various regions of the genome turn over quickly, but not all at the same time, owing to the rapid fixation of new favorable mutations. The oldest estimates of the ancestry of *P. falciparum* extend to 300,000 to 400,000 years ago (Hughes and Verra, 2001), but these are based on older and less-reliable GenBank sequences (Barry et al., 2003; Nielsen et al., 2003).

Although different regions of the nuclear genome may have different evolutionary histories, this is not the case for the 6-kb linear mitochondrial genome (mtDNA) because, in so far as is known, the mitochondrial genome is uniparentally inherited and does not undergo recombination. This feature of mtDNA was exploited by Conway et al. (2000), who argued from mtDNA polymorphisms that *P. falciparum* radiated out of Africa and then into southeast Asia and South America. Joy et al. (2003) sequenced and analyzed the complete mtDNA genomes from 100 geographically diverse isolates. All but a few of these genomes appear to have originated from the expansion of a common ancestral sequence some 10,000 years ago, but the few remaining appear to date back as far as 50,000 to 100,000 years. These dates suggest that *P. falciparum* underwent two migrations out of Africa, an early one perhaps coinciding with the Pleistocene migration of human beings (Templeton, 2002) and a later one more recently than 10,000 years ago. The findings from the nuclear genome are not contrary to this model.

P. vivax Populations: Studies at the Genome Level

Acquisition of the genome sequence of *P. vivax* has helped spark renewed interest in the population genetics of this widespread parasite in spite of its inability to be continually cultured in erythrocytes in vitro. Much previous work has focused on single loci, usually genes encoding surface antigens such as apical membrane antigen or merozoite surface proteins, often with geographically restricted samples and without reference to worldwide population structure (Figtree et al., 2000; Gutierrez et al., 2000). However, two recent studies have analyzed genetic variation that is more likely to be evolving in a nearly neutral fashion and begin to address the issue of how much genetic variation exists among natural isolates of *P. vivax*. These studies represent the beginning of genome-level studies of this parasite.

The first study (Feng et al., 2003) focused on a 100-kb region of a *P. vivax* chromosome that is syntenic with a region of chromosome 3 in *P. falciparum*. This region includes 26 annotated genes, representing a range of functions, plus their intergenic sequences. The region was sequenced from each of five culture-adapted lines from either South America or southeast Asia. An examination of genetic variation across the region revealed a level of single-nucleotide polymorphism in *P. vivax* that was about twice as high as that found in *P. falciparum*. From these data, the authors reported that the world population of *P. vivax* is highly diverse.

The second study (Leclerc et al., 2004) examined length polymorphisms in 13 microsatellite loci from 108 samples from eight localities in Asia, Africa, New Guinea, and South America. Among the microsatellites, nine loci were completely monomorphic and three were polymorphic, owing to a rare allele found in one or two samples. Only one microsatellite locus was polymorphic with multiple, common alleles. Furthermore, among the nine microsatellite loci found to be monomorphic in *P. vivax*, eight were polymorphic in several *Plasmodium* species related to *P. vivax*; however, *P. simium*, a sister species of *P. vivax* that infects New World monkeys, was genetically indistinguishable from *P. vivax* based on the microsatellite markers. The authors conclude that the dearth of microsatellite polymorphisms in *P. vivax* could reflect severe recent population bottlenecks (a severe reduction in population size followed by an expansion from this small sample) or multiple selective sweeps of strongly favorable new mutations, consistent with a world expansion of this parasite <10,000 years ago.

How to reconcile these studies is far from certain. On the one hand, the 100-kb region studied by Feng et al. (2003) could be atypical in containing genes subject to selection for heterozygosity, which would help maintain linked polymorphisms, or it could be a region that escaped the effects of recent selective sweeps and thus retained a larger than average amount of ancestral polymorphism. On the other hand, the expectation of a high level of microsatellite polymorphism in *P. vivax* is based in part on the abundant microsatellite polymorphisms in *P. falciparum* (Anderson et al., 2000) that result from their high mutation rate (Su et al., 2004). The mutation rate of microsatellites in *P. vivax* has not been estimated, but these rates can differ tremendously among organisms. For example, the microsatellite mutation rate in mammals is 10 to 1,000 times greater than that observed in *Drosophila* and depends on the particular repeating unit of the microsatellite (Schug et al., 1998). At this point, the level of selectively neutral genetic variation that is present across the genome of *P. vivax* is very uncertain, and additional studies are needed to resolve the situation.

CONCLUSION: GENOME RESEARCH AND PARASITE EVOLUTION

For many years questions about the evolution of human malaria were beyond definitive resolution. Many studies led to intriguing but fragmentary conclusions, each only able to tell a small part of the story. The result was strong disagreements in the literature and theories continually contested and revised. Genome-level research has made it possible to explore malaria evolution in greater depth and rigor to help complete the picture. These changes have come through malaria parasite genome projects (for *P. falciparum*, *P. vivax*, *P. yoelii*, and *P. reichenowi*) and their associated databases, access to automated DNA sequencers in most modern laboratories, an increasingly sophisticated understanding of the unusual molecular biology and evolution of the parasites, and improvements in methods of statistical inference from DNA sequence differences within and among populations. The tools are in place, and the field is full of important questions that need to be addressed and paradoxes that are ready to be resolved. The potential payoff is in the form of significant advances in our understanding of the evolution and control of drug resistance, immune evasion, human defenses, and enhanced use of model systems to study malaria.

REFERENCES

Anderson, T. J. C., B. Haubold, J. T. Williams, J. G. Estrada-Franco, L. Richardson, R. Mollinedo, M. Bockarie, J. Mokili, S. Mharakurwa, N. French, J. Whitworth, I. D. Velez, A. H. Brockman, F. Nosten, M. U. Ferreira, and K. P. Day. 2000. Microsatellite markers reveal a spectrum of population structures in the malaria parasite *Plasmodium falciparum*. *Mol. Biol. Evol.* **17:**1467–1482.

Baldauf, S. L., A. J. Roger, I. Wenk-Siefert, and W. F. Doolittle. 2000. A kingdom-level phylogeny of eukaryotes on combined protein data. *Science* **290:**972–977.

Barry, A. E., A. Leliwa, M. Choi, K. M. Nielsen, D. L. Hartl, and K. P. Day. 2003. DNA sequence artifacts and the estimation of time to the most recent common ancestor (TMRCA) of *Plasmodium falciparum*. *Mol. Biochem. Parasitol.* **130:**143–147.

Bastien, O., S. Lespinats, S. Roy, K. Metayer, B. Fertil, J. J. Codani, and E. Marechal. 2004. Analysis of the compositional biases in Plasmodium falciparum genome and proteome using Arabidopsis thaliana as a reference. *Gene* **336:**163–173.

Cavalier-Smith, T. 1991. Cell diversification in heterotrophic flagellates, p. 113–131. *In* D. J. Patterson and J. Larsen (ed.), *The Biology of Freeliving Heterotrophic Flagellates*. Oxford University Press, New York, N.Y.

Cavalier-Smith, T. 2004. Only six kingdoms of life. *Proc. R. Soc. Lond. B Biol. Sci.* **271:**1251–1262.

Cavalier-Smith, T., and E. E.-Y. Chao. 2003. Phylogeny of choanozoa, apusozoa, and other protozoa and early eukaryote megaevolution. *J. Mol. Evol.* **56:**540–563.

Coatney, G. R., W. E. Collins, M. Warren, and P. G. Contacos. 1971. *The Primate Malarias*. U.S. Department of Health, Education and Welfare, Bethesda, Md.

Conway, D. J. 1997. Natural selection on polymorphic malaria antigens and the search for a vaccine. *Parasitol. Today* **13:**26–29.

Conway, D. J., C. Roper, A. M. Oduola, D. E. Arnot, P. G. Kremsner, M. P. Grobusch, C. F. Curtis, and B. M. Greenwood. 1999. High recombination rate in natural populations of Plasmodium falciparum. *Proc. Natl. Acad. Sci. USA* **96:**4506–4511.

Conway, D. J., C. Fanello, J. M. Lloyd, B. M. Al-Joubori, A. H. Baloch, S. D. Somanath, C. Roper, A. M. Oduola, B. Mulder, M. M. Povoa, B. Singh, and A. W. Thomas. 2000. Origin of *Plasmodium falciparum* malaria is traced by mitochondrial DNA. *Mol. Biochem. Parasitol.* **111:**163–171.

Corredor, V., and V. Enea. 1993. Plasmodial ribosomal RNA as phylogenetic probe: a cautionary note. *Mol. Biol. Evol.* **10:**924–926.

Coulson, R. M., N. Hall, and C. A. Ouzounis. 2004. Comparative genomics of transcriptional control in the human malaria parasite Plasmodium falciparum. *Genome Res.* **14:**1548–1554.

Drakeley C. J., M. T. Duraisingh, M. Povoa, D. J. Conway, G. A. Targett, and D. A. Baker. 1996. Geographical distribution of a variant epitope of Pfs48/45, a *Plasmodium falciparum* transmission-blocking vaccine candidate. *Mol. Biochem. Parasitol.* **81:**253–257.

Escalante, A. A., and F. J. Ayala. 1994. Phylogeny of the malarial genus *Plasmodium*, derived from rRNA gene sequences. *Proc. Natl. Acad. Sci. USA* **91:**11373–11377.

Escalante, A. A., and F. J. Ayala. 1995. Evolutionary origin of *Plasmodium* and other apicomplexa based on rRNA genes. *Proc. Natl. Acad. Sci. USA* **92:**5793–5799.

Escalante, A. A., E. Barrio, and F. J. Ayala. 1995. Evolutionary origin of human and primate malarias: evidence from the circumsporozoite protein. *Mol. Biol. Evol.* **12:**616–626.

Escalante, A. A., D. E. Freeland, W. E. Collins, and A. A. Lal. 1998. The evolution of primate malaria parasites based on the gene encoding cytochrome *b* from the linear mitochondrial genome. *Proc. Natl. Acad. Sci. USA* **95:**8124–8129.

Escalante, A. A., I. F. Goldman, P. De Rijk, R. De Wachter, W. E. Collins, S. H. Qari, and A. A. Lal. 1997. Phylogenetic study of the genus *Plasmodium* based on the secondary structure-based alignment of the small subunit ribosomal RNA. *Mol. Biochem. Parasitol.* **90:**317–321.

Feng, X., J. M. Carlton, D. A. Joy, J. Mu, T. Furuya, B. Suh, Y. Wang, J. W. Barnwell, and X.-Z. Su. 2003. Single nucleotide polymorphisms and genome diversity in *Plasmodium vivax*. *Proc. Natl. Acad. Sci. USA* **100:**8502–8507.

Figtree, M., C. J. Pasay, R. Slade, Q. Cheng, N. Cloonan, J. Walker, and A. Saul. 2000. Plasmodium vivax synonymous substitution frequencies, evolution and population structure deduced from diversity in AMA 1 and MSP 1 genes. *Mol. Biochem. Parasitol.* **108:**53–66.

Gardner, M. J., N. Hall, E. Fung, O. White, M. Berriman, R. W. Hyman, J. M. Carlton, A. Pain, K. E. Nelson, S. Bowman, I. T. Paulsen, K. James, J. A. Eisen, K. Rutherford, S. L. Salzberg, A. Craig, S. Kyes, M. S. Chan, V. Nene, S. J. Shallom, B. Suh, J. Peterson, S. Angiuoli, M. Pertea, J. Allen, J. Selengut, D. Haft, M. W. Mather, A. B. Vaidya, D. M. Martin, A. H. Fairlamb, M. J. Fraunholz, D. S. Roos, S. A. Ralph, G. I. McFadden, L. M. Cummings, G. M. Subramanian, C. Mungall, J. C. Venter, D. J. Carucci, S. L. Hoffman, C. Newbold, R. W. Davis, C. M. Fraser, and B. Barrell.

2002. Genome sequence of the human malaria parasite Plasmodium falciparum. *Nature* **419**:498–511.

Gutierrez, A., J. Vicini, M. E. Patarroyo, L. A. Murillo, and M. A. Patarroyo. 2000. *Plasmodium vivax*: polymorphism in the merozoite surface protein 1 gene from wild Colombian isolates. *Exp. Parasitol.* **95**:215–219.

Hartl, D. L., and A. G. Clark. 1997. *Principles of Population Genetics*, 3rd ed. Sinauer Associates, Sunderland, Mass.

Hedges, S. B. 2002. The origin and evolution of model organisms. *Nat. Rev. Genet.* **3**:838–849.

Hughes, A. L., and F. Verra. 2001. Very large long-term effective population size in the virulent human malaria parasite *Plasmodium falciparum*. *Proc. R. Soc. Lond. B Biol. Sci.* **268**:1855–1860.

Ingman, M., H. Kaessmann, S. Pääbo, and U. Glyllensten. 2000. Mitochondrial genome variation and the origin of modern humans. *Nature* **408**:708–713.

Iwamoto, S., J. Li, N. Sugimoto, H. Okuda, and E. Kajii. 1996. Characterization of the Duffy gene promoter: evidence for tissue-specific abolishment of expression in Fy(a-b-) of black individuals. *Biochem. Biophys. Res. Commun.* **222**:852–859.

Joy, D. A., X. Feng, J. Mu, T. Furuya, K. Chotivanich, A. U. Krettli, M. Ho, A. Wang, N. J. White, E. Suh, P. Beerli, and X. Su. 2003. Early origin and recent expansion of *Plasmodium falciparum*. *Science* **300**:318–321.

Kedzierski, L., A. A. Escalante, R. Isea, C. G. Black, J. W. Barnwell, and R. L. Coppel. 2002. Phylogenetic analysis of the genus *Plasmodium* based on the gene encoding adenylosuccinate lyase. *Infect. Genet. Evol.* **1**:297–301.

Kimura, M. 1968. Evolutionary rate at the molecular level. *Nature* **217**:624–626.

Kissinger, J. C., W. E. Collins, J. Li, and T. F. McCutchan. 1998. *Plasmodium inui* is not closely related to other quartan *Plasmodium* species. *J. Parasitol.* **84**:278–282.

Kumar, S., and B. S. Hedges. 1998. A molecular timescale for vertebrate evolution. *Nature* **392**:917–920.

Kurvadina, O. N., B. S. Leander, V. V. Aleshin, A. P. Myl'Nikov, P. J. Keeling, and T. G. Simdyanov. 2002. The phylogeny of colopodellids (Alveolata) using small subunit rRNA gene sequences suggests they are the free-living sister group to apicomplexans. *J. Eukaryot. Microbiol.* **49**:498–504.

Leander, B. S., O. N. Kuvardina, V. V. Aleshin, A. P. Mylnikov, and P. J. Keeling. 2003. Molecular phylogeny and surface morphology of *Colpodella edax* (Alveolata): insights into the phagotrophic ancestry of apicomplexans. *J. Eukaryot. Microbiol.* **50**:334–340.

Leclerc, M. C., P. Durand, C. Gauthier, S. Patot, N. Billotte, M. Menegon, C. Severini, F. J. Ayala, and F. Renaud. 2004. Meager genetic variability of the human malaria agent Plasmodium vivax. *Proc. Natl. Acad. Sci. USA* **101**:14455–14460.

Li, J., W. E. Collins, R. A. Wirtz, D. Rathore, A. Lal, and T. F. McCutchan. 2001. Geographic subdivision of the range of the malaria parasite Plasmodium vivax. *Emerg. Infect. Dis.* **7**:35–42.

Livingstone, F. B. 1958. Anthropological implications of sickle cell gene distribution in West Africa. *Am. Anthropol.* **60**:533–562.

McCutchan, T. F., J. B. Dame, L. H. Miller, and J. Barnwell. 1984. Evolutionary relatedness of Plasmodium species as determined by the structure of DNA. *Science* **225**:808–811.

McCutchan, T. F., J. C. Kissinger, M. G. Touray, J. M. Rogers, J. Li, M. Sullivan, E. M. Braga, A. U. Krettli, and L. H. Miller. 1996. Comparison of circumsporozoite proteins from avian and mammalian malarias: biological and phylogenetic implications. *Proc. Natl. Acad. Sci. USA* **93**:11889–11894.

Miller, L. H., S. J. Mason, D. F. Clyde, and M. H. McGinnis. 1976. The resistance factor to *Plasmodium vivax* in blacks: the Duffy blood group genotype, FyFy. *New Engl. J. Med.* **295**:302–304.

Mu, J., J. Duan, K. D. Makova, D. A. Joy, C. Q. Huynh, O. H. Branch, W. H. Li, and X. Z. Su. 2002. Chromosome-wide SNPs reveal an ancient origin for *Plasmodium falciparum*. *Nature* **418**:323–326.

Nielsen, K. M., J. Kasper, M. Choi, T. Bedford, K. Kristiansen, D. F. Wirth, S. K. Volkman, E. R. Lozovsky, and D. L. Hartl. 2003. Gene conversion as a source of nucleotide diversity in Plasmodium falciparum. *Mol. Biol. Evol.* **20**:726–734.

Pace, N. R. 1997. A molecular view of microbial diversity and the biosphere. *Science* **276**:734–740.

Paul, R. E. L., F. Ariey, and V. Robert. 2003. The evolutionary ecology of *Plasmodium*. *Ecol. Lett.* **6**:866–880.

Perkins, S. L., and J. J. Schall. 2002. A molecular phylogeny of malaria parasites recovered from cytochrome b gene sequences. *J. Parasitol.* **88**:972–978.

Philippe, H. 2000. Long branch attraction and protist phylogeny. *Protist* **151**:307–316.

Qari, S. H., Y. P. Shi, N. J. Pieniazek, W. E. Collins, and A. A. Lal. 1996. Phylogenetic relationship among the malaria parasites based on small subunit rRNA gene sequences: monophyletic nature of the human malaria parasite, Plasmodium falciparum. *Mol. Phylogenet. Evol.* **6**:157–165.

Rasti, N., M. Wahlgren, and Q. Chen. 2004. Molecular aspects of malaria pathogenesis. *FEMS Immunol. Med. Microbiol.* **41**:9–26.

Remsen, J., and P. O'Grady. 2002. Phylogeny of Drosophilinae (Diptera: Drosophilidae), with comments on combined analysis and character support. *Mol. Phylogenet. Evol.* **24:**249–264.

Rich, S. M., and F. J. Ayala. 2000. Population structure and recent evolution of *Plasmodium falciparum*. *Proc. Natl. Acad. Sci. USA* **97:**6994–70001.

Rich, S. M., R. R. Hudson, and F. J. Ayala. 1997. Plasmodium falciparum antigenic diversity: evidence of clonal population structure. *Proc. Natl. Acad. Sci. USA* **94:**13040–13045.

Rich, S. M., M. C. Licht, R. R. Hudson, and F. J. Ayala. 1998. Malaria's Eve: evidence of a recent population bottleneck throughout the world populations of Plasmodium falciparum. *Proc. Natl. Acad. Sci. USA* **95:**4425–4430.

Rokas, A., B. L. Williams, N. King, and S. B. Carroll. 2003. Genome-scale approaches to resolving incongruence in molecular phylogenies. *Nature* **425:**798–804.

Russell, P. F., L. S. West, R. D. Manwell, and G. M. MacDonald. 1963. *Practical Malariology*, 2nd ed. Oxford University Press, London, United Kingdom.

Saunders, M. A., M. F. Hammer, and M. W. Nachman. 2002. Nucleotide variability at G6pd and the signature of malarial selection in humans. *Genetics* **162:**1849–1861.

Schug, M. D., C. M. Hutter, M. A. Noor, and C. F. Aquadro. 1998. Mutation and evolution of microsatellites in Drosophila melanogaster. *Genetica* **102–103:**359–356.

Siddall, M. E., and J. R. Barta. 1992. Phylogeny of Plasmodium species: estimation and inference. *J. Parasitol.* **78:**567–568.

Su, X.-Z., and J. C. Wootton. 2004. Genetic mapping in the human malaria parasite Plasmodium falciparum. *Mol. Microbiol.* **53:**1573–1582.

Templeton, A. 2002. Out of Africa again and again. *Nature* **416:**45–51.

Tishkoff, S. A., R. Varkonyi, N. Cahinhinan, S. Abbes, G. Argyropoulos, G. Destro-Bisol, A. Drousiotou, B. Dangerfield, G. Lefranc, J. Loiselet, A. Piro, M. Stoneking, A. Tagarelli, G. Tagarelli, H. Elias. E. H. Touma, S. M. Williams, and A. G. Clark. 2001. Haplotype diversity and linkage disequilibrium at human G6PD: recent origin of alleles that confer malarial resistance. *Science* **293:**455–482.

Vaidya, A. B., M. S. Lashgari, L. G. Pologe, and J. Morrisey. 1993. Structural features of *Plasmodium* cytochrome-b that may underlie susceptibility to 8-aminoquinolines and hydroxynaphthoquinones. *Mol. Biochem. Parasitol.* **58:**33–42.

Vargas-Serrato, E., V. Corredor, and M. R. Galinski. 2003. Phylogenetic analysis of CSP and MSP-9 gene sequences demonstrates the close relationship of Plasmodium coatneyi to Plasmodium knowlesi. *Infect. Genet. Evol.* **3:**67–73.

Volkman, S. K., A. E. Barry, E. J. Lyons, K. M. Nielsen, S. M. Thomas, M. Choi, S. S. Thakore, K. P. Day, D. F. Wirth, and D. L. Hartl. 2001. Recent origin of *Plasmodium falciparum* from a single progenitor. *Science* **293:**482–484.

Waters, A. P., D. G. Higgins, and T. F. McCutchan. 1991. *Plasmodium falciparum* appears to have arisen as a result of a lateral transfer between avian and human hosts. *Proc. Natl. Acad. Sci. USA* **88:**3140–3144.

Waters, A. P., D. G. Higgins, and T. F. McCutchan. 1993. Evolutionary relatedness of some primate models of Plasmodium. *Mol. Biol. Evol.* **10:**914–923.

White, N. J. 2003. Malaria, p. 1205–1296. *In* G. C. Cook and A. Zumla (ed.), *Manson's Tropical Diseases*, 21st ed. W. B. Saunders, Philadelphia, Pa.

Woese, C. R., O. Kandler, and M. L. Wheelis. 1990. Towards a natural system of organisms: proposal for the domains Archaea, Bacteria, and Eucarya. *Proc. Natl. Acad. Sci. USA* **87:**4576–4579.

Zilversmit, M., P. M. O'Grady, and R. DeSalle. 2002. Shallow genomics, phylogenetics, and evolution in the family Drosophilidae, p. 512–523. *In* R. Altman, A. K. Dunker, L. Hunter, K. Lauderdale, and T. Klein, (ed.) *Pacific Symposium on Biocomputing*. World Scientific Publishing Co. Pte. Ltd., Riveredge, N.J.

Zimmerman, P. A., I. Woolley, G. L. Masinde, S. M. Miller, D. T. McNamara, F. Hazlett, C. S. Mgone, M. P. Alpers, B. Genton, B. A. Boatini, and J. W. Kazura. 1999. Emergence of FY*A null in a *Plasmodium vivax*-endemic region of Papua New Guinea. *Proc. Natl. Acad. Sci. USA* **96:**13973–13977.

INVASION AND GAMETE FORMATION

II

A MECHANISTIC APPROACH TO MEROZOITE INVASION OF RED BLOOD CELLS: MEROZOITE BIOGENESIS, RUPTURE, AND INVASION OF ERYTHROCYTES

Mary R. Galinski, Anton R. Dluzewski, and John W. Barnwell

8

Malaria merozoites are initially formed during schizogony in hepatocytes, late in the parasite's development in the liver, and subsequently in red blood cells (RBCs) during the cyclical stages of erythrocytic development. If even one hepatic schizont survives a natural or vaccine-induced liver-stage immune response, it will release its many thousands of merozoites into the bloodstream, and a blood-stage infection with all its varied clinical symptoms and complications will ensue (Mackintosh et al., 2004; Maitland and Marsh, 2004). This capability illustrates the power of the merozoite and emphasizes the need to continue to unravel the biochemical and biological processes that underpin the existential basis of its invasiveness and to seek means to target potential vulnerable steps in this stage of the parasite's life cycle.

Relatively little has been revealed about the precise makeup of liver-stage schizonts and formed merozoites, since it is extremely challenging to obtain and study infected hepatocytes (Badell et al., 1995; Gruner et al., 2003). Nevertheless, transcriptional studies have been initiated to further define the range of proteins expressed in liver-stage parasites (Wang et al., 2004). These studies and knowledge from earlier investigations (Aley et al., 1987; Szarfman et al., 1988a; Szarfman et al., 1988b; Preiser et al., 2002) support the likelihood that merozoites formed in the hepatocytes are similar (or identical) in structure and function to those that are subsequently formed in the host's RBCs. To prevent the successful development of mature infective merozoites, their exit from host RBCs, or their successful invasion of new host erythrocytes, proteins from the merozoite stage have been considered targets for vaccination or therapeutic intervention.

Here, we provide an overview of the dynamic nature of malaria from the perspective of the development of merozoites within an infected erythrocyte and the release of infectious merozoites, through the initiation and completion of the reinvasion process. This review encompasses discoveries or observations obtained

---

Mary R. Galinski, Emory Vaccine Center at the Yerkes National Primate Research Center and Division of Infectious Diseases, Department of Medicine, Emory University, 954 Gatewood Rd., Atlanta, GA 30329. *Anton R. Dluzewski,* Department of Immunobiology, Guy's, King's and St. Thomas's School of Medicine, 2nd Floor, New Guy's House, King's College London, Guy's Hospital, London Bridge, London SE1 9RT, United Kingdom. *John W. Barnwell,* Malaria Branch, Division of Parasitic Diseases, Centers for Disease Control and Prevention, Mailstop F-36, Bldg. 109, Room 1121, 4770 Buford Highway, NE, Atlanta, GA 30341.

Molecular Approaches to Malaria, Edited by Irwin W. Sherman
© 2005 ASM Press, Washington, D.C.

through studies of different species of *Plasmodium*, which together have greatly aided and refined our understanding of these events. These species include not only the human malarias *Plasmodium falciparum* and *Plasmodium vivax*, but also the simian malaria *Plasmodium knowlesi*, the chimpanzee malaria *Plasmodium reichenowi*, bird malarias such as *Plasmodium elongatum* and *Plasmodium gallinaceum*, and the rodent malarias, principally *Plasmodium yoelii* and *Plasmodium berghei*. We note key discoveries from 25 years ago and onwards to clarify and emphasize recent discoveries, which have picked up pace in the past 5 to 7 years. Especially given the availability of malaria genome data, advanced microscopy capabilities, and transfection technologies using human, simian, and rodent malaria parasites (for reviews, see Crabb et al., 2004; Wel et al., 2004; and Carvalho and Menard, 2005) there is great potential now to gain a much improved understanding of merozoite biogenesis, egress, and invasion and the propagation of the disease.

Over the past 4 decades, excellent papers with a focus on the ultrastructure of merozoite biogenesis, schizogony, and merozoite invasion of RBCs have been published (Aikawa, 1966; Aikawa et al., 1967; Ladda et al., 1969; Aikawa, 1971; Aikawa and Sterling, 1974; Bannister et al., 1975; Aikawa et al., 1978; Langreth et al., 1978;

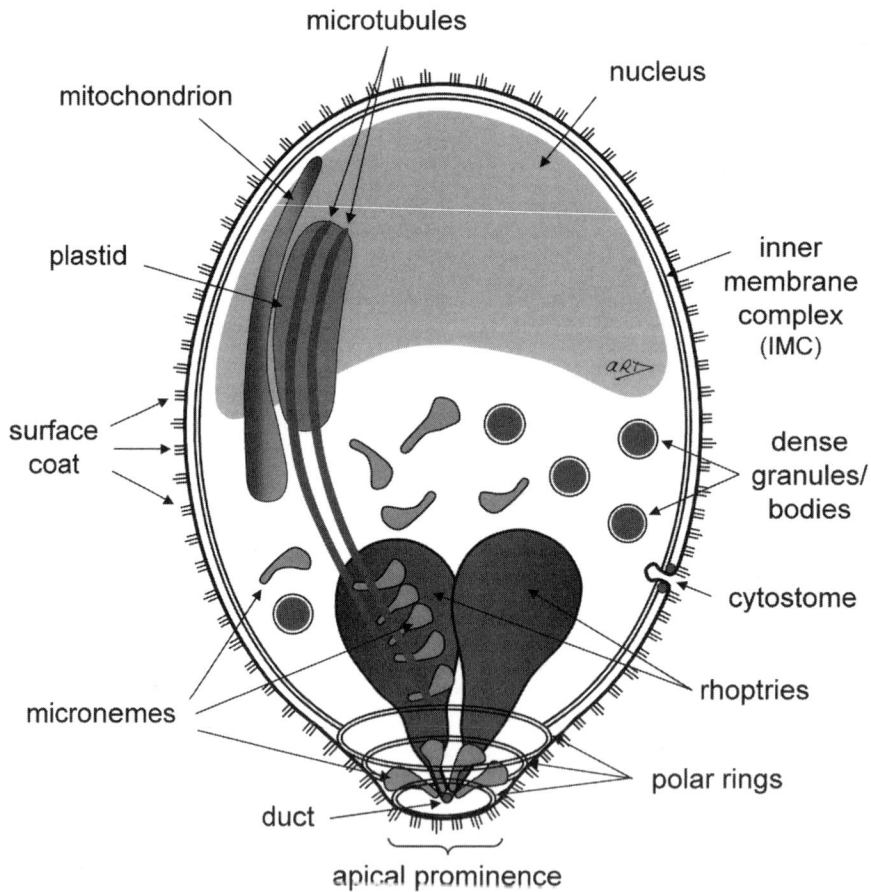

FIGURE 1 Generalized diagrammatic representation of a merozoite showing the main structural features and organelles and also depicting the process by which newly formed micronemes are translocated along the microtubules to their placement at the apical pole (Bannister et al., 2003).

McLaren et al., 1979; Miller et al., 1979; Aikawa et al., 1981; Aikawa and Miller, 1983; Bannister et al., 1986b; Mitchell and Bannister, 1988; Bannister and Mitchell, 1989; Dluzewski et al., 1989; Torii et al., 1989; Bannister and Dluzewski, 1990; Gratzer and Dluzewski, 1993; Jaikaria et al., 1993; Bannister and Mitchell, 1995; Hopkins et al., 1999; Bannister et al., 2000b; Bannister, 2001; Bannister et al., 2003; Margos et al., 2004). The original elegant, detailed morphological discoveries provide a visual picture and broad conceptual understanding of the very intricate and complex blood-stage parasite with its unique structures and organelles and its cycle of invasion and development in RBCs. These studies are worth a careful review, as they form the scaffold on which to structure details coming from ongoing advanced molecular and cell biological studies (see Chapter 3).

CELLULAR STRUCTURE OF THE MEROZOITE

Malarial merozoites are small, ovoid, atypical eukaryotic cells that are broader at the posterior end than at the anterior apical pole and measure 1.5 to 2.5 μm long and 1.0 to 2.0 μm wide, depending upon the species (Fig. 1 and 2). *P. vivax* merozoites, for example, are about twice as large as *P. falciparum* merozoites (Galinski and Barnwell, 1996). Malaria merozoites have a plasma membrane and the basic cellular machinery of typical eukaryotic cells, including a nucleus, endoplasmic reticulum, Golgi network, ribosomes, and mitochondria (Bannister et al., 2000a) (Fig. 1). In addition, they have a plastid structure (Hopkins et al., 1999) and evidence of an ancient endosymbiotic acquisition of an alga (Fast et al., 2001; Williamson et al., 2002; Waller and McFadden, 2005). Pertinent to the focus of this review, merozoites also have a specialized apical pole with its distinctive structure with three prominent polar rings composed of cytoskeletal matrix-like material surrounding the area from which internal parasite molecules are expelled during the early phases of invasion. Additionally, two or three microtubules are localized to one side of the parasite, and secretory organelles (Bannister et al., 2000a), which are important for the parasite's successful invasion of RBCs and possibly also its timely and effective egress, are present.

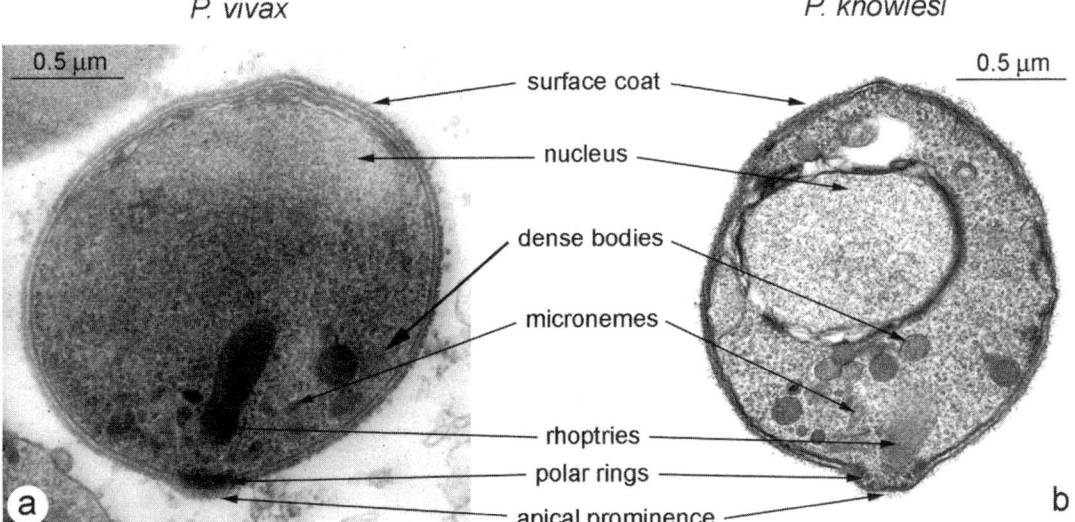

FIGURE 2 Electron micrographs of *P. vivax* (a) and *P. knowlesi* (b) merozoites, showing the main structural features and organelles, as also detailed in the schematic shown in Fig. 1. The *P. knowlesi* parasite was obtained from an infected rhesus macaque and prepared for EM morphology fixation at the Emory Vaccine Center at the Yerkes National Primate Research Center. (The photograph of *P. vivax* was produced with assistance from Michael J. Stewart and is reprinted from Galinski and Barnwell [1996], with permission from the publisher.)

The merozoite's plasma membrane and organellar membranes contain numerous intramembranous particles (IMPs) of different sizes, identified by freeze-fracture studies, suggesting the presence of as-yet-undefined functional structures in these membranes (McLaren et al., 1979; Aikawa et al., 1981; Mitchell and Bannister, 1988; Bannister et al., 2000b; Bannister et al., 2003). Under the plasma membrane beyond the apical pole region, two other membranes form a flattened cisterna structure. This triple-membrane structure forms the pellicle of the merozoite, which appears to be connected by intermittently spaced filaments. The microtubules, which extend about one-half to two-thirds the length of the merozoite, are attached to the basal member of the apical polar rings and appear to be connected to the pellicle by dispersed filaments.

A structured surface coat overlies the merozoite plasma membrane. Over the last 25 years, this surface covering of the merozoite has been described with various qualifiers, including fuzzy, amorphous, fibrillar, and spiked. Bannister and colleagues described in exquisite detail the organized and structured nature of the merozoite surface coat of *P. knowlesi*, with its clusters of tufted fibrils interspersed with amorphous proteinaceous material (Bannister et al., 1986b). The fibrillar tufts were shown to be closely spaced in a reticulated pattern, with each cluster consisting of 5 to 10 parallel filaments that are 2 to 3 nm thick and 18 to 22 nm long, with apparent thickened or bifurcated tips. Fewer interspersed thinner filaments, up to ~400 nm or more in length, were also observed extending upward or appearing parallel to the surface of the merozoite (Fig. 3b). The electron micrographs (EMs) from these studies confirmed that this coat covers the entire surface of the merozoite, including the apical pole region. The surface coat ultrastructure has also been described in somewhat less detail for *P. falciparum* (Langreth et al., 1978) and *P. vivax* (Galinski and Barnwell, 1996) merozoites with the fibrillar clusters of the latter parasite quite closely apposed to each other to give a palisaded appearance of spike-like projections (Fig. 2 and 3a). The actual appearance and precise interpretation of the merozoite surface coat may very well depend on the technical quality of the EMs obtained in different preparations and different fixation procedures and with merozoites of different species. Thus, while individual EMs may portray differences (e.g., in the degree or clarity of the spiked appearance), the natural appearance of the surface coat may in fact be similar for each species; at present, this is not known for certain.

Merozoites also have several specialized secretory organelles that are specialized adaptations for efficient invasion of erythrocytes (Fig. 1 and 2). These organelles include two large pear-shaped rhoptries, up to 40 smaller tubular shaped micronemes, and an estimated 12 or so similarly sized but rounded membrane-bound microspheres (also called dense bodies or dense granules). All of these organelles are electron dense, with distinctive granular-appearing material inside, the nature of which can vary at different locations within each, suggesting a nonuniform composition (Bannister et al., 2000a). The rhoptries, micronemes, and polar ring structure, comprising the apical complex, are typical of the phylum Apicomplexa, to which the 100 or so known species of *Plasmodium* belong (Garnham, 1966). The paired rhoptries are connected by a bifurcated duct to a common opening at the apical prominence of the merozoite. In matured merozoites, the microneme organelles surround the common duct of the rhoptries. They can appear physically continuous with the rhoptry duct, although whether they actually share a common meeting point and opening at the apical pole or not remains in question. As depicted in the schematic of a merozoite shown here (Fig. 1), published EMs have clearly shown that micronemes are formed from a Golgi-like cisterna and then are translocated along the microtubules to their position at the apical pole (Bannister et al., 2003). A number of proteins have been identified and characterized that are located in the micronemes and rhoptries (Adams et al., 2001; Sam-Yellowe et al., 2004) and that have presumed or experimentally defined roles in the process of RBC invasion. The so-called

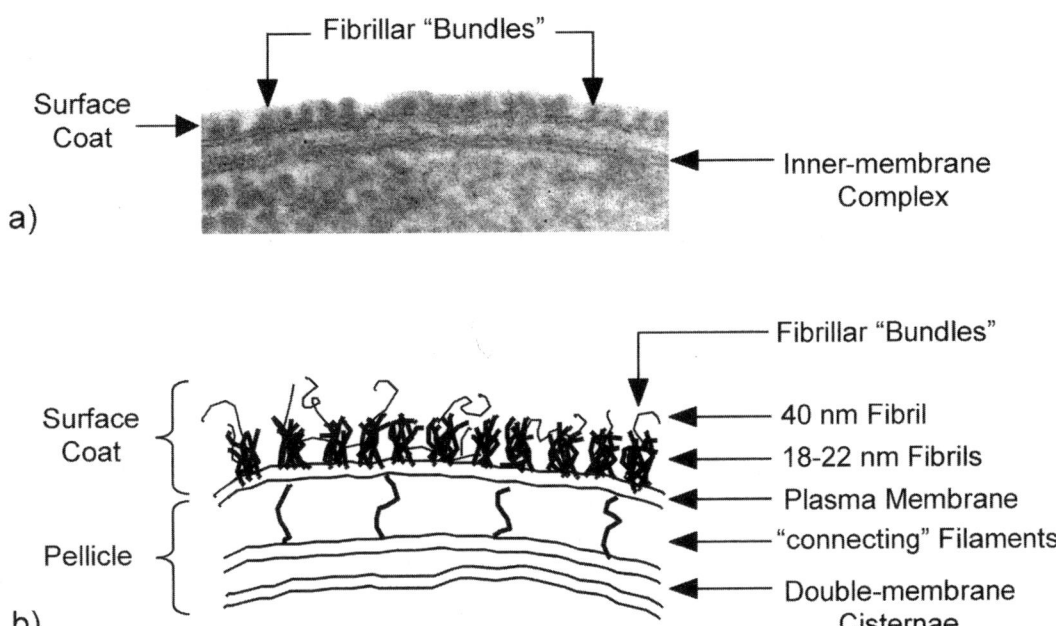

FIGURE 3 (a) EM featuring the triple membranes and surface coat of the *P. vivax* merozoite shown in Fig. 2 with a tufted or spiked appearance. (b) Schematic representation of the merozoite membrane pellicle, consisting of the inner membrane complex, plasma membrane with connecting filaments, and a fibril bundled surface coat, as observed in Bannister et al. (1986b).

dense bodies, dense granules, or microspheres are predominantly seen in the apical half of the merozoite (as the nucleus occupies a majority of the posterior portion of the cell), and these are secreted from the merozoite soon after entry is completed, with their contents playing an apparent role(s) in the differentiation of the parasite and its host cell into a growing trophozoite (Aikawa, 1971; Bannister et al., 1975; Bannister and Mitchell, 1989; Torii et al., 1989).

MEROZOITE BIOGENESIS

Merozoite biogenesis has gained increasing attention in the last few years (Bannister et al., 2000b; Bannister et al., 2003; Margos et al., 2004; Topolska et al., 2004a). With the availability of genome sequence information for multiple species of *Plasmodium* (http://plasmodb.org), the prospects for thoroughly understanding this process are greatly increasing. EMs provided the first clear evidence that malaria parasites develop within a parasitophorous vacuole (PV) (Ladda et al., 1969). As a merozoite begins to invade an RBC, an internal membrane-lined invasion pit develops (Ladda et al., 1969; Bannister et al., 1975; Langreth et al., 1978; Miller et al., 1979; Aikawa and Miller, 1983; Bannister et al., 1986a; Ward et al., 1993; Dluzewski et al., 1995). As entry proceeds, this internal space and the surrounding membrane, the parasitophorous vacuolar membrane (PVM), enlarge, and the parasite becomes contained within the PVM, separated from the erythrocyte cytosol and the erythrocyte membrane. Once inside, the parasite immediately becomes transformed into a ring-stage form (young trophozoite), contained within three membranes: the parasite plasma membrane, the PVM, and the erythrocyte membrane.

As the parasite transforms in the RBC from its ring stage of development to become an ameboid, actively feeding, and growing trophozoite form, it consumes hemoglobin with the aid of a cytostome structure (Aikawa et al., 1966) and enzymes located in a specialized food vacuole (Rosenthal et al., 2002). Amino acids are generated along with an insoluble pigment called hemozoin that eventually coalesces into a residual body in the matured parasite. Once the nucleus divides, the parasite is called a schizont and it enters the phase known as schizogony (and ultimately merogony), which results in the production of 6 to 32 new merozoites, the typical minimal and maximal number being different for each species (Fig. 4; Color Plate 6a). As they develop, the merozoite apical poles face outward towards the erythrocyte membrane, and the posterior pole of the matured merozoites is connected to the centrally located residual body via a stalk structure (Aikawa, 1966). Prior to rupture of an infected cell, the merozoites become detached from this structure and the pigmented residual body. The transition from an invading merozoite to the development of a mature schizont with new merozoites that largely fill the RBC takes approximately 24, 48, or 72 h, depending on the species (Coatney et al., 1971). The overall series of biological processes that comprise schizont development and merogony seem to be similar in each species, but for reasons that are unknown, the timing can differ. The final growth phase leads to the release of mature infective merozoites.

During the merozoite developmental process, fundamental observations have been made on the sequential production of the known merozoite organellar and membrane structures. Interestingly, the merozoite surface coat structure is detected as early as the two-nucleated schizont stage of development, well before the segmentation process occurs to form individual merozoites (Bannister et al., 1986b). The rhoptries and three apical polar rings form early in schizogony coincident with the microtubules, while the micronemes and dense bodies form much later, just prior to segmentation of the schizont into

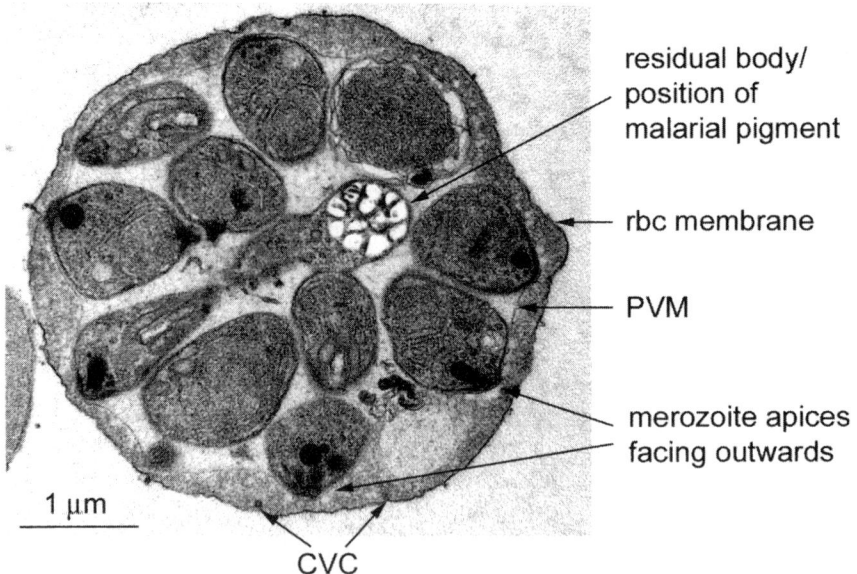

FIGURE 4 Electron micrograph of a mature *P. knowlesi* schizont with merozoites surrounding the residual body and their apical ends facing outwards. Note that the PVM still appears intact at this late stage of development. Caveola vesicle complex structures (CVC; the basis of Schüffner's stippling) (Coatney et al., 1971) can also be observed at the surface of the infected red blood cell membrane. The *P. knowlesi* parasite was obtained from an infected rhesus macaque and prepared for EM morphology fixation at the Emory Vaccine Center at Yerkes.

visible merozoites (Bannister et al., 2000b; Bannister et al., 2003). The fine ultrastructural analysis of serial sections shows that the apical pole of each merozoite forms close to the ends of endomitotic spindles coming from the nuclei, with an adjacent Golgi cisterna present (Bannister et al., 2000b). All three types of secretory organelles are generated from the nuclear envelope via a coated vesicle budding process (Jaikaria et al., 1993; Bannister et al., 2000b; Bannister et al., 2003). The rhoptry and microneme buds each have a characteristic bristled surface; contrary to the dense granules, they are directed via the apparent Golgi cisterna secretory pathway (Bannister et al., 2000b; Bannister et al., 2003). The club-shaped micronemes alone are then directed to the microtubules, where they are translocated for a distance up to 1 μm to their final destination within the polar ring region and surrounding the rhoptry ducts at the merozoite's apex (Fig. 1) (Bannister et al., 2003). The granular nature of the neck and bulbous areas in each of these organelles differ, consistent with the apparent differing destinations of the various rhoptry and microneme proteins (Sam-Yellowe et al., 1995; Bannister et al., 2000b; Preiser et al., 2000; Bannister et al., 2003; Topolska et al., 2004a). Importantly, the recent investigations in this area may explain how microtubule-disrupting agents such as colchicine function to inhibit merozoite invasion. In essence, this would happen by disruption of the trafficking and proper placement of the microneme organelles (Bannister et al., 2003), which are known to contain proteins necessary for adhesion to RBC receptors (Adams et al., 2001).

Step-by-step characteristics of the development of merozoite structures and merozoite protein expression have recently been detailed experimentally for two isolates of *P. falciparum* (Margos et al., 2004). This study, utilizing light microscopy, the analysis of serial sections by transmission EM, and comparative immunofluorescence, supports early findings reported for various individual antigens as well as today's global studies of transcription during schizogony and merogony (Bozdech et al., 2003; Florens et al., 2004; Le Roch et al., 2004) by showing that the expression of merozoite-specific proteins ensues in a relatively ordered, coordinated, and regulated fashion, consistent with the development of specific merozoite structures and organelles. For example, proteins that form the merozoite surface coat are produced in late-stage trophozoites and are present through schizogony, rhoptry proteins are produced during early schizogony, and other apically located proteins are not produced until very late in schizogony. Original immunochemical and biological studies show that some proteins that are critical to the invasion process are optimally expressed at the time of segmentation to form individual merozoites.

PHASES OF MEROZOITE INVASION AND ASSOCIATED EVENTS

The whole process of merozoite invasion can be divided into three or four distinct phases with a number of ultrastructural alterations and molecular events attributed to each phase, with an untold number of others likely to be discovered in the future (Table 1). Video microscopy recordings of *P. knowlesi* merozoites invading erythrocytes (Dvorak et al., 1975) and observations from EM (Ladda et al., 1969; Bannister et al., 1975; Aikawa et al., 1978; Miller et al., 1979) form the visual structure to which molecular characterized actions are increasingly assembled. The basic features of invasion as observed by light microscopy are captured in Color Plate 6. The first phase of erythrocyte invasion involves merozoite release and identification of a potential host cell. The second phase involves the proper reorientation of merozoites adhered to erythrocytes via reversible merozoite surface coat attachments and their successful requisite irreversible apical attachment. This is followed by the entry phase (Fig. 5; Color Plate 7), which utilizes motor proteins and positions the merozoite inside a vacuole within the cytoplasm of the erythrocyte. The final phase is characterized by events that begin to modify the environment of the parasite and transform it into a feeding, growing trophozoite stage. Once a host cell is identified, the entire process of erythrocyte invasion takes about 30 to 60 s and only

TABLE 1 Generalized time scale of events for merozoite invasion of RBCs[a]

Approx time from merozoite release	Morphology	RBC	Merozoite surface	Pellicle-MTs	Rhoptries	Micronemes	Dense bodies
0–1 min	Schizont rupture and subsequent initial RBC interactions	RBC-PVM membrane disruption-breakdown	Proteolytic processing of surface proteins	MTs required for final location of micronemes	Secretion?	Contents processed? or secreted	?
~15 s	Reorientation, apical attachment to RBC	Membrane cytoskeleton rearrangement	?	?	Secretion ✓	Processing? or secretion ✓	Secretion?
~10 s	Junction formation, RBC interaction, PVM formation	Membrane cytoskeleton rearrangement	Proteolytic processing of surface proteins	Activation of invasion motor proteins	Secretion ✓; aids PVM formation	Processing? or secretion ✓	Secretion?
~45 s	Entry-movement through junction	PVM formation	Continued processing	Distortion of pellicle during invasion	Continued, secretion-PVM formation	Continued secretion?	Secretion into vacuolar space?
~10 s	Resealing	Reestablishment of cytoskeleton continuity; RBC membrane fusion-closure	?	?	Emptied	?	?
~5 min?	Internalized merozoite differentiation		Endocytotic clearance of merozoite surface proteins	Disintegration of pellicle and MTs	Disappearance	Disappearance	Secretion ✓

[a] Based primarily on data obtained from invasion studies with *P. knowlesi* (Dvorak et al., 1975; Bannister et al., 1975; Aikawa et al., 1978; Bannister and Mitchell, 1989). *P. knowlesi* merozoites are notably more stable than the merozoites of other *Plasmodium* species, enabling such detailed, real-time studies of the *Plasmodium* merozoite invasion of RBCs. MTs, microtubules.

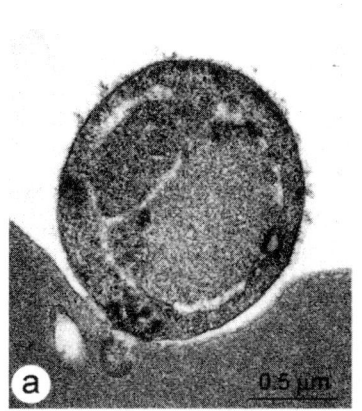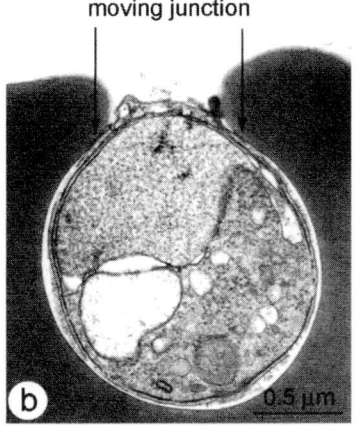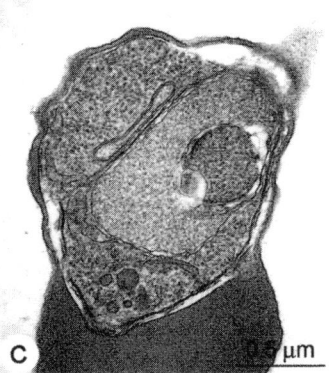

FIGURE 5 EM invasion sequence of *P. knowlesi* merozoites. (a) Apically attached merozoite (cytochalasin B treated). (b) Invading merozoite with moving junction indicated. (c) Newly invaded, fully enveloped merozoite. The parasite preparations for panels a and c were generated for EM morphology fixation at the Emory Vaccine Center at the Yerkes National Primate Research Center. Figure 5b was contributed by Lawrence H. Bannister.

about another 10 to 20 min to transform into an intracellular ring-stage trophozoite (Table 1) (Dvorak et al., 1975; Mitchell and Bannister, 1988). Individual molecules and protein families perform distinct functions simultaneously or sequentially in a processive manner that has been described as a cascade of molecular events.

PHASE I: MEROZOITE RELEASE AND INITIAL ATTACHMENT TO THE ERYTHROCYTE

Erythrocyte Rupture and Merozoite Release

Merozoites must first be released from the worn-out, hemoglobin-depleted, and extensively altered erythrocyte that hosted their development (Cooke et al., 2004; Galinski and Corredor, 2004). Extracellular merozoites are intrinsically short lived (Johnson et al., 1980); thus, as merozoites exit from their host cells and enter the fast-moving world of circulating blood, they must be physically competent to readily identify, latch onto, and invade new, appropriate host erythrocytes. It follows that the schizont-containing RBC ought not to release its merozoites prematurely. They must be fully formed and equipped for this task. In the case of some species, like *P. vivax*, only the young RBCs are suitable for invasion (Mons, 1990; Galinski et al., 1992). They therefore must discriminate between the much rarer young reticulocyte host cells from the vast majority of the more-mature erythrocytes. The apical pole can be viewed as the business end of the merozoite, which provides entry into the RBC through the engagement of its receptors, proteases, and other biochemical processes, as detailed below. Attention has increasingly become directed toward understanding molecular details surrounding the parasite's exit from the RBC (Winograd et al., 1999; Salmon et al., 2001; Li et al., 2002a; Wickham et al., 2003; Margos et al., 2004; Palacpac et al., 2004b), yet many questions still remain. The PVM and the infected erythrocyte host cell membrane with its cytoskeletal network are formidable barriers (Bennett and Lambert, 1991), which protect the developing parasites from external host immune responses. How newly formed merozoites become released from this sanctuary may enlighten us about the full scope of functional roles played by the merozoite.

Several schizont rupture scenarios have been put forth, based on video microscopic and cell biological investigations with *P. falciparum* schizonts matured in culture (Winograd et al., 1999;

Salmon et al., 2001; Wickham et al., 2003). The parasites must exit from both the PVM housing the developing merozoites and the erythrocyte membrane. Winograd and colleagues (1999) first suggested from video microscopic analyses that the merozoites, along with the hemozoin-containing residual body, exit in about 1 s through a single estimated 2.5-μm opening in the RBC membrane, leaving the RBC and PVM intact. Two more recent studies, however, using protease inhibitors and defined immunochemical reagents and microscopic analyses to distinguish the PVM and the RBC membrane suggest alternative multistep processes. These studies suggest that several different parasite-encoded proteases function in a possible series of proteolytic events that lead to the egress of the parasites with the apparent rupturing of these membranes. Salmon and colleagues (2001) found that schizonts incubated with the cysteine protease inhibitor E64 resulted in the accumulation of so-called PVM-enclosed merozoite structures (PEMS). These studies concluded that matured merozoites, still contained in the PVM (i.e., the PEMS), are released from infected erythrocytes and that the merozoites are then dispersed, following a subsequent proteolytic degradation of the PVM (Salmon et al., 2001). Wickham and colleagues (2003) combined the use of protease inhibitors and green fluorescent protein-tagged chimeric protein constructs targeted to the PVM or the RBC cytosol. These experiments alternatively concluded that the PVM lyses while still within the infected RBC, leaving the merozoites free in the erythrocyte cytosol, and that this step is followed by the induced rupture of the erythrocyte membrane (Wickham et al., 2003). The ordering of events remains controversial, while corroborating experimentation along these lines continues. As shown in Fig. 4, the PVM is clearly present in a mature segmented *P. knowlesi* schizont. If the PVM in fact lyses first with matured merozoites then free in the erythrocyte cytosol, this must be for a very brief period. Additional studies to expand upon and confirm each of these original findings and interpretations would be welcomed.

The involvement of proteases in the rupture of the mature infected RBCs has been suspected for some time, since experiments performed in the 1980s showed that merozoite invasion, as well as rupture, could be inhibited with protease inhibitors that are specific for serine and cysteine proteases (Hadley et al., 1983; Lyon et al., 1986). A number of parasite proteases have been identified to date that are able to cleave RBC cytoskeleton and membrane proteins. These include a 37-kDa cathepsin D protease from *Plasmodium lophurae*, which was shown to digest spectrin, ankyrin, band 2.6, and band 3 (Sherman and Tanigoshi, 1983); a 76-kDa serine protease from *P. falciparum*, shown to cleave band 3 (Braun-Breton and Pereira da Silva, 1988; Roggwiller et al., 1996); a 37-kDa protease from *P. falciparum* and *P. berghei*, shown to degrade spectrin and protein 4.1 (Deguercy et al., 1990); the aspartic protease plasmepsin II, which can cleave spectrin, actin, and protein 4.1 (Le Bonniec et al., 1999); and the falcipain-2 cysteine protease, which cleaves host membrane proteins ankyrin and 4.1 (Raphael et al., 2000; Dua et al., 2001; Hanspal et al., 2002). What actual roles, if any, each of these proteases may play in erythrocyte rupture or merozoite invasion remain to be determined.

Other candidate proteases being considered potentially relevant in the rupture-release process include those associated with the members of the serine repeat antigen (SERA) family of proteins (Mercereau-Puijalon et al., 2002). SERA was originally identified in *P. falciparum* as a 126-kDa soluble protein that is secreted into the PV during trophozoite development and weakly associated with merozoites (Delplace et al., 1987); it has a signal peptide but no transmembrane domain (Bzik et al., 1988; Knapp et al., 1989). To date, there are nine *sera* paralogs identified in *P. falciparum* (Gardner et al., 2002), five in *P. vivax* (Kiefer et al., 1996), three in *Plasmodium vinkei* (Gor et al., 1998), and four in *P. yoelii* (Carlton et al., 2002). In *P. falciparum*, eight of the nine genes are clustered on chromosome 2 (Gardner et al., 2002) and *sera 4–6* and *sera 9* (chromosome 9) (Gardner et al., 2002) appear to be up-regulated in blood-stage parasites (Aoki et al., 2002; Miller et al., 2002b). Moreover, *sera 4-6* genes could not be disrupted, suggesting that they have essential roles in the life cycle of the parasite (Aoki et al., 2002; Miller et al., 2002b). *sera 5* encodes the first SERA identified, which was also called the serine protein (SERP) and shown to undergo multiple processing steps (Delplace et al., 1987; Debrabant et al., 1992). This protein is proteolytically processed into 47-kDa N-terminal, 56-kDa central, and 18-kDa C-terminal fragments. The 56-kDa fragment is then cleaved, resulting in a 50-kDa fragment (Delplace et al., 1987; Debrabant et al., 1992), and the 47-kDa fragment associates with the 18-kDa fragment via a cysteine bond (Delplace et al., 1987; Debrabant et al., 1992). The 47-kDa fragment from some alleles is further processed into 25-kDa fragments which associate with the merozoite surface via disulfide bonds to the 18-kDa fragment (Li et al., 2002b). Li and colleagues (2002b) demonstrated that SERA processing occurs just prior to rupture of mature schizonts. The 50-kDa fragment shares similarities with the catalytic sites of known papain family cysteine proteases (Higgins et al., 1990; Eakin et al., 1992), and this fragment is shed into culture medium and not found associated with the merozoite (Li et al., 2002b).

Because both the processing of SERA and the rupture of mature schizont-infected RBCs are inhibited by leupeptin, the suggestion that SERA might be involved in merozoite release was put forward (Delplace et al., 1988). Several protease inhibitors have since been shown to block the processing of SERA, and they also inhibit schizont rupture and lead to the accumulation of PEMS (Li et al., 2002a), thus adding to speculation regarding both multiple protease activities being present at this stage and the possible role SERA may have in the rupturing process. As an added point of interest, growth inhibitory antibodies against SERA5 were shown to prevent the dispersal of merozoites released from matured schizonts (Pang et al., 1999). Phylogenetic analyses show that SERA family members group in multiple *Plasmodium* species as either serine or cysteine proteases (Hodder et al., 2003) and that SERA5 can retain the autocatalytic activity of its central domain despite having a serine at the catalytic site (Higgins et al., 1989; Hodder et al., 2003). As suggested by Hodder et al. (2003), it is possible, although not yet experimentally investigated, that SERA proteins with active cysteine or serine catalytic sites function at different steps in the egress of merozoites from their host cells.

It seems that the parasite's method of exiting will increasingly be viewed as a programmed multistep process, as is the merozoite invasion cascade, in part involving specific protein processing and degradation steps. Furthermore, recent studies of the production and trafficking of lipid bodies in the asexual blood-stage parasites of *P. falciparum* suggest that free fatty acids resulting from the progressive degradation of triacylglycerol during schizogony play a role in the rupturing process (Palacpac et al., 2004a). These data suggested the presence of lipase and possibly phospholipase activities and their potential role in the rupturing process, aiding the breakdown of the PVM and the infected RBC membranes (Palacpac et al., 2004a). In this regard, it is interesting that both Seed (1976) and Bannister and colleagues (1986b) suggested that the contents of the rhoptries could also participate in the egress of merozoites from the matured schizont. Ultrastructural studies showing the spewing of rhoptry contents from within a matured schizont, along with associated signs of vesiculation and disruption of both the PVM and the erythrocyte membrane, support this possibility (Bannister et al., 1986a). Rhoptries are known to contain lipids in the form of membrane whorls (Bannister et al., 1986a; Stewart et al., 1986).

Initial Host Cell Contact and the Merozoite Surface

Once released, the initial adhesive contact between the merozoite and the erythrocyte may occur at any point on the merozoite surface and involves the merozoite surface coat filaments (Table 1; Fig. 3) (Dvorak et al., 1975). The videomicroscopic and EM observations show that these contacts coincide with dynamic heaving perturbations of the RBC membrane. The filament contacts have been observed at distances of 15 to 20 nm, and over longer distances of 40 nm or more (Bannister et al., 1986b). At this stage, the merozoite can become detached, in which case it can then readhere to another erythrocyte. These initial interactions suggest that the contacts are most likely molecularly mediated between specific receptors in the erythrocyte membrane and that ligands of the merozoite. *P. knowlesi* merozoites in vitro will specifically form initial attachments to a variety of primate erythrocytes but do not attach to prosimian, dog, or rodent RBCs (Butcher et al., 1973). While the initial adhesive contacts involve elements of the merozoite's proteinaceous fibrillar projections, and a number of these proteins have been identified, as described below, the specific identities of the molecular constituents responsible for these contacts remain in most instances to be established.

Over the last 25 years, as noted above, we have come to view the surface of the merozoite as either a fuzzy and amorphous structure or as a more complex coat structure with organized fibrillar bundles, sometimes with a spiked appearance. This period has also witnessed the discovery of at least 10 distinct proteins known to be associated with the merozoite surface and classified as merozoite surface proteins (MSPs) with designations of MSP-1 through MSP-10 and counting (Topolska et al., 2003). All 10 MSPs are present in *P. falciparum*, but this full set is not necessarily present in all *Plasmodium* species. It was noticed first from immunochemical and synteny studies and later by genome sequencing that MSP-2 is not present in *P. vivax*, the simian parasite *P. knowlesi*, or the rodent malaria parasite *Plasmodium chabaudi* (Marshall et al., 1998; Black et al., 1999; Black et al., 2002; Black et al., 2004). Conversely, MSP-3, shown in *P. vivax* to comprise a family of genetically and structurally related and coexpressed surface coat proteins (Galinski et al., 1999; Galinski et al., 2001), is now known to be part of a locus in *P. falciparum* that includes related genes encoding several MSPs, including MSP-6 (Trucco et al., 2001; Pearce et al., 2005; S. Singh et al., unpublished data). Currently, the malaria genome databases (http://plasmodb.org) contain an apparent 6 or 7 paralogs of *P. falciparum msp-3*, 2 *msp-3* genes for *P. knowlesi*, and a locus of 12 *P. vivax msp-3* paralogs, which are given Greek subdesignations (α, β, γ, etc.) as originally described for the *msp-3* family in *P. vivax* (Galinski et al., 1999; Galinski et al., 2001; and J. W. Barnwell and M. R. Galinski, unpublished data).

Certain MSPs (e.g., MSP-1, MSP-2, MSP-4, MSP-5, MSP-8, and MSP-10) are anchored into the plasma membrane of the merozoite via a C-terminal glycosylphosphatidylinositol (GPI)-lipid anchor, while other MSPs that do not have a GPI anchor or transmembrane domain (e.g., MSP-3, MSP-6, MSP-7, and MSP-9) are predicted to be associated with the surface through noncovalent interactions with the anchored proteins and with each other (McColl et al., 1994; Mulhern et al., 1995; Galinski et al., 1999; Galinski et al., 2001; Pachebat et al., 2001; Trucco et al., 2001; Vargas-Serrato et al., 2002; Pearce et al., 2004). MSP-9, when identified and characterized in *P. vivax*, *Plasmodium cynomolgi*, and *P. knowlesi* (Barnwell et al., 1999; Vargas-Serrato et al., 2002), was found to be related to the *P. falciparum* antigen known as ABRA or p101 (Stahl et al., 1986; Weber et al., 1988). While some find it regrettable that such nomenclature is updated, noting it can be confusing (Topolska et al., 2003), recognition of ABRA's relationship to MSP-9 aids our comprehensive understanding of the shared biology of the different species (Vargas-Serrato et al., 2002). In each species, MSP-9 is apparently located at the surface of merozoites and in the PV space. Its surface localization has been shown by immunofluorescence assays (IFA) and immuno-EM on individual merozoites (Barnwell et al., 1999), and by immuno-EM it can appear to "bathe" the merozoites contained within matured schizonts (unpublished data).

MSP-1 (Holder and Freeman, 1984b; Miller et al., 1993) is a large MSP expressed in all species of *Plasmodium* with a relative mobility (relative molecular mass) by sodium dodecyl sulfate-polyacrylamide gel electrophoresis (SDS-PAGE) of 180 to 250 kDa. MSP-1 was the first *Plasmodium* MSP to be discovered and has been most thoroughly investigated. Its primary structure is complex, with areas of amino acid conservation interspersed with other regions that are moderately or highly polymorphic within a species (Mackay et al., 1985; Tanabe et al., 1987; Peterson et al., 1988; Gibson et al., 1992; Miller et al., 1993; Premawansa et al., 1993; Putaporntip et al., 2002). Much of the dimorphic diversity probably results from genetic recombination occurring during meiosis, although other mechanisms, including gene conversion, also contribute to its mosaic structure (Tanabe et al., 1987; Miller et al., 1993; Kerr et al., 1994; Kaneko et al., 1997; Putaporntip et al., 1997; Putaporntip et al., 2002).

It is now well established that several MSPs undergo defined proteolytic processing events. Foremost among these, the processing of MSP-1 has been most extensively studied (Blackman et al., 1991; Blackman et al., 1996; Blackman, 2000; Kauth et al., 2003). MSP-1 undergoes extensive specific proteolytic processing either very late in schizogony or at some time after merozoite release, with the result that the molecule is cleaved into four fragments (David et al., 1984; Holder and Freeman, 1984b; Lyon and Haynes, 1986; Holder et al., 1987) bound together as a noncovalent complex on the merozoite surface (McBride and Heidrich, 1987). The fragment sizes in *P. falciparum* are 83 kDa (N terminus), 30 and 38 kDa (central portions), and 42 kDa (C terminus) (Holder et al., 1987; Stafford et al., 1996), with the latter fragment bound to the plasma membrane by a GPI anchor (Haldar et al., 1985; Gerold et al., 1994). Two other proteins that are not part of MSP-1 are coprecipitated with the MSP-1 complex of fragments in *P. falciparum* (Heidrich et al., 1983; McBride and Heidrich, 1987; Stafford et al., 1996), which are now known to be the 36- and 22-kDa processed fragments, respectively, of MSP-6 (Trucco et al., 2001) and MSP-7 (Pachebat et al., 2001). Moreover, it has recently been shown that in *P. falciparum* MSP-3 undergoes processing during schizogony and that the cleavage site motif has characteristics similar to those identified for MSP-1, MSP-6, and MSP-7 (Pearce et al., 2004). This suggests that the same or similar protease(s) may be involved in the processing of all these surface proteins.

P. falciparum MSP-2 is a small (relative molecular mass, 45 to 56 kDa) and highly polymorphic protein, characterized by a diverse central region that includes allelic versions of tandemly repeated amino acid motifs. This central variable domain is flanked by small conserved terminal domains (Smythe et al., 1990; Fenton et al., 1991; Smythe et al., 1991; Irion et al., 1998). No potential biological structure-function roles in invasion have been ascribed to MSP-2 of *P. falciparum*, although antibodies directed towards MSP-2 inhibit merozoite invasion in vitro (Clark et al., 1989). However, the failure to successfully knock out this gene by homologous recombination suggests that MSP-2 performs an essential function for the merozoite (Cowman et al., 2000).

P. falciparum MSP-3 (PfMSP-3) (McColl et al., 1994; Oeuvray et al., 1994) is a slightly diverse $\sim$43-kDa protein that has three contiguous regions consisting of four heptad amino acid repeats, with the hydrophobic amino acid alanine in the first and fourth positions of each heptad. It was reasoned that these regions could form coiled-coil alpha helices (McColl et al., 1994) and might exhibit the molecular interactions characteristic of this type of tertiary structure (Burkhard et al., 2001). The structure of a peptide containing one of the blocks of four heptads was evaluated by physical experimentation, which indicated it formed an α-helix, but the peptide did not form multimers with itself in solution (Mulhern et al., 1995).

The *P. vivax* MSP-3 family, with the individual single-copy gene members reported to date designated *msp-3*α, *msp-3*β, and *msp-3*γ (Galinski et al., 1999; Galinski et al., 2001), is related to *P. falciparum* MSP-3, although there is only 30% overall identity between the amino acid sequences of the three *P. vivax* MSP-3s characterized and PfMSP-3. Their unifying molecular characteristic, like PfMSP-3, is a large central domain that is high in alanine content (20 to 30%) and predicted to form a tertiary structure composed of several coiled-coil α-helices. Unlike the putative coiled-coil blocks of

PfMSP-3, the coiled-coil regions of the *P. vivax* family are more extensive than the minimum of four heptads required to form coiled-coil helices. The calculated molecular masses of the PvMSP-3s range from 75 to 105 kDa, while their relative molecular mass by SDS-PAGE ranges from 120 to >180 kDa. There is a very high degree of diversity in natural isolates, primarily within their central alanine-rich coiled-coil domains (Rayner et al., 2002; Rayner et al., 2004b). It remains to be determined how MSP-3 molecules interact with themselves and other proteins at the surface of the merozoite.

P. falciparum MSP-4 (relative molecular mass, ~40 kDa) has a single epidermal growth factor (EGF)-like domain at its carboxy terminus and a GPI-lipid anchor (Marshall et al., 1997). *P. falciparum msp-5* is similar to *msp-4* in size and two-exon structure with a single encoded C-terminal EGF domain, and it is located in the same genomic locus that contains *msp-4*, *msp-2*, and the adenylosuccinate lyase gene (*asl*) (Marshall et al., 1998). In *P. vivax*, the syntenic locus does not contain *msp-2*, and the *msp-5* gene is considerably larger than *msp-4*, although both retain the two-exon structure. Their protein products migrate as molecular masses of 62 and 86 kDa, respectively (Black et al., 2002). Furthermore, in *P. vivax* MSP-4 is localized by immunofluorescence circumferentially around the merozoite surface, whereas MSP-5 is located only as a patch of fluorescence at the apical end of the merozoite. This is also consistent with a different location and solubility in TX-114 of MSP-4 and MSP-5 in *P. knowlesi* (Black et al., 2004). In the rodent malarias, *msp-4* and *msp-5* are combined into a single gene; an *msp-2* ortholog is not present (Black et al., 1999; Kedzierski et al., 2000). Preliminary indications from the gene sequences of MSP-4 and MSP-5 from a few *P. falciparum* laboratory and field isolates are that they have limited diversity (Marshall et al., 1997; Wu et al., 1999; Benet et al., 2004). *P. falciparum* MSP-8, like MSP-1, contains two C-terminal EGF-like domains and undergoes proteolytic processing. It migrates in SDS-PAGE around a relative molecular mass of 98 kDa and is localized at the surface of trophozoites, schizonts, and merozoites via a GPI anchor, similar to pAg-2, an ortholog in *P. yoelii* (Burns et al., 2000; Black et al., 2001). An ortholog of *msp-8* in *P. vivax* has also been identified (Perez-Leal et al., 2004). Another MSP gene encoding two EGF-like domains was designated as *msp-10* (Black et al., 2003). Judging by immunofluorescence, this MSP is both at the surface around the merozoite and, additionally, concentrated as a patch of fluorescence at the apical end of the merozoite. The latest *msp* declared for *P. falciparum* was designated *msp-11* and localized within the *msp-3* genomic locus (Pearce et al., 2005). Interestingly, another *P. vivax* gene of similar size to *msp-1*, also with encoded double C-terminal EGF-like domains, is located in the *P. vivax* genome next to *msp-1*. This gene is also present in the *P. knowlesi* genome, although it is not present in the *P. falciparum* genome (E. V.-S. Meyer et al., unpublished data).

How are these molecules arranged on the surface of the merozoite? As noted above, MSP-1 and several other MSPs are anchored in the plasma membrane of the merozoite through the posttranslational addition of a GPI lipid anchor to the C termini of the proteins. Additionally, in each species, the MSP-3 proteins and others (MSP-6, MSP-7, and MSP-9) have neither a classical hydrophobic transmembrane domain nor a GPI-lipid modification to anchor them to the merozoite plasma membrane. Rather, the MSP-3 proteins have been described as being intimately but peripherally associated with the merozoite surface; MSP-6 and MSP-7 are known to noncovalently interact with MSP-1. In this regard, it is important to keep in mind that the EM-visible surface coat structure is relatively easy to strip from merozoites (Color Plate 7) (Aikawa et al., 1978; Mitchell and Bannister, 1988; Bannister and Dluzewski, 1990). Members of the MSP-3 family and MSP-9 could be noncovalently harnessed to the surface both by electrostatic forces and through protein-protein interactions with themselves and to other membrane-bound MSPs, with coiled-coil α-helices becoming entwined and contributing to the plush appearance of the organized fibrillar bundled clusters observed at the surface (Fig. 2 and 3). Proteins with coiled-coil α-helices have the potential to form homotypic or heterotypic multimeric protein bundles (Burkhard et al., 2001), which would be consistent with the 18- to 20-nm bundles of fibrils extending from the surface (Bannister et al., 1986b). Some other recent novel insights may further change how we view the merozoite surface coat structure. Wang and colleagues (2003) demonstrated that the GPI membrane-anchored proteins MSP-1, MSP-2, MSP-4, MSP-5, and MSP-8 are partially TX-100 insoluble at 4°C but completely soluble at 37°C, which denotes solubility properties considered to be characteristic of raft-associated proteins (Brown and Rose, 1992; Zhang and Thompson, 1997; London and Brown, 2000). Isolated detergent-resistant membranes (DRMs) were shown to contain most of the GPI-anchored MSPs, primarily those that have been shown to localize over the entire surface of merozoites, but apparently not those restricted to the apical region as is MSP-10 in *P. falciparum* and MSP-5 in *P. vivax* and *P. knowlesi*. What may be the benefit of the parasite's MSPs localizing to raft microdomains? As suggested by these authors, a speculative answer is that such localization provides a means to concentrate these molecules and this may add flexibility to the surface-exposed structures to enhance the effectiveness of initial parasite surface and erythrocyte surface interactions. Whether this new line of research will garner additional support for this notion remains to be seen. Similarly, whether one or more of these proteins has erythrocyte-adhesive properties and thus is potentially responsible for the 18- to 20-nm fibril adhesions between the merozoite and erythrocyte remains to be determined.

Usually, once a protein has been confirmed to be associated with the surface of the merozoite, it becomes a candidate for (i) possible adhesion to the RBC surface, and, thus, potentially a ligand for invasion and (ii) potential inclusion in a malaria vaccine. Extensive investigations have followed on MSP-1 (the first discovered) over the past 25 years (Blackman, 2000) and since, with the other MSPs following suit, to evaluate whether any bind to RBCs. Only two MSPs from *P. falciparum* have been associated with binding to erythrocytes to date, MSP-1 and MSP-9. Does MSP-1 participate in the initial surface interaction between the merozoite and the erythrocyte? Given its location and large size, MSP-1 could correspond to the long fibrils that tether merozoites to the erythrocyte at distances of 40 nm or greater or comprise the 20-nm fibrillar bundles (Bannister et al., 1986b). *P. falciparum* MSP-1 was first demonstrated to bind to human and primate erythrocytes by in vitro erythrocyte-binding assays and biosynthetically radiolabeled parasite proteins in culture supernatants (Perkins and Rocco, 1988). In these instances, the binding capacity of MSP-1 was dependent on terminal sialic acid residues present on erythrocyte membrane glycoproteins. However, in these studies MSP-1 was binding only as an intact full-length molecule and as such was at odds with the experimental data mentioned above showing that MSP-1 is processed by a protease(s) into four major fragments at about the time merozoites are released from infected erythrocytes. The MSP-1-processed fragments are held together in a noncovalent complex on the merozoite surface, and it is difficult to reconcile this processed structure of MSP-1 with the binding data; yet the whole protein as well as the processed fragments could conceivably retain similarly functional adhesive regions. The issue of whether MSP-1 participates in early initial merozoite adhesion has been addressed previously and remains an intriguing question (Barnwell and Galinski, 1991; Holder, 1994; Holder and Blackman, 1994; Galinski and Barnwell, 1996; Blackman, 2000). In any case, antibodies to the C-terminal EGF-like region inhibit invasion in vitro (Blackman et al., 1990), and transfection studies indicate this portion of the protein is functionally required for merozoite survival and invasion (O'Donnell et al., 2000). As such, it has become a prime malaria vaccine candidate (Mahanty et al., 2003). It has also been suggested that the primary processing of MSP-1, if it occurs during the initial attachment, might facilitate reorientation. More recently, it was suggested that the p42 fragment of MSP-1 bound to two external loops of erythrocyte band 3 through a sialic acid-independent interaction with the EGF-like domains (Goel et al., 2003). Earlier, it was shown that the N terminus of recombinant MSP-9 also bound to band 3 (Kushwaha et al., 2002), and recent data suggest that the p42 fragments of MSP-1 and MSP-9 directly interact as a coligand complex for binding to band 3 (Li et al., 2004). This is an intriguing possibility for an initial adhesion event between the RBC and the free merozoite. However, whether the proposed interaction actually occurs as an initial binding interaction or later on in invasion will need further study to clarify if the EGF domain of MSP-1 is most likely buried near the surface of the parasite membrane and not exposed close to the periphery of the surface coat. Furthermore, truncation of the C terminus of MSP-3 by genetic manipulation negatively affects the transport of MSP-9 to the merozoite surface and the PV (Mills et al., 2002). Although merozoite invasion was affected in these parasites, that invasion was 80% of normal indicates that MSP-3 and MSP-9 do not play roles that are directly necessary for invasion.

PHASE II: MEROZOITE REORIENTATION AND IRREVERSIBLE ATTACHMENT THROUGH TIGHT JUNCTION FORMATION

After initial attachment, if not already positioned for entry, the merozoite must become reoriented so that the apical prominence is facing the erythrocyte membrane (Fig. 5a). The rhythmic oscillations of the erythrocyte membrane that occur on initial contact, at times extending partially around the merozoite, seem to facilitate reorientation. Consistent with the notion that adhesive interactions and RBC membrane warping assist in reorientation of the merozoite are transmission EMs of erythrocytes wrapped around merozoites with fibrillar extensions connecting across 15- to 20-nm distances to the RBC surface (Mitchell and Bannister, 1988; O'Donnell et al., 2000). What adhesive and contractile forces of the merozoite or other molecular means induce membrane deformations remain unknown. Nevertheless, the tight attachment of the merozoite apex against the erythrocyte membrane is an essential step for subsequent host cell entry to occur (Fig. 5a). Apposition of the apical prominence directly with the erythrocyte membrane results in a close 4-nm membrane-to-membrane interaction, known as a tight junction between the parasite and the host cell (Bannister et al., 1975; Aikawa et al., 1978). At this site of contact, there is an increased electron-dense thickening under the erythrocyte membrane and a slight indentation of the erythrocyte membrane. The formation of the tight junction is irreversible, and the merozoite is now set to enter the designated host cell. These observations, of course, raise questions relating to the molecular components of both the parasite and the host that participate to bring forth these events and interactions. The specific adhesive and other molecular interactions that occur to bring about this phase of erythrocyte invasion, while becoming increasingly known, are also becoming increasingly more complex; a large body of work remains to be analyzed to gain a coherent understanding.

Erythrocyte Invasion Receptors

Investigations initiated in the 1970s have established that the erythrocyte membrane glycoprotein carrying the Duffy blood group antigen epitopes (Fy^a, Fy^b, and Fy^6), also known as glycoprotein D or Duffy-associated chemokine receptor (DARC) (Hadley, 1986; Nichols et al., 1987), is an erythrocyte determinant essential for the invasion of human RBCs by both *P. vivax* and *P. knowlesi* merozoites (Miller et al., 1975; Miller et al., 1976; Barnwell et al., 1989). The structure of the DARC deduced from the sequence of this single exon gene indicates it is largely buried in the erythrocyte membrane as a member of a family of eight transmembrane domain proteins (Chaudhuri et al., 1993). A 66-amino-acid N-terminal extracellular domain bears the Duffy epitopes (Chaudhuri et al.,

1993; Iwamoto et al., 1995; Wasniowska et al., 2004) and binds proinflammatory chemokines such as RANTES, interleukin 8, and melanoma growth stimulatory activity, as well as a merozoite ligand (Horuk et al., 1993; Chaudhuri et al., 1994). P. vivax and P. knowlesi merozoites cannot invade human erythrocytes that lack the Duffy glycoprotein through genetic mutation (Duffy negative; $FyFy$). Similarly, removal of the Fy^a, Fy^b, and Fy^6 epitopes from DARC by chymotrypsin treatment of Duffy-positive human erythrocytes also prevents invasion of P. knowlesi or P. vivax into human erythrocytes. Treatment with trypsin, on the other hand, does not remove the Duffy determinants and does not inhibit invasion (Miller et al., 1975; Barnwell and Wertheimer, 1989).

Further experiments utilizing the monkey malaria parasite P. knowlesi went on to implicate the interaction between a merozoite component and the erythrocyte Duffy glycoprotein as a key element in initiating the electron-dense tight junction (Miller et al., 1979). Initial adhesion and apical orientation with development of a tight junction will occur if invasive merozoites are treated with cytochalasin B (Fig. 5a); however, any further parasite entry into the erythrocyte is inhibited. In contrast, when P. knowlesi merozoites interact with Duffy-negative erythrocytes in the presence of cytochalasin B, attachment and reorientation occur, but no tight junction forms. Instead of forming a tight junction, adherence to the erythrocyte is maintained at a distance of 100 to 150 nm by thin fibrils that extend from the apical prominence (Miller et al., 1979). These studies implicated the Duffy antigen and the complementary merozoite ligand in initiating or establishing the tight junction. Later reports confirmed that the interaction of P. vivax (or P. knowlesi) merozoite ligands with the Duffy glycoprotein receptor is not dependent upon sialic acid or N-linked oligosaccharides present on the Duffy antigen (Mason et al., 1977; Barnwell et al., 1989; Barnwell and Galinski, 1991; Chitnis et al., 1996).

In contrast to P. vivax and P. knowlesi, P. falciparum merozoites use terminal sialic acid residues on erythrocyte membrane glycophorins as a major receptor (Miller et al., 1977a; Pasvol et al., 1982a; Pasvol et al., 1982b; Cartron et al., 1983; Facer, 1983; Perkins, 1984b; Perkins, 1984a; Perkins and Rocco, 1988), rather than the Duffy glycoprotein, and will invade erythrocytes regardless of whether they are Duffy positive or Duffy negative. The basic role of sialic acid and glycophorins in P. falciparum merozoite invasion has been established using mutant erythrocytes, enzymatic treatments, and antibodies. There are four major glycophorin polypeptides (A, B, C, and D) in the erythrocyte membrane, each with a single transmembrane domain and a variable number of N-terminal O-linked sialotetrasaccharides. Glycophorin A also has one N-linked oligosaccharide chain and along with glycophorin C is cleaved by trypsin, whereas glycophorin B is not. Depending upon the strain of P. falciparum, it was found that trypsinization of RBCs would reduce, abolish, or have little effect on merozoite invasion, indicating the possibility of different invasion determinants (see below).

Microneme Proteins: the DBL-EBL Adhesive Invasion Ligands and Their Erythrocyte Receptors

Camus and Hadley (1985) were the first to identify a potential invasion ligand by showing that a 175-kDa P. falciparum merozoite protein would bind to normal human RBC but not mutant transposon erythrocytes deficient in sialic acid or cells treated with neuraminidase to remove sialic acid; this protein was thus termed erythrocyte-binding antigen 175, or EBA-175 (Camus and Hadley, 1985). Similarly, a 135-kDa P. knowlesi protein and a 140-kDa P. vivax protein (Haynes et al., 1988; Wertheimer and Barnwell, 1989) were shown to bind specifically to Duffy-positive, but not to Duffy-negative, human erythrocytes; this protein has since become known as the Duffy-binding protein, or DBP. The P. knowlesi 135-kDa and P. falciparum 175-kDa ligand proteins, which were presumably mediating merozoite invasion to complementary known erythrocyte receptors, were subsequently shown by immuno-EM to be localized in the merozoite's microneme organelles (Adams et al., 1990; Sim et al., 1992). This has given rise to the notion that these apical tubular organelles are at least in part a repository for parasite ligands involved in the receptor interactions mediating invasion.

The 175-kDa *P. falciparum* protein, EBA-175, binds via sialic acid residues that are in an α-2,3 conformation (Orlandi et al., 1990; Klotz et al., 1992). In addition, this *P. falciparum* invasion ligand specifically binds to glycophorin A and not glycophorin B or glycophorin C/D, although these glycophorins also have terminal sialyl carbohydrate residues on their O-linked oligosaccharides (Dolan et al., 1994; Sim et al., 1994). Native and recombinantly expressed EBA-175 does not bind to glycophorin A-deficient human En(a−) RBC. The interaction between EBA-175 and glycophorin A thus apparently requires in part the peptide structure of glycophorin A, as well as sialic acid in an α-2,3 linkage (Orlandi et al., 1992; Sim et al., 1994). The N-linked oligosaccharide of glycophorin A is not involved.

The adhesion of the *P. knowlesi* and *P. vivax* 135- or 140-kDa DBPs to the Duffy glycoprotein generally correlates with the susceptibility of different primate erythrocytes to invasion (Miller et al., 1988; Wertheimer and Barnwell, 1989). One notable exception is that the DBP of *P. vivax* does not bind to squirrel monkey erythrocytes as expected, although squirrel monkeys are susceptible to infection with *P. vivax*, suggesting that in this simian host an alternative receptor may be used. Conversely, the DBPs of *P. vivax* and *P. knowlesi* do bind to *Cebus* monkey erythrocytes, but in this case, this species of monkey is not susceptible to infection with either malaria species. Apparently, the merozoites are not able to initiate early attachment to the *Cebus* erythrocytes (Miller et al., 1977b), clearly illustrating that more than one specific receptor-ligand interaction is necessary for attachment and invasion. This is also the case for EBA-175 and invasion of New World monkey RBCs, where it is evident that EBA-175 does not bind to *Aotus* or *Saimiri* monkey erythrocytes. Recently, an alternative erythrocyte-binding assay based on flow cytometry has been developed as another means to further identify and study these and other parasite adhesive proteins (Tran et al., 2005). Importantly, however, it must be remembered that the rigorous demonstration of the adhesion of native proteins (or their naturally processed fragments) is the ultimate test of specificity and function.

When the genes encoding the *P. knowlesi* and *P. vivax* DBPs and *P. falciparum* EBA-175 were first identified (Adams et al., 1990; Sim et al., 1990; Fang et al., 1991), comparative analyses showed that they were all related and constituted a family of merozoite invasion protein genes (Adams et al., 1992). The genes of this family, represented by the DBPs and EBA-175 and other paralogs, are now referred to as the erythrocyte binding-like (EBL) or, more apropos, to differentiate them from other classes of erythrocyte-adhesive proteins, the Duffy-binding-like (DBL) family (Miller et al., 2002a). The *dbl/ebl* genes have similar exon-intron structures with a primary central exon, one small 5′ intron, and three small 3′ introns (Fig. 6). Each gene encodes a protein with two cysteine-rich regions in its putative extracellular domains that share significant amino acid homologies and a hydrophobic transmembrane domain near the C terminus. One cysteine-rich region near the N terminus constitutes the adhesive or DBL domain with specific affinity for a particular erythrocyte receptor, while the other lies in the C terminus just before the transmembrane domain. In *P. falciparum* and *P. reichenowi*, the N-terminal domain is a duplication giving rise to DBL regions F1 and F2, but in all other *Plasmodium* species examined there is apparently a single N-terminal DBL domain (Adams et al., 1992; Adams et al., 2001; Ozwara et al., 2001; Rayner et al., 2004a). Of note, the relationship between the DBP interaction with the Duffy receptor and tight junction formation has only been shown with *P. knowlesi* (Miller et al., 1979). By inference, although not proven, it is thought that orthologous and paralogous merozoite proteins interacting with the Duffy glycoprotein, glycophorin A, or other erythrocyte receptors serve the same function for *P. vivax* and *P. falciparum* invasion.

While *P. vivax* apparently has only one member of the *dbl* family in its genome, other human

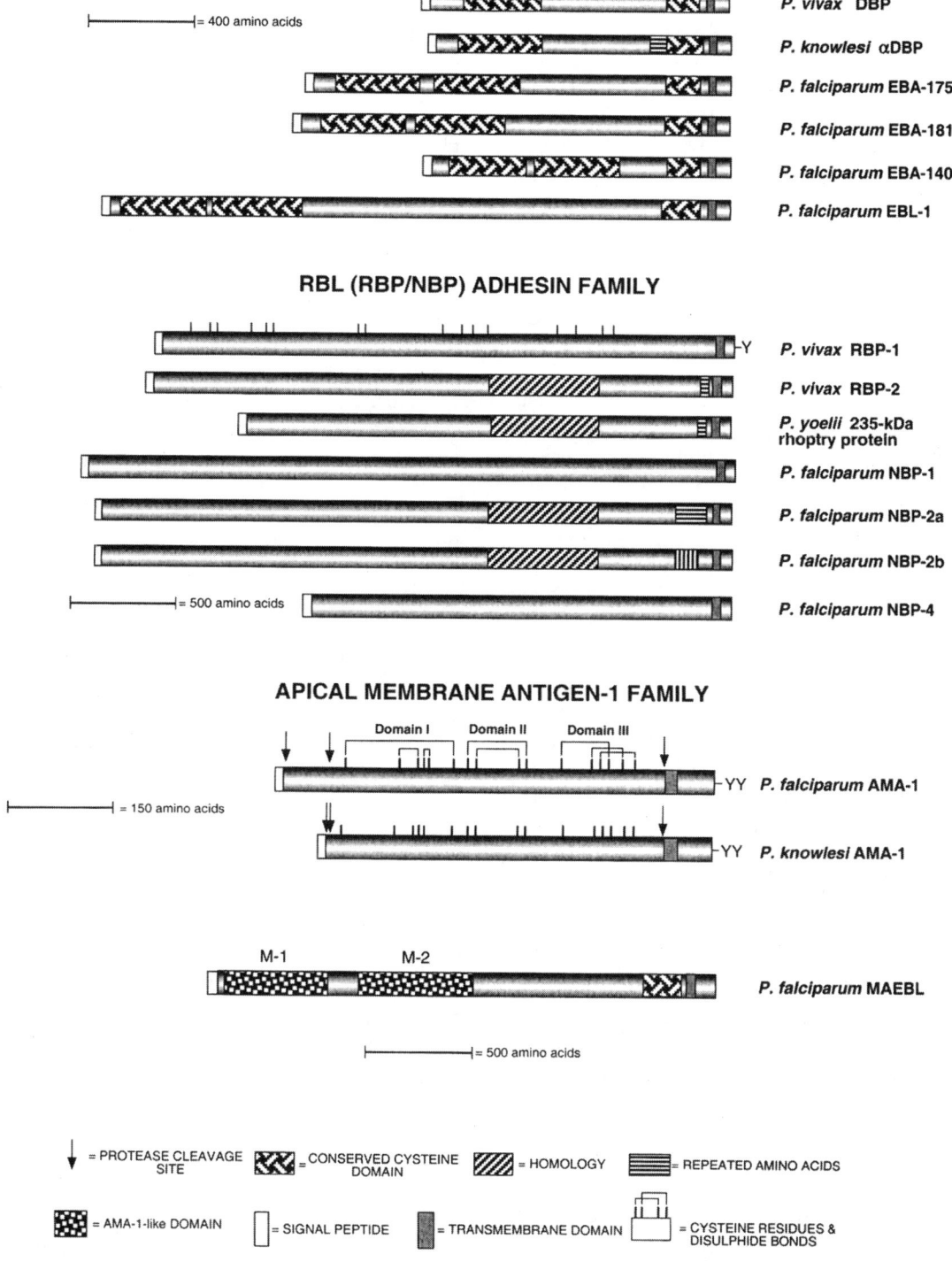

FIGURE 6 Schematic representing the DBL, RBL, and AMA-1 protein families. Several members of each family are depicted with their predominant features highlighted, as indicated by the key at the bottom of the figure.

and simian malaria species have several paralogous genes. *P. knowlesi* has three *dbl* genes termed the α-, β-, and γ-*dbp* genes, yet only the α gene product binds the Duffy glycoprotein receptor (Adams et al., 1990; Adams et al., 1992; Chitnis and Miller, 1994); DBP-β binds sialic acid residues and the erythrocyte receptor for the DBP-γ is unknown (Ranjan and Chitnis, 1999). Besides *eba-175*, four or five other paralogs of the *dbl/ebl* gene family have been identified in the *P. falciparum* genome (Adams et al., 2001). These include *ebl-1* (Peterson et al., 1995; Peterson and Wellems, 2000), *baebl/eba-140* (Mayer et al., 2001; Thompson et al., 2001; Narum et al., 2002), *pebl/eba-165* (Triglia et al., 2001b), *jsebl/eba-175* (Mayer et al., 2001; Gilberger et al., 2003b), and *maebl* (Kappe et al., 1998; Blair et al., 2002a) (Fig. 6). As noted previously, each *P. falciparum* paralog gene encodes two duplicated cysteine-rich encoded domains termed F1 and F2, except for *maebl*. This latter *dbl* family member is a chimeric gene found in many if not all other *Plasmodium* species (Kappe et al., 1998; Blair et al., 2002a; Singh et al., 2004). The N terminus of MAEBL has an extracellular cysteine domain structure similar to apical membrane antigen 1 (AMA-1) (described below) and a C-terminal region structurally similar to that of the DBL/EBL family. MAEBL is expressed in the sporozoite and appears to be critically important in sporozoite invasion biology (Kariu et al., 2002; Preiser et al., 2004). During hepatic schizogony, *maebl* transcripts are produced, albeit at low levels (Blair et al., 2002a; Blair et al., 2002b; Singh et al., 2004). In merozoites, MAEBL is apparently located in the rhoptries (Noe and Adams, 1998; Blair et al., 2002a) and not in the micronemes where DBL family members have generally been localized. The *P. falciparum* MAEBL N-terminal domain designated M2 has been reported to bind human erythrocytes independent of the Duffy receptor and the sialyl glycoproteins (Ghai et al., 2002).

The receptor-binding domains have been defined within the first cysteine-rich region for several of these DBL/EBL family members. Transfected COS cells expressing DBP and EBA-175 on their surface as a chimeric protein of region II and herpesvirus glycoprotein D bound erythrocytes according to the same specificity demonstrated by the native proteins (Chitnis and Miller, 1994; Sim et al., 1994; Ranjan and Chitnis, 1999). A 35-amino-acid peptide from the extracellular 66 amino acids of the N terminus of the human Duffy glycoprotein inhibits the binding of Duffy-positive erythrocytes to COS cells expressing the cysteine-rich-binding domain of both the *P. vivax* and *P. knowlesi* DBPs (Chitnis et al., 1996). In EBA-175, the duplicated F2 region demonstrated erythrocyte binding that was specifically sialic acid and glycophorin A dependent (Sim et al., 1994).

BAEBL/EBA-140 (Narum et al., 2002) was shown to bind human erythrocytes in a trypsin- and neuraminidase-sensitive manner, much like EBA-175 (Mayer et al., 2001). Unlike EBA-175, EBA-140 also bound En(a−) erythrocytes, which lack glycophorin A, and also bound S−s−U− erythrocytes, which lack glycophorin B. EBA-140 did not show strong adhesion to Gerbich and Yus mutant erythrocytes, which have altered glycophorin C and lack glycophorin D, implicating these sialyl glycoproteins as receptors. Thompson et al. (2001) also found that EBA-140/BAEBL binding is sensitive to neuraminidase but resistant to trypsin and other proteases. Others (Narum et al., 2002) found trypsinization only reduced binding by 50%; more recently, it has been reported that the sialic acid-dependent, trypsin-sensitive binding was due to glycophorin C in the region expressed by exon 2 (Lobo et al., 2003). In light of the above varied results, it was reported that four nonsynonymous nucleotide polymorphisms found in various *P. falciparum* strains in the F1 domain of the *baebl/eba-140* gene effectively changed binding profiles (Mayer et al., 2002), which suggested at least four different erythrocyte receptors could be involved in EBA-140/BAEBL adhesion, including glycophorin C and D.

Recently, JSEBL/EBA-181 was shown to exhibit sialic acid-dependent adhesion to erythrocytes that was resistant to trypsinization and sensitive to chymotrypsin (Gilberger et al., 2003b). However, S−s−U− cells bound EBA-181, indicating that the receptor was not glycophorin B. The unknown erythrocyte receptor has been designated E. From a comparison of only two strains it was suggested that a difference in expression levels of JSEBL/EBA-181 could be seen. Like EBA-140 (Mayer et al., 2001; Thompson et al., 2001; Narum et al., 2002), JSEBL/EBA-181, by comparative confocal immunofluorescent microscopy, appears to be located in the microneme organelles (Gilberger et al., 2003b; Mayer et al., 2004). Similar to BAEBL, polymorphism in both F1 and F2 domains of the JSEBL region II also appears to alter receptor specificity between different isolates of *P. falciparum*. In light of this, it is noted that although EBA-175 has a greater extent of polymorphism in both the F1 and F2 domains, an alteration of receptor specificity between different isolates has not yet been reported.

A specific erythrocyte receptor or role in merozoite invasion for EBL-1 has not been defined, although the *ebl-1* gene has been genetically linked to a parasite phenotype exhibiting increased proliferation (multiplication) (Peterson et al., 1995; Peterson and Wellems, 2000). Furthermore, the *pebl/eba-165* gene from 13 different strains of *P. falciparum* contains one or two frame-shift mutations (Triglia et al., 2001a; Rayner et al., 2004a). As such, it is transcribed but not translated and thus likely is a pseudogene, and it is thus designated ψ*eba140*/ψ*pebl* (Triglia et al., 2001a). In the ape malaria parasite *P. reichenowi*, *pebl/eba-165* has a complete open reading frame, as do the other *dbl* paralogs; thus presumably it is both transcribed, translated, and participating in invasion of chimpanzee erythrocytes (Rayner et al., 2004a). *P. reichenowi* cannot invade human erythrocytes and is unable to infect humans, although *P. falciparum* can infect chimpanzees and invade ape red blood cells. Some of the diversity and resultant changes in the phenotypes of receptor specificity observed between the DBL paralogs in these two closely related malaria parasites might explain this lack of cross infectivity in *P. reichenowi* (Rayner et al., 2004a; Martin et al., in press).

The RBL Family

A second family of apically located adhesive merozoite proteins implicated in the adhesion or interaction of the merozoite's apical pole to the erythrocyte is the reticulocyte-binding protein-like (RBL) family of ligands (Fig. 6). The precise localization and functional actions of the RBLs in each species of *Plasmodium* where they have been identified to date and their timing in the invasion cascade are still open questions. They may function, as previously proposed (Galinski et al., 1992; Galinski and Barnwell, 1996), prior to the binding interactions of DBP, EBA-175, or other DBL ligands. It has also been suggested recently that RBLs perform functions similar to or in parallel with those of the DBL ligands (Triglia et al., 2005). A role in the targeting of particular erythrocyte populations for invasion is also hypothesized for the RBLs, leading to the commitment of a merozoite to invade a particular host cell (Galinski et al., 1992; Galinski and Barnwell, 1996).

Two native *P. vivax* merozoite proteins were shown to bind specifically to human RBCs enriched for reticulocytes and thus termed *P. vivax* reticulocyte-binding protein 1 (PvRBP-1) and PvRBP-2 (Galinski et al., 1992). Hence, the RBL terminology was adopted once homologous invasion ligands were identified in the genomes of several *Plasmodium* species (Miller et al., 2002a). One member, first recognized to be within this class, is a 235-kDa protein in the rodent parasite *P. yoelii* that is located in the rhoptry organelles (Holder and Freeman, 1984a; Oka et al., 1984; Holder et al., 1985b). Passive transfer experiments with a monoclonal antibody (MAb) specific for the 235-kDa rhoptry protein indicated that this protein facilitated the invasion of mature mouse erythrocytes (normocytes). Mice infected with the lethal *P. yoelii* strain YM, which normally infects both young and mature erythrocytes, were protected by passively injected MAbs by restricting RBC invasion to reticulocytes (Freeman et al., 1980). *P. vivax*, like nonlethal *P. yoelii* strains or *P. berghei* in the rat, only or mostly invades reticulocytes (Kitchen, 1938; Mons, 1990). Unlike the PvRBPs, however, the *P. yoelii* 235-kDa (Py235) rhoptry protein binds to mature mouse erythrocytes and not specifically to reticulocytes (Ogun and Holder, 1996), a finding that is consistent with the data showing that antibodies block invasion of mature erythrocytes but not reticulocytes (Freeman et al., 1980). PvRBP-2, more so than PvRBP-1, was determined to share a limited but significant level of homology with Py235 rhoptry protein (Keen et al., 1994; Galinski et al., 2000), which is now known to belong to a family with 14 confirmed paralog genes (Keen et al., 1990; Carlton et al., 2002).

The adhesion of native PvRBP-1 and PvRBP-2 to reticulocytes, predicted experimentally to bind as a complex, is independent of the Duffy phenotype of the RBC, indicating a novel receptor, which remains unknown. The characterization of the two *P. vivax* genes and their encoded proteins shows some intriguing structural properties. *pvrbp-1*, *pvrbp-2*, and in fact all subsequently identified members of the *rbl* gene family have a similar two-exon structure, with the first exon encoding the signal peptide followed by a short intron and a large 6- to 9-kb exon (Fig. 6) (Galinski et al., 1992; Galinski et al., 2000; Rayner et al., 2000; Narum et al., 2001; Rayner et al., 2001; Miller et al., 2002a). Both PvRBPs are quite large with calculated masses of 325 and 330 kDa, respectively. A potential hydrophobic transmembrane motif present in each at their C termini indicates they are type I membrane-anchored proteins with short putative cytoplasmic domains. PvRBP-1 in native immunoprecipitations behaves as a dimer covalently associated via one or more disulfide bonds, suggesting they may form a complex with PvRBP-2 through noncovalent interactions (Galinski et al., 1992; Galinski and Barnwell, 1996). Immunofluorescence analysis colocalizes the PvRBPs to the apical region of the merozoite overlapping with the signal from anti-DBP antibodies. Additionally, the fluorescent signal is in a crescent-shaped pattern seemingly capping the apex of the merozoite, a pattern that is not consistent with a location entirely within the micronemes and certainly not with a classic double dot rhoptry location, despite their paralogous relationship to the Py235 rhoptry proteins (Oka et al., 1984).

Prior to the advent of the genome sequence databases for various *Plasmodium* species, antibody cross-reactivity to a few regions and low-stringency DNA hybridization had indicated that proteins and genes sharing homology to PvRBP-1 and PvRBP-2 also existed in *P. knowlesi* and *P. falciparum*, two species that invade erythrocytes regardless of their maturity. Initially, fragments of two distinct genes showing highest homology to PvRBP-2 were isolated from *P. falciparum* (Rayner et al., 2000). As the *P. yoelii* 235-kDa rhoptry protein *rbl* gene was already known to be a member of a multigene family (Keen et al., 1990; Borre et al., 1995; Sinha et al., 1996; Carlton et al., 2002), it was not entirely surprising to find two *pvrbp2*-related genes in *P. falciparum*. The sequencing of the *P. falciparum* genome and long-range PCR amplification showed these two RBP-2-like *rbl* genes are nearly identical in sequence over a large 5′ portion of their coding region. Their sequences become unique only at the 3′ ends of the genes (Rayner et al., 2000; Triglia et al., 2001a). Subsequent searches of the *P. falciparum* genome databases revealed three other related genes: the five currently known *P. falciparum* RBLs have been noted as reticulocyte-binding protein homologs (Rh), as well as normocyte-binding proteins (NBPs) (Galinski and Barnwell, 1996), to distinguish them in relationship to their anticipated binding specificity from that of the RBPs. Thus, species that invade reticulocytes as a main host cell would have RBPs, while those generally invading all RBCs would have NBPs. The first discovered *P. falciparum* RBLs are thus known as PfNBP2a and -2b (as well as Rh2a and Rh2b) (Rayner et al., 2000; Triglia et al., 2001a). *P. falciparum nbp1* (*rh1*) was then identified and shown to exhibit a stronger homology to *pvrbp1* than *pvrbp2* (Rayner et al., 2001). *nbp3* (*rh3*) was shown to be a pseudogene with frameshift mutations that are transcribed but not translated (Taylor et al., 2001); *nbp4* (*rh4*) encodes a much smaller protein (~205 kDa) than the other four (>325 kDa) but falls within the *rbp1* subgroup of the *rbl* family (Kaneko et al., 2002). Only NBP1 (Rh1) has been shown so far to bind erythrocytes and with a specificity that was not age dependent or sensitive to trypsin, but sialic acid dependent (Rayner et al., 2001). The adhesion of PfNBP1 to human erythrocytes, however, did not involve glycophorin B, and the unidentified receptor was designated Y. Some studies indicate that NBP1, NBP2a, and NBP2b are differentially expressed between strains (Taylor et al., 2002; Duraisingh et al., 2003; Triglia et al., 2005), but not in another case (Rayner et al., 2001).

PfNBP-1 by immunofluorescence appears to colocalize with EBA-175 and like the RBPs of *P. vivax* to also form a crescent pattern over the apex of the merozoite (Rayner et al., 2001). PfNBP2a and PfNBP2b, on the other hand, occasionally show a double dot of apical fluorescence, and they colocalize with Pf240/Pf225 (Roger et al., 1988) and less so with Pf140 and rhoptry-associated protein 1 (RAP-1), which by immuno-EM are located in the peduncle (neck) and bulbous body of the rhoptries, respectively (Rayner et al., 2000; Rayner et al., 2001; Triglia et al., 2001a). Immuno-EM experiments have further suggested a rhoptry peduncle and body localization for PfNBP2 (Duraisingh et al., 2003). PfNBP4 (Rh4) antibodies produce a punctate apical fluorescent pattern that colocalizes with anti-EBA-175 fluorescence, similar to that described for PfNBP1, with a crescent pattern like that of PvRBP1 (Kaneko et al., 2002).

Genome databases and genetic studies of other species of *Plasmodium* have revealed a number of other interesting insights about the RBL ligand family. *rbl*-type genes are probably present in all *Plasmodium* species, including avian malaria parasites as represented by *P. gallinaceum*. In the genomes of two other rodent malaria parasites, *P. berghei* and *P. chabaudi*, there are families of genes that are highly related to the *P. yoelii* 235-kDa rhoptry protein genes and, like *py235*, fall into the RBP2 subgroup of the RBL family. In *P. cynomolgi* two expressed *rbl* genes exist (Okenu et al., in press) that are highly homologous to *pvrbp1* and *pvrbp2* with ~85% identity, as expected given the close genetic kinship between these two parasite species (Escalante et al., 2001). In contrast, *P. knowlesi*, also a member of the simian malaria clade and related to *P. vivax*, has no detectable *rbp2*-like gene by DNA hybridization or BLAST search of genome databases. A highly homologous *pvrbp1*-like sequence is detectable by Southern blot hybridization and BLAST search, but this is a pseudogene that has numerous frame-shift mutations and deletions, such that most of the gene has disappeared from the genome (E. V.-S. Meyer et al., submitted for publication). Conversely, there are two other *rbl* genes in the *P. knowlesi* genome that show very low homology to *pvrbp1* and *pvrbp2* but high homology to two other genes detected by searching with these sequences in the *P. vivax* genome; we have termed these new *rbl* genes the *PkNBPXa/PvRBPXa* and *PkNBPXb/PvRBPXb* genes (Meyer et al., submitted). Interestingly, in *Plasmodium coatneyi*, which is very closely related to *P. knowlesi* (Vargas-Serrato et al., 2003), *pvrbp1* and *pvrbp2* orthologs, as well as *nbpx* genes, are present (Meyer et al., submitted). In *P. reichenowi*, the *nbp1* gene is a deteriorating pseudogene, but the *nbp3/rh3* gene has a complete open reading frame, unlike in *P. falciparum*. The other *P. falciparum rbl* orthologs in *P. reichenowi* also have complete open reading frames and are presumably expressed (Rayner et al., 2004c).

Based upon the reticulocyte-binding specificity of the PvRBP complex, its apical location, and the known biology of the *P. vivax* merozoite invasion, this receptor complex could function to aid *P. vivax* merozoites in specifically targeting reticulocytes for invasion (Galinski et al., 1992; Galinski and Barnwell, 1996). Intuitively, this selective interaction would have to occur prior to any adhesive step(s) leading irreversibly to merozoite-erythrocyte tight junction formation, since *P. vivax* merozoites must come into contact with numerous normocytes before necessarily encountering reticulocytes, which comprise a minority (<1%) of the total RBC population. The Duffy glycoprotein is expressed on all circulating erythrocytes; premature, incidental interaction between the parasite's DBP and the Duffy receptor prior to an RBP interaction with a reticulocyte could be detrimental to *P. vivax* survival, regardless of whether the DBP initiates or is an integral part of the nonreversible tight junction. The DBP/EBA-175 proteins are stored in the microneme organelles (Adams et al., 1990; Sim et al., 1992); but unlike AMA-1, there is no evidence for the DBL ligands being exposed at the surface prior to the initiation and formation of the tight junction. This suggests another possible more general function for the PvRBPs and, perhaps generally from a biological perspective, for the RBL class of merozoite proteins. That is, binding of these proteins may be necessary to trigger (signal) the timely release of the sequestered microneme proteins, so that tight junction formation and entry can proceed (Galinski et al., 1992; Galinski and Barnwell, 1996). Redundancy with different gene copy numbers in each species and diversity among members may be the means of preserving a unique role of the RBL family.

The Py235 protein is associated with the invasion of *P. yoelii* into mature mouse erythrocytes; neither *P. knowlesi* nor *P. falciparum* has a strong biological preference for invasion of one erythrocyte population over another. In the case of these three species, there should be no need to target a subpopulation of erythrocytes such as reticulocytes. We also know that in the absence of the development of a tight junction, as shown for *P. knowlesi*, long-range fibrillar connections tether the apical end of the merozoite to the erythrocyte (Miller et al., 1979; Aikawa and Miller, 1983). This suggests that adhesive interactions occur at the apex of the merozoite in the absence of a DBL ligand interaction. Furthermore, apical fibrils can be observed as the merozoite is entering the PV (Bannister et al., 1975; Mitchell and Bannister, 1988; Bannister and Dluzewski, 1990). These observations add emphasis to the positioning of the merozoite apical pole prior to entry (as well as subsequently) and from a structural point of view invite speculation that the RBL ligands might represent the initially observed fibrillar adhesins. In any case, regardless of their binding specificity, RBLs could aid the positioning of the apical pole of the merozoite. Quite plausibly, this or some other early contact could initiate a signal transduction event(s) to induce the release of micronemal DBL proteins at a time when the merozoite is in a position to commit to junction formation.

AMA-1

Another merozoite protein thought to have an adhesive or receptor-binding function in RBC invasion and perhaps to play a unique role is AMA-1 (Fig. 6). While initially localized by immuno-EM in the neck of each rhoptry (Crewther et al., 1990), AMA-1 has recently been shown to be stored in the micronemes from the time these vesicular organelles are transported from the Golgi cisterni along microtubules to the apical prominence of the merozoite until released to the merozoite surface (Healer et al., 2002; Bannister et al., 2003). Evidence for a central role of AMA-1 in invasion comes from in vitro invasion inhibition and genetic studies. Fab fragments of an immunoglobulin G MAb inhibited invasion of *P. knowlesi* merozoites as well as the intact immunoglobulin G (Deans et al., 1982; Thomas et al., 1984), and recently antibodies against *P. falciparum* AMA-1 were also shown to inhibit invasion (Dutta et al., 2003; Healer et al., 2004). Additionally, attempts to knock out AMA-1 by gene disruption experiments have not been possible, even though it was shown that *ama-1* could be targeted by homologous recombination, suggesting a critical function of AMA-1 in invasion (Triglia et al., 2000). Nevertheless, a well-defined receptor-ligand interaction for AMA-1, whether it involves a component of the erythrocyte or another merozoite protein, has not been conclusively demonstrated. It has been shown, however, that *P. yoelii* AMA-1 expressed on the surface of COS cells will bind rat and mouse, but not human, RBCs (Fraser et al., 2001). The reverse experiment of expressing *P. falciparum*

AMA-1 and demonstrating binding of human RBCs has not been reported. In this context, complementation of *P. falciparum* AMA-1 (PfAMA-1) with AMA-1 of the rodent malaria parasite *P. chabaudi* allows invasion at 35% efficiency of normal and increases the efficiency of *P. falciparum* invasion of mouse erythrocytes (Triglia et al., 2000), suggesting a role for AMA-1 in invasion with some degree of host specificity.

The gene encoding AMA-1 has been characterized from many species of *Plasmodium*, including human and ape malarias (Peterson et al., 1989; Cheng and Saul, 1994; Kocken et al., 2000), the simian malarias (Peterson et al., 1990; Waters et al., 1990; Dutta et al., 1995), and rodent malarias (Marshall et al., 1989; Kappe and Adams, 1996). AMA-1 is conserved in structure (Fig. 6) and function even in genetically distant Apicomplexan parasites such as *Toxoplasma gondii* and *Babesia bovis* (Donahue et al., 2000; Michon et al., 2002; Gaffar et al., 2004). AMA-1 is synthesized as an 83-kDa type I integral transmembrane protein in *P. falciparum*, but in all other species besides *P. reichenowi* it is approximately 66 kDa in mass with the extra mass due to a much larger propeptide sequence after the signal peptide. Located in the apical micronemes at or near the time of merozoite release from the infected RBC (Healer et al., 2002; Bannister et al., 2003), AMA-1 becomes translocated to and distributed over the entire surface of the merozoite but perhaps in greatest concentration at the apical pole (Thomas et al., 1990; Narum and Thomas, 1994). There are 16 conserved cysteine residues in the extracellular part of the protein, and the cysteine-bonding pattern indicates that there are three disulfide-bonded domains formed (Hodder et al., 1996). AMA-1, like MSP-1 and other merozoite proteins, undergoes extensive proteolytic processing, with initial clipping of the propeptide sequence signaling the translocation of the 66-kDa form from the micronemes to the merozoite surface. Further protease-mediated processing at the surface removes cysteine domains 1, 2, and 3 by cleavage at a threonine 29 residues N-terminal of the transmembrane domain at release or during invasion to create the soluble 44-, 42-, 48-, and 44-kDa fragments (Howell et al., 2001; Howell et al., 2003). Since the merozoite surface coat architecture is formed during schizogony prior to redistribution of AMA-1 from the micronemes, AMA-1 is certainly not one of the initial molecular building blocks of the coat. There is no evidence that AMA-1 participates in the initial reversible adhesion of merozoites to RBCs, although mammalian cell expressed AMA-1 binds rodent erythrocytes (Fraser et al., 2001). However, a recent report taking advantage of the invasiveness of free *P. knowlesi* merozoites and observations by EM have suggested that AMA-1 might function in apical reorientation or in promoting close apical attachment, leading to tight junction formation (Mitchell et al., 2004). Interestingly, AMA-1 is also expressed in sporozoites and probably has a role in the invasion of hepatocytes (Silvie et al., 2004; Singh et al., 2004).

Alternate Pathways for Invasion

The multiple members of the parasite's ligand families between and within species suggest that malarial merozoites must possess different sets of "locks" and "keys" to gain access into an erythrocyte. These may bind in a hierarchical sequential fashion or may act as substitutes or backups when alternative mechanisms for entry must be utilized. For example, early invasion studies using a variety of mutant human erythrocytes and enzymatic treatments showed that different isolates of *P. falciparum* were not totally dependent on sialic acid or just glycophorin A for mediating invasion and that, in general, there were sialic acid-dependent and -independent pathways (Mitchell et al., 1986; Hadley et al., 1987; Perkins and Holt, 1988; Holt et al., 1989; Dolan et al., 1990; Dolan et al., 1994; Sim et al., 1994). At least three pathways using different erythrocyte receptors for entry of *P. falciparum* merozoites into RBCs became evident from subsequent investigations (Dolan et al., 1994; Sim et al., 1994). In addition to the sialoglycophorin A/EBA-175 pathway, a pathway utilizing specifically sialic acid and glycophorin B was identified, as well as a sialic acid-independent, trypsin-sensitive receptor pathway; as yet, the RBC receptor is still uncharacterized. The merozoite proteins that would potentially interact with either glycophorin B or the sialic acid–independent, trypsin-sensitive invasion routes have not been identified, despite the characterization of all the EBL (and RBL) ligand genes known in *P. falciparum*. Studies of other cultured strains of *P. falciparum* have indicated additional specificities for invasion receptor-ligand interactions in *P. falciparum* (Gaur et al., 2003; Cortes et al., 2004). Earlier, glycophorin C was also believed to participate in *P. falciparum* invasion (Pasvol et al., 1984) and just recently it was shown that some alleles of BAEBL/EBA-140 do interact with this glycophorin (Mayer et al., 2001; Lobo et al., 2003; Maier et al., 2003). JSEBL, for at least one strain (Dd2), also binds a neuraminidase-sensitive, trypsin-resistant receptor that is not glycophorin B (Gilberger et al., 2003b). In fact, studies in South America, Africa, and Asia indicate many or most strains of *P. falciparum* are capable of using pathways that are alternative to the classic sialic acid-dependent or glycophorin A/EBA-175 route of invasion (Okoyeh et al., 1999; Baum et al., 2003; Lobo et al., 2004). However, the polymorphisms in the binding domains of BAEBL and JSEBL alone provide eight alternative receptor mechanisms for invasion if BAEBL and JSEBL are functionally equivalent to EBA-175 (Mayer et al., 2002; Mayer et al., 2004). In fact, genetic manipulation of *P. falciparum* that either C-terminally truncated EBA-175 or deleted *eba-175*, *baebl*, or *jsebl* in either sialic acid-dependent or -independent strains has not shown changes in the efficacy of merozoite invasion much different from wild-type parasites without alteration of the host cell (Reed et al., 2000; Duraisingh et al., 2003; Gilberger et al., 2003a; Maier et al., 2003). Surprisingly, EBA-175, which has many more polymorphic sites in the F1 and F2 domains of regions II than BAEBL or JSEBL, does not appear to have receptor specificities other than for glycophorin A (Binks and Conway, 1999). So, why does *P. falciparum* make use of an apparent myriad of possible receptor-ligand specificities with the potential to effect successful invasion of host cells, when *P. vivax*, an equally successful human parasite, seems to make do with only two known specific interactions at critical junctures in the invasion process—reticulocyte selection via RBPs and junction formation via DBP?

It is obvious that the use of alternative receptors in invasion pathways should provide a survival advantage for *P. falciparum* in the face of genetic mutations that might alter the structure or expression of essential erythrocyte receptors. Mutations altering the expression and structure of erythrocyte glycophorins do occur with some frequency in human populations, often without severe effects in the affected individuals in areas where malaria is highly endemic. In this regard, the African continent remains an area where *P. falciparum* is highly endemic, as is south Asia and New Guinea. In Africa, there appears to be no high rate of fixation in the populations of glycophorin mutations that have been shown to affect merozoite invasion in laboratory investigations. In the New Guinea lowlands, the Gerbich mutation in glycophorin C occurs at a high rate (47%). Although some evidence suggests this mutation affects *P. falciparum* invasion through a glycophorin C pathway (Maier et al., 2003), there is little other evidence for this mutation affecting parasite success or protecting against pathological consequences of infection (Patel et al., 2001). In the face of developing naturally acquired protective immune responses against EBA-175 or other paralogous DBL/EBL ligands, the availability of alternative invasion pathways would provide a selective advantage to overcome immune attack. If vaccines based upon blocking merozoite erythrocyte receptor interactions are to be developed and effectively employed, then it would be important to identify and employ all the alternative ligands to impart effective immunity. In this regard, though, the testing of EBA-175-based vaccines may also be problematic, since the *Aotus* and *Saimiri* monkey small-primate models used in preclinical testing of malaria vaccines apparently do not utilize an EBA-175 as the principal invasion pathway, as their RBCs do not bind EBA-175 in erythrocyte-binding assays (Sim et al., 1994).

In contrast to the epidemiological profile of *P. falciparum* in much of western and central sub-Saharan Africa where this parasite is abundant, *P. vivax* is nearly absent. This is because the vast majority of the indigenous human residents in these areas are Duffy negative (Miller and Carter, 1976). In this case, however, the genetic mutation was probably responsible for a negative selection on the parasite and not a positive selection by the parasite for the mutation, since *P. vivax* likely arose as a human pathogen in southeast Asia and not in Africa (Escalante et al., 2005; Mu et al., in press). It is believed that *P. vivax* merozoite invasion absolutely requires the Duffy receptor, as studies have not yet turned up evidence for alternative or multiple DBL pathways used by this parasite in human populations; in contrast, *P. knowlesi* has three DBL family members, each with a different erythrocyte receptor. Nonetheless, some data suggest the potential for alternative entry pathways for *P. vivax* at this stage in the invasion cascade. *Saimiri* monkey erythrocytes do not bind DBP, and this primate species is readily infected with *P. vivax* (Wertheimer and Barnwell, 1989; Chitnis et al., 1996). Although not readily apparent in human populations, this at least suggests that the potential for utilizing alternative receptors in place of DARC exists in *P. vivax*. Novel proteins that by immuno-EM are found to localize specifically to the apical end of the merozoite and bind to *Saimiri* monkey RBCs may provide clues to which protein(s) might function as a backup for the DBP of *P. vivax*. Such studies will also help define the relative roles of RBLs, DBLs, and other apically localized adhesive proteins in the invasion cascade.

The ability to genetically transform *Plasmodium* to eliminate the expression or alter the structure of an expressed protein has advanced our understanding of invasion mechanisms and the ability to probe the molecular redundancies and intricacies of invasion. However, there are limitations in current methods that affect our ability to interpret results. Additional methods such as inducible expression systems and better-defined targets on host cells are required. For example, certain strains of *P. falciparum* are used more frequently in such studies than others that might offer more appropriate phenotypes. In some strains, the knockout of a particular gene may not produce a viable phenotype, while in another strain viable parasites are recovered, but no interpretable phenotype is observed, probably reflecting differences of reliance on a particular pathway or some biological advantage (Gilberger et al., 2003b). Furthermore, W2mef and a sialic acid-independent, trypsin-resistant strain of 3D7 are most often used in transformation studies (Kaneko et al., 2000; Reeder et al., 2000; Duraisingh et al., 2003; Gilberger et al., 2003a). However, the latter strain intrinsically expresses ligands that can use non-glycophorin A, non-sialic acid receptor interactions efficiently for invasion, whereas the former (W2mef) is very dependent upon sialic acid but relatively trypsin resistant. However, W2mef is the parent clone of Dd2 (all derived from Indochina III), which has a biology as yet not observed in other strains highly dependent on sialic acid, including Malayan Camp and FCR3 (Mitchell et al., 1986; Dolan et al., 1990). Remarkably, the *P. falciparum* clone Dd2 and subclones of Dd2, which nominally use sialic acid-dependent receptors, can switch at a frequency of 10^{-4} to a sialic acid-independent pathway, and this switch is genetically stable (Dolan et al., 1990).

Furthermore, reliance on enzymatic treatment of erythrocytes and antibody-mediated inhibition of invasion is problematic in regards to the interpretation of invasion phenotypes resulting from genetic transformation experiments. Invasion is a procession of different steps and molecular interactions defined by an ordered cascade of required events. While proteases relevant to invasion may have specific cleavage sites, they most certainly target more than one protein exposed on the surface of erythrocytes, and therefore such treatments potentially would affect more than one receptor, which may be required for different molecular steps in the invasion cascade. One example of this type of unintended consequence comes from experiments on *P. knowlesi* invasion. *P. knowlesi*, enigmatically, while not able to invade

Duffy-negative human erythrocytes, can invade these refractory cells if they have been treated with trypsin (Miller et al., 1979). Then, invasion does not seem to involve the *P. knowlesi* β and γ *dbp* gene products, since trypsinized human Duffy-negative erythrocytes do not adhere to their predicted binding domains (Chitnis and Miller, 1994). Presumably, a cryptic erythrocyte receptor and another merozoite molecule, both as yet unidentified, can interact with and assume the invasion role normally accomplished by the interaction of the DBP and the Duffy glycoprotein. Antibody-mediated inhibition of invasion targeting EBA-175 can also have nonspecific effects, since significant inhibition against strains that are not dependent upon this invasion pathway is often exhibited (Narum et al., 2000; Pandey et al., 2002). Clearly, there is much to be discovered yet with regards to the various molecules involved both as ligands and receptors and the ordering of events in these early critical adhesion steps that position the apex of the merozoite flush with its chosen host cell

PHASE III: ERYTHROCYTE ENTRY AND FORMATION OF THE PV

Prior to actual entry into the erythrocyte, through the presumed actions of the RBL and DBL ligands, AMA-1, and probably other as-yet-unknown molecular players, the merozoite apex becomes closely positioned within 4 nm of the RBC surface, and an ultrastructural dense region appears beneath the area of contact (Aikawa et al., 1978). The molecular composition of this junctional zone and whether the known merozoite ligands such as the RBLs, DBLs, or AMA-1 are engaged with respective receptors still remain unclear. However, freeze-fracture studies of invading *P. knowlesi* merozoites indicate an increased density of intramembrane particles (IMPs) in this region of contact, the inference being that integral transmembrane proteins such as band 3, glycophorins, and DARC are clustered at this point of contact, with these in turn being connected to one or more of the merozoite ligands previously described or, for that matter, other unknown connectors (Color Plate 7a).

The entry phase of merozoite invasion into RBCs is a highly dynamic and rapid process that, after apical attachment and formation of the tight junction, involves a number of molecular events that allow the merozoite to gain entry into the erythrocyte and become positioned within a vacuole inside the erythrocyte. These active processes are largely, if not wholly, initiated by the merozoite, since the mature erythrocyte is not capable of either phagocytosis or receptor-mediated endocytosis. Biologically, this requires movement of the junction zone posteriorly over the merozoite; at the same time, a PVM is produced into which the merozoite enters. Molecularly, this involves (i) a coordinated cascade of signaling events (Sibley, 2004) that leads to proteolytic processing of essential invasion-related molecules (Blackman, 2004), (ii) the activation of motor proteins (Pinder et al., 2000), (iii) the secretory release of proteins and lipids from the rhoptry organelles (Blackman and Bannister, 2001), (iv) localized alteration of the erythrocyte membrane architecture with general membrane remodeling, and (v) closure and "sealing" of the erythrocyte and PVMs.

The Ultrastructure of Invasion

EMs show that upon initiation of merozoite entry an indentation of the erythrocyte membrane occurs at the site where the tight junction has formed (Aikawa et al., 1978). The junction, in the form of a narrow band which retains its suberythrocyte electron density, begins to move posteriorly over the surface of the merozoite (Aikawa et al., 1978). As this is happening, a membrane-lined invasion pit, the PV, begins to form between the merozoite and the erythrocyte interiorly beyond the boundaries of the junctional band. The PVM progressively expands inward as the junctional band continues to encircle and be driven posteriorly over the merozoite, pushing it deeper into the enlarging PV (Fig. 5b). The structured merozoite surface coat is absent from the portion of the parasite within the confines of the PV, as it has apparently been stripped away as the junction moves over the surface of the parasite (Ladda et al., 1969; Aikawa et al., 1978). As these events transpire, it appears that contents of the rhoptries are discharged into the PV (Bannister et al., 1986a; Bannister and Mitchell, 1989). These ultrastructural observations, together with proteomic data (Sam-Yellowe et al., 2004), indicate that the rhoptries contain a mixture of proteinaceous and lipid material that assists and perhaps induces the invagination of the erythrocyte membrane and its restructuring, which results in formation of the PVM and the PV.

Rhoptry Proteins

Mostly described from studies of *P. falciparum,* a number of malarial proteins are transported to and stored in the rhoptry organelles, whereas other polypeptides are present as the rhoptries are matured from the Golgi membrane (Bannister et al., 2000b; Topolska et al., 2004b). One such protein is a 240-kDa protein, which is processed to a 225-kDa protein, reportedly localized in the neck of each rhoptry (Roger et al., 1988). We now know that this 240- to 225-kDa (designated 240/225-kDa) protein is identical to the RBL ligands NBP-2a and NBP-2b discussed above (J. C. Rayner et al., submitted for publication). Two separate noncovalently bound protein complexes are known to be present in the bulbous bodies of the rhoptries. One complex, termed the RhopH or high-molecular-mass rhoptry protein complex, comprises three distinct proteins of 155/140, 140/130, and 110/105 kDa, designated RhopH-1, RhopH-2, and RhopH-3, respectively (Holder et al., 1985a; Cooper et al., 1988). The genes for all three proteins have been characterized (Brown and Coppel, 1991; Kaneko et al., 2001; Ling et al., 2003). *rhoph1* was identified as a member of the cytoadherence-linked asexual gene (*clag*) family, so named because one member (*clag9*) was associated with the sequestration of infected erythrocytes (Holt et al., 1999). *rhoph2* and *rhoph3* are single-copy genes with six and nine introns, respectively (Brown and Coppel, 1991; Ling et al., 2003). More recently, *clag9* was localized by immunofluorescence (Gardiner et al., 2004) and immuno-EM unequivocally to the rhoptry organelles, exclusively to the basal bulb region rather than to the peduncle. The Clag9-specific antibodies immunoprecipitated the RhopH complex, and Clag 9 was detected in the complex purified by antibodies to RhopH2, indicating that the RhopH complex could be a mix of *clag* components (Ling et al., 2004).

The second rhoptry protein complex is composed of three lower-molecular-mass proteins of 86 and 42/39 kDa called RAP-1 and RAP-2 (Howard and Reese, 1984; Bushell et al., 1986; Schofield et al., 1986; Clark et al., 1987; Ridley et al., 1990), and RAP-3, a third protein of 37 kDa (Howard, 1990; Howard et al., 1998). RAP-1 is rapidly processed from an 86-kDa precursor to 82 kDa, which is the predominant form found in free merozoites and ring-stage parasites (Bushell et al., 1986; Howard et al., 1998). The genes encoding RAP-1, RAP-2, and RAP-3 (Ridley et al., 1990; Baldi et al., 2002) encode proteins that have neither transmembrane domains nor repeated amino acid motifs. How these three proteins form a complex together is not known and is not evident from structure predictions based on their primary sequences. *rap-3* is a paralog of *rap-2*, and these genes are arranged on chromosome 5 in a head-to-tail fashion. They are conserved across species, but disruption of the *rap-3* gene did not affect RBC invasion efficiency (Baldi et al., 2002).

The function(s) of the RhopH and RAP complexes in the invasion process is not known. The 110-kDa and other associated members of the RhopH complex interact with inner leaflet lipids of the erythrocyte membrane in vitro (Sam-Yellowe and Perkins, 1991). During or shortly after invasion, the complex is localized to the plasma membrane of the newly intracellular parasite or its PVM (Lustigman et al., 1988; Sam-Yellowe et al., 1988; Ling et al., 2003). The role of the RhopH complex could be in remodeling the infected erythrocyte during or after invasion, and it might have different functions dependent on the molecular composition of the RhopH complex with respect to the Clags (Ling et al., 2004). The RAP-1/RAP-2 complex has been associated with the lamellar membrane-like structures released from the rhoptries of free merozoites that have not invaded erythrocytes (Bannister et al., 1986a). After invasion, this complex is associated with newly formed ring-stage parasites, apparently incorporated into the PVM and/or the parasite plasma membrane. However, truncation of the C-terminal half of RAP-1 disrupted the formation of the RAP complex with a failure of RAP-2 being transported to the rhoptries in merozoites, although the truncated RAP-1 was present in these organelles (Baldi et al., 2000). In the absence of an intact RAP-1 and no rhoptry-localized RAP-2, the mutant merozoites were still able to invade erythrocytes efficiently. Due to a lack of anti-RAP-3-specific antibodies, unfortunately, it was not determined if RAP-3 was located in the rhoptries of RAP-1 mutant parasites.

The rhoptry-associated membrane antigen (RAMA) is yet another rhoptry protein characterized to date, which is proteolytically processed from a 170-kDa precursor to a 60-kDa form anchored in the interior side of the rhoptry membrane via a GPI anchor (Topolska et al., 2004a). The 170-kDa form is translated in early trophozoites, where it is transiently localized within regions of the Golgi membrane that bud and mature into rhoptries. During invasion, RAMA is discharged from the rhoptries, apparently binds to the erythrocyte membrane, and in early ring stages is found in the PV or PVM. Other proteins that have been initially ascribed a location in the rhoptry organelles are a 52-kDa molecule (Storey, 1992), a 55-kDa GPI-linked protein (Smythe et al., 1988) that is the processed fragment of RAMA discussed above, a serine protease (Braun-Breton and Pereira da Silva, 1988), and a novel *P. falciparum* rhoptry protein called PfRhop148 that is highly rich in Asn residues (Lobo et al., 2003). Yet another protein newly localized to the paired rhoptry organelles is a *P. falciparum* ortholog of stomatin, an oligomeric integral membrane protein that is apparently inserted into the newly formed PVM as the parasite invades (Hiller et al., 2003). In fact, there are likely to be many other as-yet-undiscovered proteins that function in rhoptries and invasion, as freeze-fracture studies of rhoptry membranes show a diversity, both in size and number, of intramembranous particles on the internal and external fracture faces (Bannister et al., 1986a), and the proteomic profile to date supports this likelihood (Sam-Yellowe et al., 2004).

The Merozoite Actomyosin Motor and Junctional Movement

Mechanistically, the process of a merozoite entering its host cell has intrigued malariologists for over 25 years, ever since real-time video microscopy (Dvorak et al., 1975) showed this to be a highly dynamic, albeit brief, event (summarized in Table 1). Recent advances from many biochemical studies on *Plasmodium* and related Apicomplexa indicate that movement into the host cell is based on an idiosyncratic actomyosin motor which together with accessory proteins is connected internally and externally to the cytoskeleton of the merozoite and RBC, respectively. Merozoite entry into an erythrocyte is clearly not a passive process; it requires motility and the expenditure of energy. Since the mature erythrocyte alone displays no phagocytic or receptor-mediated endocytic activities, most of the driving force and energy is probably derived from the parasite. However, erythrocytes devoid of ATP are not invaded (Dluzewski et al., 1983b). This requirement is thought to be linked to phosphorylation of RBC membrane cytoskeletal proteins and/or the maintenance of RBC membrane phospholipid asymmetry (Rangachari et al., 1986). Treatment of *P. knowlesi* and *P. falciparum* merozoites with cytochalasins prior to invasion prevents invasion, whereas prior treatment of erythrocytes does not (Miller et al., 1979; Field et al., 1993). Merozoites subjected to cytochalasin treatment are able to attach to RBCs and form a typical tight junction, but movement of the junction over the merozoite surface, and thus entry, is arrested. Cytochalasins are fungal products that bind actin and prevent its polymerization, thus preventing actin-linked motor function (MacLean-Fletcher and Pollard, 1980; Cooper, 1987). This strongly suggests that an actin-linked parasite molecular motor functions to drive the movement of the junction (Color Plate 7). In the related apicomplexan *T. gondii*, which in contrast invades host cells that have an actin-based motility system, it was clearly shown that host cell invasion and junctional movement are dependent upon parasite actin and not host cell actin (Dobrowolski and Sibley, 1996). Thus, mutations in the only actin gene of *T. gondii* resulted in phenotypes refractory to cytochalasin-mediated inhibition of invasion. *P. falciparum* was initially shown to have two divergent actin genes, *pf-actin I*, expressed throughout the life cycle, and *pf-actin II*, expressed only in the sexual stages (Wesseling et al., 1988; Wesseling et al., 1989), but there are now known to be seven actin or actin-like genes as judged by database analysis of the *P. falciparum* genome (http://plasmodb.org) and microarray analysis (Ben Mamoun et al., 2001; Bozdech et al., 2003; Le Roch et al., 2003).

Actin microfilaments per se at the junction zone are widely predicted but as yet have not been visualized by EM in the merozoite on the basis of their characteristic ultrastructure. However, based on indirect evidence, actin is present in the merozoites, with approximately one-third as soluble actin (G-actin, most of which appears to be ubiquitinated) and the remainder in the filamentous form (F-actin) (Field et al., 1993). More recently, F-actin distribution in the merozoite was demonstrated by the use of fluoresceinated derivatives of phalloidin, a fungal metabolite that generally binds specifically to F-actin (Cooper, 1987) and actin-specific antibodies (Webb et al., 1996). Fluorescence from the bound phalloidin and antibodies was confined to the poles and periphery of merozoites.

Given the strong evidence for actin-linked motility and the attractive proposition of a linear cellular motor operating during invasion (King, 1988), the obvious counterpart to look for to provide the motive force is myosin. Early attempts to recognize apicomplexan myosins were limited and sought antibody cross-reactivity with known myosins (Schwartzman and Pfefferkorn, 1983; Webb et al., 1996). This expanded rapidly with the application of molecular biology techniques and led to the identification of both *Toxoplasma* and *Plasmodium* myosins (Heintzelman and Schwartzman, 1997; Pinder et al., 1998; Hettmann et al., 2000), which constituted a new class of myosins (class XIV). Most interestingly, one of the smallest of the myosins, Pfmyo-A (Pfmyo-1), is implicated in the merozoite invasion process (Pinder et al., 1998). Protein expression analysis, immunofluorescence, and immuno-EM localization studies using antibodies specific for Pfmyo-A show this myosin to be expressed only in mature merozoites and located in the region of the plasma membrane and inner membrane complex (IMC). Furthermore, the relatively specific myosin ATPase inhibitor, BDM, inhibited invasion. Perhaps the most convincing evidence that myo-A is the invasion motor comes by inference from Meissner and colleagues (2002). An inducible protein expression system was generated in *T. gondii*, and it was shown that TgMyoA was essential both for host cell invasion and for gliding. Conservation of this motor protein applies equally between motile stages (i.e., merozoites, ookinetes [Margos et al., 2000], and sporozoites [Matuschewski et al., 2001]) and between species such that several elegant details describing it and its associated proteins have been advanced. These are largely from studies of *T. gondii* tachyzoites (Keeley and Soldati, 2004; Sibley, 2004; Soldati and Meissner, 2004), and *Plasmodium* sporozoites (Bergman et al., 2003), known for their gliding motility (Stewart and Vanderberg, 1988). Of the six myosins that *P. falciparum* is now known to have (Chaparro-Olaya et al., 2005), preliminary characterization of two of these shows that the smallest myosin, Pfmyo-B (relative molecular mass, 88 kDa), is expressed anteriorly only in mature merozoites (Chaparro et al., 2003), whereas the largest, Pfmyo-D (relative molecular mass, 245 kDa), is expressed in the trophozoite and/or schizont stages but not abundantly in mature merozoites. Whether either is involved in the invasion process is unclear at present.

The Moving Junction

A generalized scheme indicating how the merozoite generates a specialized junction zone to effect movement into the host RBC can be envisioned as illustrated by the diagrammatic representation of the molecular constituents of a moving junction (Color Plate 7a). It is assumed that the moving junction initially forms as an annulus at the extreme apical end of the IMC where it abuts with the uppermost polar ring. Myosin-A, present in the IMC, is a single-headed nonprocessive myosin that moves toward the positive ends (the growing or barbed end) of actin filaments (all myosins move in this direction on actin filaments with the exception of myosin VI) (Lister et al., 2004). Its tail associates with a connector, myosin A tail domain-interacting protein (Bergman et al., 2003), which is hypothesized to be linked to the IMC, perhaps via the recently discovered gliding-associated proteins (designated GAP50/GAP45), identified in *T. gondii* tachyzoites and for which homologs exist in *P. falciparum* (Gaskins et al., 2004). The motor end of myosin-A, the head, binds to actin, the latter almost certainly existing in the form of short filaments. Estimates indicate that there is enough actin to make $\sim$20 μm of F-actin (Field et al., 1993). If the junction is $\sim$100-nm deep and most of the G-actin is assembled into short filaments of this length, then $\sim$200 such filaments could exist longitudinally oriented around the circumferential junction zone. Actin also binds the glycolytic enzyme aldolase (Jewett and Sibley, 2003), and this is the proposed link to the cytoplasmic tail of a transmembrane parasite receptor molecule. For *P. falciparum* sporozoites, it is PfTRAP (thrombospondin-related anonymous protein) (Buscaglia et al., 2003); in *T. gondii*, it is TgMIC2 (Jewett and Sibley, 2003), a homolog of TRAP. Both are micronemal proteins. It remains to be determined if it is a micronemal adhesion protein(s) in malaria merozoites that binds to aldolase and potentially connects to the actomyosin motor; TRAP homologs have not been localized in merozoites with certainty. However, this ligand in merozoites could be EBL, AMA-1, or other micronemal proteins unidentified or recently characterized, such as *Plasmodium* thrombospondin-related apical merozoite protein (PTRAMP) (Thompson et al., 2001). Rather surprisingly, it has been found that the cytoplasmic domain of EBA-175 could be replaced by that of TRAP in *P. falciparum* and apparently remain fully functional in merozoite invasion (Gilberger et al., 2003a). Also of interest is a 60-kDa *P. falciparum* merozoite protein described 15 years ago. By immunofluorescent detection, the protein appeared to exhibit a mobile pattern during invasion (Klotz et al., 1989) and seemed to mimic the position of the electron-dense junction when viewed by EM. The protein was named merozoite-capping protein 1 (MCP-1), and the data suggest that MCP-1 could be a component of the junction or facilitate junctional movement. The putative *mcp-1* gene encodes a cytoplasmic protein, which has a sub-membrane localization in merozoites, with a domain homologous to oxidoreductases, a negatively charged domain, and a C-terminal positively charged domain, which could potentially mediate interactions with merozoite cytoskeletal elements, perhaps like aldolase (Hudson-Taylor et al., 1995). The parasite ligand would then be bound to a host cell receptor such as one of the glycophorins or band 3, which would be cleared of the zone of the developing PVM. The non-micronemal surface proteins, such as MSP-1 (or MSP-8), could also be involved through their C-terminal EGF domains with the moving junction (Goel et al., 2003).

The junction thus appears to be an annular structure fixed relative to the RBC, but one which can expand and contract circumferentially so that the merozoite can move through it. The power stroke of myosin can be envisaged to shift the merozoite IMC inwards by virtue of its associated complex of connecting and anchoring proteins, but how does the plasma membrane move past the junction into the incipient PV? It could be synchronously detached at the corresponding receptor-ligand attachment site after each power stroke, probably involving a proteolytic event or events (see the discussion below on proteases). The merozoite plasma membrane with clipped ligands could then flow past, allowing new ligands to occupy the now-vacant receptors. Alternatively, the receptor-ligand interactions could remain largely intact at the junction zone, as do the actin filaments. The myo-A power stroke would pull the IMC inwards; at the same time, the plasma membrane would filter through the junction zone. Some MSPs would have to be processed and/or sieved at the junctional zone. At present, though, we have no direct knowledge of what the connector(s) may be or, for that matter, of the actual composition of the moving junction. Although the binding of the EBLs through PkDBP or AMA-1 (in *P. knowlesi*) with their respective RBC receptors has been associated with junction information, it is not known if these molecules are actually present in the junction when it forms and moves (as alluded to above).

Proteolysis and Invasion

Parasite proteases are essential for merozoite invasion of erythrocytes (Blackman, 2000; Blackman, 2004). In addition to their apparent role in the rupture of PVM and erythrocyte membranes to release merozoites (reviewed in Bannister, 2001, and Lew, 2001, and discussed above), parasite proteases are essential for processing certain merozoite proteins to initiate invasion and to complete their entry into a new host cell. The first indication that proteases were directly necessary for invasion came from a study showing that chymostatin, an inhibitor of chymotrypsin-like proteases, but not leupeptin, an inhibitor of some serine and cysteine proteases, inhibited invasion by isolated invasive *P. knowlesi* merozoites (Hadley et al., 1983). The chymostatin-inhibited merozoites could initiate apical junctional formation, but further entry into the erythrocyte was prohibited. Interestingly, it could also be shown with *P. knowlesi* merozoites that other protease inhibitors prevented merozoite attachment. In *P. falciparum* merozoite invasion, both chymostatin and leupeptin were inhibitory, and inhibition by chymostatin could be partially reversed by pretreating erythrocytes with chymotrypsin (Dluzewski et al., 1986a). This suggested that the target of a chymostatin-sensitive parasite protease might be one or more of the membrane proteins exposed extracellularly at the erythrocyte surface. In contrast, inhibition of *P. falciparum* merozoite invasion by leupeptin was not reversed by treating the erythrocytes with chymotrypsin, suggesting that leupeptin may inhibit a different activity. Inhibitors of calpain I and calpain II, which are proteases that are calcium dependent, also appear to be potent inhibitors of merozoite invasion (Olaya and Wasserman, 1991). Interestingly, loading of erythrocyte ghosts with protease inhibitors also reduces invasion, thus suggesting proteolytic cleavage of internal cytoskeletal interactions may also be necessary for merozoite entry (Dluzewski et al., 1986b; Olaya and Wasserman, 1991). However, the inability to obtain free invasive merozoites from *P. falciparum* complicates direct corroboration of the observed effects, since these same inhibitors also inhibit merozoite release and perhaps development.

The experimental conditions and experiences with protease inhibitors described above would suggest that various proteases in the apical organelles play different roles during invasion, processing other proteins for release and activation such as AMA-1 (Narum and Thomas, 1994; Howell et al., 2001) or altering the erythrocyte architecture. One protease, called gp76, was characterized as an 83-Da protein, which was processed to a 76-kDa protein during schizogony (Braun-Breton and Pereira da Silva, 1988). Based upon an IFA pattern exhibited by an anti-gp76 MAb, it was believed to be localized in the rhoptries and activated with PI-specific phospholipase C (Braun-Breton and Pereira da Silva, 1988). Characterization of the specificity of the activities and inhibitor profiles of the gp76/gp68 protease indicated a chymotrypsin-like neutral serine protease. Consistent with chymotrypsin treatment of erythrocytes partially restoring invasion in the presence of chymostatin, the gp76 protease could cleave an outer loop of erythrocyte band 3 in intact erythrocytes, producing two fragments of 65 and 85 kDa. Importantly, this proteolytic activity partially reversed the chymostatin-induced inhibition of invasion (Braun-Breton et al., 1994; Roggwiller et al., 1996). Additionally, chymotrypsin or gp76 cleavage of band 3 induced erythrocyte membrane internalization and vacuolization (Tilley et al., 1990; Roggwiller et al., 1996), which suggests that gp76-mediated proteolysis of band 3 is a necessary process in invasion, to facilitate erythrocyte membrane invagination and PVM formation. In the context of the importance of band 3 for merozoite invasion, it is of interest that erythrocytes made deficient in band 3 become spheroidal in shape and are vulnerable to membrane blebbing and vesiculation, even though the cytoskeletal matrix remains essentially intact (Jay, 1996; Peters et al., 1996).

Antisera localized by IFA on merozoites a *P. berghei* p68 protease, which is different from the gp76/gp68 protease noted above, as an apical polar patch of fluorescence (Bernard et al., 1987). The p68 protease, which is inhibited by leupeptin and antipain, has been purified from both *P. berghei* and *P. falciparum* (Bernard et al., 1987; Grellier et al., 1989). Specific peptides capable of inhibiting its proteolytic activity were found to inhibit *P. falciparum* merozoite invasion in vitro (Mayer et al., 1991). The gene for p68 was recently characterized and was found to be an M1 zinc aminopeptidase, expressed as a 122-kDa protein diffusely distributed in trophozoites and processed to 90 and 68 kDa in schizonts, where in merozoites it becomes organized into a vesicle or single spot (Allary et al., 2002).

Many of the surface proteins are swept away from the surface of the merozoite at the point of contact with the moving junction as it advances over the merozoite surface (Ladda et al., 1969; Bannister et al., 1975; Aikawa et al., 1978; Mitchell and Bannister, 1988), a process facilitated by specific protease activity at the surface of the entering merozoite, which is necessary in merozoite invasion. The most comprehensive research into the biochemical basis for the clearance has focused on MSP-1 and the secondary proteolytic processing of its 42-kDa fragment of MSP-1 into p19 and p33 fragments, which is necessary for invasion to be completed (Holder, 1988; Holder et al., 1994). The GPI-anchored p19 fragment is carried into the erythrocyte with the invading merozoite, while the p33 fragment, along with other MSP-1 fragments and other MSPs of the complex, is shed from the surface (Blackman et al., 1990; Blackman et al., 1991; Blackman and Holder, 1992; Blackman et al., 1996; Stafford et al., 1996). The processing of AMA-1 by proteases and the relationship of the processing steps to the movement of this ligand from the micronemes to the surface of the merozoite have also been well studied (Narum and Thomas, 1994; Howell et al., 2001; Dutta et al., 2003; Howell et al., 2003). While processing of MSP-1 and AMA-1 proteins at the surface of merozoites is well characterized, the exact timing and mechanism of the primary and secondary processing are still to be evaluated (Blackman, 2004).

However, recent research indicates that a common "sheddase" enzyme operates to secondarily cleave both MSP1 and AMA1 (Howell et al., 2003), leaving the C-terminal fragments secured in the merozoite membrane to be internalized and releasing the majority of the corresponding proteins and protein complexes. A subtilisin-like serine protease (SUB-2), shown to be important during invasion (Barale et al., 1999; Hackett et al., 1999) and essential for the parasite (Uzureau et al., 2004), has been proposed for this function, possibly acting in concert with the moving junction (Blackman, 2004). However, SUB-2 localization in the dense granules is problematic in this regard unless it can be shown that some of these organelles release their contents onto the merozoite surface prior to completing invasion. Another possibility for a sheddase is one of the *Plasmodium* membrane-bound rhomboid proteases, which in *T. gondii* have been shown to clip off MIC-2 during invasion (Urban and Freeman, 2003).

What is not clear at this time is whether the processing is important for attachment or if processing mainly acts to facilitate movement of the junction and merozoite entry. One proposal is that this processing allows the EGF-like domains, which are known in other proteins to participate in molecular adhesion, to interact with a host or some other parasite molecule (Holder et al., 1994). An alternative possibility is that the processing allows the disengagement of an interaction between the 42-kDa fragment and another molecule. EGF-like domains in adhesins, such as the selectins, often act in cooperation with other nearby peptide domains to facilitate specific receptor-ligand interactions. The separation of two such domains could abolish or decrease an interaction. Since the same protease is involved in the specific processing of AMA-1, a similar action may also occur for this invasion ligand and perhaps for other MSPs such as MSP-4, MSP-5, and MSP-8 that have EGF domains.

Apical Organelle Secretion—Composition and Formation of the PV

As is evident from the preceding commentary, the specialized secretory organelles that merozoites have evolved at their anterior end (Fig. 1 and 2; Color Plate 7b) are discharged sequentially during the course of invasion (Sibley, 2003). It is noteworthy that while cytochalasins inhibit invasion by interfering with movement of the tight junction but not its formation, these agents do not inhibit microneme or rhoptry secretion (Fig. 5a). Micronemes are the smallest of the organelles, and the cargo of these tubular vesicles as discussed above includes AMA-1, DBP, EBA-175, and other EBL family proteins, as well as PTRAMPs, perhaps some RBL family members, and likely other unknown coligands. In this context, it is a matter of debate as to whether all micronemes contain the complete repertoire of micronemal proteins or whether there are subsets of micronemes with specific contents. This question arises both from localization studies and from a consideration of the biological characteristics of the micronemal proteins, notably AMA-1, DBP, and EBA-175. When released from schizonts, some free merozoites and mature merozoites of segmented schizonts display AMA-1 around the cell surface (Narum and Thomas, 1994), while in others it is apparent that AMA-1 is retained in the micronemes, indicating that at least some micronemes have secreted their contents. However, under these same conditions neither DBP nor EBA-175 is observed over the surface or apparently capping the merozoite apex. This suggests there is a mechanism for differential secretion of compartments within a microneme or that different micronemes have different contents that are induced to secrete by separate signals. In *T. gondi* and *Cryptosporidium parvum*, microneme secretion and invasion are effected by elevation of cytosolic calcium (Lovett and Sibley, 2003; Chen et al., 2004). However, when AMA-1 appears on the surface of free merozoites, it is not known if this is a normal viable state or whether the microneme contents have been secreted prematurely. It is clear that micronemes secrete their contents from the apical prominence of the merozoite, but whether they achieve this by fusing with the rhoptry duct (Mitchell and Bannister, 1988) or are open individually at the apical end is also still a matter of debate. As noted in numerous ultrastructural studies of invasion described above, rhoptries appear to secrete their contents when close apical contact is made, presumably secondarily to microneme secretion, at or just after the initiation of the tight junction and as the junction moves back (relatively) over the merozoite. Some EMs show lamellar-like material deposited in the developing vacuole, while others show a filmy band extending from the apical opening of the rhoptry duct that connects the merozoite to the enlarging PV membrane when the junction has moved posteriorly over the merozoite and away from the apical prominence. This filmy band seems to represent material emanating from the rhoptries (Bannister et al., 1975; Aikawa et al., 1978; Aikawa and Miller, 1983; Bannister et al., 1986a; Bannister and Mitchell, 1989), giving rise to the notion that rhoptries contribute to PVM formation by providing both membrane remodeling proteins and membrane lipid material.

Among many past controversial points regarding merozoite invasion, the composition of the PVM and how it comes about are two interrelated questions that remain, but recent studies of both *Plasmodium* and *Toxoplasma* are providing some answers. One hypothesis contends that the PVM is essentially formed from the invagination of the erythrocyte membrane. An alternative hypothesis asserts that the PVM is primarily formed by the merozoite's release of membranous secretions. The truth lies somewhere in the middle.

The developing vacuolar membrane of invading merozoites, as opposed to the uninvolved RBC membrane beyond the junction, is cleared of most IMPs (McLaren et al., 1979; Aikawa et al., 1981). The major erythrocyte membrane proteins, band 3 and glycophorin A, and cytoskeletal proteins such as spectrin are absent or much reduced in the PVM, which is consistent with a lower IMP content (Atkinson et al., 1988; Dluzewski et al., 1988; Dluzewski et al., 1989; Haldar, 1992; Ward et al., 1993). An increased local concentration of IMP is thought to be a major contributor to the appearance of the thickened electron-dense region under the erythrocyte membrane at the site of the junction (Aikawa et al., 1981; Mitchell and Bannister, 1988). The implication from these observations is that the moving junction is able to clear the invaginating erythrocyte membrane of surface proteins as it traverses the merozoite surface, acting as a weir and allowing host lipid to flow into the forming PVM. To allow this action, cytoskeletal protein interactions and attachments of certain membrane proteins to the cytoskeleton proteins would have to be disrupted. Interestingly, mechanically induced deformation of normal RBC membranes results in vesiculation, uncoupling the cytoskeleton from the lipid bilayer; as the cytoskeleton retracts from the area of deformation and vesiculation, band 3 and glycophorin A are depleted (Knowles et al., 1997). Concomitantly, in the regions of membrane deformation and vesiculation, phosphatidylethanolamine and GPI-linked CD59 become enriched (Knowles et al., 1997).

Other evidence for a host cell origin of the PVM comes from experiments using free invasive *P. knowlesi* merozoites and fluorescent lipids incorporated into the membranes of targeted erythrocytes (Ward et al., 1993). As merozoites invaded the cells, no difference was found in fluorescent intensity between the erythrocyte membrane and the developing PVM, suggesting that the PVM was composed of host-derived lipid. Another study also showed that *P. falciparum* parasites are surrounded by fluorescent lipid after invasion of tagged erythrocytes (Haldar and Uyetake, 1992), suggesting that host cell membrane lipids are substantially involved in PVM formation. The notion that PVM formation is mostly of parasite origination implies that the area of the PVM increases because lipid membranous material is secreted by the parasite, with the junction remaining in place at a localized site on the erythrocyte membrane and the motive force at the junction ratcheting the merozoite forward into the expanding PV. Most observations giving rise to the suggestion that the rhoptries may store and extrude lipid membranes come from EMs. By EM, fixation in tannic acid of free viable merozoites shows membranous whorls in the rhoptry and whorled material extruding from the apical duct (Bannister et al., 1986a; Stewart et al., 1986). Membranous vacuoles and tubular structures have also been seen in the erythrocyte cytoplasm extending from the site of junction and nascent PV formation, as well as transparent, amorphous connections between the nascent enlarging vacuole membrane and the rhoptry duct of the invading merozoite (Aikawa et al., 1978; Koontz et al., 1979; Bannister and Mitchell, 1989). However, direct evidence of the lipidic content of the whorls and lamellar structures has not yet been reported.

Similarly, we know individual rhoptry proteins and protein complexes, as discussed earlier, which appear to be associated with membrane-like secretions, bind to erythrocyte membrane and become incorporated into the PV-PVM. Furthermore, NBD-phosphatidylethanolamine phospholipid incorporated into the membrane of erythrocytes and visualized by immuno-EM was largely excluded from the PVM surrounding newly invaded *P. falciparum* parasites and from the zone of attachment of *P. knowlesi* merozoites (Mikkelsen et al., 1988; Dluzewski et al., 1992), which is in contrast to the experiments mentioned above. In a reverse experiment, *P. falciparum* merozoites were labeled with fluorescent fatty acids with the fluorescence concentrated at the apical ends of the merozoites. Within the limitations of the resolution of microscopic images of fixed invading merozoites, the parasite-derived lipid appeared to be inserted during invasion into the erythrocyte membrane at the point of contact, and fluorescence was noted in the PVM surrounding newly invaded parasites (Mikkelsen et al., 1988). Measurement of the physical area of the membrane of infected erythrocytes versus that of normal erythrocytes found no demonstrable differences, even with multiple invasions (Dluzewski et al., 1995). These sets of observations tend to support the idea that the PVM is also formed by, or at least modified by, parasite-derived lipid and protein.

If one assumes that host cell lipid and parasite rhoptry components ultimately comprise the PV, where and how do the two come together? There could be membrane fusion as the rhoptry contents exit at the initial junction point and continue to flow as invasion proceeds, or the rhoptries could insert micelles, which fuse back with the host lipid, or a combination of both could occur. Support for the former mechanism is evident for *T. gondii*, in which cytochalasin D-treated tachyzoite rhoptry-derived vacuoles have been shown to fuse back with the PVM (Hakansson et al., 2001). Host cell cholesterol is also important for parasite invasion and vacuole formation, since host cells depleted of cholesterol are refractory to *T. gondii* tachyzoite invasion, whereas those with cholesterol added back were not (Coppens and Joiner, 2003). Detailed examination of the modulatory effect of cholesterol on parasite invasion showed that only host cholesterol and not parasite rhoptry cholesterol was recruited to the PVM. If the requirements for vacuole formation are similar for *P. falciparum* merozoites, then this would be the simplest explanation for an earlier finding of an observed inhibition of invasion when cholesterol-depleted RBCs were employed (Dluzewski et al., 1985). More recent investigations support the role of cholesterol in merozoite invasion and PVM formation. Cholesterol-rich DRM lipids or rafts have been implicated as a factor in the selective recruitment of 10 membrane proteins into the developing and expanding PVM and exclusion of other erythrocyte membrane proteins such as band 3, band 7, and glycophorin A (Lauer et al., 2000; Samuel et al., 2001; Murphy et al., 2004). However, why would a host cell DRM-associated stomatin, band 7, be excluded from the developing PVM while a parasite stomatin in rhoptry DRM rafts and the RhopH complex be inserted into the forming PVM (Hiller et al., 2003)?

Signaling the Invasion Cascade

Given the molecular complexity of merozoite invasion, a question that needs to be addressed is how the molecular events are coordinated. While it must be assumed that signals are provided that cause microneme or rhoptry release at the appropriate moment, little is actually known about the nature of those proposed signals. Similarly, are there a series of biochemical events that trigger a switch that activates the actomyosin motor to move the tight junction? Most of these questions remain open to future investigations relating to merozoite invasion. Staurosporine, an inhibitor of serine/threonine kinases, profoundly inhibited invasion by *P. knowlesi* merozoites but did not prevent attachment and initial organization of the junction (Ward et al., 1994). Based upon the known effects of staurosporine on the phosphorylation of actin and actin motor organization, it was proposed that a protein kinase C is involved in junctional movement. Okadaic acid, an inhibitor of serine/threonine phosphatases 1 and 2A, had variable effects on *P. knowlesi* invasion but inhibited *P. falciparum* merozoite invasion. The fact that the two inhibitors can also affect intracellular growth of *P. falciparum* complicates the experimental results (Dluzewski and Garcia, 1996).

Between 65 and 96 kinase genes have been identified in the *P. falciparum* genome, with Ca^{2+}-dependent kinase family members being well represented (Ward et al., 1994; Anamika et al., 2005). One form of a *P. falciparum* kinase, PfPK4, has been found to have a punctate distribution in merozoites, colocalizing with AMA-1 but not with RAP-1 (Mohrle et al., 1997). Although the localization would suggest a micronemal location, we do not know if PK4 is operational in the merozoite in the context of invasion. It has been known for some time now that phosphorylation of erythrocyte membrane cytoskeletal proteins and ATP levels in the RBC can profoundly influence RBC invasion by merozoites and that a membrane-bound cyclic AMP-independent kinase-targeting spectrin was critical (Dluzewski et al., 1983a; Rangachari et al., 1986). Most recently, evidence was presented that indicated signaling via the RBC β2-adrenergic receptor and heterotrimeric guanine-binding protein influenced the ability of *P. falciparum* merozoites to invade erythrocytes (Harrison et al., 2003). Agonistic stimulation of increased cyclic AMP levels led to increased invasion; such induced signaling may be behind the rapid reorganization of the RBC membrane at the point of invasion.

If Ca^{2+} is chelated, then RBC invasion by isolated free and invasive *P. knowlesi* merozoites is inhibited (Johnson et al., 1981). An investigation on *P. falciparum* merozoite invasion concluded that the requirement for Ca^{2+} was extracellular and not intracellular (McCallum-Deighton and Holder, 1992). However, this could be relevent to a Ca^{2+}-dependent serine protease on the merozoite surface that processes MSP-1 (Blackman and Holder, 1992). It follows that since this processing is required for invasion to proceed, this Ca^{2+} dependence would be external. Some studies have suggested intracellular Ca^{2+} and modulators of Ca^{2+} levels such as calmodulin are involved in *P. falciparum* merozoite invasion (Matsumoto et al., 1987; Wasserman, 1990; Wasserman and Chaparro, 1996). However, more-detailed studies with specific inhibitors of intracellular Ca^{2+} flux and mediators of signal transduction reveal a role and requirement for intracellular calcium-mediated signal transduction in invasion by other apicomplexan parasites. Intracellular Ca^{2+} and activation of signal transduction pathways occur in *Theileria parva* sporozoites during invasion of a host cell (Shaw, 1997). Several studies have shown that intracellular levels and mobilization of parasite Ca^{2+} influence host cell invasion and stimulate the discharge of microneme proteins such as AMA-1 in *T. gondii* tachyzoites and *C. parvum* sporozoites (Carruthers et al., 1999; Donahue et al., 2000; Vieira and Moreno, 2000; Lovett and Sibley, 2003; Chen et al., 2004). Whether transduced signals mediate a coordinated release of the contents of malaria merozoite rhoptry and microneme organelles remains unanswered; similarly, the type of interactions that could induce the proposed signal transduction processes necessary to initiate and coordinate a molecular cascade for invasion is an open question.

Completion of Entry—PV and Erythrocyte Membrane Closure

When the merozoite is inside its new host cell, it is entirely enclosed by the PVM, with the junction remaining for a period of time at the posterior end of the merozoite where it fuses to close any connection between the PVM and the erythrocyte membrane (Aikawa et al., 1978; Aikawa and Miller, 1983). At this stage, the microspheres (dense bodies), the third set of secretory organelles, discharge their contents. Ultrastructural evidence has most often indicated that in the immediate postentry period, these rounded bodies move to the periphery of the merozoite, fuse with the plasmalemma, and secrete their granular contents externally into the PV (Bannister et al., 1975; Bannister and Mitchell, 1989; Torii et al., 1989). This fusion and release often distort the PVM in the process (Bannister et al., 1975; Torii et al., 1989). If dense granules are to play a role during invasion and not just afterward, this membrane fusion and release would have to occur before the merozoite is fully internalized, which has not yet been observed. However, perhaps one indication that this phenomenon may occur during invasion comes from studies on the first dense granule protein to be identified, the ring-infected erythrocyte surface antigen (RESA) (Aikawa et al., 1990; Culvenor et al., 1991), in which anti-RESA antibodies were found to inhibit invasion (Anders et al., 1987). The function of this protein is unknown, but ultimately it ends up in association with the cytoskeleton on the cytoplasmic surface of the RBC after release.

Although many proteins have been identified in the spheroidal dense bodies of *T. gondii*, only a few proteins besides RESA have been located in the dense bodies of *Plasmodium*. One, which has been termed the ring membrane antigen, is a 14-kDa protein that remains localized to the early trophozoite plasma membrane (Trager et al., 1992). The only other dense granule proteins are two subtilisin serine proteases, SUB-1 and SUB-2, the latter structurally also a type I integral membrane protein (Blackman et al., 1998; Barale et al., 1999; Hackett et al., 1999). One of the postulated functions of these proteases, especially of SUB-2, is to be the surface exposed sheddase protease responsible for processing of AMA-1, MSP-1, and perhaps other surface coat proteins (Withers-Martinez et al., 2004). However, to perform this role they would have to be released at the beginning of invasion and not at the end when dense granule discharge has been observed (Torii et al., 1989; Culvenor et al., 1991; Trager et al., 1992). There are very likely other proteins and chemicals yet to be found in the *Plasmodium* dense granules; at present, these organelles appear to function in altering the environment of the internalized merozoite and assisting in the transition to a growing trophozoite.

ACKNOWLEDGMENTS

Special thanks are extended to Lawrie H. Bannister, John M. Hopkins, Gabriele Margos (KCL), Esmeralda V.-S. Meyer (EVC), and members of the malaria research laboratories at the Emory Vaccine Center at Yerkes and King's College London for their generous contributions toward the preparation of the electron and light microscopy images. Dan Osborn, Simon Hughes (KCL), and Bill Jarra (National Institute for Medical Research, London, United Kingdom) provided assistance in generating the images shown in Color Plate 6.

A.R.D. received financial support from The Wellcome Trust (grant no. 069515), and M.R.G. is supported by NIH grant no. AI55994, AI24710, AI52371, and AI35804 and the Yerkes National Primate Research Center base grant no. RR-00165 awarded by the Comparative Medicine Program, National Center for Research Resources, National Institutes of Health.

REFERENCES

Adams, J. H., P. L. Blair, O. Kaneko, and D. S. Peterson. 2001. An expanding *ebl* family of *Plasmodium falciparum*. *Trends Parasitol.* **17:**297–299.

Adams, J. H., D. E. Hudson, M. Torii, G. E. Ward, T. E. Wellems, M. Aikawa, and L. H. Miller. 1990. The Duffy receptor family of *Plasmodium knowlesi* is located within the micronemes of invasive malaria merozoites. *Cell* **63:**141–153.

Adams, J. H., B. K. Sim, S. A. Dolan, X. Fang, D. C. Kaslow, and L. H. Miller. 1992. A family of erythrocyte binding proteins of malaria parasites. *Proc. Natl. Acad. Sci. USA* **89:**7085–7089.

Aikawa, M. 1966. The fine structure of the erythrocytic stages of three avian malarial parasites, *Plasmodium fallax*, *P. lophurae*, and *P. cathemerium*. *Am. J. Trop. Med. Hyg.* **15:**449–471.

Aikawa, M. 1971. Parasitological review. *Plasmodium*: the fine structure of malarial parasites. *Exp. Parasitol.* **30:**284–320.

Aikawa, M., P. K. Hepler, C. G. Huff, and H. Sprinz. 1966. The feeding mechanism of avian malarial parasites. *J. Cell Biol.* **28:**355–373.

Aikawa, M., C. G. Huff, and H. Sprinz. 1967. Fine structure of the asexual stages of *Plasmodium elongatum*. *J. Cell Biol.* **34:**229–249.

Aikawa, M., and L. H. Miller. 1983. Structural alteration of the erythrocyte membrane during malarial parasite invasion and intraerythrocytic development. *Ciba Found. Symp.* **94:**45–63.

Aikawa, M., L. H. Miller, J. Johnson, and J. Rabbege. 1978. Erythrocyte entry by malarial parasites. A moving junction between erythrocyte and parasite. *J. Cell Biol.* **77:**72–82.

Aikawa, M., L. H. Miller, J. R. Rabbege, and N. Epstein. 1981. Freeze-fracture study on the erythrocyte membrane during malarial parasite invasion. *J. Cell Biol.* **91:**55–62.

Aikawa, M., and C. Sterling. 1974. High voltage electron microscopy on microgametogenesis of *Haemoproteus columbae*. *Z. Zellforsch. Mikrosk. Anat.* **147:**353–360.

Aikawa, M., M. Torii, A. Sjolander, K. Berzins, P. Perlmann, and L. H. Miller. 1990. Pf155/RESA antigen is localized in dense granules of *Plasmodium falciparum* merozoites. *Exp. Parasitol.* **71:**326–329.

Aley, S. B., C. T. Atkinson, M. Aikawa, W. L. Maloy, and M. R. Hollingdale. 1987. Ultrastructural localization of *Plasmodium falciparum* circumsporozoite protein in newly invaded hepatoma cells. *J. Parasitol.* **73:**1241–1245.

Allary, M., J. Schrevel, and I. Florent. 2002. Properties, stage-dependent expression and localization of *Plasmodium falciparum* M1 family zinc-aminopeptidase. *Parasitology* **125:**1–10.

Anamika, N. Srinivasan, and A. Krupa. 2005. A genomic perspective of protein kinases in *Plasmodium falciparum*. *Proteins* **58:**180–189.

Anders, R. F., L. J. Murray, L. M. Thomas, K. M. Davern, G. V. Brown, and D. J. Kemp. 1987. Structure and function of candidate vaccine antigens in *Plasmodium falciparum*. *Biochem. Soc. Symp.* **53:**103–114.

Aoki, S., J. Li, S. Itagaki, B. A. Okech, T. G. Egwang, H. Matsuoka, N. M. Palacpac, T. Mitamura, and T. Horii. 2002. Serine repeat antigen

(SERA5) is predominantly expressed among the SERA multigene family of *Plasmodium falciparum*, and the acquired antibody titers correlate with serum inhibition of the parasite growth. *J. Biol. Chem.* **277**:47533–47540.

Badell, E., V. Pasquetto, W. Eling, A. Thomas, and P. Druilhe. 1995. Human *Plasmodium* liver stages in SCID mice: a feasible model? *Parasitol. Today* **11**:169–171.

Baldi, D. L., K. T. Andrews, R. F. Waller, D. S. Roos, R. F. Howard, B. S. Crabb, and A. F. Cowman. 2000. RAP1 controls rhoptry targeting of RAP2 in the malaria parasite *Plasmodium falciparum*. *EMBO J.* **19**:2435–2443.

Baldi, D. L., R. Good, M. T. Duraisingh, B. S. Crabb, and A. F. Cowman. 2002. Identification and disruption of the gene encoding the third member of the low-molecular-mass rhoptry complex in *Plasmodium falciparum*. *Infect. Immun.* **70**:5236–5245.

Bannister, L. H. 2001. Looking for the exit: how do malaria parasites escape from red blood cells? *Proc. Natl. Acad. Sci. USA* **98**:383–384.

Bannister, L. H., G. A. Butcher, E. D. Dennis, and G. H. Mitchell. 1975. Structure and invasive behaviour of *Plasmodium knowlesi* merozoites in vitro. *Parasitology* **71**:483–491.

Bannister, L. H., and A. R. Dluzewski. 1990. The ultrastructure of red cell invasion in malaria infections: a review. *Blood Cells* **16**:257–292.

Bannister, L. H., J. M. Hopkins, A. R. Dluzewski, G. Margos, I. T. Williams, M. J. Blackman, C. H. Kocken, A. W. Thomas, and G. H. Mitchell. 2003. *Plasmodium falciparum* apical membrane antigen 1 (PfAMA-1) is translocated within micronemes along subpellicular microtubules during merozoite development. *J. Cell Sci.* **116**:3825–3834.

Bannister, L. H., J. M. Hopkins, R. E. Fowler, S. Krishna, and G. H. Mitchell. 2000a. A brief illustrated guide to the ultrastructure of *Plasmodium falciparum* asexual blood stages. *Parasitol. Today* **16**:427–433.

Bannister, L. H., J. M. Hopkins, R. E. Fowler, S. Krishna, and G. H. Mitchell. 2000b. Ultrastructure of rhoptry development in *Plasmodium falciparum* erythrocytic schizonts. *Parasitology* **121**:273–287.

Bannister, L. H., and G. H. Mitchell. 1989. The fine structure of secretion by *Plasmodium knowlesi* merozoites during red cell invasion. *J. Protozool.* **36**:362–367.

Bannister, L. H., and G. H. Mitchell. 1995. The role of the cytoskeleton in *Plasmodium falciparum* merozoite biology: an electron-microscopic view. *Ann. Trop. Med. Parasitol.* **89**:105–111.

Bannister, L. H., G. H. Mitchell, G. A. Butcher, and E. D. Dennis. 1986a. Lamellar membranes associated with rhoptries in erythrocytic merozoites of *Plasmodium knowlesi*: a clue to the mechanism of invasion. *Parasitology* **92**:291–303.

Bannister, L. H., G. H. Mitchell, G. A. Butcher, E. D. Dennis, and S. Cohen. 1986b. Structure and development of the surface coat of erythrocytic merozoites of *Plasmodium knowlesi*. *Cell Tissue Res.* **245**:281–290.

Barale, J. C., T. Blisnick, H. Fujioka, P. M. Alzari, M. Aikawa, C. Braun-Breton, and G. Langsley. 1999. *Plasmodium falciparum* subtilisin-like protease 2, a merozoite candidate for the merozoite surface protein 1–42 maturase. *Proc. Natl. Acad. Sci. USA* **96**:6445–6450.

Barnwell, J. W., and M. R. Galinski. 1991. The adhesion of malaria merozoite proteins to erythrocytes: a reflection of function? *Res. Immunol.* **142**:666–672.

Barnwell, J. W., M. R. Galinski, S. G. DeSimone, F. Perler, and P. Ingravallo. 1999. *Plasmodium vivax*, *P. cynomolgi*, and *P. knowlesi*: identification of homologue proteins associated with the surface of merozoites. *Exp. Parasitol.* **91**:238–249.

Barnwell, J. W., M. E. Nichols, and P. Rubinstein. 1989. In vitro evaluation of the role of the Duffy blood group in erythrocyte invasion by *Plasmodium vivax*. *J. Exp. Med.* **169**:1795–1802.

Barnwell, J. W., and S. P. Wertheimer. 1989. *Plasmodium vivax*: merozoite antigens, the Duffy blood group, and erythrocyte invasion. *Prog. Clin. Biol. Res.* **313**:1–11.

Baum, J., M. Pinder, and D. J. Conway. 2003. Erythrocyte invasion phenotypes of *Plasmodium falciparum* in The Gambia. *Infect. Immun.* **71**:1856–1863.

Ben Mamoun, C., I. Y. Gluzman, C. Hott, S. K. MacMillan, A. S. Amarakone, D. L. Anderson, J. M. Carlton, J. B. Dame, D. Chakrabarti, R. K. Martin, B. H. Brownstein, and D. E. Goldberg. 2001. Co-ordinated programme of gene expression during asexual intraerythrocytic development of the human malaria parasite *Plasmodium falciparum* revealed by microarray analysis. *Mol. Microbiol.* **39**:26–36.

Benet, A., L. Tavul, J. C. Reeder, and A. Cortes. 2004. Diversity of *Plasmodium falciparum* vaccine candidate merozoite surface protein 4 (MSP4) in a natural population. *Mol. Biochem. Parasitol.* **134**:275–280.

Bennett, V., and S. Lambert. 1991. The spectrin skeleton: from red cells to brain. *J. Clin. Investig.* **87**:1483–1489.

Bergman, L. W., K. Kaiser, H. Fujioka, I. Coppens, T. M. Daly, S. Fox, K. Matuschewski, V. Nussenzweig, and S. H. Kappe. 2003. Myosin A tail domain interacting protein (MTIP) localizes to the inner membrane complex of *Plasmodium* sporozoites. *J. Cell Sci.* **116**:39–49.

Bernard, F., R. Mayer, I. Picard, A. Deguercy, M. Monsigny, and J. Schrevel. 1987. *Plasmodium berghei* and *Plasmodium chabaudi*: a neutral endopeptidase in parasite extracts and plasma of infected animals. *Exp. Parasitol.* **64:**95–103.

Binks, R. H., and D. J. Conway. 1999. The major allelic dimorphisms in four *Plasmodium falciparum* merozoite proteins are not associated with alternative pathways of erythrocyte invasion. *Mol. Biochem. Parasitol.* **103:**123–127.

Black, C. G., J. W. Barnwell, C. S. Huber, M. R. Galinski, and R. L. Coppel. 2002. The *Plasmodium vivax* homologues of merozoite surface proteins 4 and 5 from Plasmodium falciparum are expressed at different locations in the merozoite. *Mol. Biochem. Parasitol.* **120:**215–224.

Black, C. G., L. Wang, A. R. Hibbs, E. Werner, and R. L. Coppel. 1999. Identification of the *Plasmodium chabaudi* homologue of merozoite surface proteins 4 and 5 of *Plasmodium falciparum*. *Infect. Immun.* **67:**2075–2081.

Black, C. G., L. Wang, A. E. Topolska, D. I. Finkelstein, M. K. Horne, A. W. Thomas, N. Mohandas, and R. L. Coppel. 2004. Merozoite surface proteins 4 and 5 of *Plasmodium knowlesi* have differing cellular localisation and association with lipid rafts. *Mol. Biochem. Parasitol.* **138:**153–158.

Black, C. G., L. Wang, T. Wu, and R. L. Coppel. 2003. Apical location of a novel EGF-like domain-containing protein of *Plasmodium falciparum*. *Mol. Biochem. Parasitol.* **127:**59–68.

Black, C. G., T. Wu, L. Wang, A. R. Hibbs, and R. L. Coppel. 2001. Merozoite surface protein 8 of *Plasmodium falciparum* contains two epidermal growth factor-like domains. *Mol. Biochem. Parasitol.* **114:**217–226.

Blackman, M. J. 2004. Proteases in host cell invasion by the malaria parasite. *Cell. Microbiol.* **6:**893–903.

Blackman, M. J. 2000. Proteases involved in erythrocyte invasion by the malaria parasite: function and potential as chemotherapeutic targets. *Curr. Drug Targets* **1:**59–83.

Blackman, M. J., and L. H. Bannister. 2001. Apical organelles of Apicomplexa: biology and isolation by subcellular fractionation. *Mol. Biochem. Parasitol.* **117:**11–25.

Blackman, M. J., E. D. Dennis, E. M. Hirst, C. H. Kocken, T. J. Scott-Finnigan, and A. W. Thomas. 1996. *Plasmodium knowlesi*: secondary processing of the malaria merozoite surface protein-1. *Exp. Parasitol.* **83:**229–239.

Blackman, M. J., H. Fujioka, W. H. Stafford, M. Sajid, B. Clough, S. L. Fleck, M. Aikawa, M. Grainger, and F. Hackett. 1998. A subtilisin-like protein in secretory organelles of *Plasmodium falciparum* merozoites. *J. Biol. Chem.* **273:**23398–23409.

Blackman, M. J., H. G. Heidrich, S. Donachie, J. S. McBride, and A. A. Holder. 1990. A single fragment of a malaria merozoite surface protein remains on the parasite during red cell invasion and is the target of invasion-inhibiting antibodies. *J. Exp. Med.* **172:**379–382.

Blackman, M. J., and A. A. Holder. 1992. Secondary processing of the *Plasmodium falciparum* merozoite surface protein-1 (MSP1) by a calcium-dependent membrane-bound serine protease: shedding of MSP133 as a noncovalently associated complex with other fragments of the MSP1. *Mol. Biochem. Parasitol.* **50:**307–315.

Blackman, M. J., H. Whittle, and A. A. Holder. 1991. Processing of the *Plasmodium falciparum* major merozoite surface protein-1: identification of a 33-kilodalton secondary processing product which is shed prior to erythrocyte invasion. *Mol. Biochem. Parasitol.* **49:**35–44.

Blair, P. L., S. H. Kappe, J. E. Maciel, B. Balu, and J. H. Adams. 2002a. *Plasmodium falciparum* MAEBL is a unique member of the *ebl* family. *Mol. Biochem. Parasitol.* **122:**35–44.

Blair, P. L., A. Witney, J. D. Haynes, J. K. Moch, D. J. Carucci, and J. H. Adams. 2002b. Transcripts of developmentally regulated *Plasmodium falciparum* genes quantified by real-time RT-PCR. *Nucleic Acids Res.* **30:**2224–2231.

Borre, M. B., C. A. Owen, J. K. Keen, K. A. Sinha, and A. A. Holder. 1995. Multiple genes code for high-molecular-mass rhoptry proteins of *Plasmodium yoelii*. *Mol. Biochem. Parasitol.* **70:**149–155.

Bozdech, Z., J. Zhu, M. P. Joachimiak, F. E. Cohen, B. Pulliam, and J. L. DeRisi. 2003. Expression profiling of the schizont and trophozoite stages of *Plasmodium falciparum* with a long-oligonucleotide microarray. *Genome Biol.* **4:**R9.

Braun-Breton, C., T. Blisnick, M. E. Morales-Betoulle, J. C. Barale, and G. Langsley. 1994. Malaria parasites: enzymes involved in red blood cell invasion. *Braz. J. Med. Biol. Res.* **27:**363–367.

Braun-Breton, C., and L. Pereira da Silva. 1988. Activation of a *Plasmodium falciparum* protease correlated with merozoite maturation and erythrocyte invasion. *Biol. Cell* **64:**223–231.

Brown, D. A., and J. K. Rose. 1992. Sorting of GPI-anchored proteins to glycolipid-enriched membrane subdomains during transport to the apical cell surface. *Cell* **68:**533–544.

Brown, H. J., and R. L. Coppel. 1991. Primary structure of a *Plasmodium falciparum* rhoptry antigen. *Mol. Biochem. Parasitol.* **49:**99–110.

Burkhard, P., J. Stetefeld, and S. V. Strelkov. 2001. Coiled coils: a highly versatile protein folding motif. *Trends Cell Biol.* **11:**82–88.

Burns, J. M., E. K. Adeeku, C. C. Belk, and P. D. Dunn. 2000. An unusual tryptophan-rich domain characterizes two secreted antigens of *Plasmodium yoelii*-infected erythrocytes. *Mol. Biochem. Parasitol.* **110:**11–21.

Buscaglia, C. A., I. Coppens, W. G. Hol, and V. Nussenzweig. 2003. Sites of interaction between aldolase and thrombospondin-related anonymous protein in *Plasmodium*. *Mol. Biol. Cell* **14**:4947–4957.

Bushell, G. R., J. A. Cooper, L. T. Ingram, L. Schofield, A. Saul, R. J. Epping, S. Chiu, S. Jelacic, and J. A. Upcroft. 1986. Identification of key antigens of *Plasmodium falciparum* as vaccine candidates. *P. N. G. Med. J.* **29**:69–73.

Butcher, G. A., G. H. Mitchell, and S. Cohen. 1973. Mechanism of host specificity in malarial infection. *Nature* **244**:40–41. (Letter.)

Bzik, D. J., W. B. Li, T. Horii, and J. Inselburg. 1988. Amino acid sequence of the serine-repeat antigen (SERA) of *Plasmodium falciparum* determined from cloned cDNA. *Mol. Biochem. Parasitol.* **30**:279–288.

Camus, D., and T. J. Hadley. 1985. A *Plasmodium falciparum* antigen that binds to host erythrocytes and merozoites. *Science* **230**:553–556.

Carlton, J. M., S. V. Angiuoli, B. B. Suh, T. W. Kooij, M. Pertea, J. C. Silva, M. D. Ermolaeva, J. E. Allen, J. D. Selengut, H. L. Koo, J. D. Peterson, M. Pop, D. S. Kosack, M. F. Shumway, S. L. Bidwell, S. J. Shallom, S. E. van Aken, S. B. Riedmuller, T. V. Feldblyum, J. K. Cho, J. Quackenbush, M. Sedegah, A. Shoaibi, L. M. Cummings, L. Florens, J. R. Yates, J. D. Raine, R. E. Sinden, M. A. Harris, D. A. Cunningham, P. R. Preiser, L. W. Bergman, A. B. Vaidya, L. H. van Lin, C. J. Janse, A. P. Waters, H. O. Smith, O. R. White, S. L. Salzberg, J. C. Venter, C. M. Fraser, S. L. Hoffman, M. J. Gardner, and D. J. Carucci. 2002. Genome sequence and comparative analysis of the model rodent malaria parasite Plasmodium yoelii yoelii. *Nature* **419**:512–519.

Carruthers, V. B., O. K. Giddings, and L. D. Sibley. 1999. Secretion of micronemal proteins is associated with toxoplasma invasion of host cells. *Cell. Microbiol.* **1**:225–235.

Cartron, J. P., O. Prou, M. Luilier, and J. P. Soulier. 1983. Susceptibility to invasion by *Plasmodium falciparum* of some human erythrocytes carrying rare blood group antigens. *Br. J. Haematol.* **55**:639–647.

Carvalho, T. G., and R. Menard. 2005. Manipulating the *Plasmodium* genome. *Curr. Issues Mol. Biol.* **7**:39–55.

Chaparro, J., A. R. Dluzewski, G. Margos, M. M. Wasserman, G. H. Mitchell, L. H. Bannister, and J. C. Pinder. 2003. The multiple myosins of malaria: the smallest malaria myosin, *Plasmodium falciparum* myosin-B (Pfmyo-B) is expressed in late-stage schizonts and merozoites. *Eur. J. Protistol.* **39**:423–427.

Chaparro-Olaya, J., G. Margos, D. J. Coles, A. R. Dluzewski, G. H. Mitchell, M. M. Wasserman, and J. C. Pinder. 2005. *Plasmodium falciparum* myosins: transcription and translation during asexual parasite development. *Cell Motil. Cytoskeleton* **60**:200–213.

Chaudhuri, A., J. Polyakova, V. Zbrzezna, K. Williams, S. Gulati, and A. O. Pogo. 1993. Cloning of glycoprotein D cDNA, which encodes the major subunit of the Duffy blood group system and the receptor for the *Plasmodium vivax* malaria parasite. *Proc. Natl. Acad. Sci. USA* **90**:10793–10797.

Chaudhuri, A., V. Zbrzezna, J. Polyakova, A. O. Pogo, J. Hesselgesser, and R. Horuk. 1994. Expression of the Duffy antigen in K562 cells. Evidence that it is the human erythrocyte chemokine receptor. *J. Biol. Chem.* **269**:7835–7838.

Chen, X. M., S. P. O'Hara, B. Q. Huang, J. B. Nelson, J. J. Lin, G. Zhu, H. D. Ward, and N. F. LaRusso. 2004. Apical organelle discharge by *Cryptosporidium parvum* is temperature, cytoskeleton, and intracellular calcium dependent and required for host cell invasion. *Infect. Immun.* **72**:6806–6816.

Cheng, Q., and A. Saul. 1994. Sequence analysis of the apical membrane antigen I (AMA-1) of *Plasmodium vivax*. *Mol. Biochem. Parasitol.* **65**:183–187.

Chitnis, C. E., A. Chaudhuri, R. Horuk, A. O. Pogo, and L. H. Miller. 1996. The domain on the Duffy blood group antigen for binding *Plasmodium vivax* and *P. knowlesi* malarial parasites to erythrocytes. *J. Exp. Med.* **184**:1531–1536.

Chitnis, C. E., and L. H. Miller. 1994. Identification of the erythrocyte binding domains of *Plasmodium vivax* and *Plasmodium knowlesi* proteins involved in erythrocyte invasion. *J. Exp. Med.* **180**:497–506.

Clark, I. A., N. H. Hunt, G. A. Butcher, and W. B. Cowden. 1987. Inhibition of murine malaria (*Plasmodium chabaudi*) in vivo by recombinant interferon-gamma or tumor necrosis factor, and its enhancement by butylated hydroxyanisole. *J. Immunol.* **139**:3493–3496.

Clark, J. T., S. Donachie, R. Anand, C. F. Wilson, H. G. Heidrich, and J. S. McBride. 1989. 46–53 kilodalton glycoprotein from the surface of *Plasmodium falciparum* merozoites. *Mol. Biochem. Parasitol.* **32**:15–24.

Coatney, R. G., W. E. Collins, M. Warren, and P. G. Contacos. 1971. *The Primate Malarias*. U.S. Government Printing Office, Washington, D.C.

Cooke, B. M., K. Lingelbach, L. H. Bannister, and L. Tilley. 2004. Protein trafficking in *Plasmodium falciparum*-infected red blood cells. *Trends Parasitol.* **20**:581–589.

Cooper, J. A. 1987. Effects of cytochalasin and phalloidin on actin. *J. Cell Biol.* **105**:1473–1478.

Cooper, J. A., L. T. Ingram, G. R. Bushell, C. A. Fardoulys, D. Stenzel, L. Schofield, and A. J. Saul. 1988. The 140/130/105 kilodalton protein complex in the rhoptries of *Plasmodium falciparum* consists of discrete polypeptides. *Mol. Biochem. Parasitol.* **29**:251–260.

Coppens, I., and K. A. Joiner. 2003. Host but not parasite cholesterol controls *Toxoplasma* cell entry by modulating organelle discharge. *Mol. Biol. Cell* **14:**3804–3820.

Cortes, A., A. Benet, B. M. Cooke, J. W. Barnwell, and J. C. Reeder. 2004. Ability of *Plasmodium falciparum* to invade southeast Asian ovalocytes varies between parasite lines. *Blood* **104:**2961–2966.

Cowman, A. F., D. L. Baldi, J. Healer, K. E. Mills, R. A. O'Donnell, M. B. Reed, T. Triglia, M. E. Wickham, and B. S. Crabb. 2000. Functional analysis of proteins involved in *Plasmodium falciparum* merozoite invasion of red blood cells. *FEBS Lett.* **476:**84–88.

Crabb, B. S., M. Rug, T. W. Gilberger, J. K. Thompson, T. Triglia, A. G. Maier, and A. F. Cowman. 2004. Transfection of the human malaria parasite *Plasmodium falciparum*. *Methods Mol. Biol.* **270:**263–276.

Crewther, P. E., J. G. Culvenor, A. Silva, J. A. Cooper, and R. F. Anders. 1990. *Plasmodium falciparum*: two antigens of similar size are located in different compartments of the rhoptry. *Exp. Parasitol.* **70:**193–206.

Culvenor, J. G., K. P. Day, and R. F. Anders. 1991. *Plasmodium falciparum* ring-infected erythrocyte surface antigen is released from merozoite dense granules after erythrocyte invasion. *Infect. Immun.* **59:**1183–1187.

David, P. H., T. J. Hadley, M. Aikawa, and L. H. Miller. 1984. Processing of a major parasite surface glycoprotein during the ultimate stages of differentiation in *Plasmodium knowlesi*. *Mol. Biochem. Parasitol.* **11:**267–282.

Deans, J. A., T. Alderson, A. W. Thomas, G. H. Mitchell, E. S. Lennox, and S. Cohen. 1982. Rat monoclonal antibodies which inhibit the in vitro multiplication of *Plasmodium knowlesi*. *Clin. Exp. Immunol.* **49:**297–309.

Debrabant, A., P. Maes, P. Delplace, J. F. Dubremetz, A. Tartar, and D. Camus. 1992. Intramolecular mapping of *Plasmodium falciparum* P126 proteolytic fragments by N-terminal amino acid sequencing. *Mol. Biochem. Parasitol.* **53:**89–95.

Deguercy, A., M. Hommel, and J. Schrevel. 1990. Purification and characterization of 37-kilodalton proteases from *Plasmodium falciparum* and *Plasmodium berghei* which cleave erythrocyte cytoskeletal components. *Mol. Biochem. Parasitol.* **38:**233–244.

Delplace, P., A. Bhatia, M. Cagnard, D. Camus, G. Colombet, A. Debrabant, J. F. Dubremetz, N. Dubreuil, G. Prensier, B. Fortier, et al. 1988. Protein p126: a parasitophorous vacuole antigen associated with the release of *Plasmodium falciparum* merozoites. *Biol. Cell* **64:**215–221.

Delplace, P., B. Fortier, G. Tronchin, J. F. Dubremetz, and A. Vernes. 1987. Localization, biosynthesis, processing and isolation of a major 126 kDa antigen of the parasitophorous vacuole of *Plasmodium falciparum*. *Mol. Biochem. Parasitol.* **23:**193–201.

Dluzewski, A. R., P. R. Fryer, S. Griffiths, R. J. Wilson, and W. B. Gratzer. 1989. Red cell membrane protein distribution during malarial invasion. *J. Cell Sci.* **92:**691–699.

Dluzewski, A. R., and C. R. Garcia. 1996. Inhibition of invasion and intraerythrocytic development of *Plasmodium falciparum* by kinase inhibitors. *Experientia* **52:**621–623.

Dluzewski, A. R., G. H. Mitchell, P. R. Fryer, S. Griffiths, R. J. Wilson, and W. B. Gratzer. 1992. Origins of the parasitophorous vacuole membrane of the malaria parasite, *Plasmodium falciparum*, in human red blood cells. *J. Cell Sci.* **102:**527–532.

Dluzewski, A. R., K. Rangachari, W. B. Gratzer, and R. J. Wilson. 1983a. Inhibition of malarial invasion of red cells by chemical and immunochemical linking of spectrin molecules. *Br. J. Haematol.* **55:**629–637.

Dluzewski, A. R., K. Rangachari, R. J. Wilson, and W. B. Gratzer. 1983b. Properties of red cell ghost preparations susceptible to invasion by malaria parasites. *Parasitology* **87:**429–438.

Dluzewski, A. R., K. Rangachari, M. J. Tanner, D. J. Anstee, R. J. Wilson, and W. B. Gratzer. 1986a. Inhibition of malarial invasion by intracellular antibodies against intrinsic membrane proteins in the red cell. *Parasitology* **93:**427–431.

Dluzewski, A. R., K. Rangachari, R. J. Wilson, and W. B. Gratzer. 1986b. *Plasmodium falciparum*: protease inhibitors and inhibition of erythrocyte invasion. *Exp. Parasitol.* **62:**416–422.

Dluzewski, A. R., K. Rangachari, R. J. Wilson, and W. B. Gratzer. 1985. Relation of red cell membrane properties to invasion by *Plasmodium falciparum*. *Parasitology* **91:**273–280.

Dluzewski, A. R., D. Zicha, G. A. Dunn, and W. B. Gratzer. 1995. Origins of the parasitophorous vacuole membrane of the malaria parasite: surface area of the parasitized red cell. *Eur. J. Cell Biol.* **68:**446–449.

Dobrowolski, J. M., and L. D. Sibley. 1996. *Toxoplasma* invasion of mammalian cells is powered by the actin cytoskeleton of the parasite. *Cell* **84:**933–939.

Dolan, S. A., L. H. Miller, and T. E. Wellems. 1990. Evidence for a switching mechanism in the invasion of erythrocytes by *Plasmodium falciparum*. *J. Clin. Investig.* **86:**618–624.

Dolan, S. A., J. L. Proctor, D. W. Alling, Y. Okubo, T. E. Wellems, and L. H. Miller. 1994. Glycophorin B as an EBA-175 independent *Plasmodium falciparum* receptor of human erythrocytes. *Mol. Biochem. Parasitol.* **64:**55–63.

Donahue, C. G., V. B. Carruthers, S. D. Gilk, and G. E. Ward. 2000. The *Toxoplasma* homolog of *Plasmodium* apical membrane antigen-1 (AMA-1) is a microneme protein secreted in response to elevated intracellular calcium levels. *Mol. Biochem. Parasitol.* **111**:15–30.

Dua, M., P. Raphael, P. S. Sijwali, P. J. Rosenthal, and M. Hanspal. 2001. Recombinant falcipain-2 cleaves erythrocyte membrane ankyrin and protein 4.1. *Mol. Biochem. Parasitol.* **116**:95–99.

Duraisingh, M. T., A. G. Maier, T. Triglia, and A. F. Cowman. 2003. Erythrocyte-binding antigen 175 mediates invasion in *Plasmodium falciparum* utilizing sialic acid-dependent and -independent pathways. *Proc. Natl. Acad. Sci. USA* **100**:4796–4801.

Dutta, S., J. D. Haynes, J. K. Moch, A. Barbosa, and D. E. Lanar. 2003. Invasion-inhibitory antibodies inhibit proteolytic processing of apical membrane antigen 1 of *Plasmodium falciparum* merozoites. *Proc. Natl. Acad. Sci. USA* **100**:12295–12300.

Dutta, S., P. Malhotra, and V. S. Chauhan. 1995. Sequence analysis of apical membrane antigen 1 (AMA-1) of Plasmodium cynomolgi bastianelli. *Mol. Biochem. Parasitol.* **73**:267–270.

Dvorak, J. A., L. H. Miller, W. C. Whitehouse, and T. Shiroishi. 1975. Invasion of erythrocytes by malaria merozoites. *Science* **187**:748–750.

Eakin, A. E., A. A. Mills, G. Harth, J. H. McKerrow, and C. S. Craik. 1992. The sequence, organization, and expression of the major cysteine protease (cruzain) from *Trypanosoma cruzi. J. Biol. Chem.* **267**:7411–7420.

Escalante, A. A., O. E. Cornejo, D. E. Freeland, A. C. Poe, E. Durrego, W. E. Collins, and A. A. Lal. 2005. A monkey's tale: the origin of *Plasmodium vivax* as a human malaria parasite. *Proc. Natl. Acad. Sci. USA* **102**:1980–1985.

Escalante, A. A., H. M. Grebert, S. C. Chaiyaroj, M. Magris, S. Biswas, B. L. Nahlen, and A. A. Lal. 2001. Polymorphism in the gene encoding the apical membrane antigen-1 (AMA-1) of *Plasmodium falciparum*. X. Asembo Bay Cohort Project. *Mol. Biochem. Parasitol.* **113**:279–287.

Facer, C. A. 1983. Erythrocyte sialoglycoproteins and *Plasmodium falciparum* invasion. *Trans. R. Soc. Trop. Med. Hyg.* **77**:524–530.

Fang, X. D., D. C. Kaslow, J. H. Adams, and L. H. Miller. 1991. Cloning of the *Plasmodium vivax* Duffy receptor. *Mol. Biochem. Parasitol.* **44**:125–132.

Fast, N. M., J. C. Kissinger, D. S. Roos, and P. J. Keeling. 2001. Nuclear-encoded, plastid-targeted genes suggest a single common origin for apicomplexan and dinoflagellate plastids. *Mol. Biol. Evol.* **18**:418–426.

Fenton, B., J. T. Clark, C. M. Khan, J. V. Robinson, D. Walliker, R. Ridley, J. G. Scaife, and J. S. McBride. 1991. Structural and antigenic polymorphism of the 35- to 48-kilodalton merozoite surface antigen (MSA-2) of the malaria parasite *Plasmodium falciparum*. *Mol. Cell. Biol.* **11**:963–971.

Field, S. J., J. C. Pinder, B. Clough, A. R. Dluzewski, R. J. Wilson, and W. B. Gratzer. 1993. Actin in the merozoite of the malaria parasite, *Plasmodium falciparum*. *Cell Motil. Cytoskeleton* **25**:43–48.

Florens, L., X. Liu, Y. Wang, S. Yang, O. Schwartz, M. Peglar, D. J. Carucci, J. R. Yates III, and Y. Wub. 2004. Proteomics approach reveals novel proteins on the surface of malaria-infected erythrocytes. *Mol. Biochem. Parasitol.* **135**:1–11.

Fraser, T. S., S. H. Kappe, D. L. Narum, K. M. VanBuskirk, and J. H. Adams. 2001. Erythrocyte-binding activity of *Plasmodium yoelii* apical membrane antigen-1 expressed on the surface of transfected COS-7 cells. *Mol. Biochem. Parasitol.* **117**:49–59.

Freeman, R. R., A. J. Trejdosiewicz, and G. A. Cross. 1980. Protective monoclonal antibodies recognising stage-specific merozoite antigens of a rodent malaria parasite. *Nature* **284**:366–368.

Gaffar, F. R., A. P. Yatsuda, F. F. Franssen, and E. de Vries. 2004. A *Babesia bovis* merozoite protein with a domain architecture highly similar to the thrombospondin-related anonymous protein (TRAP) present in *Plasmodium* sporozoites. *Mol. Biochem. Parasitol.* **136**:25–34.

Galinski, M. R., and J. W. Barnwell. 1996. *Plasmodium vivax*: merozoites, invasion of reticulocytes and considerations for malaria vaccine development. *Parasitol. Today* **12**:20–29.

Galinski, M. R., and V. Corredor. 2004. Variant antigen expression in malaria infections: posttranscriptional gene silencing, virulence and severe pathology. *Mol. Biochem. Parasitol.* **134**:17–25.

Galinski, M. R., C. Corredor-Medina, M. Povoa, J. Crosby, P. Ingravallo, and J. W. Barnwell. 1999. *Plasmodium vivax* merozoite surface protein-3 contains coiled-coil motifs in an alanine-rich central domain. *Mol. Biochem. Parasitol.* **101**:131–147.

Galinski, M. R., P. Ingravallo, C. Corredor-Medina, B. Al-Khedery, M. Povoa, and J. W. Barnwell. 2001. *Plasmodium vivax* merozoite surface proteins-3β and -3γ share structural similarities with P. vivax merozoite surface protein-3α and define a new gene family. *Mol. Biochem. Parasitol.* **115**:41–53.

Galinski, M. R., C. C. Medina, P. Ingravallo, and J. W. Barnwell. 1992. A reticulocyte-binding protein complex of *Plasmodium vivax* merozoites. *Cell* **69**:1213–1226.

Galinski, M. R., M. Xu, and J. W. Barnwell. 2000. *Plasmodium vivax* reticulocyte binding protein-2 (PvRBP-2) shares structural features with PvRBP-

1 and the *Plasmodium yoelii* 235 kDa rhoptry protein family. *Mol. Biochem. Parasitol.* **108:**257–262.

Gardiner, D. L., T. Spielmann, M. W. Dixon, P. L. Hawthorne, M. R. Ortega, K. L. Anderson, T. S. Skinner-Adams, D. J. Kemp, and K. R. Trenholme. 2004. CLAG 9 is located in the rhoptries of *Plasmodium falciparum*. *Parasitol. Res.* **93:**64–67.

Gardner, M. J., N. Hall, E. Fung, O. White, M. Berriman, R. W. Hyman, J. M. Carlton, A. Pain, K. E. Nelson, S. Bowman, I. T. Paulsen, K. James, J. A. Eisen, K. Rutherford, S. L. Salzberg, A. Craig, S. Kyes, M. S. Chan, V. Nene, S. J. Shallom, B. Suh, J. Peterson, S. Angiuoli, M. Pertea, J. Allen, J. Selengut, D. Haft, M. W. Mather, A. B. Vaidya, D. M. Martin, A. H. Fairlamb, M. J. Fraunholz, D. S. Roos, S. A. Ralph, G. I. McFadden, L. M. Cummings, G. M. Subramanian, C. Mungall, J. C. Venter, D. J. Carucci, S. L. Hoffman, C. Newbold, R. W. Davis, C. M. Fraser, and B. Barrell. 2002. Genome sequence of the human malaria parasite *Plasmodium falciparum*. *Nature* **419:**498–511.

Garnham, P. C. 1966. *Malaria Parasites and Other Haemosporidia.* Blackwell Press, London, United Kingdom.

Gaskins, E., S. Gilk, N. DeVore, T. Mann, G. Ward, and C. Beckers. 2004. Identification of the membrane receptor of a class XIV myosin in *Toxoplasma gondii*. *J. Cell Biol.* **165:**383–393.

Gaur, D., J. R. Storry, M. E. Reid, J. W. Barnwell, and L. H. Miller. 2003. *Plasmodium falciparum* is able to invade erythrocytes through a trypsin-resistant pathway independent of glycophorin B. *Infect. Immun.* **71:**6742–6746.

Gerold, P., A. Dieckmann-Schuppert, and R. T. Schwarz. 1994. Glycosylphosphatidylinositols synthesized by asexual erythrocytic stages of the malarial parasite, *Plasmodium falciparum*. Candidates for plasmodial glycosylphosphatidylinositol membrane anchor precursors and pathogenicity factors. *J. Biol. Chem.* **269:**2597–2606.

Ghai, M., S. Dutta, T. Hall, D. Freilich, and C. F. Ockenhouse. 2002. Identification, expression, and functional characterization of MAEBL, a sporozoite and asexual blood stage chimeric erythrocyte-binding protein of *Plasmodium falciparum*. *Mol. Biochem. Parasitol.* **123:**35–45.

Gibson, H. L., J. E. Tucker, D. C. Kaslow, A. U. Krettli, W. E. Collins, M. C. Kiefer, I. C. Bathurst, and P. J. Barr. 1992. Structure and expression of the gene for Pv200, a major blood-stage surface antigen of *Plasmodium vivax*. *Mol. Biochem. Parasitol.* **50:**325–333.

Gilberger, T. W., J. K. Thompson, M. B. Reed, R. T. Good, and A. F. Cowman. 2003a. The cytoplasmic domain of the *Plasmodium falciparum* ligand EBA-175 is essential for invasion but not protein trafficking. *J. Cell Biol.* **162:**317–327.

Gilberger, T. W., J. K. Thompson, T. Triglia, R. T. Good, M. T. Duraisingh, and A. F. Cowman. 2003b. A novel erythrocyte binding antigen-175 paralogue from *Plasmodium falciparum* defines a new trypsin-resistant receptor on human erythrocytes. *J. Biol. Chem.* **278:**14480–14486.

Goel, V. K., X. Li, H. Chen, S. C. Liu, A. H. Chishti, and S. S. Oh. 2003. Band 3 is a host receptor binding merozoite surface protein 1 during the *Plasmodium falciparum* invasion of erythrocytes. *Proc. Natl. Acad. Sci. USA* **100:**5164–5169.

Gor, D. O., A. C. Li, and P. J. Rosenthal. 1998. Protective immune responses against protease-like antigens of the murine malaria parasite *Plasmodium vinckei*. *Vaccine* **16:**1193–1202.

Gratzer, W. B., and A. R. Dluzewski. 1993. The red blood cell and malaria parasite invasion. *Semin. Hematol.* **30:**232–247.

Grellier, P., I. Picard, F. Bernard, R. Mayer, H. G. Heidrich, M. Monsigny, and J. Schrevel. 1989. Purification and identification of a neutral endopeptidase in *Plasmodium falciparum* schizonts and merozoites. *Parasitol. Res.* **75:**455–460.

Gruner, A. C., G. Snounou, K. Brahimi, F. Letourneur, L. Renia, and P. Druilhe. 2003. Pre-erythrocytic antigens of *Plasmodium falciparum*: from rags to riches? *Trends Parasitol.* **19:**74–78.

Hackett, F., M. Sajid, C. Withers-Martinez, M. Grainger, and M. J. Blackman. 1999. PfSUB-2: a second subtilisin-like protein in *Plasmodium falciparum* merozoites. *Mol. Biochem. Parasitol.* **103:**183–195.

Hadley, T., M. Aikawa, and L. H. Miller. 1983. *Plasmodium knowlesi*: studies on invasion of rhesus erythrocytes by merozoites in the presence of protease inhibitors. *Exp. Parasitol.* **55:**306–311.

Hadley, T. J. 1986. Invasion of erythrocytes by malaria parasites: a cellular and molecular overview. *Annu. Rev. Microbiol.* **40:**451–477.

Hadley, T. J., F. W. Klotz, G. Pasvol, J. D. Haynes, M. H. McGinniss, Y. Okubo, and L. H. Miller. 1987. Falciparum malaria parasites invade erythrocytes that lack glycophorin A and B (MkMk). Strain differences indicate receptor heterogeneity and two pathways for invasion. *J. Clin. Investig.* **80:**1190–1193.

Hakansson, S., A. J. Charron, and L. D. Sibley. 2001. *Toxoplasma* evacuoles: a two-step process of secretion and fusion forms the parasitophorous vacuole. *EMBO J.* **20:**3132–3144.

Haldar, K., M. A. Ferguson, and G. A. Cross. 1985. Acylation of a *Plasmodium falciparum* merozoite surface antigen via sn-1,2-diacyl glycerol. *J. Biol. Chem.* **260:**4969–4974.

Haldar, K., and L. Uyetake. 1992. The movement of fluorescent endocytic tracers in *Plasmodium falciparum* infected erythrocytes. *Mol. Biochem. Parasitol.* **50:**161–177.

Hanspal, M., M. Dua, Y. Takakuwa, A. H. Chishti, and A. Mizuno. 2002. *Plasmodium falciparum* cysteine protease falcipain-2 cleaves erythrocyte membrane skeletal proteins at late stages of parasite development. *Blood* **100:**1048–1054.

Harrison, T., B. U. Samuel, T. Akompong, H. Hamm, N. Mohandas, J. W. Lomasney, and K. Haldar. 2003. Erythrocyte G protein-coupled receptor signaling in malarial infection. *Science* **301:**1734–1736.

Haynes, J. D., J. P. Dalton, F. W. Klotz, M. H. McGinniss, T. J. Hadley, D. E. Hudson, and L. H. Miller. 1988. Receptor-like specificity of a *Plasmodium knowlesi* malarial protein that binds to Duffy antigen ligands on erythrocytes. *J. Exp. Med.* **167:**1873–1881.

Healer, J., S. Crawford, S. Ralph, G. McFadden, and A. F. Cowman. 2002. Independent translocation of two micronemal proteins in developing *Plasmodium falciparum* merozoites. *Infect. Immun.* **70:**5751–5758.

Healer, J., V. Murphy, A. N. Hodder, R. Masciantonio, A. W. Gemmill, R. F. Anders, A. F. Cowman, and A. Batchelor. 2004. Allelic polymorphisms in apical membrane antigen-1 are responsible for evasion of antibody-mediated inhibition in *Plasmodium falciparum*. *Mol. Microbiol.* **52:**159–168.

Heidrich, H. G., W. Strych, and J. E. Mrema. 1983. Identification of surface and internal antigens from spontaneously released *Plasmodium falciparum* merozoites by radio-iodination and metabolic labelling. *Z. Parasitenkd.* **69:**715–725.

Heintzelman, M. B., and J. D. Schwartzman. 1997. A novel class of unconventional myosins from *Toxoplasma gondii*. *J. Mol. Biol.* **271:**139–146.

Hettmann, C., A. Herm, A. Geiter, B. Frank, E. Schwarz, T. Soldati, and D. Soldati. 2000. A dibasic motif in the tail of a class XIV apicomplexan myosin is an essential determinant of plasma membrane localization. *Mol. Biol. Cell* **11:**1385–1400.

Higgins, D. G., D. J. McConnell, and P. M. Sharp. 1989. Malarial proteinase? *Nature* **340:**604.

Higgins, D. L., M. C. Lamb, S. L. Young, D. B. Powers, and S. Anderson. 1990. The effect of the one-chain to two-chain conversion in tissue plasminogen activator: characterization of mutations at position 275. *Thromb. Res.* **57:**527–539.

Hiller, N. L., T. Akompong, J. S. Morrow, A. A. Holder, and K. Haldar. 2003. Identification of a stomatin orthologue in vacuoles induced in human erythrocytes by malaria parasites. A role for microbial raft proteins in apicomplexan vacuole biogenesis. *J. Biol. Chem.* **278:**48413–48421.

Hodder, A. N., P. E. Crewther, M. L. Matthew, G. E. Reid, R. L. Moritz, R. J. Simpson, and R. F. Anders. 1996. The disulfide bond structure of *Plasmodium* apical membrane antigen-1. *J. Biol. Chem.* **271:**29446–29452.

Hodder, A. N., D. R. Drew, V. C. Epa, M. Delorenzi, R. Bourgon, S. K. Miller, R. L. Moritz, D. F. Frecklington, R. J. Simpson, T. P. Speed, R. N. Pike, and B. S. Crabb. 2003. Enzymic, phylogenetic, and structural characterization of the unusual papain-like protease domain of *Plasmodium falciparum* SERA5. *J. Biol. Chem.* **278:**48169–48177.

Holder, A. A. 1988. The precursor to major merozoite surface antigens: structure and role in immunity. *Prog. Allergy* **41:**72–97.

Holder, A. A. 1994. Proteins on the surface of the malaria parasite and cell invasion. *Parasitology* **108(Suppl):**S5–S18.

Holder, A. A., and M. J. Blackman. 1994. What is the function of MSP-I on the malaria merozoite? *Parasitol. Today* **10:**182–184.

Holder, A. A., and R. R. Freeman. 1984a. Protective antigens of rodent and human bloodstage malaria. *Philos. Trans. R. Soc. Lond. B Biol. Sci.* **307:**171–177.

Holder, A. A., and R. R. Freeman. 1984b. The three major antigens on the surface of *Plasmodium falciparum* merozoites are derived from a single high molecular weight precursor. *J. Exp. Med.* **160:**624–629.

Holder, A. A., M. J. Blackman, M. Borre, P. A. Burghaus, J. A. Chappel, J. K. Keen, I. T. Ling, S. A. Ogun, C. A. Owen, and K. A. Sinha. 1994. Malaria parasites and erythrocyte invasion. *Biochem. Soc. Trans.* **22:**291–295.

Holder, A. A., R. R. Freeman, S. Uni, and M. Aikawa. 1985a. Isolation of a *Plasmodium falciparum* rhoptry protein. *Mol. Biochem. Parasitol.* **14:**293–303.

Holder, A. A., M. J. Lockyer, K. G. Odink, J. S. Sandhu, V. Riveros-Moreno, S. C. Nicholls, Y. Hillman, L. S. Davey, M. L. Tizard, R. T. Schwarz, et al. 1985b. Primary structure of the precursor to the three major surface antigens of *Plasmodium falciparum* merozoites. *Nature* **317:**270–273.

Holder, A. A., J. S. Sandhu, Y. Hillman, L. S. Davey, S. C. Nicholls, H. Cooper, and M. J. Lockyer. 1987. Processing of the precursor to the major merozoite surface antigens of *Plasmodium falciparum*. *Parasitology* **94:**199–208.

Holt, D. C., D. L. Gardiner, E. A. Thomas, M. Mayo, P. F. Bourke, C. J. Sutherland, R. Carter, G. Myers, D. J. Kemp, and K. R. Trenholme. 1999. The cytoadherence linked asexual gene family of *Plasmodium falciparum*: are there roles other than cytoadherence? *Int. J. Parasitol.* **29:**939–944.

Holt, E. H., M. E. Nichols, Z. Etzion, and M. E. Perkins. 1989. Erythrocyte invasion by two *Plasmodium falciparum* isolates differing in sialic acid dependency in the presence of glycophorin A antibodies. *Am. J. Trop. Med. Hyg.* **40:**245–251.

Hopkins, J., R. Fowler, S. Krishna, I. Wilson, G. Mitchell, and L. Bannister. 1999. The plastid in *Plasmodium falciparum* asexual blood stages: a three-dimensional ultrastructural analysis. *Protist* **150:** 283–295.

Horuk, R., C. E. Chitnis, W. C. Darbonne, T. J. Colby, A. Rybicki, T. J. Hadley, and L. H. Miller. 1993. A receptor for the malarial parasite *Plasmodium vivax*: the erythrocyte chemokine receptor. *Science* **261:**1182–1184.

Howard, R. F. 1990. The lower-molecular-weight protein complex (RI) of the *Plasmodium falciparum* rhoptries lacks the glycolytic enzyme aldolase. *Mol. Biochem. Parasitol.* **42:**235–240.

Howard, R. F., D. L. Narum, M. Blackman, and J. Thurman. 1998. Analysis of the processing of *Plasmodium falciparum* rhoptry-associated protein 1 and localization of Pr86 to schizont rhoptries and p67 to free merozoites. *Mol. Biochem. Parasitol.* **92:** 111–122.

Howard, R. F., and R. T. Reese. 1984. Synthesis of merozoite proteins and glycoproteins during the schizogony of *Plasmodium falciparum*. *Mol. Biochem. Parasitol.* **10:**319–334.

Howell, S. A., I. Well, S. L. Fleck, C. Kettleborough, C. R. Collins, and M. J. Blackman. 2003. A single malaria merozoite serine protease mediates shedding of multiple surface proteins by juxtamembrane cleavage. *J. Biol. Chem.* **278:**23890–23898.

Howell, S. A., C. Withers-Martinez, C. H. Kocken, A. W. Thomas, and M. J. Blackman. 2001. Proteolytic processing and primary structure of *Plasmodium falciparum* apical membrane antigen-1. *J. Biol. Chem.* **276:**31311–31320.

Hudson-Taylor, D. E., S. A. Dolan, F. W. Klotz, H. Fujioka, M. Aikawa, E. V. Koonin, and L. H. Miller. 1995. *Plasmodium falciparum* protein associated with the invasion junction contains a conserved oxidoreductase domain. *Mol. Microbiol.* **15:**463–471.

Irion, A., I. Felger, S. Abdulla, T. Smith, R. Mull, M. Tanner, C. Hatz, and H. P. Beck. 1998. Distinction of recrudescences from new infections by PCR-RFLP analysis in a comparative trial of CGP 56 697 and chloroquine in Tanzanian children. *Trop. Med. Int. Health* **3:**490–497.

Iwamoto, S., T. Omi, E. Kajii, and S. Ikemoto. 1995. Genomic organization of the glycoprotein D gene: Duffy blood group Fya/Fyb alloantigen system is associated with a polymorphism at the 44-amino acid residue. *Blood* **85:**622–626.

Jaikaria, N. S., C. Rozario, R. G. Ridley, and M. E. Perkins. 1993. Biogenesis of rhoptry organelles in *Plasmodium falciparum*. *Mol. Biochem. Parasitol.* **57:** 269–279.

Jay, D. G. 1996. Role of band 3 in homeostasis and cell shape. *Cell* **86:**853–854.

Jewett, T. J., and L. D. Sibley. 2003. Aldolase forms a bridge between cell surface adhesins and the actin cytoskeleton in apicomplexan parasites. *Mol. Cell* **11:**885–894.

Johnson, J. G., N. Epstein, T. Shiroishi, and L. H. Miller. 1980. Factors affecting the ability of isolated *Plasmodium knowlesi* merozoites to attach to and invade erythrocytes. *Parasitology* **80:**539–550.

Johnson, J. G., N. Epstein, T. Shiroishi, and L. H. Miller. 1981. Identification of surface proteins on viable *Plasmodium knowlesi* merozoites. *J. Protozool.* **28:**160–164.

Kaneko, O., D. A. Fidock, O. M. Schwartz, and L. H. Miller. 2000. Disruption of the C-terminal region of EBA-175 in the Dd2/Nm clone of *Plasmodium falciparum* does not affect erythrocyte invasion. *Mol. Biochem. Parasitol.* **110:**135–146.

Kaneko, O., M. Kimura, F. Kawamoto, M. U. Ferreira, and K. Tanabe. 1997. *Plasmodium falciparum*: allelic variation in the merozoite surface protein 1 gene in wild isolates from southern Vietnam. *Exp. Parasitol.* **86:**45–57.

Kaneko, O., J. Mu, T. Tsuboi, X. Su, and M. Torii. 2002. Gene structure and expression of a *Plasmodium falciparum* 220-kDa protein homologous to the *Plasmodium vivax* reticulocyte binding proteins. *Mol. Biochem. Parasitol.* **121:**275–278.

Kaneko, O., T. Tsuboi, I. T. Ling, S. Howell, M. Shirano, M. Tachibana, Y. M. Cao, A. A. Holder, and M. Torii. 2001. The high molecular mass rhoptry protein, RhopH1, is encoded by members of the clag multigene family in *Plasmodium falciparum* and *Plasmodium yoelii*. *Mol. Biochem. Parasitol.* **118:**223–231.

Kappe, S. H., and J. H. Adams. 1996. Sequence analysis of the apical membrane antigen-1 genes (ama-1) of *Plasmodium yoelii yoelii* and *Plasmodium berghei*. *Mol. Biochem. Parasitol.* **78:**279–283.

Kappe, S. H., A. R. Noe, T. S. Fraser, P. L. Blair, and J. H. Adams. 1998. A family of chimeric erythrocyte binding proteins of malaria parasites. *Proc. Natl. Acad. Sci. USA* **95:**1230–1235.

Kariu, T., M. Yuda, K. Yano, and Y. Chinzei. 2002. MAEBL is essential for malarial sporozoite infection of the mosquito salivary gland. *J. Exp. Med.* **195:** 1317–1323.

Kauth, C. W., C. Epp, H. Bujard, and R. Lutz. 2003. The merozoite surface protein 1 complex of human malaria parasite *Plasmodium falciparum*: interactions and arrangements of subunits. *J. Biol. Chem.* **278:**22257–22264.

Kedzierski, L., C. G. Black, and R. L. Coppel. 2000. Characterization of the merozoite surface protein 4/5 gene of *Plasmodium berghei* and *Plasmodium yoelii*. *Mol. Biochem. Parasitol.* **105:**137–147.

Keeley, A., and D. Soldati. 2004. The glideosome: a molecular machine powering motility and host-cell invasion by Apicomplexa. *Trends Cell Biol.* **14:**528–532.

Keen, J., A. Holder, J. Playfair, M. Lockyer, and A. Lewis. 1990. Identification of the gene for a *Plasmodium yoelii* rhoptry protein. Multiple copies in the parasite genome. *Mol. Biochem. Parasitol.* **42:**241–246.

Keen, J. K., K. A. Sinha, K. N. Brown, and A. A. Holder. 1994. A gene coding for a high-molecular mass rhoptry protein of *Plasmodium yoelii*. *Mol. Biochem. Parasitol.* **65:**171–177.

Kerr, P. J., L. C. Ranford-Cartwright, and D. Walliker. 1994. Proof of intragenic recombination in *Plasmodium falciparum*. *Mol. Biochem. Parasitol.* **66:**241–248.

Kiefer, M. C., K. A. Crawford, L. J. Boley, K. E. Landsberg, H. L. Gibson, D. C. Kaslow, and P. J. Barr. 1996. Identification and cloning of a locus of serine repeat antigen (sera)-related genes from *Plasmodium vivax*. *Mol. Biochem. Parasitol.* **78:**55–65.

King, C. A. 1988. Cell motility of sporozoan protozoa. *Parasitol. Today* **4:**315–319.

Kitchen, S. K. 1938. The infection of reticulocytes by *Plasmodium vivax*. *Am. J. Trop. Med. Hyg.* **18:**347–353.

Klotz, F. W., P. A. Orlandi, G. Reuter, S. J. Cohen, J. D. Haynes, R. Schauer, R. J. Howard, P. Palese, and L. H. Miller. 1992. Binding of *Plasmodium falciparum* 175-kilodalton erythrocyte binding antigen and invasion of murine erythrocytes requires N-acetylneuraminic acid but not its O-acetylated form. *Mol. Biochem. Parasitol.* **51:**49–54.

Knapp, B., E. Hundt, and H. A. Kupper. 1989. A new blood stage antigen of *Plasmodium falciparum* transported to the erythrocyte surface. *Mol. Biochem. Parasitol.* **37:**47–56.

Knowles, D. W., L. Tilley, N. Mohandas, and J. A. Chasis. 1997. Erythrocyte membrane vesiculation: model for the molecular mechanism of protein sorting. *Proc. Natl. Acad. Sci. USA* **94:**12969–12974.

Kocken, C. H., D. L. Narum, A. Massougbodji, B. Ayivi, M. A. Dubbeld, A. van der Wel, D. J. Conway, A. Sanni, and A. W. Thomas. 2000. Molecular characterisation of *Plasmodium reichenowi* apical membrane antigen-1 (AMA-1), comparison with *P. falciparum* AMA-1, and antibody-mediated inhibition of red cell invasion. *Mol. Biochem. Parasitol.* **109:**147–156.

Koontz, L. C., R. L. Jacobs, W. L. Lummis, and L. H. Miller. 1979. *Plasmodium berghei*: uptake of clindamycin and its metabolites by mouse erythrocytes with clindamycin-sensitive and clindamycin-resistant parasites. *Exp. Parasitol.* **48:**206–212.

Kushwaha, A., A. Perween, S. Mukund, S. Majumdar, D. Bhardwaj, N. R. Chowdhury, and V. S. Chauhan. 2002. Amino terminus of *Plasmodium falciparum* acidic basic repeat antigen interacts with the erythrocyte membrane through band 3 protein. *Mol. Biochem. Parasitol.* **122:**45–54.

Ladda, R., M. Aikawa, and H. Sprinz. 1969. Penetration of erythrocytes by merozoites of mammalian and avian malarial parasites. *J. Parasitol.* **55:**633–644.

Langreth, S. G., J. B. Jensen, R. T. Reese, and W. Trager. 1978. Fine structure of human malaria in vitro. *J. Protozool.* **25:**443–452.

Lauer, S., J. VanWye, T. Harrison, H. McManus, B. U. Samuel, N. L. Hiller, N. Mohandas, and K. Haldar. 2000. Vacuolar uptake of host components, and a role for cholesterol and sphingomyelin in malarial infection. *EMBO J.* **19:**3556–3564.

Le Bonniec, S., C. Deregnaucourt, V. Redeker, R. Banerjee, P. Grellier, D. E. Goldberg, and J. Schrevel. 1999. Plasmepsin II, an acidic hemoglobinase from the *Plasmodium falciparum* food vacuole, is active at neutral pH on the host erythrocyte membrane skeleton. *J. Biol. Chem.* **274:**14218–14223.

Le Roch, K. G., J. R. Johnson, L. Florens, Y. Zhou, A. Santrosyan, M. Grainger, S. F. Yan, K. C. Williamson, A. A. Holder, D. J. Carucci, J. R. Yates III, and E. A. Winzeler. 2004. Global analysis of transcript and protein levels across the *Plasmodium falciparum* life cycle. *Genome Res.* **14:**2308–2318.

Le Roch, K. G., Y. Zhou, P. L. Blair, M. Grainger, J. K. Moch, J. D. Haynes, P. De La Vega, A. A. Holder, S. Batalov, D. J. Carucci, and E. A. Winzeler. 2003. Discovery of gene function by expression profiling of the malaria parasite life cycle. *Science* **301:**1503–1508.

Li, J., H. Matsuoka, T. Mitamura, and T. Horii. 2002a. Characterization of proteases involved in the processing of *Plasmodium falciparum* serine repeat antigen (SERA). *Mol. Biochem. Parasitol.* **120:**177–186.

Li, J., T. Mitamura, B. A. Fox, D. J. Bzik, and T. Horii. 2002b. Differential localization of processed fragments of *Plasmodium falciparum* serine repeat antigen and further processing of its N-terminal 47 kDa fragment. *Parasitol. Int.* **51:**343–352.

Li, X., H. Chen, T. H. Oo, T. M. Daly, L. W. Bergman, S. C. Liu, A. H. Chishti, and S. S. Oh. 2004. A co-ligand complex anchors *Plasmodium falciparum* merozoites to the erythrocyte invasion receptor band 3. *J. Biol. Chem.* **279:**5765–5771.

Ling, I. T., L. Florens, A. R. Dluzewski, O. Kaneko, M. Grainger, B. Y. Yim Lim, T. Tsuboi, J. M. Hopkins, J. R. Johnson, M. Torii, L. H. Bannister, J. R. Yates III, A. A. Holder, and D. Mattei. 2004. The *Plasmodium falciparum* clag9 gene encodes a rhoptry protein that is transferred to the host erythrocyte upon invasion. *Mol. Microbiol.* **52:**107–118.

Ling, I. T., O. Kaneko, D. L. Narum, T. Tsuboi, S. Howell, H. M. Taylor, T. J. Scott-Finnigan, M. Torii, and A. A. Holder. 2003. Characterisation of the rhoph2 gene of *Plasmodium falciparum* and *Plasmodium yoelii*. *Mol. Biochem. Parasitol.* **127:**47–57.

Lister, I., S. Schmitz, M. Walker, J. Trinick, F. Buss, C. Veigel, and J. Kendrick-Jones. 2004. A monomeric myosin VI with a large working stroke. *EMBO J.* **23:**1729–1738.

Lobo, C. A., K. de Frazao, M. Rodriguez, M. Reid, M. Zalis, and S. Lustigman. 2004. Invasion profiles of Brazilian field isolates of *Plasmodium falciparum*: phenotypic and genotypic analyses. *Infect. Immun.* **72:**5886–5891.

Lobo, C. A., M. Rodriguez, M. Reid, and S. Lustigman. 2003. Glycophorin C is the receptor for the *Plasmodium falciparum* erythrocyte binding ligand PfEBP-2 (baebl). *Blood* **101:**4628–4631.

London, E., and D. A. Brown. 2000. Insolubility of lipids in Triton X-100: physical origin and relationship to sphingolipid/cholesterol membrane domains (rafts). *Biochim. Biophys. Acta* **1508:**182–195.

Lovett, J. L., and L. D. Sibley. 2003. Intracellular calcium stores in *Toxoplasma gondii* govern invasion of host cells. *J. Cell Sci.* **116:**3009–3016.

Lustigman, S., R. F. Anders, G. V. Brown, and R. L. Coppel. 1988. A component of an antigenic rhoptry complex of *Plasmodium falciparum* is modified after merozoite invasion. *Mol. Biochem. Parasitol.* **30:**217–224.

Lyon, J. A., and J. D. Haynes. 1986. Plasmodium falciparum antigens synthesized by schizonts and stabilized at the merozoite surface when schizonts mature in the presence of protease inhibitors. *J. Immunol.* **136:**2245–2251.

Lyon, J. A., J. D. Haynes, C. L. Diggs, J. D. Chulay, and J. M. Pratt-Rossiter. 1986. *Plasmodium falciparum* antigens synthesized by schizonts and stabilized at the merozoite surface by antibodies when schizonts mature in the presence of growth inhibitory immune serum. *J. Immunol.* **136:**2252–2258.

Mackay, M., M. Goman, N. Bone, J. E. Hyde, J. Scaife, U. Certa, H. Stunnenberg, and H. Bujard. 1985. Polymorphism of the precursor for the major surface antigens of *Plasmodium falciparum* merozoites: studies at the genetic level. *EMBO J.* **4:**3823–3829.

Mackintosh, C. L., J. G. Beeson, and K. Marsh. 2004. Clinical features and pathogenesis of severe malaria. *Trends Parasitol.* **20:**597–603.

MacLean-Fletcher, S., and T. D. Pollard. 1980. Mechanism of action of cytochalasin B on actin. *Cell* **20:**329–341.

Mahanty, S., A. Saul, and L. H. Miller. 2003. Progress in the development of recombinant and synthetic blood-stage malaria vaccines. *J. Exp. Biol.* **206:**3781–3788.

Maier, A. G., M. T. Duraisingh, J. C. Reeder, S. S. Patel, J. W. Kazura, P. A. Zimmerman, and A. F. Cowman. 2003. *Plasmodium falciparum* erythrocyte invasion through glycophorin C and selection for Gerbich negativity in human populations. *Nat. Med.* **9:**87–92.

Maitland, K., and K. Marsh. 2004. Pathophysiology of severe malaria in children. *Acta Trop.* **90:**131–140.

Margos, G., L. H. Bannister, A. R. Dluzewski, J. Hopkins, I. T. Williams, and G. H. Mitchell. 2004. Correlation of structural development and differential expression of invasion-related molecules in schizonts of *Plasmodium falciparum*. *Parasitology* **129:**273–287.

Margos, G., I. Siden-Kiamos, R. E. Fowler, T. R. Gillman, R. Spaccapelo, G. Lycett, D. Vlachou, G. Papagiannakis, W. M. Eling, G. H. Mitchell, and C. Louis. 2000. Myosin A expressions in sporogonic stages of *Plasmodium*. *Mol. Biochem. Parasitol.* **111:**465–469.

Marshall, V. M., M. G. Peterson, A. M. Lew, and D. J. Kemp. 1989. Structure of the apical membrane antigen I (AMA-1) of *Plasmodium chabaudi*. *Mol. Biochem. Parasitol.* **37:**281–283.

Marshall, V. M., A. Silva, M. Foley, S. Cranmer, L. Wang, D. J. McColl, D. J. Kemp, and R. L. Coppel. 1997. A second merozoite surface protein (MSP-4) of *Plasmodium falciparum* that contains an epidermal growth factor-like domain. *Infect. Immun.* **65:**4460–4467.

Marshall, V. M., W. Tieqiao, and R. L. Coppel. 1998. Close linkage of three merozoite surface protein genes on chromosome 2 of *Plasmodium falciparum*. *Mol. Biochem. Parasitol.* **94:**13–25.

Martin, M. J., J. C. Rayner, P. Gagneaux, J. W. Barnwell, and A. Varki. Human-chimpanzee selectivity of *Plasmodium falciparum* and *Plasmodium reichenowi* explained by a single oxygen atom difference. Submitted for publication.

Mason, S. J., L. H. Miller, T. Shiroishi, J. A. Dvorak, and M. H. McGinniss. 1977. The Duffy blood group determinants: their role in the susceptibility of human and animal erythrocytes to *Plasmodium knowlesi* malaria. *Br. J. Haematol.* **36:**327–335.

Matsumoto, Y., G. Perry, L. W. Scheibel, and M. Aikawa. 1987. Role of calmodulin in *Plasmodium falciparum*: implications for erythrocyte invasion by the merozoite. *Eur. J. Cell Biol.* **45:**36–43.

Matuschewski, K., M. M. Mota, J. C. Pinder, V. Nussenzweig, and S. H. Kappe. 2001. Identification of the class XIV myosins Pb-MyoA and Py-MyoA and expression in *Plasmodium* sporozoites. *Mol. Biochem. Parasitol.* **112:**157–161.

Mayer, D. C., O. Kaneko, D. E. Hudson-Taylor, M. E. Reid, and L. H. Miller. 2001. Characterization of a *Plasmodium falciparum* erythrocyte-binding protein paralogous to EBA-175. *Proc. Natl. Acad. Sci. USA* **98:**5222–5227.

Mayer, D. C., J. B. Mu, X. Feng, X. Z. Su, and L. H. Miller. 2002. Polymorphism in a Plasmodium falciparum erythrocyte-binding ligand changes its receptor specificity. *J. Exp. Med.* **196:**1523–1528.

Mayer, D. C., J. B. Mu, O. Kaneko, J. Duan, X. Z. Su, and L. H. Miller. 2004. Polymorphism in the *Plasmodium falciparum* erythrocyte-binding ligand JESEBL/EBA-181 alters its receptor specificity. *Proc. Natl. Acad. Sci. USA* **101:**2518–2523.

Mayer, R., I. Picard, P. Lawton, P. Grellier, C. Barrault, M. Monsigny, and J. Schrevel. 1991. Peptide derivatives specific for a *Plasmodium falciparum* proteinase inhibit the human erythrocyte invasion by merozoites. *J. Med. Chem.* **34:**3029–3035.

McBride, J. S., and H. G. Heidrich. 1987. Fragments of the polymorphic Mr 185,000 glycoprotein from the surface of isolated *Plasmodium falciparum* merozoites form an antigenic complex. *Mol. Biochem. Parasitol.* **23:**71–84.

McCallum-Deighton, N., and A. A. Holder. 1992. The role of calcium in the invasion of human erythrocytes by *Plasmodium falciparum*. *Mol. Biochem. Parasitol.* **50:**317–323.

McColl, D. J., A. Silva, M. Foley, J. F. Kun, J. M. Favaloro, J. K. Thompson, V. M. Marshall, R. L. Coppel, D. J. Kemp, and R. F. Anders. 1994. Molecular variation in a novel polymorphic antigen associated with *Plasmodium falciparum* merozoites. *Mol. Biochem. Parasitol.* **68:**53–67.

McLaren, D. J., L. H. Bannister, P. I. Trigg, and G. A. Butcher. 1979. Freeze fracture studies on the interaction between the malaria parasite and the host erythrocyte in *Plasmodium knowlesi* infections. *Parasitology* **79:**125–139.

Meissner, M., D. Schluter, and D. Soldati. 2002. Role of *Toxoplasma gondii* myosin A in powering parasite gliding and host cell invasion. *Science* **298:**837–840.

Mercereau-Puijalon, O., J. C. Barale, and E. Bischoff. 2002. Three multigene families in *Plasmodium* parasites: facts and questions. *Int. J. Parasitol.* **32:**1323–1344.

Michon, P., J. R. Stevens, O. Kaneko, and J. H. Adams. 2002. Evolutionary relationships of conserved cysteine-rich motifs in adhesive molecules of malaria parasites. *Mol. Biol. Evol.* **19:**1128–1142.

Mikkelsen, R. B., M. Kamber, K. S. Wadwa, P. S. Lin, and R. Schmidt-Ullrich. 1988. The role of lipids in *Plasmodium falciparum* invasion of erythrocytes: a coordinated biochemical and microscopic analysis. *Proc. Natl. Acad. Sci. USA* **85:**5956–5960.

Miller, L. H., and R. Carter. 1976. A review. Innate resistance in malaria. *Exp. Parasitol.* **40:**132–146.

Miller, L. H., M. Aikawa, J. G. Johnson, and T. Shiroishi. 1979. Interaction between cytochalasin B-treated malarial parasites and erythrocytes. Attachment and junction formation. *J. Exp. Med.* **149:**172–184.

Miller, L. H., D. I. Baruch, K. Marsh, and O. K. Doumbo. 2002a. The pathogenic basis of malaria. *Nature* **415:**673–679.

Miller, L. H., J. D. Haynes, F. M. McAuliffe, T. Shiroishi, J. R. Durocher, and M. H. McGinniss. 1977a. Evidence for differences in erythrocyte surface receptors for the malarial parasites, *Plasmodium falciparum* and *Plasmodium knowlesi*. *J. Exp. Med.* **146:**277–281.

Miller, L. H., F. M. McAuliffe, and S. J. Mason. 1977b. Erythrocyte receptors for malaria merozoites. *Am. J. Trop. Med. Hyg.* **26:**204–208.

Miller, L. H., D. Hudson, and J. D. Haynes. 1988. Identification of *Plasmodium knowlesi* erythrocyte binding proteins. *Mol. Biochem. Parasitol.* **31:**217–222.

Miller, L. H., S. J. Mason, D. F. Clyde, and M. H. McGinniss. 1976. The resistance factor to *Plasmodium vivax* in blacks. The Duffy-blood-group genotype, FyFy. *N. Engl. J. Med.* **295:**302–304.

Miller, L. H., S. J. Mason, J. A. Dvorak, M. H. McGinniss, and I. K. Rothman. 1975. Erythrocyte receptors for (*Plasmodium knowlesi*) malaria: Duffy blood group determinants. *Science* **189:**561–563.

Miller, L. H., T. Roberts, M. Shahabuddin, and T. F. McCutchan. 1993. Analysis of sequence diversity in the *Plasmodium falciparum* merozoite surface protein-1 (MSP-1). *Mol. Biochem. Parasitol.* **59:**1–14.

Miller, S. K., R. T. Good, D. R. Drew, M. Delorenzi, P. R. Sanders, A. N. Hodder, T. P. Speed, A. F. Cowman, T. F. de Koning-Ward, and B. S. Crabb. 2002b. A subset of *Plasmodium falciparum* SERA genes are expressed and appear to play an important role in the erythrocytic cycle. *J. Biol. Chem.* **277:**47524–47532.

Mills, K. E., J. A. Pearce, B. S. Crabb, and A. F. Cowman. 2002. Truncation of merozoite surface protein 3 disrupts its trafficking and that of acidic-basic repeat protein to the surface of *Plasmodium falciparum* merozoites. *Mol. Microbiol.* **43:**1401–1411.

Mitchell, G. H., and L. H. Bannister. 1988. Malaria parasite invasion: interactions with the red cell membrane. *Crit. Rev. Oncol. Hematol.* **8:**225–310.

Mitchell, G. H., T. J. Hadley, M. H. McGinniss, F. W. Klotz, and L. H. Miller. 1986. Invasion of erythrocytes by *Plasmodium falciparum* malaria parasites: evidence for receptor heterogeneity and two receptors. *Blood* **67:**1519–1521.

Mitchell, G. H., A. W. Thomas, G. Margos, A. R. Dluzewski, and L. H. Bannister. 2004. Apical membrane antigen 1, a major malaria vaccine candidate, mediates the close attachment of invasive merozoites to host red blood cells. *Infect. Immun.* **72:**154–158.

Mohrle, J. J., Y. Zhao, B. Wernli, R. M. Franklin, and B. Kappes. 1997. Molecular cloning, characterization and localization of PfPK4, an eIF-2α kinase-related enzyme from the malarial parasite *Plasmodium falciparum. Biochem. J.* **328:**677–687.

Mons, B. 1990. Preferential invasion of malarial merozoites into young red blood cells. *Blood Cells* **16:**299–312.

Mu, J., D. A. Joy, J. Duan, Y. Huang, J. Carlton, J. Walker, J. Barnwell, P. Beerli, M. A. Charleston, O. G. Pybus, and X. Su. Host switch leads to emergence of *Plasmodium vivax* malaria in humans. *Proc. Natl. Acad. Sci. USA,* in press.

Mulhern, T. D., G. J. Howlett, G. E. Reid, R. J. Simpson, D. J. McColl, R. F. Anders, and R. S. Norton. 1995. Solution structure of a polypeptide containing four heptad repeat units from a merozoite surface antigen of *Plasmodium falciparum. Biochemistry* **34:**3479–3491.

Murphy, S. C., B. U. Samuel, T. Harrison, K. D. Speicher, D. W. Speicher, M. E. Reid, R. Prohaska, P. S. Low, M. J. Tanner, N. Mohandas, and K. Haldar. 2004. Erythrocyte detergent-resistant membrane proteins: their characterization and selective uptake during malarial infection. *Blood* **103:**1920–1928.

Narum, D. L., S. R. Fuhrmann, T. Luu, and B. K. Sim. 2002. A novel *Plasmodium falciparum* erythrocyte binding protein-2 (EBP2/BAEBL) involved in erythrocyte receptor binding. *Mol. Biochem. Parasitol.* **119:**159–168.

Narum, D. L., J. L. Green, S. A. Ogun, and A. A. Holder. 2001. Sequence diversity and antigenic polymorphism in the *Plasmodium yoelii* p235 high molecular mass rhoptry proteins and their genes. *Mol. Biochem. Parasitol.* **112:**193–200.

Narum, D. L., J. D. Haynes, S. Fuhrmann, K. Moch, H. Liang, S. L. Hoffman, and B. K. Sim. 2000. Antibodies against the *Plasmodium falciparum* receptor binding domain of EBA-175 block invasion pathways that do not involve sialic acids. *Infect. Immun.* **68:**1964–1966.

Narum, D. L., and A. W. Thomas. 1994. Differential localization of full-length and processed forms of PF83/AMA-1 an apical membrane antigen of *Plasmodium falciparum* merozoites. *Mol. Biochem. Parasitol.* **67:**59–68.

Nichols, M. E., P. Rubinstein, J. Barnwell, S. Rodriguez de Cordoba, and R. E. Rosenfield. 1987. A new human Duffy blood group specificity defined by a murine monoclonal antibody. Immunogenetics and association with susceptibility to *Plasmodium vivax. J. Exp. Med.* **166:**776–785.

Noe, A. R., and J. H. Adams. 1998. *Plasmodium yoelii* YM MAEBL protein is coexpressed and colocalizes with rhoptry proteins. *Mol. Biochem. Parasitol.* **96:**27–35.

O'Donnell, R. A., A. Saul, A. F. Cowman, and B. S. Crabb. 2000. Functional conservation of the malaria vaccine antigen MSP-119across distantly related Plasmodium species. *Nat. Med.* **6:**91–95.

Oeuvray, C., H. Bouharoun-Tayoun, H. Gras-Masse, E. Bottius, T. Kaidoh, M. Aikawa, M. C. Filgueira, A. Tartar, and P. Druilhe. 1994. Merozoite surface protein-3: a malaria protein inducing antibodies that promote *Plasmodium falciparum* killing by cooperation with blood monocytes. *Blood* **84:**1594–1602.

Ogun, S. A., and A. A. Holder. 1996. A high molecular mass *Plasmodium yoelii* rhoptry protein binds to erythrocytes. *Mol. Biochem. Parasitol.* **76:**321–324.

Oka, M., M. Aikawa, R. R. Freeman, A. A. Holder, and E. Fine. 1984. Ultrastructural localization of protective antigens of *Plasmodium yoelii* merozoites by the use of monoclonal antibodies and ultrathin cryomicrotomy. *Am. J. Trop. Med. Hyg.* **33:**342–346.

Okenu, D. M. N., E. V.-S. Meyer, T. C. Puckett, G. Rosas-Acosta, J. W. Barnwell, and M. R. Galinski. The reticulocyte binding proteins of *Plasmodium cynomolgi*: a model system for studies of *P. vivax. Mol. Biochem. Parasitol.,* in press.

Okoyeh, J. N., C. R. Pillai, and C. E. Chitnis. 1999. *Plasmodium falciparum* field isolates commonly use erythrocyte invasion pathways that are independent of sialic acid residues of glycophorin A. *Infect. Immun.* **67:**5784–5791.

Olaya, P., and M. Wasserman. 1991. Effect of calpain inhibitors on the invasion of human erythrocytes by the parasite *Plasmodium falciparum. Biochim. Biophys. Acta* **1096:**217–221.

Orlandi, P. A., B. K. Sim, J. D. Chulay, and J. D. Haynes. 1990. Characterization of the 175-kilodalton erythrocyte binding antigen of *Plasmodium falciparum. Mol. Biochem. Parasitol.* **40:**285–294.

Ozwara, H., C. H. Kocken, D. J. Conway, J. M. Mwenda, and A. W. Thomas. 2001. Comparative analysis of *Plasmodium reichenowi* and *P. falciparum* erythrocyte-binding proteins reveals selection to maintain polymorphism in the erythrocyte-binding region of EBA-175. *Mol. Biochem. Parasitol.* **116:** 81–84.

Pachebat, J. A., I. T. Ling, M. Grainger, C. Trucco, S. Howell, D. Fernandez-Reyes, R. Gunaratne, and A. A. Holder. 2001. The 22 kDa component of the protein complex on the surface of *Plasmodium falciparum* merozoites is derived from a larger precursor, merozoite surface protein 7. *Mol. Biochem. Parasitol.* **117:**83–89.

Palacpac, N. M., Y. Hiramine, F. Mi-ichi, M. Torii, K. Kita, R. Hiramatsu, T. Horii, and T. Mitamura. 2004a. Developmental-stage-specific triacylglycerol biosynthesis, degradation and trafficking as lipid bodies in *Plasmodium falciparum*-infected erythrocytes. *J. Cell Sci.* **117**:1469–1480.

Palacpac, N. M., Y. Hiramine, S. Seto, R. Hiramatsu, T. Horii, and T. Mitamura. 2004b. Evidence that *Plasmodium falciparum* diacylglycerol acyltransferase is essential for intraerythrocytic proliferation. *Biochem. Biophys. Res. Commun.* **321**: 1062–1068.

Pandey, K. C., S. Singh, P. Pattnaik, C. R. Pillai, U. Pillai, A. Lynn, S. K. Jain, and C. E. Chitnis. 2002. Bacterially expressed and refolded receptor binding domain of *Plasmodium falciparum* EBA-175 elicits invasion inhibitory antibodies. *Mol. Biochem. Parasitol.* **123**:23–33.

Pang, X. L., T. Mitamura, and T. Horii. 1999. Antibodies reactive with the N-terminal domain of *Plasmodium falciparum* serine repeat antigen inhibit cell proliferation by agglutinating merozoites and schizonts. *Infect. Immun.* **67**:1821–1827.

Pasvol, G., D. Anstee, and M. J. Tanner. 1984. Glycophorin C and the invasion of red cells by *Plasmodium falciparum*. *Lancet* **i**:907–908.

Pasvol, G., M. Jungery, D. J. Weatherall, S. F. Parsons, D. J. Anstee, and M. J. Tanner. 1982a. Glycophorin as a possible receptor for *Plasmodium falciparum*. *Lancet* **ii**:947–950.

Pasvol, G., J. S. Wainscoat, and D. J. Weatherall. 1982b. Erythrocytes deficiency in glycophorin resist invasion by the malarial parasite *Plasmodium falciparum*. *Nature* **297**:64–66.

Patel, S. S., R. K. Mehlotra, W. Kastens, C. S. Mgone, J. W. Kazura, and P. A. Zimmerman. 2001. The association of the glycophorin C exon 3 deletion with ovalocytosis and malaria susceptibility in the Wosera, Papua New Guinea. *Blood* **98**: 3489–3491.

Pearce, J. A., A. N. Hodder, and R. F. Anders. 2004. The alanine-rich heptad repeats are intact in the processed form of *Plasmodium falciparum* MSP3. *Exp. Parasitol.* **108**:186–189.

Pearce, J. A., K. E. Mills, T. Triglia, A. F. Cowman, and R. F. Anders. 2005. Characterisation of two novel proteins from the asexual stage of *Plasmodium falciparum*, H101 and H103. *Mol. Biochem. Parasitol.* **139**:141–151.

Perez-Leal, O., A. Y. Sierra, C. A. Barrero, C. Moncada, P. Martinez, J. Cortes, Y. Lopez, E. Torres, L. M. Salazar, and M. A. Patarroyo. 2004. *Plasmodium vivax* merozoite surface protein 8 cloning, expression, and characterisation. *Biochem. Biophys. Res. Commun.* **324**:1393–1399.

Perkins, M. E. 1984a. Binding of glycophorins to *Plasmodium falciparum* merozoites. *Mol. Biochem. Parasitol.* **10**:67–78.

Perkins, M. E. 1984b. Surface proteins of *Plasmodium falciparum* merozoites binding to the erythrocyte receptor, glycophorin. *J. Exp. Med.* **160**:788–798.

Perkins, M. E., and E. H. Holt. 1988. Erythrocyte receptor recognition varies in *Plasmodium falciparum* isolates. *Mol. Biochem. Parasitol.* **27**:23–34.

Perkins, M. E., and L. J. Rocco. 1988. Sialic acid-dependent binding of *Plasmodium falciparum* merozoite surface antigen, Pf200, to human erythrocytes. *J. Immunol.* **141**:3190–3196.

Peters, L. L., R. A. Shivdasani, S. C. Liu, M. Hanspal, K. M. John, J. M. Gonzalez, C. Brugnara, B. Gwynn, N. Mohandas, S. L. Alper, S. H. Orkin, and S. E. Lux. 1996. Anion exchanger 1 (band 3) is required to prevent erythrocyte membrane surface loss but not to form the membrane skeleton. *Cell* **86**:917–927.

Peterson, D. S., L. H. Miller, and T. E. Wellems. 1995. Isolation of multiple sequences from the *Plasmodium falciparum* genome that encode conserved domains homologous to those in erythrocyte-binding proteins. *Proc. Natl. Acad. Sci. USA* **92**:7100–7104.

Peterson, D. S., and T. E. Wellems. 2000. EBL-1, a putative erythrocyte binding protein of *Plasmodium falciparum*, maps within a favored linkage group in two genetic crosses. *Mol. Biochem. Parasitol.* **105**: 105–113.

Peterson, M. G., R. L. Coppel, P. McIntyre, C. J. Langford, G. Woodrow, G. V. Brown, R. F. Anders, and D. J. Kemp. 1988. Variation in the precursor to the major merozoite surface antigens of *Plasmodium falciparum*. *Mol. Biochem. Parasitol.* **27**: 291–301.

Peterson, M. G., V. M. Marshall, J. A. Smythe, P. E. Crewther, A. Lew, A. Silva, R. F. Anders, and D. J. Kemp. 1989. Integral membrane protein located in the apical complex of *Plasmodium falciparum*. *Mol. Cell. Biol.* **9**:3151–3154.

Peterson, M. G., P. Nguyen-Dinh, V. M. Marshall, J. F. Elliott, W. E. Collins, R. F. Anders, and D. J. Kemp. 1990. Apical membrane antigen of *Plasmodium fragile*. *Mol. Biochem. Parasitol.* **39**:279–283.

Pinder, J., R. Fowler, L. Bannister, A. Dluzewski, and G. H. Mitchell. 2000. Motile systems in malaria merozoites: how is the red blood cell invaded? *Parasitol. Today* **16**:240–245.

Pinder, J. C., R. E. Fowler, A. R. Dluzewski, L. H. Bannister, F. M. Lavin, G. H. Mitchell, R. J. Wilson, and W. B. Gratzer. 1998. Actomyosin motor in the merozoite of the malaria parasite, *Plasmodium falciparum*: implications for red cell invasion. *J. Cell Sci.* **111**:1831–1839.

Preiser, P., M. Kaviratne, S. Khan, L. Bannister, and W. Jarra. 2000. The apical organelles of malaria merozoites: host cell selection, invasion, host immunity and immune evasion. *Microbes Infect.* **2**:1461–1477.

Preiser, P., L. Renia, N. Singh, B. Balu, W. Jarra, T. Voza, O. Kaneko, P. Blair, M. Torii, I. Landau, and J. H. Adams. 2004. Antibodies against MAEBL ligand domains M1 and M2 inhibit sporozoite development in vitro. *Infect. Immun.* **72:**3604–3608.

Preiser, P. R., S. Khan, F. T. Costa, W. Jarra, E. Belnoue, S. Ogun, A. A. Holder, T. Voza, I. Landau, G. Snounou, and L. Renia. 2002. Stage-specific transcription of distinct repertoires of a multigene family during *Plasmodium* life cycle. *Science* **295:**342–345.

Premawansa, S., V. A. Snewin, E. Khouri, K. N. Mendis, and P. H. David. 1993. *Plasmodium vivax*: recombination between potential allelic types of the merozoite surface protein MSP1 in parasites isolated from patients. *Exp. Parasitol.* **76:**192–199.

Putaporntip, C., S. Jongwutiwes, N. Sakihama, M. U. Ferreira, W. G. Kho, A. Kaneko, H. Kanbara, T. Hattori, and K. Tanabe. 2002. Mosaic organization and heterogeneity in frequency of allelic recombination of the Plasmodium vivax merozoite surface protein-1 locus. *Proc. Natl. Acad. Sci. USA* **99:**16348–16353.

Putaporntip, C., S. Jongwutiwes, K. Tanabe, and S. Thaithong. 1997. Interallelic recombination in the merozoite surface protein 1 (MSP-1) gene of *Plasmodium vivax* from Thai isolates. *Mol. Biochem. Parasitol.* **84:**49–56.

Rangachari, K., A. Dluzewski, R. J. Wilson, and W. B. Gratzer. 1986. Control of malarial invasion by phosphorylation of the host cell membrane cytoskeleton. *Nature* **324:**364–365.

Ranjan, A., and C. E. Chitnis. 1999. Mapping regions containing binding residues within functional domains of *Plasmodium vivax* and *Plasmodium knowlesi* erythrocyte-binding proteins. *Proc. Natl. Acad. Sci. USA* **96:**14067–14072.

Raphael, P., Y. Takakuwa, S. Manno, S. C. Liu, A. H. Chishti, and M. Hanspal. 2000. A cysteine protease activity from *Plasmodium falciparum* cleaves human erythrocyte ankyrin. *Mol. Biochem. Parasitol.* **110:**259–272.

Rayner, J. C., V. Corredor, D. Feldman, P. Ingravallo, F. Iderabdullah, M. R. Galinski, and J. W. Barnwell. 2002. Extensive polymorphism in the *Plasmodium vivax* merozoite surface coat protein MSP-3alpha is limited to specific domains. *Parasitology* **125:**393–405.

Rayner, J. C., M. R. Galinski, P. Ingravallo, and J. W. Barnwell. 2000. Two *Plasmodium falciparum* genes express merozoite proteins that are related to *Plasmodium vivax* and *Plasmodium yoelii* adhesive proteins involved in host cell selection and invasion. *Proc. Natl. Acad. Sci. USA* **97:**9648–9653.

Rayner, J. C., C. S. Huber, and J. W. Barnwell. 2004a. Conservation and divergence in erythrocyte invasion ligands: *Plasmodium reichenowi* EBL genes. *Mol. Biochem. Parasitol.* **138:**243–247.

Rayner, J. C., C. S. Huber, D. Feldman, P. Ingravallo, M. R. Galinski, and J. W. Barnwell. 2004b. *Plasmodium vivax* merozoite surface protein PvMSP-3 beta is radically polymorphic through mutation and large insertions and deletions. *Infect. Genet. Evol.* **4:**309–319.

Rayner, J. C., C. S. Huber, M. R. Galinski, and J. W. Barnwell. 2004c. Rapid evolution of an erythrocyte invasion gene family: the *Plasmodium reichenowi* reticulocyte binding like (RBL) genes. *Mol. Biochem. Parasitol.* **133:**287–296.

Rayner, J. C., E. Vargas-Serrato, C. S. Huber, M. R. Galinski, and J. W. Barnwell. 2001. A *Plasmodium falciparum* homologue of *Plasmodium vivax* reticulocyte binding protein (PvRBP1) defines a trypsin-resistant erythrocyte invasion pathway. *J. Exp. Med.* **194:**1571–1581.

Reed, M. B., S. R. Caruana, A. H. Batchelor, J. K. Thompson, B. S. Crabb, and A. F. Cowman. 2000. Targeted disruption of an erythrocyte binding antigen in *Plasmodium falciparum* is associated with a switch toward a sialic acid-independent pathway of invasion. *Proc. Natl. Acad. Sci. USA* **97:**7509–7514.

Reeder, J. C., A. N. Hodder, J. G. Beeson, and G. V. Brown. 2000. Identification of glycosaminoglycan binding domains in *Plasmodium falciparum* erythrocyte membrane protein 1 of a chondroitin sulfate A-adherent parasite. *Infect. Immun.* **68:**3923–3926.

Ridley, R. G., B. Takacs, H. W. Lahm, C. J. Delves, M. Goman, U. Certa, H. Matile, G. R. Woollett, and J. G. Scaife. 1990. Characterisation and sequence of a protective rhoptry antigen from *Plasmodium falciparum*. *Mol. Biochem. Parasitol.* **41:**125–134.

Roger, N., J. F. Dubremetz, P. Delplace, B. Fortier, G. Tronchin, and A. Vernes. 1988. Characterization of a 225 kilodalton rhoptry protein of *Plasmodium falciparum*. *Mol. Biochem. Parasitol.* **27:**135–141.

Roggwiller, E., M. E. Betoulle, T. Blisnick, and C. Braun Breton. 1996. A role for erythrocyte band 3 degradation by the parasite gp76 serine protease in the formation of the parasitophorous vacuole during invasion of erythrocytes by *Plasmodium falciparum*. *Mol. Biochem. Parasitol.* **82:**13–24.

Rosenthal, P. J., P. S. Sijwali, A. Singh, and B. R. Shenai. 2002. Cysteine proteases of malaria parasites: targets for chemotherapy. *Curr. Pharm. Des.* **8:**1659–1672.

Salmon, B. L., A. Oksman, and D. E. Goldberg. 2001. Malaria parasite exit from the host erythrocyte: a two-step process requiring extraerythrocytic proteolysis. *Proc. Natl. Acad. Sci. USA* **98:**271–276.

Samuel, B. U., N. Mohandas, T. Harrison, H. McManus, W. Rosse, M. Reid, and K. Haldar. 2001. The role of cholesterol and glycosylphosphatidylinositol-anchored proteins of erythrocyte rafts in regulating raft protein content and malarial infection. *J. Biol. Chem.* **276:**29319–29329.

Sam-Yellowe, T. Y., L. Florens, T. Wang, J. D. Raine, D. J. Carucci, R. Sinden, and J. R. Yates III. 2004. Proteome analysis of rhoptry-enriched fractions isolated from *Plasmodium* merozoites. *J. Proteome Res.* **3:**995–1001.

Sam-Yellowe, T. Y., H. Fujioka, M. Aikawa, and D. G. Messineo. 1995. *Plasmodium falciparum* rhoptry proteins of 140/130/110 kd (Rhop-H) are located in an electron lucent compartment in the neck of the rhoptries. *J. Eukaryot. Microbiol.* **42:**224–231.

Sam-Yellowe, T. Y., and M. E. Perkins. 1991. Interaction of the 140/130/110 kDa rhoptry protein complex of *Plasmodium falciparum* with the erythrocyte membrane and liposomes. *Exp. Parasitol.* **73:**161–171.

Sam-Yellowe, T. Y., H. Shio, and M. E. Perkins. 1988. Secretion of *Plasmodium falciparum* rhoptry protein into the plasma membrane of host erythrocytes. *J. Cell Biol.* **106:**1507–1513.

Schofield, L., G. R. Bushell, J. A. Cooper, A. J. Saul, J. A. Upcroft, and C. Kidson. 1986. A rhoptry antigen of *Plasmodium falciparum* contains conserved and variable epitopes recognized by inhibitory monoclonal antibodies. *Mol. Biochem. Parasitol.* **18:**183–195.

Schwartzman, J. D., and E. R. Pfefferkorn. 1983. Immunofluorescent localization of myosin at the anterior pole of the coccidian, *Toxoplasma gondii*. *J. Protozool.* **30:**657–661.

Seed, T. M. 1976. *Plasmodium simium*: ultrastructure of erythrocytic phase. *Exp. Parasitol.* **39:**262–276.

Shaw, M. K. 1997. The same but different: the biology of *Theileria* sporozoite entry into bovine cells. *Int. J. Parasitol.* **27:**457–474.

Sherman, I. W., and L. Tanigoshi. 1983. Purification of Plasmodium lophurae cathepsin D and its effects on erythrocyte membrane proteins. *Mol. Biochem. Parasitol.* **8:**207–226.

Sibley, L. D. 2004. Intracellular parasite invasion strategies. *Science* **304:**248–253.

Sibley, L. D. 2003. Toxoplasma gondii: perfecting an intracellular life style. *Traffic* **4:**581–586.

Silvie, O., J. F. Franetich, S. Charrin, M. S. Mueller, A. Siau, M. Bodescot, E. Rubinstein, L. Hannoun, Y. Charoenvit, C. H. Kocken, A. W. Thomas, G. J. Van Gemert, R. W. Sauerwein, M. J. Blackman, R. F. Anders, G. Pluschke, and D. Mazier. 2004. A role for apical membrane antigen 1 during invasion of hepatocytes by *Plasmodium falciparum* sporozoites. *J. Biol. Chem.* **279:**9490–9496.

Sim, B. K., C. E. Chitnis, K. Wasniowska, T. J. Hadley, and L. H. Miller. 1994. Receptor and ligand domains for invasion of erythrocytes by *Plasmodium falciparum*. *Science* **264:**1941–1944.

Sim, B. K., P. A. Orlandi, J. D. Haynes, F. W. Klotz, J. M. Carter, D. Camus, M. E. Zegans, and J. D. Chulay. 1990. Primary structure of the 175K *Plasmodium falciparum* erythrocyte binding antigen and identification of a peptide which elicits antibodies that inhibit malaria merozoite invasion. *J. Cell Biol.* **111:**1877–1884.

Sim, B. K., T. Toyoshima, J. D. Haynes, and M. Aikawa. 1992. Localization of the 175-kilodalton erythrocyte binding antigen in micronemes of *Plasmodium falciparum* merozoites. *Mol. Biochem. Parasitol.* **51:**157–159.

Singh, N., P. Preiser, L. Renia, B. Balu, J. Barnwell, P. Blair, W. Jarra, T. Voza, I. Landau, and J. H. Adams. 2004. Conservation and developmental control of alternative splicing in maebl among malaria parasites. *J. Mol. Biol.* **343:**589–599.

Sinha, K. A., J. K. Keen, S. A. Ogun, and A. A. Holder. 1996. Comparison of two members of a multigene family coding for high-molecular mass rhoptry proteins of *Plasmodium yoelii*. *Mol. Biochem. Parasitol.* **76:**329–332.

Smythe, J. A., R. L. Coppel, G. V. Brown, R. Ramasamy, D. J. Kemp, and R. F. Anders. 1988. Identification of two integral membrane proteins of *Plasmodium falciparum*. *Proc. Natl. Acad. Sci. USA* **85:**5195–5199.

Smythe, J. A., R. L. Coppel, K. P. Day, R. K. Martin, A. M. Oduola, D. J. Kemp, and R. F. Anders. 1991. Structural diversity in the Plasmodium falciparum merozoite surface antigen 2. *Proc. Natl. Acad. Sci. USA* **88:**1751–1755.

Smythe, J. A., M. G. Peterson, R. L. Coppel, A. J. Saul, D. J. Kemp, and R. F. Anders. 1990. Structural diversity in the 45-kilodalton merozoite surface antigen of Plasmodium falciparum. *Mol. Biochem. Parasitol.* **39:**227–234.

Soldati, D., and M. Meissner. 2004. *Toxoplasma* as a novel system for motility. *Curr. Opin. Cell Biol.* **16:**32–40.

Stafford, W. H., R. W. Stockley, S. B. Ludbrook, and A. A. Holder. 1996. Isolation, expression and characterization of the gene for an ADP-ribosylation factor from the human malaria parasite, *Plasmodium falciparum*. *Eur. J. Biochem.* **242:**104–113.

Stahl, H. D., A. E. Bianco, P. E. Crewther, R. F. Anders, A. P. Kyne, R. L. Coppel, G. F. Mitchell, D. J. Kemp, and G. V. Brown. 1986. Sorting large numbers of clones expressing *Plasmodium falciparum* antigens in *Escherichia coli* by differential antibody screening. *Mol. Biol. Med.* **3:**351–368.

Stewart, M. J., S. Schulman, and J. P. Vanderberg. 1986. Rhoptry secretion of membranous whorls by *Plasmodium falciparum* merozoites. *Am. J. Trop. Med. Hyg.* **35:**37–44.

Stewart, M. J., and J. P. Vanderberg. 1988. Malaria sporozoites leave behind trails of circumsporozoite protein during gliding motility. *J. Protozool.* **35:**389–393.

Storey, E. 1992. A polyclonal but not a monoclonal antibody to an M_r 52-kD protein responsible for a punctate fluorescence pattern in *Plasmodium falciparum* merozoites inhibits invasion in vitro. *Am. J. Trop. Med. Hyg.* **47:**663–674.

Szarfman, A., J. A. Lyon, D. Walliker, I. Quakyi, R. J. Howard, S. Sun, W. R. Ballou, K. Esser, W. T. London, R. A. Wirtz, et al. 1988a. Mature liver stages of cloned *Plasmodium falciparum* share epitopes with proteins from sporozoites and asexual blood stages. *Parasite Immunol.* **10:**339–351.

Szarfman, A., D. Walliker, J. S. McBride, J. A. Lyon, I. A. Quakyi, and R. Carter. 1988b. Allelic forms of gp195, a major blood-stage antigen of *Plasmodium falciparum*, are expressed in liver stages. *J. Exp. Med.* **167:**231–236.

Tanabe, K., M. Mackay, M. Goman, and J. G. Scaife. 1987. Allelic dimorphism in a surface antigen gene of the malaria parasite *Plasmodium falciparum*. *J. Mol. Biol.* **195:**273–287.

Taylor, H. M., M. Grainger, and A. A. Holder. 2002. Variation in the expression of a *Plasmodium falciparum* protein family implicated in erythrocyte invasion. *Infect. Immun.* **70:**5779–5789.

Taylor, H. M., T. Triglia, J. Thompson, M. Sajid, R. Fowler, M. E. Wickham, A. F. Cowman, and A. A. Holder. 2001. *Plasmodium falciparum* homologue of the genes for *Plasmodium vivax* and *Plasmodium yoelii* adhesive proteins, which is transcribed but not translated. *Infect. Immun.* **69:**3635–3645.

Thomas, A. W., L. H. Bannister, and A. P. Waters. 1990. Sixty-six kilodalton-related antigens of *Plasmodium knowlesi* are merozoite surface antigens associated with the apical prominence. *Parasite Immunol.* **12:**105–113.

Thomas, A. W., J. A. Deans, G. H. Mitchell, T. Alderson, and S. Cohen. 1984. The Fab fragments of monoclonal IgG to a merozoite surface antigen inhibit *Plasmodium knowlesi* invasion of erythrocytes. *Mol. Biochem. Parasitol.* **13:**187–199.

Thompson, J. K., T. Triglia, M. B. Reed, and A. F. Cowman. 2001. A novel ligand from *Plasmodium falciparum* that binds to a sialic acid-containing receptor on the surface of human erythrocytes. *Mol. Microbiol.* **41:**47–58.

Tilley, L., M. Foley, R. F. Anders, A. R. Dluzewski, W. B. Gratzer, G. L. Jones, and W. H. Sawyer. 1990. Rotational dynamics of the integral membrane protein, band 3, as a probe of the membrane events associated with *Plasmodium falciparum* infections of human erythrocytes. *Biochim. Biophys. Acta* **1025:**135–142.

Topolska, A. E., A. Lidgett, D. Truman, H. Fujioka, and R. L. Coppel. 2004a. Characterization of a membrane-associated rhoptry protein of *Plasmodium falciparum*. *J. Biol. Chem.* **279:**4648–4656.

Topolska, A. E., T. L. Richie, D. H. Nhan, and R. L. Coppel. 2004b. Associations between responses to the rhoptry-associated membrane antigen of *Plasmodium falciparum* and immunity to malaria infection. *Infect. Immun.* **72:**3325–3330.

Topolska, A. E., L. Wang, C. G. Black, and R. L. Coppel. 2003. Merozoite cell biology, p. 364–444. *In* A. P. Waters and C. J. Janse (ed.), *Malaria Parasites: Genomes and Molecular Biology*. Caister Academic Press, Norfolk, England.

Torii, M., J. H. Adams, L. H. Miller, and M. Aikawa. 1989. Release of merozoite dense granules during erythrocyte invasion by *Plasmodium knowlesi*. *Infect. Immun.* **57:**3230–3233.

Trager, W., C. Rozario, H. Shio, J. Williams, and M. E. Perkins. 1992. Transfer of a dense granule protein of *Plasmodium falciparum* to the membrane of ring stages and isolation of dense granules. *Infect. Immun.* **60:**4656–4661.

Tran, T. M., A. Moreno, S. S. Yazdani, C. E. Chitnis, J. W. Barnwell, and M. R. Galinski. 2005. Detection of a *Plasmodium vivax* erythrocyte binding protein by flow cytometry. *Cytometry A* **63A:**59–66.

Triglia, T., M. T. Duraisingh, R. T. Good, and A. F. Cowman. 2005. Reticulocyte-binding protein homologue 1 is required for sialic acid-dependent invasion into human erythrocytes by *Plasmodium falciparum*. *Mol. Microbiol.* **55:**162–174.

Triglia, T., J. Healer, S. R. Caruana, A. N. Hodder, R. F. Anders, B. S. Crabb, and A. F. Cowman. 2000. Apical membrane antigen 1 plays a central role in erythrocyte invasion by *Plasmodium* species. *Mol. Microbiol.* **38:**706–718.

Triglia, T., J. Thompson, S. R. Caruana, M. Delorenzi, T. Speed, and A. F. Cowman. 2001a. Identification of proteins from *Plasmodium falciparum* that are homologous to reticulocyte binding proteins in *Plasmodium vivax*. *Infect. Immun.* **69:**1084–1092.

Triglia, T., J. K. Thompson, and A. F. Cowman. 2001b. An EBA175 homologue which is transcribed but not translated in erythrocytic stages of *Plasmodium falciparum*. *Mol. Biochem. Parasitol.* **116:**55–63.

Trucco, C., D. Fernandez-Reyes, S. Howell, W. H. Stafford, T. J. Scott-Finnigan, M. Grainger, S. A. Ogun, W. R. Taylor, and A. A. Holder. 2001. The merozoite surface protein 6 gene codes for a 36 kDa protein associated with the *Plasmodium falciparum* merozoite surface protein-1 complex. *Mol. Biochem. Parasitol.* **112:**91–101.

Urban, S., and M. Freeman. 2003. Substrate specificity of rhomboid intramembrane proteases is governed by helix-breaking residues in the substrate transmembrane domain. *Mol. Cell* **11:**1425–1434.

Uzureau, P., J. C. Barale, C. J. Janse, A. P. Waters, and C. B. Breton. 2004. Gene targeting demonstrates that the *Plasmodium berghei* subtilisin PbSUB2 is essential for red cell invasion and reveals spontaneous genetic recombination events. *Cell. Microbiol.* **6:**65–78.

Vargas-Serrato, E., J. W. Barnwell, P. Ingravallo, F. B. Perler, and M. R. Galinski. 2002. Merozoite surface protein-9 of *Plasmodium vivax* and related simian malaria parasites is orthologous to p101/ABRA of *P. falciparum*. *Mol. Biochem. Parasitol.* **120:**41–52.

Vargas-Serrato, E., V. Corredor, and M. R. Galinski. 2003. Phylogenetic analysis of CSP and MSP-9 gene sequences demonstrates the close relationship of *Plasmodium coatneyi* to *Plasmodium knowlesi*. *Infect. Genet. Evol.* **3:**67–73.

Vieira, M. C., and S. N. Moreno. 2000. Mobilization of intracellular calcium upon attachment of *Toxoplasma gondii* tachyzoites to human fibroblasts is required for invasion. *Mol. Biochem. Parasitol.* **106:**157–162.

Waller, R. F., and G. I. McFadden. 2005. The apicoplast: a review of the derived plastid of apicomplexan parasites. *Curr. Issues Mol. Biol.* **7:**57–79.

Wang, L., N. Mohandas, A. Thomas, and R. L. Coppel. 2003. Detection of detergent-resistant membranes in asexual blood-stage parasites of *Plasmodium falciparum*. *Mol. Biochem. Parasitol.* **130:**149–153.

Wang, Q., S. Brown, D. S. Roos, V. Nussenzweig, and P. Bhanot. 2004. Transcriptome of axenic liver stages of *Plasmodium yoelii*. *Mol. Biochem. Parasitol.* **137:**161–168.

Ward, G. E., H. Fujioka, M. Aikawa, and L. H. Miller. 1994. Staurosporine inhibits invasion of erythrocytes by malarial merozoites. *Exp. Parasitol.* **79:**480–487.

Ward, G. E., L. H. Miller, and J. A. Dvorak. 1993. The origin of parasitophorous vacuole membrane lipids in malaria-infected erythrocytes. *J. Cell Sci.* **106:**237–248.

Wasniowska, K., E. Lisowska, G. R. Halverson, A. Chaudhuri, and M. E. Reid. 2004. The Fya, Fy6 and Fy3 epitopes of the Duffy blood group system recognized by new monoclonal antibodies: identification of a linear Fy3 epitope. *Br. J. Haematol.* **124:**118–122.

Wasserman, M. 1990. The role of calcium ions in the invasion of *Plasmodium falciparum*. *Blood Cells* **16:**450–451.

Wasserman, M., and J. Chaparro. 1996. Intraerythrocytic calcium chelators inhibit the invasion of *Plasmodium falciparum*. *Parasitol. Res.* **82:**102–107.

Waters, A. P., A. W. Thomas, J. A. Deans, G. H. Mitchell, D. E. Hudson, L. H. Miller, T. F. McCutchan, and S. Cohen. 1990. A merozoite receptor protein from *Plasmodium knowlesi* is highly conserved and distributed throughout *Plasmodium*. *J. Biol. Chem.* **265:**17974–17979.

Webb, S. E., R. E. Fowler, C. O'Shaughnessy, J. C. Pinder, A. R. Dluzewski, W. B. Gratzer, L. H. Bannister, and G. H. Mitchell. 1996. Contractile protein system in the asexual stages of the malaria parasite *Plasmodium falciparum*. *Parasitology* **112:**451–457.

Weber, J. L., J. A. Lyon, R. H. Wolff, T. Hall, G. H. Lowell, and J. D. Chulay. 1988. Primary structure of a *Plasmodium falciparum* malaria antigen located at the merozoite surface and within the parasitophorous vacuole. *J. Biol. Chem.* **263:**11421–11425.

Wel, A., C. H. Kocken, T. C. Pronk, B. Franke-Fayard, and A. W. Thomas. 2004. New selectable markers and single crossover integration for the highly versatile *Plasmodium knowlesi* transfection system. *Mol. Biochem. Parasitol.* **134:**97–104.

Wertheimer, S. P., and J. W. Barnwell. 1989. *Plasmodium vivax* interaction with the human Duffy blood group glycoprotein: identification of a parasite receptor-like protein. *Exp. Parasitol.* **69:**340–350.

Wesseling, J. G., M. A. Smits, and J. G. Schoenmakers. 1988. Extremely diverged actin proteins in *Plasmodium falciparum*. *Mol. Biochem. Parasitol.* **30:**143–153.

Wesseling, J. G., P. J. Snijders, P. van Someren, J. Jansen, M. A. Smits, and J. G. Schoenmakers. 1989. Stage-specific expression and genomic organization of the actin genes of the malaria parasite *Plasmodium falciparum*. *Mol. Biochem. Parasitol.* **35:**167–176.

Wickham, M. E., J. G. Culvenor, and A. F. Cowman. 2003. Selective inhibition of a two-step egress of malaria parasites from the host erythrocyte. *J. Biol. Chem.* **278:**37658–37663.

Williamson, D. H., P. R. Preiser, P. W. Moore, S. McCready, M. Strath, and R. J. Wilson. 2002. The plastid DNA of the malaria parasite Plasmodium falciparum is replicated by two mechanisms. *Mol. Microbiol.* **45:**533–542.

Winograd, E., C. A. Clavijo, L. Y. Bustamante, and M. Jaramillo. 1999. Release of merozoites from *Plasmodium falciparum*-infected erythrocytes could be mediated by a non-explosive event. *Parasitol. Res.* **85:**621–624.

Withers-Martinez, C., L. Jean, and M. J. Blackman. 2004. Subtilisin-like proteases of the malaria parasite. *Mol. Microbiol.* **53:**55–63.

Wu, T., C. G. Black, L. Wang, A. R. Hibbs, and R. L. Coppel. 1999. Lack of sequence diversity in the gene encoding merozoite surface protein 5 of *Plasmodium falciparum*. *Mol. Biochem. Parasitol.* **103:**243–250.

Zhang, X., and G. A. Thompson, Jr. 1997. An apparent association between glycosylphosphatidylinositol-anchored proteins and a sphingolipid in *Tetrahymena mimbres*. *Biochem. J.* **323:**197–206.

THE SPOROZOITE

R. E. Sinden and K. Matuschewski

9

The history of research into the biology of the malarial sporozoite has produced some of the most exciting, the most bizarre, and most frustrating data. The most exciting embraces the recognition that the irradiated sporozoite could induce protective immunity and that recombinant circumsporozoite protein (CSP) is a potent immunogen (Nussenzweig et al., 1967; Beaudoin et al., 1976; Nussenzweig and Zavala, 1997). The most bizarre (always in retrospect) is that the sporozoite directly invades the erythrocyte (Schaudin, 1900); still amongst the most contentious data are that the sporozoite directly invades the hepatocyte from the liver sinusoid (Shin et al., 1982). This rich history has been extensively reviewed and the reader is directed to the excellent publications of Ménard (2000), Baldacci and Ménard (2004), Frevert (2004), and Kappe et al. (2004b), among many others.

The intense interest in sporozoite biology has been richly rewarded by the progressive unravelling of the truly remarkable journey undertaken by such a simple cell. The diversity and elegance of the strategies employed are only now being revealed. The key new comprehensive technologies underpinning these advances in our understanding include the following.

Transgenic technologies. The generation of fluorescent-tagged parasites (e.g., Natarajan et al. 2001) has permitted penetrating observations in vivo on the migration of the sporozoite within both its mosquito and its mammalian hosts (Frischknecht et al., 2004; Vanderberg and Frevert, 2004).

Targeted gene disruptions. Observation of the downstream phenotypes provides important insights into potential protein function.

Microarray methods. These permit the rapid and sensitive detection of gene transcription and thus offer useful, but indirect, analyses of the developmental regulation of the parasite (LeRoch et al., 2003).

Proteomic analyses. These directly describe the expression of gene products in the whole cell or isolated organelles and thus are rapidly bridging the gap in our under-

---

R. E. Sinden, Department of Biological Sciences, Imperial College London, Sir Alexander Fleming Building, Imperial College Rd., South Kensington, London SW7 2AZ, United Kingdom. *K. Matuschewski*, Department of Parasitology, Heidelberg University School of Medicine, Im Neuenheimer Feld 324, 69120 Heidelberg, Germany.

standing of parasite structure and function. Together with transcriptome analysis, the method provides penetrating insights into the relative impact of transcription and translation control mechanisms on developmental regulation. The latter is proving to be a major player as the parasite moves between vertebrate and invertebrate hosts (in both directions).

MALARIAL INVASIVE STAGES

The sporozoite is one of three invasive stages in the malaria life cycle. The others are the ookinete and the merozoite. They have highly conserved organization, the key components of which are a surface designed to interact with the host cell, a cytoskeleton against which the actomyosin cell motor exerts its power, and secretory organelles that modify the host cell to permit entry (the micronemes) and that establish a parasitophorous vacuole in which the parasite may replicate (the rhoptries). It will therefore be particularly important to study the function of genes that are shared and those that differ between all invasive stages. For instance, one of the lead vaccine candidates against erythrocytic stages of *Plasmodium falciparum* is the apical membrane antigen 1 (AMA-1). AMA-1 localizes to merozoite micronemes (Healer et al., 2002; Bannister et al., 2003), suggesting that it plays a specific role during erythrocyte invasion, possibly during reorientation of the merozoite prior to cell entry (Mitchell et al., 2004). However, AMA-1 is again synthesized in sporozoites, and antibody inhibition assays indicate that it may contribute to hepatocyte invasion (Silvie et al., 2004). AMA-1 localizes to the surface of salivary gland sporozoites, further supporting a potential role after transmission to the mammalian host (Srinivasan et al., 2004). Because of the essential role during erythrocytic schizogony (Triglia et al., 2000), conditional mutagenesis will be required to study AMA-1 gene function in sporozoites. A recent strategy based on Flp-mediated site-specific recombination should permit functional analysis of essential genes in the haploid stages after sporogony (Carvalho et al., 2004).

SPOROZOITE FORMATION

Sporozoite formation is the final step of differentiation of the oocyst (Fig. 1). Oocyst differentiation, like that of the liver and blood schizont, essentially involves a massive expansion in cell mass and increase in nuclear number, the latter being achieved by a series of synchronous endomitoses. The fascinating disconnect between the normal eukaryotic alternation of nuclear division and cytokinesis raises very interesting questions as to the regulation of the cell cycle and its component pathways. A hypothetical model for this regulation is given in Fig. 2. A major question is what determines when cytokinesis (sporozoite formation) begins? Is it the nuclear/cytoplasmic ratio, and how does this regulation differ between the oocyst (2,000 to 8,000 daughter cells), liver schizont (30,000 to 50,000 cells), and blood schizont (8 to 32 cells)?

The process of sporozoite formation (Fig. 1), in very large part, mirrors that of the merozoite (both liver stage and erythrocytic) and that of the ookinete (where, atypically, just one daughter cell is produced following two meiotic divisions of the genome). Within the cytoplasm of the sporoblast lobes of the divided oocyst, peripherally located nuclei undergo the final mitotic division. The microtubule organizing center (MTOC) that regulates the formation of the final mitotic spindle divides, and each daughter MTOC nucleates the formation of the apical complex of the invasive stage between the nucleus and the plasma membrane (Fig 1A). The molecular mechanism positioning the MTOC to the cell cortex under the plasma membrane is unknown but may in part be regulated by CSP expression (Table 1) (Thathy et al., 2002). Whether this is achieved directly by the glycosyl phosphatidylinositol-anchored protein affecting the microdomain structure of the oocyst plasmalemma or by other transmembrane proteins in a multimolecular complex is unknown but worthy of study (see below). Following the assembly of the apical rings, the subpellicular microtubules polymerize and attach to the cytoplasmic face of the inner membrane complex (IMC). A striated connection between the spindle pole body, located in a pore of the

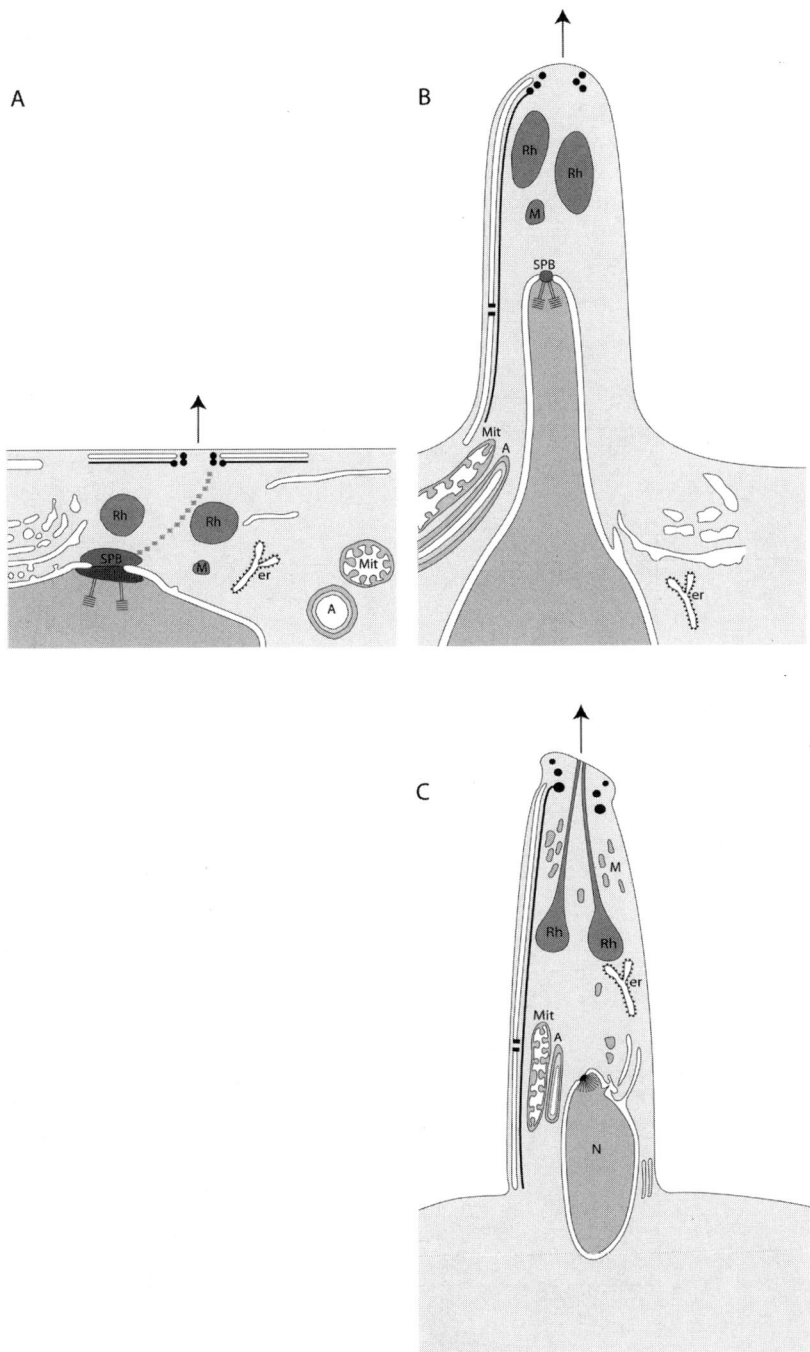

FIGURE 1 Diagram illustrating the morphogenesis of one sporozoite on the surface of the sporoblast in the malarial oocyst. Initial bud formation (*P. berghei* day 7 to 8) (A); intermediate elongation (day 9 to 10) (B); elongate but immature sporozoite (day 11) (C). N, nucleus; A, apicoplast; Mit, mitochondrion; Rh, rhoptry (a regulated secretory vesicle); M, microneme (a regulated secretory vesicle); SPB, spindle pole body; er, endoplasmic reticulum.

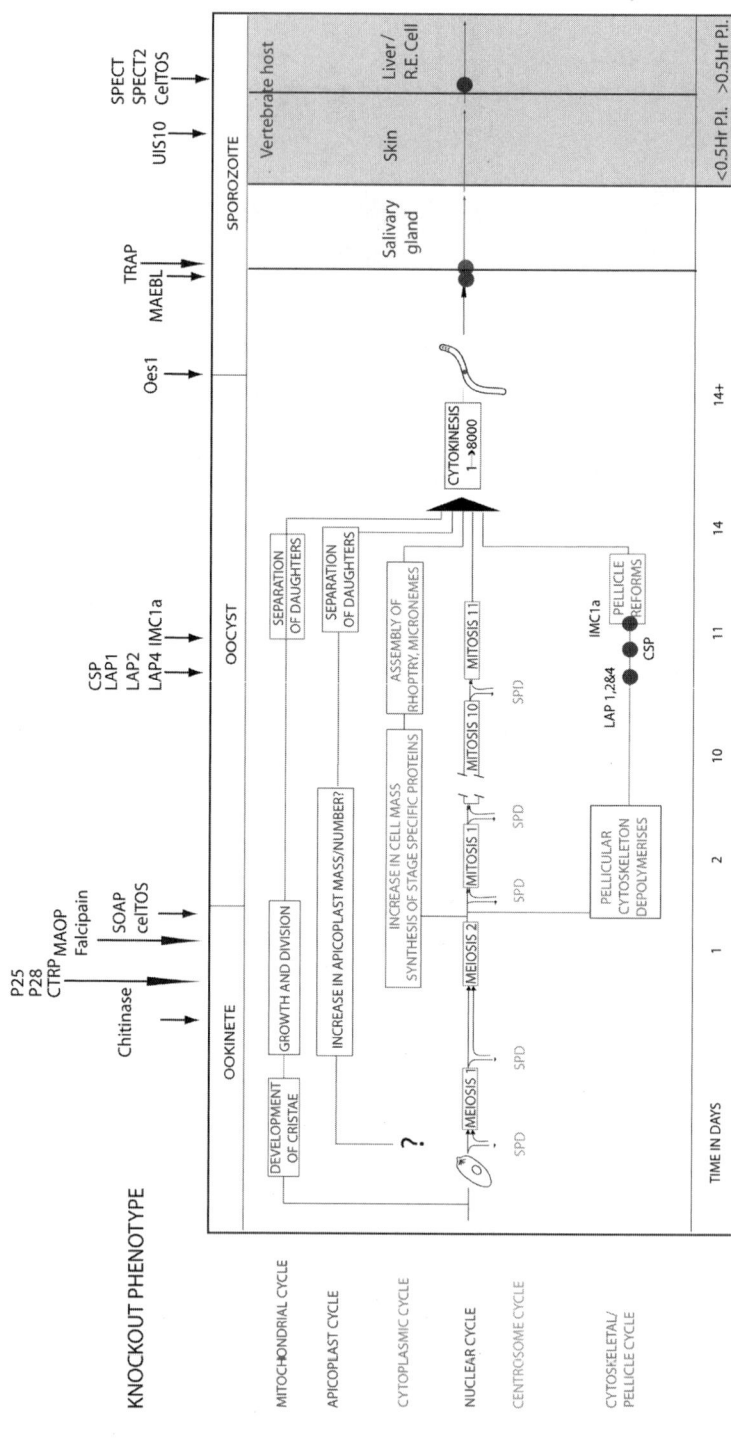

FIGURE 2 Diagram illustrating some of the separate cellular activities that make up the cell cycle of the malarial oocyst (based on a model proposed for *Saccharomyces*). The time scale is indicated at the bottom of the diagram, and the possible times at which different gene knockouts exert their impact on development are shown at the top. The diagram attempts to illustrate the difficulty in correctly assigning a point of biological impact of any gene knockout. If morphological criteria alone are used, it will depend in which subcycle the mutant acts as to when the aberrant phenotype is seen. Also noting the interdependence of the subcycles, it is perfectly feasible to ascribe a late phenotype to a mutant with an early action but with a morphological phenotype visible only following its interaction with a second and previously unaffected subcycle, for example.

TABLE 1 Characteristics of some proteins expressed in *Plasmodium* sporozoites[a]

Protein [kDa]	Primary structure	Location	Stages	Function	References
AMA-1 [83]		micronemes	sporozoite merozoite	?	Silvie et al., 2004
CSP [52]	RI RIII TSR GPI	surface	sporozoite	sporozoite formation target cell adhesion	Menard et al., 1997 Cerami et al., 1992
IMC1a [92]	Cys ... Cys	peripheral	sporozoite	sporozoite formation	Khater et al., 2004
MAEBL [220]	M1 M2 AMA-1 like Cys	micronemes	sporozoite liver stage merozoite	salivary gland adhesion liver stage development	Kariu et al., 2002 Preiser et al., 2004
MCP1 [43]	PR	cytoplasmic	sporozoite merozoite	?	Matuschewski et al., 2002b
P52 [52]	6-Cys	?	sporozoite	?	Kappe et al., 2001
PL [79]	LCAT	?	sporozoite	transmigration	Bhanot et al., 2005
PLP1/SPECT2 [88]	MACPF	micronemes	sporozoite	transmigration	Kaiser et al., 2004 Ishino et al., 2005
SPATR [25]	ATSR	surface	sporozoite	liver invasion	Kappe et al., 2001 Chattopadhyay et al., 2003
SPECT [25]		micronemes	sporozoite	transmigration	Ishino et al., 2004
SR [148]	LH2 LCCL SRCR LCCL PTX LCCL	surface	sporozoite	sporozoite formation	Claudianos et al., 2002
STARP [78]		surface	sporozoite early rings	?	Fidock et al., 1994
TRAP [80]	A-domain TSR CTD	micronemes apical tip	sporozoite	gliding motility target cell invasion	Sultan et al., 1997
UIS3 [24]		micronemes	sporozoite liver stage	liver stage development	Mueller et al., 2005a
UIS4 [22]		micronemes	sporozoite liver stage	liver stage development	Matuschewski et al., 2002b Mueller et al., 2005b

[a]Note that the table is a nonexhaustive list of sporozoite proteins. Predicted molecular weights are shown in brackets and may vary between *P. falciparum* and other *Plasmodium* species. Predicted primary structures are not to scale; stripes indicate repeat regions, grey and black boxes indicate putative cleavable signal peptides and transmembrane spans, respectively. Cellular location is typically assigned by fluorescence microscopy and stage specificity by Western and/or reverse transcription-PCR. Putative function is based on reverse genetic data (italics) or recombinant protein expression and/or antibody inhibition studies (Roman). References are listed for sporozoite expression studies only. Abbreviations: I to III, AMA-1 domains; 6-Cys, six-cysteine motif; A-domain, von Willebrand factor A domain; AMA-1, apical membrane antigen 1; ATSR, altered thrombospondin repeat; CSP, circumsporozoite protein; CTD, TRAP family cytoplasmic tail domain; Cys, cysteine-rich domain; GPI, gycosylphospatidylinositol anchor; IMC1a, inner membrane complex protein 1a; LCAT, lecithin-cholesterol acyltransferase; LCCL, *Limulus* factor C, Coch-5b2, Lgl1 domain; LH2, lipoxygenase domain; M1 and M2, MAEBL ligand domains; MACPF, membrane attack complex/perforin; MAEBL, apical membrane antigen/erythrocyte binding-like protein; MCP1, merozoite-capping protein 1; P52, six-cysteine family protein of 52 kDa; PL, phospholipase; PLP1, perforin-like protein 1; PR, peroxiredoxin; PTX, pentraxin domain; RI, region I; RIII, region III; SPATR, secreted protein with altered thrombospondin repeat; SPECT, sporozoite microneme protein essential for cell traversal; SR, scavenger receptor-like protein; SRCR, scavenger receptor cysteine rich; STARP, sporozoite threonine and asparagine-rich protein; TRAP, thrombospondin-related anonymous protein; TSR, thrombospondin type I repeat; UIS3 and UIS4, upregulated in infective sporozoites proteins 3 and 4.

mitotic nucleus, and the apical ring ensures that a haploid genome is drawn from the nucleus into the daughter cell as the sporozoite elongates. A mitochondrion and an apicoplast are carried, together with the nucleus, into each elongating sporozoite. During sporozoite elongation, the somewhat straight and inflexible cell is nonetheless motile but is restricted to an oscillation around the attachment point. Following its release from the residual sporoblast, the now more motile sporozoite (Vanderberg, 1974) assumes the typical sinusoidal or curved shape of the mature cell. This shape is believed to reflect the uneven distribution of (ca. 15 + 1) subpellicular microtubules under the IMC (Sinden and Garnham, 1973).

The molecular events of sporozoite development were very poorly understood until recently, due to the difficulties in obtaining adequate material of appropriate purity. However, the first transcriptome (LeRoch et al., 2003), transcriptional profiling (Kappe et al., 2001; Srinivasan et al., 2004), and proteome (Florens et al., 2003; Hall et al., 2005) analyses have now been published. Perhaps one of the more compelling of the biochemical strategies revealed in the sporogonic stages is the evidence that the mitochondrion is more highly developed than in the blood stages of most mammalian malarial parasites. Morphologically, the organelle develops from being (relatively) acristate to cristate and proteins and/or transcripts encoded by genes of the tricarboxylic acid oxidative phosphorylation pathways are significantly upregulated (Hall et al., 2005). This transformation should alert us to the idea that drugs targeting such parasite pathways might be more effective against the parasite in the mosquito rather than the oft-targeted mammalian hosts. Drugs of note in this context include primaquine (Coleman et al., 1994) and atovaquone (Fowler et al., 1995).

Gene knockout studies have already highlighted a number of molecules that are essential to the maturation of the oocyst (Fig. 2). Interestingly, the majority so far described (CSP, lectin adhesive-like protein 1 [LAP1], LAP2, LAP4, and IMC1a) inhibit the formation of the mature sporozoite (i.e., they result in failed or defective cytokinesis). Fortuitously, all these gene products may indeed be critically important for cytokinesis. Alternatively, these observations may forecast the probability that the unusual division cycle of *Plasmodium* is not easily analyzed by this technology. Possible sites of action of these proteins are illustrated in Fig. 2. It is immediately evident from the illustrated model that any genetic lesion in the cytoplasmic subcycle, irrespective of whether it occurs on day 1 or day 10, will only produce a division phenotype on day 10. Similarly, IMC1a, which is integral to the cytoskeletal subcycle, will also be seen to act only at the very end of the cytoplasmic cycle despite an entirely different site and time of action. The time and place of function of LAP1, -2, and -4 mutants (Claudianos et al., 2003; Trueman et al., 2004; Pradel et al., 2004) should at present remain prudently undefined, not least because they appear to result in fewer and larger nuclei. Observing a number of CSP knockout parasites, Thathy et al. (2002) noted that the formation of the IMC was atypical and not confined to discrete sporozoite buds, yet CSP itself is not confined to the plasmalemma of the bud (Hamilton et al., 1988) and the few sporozoites that form do so parallel to the sporoblast surface. This was interpreted as indicating that CSP is part of a protein complex that specifically inhibits premature assembly and/or elongation of the inner pellicle complex (comprising both membrane-vacuole and microtubules). This latter aspect of the phenotype was also noted in IMC1a knockouts (Khater et al., 2004). How is the elongation of the subpellicular vacuole regulated?

The sporozoites in the oocyst are both physiologically and biochemically distinct from those in the glands. Morphologically, the micronemes are relatively poorly developed in the oocyst, but the rhoptries are prominent, having differentiated from two large vesicles that develop immediately anterior to the nuclear spindle pole in the sporozoite bud (Fig. 1). Conversely in the salivary gland, the sporozoite micronemes are the dominant apical organelle. This differentiation parallels an increasing infectiousness to the vertebrate host and, following infection of the salivary glands, the loss of infectivity for this tissue (Vanderberg, 1975, Touray et al., 1993).

ESCAPE FROM THE OOCYST AND TRAVERSAL OF THE HEMOCOELE

The mechanism of sporozoite escape from the oocyst is very poorly understood, not least because the structure of the oocyst wall is completely unknown. The structure seen by light microscopy to be the oocyst wall or capsule is clearly bipartite. The inner layer is of parasite origin and has a granular structure; the outer layer is the basal lamina of the midgut as revealed by its typical layered organization (Sinden et al., 2004). The only molecules associated with the oocyst wall to date are transglutaminase (Adini et al., 2001) and mosquito laminin. Upon maturation of the oocyst, the basal lamina can be passively breached simply by the growth of the parasite (from 3 to 60 μm in diameter), which tears apart the laminin-collagen network (Sinden, 1974a), or it may be disrupted by sporozoite enzymes (see below). The oocyst wall is initially perforated by small holes through which sporozoites can emerge; thereafter, the weakened wall can rupture, catastrophically spilling the sporozoites into the hemocoele (Sinden, 1974a, 1974b). A central question is whether sporozoite-derived enzymes (proteases, e.g., subtilisins) are required to perforate or break down the oocyst wall. Targeted gene disruption of one stage-specific protease resulted in mature oocysts that are full of sporozoites. These oocysts do not burst unaided (A. S. I. Aly and K. Matuschewski, unpublished data). Intriguingly, in *Anopheles* mosquitoes that are infected with *ecp1*-negative (for egress cysteine protease 1) mutant parasites, the oocysts grow larger and for a longer period of time than with the wild-type parasite. Despite the long maintenance of high numbers of oocysts, the *ecp1*-negative oocyst is not recognized by the mosquito, i.e., neither oocyst melanization nor mosquito mortality increased, reminding us that the innate immune response of *Anopheles* may be more effective against the ookinete than the oocyst (Lowenberger et al., 1999). When mechanically liberated, *ecp1*-negative sporozoites are capable of performing the discontinuous gliding locomotion that is typically observed in oocyst sporozoites. Together, these data suggest that oocyst escape is an active sporozoite-mediated process that involves at least one stage-specific protease.

Sporozoites spill from the ruptured oocyst into the hemocoele, onto the midgut wall, and not infrequently into the midgut epithelial cell. In the last case, sporozoites are seen both directly in the cytoplasm and in membrane-limited compartments (Sinden and Garnham, 1971; Beaudoin et al., 1974). Sporozoite-infected epithelial cells can be expelled from the midgut wall into the lumen of the gut in a manner highly reminiscent of the time bomb theory of ookinete-midgut interaction (Han et al., 2000). Sporozoite losses in the hemocoele are significant; only 25 to 40% of those produced can be recovered from the glands. Reasons for the losses are unclear. Certainly, sporozoites are seen to circulate in all areas bathed in hemolymph, even in the feet and wing veins (Garnham, 1966). It may be assumed this is due to passive carriage in the hemocoelomic circulation. Directed sporozoite migration when attached to tissues has not been reported (R. Ménard, personal communication). Reports of sporozoite uptake by hemocytes must remain speculative. The question could be resolved using green fluorescent protein (GFP)-tagged sporozoites combined with definitive identification of hemocytes by indirect fluorescent-antibody test. Interestingly, sporozoites that lost their capacity to invade salivary glands, i.e., *trap*-negative and *maebl*-negative parasites (Sultan et al., 1997; Kariu et al., 2002), accumulate in the hemocoele, suggesting that clearance of sporozoites, however it is achieved, is not a rapid process.

RECOGNITION OF THE SALIVARY GLANDS

The salivary glands present a significant barrier to sporozoite development. This barrier may well be selective such that only developmentally advanced sporozoites gain access to this compartment. Once crossed, the further differentiation of the sporozoite means it is unable to surmount this barrier again (Touray et al., 1992). It is now widely accepted that the malaria sporozoite specifically recognizes the salivary glands in both a species- and tissue-specific manner. Sporozoites enter the salivary glands by crossing the secretory acinar cells. Salivary gland invasion is therefore a multistep transmigration process that involves interaction and penetration of the basal lamina, breaching of the basal and the apical membrane of the acinar cells, and entry into the salivary duct (Pimenta et al., 1994). Sporozoites accumulate first in the distal portions of the lateral lobe (Sterling et al., 1973; Frischknecht et al., 2004), suggesting that sporozoites recognize specific receptors on the basal side of these limited areas of the salivary gland surface. At least three sporozoite surface proteins play important roles during the various steps of salivary gland invasion.

The major surface protein is the CSP, a sporozoite-specific extracellular protein that forms a dense coat (Yoshida et al., 1981). CSP is essential for sporozoite formation (Ménard et al., 1997; Thathy, 2002). One of the hallmarks of this abundant protein, which typically makes up to 10% of metabolically labelled parasites (Yoshida et al., 1981), is characteristic proline-containing oligopeptide tandem repeats that may fold into zipper-like stacks with neighboring CSPs, thus covering the entire sporozoite surface (Godson et al., 1983, 1984). This potential structural function of the central repeat region is supported by a genetic swap experiment, where the *Plasmodium berghei* CSP repeat region was replaced by the corresponding tandem repeats of *P. falciparum* CSP without an alteration of the sporozoite phenotype (Persson et al., 2002). In contrast, replacement of the entire *P. berghei* CSP with the corresponding *P. falciparum* ortholog resulted in dramatic impairment of salivary gland invasion, suggesting a species-specific role of CSP in this process (Tewari et al., 2002). CSP is probably glycosyl phosphatidylinositol anchored and contains three additional conserved motifs, called regions I to III. By analogy with its known role in recognition and attachment to hepatocytes, it is likely that CSP similarly mediates initial contact to the basal lamina of the salivary glands. Recombinant CSP binds to the salivary glands, and this binding is successfully competed by the addition of the region 1 peptide KLKQP. Region 1A of CSP is responsible for binding to heparan sulfate (Ancsin and Kisilevsky, 2004). Competition studies using the N-terminal region or the entire CS protein also blocked sporozoite recognition of the glands (Myung et al., 2004). The binding of recombinant CSP, predominantly to the median and distal lateral lobes, closely mirrors the distribution of sporozoites within the salivary glands (Sidjanski et al., 1997). Antibodies to CSP also prevent the invasion of the salivary glands whether ingested in a blood meal (do Rosario et al., 1989), injected into the hemocoele (Warburg et al., 1992; Barreau et al., 1995), or expressed as short-chain antibody in recombinant mosquitoes (de Lara Cappuro et al., 2000). The successful application of this genetic manipulation technology in the laboratory has led to the concept that such approaches might be used to regulate parasite transmission in the field (Nirmala and James, 2003; James, 2003), but the advisability of using a technology targeting its attack upon a single protein epitope must be doubted. An attractive hypothesis is that CSP contributes to the different vector preferences of avian and mammalian malaria sporozoites. Regions I and II of *Plasmodium gallinaceum* CSP (PgCSP) differ from those of the mammalian parasites (McCutchan et al., 1996), and sporozoites of *P. gallinaceum* invade *Aedes* glands but not those of *Anopheles* (Alavi et al., 2003). Rosenberg (1985) clearly demonstrated that sporozoites of *Plasmodium knowlesi* can invade the glands of *Anopheles dirus* but not those of *Anopheles freeborni*.

Gene-targeting experiments demonstrated a clear role of the thrombospondin-related anonymous protein (TRAP) during sporozoite entry into salivary glands (Sultan et al., 1997; Kappe et al., 1999; Wengelnik et al., 1999; Matuschewski et al., 2002a). Disruption of TRAP resulted in viable sporozoites that are no longer motile and that had lost their capacity to enter the acinar cells (Sultan et al., 1997; Wengelnik et al., 1999). An extracellular recognition event is transmitted by this transmembrane protein to the parasite cytoplasm, as revealed by loss of function mutants in the conserved cytoplasmic tail domain in which sporozoite movement and invasive potential were abolished (Kappe et al., 1999). TRAP may also play a species-specific role, as replacement of the entire *P. berghei* TRAP with the corresponding *P. falciparum* ortholog resulted in severe impairment of salivary gland invasion (Wengelnik et al., 1999). Salivary gland invasion is largely mediated by an adhesion motif termed the A-domain (Matuschewski et al., 2002a), which is typically found in α-subunits of mammalian integrins. Plasmodium TRAP proteins contain a more conserved, so-called N-10 region within their amino-terminal portion (Gantt et al., 2000), which is again different in avian *Plasmodium* species (Templeton et al., 1997) and may contribute to the selectivity of salivary gland invasion.

In marked contrast to both CSP and TRAP, which continue to play important roles after sporozoite transmission to the vertebrate host, gene-targeting experiments suggest that the so-called apical membrane antigen/erythrocyte-binding-like protein (MAEBL) is exclusively required for salivary gland invasion (Kariu et al., 2002). MAEBL is predominantly expressed in micronemes of immature and/or oocyst sporozoites, and 5′ insertion mutants display a specific defect in junction formation with the basal lamina of the salivary glands. Interestingly, in vitro gliding motility and the capacity to infect hepatocytes are retained in these mutants. Systematic detection of MAEBL throughout the *Plasmodium* life cycle, however, revealed that MAEBL is resynthesized within salivary glands and may function also in hepatocyte invasion and liver stage development (Preiser et al., 2004). Differential posttranslational processing (Preiser et al., 2004) and alternative splicing of *maebl* transcripts (Singh et al., 2004) may result in multiple MAEBL variants that serve individual functions, including erythrocyte binding, as originally proposed (Kappe et al., 1998). The role of additional sporozoite molecules, including sporozoite threonine and asparagine-rich protein (designated STARP) (Fidock et al., 1994), sporozoite and liver-stage antigen (designated SALSA) (Bottius et al., 1996), sporozoite protein with an altered thrombospondin domain (designated SPATR) (Kappe et al., 2001), AMA-1 (Silvie et al., 2004), and erythrocyte-binding antigen 175 (designated EBA 175) (Grüner et al., 2001), remains to be determined. Often, these proteins are shared with blood stages, thus excluding analysis by conventional gene targeting.

Recognition and successful invasion of the salivary glands are a complex process. Host ligands would putatively contain region-1-binding motifs, sulfated glycoconjugates, and Duffy-like domains. The question remains: what are the host molecules bearing these domains and where are they? The salivary gland epithelial cells are bounded by a basal lamina (as are all tissues contacting the hemocoele). Do these host ligands lie on the lamina and/or on the basement membrane of the cell? Using a phage-display library, a synthetic 12-amino-acid peptide, termed SM1, was identified that specifically binds to the luminal side of the midgut epithelium and the distal lobes of salivary glands (Ghosh et al., 2001). Transgenic mosquitoes that express the SM1 peptide in the mosquito midgut exhibit severely impaired vector competence (Ito et al., 2002). The SM1 peptide appears to mimic a region within the A-domains of TRAP in the case of salivary gland invasion and probably those of the circumsporozoite- and TRAP-related protein CTRP during ookinete penetration of the midgut epithelium (Ghosh et al., 2004). This finding should pave the way for the isolation of the respective receptors recognized by the parasite invasins. Identification of the mosquito ligands for the sporozoite is a daunting objective, and progress has been slow. To date, the best we can say is that molecules bearing N-acetyl glucosamine groups (i.e., those that bind wheat germ agglutinin) are involved (Barreau et al., 1995), that antibodies to salivary gland proteins can block invasion, and that among the immunogenic proteins is one of 100 kDa. Monoclonal antibodies to this protein substantially inhibit sporozoite invasion (Brennan et al., 2000).

SPOROZOITE MOTILITY AND HOST CELL INVASION

Whereas sporozoite migration to the salivary glands may be a passive process, invasion of the salivary glands requires active movement by the parasite. Whether CSP and MAEBL are active participants in the invasion machinery is unclear. Certainly, the absence of MAEBL does not render the sporozoite immotile. Inhibition of sporozoite motility by anti-CSP antibodies (Stewart et al., 1986) and deposition of CSP in the trails of gliding sporozoites (Stewart and Vanderberg, 1988) strongly suggest a function in gliding motility. Employing mutagenesis in a heterologous construct, region II of CSP was found to play a crucial role in gliding motility (Tewari et al., 2002), although these findings need to be confirmed in a homologous system.

The current concept of the molecular motor, its links to the parasite receptors, and host ligands are summarized in Fig. 3. Locomotion and cell invasion are powered by an actin/myosin motor (Dobrowolski and Sibley, 1996; Sibley, 2004). Until recently, it was proposed that the scaffold for force generation is provided, as usual, by filamentous actin (Opitz and Soldati, 2002). This view was recently challenged through the localization of the myosin motor to the IMC (Bergman et al., 2003) and the isolation of the actin-bridging molecule aldolase (Jewett and Sibley, 2003), which is thought to link TRAP family invasins to actin instead of myosin, as initially assumed. The unexpected reverse orientation of the motor complex (Kappe et al., 2004a) does not challenge the early model of backward movement of the receptor-ligand complexes (e.g., TRAP/extracellular matrix) required to propel the parasite forward and into the host cell (Russell and Sinden, 1981; King, 1988). However, it leads to an important prediction for the role of actin. For an actin/myosin motor to be remain functional, the actin polymers must be short, since the motor myosins are anchored to the IMC serving as the scaffold (Bergman et al., 2003), and only independent units of actin-bound receptor/ligand complexes can be pulled backwards by a set of myosin motors and passed on to the next motor unit. This model would also explain the puzzling observation that microfilaments cannot be visualized within the parasites with fluorescent derivatives of the F-actin-binding toxin, phalloidin (Russell, 1983; Dobrowolski et al., 1997; Shaw and

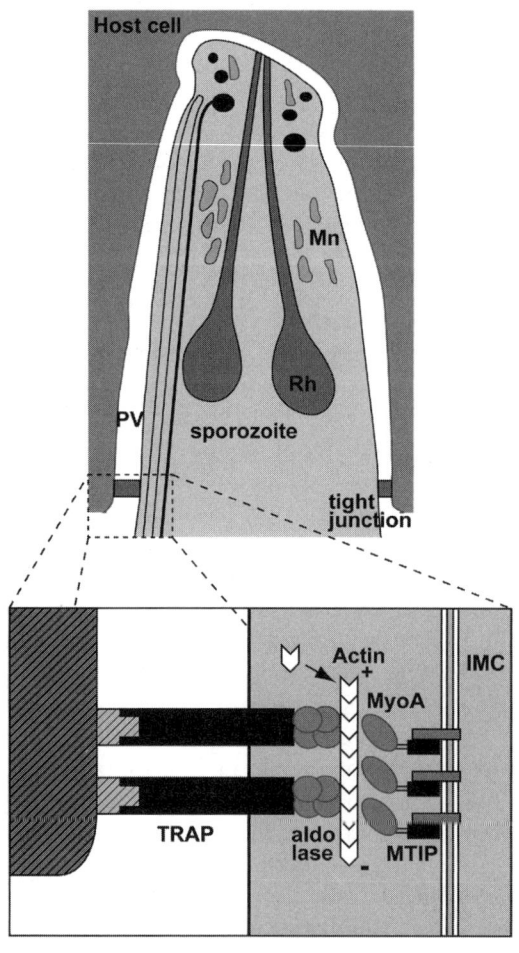

FIGURE 3 Role of the sporozoite molecular motor during host cell invasion. Sporozoites actively invade their respective host cells under simultaneous formation of a novel cellular compartment of the host cell's own making, the parasitophorous vacuole (PV). The PV membrane is remodeled through the secretion of proteins and lipids from secretory organelles, i.e., micronemes (Mn), rhoptries (Rh), and dense granules. The sporozoite invasin TRAP interacts with as-yet-unidentified cellular receptors leading to formation of a tight moving junction. This junction acts as a molecular sieve to exclude host plasma membrane proteins. Through its conserved cytoplasmic domain, TRAP is tethered via aldolase tetramers to short actin filaments. Actin polymerization is temporarily and spatially regulated and is likely to be the rate-limiting factor for parasite locomotion. Unconventional class XIV myosins (MyoA) are immobilized via myosin A tail domain interacting protein (MTIP) and other proteins to the IMC, flattened membrane stacks that are anchored to subpellicular filaments and microtubules. This structure determines the shape of sporozoites and serves as the scaffold for actin myosin-based motility. Plus-end-directed motility along the actin units results in backward translocation of the actin oligomers to the posterior end of the parasite and, thus, a net forward movement of the sporozoite.

Tilney, 1999; Gantt et al., 2000), simply because actin serves an entirely different role. Indeed, purified *Plasmodium* actin lacks the key intrinsic property to form elongated microfilaments (Schüler et al., 2005). To be correct, this model must also explain how the sporozoite binds to and recognizes the basal lamina before it gains mechanical purchase on the tissue and how TRAP directly or indirectly binds to the parasite's various environments. These sequentially change from the basal lamina to the host cell plasmalemma, host cell cytoplasm, secretory lumen, and duct.

How the sporozoite disrupts the basal lamina is unknown. Initial invasion of the salivary gland cell is achieved by the classical induction of a parasitophorous vacuole, presumably involving secretion of both micronemal and rhoptry proteins, the former producing the invagination and the latter stabilizing the vacuole. Host mitochondria associate with the vacuole membrane, but this membrane rapidly breaks down, and the sporozoite traverses the cell cytoplasm and crosses the apical membrane into the secretory vesicle.

SPOROZOITE MATURATION IN THE SALIVARY GLANDS

The notion that residence in the salivary gland provides unique and essential cues for the development of infectivity to the vertebrate is, at first sight, inconsistent with the demonstration that *P. berghei* sporozoites, which develop in vitro by coculture with *Drosophila melanogaster* S_2 cells, are infectious to the mouse (Al Olayan et al., 2002) and with the production of infectious sporozoites in the hemocoele of infected *D. melanogaster* (Schneider and Shahabuddin, 2000). Both observations suggest maturation is at least in part temporally regulated. However, acquisition of full infectivity may be triggered by receptor recognition during salivary gland transmigration or the new physiological environment of the salivary gland.

Once inside the acinar space and duct of the salivary glands, the development of the malaria parasite in the mosquito vector is complete, and sporozoites must prepare for an entirely different endeavor, the journey into the vertebrate host. Following the sporozoite's transmission by mosquito bite, the new challenges to the sporozoite include, among others, a temperature shift, new physiological and metabolic environments, a novel innate immune response, and the necessity to escape the mosquito bite site to reach its final target cell, which will permit the parasite to undergo replication. Sporozoites respond to the new tasks by switching their gene expression pattern. This developmental maturation of sporozoite infectivity was first noted when different sporogonic stages were irradiated and tested for their capacity to induce protective immunity against a sporozoite challenge (Vanderberg et al., 1972). Only sporozoites isolated from salivary glands confer significant sterile protection. Sporozoite maturation correlates with a dramatic increase of gliding motility, a unique form of substrate-dependent locomotion (Vanderberg, 1974), and infectivity to the mammalian host (Vanderberg, 1975). Despite being superficially similar, early sporozoites isolated from midgut oocysts are ~10,000-fold-less infective to the mammalian host, do not exhibit the typical continuous gliding pattern, and (probably as a consequence of low infectivity) fail to elicit protective immunity. Sporozoites that have entered salivary glands appear to be irreversibly programmed to undertake new roles in the vertebrate host. This was demonstrated by isolating salivary gland sporozoites and testing them for reinfection of salivary glands (Touray et al., 1992). Once inside salivary glands, sporozoites lost their capacity to reinvade. Therefore, oocyst sporozoites are commissioned specifically to invade the mosquito salivary gland. Subsequently, the molecular makeup of salivary gland sporozoites is remodeled to leave the mosquito inoculation site and enter their final target cell. That developmental regulation of gene expression during sporozoite maturation exists and correlates with the observed phenotypic differences was demonstrated by differential expression profiling (Matuschewski et al., 2002b). In this study, thirty genes were identified that are specifically *uis* genes (for upregulated in infective [salivary gland] sporozoites. Only one *uis* transcript, uis16/mcp-1, is shared between sporozoites and merozoites. Merozoite-capping protein 1 (MCP-1) was initially identified in the tight moving junction between the invading merozoite and the erythrocyte plasma membrane (Klotz et al., 1989; Hudson-Taylor, 1995). Remarkably, merozoites and salivary gland sporozoites are the only invasive stages of the *Plasmodium* life cycle that enter a host cell with the simultaneous formation of the parasitophorous vacuole, a newly formed replication-permissive organelle (Lingelbach and Joiner, 1998). The roles of the remaining sporozoite-specific *uis* genes are currently being examined by a systematic gene-targeting approach. It is likely that many of them function somewhere between sporozoite delivery into the mammalian host and liver stage maturation. The two known sporozoite-specific proteins, the coat protein CSP and the invasin TRAP, were initially described as differentiation antigens (Aikawa et al., 1981; Robson et al., 1995). Recent analysis showed that both proteins are constitutively transcribed and translated during sporozoite maturation (Matuschewski et al., 2002b), suggesting an additional level of regulation, such as posttranslational control of gene expression. For instance, one of the genes under the most stringent developmental control, *uis1*, encodes a *Plasmodium*-specific protein kinase (Matuschewski et al., 2002b). Other potential mechanisms of posttranslational gene regulation include ubiquitinylation and covalent attachment of ubiquitin-like molecules. A crucial question is whether translational control regulates gene expression to an extent similar to its regulation during transmission into the mosquito. As shown by the analysis of gametocyte transcriptome and gametocyte-ookinete proteomes (Hall et al., 2005), the comparison of sporozoite proteome with the transcriptome (Kappe et al., 2001; Matuschewski et al., 2002a; Le Roch et al., 2003) will allow this question to be addressed.

Sporozoite invasion (traversal) of the salivary gland cell has some unique outcomes. Each cell in the distal lateral lobes is invaded by scores, if not hundreds, of sporozoites which induce vacuole formation, invade the cytoplasm, and emerge into the apical secretory space—yet we have no evidence that these cells die as a result. Why not? The ookinete and the merozoite (schizont) destroy the host cell during their transit-exit. We have yet to disclose whether the invasion machinery (rhoptry and microneme content) differs critically between the invasive forms or in fact whether the salivary gland cells die. We should look. Sporozoites (Kaiser et al., 2004a) and ookinetes (Hall et al., 2005) both secrete perforin-like proteins, a family of micronemal molecules widely used to escape phagosomes and host cells by a range of pathogens. Ishino et al. (2004) have demonstrated that disruption of a perforin-like protein gene (named by them SPECT2, for sporozoite microneme protein essential for cell traversal 2), specifically expressed by the sporozoite, does not block invasion of the salivary glands but does prevent traversal of the sinusoidal cells (see below). Similarly, Kadota et al. (2004) have shown that disruption of a second perforin-like protein (PLP) gene (termed malaria attack ookinete protein, or MAOP) blocks disruption of the plasmalemma of the mosquito midgut epithelial cell by the ookinete. The former observation is consistent with the described induction of a parasitophorous vacuole in the salivary gland cell. Unfortunately, the authors assessed sporozoite infectivity by intravenous inoculation of knockout sporozoites and thus failed to observe whether SPECT2 is critical to the escape of the sporozoite from the salivary gland cell into the salivary duct. Such a possibility is raised because the protein is essential to the successful traversal of the Kupffer cell cytoplasm. An additional secreted phospholipase that has membrane lytic activity is encoded by UIS10/PL (Bhanot et al., 2005). Loss of function mutants display a phenotype only during natural transmission by mosquito bite, indicating a function prior to invasion of the blood vessel.

Evidence that sporozoite infection of the salivary glands affects mosquito physiology is strong; Rossignol et al. (1984, 1986) showed that infected mosquitoes probed more frequently than uninfected ones. While this has frequently been interpreted as indicating how the parasite might modify host behavior, resulting in enhanced probability of transmission, the recent evidence of Frischknecht et al. (2004) that the initial ejection contains the majority of sporozoites decreases the power of the former argument. Anderson et al. (2000) have additionally demonstrated increased feeding-associated mortality among gland-infected mosquitoes.

SPOROZOITES AND MOSQUITO IMMUNITY

Evidence that sporozoites can be melanized within the developing oocyst is well established (James, 1928; Collins et al., 1986), and current data suggest that this event follows the death of the parasite. Mosquito immune response genes are reportedly upregulated in a tissue-specific and systemic manner both at the time of sporozoite release into the hemocoele and at the time of invasion of the salivary glands (Dimopoulos et al., 1998; Gakhar and Shandilya, 2001). Sporozoite attrition during transit from the oocyst to the salivary gland in susceptible mosquitoes reportedly varies between 25 and 40% (Lowenberger et al., 1999). Direct evidence for the mechanism by which this reduction is achieved is lacking. Evidence for phagocytosis is questionable, and to date only heterologous immune peptides, e.g., frog magainin and dragonfly defensins, in high concentrations have been shown to kill the parasite (Gwadz et al., 1989; Shahabuddin et al., 1998). Nitric oxide synthase activity is upregulated both at the time of sporozoite release from the oocyst and at gland invasion (Dimopoulos et al., 1998; Luckhart et al., 1998). The sporozoite and other life stages are susceptible to the direct and/or indirect impact of nitric oxide (Motard et al., 1993; Naotunne et al., 1993). Whatever parasiticidal mechanisms are finally described, it remains to be determined whether the biological cost of mounting an effective immune response against the sporozoite is sustainable in natural populations, and thus whether this is a practical target for the design of genetically modified refractory vectors (Koella et al., 2004).

An observation that intrigues the authors is that a key immune response gene (NOS) is downregulated in the salivary glands a few days after sporozoite infection. The question emerges: is this observation related to the as-yet-unseen destruction of the infected cells or the selective downregulation of the immune response? If the latter, is this due to overall suppression of protein expression, or the direct targeting of immune response gene products? The putative role of CSP in modulating ribosome function in the infected hepatocyte (Frevert et al., 1998) is not inconsistent with the former hypothesis.

SPOROZOITE DELIVERY TO THE VERTEBRATE HOST

Transmission to the mammalian host occurs by ejection of sporozoites during the first probing phase of the mosquito bite. It typically takes repeated attempts for the proboscis to hit a blood vessel, resulting in multiple deposits of sporozoites in the avascular tissue of the dermis. The mosquito blood meal commences by either direct puncture of a blood vessel or pool feeding from hematomas formed by vessel injury. During this phase of the mosquito bite, sporozoites are ingested and can be detected in the *Anopheles* midgut, reducing the overall number of transmitted sporozoites (Beier et al., 1992). Surgical removal of the bitten area indicated a delayed migration from the bite site to the blood (Sidjanski and Vanderberg, 1997). Crucial insights into the role of active locomotion during sporozoite transmission came from the generation of transgenic *P. berghei* sporozoites that expressed GFP under temporal control of the CSP promoter (Natarajan et al., 2001). Using these parasites, it was shown that sporozoites already perform limited locomotion inside the salivary cavities and ducts (Frischknecht et al., 2004). Moreover, only a small number of sporozoites are ejected during salivation. Numbers delivered in each bite rise with increasing gland density but not in a linear relationship, i.e., a gland sporozoite count of 5,200 delivers 21 sporozoites/bite; a gland sporozoite count of 32,000 delivers 39 sporozoites (Frischknecht et al., 2004). Rarely does the number delivered exceed 200, and most estimates from experimental infections record numbers around 20 (Vanderberg, 1977; Rosenberg et al., 1990; Ponnudurai et al., 1991). Sporozoites are released during advance and retraction of the proboscis, and sporozoite locomotion starts only after a delay in the avascular tissue (Amino et al., 2004; Vanderberg and Frevert, 2004).

MIGRATION FROM THE SKIN TO THE LIVER

Once inside a new host, a parasite must typically cross a number of biological barriers en route to the final target cell (Barragan and Sibley, 2003). In the case of the sporozoite, the mechanisms employed to cross barriers (transmigration) and to achieve final cell invasion, which requires simultaneous formation of a replication-permissive parasitophorous vacuole, are clearly conserved. Actin-myosin-dependent locomotion appears to be the molecular basis for both routes. Like all apicomplexan parasites, e.g., *Eimeria* (Roberts et al., 1971; Danforth et al., 1992) and *Toxoplasma* (Speer et al., 1997), sporozoites breach virtually any nucleated cell that they encounter during locomotion in vitro (Vanderberg et al., 1990). The potential barriers to be met by the sporozoite include the capillary endothelium, the Kupffer cell, and the hepatocyte.

Sporozoites glide extensively through the avascular dermis until they reach a capillary. Having penetrated through the basal side of the endothelial cell layer, a proportion of sporozoites invade blood vessels and get carried away by the blood flow, whereas others actively enter lymph vessels or remain as residual sporozoites in the skin tissue (Amino et al., 2004). Apparently, sporozoites migrate for a relatively long time (at least 30 min) within the skin (Vanderberg and Frevert, 2004). The same authors could show that sporozoite migration and subsequent blood vessel invasion are blocked almost instantaneously in sporozoite-immunized animals. This finding highlights the often-neglected contribution of antibody-mediated immunity to sterile protection during natural infection by mosquito bite (Sinnis and Nardin, 2002).

For sporozoites of *Plasmodium* spp. that are infectious to mammals, the next potential barrier met by the sporozoite is the Kupffer cell. While there was considerable early debate as to whether this interaction had a positive (Sinden and Smith, 1982) or negative (Vreden et al., 1993) role in infection of the host and whether it was essential (Smith and Sinden, 1982) or irrelevant (Shin et al., 1982) to infection, it is now clear that if most sporozoites find themselves inside a Kupffer cell, as is probable, they have the ability to inactivate the hosts' killing mechanism (Smith and Alexander, 1986), escape the lethal clutches of this macrophage (Danforth et al., 1980; Pradel et al., 2001), and subsequently infect the hepatocytes lying on the extravascular side of the fenestrated sinusoid epithelium. Irrespective of whether the sporozoite journey has been direct or indirect, it now has to recognize and invade the hepatocyte (if a parasite of mammals) or cells of the reticuloendothelial system (if a parasite of birds or lizards).

The sequence of events was recently visualized and documented in vivo by intravital imaging of fluorescent sporozoites that express GFP infecting the liver of live animals either in sinusoidal endothelia or macrophages including Kupffer cells (U. Frevert et al., submitted for publication). Once inside the liver, sinusoid sporozoites suddenly adhere to the endothelial cell layer and commence gliding locomotion along the surface until they reach a Kupffer cell. Remarkably, gliding occurs predominantly against the bloodstream, suggesting very tight, yet transient, interactions of the sporozoite with the sinusoidal cell layer. Since sporozoites reportedly do not bind to endothelia directly, the basis of this interaction is most likely provided by extracellular matrix proteoglycans protruding from the space of Disse into the sinusoidal lumen (Pradel et al., 2002). The rate of sporozoite transmigration through a Kupffer cell is an order of magnitude slower than passage through hepatocytes or extracellular gliding. Initial recognition of the target organ, i.e., homing to the liver, is via CSP (Cerami et al., 1992; Frevert et al., 1993). Sporozoite adhesion is thought to be mediated by charged interactions between the positively charged regions I and II of CSP (Cerami et al., 1994; Sinnis et al., 1994; Ying et al., 1997) and highly negatively charged glucosaminoglycans. Both Kupffer cells and stellate cells produce glucosaminoglycans, with those from the latter protruding into the sinusoidal lumen through epithelial fenestrae (Pradel et al., 2002). Apparently, endothelial cells themselves do not contribute to either CSP- or TRAP-mediated adhesion of sporozoites. In contrast, Kupffer cells strongly bind both recombinant proteins, supporting their role as the portal for sporozoites (Pradel et al., 2002). Although not yet supported by intravital microscopy, transendothelial migration (probably through the epithelial fenestrae and the space of Disse) still remains an alternative rapid entry path for sporozoites (Shin et al., 1982). How and whether the sporozoite differentiates between initial attachment to the extracellular matrix and arrest at the surface of Kupffer cells remain an open question. Answers come from a recent study revealing that sporozoite contact with target cells results in activation of CSP processing (Coppi et al., 2005). Processing activity is sensitive to the inhibitors E64 and phenylmethylsulfonyl fluoride. These two compounds share inhibitory activity against the papain family of cysteine proteases, suggesting a role for members of this enzyme family in maturation of CSP. Indeed, pretreatment of animals with E64 leads to a dramatic reduction of infection by sporozoite transmission, presumably because sporozoites no longer productively invade hepatocytes (Coppi et al., 2005). A likely scenario is that CSP processing upon target cell contact triggers signaling events and host cell invasion. Next, moving junction formation is directed by the sporozoite invasin TRAP, largely through binding of the A-domain to as-yet-unidentified liver receptors (Sultan et al., 1997; Matuschewski et al., 2002). Antibody inhibition experiments indicated a contribution of SPATR to sporozoite invasion (Chattopadhyay et al., 2003). In addition, antibodies against AMA-1 (Silvie et al., 2004) and MAEBL (Preiser et al., 2004) result in fewer liver stages in in vitro-cultured hepatocytes. A recent study suggested that CD81, the ubiquitous founding member of the tetraspanin superfamily (Oren et al., 1990), appears to play a role in infection of some *Plasmodium* parasites (Silvie et al., 2003). However, the function of CD81 is probably indirect, consistent with a general role in tetraspanin microdomains and membrane fusion (Hemler et al., 2003).

Transmigration of hepatocytes occurs by temporary breaching of the plasma membrane, which can be revealed by membrane-impermeant fluorescent dyes (Mota et al., 2001). It was initially proposed that transmigration activates the parasite for subsequent productive entry into hepatocytes (Mota et al., 2002; Carrolo et al., 2003). The role of in vitro transmigration is likely a separate function to overcome barriers, such as entering and/or exiting a blood capillary or crossing the Kupffer cell or space of Disse en route from the hepatic sinusoid to the target hepatocyte. Important insights into the in vivo role of transmigration come from a study that employed reverse genetics in the rodent malaria *P. berghei*. Ishino et al. (2004) disrupted a gene that encodes a small micronemal protein dubbed SPECT. SPECT appears to be expressed exclusively by mature salivary gland sporozoites. SPECT-negative sporozoites are infective to the mammalian host, ruling out a vital function of SPECT. However, loss of SPECT function leads to a delay in patency, indicative of impairment at some point during the sporozoite journey to the liver. In vitro studies showed that neither gliding motility nor liver-stage development was affected. What was completely abolished by the absence of SPECT was cellular transmigration. Additional proteins, dubbed PLP1/SPECT2 (Kaiser et al., 2004; Ishino et al., 2005) and CelTOS (Yuda and Ishino, 2004), appear to play additional roles in transmigration. Hence, two functions of *Plasmodium* sporozoites, i.e., transmigration and host cell invasion, can be separated genetically. From these findings, it may be predicted that one can genetically engineer a parasite with the reverse phenotype, i.e., complete inhibition of host cell infectivity without affecting gliding motility or transmigration capacity. Staurosporine, a general protein-kinase inhibitor, induces just such a phenotype in *Plasmodium* sporozoites (Mota et al., 2002). General actin or myosin inhibitors, such as cytochalasin D or butanedione monoxime (BDM), act on every motility-related function of apicomplexan parasites (Mota et al., 2001; Mota et al., 2002,

Sibley, 2002). A hypothesis that attracts the authors is that migration only requires microneme secretion. In contrast, during host cell invasion both micronemes and rhoptries may be triggered to secrete their contents. Hence, it is possible that rhoptry secretion is differentially regulated through an additional signaling cascade that is activated from specific receptor-ligand interactions. The recognition of the final target cell would then switch the parasite's program from transmigration to productive entry by simultaneous formation of a parasitophorous vacuole. In this regard, it is interesting that the ookinete has a full motor complex and microneme system but lacks rhoptries, does not form a parasitophorous vacuole in the host cell, and (like the sporozoite) can migrate through numerous host cells before differentiating into a trophic form (Hall et al., 2005).

REFERENCES

Adini, A., M. Krugliak, H. Ginsburg, L. Li, L. Lavie, and A. Warburg. 2001. Transglutaminase in Plasmodium parasites: activity and putative role in oocysts and blood stages. *Mol. Biochem. Parasitol.* **117:**161-168.

Aikawa, M, N. Yoshida, R. S. Nussenzweig, and V. Nussenzweig. 1981. The protective antigen of malarial sporozoites (Plasmodium berghei) is a differentiation antigen. *Trans. R. Soc. Trop. Med. Hyg.* **126:**2494–2495.

Alavi, Y., M. Arai, J. Mendoza, M. Tuffet-Bayona, R. Sinha, K. Fowler, O. Billker, B. Franke-Fayard, C. J. Janse, A. Waters, and R. E. Sinden. 2003. The dynamics of interactions between Plasmodium and the mosquito: a study of the infectivity of Plasmodium berghei and Plasmodium gallinaceum, and their transmission by Anopheles stephensi, Anopheles gambiae and Aedes aegyptii. *Int. J. Parasitol.* **33:**933–943.

Al-Olayan, E. M., A. L. Beetsma, G. A. Butcher, R. E. Sinden, and H. Hurd. 2002. Complete development of mosquito phases of malaria parasite in vitro. *Science* **295:**677–679.

Amino, R., A. Genovesio, S. Celli, P. Roux, J. C. Olivo-Marin, S. L. Shorte, R. Ménard, and F. Frischknecht. 2004. From the skin to the liver: in vivo imaging of Plasmodium sporozoites, abst. 5C. Molecular Parasitology Meeting XV, Woods Hole, Mass.

Ancsin, J. B., and R. Kisilevsky. 2004. A binding site for highly sulphated heparan sulphate is identified at the N terminus of the circumsporozoite protein: significance for malarial sporozoite attachment to hepatocytes. *J. Biol. Chem.* **279:**21824–21832.

Anderson, R. A., B. G. Knols, and J. C. Koella. 2000. Plasmodium falciparum sporozoites increase feeding-association mortality of their mosquito hosts Anopheles gambiae s.l. *Parasitology* **120:**329–333.

Baldacci, P., and R. Ménard. 2004. The elusive malaria sporozoite in the mammalian host. *Mol. Microbiol.* **54:**298–306.

Bannister, L. H., J. M. Hopkins, A. R. Dluzewski, G. Margos, I. T. Williams, M. J. Blackman, C. H. Kocken, A. W. Thomas, and G. H. Mitchell. 2003. Plasmodium falciparum apical membrane antigen 1 (PfAMA-1) is translocated within micronemes along subpellicular microtubules during merozoite development. *J. Cell Sci.* **116:**3825–3834.

Barragan, A., and L. D. Sibley. 2003. Migration of Toxoplasma gondii across biological barriers. *Trends Microbiol.* **11:**426–30.

Barreau, C., M. Touray, P. F. Pimenta, L. H. Miller, and K. D. Vernick. 1995. Plasmodium gallinaceum sporozoite invasion of Aedes aegypti salivary glands is inhibited by anti-gland antibodies and by lectins. *Exp. Parasitol.* **81:**332–343.

Beaudoin, R. L., C. P. A. Strome, and T. A. Tubergen. 1974. Plasmodium berghei berghei: ectopic development of the ANKA strain in Anopheles stephensi. *Exp. Parasitol.* **36:**189–201.

Beaudoin, R. L., C. P. A. Strome, T. A. Tubergen, and F. Mitchell. 1976. Plasmodium berghei berghei: irradiated sporozoites of the ANKA strain as immunizing agents in mice. *Exp. Parasitol.* **39:**438–443.

Beier, M. S., J. R. Davis, C. B. Pumpuni, B. H. Noden, and J. C. Beier. 1992. Ingestion of Plasmodium falciparum by anopheline mosquitoes. *Am. J. Trop. Med. Hyg.* **47:**195–200.

Bergman, L. W., K. Kaiser, H. Fujioka, I. Coppens, T. M. Daly, S. Fox, K. Matuschewski, V. Nussenzweig and S. H. I. Kappe. 2003. Myosin A tail domain interacting protein (MTIP) localizes to the inner membrane complex of Plasmodium sporozoites. *J. Cell Sci.* **116:**39–49.

Bhanot, P., K. Schauer, I. Coppens, and V. Nussenzweig. 2005. A surface phospholipase is involved in the migration of plasmodium sporozoites through cells. *J. Biol. Chem.* **280:**6752–6760.

Bottius, E., L. Ben Mohamed, K. Brahimi, H. Gras, J. P. Lepers, L. Raharimalala, M. Aikawa, J. Meis, B. Slierendregt, A. Tartar, A. Thomas, and P. Druilhe. 1996. A novel Plasmodium falciparum sporozoite and liver stage antigen (SALSA) defines major B, T helper, and CTL epitopes. *J. Immunol.* **156:**2874–2884.

Brennan, J. D. G., M. Kent, R. Dhar, H. Fujioka, and N. Kumar. 2000. Anopheles gambiae salivary gland proteins as putative targets for blocking transmission of malaria parasites. *Proc. Natl. Acad. Sci. USA* **97:**13859–13864.

Carrolo, M., S. Giordano, L. Cabrita-Santos, S. Corso, A. M. Vigario, S. Silva, P. Leiriao, D. Carapau, R. Armas-Portela, P. M. Comoglio, A. Rodriguez, and M. M. Mota. 2003. Hepatocyte growth factor and its receptor are required for malaria infection. *Nat. Med.* **9:**1363–1369.

Carvalho, T. G., S. Thiberge, H. Sakamoto, and R. Ménard, R. 2004. Conditional mutagenesis using site-specific recombination in Plasmodium berghei. *Proc. Natl. Acad. Sci. USA* **101:**14931–14936.

Cerami, C., U. Frevert, P. Sinnis, B. Takacs, P. Clavijo, M. J. Santos, and V. Nussenzweig. 1992. The basolateral domain of the hepatocyte plasma membrane bears receptors for the circumsporozoite protein of Plasmodium falciparum sporozoites. *Cell* **70:**1021–1033.

Cerami, C., U. Frevert, P. Sinnis, B. Takacs, and V. Nussenzweig. 1994. Rapid clearance of malaria circumsporozoite protein (CS) by hepatocytes. *J. Exp. Med.* **179:**695–701.

Chattopadhyay, R., D. Rathore, H. Fujioka, S. Kumar, S., P. de la Vega, D. Haynes, K. Moch, D. Fryauff, R. Wang, D. J. Carucci, and S. L. Hoffman. 2003. PfSPATR, a Plasmodium falciparum protein containing an altered thrombospondin type I repeat domain is expressed at several stages of the parasite life cycle and is the target of inhibitory antibodies. *J. Biol. Chem.* **278:**25977–25981.

Claudianos, C., H. E. Trueman, J. T. Dessens, M. Arai, J. Mendoza, G. A. Butcher, T. Crompton, and R. E. Sinden. 2002. A malaria scavenger receptor-like protein essential for parasite development. *Mol. Microbiol.* **45:**1473–1484.

Coleman, R. E., A. K. Nath, I. Schneider, G. H. Song, T. A. Klein, and W. K. Milhous. 1994. Prevention of sporogony of Plasmodium falciparum and P. berghei in Anopheles stephensi mosquitoes by transmission-blocking antimalarials. *Am. J. Trop. Med. Hyg.* **50:**646–653.

Collins, F. H., R. K. Sakai, K. D. Vernick, S. Paskewitz, D. C. Seeley, L. H. Miller, W. E. Collins, C. C. Campbell, and R. W. Gwadz. 1986. Genetic selection of a plasmodium refractory strain of the malaria vector Anopheles gambiae. *Science* **234:**607–610.

Coppi, A., C. Pinzon-Ortiz, C. Hutter, and P. Sinnis. 2005. The *Plasmodium* circumsporozoite protein is proteolytically processed during cell invasion. *J. Exp. Med.* **201:**27–33.

Danforth, H., D. M. Aikawa, A. H. Cochrane, and R. S. Nussenzweig. 1980. Sporozoites of mammalian malaria: attachment to, interiorization and fate within macrophages. *J. Protozool.* **27:**193–202.

Danforth, H. D., R. Entzeroth, and B. Chobotar. 1992. Scanning and transmission electron microscopy of host cell pathology associated with penetration of Eimeria papillata sporozoites. *Parasitol. Res.* **78:**570–573.

de Lara Capurro, M., J. Coleman, B. T. Beerntsen, K. M. Myles, K. I. E. Olson, A. U. Krettli, and A. A. James. 2000. Virus-expressed, recombinant single-chain antibody blocks sporozoite infection of salivary glands in *Plasmodium gallinaceum*-infected *Aedes aegypti*. *Am. J. Trop. Med. Hyg.* **62:**427–433.

Dimopoulos, G., D. Seeley, A. Wolf, and F. C. Kafatos. 1998. Malaria infection of the mosquito Anopheles gambiae activates immune-responsive genes during critical transition stages of the parasite life cycle. *EMBO J.* **17:**6115–6123.

Dobrowolski, J. M., and L. D. Sibley. 1996. Toxoplasma invasion of mammalian cells is powered by the actin cytoskeleton of the parasite. *Cell* **84:**933–939.

do Rosario, V. E., A. Appiah, J. A. Vaughan, and M. R. Hollingdale. 1989. Plasmodium falciparum: administration of anti-sporozoite antibodies during sporogony results in production of sporozoites which are not neutralized by human anti-circumsporozoite protein vaccine sera. *Trans. R. Soc. Trop. Med. Hyg.* **83:**305–307.

Fidock, D. A., E. Bottius, K. Brahimi, I. I. M. D. Moelans, M. Aikawa, R. N. H. Konings, U. Certa, P. Olafsson, T. Kaidoh, A. Asavanich, C. Guerinmarchand, and P. Druilhe. 1994. Cloning and characterization of a novel Plasmodium falciparum sporozoite surface antigen, STARP. *Mol. Biochem. Parasitol.* **64:**219–232.

Florens, L., M. P. Washburn, J. D. Raine, R. M. Anthony, M. Grainger, J. D. Haynes, J. K. Moch, N. Muster, J. B. Sacci, D. L. Tabb, A. A. Witney, D. Wolters, Y. Wu, M. J. Gardner, A. A. Holder, R. E. Sinden, J. R. Yates III, and D. J. Carucci. 2002. A proteomic view of the Plasmodium falciparum life cycle. *Nature* **419:**520–526.

Fowler, R. E., R. E. Sinden, and M. Pudney. 1995. Inhibitory activity of the anti-malarial atovaquone (566C80) against ookinete, oocysts and sporozoites of Plasmodium berghei. *J. Parasitol.* **8:**454–458.

Frevert, U. 2004. Sneaking through the back entrance: the biology of malaria sporozoite stages. *Trends Parasitol.* **20:**417–424.

Frevert, U., M. R. Galinski, F-U. Hugel, N. Allon, H. Schreier, S. Smulevitch, M. Shakibaei, and P. Clavijo. 1998. Malaria circumsporozoite protein inhibits protein synthesis in mammalian cells. *EMBO J.* **17:**3816–3826.

Frevert, U., P. Sinnis, C. Cerami, W. Shreffler, B. Takacs, and V. Nussenzweig. 1993. Malaria circumsporozoite protein binds to heparan sulfate proteoglycans associated with the surface membrane of hepatocytes. *J. Exp. Med.* **177:**1287–1298.

Frevert, U., S. Engelmann, S. Zougbédé, J. Stange, B. Ng, K. Matuschewski, L. Liebes, and H. Yee.

2005. Intravital observation of Plasmodium berghei sporozoite infection of the liver. *PLoS Biol.* **3:**e192.

Frischknecht, F., P. Baldacci, B. Martin, C. Zimmer, S. Thiberge, J.-C. Olivo-Marin, S. L. Shorte, and R. Ménard. 2004. Imaging movement of malaria parasites during transmission by Anopheles mosquitoes. *Cell. Microbiol.* **6:**687–694.

Gakhar, S. K., and H. K. Shandilya. 2001. Midgut specific immune response of vector mosquito Anopheles stephensi to malaria parasite Plasmodium. *Ind. J. Exp. Biol.* **39:**287–290.

Gantt, S., C. Persson, K. Rose, A. J. Birkett, R. Abagayan, and V. Nussenzweig. 2000. Antibodies against the thrombospondin-related anonymous protein do not inhibit Plasmodium sporozoite infectivity in vivo. *Infect. Immun.* **68:**3667–3673.

Garnham, P. C. C. 1966. *Malaria Parasites and Other Haemosporidia.* Blackwell Scientific Publications, Oxford, United Kingdom.

Ghosh, A. K., D. Jethwaney, M. Devenport, V. Anderson, A. A. Sultan, and M. Jacobs-Lorena. 2004. Molecular dissection of sporozoite-salivary gland interactions, abstr. 111. American Society of Tropical Medicine and Hygiene (ASTMH) 53rd Annual Meeting, Miami, Fla.

Ghosh, A. K., P. E. Ribolla, and M. Jacobs-Lorena. 2001. Targeting Plasmodium ligands on mosquito salivary glands and midgut with a phage display peptide library. *Proc. Natl. Acad. Sci. USA* **98:**13278–13281.

Godson, G. N., J. Ellis, J. R. Lupski, L. S. Ozaki, and P. Svec. 1984. Structure and organization of genes for sporozoite surface antigens. *Philos. Trans. R. Soc. Lond. B Biol. Sci.* **307:**129–139.

Godson, G. N., J. Ellis, P. Svec, D. H. Schlesinger, and V. Nussenzweig. 1983. Identification and chemical synthesis of a tandemly repeated immunogenic region of Plasmodium knowlesi circumsporozoite protein. *Nature* **305:**29–33.

Grüner, A. C., K. Brahimi, F. Letourneur, L. Renia, W. Eling, G. Snounou, and P. Druilhe. 2001. Expression of the erythrocyte-binding antigen 175 in sporozoites and in liver stages of Plasmodium falciparum. *J. Infect. Dis.* **184:**892–897.

Gwadz, R. W., D. Kaslow, J. Y. Lee, W. L. Maloy, M. Zasloff, and L. H. Miller. 1989. Effects of magainins and cecropins on the sporogonic development of malaria parasites in mosquitoes. *Infect. Immun.* **57:**2628–2633.

Hall, N., M. Karras, J. D. Raine, J. M. Carlton, T. W. A. Kooij, M. Berriman, L. Florens, C. S. Janssen, A. Pain, G. K. Christophides, K. James, K. Rutherford, B. Harris, D. Harris, C. Churcher, M. A. Quail, D. Ormond, J. Doggett, H. E. Trueman, J. Mendoza, S. L. Bidwell, M.-A. Rajandream, D. J. Carucci, J. R. Yates III, F. C. Kafatos, C. J. Janse, B. Barrell, C. M. R. Turner, A. P. Waters, and R. E. Sinden. 2005. A comprehensive survey of the Plasmodium life cycle by genomic, transcriptomic, and proteomic analyses. *Science* **307:**82–86.

Hamilton, A. J., C. S. Davies, and R. E. Sinden. 1988. Expression of the circumsporozoite proteins revealed in situ in the mosquito stages of Plasmodium berghei by the Lowicryl-immunogold technique. *Parasitology* **96:**273–280.

Han, Y. S., J. Thompson, F. C. Kafatos, and C. Barillas-Mury. 2000. Molecular interactions between Anopheles stephensi midgut cells and Plasmodium berghei: the time bomb theory of ookinete invasion of mosquitoes. *EMBO J.* **19:**6030–6039.

Healer, J. F., S. Crawford, S. Ralph, G. McFadden, and A. F. Cowman. 2002. Independent translocation of two micronemal proteins in developing *Plasmodium falciparum* merozoites. *Infect. Immun.* **70:** 5751–5758.

Hemler, M. E. 2003. Tetraspanin proteins mediate cellular penetration, invasion, and fusion events and define a novel type of membrane microdomain. *Annu. Rev. Cell. Dev. Biol.* **19:**397–422.

Hillyer, J. F., S. L. Schmidt, and B. M. Christensen. 2003. Rapid phagocytosis and melanization of bacteria and Plasmodium sporozoites by hemocytes of the mosquito Aedes aegypti. *J. Parasitol.* **89:**62–69.

Hudson-Taylor, D. E., S. A. Dolan, F. W. Klotz, H. Fujioka, M. Aikawa, E. V. Koonin, and L. H. Miller. 1995. Plasmodium falciparum protein associated with the invasion junction contains a conserved oxidoreductase domain. *Mol. Microbiol.* **15:**463–471.

Ishino, T., K. Yano, Y. Chinzei, and M. Yuda. 2004. Cell-passage activity is required for the malarial parasite to cross the liver sinusoidal cell layer. *PLoS Biol.* **2:**E4.

Ishino, T., Y. Chinzei, and M. Yuda. 2005. A Plasmodium sporozoite protein with a membrane attack complex domain is required for breaching the liver sinusoidal cell layer prior to hepatocyte infection. *Cell. Microbiol.* **7:**199–208.

Ito, J., A. Ghosh, L. A. Moreira, E. A. Wimmer, and M. Jacobs-Lorena. 2002. Transgenic anopheline mosquitoes impaired in transmission of a malaria parasite. *Nature* **417:**452–455.

James, A. A. 2003. Blocking malaria parasite invasion of mosquito salivary glands. *J. Exp. Biol.* **206:**3817–3821.

James, S. P. 1928. Ross's "black spores" in the stomach and salivary glands of malaria-infected mosquitoes. *Trans. R. Soc. Trop. Med. Hyg.* **21:**258.

Jewett, T. J., and L. D. Sibley. 2003. Aldolase forms a bridge between cell surface adhesins and the actin cytoskeleton in apicomplexan parasites. *Mol. Cell* **11:**885–894.

Kadota, K., T. Ishino, T. Matsuyama, Y. Chinzei, and M. Yuda. Essential role of membrane attack protein in malarial transmission to the mosquito host. *Proc. Natl. Acad. Sci. USA* **101:**16310–16315.

Kaiser, K., N. Camargo, I. Coppens, J. M. Morrisey, A. B. Vaidya, and S. H. I. Kappe. 2004a. A member of a conserved Plasmodium protein family with membrane-attack complex/perforin (MACPF)-like domains localizes to the micronemes of sporozoites. *Mol. Biochem. Parasitol.* **133:**15–26.

Kaiser, K., K. Matuschewski, N. Camargo, J. Ross, and S. H. I. Kappe. 2004b. Differential transcriptome profiling identifies Plasmodium genes encoding pre-erythrocytic stage-specific proteins. *Mol. Microbiol.* 51:1221–1232.

Kappe, S., T. Bruderer, S. Gantt, H. Fujioka, V. Nussenzweig, and R. Ménard. 1999. Conservation of a gliding motility and cell invasion machinery in apicomplexan parasites. *J. Cell Biol.* **147:**937–944.

Kappe, S. H. I., C. A. Buscaglia, L. W. Bergman, I. Coppens, and V. Nussenzweig. 2004a. Apicomplexan gliding motility and host cell invasion: overhauling the motor model. *Trends Parasitol.* **20:**13–16.

Kappe, S. H. I., C. A. Buscaglia, and V. Nussenzweig. 2004b. Plasmodium sporozoite molecular cell biology. *Annu. Rev. Cell Dev. Biol.* **20:**29–59.

Kappe, S. H. I., M. J. Gardner, S. M. Brown, J. Ross, K. Matuschewski, J. M. Ribeiro, J. H. Adams, J. Quackenbush, J. Cho, D. J. Carucci, S. L. Hoffman, and V. Nussenzweig. 2001. Exploring the transcriptome of the malaria sporozoite stage. *Proc. Natl. Acad. Sci. USA* **98:**9895–9900.

Kappe, S. H. I., A. R. Noe, T. S. Fraser, P. L. Blair, and J. H. Adams. 1998. A family of chimeric erythrocyte binding proteins of malaria parasites. *Proc. Natl. Acad. Sci. USA* **95:**1230–1235.

Kariu, T., M. Yuda, K. Yano, and Y. Chinzei. 2002. MAEBL is essential for malarial sporozoite infection of the mosquito salivary gland. *J. Exp. Med.* **195:**1317–1323.

Khater, E. I., R. E. Sinden, and J. T. Dessens. 2004. A malaria membrane skeletal protein is essential for normal morphogenesis, motility and infectivity of sporozoites. *J. Cell Biol.* **167:**425–432.

King, C. A. 1988. Cell motility of sporozoan protozoa. *Parasitol. Today* **4:**315–319.

Klotz, F. W., T. J. Hadley, M. Aikawa, J. Leech, R. J. Howard, and L. H. Miller. 1989. A 60-kDa Plasmodium falciparum protein at the moving junction formed between merozoite and erythrocyte during invasion. *Mol. Biochem. Parasitol.* **36:**177–186.

Koella, J. C., and C. Boete. 2003. A model for the coevolution of immunity and immune evasion in vector-borne diseases with implications for the epidemiology of malaria. *Am. Nat.* **161:**698–707.

Le Roch, K. G., Y. Zhou, P. L. Blair, M. Grainger, J. K. Moch, J. D. Haynes, P. Vega, A. A. Holder, S. Batalov, D. J. Carucci, and E. A. Winzeler. 2003. Discovery of gene function by expression profiling of the malaria parasite life cycle. *Science* **12:**1487–1488.

Lingelbach, K., and K. A. Joiner. 1998. The parasitophorous vacuole membrane surrounding Plasmodium and Toxoplasma: an unusual compartment in infected cells. *J. Cell Sci.* **111:**1467–1475.

Lowenberger, C. A., S. Kamal, J. Chiles, S. Paskewitz, P. Bulet, J. A. Hoffmann, and B. M. Christensen. 1999. Mosquito-Plasmodium interactions in response to immune activation of the vector. *Exp. Parasitol.* **1:**59–69.

Luckhart, S., Y. Vodovotz, L. W. Cui, and R. Rosenberg. 1998. The mosquito Anopheles stephensi limits malaria parasite development with inducible synthesis of nitric oxide. *Proc. Natl. Acad. Sci. USA* **95:**5700–5705.

Matuschewski, K., J. Ross, S. M. Brown, K. Kaiser, V. Nussenzweig, and S. H. I. Kappe. 2000a. Infectivity-associated changes in the transcriptional repertoire of the malaria sporozoite stage. *J. Biol. Chem.* **277:**41948–41953.

Matuschewski, K., A. C. Nunes, V. Nussenzweig, and R. Ménard. 2002b. Plasmodium sporozoite invasion into insect and mammalian cells is directed by the same dual binding system. *EMBO J.* **21:**1597–1606.

McCutchan, T. F., J. C. Kissinger, M. G. Touray, M. J. Rogers, M. Li, M. Sullivan, E. M. Braga, A. U. Krettli, and L. H. Miller. 1996. Comparison of circumsporozoite proteins from avian and mammalian malarias: biological and phylogenetic implications. *Proc. Natl. Acad. Sci. USA* **93:**11889–11894.

Ménard, R. 2000. The journey of the malaria sporozoite through its hosts: two parasite proteins lead the way. *Microbes Infect.* **2:**633–642.

Ménard, R., A. A. Sultan, C. Cortes, R. Altszuler, M. R. van Dijk, C. J. Janse, A. P. Waters, R. S. Nussenzweig, and V. Nussenzweig. 1997. Circumsporozoite protein is required for development of malaria sporozoites in mosquitoes. *Nature* **385:**336–340.

Mitchell, G. H., A. W. Thomas, G. Margos, A. R. Dluzewski, and L. H. Bannister. 2004. Apical membrane antigen 1, a major malaria vaccine candidate, mediates the close attachment of invasive merozoites to host red blood cells. *Infect. Immun.* **72:**154–158.

Mota, M. M., G. Pradel, J. P. Vanderberg, J. C. R. Hafalla, U. Frevert, R. S. Nussenzweig, V. Nussenzweig, and A. Rodriguez. 2001. Migration of Plasmodium sporozoites through cells before infection. *Science* **291:**141–144.

Mota, M. M., J. C. Hafalla, and A. Rodriguez. 2002. Migration through host cells activates Plasmodium sporozoites for infection. *Nat Med.* **8:**1318–1322.

Motard, A., I. Landau, A. Nussler, G. Grau, D. Baccam, D. Mazier, and G. A. T. Targett. 1993. The role of reactive nitrogen intermediates in modulation of gametocyte infectivity of rodent malaria parasites. *Parasite Immunol.* **15:**21–26.

Mueller, A.-K., M. Labaied, S. H. I. Kappe, and K. Matuschewski. 2005a. Genetically modified Plasmodium parasites as a protective experimental malaria vaccine. *Nature* **433:**164–167.

Mueller, A.-K., N. Camargo, K. Kaiser, C. Andorfer, U. Frevert, K. Matuschewski, and S. H. I. Kappe. 2005. Plasmodium liver stage developmental arrest by depletion of a protein at the parasite-host interface. *Proc. Natl. Acad. Sci. USA* **102:**3022–3027.

Myung, J. M., P. Marshall, and P. Sinnis. 2004. The Plasmodium circumsporozoite protein is involved in most salivary gland invasion by sporozoites. *Mol. Biochem. Parasitol.* **133:**53–59.

Naotunne, T. S., N. D. Karunaweera, K. N. Mendis, and R. Carter. 1993. Cytokine-mediated inactivation of malarial gametocytes is dependent on the presence of white blood cells and involves reactive nitrogen intermediates. *Immunology* **78:**555–562.

Natarajan, R., V. Thathy, M. M. Mota, J. C. Hafala, R. Ménard, and K. D. Vernick. 2001. Fluorescent Plasmodium berghei sporozoites and pre-erythrocytic stages: a new tool to study mosquito and mammalian host interactions with malaria parasites. *Cell Microbiol.* **3:**371–379.

Nirmala, X., and A. A. James. 2003. Engineering Plasmodium-refractory phenotypes in mosquitoes. *Trends Parasitol.* **19:**384–387.

Nussenzweig, R. S., and F. Zavala. 1997. A malaria vaccine based on a sporozoite antigen. *N. Engl. J. Med.* **336:**128–130.

Nussenzweig, R. S., J. P. Vanderberg, H. Most, and C. Orton. 1967. Protective immunity produced by injection of X-irradiated sporozoites of Plasmodium berghei. *Nature* **216:**160–162.

Opitz, C., and D. Soldati. 2002. 'The glideosome': a dynamic complex powering gliding motion and host cell invasion by *Toxoplasma gondii*. *Mol. Microbiol.* **45:**597–604.

Oren, R., S. Takahashi, C. Doss, R. Levy, and S. Levy. 1990. TAPA-1, the target of the antiproliferative antibody, defines a new family of transmembrane proteins. *Mol. Cell. Biol.* **10:**4007–4015.

Persson, C., G. A. Oliveira, A. A. Sultan, P. Ghanot, V. Nussenzweig, and E. Nardin. 2002. Cutting Edge: a new tool to evaluate human pre-erythrocytic malaria vaccines: rodent parasites bearing a hybrid Plasmodium falciparum circumsporozoite protein. *J. Immunol.* **169:**6681–6685.

Pimenta, P. F., M. Touray, and L. Miller. 1994. The journey of malaria sporozoites in the mosquito salivary gland. *J. Euk. Microbiol.* **41:**608–624.

Ponnudurai, T., A. H. W. Lensen, G. J. A. Vangemert, M. G. Bolmer, and J. H. E. T. Meuwissen. 1991. Feeding behaviour and sporozoite ejection by infected Anopheles stephensi. *Trans. R. Soc. Trop. Med. Hyg.* **85:**175–180.

Pradel, G., and U. Frevert. 2001. Malaria sporozoites actively enter and pass through rat Kupffer cells prior to hepatocyte invasion. *Hepatology* **33:**1154–1165.

Pradel, G., S. Garapaty, and U. Frevert. 2002. Proteoglycans mediate malaria sporozoite targeting to the liver. *Mol. Microbiol.* **45:**637–651.

Pradel, G., K. Hayton, L. Aravind, L. M. Iyer, M. S. Abrahamsen, A. Bonawitz, C. Mejia, and T. J. Templeton. A multidomain adhesion protein family expressed in Plasmodium falciparum is essential for transmission to the mosquito. *J. Exp. Med.* **199:**1533–1544.

Preiser, P., L. Renia, N. Singh, B. Balu, W. Jarra, T. Voza, O. Kaneko, P. Blair, M. Torii, I. Landau, and J. H. Adams. 2004. Antibodies against MAEBL ligand domains M1 and M2 inhibit sporozoite development in vitro. *Infect. Immun.* **72:**3604–3608.

Roberts, W. L., C. A. Speer, and D. M. Hammond. 1971. Penetration of Eimeria larimerensis sporozoites into cultured cells as observed with the light and electron microscopes. *J. Parasitol.* **57:**615–625.

Robson, K. J. H., U. Frevert, I. Reckmann, G. Cowan, J. Beier, I. G. Scragg, K. Takehara, D. H. L. Bishop, G. Pradel, R. Sinden, S. Saccheo, H. M. Muller, and A. Crisanti. 1995. Thrombospondin-related adhesive protein (TRAP) of plasmodium falciparum: expression during sporozoite ontogeny and binding to human hepatocytes. *EMBO J.* **14:**3883–3894.

Rosenberg, R. 1992. Ejection of malaria sporozoites by feeding mosquitoes. *Trans. R. Soc. Trop. Med. Hyg.* **86:**109.

Rosenberg, R. 1985. Inability of Plasmodium knowlesi sporozoites to invade Anopheles freeborni salivary glands. *Am. J. Trop. Med. Hyg.* **34:**687–691.

Rossignol, P. A., J. M. C. Ribeiro, and A. Spielman. 1986. Increased biting rate and reduced fertility in sporozoite-infected mosquitoes. *Am. J. Trop. Med. Hyg.* **35:**277–279.

Rossignol, P. A., J. M. C. Ribeiro, and A. Spielman. 1984. Increased intradermal probing time in sporozoite-infected mosquitoes. *Am. J. Trop. Med. Hyg.* **33:**17–20.

Russell, D. G. 1983. Host cell invasion by Apicomplexa: an expression of the parasite's contractile system. *Parasitology* **87:**199–209.

Russell, D. G., and R. E. Sinden. 1981. The role of the cytoskeleton in the motility of coccidian sporozoites. *J. Cell Sci.* **50:**345–359.

Schaudin, F. 1903. Studien ueber krankheitserregende Protozoen. II. Plasmodium vivax (Grassi und Feletti), der Erreger des Tertien-Fiebers beim Menschen. *Arb. Kaiserlichen Gesundheitsamte* **19**:169–250.

Schneider, D., and M. Shahabuddin. 2000. Malaria parasite development in a Drosophila model. *Science* **288**:2376–2379.

Schüler, H., A.-K. Mueller, and K. Matuschewski. 2005. Unusual properties of Plasmodium falciparum actin: new insights into microfilament dynamics of apicomplexan parasites. *FEBS Lett.* **579**:655–660.

Shahabuddin, M., I. Fields, P. Bulet, J. A. Hoffmann, and L. H. Miller. 1998. Plasmodium gallinaceum: differential killing of some mosquito stages of the parasite by insect defensin. *Exp. Parasitol.* **89**:103–112.

Shaw, M. K., and L. G. Tilney. 1999. Induction of an acrosomal process in Toxoplasma gondii: visualization of actin filaments in a protozoan parasite. *Proc. Natl. Acad. Sci. USA* **96**:9095–9099.

Shin, S. C. J., J. P. Vanderberg, and J. A. Terzakis. 1982. Direct infection of hepatocytes by sporozoites of Plasmodium berghei. *J. Protozool.* **29**:448–454.

Sibley, L. D. 2004. Intracellular parasite invasion strategies. *Science* **304**:248–253.

Sidjanski, S., and J. P. Vanderberg. 1997. Delayed migration of Plasmodium sporozoites from the mosquito bite site to the blood. *Am. J. Trop. Med. Hyg.* **57**:426–429.

Sidjanski, S. P., J. P. Vanderberg, and P. Sinnis. 1997. Anopheles stephensi salivary glands bear receptors for region I of the circumsporozoite protein of Plasmodium falciparum. *Mol. Biochem. Parasitol.* **90**:33–41.

Silvie, O., J.-F. Franetich, S. Charrin, M. S. Mueller, A. Siau, M. Bodescot, E. Rubinstein, L. Hannoun, Y. Charoenvit, C. H. Kocken, A. W. Thomas, G. J. Van Gemert, R. W. Sauerwein, M. J. Blackman, R. F. Anders, G. Plüschke, and D. Mazier. 2004. A role for apical membrane antigen 1 during invasion of hepatocytes by Plasmodium falciparum sporozoites. *J. Biol. Chem.* **279**:9490–9496.

Silvie, O., E. Rubinstein, J. F. Franetich, M. Prenant, E. Belnoue, L. Renia, L. Hannoun, W. Eling, S. Levy, C. Boucheix, and D. Mazier. 2003. Hepatocyte CD81 is required for Plasmodium falciparum and Plasmodium yoelii sporozoite infectivity. *Nat. Med.* **9**:93–96.

Sinden, R. E. 1974a. Excystment by sporozoites of malaria parasites. *Nature* **252**:314–310.

Sinden, R. E. 1974b. The sporogonic cycle of Plasmodium yoelii nigeriensis: a scanning electron microscope study. *Protistologica* **11**:31–39.

Sinden, R. E., Y. I. H. Alavi, G. A. Butcher, J. T. Dessens, J. D. Raine, H. E. Trueman. 2004. Chapter 15. Ookinete cell biology, p. 475–500. *In* A. P. Waters and C. J. Janse (ed.), *Genomes and the Molecular Cell Biology of Malaria Parasites*. Caister Academic Press, Wymondham, United Kingdom.

Sinden, R. E., and P. C. C. Garnham. 1973. A comparative study of the ultrastructure of Plasmodium sporozoites within the oocyst and salivary glands, with particular reference to the incidence of the micropore. *Trans. R. Soc. Trop. Med. Hyg.* **67**:631–637.

Sinden, R. E., and J. E. Smith. 1982. The role of the Kupffer cell in the infection of rodents by sporozoites of Plasmodium: uptake of sporozoites by perfused liver and the establishment of infection in vivo. *Acta Trop.* **39**:11–27.

Singh, N., P. Preiser, L. Renia, B. Balu, J. Barnwell, P. Blair, W. Jarra, T. Voza, I. Landau, and J. H. Adams. 2004. Conservation and developmental control of alternative splicing in maebl among malaria parasites. *J. Mol. Biol.* **343**:589–599.

Sinnis, P., P. Clavijo, D. Fenyö, B. T. Chait, C. Cerami, and V. Nussenzweig. 1994. Structural and functional properties of region II-plus of the malaria circumsporozoite protein. *J. Exp. Med.* **180**:297–306.

Sinnis, P., and E. Nardin. 2002. Sporozoite antigens: biology and immunology of the circumsporozoite protein and thrombospondin-related anonymous protein. *Chem. Immunol.* **80**:70–96.

Smith, J. E., and J. Alexander. 1986. Evasion of macrophage microbicidal mechanisms by mature sporozoites of Plasmodium yoelii yoelii. *Parasitology* **93**:33–38.

Smith, J. E., and R. E. Sinden. 1982. On the relationship between Kupffer cell activity and the uptake and infectivity of sporozoites of Plasmodium yoelii nigeriensis, p. 437–444. *In* D. L. Knook and E. Wisse (ed.), *Sinusoidal Liver Cells*. Elsevier Biomedical Press Amsterdam, The Netherlands.

Speer, C. A., J. P. Dubey, J. A. Blixt, and K. Prokop. 1997. Time lapse video microscopy and ultrastructure of penetrating sporozoites, types 1 and 2 parasitophorous vacuoles, and the transformation of sporozoites to tachyzoites of the VEG strain of Toxoplasma gondii. *J. Parasitol.* **83**:565–574.

Srinivasan, P., E. G. Abraham, A. K. Ghosh, J. Valenzuela, J. M. Ribeiro, G. Dimopoulos, F. C. Kafatos, J. H. Adams, H. Fujioka, and M. Jacobs-Lorena. 2004. Analysis of the Plasmodium and Anopheles transcriptomes during oocyst differentiation. *J. Biol. Chem.* **279**:5581–5587.

Sterling, C. R., M. Aikawa, and J. P. Vanderberg. 1973. Passage of Plasmodium berghei sporozoites through the salivary glands of Anopheles stephensi: an electron microscope study. *J. Parasitol.* **59**:593–605.

Stewart, M. J., R. Nawrot, S. Schulman, and J. P. Vanderberg. 1986. Plasmodium berghei sporozoite invasion is blocked in vitro by sporozoite-immobilizing antibodies. *Infect. Immun.* **51**:859–864.

Stewart, M. J., and J. P. Vanderberg. 1988. Malaria sporozoites leave behind a trail of circumsporozoite protein during gliding motility. *J. Protozool.* **35:**389–393.

Sultan, A. A., V. Thathy, U. Frevert, K. J. Robson, A. Crisanti, V. Nussenzweig, R. Nussenzweig, and R. Ménard. 1997. TRAP is necessary for gliding motility and infectivity of Plasmodium sporozoites. *Cell* **90:**511–522.

Templeton, T. J., and D. C. Kaslow. 1997. Cloning and cross-species comparison of the thrombospondin-related anonymous protein (TRAP) gene from Plasmodium knowlesi, Plasmodium vivax and Plasmodium gallinaceum. *Mol. Biochem. Parasitol.* **84:**13–24.

Tewari, R., R. Spaccapelo, F. Bistoni, A. A. Holder, and A. Crisanti. 2002. Function of region I and II adhesive motifs of Plasmodium falciparum circumsporozoite protein in sporozoite motility and infectivity. *J. Biol. Chem.* **277:**47613–47618.

Thathy, V., H. Fujioka, S. Gantt R. Nussenzweig, V. Nussenzweig, and R. Ménard. 2002. Levels of circumsporozoite protein in the Plasmodium oocyst determine sporozoite morphology. *EMBO J.* **21:**1586–1596.

Touray, M. G., A. Warburg, A. Laughinghouse, A. U. Krettli, and L. H. Miller. 1992. Developmentally regulated infectivity of malaria sporozoites for mosquito salivary glands and the vertebrate host. *J. Exp. Med.* **175:**1607–1612.

Triglia, T., J. Healer, S. R. Cruana, A. N. Hodder, R. F. Anders, B. S. Crabb, and A. F. Cowman. 2000. Apical membrane antigen 1 plays a central role in erythrocyte invasion by Plasmodium species. *Mol. Microbiol.* **38:**706–718.

Trueman, H. E., J. D. Raine, L. Florens, J. T. Dessens, J. Mendoza, J. Johnson C. C. Waller, I. Delrieu, A. A. Holder, J. Langhorne, D. J. Carucci, J. R. Yates III, and R. E. Sinden. 2004. Functional characterisation of an LCCL/lectin-domain containing protein family in Plasmodium berghei. *J. Parasitol.* **90:**1062–1071.

Vanderberg, J. P. 1975. Development of infectivity by the Plasmodium berghei sporozoite. *J. Parasitol.* **61:**43–50.

Vanderberg, J. P. 1974. Studies on the motility of Plasmodium sporozoites. *J. Protozool.* **21:**527–537.

Vanderberg, J. P. 1977. Plasmodium berghei: quantitation of sporozoites injected by mosquitoes feeding on a rodent host. *Exp. Parasitol.* **42:**169–181.

Vanderberg, J. P. 1990. Plasmodium sporozoite interactions with macrophages in vitro: a videomicroscopic analysis. *J. Protozool.* **37:**528–536.

Vanderberg, J. P., R. S. Nussenzweig, Y. Sanabria, R. Nawrot, and H. Most. 1972. Stage specificity of anti-sporozoite antibodies in rodent malaria and its relationship to protective immunity. *Proc. Helminthol. Soc. Washington* **39:**514–525.

Vanderberg, J. P., and U. Frevert. 2004. Intravital microscopy demonstrating antibody-mediated immobilisation of Plasmodium berghei sporozoites injected into skin by mosquitoes. *Int. J. Parasitol.* **34:**991–996.

Vreden, S. G. S., R. W. Sauerwein, J. P. Verhave, N. Vanrooijen, J. H. E. T. Meuwissen, and M. F. Vandenbroek. 1993. Kupffer cell elimination enhances development of liver schizonts of *Plasmodium berghei* in rats. *Infect. Immun.* **61:**1936–1939.

Warburg, A., M. Touray, A. U. Krettli, and L. H. Miller. 1992. Plasmodium gallinaceum—antibodies to circumsporozoite protein prevent sporozoites from invading the salivary glands of Aedes aegypti. *Exp. Parasitol.* **75:**303–307.

Wengelnik, K., R. Spaccapelo, S. Naitza, K. J. H. Robson, C. J. Janse, F. Bistoni, A. P. Waters, and A. Crisanti. 1999. The A-domain and the thrombospondin-related motif of Plasmodium falciparum TRAP are implicated in the invasion process of mosquito salivary glands. *EMBO J.* **18:**5195–5204.

Ying, P., M. Shakibaei, M. S. Patankar, P. Clavijo, R. C. Beavis, G. F. Clark, and U. Frevert. 1997. The malaria circumsporozoite protein: interaction of the conserved regions I and II-plus with heparin-like oligosaccharides in heparan sulfate. *Exp. Parasitol.* **85:**168–182.

Yoshida, N., P. Potocnjak, V. Nussenzweig, and R. S. Nussenzweig. 1981. Biosynthesis of Pb44, the protective antigen of sporozoites of Plasmodium berghei. *J. Exp. Med.* **154:**1225–1236.

Yuda, M., and T. Ishino. 2004. Liver invasion by malarial parasites—how do malarial parasites break through the host barrier? *Cell. Microbiol.* **6:**1119–1125.

GAMETOCYTES AND GAMETES

Pietro Alano and Oliver Billker

10

Plasmodium gametes are the only parasite stages to survive and develop in the blood meal engorged by an *Anopheles* mosquito. This simple fact highlights the central role of this cell type in achieving parasite transmission and the relevance of the process of gamete formation to the spread of malaria. For malaria parasites, transmission is synonymous with sexuality. Fertilization produces the only short-lived, diploid stage of the parasite—the ookinete—in which the parental genomes are reassorted in a conventional meiosis (Sinden et al., 1985). Gametes and their precursors, the gametocytes, are the very special cell types that *Plasmodium* developed in the course of evolution to accomplish such key steps in its life cycle and are the subject of this chapter.

In the main, cellular and molecular aspects of gametocyte and gamete formation will be covered here. Readers will find literature on these and additional aspects of sexual differentiation in excellent recent reviews (Janse and Waters, 2004; Sauerwein and Eling, 2002; Smith et al., 2002; Talman et al., 2004a).

THE SWITCH TO SEXUAL DEVELOPMENT

Plasmodium gametocytes originate in the vertebrate host. Some derive directly from merozoites produced by hepatic schizonts (Alano and Carter, 1990, and references therein), similarly to the related Haemosporida blood parasites, in which exoerythrocytic forms are the only source of gametocytes (Smith et al., 2002). *Plasmodium* species are unique within the order Haemosporida for having evolved the ability to undergo asexual propagation in the bloodstream, producing at each cycle variable numbers of gametocytes. These are the major source of gametocytes in malaria infections.

Asexual multiplication in the bloodstream provides *Plasmodium* with a novel environment for its propagation but also extends the period available for sexual differentiation and transmission. The proportion of parasites differentiating into gametocytes at each generation can vary greatly during the course of a natural infection (Smalley et al., 1981), with gametocyte conversion rates changing over 50-fold, as mea-

---

Pietro Alano, Dipartimento di Malattie Infettive, Parassitarie e Immunomediate, Istituto Superiore di Sanità, Viale Regina Elena 299, 00161 Rome, Italy. *Oliver Billker,* Department of Biological Sciences, Imperial College London, London SW7 2AZ, United Kingdom.

Molecular Approaches to Malaria, Edited by Irwin W. Sherman
© 2005 ASM Press, Washington, D.C.

sured in vitro (Bruce et al., 1990). Modulation of gametocyte production is thought to provide the parasite with a flexible escape mechanism from immunological and pharmacological aggression, in marked contrast to other apicomplexans, such as *Eimeria*, in which sexual differentiation is rigidly programmed (Smith et al., 2002). Additional plasticity is given by the fact that sex ratios of gametocytes are also highly variable, both between *Plasmodium* isolates and in the course of a single infection. A high flexibility of sexual differentiation, combined with the nonlinear relationships between gametocyte densities and infectivity to the mosquito vector (Sinden et al., 1996), reduces the predictability of parasite transmission dynamics.

Albeit flexible, the developmental mechanism governing sexual differentiation is clearly inherited, since parasites can irreversibly lose the ability to undergo gametocytogenesis after prolonged asexual propagation, and different parasite isolates and clones are characterized by different efficiencies in gametocyte production (Alano and Carter, 1990, and references therein). No master regulatory gene for sexual differentiation has, however, been identified. Studies of *Plasmodium berghei* and *Plasmodium falciparum* lines in which chromosomal deletions were associated with impaired or absent gametocyte production (Janse et al., 1992; Day et al., 1993) have so far failed to point the spotlight at such genes, although analysis of chromosomal rearrangements identified loci associated with alterations in gametocyte production or maturation (Guinet and Wellems, 1997; Gardiner et al., 2005).

Commitment to Sexual Differentiation

The parasite's developmental switch from asexual cycling to gametocytogenesis has been investigated mainly in the human and the rodent parasites. In *P. falciparum*, sibling parasites derived from individual, isolated schizonts were analyzed in different studies by morphology and with specific antibodies distinguishing between gametocytes and asexual parasites or male and female gametocytes. Results suggested that a developmental decision is made during maturation of an asexual schizont that results in the production of a progeny of merozoites committed either to develop once more as asexual parasites or to all differentiate as gametocytes (Bruce et al., 1990) whose sex seems also predetermined (Silvestrini et al., 2000). Experiments with the rodent parasite *P. berghei* suggested instead that the developmental switch occurs after merozoite invasion in the first 16 h of growth of the uncommitted parasite (Mons, 1986). It is somewhat surprising that this key developmental mechanism may not be conserved between the human and the rodent parasite, given other conserved features of sexual differentiation. Identification of earlier markers of sexual differentiation will clarify this issue.

Modulation of Gametocyte Production

A vast literature exists on environmental, immunological, and chemical factors reported to induce and modulate gametocyte production (Alano and Carter, 1990; Talman et al., 2004a), although not all studies clearly distinguish between true developmental induction and differential survival or killing of asexual and sexual stages. However, the state of the art is that no specific molecular or cell-mediated mechanism other than a general parasite response to adverse growth conditions has been conclusively linked to the rate of gametocyte production in any of the above cases.

Identification of such a mechanism(s) would be relevant to explain clinical and biological observations of increased gametocyte densities following drug treatment of human and experimental infections (Price et al., 1999; Buckling et al., 1999; Bousema et al., 2003). Indications exist that inefficient treatment and increased rates of recrudescence are associated with enhanced gametocyte production (Price et al., 1999) and that chloroquine-resistant parasites are preferentially transmitted in rodent and human infections (Gautret et al., 2000; Hallett et al., 2004). These observations propose the concerning scenario that selection of drug-resistant genotypes is accompanied by increased chances for their transmission. Although the relationship between drug resistance and transmission is complex, depending on clone multiplicity of infection, transmission intensity, and the single- or multigenic nature of the resistance trait (Talisuna et al., 2003), understanding how parasites modulate gametocyte production exposed to drug treatment is highly important for rational design of future control strategies.

Modulation of Gametocyte Sex Ratio

The fact that a single, haploid malaria parasite gives rise to infections producing gametocytes of both sexes and the absence of sex chromosomes in *Plasmodium* indicate that gametocyte sex determination is governed by differential gene expression (Alano and Carter, 1990). Malaria parasites typically produce more female than male gametocytes, but they evolved the ability to modify gametocyte sex ratio during an infection. Progressive increases in the proportion of male gametocytes in infections of *P. falciparum* and other malaria parasites have been described since the first half of the 20th century (Schall, 1989 and references therein). Recent work on the avian and rodent parasites *Plasmodium gallinaceum* and *Plasmodium vinckei* showed that anemia or hypoxia and exogenous erythropoietin, but not the host's immunological response, were associated with an increased proportion of male gametocytes (Paul et al., 2000). These results suggest that not only does the mechanism inducing sexual differentiation respond to environmental signals, but the one governing sex determination does as well. Gametocyte-inducing factors such as chemotherapy in in vivo infections do not specifically affect sex ratio (Talman et al., 2004b). On the other hand, stimulation of erythropoiesis produces a male-biased sex ratio without significantly affecting the overall gametocyte conversion rate (Paul et al., 2000; Paul et al., 1999). These results suggest that induction of gametocytogenesis and sex determination are independently controlled, while the experiments on developmental commitment in *P. falciparum* instead suggest that they are intimately linked. Fundamental questions thus remain unanswered after more than a century of malaria research and well after the entrance of malariology in the postgenomic era.

GAMETOCYTOGENESIS

The length of the erythrocytic asexual cycle varies significantly between malaria species, ranging from the 18 h in *Plasmodium yoelii* to 72 h in *Plasmodium malariae*. Gametocyte formation invariably requires more time, ranging in most species from 1.1 to 2 times the length of the asexual cycle (Paul et al., 2003; and Gautret and Motard, 1999; and references therein). An exceptionally extended time of around 10 days is needed for gametocytogenesis in *P. falciparum* in humans (Thomson and Robertson, 1935), in the experimental *Aotus* monkey model (Hawking et al., 1971), and in vitro (Jensen, 1979).

Gametocyte Formation

The most detailed and multifaceted descriptions of gametocytogenesis are those available for *P. berghei* and *P. falciparum*, allowing one of the fastest and the slowest processes of sexual differentiation to be compared.

In *P. falciparum*, gametocyte development has been classified into five morphological stages (Hawking et al., 1971), characterized by cellular and molecular markers (Fig. 1).

THE EARLY EVENTS IN *P. FALCIPARUM* GAMETOCYTOGENESIS

Until 24 h after merozoite invasion, a young gametocyte appears as a small, round cell not dissimilar from a small trophozoite. Synthesis of the parasitophorous vacuole protein Pfs16 (Bruce et al., 1994) is then induced, followed by the massive production of the cytoplasmic phosphoprotein Pfg27 (Carter et al., 1989). At this stage, a bilamellar membrane sac adjacent to a few subpellicular microtubules is detected in ultrastructural analysis as diagnostic of the early sexual cell (Sinden, 1982), announcing the major cell structure changes leading to the crescent-shaped stage II gametocyte at 48 h. Pfs16 and Pfg27 are the earliest-characterized molecular markers of gametocytogenesis in *P. falciparum*. Pfs16 function was not conclusively elucidated in a gene-targeting experiment (Kongkasuriyachai et al., 2004). Disruption of the *pfg27* gene resulted instead in complete abolishment of

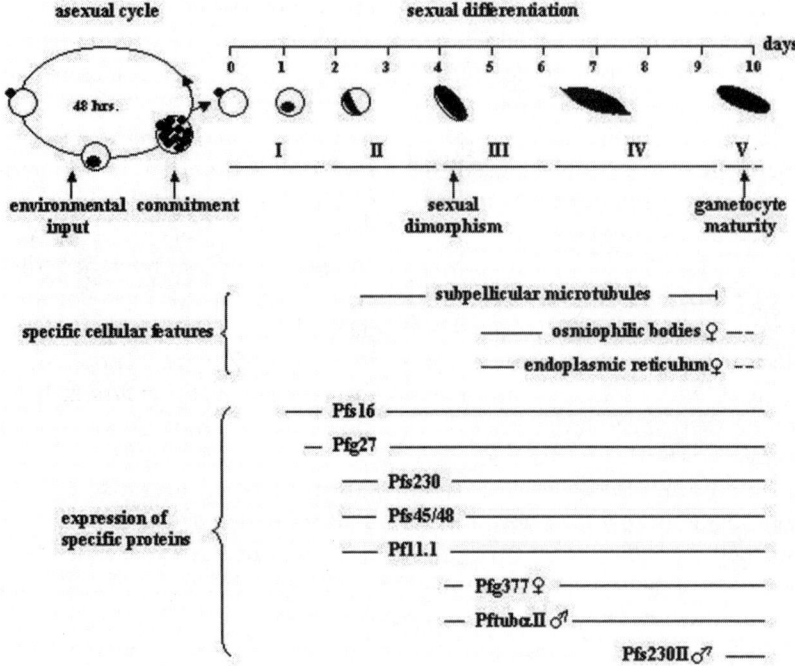

FIGURE 1 Diagram of the developmental steps in *P. falciparum* gametocytogenesis and the appearance of specific cellular features and sexual stage-specific proteins during gametocyte maturation.

gametocyte maturation and in the formation of large abnormal vacuoles in parasites identified as abortive gametocytes (Lobo et al., 1999). The three-dimensional structure of Pfg27 (Fig. 2) was the first to be resolved from the vast class of malaria proteins lacking homologues in eukaryotic databases (Sharma et al., 2003). Structure determination revealed that the two non-homologous halves of the protein are arranged in two structurally superimposable domains characterized by an all-α-helical novel fold. From structural clues, it was experimentally shown that a Pfg27 homodimer is able to bind specifically single-stranded RNA and SH3 protein motifs in vitro (Sharma et al., 2003). As classic SH3 domains are not detected in the predicted repertoire of *P. falciparum* proteins (Aravind et al., 2003), the search for Pfg27-interacting proteins promises to be particularly exciting. A key role for Pfg27, suggested by the observation that it constitutes almost 10% of the protein content of the early gametocyte (Carter et al., 1989), is thus further supported by its potential interactions with parasite proteins and RNAs. Finally, the fact that Pfg27 is phosphorylated (Kumar, 1997) opens additional questions on its functional role and on modulation of its activity by signalling pathways in this particular time of parasite development.

THE MIDDLE PHASE OF *P. FALCIPARUM* GAMETOCYTE MATURATION

Formation of stage III gametocyte around day 4 is an important physiological transition in gametocyte maturation. Sexual dimorphism clearly appears at the ultrastructural level (see Chapter 3) and it is accompanied by expression of sex-specific molecular markers. A specific α-tubulin isoform is detectable in male gametocytes and will be incorporated into the axoneme of the microgamete (Rawlings et al., 1992), while protein Pfg377 starts to be produced preferentially in female gametocytes (Severini et al., 1999),

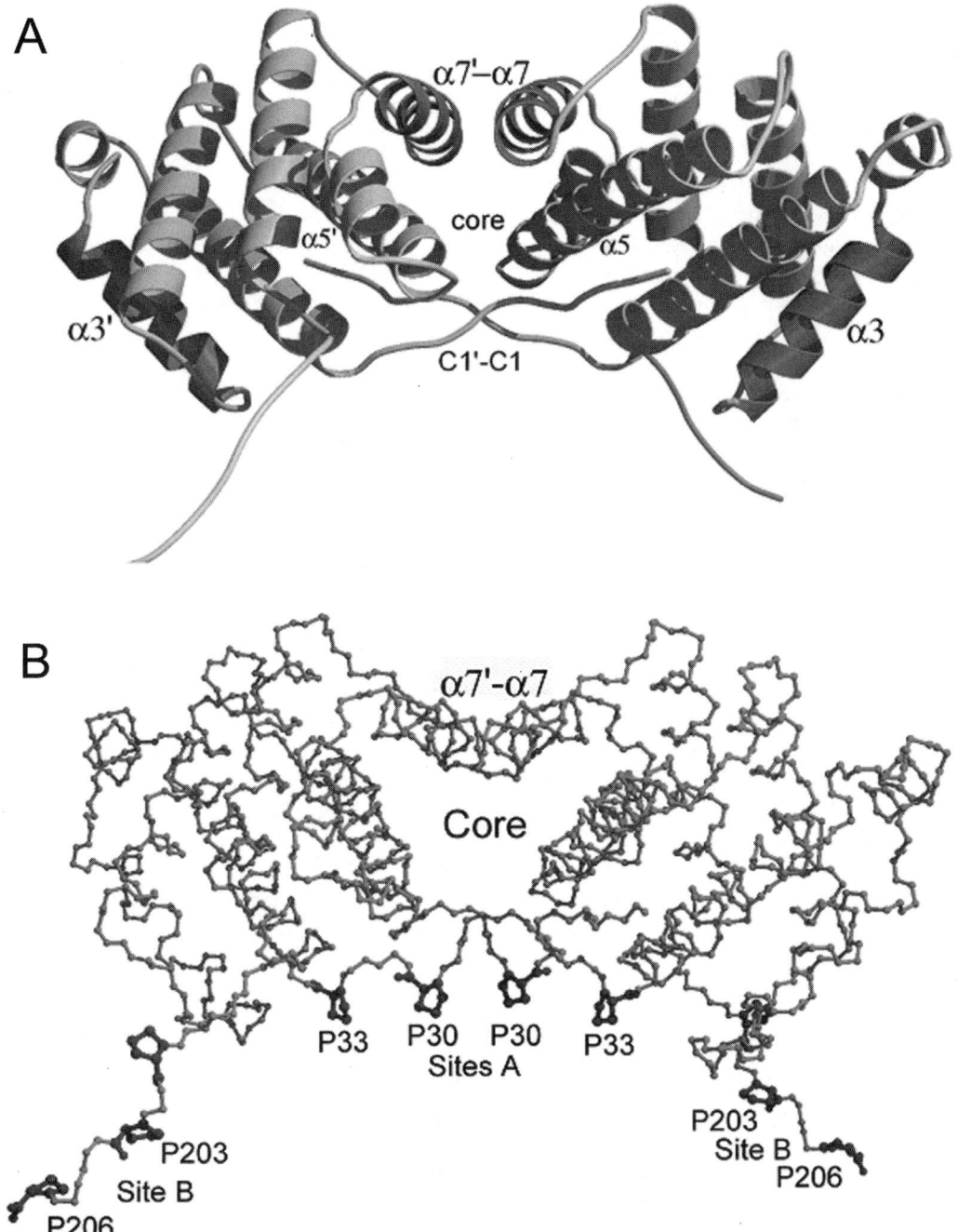

FIGURE 2 Structure of Pfg27. (A) Structure of the Pfg27 homodimer showing the helices involved in protein dimerization ($\alpha 7'$ and $\alpha 7$) and in RNA binding ($\alpha 3'$ and $\alpha 3$). (B) Positions of the proline residues of the PXXP motifs involved in protein-protein interaction and recognition of SH3 modules. Each protein monomer contains one site A (30-PLSP-33) and one site B (203-PALP-206). (Reprinted from Sharma et al., 2003, with permission of the publisher.)

localized in the osmiophilic bodies. These are membrane-limited, electron-dense organelles proposed to participate in the process of emergence of the female gametocytes from the red blood cell (Alano et al., 1995). Production of this protein, the only specific marker for osmiophilic bodies, precedes ultrastructural detection of these organelles in stage IV gametocytes (Sinden, 1982), suggesting that Pfg377 might be functionally related to their formation.

From stage III of maturation, gametocytes become insensitive to many drugs active against asexual parasites and younger sexual stages. Chloroquine, pyrimethamine, and atovaquone are not effective against late gametocytes but inhibit maturation of young sexual stages (Lang-Unnasch and Murphy, 1998). The fact that early gametocytes are generally sensitive to drugs killing asexual forms suggests that the two parasite stages share a substantially similar metabolism. The failure of pyrimethamine and atovaquone to kill late gametocytes indicates for instance that de novo pyrimidine synthesis is negligible in these stages and active in the sensitive early sexual cells (Lang-Unnasch and Murphy, 1998). Similarly, chloroquine toxicity in gametocytes was proposed to be related to the ability of young, but not late, gametocytes to digest hemoglobin (Hayward et al., 2000b). In contrast to the above cases, primaquine, the 8-aminoquinoline traditionally used to kill *Plasmodium vivax* hypnozoites and *P. falciparum* gametocytes, is active against late gametocyte but much less active against asexual forms, even though sensitivity of young gametocytes was not specifically tested (Pukrittayakamee et al., 2004). This might be related to adverse effects of primaquine on structure and functions of mitochondria (Lang-Unnasch and Murphy, 1998, and references therein), suggested to be more active in gametocytes than in asexual stages because of the increased presence of cristae and upregulation of some mitochondrial proteins (Hall et al., 2005). Information on mitochondrial function in sexual differentiation is still limited. Comparison of electron-transporting enzymes and oxygen consumption between mitochondria purified from asexual parasites and gametocytes gave conflicting results (Learngaramkul et al., 1999; Krungkrai et al., 1999), and even estimates of the number of these organelles in the gametocyte is controversial, ranging between six (Krungkrai et al., 2000) and one (Kato et al., 1990), as in asexual stages (Sato et al., 2003).

THE LATE PHASE OF *P. FALCIPARUM* GAMETOCYTE MATURATION

Metabolism in gametocytes at stages IV and V is relatively inactive (Sinden and Smalley, 1979). Stage V gametocytes remain in a quiescent state for several days, and their competence to undergo gametogenesis declines with a half-life of 2 to 3 days (Smalley and Sinden, 1977). Nevertheless, physiologically significant events take place at this time of maturation. The array of subpellicular microtubules disappears (Sinden, 1982), resulting in the smoothing of the gametocyte's edges and the curved appearance of stage V gametocytes, a character more pronounced in females. Other transformations, possibly related to the latter, are likely to occur at the surface of the gametocyte-infected red blood cell, causing them to abandon the hidden maturation sites in the bone marrow and spleen (see below). In addition, male gametocytes of stage V express a paralog of the *pfs230* gene (Eksi and Williamson, 2002). The quiescent state of stage V gametocytes is nevertheless highly sensitive to environmental signals. As we will see, these cells are fully equipped for mastering the sequence of coordinated events of gametogenesis in a very short time.

MATURATION OF *P. BERGHEI* GAMETOCYTES

In *P. berghei*, no morphological difference distinguishes a developing gametocyte from a young asexual parasite before 20 h postinvasion, while ultrastructurally the first diagnostic feature of the young gametocyte, the eccentric position of the parasite nucleus, is perceived only from 15 to 18 h postinvasion (Mons, 1986). Sex dimorphism is evident at 24 h with female gametocytes having more osmiophilic bodies (which first appear at around 22 h), an extended endoplasmic reticulum, and a smaller and more compact nucleus than male gametocytes (Mons, 1986). At the molecular level, transcripts for genes *pbs21* and *pbg377* are detectable from 20 h postinvasion, while the mRNA for the gamete fertilization surface antigen Pbs230 is the earliest marker of sexual differentiation, appearing from 16 h postinvasion (Janse et al., 2003).

Interactions with the Vertebrate Host

Gametocytes do not appear to contribute directly to the pathology of malaria in the vertebrate host. However, the immune system of infected individuals specifically detects gametocytes, produces antigen-specific antibodies (Sauerwein and Eling, 2002), and mounts cell- and cytokine-mediated responses against sexual stages (Ramsey et al., 2002; Smith et al., 2003; Naotunne et al., 1993). Intense investigations of the immune response against sexual stages are motivated by the attempt to design a vaccine able to arrest parasite transmission from an infected individual to the insect vector, based on the fact that antibodies against *Plasmodium* gametocyte and gamete surface antigens were shown to gain access to the parasite after emergence in the mosquito gut and prevent further parasite development (Kaslow, 2002).

GAMETOCYTE SEQUESTRATION

Plasmodium species evolved the strategy to confine part of their asexual multiplication away from the peripheral circulation by adhering to the vasculature endothelium of internal organs. This phenomenon has been described in *P. falciparum* (Cooke and Coppel, 1995), in simian malaria parasites *Plasmodium fragile* and *Plasmodium coatneyi* (David et al., 1988; de Sousa and Riley, 2002), and in the rodent parasites *P. berghei*, *P. yoelii*, and *Plasmodium chabaudi* (Landau and Gautret, 1998; Gilks et al., 1990). For gametocytes, much less is known about sequestration, except for *P. falciparum* where there is consolidated evidence that sexual stages normally develop in internal organs and only mature stages circulate in the peripheral blood. Detailed descriptions of the early sexual stages of *P. falciparum*, specifically sequestered in the bone marrow and secondarily in the spleen date back to the earliest years of malaria research (Bastianelli and Bignami, 1899) and were followed by a limited number of subsequent investigations (Thomson and Robertson, 1935; Smalley et al., 1980).

In *P. falciparum*, asexual parasites adhere to the endothelium of several organs through parasite-induced modifications on the surface of infected erythrocytes, called knobs (Crabb et al., 1997). The major parasite adhesin involved is encoded by members of the *PfEMP1* polymorphic gene family, which mediates interactions with host endothelial ligands (Sherman et al., 2003). Mechanisms of gametocyte cytoadhesion have been comparatively less frequently investigated, in most cases with cell lines and molecules utilized in studies of asexual cytoadherence. A summary of the information from such studies and on gametocyte-specific molecules containing conserved adhesive motifs is presented in Fig. 3.

Gametocyte cytoadhesive properties are modulated in the course of sexual differentiation. Early gametocytes were reported to cytoadhere through the host ligand CD36 with an efficiency similar to those of asexual parasites (Day et al., 1998). Young gametocytes produce and expose PfEMP1 molecules on their surfaces (Smith et al., 2003), but conflicting data exist on the reported presence of knobs on these cells (Day et al., 1998; Sinden, 1982). In contrast, in stage III and IV gametocytes CD36-mediated cytoadherence is absent or much reduced compared to asexual forms (Day et al., 1998; Rogers et al., 1996a; Rogers et al., 1996b). These stages were instead shown to bind through specific host ligands to cell lines derived from bone marrow endothelium and stroma, albeit with lower avidity than the CD36-mediated adhesion of young gametocytes (Rogers et al., 2000). No cytoadhesion was ever observed in any of the above studies for stage V gametocytes. Progressively reduced binding efficiency of *P. falciparum* gametocytes during their development is consistent with observed maturation of these cells in tissues characterized by reduced blood flow, such as bone marrow and spleen. The above observations do not, however, explain why

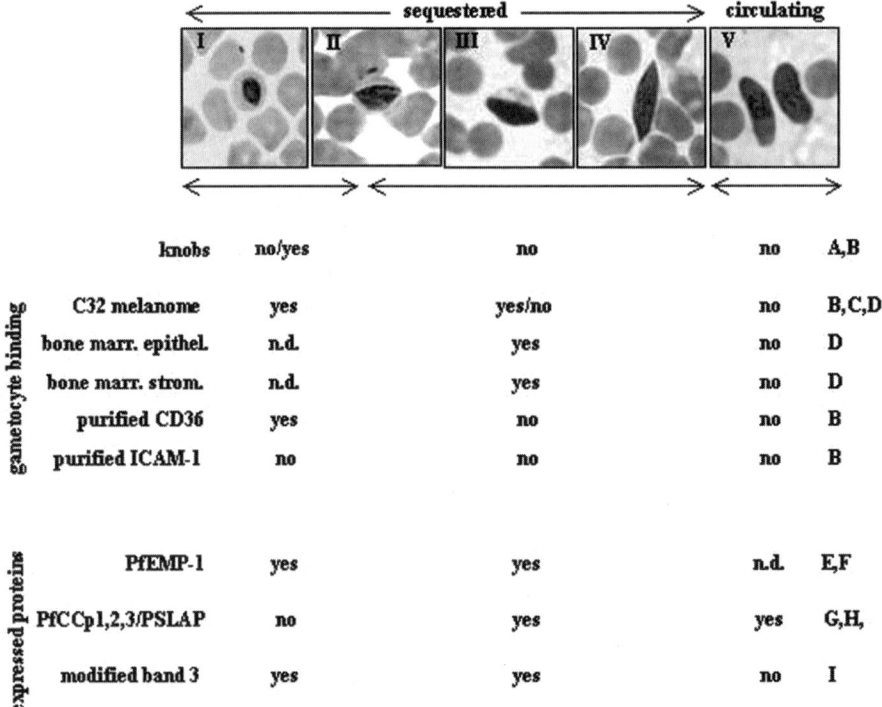

FIGURE 3 Current knowledge of cytoadhesive properties of *P. falciparum* gametocytes. Microphotographs (kindly provided by A. Olivieri) show the five stages of gametocyte maturation. Information from ultrastructure, cell-binding studies, and biochemical analysis of parasite and host proteins shown to be or possibly involved in cytoadhesion is indicated under the stages to which the images refer. The rightmost column contains the following references: Sinden, 1982 (A); Day et al., 1998 (B); Rogers et al., 1996a (C); Rogers et al., 2000 (D); Smith et al., 2003 (E); Florens et al., 2002 (F); Delrieu et al., 2002 (G); Trueman et al., 2004 (H); and Rogers et al., 1996b (I).

young gametocytes colonize only these organs, given that they share similar cytoadhesive properties with asexual parasites. Alternative hypotheses could be either that specific homing mechanisms exist for early gametocytes, mediated by specific ligands, or that the bone marrow microenvironment is particularly effective in inducing sexual development in schizonts sequestered in that compartment, and it is somehow able to retain their gametocyte progeny.

No parasite sexual stage-specific ligand involved in gametocyte cytoadhesion is described. Proteomic analysis reported that distinct PfEMP1 polypetides appear specifically in gametocyte samples (Florens et al., 2002) and that five proteins containing multiple lectin-like LCCL motifs were detected only in gametocyte and gamete samples (Lasonder et al., 2002). Further studies of this gene family conserved in human and rodent parasites showed that three of them (PfCCp1, PfCCp2, and PSLAP/PfCCp3) are associated but not exposed on the surface of *P. falciparum* gametocytes from stage III to V of maturation (Delrieu et al., 2002; Trueman et al., 2004). However, targeted gene disruption of PSLAP/PfCCp3, its *P. berghei* homolog PbSR1/LAP-1, and PfCCp2 suggests that their essential functions are not in gametocytes but at the oocyst or sporozoite stage (Pradel, 2004; Trueman et al., 2004).

GAMETOCYTE DISTRIBUTION IN THE PERIPHERAL CIRCULATION

Observations of *P. chabaudi*, *P. vinckei*, and *P. yoelii* reported a nonrandom distribution of gametocytes in the peripheral circulation, suggesting that infective stages are sequestered in a subdermal compartment of the host, more amenable for mosquito transmission (Landau and Gautret, 1998). Analogous observations of *P. falciparum* suggested that gametocytes form aggregates in the peripheral circulation (Pichon et al., 2000). Also, the most extreme case of male and female gametocyte proximity (the simultaneous presence within the same red blood cell) has been specifically investigated and discussed for its possible implications for transmission (Jovani et al., 2004). Observations of these phenomena have been so far either preliminary or mainly descriptive, and molecular characterization of underlying mechanisms is required to confirm their proposed functional significance.

Gene Expression in Gametocytogenesis

Pregenomic identification of several genes expressed specifically or preferentially in gametocytes and gametes (Table 1) suggested that the high degree of specialization of *Plasmodium* sexual cells relies on a set of specific proteins and showed that even a sexual stage-specific ribosomal RNA gene is utilized in gametocytogenesis of the four present in malaria parasites (see Chapter 18). A complement of gametocyte-specific proteins is indeed being detected by stage-specific genome-wide analyses of rodent and human malaria parasites (see Chapters 5 and 6). Transcriptome and proteome data are available for *P. falciparum* stage III and IV gametocytes and gametes, *P. berghei* gametocytes and ookinetes, and *P. yoelii* gametocytes (proteome) (Hayward et al., 2000a; Le Roche et al., 2003; Hall et al., 2005; Florens et al., 2002; Lasonder et al., 2002; Carlton et al., 2002). *P. berghei* transgenic lines that express green fluorescent protein under gender-specific promoters have been obtained (Khan and Waters, 2004) in which gametocytes can now be sorted by sex, and protein composition of males and females can be analyzed separately.

Regulatory mechanisms governing gene expression during gametocyte formation are still largely unknown, but available data indicate that they utilize transcriptional and posttranscriptional mechanisms. It is intriguing that specific regulatory transcription factors have not been identified in the *P. falciparum* genome (Aravind et al., 2003; Coulson et al., 2004). Nevertheless, functional analyses of malaria gene upstream regions revealed conventional regulation of transcription initiation in most cases, albeit mediated by malaria-specific sequence elements (Militello et al., 2004, and references therein). Regulated control of transcription initiation for sexual stage-specific genes was reported with nuclear run-on assays or with transfection experiments for *P. falciparum* genes *pfs16*, *pfg27*, and *pfs25* (Alano et al., 1996; Dechering et al., 1997; Dechering et al., 1999) and the *P. berghei* genes *pbNAP(B7)*, *pbtubulin-αII*, and *pbs21* (M. Ponzi et al., personal communication; Khan and Waters, 2004). With *P. berghei* gene *pbs21*, deletion of sequences responsible for stage-specific regulation was shown to cause constitutive gene expression in all blood stages (Margos et al., 1998).

Gametocyte-specific gene expression can involve the use of alternative promoters and differential splicing. This was suggested for mRNAs for *P. falciparum* β tubulin (Delves et al., 1990) and DNA polymerase δ (Ridley et al., 1991), which showed different sizes in asexual parasites and gametocytes, and was demonstrated for the *P. berghei* gene *NAP(B7)*. Two promoters are utilized to transcribe this gene in asexual parasites and in gametocytes, and a noncoding intron is spliced from the 5′ untranslated region only in the gametocyte-specific transcript (Pace et al., 1998). Elucidating some of the presumably unusual mechanisms of transcriptional regulation in malaria parasites will not only be critical to our understanding of sexual development but may also provide new targets for pharmacological intervention.

GAMETOGENESIS

Exflagellation is visually the most striking event in the malarial life cycle. When Alphonse Laveran (1880) observed for the first time how crescent-shaped parasites transformed into spheres, which then released vividly moving flagellar forms, he became convinced that malaria was caused by a protozoan parasite. Exflagellation has fascinated scientists ever since and much

TABLE 1 Genes specifically or predominantly expressed in gametocytes and gametes of *P. falciparum* and *P. berghei*

Gene name	PlasmoDB name	Annotation, function, localization	Expression	KO phenotype	Reference(s)
Cytoplasmic, organellar					
pfs16	PFD0310w	Parasitophorous vacuole	From 24 h p.i.	Diminished gametocyte production	Bruce et al., 1994; Baker et al., 1994; Kongkasuriyachai et al., 2004
pfg27	PF13_0011	Cytoplasm	From 30 h p.i.	Vacuolated, abortive gametocytes	Carter et al., 1989; Lobo et al., 1999; Cui et al., 2002
pfPuf1; pfPuf2	PFE0935c; PFD0825c		From stage I (mRNA)		
pfactin II	PF14_0124	Axoneme	Gametocytes (mRNA)		Wesseling et al., 1989
pftubulin αII	PFD1050w		From stage III (mRNA-protein)		Rawlings et al., 1992
Pf77	MAL6P1.213		Female gametocytes III–V		Baker et al., 1995
pf11.1	PF10_0374		Gametocytes (mRNA-proteins)		Scherf et al., 1992
pfORC1	PFL0150w	Homologous to origin recognition complex subunit 1	Gametocytes (mRNA)		Li and Cox, 2003
pfMCM4	PF13_0095	Homologous to minichromosome maintenance protein 4	Gametocytes (mRNA)		Li and Cox, 2001
pfs377	PFL2405c	Osmiophilic bodies	From stage III (protein)		Alano et al., 1995; Severini et al., 1999
pbg377					
PfPEPCK	PF13_0234	Phosphoenolpyruvate carboxykinase	From 16 h p.i. (mRNA) Stage IV–V gametes-zygotes (mRNA-enzyme activity)		Janse et al., 2003; Hayward, 2000
cytochrome b			Gametocytes (mRNA)		Petmitr and Krungkrai, 1995
Ribosomal RNAs					
pfrRNA S1	GenBank AF503869	18S ribosomal RNA	Gametocytes		Fang et al., 2004
pbrRNA C and D (S type)		18S ribosomal RNA	Oocyst	Individual KOs do not affect oocyst	van Spaendonk et al., 2001

Family/Gene	ID	Feature	Expression	Phenotype	References
Cytoadhesion–cell-cell interaction–membrane					
PfCCp1; -2; -3/PSLAP	PF14_0723, PF14_0067, PF14_0532,	LCCL cytoadhesive motifs	Stage III to V (mRNA/protein)	Sporozoites do not invade salivary glands	Delrieu et al., 2002; Lasonder et al., 2002; Pradel et al., 2004
pbSR/pLAP1		LCCL cytoadhesive motifs	RNA asexual and sexual, protein in sporozoites	Oocyst do not form sporozoites	Claudianos et al., 2002; Trueman et al., 2004
PfEMP1		Adhesins (on young gametocytes)	Specific peptides in gametocyte proteome		Florens et al., 2002
stevor			Specific mRNAs in gametocytes		Sutherland, 2001
6-Cys domain family					
pfs230	PFB0405w	Gamete surface	From stage II (mRNA/protein)	Normal production of gametocytes and gametes	Sauerwein and Eling, 2002; Eksi et al., 2002
pfs230 paralog	PFB0400w		Stage V male gametocytes (protein)		Eksi and Williamson, 2002
pfs45/48	PF13_0247	Gamete surface	From stage II (mRNA/protein)	Male gamete fertility inhibited	Sauerwein and Eling, 2002; van Dijk et al., 2001
pb230		Gamete surface	From 20 h p.i. (mRNA)		Janse et al., 2003
pb45/48		Gamete surface	From 20 h p.i. (mRNA)	Male gamete fertility inhibited	van Dijk et al., 2001
p25/p28 gene family					
Pfs25	PF10_0303	Epidermal growth factor-like domain, ookinete surface	Gametes (mRNA) ookinete (protein)		Sauerwein and Eling, 2002
Pfs28	PF10_0302	Epidermal growth factor-like domain, ookinete surface	Gametocyte (mRNA) ookinete (protein)	Double KO ookinete inhibited; single KOs mildly compromised	Paton et al., 1993; Tsuboi et al., 1997; Tomas et al., 2001
Pbs21, Pbs25					
Signaling molecules					
PbCDPK4		CPDK	Male gametocytes	Microgametocytes do not enter cell cycle upon activation	Billker et al., 2004

(*Continued*)

TABLE 1 (*Continued*)

Gene name	PlasmoDB name	Annotation, function, localization	Expression	KO phenotype	Reference(s)
Pfcrkl	PFD0865c	Cyclin-dependent kinase-like	Gametocytes (mRNA)		Doerig et al., 1995
pfmrk	PF10_0141	Cyclin-dependent kinase in vitro	Gametocytes (mRNA)		Li et al., 1996
Pfmap2	PF11_0147	MAP-like protein kinase	Gametocytes (mRNA)		Dorin et al., 1999
Pfkin	PF14_0516	Serine/threonine protein kinase	Gametocytes (mRNA)		Bracchi et al., 1996
Pflammer	PF14_0431	Homologous to *Arabidopsis thaliana* AFC3 kinase	Gametocytes (mRNA)		Li et al., 2001
PfCDPK3	PFC0420w	CPDK	Gametocytes (mRNA)		Li et al., 2000
Protein phosphatase α	PF14_0630	Putative serine threonine phosphatase	Gametocytes (mRNA)		Li and Baker, 1998
Protein phosphatase β	PF11245c	Putative serine threonine phosphatase	Gametocytes (mRNA)		Li and Baker, 1997
pf guanylyl cyclase α	PF11_0395	Parasite-parasitophorous vacuole membrane	Gametocytes		Carucci et al., 2000
pf guanylyl cyclase β	PF13_0320		Gametocytes		Carucci et al., 2000

work has been directed towards three principal areas: (i) the description of gametogenesis at the ultrastructural, cellular, and cell-cycle levels, (ii) the identification of environmental factors and signal transduction mechanisms that control activation of the quiescent gametocytes as they enter the mosquito, and (iii) the identification and functional characterization of gamete and zygote surface antigens that are targets of transmission-blocking antibodies.

The Cell Biology of Gametogenesis

Studies by light and electron microscopy that deal specifically with gametogenesis are available for *P. falciparum* (Sinden et al., 1978), *Plasmodium cynomolgi* (Garnham et al., 1967), *P. berghei* (Garnham et al., 1967; Billker et al., 2002), *Plasmodium yoelii nigeriensis* (Sinden et al., 1976; Sinden, 1975; Sinden and Croll, 1975), and *P. gallinaceum* (Aikawa et al., 1984). Gametogenesis seems highly conserved at the ultrastructural level, which is perhaps surprising in view of the fundamental species differences in gametocytogenesis and gametocyte morphology reviewed above. Figure 4 provides a summary of the constituent events of gametogenesis and their approximate timing. Some processes, such as emergence, are shared between males and females, whereas others are sex specific. The highly precommitted nature of the gametocytes and the speed with which they respond to activation preclude a need for much de novo protein

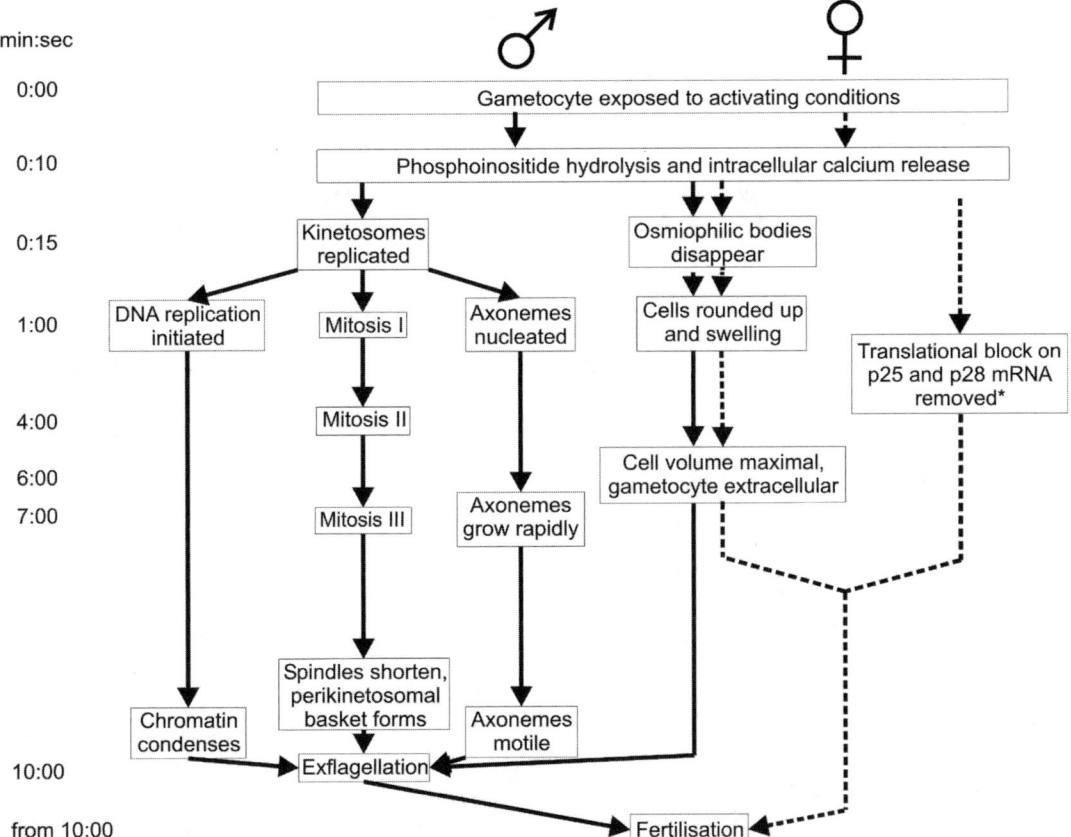

FIGURE 4 Constituent events of gametogenesis and their approximate timing, generalized from different studies of *P. falciparum* and *P. berghei*. *, Translational de-repression of p25 and p28 mRNAs occurs rapidly upon macrogametocyte activation, but neither protein is required for fertilization.

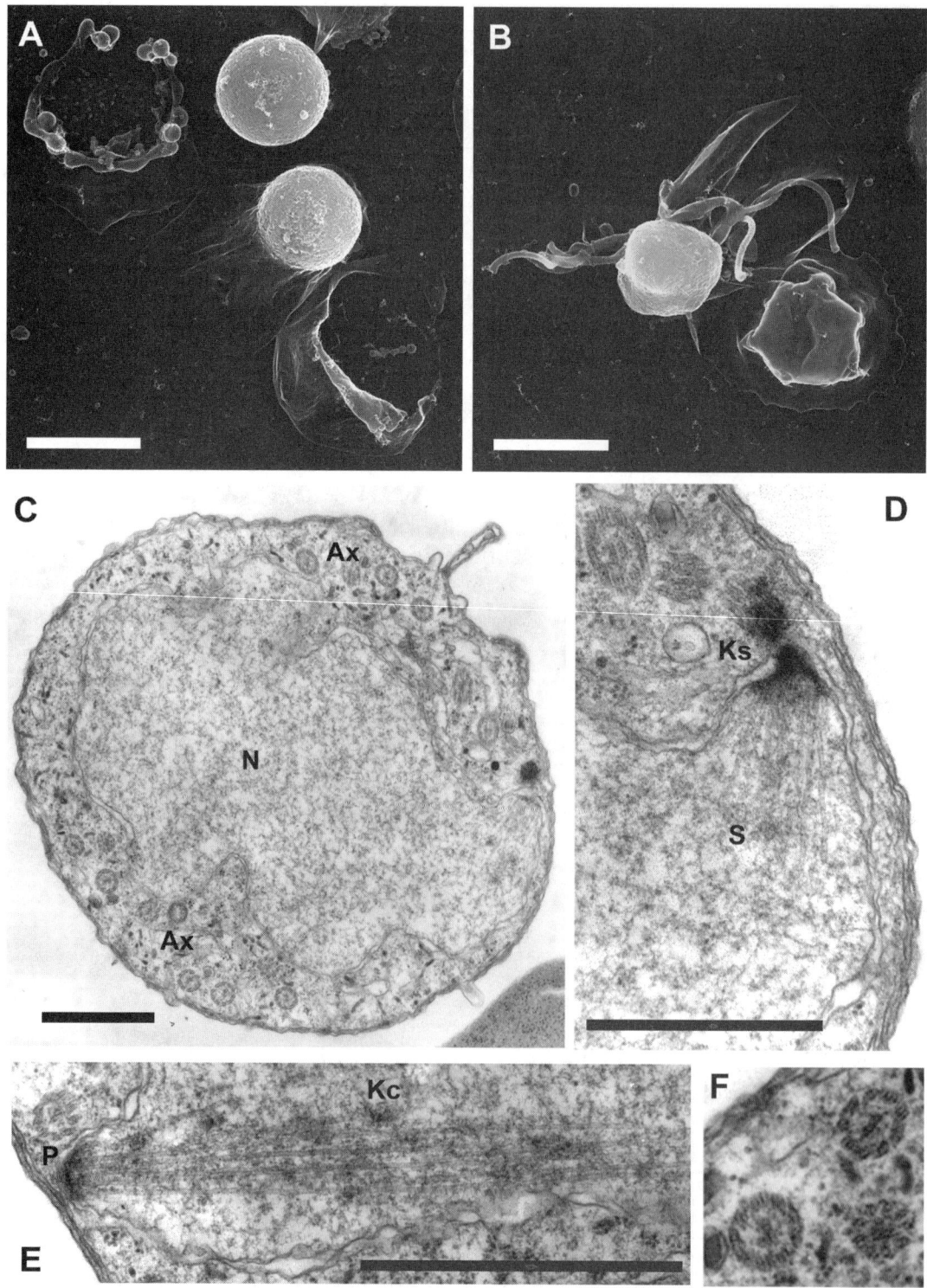

synthesis, although studies using an inhibitor of transcription, actinomycin D, suggest some may be required (Toye et al., 1977; Sinden, 1983b).

GAMETOCYTE EMERGENCE

The escape of gametocytes from their host cells is thought to involve both enzymatic and mechanical forces. Ultrastructural studies have identified secretory vesicles with an unknown osmiophilic content that are more abundant in macrogametocytes and are exocytosed within minutes of gametocyte activation, releasing their content into the parasitophorous vacuole (Sinden et al., 1976; Sinden et al., 1978). Disappearance of osmiophilic bodies correlates with the breakdown of the erythrocyte membrane, but a biochemical link between the two events remains to be established, as does the function of the osmiophilic body membrane protein Pfg377 (Severini et al., 1999; Alano et al., 1995). That microgametocytes have fewer osmiophilic bodies than females has been taken to suggest that males additionally rely on microgamete motility to shed the host cell plasma membrane (Sinden et al., 1996). A megadalton gametocyte antigen, Pf11-1, located in the erythrocyte cytoplasm, has been implicated in emergence, because upon gametocyte activation it associates with the membrane of the lysed host cell (Scherf et al., 1992). Whether gametocyte emergence requires the same types of protease activities as schizont rupture (Salmon et al., 2001; Wickham et al., 2003) is unknown.

Within a few minutes of activation gametocytes begin to expand and round up, forming perfect spheres (Fig. 5) with a centrally located nucleus. Gametocyte swelling, which may require the rapid de novo synthesis of membrane phospholipids, is thought to provide a mechanical force that facilitates emergence in both sexes. It is consistent with this notion that in *P. falciparum*, where cell volume can increase up to sixfold, the last erythrocyte remnants are shed towards the end of the expansion period (Sinden et al., 1978). In the male gametocyte, maximal expansion is followed by a period of contraction that precedes the onset of exflagellation by a few minutes.

MALE GAMETE DEVELOPMENT

At the optimal temperature, which differs somewhat between species, male gametogenesis invariably takes between 8 to 10 min in vitro until gametes begin to be released (Bishop and McConnachie, 1956; Sinden and Croll, 1975; Sinden et al., 1978). Exflagellation is always delayed at suboptimal temperatures and is totally arrested below about 10°C (Carter and Graves, 1988).

The first morphological evidence for microgametocyte activation appears within 15 s of induction, when a single amorphous microtubule organizing center, situated on the cytoplasmic face of a nuclear pore, gives rise to eight kinetosomes (Sinden et al., 1976). These are organized in two superimposed planar tetrads, within each of which four kinetosomes are orthogonally arranged. The tetrads split into two halves, which move into opposition along the nuclear envelope and between them form the spindle of mitosis I. Within 1 min of activation, most of the tubulin monomers from the nucleoplasm are recruited into this spindle, which is located at the periphery of the nucleus (Billker et al., 2002). Coincident with mitosis I, each cytosolic kinetosome nucleates one of eight axonemes, which display the conventional arrangement of nine outer doublets and two inner

FIGURE 5 Electron micrographs (EM) of gametogenesis in *P. berghei*. (A and B) Scanning EMs of gametocytes fixed 12 min after activation. Scale bar, 5 μm. (A) Two gametocytes that have rounded up and are in the process of lysing their host cells. Erythrocyte ghosts from which gametocytes have emerged are also seen. (B) Scanning EM of an exflagellating microgametocyte still associated with the ghost of the lysed host erythrocyte. (C to F) Transmission EM of activated microgametocytes. Scale bar, 1 μm. (C) At 10 min after activation cytosolic axonemes (Ax) have formed and are wrapped around the large central nucleus (N) of the round cell. (D) A cytosolic kinetosome (Ks) that has nucleated an axoneme and is associated with a nuclear spindle plaque plus hemispindle (S). (E) A long spindle of mitosis I with kinetochores and a spindle plaque (P) in a nuclear pore. (F) Axonemes in cross section.

singlet microtubules. A gametocyte-specific α-tubulin isoform II becomes incorporated into the axonemes and may play a role in microgamete motility (Rawlings et al., 1992). While axonemes grow to their full length of 22 μm in *P. falciparum* (Sinden et al., 1978), two more rounds of endomitosis occur. By 6 min, the four spindles of mitosis III have formed, one parallel pair situated on either side of a circular plane that is defined by a cytosolic ring formed by the eight axonemes. Over the following 2 min, spindles elongate while kinetochore microtubules shorten, pulling chromosomes towards the opposite poles.

Upon activation, the microgametocyte replicates its genome three times (about once every 3 min) until the octoploid value is reached (Janse et al., 1986a; Janse et al., 1988). This is extraordinarily fast if compared to other life cycle stages of *Plasmodium*. It is reminiscent of replication in animal embryos, in which the duration of S phase is reduced drastically during early divisions through the use of many additional origins of replication, resulting in smaller replicons. A similar mechanism seems to operate in microgametocytes, where key components of the origin of replication complex, ORC1 and MCM4, are strongly upregulated when compared to asexual blood stages (Li and Cox, 2001, 2003).

Chromatin condensation only sets in at the end of mitosis III, when each of eight short hemispindles anchors one haploid set of 14 chromosomes to each of the microtubule organizing centers and thereby to one of the extranuclear kinetosome/axoneme complexes. The latter are at this stage surrounded by a cytosolic basket of membranous tubovesicular structures that may provide membrane material for the forming microgamete. Approximately 10 min after induction, axonemes become motile. Within a few seconds, microgametes begin to emerge with the kinetosome first, which drags along a hemispindle with attached chromosomes and part of the nuclear envelop, which stays intact and undivided throughout mitosis. These structures become incorporated into the forming microgamete, and it is only at this stage that chromatin condensation is complete. Microgametes elongate at a remarkable speed of about 1 μm/s until they are eventually propelled away from the residual body of the gametocyte by the flagellar movement of their axoneme. They lack an obvious acrosome structure. No mitochondrion and apicoplast seem to be present (Sinden et al., 1978) and these organelles and their genomes are maternally inherited (Vaidya et al., 1993; Creasey et al., 1994).

Microgametes move by sinusoidal or helical waves. Motility can persist for about 40 min, during which periods of rapid movement become more and more disrupted by phases of slow or no motility (Sinden and Croll, 1975). Male gametes interact strongly with erythrocytes, and as a result exflagellating parasites surround themselves with characteristic rosettes of host cells, which are moved about vigorously in vitro. Red blood cell binding of *P. falciparum* involves sialic acid and glycophorin A on the red blood cell surface (Templeton et al., 1998). Binding is species specific, suggesting it may be of functional importance, but its role remains unclear.

The extraordinary speed of exflagellation incurs a high error rate. Microgametes with two axonemes are often observed, and one study found that 23% of microgametes generated by *P. falciparum* in vitro were anucleate (Sinden et al., 1978). Perhaps microgametogenesis needs to be fast because coagulation of the blood meal or its concentration by diuresis quickly restricts microgamete movement. Alternatively, or additionally, exflagellation may be an unusual example of sperm competition, where different microgametocytes compete for a limited number of females.

FEMALE DEVELOPMENT AND POSTTRANSLATIONAL GENE REGULATION

Following their emergence, macrogametes undergo little further maturation and seem to become available for fertilization without delay. The macrogametocyte is already highly preadapted to differentiate into an ookinete upon fertilization. This involves the rapid de novo synthesis of proteins using ribosomes and mRNAs that are already abundant in the macrogametocyte. For example, the messengers for two major glycosylphosphatidylinositol-anchored surface proteins of the ookinete, p25 and p28, start to accumulate in the macrogametocyte but are translationally repressed until the parasite enters the mosquito (Paton et al., 1993). p25 and p28 elicit potent transmission-blocking antibody responses and are important vaccine candidates (Kaslow, 2002). Tight translational control prevents their exposure to the vertebrate immune system, and this may be crucial to parasite transmission. The translational block on the *p25* and *p28* gene messages is lifted by macrogametocyte activation, independently of fertilization (Billker et al., 1997). This is different from translationally repressed maternal mRNAs in metazoan oocytes, in which fertilization triggers translation.

Translational repression may be used extensively by the macrogametocyte as a means of responding rapidly to the new environment of the mosquito. This has been suggested by two studies in which the comparison of stage-specific transcript and protein levels from *P. falciparum* (Le Roch et al., 2004) and *P. berghei* (Hall et al., 2005) has identified numerous gametocyte transcripts for which the corresponding protein only becomes detectable in the next parasite stage examined (the gamete or zygote for *P. falciparum* and ookinete for *P. berghei*). In other eukaryotes, translational repression is known to involve specific 3′ untranslated region-binding proteins, such as pumilio in *Drosophila melanogaster*, which contains characteristic Puf domains. Two Puf domain proteins, PfPuf1 and PfPuf2, are upregulated in gametocytes over asexual blood stages (Cui et al., 2002), but their functions remain to be investigated. A 47-bp motif enriched in putative UUGU-binding sites for Puf domain proteins has been identified in downstream flanking regions of 29 gametocyte-expressed genes (including *p25* and *p28*) that are candidates for translational repression (Hall et al., 2005).

A role for the abundant RNA-binding gametocyte protein Pfg27 (Sharma et al., 2003) in translational repression during the extended gametocyte maturation in *P. falciparum* has not been investigated but might explain the failure of *pfg27* knockout gametocytes to reach maturity (Lobo et al., 1999). It is intriguing to note that *Plasmodium reichenowi* shares the long gametocyte maturation period with *P. falciparum* and has an ortholog of Pfg27, while none has so far been found in the (still incompletely sequenced) genomes of other *Plasmodium* species.

GAMETOGENESIS AND THE CELL CYCLE

DNA replication to the octoploid level occurs only upon microgametocyte activation and is required for exflagellation (Janse et al., 1986a; Janse et al., 1988), suggesting that circulating gametocytes are arrested at a G_0-like stage of the cell cycle. However, things may be more complicated. Fluorometric data suggest that the DNA content of nonactivated gametocytes is already elevated well above the haploid value in both sexes, although the diploid level is not quite achieved (Janse et al., 1986a; Janse et al., 1988). This excess DNA remains unexplained and should be confirmed by independent methods. There is no evidence for selective gene amplification; the alternative hypothesis, that gametocytes are arrested in mid-S phase, is unsatisfactory, because the female nucleus does not need to enter S phase until after fertilization (Janse et al., 1986b). The appearance of a hemispindle in activated macrogametocytes of *P. falciparum* (Sinden et al., 1978) is another observation difficult to reconcile with conventional cell cycle models. During microgametogenesis, the repeated rounds of replication and mitosis seem to proceed independently and uncoupled from some of the usual cell cycle checkpoints; as a result, DNA synthesis goes to completion even when mitosis is blocked by nocodazole, a microtubule-destabilizing drug, or by azadirachtin, a plant limnoid that inhibits exflagellation by interfering with mitotic spindles and axonemes (Billker et al., 2002).

Regulation of Gametogenesis by Extracellular Triggers

The mosquito midgut environment provides gametocytes with unique environmental cues to coordinate the appearance of the short-lived gametes precisely in time and space. Tight control over gametocyte activation is crucial to avoid a waste of transmission potential and possibly to prevent antigens encoded by translationally repressed gametocyte mRNAs from being exposed to the vertebrate immune system. Three physiological triggers of gametogenesis have been identified to date: (i) a drop in temperature that the parasite experiences as it leaves the vertebrate blood-stream, (ii) a rise in extracellular pH that occurs readily when blood is exposed to air in vitro but that is much slower in the mosquito blood meal, and (iii) a small mosquito molecule, xanthurenic acid (XA). The only other factors required for gametocyte activation in vitro are glucose and physiological concentrations of three essential inorganic ions, Na^+, Cl^-, and HCO_3^-.

TEMPERATURE

Gametogenesis is inhibited completely at the body temperature of the host, and a drop of at least 4°C is required for gametocyte activation in vitro. The temperature optimum is higher in *P. falciparum* (28 to 34°C) and *P. gallinaceum* (24 to 30°C) (Carter and Graves, 1988) than in rodent species (19 to 21°C) (Sinden, 1983a), reflecting the different climates in which these parasites are transmitted and their different overall temperature requirements for sporogony. *P. berghei* gametocytes remain quiescent when ingested by *Anopheles stephensi* at 37°C, confirming in vivo the dominant effect of high temperature inhibition over any other activating factors the parasite encounters in the mosquito (Billker et al., 1997).

EXTRACELLULAR pH

A drop in temperature is required but not sufficient to activate gametocytes. Early experiments with *P. falciparum* (then called *Plasmodium praecox*) found that, at a permissive temperature, exflagellation was triggered in vitro when extracellular pH rose above the physiological pH 7.4 (Chorine, 1933). A sufficient pH increase occurs within a few minutes when blood is allowed to equilibrate with air (Bishop and McConnachie, 1956 and 1960), which offers a convenient but most likely artificial way of activating gametocytes in vitro. Recognizing the inverse relationship between pCO2 and pH in bicarbonate-buffered solutions such as blood (Carter and Nijhout, 1977) confirmed that pH (not reduced CO_2 tension) was responsible in vitro. They identified a narrow pH optimum of pH 7.9 to 8.1 for *P. gallinaceum*. Similar pH ranges emerged from later in vitro studies of *P. falciparum* and *P. berghei* (pH 7.8 to 8.0) (Sinden, 1983b). In vivo, the mosquito cuticle and integument impede gas exchange; as a result, the pH in the blood meal rises by only 0.12 to 0.18 pH units during the first half hour after feeding (Micks et al., 1948; Bishop and McConnachie, 1956; Billker et al., 2000), suggesting that in vivo mosquito factors contribute to gametocyte activation.

XA

Evidence for a mosquito exflagellation factor from *Aedes aegypti* and *A. stephensi* was first presented by Nijhout and Carter (1979). The activity was purified from mosquito extracts and identified as XA (Billker et al., 1998; Garcia et al., 1998). Synthetic XA triggers gametogenesis of all *Plasmodium* species examined but the 50% effective concentration values vary widely from 0.08 μM in *P. gallinaceum* to 2 μM in *P. falciparum* and 9 μM in *P. berghei* (Arai et al., 2001). In all species, XA can completely replace the pH stimulus, but evidence from *P. berghei* suggests it cannot overcome the low-temperature requirement (Billker et al., 1997). XA and pH act synergistically in *P. berghei* (Billker et al., 2000), such that raising pH in vitro from pH 7.40 to that of the blood meal, pH 7.58, renders gametocytes about three times more sensitive to XA.

XA is a product of the kynurenine pathway of tryptophan oxidation, which in mosquitoes generates the brown ommochrome pigments of the compound eyes and ocelli. Any excess of the reactive pigment precursor 3OH-kynurenine is detoxified to XA by a 3OH-KYN transaminase (Han et al., 2002) and then deposited in the eye or secreted. XA enters the alimentary canal, since it is found in the prediuresis fluid that is excreted when *A. stephensi* feeds on saline (Nijhout, 1979; Garcia et al., 1997; Billker, unpublished). However, whether the gut epithelium, the Malpighian tubules, or perhaps even the salivary glands (Hirai et al., 2001) are the main source remains to be investigated further.

A spontaneous mutant in *A. aegypti* lacks XA and eye pigmentation due to a 167-nucleotide deletion in the kynurenine hydroxylase gene (Bhalla, 1968; Han et al., 2003). Using this mutant, Arai et al. (2001) showed that mosquito-derived XA was essential for exflagellation of *P. gallinaceum* but only when gametocytes entered the midgut in a minimal medium that excluded host serum components. In contrast, exflagellation in complete infected chicken blood was only about halved in the XA-deficient mosquito. Clearly, the role of mosquito-derived XA can be partly fulfilled by a component from chicken serum in this vector-parasite combination. If and how this unknown factor (which could be chicken serum-derived XA) could contribute to limiting gametogenesis to the mosquito is unclear. Important open questions include whether XA is essential to the transmission of mammalian malaria by *Anopheles*, its source in the mosquito, and why there are such marked differences in sensitivity between parasite species.

MOSQUITO SPECIES-SPECIFIC INHIBITORS OF EXFLAGELLATION

Mosquitoes possess a highly effective innate immune response that limits malaria infections even in susceptible strains, suggesting that the basic state of a mosquito is one of refractoriness to malarial infection (Sinden et al., 2004). There are only a few examples of mosquito-parasite combinations, in which refractoriness becomes evident already at the level of gametogenesis (Micks et al., 1948; Omar, 1968; Alavi et al., 2003), and these have not been analyzed at the molecular level. Collectively, these observations are difficult to explain in terms of the known triggers of gametogenesis. They suggest some mosquitoes may produce inhibitory factors, the identification of which would be of interest.

Signaling Pathways Regulating Gametogenesis

Current evidence indicates that the still largely unknown signal transduction pathways that regulate gametogenesis fulfill three minimal requirements. (i) They enable the intracellular gametocyte to respond to physical and chemical cues from outside the host cell. (ii) They integrate these cues upstream of a single, irreversible decision point. (iii) They then execute and coordinate the parallel constituent events of gametogenesis, many of which differ between sexes. The first point involves receptor mechanisms, the other two involve secondary messengers and signal transduction cascades. Pharmacological, biochemical, and genetic studies have begun to identify the signaling pathways involved (Fig. 6). We will consider these first before briefly addressing the much more uncertain issue of parasite receptors.

CALCIUM AND CALCIUM-DEPENDENT PATHWAYS

In *P. falciparum*, gametocyte activation by pH shift triggers a phosphoinositide-specific phospholipase C (PI-PLC) activity that generates the second messengers diacylglycerol and inositol(1,4,5)trisphosphate (IP_3) (Martin et al., 1994). Combined with pharmacological studies indicating a role for intracellular calcium in the regulation of exflagellation (Martin et al., 1978; Kawamoto et al., 1993; Kawamoto et al., 1990), these data suggested that a conserved PI-PLC–IP_3–calcium signalling module exists in the gametocyte and plays a role in gametocyte activation. A transgenic *P. berghei* reporter parasite expressing the calcium-sensitive luciferase, aequorin, allowed the detection of calcium specifically in the cytosol of the intracellular gametocyte (Billker et al., 2004). At a permissive temperature, both XA and pH 8.0 were found to trigger a rise in cytosolic calcium that peaks within 15 s, is independent of extracellular calcium, is specific to the gametocyte, and is essential for the activation of male and female gametocytes.

Calcium is an important secondary messenger in the Apicomplexa. Calcium effector proteins include a family of calcium-dependent protein kinases (CDPKs). Typically, CDPKs are composed of an N-terminal serine/threonine kinase domain, a pseudosubstrate autoinhibitory domain, and a C-terminal calmodulin-like domain composed of four calcium-binding EF-hands. This domain architecture is not found in

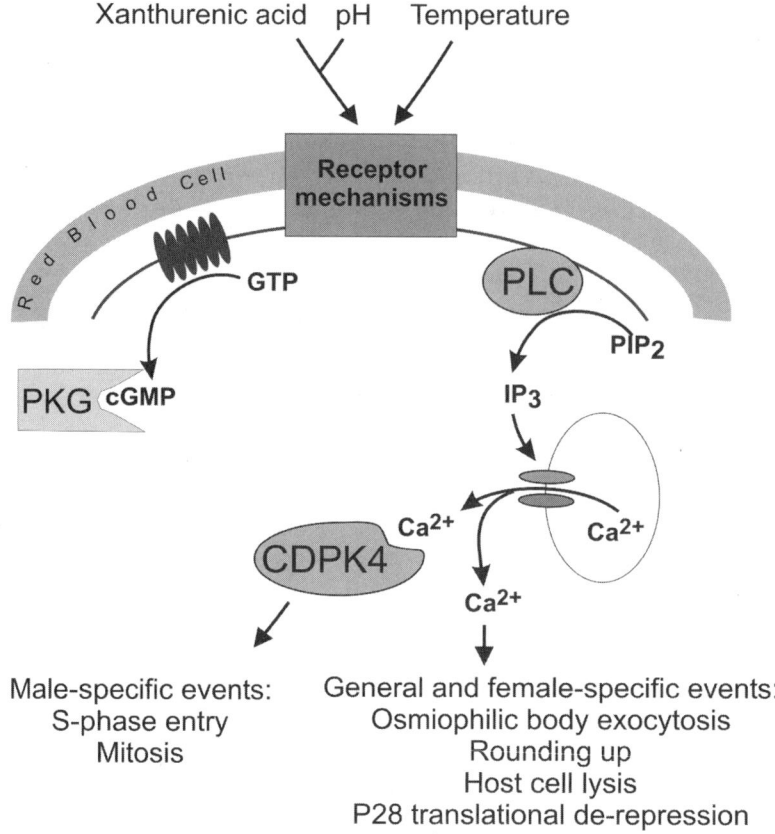

FIGURE 6 Triggers and signal transduction mechanisms regulating gametogenesis. Receptor mechanisms are unknown and may be located in either the parasite or the host cell. Signaling pathways involved in gametocyte activation include a conserved PI-PLC/IP$_3$/calcium signaling module. Cytosolic calcium triggers male-specific constituent events through the protein kinase CDPK4. A role for cGMP in the regulation of gametogenesis is likely but needs to be verified. PIP$_2$, phosphatidylinositol-4,5-bisphosphate; PKG, protein kinase G.

animal cells and seems to be limited to ciliates, the Apicomplexa, and plants. One member of this family, CDPK4, is expressed predominantly in the male gametocyte where it mediates the exclusively male-specific effects of the initial calcium signal (Billker et al., 2004). Microgametocytes with a disrupted CDPK4 gene fail to replicate their genomes upon stimulation by either XA or pH 8.0, but neither the translational de-repression of the p28 mRNA in the female nor emergence in gametocytes of either sex is affected. Transmission to the mosquito is blocked completely in CDPK4-disrupted parasites. CDPK4 thus has a key role specifically in male cell cycle regulation and other, as-yet-unidentified calcium effector pathways must operate in parallel to regulated emergence and female gametocyte differentiation.

How CDPK4 regulates male S-phase entry is unclear. Little is known about substrates and downstream effectors of CDPKs in any species. In plants, cross talk from different stress-sensing CDPKs to the same mitogen-activated protein (MAP) kinase pathway is suspected (Ludwig et al., 2003). Interestingly, there is a MAP-like kinase, PfMAP2 (Dorin et al., 1999), of unknown function expressed specifically in gametocytes. Other likely downstream effectors include cyclins and cdc2-related protein kinases that play key roles in regulating the eukaryotic cell cycle. These are well represented in the *Plasmodium* genome (Ward et al., 2004); some, such as cdc2-related protein kinase 1 (Doerig et al., 1995), are highly expressed in gametocytes versus asexual blood stages. (See also chapter 15.)

CYCLIC NUCLEOTIDE-DEPENDENT PATHWAYS

Malaria parasites possess the necessary molecular components for cyclic nucleotide signalling (Baker and Kelly, 2004). Following earlier pharmacological studies on cyclic nucleotides in exflagellation (Martin et al., 1978; Kawamoto et al., 1993; Kawamoto et al., 1990), the first tangible evidence was recently provided by Muhia et al. (2001), who found that XA enhances the generation of cyclic GMP (cGMP) in membrane fractions of gametocyte-infected erythrocytes, but not from asexual parasite stages of *P. falciparum*. Adenylyl cyclase activity was not affected. This observation raises the possibility that XA might act by direct activation of one of the two unusual guanylyl cyclases expressed in the gametocyte (Carucci et al., 2000) or perhaps by inhibition of a phosphodiesterase.

IONIC EXCHANGE MECHANISMS

At least three functions in gametogenesis can be envisaged for ion exchangers in the gametocyte and erythrocyte plasma membranes. (i) Ion exchangers in both membranes may be involved in mediating the dramatic effect of extracellular pH on the intracellular gametocyte. (ii) Related to this may be the general role of ion exchangers in regulating cytosolic pH, which often increases when cells come out of a quiescent state. (iii) Ion exchangers in the parasite's plasma membrane are almost certainly involved in regulating changes in cell volume that accompany gametogenesis.

Evidence for the type of ion exchangers involved comes from studies of *P. gallinaceum* and *P. berghei*, which have shown that complete activation of gametocytes in vitro can be achieved in a minimal medium containing no more than an energy source (such as 10 mM glucose) and three essential ions: Na^+, Cl^-, and HCO_3^- (Bishop and McConnachie, 1960; Kawamoto et al., 1992; Kawamoto et al., 1991). It is important to note that the role of bicarbonate as an essential ion is independent of the buffer system used to control pH. At pH 8.0 exflagellation in both species is still absolutely dependent on the presence of bicarbonate ions. Interestingly, mosquito exflagellation factor/XA overcomes the bicarbonate requirement in *P. gallinaceum* (Nijhout, 1979) but not in *P. berghei* (Butcher et al., 1996). In *P. berghei* acidosis caused by high parasitemias in vivo means that the concentration of HCO_3^- can become limiting for mosquito transmission (Butcher et al., 1996).

RECEPTORS AND SIGNAL INTEGRATION

The ability of the *P. gallinaceum* gametocyte to respond rapidly to nanomolar concentrations of XA is difficult to explain without assuming that an XA receptor protein exists. Current evidence is insufficient to decide whether such a receptor is derived from the host cell or the parasite and in the latter case whether it is exported to the host cell or whether XA gains access to the intracellular gametocyte. The unknown receptor mechanism can be synergistically activated by both pH and XA, has at least one gametocyte-specific component, and operates upstream of Ca^{2+} release and probably PI-PLC. The pathway may include a cGMP signaling module and a temperature-sensitive element. It is also clear that in *P. berghei*, gametocytes integrate all three environmental conditions before deciding whether to differentiate into gametes (Billker et al., 1997; Billker et al., 2004). Intriguing data by Kawamoto et al. (1991, 1993) supported an alternative model in which most constituent events of gametogenesis are regulated by temperature alone. However, these observations were not confirmed (Billker et al., 1997) and may have resulted from inadvertent partial activation of gametocytes during handling prior to experiments. This problem is difficult to avoid and may have afflicted several of the published studies on gametogenesis.

FERTILIZATION

Following the initial contact, male and female gametes quickly become closely associated. For a few seconds, the microgamete moves back and forth on the surface of the female. Eventually, the plasma membranes of the gametes fuse and with a burst of vibratory movement the axoneme and attached male nucleus enter the female cytoplasm. Nuclear fusion ensues, and over the next 3 h the zygote becomes tetraploid (Janse et al., 1986b). A synaptonemal complex and two meiotic divisions were described by Sinden et al. (1985), during which inter- and intrachromosomal recombination takes place (Walliker, 1983). No polar bodies are formed by the zygote, and all four haploid genomes therefore persist throughout the ookinete stage and replicate within the single nuclear envelope of the maturing oocyst until sporozoite budding restores the haploid state.

Members of the *p48/45* family of 10 genes, characterized by six-cysteine structural domains (Templeton and Kaslow, 1999; Thompson et al., 2001), may play important roles in gamete recognition. Several members are now known to be upregulated in gametocytes (Janse and Waters, 2004). The p48/45 gene itself encodes a conserved protein on the surface of gametes and gametocytes of either sex that is a target for transmission-blocking antibodies and a major vaccine candidate. Targeted disruption of the *p48/45* gene leads to an almost complete blockade of fertilization, resulting from a specific defect of the *p48/45* knockout microgametes to bind and fuse with macrogametes, independently of whether the latter express p48/45 or not (van Dijk et al., 2001). Population analysis of sequence variation in the *p48/45* gene led Conway et al. (2001) to predict that *p48/45* may be involved in gamete incompatibility between different regional populations of *P. falciparum*. The function of another major gamete surface antigen, P230, is yet to be demonstrated. The protein is processed upon gametocyte activation, and the two resulting forms are proposed to interact with P48/45 on the gamete surface (Sauerwein and Eling, 2002). Partial disruption of the *p230* gene in *P. falciparum* showed that gametocyte and gamete maturation proceed also in the absence of Pfs230 association to the cell surface, while fertilization efficiency was not investigated (Eksi et al., 2002).

REFERENCES

Aikawa, M., R. Carter, Y. Ito, and M. Nijhout. 1984. New observations on gametogenesis, fertilization, and zygote transformation in *Plasmodium gallinaceum*. *J. Protozool.* **31:**403–413.

Alano, P., and R. Carter. 1990. Sexual differentiation in malaria parasites. *Annu. Rev. Microbiol.* **44:**429–449.

Alano, P., D. Read, M. Bruce, M. Aikawa, T. Kaido, T. Tegoshi, S. Bhatti, D. K. Smith, C. Luo, and S. Hansra. 1995. COS cell expression cloning of Pfg377, a *Plasmodium falciparum* gametocyte antigen associated with osmiophilic bodies. *Mol. Biochem. Parasitol.* **74:**143–156.

Alano, P., F. Silvestrini, and L. Roca. 1996. Structure and polymorphism of the upstream region of the *pfg27/25* gene, transcriptionally regulated in gametocytogenesis of *Plasmodium falciparum*. *Mol. Biochem. Parasitol.* **79:**207–217.

Alavi, Y., M. Arai, J. Mendoza, M. Tufet-Bayona, R. Sinha, K. Fowler, O. Billker, B. Franke-Fayard, C. J. Janse, A. Waters, and R. E. Sinden. 2003. The dynamics of interactions between *Plasmodium* and the mosquito: a study of the infectivity of *Plasmodium berghei* and *Plasmodium gallinaceum*, and their transmission by *Anopheles stephensi*, *Anopheles gambiae* and *Aedes aegypti*. *Int. J. Parasitol.* **33:** 933–943.

Arai, M., O. Billker, H. R. Morris, M. Panico, M. Delcroix, D. Dixon, S. V. Ley, and R. E. Sinden. 2001. Both mosquito-derived xanthurenic acid and a host blood-derived factor regulate gametogenesis of *Plasmodium* in the midgut of the mosquito. *Mol. Biochem. Parasitol.* **116:**17–24.

Aravind, L., L. M. Iyer, T. E. Wellems, and L. H. Miller. 2003. Plasmodium biology: genomic gleanings. *Cell* **115:**771–785.

Baker, D. A., and J. M. Kelly. 2004. Purine nucleotide cyclases in the malaria parasite. *Trends Parasitol.* **20:**227–232.

Baker, D. A., O. Daramola, M. V. McCrossan, J. Harmer, and G. A. Targett. 1994. Subcellular localization of Pfs16, a *Plasmodium falciparum* gametocyte antigen. *Parasitology* **108:**129–137.

Baker, D. A., J. Thompson, O. O. Daramola, J. M. Carlton, and G. A. Targett. 1995. Sexual-stage-specific RNA expression of a new *Plasmodium falciparum* gene detected by in situ hybridisation. *Mol. Biochem. Parasitol.* **72:**193–201.

Bastianelli, G., and A. Bignami. 1899. Sulla struttura dei parassiti malarici, e in specie dei gameti dei parassiti estivo-autunnali. Atti della Società per gli studi della malaria **1:**1–13.

Bhalla, S. C. 1968. White Eye, a new sex-linked mutant of *Aedes aegypti. Mosquito News* **28:**380–385.

Billker, O., A. J. Miller, and R. E. Sinden. 2000. Determination of mosquito bloodmeal pH in situ by ion-selective microelectrode measurement: implications for the regulation of malarial gametogenesis. *Parasitology* **120:**547–551.

Billker, O., M. K. Shaw, G. Margos, and R. E. Sinden. 1997. The roles of temperature, pH and mosquito factors as triggers of male and female gametogenesis of *Plasmodium berghei* in vitro. *Parasitology* **115:**1–7.

Billker, O., M. K. Shaw, I. W. Jones, S. V. Ley, A. J. Mordue, and R. E. Sinden. 2002. Azadirachtin disrupts formation of organised microtubule arrays during microgametogenesis of *Plasmodium berghei. J. Eukaryot. Microbiol.* **49:**489–497.

Billker, O., S. Dechamps, R. Tewari, G. Wenig, B. Franke-Fayard, and V. Brinkmann. 2004. Calcium and a calcium-dependent protein kinase regulate gamete formation and mosquito transmission in a malaria parasite. *Cell* **117:**503–514.

Billker, O., V. Lindo, M. Panico, A. E. Etienne, T. Paxton, A. Dell, M. Rogers, R. E. Sinden, and H. R. Morris. 1998. Identification of xanthurenic acid as the putative inducer of malaria development in the mosquito. *Nature* **392:**289–292.

Bishop, A., and E. W. McConnachie. 1956. A study of the factors affecting the emergence of the gametocytes of *Plasmodium gallinaceum* from the erythrocytes and the exflagellation of the male gametocytes. *Parasitology* **46:**192–215.

Bishop, A., and E. W. McConnachie. 1960. Further observations on the in vitro development of the gametocytes of *Plasmodium gallinaceum. Parasitology* **50:**431–448.

Bousema, J. T., L. C. Gouagna, A. M. Meutstege, B. E. Okech, N. I. Akim, J. I. Githure, J. C. Beier, and R. W. Sauerwein. 2003. Treatment failure of pyrimethamine-sulphadoxine and induction of *Plasmodium falciparum* gametocytaemia in children in western Kenya. *Trop. Med. Int. Health* **8:**427–430.

Bracchi, V., G. Langsley, J. Thelu, W. Eling, and P. Ambroise-Thomas. 1996. PfKIN, an SNF1 type protein kinase of *Plasmodium falciparum* predominantly expressed in gametocytes. *Mol. Biochem. Parasitol.* **76:**299–303.

Bruce, M. C., R. N. Carter, K. Nakamura, M. Aikawa, and R. Carter. 1994. Cellular location and temporal expression of the *Plasmodium falciparum* sexual stage antigen Pfs16. *Mol. Biochem. Parasitol.* **65:**11–22.

Buckling, A., L. Crooks, and A. Read. 1999. *Plasmodium chabaudi*: effect of antimalarial drugs on gametocytogenesis. *Exp. Parasitol.* **93:**45–54.

Butcher, G. A., R. E. Sinden, and O. Billker. 1996. *Plasmodium berghei*: infectivity of mice to *Anopheles stephensi* mosquitoes. *Exp. Parasitol.* **84:**371–379.

Carlton, J. M., et al. 2002. Genome sequence and comparative analysis of the model rodent malaria parasite *Plasmodium yoelii yoelii. Nature* **419:**512–519.

Carter, R., and P. M. Graves. 1988. Gametocytes, p. 273–305. *In* W. H. Wernsdorfer and I. McGregor (ed.), *Malaria: Principles and Practice of Malariology*. Churchill Livingstone, Edinburgh, United Kingdom.

Carter, R., P. M. Graves, A. Creasey, K. D. Byrne, D. Read, P. Alano, and B. Fenton. 1989. *Plasmodium falciparum*: an abundant stage-specific protein expressed during early gametocyte development. *Exp. Parasitol.* **69:**140–149.

Carucci, D. J., A. A. Witney, D. K. Muhia, D. C. Warhurst, P. Schaap, M. Meima, J. L. Li, M. C. Taylor, J. M. Kelly, and D. A. Baker. 2000. Guanylyl cyclase activity associated with putative bifunctional integral membrane proteins in *Plasmodium falciparum. J. Biol. Chem.* **275:**22147–22156.

Chorine, V. 1933. Conditions qui régissent la fécondation de *Plasmodium praecox. Arch. Inst. Pasteur Alger.* **11:**1–8.

Conway, D. J., R. L. Machado, B. Singh, P. Dessert, Z. S. Mikes, M. M. Povoa, A. M. Oduola, and C. Roper. 2001. Extreme geographical fixation of variation in the *Plasmodium falciparum* gamete surface protein gene Pfs48/45 compared with microsatellite loci. *Mol. Biochem. Parasitol.* **115:**145–156.

Cooke, B. M., and R. L. Coppel. 1995. Cytoadhesion and falciparum malaria: going with the flow. *Parasitol. Today* **11:**282–287.

Coulson, R. M., N. Hall, and C. A. Ouzounis. 2004. Comparative genomics of transcriptional control in the human malaria parasite *Plasmodium falciparum. Genome Res.* **14:**1548–1554.

Crabb, B. S., B. M. Cooke, J. C. Reeder, R. F. Waller, S. R. Caruana, K. M. Davern, M. E. Wickham, G. V. Brown, R. L. Coppel, and A. F. Cowman. 1997. Targeted gene disruption shows that knobs enable malaria-infected red cells to cytoadhere under physiological shear stress. *Cell* **89:**287–296.

Creasey, A., K. Mendis, J. Carlton, D. Williamson, I. Wilson, and R. Carter. 1994. Maternal inheritance of extrachromosomal DNA in malaria parasites. *Mol. Biochem. Parasitol.* **65:**95–98.

Cui, L., Q. Fan, and J. Li. 2002. The malaria parasite *Plasmodium falciparum* encodes members of the Puf RNA-binding protein family with conserved RNA binding activity. *Nucleic Acids Res.* **30:**4607–4617.

David, P. H., S. M. Handunnetti, J. H. Leech, P. Gamage, and K. N. Mendis. 1988. Rosetting: a new cytoadherence property of malaria-infected erythrocytes. *Am. J. Trop. Med. Hyg.* **38:**289–297.

Day, K. P., R. E. Hayward, D. Smith, and J. G. Culvenor. 1998. CD36-dependent adhesion and knob expression of the transmission stages of *Plasmodium falciparum* is stage specific. *Mol. Biochem. Parasitol.* **93:**167–177.

Day, K. P., F. Karamalis, J. Thompson, D. A. Barnes, C. Peterson, H. Brown, G. V. Brown, and D. J. Kemp. 1993. Genes necessary for expression of a virulence determinant and for transmission of *Plasmodium falciparum* are located on a 0.3-megabase region of chromosome 9. *Proc. Natl. Acad. Sci. USA* **90:**8292–8296.

Dechering, K. J., A. M. Kaan, W. Mbacham, D. F. Wirth, W. Eling, R. N. Konings, and H. G. Stunnenberg. 1999. Isolation and functional characterization of two distinct sexual-stage-specific promoters of the human malaria parasite *Plasmodium falciparum*. *Mol. Cell. Biol.* **19:**967–978.

Dechering, K. J., J. Thompson, H. J. Dodemont, W. Eling, and R. N. Konings. 1997. Developmentally regulated expression of pfs16, a marker for sexual differentiation of the human malaria parasite *Plasmodium falciparum*. *Mol. Biochem. Parasitol.* **89:**235–244.

Delrieu, I., C. C. Waller, M. M. Mota, M. Grainger, J. Langhorne, and A. A. Holder. 2002. PSLAP, a protein with multiple adhesive motifs, is expressed in *Plasmodium falciparum* gametocytes. *Mol. Biochem. Parasitol.* **121:**11–20.

Delves, C. J., P. Alano, R. G. Ridley, M. Goman, S. P. Holloway, J. E. Hyde, and J. G. Scaife. 1990. Expression of alpha and beta tubulin genes during the asexual and sexual blood stages of *Plasmodium falciparum*. *Mol. Biochem. Parasitol.* **43:**271–278.

de Souza, J. B., and E. M. Riley. 2002. Cerebral malaria: the contribution of studies in animal models to our understanding of immunopathogenesis. *Microbes Infect.* **4:**291–300.

Doerig, C., P. Horrocks, J. Coyle, J. Carlton, A. Sultan, D. Arnot, and R. Carter. 1995. Pfcrk-1, a developmentally regulated cdc2-related protein kinase of *Plasmodium falciparum*. *Mol. Biochem. Parasitol.* **70:**167–174.

Dorin, D., P. Alano, I. Boccaccio, L. Ciceron, C. Doerig, R. Sulpice, D. Parzy, and C. Doerig. 1999. An atypical mitogen-activated protein kinase (MAPK) homologue expressed in gametocytes of the human malaria parasite *Plasmodium falciparum*. Identification of a MAPK signature. *J. Biol. Chem.* **274:**29912–29920.

Eksi, S., A. Stump, S. L. Fanning, M. I. Shenouda, H. Fujioka, and K. C. Williamson. 2002. Targeting and sequestration of truncated Pfs230 in an intraerythrocytic compartment during *Plasmodium falciparum* gametocytogenesis. *Mol. Microbiol.* **44:**1507–1516.

Eksi, S., and K. C. Williamson. 2002. Male-specific expression of the paralog of malaria transmission-blocking target antigen Pfs230, PfB0400w. *Mol. Biochem. Parasitol.* **122:**127–130.

Fang, J., M. Sullivan, and T. F. McCutchan. 2004. The effects of glucose concentration on the reciprocal regulation of rRNA promoters in *Plasmodium falciparum*. *J. Biol. Chem.* **279:**720–725.

Florens, L., M. P. Washburn, J. D. Raine, R. M. Anthony, M. Grainger, J. D. Haynes, J. K. Moch, N. Muster, J. B. Sacci, D. L. Tabb, A. A. Witney, D. Wolters, Y. Wu, M. J. Gardner, A. A. Holder, R. E. Sinden, J. R. Yates, and D. J. Carucci. 2002. A proteomic view of the *Plasmodium falciparum* life cycle. *Nature* **419:**520–526.

Garcia, G. E., R. A. Wirtz, and R. Rosenberg. 1997. Isolation of a substance from the mosquito that activates *Plasmodium* fertilization. *Mol. Biochem. Parasitol.* **88:**127–135.

Garcia, G. E., R. A. Wirtz, J. R. Barr, A. Woolfitt, and R. Rosenberg. 1998. Xanthurenic acid induces gametogenesis in *Plasmodium*, the malaria parasite. *J. Biol. Chem.* **273:**12003–12005.

Gardiner, D. L., M. W. A. Dixon, T. Spielmann, T. S. Skinner-Adams, P. L. Hawthorne, M. R. Ortega, D. J. Kemp, and K. R. Trenholme. 2005. Implication of a *Plasmodium falciparum* gene in the switch between asexual reproduction and gametocytogenesis. *Mol. Biochem. Parasitol.* **140:**153–160.

Garnham, P. C. C., R. G. Bird, and J. R. Baker. 1967. Electron microscope studies of motile stages of malaria parasites. V. Exflagellation in *Plasmodium*, hepatocystis and leucocytozoon. *Trans. R. Soc. Trop. Med. Hyg.* **61:**58–68.

Gautret, P., and A. Motard. 1999. Periodic infectivity of Plasmodium gametocytes to the vector. A review. *Parasite* **6:**103–111.

Gautret, P., I. Landau, L. Tailhardat, F. Miltgen, F. Coquelin, T. Voza, A. G. Chabaud, and J. L. Jacquemin. 2000. The effects of subcurative doses of chloroquine on *Plasmodium vinckei petteri* gametocytes and on their infectivity to mosquitoes. *Int. J. Parasitol.* **30:**1193–1198.

Gilks, C. F., D. Walliker, and C. I. Newbold. 1990. Relationships between sequestration, antigenic variation and chronic parasitism in *Plasmodium chabaudi chabaudi*—a rodent malaria model. *Parasite Immunol.* **12:**45–64.

Guinet, F., and T. E. Wellems. 1997. Physical mapping of a defect in *Plasmodium falciparum* male gametocytogenesis to an 800 kb segment of chromosome 12. *Mol. Biochem. Parasitol.* **90:**343–346.

Hall, N., M. Karras, J. D. Raine, J. M. Carlton, T. W. Kooij, M. Berriman, L. Florens, C. S. Janssen, A. Pain, G. K. Christophides, K. James,

K. Rutherford, B. Harris, D. Harris, C. Churcher, M. A. Quail, D. Ormond, J. Doggett, H. E. Trueman, J. Mendoza, S. L. Bidwell, M. A. Rajandream, D. J. Carucci, J. R. Yates III, F. C. Kafatos, C. J. Janse, B. Barrell, C. M. Turner, A. P. Waters, and R. E. Sinden. 2005. A comprehensive survey of the *Plasmodium* life cycle by genomic, transcriptomic, and proteomic analyses. *Science* **307:**82–86.

Hallett, R. L., C. J. Sutherland, N. Alexander, R. Ord, M. Jawara, C. J. Drakeley, M. Pinder, G. Walraven, G. A. T. Targett, and A. Alloueche. 2004. Combination therapy counteracts the enhanced transmission of drug resistant malaria parasites to mosquito. *Antimicrob. Agents Chemother.* **48:**3940–3943.

Han, Q., E. Calvo, O. Marinotti, J. Fang, M. Rizzi, A. A. James, and J. Li. 2003. Analysis of the wild-type and mutant genes encoding the enzyme kynurenine monooxygenase of the yellow fever mosquito, *Aedes aegypti*. *Insect Mol. Biol.* **12:**483–490.

Han, Q., J. Fang, and J. Li. 2002. 3-Hydroxykynurenine transaminase identity with alanine glyoxylate transaminase. A probable detoxification protein in *Aedes aegypti*. *J. Biol. Chem.* **277:**15781–15787.

Hawking, F., M. E. Wilson, and K. Gammage. 1971. Evidence for cyclic development and short-lived maturity in the gametocytes of *Plasmodium falciparum*. *Trans. R. Soc. Trop. Med. Hyg.* **65:**549–559.

Hayward, R. E. 2000. *Plasmodium falciparum* phosphoenolpyruvate carboxykinase is developmentally regulated in gametocytes. *Mol. Biochem. Parasitol.* **107:**227–240.

Hayward, R. E., J. L. Derisi, S. Alfadhli, D. C. Kaslow, P. O. Brown, and P. K. Rathod. 2000a. Shotgun DNA microarrays and stage-specific gene expression in *Plasmodium falciparum* malaria. *Mol. Microbiol.* **35:**6–14.

Hayward, R. E., D. J. Sullivan, and K. P. Day. 2000b. *Plasmodium falciparum*: histidine-rich protein II is expressed during gametocyte development. *Exp. Parasitol.* **96:**139–146.

Hirai, M., J. Wang, S. Yoshida, A. Ishii, and H. Matsuoka. 2001. Characterization and identification of exflagellation-inducing factor in the salivary gland of *Anopheles stephensi* (Diptera: Culicidae). *Biochem. Biophys. Res. Commun.* **287:**859–864.

Janse, C. J., and A. P. Waters. 2004. Sexual development of malaria parasites, p. 445–474. *In* A. P. Waters and C. J. Janse (ed.), *Malaria Parasites: Genomes and Molecular Biology.* Caister Academic Press, Norwich, England.

Janse, C. J., P. F. J. Klooster, H. J. Kaay, M. Ploeg, and J. P. Overdulve. 1986a. DNA synthesis in *Plasmodium berghei* during asexual and sexual development. *Mol. Biochem. Parasitol.* **20:**173–182.

Janse, C. J., P. F. van der Klooster, H. J. van der Kaay, M. Van der Ploeg, and J. P. Overdulve. 1986b. Rapid repeated DNA replication during microgametogenesis and DNA synthesis in young zygotes of *Plasmodium berghei*. *Trans. R. Soc. Trop. Med. Hyg.* **80:**154–157.

Janse, C. J., T. Ponnudurai, A. H. W. Lensen, J. H. E. T. Meuwissen, M. Van der Ploeg, and J. P. Overdulve. 1988. DNA synthesis in gametocytes of *Plasmodium falciparum*. *Parasitology* **96:**1–7.

Janse, C. J., A. Haghparast, M. A. Speranca, J. Ramesar, H. Kroeze, H. A. del Portillo, and A. P. Waters. 2003. Malaria parasites lacking eef1a have a normal S/M phase yet grow more slowly due to a longer G1 phase. *Mol. Microbiol.* **50:**1539–1551.

Janse, C. J., J. Ramesar, F. M. van der Berg, and B. Mons. 1992. *Plasmodium berghei*: in vivo generation and selection of karyotype mutants and non-gametocyte producer mutants. *Exp. Parasitol.* **74:**1–10.

Jensen, J. B. 1979. Observations on gametogenesis in *Plasmodium falciparum* from continuous culture. *J. Protozool.* **26:**129–132.

Jovani, R., L. Amo, E. Arriero, O. Krone, A. Marzal, P. Shurulinkov, G. Tomas, D. Sol, J. Hagen, P. Lopez, J. Martin, C. Navarro, and J. Torres. 2004. Double gametocyte infections in apicomplexan parasites of birds and reptiles. *Parasitol. Res.* **94:**155–157.

Kaslow, D. C. 2002. Transmission-blocking vaccines. *Chem. Immunol.* **80:**287–307.

Kato, M., K. Tanabe, A. Miki, K. Ichimori, and S. Waki. 1990. Membrane potential of *Plasmodium falciparum* gametocytes monitored with rhodamine 123. *FEMS Microbiol. Lett.* **69:**283–288.

Kawamoto, F., H. Fujioka, R. I. Murakami, Syafruddin, M. Hagiwara, T. Ishikawa, and H. Hidaka. 1993. The roles of Ca^{2+}/calmodulin- and cGMP-dependent pathways in gametogenesis of a rodent malaria parasite, *Plasmodium berghei*. *Eur. J. Cell Biol.* **60:**101–107.

Kawamoto, F., N. Kido, T. Hanaichi, M. B. A. Djamgoz, and R. E. Sinden. 1992. Gamete development in *Plasmodium berghei* regulated by ionic exchange mechanisms. *Parasitol. Res.* **78:**277–284.

Kawamoto, F., R. Alejo-Blanco, S. L. Fleck, and R. E. Sinden. 1991. *Plasmodium berghei*: ionic regulation and the induction of gametogenesis. *Exp. Parasitol.* **72:**33–42.

Kawamoto, F., R. Alejo-Blanco, S. L. Fleck, Y. Kawamoto, and R. E. Sinden. 1990. Possible roles of Ca^{2+} and cGMP as mediators of the exflagellation of *Plasmodium berghei* and *Plasmodium falciparum*. *Mol. Biochem. Parasitol.* **42:**101–108.

Khan, S. M., and A. P. Waters. 2004. Malaria parasite transmission stages: an update. *Trends Parasitol.* **20:**575–580.

Kongkasuriyachai, D., H. Fujioka, and N. Kumar. 2004. Functional analysis of *Plasmodium falciparum* parasitophorous vacuole membrane protein (Pfs16) during gametocytogenesis and gametogenesis by targeted gene disruption. *Mol. Biochem. Parasitol.* **133:**275–285.

Krungkrai, J., D. Burat, S. Kudan, S. Krungkrai, and P. Prapunwattana. 1999. Mitochondrial oxygen consumption in asexual and sexual blood stages of the human malarial parasite, *Plasmodium falciparum*. *Southeast Asian J. Trop. Med. Public Health* **30:**636–642.

Krungkrai, J., P. Prapunwattana, and S. R. Krungkrai. 2000. Ultrastructure and function of mitochondria in gametocytic stage of *Plasmodium falciparum*. *Parasite* **7:**19–26.

Kumar, N. 1997. Protein phosphorylation during sexual differentiation in the malaria parasite *Plasmodium falciparum*. *Mol. Biochem. Parasitol.* **87:**205–210.

Landau, I., and P. Gautret. 1998. Animal models: rodents, p. 401–417. *In* I. W. Sherman (ed.), *Malaria: Parasite Biology, Pathogenesis, and Protection*. ASM Press, Washington, D.C.

Lang-Unnasch, N., and A. D. Murphy. 1998. Metabolic changes of the malaria parasite during the transition from the human to the mosquito host. *Annu. Rev. Microbiol.* **52:**561–590.

Lasonder, E., Y. Ishihama, J. S. Andersen, A. M. Vermunt, A. Pain, R. W. Sauerwein, W. M. Eling, N. Hall, A. P. Waters, H. G. Stunnenberg, and M. Mann. 2002 Analysis of the *Plasmodium falciparum* proteome by high-accuracy mass spectrometry. *Nature* **419:**537–542.

Laveran, A. 1880. Note sur un nouveau parasite trouvé dans le sang de plusieurs malades atteints de fièvre palustre. *Bull. Acad. Med.* **9:**1235–1236.

Learngaramkul, P., S. Petmitr, S. R. Krungkrai, P. Prapunwattana, and J. Krungkrai. 1999. Molecular characterization of mitochondria in asexual and sexual blood stages of *Plasmodium falciparum*. *Mol. Cell. Biol. Res. Commun.* **2:**15–20.

Le Roch, K. G., J. R. Johnson, L. Florens, Y. Zhou, A. Santrosyan, M. Grainger, S. F. Yan, K. C. Williamson, A. A. Holder, D. J. Carucci, J. R. Yates III, and E. A. Winzeler. 2004. Global analysis of transcript and protein levels across the *Plasmodium falciparum* life cycle. *Genome Res.* **14:**2308–2318.

Le Roch, K. G., Y. Zhou, P. L. Blair, M. Grainger, J. K. Moch, J. D. Haynes, P. De La Vega, A. A. Holder, S. Batalov, D. J. Carucci, and E. A. Winzeler. 2003. Discovery of gene function by expression profiling of the malaria parasite life cycle. *Science* **301:**1503–1508.

Li, J. L., and L. S. Cox. 2001. Identification of an MCM4 homologue expressed specifically in the sexual stage of *Plasmodium falciparum*. *Int. J. Parasitol.* **31:**1246–1252.

Li, J. L., and L. S. Cox. 2003. Characterisation of a sexual stage-specific gene encoding ORC1 homologue in the human malaria parasite *Plasmodium falciparum*. *Parasitol. Int.* **52:**41–52.

Li, J. L., and D. A. Baker. 1997. Protein phosphatase beta, a putative type-2A protein phosphatase from the human malaria parasite *Plasmodium falciparum*. *Eur. J. Biochem.* **249:**98–106.

Li, J. L., and D. A. Baker. 1998. A putative protein serine/threonine phosphatase from *Plasmodium falciparum* contains a large N-terminal extension and five unique inserts in the catalytic domain. *Mol. Biochem. Parasitol.* **95:**287–295.

Li, J. L., D. A. Baker, and L. S. Cox. 2000. Sexual stage-specific expression of a third calcium-dependent protein kinase from *Plasmodium falciparum*. *Biochim. Biophys. Acta* **1491:**341–349.

Li, J. L., K. J. Robson, J. L. Chen, G. A. Targett, and D. A. Baker. 1996. Pfmrk, a MO15-related protein kinase from *Plasmodium falciparum*. Gene cloning, sequence, stage-specific expression and chromosome localization. *Eur. J. Biochem.* **241:**805–813.

Li, J. L., G. A. Targett, and D. A. Baker. 2001. Primary structure and sexual stage-specific expression of a LAMMER protein kinase of *Plasmodium falciparum*. *Int. J. Parasitol.* **31:**387–392.

Lobo, C. A., H. Fujioka, M. Aikawa, and N. Kumar. 1999. Disruption of the Pfg27 locus by homologous recombination leads to loss of the sexual phenotype in *P. falciparum*. *Mol. Cell* **3:**793–798.

Ludwig, A. A., T. Romeis, and J. D. G. Jones. 2003. CDPK-mediated signalling pathways: specificity and cross-talk. *J. Exp. Bot.* **55:**1–8.

Margos, G., M. R. van Dijk, J. Ramesar, C. J. Janse, A. P. Waters, and R. E Sinden. 1998. Transgenic expression of a mosquito-stage malarial protein, Pbs21, in blood stages of transformed *Plasmodium berghei* and induction of an immune response upon infection. *Infect. Immun.* **66:**3884–3891.

Martin, S. K., L. H. Miller, M. M. Nijhout, and R. Carter. 1978. *Plasmodium gallinaceum*: induction of male gametocyte exflagellation by phosphodiesterase inhibitors. *Exp. Parasitol.* **44:**239–242.

Martin, S. K., M. Jett, and I. Schneider. 1994. Correlation of phosphoinositide hydrolysis with exflagellation in the malaria microgametocyte. *J. Parasitol.* **80:**371–378.

Micks, D. W., P. F. de Caires, and L. B. Franco. 1948. The relationship of exflagellation in avian plasmodia to pH and immunity in the mosquito. *Am. J. Hyg.* **48:**182–190.

Militello, K. T., M. Dodge, L. Bethke, and D. F. Wirth. 2004. Identification of regulatory elements in the *Plasmodium falciparum* genome. *Mol. Biochem. Parasitol.* **134:**75–88.

Mons, B. 1986. Intraerythrocytic differentiation of *Plasmodium berghei*. *Acta Leidensa* **54:**1–124.

Muhia, D. K., C. A. Swales, W. Deng, J. M. Kelly, and D. A. Baker. 2001. The gametocyte-activating factor xanthurenic acid stimulates an increase in membrane-associated guanylyl cyclase activity in the human malaria parasite *Plasmodium falciparum*. *Mol. Microbiol.* **42:**553–560.

Naotunne, T. S., N. D. Karunaweera, K. N. Mendis, and R. Carter. 1993. Cytokine-mediated inactivation of malarial gametocytes is dependent on the presence of white blood cells and involves reactive nitrogen intermediates. *Immunology* **78:**555–562.

Nijhout, M. M. 1979. *Plasmodium gallinaceum*: exflagellation stimulated by a mosquito factor. *Exp. Parasitol.* **48:**75–80.

Omar, S. M. 1968. Vergleichende Beobachtungen über die Entwicklung von *Plasmodium cynomolgi bastianellii* in *Anopheles stephensi* und *Anopheles albimanus*. *Z. Tropenmed. Parasitol.* **19:**370–389.

Pace, T., C. Birago, C. J. Janse, L. Picci, and M. Ponzi. 1998. Developmental regulation of a Plasmodium gene involves the generation of stage-specific 5' untranslated sequences. *Mol. Biochem. Parasitol.* **97:**45–53.

Paton, M. G., G. C. Barker, H. Matsuoka, J. Ramesar, C. J. Janse, A. P. Waters, and R. E. Sinden. 1993. Structure and expression of a post-transcriptionally regulated malaria gene encoding a surface protein from the sexual stages of *Plasmodium berghei*. *Mol. Biochem. Parasitol.* **59:**263–276.

Paul, R. E., T. N. Coulson, A. Raibaud, and P. T. Brey. 2000. Sex determination in malaria parasites. *Science* **287:**128–131

Paul, R. E., A. Raibaud, and P. T. Brey. 1999. Sex ratio adjustment in *Plasmodium gallinaceum*. *Parassitologia* **41:**153–158.

Paul, R. E. L., F. Ariey, and V. Robert. 2003. The evolutionary ecology of Plasmodium. *Ecol. Lett.* **6:**866–880.

Petmitr, S., and J. Krungkrai. 1995. Mitochondrial cytochrome b gene in two developmental stages of human malarial parasite *Plasmodium falciparum*. *Southeast Asian J. Trop. Med. Public Health* **26:**600–605.

Pichon, G., H. P. Awono-Ambene, and V. Robert. 2000. High heterogeneity in the number of *Plasmodium falciparum* gametocytes in the bloodmeal of mosquitoes fed on the same host. *Parasitology* **121:**115–120.

Pradel, G., K. Hayton, L. Aravind, L. M. Iyer, M. S. Abrahamsen, A. Bonawitz, C. Mejia, and T. J. Templeton. 2004. A multidomain adhesion protein family expressed in *Plasmodium falciparum* is essential for transmission to the mosquito. *J. Exp. Med.* **199:**1533–1544.

Price, R., F. Nosten, J. A. Simpson, C. Luxemburger, L. Phaipun, ter F. Kuile, van M. Vugt, T. Chongsuphajaisiddhi, and N. J. White. 1999. Risk factors for gametocyte carriage in uncomplicated falciparum malaria. *Am. J. Trop. Med. Hyg.* **60:**1019–1023.

Pukrittayakamee, S., K. Chotivanich, A. Chantra, R. Clemens, S. Looareesuwan, and N. J. White. 2004. Activities of artesunate and primaquine against asexual- and sexual-stage parasites in falciparum malaria. *Antimicrob. Agents Chemother.* **48:**1329–1334.

Ramsey, J. M., A. Tello, C. O. Contreras, R. Ordonez, N. Chirino, J. Rojo, and F. Garcia. 2002. *Plasmodium falciparum* and *P. vivax* gametocyte-specific exoantigens stimulate proliferation of TCR gammadelta lymphocytes. *J. Parasitol.* **88:**59–68.

Rawlings, D. J., H. Fujioka, M. Fried, D. B. Keister, M. Aikawa, and D. C. Kaslow. 1992. α-Tubulin II is a male-specific protein in *Plasmodium falciparum*. *Mol. Biochem. Parasitol.* **56:**239–250.

Ridley, R. G., J. H. White, S. M. McAleese, M. Goman, P. Alano, E. de Vries, and B. J. Kilbey. 1991. DNA polymerase delta: gene sequences from *Plasmodium falciparum* indicate that this enzyme is more highly conserved than DNA polymerase alpha. *Nucleic Acids Res.* **19:**6731–6736.

Rogers, N. J., O. Daramola, G. A. Targett, and B. S. Hall. 1996a. CD36 and intercellular adhesion molecule 1 mediate adhesion of developing *Plasmodium falciparum* gametocytes. *Infect. Immun.* **64:**1480–1483.

Rogers, N. J., B. S. Hall, J. Obiero, G. A. Targett, and C. J. Sutherland. 2000. A model for sequestration of the transmission stages of *Plasmodium falciparum*: adhesion of gametocyte-infected erythrocytes to human bone marrow cells. *Infect. Immun.* **68:**3455–3462.

Rogers, N. J., G. A. Targett, and B. S. Hall. 1996b. *Plasmodium falciparum* gametocyte adhesion to C32 cells via CD36 is inhibited by antibodies to modified band 3. *Infect. Immun.* **64:**4261–4268.

Salmon, B. L., A. Oksman, and D. E. Goldberg. 2001. Malaria parasite exit from the host erythrocyte: a two-step process requiring extraerythrocytic proteolysis. *Proc. Natl. Acad. Sci. USA* **98:**271–276.

Sato, S., K. Rangachari, and R. J. M. Wilson. 2003. Targeting GFP to the malarial mitochondrion. *Mol. Biochem. Parasitol.* **130:**155–158.

Sauerwein, R. W., and W. M. Eling. 2002. Sexual and sporogonic stage antigens. *Chem. Immunol.* **80:**188–203.

Schall, J. J. 1989. The sex ratio of Plasmodium gametocytes. *Parasitology* **98:**343–350.

Scherf, A., R. Carter, C. Petersen, P. Alano, R. Nelson, M. Aikawa, D. Mattei, D. S. Pereira, and J. Leech. 1992. Gene inactivation of Pf11-1 of *Plasmodium falciparum* by chromosome breakage and healing: identification of a gametocyte-specific protein with a potential role in gametogenesis. *EMBO J.* **11:**2293–2301.

Severini, C., F. Silvestrini, A. Sannella, S. Barca, L. Gradoni, and P. Alano. 1999. The production of the osmiophilic body protein Pfg377 is associated with stage of maturation and sex in *Plasmodium falciparum* gametocytes. *Mol. Biochem. Parasitol.* **100:**247–252.

Sharma, A., I. Sharma, D. Kogkasuriyachai, and N. Kumar. 2003. Structure of a gametocyte protein essential for sexual development in *Plasmodium falciparum*. *Nat. Struct. Biol.* **10:**197–203.

Sherman, I. W., S. Eda, and E. Winograd. 2003. Cytoadherence and sequestration in *Plasmodium falciparum*: defining the ties that bind. *Microbes Infect.* **5:**897–909.

Silvestrini, F., P. Alano, and J. L. Williams. 2000. Commitment to the production of male and female gametocytes in the human malaria parasite *Plasmodium falciparum*. *Parasitology* **121:**465–471.

Sinden, R. E. 1975. Microgametogenesis in *Plasmodium yoelii nigeriensis*: a scanning electron microscope investigation. *Protistologica* **11:**263–268.

Sinden, R. E. 1983a. Sexual development of malarial parasites. *Adv. Parasitol.* **22:**153–216.

Sinden, R. E. 1983b. The cell biology of sexual development in *Plasmodium*. *Parasitology* **86:**7–28.

Sinden, R. E., and N. A. Croll. 1975. Cytology and kinetics of microgametogenesis and fertilization in *Plasmodium yoelii nigeriensis*. *Parasitology* **70:**53–65.

Sinden, R. E., E. U. Canning, and B. Spain. 1976. Gametogenesis and fertilization in *Plasmodium yoelii nigeriensis*: a transmission electron microscope study. *Proc. R. Soc. Lond. B Biol. Sci.* **193:**55–76.

Sinden, R. E., E. U. Canning, R. S. Bray, and M. E. Smalley. 1978. Gametocyte and gamete development in *Plasmodium falciparum*. *Proc. R. Soc. Lond. B Biol. Sci.* **201:**375–399.

Sinden, R. E., G. Butcher, O. Billker, and S. Fleck. 1996. Regulation of infectivity of *Plasmodium* to the mosquito vector. *Adv. Parasitol.* **38:**54–117.

Sinden, R. E., Y. Alavi, and J. D. Raine. 2004. Mosquito-malaria interactions: a reappraisal of the concepts of susceptibility and refractoriness. *Insect Biochem. Mol. Biol.* **34:**625–629.

Sinden, R. E. 1982. Gametocytogenesis of *Plasmodium falciparum* in vitro: an electron microscopic study. *Parasitology* **84:**1–11.

Sinden, R. E., and M. E. Smalley. 1979. Gametocytogenesis of *Plasmodium falciparum* in vitro: the cell-cycle. *Parasitology* **79:**277–296.

Sinden, R. E., R. H. Hartley, and L. Winger. 1985. The development of Plasmodium ookinetes in vitro: an ultrastructural study including a description of meiotic division. *Parasitology* **91:**227–244.

Smalley, M. E., S. Abdalla, and J. Brown. 1980. The distribution of *Plasmodium falciparum* in the peripheral blood and bone marrow of Gambian children. *Trans. R. Soc. Trop. Med. Hyg.* **75:**103–105.

Smalley, M. E., and R. E. Sinden. 1977. *Plasmodium falciparum* gametocytes: their longevity and infectivity. *Parasitology* **74:**1–8.

Smalley, M. E., J. Brown, and N. M. Bassett. 1981. The rate of production of *Plasmodium falciparum* gametocytes during natural infections. *Trans. R. Soc. Trop. Med. Hyg.* **75:**318–319.

Smith, T. G., L. Serghides, S. N. Patel, M. Febbraio, R. L. Silverstein, and K.C Kain. 2003. CD36-mediated nonopsonic phagocytosis of erythrocytes infected with stage I and IIA gametocytes of *Plasmodium falciparum*. *Infect. Immun.* **71:**393–400.

Smith, T. G., D. Walliker, and L. C. Ranford-Cartwright. 2002. Sexual differentiation and sex determination in the Apicomplexa. *Trends Parasitol.* **18:**315–323.

Sutherland, C. J. 2001. Stevor transcripts from *Plasmodium falciparum* gametocytes encode truncated polypeptides. *Mol. Biochem. Parasitol.* **113:**331–335.

Talisuna, A. O., P. Langi, T. K. Mutabingwa, E. Van Marck, N. Speybroeck, T. G. Egwang, W. W. Watkins, I. M. Hastings, and U. D'Alessandro. 2003. Intensity of transmission and spread of gene mutations linked to chloroquine and sulphadoxine-pyrimethamine resistance in falciparum malaria. *Int. J. Parasitol.* **33:**1051–1058.

Talman, A. M., O. Domarle, F. E. McKenzie, F. Ariey, and V. Robert. 2004a. Gametocytogenesis: the puberty of *Plasmodium falciparum*. *Malar. J.* **3:**24.

Talman, A. M., R. E. L. Paul, C. S. Sokhna, O. Domarle, F. Ariey, J.-F. Trape, and V. Robert. 2004b. Influence of chemotherapy on Plasmodium gametocyte sex ratio of mice and men. *Am. J. Trop. Med. Hyg.* **71:**739–744.

Templeton, T. J., and D. C. Kaslow. 1999. Identification of additional members define a *Plasmodium falciparum* gene superfamily which includes Pfs48/45 and Pfs230. *Mol. Biochem. Parasitol.* **101:**223–227.

Templeton, T. J., D. B. Keister, O. Muratova, J. L. Procter, and D. C. Kaslow. 1998. Adherence of erythrocytes during exflagellation of *Plasmodium falciparum* microgametes is dependent on erythrocyte surface sialic acid and glycophorins. *J. Exp. Med.* **187:**1599–1609.

Thompson, J., C. J. Janse, and A. P. Waters. 2001. Comparative genomics in *Plasmodium*: a tool for the identification of genes and functional analysis. *Mol. Biochem. Parasitol.* **118:**147–154.

Thomson, J. G., and A. Robertson. 1935. The structure and development of *Plasmodium falciparum* gametocytes in the internal organs and in the peripheral circulation. *Trans. R. Soc. Trop. Med. Hyg.* **29:**31–40.

Tomas, A. M., G. Margos, G. Dimopoulos, L. H. van Lin, T. F. de Koning-Ward, R. Sinha, P. Lupetti, A. L. Beetsma, M. C. Rodriguez, M.

Karras, A. Hager, J. Mendoza, G. A. Butcher, F. Kafatos, C. J. Janse, A. P. Waters, and R. E. Sinden. 2001. P25 and P28 proteins of the malaria ookinete surface have multiple and partially redundant functions. *EMBO J.* **20:**3975–3983.

Toye, P. J., R. E. Sinden, and E. U. Canning. 1977. The action of metabolic inhibitors on microgametogenesis in *Plasmodium yoelii nigeriensis*. *Z. Parasitenkd.* **53:**133–141.

Trueman, H. E., J. D. Raine, L. Florens, J. T. Dessens, J. Mendoza, J. Johnson, C. C. Waller, I. Delrieu, A. A. Holders, J. Langhorne, D. J. Carucci, J. R. Yates III, and R. E. Sinden. 2004. Functional characterization of an LCCL-lectin domain containing protein family in *Plasmodium berghei*. *J. Parasitol.* **90:**1062–1071.

Tsuboi, T., Y. M. Cao, D. C. Kaslow, K. Shiwaku, and M. Torii. 1997. Primary structure of a novel ookinete surface protein from *Plasmodium berghei*. *Mol. Biochem. Parasitol.* **85:**131–134.

Vaidya, A. B., J. Morrisey, C. V. Plowe, D. C. Kaslow, and T. E. Wellems. 1993. Unidirectional dominance of cytoplasmic inheritance in two genetic crosses of *Plasmodium falciparum*. *Mol. Cell. Biol.* **13:**7349–7357.

van Dijk, M. R., C. J. Janse, J. Thompson, A. P. Waters, J. A. Braks, H. J. Dodemont, H. G. Stunnenberg, G. J. van Gemert, R. W. Sauerwein, and W. Eling. 2001. A central role for P48/45 in malaria parasite male gamete fertility. *Cell* **104:**153–164.

vanSpaendonk, R. M., J. Ramesar, A. van Wigcheren, W. Eling, A. L. Beetsma, G. J. van Gemert, J. Hooghof, C. J. Janse, and A. P. Waters. 2001. Functional equivalence of structurally distinct ribosomes in the malaria parasite, *Plasmodium berghei*. *J. Biol. Chem.* **276:**22638–22247.

Walliker, D. 1983. The genetic basis of diversity in malaria parasites. *Adv. Parasitol.* **22:**217–259.

Ward, P., L. Equinet, J. Packer, and C. Doerig. 2004. Protein kinases of the human malaria parasite *Plasmodium falciparum*: the kinome of a divergent eukaryote. *BMC Genomics* **5:**79.

Wesseling, J. G., P. J. Snijders, P. van Someren, J. Jansen, M. A. Smits, and J. G. Schoenmakers. 1989. Stage-specific expression and genomic organization of the actin genes of the malaria parasite *Plasmodium falciparum*. *Mol. Biochem. Parasitol.* **35:**167–176.

Wickham, M. E., J. G. Culvenor, and A. F. Cowman. 2003. Selective inhibition of a two-step egress of malaria parasites from the host erythrocyte. *J. Biol. Chem.* **278:**37658–37663.

GROWTH AND METABOLISM

III

MOLECULAR APPROACHES TO MALARIA: GLYCOLYSIS IN ASEXUAL-STAGE PARASITES

Charles J. Woodrow and Sanjeev Krishna

11

In erythrocytes, *Plasmodium falciparum* has no obvious energy stores. Glucose storage forms such as amylopectin and mannitol identified in other apicomplexan parasites (Heise et al., 1999; Petry and Harris, 1999; Speer et al., 1998) are not reported in *P. falciparum*. Analysis of the completed *P. falciparum* genome sequence has not revealed genes for fructose 1,6-bisphosphatase or pyruvate carboxylase, suggesting that gluconeogenesis is absent (Fig. 1). Enzymes for the synthesis of trehalose, glycogen, and other carbohydrate stores are also apparently missing. To meet requirements for rapid growth and multiplication, parasites must continuously import and use glucose. Uninfected erythrocytes use relatively little glucose, mainly to maintain ion gradients and regulate cell volume by action of ion transporters such as the plasma membrane calcium and sodium-potassium ATPases. They metabolize glucose relatively inefficiently, making ATP by anaerobic glycolysis because they have no mitochondria. Asexual-stage *P. falciparum* has a single mitochondrion apposed to its apicoplast, and citric acid cycle enzymes are expressed, particularly in sexual stages of the parasite life cycle (Florens et al., 2002). The *P. falciparum* mitochondrion maintains a transmembrane potential gradient that is essential for survival and is a target for the antimalarial atovaquone; however, it is not used for aerobic glycolysis in asexual-stage parasites. This conclusion is confirmed by several observations. First, absence of an NADH dehydrogenase complex I in the parasite's genome implies limited ATP synthesis through oxidative phosphorylation because this enzyme is the first electrogenic step of conventional electron transfer chains. Second, there is no relationship between the mitochondrial membrane potential of asexual stage *Plasmodium yoelii* and oxidative phosphorylation (Srivastava et al., 1997). Third, respiratory complex activity and electron transport rates are lower than those observed in mammalian mitochondria (Murphy et al., 1997; Uyemura et al., 2000; Takashima et al., 2001). Oxidative phosphorylation and ATP synthesis are detected in mitochondria of *Plasmodium berghei* of digitonin-permeabilized trophozoites (Uyemura et al., 2000), but the ATP synthetic activity is limited both by the very low electron transport rate and by the

Charles J. Woodrow and Sanjeev Krishna, Department of Cellular and Molecular Medicine, St. George's Hospital Medical School, Cranmer Terrace, London SW17 0RE, United Kingdom.

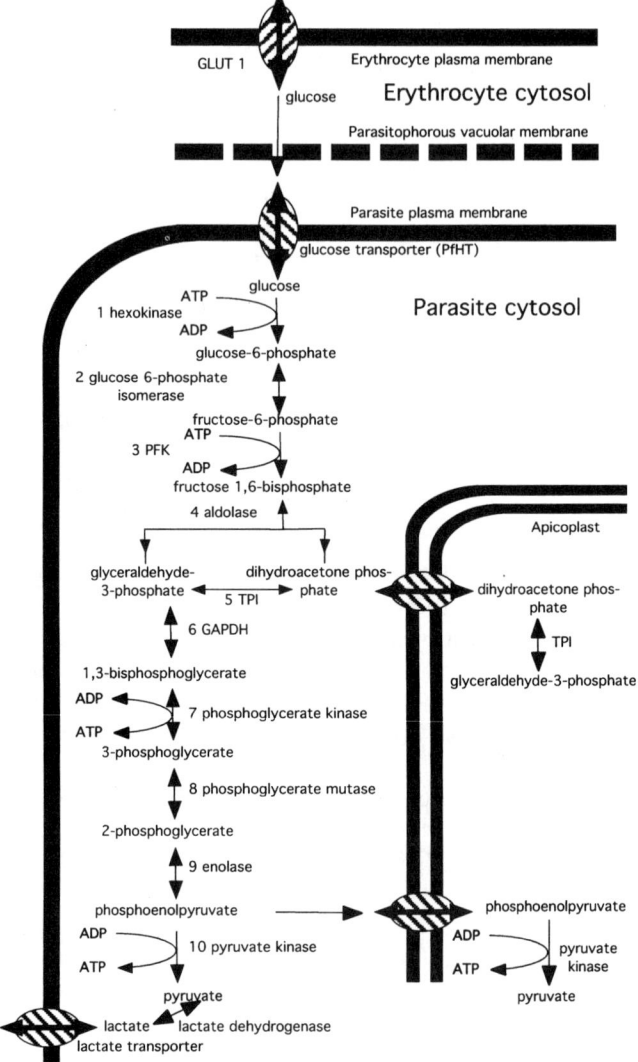

FIGURE 1 Postulated scheme for glycolytic reactions in *P. falciparum*-infected erythrocytes. Distinct cytosolic and apicoplast enzymes for TPI and pyruvate kinase are shown (see Table 1). Abbreviations: triosephosphate isomerase, TPI. For Enzyme Commission numbers, see Table 1.

activity of an uncoupling protein bypassing the ATP synthase and therefore dissipating part of the transmembrane electrochemical gradient. From these experiments, one may conclude that malarial mitochondria do not provide a significant part of the ATP required for growth during asexual intraerythrocytic stages. Asexual-stage parasites are microaerophilic homolactate fermenters (Sherman, 1998).

When parasites are maturing in erythrocytes and transforming into their most metabolically active stage of development, they can increase utilization of glucose by up to 100-fold compared with uninfected erythrocytes (Homewood, 1977). This high rate of anaerobic glycolysis is essential for parasite survival. Lack of glucose kills parasites in culture. This absolute dependence on glucose for energy supply for asexual *P. falciparum* is best shown by experiments with freed parasites, where removal of extracellular glucose results in an immediate drop (within minutes) in intraparasite pH. This acidification can be equally rapidly reversed by adding glucose (or fructose) to the extracellular medium (Saliba et al., 2004). As well as providing energy in the form of ATP, glycolysis provides synthetic precursors (e.g., for amino acids serine and alanine, nucleotides, and lipids) and reducing power in the form of NADH for oxidation-reduction reactions.

This high demand of asexual *P. falciparum* parasites for glucose may result in competition with host cells for this essential substrate, particularly in areas where there are many infected cells sequestered and blocking normal bood flow. If hypoglycemia complicates malaria, then effects of competition for metabolites may become exacerbated (Krishna and Woodrow, 1999). The major product of glycolysis by asexual-stage parasites is lactate, with small quantities of acetate and succinate also detectable (Sherman, 1979). Large quantities of lactic acid produced in the vicinity of hypoxic host tissue may impair function of host cells.

There are, therefore, two independent reasons for targeting glycolysis in the postgenome era: first, to kill parasites by identifying new inhibitors and eventually developing novel drugs, and second, to decrease use of glucose and output of lactic acid in those regions where there are many parasites that could compete with host tissues for glucose and add to the problem of disposal of lactate. Even rapidly acting conventional antimalarials (chloroquine, quinine, and artemisinin) take many hours before they begin to inhibit glycolysis significantly, indicating that they do not primarily target glycolysis (ter Kuile et al., 1993).

GLYCOLYTIC ENZYMES IN *P. FALCIPARUM*

Glycolysis has been studied in detail since the 19th century, and the importance of glucose in the metabolism of malaria parasites has been established since the early 20th century (Sherman, 1979). Systematic biochemical studies of carbohydrate metabolism confirmed the overall pattern of glucose utilization in malaria parasites to be similar to that in other eukaryotic cells. Increased metabolic activity of *Plasmodium*-infected red blood cells is accompanied by the appearance of glycolytic enzymes with properties distinct from host red blood cell enzymes (Roth et al., 1988). In the last 15 years, the glycolytic enzymes of *P. falciparum* have been studied individually by classical molecular and biochemical approaches and more recently by analysis of the *P. falciparum* genome (Fig. 1 and Table 1).

Functional Genomic Analysis of Glycolysis

Genes for certain *P. falciparum* glycolytic enzymes (reactions 1, 2, 4, 5, 6, 7, and 9) (Kaslow and Hill, 1990; Olafsson et al., 1992; Certa et al., 1988; Daubenberger et al., 2000; Hicks et al., 1991; Ranie et al., 1993; Read et al., 1994), as well as the fermentation step catalyzed by lactate dehydrogenase (LDH) (Bzik et al., 1993; Simmons et al., 1985), were identified by using molecular technology. More recently, the completed *P. falciparum* genome (Gardner et al., 2002) has confirmed and clarified these studies and identified missing or unpredicted enzymes. Genome data allow homology-based analyses, which have recently been enhanced and complemented by transcriptional analysis of highly synchronized asexual-stage *P. falciparum* parasites (Ben Mamoun et al., 2001; Bozdech et al., 2003; Le Roch et al., 2003) including merozoites, as well as gametocytes and sporozoites (Le Roch et al., 2003). Protein expression in various stages of the life cycle has also been analyzed by proteomic technology (Florens et al., 2002; Lasonder et al., 2002). All these data are freely available at PlasmoDB (www.plasmodb.org) (Kissinger et al., 2002).

We have BLAST searched PlasmoDB, using as queries the amino acid sequences of human analogs of glycolytic enzymes as well as LDH, glucose, and monocarboxylate transporters (Table 1). This method is not wholly sensitive or specific in detecting orthologs. So additionally, we searched PlasmoDB for each gene by name, relying on existing annotations of sequences. All sequences picked up by these approaches are presented in Table 1. For many enzymes, more than one gene shows significant homology to human sequences or is designated a glycolytic enzyme in PlasmoDB. We then analyzed

TABLE 1 In silico analysis of glycolytic enzymes of *P. falciparum*[a]

Protein	Step	EC-TC no.	Gene	Chr	ORF (aa)	Score (bits)	Query (ORF)	Reference(s)	Additional sequence
Glucose transporter		2.A.1.1.24	PFB0210c	2	504	147	P11166 (492)	Woodrow et al., 1999	PFI0955w[b]
Hexokinase	1	2.7.1.1	MAL6P1.189	6[c]	493[d]	208	P19367 (917)	Olafsson et al., 1992	
G6PI	2	5.3.1.9	PF14_0341	14	579[e]	354	P06744 (557)	Kaslow and Hill, 1990	
Phosphofructokinase	3	2.7.1.11	PFI0755c	9	1,418	46	P17858 (779)		PF11_0294[f]
Aldolase	4	4.1.2.13	PF14_0425	14	369	332	P04075 (363)	Certa et al., 1988; Knapp et al., 1990	
TPI	5	5.3.1.1	PF14_0378	14	248	213	P60174 (248)	Ranie et al., 1993	PFC0831w[g]
GAPDH	6	1.2.1.12	PF14_0598	14	337	441	P04406 (334)	Daubenberger et al., 2000	
Phosphoglycerate kinase	7	2.7.2.3	PFI_1105w	9	416	501	P00558 (416)	Hicks et al., 1991	
Phosphoglycerate mutase	8	5.4.2.1	PF11_0208	11	250	280	P18669 (253)		PFD0660w[h]
Enolase	9	4.2.1.11	PF10_0155	10	446	571	P06733 (433)	Read et al., 1994	PF10_0363[i]
Pyruvate kinase	10	2.7.1.40	MAL6P1.160	6	511	394	P14618 (530)	Chan and Sim, 2004	PF13_0144[j]
LDH		1.1.1.27	PF13_0141	13	316	142	P00338 (331)	Bzik et al., 1993; Simmons et al., 1985	MAL6P1.242[k]
Lactate transporter		2.A.1.13.1	None found				P53985 (500)		

[a]BLASTP searches used human enzyme orthologs (Swiss-Prot accession numbers are given) as a query sequence. Genes for which primary sequence data (e.g., low homology scores, presence of apicoplast signal peptide) or combined transcriptomic and proteomic data indicate predominantly sexual-phase expression are presumed not to mediate glycolysis in asexual parasites and are placed in the last column. Abbreviations: EC, Enzyme Commission; TC, Transporter Commission; Chr, chromosome; aa, amino acids; G6PI, glucose-6-phosphate isomerase.
[b]A 476-amino-acid protein not picked up by using GLUT1 as query, expressed in sporozoites and gametocytes.
[c]Originally described as chromosome 8 (Olafsson et al., 1992).
[d]Size typical of invertebrate hexokinases (Ureta, 1982).
[e]Previously reported as 591 (Kaslow and Hill, 1990).
[f]A total of 1,570 amino acids picked up by name search only, expressed in gametocytes.
[g]A total of 372 amino acids (score 96), low levels of expression in asexual life cycle, apicoplast-targeting signal.
[h]A total of 295 amino acids (picked up by name search only), atypical expression pattern.
[i]A total of 745 amino acids (score 137), containing an apicoplast-targeting signal.
[j]A total of 334 amino acids (score 107), expressed in sporozoites and gametocytes.
[k]A total of 313 amino acids (score 95), expressed in sporozoites and gametocytes.

transcriptional and proteomic data available via PlasmoDB to assess timing of expression and correlated these results with original, individual gene studies (Grall et al., 1992; Olafsson and Certa, 1994; Srivastava et al., 1992). These expression studies reveal that most glycolytic enzymes display tight regulation of expression, which tends to peak at the early trophozoite stage with up to 50-fold increases in transcription compared to the early ring stage (Ben Mamoun et al., 2001; Bozdech et al., 2003).

Our in silico search confirmed the presence of seven previously reported glycolytic enzyme sequences and LDH (Table 1). Searching with human triosephosphate isomerase revealed two genes: the first (PF14_0378) (Ranie et al., 1993) shows the typical glycolytic characteristic of upregulation through asexual development. In contrast, the second (PFC0831w) displays low-level expression at all life cycle stages, although it may be involved in apicoplast metabolism (Ralph et al., 2004). Only one copy of aldolase is apparent, in contrast to the genome of *P. berghei* (Meier et al., 1992). A single glyceraldehyde-3-phosphate dehydrogenase (GAPDH) protein (Daubenberger et al., 2000) localizes to the cytosol for much of the asexual cycle, suggesting that plastid-targeted GAPDH seen in other apicomplexans (Fast et al., 2001) has been lost from the *P. falciparum* genome; however, PfGAPDH does associate with the apical region of merozoites, perhaps indicating that it also mediates nonglycolytic functions (Daubenberger et al., 2003). As well as the previously studied LDH (Simmons et al., 1985; Bzik et al., 1993), two additional LDH paralogs are seen which are of uncertain significance; they appear to be expressed preferentially during gametocyte and sporozoite stages.

NOVEL SEQUENCES FROM THE GENOME PROJECT

BLAST searching with human phosphofructokinase (PFK) yields a gene (PFI0755c) that encodes a strikingly long protein, approximately double the length of other PFK enzymes. This phenomenon of an anomalously lengthened PFK is apparently restricted to *P. falciparum* and *Cryptosporidium parvum* (Abrahamsen et al., 2004). Protein matrix analysis suggests duplication in this sequence, which subsequently underwent fusion, such that in vivo the enzyme presumably consists of a homodimer rather than the heterotetramer state generally adopted by these enzymes. Phylogenetic analysis reveals that PFK genes have undergone many horizontal gene transfer events, with PFI0755c located in the long group among distantly related organisms from both eukaryotic and prokaryotic sources (Bapteste et al., 2003). Sequence alignment shows that the amino-terminal section of PFK (PFI0755c) contains a combination of critical catalytic amino acids that are associated with the use of ATP, rather than inorganic pyrophosphate, as energy phosphodonors in other protists (Claustre et al., 2002; Lopez et al., 2002). PFK (PFI0755c) therefore appears to be an ATP-PFK, which acts irreversibly in glycolysis, rather than an inorganic pyrophosphate-PFK, which may function in either glycolysis or gluconeogenesis. The carboxy-terminal half of PFI0755c lacks amino acids associated with catalysis at critical positions and may serve a regulatory function. Searching by name for PFK enzymes reveals a second double-length sequence on chromosome 11 (PF11_0294) that appears to be expressed only in gametocytes, lacks amino acids associated with catalytic activity, and is therefore presumably not involved in glycolysis.

BLAST searching using human phosphoglycerate mutase yields PF11_0208, a gene displaying an expression pattern typical of other glycolytic enzymes. Name searching reveals a second sequence (PFD0660w), but this annotation appears dubious; when used as a probe against National Center for Biotechnology Information databases, PFD0660w shows weak homology to unidentified proteins only, and its pattern of expression is atypical for glycolytic enzymes, appearing only in the latter stages of the asexual phase, as well as sporozoites and gametocytes.

Pyruvate kinase is one of the "payback" glycolytic steps that generates ATP from ADP. It is interesting, therefore, that BLAST searching with human pyruvate kinase reveals two genes. MAL6P1.160 is likely to be cytosolic, showing typical upregulation at the early trophozoite stage; it has recently been expressed in *E. coli* (Chan and Sim, 2004). A second sequence, PF10_0363, is longer, contains a potential apicoplast transit peptide, and may therefore be involved in ATP generation in the apicoplast (Ralph et al., 2004).

In this manner, we can confidently assign glycolytic enzyme activities of asexual stage *P. falciparum* to single genes in each case. The list given in Table 1 may serve as a useful guide to further studies. However, definitive studies of individual genes confirming location and pattern of expression, as well as functional characterization, remain necessary. These are already available for several enzymes described prior to the release of the genome (Daubenberger et al., 2000; Kaslow and Hill, 1990; Olafsson and Certa, 1994; Srivastava et al., 1992) as well as two new sequences, pyruvate kinase (above) and the glucose transporter (see below).

The unveiling of the sequence of the *P. falciparum* glucose transporter represents one of the most significant contributions by the genome project to our understanding of carbohydrate metabolism in *P. falciparum*. The multiple membrane systems of infected red blood cells render study of functional aspects of glucose transport into *P. falciparum* more difficult than for glycolytic enzymes. Prior to the genome project, it was determined that glucose transit through the infected red blood cell was passive (Kirk et al., 1996) with glucose uptake by parasites via a facilitative transporter on the parasite plasma membrane (Goodyer et al., 1997). We had attempted to use oligonucleotides to clone a *P. falciparum* hexose transporter, but these transporters are generally much less conserved than many glycolytic enzymes both in terms of overall identity and signature motifs that allow design of AT-biased oligonucleotides as probes to search libraries or in PCR experiments. In retrospect, without a genome sequence, it is likely that the protein could only have been cloned by expression. A single open reading frame (ORF) PFB0210c on chromosome 2, revealed in the early stage of the genome sequencing project (Gardner et al., 1998), resembled a sugar transporter from the major facilitator superfamily.

Expression of its protein product (*P. falciparum* hexose transporter [PfHT]) in asexual stages peaks earlier than that of cytoplasmic glycolytic enzymes, perhaps because it takes a relatively long time for PfHT to reach the parasite plasma membrane (Bozdech et al., 2003; Woodrow et al., 1999). As predicted, the protein localizes to the region of the parasite plasma membrane (Rager et al., 2001; Woodrow et al., 1999). PfHT was the first malaria transporter to be expressed (in *Xenopus laevis* oocytes) where it has been shown to transport both glucose and fructose in a facilitative manner.

Word searching of PlasmoDB reveals another member of the major facilitator superfamily (PFI0955w) listed as a putative sugar transporter, but homology between PfHT and this sequence, although statistically significant, is in regions not conserved across hexose transporters. This suggests that PFI0955w may have evolved from PfHT by duplication with subsequent divergence (Gardner et al., 2002). The function of PFI0955w remains an enigma, although it is interesting that it is expressed in sporozoites and gametocytes rather than the asexual phase.

In contrast to the success of the genome project in revealing the molecular basis of glucose entry, lactate export remains unassigned. At the primary sequence level, there are no genes with significant homology to human monocarboxylate transporters in the genome. The molecular basis of lactate removal from parasites (Elliott et al., 2001) is unclear, demonstrating elegantly that genome analysis may not provide an answer to every question.

TOWARDS AN INHIBITOR OF GLYCOLYSIS

A daunting number of steps are required to develop an antimalarial by the process of rational drug development (Ridley, 1997). The first is to identify specific inhibitors of important metabolic processes in *P. falciparum*. For reasons given in the introduction, glycolytic enzymes may be excellent antimalarial targets. The most obvious way for inhibitors to target the action of glycolytic enzymes is by blocking their catalytic sites, which allows substrate analogs to be used as probes. In addition, since these enzymes are allosteric multimers, it is possible to inhibit intersubunit interactions and interaction with the cytoskeleton, where relevant.

Design of specific inhibitors may be enhanced by crystal structures, which are available for aldolase (Kim et al., 1998), triosephosphate isomerase (Parthasarathy et al., 2002; Parthasarathy et al., 2003; Velanker et al., 1997), phosphoglycerate kinase (Pal et al., 2004), and LDH (Dunn et al., 1996) (Color Plate 8). Structure-based approaches have been recently reviewed (Brady and Cameron, 2004) and although studies are still limited, the glycolytic pathway is one of the areas of promise. LDH has received most attention in this context. Inhibition of LDH prevents regeneration of NAD^+ from NADH and so removes this glycolytic prerequisite, thereby stopping ATP production. LDH has been studied for decades in a variety of organisms, with its kinetic mechanism well understood. The protein thought to mediate asexual-phase LDH activity (Table 1) has been functionally studied and shown to display kinetic properties distinct from human variants (Makler et al., 1993). Crystal structure analysis of *P. falciparum* (Dunn et al., 1996) and human variants (Read et al., 2001) reveals that the NADH cofactor binding pocket is positioned differently, and there is a five-amino-acid insert in a loop region involved in the substrate binding site, also seen in other *Plasmodium* species (Brown et al., 2004) and apicomplexans (Yang and Parmley, 1997). These two distinct features produce an increase in the volume of the active site cleft in *Plasmodium* versions compared to the human enzyme (Dunn et al., 1996; Read et al., 2001; Winter et al., 2003). These findings intuitively suggest that specific inhibitors can be generated (Ridley, 1997). Studies of the cofactor-competitive gossypol compounds (Gomez et al., 1997; Royer et al., 1986) have shown submicromolar K_i values (inhibitory constants) for LDH; unfortunately, these compounds are relatively nonselective for the parasite enzyme and therefore cytotoxic to mammalian cells. Less-toxic derivatives have been examined (Deck et al., 1998; Razakantoanina et al., 2000), and these show parasiticidal activity in the concentration range of 10 to 100 μM (Razakantoanina et al., 2000).

High-throughput screening of over 500,000 compounds identified several chemical families with specific PfLDH activity, of which the azole family was investigated in more detail by crystallographic analysis of the LDH-inhibitor complex and parasite inhibition assays (Cameron et al., 2004). One compound with an enzyme K_i of just under 1 μM disappointingly had a *P. falciparum* 50% inhibitory concentration (IC_{50}) for parasites of about 20 μM. One reason for this may be poor uptake of these charged anionic compounds across the multiple membrane systems of the infected red blood cell, in comparison to uptake of other compounds such as chloroquine. This example demonstrates a significant problem associated with the rational drug development approach. Even if specific enzyme inhibitors can be generated as lead compounds, several further hurdles need to be overcome. The drug must be able to gain access to the appropriate location for an enzyme. Efficient enzyme inhibition does not necessarily equate with successful inhibition of a complete pathway. Finally, even if a parasite inhibitor is found, it has to be nontoxic and have appropriate pharmacokinetic behavior.

Many biochemistry textbooks describe certain enzymatic steps as rate limiting, but a more modern metabolic control analysis has led to reinterpretation of these statements. Such analysis has been applied to trypanosomes, where glycolytic enzymes are localized in a specific organelle, the glycosome (Opperdoes, 1987). Both computer simulation and experimental studies of glycolysis in bloodstream forms of *Trypanosoma brucei* have provided quantitative information about the prospects of decreasing flux by inhibition of any individual enzyme; both approaches lead to the conclusion that the plasma membrane glucose transporter is the most promising target from this perspective (Bakker et al., 1999; Bakker et al., 2000). If the same principle applies to *P. falciparum*, then the glucose transporter PfHT represents a promising drug target (Joet et al., 2003b). Detailed analysis of the interaction between PfHT and hexose analogs pointed to significant differences between human and *P. falciparum* transporters in terms of interaction at the O-3 position of glucose (Woodrow et al., 2000; Joet et al., 2003a). An analog-based inhibitor (compound 3361) displays K_i values for PfHT and *Plasmodium vivax* hexose transporter (expressed in *Xenopus* oocytes) of 53 and 120 μM, respectively, and IC_{50} values in *P. falciparum* and *P. vivax* cultures of 16 and 9 μM, respectively. These antiparasitic effects are specifically via inhibition of glucose permeation and led directly to a decline in ATP levels and loss of intraparasitic pH regulation (Saliba et al., 2004). The relatively potent IC_{50} compared to K_i value for this compound is consistent with the hypothesis that glucose transport is the most critical step of the whole glycolytic pathway and implies efficient uptake of the inhibitor. Inhibition of PfHT is therefore a promising area for further study, perhaps with high-throughput screening approaches.

Diagnostics

The distinct antigenic properties of *P. falciparum* LDH have allowed the development of a useful antigen detection assay that has been formulated into a field dipstick (optiMAL) (Cooke et al., 1999; Palmer et al., 1998). At present, glycolytic enzymes are not thought to be involved in immunity against malaria.

CONCLUSIONS

An improved description and understanding of the enzymes of the glycolytic pathway of *P. falciparum* should help to define further unique biochemical targets as a basis for the development of antimalarials. Single enzymes appear to mediate cytosolic glycolysis during the asexual cycle, while two distinct apicoplast-targeted counterparts may be involved in apicoplast energy metabolism. Glucose transport may be a promising target on theoretical grounds; LDH has received the most attention so far in terms of rational drug development. Glycolysis in *P. falciparum* may contribute to disease pathogenesis by competing for glucose in host tissues, lending added impetus to discovering ways of inhibiting this key pathway.

ACKNOWLEDGMENTS

In silico analysis of glycolytic enzymes used PlasmoDB (Kissinger et al., 2002).

We thank Angus Cameron for providing figures for crystal structures and the MRC (United Kingdom) and The Wellcome Trust for funding.

REFERENCES

Abrahamsen, M. S., T. J. Templeton, S. Enomoto, J. E. Abrahante, G. Zhu, C. A. Lancto, M. Deng, C. Liu, G. Widmer, S. Tzipori, G. A. Buck, P. Xu, A. T. Bankier, P. H. Dear, B. A. Konfortov, H. F. Spriggs, L. Iyer, V. Anantharaman, L. Aravind, and V. Kapur. 2004. Complete genome sequence of the apicomplexan, Cryptosporidium parvum. *Science* **304**:441–445.

Bakker, B. M., M. C. Walsh, B. H. ter Kuile, F. I. Mensonides, P. A. Michels, F. R. Opperdoes, and H. V. Westerhoff. 1999. Contribution of glucose transport to the control of the glycolytic flux in Trypanosoma brucei. *Proc. Natl. Acad. Sci. USA* **96**:10098–10103.

Bakker, B. M., H. V. Westerhoff, F. R. Opperdoes, and P. A. Michels. 2000. Metabolic control analysis of glycolysis in trypanosomes as an approach to improve selectivity and effectiveness of drugs. *Mol. Biochem. Parasitol.* **106**:1–10.

Bapteste, E., D. Moreira, and H. Philippe. 2003. Rampant horizontal gene transfer and phosphodonor change in the evolution of the phosphofructokinase. *Gene* **318**:185–191.

Ben Mamoun, C., I. Y. Gluzman, C. Hott, S. K. MacMillan, A. S. Amarakone, D. L. Anderson, J. M. Carlton, J. B. Dame, D. Chakrabarti, R. K. Martin, B. H. Brownstein, and D. E. Goldberg. 2001. Co-ordinated programme of gene expression during asexual intraerythrocytic development of the human malaria parasite Plasmodium falciparum revealed by microarray analysis. *Mol. Microbiol.* **39**:26–36.

Bozdech, Z., M. Llinas, B. L. Pulliam, E. D. Wong, J. Zhu, and J. L. DeRisi. 2003. The transcriptome of the intraerythrocytic developmental cycle of Plasmodium falciparum. *PLoS Biol.* **1**:E5.

Brady, R. L., and A. Cameron. 2004. Structure-based approaches to the development of novel antimalarials. *Curr. Drug Targets* **5**:137–149.

Brown, W. M., C. A. Yowell, A. Hoard, T. A. Vander Jagt, L. A. Hunsaker, L. M. Deck, R. E. Royer, R. C. Piper, J. B. Dame, M. T. Makler, and D. L. Vander Jagt. 2004. Comparative structural analysis and kinetic properties of lactate dehydrogenases from the four species of human malarial parasites. *Biochemistry* **43**:6219–6229.

Bzik, D. J., B. A. Fox, and K. Gonyer. 1993. Expression of Plasmodium falciparum lactate dehydrogenase in Escherichia coli. *Mol. Biochem. Parasitol.* **59:**155–166.

Cameron, A., J. Read, R. Tranter, V. J. Winter, R. B. Sessions, R. L. Brady, L. Vivas, A. Easton, H. Kendrick, S. L. Croft, D. Barros, J. L. Lavandera, J. J. Martin, F. Risco, S. Garcia-Ochoa, F. J. Gamo, L. Sanz, L. Leon, J. R. Ruiz, R. Gabarro, A. Mallo, and F. Gomez de las Heras. 2004. Identification and activity of a series of azole-based compounds with lactate dehydrogenase-directed anti-malarial activity. *J. Biol. Chem.* **279:**31429–31439.

Certa, U., P. Ghersa, H. Dobeli, H. Matile, H. P. Kocher, I. K. Shrivastava, A. R. Shaw, and L. H. Perrin. 1988. Aldolase activity of a Plasmodium falciparum protein with protective properties. *Science* **240:**1036–1038.

Chan, M., and T. S. Sim. 2004. Functional analysis, overexpression, and kinetic characterization of pyruvate kinase from Plasmodium falciparum. *Biochem. Biophys. Res. Commun.* **326:**188–196.

Claustre, S., C. Denier, F. Lakhdar-Ghazal, A. Lougare, C. Lopez, N. Chevalier, P. A. Michels, J. Perie, and M. Willson. 2002. Exploring the active site of Trypanosoma brucei phosphofructokinase by inhibition studies: specific irreversible inhibition. *Biochemistry* **41:**10183–10193.

Cooke, A. H., P. L. Chiodini, T. Doherty, A. H. Moody, J. Ries, and M. Pinder. 1999. Comparison of a parasite lactate dehydrogenase-based immunochromatographic antigen detection assay (OptiMAL) with microscopy for the detection of malaria parasites in human blood samples. *Am. J. Trop. Med. Hyg.* **60:**173–176.

Daubenberger, C. A., F. Poltl-Frank, G. Jiang, J. Lipp, U. Certa, and G. Pluschke. 2000. Identification and recombinant expression of glyceraldehyde-3-phosphate dehydrogenase of Plasmodium falciparum. *Gene* **246:**255–264.

Daubenberger, C. A., E. J. Tisdale, M. Curcic, D. Diaz, O. Silvie, D. Mazier, W. Eling, B. Bohrmann, H. Matile, and G. Pluschke. 2003. The N′-terminal domain of glyceraldehyde-3-phosphate dehydrogenase of the apicomplexan Plasmodium falciparum mediates GTPase Rab2-dependent recruitment to membranes. *Biol. Chem.* **384:**1227–1237.

Deck, L. M., R. E. Royer, B. B. Chamblee, V. M. Hernandez, R. R. Malone, J. E. Torres, L. A. Hunsaker, R. C. Piper, M. T. Makler, and D. L. Vander Jagt. 1998. Selective inhibitors of human lactate dehydrogenases and lactate dehydrogenase from the malarial parasite Plasmodium falciparum. *J. Med. Chem.* **41:**3879–3887.

Dunn, C. R., M. J. Banfield, J. J. Barker, C. W. Higham, K. M. Moreton, D. Turgut-Balik, R. L. Brady, and J. J. Holbrook. 1996. The structure of lactate dehydrogenase from Plasmodium falciparum reveals a new target for anti-malarial design. *Nat. Struct. Biol.* **3:**912–915.

Elliott, J. L., K. J. Saliba, and K. Kirk. 2001. Transport of lactate and pyruvate in the intraerythrocytic malaria parasite, Plasmodium falciparum. *Biochem. J.* **355:**733–739.

Fast, N. M., J. C. Kissinger, D. S. Roos, and P. J. Keeling. 2001. Nuclear-encoded, plastid-targeted genes suggest a single common origin for apicomplexan and dinoflagellate plastids. *Mol. Biol. Evol.* **18:**418–426.

Florens, L., M. P. Washburn, J. D. Raine, R. M. Anthony, M. Grainger, J. D. Haynes, J. K. Moch, N. Muster, J. B. Sacci, D. L. Tabb, A. A. Witney, D. Wolters, Y. Wu, M. J. Gardner, A. A. Holder, R. E. Sinden, J. R. Yates, and D. J. Carucci. 2002. A proteomic view of the Plasmodium falciparum life cycle. *Nature* **419:**520–526.

Gardner, M. J., N. Hall, E. Fung, O. White, M. Berriman, R. W. Hyman, J. M. Carlton, A. Pain, K. E. Nelson, S. Bowman, I. T. Paulsen, K. James, J. A. Eisen, K. Rutherford, S. L. Salzberg, A. Craig, S. Kyes, M. S. Chan, V. Nene, S. J. Shallom, B. Suh, J. Peterson, S. Angiuoli, M. Pertea, J. Allen, J. Selengut, D. Haft, M. W. Mather, A. B. Vaidya, D. M. Martin, A. H. Fairlamb, M. J. Fraunholz, D. S. Roos, S. A. Ralph, G. I. McFadden, L. M. Cummings, G. M. Subramanian, C. Mungall, J. C. Venter, D. J. Carucci, S. L. Hoffman, C. Newbold, R. W. Davis, C. M. Fraser, and B. Barrell. 2002. Genome sequence of the human malaria parasite Plasmodium falciparum. *Nature* **419:**498–511.

Gardner, M. J., H. Tettelin, D. J. Carucci, L. M. Cummings, L. Aravind, E. V. Koonin, S. Shallom, T. Mason, K. Yu, C. Fujii, J. Pederson, K. Shen, J. Jing, C. Aston, Z. Lai, D. C. Schwartz, M. Pertea, S. Salzberg, L. Zhou, G. G. Sutton, R. Clayton, O. White, H. O. Smith, C. M. Fraser, S. L. Hoffman, et al. 1998. Chromosome 2 sequence of the human malaria parasite Plasmodium falciparum. *Science* **282:**1126–1132.

Gomez, M. S., R. C. Piper, L. A. Hunsaker, R. E. Royer, L. M. Deck, M. T. Makler, and D. L. Vander Jagt. 1997. Substrate and cofactor specificity and selective inhibition of lactate dehydrogenase from the malarial parasite P. falciparum. *Mol. Biochem. Parasitol.* **90:**235–246.

Goodyer, I. D., D. J. Hayes, and R. Eisenthal. 1997. Efflux of 6-deoxy-D-glucose from Plasmodium falciparum-infected erythrocytes via two saturable carriers. *Mol. Biochem. Parasitol.* **84:**229–239.

Grall, M., I. K. Srivastava, M. Schmidt, A. M. Garcia, J. Mauel, and L. H. Perrin. 1992. Plasmodium falciparum: identification and purification of the phosphoglycerate kinase of the malaria parasite. *Exp. Parasitol.* **75:**10–18.

Heise, A., W. Peters, and H. Zahner. 1999. A monoclonal antibody reacts species-specifically with amylopectin granules of Eimeria bovis merozoites. *Parasitol. Res.* **85:**500–503.

Hicks, K. E., M. Read, S. P. Holloway, P. F. Sims, and J. E. Hyde. 1991. Glycolytic pathway of the human malaria parasite Plasmodium falciparum: primary sequence analysis of the gene encoding 3-phosphoglycerate kinase and chromosomal mapping studies. *Gene* **100:**123–129.

Homewood, C. A. 1977. Carbohydrate metabolism of malarial parasites. *Bull. W. H. O.* **55:**229–235.

Joet, T., U. Eckstein-Ludwig, C. Morin, and S. Krishna. 2003a. Validation of the hexose transporter of Plasmodium falciparum as a novel drug target. *Proc. Natl. Acad. Sci. USA* **100:**7476–7479.

Joet, T., C. Morin, J. Fischbarg, A. I. Louw, U. Eckstein-Ludwig, C. Woodrow, and S. Krishna. 2003b. Why is the Plasmodium falciparum hexose transporter a promising new drug target? *Expert Opin. Ther. Targets* **7:**593–602.

Kaslow, D. C., and S. Hill. 1990. Cloning metabolic pathway genes by complementation in Escherichia coli. Isolation and expression of Plasmodium falciparum glucose phosphate isomerase. *J. Biol. Chem.* **265:**12337–12341.

Kim, H., U. Certa, H. Dobeli, P. Jakob, and W. G. Hol. 1998. Crystal structure of fructose-1,6-bisphosphate aldolase from the human malaria parasite Plasmodium falciparum. *Biochemistry* **37:**4388–4396.

Kirk, K., H. A. Horner, and J. Kirk. 1996. Glucose uptake in Plasmodium falciparum-infected erythrocytes is an equilibrative not an active process. *Mol. Biochem. Parasitol.* **82:**195–205.

Kissinger, J. C., B. P. Brunk, J. Crabtree, M. J. Fraunholz, B. Gajria, A. J. Milgram, D. S. Pearson, J. Schug, A. Bahl, S. J. Diskin, H. Ginsburg, G. R. Grant, D. Gupta, P. Labo, L. Li, M. D. Mailman, S. K. McWeeney, P. Whetzel, C. J. Stoeckert, and D. S. Roos. 2002. The Plasmodium genome database. *Nature* **419:**490–492.

Knapp, B., E. Hundt, and H. A. Kupper. 1990. Plasmodium falciparum aldolase: gene structure and localization. *Mol. Biochem. Parasitol.* **40:**1–12.

Krishna, S., and C. J. Woodrow. 1999. Expression of parasite transporters in Xenopus oocytes. *Novartis Found. Symp.* **226:**126–139.

Lasonder, E., Y. Ishihama, J. S. Andersen, A. M. Vermunt, A. Pain, R. W. Sauerwein, W. M. Eling, N. Hall, A. P. Waters, H. G. Stunnenberg, and M. Mann. 2002. Analysis of the Plasmodium falciparum proteome by high-accuracy mass spectrometry. *Nature* **419:**537–542.

Le Roch, K. G., Y. Zhou, P. L. Blair, M. Grainger, J. K. Moch, J. D. Haynes, P. De La Vega, A. A. Holder, S. Batalov, D. J. Carucci, and E. A. Winzeler. 2003. Discovery of gene function by expression profiling of the malaria parasite life cycle. *Science* **301:**1503–1508.

Lopez, C., N. Chevalier, V. Hannaert, D. J. Rigden, P. A. Michels, and J. L. Ramirez. 2002. Leishmania donovani phosphofructokinase. Gene characterization, biochemical properties and structure-modeling studies. *Eur. J. Biochem.* **269:**3978–3989.

Makler, M. T., J. M. Ries, J. A. Williams, J. E. Bancroft, R. C. Piper, B. L. Gibbins, and D. J. Hinrichs. 1993. Parasite lactate dehydrogenase as an assay for Plasmodium falciparum drug sensitivity. *Am. J. Trop. Med. Hyg.* **48:**739–741.

Meier, B., H. Dobeli, and U. Certa. 1992. Stage-specific expression of aldolase isoenzymes in the rodent malaria parasite Plasmodium berghei. *Mol. Biochem. Parasitol.* **52:**15–27.

Murphy, A. D., J. E. Doeller, B. Hearn, and N. Lang-Unnasch. 1997. Plasmodium falciparum: cyanide-resistant oxygen consumption. *Exp. Parasitol.* **87:**112–120.

Olafsson, P., and U. Certa. 1994. Expression and cellular localisation of hexokinase during the blood-stage development of Plasmodium falciparum. *Mol. Biochem. Parasitol.* **63:**171–174.

Olafsson, P., H. Matile, and U. Certa. 1992. Molecular analysis of Plasmodium falciparum hexokinase. *Mol. Biochem. Parasitol.* **56:**89–101.

Opperdoes, F. R. 1987. Compartmentation of carbohydrate metabolism in trypanosomes. *Annu. Rev. Microbiol.* **41:**127–151.

Pal, B., B. Pybus, D. D. Muccio, and D. Chattopadhyay. 2004. Biochemical characterization and crystallization of recombinant 3-phosphoglycerate kinase of Plasmodium falciparum. *Biochim. Biophys. Acta* **1699:**277–280.

Palmer, C. J., J. F. Lindo, W. I. Klaskala, J. A. Quesada, R. Kaminsky, M. K. Baum, and A. L. Ager. 1998. Evaluation of the OptiMAL test for rapid diagnosis of Plasmodium vivax and Plasmodium falciparum malaria. *J. Clin. Microbiol.* **36:**203–206.

Parthasarathy, S., H. Balaram, P. Balaram, and M. R. Murthy. 2002. Structures of Plasmodium falciparum triosephosphate isomerase complexed to substrate analogues: observation of the catalytic loop in the open conformation in the ligand-bound state. *Acta Crystallogr. D Biol. Crystallogr.* **58:**1992–2000.

Parthasarathy, S., K. Eaazhisai, H. Balaram, P. Balaram, and M. R. Murthy. 2003. Structure of Plasmodium falciparum triose-phosphate isomerase-2-phosphoglycerate complex at 1.1-A resolution. *J. Biol. Chem.* **278:**52461–52470.

Petry, F., and J. R. Harris. 1999. Ultrastructure, fractionation and biochemical analysis of Cryptosporidium parvum sporozoites. *Int. J. Parasitol.* **29:**1249–1260.

Rager, N., C. B. Mamoun, N. S. Carter, D. E. Goldberg, and B. Ullman. 2001. Localization of the Plasmodium falciparum PfNT1 nucleoside transporter to the parasite plasma membrane. *J. Biol. Chem.* **276:**41095–41099.

Ralph, S. A., G. G. Van Dooren, R. F. Waller, M. J. Crawford, M. J. Fraunholz, B. J. Foth, C. J. Tonkin, D. S. Roos, and G. I. McFadden. 2004. Tropical infectious diseases: metabolic maps and functions of the Plasmodium falciparum apicoplast. *Nat. Rev. Microbiol.* **2:**203–216.

Ranie, J., V. P. Kumar, and H. Balaram. 1993. Cloning of the triosephosphate isomerase gene of Plasmodium falciparum and expression in Escherichia coli. *Mol. Biochem. Parasitol.* **61:**159–169.

Razakantoanina, V., P. P. Nguyen Kim, and G. Jaureguiberry. 2000. Antimalarial activity of new gossypol derivatives. *Parasitol. Res.* **86:**665–668.

Read, J. A., V. J. Winter, C. M. Eszes, R. B. Sessions, and R. L. Brady. 2001. Structural basis for altered activity of M- and H-isozyme forms of human lactate dehydrogenase. *Proteins* **43:**175–185.

Read, M., K. E. Hicks, P. F. Sims, and J. E. Hyde. 1994. Molecular characterisation of the enolase gene from the human malaria parasite Plasmodium falciparum. Evidence for ancestry within a photosynthetic lineage. *Eur. J. Biochem.* **220:**513–520.

Ridley, R. G. 1997. Plasmodium: drug discovery and development—an industrial perspective. *Exp. Parasitol.* **87:**293–304.

Roth, E. F., Jr., M. C. Calvin, I. Max-Audit, J. Rosa, and R. Rosa. 1988. The enzymes of the glycolytic pathway in erythrocytes infected with Plasmodium falciparum malaria parasites. *Blood* **72:**1922–1925.

Royer, R. E., L. M. Deck, N. M. Campos, L. A. Hunsaker, and D. L. Vander Jagt. 1986. Biologically active derivatives of gossypol: synthesis and antimalarial activities of peri-acylated gossylic nitriles. *J. Med. Chem.* **29:**1799–1801.

Saliba, K. J., S. Krishna, and K. Kirk. 2004. Inhibition of hexose transport and abrogation of pH homeostasis in the intraerythrocytic malaria parasite by an O-3-hexose derivative. *FEBS Lett.* **570:**93–96.

Sherman, I. W. 1979. Biochemistry of *Plasmodium* (malarial parasites). *Microbiol. Rev.* **43:**453–495.

Sherman, I. W. 1998. *Malaria*, p. 135–143. ASM Press, Washington, D.C.

Simmons, D. L., J. E. Hyde, M. Mackay, M. Goman, and J. Scaife. 1985. Cloning studies on the gene coding for L-(+)-lactate dehydrogenase of Plasmodium falciparum. *Mol. Biochem. Parasitol.* **15:**231–243.

Speer, C. A., S. Clark, and J. P. Dubey. 1998. Ultrastructure of the oocysts, sporocysts, and sporozoites of Toxoplasma gondii. *J. Parasitol.* **84:**505–512.

Srivastava, I. K., H. Rottenberg, and A. B. Vaidya. 1997. Atovaquone, a broad spectrum antiparasitic drug, collapses mitochondrial membrane potential in a malarial parasite. *J. Biol. Chem.* **272:**3961–3966.

Srivastava, I. K., M. Schmidt, M. Grall, U. Certa, A. M. Garcia, and L. H. Perrin. 1992. Identification and purification of glucose phosphate isomerase of Plasmodium falciparum. *Mol. Biochem. Parasitol.* **54:**153–164.

Takashima, E., S. Takamiya, S. Takeo, F. Mi-ichi, H. Amino, and K. Kita. 2001. Isolation of mitochondria from Plasmodium falciparum showing dihydroorotate dependent respiration. *Parasitol. Int.* **50:**273–278.

ter Kuile, F., N. J. White, P. Holloway, G. Pasvol, and S. Krishna. 1993. Plasmodium falciparum: in vitro studies of the pharmacodynamic properties of drugs used for the treatment of severe malaria. *Exp. Parasitol.* **76:**85–95.

Ureta, T. 1982. The comparative isozymology of vertebrate hexokinases. *Comp. Biochem. Physiol. B* **71:**549–555.

Uyemura, S. A., S. Luo, S. N. Moreno, and R. Docampo. 2000. Oxidative phosphorylation, Ca^{2+} transport, and fatty acid-induced uncoupling in malaria parasites mitochondria. *J. Biol. Chem.* **275:**9709–9715.

Velanker, S. S., S. S. Ray, R. S. Gokhale, S. Suma, H. Balaram, P. Balaram, and M. R. Murthy. 1997. Triosephosphate isomerase from Plasmodium falciparum: the crystal structure provides insights into antimalarial drug design. *Structure* **5:**751–761.

Winter, V. J., A. Cameron, R. Tranter, R. B. Sessions, and R. L. Brady. 2003. Crystal structure of Plasmodium berghei lactate dehydrogenase indicates the unique structural differences of these enzymes are shared across the Plasmodium genus. *Mol. Biochem. Parasitol.* **131:**1–10.

Woodrow, C. J., R. J. Burchmore, and S. Krishna. 2000. Hexose permeation pathways in Plasmodium falciparum-infected erythrocytes. *Proc. Natl. Acad. Sci. USA* **97:**9931–9936.

Woodrow, C. J., J. I. Penny, and S. Krishna. 1999. Intraerythrocytic Plasmodium falciparum expresses a high affinity facilitative hexose transporter. *J. Biol. Chem.* **274:**7272–7277.

Yang, S., and S. F. Parmley. 1997. Toxoplasma gondii expresses two distinct lactate dehydrogenase homologous genes during its life cycle in intermediate hosts. *Gene* **184:**1–12.

THE MITOCHONDRION

Akhil B. Vaidya

12

As we gather more genomic information from greater diversity of organisms, it is becoming increasingly clear that genomes are in evolutionary flux, driven as much by horizontal gene transfers as by mutational changes. Endosymbiotic origins of mitochondria and chloroplasts exemplify gene transfer events on a massive scale. Indeed, such gene transfers are now believed to be at the root of the emergent precursor of all eukaryotic cells. In recent years, the concept that the extant amitochondriate organisms such as *Entamoeba histolytica* and *Giardia lamblia* represent deep-branching primitive eukaryotes prior to their acquisition of mitochondria (Sogin, 1991) has come to be seriously questioned: rather than being examples of primitive precursors that never acquired mitochondria, these organisms are now known to have lost their mitochondrial genomes while maintaining relict mitochondrial structures and functions (Mai et al., 1999; Bui et al., 1996; Dyall and Johnson, 2000; Vanacova et al., 2003; Gray et al., 2004). Apicomplexan parasites provide excellent examples of an evolutionary process of acquisitions and mergers involving genomes of different lineages. The discovery in the late 1980s that the mitochondrial genome of malaria parasites consisted of an unusually small 6-kb DNA molecule (Vaidya and Arasu, 1987; Vaidya et al., 1989; Aldritt et al., 1989; Joseph et al., 1989) gave impetus to a reexamination of the provenance of the 35-kb circular DNA molecule that was then considered to be the mitochondrial genome (Kilejian, 1975; Gardner et al., 1988; Williamson et al., 1985). As it turned out, the 35-kb DNA molecule is a derivative of a chloroplast genome, acquired originally from an alga that formed an alliance with another eukaryote, the progenitor of phylum Apicomplexa (Wilson and Williamson, 1997; Foth and McFadden, 2003; Wilson, 2002). Initially, this alliance must have involved the coexistence of five genomes in a single organism: two nuclear genomes, two mitochondrial genomes, and one chloroplast genome. Over time, elimination of the nucleus and the mitochondrion of the algal partner, as well as reshuffling of genes originally present in these organelles while maintaining a relict chloroplast with its own genome, led to the presence of

---

Akhil B. Vaidya, Center for Molecular Parasitology, Department of Microbiology and Immunology, Drexel University College of Medicine, Philadelphia, PA 19129.

Molecular Approaches to Malaria, Edited by Irwin W. Sherman
© 2005 ASM Press, Washington, D.C.

three genomes in most extant apicomplexan parasites. Genome sequence data clearly show footprints of these events, by revealing the presence of algal-like genes in the parasite nucleus, encoding physiological processes usually associated with plants (Gardner et al., 2002).

This chapter aims to provide an overview of the mitochondrion in malaria parasites in light of what has been learned from the genome sequence. Historically, functions of the mitochondrion in malaria parasites were unclear because of its sac-like appearance and paucity of cristae. Because the organelle from sexual stages of the parasites assumed more conventional morphology, it was assumed that in erythrocytic stages the mitochondrion serves minimal functions, only to be more active in sexual and insect stages. While some of the morphological differences may result from artifacts associated with fixation in blood stages of parasites, the question of differential functions of the mitochondrion in different stages remains unsettled at this point. Genomic and proteomic approaches can address this point to a degree, but biochemical investigations are only possible with blood stages of malaria parasites, and those with some difficulty. Discussion in this chapter, by necessity, is largely restricted to the mitochondrion as studied in asexual blood stages. The reader may wish also to refer to previous reviews that provide details of earlier work on parasite biochemistry, as well as of mitochondrial functions in malaria parasites (Sherman, 1979; Vaidya, 1998, 2004; Fry, 1991; Krungkrai, 2004).

METABOLIC FUNCTIONS OF THE MITOCHONDRION

The Challenge of Biochemical Studies of Mitochondria from *Plasmodium*

As the mitochondrion is surrounded by multiple membranes in erythrocytic stages, isolation of intact mitochondria of acceptable quality has been a major obstacle in carrying out biochemical studies. Prior to successful culture of *Plasmodium falciparum*, work on mitochondrial biochemistry also suffered from contamination with host components (Sherman, 1979). The need to be aware of host mitochondrial contamination is still valid when using rodent malaria parasites as a source. *Plasmodium berghei*, for instance, prefers to grow in reticulocytes, which continue to harbor some mammalian mitochondria. Similarly, contamination with platelets with their own mitochondria can also give false impressions of parasite mitochondrial physiology. Fry and Beesley (1991) were the first to explore mitochondrial isolation from mammalian *Plasmodium* spp. for biochemical studies. They isolated mitochondria from *Plasmodium yoelii* and *P. falciparum* in relatively low yield but of acceptable quality to carry out initial characterization of components involved in the electron transport chain. Fry and coworkers (Fry, 1991; Fry and Beesley, 1991) noticed (and we have confirmed) that saponin-released parasites are surprisingly tough; even prolonged homogenization results only as a result of the starting material being broken. Although the use of nitrogen cavitation as a means to break open the parasites has provided biochemically active mitochondrial preparations, the purity of the fractions, the yield, and the robustness of their biochemical activities were not described (Takashima et al., 2001). There appears to be an inverse relationship between the efficiency of parasite breakage and the purity of the mitochondrial preparations: the more thoroughly disrupted parasite material yields mitochondria that contain larger amounts of food vacuole fragments with hemoglobin in various stages of digestion. Clearly, alternative means of mitochondrial isolation need to be explored.

To circumvent some of these problems, a flow cytometric assay was developed to measure mitochondrial membrane potential in intact parasite-infected red blood cells (RBCs) using a very low concentration of a cationic lipophilic fluorescent probe that sequesters in the parasite mitochondria with charged inner membranes (Srivastava et al., 1997). In addition, respiration by the intact parasite-infected RBC was used as a means to measure mitochondrial electron transport (Srivastava et al., 1997). These in situ methods have been useful in investigating the mechanism of action of antimalarial compounds that target mitochondrial physiology in malaria parasites. These methods need to be used with a great deal of care to avoid artifacts that may arise from using probes that have intrinsic inhibitory activity at higher concentrations or that are not fully dependent upon membrane potential for accumulation within mitochondria.

In kinetoplastid parasites, permeabilization of the plasma membrane by cholesterol-dependent detergent digitonin has been used for in situ investigations of mitochondrial reactions (Vercesi et al., 1991). The reason for the success of this approach in the kinetoplastids is that while the plasma membrane contains significant amounts of cholesterol, the mitochondrial membranes are poor in cholesterol and thus are left intact. Digitonin permeabilization of saponin-released *P. berghei*, *P. falciparum*, and *P. yoelii* has been reported to provide means for similar in situ mitochondrial biochemical studies of malaria parasites (Uyemura et al., 2000; Uyemura et al., 2004). This approach, however, is difficult to reconcile with the observation that the plasma membrane of intraerythrocytic malaria parasites is particularly poor in cholesterol (Elmendorf et al., 1992; Behari and Haldar, 1994; Elmendorf and Haldar, 1994; Lauer et al., 2000). Indeed, saponin is capable of releasing intact parasites because this cholesterol-dependent detergent forms pores in the RBC plasma membrane while sparing the cholesterol-poor parasite plasma membrane. Therefore, it is difficult to envision how digitonin is able to permeabilize parasite plasma membranes of saponin-released *Plasmodium* spp. Biochemical activities reported by using digitonin-permeabilized parasites need to be reproduced by other groups.

The Electron Transport Chain and the Generation of Proton Motive Force

Although mitochondria are called the powerhouse of cells for providing ATP, it is the generation of proton motive force across their inner membrane that justifies this moniker. It is the proton motive force that powers the metabolic processes in mitochondria. The proton motive force consists of two components: a difference in electrical charge across the inner membrane ($\Delta\psi_m$) and a difference in pH (ΔpH). This is established by the transfer of protons from the matrix to the intermembrane space against the gradient that is powered by the electron transport chain. In malaria parasites, the tricarboxylic acid (TCA) cycle, which will be discussed later, is at best a minor source of the electrons. Instead, a set of dehydrogenases localized in the inner membrane initiate the electron transport by using ubiquinone as the electron acceptor. Components of complex I, the multisubunit proton-pumping NADH dehydrogenase that transfers electrons generated by the TCA cycle to the electron transport chain, are absent in malaria parasites (Gardner et al., 2002). Instead, the gene for a single-subunit NADH dehydrogenase that reduces ubiquinone has been found (Eren, 2003). Orthologs of such alternative or rotenone-insensitive NADH dehydrogenases are often present in plants and fungi (Yagi et al., 2001). Since NADH is an excellent substrate for the electron transport chain in isolated intact mitochondria from malaria parasites (Fry and Beesley, 1991) and since mitochondria cannot import NADH, the parasite NADH dehydrogenase is likely to face the intermembrane space and use the cytosolic NADH as the substrate. Other dehydrogenases that are likely to face the cytosolic side of the inner membrane are dihydroorotate dehydrogenase (DHODH) and glycerol 3-phosphate dehydrogenase. Since malaria parasites rely solely on de novo pyrimidine biosynthesis, DHODH is a critical enzyme, and the disposal of the electrons generated by DHODH is a key function of the parasite mitochondrion (Gutteridge et al., 1979; Krungkrai et al., 1990, 1991; Baldwin et al., 2002). Succinate dehydrogenase has been identified in malaria parasites, and although two catalytic subunit genes are present within the genome, membrane-anchoring subunits could not be detected (Takeo et al., 2000). Biochemical properties of the succinate dehydrogenase suggest that the enzyme may act more as a fumarate reductase, acting mainly to synthesize succinate (Suraveratum et al., 2000; Takeo et al., 2000). Succinate would be used for generation of succinyl-coenzyme A (CoA), a substrate for heme biosynthesis.

The overall organization of various dehydrogenases within the mitochondrion suggests that the major sources of reducing equivalents needed for the electron transport chain are substrates (NADH, dihydroorotate, and glycerol 3-phosphate) located within the intermembrane space rather than the mitochondrial matrix. In this regard, the mitochondrion in malaria parasites is fundamentally different from the conventional mitochondria. This again supports the notion of the malarial mitochondrion as a highly derived organelle. A schematic depiction of the mitochondrial processes is provided by Fig. 1.

The electron acceptor of the mitochondrial dehydrogenases is ubiquinone, which in turn is reoxidized by the mutisubunit complex III, the cytochrome bc_1 complex that transfers the electrons to cytochrome c while translocating protons to the intermembrane space. Thus, the bc_1 complex is the central enzyme in electron transport and generation of proton motive force in mitochondria. In malaria parasites, the bc_1

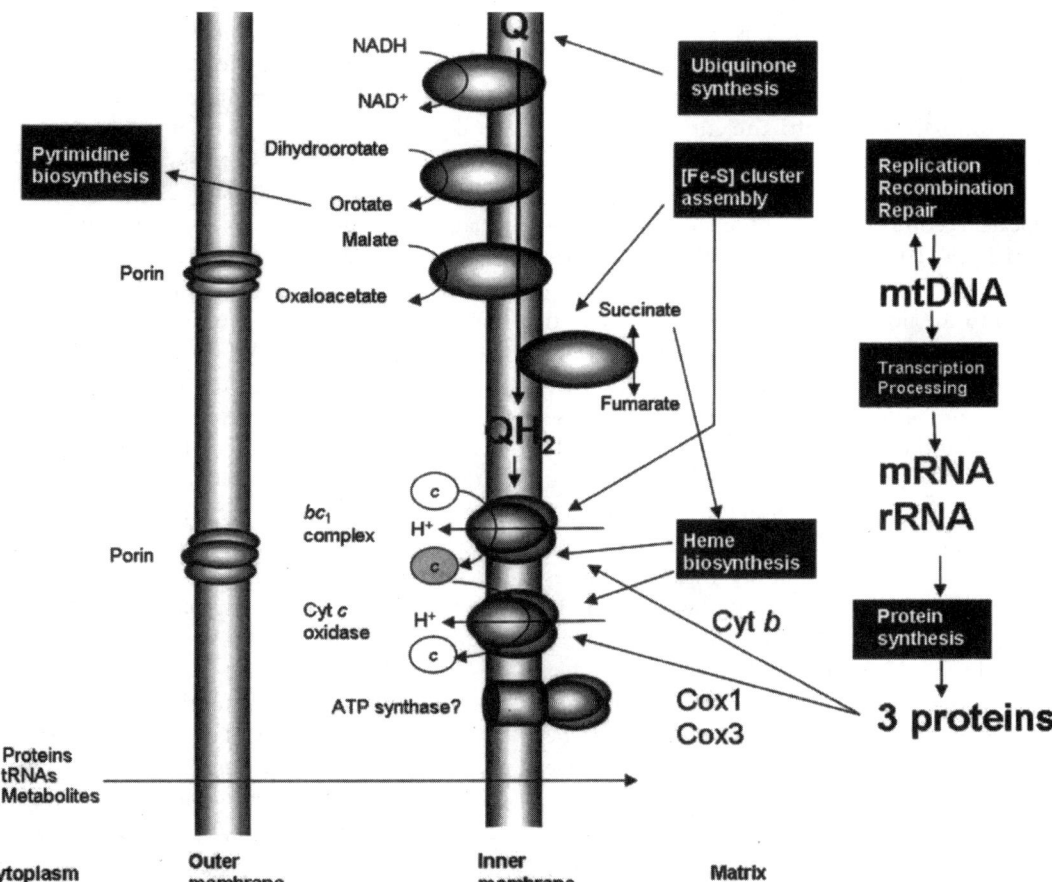

FIGURE 1 An outline of mitochondrial processes in malaria parasites. An outer membrane with a putative porin would allow molecules up to 1,500 daltons to freely diffuse to and from the intermembrane space. Transporters in the outer and inner membrane that assist mitochondrial import of proteins are not shown, nor are metabolite transporters located in the inner membrane. Major metabolic processes described in the text are shown in black boxes, and the flow of substrates and metabolites is indicated by arrows. The orientation of the dehydrogenases is assumed from biochemical studies of Fry and Beesley (1991). Glycerol 3-phosphate dehydrogenase is not shown, but is likely to face the intermembrane space.

complex appears to be composed of 7 subunits (compared to 11 in mammalian mitochondria), of which only the cytochrome *b* subunit is encoded in the mitochondrial genome; the rest of the subunits are encoded in the nucleus and will need to be imported. Structural features of cytochrome *b* revealed by sequence analysis suggested it to be a target for antimalarial drugs of the hydroxynaphthoquinone class (Vaidya et al., 1993a). This has now been confirmed by investigations on the mode of action for atovaquone (2-[*trans*-4-[4′-chlorophenyl]cyclohexyl]-3-hydroxy-1,4-naphthoquinone), as will be described later (Srivastava et al., 1997; Vaidya, 1998).

Reduced cytochrome *c* is oxidized by the multisubunit complex IV, the cytochrome *c* oxidase, which transfers the electrons to oxygen and pumps protons into the intermembrane space. Thus, this is the second complex that generates the proton motive force. The complex appears to consist of five subunits, of which subunits I and III are encoded within the mitochondrial genome. The subunit II gene is transferred to the nucleus and split into two parts in the process. Until now, this was observed only in the chlamydomonad algae family (Perez-Martinez et al., 2001), which suggests the possibility that the cytochrome *c* oxidase subunit II originated in the mitochondrion of the algal endosymbiont, rather than in the ancestral protist mitochondrion. The activity of cytochrome *c* oxidase is the chief reason that malaria parasites require oxygen for growth. A low-level alternative oxidase activity, believed to transfer electrons directly from ubiquinol to oxygen, has been reported (Murphy et al., 1997). However, the gene encoding such a protein cannot be detected in the genome, although an alternative oxidase is encoded in the *Cryptosporidium* genome (Xu et al., 2004; Abrahamsen et al., 2004; Roberts et al., 2004; Suzuki et al., 2004).

The only other potentially electrogenic enzyme possibly targeted to the mitochondrion is a proton translocating nicotinamide nucleotide transhydrogenase (Hatefi and Yamaguchi, 1996) that is encoded in the parasite genome (Eren, 2003). This protein, however, is not expressed in erythrocytic stages but in the sporozoite stages where it localizes to organelles that appear to be distinct from the mitochondrion (Eren, 2003). Thus, unlike its function in mammalian mitochondria, transhydrogenase does not appear to be a contributor to mitochondrial physiology.

The TCA Cycle?

Extensive biochemical studies done previously failed to detect many of the enzymes necessary for the TCA cycle in malaria parasites (Sherman, 1979, 1998). Furthermore, demonstration that almost all of pyruvate generated through glycolysis is converted to lactate deprives the TCA cycle of its key substrate (Vander Jagt et al., 1990). Thus, the TCA cycle was assumed to be absent in the erythrocytic stages of malaria parasites. The genome project, however, has revealed the presence of genes encoding essentially all TCA cycle enzymes necessary (Gardner et al., 2002). This has raised the question of whether malaria parasites do possess a functional TCA cycle. Gene expression profiles of blood-stage parasites revealed an apparently coordinated expression of genes for the TCA cycle enzymes (Bozdech et al., 2003). The overall level of expression, however, is quite low for these genes, resulting in enzymatic activities that were below the levels of detection in biochemical assays done earlier. Experiments involving fusion of N-terminal targeting signal sequences to the green fluorescent protein have supported mitochondrial localization of citrate synthetase, isocitrate dehydrogenase, malate-quinone oxidoreductase (MQO) (M. W. Mather, J. M. Morrisey, and A. B. Vaidya, unpublished data), and one of the dihydrolipomide dehydrogenases in malaria parasites (McMillan et al., 2005). It is interesting that MQO has until now has been detected only in prokaryotic cells. Malate dehydrogenase, which usually carries out the reaction catalyzed by MQO, is a cytosolic enzyme in malaria parasites (Lang-Unnasch, 1995) and is likely to convert oxaloacetate to malate (Tripathi et al., 2004). Pyruvate dehydrogenase complex subunits E1 and E2 and one of the E3 (the second dihydrolipoamide dehydrogenase) appear to be targeted to the apicoplast (Foth et al., 2005); acetyl-CoA generated here likely serves as the substrate for fatty acid synthesis within the apicoplast. The lack of mitochondrial pyruvate dehydrogenase eliminates the conventional generation of acetyl-CoA within the mitochondrion, which is necessary for the initiation of the TCA cycle. Although there is no direct evidence to support this, it is often suggested that the TCA cycle is more robust in insect stages of the parasite. However, the absence of a system that can provide acetyl-CoA in the mitochondrion makes this proposition difficult to envision. Therefore, the TCA cycle enzymes located within the parasite mitochondria, instead of serving as the main source of reducing equivalents for the electron transport chain, are likely to be involved in biosynthetic reactions, mainly to provide succinyl-CoA for use in heme biosynthesis.

ATP Synthesis?

Mitochondrial ATP synthesis is carried out by a multiprotein rotary enzyme, F_0F_1 ATP synthase (complex V), located within the inner membrane by utilizing the proton motive force (Boyer, 1997). Previous studies have suggested that the parasite mitochondrion contributes little to the ATP pool within malaria parasites (Fry et al., 1990). Genome sequence has now revealed the presence of genes encoding all components of the F_1 sector of F_0F_1 ATP synthase, but only subunit c of the F_0 sector (Gardner et al., 2002). Genes encoding subunits a and b, which function as a stator for the rotary enzyme, as well as a part of the proton channel, cannot be detected in the genome of any malaria parasite species. Without these subunits, the ATP synthase cannot function to generate ATP. Some highly unusual arrangement needs to be evoked for this enzyme complex to synthesize ATP. Indeed, a common belief that the parasite mitochondrion is a source of ATP in insect stages cannot be supported if we are to ascribe conventional features to the ATP synthase. Indirect evidence for ATP synthesis by digitonin-treated rodent malaria parasites is modest and subject to alternative explanations (Uyemura et al., 2000; Uyemura et al., 2004), and thus does not provide sufficient evidence for mitochondrial ATP synthesis.

Pyrimidine Biosynthesis

If the TCA cycle and ATP synthase are both disabled in the malaria parasite mitochondrion, what is left for the organelle to do? Clearly, there are several other essential functions that the mitochondrion serves. A central function, as alluded to earlier, is to serve as the location where the fourth step in pyrimidine biosynthesis occurs (Gutteridge et al., 1979). The enzyme DHODH oxidizes dihydroorotate to produce orotate, transferring electrons to ubiquinone. DHODH from malaria parasites, in both its native form partially purified from the parasite (Krungkrai, 1991, 1995; Krungkrai et al., 1991) and a recombinant form expressed in bacteria (Baldwin et al., 2002), has been characterized. DHODH from humans is the target for antiarthritis and antiproliferative compounds such as leflunomide and Redoxal (Fox, 1998; Loffler et al., 1998; Batt et al., 1998). These drugs have been shown to have inhibition profiles that are characteristic of the species from which DHODH is examined. Plasmodium DHODH is inhibited by dichoroallyl lawsone and Redoxal at concentrations >1,000-fold higher than those for human DHODH (Baldwin et al., 2002). This suggests that it should be possible to screen for compounds that preferentially inhibit the parasite DHODH.

In mammals, a multifunctional protein closely associated with the outer surface of the mitochondrion carries out the first three steps of pyrimidine biosynthesis (Evans and Guy, 2004); perimitochondrial localization of this protein can be presumed to be for efficient channeling of the substrate (dihydroorotate) to DHODH in the inner membrane facing the intermembrane space. In malaria parasites, enzymes for these steps (carbamoyl-phosphate synthetase, aspartate transcarbamylase, and dihydroorotase) are encoded by separate genes, but their localization is not clear at this point.

Heme Biosynthesis

Although malaria parasites digest hemoglobin, the liberated heme is likely not used for assembly of hemoproteins. Instead, malaria parasites seem to carry out de novo heme biosynthesis.

Heme is required for the assembly of *a*-, *b*-, and *c*-type cytochromes, almost all them localized to the mitochondrion. The story of heme biosynthesis, however, appears to be rather complicated, with both parasite-encoded and host-derived heme synthesis enzymes localized to various compartments of malaria parasites. The rate-limiting reaction in heme biosynthesis is the first step involving synthesis of δ-aminolevulinate (δ-ALA), eight molecules of which are needed for the assembly of a single heme molecule. There are two different pathways for δ-ALA synthesis in eukaryotes. The Shemin pathway, involving conjugation of succinyl-CoA and glycine by δ-ALA synthetase, is localized in mitochondria, whereas the C5 pathway involving utilization of the carbon skeleton of glutamate through the activity of three different enzymes is localized in chloroplasts. It is now well established that the parasite genome encodes δ-ALA synthetase that is localized to the mitochondrion (Wilson et al., 1996; Varadharajan et al., 2002; Sato et al., 2004). All the rest of the heme biosynthesis enzymes, except for uroporphyrinogen III synthetase, are also encoded in the parasite genome. The parasite enzymes responsible for the next steps in heme biosynthesis— δ-ALA dehydratase, porphobilinogen synthase, and hydroxymethylbilane synthase—have been shown to localize to the apicoplast, and bear the characteristic bipartite apicoplast-targeting signals (Sato and Wilson, 2002; Sato et al., 2004). Similar signal sequences are also present within other heme biosynthesis enzymes, except ferrochelatase. Since ferrochelatase carries out the final step of loading iron into the porphyrin ring, and since the mitochondrion is the chief user of heme in malaria parasites, it was suggested that this enzyme may be localized to the mitochondrion (Gardner et al., 2002). Thus, the first and the last steps in heme biosynthesis were proposed to be localized to the mitochondrion, whereas the other steps would be localized to the apicoplast. This proposal would require shuttling of substrates between the mitochondrion and the apicoplast. Varadhrajan et al., however, have provided data suggesting that ferrochelatase is localized to the apicoplast (Varadharajan et al., 2004). This raises the specter of a possible C5 pathway relegated to the apicoplast. The genomic sequence, however, does not reveal the presence of the enzymes required to carry out the C5 pathway synthesis of ALA. Surprisingly, the parasite cytosol appears to harbor host-derived heme biosynthesis enzymes that could carry out heme synthesis in vitro if provided with ALA (Varadharajan et al., 2004). Indeed, in vitro heme biosynthetic activity of the cytosolic fractions far exceeds that of the organellar fractions. The suggestion seems to be that the host enzymes imported by the parasite into its cytoplasm can carry out the complete heme assembly once the parasite mitochondrion provides ALA as the starting material. This is clearly a radical view and raises many questions as to regulation of these reactions, chaperoning of apocytochromes, maintenance of free heme prior to its import into the mitochondrion, and mechanisms of appropriate transport of heme. Another unresolved problem is the source of iron to be used by ferrochelatase, since ionic form of iron is unlikely to be present in free form within the parasite cytosol. Clearly, careful studies need to be designed to resolve these problems.

Iron-Sulfur Cluster Synthesis

Iron-sulfur ([Fe-S]) clusters are cofactors of numerous proteins that are central to cellular processes such as redox reactions, metabolic catalysis, and sensing of iron and oxygen levels (Muhlenhoff and Lill, 2000; Lill and Kispal, 2000). The assembly of [Fe-S] clusters is a complex process requiring participation of several enzymes, chaperones, and transporters. These processes emerged early in evolution, being integral to all prokaryotes. In eukaryotes, [Fe-S] cluster synthesis continues to be relegated to the organelles descendent from their prokaryotic endosymbionts, namely the mitochondria and chloroplasts, even as genes encoding components of the synthetic apparatus have mostly been transferred to the nucleus. [Fe-S] clusters synthesized in these organelles are used not only for the assembly of proteins resident in the organelles, but also are exported to be part of numerous cytosolic and nuclear proteins. Indeed, even in eukaryotes that lack mitochondrial DNA (mtDNA) (such as *Trichomonas* and *Giardia*), vestigial mitochondrion-like structures are maintained with a major function being to serve as the sites for the [Fe-S] cluster synthesis (Tachezy et al., 2001; Seeber, 2002a, 2002b). Although we know very little about the biochemical features of [Fe-S] cluster synthesis in malaria parasites, the genomic sequence reveals interesting arrangements. There seem to be two separate sets of genes encoding orthologs of proteins involved in [Fe-S] cluster synthesis: one set of proteins is likely to be targeted to the apicoplast (reviewed in Chapter 14), whereas the other set is likely to be involved in mitochondrial [Fe-S] cluster synthesis. There are at least 10 different proteins encoded in the nucleus that are likely to be targeted to the mitochondrion to participate in the complex series of transport, chaperoning, and enzymatic reactions necessary for [Fe-S] cluster formation. A review by Seeber (2002a) provides an excellent genome-based view of [Fe-S] biogenesis in apicomplexan parasites.

Other Functions

In addition to their role in bioenergetics and the processes described above, many other functions have been ascribed to mitochondria in various organisms. These include amino acid metabolism, sterol biosynthesis, calcium homeostasis, ubiquinone synthesis, and regulation of apoptosis. As to the amino acid metabolism, isolated mitochondria from both *P. falciparum* and *P. yoelii* are able to oxidize proline (Fry and Beesley, 1991), suggesting the presence of glutamate dehydrogenase within the mitochondrion. There are three different genes encoding glutamate dehydrogenase in *Plasmodium*, but none of these appear to have a mitochondrial targeting sequence. The major glutamate dehydrogenase activity appears to be restricted to the cytoplasm and likely to be a source for NADPH (Wagner et al., 1998). Malaria parasites do not synthesize sterols but do produce isoprenols, which are synthesized in the apicoplast through a nonmavelonate pathway. Ubiquinone synthesis, which requires isoprenols as substrates, is likely to be localized to the mitochondrion. Several of the enzymes necessary for ubiquinone synthesis are encoded in the genome. Calcium homeostasis is now recognized as a major function of energized mitochondria in mammals. At this point, it is not clear how active a role the mitochondrion plays in calcium homeostasis in malaria parasites, although initial investigations with calcium-sensitive fluorescent probes suggest significant uptake of calcium by the mitochondrion in malaria parasites (Gazarini and Garcia, 2004).

In metazoa, mitochondria are central to the process of programmed cell death or apoptosis, releasing several of proapoptotic molecules such as cytochrome *c* in response to a number of different apoptotic signals, including the collapse of mitochondrial membrane potential (Green and Kroemer, 2004). There is no evidence yet of a programmed cell death pathway in malaria parasites, and typical features of apoptosis are not seen in dying malaria parasites. As will be discussed later, collapse of mitochondrial membrane potential by atovaquone does not appear to initiate a typical apoptotic pathway in malaria parasites.

THE MITOCHONDRIAL GENOME: ITS EXPRESSION AND REPLICATION

The Mitochondrial Genome

Mitochondrial genomes come in wide arrays of size, structure, and gene content (Gray et al., 2004). In general, however, characteristics of mtDNA for a given phylum or order remain relatively similar. For instance, all metazoa possess circular mtDNA of about 15 to 20 kb with similar gene content and organization. Being much more diverse, protists overall demonstrate the greatest range of diversity, with each of the taxa having its characteristic mitochondrial genome architecture and gene content. For instance, all kinetoplastids have the maxicircle-minicircle organization for their mtDNA. Members of the phylum Apicomplexa, however, appear to have a greater range of mtDNA organization, as judged by data from a few members of the group. The size of mtDNA is about 6 kb, which may be present as linear molecules bounded by repeated sequences (in *Theileria* spp.) (Kairo et al., 1994) or as tandem arrays of molecules joined head to tail (in *Plasmodium* spp.) (Vaidya and Arasu, 1987). On the other hand, *Cryptosporidium* spp. have lost their mtDNA altogether (Abrahamsen et al., 2004; Xu et al., 2004), whereas *Toxoplasma* contains >100 copies of portions of mtDNA scattered throughout the nuclear genome (Ossorio et al., 1991). The overall impression is of an organellar genome in flux. This may reflect different physiological roles played by the mitochondrion in different species of apicomplexan parasites, dictated by the ecological niche of their parasitic existence.

In *Plasmodium*, estimates of copy numbers of mtDNA range from 30 to 100 per parasite, mostly arranged in head-to-tail tandem arrays of various lengths; a few unit length circles may also be present. The mtDNA has the smallest gene content of any known organellar genomes, encoding only three proteins: cytochrome *b* and subunits I and III of cytochrome *c* oxidase. Cytochrome *c* oxidase subunit I and cytochrome *b* are encoded from a contiguous region with separate mRNA for each protein; cytochrome *c* oxidase subunit III is encoded by the opposite strand. Instead of contiguous genes for ribosomal RNA, the large- and small-subunit rRNA genes are fragmented, scrambled in their order, and encoded by both strands (Vaidya et al., 1989; Vaidya et al., 1993a; Feagin et al., 1992). In addition to the three mRNAs, 20 small RNA molecules ranging from 29 to 200 nucleotides in length are encoded by the mtDNA (Feagin et al., 1992; Feagin et al., 1997; Suplick et al., 1990; McIntosh, 1995). These molecules constitute the highly conserved core sequences of the small and large subunits of rRNA that possess the ability to form appropriate hydrogen-bonded helices of conventional rRNA. The parasite mtDNA does not encode any tRNA or ribosomal protein.

In both its sequence and organization, the mtDNA is highly conserved in all malaria parasite species examined, with 90% sequence identity at the nucleotide level (McIntosh et al., 1998). Because nuclear genomes of these parasite species demonstrate significant differences in their A+T content, different evolutionary pressures are likely to be influencing nuclear and mitochondrial genomes (McIntosh et al., 1998). There do not appear to be any mtDNA sequence variations within a given clone of malaria parasite, suggesting a lack of heteroplasmy. Single nucleotide polymorphisms, however, can be detected in mtDNA from different geographical isolates (Vaidya et al., 1993b; McIntosh et al., 1998; Joy et al., 2003). Indeed, sequencing of 100 different *P. falciparum* isolates provided haplotype lineages of the parasite with clues as to the ancestral relationship between these clones and the evolutionary origins of *P. falciparum* (Joy et al., 2003; Su et al., 2003).

During male gametogenesis, mtDNA is not segregated within the flagellated male gamete (Vaidya et al., 1993b; Creasey et al., 1994). Therefore, mtDNA (as well as apicoplast DNA) is inherited from the female gamete during fertilization in the mosquito midgut (Vaidya et al., 1993b; Creasey et al., 1993). As in animals, the cytoplasmic genomes are maternally derived in all extant malaria parasites. This observation provides the means to track parental lineage of progeny parasites generated from genetic crosses (Vaidya et al., 1993b; Creasey et al., 1993).

Transcription, Translation, and Replication of mtDNA

mtDNA is transcribed into at least 20 different stable RNA molecules, only 3 of which are mRNAs; the rest constitute fragments of large and small subunits of rRNA or precursors from which such fragments are generated (Feagin et al., 1992; Feagin et al., 1997; Suplick et al., 1990; McIntosh, 1995). An RNA polymerase resembling bacteriophage T3-T7 polymerase has been identified (Li et al., 2001). Because of the high density of genes within the parasite mtDNA, these are likely to be transcribed as larger precursors and processed (Ji et al., 1996; Rehkopf et al., 2000). Almost all RNA molecules also appear to have oligoadenylation at their 3' ends (Gillespie et al., 1999). The length of these tails varies among different RNA molecules. At this point, identities of the promoter sequences and components that regulate transcription from the mtDNA are unknown.

The three proteins encoded by the mtDNA will need to be translated in the organelle. The translational machinery that carries out this process is likely to be the most unusual in its structural and biochemical features. The mitochondrial ribosomes in malaria parasites can be predicted to have unprecedented organization: multiple rRNA fragments will need to interact in *trans* to form the core small and large subunit rRNA structures, which in turn will interact with imported ribosomal proteins to form functional ribosomes (McIntosh, 1995). At present, little is known about the components and the process underlying the assembly of these bizarre ribosomes. Through a bioinformatics approach, a number of nuclearly encoded putative mitochondrial ribosomal proteins have been identified, and the N-terminal signal of one has been experimentally shown to target a fused green fluorescent protein to the mitochondrion (R. Perrault, J. M. Morrisey, and A. B. Vaidya, unpublished data). Because the mtDNA does not encode any tRNAs or tRNA synthetases, these will need to be imported from the cytoplasm. The parasite genome does not reveal any tRNA genes other than those for cytoplasmic protein synthesis and those encoded in the apicoplast genome. Thus, the tRNAs may need to serve protein synthesis both in the cytoplasm and the mitochondrion, participating in protein synthesis off both eukaryotic and (presumably highly derivatized) prokaryotic ribosomes. Furthermore, there do not appear to be sufficient tRNA synthetase genes to independently serve the three compartments—the cytoplasm, the mitochondrion, and the apicoplast—in which protein synthesis occurs in malaria parasites. Hence, some of the tRNA synthetases will have dual or triple targeting. A similar situation is likely to exist for a number of different components required for protein synthesis.

A rolling-circle mode of replication with extensive gene conversion and recombination for the parasite mtDNA was proposed by Preiser et al. (1996), based on morphological and electrophoretic mobility. The architecture of mtDNA in malaria parasites, i.e., tandem arrays of 6-kb DNA molecules in head-to-tail arrangement, may derive from this mode of replication. Also, the extensive gene conversion implied here may result in copy correction and could also explain the high degree of sequence conservation within a given clone of malaria parasite, as well as the lack of heteroplasmy. DNA replication, recombination, and repair would require participation of a large number of proteins in these complex processes. At present, little is known about these proteins in malaria parasites. A gene for DNA polymerase γ, the canonical mitochondrial DNA polymerase, cannot be detected in the genome sequence. However, two genes encoding proteins with homology to DNA polymerase I were found. One of these, PF14_0112, currently annotated as POM1, encodes a 2,016-amino-acid open reading frame with homology to DNA polymerase I, as well as domains with homology to a $3'$-to-$5'$ exonuclease and a DNA helicase-primase. The predicted protein appears to have a putative apicoplast-targeting signal at its amino terminal, and thus may be the apicoplast DNA polymerase (M. Barrett et al., personal communication). The second gene, PFF1225c, encodes a 1,444-amino-acid open reading frame with the DNA polymerase domain located at the C-terminal region and a signal sequence for mitochondrial targeting at its N-terminal.

MITOCHONDRIAL TARGETS FOR ANTIMALARIAL DRUGS

Cytochrome bc_1 Complex and Atovaquone-Proguanil Combination

Early sequence analysis of *Plasmodium* cytochrome *b* revealed unique structural features that were suggested to underlie the sensitivity of malaria parasites to hydroxynaphthoquinones (Vaidya et al., 1993a). To understand the basis for this suggestion and the mechanism of atovaquone action, an overview of the mechanism by which the cytochrome bc_1 complex transfers electrons from ubiquinol to cytochrome *c* while translocating protons from the matrix to the intermembrane space in mitochondria is necessary and is summarized in Fig. 2 (Yu et al., 1999;

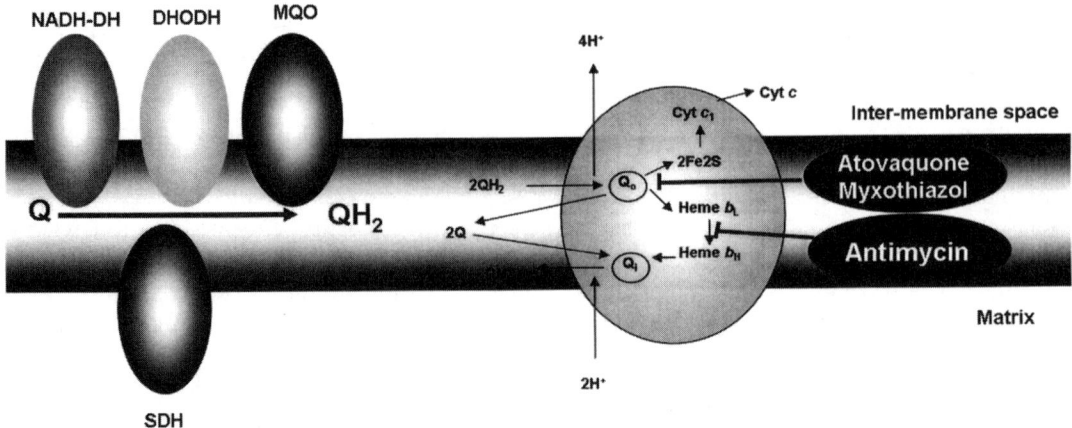

FIGURE 2 A description of the Q cycle through which ubiquinol is oxidized and protons are translocated at the cytochrome bc_1 complex. Bifurcation of electrons from QH_2 at the Q_o site is a key feature of the Q cycle. Sites at which cytochrome bc_1 inhibitors work are indicated at the right. See the text for a description.

Zhang et al., 2000; Gao et al., 2003; Hunte et al., 2003; Crofts, 2004). Structural studies of the complex have led to a consensus regarding the general features of how this process, called the Q cycle, is believed to work. Ubiquinol (QH_2), produced by various mitochondrial dehydrogenases, is oxidized to ubiquinone through a bifurcated electron transfer at a site within the bc_1 complex with concomitant translocation of $2H^+$ into the intermembrane space for each ubiquinol oxidized. During bifurcation, one of the electrons reduces the high-potential 2Fe2S cluster of the iron-sulfur protein; the other reduces the low-potential heme b_L of cytochrome b at the Q_o site. The reduced heme b_L immediately passes the electron to the high-potential heme b_H of cytochrome b at the Q_i site, where a resident quinone is reduced to semiquinone through one electron transfer. A second round of QH_2 oxidation is necessary for full reduction of semiquinone at the Q_i site. Reduction of the 2Fe2S cluster is followed by a major movement of the head domain of the iron-sulfur protein, which positions the 2Fe2S cluster proximal to the heme of cytochrome c_1, leading to the reduction of the heme c_1. This is followed by electron transfer to cytochrome c, which is docked on the intermembrane side of the cytochrome bc_1 complex. A number of inhibitors have been identified that work either at the Q_o (e.g., myxothiazol and stigmatellin) or the Q_i (e.g., antimycin and funiculosin) site. Missense mutations within cytochrome b can result in resistance to these inhibitors, and their locations within the protein initially helped to define the Q_o and Q_i sites of the complex. X-ray crystal structures of the complex in the presence of these inhibitors have provided clear definitions of these sites (Xia et al., 1997; Iwata et al., 1998; Kim et al., 1998).

Examination of the cytochrome b sequence from malaria parasites revealed features of its Q_o and Q_i regions that were associated with resistance to myxothiazol and antimycin (Vaidya et al., 1993a). This led to a suggestion that the selective toxicity of hydroxynaphthoquinones for malaria parasites may arise from such differences. Hydroxynaphthoquinones have been suspected to act as ubiquinone competitors since the 1940s (Fieser, 1948), but the first demonstration of their selective action came from the work of Fry and Pudney, who showed that atovaquone, a compound being developed as an antimalarial, inhibited the parasite mitochondrial cytochrome bc_1 complex at a 1,000-fold lower concentration than the complex from the mammalian mitochondria (Fry and Pudney, 1992). With an assay to study mitochondrial physiology in live intact malaria parasites by measuring mitochondrial membrane potential, atovaquone was found to collapse the malarial mitochondrial membrane potential without affecting mammalian mitochondria at therapeutic concentrations (Srivastava et al., 1997). As expected, since it inhibits mitochondrial electron transport, atovaquone was also found to inhibit respiration by malaria parasites (Srivastava et al., 1997).

While malaria parasites were found to be exquisitely sensitive to atovaquone both in vitro and in animal malaria models, there was a 30% treatment failure when it was used as a single agent in clinical trials to treat uncomplicated *P. falciparum* malaria, with atovaquone-resistant parasites emerging within 28 days after treatment (Looareesuwan et al., 1996; Looareesuwan et al., 1999a, 1999b). Indeed, atovaquone-resistant malaria parasites easily arose in an experimental rodent malaria system (Srivastava et al., 1999) as well as in *P. falciparum* cultures (Gassis and Rathod, 1996). Examination of the cytochrome *b* sequence from nine independently derived atovaquone-resistant *P. yoelii* lines revealed missense mutations centered around the Q_o region, especially near the highly conserved PEWY sequence (Srivastava et al., 1999). Similar mutations in atovaquone-resistant parasites from other malaria species have also been observed (Syafruddin et al., 1999; Korsinczky et al., 2000; Fivelman et al., 2002; Vaidya and Mather, 2000). The collection of mutants, therefore, provided a strong argument that atovaquone is a Q_o site inhibitor and also delineated the pocket where atovaquone appears to bind. Clearly, subtle changes in the charge and volume of this pocket are all that is required for emergence of >1,000-fold resistance of the complex to atovaquone. The mammalian bc_1 complex naturally possesses these changes and is therefore relatively impervious to the drug, thereby providing therapeutic margins.

The unacceptable rate of treatment failure almost doomed atovaquone as an antimalarial. To rescue it, however, a synergistic partner was sought. Proguanil, an antimalarial developed in the 1950s, was found to synergize with atovaquone in *P. falciparum* cultures and was chosen for further development (Canfield et al., 1995). A number of clinical trials found a fixed combination of 1,000 mg of atovaquone and 250 mg of proguanil to be highly effective in treating uncomplicated falciparum malaria, with cure rates of close to 100% (Looareesuwan et al., 1999a, 1999b). Additional clinical trials found this combination also to be effective as a prophylactic agent and against vivax malaria (Looareesuwan et al., 1999c; Ling et al., 2002; Lacy et al., 2002). The combination of atovaquone and proguanil, trademarked Malarone, is being increasingly used for prophylaxis in travelers. The extremely high cost of Malarone makes this drug essentially unavailable to most of the people in countries where malaria is endemic, even though most of the clinical trials that led to its certification were carried out in such countries. Ethical questions raised by this are beyond the scope of the present chapter.

It is interesting that Malarone does not have efficacy against the dormant (hypnozoite) forms of *Plasmodium vivax*, as judged from clinical experience with relapsing vivax malaria (Povinelli et al., 2003; Looareesuwan et al., 1999c). Although we know practically nothing about the biochemical features of the hypnozoites, this observation would suggest that the maintenance of mitochondrial membrane potential is not essential for their survival. Indeed, when atovaquone treatment is applied, parasite demise, even in asexual stages, does not seem to involve the apoptotic pathways induced in animal cells with collapsed mitochondrial membrane potential. Significant length of exposure to atovaquone in certain stages of parasites appears to be required for parasite death, an observation that needs to be explored in detail (H. Painter, J. M. Morrisey, and A. B. Vaidya, unpublished data).

Possible mechanisms for the synergy between atovaquone and proguanil have been investigated (Srivastava and Vaidya, 1999). Proguanil at therapeutic concentrations did not affect mitochondrial physiology of malarial parasites. However, when combined with atovaquone, it lowered the midpoint inhibition concentration at which atovaquone collapsed mitochondrial membrane potential by six- to eightfold. Interestingly, it did not affect the midpoint inhibition concentration at which atovaquone inhibited respiration by malaria parasites. This suggests that the electron transport inhibition and membrane potential collapse effects of atovaquone may be separable. The molecular details of how this could be achieved are unclear at this point. Proguanil as an antimalarial on its own is a prodrug that needs to be converted to cycloguanil to act as a parasite dihydrofolate reductase inhibitor (Carrington, 1951; Crowther, 1953). The synergy with atovaquone, however, was the property of the prodrug itself; conversion to cycloguanil was not required (Canfield et al., 1995; Srivastava and Vaidya, 1999). Indeed, cycloguanil as well as other dihydrofolate reductase inhibitors failed to synergize with atovaquone (Canfield et al., 1995). The fact that this is a true synergy between atovaquone and proguanil is further supported by the observation that atovaquone-resistant parasites are equally resistant to the atovaquone-proguanil combination as well (Srivastava and Vaidya, 1999).

The cytochrome bc_1 complex of malaria parasites is clearly a fully validated target for antimalarial drug development. The aim ought to be to develop significantly cheaper compounds that also target the parasite bc_1 complex and do so without the risk of rapid resistance emergence. Such endeavors are already under way with the hope that new, effective, and affordable therapeutics for the areas of endemicity will emerge from these studies.

Mitochondrial Protein Synthesis and Antibiotics

Antibiotics that inhibit prokaryotic protein synthesis have been known to possess antimalarial activity. This was assumed to be due to the prokaryotic nature of mitochondrial protein synthesis. The discovery of the apicoplast has added yet another venue where antibiotics may work as antimalarials. Mitochondrial translation machinery is necessary for the synthesis of three proteins encoded by mtDNA. Although mitochondrial ribosomes have an unorthodox architecture, it is safe to assume that these ribosomes are fundamentally prokaryotic in nature and thus are possible targets for antibiotics that inhibit protein synthesis. Tetracyclines, especially those with greater membrane permeability such as doxycycline, are well-known antimalarial antibiotics and are likely to affect mitochondrial and/or plastid protein synthesis (Clough and Wilson, 2001). The tetracycline-binding site in prokaryotes is within the small ribosomal subunit and is dependent upon a specific nucleotide position (A892 in *Escherichia coli* numbering) within the small-subunit rRNA (Cundliffe, 1990). Both plastid and mitochondrial small-subunit rRNA possesses this conserved site. It is important to note that the tetracycline-binding site within the prokaryotic ribosome appears to possess a critical structural feature, since tetracycline-resistant ribosomes with altered binding sites have not been observed for any bacteria (tetracycline resistance results through transport-efflux mechanisms). Interestingly, tetracycline-resistant malaria parasites have not been reported to date. The disadvantage of tetracyclines as antimalarials is their slow rate of action for treating acute malaria, although they work well as a prophylactic measure. Other antibiotics with antimalarial effects (e.g., clindamycin and azithromycin) are likely to affect plastid functions rather than the mitochondrion, since the mitochondrial rRNA sequence appears to bear changes that are associated with resistance to these antibiotics (Cundliffe, 1990).

Other Potential Targets

The great evolutionary divergence of mitochondrial physiological processes between malaria parasites and their host offers several opportunities to explore means by which inhibitors with differential activities can be developed. As mentioned earlier, DHODH provides one such opportunity. This mitochondrially located enzyme is essential for parasite survival and has already been shown to have responses to inhibitors that would suggest identification of compounds with a high therapeutic index (Baldwin et al., 2002). Another potential target is the single-subunit NADH dehydrogenase, an enzyme that does not have an apparent ortholog in mammals. The choice of this target, however, should await its validation as an essential enzyme in parasite physiology. Many of the other components discussed in this chapter could also be potential targets, but these too should be subjected to tests of their validity as targets, as well as an evaluation of the challenges of identifying selective inhibitors. Because many of the critical mitochondrial processes involve participation of ubiquinone, its analogs that direct more than one process may prove beneficial in that resistance may be harder to achieve if more than one molecule are targets for action.

In addition to antibiotics that inhibit prokaryotic protein synthesis, DNA gyrase inhibitors such as quinolones have also been found to have antimalarial activity (Krishna et al., 1988; Wyler, 1989). In general, this group of antibiotics is believed to act on apicoplast-targeted DNA gyrase and to inhibit apicoplast DNA replication, causing a delayed death phenotype in at least *Toxoplasma* (Fichera and Roos, 1997). A topoisomerase will clearly be necessary for mitochondrial DNA replication, but its identity is unknown at present. As in case of *Trypanosoma*, topoisomerase inhibitory compounds may provide attractive antimalarial leads.

PERSPECTIVES

On an evolutionary time scale, the mitochondrion in malaria parasites is caught in the process of being minimized. Its diminished genome, its bizarre ribosomes, and its shedding of conventional roles in cellular physiology all speak to this minimization. Its brethren in *Cryptosporidium* spp. are even further down this path, having gotten rid of the genome itself. Minimization, however, does not equal irrelevance. This chapter has attempted to provide an overview of why this "relict-in-the-making" mitochondrion is still critical for parasite physiology and a valid target for antimalarial drugs. The challenge remains to understand mitochondrial contributions in greater detail with the hope of identifying features that are ripe for further chemotherapeutic intervention.

ACKNOWLEDGMENTS

I thank my colleagues in the Center for Molecular Parasitology at Drexel University College of Medicine for providing a wonderful academic atmosphere and for stimulating discussions.

This work was supported by grants AI28398 and AI53148 from NIH.

REFERENCES

Abrahamsen, M. S., T. J. Templeton, S. Enomoto, J. E. Abrahante, G. Zhu, C. A. Lancto, M. Deng, C. Liu, G. Widmer, S. Tzipori, G. A. Buck, P. Xu, A. T. Bankier, P. H. Dear, B. A. Konfortov, H. F. Spriggs, L. Iyer, V. Anantharaman, L. Aravind, and V. Kapur. 2004. Complete genome sequence of the apicomplexan, Cryptosporidium parvum. *Science* **304:**441–445.

Aldritt, S. M., J. T. Joseph, and D. F. Wirth. 1989. Sequence identification of cytochrome *b* in Plasmodium gallinaceum. *Mol. Cell. Biol.* **9:**3614–3620.

Baldwin, J., A. M. Farajallah, N. A. Malmquist, P. K. Rathod, and M. A. Phillips. 2002. Malarial dihydroorotate dehydrogenase. Substrate and inhibitor specificity. *J. Biol. Chem.* **277:**41827–41834.

Batt, D. G., J. J. Petraitis, S. R. Sherk, R. A. Copeland, R. L. Dowling, T. L. Taylor, E. A. Jones, R. L. Magolda, and B. D. Jaffee. 1998. Heteroatom- and carbon-linked biphenyl analogs of Brequinar as immunosuppressive agents. *Bioorg. Med. Chem. Lett.* **8:**1745–750.

Behari, R., and K. Haldar. 1994. Plasmodium falciparum: protein localization along a novel, lipid-rich tubovesicular membrane network in infected erythrocytes. *Exp. Parasitol.* **79:**250–259.

Boyer, P. D. 1997. The ATP synthase—a splendid molecular machine. *Annu. Rev. Biochem.* **66:**717–749.

Bozdech, Z., M. Llinas, B. L. Pulliam, E. D. Wong, J. Zhu, and J. L. DeRisi. 2003. The transcriptome of the intraerythrocytic developmental cycle of Plasmodium falciparum. *PLoS Biol.* **1:**E5.

Bui, E. T., P. J. Bradley, and P. J. Johnson. 1996. A common evolutionary origin for mitochondria and hydrogenosomes. *Proc. Natl. Acad. Sci. USA* **93:**9651–9656.

Canfield, C. J., M. Pudney, and W. E. Gutteridge. 1995. Interactions of atovaquone with other antimalarial drugs against Plasmodium falciparum in vitro. *Exp. Parasitol.* **80:**373–381.

Carrington, H. C., A. F. Crowther, D. G. Davey, A. A. Levi, and F. L. Rose. 1951. A metabolite of "Paludrine" with high antimalarial activity. *Nature* **168:**1080.

Clough, B., and R. J. Wilson. 2001. Antibiotics and the plasmodial plastid organelle, p. 265–286. *In* P. J. Rosenthal (ed.), *Antimalarial Chemotherapy: Mechanisms of*

Action, Resistance, and New Directions in Drug Discovery. Human Press, Totowa, N.J.

Creasey, A., K. Mendis, J. Carlton, D. Williamson, I. Wilson, and R. Carter. 1994. Maternal inheritance of extrachromosomal DNA in malaria parasites. *Mol. Biochem. Parasitol.* **65:**95–98.

Creasey, A. M., L. C. Ranford-Cartwright, D. J. Moore, D. H. Williamson, R. J. Wilson, D. Walliker, and R. Carter. 1993. Uniparental inheritance of the mitochondrial gene cytochrome b in Plasmodium falciparum. *Curr. Genet.* **23:**360–364.

Crofts, A. R. 2004. The cytochrome bc1 complex: function in the context of structure. *Annu. Rev. Physiol.* **66:**689–733.

Crowther, A. F., and A. A. Levi. 1953. Proguanil—the isolation of a metabolite with high antimalarial activity. *Br. J. Pharmacol.* **8:**93–97.

Cundliffe, E. 1990. Recognition sites for antibiotics within rRNA, p. 479–490. *In* W. Hill, A. Dahlberg, R. A. Garrett, P. B. Moore, D. Schlessinger, and J. R. Warner (ed.), *The Ribosome: Structure, Function, and Evolution.* ASM Press, Washington, D.C.

Dyall, S. D., and P. J. Johnson. 2000. Origins of hydrogenosomes and mitochondria: evolution and organelle biogenesis. *Curr. Opin. Microbiol.* **3:**404–411.

Elmendorf, H. G., J. D. Bangs, and K. Haldar. 1992. Synthesis and secretion of proteins by released malarial parasites. *Mol. Biochem. Parasitol.* **52:**215–230.

Elmendorf, H. G., and K. Haldar. 1994. Plasmodium falciparum exports the Golgi marker sphingomyelin synthase into a tubovesicular network in the cytoplasm of mature erythrocytes. *J. Cell Biol.* **124:**449–462.

Eren, D. 2003. *Plasmodium falciparum* enzymes involved in redox balancing of nicotinamide nucleotides. Ph.D. thesis. Drexel University College of Medicine, Philadelphia, Pa.

Evans, D. R., and H. I. Guy. 2004. Mammalian pyrimidine biosynthesis: fresh insights into an ancient pathway. *J. Biol. Chem.* **279:**33035–33038.

Feagin, J. E., B. L. Mericle, E. Werner, and M. Morris. 1997. Identification of additional rRNA fragments encoded by the Plasmodium falciparum 6 kb element. *Nucleic Acids Res.* **25:**438–446.

Feagin, J. E., E. Werner, M. J. Gardner, D. H. Williamson, and R. J. Wilson. 1992. Homologies between the contiguous and fragmented rRNAs of the two Plasmodium falciparum extrachromosomal DNAs are limited to core sequences. *Nucleic Acids Res.* **20:**879–887.

Fichera, M. E., and D. S. Roos. 1997. A plastid organelle as a drug target in apicomplexan parasites. *Nature* **390:**407–409.

Fieser, L. F., E. Berliner, F. J. Bondhus, F. C. Chang, W. G. Dauben, M. G. Etlinger, G. Fawaz, G. Fields, M. Feiser, C. Heidelberger, H. Heymann, A. M. Seligman, W. R. Vaughan, A. G. Wilson, E. Wilson, M. I. Wu, M. T. Leffler, K. E. Hamlin, R. J. Hathaway, E. J. Matson, E. E. Moore, M. B. Moore, R. T. Rapala, and H. E. Zaugg. 1948. Naphthoquinone antimalarials. I-XVII. *J. Am. Chem. Soc.* **70:**3151–3244.

Fivelman, Q. L., G. A. Butcher, I. S. Adagu, D. C. Warhurst, and G. Pasvol. 2002. Malarone treatment failure and in vitro confirmation of resistance of Plasmodium falciparum isolate from Lagos, Nigeria. *Malar. J.* **1:**1.

Foth, B. J., and G. I. McFadden. 2003. The apicoplast: a plastid in Plasmodium falciparum and other apicomplexan parasites. *Int. Rev. Cytol.* **224:**57–110.

Foth, B. J., L. M. Stimmler, E. Handman, B. S. Crabb, A. N. Hodder, and G. I. McFadden. 2005. The malaria parasite Plasmodium falciparum has only one pyruvate dehydrogenase complex, which is located in the apicoplast. *Mol. Microbiol.* **55:**39–53.

Fox, R. I. 1998. Mechanism of action of leflunomide in rheumatoid arthritis. *J. Rheumatol. Suppl.* **53:**20–26.

Fry, M. 1991. Mitochondria of *Plasmodium*, p. 154–167. *In* G. H. N. Coombs and M. J. North (ed.), *Biochemical Protozoology.* Taylor and Francis, London, United Kingdom.

Fry, M., and J. E. Beesley. 1991. Mitochondria of mammalian Plasmodium spp. *Parasitology* **102:**17–26.

Fry, M., and M. Pudney. 1992. Site of action of the antimalarial hydroxynaphthoquinone, 2-[*trans*-4-(4′-chlorophenyl) cyclohexyl]-3-hydroxy-1,4-naphthoquinone (566C80). *Biochem. Pharmacol.* **43:**1545–1553.

Fry, M., E. Webb, and M. Pudney. 1990. Effect of mitochondrial inhibitors on adenosinetriphosphate levels in Plasmodium falciparum. *Comp. Biochem. Physiol. B* **96:**775–782.

Gao, X., X. Wen, L. Esser, B. Quinn, L. Yu, C. A. Yu, and D. Xia. 2003. Structural basis for the quinone reduction in the bc1 complex: a comparative analysis of crystal structures of mitochondrial cytochrome bc1 with bound substrate and inhibitors at the Qi site. *Biochemistry* **42:**9067–9080.

Gardner, M. J., P. A. Bates, I. T. Ling, D. J. Moore, S. McCready, M. B. Gunasekera, R. J. Wilson, and D. H. Williamson. 1988. Mitochondrial DNA of the human malarial parasite Plasmodium falciparum. *Mol. Biochem. Parasitol.* **31:**11–17.

Gardner, M. J., N. Hall, E. Fung, O. White, M. Berriman, R. W. Hyman, J. M. Carlton, A. Pain, K. E. Nelson, S. Bowman, I. T. Paulsen, K. James, J. A. Eisen, K. Rutherford, S. L. Salzberg, A. Craig, S. Kyes, M. S. Chan, V. Nene, S. J. Shallom, B. Suh, J. Peterson, S. Angiuoli, M. Pertea, J. Allen, J. Selengut, D. Haft, M. W. Mather, A. B. Vaidya, D. M. Martin, A. H. Fairlamb, M. J. Fraunholz, D. S. Roos, S. A. Ralph, G. I. McFadden, L. M. Cummings, G. M. Subramanian, C. Mungall, J. C. Venter, D. J. Carucci, S. L. Hoffman, C. Newbold, R. W. Davis, C. M. Fraser, and B. Barrell. 2002. Genome sequence of the human malaria parasite Plasmodium falciparum. *Nature* **419:**498–511.

Gassis, S., and P. K. Rathod. 1996. Frequency of drug resistance in *Plasmodium falciparum*: a nonsynergistic combination of 5-fluoroorotate and atovaquone suppresses in vitro resistance. *Antimicrob. Agents Chemother.* **40:**914–919.

Gazarini, M. L., and C. R. Garcia. 2004. The malaria parasite mitochondrion senses cytosolic Ca2+ fluctuations. *Biochem. Biophys. Res. Commun.* **321:**138–144.

Gillespie, D. E., N. A. Salazar, D. H. Rehkopf, and J. E. Feagin. 1999. The fragmented mitochondrial ribosomal RNAs of Plasmodium falciparum have short A tails. *Nucleic Acids Res.* **27:**2416–2422.

Gray, M. W., B. F. Lang, and G. Burger. 2004. Mitochondria of protists. *Annu. Rev. Genet.* **38:**477–524.

Green, D. R., and G. Kroemer. 2004. The pathophysiology of mitochondrial cell death. *Science* **305:**626–629.

Gutteridge, W. E., D. Dave, and W. H. Richards. 1979. Conversion of dihydroorotate to orotate in parasitic protozoa. *Biochim. Biophys. Acta* **582:**390–401.

Hatefi, Y., and M. Yamaguchi. 1996. Nicotinamide nucleotide transhydrogenase: a model for utilization of substrate binding energy for proton translocation. *FASEB J.* **10:**444–452.

Hunte, C., H. Palsdottir, and B. L. Trumpower. 2003. Protonmotive pathways and mechanisms in the cytochrome bc1 complex. *FEBS Lett.* **545:**39–46.

Iwata, S., J. W. Lee, K. Okada, J. K. Lee, M. Iwata, B. Rasmussen, T. A. Link, S. Ramaswamy, and B. K. Jap. 1998. Complete structure of the 11-subunit bovine mitochondrial cytochrome bc1 complex. *Science* **281:**64–71.

Ji, Y. E., B. L. Mericle, D. H. Rehkopf, J. D. Anderson, and J. E. Feagin. 1996. The Plasmodium falciparum 6 kb element is polycistronically transcribed. *Mol. Biochem. Parasitol.* **81:**211–223.

Joseph, J. T., S. M. Aldritt, T. Unnasch, O. Puijalon, and D. F. Wirth. 1989. Characterization of a conserved extrachromosomal element isolated from the avian malarial parasite Plasmodium gallinaceum. *Mol. Cell. Biol.* **9:**3621–3629.

Joy, D. A., X. Feng, J. Mu, T. Furuya, K. Chotivanich, A. U. Krettli, M. Ho, A. Wang, N. J. White, E. Suh, P. Beerli, and X. Z. Su. 2003. Early origin and recent expansion of Plasmodium falciparum. *Science* **300:**318–321.

Kairo, A., A. H. Fairlamb, E. Gobright, and V. Nene. 1994. A 7.1 kb linear DNA molecule of Theileria parva has scrambled rDNA sequences and open reading frames for mitochondrially encoded proteins. *EMBO J.* **13:**898–905.

Kilejian, A. 1975. Circular mitochondrial DNA from the avian malarial parasite Plasmodium lophurae. *Biochim. Biophys. Acta* **390:**276–284.

Kim, H., D. Xia, C. A. Yu, J. Z. Xia, A. M. Kachurin, L. Zhang, L. Yu, and J. Deisenhofer. 1998. Inhibitor binding changes domain mobility in the iron-sulfur protein of the mitochondrial bc1 complex from bovine heart. *Proc. Natl. Acad. Sci. USA* **95:**8026–8033.

Korsinczky, M., N. Chen, B. Kotecka, A. Saul, K. Rieckmann, and Q. Cheng. 2000. Mutations in *Plasmodium falciparum* cytochrome *b* that are associated with atovaquone resistance are located at a putative drug-binding site. *Antimicrob. Agents Chemother.* **44:**2100–2108.

Krishna, S., T. M. Davis, P. C. Chan, R. A. Wells, and K. J. Robson. 1988. Ciprofloxacin and malaria. *Lancet* **1:**1231–1232.

Krungkrai, J. 1991. Malarial dihydroorotate dehydrogenase mediates superoxide radical production. *Biochem. Int.* **24:**833–839.

Krungkrai, J. 2004. The multiple roles of the mitochondrion of the malarial parasite. *Parasitology* **129:**511–524.

Krungkrai, J. 1995. Purification, characterization and localization of mitochondrial dihydroorotate dehydrogenase in Plasmodium falciparum, human malaria parasite. *Biochim. Biophys. Acta* **1243:**351–360.

Krungkrai, J., A. Cerami, and G. B. Henderson. 1991. Purification and characterization of dihydroorotate dehydrogenase from the rodent malaria parasite Plasmodium berghei. *Biochemistry* **30:**1934–1939.

Krungkrai, J., A. Cerami, and G. B. Henderson. 1990. Pyrimidine biosynthesis in parasitic protozoa: purification of a monofunctional dihydroorotase from Plasmodium berghei and Crithidia fasciculata. *Biochemistry* **29:**6270–6275.

Lacy, M. D., J. D. Maguire, M. J. Barcus, J. Ling, M. J. Bangs, R. Gramzinski, H. Basri, P. Sismadi, G. B. Miller, J. D. Chulay, D. J. Fryauff, S. L. Hoffman, and J. K. Baird. 2002. Atovaquone/proguanil therapy for Plasmodium falciparum and Plasmodium vivax malaria in Indonesians who lack clinical immunity. *Clin. Infect. Dis.* **35:**e92–e95.

Lang-Unnasch, N. 1995. Plasmodium falciparum: antiserum to malate dehydrogenase. *Exp. Parasitol.* **80:**357–359.

Lauer, S., J. VanWye, T. Harrison, H. McManus, B. U. Samuel, N. L. Hiller, N. Mohandas, and K. Haldar. 2000. Vacuolar uptake of host components, and a role for cholesterol and sphingomyelin in malarial infection. *EMBO J.* **19:**3556–3564.

Li, J., J. A. Maga, N. Cermakian, R. Cedergren, and J. E. Feagin. 2001. Identification and characterization of a Plasmodium falciparum RNA polymerase gene with similarity to mitochondrial RNA polymerases. *Mol. Biochem. Parasitol.* **113:**261–269.

Lill, R., and G. Kispal. 2000. Maturation of cellular Fe-S proteins: an essential function of mitochondria. *Trends Biochem. Sci.* **25:**352–356.

Ling, J., J. K. Baird, D. J. Fryauff, P. Sismadi, M. J. Bangs, M. Lacy, M. J. Barcus, R. Gramzinski, J. D. Maguire, M. Kumusumangsih, G. B. Miller, T. R. Jones, J. D. Chulay, and S. L. Hoffman. 2002. Randomized, placebo-controlled trial of atovaquone/proguanil for the prevention of Plasmodium falciparum or Plasmodium vivax malaria among migrants to Papua, Indonesia. *Clin. Infect. Dis.* **35:**825–833.

Loffler, M., K. Grein, W. Knecht, A. Klein, and U. Bergjohann. 1998. Dihydroorotate dehydrogenase. Profile of a novel target for antiproliferative and immunosuppressive drugs. *Adv. Exp. Med. Biol.* **431:**507–513.

Looareesuwan, S., J. D. Chulay, C. J. Canfield, and D. B. Hutchinson. 1999a. Malarone (atovaquone and proguanil hydrochloride): a review of its clinical development for treatment of malaria. *Am. J. Trop. Med. Hyg.* **60:**533–541.

Looareesuwan, S., C. Viravan, H. K. Webster, D. E. Kyle, D. B. Hutchinson, and C. J. Canfield. 1996. Clinical studies of atovaquone, alone or in combination with other antimalarial drugs, for treatment of acute uncomplicated malaria in Thailand. *Am. J. Trop. Med. Hyg.* **54:**62–66.

Looareesuwan, S., P. Wilairatana, K. Chalermarut, Y. Rattanapong, C. J. Canfield, and D. B. Hutchinson. 1999b. Efficacy and safety of atovaquone/proguanil compared with mefloquine for treatment of acute Plasmodium falciparum malaria in Thailand. *Am. J. Trop. Med. Hyg.* **60:**526–532.

Looareesuwan, S., P. Wilairatana, R. Glanarongran, K. A. Indravijit, L. Supeeranontha, S. Chinnapha, T. R. Scott, and J. D. Chulay. 1999c. Atovaquone and proguanil hydrochloride followed by primaquine for treatment of Plasmodium vivax malaria in Thailand. *Trans. R. Soc. Trop. Med. Hyg.* **93:**637–640.

Mai, Z., S. Ghosh, M. Frisardi, B. Rosenthal, R. Rogers, and J. Samuelson. 1999. Hsp60 is targeted to a cryptic mitochondrion-derived organelle ("crypton") in the microaerophilic protozoan parasite Entamoeba histolytica. *Mol. Cell. Biol.* **19:**2198–2205.

McIntosh, M. T. 1995. Biological significance of discontinuous and scrambled ribosomal RNA genes in the mitochondria of malarial parasites. Ph.D. thesis. Medical College of Pennsylvania and Hahnemann University, Philadelphia.

McIntosh, M. T., R. Srivastava, and A. B. Vaidya. 1998. Divergent evolutionary constraints on mitochondrial and nuclear genomes of malaria parasites. *Mol. Biochem. Parasitol.* **95:**69–80.

McMillan, P. J., L. M. Stimmler, B. J. Foth, G. I. McFadden, and S. Muller. 2005. The human malaria parasite Plasmodium falciparum possesses two distinct dihydrolipoamide dehydrogenases. *Mol. Microbiol.* **55:**27–38.

Muhlenhoff, U., and R. Lill. 2000. Biogenesis of iron-sulfur proteins in eukaryotes: a novel task of mitochondria that is inherited from bacteria. *Biochim. Biophys. Acta* **1459:**370–382.

Murphy, A. D., J. E. Doeller, B. Hearn, and N. Lang-Unnasch. 1997. Plasmodium falciparum: cyanide-resistant oxygen consumption. *Exp. Parasitol.* **87:**112–120.

Ossorio, P. N., L. D. Sibley, and J. C. Boothroyd. 1991. Mitochondrial-like DNA sequences flanked by direct and inverted repeats in the nuclear genome of Toxoplasma gondii. *J. Mol. Biol.* **222:**525–536.

Perez-Martinez, X., A. Antaramian, M. Vazquez-Acevedo, S. Funes, E. Tolkunova, J. d'Alayer, M. G. Claros, E. Davidson, M. P. King, and D. Gonzalez-Halphen. 2001. Subunit II of cytochrome c oxidase in chlamydomonad algae is a heterodimer encoded by two independent nuclear genes. *J. Biol. Chem.* **276:**11302–11309.

Povinelli, L., T. A. Monson, B. C. Fox, M. E. Parise, J. M. Morrisey, and A. B. Vaidya. 2003. Plasmodium vivax malaria in spite of atovaquone/proguanil (malarone) prophylaxis. *J. Travel Med.* **10:**353–355.

Preiser, P. R., R. J. Wilson, P. W. Moore, S. McCready, M. A. Hajibagheri, K. J. Blight, M. Strath, and D. H. Williamson. 1996. Recombination associated with replication of malarial mitochondrial DNA. *EMBO J.* **15:**684–693.

Rehkopf, D. H., D. E. Gillespie, M. I. Harrell, and J. E. Feagin. 2000. Transcriptional mapping and RNA processing of the Plasmodium falciparum mitochondrial mRNAs. *Mol. Biochem. Parasitol.* **105:**91–103.

Roberts, C. W., F. Roberts, F. L. Henriquez, D. Akiyoshi, B. U. Samuel, T. A. Richards, W. Milhous, D. Kyle, L. McIntosh, G. C. Hill, M. Chaudhuri, S. Tzipori, and R. McLeod. 2004. Evidence for mitochondrial-derived alternative oxidase in the apicomplexan parasite Cryptosporidium parvum: a potential anti-microbial agent target. *Int. J. Parasitol.* **34:**297–308.

Sato, S., B. Clough, L. Coates, and R. J. Wilson. 2004. Enzymes for heme biosynthesis are found in both the mitochondrion and plastid of the malaria parasite Plasmodium falciparum. *Protist* **155:**117–125.

Sato, S., and R. J. Wilson. 2002. The genome of Plasmodium falciparum encodes an active delta-aminolevulinic acid dehydratase. *Curr. Genet.* **40**: 391–398.

Seeber, F. 2002a. Biogenesis of iron-sulphur clusters in amitochondriate and apicomplexan protists. *Int. J. Parasitol.* **32**:1207–1217.

Seeber, F. 2002b. Eukaryotic genomes contain a [2Fez.sbnd;2S] ferredoxin isoform with a conserved C-terminal sequence motif. *Trends Biochem. Sci.* **27**:545–547.

Sherman, I. W. 1979. Biochemistry of *Plasmodium* (malarial parasites). *Microbiol. Rev.* **43**:453–495.

Sherman, I. W. 1998. Carbohydrate metabolism of asexual stages, p. 135–144. *In* I. W. Sherman (ed.), *Malaria: Parasite Biology, Pathogenesis, and Protection.* ASM Press, Washington, D.C.

Sogin, M. L. 1991. Early evolution and the origin of eukaryotes. *Curr. Opin. Genet. Dev.* **1**:457–463.

Srivastava, I. K., J. M. Morrisey, E. Darrouzet, F. Daldal, and A. B. Vaidya. 1999. Resistance mutations reveal the atovaquone-binding domain of cytochrome b in malaria parasites. *Mol. Microbiol.* **33**:704–711.

Srivastava, I. K., H. Rottenberg, and A. B. Vaidya. 1997. Atovaquone, a broad spectrum antiparasitic drug, collapses mitochondrial membrane potential in a malarial parasite. *J. Biol. Chem.* **272**:3961–3966.

Srivastava, I. K., and A. B. Vaidya. 1999. A mechanism for the synergistic antimalarial action of atovaquone and proguanil. *Antimicrob. Agents Chemother.* **43**:1334–1339.

Su, X. Z., J. Mu, and D. A. Joy. 2003. The "Malaria's Eve" hypothesis and the debate concerning the origin of the human malaria parasite Plasmodium falciparum. *Microbes Infect.* **5**:891–896.

Suplick, K., J. Morrisey, and A. B. Vaidya. 1990. Complex transcription from the extrachromosomal DNA encoding mitochondrial functions of *Plasmodium yoelii*. *Mol. Cell. Biol.* **10**:6381–6388.

Suraveratum, N., S. R. Krungkrai, P. Leangaramgul, P. Prapunwattana, and J. Krungkrai. 2000. Purification and characterization of Plasmodium falciparum succinate dehydrogenase. *Mol. Biochem. Parasitol.* **105**:215–222.

Suzuki, T., T. Hashimoto, Y. Yabu, Y. Kido, K. Sakamoto, C. Nihei, M. Hato, S. Suzuki, Y. Amano, K. Nagai, T. Hosokawa, N. Minagawa, N. Ohta, and K. Kita. 2004. Direct evidence for cyanide-insensitive quinol oxidase (alternative oxidase) in apicomplexan parasite Cryptosporidium parvum: phylogenetic and therapeutic implications. *Biochem. Biophys. Res. Commun.* **313**:1044–1052.

Syafruddin, D., J. E. Siregar, and S. Marzuki. 1999. Mutations in the cytochrome b gene of Plasmodium berghei conferring resistance to atovaquone. *Mol. Biochem. Parasitol.* **104**:185–194.

Tachezy, J., L. B. Sanchez, and M. Muller. 2001. Mitochondrial type iron-sulfur cluster assembly in the amitochondriate eukaryotes Trichomonas vaginalis and Giardia intestinalis, as indicated by the phylogeny of IscS. *Mol. Biol. Evol.* **18**:1919–1928.

Takashima, E., S. Takamiya, S. Takeo, F. Mi-ichi, H. Amino, and K. Kita. 2001. Isolation of mitochondria from Plasmodium falciparum showing dihydroorotate dependent respiration. *Parasitol. Int.* **50**:273–278.

Takeo, S., A. Kokaze, C. S. Ng, D. Mizuchi, J. I. Watanabe, K. Tanabe, S. Kojima, and K. Kita. 2000. Succinate dehydrogenase in Plasmodium falciparum mitochondria: molecular characterization of the SDHA and SDHB genes for the catalytic subunits, the flavoprotein (Fp) and iron-sulfur (Ip) subunits. *Mol. Biochem. Parasitol.* **107**:191–205.

Tripathi, A. K., P. V. Desai, A. Pradhan, S. I. Khan, M. A. Avery, L. A. Walker, and B. L. Tekwani. 2004. An alpha-proteobacterial type malate dehydrogenase may complement LDH function in Plasmodium falciparum. Cloning and biochemical characterization of the enzyme. *Eur. J. Biochem.* **271**: 3488–3502.

Uyemura, S. A., S. Luo, S. N. Moreno, and R. Docampo. 2000. Oxidative phosphorylation, Ca^{2+} transport, and fatty acid-induced uncoupling in malaria parasites mitochondria. *J. Biol. Chem.* **275**: 9709–9715.

Uyemura, S. A., S. Luo, M. Vieira, S. N. Moreno, and R. Docampo. 2004. Oxidative phosphorylation and rotenone-insensitive malate- and NADH-quinone oxidoreductases in Plasmodium yoelii yoelii mitochondria in situ. *J. Biol. Chem.* **279**:385–393.

Vaidya, A. B. 2004. Mitochondrial and plastid functions as antimalarial drug targets. *Curr. Drug Targets Infect. Disord.* **4**:11–23.

Vaidya, A. B. 1998. Mitochondrial physiology as a target for atovaquone and other antimalarials, p. 355–368. *In* I. W. Sherman (ed.), *Malaria: Parasite Biology, Pathogenesis, and Protection.* ASM Press, Washington, D.C.

Vaidya, A. B., R. Akella, and K. Suplick. 1989. Sequences similar to genes for two mitochondrial proteins and portions of ribosomal RNA in tandemly arrayed 6-kilobase-pair DNA of a malarial parasite. *Mol. Biochem. Parasitol.* **35**:97–107.

Vaidya, A. B., and P. Arasu. 1987. Tandemly arranged gene clusters of malarial parasites that are highly conserved and transcribed. *Mol. Biochem. Parasitol.* **22**:249–257.

Vaidya, A. B., and M. W. Mather. 2000. Atovaquone resistance in malaria parasites. *Drug Resist Updates* **3**:283–287.

Vaidya, A. B., M. S. Lashgari, L. G. Pologe, and J. Morrisey. 1993a. Structural features of Plasmodium cytochrome b that may underlie susceptibility to 8-aminoquinolines and hydroxynaphthoquinones. *Mol. Biochem. Parasitol.* **58**:33–42.

Vaidya, A. B., J. Morrisey, C. V. Plowe, D. C. Kaslow, and T. E. Wellems. 1993b. Unidirectional dominance of cytoplasmic inheritance in two genetic crosses of Plasmodium falciparum. *Mol. Cell. Biol.* **13:**7349–7357.

Vanacova, S., D. R. Liston, J. Tachezy, and P. J. Johnson. 2003. Molecular biology of the amitochondriate parasites, Giardia intestinalis, Entamoeba histolytica and Trichomonas vaginalis. *Int. J. Parasitol.* **33:**235–255.

Vander Jagt, D. L., L. A. Hunsaker, N. M. Campos, and B. R. Baack. 1990. D-Lactate production in erythrocytes infected with Plasmodium falciparum. *Mol. Biochem. Parasitol.* **42:**277–284.

Varadharajan, S., S. Dhanasekaran, Z. Q. Bonday, P. N. Rangarajan, and G. Padmanaban. 2002. Involvement of delta-aminolaevulinate synthase encoded by the parasite gene in de novo haem synthesis by Plasmodium falciparum. *Biochem. J.* **367:**321–327.

Varadharajan, S., B. K. Sagar, P. N. Rangarajan, and G. Padmanaban. 2004. Localization of ferrochelatase in Plasmodium falciparum. *Biochem. J.* **384:**429–436.

Vercesi, A. E., C. F. Bernardes, M. E. Hoffmann, F. R. Gadelha, and R. Docampo. 1991. Digitonin permeabilization does not affect mitochondrial function and allows the determination of the mitochondrial membrane potential of Trypanosoma cruzi in situ. *J. Biol. Chem.* **266:**14431–14434.

Wagner, J. T., H. Ludemann, P. M. Farber, F. Lottspeich, and R. L. Krauth-Siegel. 1998. Glutamate dehydrogenase, the marker protein of Plasmodium falciparum—cloning, expression and characterization of the malarial enzyme. *Eur. J. Biochem.* **258:**813–819.

Williamson, D. H., R. J. Wilson, P. A. Bates, S. Mc-Cready, F. Perler, and B. U. Qiang. 1985. Nuclear and mitochondrial DNA of the primate malarial parasite Plasmodium knowlesi. *Mol. Biochem. Parasitol.* **14:**199–209.

Wilson, C. M., A. B. Smith, and R. V. Baylon. 1996. Characterization of the delta-aminolevulinate synthase gene homologue in P. falciparum. *Mol. Biochem. Parasitol.* **75:**271–276.

Wilson, R. J. 2002. Progress with parasite plastids. *J. Mol. Biol.* **319:**257–274.

Wilson, R. J., and D. H. Williamson. 1997. Extrachromosomal DNA in the Apicomplexa. *Microbiol. Mol. Biol. Rev.* **61:**1–16.

Wyler, D. J. 1989. Fluoroquinolones for malaria: the newest kid on the block? *Ann. Intern. Med.* **111:**269–271.

Xia, D., C. A. Yu, H. Kim, J. Z. Xia, A. M. Kachurin, L. Zhang, L. Yu, and J. Deisenhofer. 1997. Crystal structure of the cytochrome bc1 complex from bovine heart mitochondria. *Science* **277:**60–66.

Xu, P., G. Widmer, Y. Wang, L. S. Ozaki, J. M. Alves, M. G. Serrano, D. Puiu, P. Manque, D. Akiyoshi, A. J. Mackey, W. R. Pearson, P. H. Dear, A. T. Bankier, D. L. Peterson, M. S. Abrahamsen, V. Kapur, S. Tzipori, and G. A. Buck. 2004. The genome of Cryptosporidium hominis. *Nature* **431:**1107–1112.

Yagi, T., B. B. Seo, S. Di Bernardo, E. Nakamaru-Ogiso, M. C. Kao, and A. Matsuno-Yagi. 2001. NADH dehydrogenases: from basic science to biomedicine. *J. Bioenerg. Biomembr.* **33:**233–242.

Yu, C. A., H. Tian, L. Zhang, K. P. Deng, S. K. Shenoy, L. Yu, D. Xia, H. Kim, and J. Deisenhofer. 1999. Structural basis of multifunctional bovine mitochondrial cytochrome bc1 complex. *J. Bioenerg. Biomembr.* **31:**191–199.

Zhang, Z., E. A. Berry, L. S. Huang, and S. H. Kim. 2000. Mitochondrial cytochrome bc1 complex. *Subcell. Biochem.* **35:**541–580.

TRAFFICKING AND THE TUBULOVESICULAR MEMBRANE NETWORK[†]

Kasturi Haldar, Narla Mohandas, Souvik Bhattacharjee, Travis Harrison, N. Luisa Hiller, Konstantinos Liolios, Sean Murphy, Pamela Tamez, and Christiaan van Ooij

13

The blood stages of malarial infection are responsible for all disease symptoms and pathologies as well as death due to malaria. In these stages, malaria parasites infect reticulocytes and erythrocytes. The most virulent of human malaria parasites, *Plasmodium falciparum*, infects mature erythrocytes. Erythrocytic infection is important to the virulence of *P. falciparum* relative to *Plasmodium vivax*, another major human malaria parasite that is widespread but less virulent and restricted to reticulocytes. Masamichi Aikawa initiated studies of a wide range of *Plasmodium* spp. in the 1960s and made seminal contributions to our understanding of the ultrastructure of malarial parasite entry into the erythrocytes as well as intracellular infection. After William Trager and colleagues established an in vitro system for culturing *P. falciparum*, Aikawa and his collaborators investigated the complexities of antigenic and morphological changes inflicted by this virulent parasite on terminally differentiated mature human erythrocyte.

The mature erythrocyte is a specialized cell whose major function is to deliver oxygen to tissues and, in the process, survive transport through the capillaries. The high concentrations of hemoglobin, prominent solute and ion transport systems, and a well-developed and highly deformable submembrane cytoskeleton all reflect the erythrocyte's specialized functions. The mature human erythrocyte is also a terminally differentiated host cell that lacks a nucleus and other intracellular organelles, is incapable of de novo protein or lipid biosynthesis, and is apparently devoid of endocytic machinery (Chasis et al., 1989; Schrier, 1985). Thus, entry and intracellular survival of a pathogen in the erythrocyte presumably require the activation of membrane trafficking pathways in this host (in marked contrast to infection of macrophages

---

Kasturi Haldar, Souvik Bhattacharjee, Travis Harrison, N. Luisa Hiller, Konstantinos Liolios, Sean Murphy, Pamela Tamez, and Christiaan van Ooij, Departments of Pathology and Microbiology-Immunology, Northwestern University, 303 E. Chicago Avenue, Chicago IL 60611. *Narla Mohandas,* New York Blood Center, 310 East 67th Street, New York, NY 10021.

[†]This chapter is dedicated to the memory of Masamichi Aikawa for his seminal contributions to our understanding of the ultrastructure of malarial parasite entry into the erythrocytes as well as intracellular infection which guided early concepts and continue to influence our thinking of this unique host-pathogen niche. He remained active until his death on 13 April 2004, which brought great sadness and loss to the entire parasitology community worldwide.

Molecular Approaches to Malaria, Edited by Irwin W. Sherman
© 2005 ASM Press, Washington, D.C.

where avoidance of fusion with degradative, phagolysosomal pathways in general is critical for the pathogen's survival) (Haas, 1998).

Not unexpectedly then, erythrocytes are infected by few pathogens. Recently, *Bartonella* has emerged as a powerful system for identifying prokaryotic virulence determinants that mediate red blood cell infection (Schulein and Dehio, 2002). However, the best understanding of the mechanisms of erythrocytic infection still comes from *P. falciparum*. As elaborated elsewhere in this volume, this organism causes the most virulent form of human malaria, a disease that afflicts almost 300 million people worldwide and kills over 2 million children each year. *P. falciparum* has a robust in vitro culture system, and its genome is sequenced. Although classical genetic approaches are still limited, reverse genetics can readily be applied to this microorganism.

A critical step in plasmodial entry and subsequent survival is the establishment of a parasitophorous vacuolar membrane (PVM), which is initiated upon invasion, surrounds the parasite during intracellular growth, and manifests complex structural and transport functions at different intraerythrocytic stages (Aikawa et al., 1978; Atkinson and Aikawa, 1990; Bannister and Dluzewski, 1990). This chapter will focus on our current understanding of (i) erythrocyte signaling mechanisms and trafficking of host receptors and parasite ligands implicated in PVM formation, (ii) transport functions of a tubulovesicular membrane network (TVN) extending from the PVM to the erythrocyte, and (iii) a secretome of ~300 parasite proteins exported to the erythrocyte, which indicates vast complexity in structural and antigenic remodeling of the host milieu required for virulence and parasite survival in the erythrocyte.

ERYTHROCYTE DETERGENT-RESISTANT MEMBRANE RAFTS AND G PROTEIN SIGNALING IN MALARIAL INFECTION

Plasmodial entry into erythrocytes is a complex, dynamic process (Dvorak et al., 1975). The apically oriented, invading merozoite-stage parasite interacts with the erythrocyte to form a parasite-host cell junction. Invagination of the erythrocyte membrane results in engulfment of the parasite and establishment of the intracellular ring-stage parasite surrounded by the PVM (Aikawa et al., 1978; Bannister et al., 2000). It is well established that entry consists of multiple steps mediated by parasite proteins that reside on the surface of the merozoite and within its apical organelles; some adhere to erythrocytes upon secretion from the parasite (Chitnis, 2001; Cowman and Crabb, 2002; Holder et al., 1994). In contrast, little is known of how the erythrocyte regulates infection even as it undergoes active deformation, convolution, and involution (Dvorak et al., 1975). A significant hurdle to a mechanistic understanding of the process was that major erythrocyte membrane proteins failed to be internalized into the nascent or developing PVM. However, our studies (Harrison et al., 2003; Lauer et al., 2000; Samuel et al., 2001; Murphy et al., 2004) and those of Dvorak's group (Nagao et al., 2002) have recently shown that although major erythrocyte proteins are excluded, proteins resident in erythrocyte detergent-resistant membrane (DRM) rafts are trafficked to the malarial vacuole. Rafts are lateral heterogeneities in the bilayer, frequently conceived of as clusters or floating islands of proteins and lipids held together by a cholesterol-rich microenvironment (Brown and London, 1998a; Friedrichson and Kurzchalia, 1998; Simons and Ikonen, 1997; Varma and Mayor, 1998). At least a subset of raft proteins can be isolated as highly buoyant and cholesterol-rich DRMs, which can be separated from other detergent-resistant cytoskeletal components by density gradient centrifugation in a 5 to 40% sucrose gradient. While some authors question the significance of DRMs, multiple studies (summarized by Brown and London, 1998b) indicate that DRMs do contain aggregated rafts. Moreover, nonraft proteins and lipids are not recruited into DRMs during their isolation (Ahmed et al., 1997). Therefore, proteins in DRMs must have specialized requirements for the cholesterol-rich raft environment. Thus, DRMs have provided a powerful tool to identify as many as

10 erythrocyte proteins that are recruited to the plasmodial vacuole (Color Plate 9) (summarized in Murphy et al., 2004).

Proteomics of isolated erythrocyte DRMs indicate that these complexes contain only a few percent of the total red blood cell membrane protein (Samuel et al., 2001; Murphy et al., 2004). This may explain why earlier efforts failed to detect host proteins in the malarial vacuole. Initial studies investigating the role of host rafts in malarial infection showed that depletion of modest amounts of erythrocyte cholesterol (~23%) led to abrogation of DRM rafts and parasite entry. In addition, four resident proteins of erythrocyte DRM rafts were detected in the PVM. Whether or not this vacuolar transport reflected selective uptake of host DRM proteins was unknown. A further complication was that DRMs of vastly different protein and cholesterol contents have been isolated from erythrocytes. We limited our analysis to isolated DRMs containing the highest cholesterol-to-protein ratio with low protein mass comprising only <0.1% of the total erythrocyte membrane protein. Structural studies indicated that the major resident proteins of these complexes were band 3, flotillin 1 and flotillin 2, and stomatin (Table 1) and in conjunction with antibody studies revealed an additional 12 minor DRM proteins (Table 2). Microscopy studies showed that band 3 and stomatin, which reflect at least half the bulk mass of erythrocyte DRM proteins and all non-DRM proteins, were excluded from the *P. falciparum* vacuole (Table 2). In contrast, flotillins 1 and 2 and as many as eight minor host DRM proteins are recruited to the vacuolar parasite (Table 2). Thus, DRM association appeared necessary but not sufficient for vacuolar recruitment. Moreover, there is active vacuolar uptake of a small mass fraction of host DRM proteins and heterogeneity in molecular interactions of host DRM raft proteins during parasite infection. As indicated earlier, 10 internalized DRM proteins showed varied lipid and

TABLE 1 Erythrocyte raft proteins identified by LC-MS/MS[a]

Gel band	High-confidence assignments		Tentative assignment	
	Protein	No. of peptides	Protein	No. of peptides
250	Band 3	14	Glucose transporter 1	2
120	Band 3	21	Protein 4.2 (pallidin)	1
			S100β	1
			Peroxiredoxin 2	1
50	Flotillin 1	16	S100β	1
	Flotillin 2	11	Peroxiredoxin 2	1
	Glucose transporter 1	3		
45	Peroxiredoxin 2	7	GAPDH	2
	Band 3	6	Stomatin	1
32	Stomatin	13	S100β	2
30	Stomatin	5	Carbonic anhydrase II	2
	Carbonic anhydrase I	4	S100β	1
			Peroxiredoxin 2	1
28	Flotillin 1	4	Glucose transporter 1	2
	Stomatin	4	Flotillin 2	2
	S100β	3	Aquaporin 1	1
	Peroxiredoxin 2	3		
15	α-Globin	6		
	β-Globin	14		

[a]In addition to the reported protein identifications, β-globin and sometimes α-globin were detected in all bands analyzed. For simplicity, these results were omitted from the table. Protein identifications were considered high-confidence assignments if the TurboSEQUEST browser report matched three or more tryptic peptides with a cross-correlation score of ≥2.5. Identifications were considered tentative if only one or two peptides were matched with cross-correlation scores of ≥2.5. The number of peptides meeting these criteria is shown. GAPDH, glyceraldehyde-3-phosphate dehydrogenase.

TABLE 2 Summary of erythrocyte lipid raft proteins[a]

Protein	Molecular weight (kDa)	Membrane association	Remarks	Internalized to the PVM?
α-Globin	15	Cytoplasmic	Hemoglobin complex	ND
β-Globin	16	Cytoplasmic	Hemoglobin complex	ND
GAPDH	36	Cytoplasmic	Conversion of G3P to 1,3-BPG	Not internalized
Peroxiredoxin 2	21.9	Cytoplasmic	Eliminates peroxides; signaling?	ND
S100β	10.5	Cytoplasmic	Dimer with α chain; binds p53, tubulin, Ca^{2+}; is assembly of microtubules, filaments	ND
CA-I		Cytoplasmic	Reversible hydration of CO_2	ND
Flotillin 1	47	Endofacing hairpin loop	Organization of caveolae and/or lipid rafts	Internalized
Flotillin 2	45	Endofacing hairpin loop	High-order fotillin oligomers; raft scaffolding component	Internalized
Stomatin	31	Endofacing hairpin loop	Associates with Glu1; cation transport	Not internalized
Gαs	44	Endofacing	GPCR activation of adenylate cyclase	Internalized
CD55	55–70	GPI linked	Decay accelerating factor	Internalized
CD58	64–73	GPI linked	Unknown in erythrocytes	Internalized
CD59	20–40	GPI linked	Membrane inhibitor of reactive complement lysis	Internalized
Glu11	54	Multipass (12)	May bind stomatin; passive glucose transport	Not internalized
Band 3	101	Multipass (14)	Binds protein 42 and ankyrin; Cl^- shift	Not internalized
Aquaporin 1	28	Multipass (6)	Water channel protein for erythrocytes and renal PCT	Internalized
β$_2$AR	65	Multipass (7)	Gαs-coupled receptor	Internalized
Duffy	35–43	Multipass (7)	Chemokine and P. vivax receptor; GPCR-like	Internalized
Scramblase	35	Single pass	Movement of membrane phospholipids	Internalized

[a]Nineteen proteins were reliably identified in the floating fraction of Tx-100-resistant erythrocyte membranes, but only a subset of these proteins enters the malarial vacuole. ND, no data; G3P, glyceraldehyde-3-phosphate; 1,3-BPG, 1-3, bisphosphoglycerate; GPCR, G protein-coupled receptor; PCT, proximal convoluted tubule; GAPDH, glyceraldehyde-3-phosphate dehydrogenase; β$_2$-AR, β2-adrenergic receptor. Numbers in parentheses represent the number of transmembrane domains in multipass proteins.

peptidic anchors, indicating that multiple proteins with transmembrane domains were internalized into the vacuole. Thus, contrary to earlier proposed models of apicomplexan vacuole formation (Mordue et al., 1999), DRM association, rather than lipid anchors, provides the preferred criteria for protein recruitment into the P. falciparum vacuole.

The recent explosive interest in DRM rafts has been due to finding that they contain heterotrimeric G proteins and tyrosine kinases on their cytoplasmic face; these proteins associate with rafts because they tend to be acylated (Moffett et al., 2000; Sargiacomo et al., 1993). Although erythrocytes are not highly active in signaling, they do possess signaling molecules (Minetti and Low, 1997), including heterotrimeric G_s, $G_{i/o}$, and G_q (Gilman, 1987). Only G_s is recruited to the malarial vacuole (Fig. 1) (Lauer et al., 2000), and peptides that disrupt G_s interaction with its receptor block malarial infection (Harrison et al., 2003; Lauer et al., 2000;

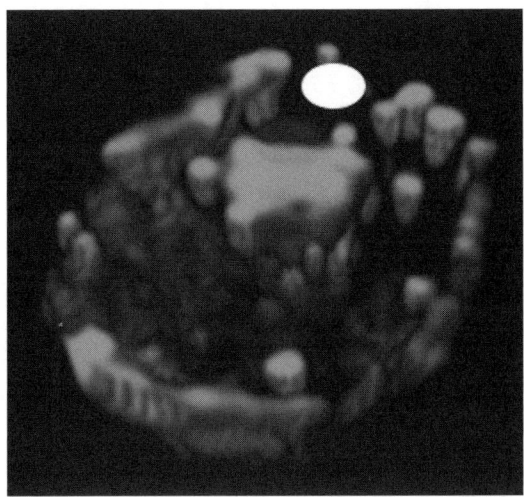

FIGURE 1 The image depicts a cross-sectional view through a three-dimensional reconstruction of signaling in a human erythrocyte upon infection by the malaria parasite *P. falciparum*. Distribution of host signaling heterotimeric Gαs protein is pseudocolored in light gray, and the parasite is represented by a white sphere.

Samuel et al., 2001). Further, agonists of G_s-coupled receptors, such as the β_2-adrenergic receptor and adenosine receptor, stimulate production of cyclic AMP (cAMP), showing that erythrocytes are capable of G-protein-coupled receptor (GPCR) signaling. Agonists stimulate malarial infection, while antagonists block both agonist-mediated enhancement of infection and production of signaling intermediates.

GPCR signaling regulates cytoskeletal reorganizations in many cell types (Etienne-Manneville and Hall, 2002; Vanhauwe et al., 2002). In malarial infection, erythrocyte G proteins may assist in clearing the erythrocyte skeleton, which could affect many erythrocyte membrane processes, including deformation and invagination, to form the vacuole during parasite entry. Activation of erythrocyte GPCRs that stimulate infection could occur in multiple ways. Bioinformatic studies failed to reveal the presence of parasite-encoded pathways for production of catecholamine-like stimulatory molecules (S. Murphy and K. Haldar, unpublished data). It is possible that the sympathetic response during malarial fevers may raise host catecholamines (epinephrine and norepinephrine) that activate host GPCRs to essentially "invite the parasite in." Regardless of how signaling is activated, drugs that dampen host signaling and reduce downstream vacuolar trafficking may protect against malarial infection by blocking vacuole formation.

Finally, although catecholamines activate the β_2-adrenergic receptor signaling to increase in cAMP in erythrocytes, this alone does not induce endovacuolar uptake or clustering of rafts in the erythrocyte membrane (B. Myles and K. Haldar, unpublished data). This latter step requires a stimulus from the parasite as well. Our working model proposes that parasite adhesive proteins couple to G_s-containing host raft complexes, all of which are internalized to the vacuole (Fig. 2).

In support of this model, we characterized a *P. falciparum* ortholog of stomatin (Pfstomatin), an important raft family protein in eukaryotes that belongs to the superfamily of flotillins (Hiller et al., 2003). Caveolins, which are the most actively studied of specialized membrane domain-forming proteins, are not present in either the host erythrocyte or the parasite. Nonetheless, by virtue of their ability to partition into cholesterol-rich domains and oligomerize, stomatins and flotillins are expected to concentrate microrafts into larger domains or platforms, thereby stabilizing and increasing the lifetime of signaling interactions on the cytoplasmic face of the platform (Anderson and Jacobson, 2002). Pfstomatin and RhopH2 (another parasite-encoded protein) reside in rhoptries, are shown to be DRM raft proteins, and insert into the newly forming host raft-enriched (and G_s-enriched) vacuole (Hiller et al., 2003). Since rafts show self-associative properties and form lipid shells (or platforms) that signal (Anderson and Jacobson, 2002), we think that interactions of parasite rafts and host G_s-containing rafts influence signaling, at least to the extent of concentrating signaling at the junction of invagination and the nascent vacuole (Fig. 2). In vitro studies suggest that catecholamines (which are abundant during malarial fevers, the time of parasite emergence

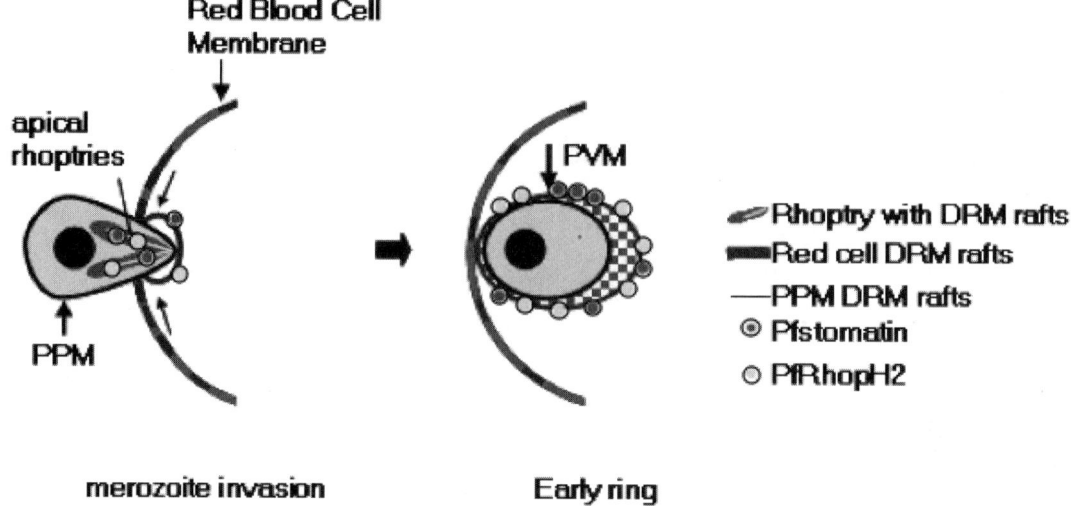

FIGURE 2 Proposed model for recruitment of parasite and host DRM rafts during PVM formation. The PVM contains proteins originating from rhoptry DRM rafts, as well as from G_s-containing erythrocyte DRM rafts. Moreover, Pfstomatin localizes to the cytoplasmic face of the vacuolar DRM rafts, with a topology consistent for that required for raft-induced plasma membrane invagination in cells. The cytoplasmic location of RhopH2 suggests that additional parasite protein complexes may contribute to raft-based vacuole biogenesis in infected erythrocytes.

and reinvasion into red blood cells) induce cAMP production and increase red blood cell flexibility, filterability, and deformation (Oonishi et al., 1997; Tuvia et al., 1998; Tuvia et al., 1999). In theory, the high catecholamine window of malarial fevers along with parasite ligands may alter membrane functions of the target erythrocyte and contribute to infection.

ROLE FOR THE TVN IN THE UPTAKE OF ERYTHROCYTE RAFT PROTEINS; EXPORT OF PARASITE PROTEINS INTO THE ERYTHROCYTE

In eukaryotes, extracellular and cell surface components are rapidly internalized by the process of endocytosis. Here, mechanistic insights have come largely from studies in mammalian cells and have historically emphasized the concept of an endocytic vesicle. These vesicles form from invagination of the host cell membrane and deliver membrane-bound receptors and extracellular ligands to intracellular destinations: the best-studied are clathrin-coated vesicles involved in endocytosis for the transferrin and low-density lipoprotein cholesterol receptors (Wakeham et al., 2000; Ybe et al., 2000). However, uncoated processes are also known to contribute to import pathways (Nichols and Lippincott-Schwartz, 2001). Most prominent among these are caveolae (containing the protein caveolin), which are flask-shaped invaginations that have been implicated in the uptake of receptors, nutrients, and infectious agents (Conner and Schmid, 2003). Membrane and luminal components of clathrin-coated vesicles and caveolae and other noncoated vesicles internalized into cells subsequently interact with prominent tubulovesicular compartments such as recycling endosomes or late endosomes, to be either recycled back to the plasma membrane or further delivered to target intracellular destinations.

The *P. falciparum*-infected erythrocyte cytoplasm is rich in tubulovesicular intermediates. This is in contrast to uninfected erythrocytes that are devoid of membranous organelles, indicating that the endovacuolar membranes developing in the intraerythrocytic space are induced by the parasite. Yet coated vesicles and their component proteins are not prominent in the intraerythrocytic milieu. Rather, uncoated tubules emerging from the PVM into the erythrocyte have been detected by advanced digitized fluorescence microscopy (Fig. 3A) (Elmendorf and Haldar, 1994; Grellier et al., 1991; Haldar et al., 1989). Large, smooth vesicular structures are also seen and a general consensus has now evolved for the existence of an interconnected network of tubular and vesicular membranes (or TVN). The TVN emerges from the PVM and extends into the erythrocyte cytoplasm. While it clearly grows during asexual parasite development in the red blood cell, the major structural elements of interconnected tubules and vesicles do not undergo rapid movement within seconds or minutes. Some studies have reported the presence of rapidly mobile vesicular elements in the intraerythocytic space (Hibbs and Saul, 1994), but their relationship to the TVN is not known.

Despite their low mobility, TVN tubules contain erythrocyte raft proteins whose import is restricted if tubule development is arrested by blocking a parasite-encoded resident sphingomyelin synthase in the network (Fig. 3B) (Lauer et al., 1997), first detected in early studies of plasmodial Golgi dynamics (Elmendorf and Haldar, 1994). Using a combination of bioinformatics and complementation in yeast, mammalian genes for sphinogmyelin synthase were recently identified (Huitema et al., 2004). Database analyses revealed two plasmodial orthologs of sphingomyelin synthase, both of which are predicted to be expressed in blood stages (Bozdech et al., 2003; Le Roch et al., 2003). This is consistent with early biochemical data on the presence of two plasmodial synthases, one of which is released into intraerythrocytic membranes (Elmendorf and Haldar, 1994) and inhibited by low (0.5 to 5 μM) concentrations of D-threo sphingolipid analogs PPMP (1-phenyl-2-palmitoylamino-3-morpholino-1-propanol) and PDMP (1-phenyl-2-decanoylamino-3-morpholino-1-propanol) (Lauer et al., 1995), which in higher eukaryotes block glucosylceramide synthase. Inhibiting this parasite activity with concentrations of 1 to 5 μM PPMP retarded development of intraerythrocytic tubules and decreased accumulation of erythrocyte raft

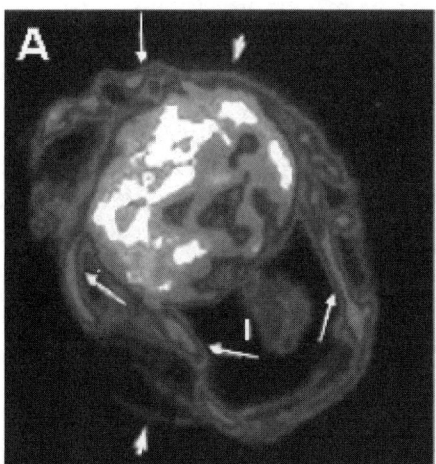 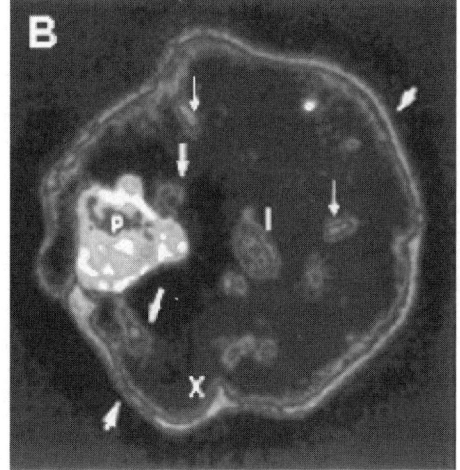

FIGURE 3 Development of the *P. falciparum* PVM-TVN and its inhibition by treatment with PPMP (Lauer et al., 1997). Grayscale images of control infected (A) and PPMP-treated infected (B) cells were labeled with Bodipy-ceramide (shown). Large arrowheads indicate periphery of the red blood cell; p, parasite; l, large vesicle or loop. (A) Long thin arrows show interconnected tubules of a normal TVN. (B) Arrested TVN. Short thin arrows indicate isolated rods; short thick arrows indicate vesicular structures attached to the parasite; x indicates TVN elements attached to the red blood cell.

proteins during intraerythrocytic growth (Lauer et al., 2000). This treatment also blocked uptake of nutrient solutes into the TVN (Lauer et al., 1997), suggesting that the network provides a transport conduit for host cell surface proteins, as well as extracellular solutes. Subsequent studies should reveal transport intermediates at junctions of interaction between the TVN and the erythrocyte membrane and the mechanisms by which they regulate import of host raft proteins, as well as nutrient solutes, into the infected erythrocyte.

Although blocking tubular development in the TVN retards solute and host protein import, it does not inhibit the export of endogenous parasite proteins to the erythrocyte (Lauer et al., 1997). This suggests that the TVN may not be essential for protein delivery to the red blood cell and that mechanisms for importing erythrocyte proteins and extracellular nutrients are distinct from those that underlie parasite protein export to the red blood cell. However, recent studies show that antigens such as a parasite histidine-rich protein II (PfHRPII) tagged with green fluorescent protein (GFP) associate with complex tubulovesicular structures in the erythrocyte cytoplasm (Color Plate 10A) (Lopez-Estrano et al., 2003). These studies reveal an unexpected complexity in TVN membranes, as they mediate both import and export pathways across the intraerythrocytic space. To better understand how they relate to intraerythrocytic structures defined by electron microscopy, we have reviewed in brief below how the assignment of function of malarial intraerythrocytic structures has evolved over the last 3 decades.

By definition, intraerythrocytic structures are referred to as membrane-bound elements that are detected in infected erythrocytes and induced by malarial infection. They were first detected by transmission electron microscopy (Atkinson and Aikawa, 1990; Bannister and Dluzewski, 1990; Langreth et al., 1978), and this was greatly facilitated (as was all of malaria research) by the establishment of an in vitro culture system for *P. falciparum* (Trager and Jensen, 1976). Prominent among intraeythrocytic structures were flattened lamellae called Maurer's clefts (also called short clefts) and large circular double-membrane loops (also referred to as circular clefts and antigenically distinct from clefts) extending from the PVM into the erythrocyte cytoplasm of early ring-infected erythrocytes (Atkinson and Aikawa, 1990; Langreth et al., 1978). Both Maurer's clefts and loops persist through the trophozoite stage when electron-dense knobs that mediate cytoadherence to endothelial cells appear at the infected red blood cell membrane. Further structural and transport links between the vacuolar loops, Maurer's clefts, and smaller intraerythrocytic vesicles (which are also seen) remain woefully obscure. Nonetheless, each intraerythrocytic structure detected by electron microscopy shows distinctive structural and transport characteristics (as described below), and their relationship to tubules and vesicles of the TVN as imaged by high-resolution fluorescence microscopy is becoming less enigmatic.

Maurer's clefts resemble isolated Golgi cisternae, share antigenic determinants with mammalian Golgi (Blisnick et al., 2000; Li et al., 1995), are located in the periphery of the erythrocyte (frequently apposed to the host cell membrane), and are surrounded by an electron-dense coat. Multiple parasite proteins destined for release into the erythrocyte associate with clefts, suggesting that they are intermediates in transport to the red blood cell (Color Plate 10A). But whether Maurer's clefts sort proteins to different destinations in the red blood cell is still not known. A truncated form of parasite proteins is reported to associate with Maurer's clefts (Waterkyn et al., 2000): *P. falciparum* knob-associated HRP (PfKAHRP) or *P. falciparum* HRP-1, a variant (VAR) antigen (also known as *P. falciparum* erythrocyte membrane protein 1 or PfEMP1), and a third protein, PfEMP3 (Waterkyn et al., 1999). Since PfEMP1 is a ligand for cytoadherence and clustered in knobs containing PfKAHRP, this led to the suggestion that the cytoadherence complex may assemble in clefts. However, knockouts in PfEMP3 and KAHRP continue to transport PfEMP1 to the red blood cell surface (Wickham et al., 2001), suggesting that assembly of such a complex is not critical for delivery of PfEMP1 to the red blood cell. Alternatively, protein association with clefts may reflect their presence within the lumen of clefts from where they are released to the erythrocyte. Thus, clefts may provide a portal for protein exit proximal to the erythrocyte membrane, the target of many exported antigens. Recent studies with live cells demonstrate that truncated forms of parasite proteins tagged to green fluorescent protein can be delivered from the parasitophorous vacuole (PV) to the clefts without release into the erythrocyte cytoplasm, suggesting that there is a pathway of protein export from the lumen of the vacuole to the lumen of the clefts (Color Plate 10B) (Lopez-Estrano et al., 2003). Distribution of green fluorescence further suggests the presence of tubulovesicular intermediates that mediate transport from the vacuole to the clefts. By virtue of their protein cargo, it should be possible to detect these transport intermediates by electron microscopy and define a pathway from the vacuole to the clefts. However, comprehensive characterization of this pathway will require the identification of resident proteins of these structures that influence their morphology and transport properties, as well as their cargo. Transport structures that mediate protein export from the Maurer's clefts to the erythrocyte membrane also need definition at the structural and molecular levels.

Intraerythrocytic loops (or circular clefts) emerging from the PVM have received considerably less attention than the Maurer's clefts. Two major resident loop proteins are *P. falciparum* exported protein 1 (PfEXP1) and PfEXP2 (Fischer et al., 1998; Günther et al., 1991). They are also detected on the PVM, and PfEXP1 is a resident DRM raft protein, suggesting that lipid-raft-based interactions may be important to formation and/or function of these loops. But what their function may be remains elusive. To date, there is no strong electron microscopy evidence that transport intermediates connect loops to Maurer's clefts and thereby contribute in protein export to multiple destinations in the erythrocyte. One possible function for PfEXP1 is stabilization of the PVM by interaction with erythrocyte raft components, as well as possibly rhoptry raft proteins deposited into the newly formed vacuolar membrane. Paradoxically, while electron microscopy studies have pioneered high-resolution structure of intraerythrocytic Maurer's cleft and loop membranes, they have failed to reveal the presence of interconnected networks of tubulovesicular membranes. One reason for this is that membrane networks may not survive the extensive processing needed to embed samples in the resins used for electron microscopy. Thus, cryoimmunoelectron microscopy may provide better resolution of hitherto missing intermediates. In addition, the undulating nature of networks and the high concentration of the electron dense protein hemoglobin may require serial reconstructions and technical innovations to define subtle (and not so subtle) network connection in the infected erythrocyte by electron microscopy. Currently, a combination of fluorescence data and electron microscopy data is needed for maximal insights into structure/function of intraerythrocytic membranes. Protein in Maurer's clefts frequently colocalizes at the tips and periphery of labeled tubules (Color Plate 10A) (Lopez-Estrano et al., 2003). This suggests that loops and clefts may be domains of a TVN, that chimeric GFP proteins released in the PV are accessible to loops and vesicles, but that their movement to the Maurer's clefts requires additional transport signals present on parasite protein cargo released into the erythrocyte. The reasons for these selective barriers in transport are not known. But it reinforces the idea that tubulovesicular networks have multiple membrane domains and, like other eukaryotic organelles, display extensive domain structures (such as the trans-Golgi network or the endoplasmic reticulum [ER]) with significant functional separation of lumenal domains.

A SECRETOME OF VIRULENCE DETERMINANTS AND EXPORTED PARASITE PROTEINS

This structure predicts a major host-targeting pathway and high complexity in erythrocyte modification during malarial infection (Hiller et al., 2004). Proteins secreted from the malaria parasite to the host erythrocyte are responsible for many disease pathologies (including death) (Miller et al., 2002) and underlie structural and transport changes in the erythrocyte required for parasite survival (Aravind et al., 2003; Haldar et al., 2002). The most prominent exported proteins belong to antigen families such as subtelomeric variable open reading frame (STEVOR) and repetitive interspersed family (RIFIN) concentrated in Maurer's clefts. The family of PfEMP1s are responsible for both antigenic and adhesive changes in the infected erythrocyte (Rasti et al., 2004) and are virulence determinants in cerebral and placental malaria (Fried and Duffy, 1996; Rasti et al., 2004). In this regard, exported proteins are analogous to microbial effectors secreted to the host cell. In many prokaryotic pathogens, release of effector molecules into host cells is known to be mediated by highly developed secretion systems (Galan, 2001; Schulein and Dehio, 2002). These systems utilize specific signals and their cognate receptors and/or translocators to deliver effectors to target destinations in the host cell. They also facilitate identification of unknown effectors by virtue of their associated transport signals and describe comprehensive sets of a defined number of pathogenic effectors for a given secretion system. In contrast, plasmodial proteins that underlie multiple phenotypic modifications in the erythrocyte, as well as the number of exported proteins, remain unknown (Schulein and Dehio, 2002). Host-targeted proteins of *Toxoplasma gondii*, a related intravacuolar apicomplexan that can serve as a model for *Plasmodium*, also remain poorly defined. Thus, a major question in plasmodium-host interactions has been whether conserved transport signals target proteins from the parasite to the erythrocyte as part of a major host-targeting pathway and enable recognition of a wide range of proteins (a secretome) that present high-value candidate effectors of disease and infection.

Plasmodium is a eukaryote, and several studies have established that a cleavable ER-type signal sequence (SS) is necessary and sufficient for protein recruitment into the secretory pathway within the parasite, as well as release at the parasite plasma membrane and into the lumen of the PV (Cheresh et al., 2002; Waller et al., 2000; Wickham et al., 2001). Further, for the KAHRP (HRPI), as well as HRPII, cleavage of the ER-type signal sequence reveals a vacuolar transport sequence (VTS) that resides within the next 40 amino acids of each protein (Lopez-Estrano et al., 2003). The VTS is required to export a reporter like GFP from the lumen of the PV to the erythrocyte cytoplasm and must be exposed at the N terminus (Lopez-Estrano et al., 2003). Additional non-histidine-rich proteins exported to the erythrocyte also contain VTSs located within 60 amino acids downstream of the SS cleavage site, suggesting that VTSs from five different exported proteins, apparently unrelated in sequence, contain equivalent information with respect to protein export from the PV into the erythrocyte.

Utilization of an in silico pattern-finding program (Multiple Expectation Maximization for Motif Elucidation [MEME]) revealed the presence of a conserved amino acid signal in these five experimentally validated VTSs. Use of this MEME pattern in the motif alignment and search tool (MAST), followed by additional data optimization, revealed that 251 sequences in the *P. falciparum* database shared an amino acid pattern (linearly represented as $Rx_1SRI\text{-}LAEx_2x_3x_4$) immediately downstream of an ER-type SS. Positions 2 and 9 to 11 allow multiple amino acids but are not completely random and thus in the linearized motif are represented as x1-4 (see the logo in Fig. 4A). The output proteins containing this motif included a surprisingly large number of hypothetical proteins (Table 3) and known exported antigens (Haeggstrom et al., 2004; Kaviratne et al., 2002; Sam-Yellowe et al., 2004) not in the original input of five, such as the 119 membrane-bound RIFINs and 22 membrane-bound STEVORs, respectively. Finding RIFINs and STEVORs in the output was surprising, since the original input sequences came only from soluble exported proteins. Although a full-length STEVOR was

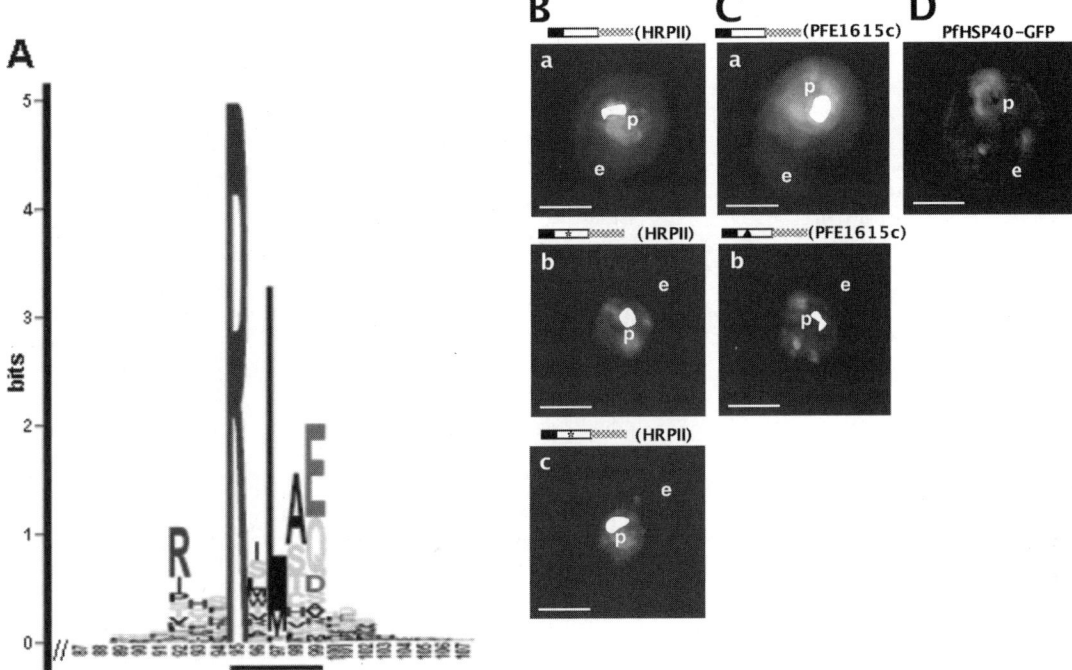

FIGURE 4 The VTS motif is a signal for vacuolar protein export and shows high value in identification of unknown parasite proteins exported to the erythrocyte. (A) Sequence logo derived from the predicted secretome. The height of the amino acids is proportional to the fraction of the observed frequency relative to the expected frequency in *P. falciparum* proteins. (B) Projections (0°) of live cells expressing GFP chimeras of SSHRPIIVTS with no change (a) or motif point mutants R4A (b), and L6A (c), where 4 and 6 indicate position 11 in the amino acid signal shown in the sequence logo; note the export of green fluorescence (GFP) (pseudocolored in gray) to erythrocyte (e) is abrogated in both point mutants. (C) Projections (0°) of live cells expressing GFP chimeras of SSVTSPFE1615c with intact motif (a), or replacement of motif sequences RILKQLE with LNAKALA (b); note that the replacement of motif residues blocks export of green fluorescence (pseudocolored in gray) to the erythrocyte. (D) Export of a secretome-predicted parasite HSP40 protein (PlasmoDB identification no. PFE0055C; GenBank no. NP_703357)-GFP chimera to the erythrocyte. Projection (0°) of a live cell expressing PfHSP40GFP. Images in panels B, C, and D were detected by digitized fluorescence microscopy. Parasite (p) nucleus is Hoechst stained (pseudocolored in white). The schematic above the panels indicates constructs containing signal sequence (black), VTS (white) of the indicated proteins, and GFP (gray). *, single amino acid substitutions; ▲, seven-residue replacement. Scale bar, 5 μm.

targeted to punctate domains in the erythrocyte (consistent with the location of STEVOR in Maurer's clefts), the first 60 amino acids downstream of STEVOR SS (but lacking the transmembrane domains) indeed functioned as a VTS and exported a GFP reporter to the erythrocyte cytosol. Hence, the motif showed equivalent function in both soluble and membrane antigens. Moreover, additional signals (possibly in the transmembrane domains) must localize full-length STEVORs to punctate cleft domains in the erythrocyte.

Representation of the 11-amino-acid MEME motif and its surrounding region as a logo (Gorodkin et al., 1997) provides pictographic depiction of the relative information content of each amino acid position in the motif. R in position 4 and L in position 6 are the most highly conserved residues. The lower but finite positional value of the three C-terminal residues

TABLE 3 Summary of predicted secretome (Hiller et al., 2004)

Category	No. of proteins	Significant characteristic(s)	Annotation source
Unknown	91	No annotation	None
Protein families	8	Range between 2 and 4 sequences from predicted secretome	This chapter
Protein rich in internal	10	Repeats	This chapter
Proteins highly enriched in ≥1 or more amino acids	11	At least 1 residue in highly enriched, overexpected frequencies	This chapter
RIFINS	119	Variant antigens	PlasmoDB
STEVORS	22	Variant antigens	PlasmoDB
Candidate phosphatases	3	Aminopeptodase, α, β-hydrolase	PlasmoDB; this chapter
		α, β-hydrolase	PlasmoDB; this chapter
		Calcineurin-like phosphoesterase	This chapter
Candidate serine-threonine kinases	3	3.8 protein	PlasmoDB
		Kinases	This chapter
		Kinases	This chapter
Predicted heat shock proteins	3	hsp40 homolog	PlasmoDB; this chapter
		DNAJ	PlasmoDB; this chapter
		DNAJ domain	PlasmoDB; this chapter
Predicted ADP-ribosylation factor	1	Missing 1 domain critical for GTP binding	PlasmoDB
Presence of glycophorin-binding repeats	3	GBP 130	PlasmoDB
		5 Repeats	PlasmoDB
		7 Repeats	PlasmoDB
Putative ABC transporter	1	Transporter	PlasmoDB
Proteins in the input	5	GBP 130	PlasmoDB
		HRPI	PlasmoDB
		HRPII	NCBI
		EMP2	PlasmoDB
		EMP3	PlasmoDB
PfEMP1	59	Absence of N-terminal SS; one sequence is a fragment	PlasmoDB

represented as xxx in the linearized motif RxSRILAExxx are also clearly seen in the logo (Fig. 4A). Replacing either position 4 or position 6 in the PfHRPII VTS with an A blocked export of GFP to the erythrocyte (Fig. 4B), indicating that the motif does indeed provide a signal for protein export from the vacuole to the erythrocyte. Inhibition of export upon single amino acid replacement implicates recognition of the signal by a receptor/transporter.

The predictive power of the motif in identifying unknown exported proteins was convincingly proven using a hypothetical protein in the MAST output that contained a functional export signal. The protein of choice (PFE1615c) was expressed during blood-stage infection (as determined by transcriptome analysis) (Bozdech et al., 2003; Le Roch et al., 2003) but contained no National Center for Biotechnology Information (NCBI) homologs or Pfam pattern recognized in silico; its motif, THSRILKQLEF, was distinct from LNKRLLHETQA, which was present and experimentally verified in PfHRPII. Nonetheless, PFE1615c was indeed found to contain a vacuolar translocation signal that can export GFP to the erythrocyte (Fig. 4Ca) and this export is blocked by replacement of RILKQLE in its motif with LNAKALA (Fig. 4Cb), confirming that the motif showed high predictive value in identifying hitherto-unknown parasite proteins exported to the erythrocyte.

Among the exported parasite protein families, the PfEMP1s have drawn the most attention, due to their involvement in cerebral and placental malaria and the phenomenon of rosetting (Miller et al., 2002). Instead of a leader SS, they contain an internal transmembrane domain, which presumably results in protein recruitment to the endoplasmic reticulum and serves as a membrane anchor (Su et al., 1995). Because they lack an SS, PfEMP1s were not included in the initial analysis of the motif. However, examination of the MEME motif in a data set that contained *P. falciparum* proteins lacking a predicted N-terminal ER-type signal sequence revealed two motifs, QFFRWFSEWSE and IGKRVHAQVQN, which are present in PfEMP1 sequences. IGKRVHAQVQN was present in only two proteins, while QFFRWFSEWSE was highly conserved in all PfEMP1s. Since a conserved export motif should be preserved in the family, subsequent analysis focused on QFFRWFSEWSE. Here again, it is useful to note that although QFFRWFSEWSE seems to differ from RxSRILAExxx, the latter is a linearized representation of a matrix motif that allows the residues in QFFRWFSEWSE at finite frequencies.

Unfortunately, due to their mass (250 to 350 kDa), full-length PfEMP1s cannot be cloned and expressed as transgenes. However, it was possible to synthesize and express a minitransgene containing the motif (which is present in full-length PfEMP1s beyond amino acid 200 but before the CIDRα domain) (Fig. 5), the conserved transmembrane, and C-terminal cytoplasmic domains (amino acids 1521 to 2042 of PfEMP1 AAB09769.1), but where most of the PfEMP1 adhesive domains were replaced by GFP (Fig. 5A). This minigene-GFP product was exported to spots (possibly reflecting Maurer's clefts) and diffuse areas in the periphery of the erythrocyte (Fig. 5Ba), but when FFRWFSEWS in the motif was replaced by AASTDIAST (Fig. 5Bb) or the N-terminal fragment was deleted (Fig. 5Bc), green fluorescence remained with the parasite. This identified the first, conserved sequence in

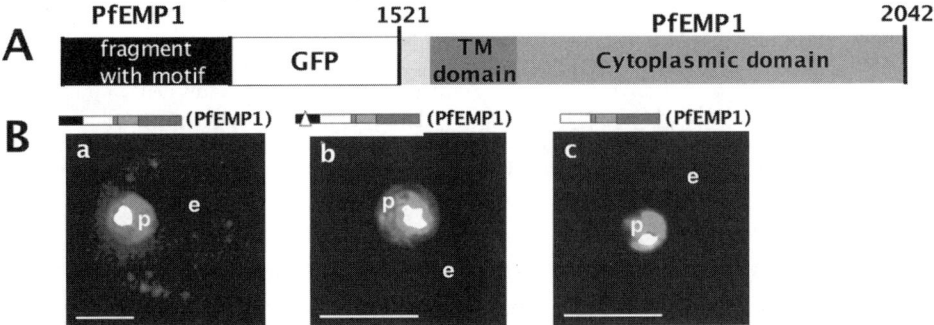

FIGURE 5 A host-targeting signal is detected in the virulence membrane protein PfEMP1. (A) Schematic representation of a PfEMP1 minigene-GFP chimera containing PfEMP1 domain (black) with the MEME motif, as well as transmembrane and cytosolic domain (amino acids 1521 to 2042 from PfEMP1 AAB09769.1) (grays). (B) Projections (0°) of live cells expressing the PfEMP1 minigene-GFP chimera with no change (a), replacement (open triangles) of FFRWFSEWS in signal by AASTDIAST (b), and deletion of the fragment spanning amino acids 203 to 321 (c), as detected by digitized fluorescence microscopy. Note the export of fluorescence (pseudocolored gray) to spots (which may correspond to erythrocyte clefts) and diffuse regions in the periphery of the erythrocyte (a). This export is blocked by either replacement of indicated residues in the predicted signal or loss of the fragment (containing the motif). Parasite (p) nucleus is Hoecht stained (pseudocolored in white). Open triangles, nine-residue replacement. Scale bar, 5 μm. Identifications shown are from PlasmoDB.

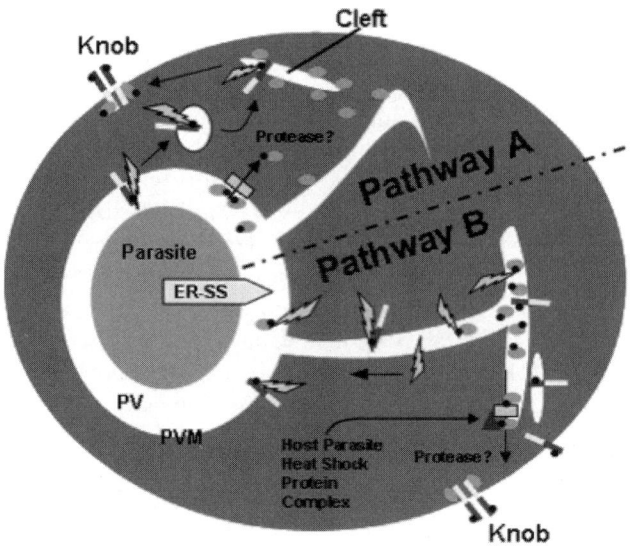

FIGURE 6 Pathways of protein export to the erythrocyte. Pathway A proposes that soluble proteins (gray oval) like PfHRPI, PfEMP3, or PfHRPII cross the PVM via a transporter (rectangle) and access the cytoplasmic face of clefts. Vesicles bearing PfEMP1 (white and gray bar) bind to a receptor (lightning bolt) bud from the PVM and are transported to clefts, where they assemble into complexes with PfHRPI and are delivered to knobs (Wickham et al., 2001), which are structures that mediate cytoadherence. A shortcoming of pathway A is that a distinct transporter and a receptor recognize the same export motif on soluble and transmembrane proteins (black dot): thus, soluble proteins could missort into vesicles, while PfEMP1 could block the transporter in the PVM. Further loss of HRPI or PfEMP3 has no effect on PfEMP1 export to the erythrocyte surface (Waterkeyn et al., 2000), and thus complex formation (of PfEMP1 and soluble proteins) in clefts is not required for cell surface expression. Pathway B proposes that a common receptor (lightning bolt) recognizes a host-targeting (HT) motif (black dot) on soluble proteins such as HRPI and HRPII (gray oval), as well as on membrane proteins like PfEMP1 (white and gray bar), and targets them to clefts. Here, soluble proteins exit (via a transporter shown as a rectangle) to the host cytosol proximal to the skeleton, and membrane proteins like PfEMP1 move to the erythrocyte surface (while others like STEVORs remain in clefts; not shown). Limitations of this pathway are as follows. (i) Membrane and soluble proteins share the same vacuolar export step, but experimental evidence shows that HT signals are equivalent in both (Hiller et al., 2004). (ii) For soluble proteins, the HT signal must target to clefts, as well as contribute to translocation from clefts to the erythrocyte cytoplasm (via a transporter), and thereby function in two distinct steps of vacuolar export. Heat shock proteins (dark triangle) and/or protease processing of leaders may greatly facilitate protein exit across the PV or clefts to red blood cell membrane or cytosol.

the virulence membrane protein PfEMP1 that can signal protein export to the erythrocyte, which is expected to occur without exporting the membrane protein into the erythrocyte cytosol. Most of the PfEMP1 adhesive domains do not appear to be essential for secretion from the vacuole. However, redundant signals may exist in these domains, well as the first 200 amino acids. The exact step of vacuolar export catalyzed in transmembrane proteins is not yet delineated, but the motif indicates a host-erythrocyte targeting signal in PfEMP1 transport. Thus, at a minimum the motif is a host-targeting signal (HTS) for both soluble and membrane proteins. Possible pathways for its action in export are discussed for the model shown in Fig. 6.

The presence of a functional signal motif in PfEMP1 suggests that the positional requirement that restricts the signal to the first 100 amino acids after the SS cleavage site may be unnecessarily restrictive. Further, that F and not L is encountered in position 6 of the export signal in PfEMP1 suggests the requirement for a large hydrophobic residue in position 6 of the signal. The dominant presence of L6 in the optimized MEME motif may reflect restriction in our initial small training set. The experimentally validated PfEMP1 signal QFFRWFSEWSE was incorporated into the MEME-MAST analysis to reiteratively and incrementally optimize the plasmodial host targeting tool and its associated pattern.

The HTS is detected in putative parasite phosphatases and kinases expressed during blood-stage infection (Table 3). This is consistent with previous data, suggesting export of these parasite activities to the erythrocyte (Kun et al., 1997). A putative parasite ADP-ribosylation factor (or ADP-ribosylation factor-like) protein is also a secretome effector, functional evaluation of which may reveal whether the parasite exports machinery underlying erythrocyte modification. Remarkably, a parasite gene-encoded heat shock protein (also identified by the plasmodial host-targeting tool) is exported to the erythrocyte (Fig. 4D), suggesting a role for these parasite chaperones in either the process of protein export and/or erythrocyte modification. The HTS is also present in 91 proteins that have no significant homologs in NCBI databases and lack all in silico annotation (Table 3). Some of these proteins represent small families, many are rich in repeats as well as specific amino acids, but the functional significance of these features remains speculative. Conservative estimates project that at least 50 of these unknown proteins are expressed during blood-stage infection, suggesting the possibility that the parasite induces significant, complex molecular changes in the erythrocyte of which we remain largely ignorant.

To further dwell on the significance of these results we present a comparative analysis in Fig. 7. Figure 7A shows an intracellular ring immediately after entry, as seen by transmission electron microscopy (the image shows an intracellular *P. knowlesi* parasite) (Aikawa et al., 1978). Figure 7B shows a trophozoite-infected erythrocyte decorated to indicate known exported antigens, as of last year. However, on the basis of current data, Fig. 7C is likely the best representation of virulence export and erythrocyte remodeling induced by malaria parasites! Importantly, the multitude of putative parasite-exported proteins provide high-value candidate effectors; future challenges lie in determining their function in virulence and disease associated with blood-stage malarial infection, as well as the ability of the parasite to survive within the erythrocyte.

The defined plasmodial HTS appears to be unique and distinct from signals currently known to drive export of proteins from prokaryotes or eukaryotes or required for protein translocation across organellar membranes (Schatz and Dobberstein, 1996). Nonetheless, plasmodial proteins containing this signal may at some low frequency be recognized by in silico programs designed to recognize known targeting sequences. PFE1615c and a parasite HSP40 homolog (Table 3) are also recognized by PlasmoAP (Foth et al., 2003); which predicts proteins targeted to the parasite apicoplast, but experimental validation reveals that both are exported to the erythrocyte (Fig. 4C and D). Additionally, there are known parasite proteins exported to the erythrocyte that do not contain an in silico recognizable export signal (Blisnick et al., 2000). Nonetheless, by virtue of the size of the motif's predicted secretome, presence in major virulence and antigenic proteins, and conservation in other plasmodial species (a very similar HTS was also identified in *Plasmodium yoelii*), it is likely to be a major export signal. The presence of a critical export signal implies its recognition by a dedicated receptor/transporter in delivering both soluble and membrane proteins from the vacuole to the erythrocyte. Future studies should reveal the nature of HTS recognition proteins, their location, and function relative to intraerythrocytic structures such as clefts and raft-rich TVN and PVM. The availability of a wide range of both host and parasite markers should enable development of a comprehensive analysis of mechanisms of membrane trafficking induced in erythrocytes upon malarial infection.

ACKNOWLEDGMENTS

Work in our laboratories was supported by NIH grants (AI39071 and HL69630 to K.H., and DK32094 to N.M.), the WHO TDR (K.H.), and the American Heart Association (N.L.H., S.M., and P.T.).

The illustration in Fig. 7C is by Kate Hiller.

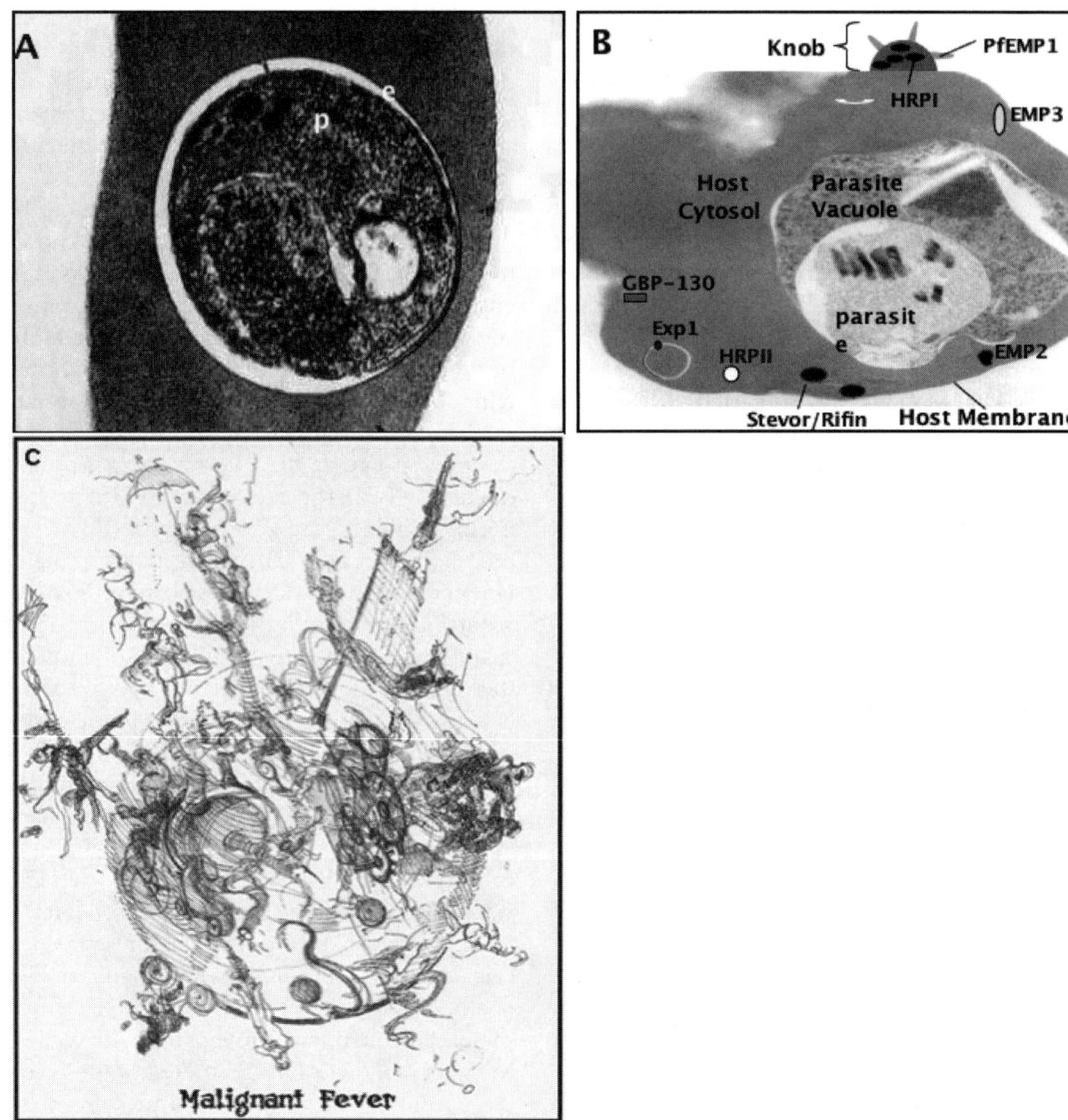

FIGURE 7 Evolving perspective on malaria parasite-host interactions in blood-stage infection. (A) Transmission electron micrograph of an intracellular ring-stage parasite immediately after merozoite entry. (B) Transmission electron micrograph of a trophozoite-infected erythrocyte decorated to indicate intraerythrocytic destinations of some known secreted effectors. (C) A bioinformatic prediction of the complexity of host-pathogen interactions in malarial infection. The illustration displays a human erythrocyte (large globe) infected by the malaria parasite *P. falciparum* (sphere within) releasing hundreds of proteins (demonic figures) to remodel the host cell and cause disease. (Fig. 7A is from Aikawa et al. [1978].)

REFERENCES

Ahmed, S. N., D. A. Brown, and E. London. 1997. On the origin of sphingolipid/cholesterol-rich detergent-insoluble cell membranes: physiological concentrations of cholesterol and sphingolipid induce formation of a detergent-insoluble, liquid-ordered lipid phase in model membranes. *Biochemistry* **36:**10944–10953.

Aikawa, M., L. H. Miller, J. Johnson, and J. Rabbege. 1978. Erythrocyte entry by malarial parasites. A moving junction between erythrocyte and parasite. *J. Cell Biol.* **77:**72–82.

Anderson, R. G., and K. Jacobson. 2002. A role for lipid shells in targeting proteins to caveolae, rafts, and other lipid domains. *Science* **296:**1821–1825.

Aravind, L., L. M. Iyer, T. E. Wellems, and L. H. Miller. 2003. Plasmodium biology: genomic gleanings. *Cell* **115:**771–785.

Atkinson, C. T., and M. Aikawa. 1990. Ultrastructure of malaria-infected erythrocytes. *Blood Cells* **16:**351–368.

Bannister, L. H., and A. R. Dluzewski. 1990. The ultrastructure of red cell invasion in malaria infections: a review. *Blood Cells* **16:**257–292.

Bannister, L. H., J. M. Hopkins, R. E. Fowler, S. Krishna, and G. H. Mitchell. 2000. A brief illustrated guide to the ultrastructure of Plasmodium falciparum asexual blood stages. *Parasitol. Today* **16:**427–433.

Blisnick, T., M. E. Morales Betoulle, J. C. Barale, P. Uzureau, L. Berry, S. Desroses, H. Fujioka, D. Mattei, and C. Braun Breton. 2000. Pfsbp1, a Maurer's cleft Plasmodium falciparum protein, is associated with the erythrocyte skeleton. *Mol. Biochem. Parasitol.* **111:**107–121.

Bozdech, Z., M. Llinas, B. L. Pulliam, E. D. Wong, J. Zhu, and J. L. DeRisi. 2003. The transcriptome of the intraerythrocytic developmental cycle of Plasmodium falciparum. *PLoS Biol.* **1:**E5.

Brown, D., and E. London. 1998a. Functions of lipid rafts in biological membranes. *Annu. Rev. Cell Dev. Biol.* **14:**111–136.

Brown, D. A., and E. London. 1998b. Structure and origin of ordered lipid domains in biological membranes. *J. Membr. Biol.* **164:**103–114.

Chasis, J. A., M. Prenant, A. Leung, and N. Mohandas. 1989. Membrane assembly and remodelling during reticulocyte maturation. *Blood* **74:**1112–1120.

Cheresh, P., T. Harrison, H. Fujioka, and K. Haldar. 2002. Targeting the malarial plastid via the parasitophorous vacuole. *J. Biol. Chem.* **277:**16265–16277.

Chitnis, C. E. 2001. Molecular insights into receptors used by malaria parasites for erythrocyte invasion. *Curr. Opin. Hematol.* **8:**85–91.

Conner, S. D., and S. L. Schmid. 2003. Regulated portals of entry into the cell. *Nature* **422:**37–44.

Cowman, A. F., and B. Crabb. 2002. The Plasmodium falciparum genome: a blueprint for erythrocyte adhesion. *Science* **298:**126–128.

Dvorak, J. A., L. H. Miller, W. C. Whitehouse, and T. Shiroshi. 1975. Invasion of erythrocytes by malaria merozoites. *Science* **187:**748–750.

Elmendorf, H. G., and K. Haldar. 1994. Plasmodium falciparum exports the Golgi marker sphingomyelin synthase into a tubovesicular network in the cytoplasm of mature erythrocytes. *J. Cell Biol.* **124:**449–462.

Etienne-Manneville, S., and A. Hall. 2002. Rho GTPases in cell biology. *Nature* **240:**629–635.

Fischer, K., T. Marti, B. Rick, D. Johnson, J. Benting, S. Baumeister, C. Helmbrecht, M. Lanzer, and K. Lingelbach. 1998. Characterization and cloning of the gene encoding the vacuolar membrane protein EXP-2 from Plasmodium falciparum. *Mol. Biochem. Parasitol.* **92:**47–57.

Foth, B. J., S. A. Ralph, C. J. Tonkin, N. S. Struck, M. Fraunholz, D. S. Roos, A. F. Cowman, and G. I. McFadden. 2003. Dissecting apicoplast targeting in the malaria parasite Plasmodium falciparum. *Science* **299:**705–708.

Fried, M., and P. E. Duffy. 1996. Adherence of Plasmodium falciparum to chondroitin sulfate A in the human placenta. *Science* **272:**1502–1504.

Friedrichson, T., and T. V. Kurzchalia. 1998. Microdomains of GPI-anchored proteins in living cells revealed by crosslinking. *Nature* **394:**802–804.

Galan, J. E. 2001. Salmonella interactions with host cells: type III secretion at work. *Annu. Rev. Cell Dev. Biol.* **17:**53–86.

Gilman, A. G. 1987. G protein transducers of receptor generated signals. *Annu. Rev. Biochem.* **56:**615–649.

Gorodkin, J., L. J. Heyer, S. Brunak, and G. D. Stormo. 1997. Displaying the information contents of structural RNA alignments: the structure logos. *Comput. Appl. Biosci.* **13:**583–586.

Grellier, P., D. Rigomier, V. Clavey, J.-C. Fruchart, and J. Schrevel. 1991. Lipid traffic between high-density lipoproteins and Plasmodium falciparum-infected red blood cells. *J. Cell Biol.* **112:**267–277.

Günther, K., M. Tümmler, H.-H. Arnold, R. Ridley, M. Gorman, J. G. Scaife, and K. Lingelbach. 1991. An exported protein of Plasmodium falciparum is synthesized as an integral membrane protein. *Mol. Biochem. Parasitol.* **46:**149–158.

Haas, A. 1998. Reprogramming the phagocytic pathway—intracellular pathogens and their vacuoles (review). *Mol. Membr. Biol.* **15:**103–121.

Haeggstrom, M., F. Kironde, K. Berzins, Q. Chen, M. Wahlgren, and V. Fernandez. 2004. Common trafficking pathway for variant antigens destined for the surface of the Plasmodium falciparum-infected erythrocyte. *Mol. Biochem. Parasitol.* **133:**1–14.

Haldar, K., A. F. de Amorin, and G. A. M. Cross. 1989. Transport of fluorescent phospholipid analogues from the erythrocyte membrane to the parasite in Plasmodium falciparum-infected red cells. *J. Cell Biol.* **108:**2183–2192.

Haldar, K., N. Mohandas, B. U. Samuel, T. Harrison, N. L. Hiller, T. Akompong, and P. Cheresh. 2002. Protein and lipid trafficking induced in erythrocytes infected by malaria parasites. *Cell. Microbiol.* **4:**383–395.

Harrison, T., B. U. Samuel, T. Akompong, H. Hamm, M. Narla, J. W. Lomasney, and K. Haldar. 2003. Erythrocyte G protein coupled receptor signaling in malaria infection. *Science* **301:**1734–1736.

Hibbs, A. R., and A. J. Saul. 1994. Plasmodium falciparum: highly mobile small vesicles in the malaria-infected red blood cell cytoplasm. *Exp. Parasitol.* **79:**260–269.

Hiller, N. L., T. Akompong, J. S. Morrow, A. A. Holder, and K. Haldar. 2003. Identification of a stomatin orthologue in vacuoles induced in human erythrocytes by malaria parasites. A role for microbial raft proteins in apicomplexan vacuole biogenesis. *J. Biol. Chem.* **278:**48413–48421.

Hiller, N. L., S. Bhattacharjee, C. van Ooij, K. Liolios, T. Harrison, C. Lopez-Estrano, and K. Haldar. 2004. A host-targeting signal in virulence proteins reveals a secretome in malarial infection. *Science* **306:**1934–1937.

Holder, A. A., M. J. Blackman, M. Borre, P. A. Burghaus, J. A. Chappel, J. K. Keen, I. T. Ling, S. A. Ogun, C. A. Owen, and K. A. Sinha. 1994. Malaria parasites and erythrocyte invasion. *Biochem. Soc. Trans.* **22:**291–295.

Huitema, K., J. van den Dikkenberg, J. F. Brouwers, and J. C. Holthuis. 2004. Identification of a family of animal sphingomyelin synthases. *EMBO J.* **23:**33–44.

Kaviratne, M., S. M. Khan, W. Jarra, and P. R. Preiser. 2002. Small variant STEVOR antigen is uniquely located within Maurer's clefts in Plasmodium falciparum-infected red blood cells. *Eukaryot. Cell* **1:**926–935.

Kun, J. F., A. R. Hibbs, A. Saul, D. J. McColl, R. L. Coppel, and R. F. Anders. 1997. A putative Plasmodium falciparum exported serinine/threonine kinase. *Mol. Biochem. Parasitol.* **85:**41–51.

Langreth, S. G., J. B. Jensen, R. T. Reese, and W. Trager. 1978. Fine structure of human malaria in vitro. *J. Protozool.* **25:**443–452.

Lauer, S., N. Ghori, and K. Haldar. 1995. Sphingolipid synthesis as a novel target for chemotherapy against malaria parasites. *Proc. Natl. Acad. Sci. USA* **92:**9181–9185.

Lauer, S. A., P. K. Rathod, N. Ghori, and K. Haldar. 1997. A membrane network for nutrient import in red cells infected with the malaria parasite. *Science* **276:**1122–1125.

Lauer, S. A., J. VanWye, T. Harrison, H. McManus, B. U. Samuel, N. L. Hiller, N. Mohandas, and K. Haldar. 2000. Vacuolar uptake of host components, and a role for cholesterol and sphingomyelin in malarial infection. *EMBO J.* **19:**3556–3564.

Le Roch, K. G., Y. Zhou, P. L. Blair, M. Grainger, J. K. Moch, J. D. Haynes, P. De La Vega, A. A. Holder, S. Batalov, D. J. Carucci, and E. A. Winzeler. 2003. Discovery of gene function by expression profiling of the malaria parasite life cycle. *Science* **301:**1503–1508.

Li, W., G. A. Keller, and K. Haldar. 1995. Recognition of a 170 kD protein in mammalian Golgi complexes by an antibody against malarial intraerythrocytic lamellae. *Tissue Cell* **27:**355–367.

Lopez-Estrano, C., S. Bhattacharjee, T. Harrison, and K. Haldar. 2003. Cooperative domains define a unique host cell-targeting signal in Plasmodium falciparum-infected erythrocytes. *Proc. Natl. Acad. Sci. USA* **100:**12402–12407.

Miller, L. H., D. I. Baruch, K. Marsh, and O. K. Doumbo. 2002. The pathogenic basis of malaria. *Nature* **415:**673–679.

Minetti, G., and P. S. Low. 1997. Erythrocyte signal transduction pathways and their possible functions. *Curr. Opin. Hematol.* **4:**116–121.

Moffett, S., D. A. Brown, and M. E. Linder. 2000. Lipid-dependent targeting of G proteins into rafts. *J. Biol. Chem.* **275:**2191–2198.

Mordue, D. G., N. Desai, M. Dustin, and L. D. Sibley. 1999. Invasion of Toxoplasma gondii establishes a moving junction that selectively excludes host cell plasma membrane proteins on the basis of their membrane anchoring. *J. Exp. Med.* **190:**173–192.

Nagao, E., K. B. Seydel, and J. A. Dvorak. 2002. Detergent-resistant erythrocyte membrane rafts are modified by a Plasmodium falciparum infection. *Exp. Parasitol.* **102:**57–59.

Nichols, B. J., and J. Lippincott-Schwartz. 2001. Endocytosis without clathrin coats. *Trends Cell Biol.* **11:**406–412.

Oonishi, T., K. Sakashita, and N. Uyesaka. 1997. Regulation of red blood cell filterability by Ca2+ influx and cAMP-mediated signaling pathways. *Am. J. Physiol.* **273:**C1828–C1834.

Rasti, N., M. Wahlgren, and Q. Chen. 2004. Molecular aspects of malaria pathogenesis. *FEMS Immunol. Med. Microbiol.* **41:**9–26.

Samuel, B. U., N. Mohandas, T. Harrison, H. McManus, W. Rosse, M. Reid, and K. Haldar. 2001. The role of cholesterol and glycosylphosphatidylinositol-anchored proteins of erythrocyte rafts in regulating raft protein content and malarial infection. *J. Biol. Chem.* **276:**29319–29329.

Sam-Yellowe, T. Y., L. Florens, J. R. Johnson, T. Wang, J. A. Drazba, K. G. Le Roch, Y. Zhou, S. Batalov, D. J. Carucci, E. A. Winzeler, and J. R. Yates III. 2004. A Plasmodium gene family encoding Maurer's cleft membrane proteins: structural properties and expression profiling. *Genome Res.* **14:**1052–1059.

Sargiacomo, M., M. Sudol, Z. Tang, and M. P. Lisanti. 1993. Signal transducing molecules and glycosyl-phosphatidylinositol-linked proteins form a caveolin-rich insoluble complex in MDCK cells. *J. Cell Biol.* **122:**789–807.

Schatz, G., and B. Dobberstein. 1996. Common principles of protein translocation across membranes. *Science* **271:**1519–1526.

Schrier, S. L. 1985. Red cell membrane biology—introduction. *Clin. Haematol.* **14:**1–12.

Schulein, R., and C. Dehio. 2002. The VirB/VirD4 type IV secretion system of Bartonella is essential for establishing intraerythrocytic infection. *Mol. Microbiol.* **46:**1053–1067.

Simons, K., and E. Ikonen. 1997. Functional rafts in cell membranes. *Nature* **387:**569–572.

Su, X.-Z., V. M. Heatwole, S. P. Wertheimer, F. Guinet, J. A. Herrfeldt, D. S. Peterson, J. A. Ravetch, and T. E. Wellems. 1995. The large diverse gene family var encodes proteins involved in cytoadherence and antigenic variation of Plasmodium falciparum-infected erythrocytes. *Cell* **82:**89–100.

Trager, W., and J. B. Jensen. 1976. Human malaria in continuous culture. *Science* **193:**673–675.

Tuvia, S., S. Levin, A. Bitler, and R. Korenstein. 1998. Mechanical fluctuations of the membrane-skeleton are dependent on F-actin ATPase in human erythrocytes. *J. Cell Biol.* **141:**1551–1561.

Tuvia, S., A. Moses, N. Gulayev, S. Levin, and R. Korenstein. 1999. Beta-adrenergic agonists regulate cell membrane fluctuations of human erythrocytes. *J. Physiol.* **516:**781–792.

Vanhauwe, J. F., T. O. Thomas, R. D. Minshall, C. Tiruppathi, A. Li, A. Gilchrist, E. J. Yoon, A. B. Malik, and H. E. Hamm. 2002. Thrombin receptors activate G_o proteins in endothelial cells to regulate intracellular calcium and cell shape changes. *J. Biol. Chem.* **277:**34143–34149.

Varma, R., and S. Mayor. 1998. GPI-anchored proteins are organized in submicron domains at the cell surface. *Nature* **394:**798–801.

Wakeham, D. E., J. A. Ybe, F. M. Brodsky, and P. K. Hwang. 2000. Molecular structures of proteins involved in vesicle coat formation. *Traffic* **1:**393–398.

Waller, R. F., M. B. Reed, A. F. Cowman, and G. I. McFadden. 2000. Protein trafficking to the plastid of Plasmodium falciparum is via the secretory pathway. *EMBO J.* **19:**1794–1802.

Waterkyn, J. G., B. S. Crabb, and A. F. Cowman. 1999. Transfection of the human malaria parasite Plasmodium falciparum. *Int. J. Parasitol.* **29:**945–955.

Wickham, M. E., M. Rug, S. A. Ralph, N. Klonis, G. I. McFadden, L. Tilley, and A. F. Cowman. 2001. Trafficking and assembly of the cytoadherence complex in Plasmodium falciparum-infected human erythrocytes. *EMBO J.* **20:**5636–5649.

Ybe, J. A., D. E. Wakeham, F. M. Brodsky, and P. K. Hwang. 2000. Molecular structures of proteins involved in vesicle fusion. *Traffic* **1:**474–479.

THE APICOPLAST

Stuart A. Ralph

14

The discovery and characterization of the apicoplast has been one of the success stories for the growing union of molecular, cellular, and genomic biology in parasitology. The combination of these three disciplines in a short space of time has shed much light on the origin, structure, biogenesis, and metabolism of the apicoplast.

THE APICOPLAST GENOME

The existence of a plastid in apicomplexans had previously been suggested by the discovery of a 35-kb plastid-like genome (Wilson et al., 1991), but the first link between genome and organelle was made through in situ electron microscopy (McFadden et al., 1996). The apicoplast had been observed by electron microscopy in many apicomplexa for decades, but its identity had remained anonymous (reviewed in McFadden et al., 1997). The apicoplast contains visible ribosomes, distinctly smaller than the eukaryotic cytosolic ribosomes, and is surrounded by multiple membranes (McFadden et al., 1996). While some controversy continues regarding the number of membranes surrounding the apicoplast, the best-preserved fixations of *Toxoplasma gondii* (Köhler et al., 1997) and *Garnia gonadati* (Diniz et al., 2000) show four limiting membranes.

Although the basic structure of the apicoplast genome is highly reminiscent of algal plastid genomes, its gene content tells us little about the function of the apicoplast. Of the 30 proteins encoded by the apicoplast genome, most are narcissistically involved with their own transcription and translation (Wilson et al., 1996b). A notable exception pointed the way to further understanding of the apicoplast. One of the apicoplast-encoded proteins, hsp93 (previously known as hsp100 or ClpC), is known to be involved in the import of nucleus-encoded proteins into plant plastids. Plants and algae import the great majority of their required plastid proteins from the cytosol, so was this also true for the Apicomplexa?

APICOPLAST-TARGETED PROTEINS

Shortly after the discovery of the apicoplast, preliminary sequences became available for both the *Plasmodium falciparum* genome and *T. gondii* expressed sequence tags (ESTs). A search by

---

Stuart A. Ralph, Biology of Host Parasite Interactions Unit, Institut Pasteur, Batiment Nicolle, 25 rue du Docteur Roux, 75724 Paris Cedex 15, France.

Molecular Approaches to Malaria, Edited by Irwin W. Sherman
© 2005 ASM Press, Washington, D.C.

Waller and colleagues revealed several nucleus-encoded genes whose products were known to be plastid located in plant and algal plastids. Examination of the N termini of each predicted protein revealed an extension of approximately 60 to 130 amino acids long not matching any other proteins. Attachment of this leader to a green fluorescent protein (GFP) reporter showed targeting to the apicoplast, and antibodies to nucleus-encoded proteins localized to the apicoplast by indirect fluorescence assays and immunoelectron microscopy (Waller et al., 1998). This experiment established the precept that nucleus-encoded genes held the key to understanding apicoplast function.

Dissection of the leader sequence revealed that it consisted of two distinct domains. The first resembles a classical signal peptide and is sufficient to direct a reporter protein into the secretory system where the signal peptide is cleaved. The second is similar in some aspects to the transit peptides used to target proteins to plant plastids. Attached to a reporter protein alone, this segment fails to bring about insertion into the endomembrane system and accumulates in the parasite cytosol (Waller et al., 2000). These results indicate that apicoplast targeting is via the secretory system. Equivalent experiments with *T. gondii* confirm these results (DeRocher et al., 2000; Yung et al., 2001). In a model based on these data (van Dooren et al., 2000), apicoplast proteins are cotranslationally inserted into the endomembrane system, which effects passage across the outermost membrane of the apicoplast. Here, a hypothetical Toc receptor, homologous to the Toc receptors in the outer membranes of plant chloroplasts, recognizes the transit peptide segment of the leader and initiates insertion through a series of Toc and Tic (homologous to inner chloroplast membrane proteins) components. Once inside the lumen of the apicoplast, the transit peptide is removed by a stromal processing peptidase (SPP) (van Dooren et al., 2002). Synthetic transit peptides and analysis of known chloroplast transit cleavages have identified consensus cleavage sites for this SPP in plant plastids. Despite experimental confirmation of several apicoplast transit peptide cleavage sites (Surolia and Surolia, 2001; van Dooren et al., 2002), no consensus is yet obvious in Apicomplexa. Experimental evidence from Harb and colleagues suggests biological recognition of this motif may not be very stringent. When the native cleavage site was deleted in reporter constructs, the targeted protein was instead cleaved at a number of other similarly redundant positions (Harb et al., 2004). Nonetheless, selective pressure presumably selects against strong recognition of cleavage sites within mature proteins.

The early characterization of the elements required for apicoplast targeting made possible the development of two bioinformatics tools to help identify additional possible apicoplast proteins. The first, called PATS (http://gecco.org.chemie.uni-frankfurt.de/pats/pats-index.php), is a neural network that was trained to blindly distinguish between presumed apicoplast proteins and nonapicoplast proteins (Zuegge et al., 2001). The second, named PlasmoAP (http://plasmodb.org/restricted/PlasmoAPcgi.shtml), is a rule-based system based on the characteristics of apicoplast transit peptides. The assumptions underlying these rules were tested through transfection of parasites with a series of GFP-targeting reporters and the results were used to improve the accuracy of the prediction tool. These experiments confirmed that the essential features of apicoplast transit peptides are enrichment in basic residues and depletion of acidic residues, particularly in the region immediately following the signal sequence (Foth et al., 2003). Mutational and bioinformatic analyses, as well as pull-down experiments, also support a role for chaperonin-binding sites in apicoplast transit peptides (Foth et al., 2003; Yung et al., 2003). The relatively low number of verified apicoplast proteins prevents us from estimating the predictive accuracy of the PlasmoAP or PATS tool, but both tools have been useful for screening the complete plasmodium genome for putative apicoplast proteins. A combination of these bioinformatics resources has been used to predict a putative apicoplast proteome of approximately 545 proteins, perhaps half the size of estimated nonredundant proteomes for photosynthetic plastids (Ralph et al., 2004).

APICOPLAST DIVISION

Organelle Division

The dissection of apicoplast targeting by Waller using green fluorescent protein, alongside similar constructs made in the laboratory of David Roos, inspired an ongoing series of broader targeting experiments in both *Toxoplasma* and *Plasmodium* that have revolutionized our understanding of intra- and extracellular trafficking in apicomplexans. In the shorter term, however, glowing green plastids made possible the easy observation of these organelles throughout *Plasmodium*'s intraerythrocytic life stages. The observation of apicoplasts in live cells showed that in *T. gondii* and *P. falciparum* at least, only one plastid is present per parasite, but every newly formed daughter cell faithfully obtains a plastid (Striepen et al., 2000; Waller et al., 2000). During division of *P. falciparum*, the apicoplast elongates and branches, assuming an exquisitely reticulated form that ramifies through the cell, eventually dividing to provide each daughter cell with its own apicoplast (Waller et al., 2000). Costaining with a live mitochondrial tracker showed that the apicoplast is in close apposition to the mitochondrion. In early stages, the mitochondrion is linked all along the side of the apicoplast, but in later stages the highly branched organelles form only several points of contact. This association is also visible in electron microscopy of chemically fixed parasites (Hopkins et al., 1999).

Given the complex division of *P. falciparum* asexual stages, it is noteworthy that each merozoite reliably receives a plastid. The mechanism behind this segregation has now been investigated by Striepen and colleagues with several apicomplexan species (Striepen et al., 2000). Using *T. gondii* parasites with apicoplast-localized GFP and a nuclear protein attached to red fluorescent protein (RFP), they showed that apicoplasts divide in association with nuclear division. Prior to division, the apicoplast genome divides and forms two nucleoid structures that move to opposite ends of the apicoplast. These ends are closely associated with the centrosomes (as observed with anti-centrin antibodies), and the apicoplast divides synchronously with the nucleus, using the mitotic spindle. The authors speculate that the stretching forces of the mitotic spindle, drawing the plastid over the newly forming pellicle, are sufficient to divide the apicoplast in two. Throughout the process, a nucleoid is maintained at each extremity of the apicoplast, close to the centrosomes, allowing reliable distribution of the apicoplast genome into each daughter apicoplast. The direct mechanism that positions the nucleoid is unclear (Striepen et al., 2000). An extrapolation of the same model could explain apicoplast division in *P. falciparum*, with the reticulate apicoplast associating with each of the mitotic spindles in the dividing parasite and at least one apicoplast genome associated with each centrosome. The complexity and small size of dividing *P. falciparum* parasites have so far made the visualization of such a mechanism difficult, but preliminary microscopic examination suggests that the dividing apicoplast is not strictly aligned with dividing daughter nuclei in early schizonts (Color Plate 11).

In contrast to this model, Matsuzaki and colleagues (2001) suggest that a plastid division ring divides the organelle, as is the case in plants and many algae. These authors present micrographs of apicoplasts with visible constriction zones, but oddly, the suggested constriction ring that is depicted is parallel to rather than perpendicular to the plane of division. The protein FtsZ forms the ring responsible for division in plants and algae, but no such gene is apparent in any apicomplexan genome. Several dynamin genes are present in the *P. falciparum* genome, and dynamin is known to be involved in mitochondrial division in animals, fungi, and land plants. Perhaps dynamins are also utilized for division of apicoplasts.

Apicoplast Genome Replication

Linear and circular forms of the apicoplast genome have been observed with both *P. falciparum* and *T. gondii* (Williamson et al., 2001; Williamson et al., 2002). In both organisms, the apicoplast genome is replicated by a rolling-circle mechanism, although curiously, the initiation site differs between the two species. The termination of this process is inefficient, leading to linear concatemers several genomes in length. In *P. falciparum*, an additional replication mechanism exists, initiating at twin displacement loops (Williamson et al., 2002). Such a mechanism would require resolution by the enzyme DNA gyrase, which works by cutting a double-stranded DNA molecule, passing the tangled lengths of DNA through the cut, and then resealing the cut to restore the circular architecture of the DNA. The fluoroquinolone compound ciprofloxacin interferes with the resealing step and results in linearization of the circular DNA, and ciprofloxacin does indeed inhibit displacement-loop replication in *P. falciparum* (Weissig et al., 1997; Williamson et al., 2002). Ciprofloxacin also inhibits replication of the *T. gondii* apicoplast genome (Fichera and Roos, 1997), but exactly how ciprofloxacin inhibits rolling-circle replication is less obvious.

Where Did the Apicoplast Come from?

From the earliest analyses of the apicoplast genome, there was speculation regarding its origin. When the circular 35-kb apicoplast genome was first noticed as an entity distinct from the nuclear DNA, it was naturally assumed to be the mitochondrial genome. However, molecular analyses soon showed that the genome's structure was similar to those of bacteria and chloroplasts, and phylogenetic analyses placed the origin of the genome among plastids (Gardner et al., 1991; Howe, 1992; Gozar and Bagnara, 1993). Plastids are organelles derived from the symbiosis of a cyanobacterium by a heterotrophic eukaryote, and all known examples retain a reduced cyanobacterial genome. Most of the cyanobacterial genes have moved to the eukaryotic nucleus, and their protein products must be redirected to the plastid, generally by means of an N-terminal transit peptide (Fig. 1). The best-known plastids are the double-membraned chloroplasts of land plants, but in fact, far more diverse are the plastids generated by secondary endosymbiosis—a process whereby a plastid-possessing eukaryote is engulfed and then retained by another eukaryote (Fig. 1).

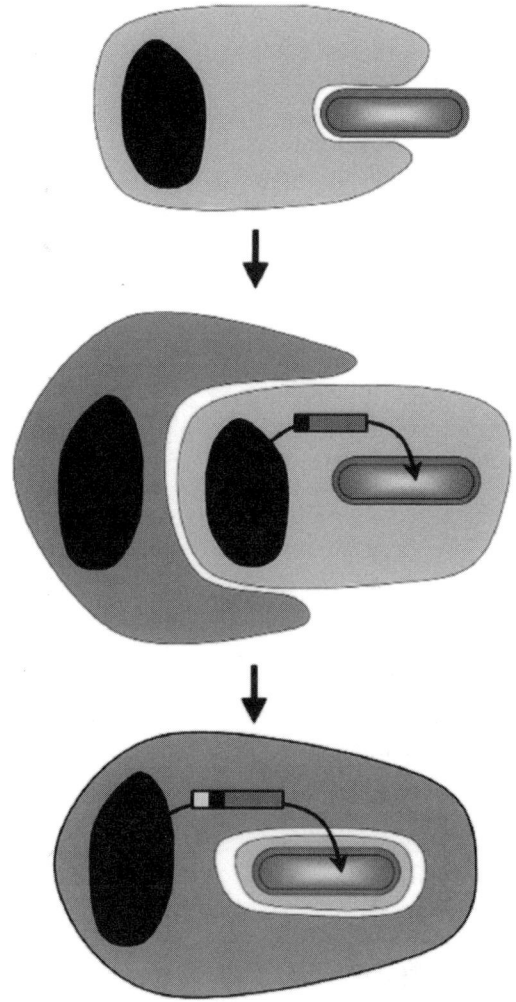

FIGURE 1 Apicoplasts arise through secondary endosymbiosis. In the primary endosymbiosis that created plastids, a cyanobacterium is engulfed by a heterotrophic eukaryote and retained in the eukaryotic cytosol. The transfer of genes from bacterium to host nucleus reinforces the dependence of the plastid on its new host. Most of the gene products that derive from these transferred genes are targeted back to the plastid with an N-terminal transit peptide. In secondary endosymbiosis, a plastid-bearing organism is itself engulfed and enslaved by an another eukaryote. As with primary endosymbionts, secondary endosymbiosis is characterized by large-scale gene transfer from endosymbiont to host. In most extant secondary endosymbionts, the nucleus of the engulfed eukaryote has completely disappeared through gene transfer and loss. The products of transferred genes are retargeted to the plastid with a bipartite leader consisting of signal sequence and transit peptide. The prey item that eventually became the apicoplast is unknown but is probably a red alga.

Genes residing in the nucleus of the engulfed eukaryote are gradually eliminated or relocated to the nucleus of the second. The protein products of these relocated genes must once again be redirected, and because the engulfed eukaryote lies topologically outside the predatory cell, such proteins are typically redirected by employing a secretory signal sequence in front of a transit peptide (Fig. 1). As we noted above, this is exactly the method employed by apicoplast proteins. The other telltale artifacts of secondary endosymbiosis are the additional membranes surrounding the plastid, and these were apparent from the earliest discoveries of the apicoplast. However, the algal cell that gave rise to the apicoplast remained to be determined, and opinions diverged between red algae (Williamson et al., 1994; McFadden et al., 1997) and green algae (Köhler et al., 1997). The high AT bias of the apicoplast genome gives rise to a highly divergent nucleotide sequence, which may have confounded some of these analyses (Blanchard and Hicks, 1999). One compelling analysis demonstrated that the operon organization of the apicoplast genome is highly reminiscent of red algal plastid genomes (McFadden and Waller, 1997).

The discovery by Waller of nuclear-encoded apicoplast genes opened a new avenue to address the question of apicoplast origins. The first such genes discovered show clear plastidic origin, but they do not group strongly with any specific plastid lineage (Waller et al., 1998). Another gene, glyceraldehyde-3-phosphate dehydrogenase (GAPDH), suggests a very interesting history that has gained increasing acceptance. Plastid-bearing organisms have two GAPDH genes, one cytosolic and the other servicing the plastid. Fast and colleagues noticed that in the Apicomplexa the plastid-targeted version is a duplicate of the cytosolic copy, well supported by highly conserved insertions (Fast et al., 2001). This character was also present in dinoflagellates, the sister group of the Apicomplexa (Wolters, 1991; Van de Peer and De Wachter, 1997), but was absent in plants, green algae, and red algae where the plastid version is evidently derived from the cyanobacterial ancestor of all plastids.

It has been established that dinoflagellates obtained their plastid from secondary endosymbiosis of a red alga, and the GAPDH data indicate that this acquisition preceded the divergence of the apicomplexans and dinoflagellates. A previous analysis of plastid 23S rRNA had suggested monophyly of the dinoflagellate and apicomplexan plastid, but with the caveat that the grouping may have been an artifact of long-branch attraction (Zhang et al., 2000). The duplicated GAPDH gene is also present in several other algal groups: heterokonts and cryptomonads. This supports the hypothesis of Cavalier-Smith (1999), who suggested that a single secondary endosymbiosis of a red alga gave rise to an enormous group that he named Chromalveolata. Chromalveolata unifies the group Chromista (heterokonts, cryptomonads, and haptophytes) with Alveolata (apicomplexans, dinoflagellates, and ciliates). A phylogeny based on four concatenated proteins supports the chromalveolate alliance (Baldauf et al., 2000), and other analyses have since supported the plastid monophyly of the chromista (Yoon et al., 2002; Harper and Keeling, 2003). However, a concatenated data set from 41 plastid proteins questions this monophyly, indicating separate plastid origins for two of the groups hypothesized to belong to the Chromalveolata, heterokonts and cryptomonads (Martin et al., 2002). More-extensive taxon sampling from within red algae as well as the hypothesized Chromalveolata might help to resolve this issue. The completion of a red algal genome (*Cyanidioschyzon merolae*) (Matsuzaki et al., 2004) also makes available a large data set to further address the question. The grouping of ciliates within Alveolata is robust, but despite the ciliates being well characterized (for example, *Paramecium* spp.), no plastid has yet been observed. If the chromalveolate hypothesis holds, ciliates may have secondarily lost their plastids.

There also appear to have been secondary plastid losses within the Apicomplexa. The most compelling case is for *Cryptosporidium*. *Cryptosporidium*'s phylogenetic placement has been debated, but it is likely an early branching apicomplexan. Tetley and colleagues reported a plastid-like organelle in electron micrographs, but the indicated organelles in the published plates are quite indistinct. No plastid membranes are apparent in the cryofixation, but the authors reported that the organelle is surrounded by a "double-membrane-sized halo" (Tetley et al., 1998). All apicoplasts observed to date possess at least three membranes, and probably four. An extensive effort to retrieve conserved regions of the apicoplast genome by PCR yielded nothing, and apicoplast-derived radiolabeled probes that hybridized to DNA of other apicomplexans were negative in *Cryptosporidium* (Zhu et al., 2000). Furthermore, *Cryptosporidium* is comparatively insensitive to numerous drugs that inhibit apicoplast function in other species. The complete sequence of the *Cryptosporidium* genome revealed no genes for apicoplast-targeted proteins (Abrahamsen et al., 2004; Xu et al., 2004) but did unearth several genes that appear to be of apicoplast origin, indicating that the organelle has been secondarily lost (Huang et al., 2004). The presence or absence of the apicoplast is unclear for several other apicomplexan species, but phylogenies support the plastid being an ancestral and widespread character in the Apicomplexa (Denny et al., 1998; Lang-Unnasch et al., 1998).

Wandering Plastid Genes

Most of the genes in the *P. falciparum* nuclear genome with strongest similarities to algal genes are hypothesized to encode apicoplast-targeted proteins. However, in several cases genes of endosymbiotic origin appear to have supplanted the original nuclear genes. Two apicomplexan glycolytic genes are alleged to have an algal ancestry, glucose-6-phosphate and enolase (Read et al., 1994; Dzierszinski et al., 1999). The argument for enolase rests mainly on a conserved five-amino-acid insertion that is shared by land plants. Keeling and Palmer (2001) point out that the distribution of the shared insertion is not congruent with a phylogeny of the rest of the gene. They argue instead that the insertion derives from subgenic recombination between fragments of the land plant and apicomplexan genes, resulting in a mosaic gene. The implication is that insertion and deletion data should be approached with care and considered in the context of whole-gene phylogeny.

IS THE APICOPLAST ESSENTIAL?

Apicoplast Inhibitors

Two different approaches have confirmed the dependence of apicomplexan parasites on the apicoplast. The first was the use of drugs that inhibit maintenance of the plastid. As the housekeeping functions of the apicoplast are bacterial in nature, many are susceptible to common antibiotics. Antibiotics such as ciprofloxacin inhibit bacterial or plastid DNA replication, while other antibiotics affect transcription, translation, and posttranslational modification. As mentioned above, ciprofloxacin immediately stops the apicoplast DNA from replicating and, importantly, causes parasite death (Divo et al., 1988). Even more interesting is the mode of death: after treatment, the parasite appears to grow healthily while inside the initial parasitophorous vacuole. Only after egress and then invasion of another host cell is parasite replication inhibited (Fichera and Roos, 1997). At a concentration of 20 nM, clindamycin kills 50% of parasites after 96 h (Pfefferkorn and Borotz, 1994). However, even at concentrations 1,000 higher than this 50% inhibitory concentration, parasite growth is unaffected in the first parasitophorous vacuole (Fichera et al., 1995). This lag in parasite response has been labeled delayed death and has been observed with several other antibiotics for both *P. falciparum* (Geary and Jensen, 1983; Sullivan et al., 2000) and *T. gondii* (Pfefferkorn and Borotz, 1994; Fichera et al., 1995) (Fig. 2).

Curiously, not all drugs that target the apicoplast exhibit the delayed death kinetic. One drug that does not is rifampicin, an inhibitor of plastid-type transcription. Plastid transcription is apparently derived from the cyanobacterial ancestor of plastids. It uses a DNA-dependent RNA polymerase homologous to the cyanobacterial polymerase called the α_2,β,β' DNA-dependent RNA polymerase. This polymerase recognizes -10 and -35 promoters courtesy of the sigma factor and transcribes polycistronic

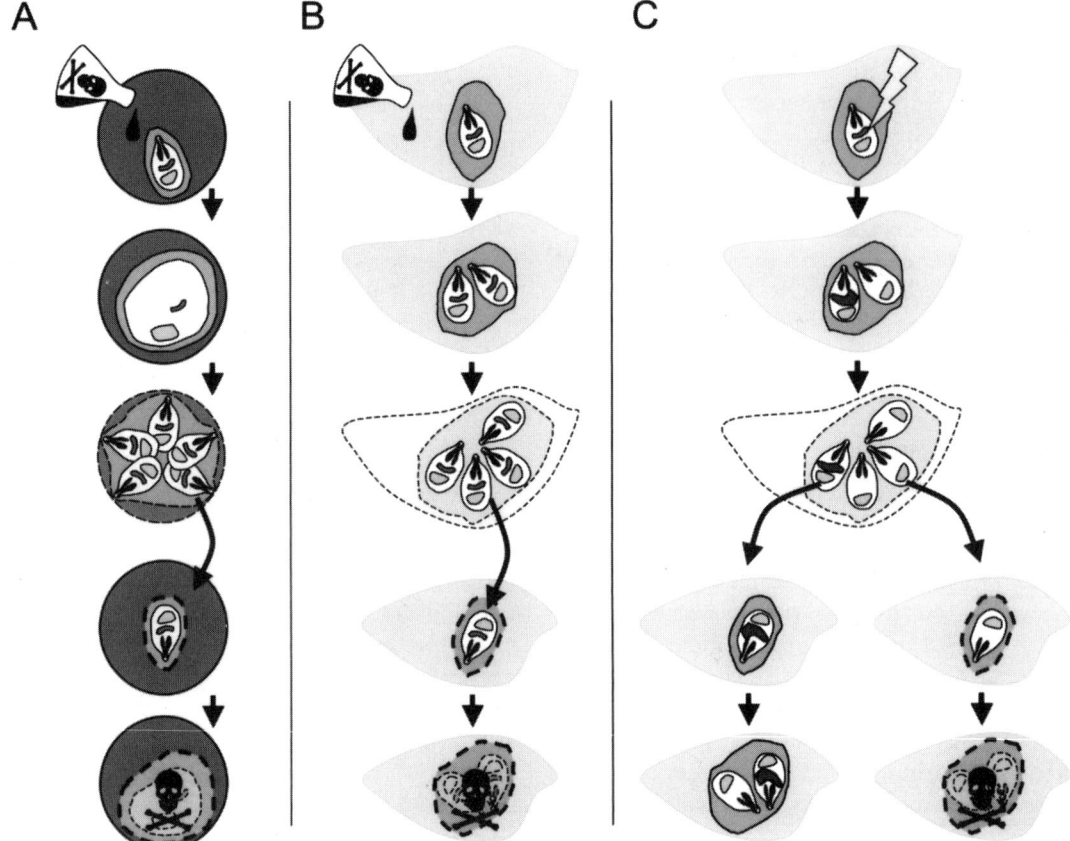

FIGURE 2 The apicoplast delayed-death phenomenon. Some of the inhibitors that are thought to target the apicoplast have delayed effects on parasite growth. Compounds such as ciprofloxacin do not appear to inhibit growth within the initial parasitophorous vacuole, but after the parasite reinvades, parasites cease to divide even when drug is removed. These reinvading parasites are not microscopically distinguishable from untreated parasites but are evidently phenotypically different. This inhibition kinetic is shared at least between *Plasmodium* (A) and *Toxoplasma* (B). In *Toxoplasma*, parasites with defects in apicoplast segregation mimic the drug-induced delayed death phenotype, with apicoplast-lacking parasites apparently growing normally until after reinvasion (C). One explanation for delayed death, consistent with known apicoplast metabolic functions, is that apicoplasts create a molecule that is needed for appropriate establishment or development of the parasitophorous vacuole membrane. In the absence of this molecule, a nonfunctional or simply nonexpanding parasitophorous vacuole restricts further growth.

RNAs from plastid DNA operons. Two subunits of this polymerase, β and β', are encoded by the apicoplast genome; the α_2 subunit and sigma factor are nucleus-encoded genes targeted to the apicoplast, strongly suggesting that the apicoplast uses a similar transcription system. The β subunit of the equivalent bacterial and plastidic polymerase is highly sensitive to rifampicin, and *Plasmodium* growth is inhibited by rifampicin. Most mitochondria use a different type of RNA polymerase, similar to that found in phage, and this phage-like RNA polymerase has been identified in *Plasmodium falciparum* (Li et al., 2001). The phage-like polymerase is rifampicin insensitive, so rifampicin very likely inhibits apicoplast transcription. Furthermore, rifampicin has been shown to specifically inhibit the formation of plastid transcripts (McConkey et al., 1997). Unlike ciprofloxacin, rifampicin kills parasites quickly, inhibiting parasite [^{3}H]hypoxanthine uptake within hours (Strath et al., 1993). An appealing explanation for this discrepancy is that ciprofloxacin prevents replication of the apicoplast genome but still allows active expression of apicoplast genes for a certain period, whereas rifampicin immediately prevents gene expression through a transcriptional block.

Unfortunately, such an explanation does not hold up to scrutiny when we consider inhibitors of organellar translation. Like DNA replication and transcription, bacterial and organellar translation is fundamentally different from eukaryotic translation. Many antibiotics inhibit bacterial translation at concentrations much lower than those required to effect eukaryotic translation. Several of these compounds specifically interact with the subunits of the RNA components of the prokaryotic ribosomes. These antibiotics have long been known to kill *P. falciparum* (Geary and Jensen, 1983), including the drug doxycycline, which likely interacts with the small (16S) subunit and is prescribed as an antimalarial. Similarly, clindamycin and thiostrepton, which bind the bacterial large (23S) subunit, are also parasiticidal (Fichera et al., 1995; Clough et al., 1997; Rogers et al., 1997). Thiostrepton inhibits the formation of new apicoplast transcripts within 6 h (McConkey et al., 1997), perhaps by inhibiting translation of the RNA-polymerase subunits, but shares the delayed death effect with ciprofloxacin (Sullivan et al., 2000). This negates the possibility that delayed death is merely due to the continued expression of genes from a nonreplicating genome. To add to the confusion, some organellar translational inhibitors lead to delayed death, while others seem to have an immediate effect (Geary and Jensen, 1983; Fichera et al., 1995). Methods of determining drug efficacy also vary from study to study, further confounding direct comparisons. Some of the drugs that cause immediate inhibition of parasite growth (including doxycycline) have possible mitochondrial targets (Beckers et al., 1995), but the pattern of kinetics is still far from clear. Conceivably, those drugs that cause immediate parasite death may have additional, nonapicoplast targets.

In sexually reproducing organisms, cytoplasmic elements such as plastids are generally maternally inherited, and the apicoplast is no exception. In *Plasmodium* spp., gametocytes form in the vertebrate host and produce male and female gametes in the mosquito midgut. Using DNA hybridization techniques, Creasey and colleagues showed that the 35-kb apicoplast genome is passed to the large female gametes but not the tiny male gametes (Creasey et al., 1994). Another study showed that thiostrepton interrupted the formation of mature parasites through sexual reproduction, although it is quite unclear at which stage the thiostrepton acts (Sullivan et al., 2000). Interestingly, the same study found that thiostrepton did not affect the infection of hepatic cells by sporozoites, implying that the apicoplast's role is reduced or redundant at this stage. Mice pretreated with thiostrepton produced no erythrocytic infection after injection of untreated sporozoites, suggesting that the apicoplast is essential for initiation of the intraerythrocytic stage, though it was not clearly shown that a primary hepatic infection was in fact established (Sullivan et al., 2000). Despite the confusing phenotypes produced by antiapicoplast compounds, several antibiotics do kill parasites both in vitro and in vivo and therefore have potential as antimalarial drugs (McFadden and Roos, 1999). Nevertheless, the delayed death result seen for some of the antiapicoplast drugs may make some of these compounds unsuitable for fast clearance of parasite burden.

Apicoplast-Lacking Mutants

The second approach that has shed light on the importance of the apicoplast is the generation of apicoplast-lacking mutants in *T. gondii*. He and colleagues serendipitously created an inadvertent fusion between an apicoplast protein, GFP, and a protein normally targeted to the rhoptry (a secretory organelle associated with host-cell invasion) (He et al., 2001a). An elegant analysis showed that the fusion protein is apparently trapped in the apicoplast protein-translocation machinery and somehow prevents correct division and segregation of the apicoplast. Within the initial parasitophorous vacuole the parasite divides normally, but the apicoplast fails to divide, yielding two parasites (called tachyzoites), one without an apicoplast (a dark parasite) and one with a large GFP-containing apicoplast (a bright parasite). This process continues, producing up to 128 tachyzoites; 127 of them lack an apicoplast. When the parasitophorous vacuole bursts, the tachyzoites can be recovered, separated by a fluorescence-activated cell sorter (FACS), and used to infect new host cells. Both bright and dark parasites are able to invade the host cell, but dark parasites divide only once or twice in the new parasitophorous vacuole, never egressing (Fig. 2) (He et al., 2001a). The result is an independent confirmation of the drug-induced delayed death phenotype. Apparently the apicoplast can be lost with no obvious short-term consequence, but over a longer period it plays an essential role. This result has led others to speculate that the function of the apicoplast is involved in membrane formation (Sullivan et al., 2000), although the dark, apicoplast-lacking parasites do at least superficially appear to establish a normal parasitophorous vacuole. Bradyzoites, the encysted life stage of *Toxoplasma* parasites, have also been occasionally observed to naturally lose their apicoplasts. No growth defect has been described for these apicoplast-lacking bradyzoites, but it is unknown whether they can redifferentiate into proliferative tachyzoites (Dzierszinski et al., 2004).

Several outstanding issues remain regarding apicoplast-lacking parasites. First is the ambiguity over how many times a dark parasite can divide: do they have a finite number of divisions after their initial formation, or is the limiting factor the formation of a new parasitophorous vacuole? Second, and more difficult to establish, does the generational distance since division from an apicoplast-bearing parasite (potentially between one and seven generations) affect the phenotype of the dark parasite? Clarification of these issues might discriminate between several models suggested by the phenotypes of the dark plastids, as follows.

1. The apicoplast produces a metabolite or macromolecule needed for ongoing parasite growth. Apicoplast loss or perturbation results in a dwindling of this mystery compound. Normal growth occurs until the stockpile runs out, and the parasite can no longer divide. Under this model, dark parasites forcibly removed from the initial parasitophorous vacuole after only one division should undergo more divisions in the new parasitophorous vacuole than if they had already divided several times. Another predicted consequence of this model is that those dark parasites separated by only one generation from an apicoplast-bearing parasite (i.e., from the final division prior to tachyzoites egress) should be capable of at least another six rounds of division, producing at least 64 tachyzoites. He and colleagues reported that dark parasites never produce more than 16 new tachyzoites (He et al., 2001a). The stockpile model is therefore inconsistent with existing data.

2. The apicoplast produces a metabolite or macromolecule needed in small amounts for ongoing parasite growth, but it is freely diffusible between parasites in the same parasitophorous vacuole. Dark parasites are sustained by the excess compound produced by the single enlarged apicoplast in the initial infection but cannot replicate independently in the subsequent parasitophorous vacuole. This model is difficult to reconcile with the drug-induced delayed death phenotype.

3. The apicoplast produces a metabolite or macromolecule integral to host invasion. If the apicoplast and mystery compound are missing, invasion is aberrant, and the new vacuole does not sustain ongoing growth, despite formation of an outwardly normal parasitophorous vacuole. This model is consistent with the essential apicoplast product being a membrane component, such as a lipid (Fig. 2).

WHAT IS THE APICOPLAST'S FUNCTION?

Chloroplasts are chlorophyll-containing organelles found in plants and algae. Their key function is photosynthesis, and they come in red, brown, and even colorless, nonphotosynthetic versions. Strictly speaking, the term chloroplast should only be used to describe green versions of the organelle; plastid is the more generic term. Plastids (of all colors) originate from endosymbiotic cyanobacteria (Cavalier-Smith, 1982; Delwiche and Palmer, 1997; Delwiche, 1999). In other words, a plastid is a modified photosynthetic bacterium living inside a eukaryotic cell. These endosymbionts are semiautonomous, living (to a limited extent) a life of their own within the host. The original acquisition of plastids presumably provided the host with the ability to photosynthesize, creating food from CO_2 and sunlight. In most apicomplexans, however, the apicoplast is never exposed to light and has no opportunity to photosynthesize. Energy requirements are presumably fulfilled by scavenging macromolecules from the host (specifically, hemoglobin in the vertebrate stages of *Plasmodium* spp.). Neither thylakoid membranes nor photosynthetic pigments are apparent by fluorescence or electron microscopy (McFadden et al., 1996; Hopkins et al., 1999). It is entirely implausible that a useless organelle could be maintained for several hundred million years. Since apicoplast loss leads to parasite death and if the plastid is not retained for its photosynthetic capacity, what does it do?

Fatty Acid Synthesis

Practically all of the advances made so far in our understanding of apicoplast function have been based on information generated by genome and EST projects for several apicomplexans. The first genes for apicoplast-targeted proteins discovered by Waller and colleagues (1998) included the enzymes acyl carrier protein (ACP), β-ketoacyl-acyl carrier protein synthase III (also known as FabH), and β-hydroxyacyl-ACP dehydratase (also referred to as FabZ). All three of these enzymes belong to a multienzyme fatty acid synthase (FAS) complex found in bacteria and plastids. The plastid FAS is referred to as type II FAS and is distinct from the type I FAS found in animals and fungi. The type I FAS appears to be derived from a fusion of the ancestral type II subunits. Plants and algae lack a type I FAS, instead relying on their plastidic type II FAS to provide requisite fatty acids (Harwood, 1996). No organisms are known to possess both type I and type II FAS. The presence of type II FAS genes in *P. falciparum* and *T. gondii* strongly implied active de novo synthesis of fatty acids in the plastid (Waller et al., 1998). In confirmation of this, several groups have now demonstrated enzymatic activity for most members of the apicoplast type II FAS pathway (Surolia and Surolia, 2001; Waters et al., 2002; Prigge et al., 2003), and *P. falciparum* has been shown to incorporate radiolabeled carbon precursors into simple fatty acids (Surolia and Surolia, 2001). An entire pathway for the import and processing of precursors for fatty acid synthesis has now been predicted from bioinformatic observations (Ralph et al., 2004). Simple fatty acids are almost certainly required within the apicoplast for the cofactor lipoic acid, but extraplastidic uses are less clear. Possible destinations include lipids of the plasma membrane or the parasitophorous vacuole. Fatty acids are usually exported from plastids using an acetyl-coenzyme A (CoA) synthase that allows the fatty acid to transfer to the endoplasmic reticulum (ER), but the presence of four apicoplast membranes presents some challenges to our understanding of how this might occur in *Plasmodium*.

Inhibitors of type II fatty acid synthase have added indirect support to the presence of active plastidic fatty acid synthesis without greatly elucidating the role of apicoplast fatty acids. Thiolactomycin, an inhibitor of the type II FAS-condensing enzymes FabB, FabF, and FabH, kills asexually cultured *P. falciparum* (Waller et al., 1998). Triclosan, an inhibitor of FabI, kills parasites in culture, as well as clearing a rodent malaria, *Plasmodium berghei*, from infected mice (Surolia and Surolia, 2001). Triclosan was also shown to specifically bind and inhibit *P. falciparum* FabI activity (Surolia and Surolia, 2001). Unlike the delayed death kinetics seen with some inhibitors of apicoplast housekeeping, both thiolactomycin and triclosan led to immediate inhibition of parasite growth. This discrepancy led Surolia and colleagues to suggest that the delayed death phenotype applies only when a drug inhibits apicoplast housekeeping function, such as transcription and translation, but leaves general metabolism unaffected (Surolia et al., 2002). They inferred that the remainder of the apicoplast's functions are not immediately affected by interruption of expression of the apicoplast-encoded genes. Only drugs that inhibit essential metabolism (they suggest fatty acid synthesis) bring about immediate death. If the dichotomy between housekeeping inhibitors and metabolic inhibitors continues to be supported (although there are already exceptions), such a hypothesis may be attractive, but it fails to explain the delayed death seen in apicoplast-lacking mutants (He et al., 2001a). Predictably, proteins normally targeted by the apicoplast are not processed to their mature versions in dark parasites (He et al., 2001b), so apicoplast metabolism could not feasibly be carried on in the absence of an apicoplast. The delayed death phenomenon is unfortunately not so easily explained.

Isopentenyl Diphosphate Synthesis

Another essential role of plastids is the synthesis of isopentenyl diphosphate (IPP), the building block of isoprenoids. It was originally thought that IPP in all organisms was synthesized via the intermediate mevalonate (for a review, see Lichtenthaler et al., 1997), but Rohmer and colleagues (1993) discovered an alternate bacterial pathway for IPP synthesis that proceeds via the intermediate 1-deoxy-D-xylulose 5-phosphate (DOXP). The alternate pathway was later discovered to be widely distributed through bacteria and the plastids of algae and land plants. The phylogenetic distribution of the pathway is incompatible with vertical transmission and must have been spread by lateral gene transfer, presumably with the plastid (Boucher and Doolittle, 2000). It is likely that only those eukaryotes with plastids contain a DOXP pathway. Jomaa and colleagues (1999) found two enzymes of the DOXP pathway, DOXP synthase and DOXP reductoisomerase, in the unfinished genome sequence of *P. falciparum*. The N-terminal leader of the DOXP reductoisomerase was sufficient to target GFP to the apicoplast of *T. gondii*, consistent with the pathway's plastid localization in plants and algae.

DOXP reductoisomerase is susceptible to a specific inhibitor, fosmidomycin, and Jomaa and colleagues showed that fosmidomycin and a derivative analog kill *P. falciparum* in culture (Jomaa et al., 1999). As with the inhibitors of fatty acid synthesis, fosmidomycin has an immediate effect on parasites, and no delayed death has been described. A combination of fosmidomycin with the antitranslational antibiotic clindamycin also has a synergistic effect on *P. falciparum* both in vitro and in vivo (Wiesner et al., 2002). Analogs of fosmidomycin are promising antimalarials with some analogs clearing mice of malaria after oral administration (Reichenberg et al., 2001), and the crystal structure of the *P. falciparum* DOXP reductoisomerase promises further analog improvement (Reuter et al., 2002). Initial human field trials with fosmidomycin show encouraging parasite clearance and low levels of toxicity with humans (Missinou et al., 2002; Borrmann et al., 2004). As with fatty acid synthesis, a complete apicoplast-based pathway for the synthesis of IPP from simple precursors is apparent by bioinformatic analysis. One of the endpoints of the pathway, dimethylallyl diphosphate, appears to be used within the apicoplast to modify the anticodons of several tRNAs. This modification, called i6a, may be used to suppress frameshifts and premature stop codons within the apicoplast genomes. Products of the IPP pathway are presumably also used by mitochondrial ubiquinones, by dolichol in the ER and Golgi, and to prenylate proteins within the parasite's endomembrane system. The movement of isoprenes between organelles has not been well characterized in any organism, but the apicoplast presumably has a specific mechanism to export the products of the IPP pathway.

Aromatic Amino Acids

Plants synthesize all required aromatic amino acids in their plastids. These plastids create aromatic amino acids via the shikimate pathway, also found in bacteria and fungi but not vertebrates (Kishore and Shah, 1988). A shikimate pathway enzyme, 5-enolpyruvyl shikimate 3-phosphate (EPSP) synthase, is the target for the herbicide glyphosate (Boocock and Coggins, 1983), and Roberts and colleagues (1998) made the promising discovery that glyphosate inhibits the activity of this enzyme in *T. gondii* extracts. The authors suggested that the enzyme was apicoplast targeted. However, the enzyme contains no apicoplast-targeting leader and groups phylogenetically with fungal cytosolic homologs rather than with plastid homologs, implying that this enzyme is not found in the *P. falciparum* apicoplast (Keeling et al., 1998). Another key shikimate enzyme has since been shown to be cytosolic in *P. falciparum* (Fitzpatrick et al., 2001), and the genome project shows that no shikimate enzymes have apicoplast-targeting leaders (Gardner et al., 2002).

Heme Biosynthesis

Heme is a prosthetic group that plays an essential role in the oxygen- and electron-carrying abilities of hemoglobin and cytochromes, respectively. Several cytochromes are present in *Plasmodium* spp., and these proteins must have heme moieties attached. Despite swimming in a sea of heme released by the breakdown of vertebrate hemoglobin, *Plasmodium* relies on its own heme synthesis to survive (Surolia and Pasmanaban, 1992). Most plants and algae depend on plastid-located heme synthesis, which initiates with glutamate and a cofactor tRNA^{Glu}. This is distinct from the Shemin pathway found in the mitochondria of fungi and animals, which creates the heme precursor δ-aminolevulinate from glycine and succinyl co-A. Radiolabeled glycine is incorporated into *P. falciparum* heme (Surolia and Pasmanaban, 1992); the responsible enzyme, δ-aminolevulinate synthetase (ALAS), has been characterized in *P. falciparum* (Wilson et al., 1996a) and appears to be mitochondrial. The subsequent step of heme synthesis is the dehydration of δ-aminolevulinate by δ-aminolevulinate dehydratase (ALAD). Data from Bonday and colleagues (Bonday et al., 1997; Bonday et al., 2000) suggest that *Plasmodium* uses ALAD imported from the vertebrate host, but an active ALAD is also encoded by the parasite's own genome (Sato et al., 2000; Dhanasekaran et al., 2004). This ALAD complements an *Escherichia coli* ALAD null mutant (Sato and Wilson, 2002) and is transcribed in blood-stage parasites (van Dooren et al., 2002). The parasite ALAD possesses an apicoplast-targeting leader, suggesting the pathway may be split between mitochondria and apicoplast (van Dooren et al., 2002). Further analyses indicate that steps of this pathway are shared between the mitochondria and apicoplast (Ralph et al., 2004; Sato et al., 2004), perhaps explaining the close contact observed between these two compartments (Hopkins et al., 1999; Waller et al., 2000). The heme pathway is the target of diphenyl ether herbicides (Prasad and Dailey, 1995), so elucidation of heme synthesis in *P. falciparum* is potentially important for malarial drug development.

Iron-Sulfur Cluster Assembly

Iron-sulfur (Fe-S) clusters are prosthetic groups required as cofactors for many proteins involved in electron transport and redox control. Their assembly is catalyzed by at least three alternative pathways that are to some extent interchangeable, designated the NIF, SUF, and ISC sytems. The functional interchangeability of components between these systems has occasionally made the nomenclature of this field unclear. In bacterial systems, the genes for these pathways are often grouped in their own operons. Iron cluster assembly also occurs in eukaryotes and can be cytosolic, mitochondrial, or plastidic. Mitochondrial ISC Fe-S assembly systems are probably descended from ISC α-proteobacterial versions; likewise, plastidic SUF versions appear to come from the *suf* operon of the cyanobacterial endosymbiont. Common elements of the assembly pathways are cysteine desulferases (NifS, IscS, or SufS) that provide the pathway with sulfur liberated from cysteine, scaffolds for intermediate steps of the assembly, and several other proteins that burn ATP for the energy needed to acquire iron and to attach Fe-S clusters to proteins. The exact roles of several NIF, SUF, and ISC members are unknown. The apicoplast genome, consistent with its ultimate cyanobacterial origin, still encodes a *sufB*, and homologs of other members of the *suf* operon (SufC, SufD, SufS) appear to be apicoplast targeted (Ellis et al., 2001; Seeber, 2002; Ralph et al., 2004). The assembly scaffold used by the SUF system is not definitively known (*sufA* has been proposed), but the NifU scaffold protein does appears to be apicoplast targeted. The pair of apicoplast-targeted proteins ferredoxin and ferredoxin:NADP reductase are probably involved in the redox regulation of iron-sulfur cluster assembly; in an interesting twist, ferredoxin itself is an Fe-S protein. It is not known whether plastid-assembled Fe-S is used within the apicoplast or elsewhere, but Fe-S assembled in mitochondria at least is thought to be exported to the cytosol by a specific transporter.

Anabolic Precursors

In plants, plastids derive both the precursors they need for anabolism and the energy requirements to drive synthetic processes from carbon compounds fixed during photosynthesis. Because the apicoplast is nonphotosynthetic, these demands are presumably fulfilled by extraplastidic sources in *Plasmodium*. The way apicoplasts solve this problem is probably by acquiring three-carbon molecules such as triosephosphates or phosphoenolpyruvate (PEP) from the parasite cytosol. Plants cells that do not photosynthesize (e.g., root cells) use specific transporters to actively import these molecules, and *Plasmodium* possesses orthologs of phosphate transporters that likely fill this role. Using these transporters, the apicoplast could feed imported dihydroxyacetone directly into IPP synthesis, while imported PEP could be converted to pyruvate to feed both the IPP and fatty acid synthesis pathways (Fig. 3). The enzyme that carries out this conversion is the huge multiprotein complex pyruvate dehydrogenase. The subunits that make up this protein in *Plasmodium* have now been characterized and shown to be both enzymatically active and apicoplast located

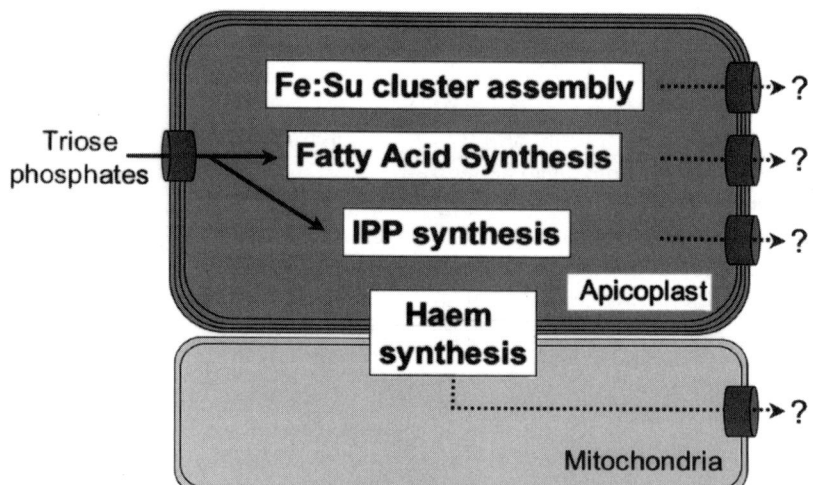

FIGURE 3 Major apicoplast anabolic pathways. Bioinformatic analyses and confirmation by biochemical methods suggest that the apicoplast is responsible for at least four major anabolic pathways, fatty acid synthesis, isopentenyl diphosphate synthesis, iron-sulfur cluster assembly, and heme synthesis, in conjunction with the mitochondrion. Plastid-specific transporters probably import triose phosphates and/or phosphoenolpyruvate to be used as the carbon building blocks for synthesis, as well as generating some energy for synthetic pathways. Each pathway may have organellar uses, but the indispensability of the apicoplast indicates there are also extraplastidic endpoints for at least one of these anabolic pathways.

(Foth et al., 2005; McMillan et al., 2005). Pyruvate dehydrogenase also yields ATP from the conversion of PEP to pyruvate, so some energy to run apicoplast processes may derive from this reaction.

CONCLUDING REMARKS

The past decade has seen great strides forward in our understanding of the apicoplast. Significant developments include the elucidation of the apicoplast targeting signal, the recognition and development of the apicoplast as a drug target, a widening comprehension of complex plastid origins, and a growing understanding of apicoplast metabolism and biochemistry. Despite these rapid advances, we still have no good answer for the fundamental question: what does the apicoplast do? What crucial service does it supply to the parasite that has made it indispensable over hundreds of millions of years? Answering this question will be the central challenge of apicoplast biology in the years ahead.

REFERENCES

Abrahamsen, M. S., T. J. Templeton, S. Enomoto, J. E. Abrahante, G. Zhu, C. A. Lancto, M. Deng, C. Liu, G. Widmer, S. Tzipori, G. A. Buck, P. Xu, A. T. Bankier, P. H. Dear, B. A. Konfortov, H. F. Spriggs, L. Iyer, V. Anantharaman, L. Aravind, and V. Kapur. 2004. Complete genome sequence of the apicomplexan, Cryptosporidium parvum. *Science* **304:**441–445.

Baldauf, S. L., A. J. Roger, I. Wenk-Siefert, and W. F. Doolittle. 2000. A kingdom-level phylogeny of eukaryotes based on combined protein data. *Science* **290:**972–977.

Beckers, C. J. M., D. S. Roos, R. G. K. Donald, B. J. Luft, J. C. Schwab, Y. Cao, and K. A. Joiner. 1995. Inhibition of cytoplasmic and organellar protein synthesis in toxoplasma gondii—implications for the target of macrolide antibiotics. *J. Clin. Investig.* **95:**367–376.

Blanchard, J., and J. S. Hicks. 1999. The non-photosynthetic plastid in malarial parasites and other apicomplexans is derived from outside the green plastid lineage. *J. Eukaryot. Microbiol.* **46:**367–375.

Bonday, Z. Q., S. Dhanasekaran, P. N. Rangarajan, and G. Padmanaban. 2000. Import of host delta-aminolevulinate dehydratase into the malarial parasite: identification of a new drug target. *Nat. Med.* **6:**898–903.

Bonday, Z. Q., S. Taketani, P. D. Gupta, and G. Padmanaban. 1997. Heme biosynthesis by the malarial parasite. Import of delta-aminolevulinate dehydrase from the host red cell. *J. Biol. Chem.* **272:**21839–21846.

Boocock, M. R., and J. R. Coggins. 1983. Kinetics of 5-enolpyruvylshikimate-3-phosphate synthase inhibition by glyphosate. *FEBS Lett.* **154:**127–133.

Borrmann, S., S. Issifou, G. Esser, A. A. Adegnika, M. Ramharter, P. B. Matsiegui, S. Oyakhirome, D. P. Mawili-Mboumba, M. A. Missinou, J. F. Kun, H. Jomaa, and P. G. Kremsner. 2004. Fosmidomycin-clindamycin for the treatment of Plasmodium falciparum malaria. *J. Infect. Dis.* **190:**1534–1540.

Boucher, Y., and W. F. Doolittle. 2000. The role of lateral gene transfer in the evolution of isoprenoid biosynthesis pathways. *Mol. Microbiol.* **37:**703–716.

Cavalier-Smith, T. 1982. The origins of plastids. *Biol. J. Linn. Soc. Lond.* **17:**289–306.

Cavalier-Smith, T. 1999. Principles of protein and lipid targeting in secondary symbiogenesis: euglenoid, dinoflagellate, and sporozoan plastid origins and the eukaryote family tree. *J. Eukaryot. Microbiol.* **46:**347–366.

Clough, B., M. Strath, P. Preiser, P. Denny, and R. Wilson. 1997. Thiostrepton binds to malarial plastid rRNA. *FEBS Lett.* **406:**123–125.

Creasey, A., K. Mendis, J. Carlton, D. Williamson, I. Wilson, and R. Carter. 1994. Maternal inheritance of extrachromosomal DNA in malaria parasites. *Mol. Biochem. Parasitol.* **65:**95–98.

Delwiche, C. 1999. Tracing the tread of plastid diversity through the tapestry of life. *Am. Nat.* **154:**S164–S177.

Delwiche, C. F., and J. D. Palmer. 1997. The origin of plastids and their spread via secondary endosymbiosis. *Plant Syst. Evol.* **11**(Suppl.)**:**51–86.

Denny, P., P. Preisser, D. Williamson, and I. Wilson. 1998. Evidence for a single origin of the 35kb plastid DNA in apicomplexans. *Protist* **149:**51–59.

DeRocher, A., C. B. Hagen, J. E. Froehlich, J. E. Feagin, and M. Parsons. 2000. Analysis of targeting sequences demonstrates that trafficking to the Toxoplasma gondii plastid branches off the secretory system. *J. Cell Sci.* **113:**3969–3977.

Dhanasekaran, S., N. R. Chandra, B. K. Chandrasekhar Sagar, P. N. Rangarajan, and G. Padmanaban. 2004. Delta-aminolevulinic acid dehydratase from Plasmodium falciparum: indigenous versus imported. *J. Biol. Chem.* **279:**6934–6942.

Diniz, J. A. P., E. O. Silva, R. Lainson, and W. de Souza. 2000. The fine structure of Garnia gonadati and its association with the host cell. *Parasitol. Res.* **86:**971–977.

Divo, A., A. Sartorelli, C. Patton, and F. Bia. 1988. Activity of fluoroqinolone antibiotics against *Plasmodium falciparum* in vitro. *Antimicrob. Agents Chemother.* **32:**1182–1186.

Dzierszinski, F., M. Nishi, L. Ouko, and D. S. Roos. 2004. Dynamics of *Toxoplasma gondii* differentiation. *Eukaryot. Cell* **3**:992–1003.

Dzierszinski, F., O. Popescu, C. Toursel, C. Slomianny, B. Yahiaoui, and S. Tomavo. 1999. The protozoan parasite Toxoplasma gondii expresses two functional plant-like glycolytic enzymes—implications for evolutionary origin of apicomplexans. *J. Biol. Chem.* **274**:24888–24895.

Ellis, K. E., B. Clough, J. W. Saldanha, and R. J. Wilson. 2001. Nifs and Sufs in malaria. *Mol. Microbiol.* **41**:973–981.

Fast, N. M., J. C. Kissinger, D. S. Roos, and P. J. Keeling. 2001. Nuclear-encoded, plastid-targeted genes suggest a single common origin for apicomplexan and dinoflagellate plastids. *Mol. Biol. Evol.* **18**:418–426.

Fichera, M. E., M. K. Bhopale, and D. S. Roos. 1995. In vitro assays elucidate peculiar kinetics of clindamycin action against *Toxoplasma gondii*. *Antimicrob. Agents Chemother.* **39**:1530–1537.

Fichera, M. E., and D. S. Roos. 1997. A plastid organelle as a drug target in apicomplexan parasites. *Nature* **390**:407–409.

Fitzpatrick, T., S. Ricken, M. Lanzer, N. Amrhein, P. MacHeroux, and B. Kappes. 2001. Subcellular localization and characterization of chorismate synthase in the apicomplexan Plasmodium falciparum. *Mol. Microbiol.* **40**:65–75.

Foth, B. J., S. A. Ralph, C. J. Tonkin, N. S. Struck, M. Fraunholz, D. S. Roos, A. F. Cowman, and G. I. McFadden. 2003. Dissecting apicoplast targeting in the malaria parasite Plasmodium falciparum. *Science* **299**:705–708.

Foth, B. J., L. M. Stimmler, E. Handman, B. S. Crabb, A. N. Hodder, and G. I. McFadden. 2005. The malaria parasite Plasmodium falciparum has only one pyruvate dehydrogenase complex, which is located in the apicoplast. *Mol. Microbiol.* **55**:39–53.

Gardner, M. J., N. Hall, E. Fung, O. White, M. Berriman, R. W. Hyman, J. M. Carlton, A. Pain, K. E. Nelson, S. Bowman, I. T. Paulsen, K. James, J. A. Eisen, K. Rutherford, S. L. Salzberg, A. Craig, S. Kyes, M. S. Chan, V. Nene, S. J. Shallom, B. Suh, J. Peterson, S. Angiuoli, M. Pertea, J. Allen, J. Selengut, D. Haft, M. W. Mather, A. B. Vaidya, D. M. Martin, A. H. Fairlamb, M. J. Fraunholz, D. S. Roos, S. A. Ralph, G. I. McFadden, L. M. Cummings, G. M. Subramanian, C. Mungall, J. C. Venter, D. J. Carucci, S. L. Hoffman, C. Newbold, R. W. Davis, C. M. Fraser, and B. Barrell. 2002. Genome sequence of the human malaria parasite Plasmodium falciparum. *Nature* **419**:498–511.

Gardner, M. J., D. H. Williamson, and R. J. M. Wilson. 1991. A circular DNA in malaria parasites encodes an RNA polymerase like that of prokaryotes and chloroplasts. *Mol. Biochem. Parasitol.* **44**:115–23.

Geary, T. G., and J. B. Jensen. 1983. Effects of antibiotics on Plasmodium falciparum in vitro. *Am. J. Trop. Med. Hyg.* **32**:221–225.

Gozar, M. M., and A. S. Bagnara. 1993. Identification of a Babesia bovis gene with homology to the small subunit ribosomal RNA gene from the 35-kilobase circular DNA of Plasmodium falciparum. *Int. J. Parasitol.* **23**:145–148.

Harb, O. S., B. Chatterjee, M. J. Fraunholz, M. J. Crawford, M. Nishi, and D. S. Roos. 2004. Multiple functionally redundant signals mediate targeting to the apicoplast in the apicomplexan parasite *Toxoplasma gondii*. *Eukaryot. Cell* **3**:663–674.

Harper, J. T., and P. J. Keeling. 2003. Nucleus-encoded, plastid-targeted glyceraldehyde-3-phosphate dehydrogenase (GAPDH) indicates a single origin for chromalveolate plastids. *Mol. Biol. Evol.* **20**:1730–1735.

Harwood, J. 1996. Recent advances in the biosynthesis of plant fatty-acids. *Biochim. Biophys. Acta* **1301**:7–56.

He, C. Y., M. K. Shaw, C. H. Pletcher, B. Striepen, L. G. Tilney, and D. S. Roos. 2001a. A plastid segregation defect in the protozoan parasite Toxoplasma gondii. *EMBO J.* **20**:330–339.

He, C. Y., B. Striepen, C. H. Pletcher, J. M. Murray, and D. S. Roos. 2001b. Targeting and processing of nuclear-encoded apicoplast proteins in plastid segregation mutants of Toxoplasma gondii. *J. Biol. Chem.* **276**:28436–28442.

Hopkins, J., R. Fowler, S. Krishna, I. Wilson, G. Mitchell, and L. Bannister. 1999. The plastid in Plasmodium falciparum asexual blood stages: a three-dimensional ultrastructural analysis. *Protist* **150**:283–295.

Howe, C. J. 1992. Plastid origin of an extrachromosomal DNA molecule from *Plasmodium*, the causative agent of malaria. *J. Theor. Biol.* **158**:199–205.

Huang, J., N. Mullapudi, C. A. Lancto, M. Scott, M. S. Abrahamsen, and J. C. Kissinger. 2004. Phylogenomic evidence supports past endosymbiosis, intracellular and horizontal gene transfer in Cryptosporidium parvum. *Genome Biol.* **5**:R88.

Jomaa, H., J. Wiesner, S. Sanderbrand, B. Altincicek, C. Weidemeyer, M. Hintz, I. Turbachova, M. Eberl, J. Zeidler, H. K. Lichtenthaler, D. Soldati, and E. Beck. 1999. Inhibitors of the nonmevalonate pathway of isoprenoid biosynthesis as antimalarial drugs. *Science* **285**:1573–1576.

Keeling, P. J., and J. D. Palmer. 2001. Lateral transfer at the gene and subgenic levels in the evolution of eukaryotic enolase. *Proc. Natl. Acad. Sci.* **98**:10745–10750.

Keeling, P. J., J. D. Palmer, R. G. K. Donald, D. S. Roos, R. F. Waller, and G. I. McFadden. 1998.

Shikimate pathway in apicomplexan parasites. *Nature* **397**:219–220.

Kishore, G. M., and D. M. Shah. 1988. Amino acid biosynthesis inhibitors as herbicides. *Annu. Rev. Biochem.* **57**:627–663.

Köhler, S., C. F. Delwiche, P. W. Denny, L. G. Tilney, P. Webster, R. J. M. Wilson, J. D. Palmer, and D. S. Roos. 1997. A plastid of probable green algal origin in apicomplexan parasites. *Science* **275**:1485–1488.

Lang-Unnasch, N., M. E. Reith, J. Munholland, and J. R. Barta. 1998. Plastids are widespread and ancient in parasites of the phylum Apicomplexa. *Int. J. Parasitol.* **28**:1743–1754.

Li, J. N., J. A. Maga, N. Cermakian, R. Cedergren, and J. E. Feagin. 2001. Identification and characterization of a Plasmodium falciparum RNA polymerase gene with similarity to mitochondrial RNA polymerases. *Mol. Biochem. Parasitol.* **113**:261–269.

Lichtenthaler, H. K., M. Rohmer, and J. Schwender. 1997. Two independent biochemical pathways for isopentenyl diphosphate and isoprenoid biosynthesis in higher plants. *Physiol. Plantarum* **101**:643–652.

Martin, W., T. Rujan, E. Richly, A. Hansen, S. Cornelsen, T. Lins, D. Leister, B. Stoebe, M. Hasegawa, and D. Penny. 2002. Evolutionary analysis of Arabidopsis, cyanobacterial, and chloroplast genomes reveals plastid phylogeny and thousands of cyanobacterial genes in the nucleus. *Proc. Natl. Acad. Sci. USA* **99**:12246–12251.

Matsuzaki, M., T. Kikuchi, K. Kita, S. Kojima, and T. Kuroiwa. 2001. Large amounts of apicoplast nucleoid DNA and its segregation in Toxoplasma gondii. *Protoplasma* **218**:180–191.

Matsuzaki, M., O. Misumi, I. T. Shin, S. Maruyama, M. Takahara, S. Y. Miyagishima, T. Mori, K. Nishida, F. Yagisawa, Y. Yoshida, Y. Nishimura, S. Nakao, T. Kobayashi, Y. Momoyama, T. Higashiyama, A. Minoda, M. Sano, H. Nomoto, K. Oishi, H. Hayashi, F. Ohta, S. Nishizaka, S. Haga, S. Miura, T. Morishita, Y. Kabeya, K. Terasawa, Y. Suzuki, Y. Ishii, S. Asakawa, H. Takano, N. Ohta, H. Kuroiwa, K. Tanaka, N. Shimizu, S. Sugano, N. Sato, H. Nozaki, N. Ogasawara, Y. Kohara, and T. Kuroiwa. 2004. Genome sequence of the ultrasmall unicellular red alga Cyanidioschyzon merolae 10D. *Nature* **428**:653–657.

McConkey, G. A., M. J. Rogers, and T. F. McCutchan. 1997. Inhibition of *Plasmodium falciparum* protein synthesis: targeting the plastid-like organelle with thiostrepton. *J. Biol. Chem.* **272**:2046–2049.

McFadden, G. I., M. Reith, J. Munholland, and N. Lang-Unnasch. 1996. Plastid in human parasites. *Nature* **381**:482.

McFadden, G. I., and D. S. Roos. 1999. Apicomplexan plastids as drug targets. *Trends Microbiol.* **6**:328–333.

McFadden, G. I., and R. F. Waller. 1997. Plastids in parasites of humans. *Bioessays* **19**:1033–1040.

McFadden, G. I., R. F. Waller, M. Reith, J. Munholland, and N. Lang-Unnasch. 1997. Plastids in apicomplexan parasites. *Plant Syst. Evol.* **11**(Suppl.):261–287.

McMillan, P. J., L. M. Stimmler, B. J. Foth, G. I. McFadden, and S. Müller. 2005. The human malaria parasite Plasmodium falciparum possesses two distinct dihydrolipoamide dehydrogenases. *Mol. Microbiol.* **55**:27–38.

Missinou, M. A., S. Borrmann, A. Schindler, S. Issifou, A. A. Adegnika, P. B. Matsiegui, R. Binder, B. Lell, J. Wiesner, T. Baranek, H. Jomaa, and P. G. Kremsner. 2002. Fosmidomycin for malaria. *Lancet* **360**:1941–1942.

Pfefferkorn, E. R., and S. E. Borotz. 1994. Comparison of mutants of Toxoplasma gondii selected for resistance to azithromycin, spiramycin, or clindamycin. *Antimicrob. Agents Chemother.* **338**:31–37.

Prasad, A. R. K., and H. A. Dailey. 1995. Generation of resistance to the diphenyl ether herbicide acifluorfen by mel cells. *Biochem. Biophys. Res. Commun.* **215**:186–191.

Prigge, S. T., X. He, L. Gerena, N. C. Waters, and K. A. Reynolds. 2003. The initiating steps of a type II fatty acid synthase in Plasmodium falciparum are catalyzed by pfACP, pfMCAT, and pfKASIII. *Biochemistry* **42**:1160–1169.

Ralph, S. A., G. G. Van Dooren, R. F. Waller, M. J. Crawford, M. J. Fraunholz, B. J. Foth, C. J. Tonkin, D. S. Roos, and G. I. McFadden. 2004. Tropical infectious diseases: metabolic maps and functions of the Plasmodium falciparum apicoplast. *Nat. Rev. Microbiol.* **2**:203–216.

Read, M., K. E. Hicks, P. F. Sims, and J. E. Hyde. 1994. Molecular characterisation of the enolase gene from the human malaria parasite Plasmodium falciparum. Evidence for ancestry within a photosynthetic lineage. *Eur. J. Biochem.* **220**:513–520.

Reichenberg, A., J. Wiesner, C. Weidemeyer, E. Dreiseidler, S. Sanderbrand, B. Altincicek, E. Beck, M. Schlitzer, and H. Jomaa. 2001. Diaryl ester prodrugs of FR900098 with improved in vivo antimalarial activity. *Bioorg. Med. Chem. Lett.* **11**:833–835.

Reuter, K., S. Sanderbrand, H. Jomaa, J. Wiesner, I. Steinbrecher, E. Beck, M. Hintz, G. Klebe, and M. T. Stubbs. 2002. Crystal structure of 1-deoxy-D-xylulose-5-phosphate reductoisomerase, a crucial enzyme in the non-mevalonate pathway of isoprenoid biosynthesis. *J. Biol. Chem.* **277**:5378–5384.

Roberts, F., C. Roberts, J. Johnson, D. Kyle, T. Krell, J. Coggins, G. Coombs, W. Milhous, S. Tzipori, D. Ferguson, D. Chakrabarti, and R. McLeod. 1998. Evidence for the shikimate pathway apicomplexan parasites. *Nature* **393**:801–806.

Rogers, M. J., Y. V. Burkham, T. F. McCutchan, and D. E. Draper. 1997. Interaction of thiostrepton with an RNA fragment derived from the plastid-encoded ribosomal RNA of the malaria parasite. *RNA* **3**:815–820.

Rohmer, M., M. Knani, P. Simonin, B. Sutter, and H. Sahm. 1993. Isoprenoid biosynthesis in bacteria: a novel pathway for the early steps leading to isopentenyl diphosphate. *Biochem. J.* **295**:517–524.

Sato, S., B. Clough, L. Coates, and R. J. Wilson. 2004. Enzymes for heme biosynthesis are found in both the mitochondrion and plastid of the malaria parasite Plasmodium falciparum. *Protist* **155**:117–125.

Sato, S., I. Tews, and R. J. M. Wilson. 2000. Impact of a plastid-bearing endocytobiont on apicomplexan genomes. *Int. J. Parasitol.* **30**:427–439.

Sato, S., and R. J. Wilson. 2002. The genome of Plasmodium falciparum encodes an active delta-aminolevulinic acid dehydratase. *Curr. Genet.* **40**:391–398.

Seeber, F. 2002. Biogenesis of iron-sulphur clusters in amitochondriate and apicomplexan protists. *Int. J. Parasitol.* **32**:1207–1217.

Strath, M., F. T. Scott, M. Gardner, D. Williamson, and I. Wilson. 1993. Antimalarial activity of rifampicin in vitro and in rodent models. *Trans. R. Soc. Trop. Med. Hyg.* **87**:211–216.

Striepen, B., M. J. Crawford, M. K. Shaw, L. G. Tilney, F. Seeber, and D. S. Roos. 2000. The plastid of Toxoplasma gondii is divided by association with the centrosomes. *J. Cell Biol.* **151**:1423–1434.

Sullivan, M., J. Li, S. Kumar, M. J. Rogers, and T. F. McCutchan. 2000. Effects of interruption of apicoplast function on malaria infection, development, and transmission. *Mol. Biochem. Parasitol.* **109**:17–23.

Surolia, N., and G. Pasmanaban. 1992. De novo biosynthesis of heme offers a new chemotherapeutic target in the human malarial parasite. *Biochem. Biophys. Res Commun.* **187**:744–750.

Surolia, N., S. P. RamachandraRao, and A. Surolia. 2002. Paradigm shifts in malaria parasite biochemistry and anti-malarial chemotherapy. *Bioessays* **24**:192–196.

Surolia, N., and A. Surolia. 2001. Triclosan offers protection against blood stages of malaria by inhibiting enoyl-ACP reductase of Plasmodium falciparum. *Nat. Med.* **7**:167–173.

Tetley, L., S. M. Brown, V. McDonald, and G. H. Coombs. 1998. Ultrastructural analysis of the sporozoite of Cryptosporidium parvum. *Microbiology* **144**:3249–3255.

Van de Peer, Y., and R. De Wachter. 1997. Evolutionary relationships among the eukaryotic crown taxa taking into account site-to-site rate variation in 18S rRNA. *J. Mol. Evol.* **45**:619–630.

van Dooren, G. G., V. Su, M. C. DiOmbrain, and G. I. McFadden. 2002. Processing of an apicoplast leader sequence in Plasmodium falciparum, and the identification of a putative leader cleavage enzyme. *J. Biol. Chem.* **277**:23612–23619.

van Dooren, G. G., R. F. Walker, K. A. Joiner, D. S. Roos, and G. I. McFadden. 2000. Traffic jams: protein transport in Plasmodium falciparum. *Parasitol. Today* **16**:421–427.

Waller, R. F., P. J. Keeling, R. G. K. Donald, B. Striepen, E. Handman, N. Lang-Unnasch, A. F. Cowman, G. S. Besra, D. S. Roos, and G. I. McFadden. 1998. Nuclear-encoded proteins target to the plastid in Toxoplasma gondii and Plasmodium falciparum. *Proc. Natl. Acad. Sci. USA* **95**:12352–12357.

Waller, R. F., M. B. Reed, A. F. Cowman, and G. I. McFadden. 2000. Protein trafficking to the plastid of Plasmodium falciparum is via the secretory pathway. *EMBO J.* **19**:1794–1802.

Waters, N. C., K. M. Kopydlowski, T. Guszczynski, L. Wei, P. Sellers, J. T. Ferlan, P. J. Lee, Z. Li, C. L. Woodard, S. Shallom, M. J. Gardner, and S. T. Prigge. 2002. Functional characterization of the acyl carrier protein (PfACP) and beta-ketoacyl ACP synthase III (PfKASIII) from Plasmodium falciparum. *Mol. Biochem. Parasitol.* **123**:85–94.

Weissig, V., T. Vetro-Widenhouse, and T. Rowe. 1997. Topoisomerase II inhibitors induce cleavage of nuclear and 35-kb plastid DNAs in the malarial parasite Plasmodium falciparum. *DNA Cell Biol.* **16**:1483–1492.

Wiesner, J., D. Henschker, D. B. Hutchinson, E. Beck, and H. Jomaa. 2002. In vitro and in vivo synergy of fosmidomycin, a novel antimalarial drug, with clindamycin. *Antimicrob. Agents Chemother.* **46**:2889–2894.

Williamson, D. H., P. W. Denny, P. W. Moore, S. Sato, S. McCready, and R. J. Wilson. 2001. The in vivo conformation of the plastid DNA of Toxoplasma gondii: implications for replication. *J. Mol. Biol.* **306**:159–168.

Williamson, D. H., M. J. Gardner, P. Preiser, D. J. Moore, K. Rangachari, and R. J. Wilson. 1994. The evolutionary origin of the 35 kb circular DNA of Plasmodium falciparum: new evidence supports a possible rhodophyte ancestry. *Mol. Gen. Genet.* **243**:249–252.

Williamson, D. H., P. R. Preiser, P. W. Moore, S. McCready, M. Strath, and R. J. Wilson. 2002. The plastid DNA of the malaria parasite Plasmodium falciparum is replicated by two mechanisms. *Mol. Microbiol.* **45**:533–542.

Wilson, C. M., A. B. Smith, and R. V. Baylon. 1996a. Characterization of the delta-aminolevulinate synthase gene homologue in P. falciparum. *Mol. Biochem. Parasitol.* **75:**271–276.

Wilson, R. J. M., P. W. Denny, P. R. Preiser, K. Rangachari, K. Roberts, A. Roy, A. Whyte, M. Strath, D. J. Moore, P. W. Moore, and D. H. Williamson. 1996b. Complete gene map of the plastid-like DNA of the malaria parasite *Plasmodium falciparum*. *J. Mol. Biol.* **261:**155–172.

Wilson, R. J. M., M. J. Gardner, J. E. Feagin, and D. H. Williamson. 1991. Have malaria parasites three genomes? *Parasitol. Today* **7:**134–136.

Wolters, J. 1991. The troublesome parasites: molecular and morphological evidence that Apicomplexa belong to the dinoflagellate-ciliate clade. *Biosystems* **25:**75–84.

Xu, P., G. Widmer, Y. Wang, L. S. Ozaki, J. M. Alves, M. G. Serrano, D. Puiu, P. Manque, D. Akiyoshi, A. J. Mackey, W. R. Pearson, P. H. Dear, A. T. Bankier, D. L. Peterson, M. S. Abrahamsen, V. Kapur, S. Tzipori, and G. A. Buck. 2004. The genome of Cryptosporidium hominis. *Nature* **431:**1107–1112.

Yoon, H. S., J. D. Hackett, G. Pinto, and D. Bhattacharya. 2002. The single, ancient origin of chromist plastids. *Proc. Natl. Acad. Sci. USA* **99:**15507–15512.

Yung, S., T. R. Unnasch, and N. Lang-Unnasch. 2001. Analysis of apicoplast targeting and transit peptide processing in Toxoplasma gondii by deletional and insertional mutagenesis. *Mol. Biochem. Parasitol.* **118:**11–21.

Yung, S. C., T. R. Unnasch, and N. Lang-Unnasch. 2003. *cis* and *trans* factors involved in apicoplast targeting in Toxoplasma gondii. *J. Parasitol.* **89:**767–776.

Zhang, Z., B. R. Green, and T. Cavalier-Smith. 2000. Phylogeny of ultra-rapidly evolving dinoflagellate chloroplast genes: a possible common origin for sporozoan and dinoflagellate plastids. *J. Mol. Evol.* **51:**26–40.

Zhu, G., M. J. Marchewka, and J. S. Keithly. 2000. Cryptosporidium parvum appears to lack a plastid genome. *Microbiology* **146:**315–321.

Zuegge, J., S. Ralph, M. Schmuker, G. I. McFadden, and G. Schneider. 2001. Deciphering apicoplast targeting signals—feature extraction from nuclear-encoded precursors of Plasmodium falciparum apicoplast proteins. *Gene* **280:**19–26.

PROTEIN KINASES REGULATING *PLASMODIUM* PROLIFERATION AND DEVELOPMENT

Christian Doerig

15

INTRODUCTION

The Eukaryotic Cell Cycle

The last decades of the 20th century have witnessed tremendous progress in our understanding of the molecular mechanisms pertaining to that crucial problem faced by eukaryotic cells: how to achieve cell division, ensuring that genetic information is accurately passed on to daughter cells. Most of this progress has been made through the use of yeast models, which are well suited to genetic approaches aiming at identifying genes involved in a given cellular process. Such approaches led to the discovery of cyclin-dependent protein kinases (CDKs) and associated proteins as major players in the regulation of cell cycle progression and to the characterization of upstream signal transduction pathways affecting the activity of the cell cycle machinery and hence the proliferation status of the cell. Subsequent investigations on the homologs of yeast cell cycle regulators in other species revealed that the mechanisms involved in cell division are largely conserved from yeast to metazoans. A thorough description of cell cycle regulation as currently understood in the yeast or mammalian cell systems is outside the scope of this chapter but can be found in recent reviews (McGowan, 2003; Wittenberg and La Valle, 2003). We will, however, briefly consider the major features of cell cycle control, to provide a general context for our discussion of *Plasmodium* cell proliferation regulators.

CELL CYCLE PHASES

The classical cell cycle is composed of four phases: G_1, in which the cell grows and accumulates the biochemical resources necessary for genome duplication; S (for DNA synthesis), in which the genome is replicated once; G_2, in which the cell prepares for cell division; and M (for mitosis), in which the two genomes are segregated and which leads to cytokinesis. Various checkpoints ensure that a given phase is not initiated prior to completion of the previous phase.

THE CYCLIN-DEPENDENT KINASES AND ASSOCIATED PROTEINS

Progression through the cell cycle phases is controlled by the cyclin-dependent protein kinases (Morgan, 1997). These enzymes phosphorylate

---

Christian Doerig, INSERM U 609, Wellcome Centre for Molecular Parasitology, University of Glasgow, 56 Dumbarton Rd., Glasgow G11 6NU, Scotland, United Kingdom.

Molecular Approaches to Malaria, Edited by Irwin W. Sherman
© 2005 ASM Press, Washington, D.C.

a number of substrates involved in processes such as in initiation of DNA synthesis or chromosome segregation (Ubersax et al., 2003). CDKs are inactive as monomers. Cyclin binding causes a rearrangement of the tertiary structure of the kinase subunit, allowing access of substrates to the catalytic cleft, thereby activating the enzyme. Any given cyclin is present transiently only within specific time windows during the cycle and therefore confers activity to its cognate CDK subunit(s) (whose expression is not in general as stringently time regulated as that of cyclins) only at the appropriate time. As an example, a major mechanism for the transition of the G_0 resting state to G_1 (and therefore entry into the cell cycle) is the induction of transcription of cyclin D. Cyclin D then stimulates CDK4/CDK6 activity and thereby promotes entry into the cell cycle through phosphorylation of the retinoblastoma protein, which indirectly triggers the transcription of proteins required for G_1 and S phases (McGowan, 2003). A number of CDKs are found in mammalian cells, although only a subset of those (CDK1, CDK2, CDK3, CDK4, and CDK6) are directly involved in cell cycle control, while the others have roles in the regulation of transcription (CDK8, CDK9, CDK10, and CDK11) or neuronal functions (CDK5). The CDK7-cyclin H pair has been implicated both in cell cycle control, as it is able to phosphorylate other CDKs and thereby enhance their activity, and in transcriptional regulation. Likewise, many cyclins have been identified. Each cyclin interacts only with a limited subset of CDKs (in some cases with only one) and vice versa. CDK activity is also regulated by other means in addition to cyclin binding: some CDKs are further stimulated by phosphorylation on a conserved threonine residue (Thr161 in CDK1) by a CDK-activating kinase (CAK). As mentioned above, CDK7 is a CAK, but there are other CAKs which are not CDKs themselves. Inversely, phosphorylation on conserved residues in the ATP-binding pocket (Thr14 and Tyr15 in CDK2) by the Myt1 and Wee1 kinases inhibits CDK activity, a block that can be released by phosphatases of the CDC25 family (Solomon and Kaldis, 1998).

Another way of negatively regulating CDKs is through the binding of specific inhibitory proteins of the p16 (targeting CDK4 and CDK6) and p21 (targeting CDK1 and 2) families (Sherr and Roberts, 1999).

Upstream Signaling Pathways

The molecular machinery controlling cell cycle progression is in essence a mere effector of signaling pathways, which are activated by a variety of intra- or extracellular stimuli (Boonstra and van Rossuml, 2003; Kahl and Meansl, 2003; Ryves and Harwood, 2003). Most of these pathways implicate phosphorylation cascades, which lead to the modulation of the activity of cell cycle regulators, either directly (e.g., mitogen-activated protein kinase [MAPK]-dependent cell cycle arrest during yeast mating type differentiation) (Gustin et al., 1998) or indirectly through the phosphorylation-dependent activation of transcription factors which in turn express cell cycle regulators (e.g., expression of cyclin D, causing entry of resting cells into G_1 as outlined above) (McGowan, 2003).

A well-characterized example of pathways linking extra- or intracellular signals to cell cycle effectors is that of the MAPK pathways. The core of the MAPK pathways is composed of a three-component module containing the MAPK, a MAPK kinase or MAPKK, and a MAPKKK. The latter can be stimulated in a number of ways, documented upstream signaling steps including ligand binding, heterotrimeric G-protein dissociation, *ras* activation, or cross talk from another pathway. Eukaryotic cells coexpress several MAPK pathways, each of which is responsible for the transduction of a particular type of signal. Hence, in *Saccharomyces cerevisiae*, five different MAPK modules regulate mating-type differentiation, filamentation, high-osmolarity responses, cell wall remodeling, and sporulation (Gustin et al., 1998). Similarly, various mammalian MAPK modules regulate cell growth, proliferation, differentiation, stress responses, and apoptosis (Schaeffer and Weber, 1999). Several other signal transduction pathways involved in the control of cell proliferation operate in the cell, and extensive cross talk between these has been documented. These include cyclic nucleotide pathways and calcium and phospholipid signaling. It is outside the scope of this chapter to review this important field, and the interested reader is directed to specific reviews (Stork and Schmitt, 2002; Boonstra and van Rossum, 2003; Kahl and Means, 2003; Ryves and Harwood, 2003; Wittenberg and La Valle, 2003).

Atypical Features of Cell Cycle Organization in Malaria Parasites

The life cycle of malaria parasites is an alternation of developmental stages where the parasite is cell cycle arrested, and stages undergoing intense cell division. Erythrocytic schizogony is the stage that has received the most attention from researchers, because (i) it is the most amenable to the provision of biological material, and (ii) it is responsible for malaria pathogenesis. The organization of the cell cycle during erythrocytic schizogony of malaria parasites deviates considerably from that in the yeast or mammalian cell systems, a likely consequence of the early divergence of the Alveolata (which include the Apicomplexa) from the Opisthokonta (which include both metazoans and fungi, explaining the close similarity between the cell cycle control features between these two groups) (Baldauf, 2003). The merozoite and ring stages are likely to be in a state analogous to the G_1 phase, with S phase being initiated some 18 h postinvasion (Arnot and Gull, 1998). Microfluorometry measurements with *Plasmodium berghei* indicated that 18 h postinvasion the haploid amount of DNA is multiplied 16 to 20 times within 4 to 6 h (Janse et al., 1986). The nuclear multiplication following entry into S phase is difficult to describe in terms of classical cell cycle phases: nuclear divisions appear to be asynchronous within a given schizont, as evidenced by the almost Gaussian distribution of nuclear body numbers per schizont and the presence of mitotic spindles at different stages of development within a single schizont (Read et al., 1993). Parasites with three nuclei or triploid DNA content also occur in *Toxoplasma gondii* (Hu et al., 2002), suggesting that asynchronous nuclear division may be widespread in Apicomplexa. This makes biochemical investigations of cell cycle phase transitions very difficult (or impossible) to achieve, even with synchronized parasite populations. How is asynchronous nuclear division mediated, considering that all nuclei are subjected (presumably) to the same cytoplasmic environment? Is this linked to the fact that the nuclear membrane remains intact during nuclear divisions? Is the genome duplicated once in each dividing nucleus prior to nuclear mitosis, or is DNA replication a more continuous process, with the ontogeny of individual nuclei occurring from a pool of available replicated genomes? Such very basic questions still remain to be answered. Flow cytometry of parasite cultures using DNA stains allowed detection of erythrocytes containing more than one haploid parasite DNA amount, but most of these may represent multiply infected cells (Janse et al., 1987; Janse and Van Vianen, 1994). Preliminary data obtained by fluorescence-activated cell sorter analysis of early schizonts released from erythrocytes by saponin lysis (thereby excluding that multiple N DNA content arises from red blood cells infected with more than one merozoite) indicate that parasites distribute in discrete peaks with 1, 2, or 3 N DNA content, suggesting that genomes are replicated one at a time, at least in the early stages of schizogony, but this has to be ascertained by further experiments (K. Grant, C. M. Doerig, and C. Doerig, unpublished data).

Hence, *Plasmodium* cell division diverges in fundamental ways from the binary division observed with mammalian cells. As will be discussed below, this is reflected in peculiarities in components of the cell cycle machinery itself.

Control of Malaria Parasite Multiplication in Response to Environmental Factors: the Example of Male Gametogenesis

The alternation of actively dividing and cell cycle-arrested developmental stages during the life cycle of malaria parasites must be associated with an efficient and versatile cell cycle control machinery, whose activity needs to be integrated with specific cell development programs. By analogy with other experimental systems where the links between cell differentiation and cell cycle regulation are relatively well understood, it is likely that at stage transition, environmental factors trigger specific signal transduction pathways and that the cell cycle control machinery is one of the effectors of such pathways. That this is indeed the case has, to our knowledge, been well documented only for one stage transition in the malaria parasite life cycle: the development from male gametocyte to male gamete (for a full discussion of this phenomenon, see Chapter 10). Briefly, this process, called exflagellation, is characterized by very rapid (approximately 15 min) and profound morphological changes, with the formation of eight flagellated cells from each microgametocyte (Sinden et al., 1996). Exflagellation occurs upon ingestion of the male gametocyte by the mosquito, and the parasite uses a drop in temperature (from 37°C in the bloodstream of the vertebrate host to 20 to 25°C in the mosquito), as well as the presence of xanthurenic acid (a by-product of insect eye pigment synthesis) as signals to trigger gametogenesis (Billker et al., 1998; Garcia et al., 1998). Early studies have demonstrated that exflagellation is accompanied by the release of cyclic GMP (cGMP) and diacylglycerol, implicating cyclic nucleotide and phospholipid/calcium signaling pathways in the process (Kawamoto et al., 1990; Martin et al., 1994); more recently, evidence has been gathered for the release of calcium upon xanthurenic treatment (Billker et al., 2004). The implication of cyclic nucleotide and calcium signaling in exflagellation has been corroborated on one hand by the observation that xanthurenic acid is able to activate a guanylate cyclase activity located in the gametocyte-derived membrane fraction (Muhia et al., 2001), and on the other hand by a reverse genetics study demonstrating that a *Plasmodium falciparum* calcium-dependent protein kinase, PfCDPK4, is required for *P. berghei* exflagellation to occur (Billker et al., 2004). Another reverse genetics study identified a crucial role for a gametocyte-specific MAPK, Pbmap-2 (the ortholog of previously characterized Pfmap-2) (Dorin et al., 1999), in exflagellation (Rangajaran et al., 2005).

Hence, release from cell cycle arrest in the gametocyte and further development in gametes require the coordinated activation of several transduction pathways. The molecular mechanisms through which pathway stimulation reactivates the cell cycle machinery remain to be elucidated. It is likely that phosphorylation/dephosphorylation of key elements such as CDKs and associated proteins will emerge as a major feature of such mechanisms.

Although gametogenesis is to our knowledge the best-understood example of the parasite's cell cycle response to well-defined signals, it is to be expected that similar phenomena occur throughout the *Plasmodium* life cycle. Indeed, both phenomenological observations and identification of specific genes provide evidence that the parasite has developed many transduction elements belonging to the cyclic AMP (cAMP)-, cGMP- and MAPK-related pathways, as well as to calcium and phospholipid signaling. It is, however, outside the scope of this chapter to review these data in detail; see Doerig, 1997; Doerig and Chakrabarti, 2004; and "Upstream Signal Transduction," below, for a short discussion of selected database mining results relevant to these pathways.

The Search for Cell Cycle Control Proteins in Malaria Parasites

The underlying principles of cell division control at the molecular level and the identity of key players in this process such as CDKs, cyclins, and CDK inhibitors (CKIs) have been elucidated mostly through genetic analysis in yeast. This powerful approach could not be implemented to conduct similar studies with malaria parasites, mostly because the complexity of the life cycle makes classical genetic crosses difficult to achieve in sufficient numbers; furthermore, no system to obtain conditional mutants has yet been developed for *Plasmodium* species. Therefore, the strategy to identify cell cycle control elements that was adopted until the advent of the genomic database was to PCR amplify putative homologs from parasite cDNA using degenerate primers targeting conserved regions. This led to the identification of several protein

TABLE 1 The plasmodial CMGC (CDK, MAPK, GSK3 and CDK-like) protein kinases and associated proteins[a]

Name	Family	PlasmoDB identifier	Max (h)	Min (h)	Reference(s)
Pfmrk	CDK	PF10_0141	10	37	Li et al., 1996; Le Roch et al., 2000; Waters et al., 2000
Pfcrk-3	CDK	PFD0740w	12	34	
Pfmap-2	MAPK	PF11_0147	13	41	Dorin et al., 1999; Dorin et al., 2001
Pflammer	CLK	PF14_0431	13	32	Li et al., 2001a
	CLK	PFC0105w	14	40	
Pfmap-1	MAPK	PF14_0294	24	8/43	Doerig et al., 1996; Lin et al., 1996; Graeser et al., 1997
PfCK2c	CKII	PF11_0096	6–25	41	
Pfcrk-1	CDK	PFD0865c	28	6	Doerig et al., 1995
PfPK5	CDK	MAL13P1.279	28	8	Ross-Macdonald et al., 1994; Le Roch et al., 2000
	CMGC related	MAL13P1.196	30	11	
PfGSK3	GSK3	PFC0525c	33	9	Droucheau et al., 2004
Pfcrk-5	CMGC related	MAL6P1.271	33	10	
Pfcrk-4	CMGC related	PFC0755c	37	11	
	CLK	PF14_0408	38	15	
	CLK	PF11_0156	39	17	
PFPK1	GSK3	PF08_0044	45/5	30	Kappes et al., 1995
	GSK3	MAL13P1.84			
PfPK6	CMGC related	PF13_0206	Trophs		Bracchi-Ricard et al., 2000; Le Roch et al., 2000; Li et al., 2001
Pfcyc-1	Cyclin (H)	PF14_0605	11	41	Le Roch et al., 2000
Pfcyc-3	Cyclin	PFE0920c	6–25	40	Merckx et al., 2003
Pfcyc-4	Cyclin (L)	PF13_0022	27	8/40	Merckx et al., 2003
Pfcyc-2	Cyclin (mitotic)	PFL1330c	30	18	Merckx et al., 2003
PfCKIIR1	CKII regulatory	PF11_0048	6–25	40	
PfCKIIR2	CKII regulatory	PF13_0232	10–25	41	

[a]The entries are sorted according to the timing of their expression during schizogony, as determined by Bozdech et al. (2003) and detailed in Fig. 2. The times of maximal and minimal relative expression are indicated. Two values separated by a slash (/) are given for those genes whose pattern of expression show two maxima or minima over one cycle (e.g., Pfmap-1); two values separated by a hyphen indicate a plateau with no well-defined peaks. Trophs, maximal PfPK6 Northern blot signal in trophozoites (Bracchi-Ricard et al., 2000). No microarray data are available for this gene in the Bozdech et al. study.

kinases, including the CDK-related kinases PfPK5 (Ross-Macdonald et al., 1994), PfPK6 (Bracchi-Ricard et al., 2000), Pfmrk (Li et al., 1996), and Pfcrk-1 (Doerig et al., 1995), as well as the MAPK homologs Pfmap-1 and Pfmap-2 (Doerig et al., 1996; Lin et al., 1996; Dorin et al., 1999) and two members of the glycogen synthase kinase 3 (GSK3) family (PfPK1 and PfGSK3) (Kappes et al., 1995; Droucheau et al., 2004). Despite attempts from several laboratories, neither cyclin- nor CKI-encoding genes were found by this approach. However, thanks to its policy of making sequence data available on the Web as they were being generated (instead of waiting for a complete genomic database, which became available at the end of 2002), the *Plasmodium falciparum* Genome Consortium allowed fast progress to be made from the late 1990s onwards: the first cyclin was cloned and characterized in 2000 (Le Roch et al., 2000), three more cyclins were reported in 2003 (Merckx et al., 2003), and a recent exhaustive study of the complement of protein kinase-encoding genes (Ward et al., 2004) led to the identification of a novel CDK-related enzyme (Pfcrk-5). Table 1 summarizes the catalogue of CMGC (for CDK,

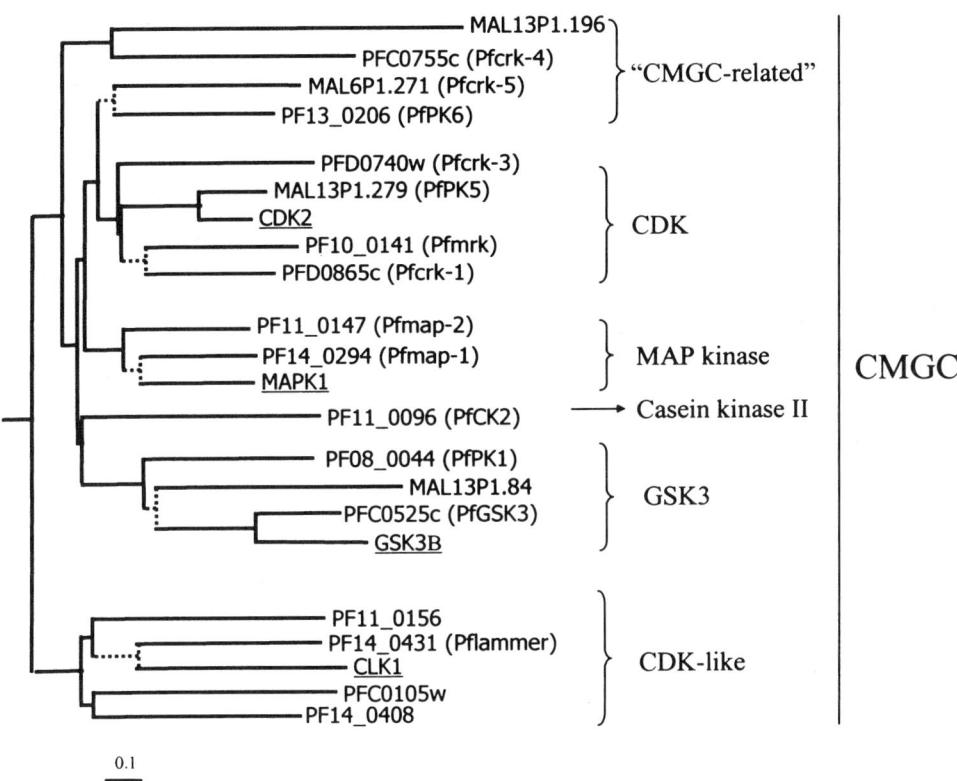

FIGURE 1 Plasmodial protein kinases clustering within the CMGC family. The figure depicts only the CMGC branch of a larger tree constructed (by J. Packer, Abbott Laboratories) from a Hidden Markov Model-derived alignment of all protein kinases in the *P. falciparum* genome. Four human protein kinases (underlined) representing the four major families in the CMGC group (CDKs, MAPKs, GSK3, and CDK-like) were included in the alignment to anchor the position of these families in the tree. Branches with bootstrap values >40 are shown with dashes. The scale bar represents 0.1 mutational changes per residue (10 PAM units). Although some plasmodial sequences clearly cluster with established families (e.g., PfPK5 with CDK1/2 or Pfmap-1 and Pfmap-2 with the MAPKs), others (e.g., those labeled "CMGC-like" on the figure, because their branch originates near the base of the branch containing the CDK, MAPK and GSK3 groups) are much more difficult to classify. Adapted from Ward et al., 2004, which should be consulted for details, including bootstrap values.

MAPK, GSK3, and CDK-like) kinases, cyclins, and potentially associated proteins present in the *P. falciparum* genome; Fig. 1 gives an indication of their phylogenetic relationships.

CDK- AND CYCLIN-RELATED PROTEINS IN ORDER OF APPEARANCE DURING ERYTHROCYTIC SCHIZOGONY

No experimental functional data are available that assign clear roles for the CDK-related kinases and associated proteins in the process of parasite multiplication. However, microarray and proteomics data (available on PlasmoDB) (Bahl et al., 2002) can provide some basis for speculation on the possible function these proteins. Rather than providing an exhaustive list of the biochemical properties of CDK-related kinases and cyclins (which has been done previously) (Doerig et al., 2002), we will focus on considering their possible role in cell division on the basis of the timing of their expression during erythrocytic schizogony, which is summarized in Table 1 and Fig. 2. Among the 18 plasmodial

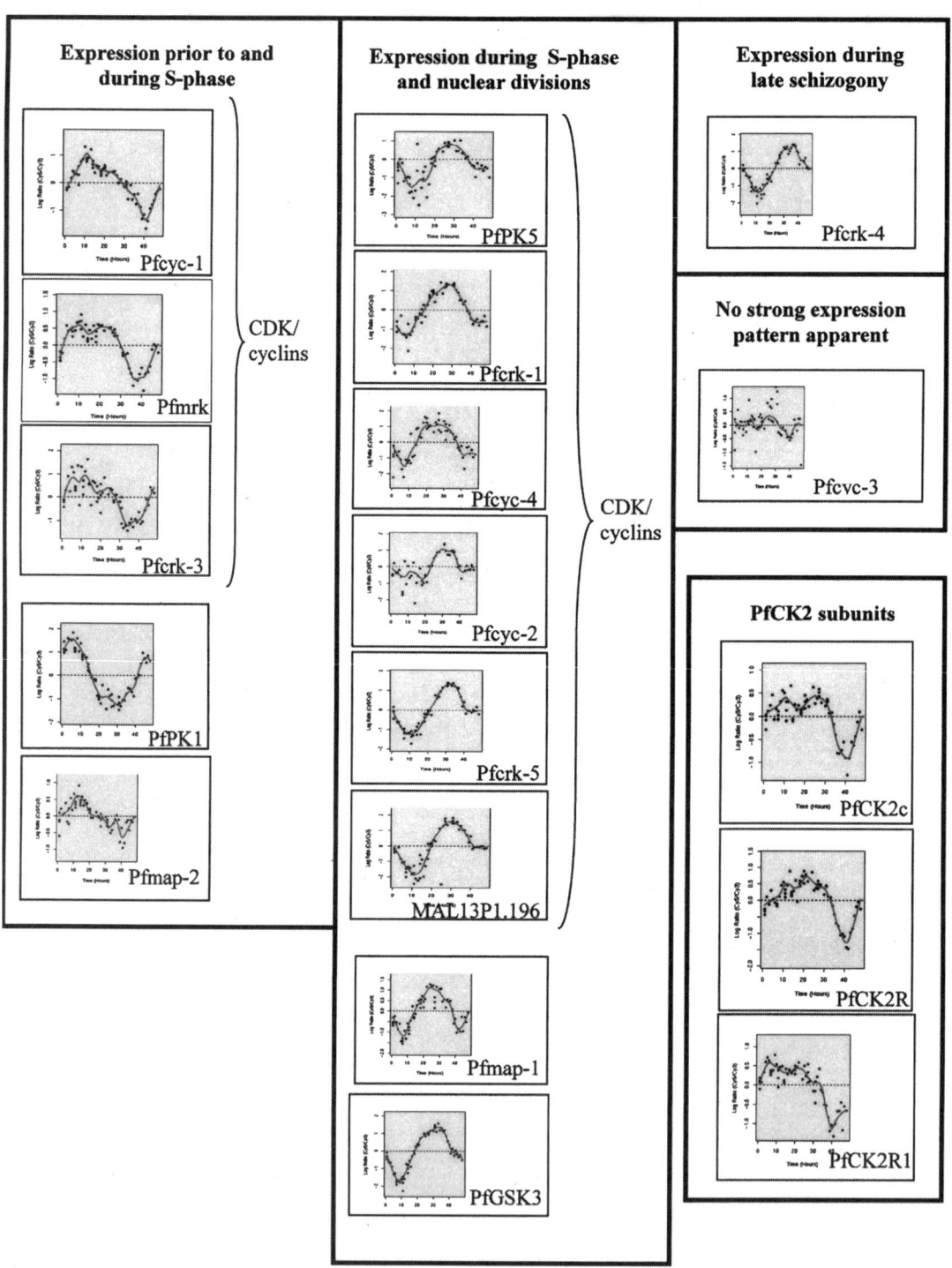

FIGURE 2 Raw microarray expression data for the CDK-related protein kinases and cylins compiled from the files from the Bozdech et al. study and made available on PlasmoDB (www.plasmodb.org). See Bozdech et al., 2003, and the PlasmoDB website for details.

protein kinases that cluster with the CMGC family (Table 1), we will consider in the next section only those which cluster within the CDK- and CMGC-related branches (Fig. 1) and for which at least some biochemical data are available, as well as the four plasmodial cyclin-related proteins characterized so far.

CDK- and Cyclin-Related Proteins Whose mRNA Is Expressed prior to the Onset of DNA Synthesis

As mentioned above, merozoites, rings, and early trophozoites are thought to be in a state equivalent to the G_1 phase of higher eukaryotes. Two CDK-related kinases, Pfmrk and Pfcrk-3, and one of the four cyclins characterized (Pfcyc-1) are expressed in these stages at the mRNA level.

Pfmrk

Pfmrk (MO15-related kinase) was identified during a PCR-based search for genes encoding CDK-related proteins (Li et al., 1996). Pfmrk shares 46% identity with the human CDK-activating kinase MO15, also known as CDK7, in comparison to 34 to 43% identity with other human CDKs. In mammalian cells, the CDK7-cyclin H pair is involved both in cell cycle control (through its ability to activate other CDKs on a conserved Thr residue in the so-called T-loop) and in the regulation of transcription (through its ability to phosphorylate the repeated heptapeptide in the carboxy-terminal domain [CTD] of RNA polymerase II) (Kaldis, 1999; Stiller and Hall, 2002). Pfmrk is one of the first CDK-related kinases to be expressed during schizogony at a stage where (according to the microarray data available on PlasmoDB) (Bozdech et al., 2003) the only other CDK-related kinase to be expressed is Pfcrk-3 (see below). This suggests that the latter enzyme may be a substrate for Pfmrk, a proposition that should be easily testable with recombinant enzymes. Since large-scale mRNA synthesis starts prior to S phase (Arnot and Gull, 1998), it is possible that Pfmrk also plays a role in transcription. Indeed, in the presence of the Pfcyc-1 cyclin (see below), recombinant Pfmrk is able to phosphorylate the CTD of mammalian RNA polymerase II (Li et al., 2001b). It is worthwhile mentioning here that the *P. falciparum* RNA polymerase II (PlasmoDB gene identifier PFC0805w) possesses a CTD, albeit with a non-canonical heptapeptide repeated motif which nevertheless contains potential CDK phosphorylation sites (Stiller and Hall, 2002). The two functions, transcription and CDK activation, are of course not mutually exclusive, as is well established by the dual role of mammalian CDK7. Obviously, Pfmrk protein can be maintained until later stages of parasite development and hence may have activity on additional CDK-related enzymes, even though its mRNA levels appear to decrease from 30 h onwards. However, no effect of Pfmrk–Pfcyc-1 (either phosphorylation or activation) was observed in vitro on recombinant PfPK5, the putative CDK1 malarial homolog (Le Roch et al., 2000). This may reflect a true absence of CAK activity or be a consequence of a lack of an essential cofactor (such as a functional homolog of MAT1, a protein required for full CDK7 activity) in this in vitro assay.

Pfcrk-3

This kinase is related to CDKs over its catalytic domain (40% identity at the amino acid level with *Schizosaccharomyces pombe* p34^{cdc}, excluding the insertions in the Pfcrk-3 catalytic domain) and clusters within this family on a phylogentic tree (Fig. 1). However, it displays very atypical features such as a large extension at its N terminus and large insertions within the catalytic domain itself, and it is difficult to predict a function on the basis of its sequence. The catalytic domain has been expressed in *Escherichia coli*, but the recombinant protein lacks kinase activity even in the presence of human or malarial cyclins (see below). However, immunoprecipitation from asexual parasite extracts using immunopurified antibodies against Pfcrk-3 peptides allows the recovery of histone H1 kinase activity, indicating that the protein either possesses or at least is associated with protein kinase activity (L. Equinet, K. Le Roch, and C. Doerig, unpublished data). The pattern of Pfcrk-3 mRNA levels during the asexual cycle is similar to that of Pfmrk transcripts (Fig. 2).

Pfcyc-1

Pfcyc-1 is the only cyclin whose mRNA is detectable prior to S phase. Interestingly, the highest scores obtained in BLASTP analysis of generalist databases are cyclin H homologs from a variety of organisms. Since in mammalian cells cyclin H is the cognate activator of CDK7, it is tempting to speculate that Pfcyc-1 might be a partner for Pfmrk, because the latter shows highest homology to CDK7/MO15 (see above) and appears from microarray data to be coregulated with Pfcyc-1. Recombinant Pfcyc-1 is indeed able to stimulate Pfmrk-dependent phosphorylation of mammalian CTD or histone H1 in vitro (Li et al., 2001), and it would seem at first sight that this nicely mimics the mammalian CDK7-cyclin H pair. However, things are not that simple, as Pfcyc-1 has unexpectedly been shown to be a much better activator of PfPK5 (a CDK1 putative homolog, see below) than of Pfmrk (Le Roch et al., 2000)! This illustrates the difficulty of predicting specific functions (biochemical or biological) for *Plasmodium* cell cycle elements on the sole basis of their sequence. Furthermore, although some Pfcyc-1 protein is detectable in rings, in line with the microarray data, protein levels peak at the trophozoite stage and appear to mimic those of PfPK5 protein (Merckx et al., 2003), suggesting that these two polypeptides may indeed function as a complex in vivo. Both pull-down experiments and immunoprecipitation using immunopurified antibodies directed against a Pfcyc-1-derived peptide demonstrate that this protein in able to associate with a histone H1 kinase activity in parasite extracts (Merckx et al., 2003), but the identity of the kinase subunit responsible for this activity remains to be determined.

CDK- and Cyclin-Related Proteins Whose mRNA Peaks during S Phase and Nuclear Division

Because of the asynchrony of nuclear division and of the fact that DNA synthesis is still ongoing when nuclear mitoses begin, it is impossible to distinguish between gene products expressed in the plasmodial equivalents of S and M (and G_2) phases (provided true such equivalents exist, which is not established). Nevertheless, examination of microarray expression profiles indicates that several putative cell cycle regulators are not expressed in early stages but that their mRNA levels become detectable in trophozoites, at the time of S-phase onset and then drop again before schizont segmentation (Bozdech et al., 2003; Le Roch et al., 2003). These include PfPK5, the malarial CDK-related kinase that is the most closely related to CDK1 of mammalian and yeast cells, Pfcrk-5 and Pfcrk-1, as well as two cyclins, Pfcyc-2 and Pfcyc-4. PfPK6, although not associated with data in the microarray study of Bozdech et al. (2003), was found in the study by Le Roch et al. (2003) to be expressed in trophozoites and early schizonts.

PfPK6

This enzyme was identified in the context of a search for genes that are expressed at specific stages of intraerythrocytic asexual development (Bracchi-Ricard et al., 2000). Reverse-transcriptase-mediated PCR-based differential display allowed the identification of the PfPK6 mRNA in parasites undergoing the transition from ring to schizont stages. Independently, the same gene was cloned, following a homology-based PCR approach with degenerate primers targeting regions conserved in members of the CDK family. PfPK6 displays similar levels of homology to CDKs and MAPKs in BLASTP analysis and in a phylogenetic analysis appears to cluster with Pfcrk-5 in a distinct branch originating at the base of the CDK family, within the CMGC group (Fig. 1). Although mammalian cells possess protein kinases, such as the MOK enzyme, that also show some relatedness to both MAPKs and CDKs (Miyata et al., 1999), PfPK6 is not closer to these than to either CDKs or MAPKs, indicating that they evolved independently. Expression during trophozoite maturation suggested by the reverse-transcriptase-mediated PCR-based differential display results was verified by Northern and Western blot analysis, and more recently this was confirmed by the Le Roch et al. (2003) microarray analysis, in which PfPK6 mRNA was detected in trophozoites and early schizonts. This suggests that PfPK6 may be involved in the control of S phase; its ability to phosphorylate in vitro a plasmodial subunit of ribonucleotide reductase, an enzyme that is required for the synthesis of DNA precursors, is consistent with this idea (Bracchi-Ricard et al., 2000).

Recombinant PfPK6 shows robust kinase activity in the absence of cyclin, which does not necessarily imply that the full activity is cyclin independent under physiological conditions. In contrast to that of other plasmodial CDK-related enzymes, the primary structure of PfPK6 does not contain any sequence that is related to a cyclin-binding site. PfPK5, for example, does contain a PSTAIRE-like sequence (PSTTIRE); Pfcrk-1 contains an AMTSLRE motif similar to PITSLRE in CDK11, and Pfmrk has an NFVLLRE motif reminiscent of the NRTALRE motif found in CDK7.

PfPK5

Of the six malarial CDK-related kinases, PfPK5 is that which displays the closest similarity to CDK1 homologs of higher eukaryotes. It shows 60% identity to both CDK1 and CDK5, but clusters within the CDK1/2/3 group (by opposition to the CDK5 group) in a phylogenetic study (Fig. 1) (Ward et al., 2004). It is probably as a consequence of its high level of homology to CDK1 that this enzyme was the first CDK-related enzyme to be identified (Ross-Macdonald et al., 1994). Although histone H1 kinase activity was detected in immunoprecipitates from parasite extracts, enzymatic activity of recombinant PfPK5 has very little activity as a monomer. This is expected, as CDKs are notoriously inactive in the absence of their cognate cyclin. Unexpectedly, however, histone H1 kinase activity can be stimulated by a factor of up to 4 orders of magnitude by a variety of cyclins and cyclin-related proteins, some of which, in other systems, are specific for non-CDK1/2 kinase subunits. The first piece of evidence for the promiscuity of PfPK5 with respect to cyclins came from a series of experiments aimed at determining whether PfPK5 could be phosphorylated (and activated) by a Pfmrk-human cyclin H complex: not only did coincubation of PfPK5 with the complex stimulate histone H1 activity by at least 3 orders of magnitude, it was also shown that this effect occurred in the absence of Pfmrk, human cyclin H being sufficient to stimulate PfPK5 activity. This unexpected result was extended by the demonstration that Pfcyc-1, p25 (a mammalian activator which is otherwise stringently specific for CDK5), bovine cyclin A, RINGO (an activator of CDK2 in vertebrates that shows no sequence similarity to cyclins), and Pfcyc-3 (see below) were all able to activate PfPK5 (Le Roch et al., 2000; Holton et al., 2003; Merckx et al., 2003). Structural data on PfPK5 complexed with small inhibitors (Holton et al., 2003) indicate that the protein is prone to adopt the active conformation, which is consistent with the cyclin promiscuity it displays. The possible physiological significance of this unexpected property of PfPK5 remains to be established.

The peak of PfPK5 mRNA levels in late trophozoites and schizonts is mirrored by the protein levels as detected by Western blot analysis (Merckx et al., 2003). This pattern of expression, together with the close relatedness to the CDK1/2/3 family, makes PfPK5 a good candidate for a major cell cycle regulator.

Pfcrk-1

Pfcrk-1 was identified in a PCR-based search for CDK1 homologs (Doerig et al., 1995). In addition to its catalytic domain, Pfcrk-1 has a large N-terminal extension with repeated amino acid motifs, which does not show homology to any known protein. Over the catalytic domain, however, Pfcrk-1 displays maximal homology to members of the PITSLRE family (Lahti et al., 1995), whose name is derived from the residues in the cyclin-binding region (in Pfcrk-1, this sequence reads AMTSRLE). Mammalian cells express several isoforms of PITSLRE enzymes, which are distributed in two groups, p100 and p58. Upon identification of cyclin L as a putative activator of the mammalian PITSLRE, they have been renamed CDK11 (Dickinson et al., 2002; Hu et al., 2003). The CDK11p110/p58 protein kinase isoforms in humans are encoded by two distinct but closely related genes (*cdc2L1* and *cdc2L2*), whereas a single CDK11p110/p58 gene (cdc2l) is found in the mouse genome (Mock et al., 1994; Gururajan et al., 1998). All three genes express 110-kDa isoforms, as well as p58 isoforms originating from an internal ribosome entry site sequence (Cornelis et al., 2000). The p110^{PITSLRE} isoforms are involved in the control of RNA splicing (Dickinson et al., 2002; Hu et al., 2003), whereas the p58^{PITSLRE} isoform is able to bind an S-phase cyclin and appears to participate more directly to cell cycle control (Zhang et al., 2002). In mice, PITSLRE null mutants are not viable, the mutation being lethal in homozygotes at the blastocyte stage (Li et al., 2004). Interestingly, Northern blot analysis of Pfcrk-1 mRNA isolated from gametocytes detects two distinct bands (Doerig et al., 1995), but whether or not these encode two isoforms with distinct functions remains to be investigated.

Pfcrk-1 was first found to be expressed preferentially in gametocytes, with only marginal amounts of mRNA detectable in asexual parasites (Doerig et al., 1995). Nevertheless, microarray data confirm expression in late trophozoites and early schizonts (Bozdech et al., 2003; Le Roch et al., 2003). This is consistent with the recent observation that Pbcrk-1, the Pfcrk-1 ortholog in *P. berghei*, is essential for asexual multiplication (Rangajaran et al., unpublished). In view of the (at least) dual function of members of the CDK11^{PITSLRE} family in mammalian cells, it would be interesting to see if distinct Pfcrk-1 isoforms are expressed at the different developmental stages where the gene is expressed, i.e., during schizogony and gametocytogenesis, respectively. As mentioned above, two cyclins (cyclin L and cyclin D3) are known to interact with CDK11 (the former with p110^{PITSLRE} isoforms, the latter with p58^{PITSLRE} isoforms). No cyclin D homologs have been found in the *Plasmodium* genome, but the highest BLASTP scores for Pfcyc-4 (see below) correspond to cyclin L homologs. Despite this promising observation, recombinant Pfcyc-4 was unable to activate the recombinant catalytic domain of Pfcrk-1 in vitro (L. Equinet and C. Doerig, unpublished data); this may be due to the absence of the Pfcrk-1 N-terminal extension in this experiment (after all, cyclin L associates with the p100 isoforms in mammals) or to the lack of additional essential cofactors.

Pfcrk-5

Unlike most CDK-related kinases discussed in this chapter, Pfcrk-5 escaped identification in PCR-based searches. Instead, its discovery had to await an exhaustive analysis of the *P. falciparum* kinome (Ward et al., 2004). Pfcrk-5 appears to be related to the CDK1/2/3 group according to BLASTP analysis but clusters with PfPK6 at a branch close to the base of the CDK group in a more refined phylogenetic analysis (Fig. 1). Like Pfcrk-3 and Pfcrk-4, Pfcrk-5 carries a large insertion in the catalytic domain, but in contrast to these kinases it does not display a large N-terminal extension. Immunoprecipitates from asexual parasites using immunopurified immunoglobuin Ys (IgYs) directed against a Pfcrk-5-derived peptide are associated with strong histone H1 kinase activity (M. P. Nivez, D. Goldring, and C. Doerig, unpublished data), but currently nothing is known of the possible partnership of this enzyme with cyclins.

Pfcyc-2

Microarray data indicate that the mRNAs from two genes encoding cyclin-related proteins, Pfcyc-2 and Pfcyc-4, peak in late trophozoites and schizonts. The structure of the Pfcyc-2 open reading frame (ORF) is very unusual among cyclin-encoding genes: the 390-amino-acid region with low homology to mitotic (A and B) cyclins, and which contains the cyclin signature as defined in the Pfam conserved domain database, is embedded within a large ORF potentially encoding a polypeptide of approximately 2,300 amino acids. However, in Western blot analysis of synchronized asexual parasites using immunopurified IgY directed against a peptide from the 390-residue cyclin-like region, large polypeptides represent only a minor fraction of the total signals. Indeed, the major protein species has an apparent molecular mass of 30 kDa only, pointing to a possible proteolytic maturation resulting in a separate polypeptide carrying only the cyclin-like region (this hypothesis remains to be investigated experimentally) (Merckx et al., 2003). Immunopurified anti-Pfcyc-2 IgYs allowed the immunoprecipitation of a weak histone H1 kinase activity; however, considering that pull-down experiments with a recombinant protein containing the 390-residue kinase-like region did not provide evidence for association with a kinase activity, it remains unclear whether or not Pfcyc-2 functions as a cyclin. In line with the microarray data, the Pfcyc-2 30-kDa protein does peak at the trophozoite/schizont stages.

Pfcyc-4

This protein displays maximal homology to cyclins of the AniA/cyclin L family, which are characterized by the presence of two cyclin boxes (a feature shared by Pfcyc-4). As mentioned above, in mammalian cells cyclin L is a partner for $p110^{PITSLRE}$, a member of the CDK11 family (Dickinson et al., 2002; Hu et al., 2003); the relatedness of Pfcrk-1 to this family points to the possibility that Pfcrk-1 and Pfcyc-4 form a functional pair, but this has not been confirmed experimentally. It has been shown, however, that Pfcyc-4 associates with a histone H1 kinase activity in parasite extracts, both by pull-down and immunoprecipitation experiments (Merckx et al., 2003), but the identity of the kinase subunit remains to be determined. Interestingly, immunopurified IgYs against an N-terminal peptide derived from Pfcyc-4 (calculated molecular mass, 31.1 kDa) detect two major bands with distinct expression patterns: a 31-kDa band peaks in trophozoites and a 39-kDa species that clearly shows maximum expression in segmenters, with high levels maintained in rings (Merckx et al., 2003).

CDK- and Cyclin-Related Proteins Whose mRNA Peaks in Late Schizonts and Segmenters

Pfcrk-4

Very little is known about this protein. Pfcrk-4 is very similar to Pfcrk-3 in terms of primary structure: the kinase catalytic domain is preceded by a large insertion at its N terminus, and large insertions are present within the catalytic domain itself. BLASTP analysis indicates that Pfcrk-4 is related to CDKs and MAPKs, like PfPK6. As for PfPK6, phylogenetic analysis indicates that Pfcrk-4 is relatively distantly related to CDKs and MAPKs, forming a distinct branch that also includes uncharacterized MAL13P1.196 and which is located at the base of the major branch containing the CDK, MAPK, and GSK3 groups (Fig. 1). Western blots using immunopurified IgYs against a peptide derived from Pfcrk-4 indicate that the protein is expressed in blood stages, and a weak histone H1 kinase activity has been recovered in immunoprecipitates (L. Equinet, D. Goldring, and C. Doerig, unpublished data). In the absence of additional data, it is not clear at this stage whether this activity originates in Pfcrk-4 itself or in another coprecipitating enzyme.

CDK- and Cyclin-Related Proteins Whose mRNA Does Not Show Strong Regulation during Schizogony

Pfcyc-3

This protein is most similar to a *Trypanosoma* cyclin that was cloned by complementation in a yeast mutant deficient in G_1 cyclin, but it does not show particular relatedness to any specific class of mammalian cyclins. Could it be that Pfcyc-3 represents a functional homolog of G_1/S-phase cyclins D or E, for which no sequence homologs are apparently present in the *P. falciparum* genome? As a recombinant protein, it is able to activate PfPK5, and immunopurified IgYs against a Pfcyc-3-derived peptide allow the immunoprecipitation of a histone H1 kinase activity. Unfortunately, these antibodies did not work in Western blots, precluding analysis of the expression pattern of the protein during parasite development until other antibodies become available (Merckx et al., 2003).

Potential Substrates of the CDK-Related Kinases

To our knowledge, the only documented evidence for a plasmodial protein acting as a substrate for a CDK-related kinase is that concerning the in vitro phosphorylation of ribonucleotide reductase by PfPK6 (see above). However, the genomic database allows the identification of many proteins that are likely to be

CDK substrates. To cite a few examples (Fig. 3), several polypeptides with high homology to proteins directly involved in the control of DNA synthesis initiation, including components of the minichromosome maintenance complex (Forsburg, 2004) and origin recognition complex (ORC) (Dutta and Bell, 1997; Kelly and Brown, 2000), have been identified; expression of these elements peaks between 25 and 30 h postinvasion, corresponding as expected to the time of DNA synthesis. By analogy with the mechanisms of DNA synthesis initiation in other eukaryotes, it is likely that some of these proteins are regulated by CDK-dependent phosphorylation. It is noteworthy that the complement of ORC components appears to be reduced in *Plasmodium* in comparison to that in higher eukaryotes, with three ORC family members instead of six in mammalian cells. In contrast, all six subunits of minichromosome maintenance complex have been identified. Other key regulators of prereplication complex formation (Cdt1, Dbf4p-dependent kinase, and Cdc45) (Dutta and Bell 1997; Kelly and Brown 2000) appear to be absent. Thus, it appears that in comparison to the eukaryotic model, the malaria parasite may have a minimal prereplication complex for the onset of S phase, only slightly more complex than the complex in archeae (D. Chakrabarti, personal communication).

Likewise, putative homologs of components of the so-called mitotic exit network (McCollum and Gould, 2001) have been identified, some of which may represent substrates for CDKs (Fig. 3). A last example is that of the NIMA-related protein kinases (NIMA stands for never in mitosis/*Aspergillus*, because null mutants in this gene, first identified in the fungus *Aspergillus nidulans*, do not enter mitosis) (O'Connell et al., 2003). In other systems, members of this family play major roles in cell division control, for example through regulating centrosome function in a CDK-dependent way. Five NIMA-related kinases have been identified in the malarial genome, including the previously described Pfnek-1 (see below). Hence, the availability of a genomic database allows the preparation of lists of potential CDK substrates; algorithms predicting substrate peptides for a given kinase (Brinkworth et al., 2003) can help narrow the search before experimental confirmation is sought. The next several years will undoubtedly witness significant progress in the elucidation of the structure of the phosphorylation-dependent regulatory networks controlling parasite proliferation.

UPSTREAM SIGNAL TRANSDUCTION PATHWAYS WITH A POTENTIAL ROLE IN THE REGULATION OF THE CELL CYCLE MACHINERY

Signal transduction pathway elements likely to play a role in the control of parasite proliferation and/or development are listed in Fig. 3. It is outside the scope of this chapter to review each pathway in detail, but the essential components of pathways for which presence in malaria parasites has been documented, and which have been identified in PlasmoDB, are briefly alluded to in this section.

MAPK Pathways

Two *P. falciparum* genes related to MAPKs have been characterized: Pfmap-1, expressed in asexual parasites and in gametocytes (Doerig et al., 1996; Lin et al., 1996; Graeser et al., 1997), and Pfmap-2, which is detectable by Northern and Western blots only in gametocytes (although microarray data suggest it is expressed in asexual parasites as well) (Fig. 2) and possesses an atypical activation site (TSH instead of the TXY motif conserved in MAPKs in other eukaryotes) (Dorin et al., 1999). Both enzymes were originally grouped within the extracellular signal-regulated kinase 1 (ERK1)/ERK2 family of MAPKs. However, a recent phylogenetic analysis (Ward et al., 2004) of the entire complement of human protein kinase sequences (which was not available at the time Pfmap-1 and Pfmap-2 were characterized) indicates that Pfmap-1 is clearly closer to the recently described ERK7/8 (Abe et al., 2001; Abe et al., 2002), which appear to be regulated by their C-terminal extension. Interestingly, Pfmap-1 also possesses a large extension at its C terminus. This analysis also indicates that Pfmap-2 cannot be unambiguously associated with a particular MAPK family. Furthermore, the same study revealed that no mem-

15. PROTEIN KINASES REGULATING *PLASMODIUM* PROLIFERATION ■ 303

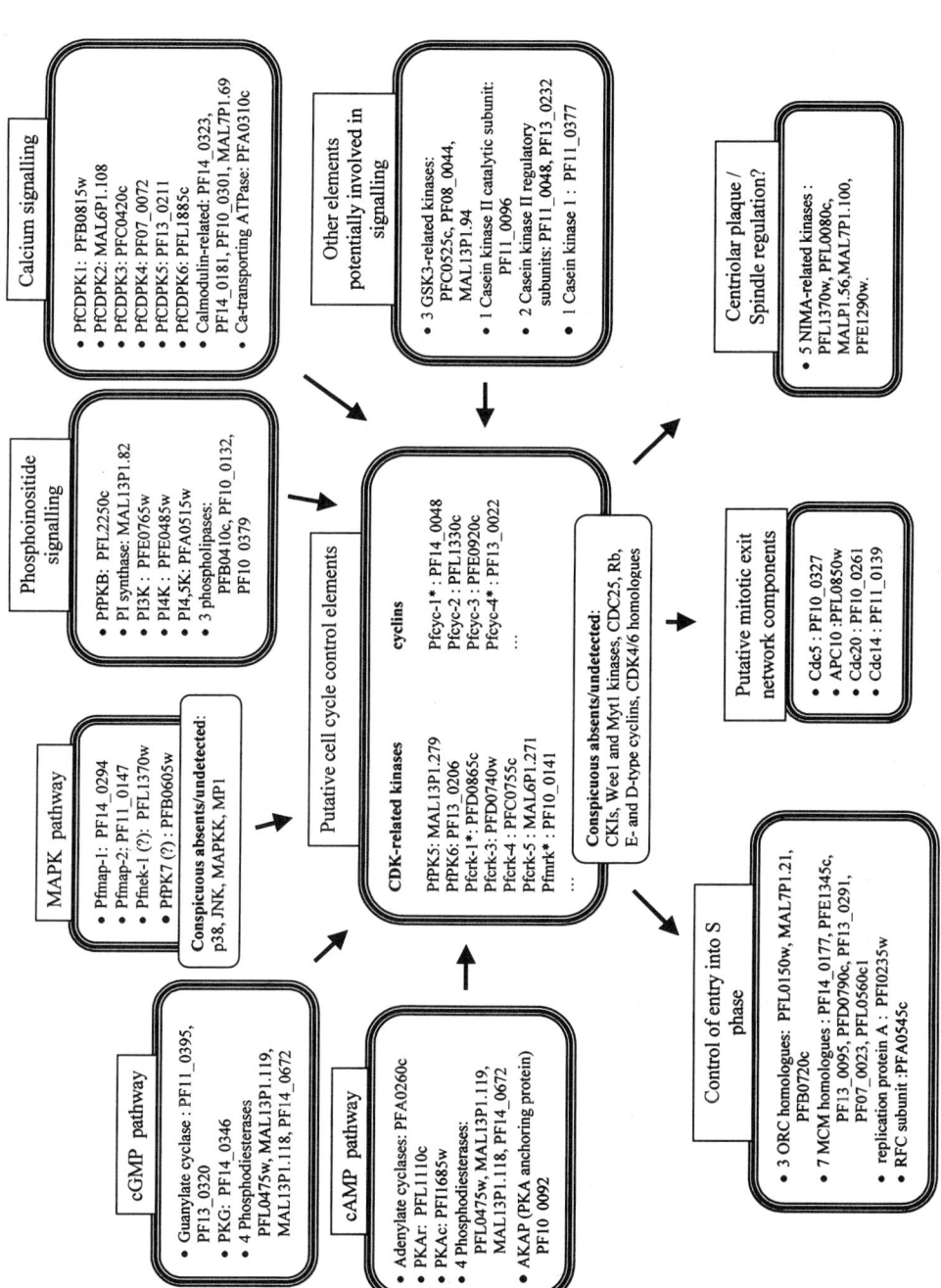

FIGURE 3 Plasmodial CDK-related kinases and cyclins, with potential upstream regulators and substrates. The figure shows selected elements, with their PlasmoDB identifier, of each pathway known to operate in *P. falciparum*. The lists given in each box do not pretend to be exhaustive, as only a subset of relevant elements was included. Furthermore, extensive cross talk most likely operates between pathways, but no attempt was made at illustrating this. Likewise, for the sake of clarity and concision, only a small number of potential substrate groups are shown. Asterisks in the central box indicate those elements for which putative homologs in other systems are involved in transcriptional regulation in addition to or instead of direct cell cycle control.

bers of the MAPKK family are found in the *Plasmodium* genome. This raises the question of the regulation of the plasmodial MAPKs. Two potential regulators of MAPK activity have been identified in PlasmoDB: Pfnek-1, which possesses a typical MAPKK activation site, but whose maximal homology is to kinases of the NIMA family (see above). Pfnek-1 is able to specifically phosphorylate Pfmap-2 (but no other MAPKs, including Pfmap-1) and acts synergistically with Pfmap-2 towards exogenous substrate phosphorylation (Dorin et al., 2001). The second potential MAPK regulator is PfPK7, which displays maximal homology to the mammalian MAPKK3/6 family (restricted to the C-terminal part of the catalytic domain) but does not possess any recognizable activation site, presents other strongly atypical features, and is unable to phosphorylate Pfmap-1 and Pfmap-2 in vitro (Dorin et al., 2005). It therefore appears that the *P. falciparum* genome does not contain typical MAPKKs as those found in other eukaryotes. It also noteworthy that the parasite does not possess modules implicating p38 or Jun N-terminal protein kinase MAP kinases, unlike metazoans and fungi in which these modules are implicated in stress response.

Cyclic Nucleotide Pathways

The cAMP-dependent kinase (protein kinase A [PKA]) exists in eukaryotic cells as an inactive complex of two regulatory (PKAr) and two catalytic (PKAc) subunits. The cAMP synthesized by adenylate cyclase in response to certain stimuli binds to the PKAr subunits, causing dissociation of the complex and the release of active PKAc subunits (Skalhegg and Tasken, 2000), which then phosphorylate effectors such as transcription factors. In the similar cGMP pathway, the cyclic nucleotide-binding regulatory domain is fused to the kinase catalytic domain. Nucleotide cyclases (Carucci et al., 2000; Muhia et al., 2003), PKAc (Syin et al., 2001), and cGMP-dependent protein kinase (Deng and Baker, 2002) homologs in *P. falciparum* have been characterized, and a PKAr subunit is under investigation (A. Merckx, S. Egée, and C. Doerig, unpublished data). Strong evidence demonstrates that the cGMP pathway is crucial for the maturation of male gametocytes to gametes (see "Control of Malaria Parasite Multiplication," above). The cAMP pathway has been implicated in gametocytogenesis (Kaushal et al., 1980), and both the catalytic and the regulatory subunits are expressed during schizogony (Syin et al., 2001; Bozdech et al., 2003; Le Roch et al., 2003). The fact that H89, a widely used PKA inhibitor, kills parasites in cultures with a low 50% inhibitory concentration (Syin et al., 2001) must be viewed with caution, in light of data showing that H89 (like most "specific" kinase inhibitors!) is actually able to inhibit other protein kinases as efficiently as PKA (Davies et al., 2000).

Calcium Signaling and Phosphoinositide Pathways

The *P. falciparum* genome encodes a number of proteins putatively involved in calcium signaling, including calmodulin-related proteins, a calcium-transporting ATPase, and a family of CDPKs, which are composed of a protein kinase catalytic domain fused to a calcium-binding domain (Zhao et al., 1994; Farber et al., 1997; Li et al., 2000; Billker et al., 2004). Enzymes with this architecture are found in plants and ciliates but not in metazoans. Although a number of protein kinases of the AGC group in *P. falciparum* have been characterized (Syin et al., 2001; Deng and Baker, 2002; Kumar et al., 2004), there is apparently no clear ortholog of PKC, a major calcium-responsive kinase in higher eukaryotes. It is therefore possible that calcium-dependent phosphorylation in malaria parasites is mediated mostly or exclusively through the CDPKs. The phosphoinositide pathway is also present in malaria parasite, as evidenced by studies implicating second messengers in gametogenesis (Martin et al., 1994) and by the presence of homologs of several components of this pathway in the plasmodial genome (Fig. 3). Intriguingly, although the malarial genome appears not to encode an ortholog of protein kinase C (Ward et al., 2004), *P. falciparum* expresses a protein with high (60% identity) homology to the receptor for activated C kinase, designated RACK (Madeira et al., 2003). The role that this protein may play in the control of cell proliferation remains to be investigated, and its kinase partner (if any) to be identified.

Other Potential Regulators
Among the 65 eukaryotic protein kinases and the few atypical protein kinases identified in the P. falciparum genome (Ward et al., 2004), several are orthologs of enzymes known to act on cell proliferation in other systems. These include (i) three enzymes related to the GSK3 family, which is involved in the response to insulin and other stimuli in mammalian cells (Grimes and Jope, 2001); two of these GSK3-related proteins have been characterized (Kappes et al., 1995; Droucheau et al., 2004); and (ii) homologs of CKI (Barik et al., 1997) and CKII (Russo et al., 2000), both of which are connected to cell cycle regulation in other eukaryotes (Fig. 3). Obviously, many of the kinases and phosphatases to which no orthology can be assigned on the basis of their sequence may well turn out to represent cell division regulators.

CONCLUSION: A PARALLEL WITH THE CELL CYCLE MACHINERY IN YEAST AND MAMMALIAN CELLS?
Considered together, the data summarized above indicate that many features of the major pathways and regulatory networks pertaining to the control of cell proliferation, as described in model organisms, are valid for malaria parasites. In particular, the fact that *P. falciparum* possesses CDK-related kinases and cyclins strongly suggests that at least some of the principal underlying cell cycle control at the molecular level in higher eukaryotes is shared by this organism. A global consideration of all available data concerning these elements, however, reveals considerable difficulties in drawing a parallel between the cell cycle machinery of *Plasmodium* with that operating in Opisthokonta (which include the yeast and mammalian cells used as models for cell cycle studies).

1. A first obvious difference is that in *P. falciparum* CDK-related kinases outnumber cyclins, in contrast with the situation in yeast and mammalian cells. Does this mean that cryptic CDK positive regulators are present among the 60% of *Plasmodium* open reading frames with unknown function? In favor of this hypothesis, it was shown that PfPK5 can be activated with p25 and RINGO, two proteins from vertebrates with no sequence homology to cyclins but nevertheless able to specifically activate a subset of mammalian CDKs (Le Roch et al., 2000; Merckx et al., 2003). Although neither p25 nor RINGO has sequence homologs in the *Plasmodium* genome, it cannot be excluded that malarial proteins with cyclin-unrelated sequences may possess cyclin activity.

2. Another observation is that with the exceptions of PfPK5, which clearly clusters with CDK1/2, and of Pfcrk-1, which clusters with CDK11, no plasmodial CDK or cyclin can be assigned clear orthology to specific elements in the yeast or mammalian machinery. Hence, none of the plasmodial CDK-related kinases appears related to mammalian CDK4 and CDK6, which are responsible for progression through G_1. Likewise, no genes encoding G_1- and S-phase D- or E-type cyclins are found in the *Plasmodium* genome, and the four malarial cyclins identified so far have only low levels of homology (typically around 20% identity over the cyclin box) to other cyclin families: by BLASTP analysis, the short cyclin-like region of the huge polypeptide potentially encoded by the Pfcyc-2 ORF (see above) seems to be related to mitotic (A- and B-type) cyclins (Merckx et al., 2003), and Pfcyc-1 seems to be related to cyclin H (Le Roch et al., 2000). Experience has proven, however, that such sequence homologies have to be interpreted with utmost caution with respect to function prediction: for example, although Pfcyc-1 (a putative cyclin H homolog) would be expected to activate Pfmrk (a putative CDK7 homolog) in vitro, it turns out that it activates PfPK5 (a putative CDK1 homolog, which would be expected to be insensitive to a cyclin H homolog (see above) (Le Roch et al., 2000). Obviously, this may be due to artifactual in vitro assay conditions (see above). Nevertheless, this exemplifies the difficulty of assigning biochemical (and henceforth biological) functions on the basis of sequence homology, and it is becoming clear that experimental determination of function is required for each putative component of the cell cycle control machinery.

3. A third discrepancy between the *Plasmodium* cell cycle machinery (as defined by database mining) and that of yeast or mammalian cells is the apparent lack of otherwise conserved regulators. Although CKIs play crucial roles in the control of cell division and differentiation in other systems and even though mammalian p21^{CIP1} (but not p16^{INK4}) is able to inhibit the activity of PfPK5 and Pfmrk complexed to Pfcyc-1, indicating that at least some plasmodial CDKs respond to CKIs (Li et al., 2001b), no genes encoding proteins of the p16 or p21 families appear to be present in the *Plasmodium* genome. Likewise, the *P. falciparum* kinome does not include kinases that can be unambiguously considered as orthologs of Myt1 and Wee1, the two enzymes that inactivate CDKs by phosphorylation of the Thr and Tyr (Thr$_{14}$ and Tyr$_{15}$ in CDK1) residues located in the ATP-binding pocket of the kinase (see above). A clear ortholog of the CDC25 phosphatase, which in other systems serves as an activator of CDK through removal of the phosphate groups on Thr$_{14}$ and Tyr$_{15}$, is also apparently missing. Interestingly, however, these Thr and Tyr residues are conserved in many of the CDK-related kinases discussed above (PfPK5, PfPK6, Pfcrk-1; in Pfcrk-3, Pfcrk-4 and Pfmrk, the Tyr residue is present but the Thr is replaced by Ala, Val, and Ser, respectively). This strongly suggests that such residues may actually function as regulatory sites, but that the kinase(s) and phosphatase(s) involved are too divergent to allow unambiguous sequence-based identification. Other polypeptides which play an important role in cell cycle control in mammalian cells, such as the retinoblastoma protein, are apparently absent in malaria parasites.

The apparent conservation of many elements of the basic molecular equipment for the regulation of cell multiplication in malaria parasites is thus accompanied by important divergences in the composition and properties of the machinery and therefore, presumably, of the mechanisms involved in the control of cell proliferation. This observation is true for both the cell cycle control machinery itself (CDKs and cyclins), as well as for upstream transduction pathways. Consequently, understanding in detail the regulation of cell proliferation in malaria parasites cannot rely on sequence homology, but requires experimental demonstration (i) of the function of the elements involved in these processes and (ii) of the organization of the regulatory networks and transduction pathways. Recent advances in *Plasmodium* reverse genetic approaches on one hand, and on proteomic techniques on the other hand, will undoubtedly be instrumental in progress in this field over the next several years. In view of the unexpected findings reported so far with respect to the properties of individual cell cycle control regulators or those of the structure of some of the signal transduction pathways, there is little doubt that many surprises are expecting the investigators who will launch themselves into this enterprise.

ACKNOWLEDGMENTS

Most of the work reviewed here, performed in a number of laboratories, was made possible by the availability of the *P. falciparum* genome database PlasmoDB. The community is indebted to all members of the teams which contributed to the development of this database, an invaluable tool for molecular research on malaria. We thank the authors of the Le Roch et al. and Bozdech et al. microarray studies for making their raw data available for compilation on the PlasmoDB website. I extend my apologies to authors whose work is not cited here, due to space constraints or to involuntary oversight on my part. I thank Debopam Chakrabarti for critical reading of the manuscript and for many years of fruitful and enjoyable collaboration on cell cycle control in malaria and also all members of my team for their dedication and for agreeing to mention unpublished work in this review.

Work in my laboratory is supported by the French Ministère de la Défense (Délégation Générale pour l'Armement [DGA]), by the French-South African joint program on Science and Technology, by the European Commission, and by INSERM. I am grateful to the Scientific Department of the French Embassy in London (and particularly to Jacques Chevalier) for continuing support of our INSERM laboratory in the United Kingdom.

REFERENCES

Abe, M. K., K. T. Kahle, M. P. Saelzler, K. Orth, J. E. Dixon, and M. R. Rosner. 2001. ERK7 is an autoactivated member of the MAPK family. *J. Biol. Chem.* **276:**21272–21279.

Abe, M. K., M. P. Saelzler, R. Espinosa III, K. T. Kahle, M. B. Hershenson, M. M. Le Beau, and

M. R. Rosner. 2002. ERK8, a new member of the mitogen-activated protein kinase family. *J. Biol. Chem.* **277:**16733–16743.

Arnot, D. E., and K. Gull. 1998. The Plasmodium cell-cycle: facts and questions. *Ann. Trop. Med. Parasitol.* **92:**361–365.

Bahl, A., B. Brunk, R. L. Coppel, J. Crabtree, S. J. Diskin, M. J. Fraunholz, G. R. Grant, D. Gupta, R. L. Huestis, J. C. Kissinger, P. Labo, L. Li, S. K. McWeeney, A. J. Milgram, D. S. Roos, J. Schug, and C. J. Stoeckert, Jr. 2002. PlasmoDB: the Plasmodium genome resource. An integrated database providing tools for accessing, analyzing and mapping expression and sequence data (both finished and unfinished). *Nucleic Acids Res.* **30:**87–90.

Baldauf, S. L. 2003. The deep roots of eukaryotes. *Science* **300:**1703–1706.

Barik, S., R. E. Taylor, and D. Chakrabarti. 1997. Identification, cloning, and mutational analysis of the casein kinase 1 cDNA of the malaria parasite, Plasmodium falciparum. Stage-specific expression of the gene. *J. Biol. Chem.* **272:**26132–26138.

Billker, O., S. Dechamps, R. Tewari, G. Wenig, B. Franke-Fayard, and V. Brinkmann. 2004. Calcium and a calcium-dependent protein kinase regulate gamete formation and mosquito transmission in a malaria parasite. *Cell* **117:**503–514.

Billker, O., V. Lindo, M. Panico, A. E. Etienne, T. Paxton, A. Dell, M. Rogers, R. E. Sinden, and H. R. Morris. 1998. Identification of xanthurenic acid as the putative inducer of malaria development in the mosquito. *Nature* **392:**289–292.

Boonstra, J., and G. S. van Rossum. 2003. The role of cytosolic phospholipase A2 in cell cycle progression. *Prog. Cell Cycle Res.* **5:**181–190.

Bozdech, Z., M. Llinas, B. L. Pulliam, E. D. Wong, J. Zhu, and J. L. DeRisi. 2003. The transcriptome of the intraerythrocytic developmental cycle of Plasmodium falciparum. *PLoS Biol.* **1:**E5.

Bracchi-Ricard, V., S. Barik, C. Delvecchio, C. Doerig, R. Chakrabarti, and D. Chakrabarti. 2000. PfPK6, a novel cyclin-dependent kinase/mitogen-activated protein kinase-related protein kinase from Plasmodium falciparum. *Biochem. J.* **347:**255–263.

Brinkworth, R. I., R. A. Breinl, and B. Kobe. 2003. Structural basis and prediction of substrate specificity in protein serine/threonine kinases. *Proc. Natl. Acad. Sci. USA* **100:**74–79.

Carucci, D. J., A. A. Witney, D. K. Muhia, D. C. Warhurst, P. Schaap, M. Meima, J. L. Li, M. C. Taylor, J. M. Kelly, and D. A. Baker. 2000. Guanylyl cyclase activity associated with putative bifunctional integral membrane proteins in Plasmodium falciparum. *J. Biol. Chem.* **275:**22147–22156.

Cornelis, S., Y. Bruynooghe, G. Denecker, S. Van Huffel, S. Tinton, and R. Beyaert. 2000. Identification and characterization of a novel cell cycle-regulated internal ribosome entry site. *Mol. Cell* **5:**597–605.

Davies, S. P., H. Reddy, M. Caivano, and P. Cohen. 2000. Specificity and mechanism of action of some commonly used protein kinase inhibitors. *Biochem. J.* **351:**95–105.

Deng, W., and D. A. Baker. 2002. A novel cyclic GMP-dependent protein kinase is expressed in the ring stage of the Plasmodium falciparum life cycle. *Mol. Microbiol.* **44:**1141–1151.

Dickinson, L. A., A. J. Edgar, J. Ehley, and J. M. Gottesfeld. 2002. Cyclin L is an RS domain protein involved in pre-mRNA splicing. *J. Biol. Chem.* **277:**25465–25473.

Doerig, C. 1997. Signal transduction in malaria parasites. *Parasitol. Today* **13:**307–313.

Doerig, C., and D. Chakrabarti. 2004. Cell cycle control in *Plasmodium falciparum*: a genomics perspective, p. 249–287. *In* A. P. Waters and C. J. Janse (ed.), *Malaria Parasites: Genomes and Molecular Biology*. Caister Academic Press, Wymondham, United Kingdom.

Doerig, C., J. Endicott, and D. Chakrabarti. 2002. Cyclin-dependent kinase homologues of Plasmodium falciparum. *Int. J. Parasitol.* **32:**1575–1585.

Doerig, C., P. Horrocks, J. Coyle, J. Carlton, A. Sultan, D. Arnot, and R. Carter. 1995. Pfcrk-1, a developmentally regulated cdc2-related protein kinase of Plasmodium falciparum. *Mol. Biochem. Parasitol.* **70:**167–174.

Doerig, C. M., D. Parzy, G. Langsley, P. Horrocks, R. Carter, and C. D. Doerig. 1996. A MAP kinase homologue from the human malaria parasite, Plasmodium falciparum. *Gene* **177:**1–6.

Dorin, D., P. Alano, I. Boccaccio, L. Ciceron, C. Doerig, R. Sulpice, and D. Parzy. 1999. An atypical mitogen-activated protein kinase (MAPK) homologue expressed in gametocytes of the human malaria parasite Plasmodium falciparum. Identification of a MAPK signature. *J. Biol. Chem.* **274:**29912–29920.

Dorin, D., K. Le Roch, P. Sallicandro, P. Alano, D. Parzy, P. Poullet, L. Meijer, and C. Doerig. 2001. Pfnek-1, a NIMA-related kinase from the human malaria parasite Plasmodium falciparum. Biochemical properties and possible involvement in MAPK regulation. *Eur. J. Biochem.* **268:**2600–2608.

Dorin, D., J. P. Semblat, P. Poullet, P. Alano, D. Goldring, C. Whittle, S. Patterson, C. Whittle, D. Chakrabarti, and C. Doerig. 2005. PfPK7, an atypical MEK-related protein kinase, reflects the absence of typical three-component MAP kinase pathways in the human malaria parasite Plasmodium falciparum. *Mol. Microbiol.* **55:**184–196.

Droucheau, E., A. Primot, V. Thomas, D. Mattei, M. Knockaert, C. Richardson, P. Sallicandro, P. Alano, A. Jafarshad, B. Baratte, C. Kunick, D. Parzy, L. Pearl, C. Doerig, and L. Meijer.

2004. Plasmodium falciparum glycogen synthase kinase-3: molecular model, expression, intracellular localisation and selective inhibitors. *Biochim. Biophys. Acta* **1697:**181–196.

Dutta, A., and S. P. Bell. 1997. Initiation of DNA replication in eukaryotic cells. *Annu. Rev. Cell Dev. Biol.* **13:**293–332.

Farber, P. M., R. Graeser, R. M. Franklin, and B. Kappes. 1997. Molecular cloning and characterization of a second calcium-dependent protein kinase of Plasmodium falciparum. *Mol. Biochem. Parasitol.* **87:**211–216.

Forsburg, S. L. 2004. Eukaryotic MCM proteins: beyond replication initiation. *Microbiol. Mol. Biol. Rev.* **68:**109–131.

Garcia, G. E., R. A. Wirtz, J. R. Barr, A. Woolfitt, and R. Rosenberg. 1998. Xanthurenic acid induces gametogenesis in Plasmodium, the malaria parasite. *J. Biol. Chem.* **273:**12003–12005.

Graeser, R., P. Kury, R. M. Franklin, and B. Kappes. 1997. Characterization of a mitogen-activated protein (MAP) kinase from Plasmodium falciparum. *Mol. Microbiol.* **23:**151–159.

Grimes, C. A., and R. S. Jope. 2001. The multifaceted roles of glycogen synthase kinase 3β in cellular signaling. *Prog. Neurobiol.* **65:**391–426.

Gururajan, R., J. M. Lahti, J. Grenet, J. Easton, I. Gruber, P. F. Ambros, and V. J. Kidd. 1998. Duplication of a genomic region containing the Cdc2L1-2 and MMP21-22 genes on human chromosome 1p36.3 and their linkage to D1Z2. *Genome Res.* **8:**929–939.

Gustin, M. C., J. Albertyn, M. Alexander, and K. Davenport. 1998. MAP kinase pathways in the yeast Saccharomyces cerevisiae. *Microbiol. Mol. Biol. Rev.* **62:**1264–1300.

Holton, S., A. Merckx, D. Burgess, C. Doerig, M. Noble, and J. Endicott. 2003. Structures of P. falciparum PfPK5 test the CDK regulation paradigm and suggest mechanisms of small molecule inhibition. *Structure* (Cambridge) **11:**1329–1337.

Hu, D., A. Mayeda, J. H. Trembley, J. M. Lahti, and V. J. Kidd. 2003. CDK11 complexes promote pre-mRNA splicing. *J. Biol. Chem.* **278:**8623–8629.

Hu, K., T. Mann, B. Striepen, C. J. Beckers, D. S. Roos, and J. M. Murray. 2002. Daughter cell assembly in the protozoan parasite Toxoplasma gondii. *Mol. Biol. Cell* **13:**593–606.

Janse, C. J., P. F. van der Klooster, H. J. van der Kaay, M. van der Ploeg, and J. P. Overdulve. 1986. DNA synthesis in Plasmodium berghei during asexual and sexual development. *Mol. Biochem. Parasitol.* **20:**173–182.

Janse, C. J., and P. H. Van Vianen. 1994. Flow cytometry in malaria detection. *Methods Cell Biol.* **42:**295–318.

Janse, C. J., P. H. van Vianen, H. J. Tanke, B. Mons, T. Ponnudurai, and J. P. Overdulve. 1987. Plasmodium species: flow cytometry and microfluorometry assessments of DNA content and synthesis. *Exp. Parasitol.* **64:**88–94.

Kahl, C. R., and A. R. Means. 2003. Regulation of cell cycle progression by calcium/calmodulin-dependent pathways. *Endocr. Rev.* **24:**719–736.

Kaldis, P. 1999. The cdk-activating kinase (CAK): from yeast to mammals. *Cell. Mol. Life Sci.* **55:**284–296.

Kappes, B., J. Yang, B. W. Suetterlin, K. Rathgeb-Szabo, M. J. Lindt, and R. M. Franklin. 1995. A Plasmodium falciparum protein kinase with two unusually large kinase inserts. *Mol. Biochem. Parasitol.* **72:**163–178.

Kaushal, D. C., R. Carter, L. H. Miller, and G. Krishna. 1980. Gametocytogenesis by malaria parasites in continuous culture. *Nature* **286:**490–492.

Kawamoto, F., R. Alejo-Blanco, S. L. Fleck, Y. Kawamoto, and R. E. Sinden. 1990. Possible roles of Ca^{2+} and cGMP as mediators of the exflagellation of Plasmodium berghei and Plasmodium falciparum. *Mol. Biochem. Parasitol.* **42:**101–108.

Kelly, T. J., and G. W. Brown. 2000. Regulation of chromosome replication. *Annu. Rev. Biochem.* **69:**829–880.

Kumar, A., A. Vaid, C. Syin, and P. Sharma. 2004. PfPKB, a novel protein kinase B like enzyme from plasmodium falciparum. I. Identification, characterization and possible role in parasite development. *J. Biol. Chem.* **279:**24255–24264.

Lahti, J. M., J. Xiang, and V. J. Kidd. 1995. The PITSLRE protein kinase family. *Prog. Cell Cycle Res.* **1:**329–338.

Le Roch, K., C. Sestier, D. Dorin, N. Waters, B. Kappes, D. Chakrabarti, L. Meijer, and C. Doerig. 2000. Activation of a Plasmodium falciparum cdc2-related kinase by heterologous p25 and cyclin H. Functional characterization of a P. falciparum cyclin homologue. *J. Biol. Chem.* **275:**8952–8958.

Le Roch, K. G., Y. Zhou, P. L. Blair, M. Grainger, J. K. Moch, J. D. Haynes, P. De La Vega, A. A. Holder, S. Batalov, D. J. Carucci, and E. A. Winzeler. 2003. Discovery of gene function by expression profiling of the malaria parasite life cycle. *Science* **301:**1503–1508.

Li, J. L., D. A. Baker, and L. S. Cox. 2000. Sexual stage-specific expression of a third calcium-dependent protein kinase from Plasmodium falciparum. *Biochim. Biophys. Acta* **1491:**341–349.

Li, J. L., K. J. Robson, J. L. Chen, G. A. Targett, and D. A. Baker. 1996. Pfmrk, a MO15-related protein kinase from Plasmodium falciparum. Gene cloning, sequence, stage-specific expression and chromosome localization. *Eur. J. Biochem.* **241:**805–813.

Li, J. L., G. A. Targett, and D. A. Baker. 2001a. Primary structure and sexual stage-specific expression of a LAMMER protein kinase of Plasmodium falciparum. *Int. J. Parasitol.* **31:**387–392.

Li, T., A. Inoue, J. M. Lahti, and V. J. Kidd. 2004. Failure to proliferate and mitotic arrest of CDK11 (p110/p58)-null mutant mice at the blastocyst stage of embryonic cell development. *Mol. Cell. Biol.* **24:**3188–3197.

Li, Z., K. Le Roch, J. A. Geyer, C. L. Woodard, S. T. Prigge, J. Koh, C. Doerig, and N. C. Waters. 2001b. Influence of human p16(INK4) and p21(CIP1) on the in vitro activity of recombinant Plasmodium falciparum cyclin-dependent protein kinases. *Biochem. Biophys. Res. Commun.* **288:**1207–1211.

Lin, D. T., N. D. Goldman, and C. Syin. 1996. Stage-specific expression of a Plasmodium falciparum protein related to the eukaryotic mitogen-activated protein kinases. *Mol. Biochem. Parasitol.* **78:**67–77.

Madeira, L., R. DeMarco, M. L. Gazarini, S. Verjovski-Almeida, and C. R. Garcia. 2003. Human malaria parasites display a receptor for activated C kinase ortholog. *Biochem. Biophys. Res. Commun.* **306:**995–1001.

Martin, S. K., M. Jett, and I. Schneider. 1994. Correlation of phosphoinositide hydrolysis with exflagellation in the malaria microgametocyte. *J. Parasitol.* **80:**371–378.

McCollum, D., and K. L. Gould. 2001. Timing is everything: regulation of mitotic exit and cytokinesis by the MEN and SIN. *Trends Cell Biol.* **11:**89–95.

McGowan, C. H. 2003. Regulation of the eukaryotic cell cycle. *Prog. Cell Cycle Res.* **5:**1–4.

Merckx, A., K. Le Roch, M. P. Nivez, D. Dorin, P. Alano, G. J. Gutierrez, A. R. Nebreda, D. Goldring, C. Whittle, S. Patterson, D. Chakrabarti, and C. Doerig. 2003. Identification and initial characterization of three novel cyclin-related proteins of the human malaria parasite Plasmodium falciparum. *J. Biol. Chem.* **278:**39839–39850.

Miyata, Y., M. Akashi, and E. Nishida. 1999. Molecular cloning and characterization of a novel member of the MAP kinase superfamily. *Genes Cells* **4:**299–309.

Mock, B. A., C. Padlan, C. A. Kozak, and V. Kidd. 1994. The gene for mouse p58cdc2L1(Cdc2l1) protein kinase maps to distal mouse chromosome 4. *Mamm. Genome* **5:**191–192.

Morgan, D. O. 1997. Cyclin-dependent kinases: engines, clocks, and microprocessors. *Annu. Rev. Cell Dev. Biol.* **13:**261–291.

Muhia, D. K., C. A. Swales, W. Deng, J. M. Kelly, and D. A. Baker. 2001. The gametocyte-activating factor xanthurenic acid stimulates an increase in membrane-associated guanylyl cyclase activity in the human malaria parasite Plasmodium falciparum. *Mol. Microbiol.* **42:**553–560.

Muhia, D. K., C. A. Swales, U. Eckstein-Ludwig, S. Saran, S. D. Polley, J. M. Kelly, P. Schaap, S. Krishna, and D. A. Baker. 2003. Multiple splice variants encode a novel adenylyl cyclase of possible plastid origin expressed in the sexual stage of the malaria parasite Plasmodium falciparum. *J. Biol. Chem.* **278:**22014–22022.

O'Connell, M. J., M. J. Krien, and T. Hunter. 2003. Never say never. The NIMA-related protein kinases in mitotic control. *Trends Cell Biol.* **13:**221–228.

Rangajaran, R., K. Bei, D. Jethawaney, P. Maldonado, D. Dorin, A. Sultan, and C. Doerig. 2005. A MAP kinase regulates gametogenesis and transmission of the malaria parasite *Plasmodium berghei*. *EMBO Rep.* **6:**464–469.

Read, M., T. Sherwin, S. P. Holloway, K. Gull, and J. E. Hyde. 1993. Microtubular organization visualized by immunofluorescence microscopy during erythrocytic schizogony in Plasmodium falciparum and investigation of post-translational modifications of parasite tubulin. *Parasitology* **106:**223–232.

Ross-Macdonald, P. B., R. Graeser, B. Kappes, R. Franklin, and D. H. Williamson. 1994. Isolation and expression of a gene specifying a cdc2-like protein kinase from the human malaria parasite Plasmodium falciparum. *Eur. J. Biochem.* **220:**693–701.

Russo, G. L., C. van den Bos, A. Sutton, P. Coccetti, M. D. Baroni, L. Alberghina, and D. R. Marshak. 2000. Phosphorylation of Cdc28 and regulation of cell size by the protein kinase CKII in Saccharomyces cerevisiae. *Biochem. J.* **351:**143–150.

Ryves, W. J., and A. J. Harwood. 2003. The interaction of glycogen synthase kinase-3 (GSK-3) with the cell cycle. *Prog. Cell Cycle Res.* **5:**489–495.

Schaeffer, H. J., and M. J. Weber. 1999. Mitogen-activated protein kinases: specific messages from ubiquitous messengers. *Mol. Cell. Biol.* **19:**2435–2444.

Sherr, C. J., and J. M. Roberts. 1999. CDK inhibitors: positive and negative regulators of G1-phase progression. *Genes Dev.* **13:**1501–1512.

Sinden, R. E., G. A. Butcher, O. Billker, and S. L. Fleck. 1996. Regulation of infectivity of Plasmodium to the mosquito vector. *Adv. Parasitol.* **38:**53–117.

Skalhegg, B. S., and K. Tasken. 2000. Specificity in the cAMP/PKA signaling pathway. Differential expression, regulation, and subcellular localization of subunits of PKA. *Front. Biosci.* **5:**D678–D693.

Solomon, M. J., and P. Kaldis. 1998. Regulation of CDKs by phosphorylation. *Results Probl. Cell Differ.* **22:**79–109.

Stiller, J. W., and B. D. Hall. 2002. Evolution of the RNA polymerase II C-terminal domain. *Proc. Natl. Acad. Sci. USA* **99:**6091–6096.

Stork, P. J., and J. M. Schmitt. 2002. Crosstalk between cAMP and MAP kinase signaling in the regulation of cell proliferation. *Trends Cell Biol.* **12:**258–266.

Syin, C., D. Parzy, F. Traincard, I. Boccaccio, M. B. Joshi, D. T. Lin, X. M. Yang, K. Assemat, C. Doerig, and G. Langsley. 2001. The H89 cAMP-dependent protein kinase inhibitor blocks Plasmodium falciparum development in infected erythrocytes. *Eur. J. Biochem.* **268:**4842–4849.

Ubersax, J. A., E. L. Woodbury, P. N. Quang, M. Paraz, J. D. Blethrow, K. Shah, K. M. Shokat, and D. O. Morgan. 2003. Targets of the cyclin-dependent kinase Cdk1. *Nature* **425:**859–864.

Ward, P., L. Equinet, J. Packer, and C. Doerig. 2004. Protein kinases of the human malaria parasite Plasmodium falciparum: the kinome of a divergent eukaryote. *BMC Genomics* **5:**79.

Waters, N. C., C. L. Woodard, and S. T. Prigge. 2000. Cyclin H activation and drug susceptibility of the Pfmrk cyclin dependent protein kinase from Plasmodium falciparum. *Mol. Biochem. Parasitol.* **107:**45–55.

Wittenberg, C., and R. La Valle. 2003. Cell-cycle-regulatory elements and the control of cell differentiation in the budding yeast. *Bioessays* **25:**856–867.

Zhang, S., M. Cai, S. Xu, S. Chen, X. Chen, C. Chen, and J. Gu. 2002. Interaction of p58(PITSLRE), a G2/M-specific protein kinase, with cyclin D3. *J. Biol. Chem.* **277:**35314–35322.

Zhao, Y., S. Pokutta, P. Maurer, M. Lindt, R. M. Franklin, and B. Kappes. 1994. Calcium-binding properties of a calcium-dependent protein kinase from Plasmodium falciparum and the significance of individual calcium-binding sites for kinase activation. *Biochemistry* **33:**3714–3721.

PROTEASES AND HEMOGLOBIN DEGRADATION

Philip J. Rosenthal

16

Erythrocytic malaria parasites take up and degrade large quantities of hemoglobin. Hemoglobin processing comes at considerable cost, as it requires a complex hydrolytic machinery and necessitates the efficient detoxification of free heme molecules. It has generally been assumed that the major reason for hemoglobin hydrolysis is the provision of amino acids for parasite protein synthesis. However, the degree to which parasites rely on amino acids from hemoglobin remains uncertain, and other explanations for hemoglobin hydrolysis have been offered. In any event, hemoglobin hydrolysis is clearly an essential parasite function, as protease inhibitors that block this process also block parasite development. Recent advances have led to the identification of multiple proteolytic enzymes that appear to contribute to hemoglobin hydrolysis, although the detailed pathway of hemoglobin processing remains uncertain. This chapter will review our current understanding of the reasons for hemoglobin degradation, the mechanism of hemoglobin breakdown, the proteases that contribute to this process, and the potential for new antimalarial therapies that block hemoglobin hydrolysis.

THE NEED FOR HEMOGLOBIN HYDROLYSIS

Supply of Parasite Amino Acids

A large body of work suggests that a principal source of amino acids for erythrocytic parasites is the hydrolysis of globin. Older studies, mostly of nonhuman malaria parasites, showed that the hemoglobin content of infected erythrocytes decreased markedly (generally estimated at about 75%) during the parasite life cycle, the concentration of free amino acids was greater in infected than uninfected erythrocytes, the composition of the amino acid pool of infected erythrocytes was similar to the composition of hemoglobin, and the infection of erythrocytes containing radiolabeled hemoglobin was followed by the appearance of the label in parasite proteins (reviewed in McKerrow et al., 1993). Only five amino acids, most of which are in limited supply in human hemoglobin, are required to support normal growth of cultured parasites (Divo et al., 1985; Francis et al., 1994).

Philip J. Rosenthal, Department of Medicine, San Francisco General Hospital, University of California, San Francisco, CA 94143.

Molecular Approaches to Malaria, Edited by Irwin W. Sherman
© 2005 ASM Press, Washington, D.C.

In addition, protease inhibitors that block hemoglobin degradation also block the development of cultured parasites (Francis et al., 1994; Rosenthal et al., 1988).

Other Sources of Amino Acids

Despite strong evidence for the utilization of amino acids from hemoglobin by malaria parasites, it remains uncertain just how much amino acids acquired from hemoglobin contribute to the makeup of parasite proteins. Indeed, it has been argued recently that the incorporation of amino acids from the bloodstream may play a greater role than previously recognized. Supporting this conclusion, cultured *Plasmodium falciparum* parasites were shown to utilize only about 16% of the amino acids obtained from globin hydrolysis, with excess amino acids expelled through parasite transport pathways (Krugliak et al., 2002). Other sources of amino acids are de novo parasite synthesis and transport into the parasite.

De novo synthesis of amino acids appears to play only a small role in supplying parasite amino acids. Parasites can metabolize glucose, pyruvate, or acetate and CO_2 into glutamate, aspartate, alanine, and leucine (Scheibel and Sherman, 1988). Glutamate dehydrogenase activity, which can catalyze the synthesis of glutamate, has been identified in *P. falciparum*. Additional amino acids can be synthesized when specific precursors are available. Thus, in *P. falciparum* (Asawamahasakda and Yuthavong, 1993), methionine is synthesized from homocysteine and serine (in part to recycle S-adenosyl methionine), and in *Plasmodium knowlesi* cysteine can be synthesized from glutathione or methionine (Scheibel and Sherman, 1988).

Probably playing a more important role than de novo synthesis, erythrocytic parasites can take up exogenous amino acids. In studies with nonhuman parasites, many amino acids were taken up, and the uptake of isoleucine, which is not present in hemoglobin, was consistently high (Sherman, 1977). In studies with cultured *P. falciparum*, parasites grown in a minimal medium required supplementation with five amino acids (isoleucine, methionine, glutamine, glutamate, and cysteine) to achieve growth rates approaching those of parasites grown in standard complete culture medium (Divo et al., 1985; Francis et al., 1994). Most of the amino acids required for growth in culture are poorly represented (cysteine, methionine, and glutamine) or absent (isoleucine) in human hemoglobin. Thus, available results suggest that parasites import those amino acids not provided by the hydrolysis of globin but also indicate that they have the ability to incorporate other needed amino acids. Our understanding of uptake mechanisms is incomplete, but transport is likely via both erythrocyte amino acid transporters and parasite-derived permeation pathways (Kirk, 2001; Chapter 20). For many amino acids, normal erythrocyte transporters may be sufficient to provide parasite needs, but the relative contributions of erythrocyte and parasite-derived transporters are unknown. For glutamate, which is not transported into uninfected erythrocytes (Elford et al., 1995), transport to parasites appears to be mediated by parasite-specific permeation pathways (Lauer et al., 1997).

Additional Explanations for Hemoglobin Hydrolysis

If amino acids are available from sources other than hemoglobin and if malaria parasites utilize only a minority of the amino acids generated by hemoglobin hydrolysis, why do they expend the energy and risk the toxicity engendered by this process? Hemoglobin hydrolysis might be required to obtain one or more amino acids that are difficult to acquire by other means. In this case, even if most amino acids from hemoglobin are not utilized, the need for certain essential amino acids may obligate parasites to hydrolyze large quantities of hemoglobin. However, there is no clear evidence identifying any single amino acid as an essential product of hemoglobin hydrolysis.

Another explanation for hemoglobin hydrolysis far in excess of parasite amino acid needs is that sufficient space for the growing malaria parasite must be provided by hemoglobin degradation. The hydrolysis of hemoglobin, with transport from the infected erythrocyte of most free amino acids, might free intraerythrocytic space for the parasite.

It has recently been suggested that hemoglobin hydrolysis is required to maintain the osmotic stability of the growing intraerythrocytic parasite (Lew et al., 2003). Mathematical modeling was used to predict the means by which parasites maintain homeostasis in the face of the dramatic alterations in cell permeability that accompany erythrocyte infection. Models predicted stage-specific alterations in the volumes of infected erythrocytes that were supported by experimental evaluations (Lew et al., 2004; Lew et al., 2003). It was predicted that the osmotic stability of erythrocytes infected with mature parasites could only be maintained by large reductions in hemoglobin concentration. Thus, this new hypothesis asserts that hemoglobin hydrolysis is a necessary function of erythrocytic parasites to avoid osmotic lysis of infected erythrocytes.

MECHANISM OF HEMOGLOBIN DEGRADATION

Hemoglobin processing is a complex process, including transport of hemoglobin from erythrocyte cytosol to the parasite food vacuole, disruption of hemoglobin tetramers, removal of heme from hemoglobin or globin, detoxification of heme with the formation of hemozoin, and the hydrolysis of globin by a number of proteases, eventually to individual amino acids. Our understanding of each of these processes is improving.

Hemoglobin Uptake

In *P. falciparum*, ring-stage parasites pinocytose the hemoglobin-rich erythrocyte cytosol into small vesicles (Olliaro and Goldberg, 1995; Slomianny, 1990). As parasites develop, hemoglobin uptake increases, and trophozoites develop two structures that are apparently specialized for hemoglobin transport and processing, the food vacuole and the cytostome. The single large food vacuole appears at the trophozoite stage, probably after the fusion of small vesicles. Cytostomes also appear in trophozoites. These are peripheral tubular structures that contain apposed parasite and parasitophorous vacuole membranes and function to efficiently take up erythrocyte cytosol. Vesicles containing erythrocyte cytosol bud from the cytostomes, travel to the food vacuole, and coalesce with this structure. Many other species of malaria parasites differ from *P. falciparum* in that they contain multiple food vacuoles and transport erythrocyte cytosol to their food vacuoles via a tubular network rather than vesicular transport (Slomianny, 1990). The *P. falciparum* food vacuole appears to be a large lysosome, as it is an acidic compartment containing hydrolytic enzymes. The acidic pH of the food vacuole is maintained by two proton-pumping mechanisms, a H^+-ATPase and a H^+-pyrophosphatase (Saliba et al., 2003). Functions of the food vacuole other than the processing and detoxification of hemoglobin components remain uncertain. Hydrolytic activities typical of lysosomes other than proteases were not identified in purified food vacuoles (Goldberg et al., 1990). A recent report suggested that the food vacuole also acts as an intraparasitic calcium store (Biagini et al., 2003). It remains uncertain whether hemoglobin breakdown is initiated in the food vacuole in *P. falciparum* or whether the process begins earlier, in transport vesicles being directed to the food vacuole. Supporting at least some hemoglobin breakdown in transport vesicles is the identification of malarial pigment in these structures in ultrastructural studies (Langreth et al., 1978; Slomianny, 1990).

Separation of Hemoglobin Components

Upon its delivery to the *P. falciparum* food vacuole, and perhaps during vesicular transit to this organelle, hemoglobin is subjected to an acidic pH (Spiller et al., 2002). Sufficiently acidic pH denatures hemoglobin and releases heme, but this denaturation is very slow at the pH of the food vacuole, as demonstrated by spectrophotometric (Gabay and Ginsburg, 1993; Shenai and Rosenthal, 2002) and electrophoretic (Gamboa de Dominguez and Rosenthal, 1996) methods. It appears that parasite protease activity, which has a clear role in globin hydrolysis (see below), is required to efficiently separate heme from globin in the food vacuole, as follows. In studies utilizing purified hemoglobin, protease K-accelerated hemoglobin denaturation and a mixture of protease inhibitors slowed denaturation under conditions similar to those in the food vacuole (Gabay and Ginsburg, 1993). In studies with cultured *P. falciparum*, the proteolytic processing of hemoglobin began prior to heme release, as cultured parasites contained α- and β-globin fragments that lacked 15 to 25 carboxy-terminal amino acids but remained bound to heme (Kamchonwongpaisan et al., 1997). Suppporting a role for cysteine proteases in hemoglobin dissociation, the cysteine protease inhibitor E-64 (but not the aspartic protease inhibitor pepstatin) inhibited the dissociation of the hemoglobin tetramer (Gamboa de Dominguez and Rosenthal, 1996), the release of heme from globin (Gamboa de Dominguez and Rosenthal, 1996), the initial processing of α- and β-globin (Kamchonwongpaisan et al., 1997), and the formation of hemozoin (Asawamahasakda et al., 1994). Another study supported a role for aspartic proteases in hemoglobin dissociation, as in this analysis plasmepsin inhibitors were much more active than cysteine protease inhibitors in blocking the formation of hemozoin (Bray et al., 1999). Considering available results, it appears that both aspartic and cysteine protease activities contribute to the separation of heme from globin. This dissociation likely facilitates the efficient hydrolysis of globin and detoxification of heme that will be discussed below.

Hydrolysis of Hemoglobin

Proteases of multiple catalytic classes appear to contribute to hemoglobin degradation. Biochemical characterizations of food vacuole aspartic and cysteine proteases and metalloproteases have shown that these enzymes hydrolyze hemoglobin or globin in vitro, supporting roles in hemoglobin hydrolysis. Studies with protease inhibitors have helped to identify specific functions. In particular, inhibitors of both cysteine and aspartic proteases exert potent antiparasitic effects, although only cysteine protease inhibitors appear to directly block hemoglobin hydrolysis.

It appears that much of the hydrolysis of hemoglobin occurs in the food vacuole but that ultimate steps in hemoglobin hydrolysis take place in the parasite cytosol after small hemoglobin peptides are transported from the food vacuole. Supporting this conclusion, incubation of food vacuole lysates with hemoglobin generated numerous small peptides, averaging about eight amino acids in length (Kolakovich et al., 1997). It was proposed that small globin-derived peptides are transported from the food vacuole to parasite cytosol for completion of hemoglobin hydrolysis by neutral cytosolic proteases.

PROTEASES RESPONSIBLE FOR HEMOGLOBIN HYDROLYSIS

Important recent advances in our understanding of hemoglobin hydrolysis have come from the characterization of multiple plasmodial proteases. This characterization has been facilitated by the release of the *P. falciparum* genome sequence; the heterologous expression, purification, and biochemical characterization of multiple enzymes; and the use of transfection techniques to disrupt protease genes. It is now clear that multiple proteases reside in the *P. falciparum* food vacuole and likely contribute to hemoglobin hydrolysis. These enzymes include members of the aspartic and cysteine protease and metalloprotease classes (Table 1). In addition, at least one cytosolic exopeptidase is also probably required for the complete hydrolysis of hemoglobin. The specific enzymes that appear to be responsible for hemoglobin hydrolysis will be described below.

The Protease Repertoire of Malaria Parasites

It is now clear that there are a great many proteases encoded by malaria parasites. A recent analysis used the available *P. falciparum* genome sequence and a detailed sequence similarity search to identify 92 putative *P. falciparum* protease genes (Wu et al., 2003). These sequences include all previously identified *P. falciparum* protease sequences and putative members of many other protease families. Recent genomic studies have shown that most of the identified putative

TABLE 1 Features of proteases that may play roles in hemoglobin hydrolysis[a]

Protease	Class	Stage specificity transcription	Stage specificity protein expression	Predicted role
Falcipain-1	Cysteine	R	R/T/S	Unknown; principal role in nonerythrocytic parasites?
Falcipain-2	Cysteine	T	T	Early and mid-step Hb hyrolysis in FV, especially in trophozoites
Falcipain-2'	Cysteine	R/T	R/T	Hb hydrolysis in trophozoites
Falcipain-3	Cysteine	T/S	T/S	Hb hydrolysis in FV, especially in late trophozoites and schizonts
Dipeptidyl peptidase I	Cysteine	R/T	R/T	Hydrolysis of Hb peptides in FV
Plasmepsin I	Aspartic	R	R/T/S	Early and mid-step Hb hydrolysis in FV
Plasmepsin II	Aspartic	T	R/T/S	Early and mid-step Hb hydrolysis in FV
Plasmepsin III (HAP)	Aspartic	R/T	T/S	Early and mid-step Hb hydrolysis in FV
Plasmepsin IV	Aspartic	R	T/S	Early and mid-step Hb hydrolysis in FV
Falcilysin	Metallo	T	T	Hydrolysis of Hb peptides in FV
Aminopeptidase	Metallo	R/T	T	Hydrolysis of Hb peptides in parasite cytosol

[a]Summary of identified proteases with predicted roles in hemoglobin degradation. Stage-specificity data refer only to erythrocytic-stage parasites and are from primary literature and genomic and proteomic screens summarized on the PlasmoDB website (http://plasmodb.org). Abbreviations: R, ring; T, trophozoite; S, schizont; Hb, hemoglobin; FV, food vacuole; metallo, metalloprotease.

protease genes are transcribed, but biochemical and biological evaluations of most of the gene products are not yet available.

Plasmepsin Aspartic Proteases

Ten aspartic proteases, all named plasmepsins, are encoded by *P. falciparum*. The first enzymes characterized were plasmepsin I and plasmepsin II, which were purified from food vacuoles and heterologously expressed (Dame et al., 1994; Francis et al., 1994; Gluzman et al., 1994; Hill et al., 1994; Moon et al., 1997). Both of these enzymes were shown to hydrolyze native hemoglobin, although plasmepsin I was most active against this substrate, and plasmepsin II appeared to prefer denatured globin as a substrate (Gluzman et al., 1994). Plasmepsin I is also synthesized and processed earlier in the erythrocytic life cycle than is plasmepsin II (Francis et al., 1997). In in vitro studies, both of these plasmepsins cleaved hemoglobin at multiple sites, among them a peptide bond of α-globin (Phe_{33}-Leu_{34}) in a hinge region of the molecule (Gluzman et al., 1994; Goldberg et al., 1991). It was suggested that this cleavage begins to unravel hemoglobin, exposing additional peptide bonds to the action of plasmepsins and other proteases.

A more recent analysis identified plasmepsin I, plasmepsin II, plasmepsin III (also known as histoaspartic protease), and plasmepsin IV as likely food vacuole hemoglobinases. Each of these enzymes is expressed in erythrocytic parasites, localizes to the food vacuole, and hydrolyzes hemoglobin at acidic pH (Banerjee et al., 2002). Each of the genes encoding these plasmepsins is predicted to encode a 51-kDa proenzyme with a hydrophobic membrane-spanning domain within the proregion. The four plasmepsins are all processed at a conserved site to 37-kDa mature proteases, with processing mediated by an as yet unidentified processing enzyme (Banerjee et al., 2003). Studies of trafficking of plasmepsin II suggest that it is transported through the secretory pathway, but then carried in cytostomes to the food vacuole, potentially allowing action against hemoglobin before the substrate reaches the food vacuole (Francis et al., 1997; Klemba et al., 2004a).

The genes encoding the four food vacuole plasmepsins are in a tandem array on chromosome 14 (www.plasmodb.org). Their sequences show about 50 to 70% identity, which is much greater than that between these and other plasmodial aspartic proteases. The localization of the genes and their similarities suggest a common ancestor and similar biological functions. However, plasmepsin III is unique among the plasmepsins and among known aspartic proteases, as one of the two canonical aspartates is replaced by a histidine in this protease (Banerjee et al., 2002; Berry et al., 1999). This enzyme also has a pH optimum somewhat higher than that of the other food vacuole plasmepsins (Banerjee et al., 2002). Considering the presumed natural substrate of the food vacuole plasmepsins, plasmepsin III and plasmepsin IV are more active against denatured globin but less active against native hemoglobin than plasmepsin I and plasmepsin II (Banerjee et al., 2002). Combinations of the plasmepsins elicit much more rapid hemoglobin hydrolysis than do the individual enzymes. Recent gene disruption studies showed that knockout of any single food vacuole plasmepsin gene was not lethal for erythrocytic parasites and did not cause obvious abnormal morphology, although some knockouts led to diminished growth rates. More specifically, in one study, knockout of the plasmepsin II gene or a combined knockout of the plasmepsin I and plasmepsin IV gene caused decreased growth (Liu et al., 2005); in another study, knockout of either the plasmepsin I or plasmepsin IV gene had a similar effect (Omara-Opyene et al., 2004). In both cases, knockout of the plasmepsin III gene had no discernible effect on cultured parasites. Results for plasmepsin I and plasmepsin II were somewhat inconsistent, but results for plasmepsin IV were consistent, with each study showing that loss of this protease led to decreased parasite growth. In summary, available results suggest similar but not identical substrate specificities and functions for the four food vacuole plasmepsins and thus complementary roles for these proteases in hemoglobin hydrolysis.

Falcipain Cysteine Proteases

A total of 33 cysteine protease-like genes were identified in the *P. falciparum* genome (Wu et al., 2003), but these include members of quite distant families (reviewed in Rosenthal, 2004). Four enzymes are rather typical members of the papain family and are referred to as falcipains. Evidence for a role for cysteine proteases in hemoglobin hydrolysis came from the observation that cysteine protease inhibitors cause a dramatic morphological abnormality in *P. falciparum* trophozoites, whereby food vacuoles swell and fill with undegraded hemoglobin (Bailly et al., 1992; Dluzewski et al., 1986; Gamboa de Dominguez and Rosenthal, 1996; Rosenthal et al., 1988). The principal cysteine protease activities of *P. falciparum* trophozoites are falcipain-2 (Shenai et al., 2000) and falcipain-3 (Sijwali et al., 2001). Both falcipain-2 and falcipain-3 localize to the food vacuole and hydrolyze peptides, globin, and native hemoglobin at acidic pH, with preference for substrates with Leu at the P_2 position (Shenai et al., 2000; Sijwali et al., 2001; Dahl and Rosenthal, 2005). As is typical for cysteine proteases, the falcipains require reducing conditions for maximal activity, but they readily hydrolyze native hemoglobin at physiological concentrations of the cellular reductant glutathione (Shenai and Rosenthal, 2002). Although falcipain-2 is much more effective at cleaving small synthetic substrates, both falcipain-2 and falcipain-3 hydrolyze hemoglobin and globin with similar kinetics. The first identified falcipain, now termed falcipain-1, is also expressed in erythrocytic-stage parasites (Sijwali and Rosenthal, 2004) but has been difficult to characterize biochemically. Recent studies of falcipain-1 suggested a role for the enzyme in erythrocyte invasion (Greenbaum et al., 2002), but other studies have shown that the enzyme is not essential for erythrocytic parasites (Eksi et al., 2004; Sijwali et al., 2004). A fourth falcipain gene encodes a protease that is nearly identical to falcipain-2 in the catalytic domain and is named falcipain-2'. This protease is also expressed in erythrocytic parasites but has not yet been biochemically characterized. Falcipain-2, falcipain-2', and falcipain-3 are encoded by nearly contiguous genes on chromosome 11; the falcipain-1 gene is on chromosome 14.

Falcipain-2, falcipain-2', and falcipain-3 are much more similar in sequence to each other than they are to falcipain-1, and they have some unusual features for papain family proteases. First, these falcipains have unusually long prodomains, including putative type II membrane-spanning domains. This architecture suggests trafficking to the food vacuole via the secretory pathway, as demonstrated for the plasmepsins (Klemba et al., 2004a). Second, these three falcipains lack typical processing sites but rather contain a unique 15- to 22-residue amino-terminal extension on the mature protease. Uniquely among studied papain family proteases, the falcipains require this amino-terminal extension but not the prodomain for folding of the catalytic domain to an active enzyme (Pandey et al., 2004; Sijwali et al., 2002). Third, all four falcipains contain an unusual carboxy-terminal insertion of 15 residues (30 residues in falcipain-1) between the catalytic His and Asn (Rosenthal et al., 2002). For falcipain-2, this insertion encodes a distinct motif that protrudes from the globular protein and mediates binding between the protease and hemoglobin independent of the active site. The unique motif is not required for hydrolysis of a number of peptide and protein substrates but is required for the hydrolysis of globin and hemoglobin (Pandey et al., in press). Removal of the motif does not impact upon inhibition of falcipain-2 by peptide or protein inhibitors, but markedly decreases autoinhibition by the falcipain-2 prodomain (Pandey et al., in press). Thus, the unique motif appears to specifically mediate interaction between falcipains and their biologically relevant substrate (hemoglobin) and inhibitor (the prodomain).

Additional insights into the roles of falcipains in hemoglobin hydrolysis have come from the disruption of falcipain genes. Disruption of the falcipain-2 gene caused a block in hemoglobin hydrolysis in early trophozoites, but more mature parasites regained normal morphologies and completed normal development (Sijwali and Rosenthal, 2004). The falcipain-2 knockout parasites had increased expression of falcipain-2′ and falcipain-3, were more sensitive than the wild type to cysteine protease inhibitors, and were much more sensitive to aspartic protease inhibitors. These results suggest that the loss of falcipain-2 was complemented by expression of other falcipains, and they highlight apparent cooperative roles of cysteine and aspartic proteases in hemoglobin hydrolysis. Disruption of falcipain-3 has not yet been successful, suggesting that the product of this single-copy gene may be essential for erythrocytic parasites. It remains unclear if falcipain-1 also plays a role in hemoglobin hydrolysis, but as noted above erythrocytic falcipain-1 knockout parasites developed normally (Eksi et al., 2004; Sijwali et al., 2004). However, these parasites did show a marked drop in oocyst production, suggesting a role for this protease in nonerythrocytic parasites (Eksi et al., 2004).

Falcilysin

A single metalloprotease appears to contribute to hemoglobin hydrolysis in the food vacuole. Falcilysin is a large enzyme (130 kDa) that is a member of the M16 family of oligoendopeptidase metalloproteases (Eggleson et al., 1999). It is located in the food vacuole, but unlike food vacuole aspartic and cysteine proteases, it does not cleave native hemoglobin or denatured globin. Rather, falcilysin hydrolyzes smaller peptide fragments of hemoglobin. The enzyme shows a preference for cleavage at charged or polar amino acids. Unlike many other proteases, including the food vacuole aspartic and cysteine proteases, falcilysin has no inactive proform precursor (Murata and Goldberg, 2003b). The enzyme also does not contain an obvious signal peptide, and so its mode of targeting to the food vacuole appears to differ from that of the food vacuole aspartic and cysteine proteases. As is the case with other food vacuole proteases, falcilysin is active against hemoglobin peptides at acid pH. However, it is also active at neutral pH, albeit with quite different substrate specificity (Murata and Goldberg, 2003a). These data suggest that, as has been proposed for both plasmepsin II (Le Bonniec et al., 1999) and falcipain-2 (Dua et al., 2001), falcilysin may act against substrates in addition to hemoglobin, with acid activity optimized for hemoglobinolysis in the food vacuole and neutral activity optima against other substrates.

Aminopeptidases

Additional cleavages of hemoglobin peptides appear to be performed by at least two aminopeptidases. A food vacuole aminopeptidase has recently been identified (Klemba et al., 2004b). This cysteine protease, named dipeptidyl aminopeptidase I, is the product of one of three *P. falciparum* genes encoding homologs of dipeptidyl peptidases in other organisms (Wu et al., 2003). These genes predict proteases that are members of the papain family but share features with other dipeptidyl peptidases, including a large exclusion domain that prevents endopeptidase activity (Turk et al., 2001). The gene for dipeptidyl aminopeptidase I is located on the same chromosome and about 36 kb removed from the region that encodes the three closely related falcipains discussed above. Studies with labelled enzyme suggested trafficking through the parasitophorous vacuole to the food vacuole. Dipeptidyl aminopeptidase I cleaved dipeptide substrates under acidic conditions, with greatest activity against Pro-Arg. Disruption of the dipeptidyl aminopeptidase I gene was unsuccessful, suggesting that the protease is essential for erythrocytic parasites. The products of the two other putative *P. falciparum* dipeptidyl peptidase genes have not yet been characterized.

The processing of hemoglobin peptides in the cytosol is probably performed, at least in part, by a neutral metalloaminopeptidase. This M1 family zinc endopeptidase has been identified in *P. falciparum* (Curley et al., 1994; Florent et al., 1998; Kolakovich et al., 1997). The protease is processed from a 120-kDa precursor to an active enzyme and localizes diffusely in parasite cytosol (Allary et al., 2002). Aminopeptidase activity was greatest in trophozoites, the stage at which hemoglobin hydrolysis is most active and acted against synthetic peptides based on known hemoglobin hydrolysis fragments (Gavigan et al., 2001). It is not clear if this single aminopeptidase is responsible for all cytosolic cleavages of hemoglobin peptides, and it remains likely that other plasmodial proteases contribute to the hydrolysis of hemoglobin-derived peptides after they are transported from the food vacuole.

Cooperative Activity of Food Vacuole Proteases

While it has been difficult to identify a specific pathway of hemoglobin hydrolysis, it appears that multiple proteases act cooperatively to hydrolyze this substrate. Support for the cooperative action of cysteine and aspartic proteases comes from the demonstration of synergistic activity against hemoglobin of cysteine and aspartic proteases (Gluzman et al., 1994) and synergistic in vitro (Bailly et al., 1992; Gluzman et al., 1994; Semenov et al., 1998) and in vivo (Semenov et al., 1998) antimalarial activities of cysteine and aspartic protease inhibitors. Also, parasites with a disrupted falcipain-2 gene were not only more sensitive to cysteine protease inhibitors but also much more sensitive to the aspartic protease inhibitor pepstatin (Sijwali and Rosenthal, 2004). These data do not prove that falcipains and plasmepsins act cooperatively against hemoglobin, as an alternative explanation might be that one class of proteases is responsible for the activation of another. Indeed, falcipains appear to activate by autohydrolysis (Shenai et al., 2000; Sijwali et al., 2001), but the four food vacuole plasmepsins require the activity of another protease for activation. The identity of this processing protease is uncertain. The activity is blocked by some, but not other, cysteine protease inhibitors (Banerjee et al., 2003; Francis et al., 1997). This result has been interpreted as suggesting plasmepsin processing activity by an as yet unidentified protease, perhaps a predicted *P. falciparum* calpain homolog (as the effective inhibitors also inhibit calpains), but an alternative explanation is that the plasmepsins are activated by falcipains, but that differential activities of cysteine protease inhibitors are due to differential access to the proteases. In any event, results with protease inhibitors and knockout parasites offer hope that combinations of cysteine and aspartic protease inhibitors may be potent antimalarial drugs.

Other Proteases and Hemoglobin Hydrolysis

As noted above, many other proteases are predicted by the *P. falciparum* genome sequence. Some of these enzymes are quite well characterized but probably are not involved in hemoglobin hydrolysis. For example, the subtilisins, which are serine proteases (Withers-Martinez et al., 2004), and SERA family proteins, which have features of serine and cysteine proteases (Hodder et al., 2003), appear to have roles in erythrocyte rupture and invasion, based on their intracellular localization and their timing of expression.

Proteases of Other Plasmodial Species

Homologs of *P. falciparum* proteases have been identified in other plasmodial species. The best-characterized enzymes have been plasmepsin and falcipain homologs. Considering plasmepsin homologs in the three other human malaria parasites and some primate and rodent parasites, a homolog most closely related to plasmepsin IV was consistently identified (Dame et al., 2003), but homologs of the other food vacuole plasmepsins were not seen. Considering falcipain homologs, all studied species appear to contain a single falcipain-1 homolog and a varied number of homologs of falcipain-2 (including falcipain-2' and falcipain-3 in *P. falciparum*) (Rosenthal et al., 2002). *Plasmodium vivax* appears to express three falcipain-2 homologs (Na et al., 2004; B. K. Na, unpublished data). For both plasmepsins and falcipains, activities of homologous proteases from different human malaria parasites are very similar (Li et al., 2005; Na et al., 2004). Proteases from rodent malaria parasites are also similar to their *P. falciparum* homologs, but specific biochemical differences may complicate drug discovery efforts that utilize rodent malaria models (Humphreys et al., 1999; Singh et al., 2002).

FATE OF HEMOGLOBIN DEGRADATION PRODUCTS

Hemoglobin hydrolysis necessitates parasite mechanisms for dealing with breakdown products, including amino acids, iron, and heme. Amino acids may be essential for parasite protein synthesis and must be delivered to the cytosol. Hemoglobin-derived iron and heme may both be required for parasite metabolism, but both can also lead to toxicity and so require appropriate sequestration.

Amino Acids

As discussed above, free amino acids may be generated in the food vacuole, but available evidence suggests that ultimate steps in globin processing occur in the cytosol (Kolakovich et al., 1997). Mechanisms by which globin breakdown products are transported from the food vacuole to the cytosol are undefined. Free amino acids and some dipeptides can diffuse through lysosomal membranes (Bohley and Seglen, 1992), and so some amino acids may diffuse from the food vacuole to the cytosol. In addition, multiple putative plasmodial transporters have been identified (Mu et al., 2003), and it is likely that specific transport pathways exist for the transit of free amino acids and/or globin-derived peptides from the food vacuole to the cytosol. Excess amino acids are expelled from infected erythrocytes, presumably via parasite-derived permeation pathways (Kirk, 2001; Zarchin et al., 1986).

Iron and Parasite-Synthesized Heme

Malaria parasites require iron, and iron chelators demonstrate antimalarial activity (Mabeza et al., 1999). The means by which parasites acquire iron are unclear. Potential mechanisms include uptake of transferrin, acquisition of iron from ferritin ingested during hemoglobin accumulation, and the utilization of free erythrocyte iron, but an additional obvious source is hemoglobin (Mabeza et al., 1999). As a great deal of heme is liberated during the hydrolysis of hemoglobin, it is plausible that some iron is released from heme and utilized for parasite needs. However, experimental evidence in support of this possibility is not available. Curiously, even though hemoglobin hydrolysis leaves parasites with a massive heme detoxification problem, they also synthesize heme, with contributions from enzymes synthesized both by the host and the parasite (Bonday et al., 1997; Dhanasekaran et al., 2004). This capability suggests that hemoglobin-derived iron may not be required for parasite heme synthesis, but whether it is required for the synthesis of plasmodial iron-dependent enzymes is unknown.

Hemozoin Formation

The hydrolysis of hemoglobin by erythrocytic parasites liberates large amounts of heme. Free heme lyses parasites (Orjih et al., 1981), and as malaria parasites have not been shown to degrade heme enzymatically, they apparently incorporate heme molecules into particulate hemozoin as a detoxification mechanism. The structure of hemozoin remained uncertain for a long period, but it is now well characterized (Sullivan, 2002). Hemozoin consists of complexes of pure heme, without associated proteins (Fitch and Kanjananggulpan, 1987; Slater et al., 1991). Based on multiple means of assessment, the structure of hemozoin appears to be identical to synthetic β-hematin (Bohle et al., 1997). According to X-ray diffraction data, hemozoin is not an iron-carboxylate-linked polymer, as previously suggested (Slater et al., 1991), but rather a collection of heme dimers, with hydrogen bonding linking dimers into large insoluble aggregates (Pagola et al., 2000).

Hemozoin formation was earlier attributed to an enzymatic activity, termed heme polymerase (Slater and Cerami, 1992), but this enzyme has not been identified, and it is now clear that β-hematin forms chemically without the presence of any enzyme (Dorn et al., 1995; Egan et al., 1994). The process of hemozoin formation has been described as a biomineralization (or biocrystallization) process, with rapid precipitation of amorphous hematin followed by crystallization into hemozoin (Egan et al., 2001). The specific mechanism of hemozoin biocrystallization remains uncertain and somewhat controversial. It appears that a nucleation center is required to initiate the process, as is typical for biocrystallization reactions (Hempelmann and Egan, 2002). The best-characterized protein initiator of hemozoin formation is *P. falciparum* histidine-rich protein II. Native and recombinant histidine-rich protein II promoted the formation of hemozoin, and this process was inhibited by chloroquine, which inhibits the formation of hemozoin in intact parasites (Lynn et al., 1999; Sullivan et al., 1996). Other histidine-rich *P. falciparum* proteins can also initiate hemozoin formation (Sullivan, 2002). Lipids have also been shown to initiate this process (Bendrat et al., 1995; Dorn et al., 1995). Considering these data, a new model has been proposed whereby hemoglobin degradation begins in the transport vesicles that carry erythrocyte cytosol from the cytostome to the food vacuole and heme released by this process accumulates within the inner membrane of transport vesicles where membrane lipids initiate hemozoin formation (Hempelmann et al., 2003), although evidence to support this model is incomplete. In any event, it is becoming increasingly clear that malaria parasites utilize a unique biocrystallization strategy to detoxify free heme as they hydrolyze hemoglobin. As has also become clear in recent years, the hemozoin formation process is a productive target for antimalarial chemotherapy (Foley and Tilley, 1998; Stocks et al., 2001).

POTENTIAL NEW ANTIMALARIAL THERAPIES BASED ON INHIBITION OF HEMOGLOBIN HYDROLYSIS

A number of antimalarial drugs appear to act by interaction with products of hemoglobin processing. Notably, chloroquine and other aminoquinoline antimalarials are believed to disrupt the formation of hemozoin pigment, thus eliciting parasite toxicity from the buildup of free heme (Tilley et al., 2001). In addition, artemisinin antimalarials may act, at least in part, by the heme-mediated production of free radicals in the food vacuole (Meshnick, 2002). The mechanisms of action of these drugs will be further discussed in Chapter 23. This section will highlight new potential means of antimalarial therapy based on inhibition of proteases that hydrolyze hemoglobin.

As noted above, important initial insight into mechanisms of hemoglobin hydrolysis came from the observation that cysteine protease inhibitors block the process, leading to the accumulation of undegraded hemoglobin in the food vacuole and a block in parasite development (Bailly et al., 1992; Dluzewski et al., 1986; Rosenthal et al., 1988). Small molecule cysteine protease inhibitors exert potent antimalarial effects in vitro (Dominguez et al., 1997; Lee et al., 2003; Ring et al., 1993; Rockett et al., 1990; Rosenthal et al., 1996; Rosenthal et al., 1991; Shenai et al., 2003). Antimalarial activity is consistently accompanied by the appearance of hemoglobin-filled food vacuoles. Cysteine protease inhibitors also exert antimalarial activities in vivo against murine malaria models (Olson et al., 1999; Rosenthal et al., 1993). Extensive efforts to develop cysteine protease inhibitors as antimalarial drugs are under way.

Aspartic protease inhibitors are also under study as potential antimalarials. Interestingly, aspartic protease inhibitors do not appear to cause a specific block in hemoglobin hydrolysis, but they nevertheless exert potent antimalarial effects (Bailly et al., 1992; Rosenthal, 1995). A number of classes of potent plasmepsin inhibitors have been reported in recent years (reviewed in Boss et al., 2003). Some of these compounds have also inhibited the development of cultured parasites (Boss et al., 2003; Ersmark et al., 2004; Francis et al., 1994; Haque et al., 1999; Jiang et al., 2001; Johansson et al., 2004; Moon et al., 1997; Nezami et al., 2002; Noteberg et al., 2003; Silva et al., 1996).

Drug development efforts are expected to benefit from the determination of the structures of plasmepsin II (Silva et al., 1996), a plasmepsin homolog from *P. vivax* (Bernstein et al., 2003) and, very recently, falcipain-2 (S. Wang, unpublished data).

As noted above, evidence for the cooperative action of cysteine and aspartic proteases against hemoglobin includes demonstration that the two classes of proteases acted synergistically to degrade hemoglobin in vitro (Francis et al., 1994), evidence of the synergistic antimalarial activity of cysteine and aspartic protease inhibitors (Bailly et al., 1992; Semenov et al., 1998), and the demonstration that falcipain-2 knockout parasites have markedly increased sensitivity to an aspartic protease inhibitor (Sijwali and Rosenthal, 2004). These results suggest that an optimal therapeutic strategy may be to inhibit both classes of proteases. This strategy should provide increased antiparasitic activity over that achieved by inhibitors of either class and also help to prevent the selection of drug-resistant parasites. Considering the selection of drug resistance, a recent study showed that parasites resistant to vinyl sulfone cysteine protease inhibitors could be selected by the incubation of cultured parasites with stepwise increases in concentrations of inhibitor (Singh and Rosenthal, 2004). However, resistance developed only gradually and the mechanism of resistance was complex, including changes in falcipain gene copy number and inhibitor accumulation, but not alterations in falcipain sequences. The clinical significance of these findings is uncertain, but they further support the strategy of combination antimalarial therapy to prevent the selection of resistant parasites, as is now widely advocated (White, 2004).

REFERENCES

Allary, M., J. Schrevel, and I. Florent. 2002. Properties, stage-dependent expression and localization of *Plasmodium falciparum* M1 family zinc-aminopeptidase. *Parasitology* **125**:1–10.

Asawamahasakda, W., I. Ittarat, C. C. Chang, P. McElroy, and S. R. Meshnick. 1994. Effects of antimalarials and protease inhibitors on plasmodial hemozoin production. *Mol. Biochem. Parasitol.* **67**:183–191.

Asawamahasakda, W., and Y. Yuthavong. 1993. The methionine synthesis cycle and salvage of methyltetrahydrofolate from host red cells in the malaria parasite (*Plasmodium falciparum*). *Parasitology* **107**:1–10.

Bailly, E., R. Jambou, J. Savel, and G. Jaureguiberry. 1992. *Plasmodium falciparum*: differential sensitivity in vitro to E-64 (cysteine protease inhibitor) and pepstatin A (aspartyl protease inhibitor). *J. Protozool.* **39**:593–599.

Banerjee, R., S. E. Francis, and D. E. Goldberg. 2003. Food vacuole plasmepsins are processed at a conserved site by an acidic convertase activity in *Plasmodium falciparum*. *Mol. Biochem. Parasitol.* **129**:157–165.

Banerjee, R., J. Liu, W. Beatty, L. Pelosof, M. Klemba, and D. E. Goldberg. 2002. Four plasmepsins are active in the *Plasmodium falciparum* food vacuole, including a protease with an active-site histidine. *Proc. Natl. Acad. Sci. USA* **99**:990–995.

Bendrat, K., B. J. Berger, and A. Cerami. 1995. Haem polymerization in malaria. *Nature* **378**:138–139.

Bernstein, N. K., M. M. Cherney, C. A. Yowell, J. B. Dame, and M. N. James. 2003. Structural insights into the activation of *P. vivax* plasmepsin. *J. Mol. Biol.* **329**:505–524.

Berry, C., M. J. Humphreys, P. Matharu, R. Granger, P. Horrocks, R. P. Moon, U. Certa, R. G. Ridley, D. Bur, and J. Kay. 1999. A distinct member of the aspartic proteinase gene family from the human malaria parasite *Plasmodium falciparum*. *FEBS Lett.* **447**:149–154.

Biagini, G. A., P. G. Bray, D. G. Spiller, M. R. White, and S. A. Ward. 2003. The digestive food vacuole of the malaria parasite is a dynamic intracellular Ca2+ store. *J. Biol. Chem.* **278**:27910–27915.

Bohle, D. S., R. E. Dinnebier, S. K. Madsen, and P. W. Stephens. 1997. Characterization of the products of the heme detoxification pathway in malarial late trophozoites by X-ray diffraction. *J. Biol. Chem.* **272**:713–716.

Bohley, P., and P. O. Seglen. 1992. Proteases and proteolysis in the lysosome. *Experientia* **48**:151–157.

Bonday, Z. Q., S. Taketani, P. D. Gupta, and G. Padmanaban. 1997. Heme biosynthesis by the malarial parasite. Import of delta-aminolevulinate dehydrase from the host red cell. *J. Biol. Chem.* **272**:21839–21846.

Boss, C., S. Richard-Bildstein, T. Weller, W. Fischli, S. Meyer, and C. Binkert. 2003. Inhibitors of the *Plasmodium falciparum* parasite aspartic protease plasmepsin II as potential antimalarial agents. *Curr. Med. Chem.* **10**:883–907.

Bray, P. G., O. Janneh, K. J. Raynes, M. Mungthin, H. Ginsburg, and S. A. Ward. 1999. Cellular uptake of chloroquine is dependent on binding to ferriprotoporphyrin IX and is independent of NHE activity in *Plasmodium falciparum*. *J. Cell Biol.* **145**:363–376.

Curley, G. P., S. M. O'Donovan, J. McNally, M. Mullally, H. O'Hara, A. Troy, S. A. O'Callaghan, and J. P. Dalton. 1994. Aminopeptidases from *Plasmodium falciparum*, *Plasmodium chabaudi chabaudi* and *Plasmodium berghei*. *J. Eukaryot. Microbiol.* **41**:119–123.

Dahl, E. L., and P. J. Rosenthal. 2005. Biosynthesis, localization, and processing of falcipain cysteine proteases of *Plasmodium falciparum*. *Mol. Biochem. Parasitol.* **139**:205–212.

Dame, J. B., G. R. Reddy, C. A. Yowell, B. M. Dunn, J. Kay, and C. Berry. 1994. Sequence, expression and modeled structure of an aspartic proteinase from the human malaria parasite *Plasmodium falciparum*. *Mol. Biochem. Parasitol.* **64**:177–190.

Dame, J. B., C. A. Yowell, L. Omara-Opyene, J. M. Carlton, R. A. Cooper, and T. Li. 2003. Plasmepsin 4, the food vacuole aspartic proteinase found in all *Plasmodium* spp. infecting man. *Mol. Biochem. Parasitol.* **130**:1–12.

Dhanasekaran, S., N. R. Chandra, B. K. Chandrasekhar Sagar, P. N. Rangarajan, and G. Padmanaban. 2004. Delta-aminolevulinic acid dehydratase from *Plasmodium falciparum*: indigenous versus imported. *J. Biol. Chem.* **279**:6934–6942.

Divo, A. A., T. G. Geary, N. L. Davis, and J. B. Jensen. 1985. Nutritional requirements of *Plasmodium falciparum* in culture. I. Exogenously supplied dialyzable components necessary for continuous growth. *J. Protozool.* **32**:59–64.

Dluzewski, A. R., K. Rangachari, R. J. Wilson, and W. B. Gratzer. 1986. *Plasmodium falciparum*: protease inhibitors and inhibition of erythrocyte invasion. *Exp. Parasitol.* **62**:416–422.

Dominguez, J. N., S. Lopez, J. Charris, L. Iarruso, G. Lobo, A. Semenov, J. E. Olson, and P. J. Rosenthal. 1997. Synthesis and antimalarial effects of phenothiazine inhibitors of a *Plasmodium falciparum* cysteine protease. *J. Med. Chem.* **40**:2726–2732.

Dorn, A., R. Stoffel, H. Matile, A. Bubendorf, and R. G. Ridley. 1995. Malarial haemozoin/ beta-haematin supports haem polymerization in the absence of protein. *Nature* **374**:269–271.

Dua, M., P. Raphael, P. S. Sijwali, P. J. Rosenthal, and M. Hanspal. 2001. Recombinant falcipain-2 cleaves erythrocyte membrane ankyrin and protein 4.1. *Mol. Biochem. Parasitol.* **116**:95–99.

Egan, T. J., W. W. Mavuso, and K. K. Ncokazi. 2001. The mechanism of beta-hematin formation in acetate solution. Parallels between hemozoin formation and biomineralization processes. *Biochemistry* **40**:204–213.

Egan, T. J., D. C. Ross, and P. A. Adams. 1994. Quinoline anti-malarial drugs inhibit spontaneous formation of beta-haematin (malaria pigment). *FEBS Lett.* **352**:54–57.

Eggleson, K. K., K. L. Duffin, and D. E. Goldberg. 1999. Identification and characterization of falcilysin, a metallopeptidase involved in hemoglobin catabolism within the malaria parasite *Plasmodium falciparum*. *J. Biol. Chem.* **274**:32411–32417.

Eksi, S., B. Czesny, D. C. Greenbaum, M. Bogyo, and K. C. Williamson. 2004. Targeted disruption of *Plasmodium falciparum* cysteine protease, falcipain 1, reduces oocyst production, not erythrocytic stage growth. *Mol. Microbiol.* **53**:243–250.

Elford, B. C., G. M. Cowan, and D. J. Ferguson. 1995. Parasite-regulated membrane transport processes and metabolic control in malaria-infected erythrocytes. *Biochem. J.* **308**:361–374.

Ersmark, K., I. Feierberg, S. Bjelic, E. Hamelink, F. Hackett, M. J. Blackman, J. Hulten, B. Samuelsson, J. Aqvist, and A. Hallberg. 2004. Potent inhibitors of the *Plasmodium falciparum* enzymes plasmepsin I and II devoid of cathepsin D inhibitory activity. *J. Med. Chem.* **47**:110–122.

Fitch, C. D., and P. Kanjananggulpan. 1987. The state of ferriprotoporphyrin IX in malaria pigment. *J. Biol. Chem.* **262**:15552–15555.

Florent, I., Z. Derhy, M. Allary, M. Monsigny, R. Mayer, and J. Schrevel. 1998. A *Plasmodium falciparum* aminopeptidase gene belonging to the M1 family of zinc-metallopeptidases is expressed in erythrocytic stages. *Mol. Biochem. Parasitol.* **97**:149–160.

Foley, M., and L. Tilley. 1998. Quinoline antimalarials: mechanisms of action and resistance and prospects for new agents. *Pharmacol. Ther.* **79**:55–87.

Francis, S. E., R. Banerjee, and D. E. Goldberg. 1997. Biosynthesis and maturation of the malaria aspartic hemoglobinases plasmepsins I and II. *J. Biol. Chem.* **272**:14961–14968.

Francis, S. E., I. Y. Gluzman, A. Oksman, A. Knickerbocker, R. Mueller, M. L. Bryant, D. R. Sherman, D. G. Russell, and D. E. Goldberg. 1994. Molecular characterization and inhibition of a *Plasmodium falciparum* aspartic hemoglobinase. *EMBO J.* **13**:306–317.

Gabay, T., and H. Ginsburg. 1993. Hemoglobin denaturation and iron release in acidified red blood cell lysate—a possible source of iron for intraerythrocytic malaria parasites. *Exp. Parasitol.* **77**:261–272.

Gamboa de Dominguez, N. D., and P. J. Rosenthal. 1996. Cysteine proteinase inhibitors block early steps in hemoglobin degradation by cultured malaria parasites. *Blood* **87**:4448–4454.

Gavigan, C. S., J. P. Dalton, and A. Bell. 2001. The role of aminopeptidases in haemoglobin degradation in *Plasmodium falciparum*-infected erythrocytes. *Mol. Biochem. Parasitol.* **117**:37–48.

Gluzman, I. Y., S. E. Francis, A. Oksman, C. E. Smith, K. L. Duffin, and D. E. Goldberg. 1994. Order and specificity of the *Plasmodium falciparum* hemoglobin degradation pathway. *J. Clin Investig.* **93**:1602–1608.

Goldberg, D. E., A. F. Slater, R. Beavis, B. Chait, A. Cerami, and G. B. Henderson. 1991. Hemoglobin degradation in the human malaria pathogen *Plasmodium falciparum*: a catabolic pathway initiated by a specific aspartic protease. *J. Exp. Med.* **173**:961–969.

Goldberg, D. E., A. F. Slater, A. Cerami, and G. B. Henderson. 1990. Hemoglobin degradation in the malaria parasite *Plasmodium falciparum*: an ordered process in a unique organelle. *Proc. Natl. Acad. Sci. USA* **87**:2931–2935.

Greenbaum, D. C., A. Baruch, M. Grainger, Z. Bozdech, K. F. Medzihradszky, J. Engel, J. DeRisi, A. A. Holder, and M. Bogyo. 2002. A role for the protease falcipain 1 in host cell invasion by the human malaria parasite. *Science* **298**:2002–2006.

Haque, T. S., A. G. Skillman, C. E. Lee, H. Habashita, I. Y. Gluzman, T. J. Ewing, D. E. Goldberg, I. D. Kuntz, and J. A. Ellman. 1999. Potent, low-molecular-weight non-peptide inhibitors of malarial aspartyl protease plasmepsin II. *J. Med. Chem.* **42:**1428–1440.

Hempelmann, E., and T. J. Egan. 2002. Pigment biocrystallization in Plasmodium falciparum. *Trends Parasitol.* **18:**11.

Hempelmann, E., C. Motta, R. Hughes, S. A. Ward, and P. G. Bray. 2003. Plasmodium falciparum: sacrificing membrane to grow crystals? *Trends Parasitol.* **19:**23–26.

Hill, J., L. Tyas, L. H. Phylip, J. Kay, B. M. Dunn, and C. Berry. 1994. High level expression and characterisation of plasmepsin II, an aspartic proteinase from Plasmodium falciparum. *FEBS Lett.* **352:**155–158.

Hodder, A. N., D. R. Drew, V. C. Epa, M. Delorenzi, R. Bourgon, S. K. Miller, R. L. Moritz, D. F. Frecklington, R. J. Simpson, T. P. Speed, R. N. Pike, and B. S. Crabb. 2003. Enzymic, phylogenetic, and structural characterization of the unusual papain-like protease domain of Plasmodium falciparum SERA5. *J. Biol. Chem.* **278:**48169–48177.

Humphreys, M. J., R. P. Moon, A. Klinder, S. D. Fowler, K. Rupp, D. Bur, R. G. Ridley, and C. Berry. 1999. The aspartic proteinase from the rodent parasite Plasmodium berghei as a potential model for plasmepsins from the human malaria parasite, Plasmodium falciparum. *FEBS Lett.* **463:**43–48.

Jiang, S., S. T. Prigge, L. Wei, Y. Gao, T. H. Hudson, L. Gerena, J. B. Dame, and D. E. Kyle. 2001. New class of small nonpeptidyl compounds blocks Plasmodium falciparum development in vitro by inhibiting plasmepsins. *Antimicrob. Agents Chemother.* **45:**2577–2584.

Johansson, P. O., Y. Chen, A. K. Belfrage, M. J. Blackman, I. Kvarnstrom, K. Jansson, L. Vrang, E. Hamelink, A. Hallberg, A. Rosenquist, and B. Samuelsson. 2004. Design and synthesis of potent inhibitors of the malaria aspartyl proteases plasmepsin I and II. Use of solid-phase synthesis to explore novel statine motifs. *J. Med. Chem.* **47:**3353–3366.

Kamchonwongpaisan, S., E. Samoff, and S. R. Meshnick. 1997. Identification of hemoglobin degradation products in Plasmodium falciparum. *Mol. Biochem. Parasitol.* **86:**179–186.

Kirk, K. 2001. Membrane transport in the malaria-infected erythrocyte. *Physiol. Rev.* **81:**495–537.

Klemba, M., W. Beatty, I. Gluzman, and D. E. Goldberg. 2004a. Trafficking of plasmepsin II to the food vacuole of the malaria parasite Plasmodium falciparum. *J. Cell Biol.* **164:**47–56.

Klemba, M., I. Gluzman, and D. E. Goldberg. 2004b. A Plasmodium falciparum dipeptidyl aminopeptidase I participates in vacuolar hemoglobin degradation. *J. Biol. Chem.* **279:**43000–43007.

Kolakovich, K. A., I. Y. Gluzman, K. L. Duffin, and D. E. Goldberg. 1997. Generation of hemoglobin peptides in the acidic digestive vacuole of Plasmodium falciparum implicates peptide transport in amino acid production. *Mol. Biochem. Parasitol.* **87:**123–135.

Krugliak, M., J. Zhang, and H. Ginsburg. 2002. Intraerythrocytic Plasmodium falciparum utilizes only a fraction of the amino acids derived from the digestion of host cell cytosol for the biosynthesis of its proteins. *Mol. Biochem. Parasitol.* **119:**249–256.

Langreth, S. G., J. B. Jensen, R. T. Reese, and W. Trager. 1978. Fine structure of human malaria in vitro. *J. Protozool.* **25:**443–452.

Lauer, S. A., P. K. Rathod, N. Ghori, and K. Haldar. 1997. A membrane network for nutrient import in red cells infected with the malaria parasite. *Science* **276:**1122–1125.

Le Bonniec, S., C. Deregnaucourt, V. Redeker, R. Banerjee, P. Grellier, D. E. Goldberg, and J. Schrevel. 1999. Plasmepsin II, an acidic hemoglobinase from the Plasmodium falciparum food vacuole, is active at neutral pH on the host erythrocyte membrane skeleton. *J. Biol. Chem.* **274:**14218–14223.

Lee, B. J., A. Singh, P. Chiang, S. J. Kemp, E. A. Goldman, M. I. Weinhouse, G. P. Vlasuk, and P. J. Rosenthal. 2003. Antimalarial activities of novel synthetic cysteine protease inhibitors. *Antimicrob. Agents Chemother.* **47:**3810–3814.

Lew, V. L., L. Macdonald, H. Ginsburg, M. Krugliak, and T. Tiffert. 2004. Excess haemoglobin digestion by malaria parasites: a strategy to prevent premature host cell lysis. *Blood Cells Mol. Dis.* **32:**353–359.

Lew, V. L., T. Tiffert, and H. Ginsburg. 2003. Excess hemoglobin digestion and the osmotic stability of Plasmodium falciparum-infected red blood cells. *Blood* **101:**4189–4194.

Li, T., C. A. Yowell, B. B. Beyer, S. H. Hung, J. Westling, M. T. Lam, B. M. Dunn, and J. B. Dame. 2004. Recombinant expression and enzymatic subsite characterization of plasmepsin 4 from the four Plasmodium species infecting man. *Mol. Biochem. Parasitol.* **135:**101–109.

Liu, J., I. Y. Gluzman, M. E. Drew, and D. E. Goldberg. 2005. The role of Plasmodium falciparum food vacuole plasmepsins. *J. Biol. Chem.* **280:**1432–1437.

Lynn, A., S. Chandra, P. Malhotra, and V. S. Chauhan. 1999. Heme binding and polymerization by Plasmodium falciparum histidine rich protein II: influence of pH on activity and conformation. *FEBS Lett.* **459:**267–271.

Mabeza, G. F., M. Loyevsky, V. R. Gordeuk, and G. Weiss. 1999. Iron chelation therapy for malaria: a review. *Pharmacol. Ther.* **81:**53–75.

McKerrow, J. H., E. Sun, P. J. Rosenthal, and J. Bouvier. 1993. The proteases and pathogenicity of parasitic protozoa. *Annu. Rev. Microbiol.* **47:**821–853.

Meshnick, S. R. 2002. Artemisinin: mechanisms of action, resistance and toxicity. *Int. J. Parasitol.* **32:** 1655–1660.

Moon, R. P., L. Tyas, U. Certa, K. Rupp, D. Bur, C. Jacquet, H. Matile, H. Loetscher, F. Grueninger-Leitch, J. Kay, B. M. Dunn, C. Berry, and R. G. Ridley. 1997. Expression and characterisation of plasmepsin I from *Plasmodium falciparum*. *Eur. J. Biochem.* **244:**552–560.

Mu, J., M. T. Ferdig, X. Feng, D. A. Joy, J. Duan, T. Furuya, G. Subramanian, L. Aravind, R. A. Cooper, J. C. Wootton, M. Xiong, and X. Z. Su. 2003. Multiple transporters associated with malaria parasite responses to chloroquine and quinine. *Mol. Microbiol.* **49:**977–989.

Murata, C. E., and D. E. Goldberg. 2003a. *Plasmodium falciparum* falcilysin: a metalloprotease with dual specificity. *J. Biol. Chem.* **278:**38022–38028.

Murata, C. E., and D. E. Goldberg. 2003b. *Plasmodium falciparum* falcilysin: an unprocessed food vacuole enzyme. *Mol. Biochem. Parasitol.* **129:**123–126.

Na, B. K., B. R. Shenai, P. S. Sijwali, Y. Choe, K. C. Pandey, A. Singh, C. S. Craik, and P. J. Rosenthal. 2004. Identification and biochemical characterization of vivapains, cysteine proteases of the malaria parasite Plasmodium vivax. *Biochem. J.* **378:** 529–538.

Nezami, A., I. Luque, T. Kimura, Y. Kiso, and E. Freire. 2002. Identification and characterization of allophenylnorstatine-based inhibitors of plasmepsin II, an antimalarial target. *Biochemistry* **41:**2273–2280.

Noteberg, D., E. Hamelink, J. Hulten, M. Wahlgren, L. Vrang, B. Samuelsson, and A. Hallberg. 2003. Design and synthesis of plasmepsin I and plasmepsin II inhibitors with activity in *Plasmodium falciparum*-infected cultured human erythrocytes. *J. Med. Chem.* **46:**734–746.

Olliaro, P. L., and D. E. Goldberg. 1995. The *Plasmodium falciparum* digestive vacuole: metabolic headquarters and choice drug target. *Parasitol. Today* **11:** 294–297.

Olson, J. E., G. K. Lee, A. Semenov, and P. J. Rosenthal. 1999. Antimalarial effects in mice of orally administered peptidyl cysteine protease inhibitors. *Bioorg. Med. Chem.* **7:**633–638.

Omara-Opyene, A. L., P. A. Moura, C. R. Sulsona, J. A. Bonilla, C. A. Yowell, H. Fujioka, D. A. Fidock, and J. B. Dame. 2004. Genetic disruption of the *Plasmodium falciparum* digestive vacuole plasmepsins demonstrates their functional redundancy. *J. Biol. Chem.* **279:**54088–54096.

Orjih, A. U., H. S. Banyal, R. Chevli, and C. D. Fitch. 1981. Hemin lyses malaria parasites. *Science* **214:**667–669.

Pagola, S., P. W. Stephens, D. S. Bohle, A. D. Kosar, and S. K. Madsen. 2000. The structure of malaria pigment beta-haematin. *Nature* **404:**307–310.

Pandey, K. C., P. S. Sijwali, A. Singh, B. K. Na, and P. J. Rosenthal. 2004. Independent intramolecular mediators of folding, activity, and inhibition for the *Plasmodium falciparum* cysteine protease falcipain-2. *J. Biol. Chem.* **279:**3484–3491.

Pandey, K. C., S. X. Wang, P. S. Sijwali, A. L. Lau, J. H. McKerrow, and P. J. Rosenthal. The *Plasmodium falciparum* cysteine protease falcipain-2 captures its substrate, hemoglobin, via a unique motif. *Proc. Natl. Acad. Sci. USA*, in press.

Ring, C. S., E. Sun, J. H. McKerrow, G. K. Lee, P. J. Rosenthal, I. D. Kuntz, and F. E. Cohen. 1993. Structure-based inhibitor design by using protein models for the development of antiparasitic agents. *Proc. Natl. Acad. Sci. USA* **90:**3583–3587.

Rockett, K. A., J. H. Playfair, F. Ashall, G. A. Targett, H. Angliker, and E. Shaw. 1990. Inhibition of intraerythrocytic development of *Plasmodium falciparum* by proteinase inhibitors. *FEBS Lett.* **259:**257–259.

Rosenthal, P. J. 1995. *Plasmodium falciparum*: effects of proteinase inhibitors on globin hydrolysis by cultured malaria parasites. *Exp. Parasitol.* **80:**272–281.

Rosenthal, P. J. 2004. Cysteine proteases of malaria parasites. *Int. J. Parasitol.* **34:**1489–1499.

Rosenthal, P. J., G. K. Lee, and R. E. Smith. 1993. Inhibition of a Plasmodium vinckei cysteine proteinase cures murine malaria. *J. Clin. Investig.* **91:**1052–1056.

Rosenthal, P. J., J. H. McKerrow, M. Aikawa, H. Nagasawa, and J. H. Leech. 1988. A malarial cysteine proteinase is necessary for hemoglobin degradation by *Plasmodium falciparum*. *J. Clin. Investig.* **82:**1560–1566.

Rosenthal, P. J., J. E. Olson, G. K. Lee, J. T. Palmer, J. L. Klaus, and D. Rasnick. 1996. Antimalarial effects of vinyl sulfone cysteine proteinase inhibitors. *Antimicrob. Agents Chemother.* **40:**1600–1603.

Rosenthal, P. J., P. S. Sijwali, A. Singh, and B. R. Shenai. 2002. Cysteine proteases of malaria parasites: targets for chemotherapy. *Curr. Pharm. Des.* **8:**1659–1672.

Rosenthal, P. J., W. S. Wollish, J. T. Palmer, and D. Rasnick. 1991. Antimalarial effects of peptide inhibitors of a *Plasmodium falciparum* cysteine proteinase. *J. Clin Investig.* **88:**1467–1472.

Saliba, K. J., R. J. Allen, S. Zissis, P. G. Bray, S. A. Ward, and K. Kirk. 2003. Acidification of the malaria parasite's digestive vacuole by a H+-ATPase and a H+-pyrophosphatase. *J. Biol. Chem.* **278:**5605–5612.

Scheibel, L. W., and I. W. Sherman. 1988. Plasmodial metabolism and related organellar function during various stages of the life-cycle: proteins, lipids, nucleic acids and vitamins, p. 219–252. *In* W. H. Wernsdorfer and I. McGregor (ed.), *Malaria: Principles and Practice of Malariology*. Churchill Livingstone, Ltd., Edinburgh, United Kingdom.

Semenov, A., J. E. Olson, and P. J. Rosenthal. 1998. Antimalarial synergy of cysteine and aspartic protease inhibitors. *Antimicrob. Agents Chemother.* **42:**2254–2258.

Shenai, B. R., B. J. Lee, A. Alvarez-Hernandez, P. Y. Chong, C. D. Emal, R. J. Neitz, W. R. Roush, and P. J. Rosenthal. 2003. Structure-activity relationships for inhibition of cysteine protease activity and development of *Plasmodium falciparum* by peptidyl vinyl sulfones. *Antimicrob. Agents Chemother.* **47:**154–160.

Shenai, B. R., and P. J. Rosenthal. 2002. Reducing requirements for hemoglobin hydrolysis by *Plasmodium falciparum* cysteine proteases. *Mol. Biochem. Parasitol.* **122:**99–104.

Shenai, B. R., P. S. Sijwali, A. Singh, and P. J. Rosenthal. 2000. Characterization of native and recombinant falcipain-2, a principal trophozoite cysteine protease and essential hemoglobinase of *Plasmodium falciparum*. *J. Biol. Chem.* **275:**29000–29010.

Sherman, I. W. 1977. Transport of amino acids and nucleic acid precursors in malarial parasites. *Bull. W. H. O.* **55:**211–225.

Sijwali, P. S., K. Kato, K. B. Seydel, J. Gut, J. Lehman, M. Klemba, D. E. Goldberg, L. H. Miller, and P. J. Rosenthal. 2004. *Plasmodium falciparum* cysteine protease falcipain-1 is not essential in erythrocytic stage malaria parasites. *Proc. Natl. Acad. Sci. USA* **101:**8721–8726.

Sijwali, P. S., and P. J. Rosenthal. 2004. Gene disruption confirms a critical role for the cysteine protease falcipain-2 in hemoglobin hydrolysis by *Plasmodium falciparum*. *Proc. Natl. Acad. Sci. USA* **101:**4384–4389.

Sijwali, P. S., B. R. Shenai, J. Gut, A. Singh, and P. J. Rosenthal. 2001. Expression and characterization of the *Plasmodium falciparum* haemoglobinase falcipain-3. *Biochem. J.* **360:**481–489.

Sijwali, P. S., B. R. Shenai, and P. J. Rosenthal. 2002. Folding of the *Plasmodium falciparum* cysteine protease falcipain-2 is mediated by a chaperone-like peptide and not the prodomain. *J. Biol. Chem.* **277:**14910–14915.

Silva, A. M., A. Y. Lee, S. V. Gulnik, P. Maier, J. Collins, T. N. Bhat, P. J. Collins, R. E. Cachau, K. E. Luker, I. Y. Gluzman, S. E. Francis, A. Oksman, D. E. Goldberg, and J. W. Erickson. 1996. Structure and inhibition of plasmepsin II, a hemoglobin-degrading enzyme from *Plasmodium falciparum*. *Proc. Natl. Acad. Sci. USA* **93:**10034–10039.

Singh, A., and P. J. Rosenthal. 2004. Selection of cysteine protease inhibitor-resistant malaria parasites is accompanied by amplification of falcipain genes and alteration in inhibitor transport. *J. Biol. Chem.* **279:**35236–35241.

Singh, A., B. R. Shenai, Y. Choe, J. Gut, P. S. Sijwali, C. S. Craik, and P. J. Rosenthal. 2002. Critical role of amino acid 23 in mediating activity and specificity of vinckepain-2, a papain-family cysteine protease of rodent malaria parasites. *Biochem. J.* **368:**273–281.

Slater, A. F., and A. Cerami. 1992. Inhibition by chloroquine of a novel haem polymerase enzyme activity in malaria trophozoites. *Nature* **355:**167–169.

Slater, A. F., W. J. Swiggard, B. R. Orton, W. D. Flitter, D. E. Goldberg, A. Cerami, and G. B. Henderson. 1991. An iron-carboxylate bond links the heme units of malaria pigment. *Proc. Natl. Acad. Sci. USA* **88:**325–329.

Slomianny, C. 1990. Three-dimensional reconstruction of the feeding process of the malaria parasite. *Blood Cells* **16:**369–378.

Spiller, D. G., P. G. Bray, R. H. Hughes, S. A. Ward, and M. R. White. 2002. The pH of the *Plasmodium falciparum* digestive vacuole: holy grail or dead-end trail? *Trends Parasitol.* **18:**441–444.

Stocks, P. A., K. J. Raynes, and S. A. Ward. 2001. Novel quinoline antimalarials, p. 235–253. *In* P. J. Rosenthal (ed.), *Antimalarial Chemotherapy: Mechanisms of Action, Resistance, and New Directions in Drug Discovery*. Humana Press, Totowa, N.J.

Sullivan, D. J. 2002. Theories on malarial pigment formation and quinoline action. *Int. J. Parasitol.* **32:**1645–1653.

Sullivan, D. J., Jr., I. Y. Gluzman, and D. E. Goldberg. 1996. Plasmodium hemozoin formation mediated by histidine-rich proteins. *Science* **271:**219–222.

Tilley, L., P. Loria, and M. Foley. 2001. Chloroquine and other quinoline antimalarials, p. 87–121. *In* P. J. Rosenthal (ed.), *Antimalarial Chemotherapy: Mechanisms of Action, Resistance, and New Directions in Drug Discovery*. Humana Press, Totowa, N.J.

Turk, D., V. Janjic, I. Stern, M. Podobnik, D. Lamba, S. W. Dahl, C. Lauritzen, J. Pedersen, V. Turk, and B. Turk. 2001. Structure of human dipeptidyl peptidase I (cathepsin C): exclusion domain added to an endopeptidase framework creates the machine for activation of granular serine proteases. *EMBO J.* **20:**6570–6582.

White, N. J. 2004. Antimalarial drug resistance. *J. Clin. Investig.* **113:**1084–1092.

Withers-Martinez, C., L. Jean, and M. J. Blackman. 2004. Subtilisin-like proteases of the malaria parasite. *Mol Microbiol.* **53:**55–63.

Wu, Y., X. Wang, X. Liu, and Y. Wang. 2003. Data-mining approaches reveal hidden families of proteases in the genome of malaria parasite. *Genome Res.* **13:**601–616.

Zarchin, S., M. Krugliak, and H. Ginsburg. 1986. Digestion of the host erythrocyte by malaria parasites is the primary target for quinoline-containing antimalarials. *Biochem. Pharmacol.* **35:**2435–2442.

PLASMODIUM LIPIDS: METABOLISM AND FUNCTION

Henri J. Vial and Choukri Ben Mamoun

17

In this chapter, we review recent advances in understanding the pathways for membrane biogenesis in *Plasmodium*, present new information in this field learned from the available *Plasmodium* genome sequence and its annotation, and discuss progress in lipid-based antimalarial chemotherapy. The present review will focus on the pathways of synthesis of phospholipids (PLs) and neutral lipids and their importance in parasite physiology, intracellular localization and trafficking of lipids, and newly identified pharmacological targets. Several comprehensive reviews have covered lipids of the malarial parasite (Holz, 1977; Sherman, 1984; Vial et al., 1990; Vial et al., 1992). Therefore, aspects of *Plasmodium* lipid metabolism that have received detailed coverage in previous reports will not be addressed in this chapter. For example, plasmatic lipids, phosphoinositide metabolism and signaling (Vial et al., 2003), fatty acid (FA) biosynthesis in the apicoplast (Ralph et al., 2004; Surolia et al., 2004), glycosylphosphatidylinositol, and galactolipids (Marechal et al., 2002; Vial et al., 2003) will not be detailed here.

In human red blood cells, one single *P. falciparum* parasite can grow and divide to generate up to 36 new daughter parasites in a 40- to 48-h cycle. Successful growth and multiplication of the parasite require an active and temporally controlled metabolic program leading to the duplication of the structural components (Ben Mamoun et al., 2001; Bozdech et al., 2003; Le Roch et al., 2003). Lipids are among the most critical components to be duplicated. Indeed, there is considerable evidence supporting the role of lipids and the enzymes involved in their synthesis in parasite growth, differentiation, and various other cellular events such as signaling and hemozoin formation

Lipid biogenesis in *Plasmodium* includes all the metabolic processes that are responsible for the synthesis of the membranes that serve as permeability barriers between the parasite and its host, the membranes of the subcellular organelles, the membrane networks in the cytoplasm of infected erythrocytes, and the lipid-derived signaling molecules that regulate parasite

---

Henri J. Vial, Dynamique Moléculaire des Interactions Membranaires, UMR 5539 CNRS/Université Montpellier II, case 107, Place Eugène Bataillon, F-34095 Montpellier Cedex 5, France. *Choukri Ben Mamoun,* Center for Microbial Pathogenesis, E7041, Department of Genetics and Development Biology, University of Connecticut Health Center, 263 Farmington Ave., Farmington, CT 06030-3710.

Molecular Approaches to Malaria, Edited by Irwin W. Sherman
© 2005 ASM Press, Washington, D.C.

development and proliferation. *Plasmodium* membranes are composed mainly of glycerophospholipids and sphingolipids but contain little or no cholesterol. Phospholipids generally contain unsaturated acyl chains and therefore tend to be loosely packaged in membrane bilayers, forming liquid-disordered membranes that allow rapid lateral and rotational movement of lipids. By contrast, sphingolipids contain long, saturated acyl chains that pack tightly, forming quasi-gel membranes with reduced movement. It is thought that sphingolipids are enriched in the external leaflet of the bilayer and are connected to phospholipids that contain saturated fatty acids in the inner leaflet. Cholesterol partitions preferentially in bilayers that contain saturated fatty acids, promoting the transition from the quasi-gel phase to a more fluid liquid-ordered conformation, which allows substantial lateral mobility.

The intracellular malarial parasites meet their demand for the necessary lipid species by active synthesis through de novo pathways using precursors that are actively transported from the host cytoplasm. The parasites have evolved unique features in lipid biogenesis to survive within the intraerythrocytic environment, thus prompting a growing interest in lipid metabolism to identify rational targets for malaria chemotherapy.

Most of our recent knowledge about *Plasmodium* lipid metabolism has been gained through biochemical and molecular characterization of *Plasmodium falciparum* metabolic pathways and lipid biosynthetic genes. These analyses have mostly focused on the erythrocytic stages of the parasite due to their amenability to in vitro culturing and to the fact that lipid metabolism is nonfunctional in uninfected mature erythrocytes (Van Deenen and De Gier, 1975). On the other hand, our knowledge about the lipid metabolic machineries during the exoerythrocytic development of the parasite is very limited. This is mainly due to the absence of available procedures for culturing the liver stages of the parasite and foreseeable difficulties in obtaining isolated infected hepatocytes.

Thanks to the excellent effort by the malarial genome sequencing consortium and the PlasmoDB team, several *Plasmodium* genes encoding enzymes of lipid metabolism have been identified. In most cases, the gene function can be predicted from the protein sequence by comparison with proteins of known function or by searching for the presence of specific catalytic and substrate binding motifs. This analysis, however, should be supported by biochemical and (when possible) genetic evidence before a gene-enzyme relationship can be assigned.

LIPID COMPOSITION OF *PLASMODIUM*-INFECTED ERYTHROCYTES

The lipid composition of *Plasmodium*-infected erythrocytes has been covered in several older reviews (Holz, 1977; Sherman, 1979; Vial and Ancelin, 1992; Vial et al., 1990) and will only be briefly summarized here. The predominant lipids have been thoroughly characterized at the biochemical level. On the other hand, the lack of very sensitive analytic procedures has made it difficult to learn more about the lipids present in trace amounts, such as neutral lipids, phosphatidic acid (PA), and phosphatidylglycerol (PG).

In the normal uninfected erythrocyte, cholesterol and PL constitute the major lipids present at a molar ratio around 1. The main PLs are phosphatidylcholine (PC) (35 to 40%), phosphatidylethanolamine (PE) (30 to 35%), sphingomyelin (SM) (15%), and phosphatidylserine (PS) (10%). Phosphatidylinositol (PI) and other PLs, such as PA, cardiolipid, and lyso-PL, account for <3% of total PL (Van Deenen and De Gier, 1975), and fatty acids (FA) and neutral lipids are barely detectable.

Following *Plasmodium* infection, a major increase in the PL of erythrocytes is observed with up to sixfold increase in PL content detected during the trophozoite stage. Furthermore, biochemical analyses reveal striking differences in lipid composition when comparing infected and noninfected erythrocytes, with large increases in PC, PE, and PA. Analysis of the lipid content of purified parasites identifies PC and PE as the two major PLs in the parasite membranes. PC represents 40 to 50% of the total phospholipid content, whereas PE represents

35 to 45%, which is unusually high for a eukaryotic organism. PI represents 4 to 11% of the total parasite PL content, whereas SM and PS account for <5%.

Another interesting feature of *Plasmodium* lipid metabolism is the inability of the parasite to synthesize sterols at any given intraerythrocytic stage. Although changes in the cholesterol content of *Plasmodium*-infected erythrocytes have been reported, the erythrocyte cholesterol is in a dynamic equilibrium with the cholesterol in plasma lipoproteins (see reviews in Holz, 1977; Nawabi et al., 2003; Palacpac et al., 2004; and Vial et al., 1990), and any factors that could modify the exchange of cholesterol between the erythrocyte and the plasma, such as parasite invasion or maturation, could thus account for the reported changes (Maguire and Sherman, 1990; Vial et al., 1990). The cholesterol/PL ratio in the free parasite is around 0.1, due to the near absence of cholesterol in the parasite.

Neutral lipids are found in trace amounts in normal erythrocytes. However, following *Plasmodium* infection, a major increase in the total pool of neutral lipids (including FA, diacylglycerol [DAG], and triacylglycerol [TAG]) occurs (Holz, 1977; Sherman, 1979; Vial and Ancelin, 1992; Vial et al., 1990). The increase, expressed as a percentage, is dramatically high, but the final amounts remain very low compared to other glycerolipids. This increase is most likely due to an active de novo synthesis of neutral lipids by the parasite enzymatic machinery. (See the discussion of neutral lipid biosynthesis below.) Interestingly, high levels of DAG and TAG are detected during the last step of parasite schizogony (Nawabi et al., 2003; Palacpac et al., 2004), suggesting that neutral lipids might play a role during merozoite release and/or invasion.

Plasmodium-Mediated Changes in the Composition and Structure of Erythrocytes

Studies with *P. falciparum* parasites grown in vitro or isolated from patients with malaria have revealed profound changes in the membrane composition and structure of surrounding uninfected red blood cells. Malarial infection seems to accelerate aging of uninfected erythrocytes with apparent reduction of PL and cholesterol, changes in the PL saturation, and increases in lipid peroxidation (Omodeo-Sale et al., 2003). A specific large (40%) increase in the linoleic acid level of uninfected erythrocytes was observed, which was recovered mostly in neutral lipids. Interestingly, this increase in the linoleate level was reproduced in vitro and was also localized in the neutral lipid fraction, especially in triacylglycerols (Beaumelle and Vial, 1988b). It is not yet clear how these modifications occur and what function they might play during infection. One possible mechanism could be that some parasite proteins are secreted and cause changes in the composition and structure of uninfected red blood cells, a mechanism that could facilitate subsequent invasion. Alternatively, it could be that the high nutritional demand of the parasite results in profound modifications in serum composition that subsequently result in changes in the lipid composition of uninfected erythrocytes.

GLYCEROLIPID BIOSYNTHETIC PATHWAYS

Initial Steps of Glycerolipid Metabolism

Glycerolipid metabolism in various organisms initiates with the acylation of glycerol-3-phosphate, which can be produced by the phosphorylation of glycerol by glycerokinase (EC 2.7.1.30) or the reduction of the glycolytic intermediate dihydroxyacetone-3-phosphate by dihydroxyacetone-3-phosphate dehydrogenase (EC 1.1.1.94). This acylation reaction represents the committed step in the biosynthesis of glycerolipids and is catalyzed by acyl-coenzyme A (CoA)–glycerol-*sn*-3-phosphate acyltransferases (GPAT) (EC 2.3.1.15), resulting in the production of 1-acyl-*sn*-glycerol-3-phosphate (lysophosphatidate or LPA). LPA formed then serves as a substrate for a second acylation reaction catalyzed by acyl-CoA–1-acylglycerol-*sn*-3-phosphate acyltransferase (AGPAT; also called lysophosphatidate transferase) resulting in the formation of PA. The phosphatidate formed plays critical structural and regulatory roles in lipid biosynthetic pathways. It can serve as a substrate for phosphatidic acid phosphatase (PAP) to produce diacylglycerol, which provides the fatty acid moiety for PE, PC, and TAG. Alternatively, PA can serve as a substrate for phosphatidate cytidylyltransferase (or CDP-DAG synthase) to yield CDP-DAG, which then serves as a precursor for the synthesis of PS, PI, PG, and cardiolipin.

SYNTHESIS OF LYSOPHOSPHATIDIC ACID

The first acylation step of glycerol-3-phosphate at the $sn1$ position in *P. falciparum* has only recently been characterized (Santiago et al., 2004). This activity is specific to the parasite and is developmentally regulated, with higher activity detected during the trophozoite and schizont stages of the parasite (Santiago et al., 2004). Two GPAT genes, *PfGAT* and *PfPlsB*, have been identified in *P. falciparum*. *PfGAT* encodes a 583-amino-acid polypeptide that shares homology with the yeast GPAT enzymes Gat1p and Gat2p (Santiago et al., 2004). The four motifs that define the glycerolipid acyltransferase signature sequence are present in PfGatp. These motifs are highly conserved between PfGat and yeast GPAT, especially at motifs I and III, but are divergent from mammalian, plant, and bacterial GPATs. One additional feature that distinguishes PfGat from mammalian GPATs is the spacing between motifs II and III, which in PfGat is 67 amino acid residues larger. This feature is also conserved between PfGat and yeast GPATs and might be useful in the design of specific PfGat inhibitors. Hydrophobicity analysis of PfGat predicts three transmembrane domains with the catalytic site (motifs I to IV) located within the N-terminal domain oriented toward the cytoplasm. Cellular fractionation and localization studies using PfGat-specific antibodies revealed that it is a membrane protein of the endoplasmic reticulum (Santiago et al., 2004). Evidence for a gene-enzyme relationship was demonstrated by expression of a codon-optimized version of *PfGAT* (73.4% A+T rich), $PfGAT_{CO}$ (65.5% A+T rich), in yeast. Expression of $PfGAT_{CO}$ resulted in a full suppression of the lethality of a yeast mutant, $gat1\Delta gat2\Delta$, lacking *GAT1* and *GAT2* genes (Santiago et al., 2004). PfGat has a major preference for C16:0 and C16:1 and a low preference for C14:0, C18:0, C18:1, and C12:O acyl-CoA substrates (Santiago et al., 2004). Whereas yeast GPATs have been shown to catalyze the acylation of both glycerol-3-phosphate and dihydroxyacetone phosphate (DHAP) (DHAPAT activity) substrates, PfGat has very low preference for DHAP, with its DHAPAT activity representing only 0.5 to 2.5% of its GPAT activity. Attempts to inactivate the *PfGAT* gene have failed, suggesting that this gene might be essential.

The second GPAT gene, *PfPlsB*, encodes a 400-amino-acid polypeptide with an apicoplast-targeting sequence, suggesting that it is localized to the apicoplast and probably uses acyl-acyl carrier protein produced by this organelle's type II fatty acid (FASII) biosynthetic pathway as a substrate. The mature part of the enzyme does not share any homology with PfGat but is highly similar to plant GPAT enzymes. The gene-enzyme relationship and cellular localization of PfPlsB have not yet been demonstrated.

PfGat and PfPlsB seem to be the only GPAT enzymes of *P. falciparum*. This suggests that the endoplasmic reticulum and apicoplast play a critical role in the initial steps of synthesis of malarial glycerolipids. Noteworthy, the low DHAPAT activity of PfGat and the fact that no DHAPAT and alkyldihydroxyacetone phosphate synthase genes could be detected in the finished genome sequence of *P. falciparum* suggest that the parasite might not synthesize ether lipids.

SYNTHESIS OF PHOSPHATIDIC ACID

The second step in glycerolipid metabolism is the acylation of LPA into phosphatidic acid, a reaction catalyzed by AGPAT. Only one AGPAT gene (expressing EC 2.3.1.51), *PfPlsC*, has thus far been found in the *P. falciparum* genome; like PfPlsB, it is also predicted to localize to the apicoplast. Although further studies are needed to confirm the localization of PfPlsB and PfPlsC to the apicoplast, it seems that the apicoplast has the necessary enzymes to synthesize PA.

Besides de novo synthesis via GPAT and AGPAT enzymes, available data suggest that *P. falciparum* might also synthesize PA via alternative routes. The available *Plasmodium* genome database revealed two proteins (PFI1485 and PF14_0681) similar to DAG kinases. Although further studies are still needed to examine the biochemical properties of these two enzymes, the available information suggests that DAG might be used as a substrate for PA synthesis in *Plasmodium*. Finally, no phospholipase D-like activity could be measured in *Plasmodium* and no PLD-like gene could be identified in the available *Plasmodium* genome database, suggesting that the alternative pathway for PA synthesis via hydrolysis of preformed or transported phospholipids is absent in these parasites.

SYNTHESIS OF CDP-DAG

CDP-DAG is an obligatory intermediate in phospholipid metabolism synthesized by CDP-DAG synthase (CDS) both in prokaryotes and eukaryotes. CDS sits at a branch point in the biosynthetic pathways, where PA is partitioned between CDP-DAG and DAG (Fig. 1). In bacteria, CDP-DAG is the precursor of the major anionic PL, PG, cardiolipids, the weakly represented PS, and the major zwitterionic PE (Dowhan, 1997; Kent, 1995). In eukaryotic cells, CDP-DAG gives rise to the mitochondrial PL, PG, and cardiolipids, but also to PI and PS (Kent, 1995).

The *P. falciparum* endogenous CDS activity was characterized and shown to have an apparent K_m of 0.9 mM for PA and an apparent maximal velocity (V_{max}) of 8 nmol/10^{10} cell/h (Martin et al., 2000). The *P. falciparum* CDP-diacylglycerol synthase (*PfCDS*) gene is present as a single copy on chromosome XIV (Martin et al., 2000). *PfCDS* encodes a protein of 667 amino acids with a predicted molecular mass of 78 kDa. However, a protein of this size was never detected by Western blot analysis. Instead, 28- and 51-kDa proteins, which correspond to the N- and C-terminal regions, respectively, were detected, indicating that the enzyme undergoes a rapid proteolytic processing in the parasite. The expression pattern of the 28- and 51-kDa polypeptides during the intraerythrocytic development of the parasite followed the same pattern as that of the transcripts and the enzyme activity (Martin et al., 2000). The 245-amino-acid N-terminal region is not conserved among other CDS enzymes, is hydrophilic, and contains asparagine-rich and repetitive sequences. The 422-amino-acid C-terminal region is very hydrophobic and contains the catalytic core and seven putative membrane-spanning domains. Only the conserved C-terminal region of PfCds could be expressed in a heterologous system (COS-7 cells) and its CDP-DAG synthase activity determined. As the N-terminal fragment appears to be stable in *P. falciparum*, it will be interesting to determine whether this domain could play a regulatory or stabilizing role in CDS activity or whether it might have an independent function.

SYNTHESIS OF DAG

In most organisms, DAG is produced primarily through dephosphorylation of PA by 3-*sn*-phosphatidate phosphohydrolase (PAP). Two forms of PAPs, Mg^{2+}-dependent PAP (PAP-1) and Mg^{2+}-independent PAP (PAP-2), have been reported with mammalian tissues and yeast cells. PAP activity in *Plasmodium* has not been characterized and thus it is not yet clear whether the parasite possesses both PAP-1- and PAP-2-like activities. On the other hand, the finished genomic sequence of *P. falciparum* revealed the presence of a gene, *PfPAP2* (PFF1210w), on chromosome VI encoding a protein of 461 amino acids containing a PAP-2 signature domain, but it lacks an apicoplast-targeting sequence. The prediction of transmembrane domains in the protein suggests that it is membrane associated. In mammals, DAG can also be synthesized via the monoacylglycerol pathway by acyl-CoA:monoacylglycerol acyltransferase (MGAT). This reaction utilizes *sn*-2-monoacylglycerol released after hydrolysis of the *sn*-1 and *sn*-3 ester bonds of TAG by lipases. *P. falciparum* lacks genes homologous to mammalian MGATs, suggesting that the parasite lacks

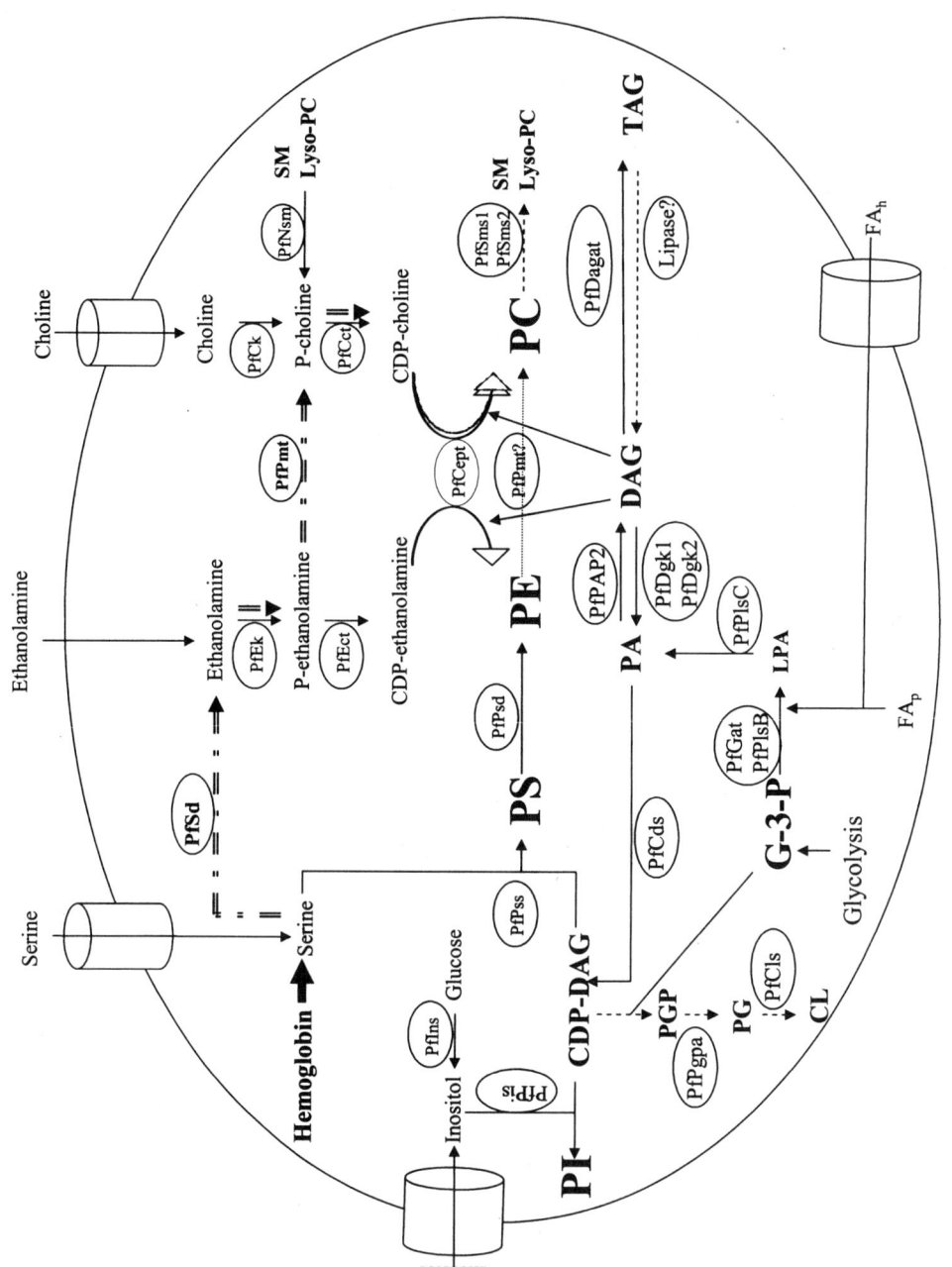

FIGURE 1 Pathways for the synthesis of phospholipids and neutral lipids in *P. falciparum*. The pathways shown include the relevant steps discussed in the text and the enzymes involved in these pathways are described in Table 1. CL, cardiolipin; PGP, phosphatidylglycerol phosphate.

TABLE 1 Known and putative phospholipid and neutral lipid enzymes of *P. falciparum*

Enzyme	Open reading frame	Chromosome	Enzyme
PfPlsB	PF13_0100	XIII	Glycerol-3-phosphate acyltransferase
PfPlsC	PF10695c	IX	1-Acyl-*sn*-glycerol-3-phosphate acyltransferase
PfGat	PFL0620c	XII	Glycerol-3-phosphate acyltransferase
PfGk	PF13_0269	XIII	Glycerol kinase
PfCept	MAL6P1.145	VI	Choline–ethanolamine-phosphate transferase
PfCk	PF14_0020	XIV	Choline kinase
PfEk	PF11_0257	XI	Ethanolamine kinase
PfEct	PF13_0253	XIII	CTP:phosphoethanolamine cytidylyltransferase
PfPsd	PFI1370c	IX	Phosphatidylserine decarboxylase
PfPssbe	MAL13P1.335	XIII	Phosphatidylserine synthase base exchange
PfPmt	MAL13P1.214	XIII	Phosphoethanolamine *N*-methyltransferase
PfPss	MAL8P1.58	VIII	Phosphatidylserine synthase; CDP-DAG dependent
PfDgk1	PF11485c	IX	Diacylglycerol kinase
PfCds	PF14_0097	XIV	Cytidine diphosphate-diacylglycerol synthase
PfPap2	PFF1210w	VI	Phosphatidic acid phosphatase
PfDgk2	PF14_0681	XIV	Diacylglycerol kinase
PfCct	PF13_0092	XIII	CTP:phosphocholine cytidylyltransferase
PfSms1	MAL6P1.178	VI	Sphingomyelin synthase
PfSms2	MAL16P1.177	VI	Sphingomyelin synthase
PfPis	MAL13P1.82	XIII	Phosphatidylinositol synthase
PfDgat1	Chr3.phat_245	III	Acyl-CoA:diacylglycerol acyltransferase
PfIns	PFE0585c	V	Myoinositol 1-phosphate synthase
PfItr	PF10955w	IX	Myoinositol transporter
PfCls	MAL6P1.97	VI	Cardiolipin synthetase
PfSpt	Chr14.phat_153	XIV	Serine palmitoyltransferase
PfSna	Chr14.phat_34	XIV	Ceramide synthase

MGAT activity or that if this activity exists it is encoded by a MGAT enzyme that shares no homology to mammalian MGATs. A third alternative route for DAG synthesis in *Plasmodium* might be via hydrolysis of glycerophospholipids through the action of the phospholipase C (PLC; sphingomyelinase, or SMase) (see "Sphingomyelinase," below).

SYNTHESIS OF TAG

Numerous studies have documented the lack of production of cholesterol and cholesterol ester in the malarial parasite (Holz, 1977; Nawabi et al., 2003; Vial and Ancelin, 1992; Vial et al., 1984a). On the other hand, biochemical studies using radioactive tracers demonstrated that the parasite possesses the enzymatic machinery necessary for the synthesis of acylglycerols, including TAG. This synthesis is stage dependent, with higher activity detected during the schizont stage (Nawabi et al., 2003; Palacpac et al., 2004; Vial et al., 1982b; Vielemeyer et al., 2004). The synthesis of TAG can be catalyzed by various enzymes such as diacylglycerol:acyl-CoA acyltransferase (DGAT) (EC 2.3.1.20), which acylates *sn*-1,2-DAG (or 2,3-DAG), *sn*-1,2(2,3)-diacylglycerol transacylase, wax ester-DGAT, or lecithin-DAG transacylase (Coleman and Lee, 2004).

The available *P. falciparum* genome has revealed the presence of only one putative acyl-CoA diacylglycerol acyltransferase gene named *PfDGAT1*. This gene encodes a polypeptide with a molecular mass of 78.1 kDa with a broad acyl-CoA specificity, localized to the microsomes (based on localization in heterologous system) (Vielemeyer et al., 2004). Although the identification of the activity using recombinant enzyme has not been formally accomplished, PfDgat11 appears to be the most likely candidate for acyl-CoA diacylglycerol acyltransferase.

The *P. falciparum* genome also revealed a putative phosphatidylcholine-sterol acyltransferase (LCAT) (GenBank NP_703950; EC 2.3.1.43). LCAT enzymes catalyze the transfer of an acyl chain from PC to the 3-hydroxyl group of cholesterol, producing lysophosphatidylcholine (lysoPC) and cholesterol ester. The absence of cholesterol esters in *P. falciparum* membranes during parasite intraerythrocytic development suggests that this activity and therefore cholesterol esters might be needed during parasite development in the liver or in the mosquito. Alternatively, the LCAT gene might encode a phospholipid:diacylglycerol acyltransferase that catalyzes the acyl-CoA-independent formation of TAG as in yeast and plants (Stahl et al., 2004). A homolog of this gene has recently been characterized in *Plasmodium berghei* and encodes a 750-amino-acid polypeptide localized to the surface of sporozoites but not expressed during the erythrocytic stage. Disruption of this gene resulted in ~90% decrease of sporozoites infectivity in the liver (Wang et al., 2004).

TAGs are contained in lipid droplets in every cell in which they have been characterized (Coleman and Lee, 2004). Electron and fluorescence microscopic studies have documented the presence of lipid bodies of 1 to 2 μm in diameter in the intraerythrocytic stage of *P. falciparum*. Their biogenesis is not affected upon incubation with excess fatty acids or lipoproteins, attesting that the formation of lipid bodies in *P. falciparum* is not regulated by extracellular lipids. Interestingly, lipid bodies are enriched during the trophozoite stage and are sensitive to brefeldin A. Initial attempts to localize the lipid bodies within the parasite using the Nile Red dye suggested an association of these particles with the parasitophorous vacuole (Nawabi et al., 2003; Palacpac et al., 2004; Vielemeyer et al., 2004). Jackson and colleagues (2004) recently reexamined this observation by taking into consideration the specific properties of excitation/emission spectrum of Nile Red fluorophores. Because the spectrum of these fluorophores changes depending on the environment and the position of the probe along the membrane plane, the authors were able to differentiate between lipid stains from lipid bodies and those from membranes. This analysis allowed the authors to identify structures rich in neutral lipids closely associated with the parasite food vacuole. This localization was further confirmed by electron microscopic and analysis of lipids from purified food vacuole.

The function of plasmodium lipid bodies is still unclear. TAG may be utilized to provide acyl groups for synthesis of lipids when local resources are scarce or might be needed when specific kinds of fatty acids or lipid precursors are required. TAG degradation might become active during the transition from schizont to segmented schizont with subsequent release of FA into the medium during schizont rupture and/or merozoite release (Palacpac et al., 2004). Thus, lipid bodies and the associated TAG might constitute dynamic cellular structures and/or components participating in the schizont structure and/or merozoite release. The enzyme(s) mediating TAG degradation remains to be elucidated.

The CDP-Choline and CDP-Ethanolamine Pathways

The de novo biosynthetic pathways of PE and PC initiate with the phosphorylation of ethanolamine and choline, conversion of the phosphoethanolamine and phosphocholine formed into CDP-ethanolamine and CDP-choline, and DAG-dependent acylation of the latter products into PE and PC, respectively. In the yeast *Saccharomyces cerevisiae*, the enzymes involved in all these steps have been identified and characterized (reviewed in Carman and Henry, 1999). It was found that two kinases, Ek1 and Ck1, are involved in the phosphorylation of ethanolamine and choline, respectively. Although the preferred substrates of Ck1 and Ek1 are choline and ethanolamine, Ck1 also exhibits ethanolamine kinase activity and Ek1 also exhibits choline kinase activity. The second steps of the CDP-choline and CDP-ethanolamine pathways are catalyzed by two enzymes, Pct1 and Ect1, respectively. Recombinant Pct1 produced in *Escherichia coli* exhibited strong preference for cholinephosphate and *N,N*-dimethylethanolaminephosphate and little or no preference for *N*-methylethanolaminephosphate and ethanolaminephosphate substrates. Furthermore, expression of a thermosensitive allele of *PCT1* in cells lacking PE methylation resulted in a major growth defect upon shift to high temperature, suggesting that loss of Pct1 activity cannot be fully complemented by Ect1. Similarly, a loss-of-function mutation of *ECT1* in cells lacking PE methylation resulted in a major reduction in PE synthesis. The final steps of the CDP-choline and CDP-ethanolamine pathways are catalyzed by two enzymes, Cpt1 and Ept1, respectively. The two enzymes share a high degree of similarity and can exhibit specificity for both substrates in vitro.

Biochemical studies of *Plasmodium* strongly support the existence of CDP-choline and CDP-ethanolamine pathways for the synthesis of PC and PE. All the enzymatic activities have been characterized, but only a few of the corresponding genes of these pathways have been identified and characterized to date.

THE CDP-CHOLINE PATHWAY

Choline phosphorylation, the first and committed step in the CDP-choline pathway, is catalyzed by choline kinase. Biochemical analysis in *P. falciparum* revealed that the choline kinase of the parasite is very specific and has little preference for ethanolamine. A gene tentatively named *PfCK* (PF14_0020) encoding a polypeptide that shares strong similarity with yeast Ck1 has been identified in the *Plasmodium* genome database. *PfCK* encodes a polypeptide of 440 amino acids predicted to be cytoplasmic. Biochemical characterization of PfCk is required before a gene-enzyme relationship can be established. It is important to note that the phosphocholine intermediate of the CDP-choline pathway can also be obtained from sphingomyelin via the activity of phospholipase C (sphingomyelinase) and from phosphoethanolamine by direct methylation via the activity of the phosphoethanolamine methyltransferase PfPmt (Pessi et al., 2004).

The second and limiting step in the CDP-choline pathway is catalyzed by CTP:phosphocholine cytidylyltransferase (CCT; EC 2.7.7.15). Studies involving time course conversion of choline into PC showed that CCT was a rate-limiting step of this essential pathway at the erythrocytic stage of *P. falciparum* and *Plasmodium knowlesi*-infected erythrocytes. The enzyme exists in an inactive soluble form and an active membrane-bound form (Ancelin and Vial, 1989). The *PfCCT* gene was cloned from a *P. falciparum* genomic library by a PCR-based approach. This gene was found to encode a polypeptide of 370 amino acids predicted to be the most acidic of all CCT identified so far, with an isoelectrical point of 5.2 (Yeo et al., 1995). Purified recombinant PfCct enzyme is activated by anionic lipids and to a slight degree by neutral lipids (Yeo et al., 1997). PfCct is a sensor of the membrane composition of the parasite and interacts with anionic PC-deficient membranes through electrostatic interaction with cationic residues present in two strongly amphipathic α-helices (Larvor et al., 2003). The available genome sequence and annotation of *P. falciparum* genome revealed that the initial *PfCCT* sequence represented a truncated version of the full-length gene. The full-length *PfCCT*, which was further confirmed by cloning and sequence of a cDNA clone, encodes a polypeptide of 896 amino acids containing two domains, each with a CCT signature catalytic motif, and sharing 85% identity in their primary sequence. Similar structure with two CCT domains is found in homologs of PfCCT in *P. knowlesi*, *Plasmodium yoelii*, and *P. berghei* (Vial et al., 2004).

PC synthesis is achieved through reaction of CDP-choline with DAG catalyzed by CDP-choline:1,2 DAG choline phosphotransferase (EC 2.7.8.2). An interesting feature that emerged from early biochemical analysis of the CDP-choline and CD-ethanolamine pathways of *P. falciparum* and further supported by the available

Plasmodium genome database is that the last steps of the two pathways in *P. falciparum* seem to involve only one enzyme. The choline phosphotransferase and ethanolamine phosphotransferase activities were found to share a similar pattern of heat inactivation and dependence on Mn^{2+} and Mg^{2+}. Furthermore, the competitive inhibition of the ethanolamine phosphotransferase by CDP-choline was similar to that of the choline phosphotransferase by CDP-ethanolamine. On the other hand, the specificity of ethanolamine phosphotransferase with respect to the molecular species of DAG is distinct from choline phosphotransferase (Vial et al., 1984b). This difference could be due to the difference in the CDP-polar head moiety cosubstrate. Accordingly, a single gene (MAL6P1.145) tentatively named *PfCEPT* (choline-ethanolamine phosphotransferase) exists in the *Plasmodium* genome database. This gene is localized on chromosome VI and encodes a polypeptide of 390 amino acids with eight putative transmembrane domains. Substrate specificity and subcellular localization analyses are needed before a gene-enzyme relationship can be established.

CDP-ETHANOLAMINE PATHWAY

Compared to most eukaryotic organisms, PE content of *P. falciparum* membrane is particularly high. An approximate 60% of parasite PE has been proposed to come through the CDP-ethanolamine pathway and 40% through decarboxylation of PS. Ethanolamine phosphorylation, the first and committed step in the CDP-ethanolamine pathways, is catalyzed by ethanolamine kinase. A gene tentatively named *PfEK* (PF11_0257) encoding a polypeptide that shares similarity with yeast Ek1 has been identified in the *Plasmodium* genome database. *PfEK* encodes a polypeptide of 423 amino acids predicted to be cytoplasmic. The second and limiting step in the CDP-ethanolamine is catalyzed by CTP:phosphoethanolamine cytidylyltransferase. A putative *P. falciparum* gene encoding this activity, *PfECT* (PF13_0253), was identified on chromosome XIII and encodes a putative polypeptide of 573 amino acids expressed throughout the intraerythrocytic life of the parasite as well in gametocytes and sporozoites. The final step of the CDP-ethanolamine pathway might be catalyzed by the same enzyme choline-ethanolamine phosphotransferase, also involved in the final step of the CDP-choline pathway (see above). Although all the genes encoding enzymes of the CDP-ethanolamine pathway could be identified in the available genome database, biochemical studies are needed to determine their substrates specificities and kinetic parameters.

The SDPM Pathway

Early indications that *P. falciparum* might utilize an alternative pathway for PC biosynthesis came indirectly from studies by Divo and colleagues and Mitamura and colleagues, who showed that *P. falciparum* can grow in the absence of choline in the medium (Divo et al., 1985; Mitamura et al., 2000). Direct evidence for this pathway, however, was demonstrated by labeling studies in *P. knowlesi* and *P. falciparum* using serine and ethanolamine as phospholipid precursors (Elabbadi et al., 1997; Pessi et al., 2004). This pathway consists of five enzymes, three of which (ethanolamine kinase, CTP:phosphocholine/phosphoethanolamine cytidylyltransferase, and choline phosphotransferase) are components of the CDP-choline pathway, and two (serine decarboxylase and phosphoethanolamine methyltransferase) catalyze two plant-like reactions that generate phosphocholine, which then serves as a precursor for PC biosynthesis (Pessi et al., 2004).

SERINE DECARBOXYLATION

The reaction sequence serine $\rightarrow$ PS $\rightarrow$ PE plays a major role in the synthesis of PE in bacteria and yeast (Carman and Henry, 1999; Shibuya, 1992). Direct evidence for decarboxylation of free serine into ethanolamine in *Plasmodium* was demonstrated using radiolabeled serine and thorough characterization of its metabolic products by extraction of labeled metabolites and subsequent identification by reverse-phase high-pressure liquid chromatography and/or three thin-layer chromatography methods (Elabbadi et al., 1997). These studies revealed that ethanolamine and phosphoethanolamine are intensively formed from serine. Because of the lack of significant amounts of labeled sphingomyelin (or any metabolic intermediate of the sphingolipid pathway) following labeling of *Plasmodium* with [^{3}H]serine or [^{3}H]palmitate, it is unlikely that the phosphoethanolamine formed from serine could be produced via the sphingolipid pathway. In vitro biochemical studies using parasite extracts confirmed the presence of a serine decarboxylase activity that is dependent on pyridoxal phosphate. The kinetic parameters of this reaction are as follows: V_{max}, ~270 nmol of ethanolamine/10^{10} cells/h; K_m value for L-serine, ~160 μM; and K_m value for pyridoxal phosphate, ~1 μM (F. Baunaure and H. J. Vial, unpublished data). Serine decarboxylase activity has also been demonstrated in plants, and the gene encoding this activity and requiring pyridoxal phosphate for activity has been identified and characterized (Rontein et al., 2001). Interestingly, no serine decarboxylase activity has been found in mammalian cells, suggesting that if this activity plays an important role in *P. falciparum* development and survival, it could be a good target for chemotherapy. Thus far, no homologs of plant serine decarboxylases could be found in the available *Plasmodium* genome database, suggesting either that the parasite uses a novel serine decarboxylase or that this activity is catalyzed by a different amino acid decarboxylase with broad substrate specificity. Purification of this enzyme by affinity chromatography from *P. falciparum* is currently in progress.

PHOSPHOETHANOLAMINE TRANSMETHYLATION

The phosphoethanolamine methyltransferase (PEAMT) activity of *P. falciparum* has been demonstrated by pulse-chase and steady-state labeling experiments using [^{14}C]ethanolamine as a phospholipid precursor (Pessi et al., 2004). Thin-layer chromatography analysis indicated that PC and phosphocholine, but not choline, could be formed from ethanolamine. This activity was also measured from parasite lysate and was found to be specific for phosphoethanolamine (Pessi et al., 2004). The gene *PfPMT* encoding this activity was further identified based on the presence of an S-adenosylmethionine (SAM) binding domain and homology to plant phosphoethanolamine methyltransferases. *PfPMT* encodes a polypeptide of 266 amino acids with a molecular mass of 31 kDa. Unlike plant PEAMT, which has a bipartite structure with two SAM-dependent catalytic motifs important for catalysis, PfPmt is a monomeric enzyme containing a single catalytic motif responsible for the three-step methylation of phosphoethanolamine into phosphocholine. Analysis of the kinetics of PfPmt activity revealed K_m values of ~79 μM and 153 μM for phosphoethanolamine and SAM, respectively, and a V_{max} of 1.2 nmol · mg^{-1} · min^{-1} for both substrates (Pessi et al., 2004). Consistent with its requirement for SAM, PfPmt activity is inhibited by the SAM analog adenosylhomocysteine. Interestingly, PfPmt is also inhibited by its product with 50% inhibition at 50 μM of phosphocholine. This property allowed identification of the phosphocholine analog, hexadecylphosphocholine (miltefosine) as an inhibitor of Pfpmt. This compound was later found to inhibit *P. falciparum* growth with a 50% inhibitory concentration (IC$_{50}$) value of ~80 μM (Pessi et al., 2004). Although PfPmt has no specificity for PE in vitro, in vivo studies are in progress to determine whether in addition to its contribution to the synthesis of PC via the SDPM pathway, PfPmt might also convert PE into PC. Indeed, although no gene encoding a conventional PE N-methylase can be found in the *Plasmodium* genome, PE is intensively methylated into PC when introduced into the membrane of a *P. knowlesi*-infected erythrocyte, using phospholipid exchange proteins (Moll et al., 1988).

PS and Lipid Metabolism of Serine

PS SYNTHESIS

In mammalian cells, PS biosynthesis is generally thought to occur solely by Ca^{2+}-dependent, enzyme-catalyzed interchange of L-serine with

PC and/or PE head groups (Kent, 1995). This situation sharply contrasts with pathways in yeast (Carman and Henry, 1999) and bacteria (Sohlenkamp et al., 2003), where de novo biosynthesis of PS occurs by condensation of L-serine with CDP-DAG catalyzed by PS synthase (PSS; EC 2.7.8.8). Few studies of PL metabolism in protozoan cells have shown that PS might also be formed by a Ca^{2+}-dependent base exchange reaction (Smith, 1993).

PS is present at a very low level in *Plasmodium* but is a key intermediate in PE synthesis (through decarboxylation). Indeed, when introduced in the infected erythrocyte using a PL transfer protein, PS is directed to the intraerythrocytic parasite where it is largely transformed into PE (Moll et al., 1988). Initial characterization of *Plasmodium* PSS activity was performed using *P. knowlesi*-infected erythrocytes obtained from *Macaca fascicularis* monkeys. PSS activity was shown to be Ca^{2+} dependent but did not require addition of phospholipids. Labeling studies using [^{14}C]serine showed that addition of CDP-DAG significantly stimulates the incorporation of this precursor into PS and PE, whereas phospholipids such as PE or PC were ineffective. Under optimal conditions, i.e., in the presence of calcium and CDP-diacylglycerol, *Plasmodium* PSS specific activity was 180 nmol/10^{10} cells/h and the apparent K_m for L-serine was 22 μM. Interestingly, this activity was found to be directly inhibited by ethanolamine (N. Elabbadi and H. J. Vial, unpublished data), suggesting that the pathways for synthesis of PS and PE might be coregulated. Similar experiments using *P. falciparum*-infected erythrocytes revealed PSS activity that uses calcium or manganese (~200 nmol PS formed/10^{10} cells/h), but only the Mn-dependent activity was significantly enhanced by CDP-DAG.

Two genes, *PfPSS* (MAL8P1.58) and *PfPSS$_{be}$* (MAL13P1.335), encoding possible PSS activities have been identified in the *P. falciparum* genome database. *PfPSS* encodes a putative CDP-DAG-dependent PS synthase (PfPss) localized in the endoplasmic reticulum. The PS synthase activity of the recombinant protein has been demonstrated and shows a K_m of 105 ± 24 μM for serine (F. Baunaure and H. J. Vial, submitted for publication). *PfPSS$_{be}$* is actively transcribed during the ring and trophozoite stages, but the catalysis of PS biosynthesis from preexisting PC and/or PE remains to be confirmed.

PS DECARBOXYLATION

In *P. knowlesi*-infected erythrocytes, PS decarboxylase (PSD) activity was demonstrated by using 3-^{14}C-serine-labeled PS as a substrate. The apparent K_m and V_{max} values for this activity were estimated to be 280 μM and 2,160 nmol/10^{10} infected cells/h, respectively. The gene *PfPSD* encoding PSD activity in *P. falciparum* was recently identified and found to encode a hydroxylamine-sensitive membrane-associated enzyme of the type I PSD family. Using PfPsd-specific antibodies, it was found that this enzyme is an endoplasmic reticulum (ER)-resident protein. It exists as a 42.6-kDa proenzyme that is posttranslationally processed at an LGST motif near its carboxyl terminus, resulting in a 5.4-kDa pyruvoyl-containing β-subunit and a 37.2-kDa membrane-bound β-subunit (Baunaure et al., 2004). Both subunits are essential for catalysis. Purified recombinant PfPsd showed an apparent K_m for PS of 63 ± 19 μM. Site-directed mutagenesis of the processing site revealed the crucial role played by each residue in enzyme processing and activity. The structural and catalytic similarities between PfPsd and its human counterpart make it unlikely that *Plasmodium* PSD activity could be a good pharmacological target.

Until now, only type-I PSD has been found in mammals. This enzyme is located in the inner mitochondrial membrane. The PS that is synthesized by PS synthase in the ER is exported to the mitochondria through the ER and mitochondrial microdomain contact sites, where PS decarboxylation takes place. PE formed in this compartment is then transferred back to the ER where it is methylated to form PC (Ardail et al., 1993). On the other hand, the *P. falciparum* PSD appears predominantly localized in the ER, suggesting that this back-and-forth movement of PS and PE is unnecessary in *P. falciparum*, since all the required enzymes (PSS and PSD) are present in the same compartment (Baunaure et al., 2004). Hence, the present ER localization appears to be a unique characteristic of this *Plasmodium* type I-related PSD.

SPHINGOLIPID METABOLISM

Sphingolipids, which are widely distributed in eukaryotes, play important roles in the growth of cells. SM is the most abundant of the mammalian sphingolipids, and degradation of SM to produce ceramide appears to be responsible for the modulation of various cellular events including proliferation, differentiation, and apoptosis.

Sphingomyelin Synthase (SMS)

In eukaryotes, de novo biosynthesis of ceramide is initiated by condensation of serine and palmitoyl-CoA, resulting in the formation of 3-ketosphinganine, which is subsequently reduced to sphinganin. Sphinganin can then be converted either to sphingosin or dihydroceramide. The former can be phosphorylated into sphingosin-1-phosphate and then converted via sphingosin-1-phosphate lyase into hexadecanal (which produces FA) and phosphoethanolamine. Dihydroceramide is transformed into ceramide, which is a precursor for complex sphingolipids: e.g., acyl-, glucosyl- (leading to glycolipids and gangliosides) or galactosyl-ceramides (leading to sulfatides). Biosynthesis of SM involves addition of phosphocholine head groups (originating from PC) to ceramide, essentially through the action of PC:ceramide choline phosphotransferase (SM synthase). The breakdown of SM into ceramides via sphingomyelinase ends the SM cycle (Fig. 1). Enzymes involved in sphinganin synthesis reside in the ER, whereas incorporation of ceramide into complex sphingolipids (including SM) occurs in the Golgi apparatus (Hannun, 1994).

Numerous studies using the erythrocytic *P. falciparum* and *P. knowlesi* stages have demonstrated a lack of incorporation of appreciable amounts of radioactive choline, fatty acids (including palmitic acid), and serine into SM in infected erythrocytes, which suggests that the parasite does not synthesize SM. This finding is in complete agreement with previous data indicating the presence of very low amounts of SM in the parasite fraction (Vial and Ancelin, 1992). Nevertheless, subsequent studies using fluorescent or radiolabeled ceramides showed their conversion into SM in infected erythrocytes, suggesting the presence of a parasite SM synthase (Ansorge et al., 1995; Haldar, 1996). This SM synthase activity was proposed to use ceramide obtained through hydrolysis of host SM and exists in two distinct isoforms present in the parasite Golgi and in tubovesicular membranes (Haldar, 1996). However, because these assays were performed in the presence of high concentrations of unnatural, more polar, substrates, it is possible that the SM formed under those conditions could have been synthesized through the action of the choline phosphotransferase.

The recent discovery by Huitema and colleagues (2004) of a new class of sphingomyelin synthases (SMSs), using a bioinformatics approach based on the presence of a sequence motif, H-[YFWH]-X2-[GA]-X3-[GSTA], has allowed the identification in *P. falciparum* genome of two genes that we named *PfSMS1* and *PfSMS2* localized in a tandem structure on chromosome VI. *PfSMS1* is expressed throughout the intraeythrocytic life cycle, as well as during the gametocyte and sporozoite stages. It encodes a protein of 461 amino acids with six predicted transmembrane domains, suggesting membrane localization. *PfSMS2* shows low expression levels in sporozoites and during intraerythrocytic development but is mostly expressed during the gametocyte stage. *PfSMS2* encodes a polypeptide of 431 amino acids with six transmembrane domains. PfSms1 and PfSms2 share 28% identity and 47% similarity in their primary polypeptide sequences. Both *PfSMS1* and *PfSMS2* genes have been cloned and expressed, and antibodies against their encoded proteins have been raised (R. Zufferey et al., unpublished data). Biochemical characterization to establish a gene-enzyme relationship and cellular localization of both proteins are in progress.

The completed genome of *P. falciparum* has also revealed the presence of two genes tentatively named *PfSPT* (chromosome XIV; phat_153) and *PfSNA* (chromosome XIV;

phat_34) encoding putative proteins that share homology with yeast serine palmitoyltransferase and ceramide synthase, respectively. *PfSPT* encodes a polypeptide of 573 amino acid residues predicted to be membrane bound. Global microarray analysis revealed that this gene is expressed throughout the erythrocytic life cycle of the parasite, albeit at very low levels. *PfSNA* encodes a putative polypeptide of 331 amino acid residues with seven putative transmembrane domains, suggesting possible membrane localization. No transcript of this gene could be detected by any of the global microarray studies. The presence of *PfSPT* and *PfSNA* genes in *P. falciparum* suggests that the parasite might be able to synthesize sphingolipids. However, the absence or low expression of these genes during the intraerythrocytic life cycle along with the failure to detect sphingolipids following serine labeling suggests that if a de novo pathway for sphingolipid biosynthesis exists in the parasite it might play little or no role during the intraerythrocytic cycle. This pathway, however, might be utilized by the parasite during differentiation and sexual development in the mosquito.

Sphingomyelinase

P. falciparum-infected erythrocytes but not uninfected erythrocytes retain the activity of a membrane-bound and Mg^{2+}-dependent neutral SMase with an optimum pH at around 7.5 (Hanada et al., 2000) involved in SM catabolism. This enzyme also has substantial PLC activity toward lysoPC and lysoplatelet-activating factor (Hanada et al., 2002). A *P. falciparum* gene, *PfNSM*, encoding a polypeptide of 393 amino acid residues with SMase activity, has been cloned and characterized (Hanada et al., 2002). Purified recombinant PfNsm exhibited biochemical parameters that were similar to those measured in *P. falciparum*-infected erythrocytes and isolated parasites. PfNsm is largely bound to the membrane fraction of the parasites, and its activity is markedly enhanced by anionic phospholipids but differs from the mammalian neutral SMase because Mn^{2+} could not substitute for Mg^{2+} for its activation. The physiological role of the SM–lysoPL-PLC in *P. falciparum* remains, however, to be determined. The plasmodial SM–lysoPL-PLC might degrade host-derived SM to supply the parasite with ceramide that could be used to modulate the cell cycle progression of intraerythrocytic parasites and/or could be used for synthesis of SM within parasitized erythrocytes. Hanada and colleagues hypothesized that host SM is accessible to the plasmodial SM–lysoPL-PLC through the tubulovesicular membrane network (Hanada et al., 2002). The plasmodial SM–lysoPL-PLC might also degrade host-derived lysoPC to supply the parasite with phosphocholine and/or monoacylglycerol for their efficient intraerythrocytic growth. Alternatively, PfNsm might be used to detoxify the potentially harmful lysoPC (unpublished data), which is abundant in the plasma and readily taken up by the infected erythrocytes (Vial et al., 1989).

MECHANISMS OF ACQUISITION OF LIPID PRECURSORS IN *PLASMODIUM*

Plasmodium Fatty Acids

The considerable increase in the phospholipid and acylglycerol content of *Plasmodium*-infected erythrocytes is associated with a comparable increase in the amounts of fatty acids bound to the glycerol backbone. Biochemical studies revealed substantial capacity for transbilayer movement of PL in the plasma membrane of infected erythrocytes (Beaumelle et al., 1988), as well as minimal uptake of vesicular PL (Moll et al., 1988), which is followed by considerable dynamic movements of these molecules within infected erythrocytes. Therefore, the transfer of significant amounts of PC from high-density lipoproteins (Grellier et al., 1991) to cover the high needs of the parasite for this major PL may be questioned, since results may only reflect the presence of phospholipase A activity in the lipoprotein fraction (Shohet, 1971). In agreement with that, serum lipoproteins have recently been shown to have detrimental effects on parasite growth and survival (Mitamura et al., 2000; Vielemeyer et al., 2004). Although it has long been known that the development and survival of *Plasmodium* within red blood cells require the presence of external lipids, it is only recently that the specific essential lipids have been identified. Research by Mitamura and colleagues indicated that fatty acids but not phospholipids or neutral lipids are essential for parasite survival (Mitamura et al., 2000; Vielemeyer et al., 2004).

Hence, the parasites incorporate fatty acids from host serum to synthesize various phospholipids, neutral lipids, and glycosylphosphatidylinositol. Interestingly, the sequencing of the genomes of various *Plasmodium* spp. revealed that these organisms also express FASII biosynthetic machinery in the apicoplast. This suggests that in addition to its ability to acquire fatty acids from the host, the parasite can also synthesize fatty acids de novo. Inhibitors of the FASII pathway have been shown to inhibit parasite growth. This suggests that this pathway might be essential for the parasite (see below). Available data suggest, however, that the bulk of the intense phospholipid and neutral lipid biosynthesis relies on the uptake of plasmatic fatty acids or fatty acids obtained via hydrolysis of some lipids such as lysoPC (Krishnegowda and Gowda, 2003; Mitamura et al., 2000; Vial et al., 1990; Vial et al., 1982a; Vielemeyer et al., 2004; Vial and Ancelin, 1992).

The activation of FA to acyl-CoA thioesters by acyl-CoA synthetase (EC 6.2.1.3) is required for most cellular reactions involving FA. Biochemical experiments have shown that the *P. knowlesi* parasite has distinct enzymes and enzymatic sites, with differential specificity for saturated and unsaturated FA (Beaumelle and Vial, 1988a). The *Plasmodium* sequencing project revealed 11 genes encoding putative acyl-CoA synthetase enzymes (PfACS) distributed in eight different chromosomes of *P. falciparum* (PF14_0761, PFC0050c, PFL2570w, AL049183, PF07_0129, PFD0085c, AC006278, PFB0695c, PFB0685c, PFE1250w, and PFL1880w). Global microarray analyses revealed that all the genes are expressed during the intraerythrocytic stage of the parasite. Interestingly, the majority of the *PfACS* genes are located in subtelomeric regions. The differential expression of these genes suggests that different enzymes might be used at different stages of the parasite life cycle. For example, although PfAcs1 and PfAcs3 are very similar, with identical carboxyl-terminal domains, *PfACS1* is strongly induced during late schizogony and soon after erythrocyte invasion, whereas *PfACS3* is mainly detected at ring stages and is undetectable in schizonts. PfACS1 might be critical for the development of membranous structures produced in the infected erythrocytes early on after invasion (Matesanz et al., 2003).

It is possible that, as in yeast, PfACS might also function as a component of the fatty acid import system, thus linking import and activation of exogenous fatty acids to intracellular utilization and signaling (Faergeman et al., 2001). PfACS1 and PfACS3 are transported in vesicle-like structures toward the host erythrocyte cytoplasm where they interact with the cytoskeletal protein ankyrin. Interestingly, coimmunoprecipitation of ankyrin and PfACS1/3 indicated that at least a fraction of these proteins is physically associated with the membrane of infected erythrocytes and suggests that such a binding may bring these enzymes closer to the host erythrocyte membrane where exogenous fatty acids are available (Tellez et al., 2003).

Interestingly, *P. falciparum* also expresses a family of three genes encoding acyl-CoA binding proteins in the subtelomeric regions of chromosomes X and XIV (NP_702638, NP_700490, and NP_700489). It is not yet clear why *P. falciparum* requires such a large number of genes encoding ACS and acyl-CoA binding proteins and why these genes are localized to the subtelomeric regions, when only a few genes are found in other protozoan parasites, including other *Plasmodium* species, and are localized in internal regions of the chromosomes. This feature seems to be unique to *P. falciparum* and suggests that this pathway might possibly be under evolutionary selection.

Malarial parasites exhibit specificity with regard to both the chain length and level of saturation of the FA acquired from the host for the synthesis of its lipids. Fatty acids with chain lengths shorter than C10 are not incorporated. In *P. knowlesi*, linoleic acid, which represents <3% of the total FA fraction, is specifically incorporated into phospholipids and accounts for ~10% of FA in PL and 18 to 22% in DAG and TAG. This occurs particularly at the expense of myristic acid, which is abundant in these lipids in the original cells (Beaumelle and Vial, 1986). Saturated and unsaturated FA are preferentially incorporated at the sn-1 and sn-2 positions of phospholipids (Krishnegowda and Gowda, 2003; Vial and Ancelin, 1992) and glycosylphosphatidylinositol (Naik et al., 2000) of *P. falciparum*. The involved FA species have to come in a pair of one saturated and one unsaturated to support *P. falciparum* growth, with palmitic and oleic acid as the best combinations (Mitamura et al., 2000). FA are initially introduced into PL during de novo synthesis, and molecular species of PL are further strictly controlled by choline and ethanolamine phosphotransferase (Vial et al., 1984b).

Although available biochemical data suggest that during its intraerythrocytic development, the parasite is unable to elongate or modify to a detectable extent the scavenged FA that are used for lipid biosynthesis, the available genome sequence of *P. falciparum* revealed two genes encoding polypeptides that share homology with long chain fatty acid elongases (MAL6P1.62) and FA desaturases (PFE0555w). The function of these putative enzymes remains to be elucidated.

Biochemical studies showed that transacylation reactions or release of the FA moiety of intact PL molecules (PC, PS, and PE) does not occur in *Plasmodium*-infected erythrocytes (Beaumelle, 1987; Moll et al., 1988). Furthermore, it appeared that fluorescent lipids are not deacylated when they are introduced into the infected erythrocyte, although these PL molecules reach the intracellular parasite (Haldar et al., 1989). Krugliak and colleagues reported an acid, Ca^{2+}-independent phospholipase activity in *P. falciparum* cytosol using a fluorescent PL (Krugliak et al., 1987). However, our attempts to identify intracellular phospholipaseA_2 activity in homogenates of *P. knowlesi*-infected erythrocytes using palmitoyl-(^{14}C)oleyl-PC failed to detect such an activity under all ionic and pH conditions tested.

Finally, the initial step of FA biosynthesis by the FASII apicoplast pathway requires that the FASII substrate acetyl-CoA be available within this organelle. Acetyl-CoA can be synthesized either from pyruvate by a pyruvate dehydrogenase complex or from acetate by an acetyl-CoA synthetase. Four genes encoding the putative four subunits of the pyruvate dehydrogenase complex and three genes encoding putative acetyl-CoA synthetases have been identified in the *Plasmodium* genome database. Interestingly, three of the four subunits of the pyruvate dehydrogenase complex have a potential apicoplast transit peptide, whereas none of the acetyl-CoA synthetases has an apicoplast-targeting sequence. This suggests that the most likely source for acetyl-CoA in the apicoplast is via conversion of pyruvate. The FASII pathway, although not a major contributor to the total lipids of the parasite, is critical to the parasite survival. The FA produced by this pathway may be utilized in a discrete cellular compartment or contribute to the synthesis of a specific but crucial type of lipid (predominantly C10 to C14, as shown by Surolia and Surolia [2001]). Alternatively, the FA synthesized in the apicoplast might be needed for the synthesis of the membranes of the apicoplast. Nevertheless, the FA produced in the apicoplast must be transported into the ER in order to be incorporated into glycerolipids.

Choline Transport

Choline is a major phospholipid precursor required for the initial step of synthesis of PC via the CDP-choline pathway. At the physiological concentration of 10 to 40 μM (Das et al., 1986), choline enters *Plasmodium*-infected erythrocytes via a carrier-mediated process, and it is likely that parasite-induced permeability pathways functionally and quantitatively operate at higher concentrations. The reported capacity of the carrier-mediated entry of choline is 170 and 95 nmol/10^{10} cells/h for *P. knowlesi*- and *Plasmodium vinckei*-infected cells (Ancelin et al., 1991; Staines and Kirk, 1998), representing a 10- to 20-fold increase in choline transport compared to uninfected erythrocytes. This increase in choline transport activity across the red blood cell membrane might be due to changes in the transport properties of the endogenous choline transporter and/or the expression of a new carrier expressed by the parasite.

Choline entry across the parasite plasma membrane has only recently been demonstrated in two independent studies using saponin-free parasites and shown to be carrier mediated (Biagini et al., 2004; Lehane et al., 2004). The transporter has an apparent K_m for choline of 25.0 ± 3.5 µM and a V_{max} of 2,800 nmol/10^{10} cells/h (Biagini et al., 2004) and an apparent K_m of 86 ± 10 µM and a V_{max} of 7,200 nmol/10^{10} cells/h (Lehane et al., 2004). Choline transport across this carrier is inhibited by choline analogs and quinine, is temperature dependent, is Na^+ independent, and is driven by the proton motive force. This transport system is likely to be the major route of entry of biscationic drugs into the intracellular parasite (Biagini et al., 2004). Available data suggest that specialized choline carriers cannot transport these large biscationic molecules. The properties of the choline transporter identified in the plasma membrane of the parasite suggest that it might function as a polyspecific organic cation transporter functionally distinct from the known dedicated eukaryotic choline carriers. Such a transporter might also be involved in the transport of other metabolites required by the intracellular parasite.

Finally, the transport capacity of the parasite plasma membrane carrier appears much higher than the choline flux through the erythrocytic membrane (combined choline flux through the endogenous and induced host cell transporters). Therefore, the rate-limiting step of choline transport in intact infected erythrocytes occurs at the host cell membrane rather than at the parasite plasma membrane. The enhanced transport rate of the parasite plasma membrane carrier relative to the host cell also means that at steady state, there will be a reduced choline concentration in the host cell relative to the medium and the parasite plasma membrane carrier will act like a choline vacuum cleaner for the host cell.

CELLULAR LOCALIZATION OF LIPID BIOSYNTHETIC ENZYMES AND LIPID TRAFFICKING

The *Plasmodium*-infected erythrocyte has a complex membranous system. No detailed analysis of its subcellular composition has been conducted, owing to the lack of efficient fractionation procedures. Unlike proteins, which can be localized using specific antibodies, and with the exception of lipid particles, which can be revealed using specific fluorescence dyes, the subcellular localization of most lipids cannot be easily determined. Although experiments using exogenous fluorescently labeled lipids can be used as a way to determine the subcellular compartments in which a certain lipid can be metabolized (Haldar, 1996), the nonnatural property of those lipids does not necessarily reflect the natural lipid composition and can lead to false interpretations. To date, there is very little or no information concerning the lipid composition of the parasitophorous vacuolar (PV) membrane or individual intraparasitic membranes (food vacuole, mitochondrion, and plastid).

Structural Organization of Lipids in the Host Cell Membrane

Erythrocytes infected with malarial parasites undergo major changes, primarily in their membrane composition. During its intraerythrocytic maturation, the parasite induces a major reorganization of the host cell membrane by inserting newly synthesized proteins, forming knob-like electron-dense structures, and by remodeling the molecular species of the membrane phospholipids. These changes can result in biophysical alterations of parasitized erythrocytes such as altered lipid packing, higher membrane fluidity, decreased microviscosity, modifications in membrane rheology, and functional modifications such as new permeability pathways.

Analysis of the lipid composition of the plasma membrane of erythrocytes after *Plasmodium* infection revealed no substantial modification with respect to lipid content and class distribution, irrespective of the mammalian host, and a possible loss of cholesterol and sphingomyelin (Maguire and Sherman, 1990; Vial and Ancelin, 1992). By contrast, the pair of fatty acids attached to the PL glycerol backbone (i.e., the PL molecular species) is changed. Fatty acids grafted on the glycerol backbone of PC and PE are modified in length and degree of unsaturation [e.g., palmitoyl (16:0)-linoleyl (18:2) and (16:0)-oleyl (18:1)-PE are increased, whereas (16:0)-arachidonyl (20:4)-, stearyl (18:0)-(20:4)- and (18:1)-(20:4)-PE are decreased] (Simões et al., 1993; Simões et al., 1990). This makes the composition very similar to that of the parasite. Importantly, this attests that the parasite is capable of fine modulation of the plasma membrane of its host and indicates intense dynamic PL traffic between the erythrocyte membrane and those of the intracellular parasite.

One interesting feature of *Plasmodium* infection relates to the location of phosphatidylserine in the membrane of infected erythrocytes. Indeed, in intact cells, PS molecules are exclusively localized in the inner leaflet of the lipid bilayer, but are flipped to the outer surface of mammalian cells under some pathological conditions. Some studies have shown that malarial infection increases the number of erythrocytes that expose PS at the outer membrane leaflet (Brand et al., 2003; Eda and Sherman, 2002; Joshi et al., 1987; Maguire et al., 1991; Sherman et al., 2003; Sherman et al., 1997), while other studies could not detect a breakdown of the phospholipid symmetry (Moll et al., 1988; Moll et al., 1990; Van der Schaft et al., 1987). Flow cytometry studies using annexin V, which specifically binds to cells that expose PS in their outer leaflet, have now confirmed that *Plasmodium* infection is accompanied by a breakdown in PS asymmetry. Upon increase of cytosolic calcium, for instance, erythrocytes undergo shrinkage, membrane blebbing, and breakdown of cell membrane PS asymmetry, all typical features of apoptosis in nucleated cells. The entry of calcium stimulates an intraerythrocytic scramblase that facilitates bidirectional PL migration across the bilayer, which in turn leads to PS exposure at the outer membrane leaflet (Lang et al., 2004). *Plasmodium*-infected erythrocytes display significantly increased annexin binding, as the intracellular parasite matures to trophozoite and schizont stages; the loss of PL asymmetry of the infected erythrocyte membrane increased with the percentage of parasitemia in the in vitro culture (Brand et al., 2003; Eda and Sherman, 2002). Increase in erythrocytic cytosolic calcium may account for the simulation of the scramblase. Loss of PS asymmetry is of great importance, since exposure of this phospholipid at the outer surface of the cell membrane is presumably followed by binding to PS receptor on macrophages and subsequent phagocytosis of the affected erythrocyte (Lang et al., 2004). Also, PS exposed to the surface of malaria-infected cells contributes to the cytoadherence of *P. falciparum*-parasitized erythrocyte (Eda and Sherman, 2002).

Membrane Raft and Malarial Growth

Despite the absence of parasite genes for cholesterol biosynthesis and its quasi-absence in the intracellular parasite, this neutral lipid appears to be essential for the establishment of infection and critical to the stability of the trophozoite-stage-infected erythrocyte (Lauer et al., 2000). The parasite membrane network is greatly depleted of cholesterol as shown by the isolation of free parasites (Vial and Ancelin, 1992) and fluorescent microscopy (Jackson et al., 2004). It is likely that some host components are incorporated into the PV membrane during invasion, and there is evidence for trafficking of raft components (cholesterol- and sphingolipid-rich domains, i.e., sphingomyelin and glycosphingolipids) between the host cell and PV membranes (Haldar et al., 2002).

Interestingly, some plasmodial proteins that reside in cholesterol-rich detergent-resistant membrane (DRM) and that are characteristic of microdomains in host cell membranes are recruited into the parasitophorous vacuole membrane (Lauer et al., 2000; Nagao et al., 2002), as if *P. falciparum* infection disrupted the association of some raft-associated proteins with erythrocyte DRMs and directed some to the vacuole, while host non-raft proteins remained on the plasma membrane (Murphy et al., 2004). Haldar et al. (2002) suggested that the erythrocyte membrane possesses a dynamic mix of cholesterol-rich microdomains and that distinct raft populations exist. Upon malarial invasion, a subset of the major raft proteins (flotilin-2 and others) and minor raft proteins (band 3, GAPDH, and others) enters the parasitophorous vacuolar membrane while other proteins are excluded (Murphy et al., 2004). Many of the cell surface merozoite proteins are themselves arranged within lipid rafts on the merozoite surface (Wang et al., 2003). Thus, the presence of raft-associated proteins in the parasitophorous membrane suggests a role for cholesterol-rich rafts in the transport of macromolecules into malaria-infected cells. Rafts also appear to be critical for parasite invasion, since red blood cells depleted of lipid rafts become resistant to invasion with malaria parasites (Samuel et al., 2001).

Thus, it appears that although cholesterol and sphingomyelin are not synthesized de novo by *Plasmodium*, mechanisms regulated by the DRM lipids, sphingomyelin and cholesterol, likely mediate (i) malarial invasion, (ii) the uptake of host DRM proteins, and (iii) maintenance of the intracellular vacuole in the nonendocytic red blood cell, which may have implications for intracellular parasitism and pathogenesis.

Lipids of the Food Vacuole and Hemozoin Formation

The food vacuole is the site of degradation of hemoglobin. In the process, ferriprotoporphyrin IX (ferric heme; FP) is formed, dimerized, and then sequestered as hemozoin, the characteristic malarial pigment. Under acidic conditions, unsaturated fatty acids and their mono- and diacylglycerols and other lipid mixtures can enhance β-hematin formation in vitro (Fitch, 2004; Pandey et al., 2003). Sonicated TAG suspensions also promote hematin crystallization, but only at low lipid-to-hematin ratios (Jackson et al., 2004).

Lipids could serve to concentrate FP and keep it in a state favorable for dimerization (Fitch, 2004).

Fitch (2004) hypothesizes that the parasite degrades the inner membrane of the endocytic vesicles by engulfing host cytosolic hemoglobin (the mechanism by which this membrane is processed is unknown) and that the released unsaturated lipids from this inner membrane are tightly coupled to FP release and dimerization in the malaria parasite. The recently shown association of lipid bodies and substantial amounts of DAG and TAG with the food vacuole suggests that these neutral lipids and their precursors might assist in the crystallization of hematin and play a role in heme detoxification during the early trophozoite stage of infection in the food vacuole before their incorporation into the neutral lipid bodies (Jackson et al., 2004). Interestingly, lipid body structures are most prominent in more mature-stage parasites, and DAG levels have been reported to be maximal in the trophozoite stages but are apparently converted to TAG by the schizont stage (Nawabi et al., 2003).

LIPID-BASED ANTIMALARIAL CHEMOTHERAPY

The available genome sequence of various *Plasmodium* spp. and the availability of strategies to genetically manipulate the parasite and assess the importance of specific genes in parasite development and survival have made it possible to identify various enzymes and metabolic machineries that could potentially constitute good malarial therapeutic targets. Furthermore, postgenomic studies have made it possible to determine the expression profile of those genes not only during the intraerythrocytic development but also during other developmental stages.

The present review shows that great strides have been made in understanding membrane biogenesis in *Plasmodium* and the importance of this process in parasite development and multiplication, but more still has to be learned. Nevertheless, because of the critical role that lipids play in parasite development and survival and the presence of major catalytic and structural differences between the lipid metabolic pathways of the parasite and those of humans, a growing interest in lipid-based therapeutic strategies has emerged in the past few years, and new classes of potent antimalarial drugs have been designed.

Quaternary Ammonium as Choline Analogs

Because of the importance of PL in *Plasmodium* development and survival, we initiated a new antimalarial pharmacological approach based on inhibition of PL metabolism using analogs of phospholipid precursors. Several compounds that mimic the structure of choline were rationally designed and optimized for their ability to inhibit de novo PC biosynthesis and *Plasmodium* growth. Bis-quaternary ammonium salts showed potent antimalarial activity in vitro and in vivo, with IC_{50} in the low nanomolar range. To remedy the low oral absorption of these first-generation compounds, a second generation consisting of G25 bioisosteric analogs (bis-amidines) was designed and found to exhibit similar potent antimalarial activities but with increased bioabsorption indices (Ancelin et al., 2003; Salom-Roig et al., 2005; Vial and Calas, 2001). Finally, we recently designed a third generation consisting of neutral prodrugs that, in the presence of plasmatic esterase, are transformed into the biscationic thiazolium drugs (Vial et al., 2004). The compounds are effective against sensitive and multidrug-resistant strains of *P. falciparum*. These molecules exert a very rapid cytotoxic effect against malarial parasites in the very low nanomolar range and are active in vivo against *P. vinckei*-infected mice, with a 50% effective dose of <0.2 mg/kg. They are able to cure highly infected mice and retain full activity after a single injection. They also retain full activity against *P. falciparum* and *Plasmodium cynomolgi* in primate models with no recrudescence and at lower doses. One interesting feature of these compounds is that they specifically accumulate in *Plasmodium*-infected erythrocytes but not in host cells to millimolar range. This property probably plays a critical role in the potency, specificity, and low toxicity of these compounds (Wengelnik et al., 2002). Biochemical and genetic analyses designed to understand the mode of action of these compounds revealed that they are potent inhibitors of the de novo PC biosynthetic pathway, are competitive inhibitors of choline entry across the parasite plasma membrane via the organic cation transporter which mediates choline entry (Biagini et al., 2004), and can also interfere with other PL metabolic pathways such as the synthesis of PE from PS (Roggero et al., 2004). Moreover, these compounds were recently found to interact with hemozoin, which enhances their antimalarial effect (Biagini et al., 2003). Thus, they appear to be dual molecules, exerting their antimalarial activity via two simultaneous toxic effects on the intracellular intraerythrocytic parasites.

Phosphocholine Analogs

The finding that the alkylphosphocholine analog, hexadecylphosphocholine (miltefosine), inhibits the activity of *Plasmodium* phosphoethanolamine methyltransferase and the growth of the parasite within human red blood cells (Pessi et al., 2004) has stimulated efforts to seek the possible use of this class of compounds in malaria chemotherapy. Alkylphosphocholines represent a novel class of lipophilic ether lipids derived from the alkyllysophospholipids that are related to natural lysophospholipids. These compounds have been demonstrated to have promising antitumor and antileishmanial activities in vitro and in vivo. Structure-activity relationship analysis of these compounds has shown that their potency is dependent on the length of their alkyl chain and the presence of a double bond. Miltefosine is the only drug within this class of compounds that has been registered for clinical purposes. Approaches to increase the entry of these compounds into *Plasmodium*-infected erythrocytes are needed before they can also be considered for treatment of malaria.

Sphingolipid Inhibitors

Available biochemical and cell biological data suggest that an SM cycle involving two possible SM synthases and one neutral Mg^{2+}-dependent SMase might exist in *P. falciparum*. Furthermore, pharmacological data using drugs known to inhibit similar activities in mammalian cells suggest that if this cycle exists it might play a critical role in the parasite life cycle. Hanada and colleagues showed that PfNms activity was sensitive to scyphostatin, an inhibitor of mammalian neutral SMase with an IC_{50} of ~3 μM (Hanada et al., 2000). These authors showed that this compound specifically impaired the maturation of trophozoites into schizonts and blocked *P. falciparum* proliferation with an IC_{50} similar to that of the enzyme (Hanada et al., 2000). Gerold and Schwarz (2001) showed that inhibitors such as fumonisin B1, cycloserine, and myriocin, known to block de novo ceramide biosynthesis in mammalian cells, inhibit the intraerythrocytic proliferation of *P. falciparum* in vitro. These compounds, however, are not very potent against the parasite, most likely because of the lack of a de novo ceramide biosynthetic pathway in this organism. Another evidence that sphingolipid inhibitors can alter parasite proliferation came from studies by Haldar and colleagues that showed that two analogs, DL-threo-1-phenyl-2-palmitolyamino-3-morpholino-1-propanol (PPMP) and DL-threo-1-phenyl-2-decanoyl-amino-3-morpholino-1-propanol, known to inhibit the synthesis of glucosylceramide in mammalian cells, block the intraerythrocytic proliferation of *P. falciparum* with an IC_{50} of 6 μM (Hanada et al., 2000; Lauer et al., 1995). These two compounds have been proposed to inhibit one of two parasitic SM synthase activities and prevent the formation of the tubulovesicular network. It remains, however, to be established whether PfSms1 and PfSms2 catalyze the synthesis of SM in the parasite, whether SM synthase activity plays an important role in the parasite life cycle, and whether their activity is the primary target of sphingolipid analogs. Nevertheless, the antimalarial activity of these drugs suggests that they can be considered as lead compounds for further chemical optimization. This strategy has been recently successfully used by Grellier and colleagues who synthesized analogs of ceramide in which the sphingosine acyl chain on carbon 3 was replaced by a phenyl group, as is also the case for PPMP, to which were attached nitro, amino, or carbon chains of various chain lengths. These studies showed that analogs containing an alkyl linkage between the fatty acid and the sphingoid core have considerably higher antimalarial activity than PPMP or ceramide analogs with an amide linkage (P. Grellier, personal communication). Unlike PPMP, whose effect against the parasite is cytostatic, the alkyl derivatives of ceramide block irreversibly the *P. falciparum* proliferation with IC_{50}s in the low nanomolar range and have no effect on parasite SM synthase activity.

Inhibitors of the FASII Pathway

Several lines of evidence suggest that the malarial FASII pathway in the apicoplast is an attractive drug target. Inhibitors of various enzymes of the FASII pathway, known to have antibacterial activity, have also now been shown to inhibit *P. falciparum* proliferation in vitro. Of these, triclosan, which inhibits the rate-limiting step in FA synthesis catalyzed by enoyl-acyl carrier protein reductase, stands as one of the most potent drugs with IC_{50}s ranging from 0.2 to 1.2 μM for different strains, targeting primarily the early stages of parasite development within red blood cells (Surolia and Surolia, 2001). Interestingly, Surolia and colleagues have shown that unlike clindamycin and chloramphenicol, which cause delayed death in the cultures of *P. falciparum* (parasitemia decreasing only after 84 h), triclosan showed a rapid antiplasmodial activity blocking progress into trophozoites within 24 h of incubation (Surolia et al., 2004). This suggests that FA synthesis in the apicoplast might be essential for parasite survival.

CONCLUSION

In the past few years, major progress has been made in understating membrane biogenesis in *Plasmodium*, in part due to the advances in strategies to express malarial genes in heterologous systems and the development of new tools for genetical manipulation of the parasite. As the number of characterized lipid biosynthetic genes grows, so does the list of possible pathways that the parasite utilizes to synthesize its lipids. However, several central questions regarding the function and importance of the different lipid pathways in parasite development and survival remain to be answered.

It is clear that appropriate lipids are required at the right place for *Plasmodium* growth at its different stages. Subcellular fractionation of the malarial parasites remains a difficult task, and cellular localization of the various lipids and mechanisms mediating their intracellular trafficking remains to be elucidated. In recent years, specific tools such as the use of fluorescent lipids, fluorescence-based complex microscopy (Jackson et al., 2004), and transfection studies using green fluorescent protein chimeras have shed new light on the subcellular location of various lipid biosynthetic enzymes. These studies have increased our understanding of the possible routes that lipid molecules can take to travel from one cellular site to another. However, many lipids persist in minute amounts and for a finite lifetime in cells, cannot be accurately detected, and require improved analytic procedures before they can be thoroughly characterized. Mass spectrometry should be of great help, but application to malarial lipids is just beginning (Enjalbal et al., 2004). Mass spectrometry could also help unravel the subtle diversity of lipids (e.g., a molecular species of PL, or acylglycerols) inside subcellular compartments. Indeed, each lipid class described in this review includes a diverse group of molecules with respect to their fatty acyl chains attached to the glycerol or sphingosine backbone. Lipid diversity appears to be very important for cells, considering the considerable amount of energy invested in producing and maintaining it.

Finally, over the past few years, new plant- and bacterial-like enzymes that are not present in animal cells and other enzymes that, although present in animal cells, exhibit biochemical properties different enough from their animal counterparts have been found in *Plasmodium*. These enzymes are likely to be excellent targets for the design of new antimalarials. However, the catalytic mechanisms and structural biology of most of the enzymes and genetic studies to validate their essential role are still to be investigated.

ACKNOWLEDGMENTS

H.J.V.'s studies were supported by the Centre National de la Recherche Scientifique (CNRS), the European Community (QLK2-CT-2000-01166, BioMalPar, LSHP-CT-2004-503578), and the Ministère de l'Education Nationale et Recherche Scientifique (PAL+). C.B.M.'s research was supported by funds from NIH and DOD.

REFERENCES

Ancelin, M. L., M. Calas, A. Bonhoure, S. Herbute, and H. J. Vial. 2003. In vivo antimalarial activities of mono- and bis quaternary ammonium salts interfering with *Plasmodium* phospholipid metabolism. *Antimicrob. Agents Chemother.* **47:**2598–2605.

Ancelin, M. L., M. Parant, M. J. Thuet, J. R. Philippot, and H. J. Vial. 1991. Increased permeability to choline in simian erythrocytes after *Plasmodium knowlesi* infection. *Biochem. J.* **273:**701–709.

Ancelin, M. L., and H. J. Vial. 1989. Regulation of phosphatidylcholine biosynthesis in *Plasmodium*-infected erythrocytes. *Biochim. Biophys. Acta* **1001:**82–89.

Ansorge, I., D. Jeckel, F. Wieland, and K. Lingelbach. 1995. *Plasmodium falciparum*-infected erythrocytes utilize a synthetic truncated ceramide precursor for synthesis and secretion of truncated sphingomyelin. *Biochem. J.* **308:**335–341.

Ardail, D., F. Gasnier, F. Lerme, C. Simonot, P. Louisot, and O. Gateau-Roesch. 1993. Involvement of mitochondrial contact sites in the subcellular compartmentalization of phospholipid biosynthetic enzymes. *J. Biol. Chem.* **268:**25985–25992.

Baunaure, F., P. Eldin, A.-M. Cathiard, and H. Vial. 2004. Characterization of a non-mitochondrial type I phosphatidylserine decarboxylase in Plasmodium falciparum. *Mol. Microbiol.* **51:**33–46.

Beaumelle, B. D. 1987. Métabolisme des acides gras et dynamique des phospholipides dans l'érythrocyte infecté par le parasite du paludisme. Ph.D. Thesis. Université de Montpellier II, Montpellier, France.

Beaumelle, B. D., and H. J. Vial. 1988a. Acyl-CoA synthetase activity in Plasmodium knowlesi-infected erythrocytes displays peculiar substrate specificities. *Biochim. Biophys. Acta* **958:**1–9.

Beaumelle, B. D., and H. J. Vial. 1988b. Uninfected red cells from malaria-infected blood: alteration of fatty acid composition involving a serum protein: an in vivo and in vitro study. *In Vitro Cell. Dev. Biol.* **24:**711–718.

Beaumelle, B. D., H. J. Vial, and A. Bienvenüe. 1988. Enhanced transbilayer mobility of phospholipids in malaria-infected monkey erythrocytes: a spin-label study. *J. Cell. Physiol.* **135:**94–100.

Ben Mamoun, C., I. Y. Gluzman, C. Hott, S. K. MacMillan, A. S. Amarakone, D. L. Anderson,

J. M. Carlton, J. B. Dame, D. Chakrabarti, R. K. Martin, B. H. Brownstein, and D. E. Goldberg. 2001. Co-ordinated programme of gene expression during asexual intraerythrocytic development of the human malaria parasite Plasmodium falciparum revealed by microarray analysis. *Mol. Microbiol.* **39:**26–36.

Biagini, G. A., E. M. Pasini, R. Hughes, H. P. De Koning, H. J. Vial, P. M. O'Neill, S. A. Ward, and P. G. Bray. 2004. Characterization of the choline carrier of Plasmodium falciparum: a route for the selective delivery of novel antimalarial drugs. *Blood* **104:**3372–3377.

Biagini, G. A., E. Richier, P. G. Bray, M. Calas, H. Vial, and S. A. Ward. 2003. Heme binding contributes to antimalarial activity of bis-quaternary ammoniums. *Antimicrob. Agents Chemother.* **47:**2584–2589.

Bozdech, Z., M. Llinas, B. L. Pulliam, E. D. Wong, J. Zhu, and J. L. DeRisi. 2003. The transcriptome of the intraerythrocytic developmental cycle of Plasmodium falciparum. *PLoS Biol.* **1:**E5.

Brand, V., C. Sandu, C. Duranton, V. Tanneur, K. Lang, S. Huber, and F. Lang. 2003. Dependence of Plasmodium falciparum in vitro growth on the cation permeability of the human host erythrocyte. *Cell. Physiol. Biochem.* **13:**347–356.

Carman, G. M., and S. A. Henry. 1999. Phospholipid biosynthesis in the yeast Saccharomyces cerevisiae and interrelationship with other metabolic processes. *Prog. Lipid Res.* **38:**361–399.

Coleman, R. A., and D. P. Lee. 2004. Enzymes of triacylglycerol synthesis and their regulation. *Prog. Lipid Res.* **43:**134–176.

Das, I., J. De Belleroche, C. J. Moore, and F. C. Rose. 1986. Determination of free choline in plasma and erythrocyte samples and choline derived from membrane phosphatidylcholine by a chemioluminescent method. *Anal. Biochem.* **152:** 178–182.

Divo, A. A., T. G. Geary, N. L. Davis, and J. B. Jensen. 1985. Nutritional requirements of Plasmodium falciparum in culture. I. Exogenously supplied dialyzable components necessary for continuous growth. *J. Protozool.* **32:**59–64.

Dowhan, W. 1997. Molecular basis for membrane phospholipid diversity: why are there so many lipids? *Annu. Rev. Biochem.* **66:**199–232.

Eda, S., and I. W. Sherman. 2002. Cytoadherence of malaria-infected red blood cells involves exposure of phosphatidylserine. *Cell. Physiol. Biochem.* **12:**373–384.

Elabbadi, N., M. L. Ancelin, and H. J. Vial. 1997. Phospholipid metabolism of serine in Plasmodium-infected erythrocytes involves phosphatidylserine and direct serine decarboxylation. *Biochem. J.* **324:**435–445.

Enjalbal, C., R. Roggero, R. Cerdan, J. Martinez, H. Vial, and J. L. Aubagnac. 2004. Automated monitoring of phosphatidylcholine biosyntheses in Plasmodium falciparum by electrospray ionization mass spectrometry through stable isotope labeling experiments. *Anal. Chem.* **76:**4515–4521.

Faergeman, N. J., P. N. Black, X. D. Zhao, J. Knudsen, and C. C. DiRusso. 2001. The acyl-CoA synthetases encoded within FAA1 and FAA4 in Saccharomyces cerevisiae function as components of the fatty acid transport system linking import, activation, and intracellular utilization. *J. Biol. Chem.* **276:**37051–37059.

Fitch, C. D. 2004. Ferriprotoporphyrin IX, phospholipids, and the antimalarial actions of quinoline drugs. *Life Sci.* **74:**1957–1972.

Gerold, P., and R. T. Schwarz. 2001. Biosynthesis of glycosphingolipids de-novo by the human malaria parasite Plasmodium falciparum. *Mol. Biochem. Parasitol.* **112:**29–37.

Grellier, P., D. Rigomier, V. Clavey, J. C. Fruchart, and J. Schrével. 1991. Lipid traffic between high density lipoproteins and Plasmodium falciparum-infected red blood cells. *J. Cell Biol.* **112:**267–277.

Haldar, K. 1996. Sphingolipid synthesis and membrane formation by Plasmodium. *Trends Cell Biol.* **6:**398–405.

Haldar, K., A. F. De Amorim, and G. A. M. Cross. 1989. Transport of fluorescent phospholipid analogues from the erythrocyte membrane to the parasite in Plasmodium falciparum-infected cells. *J. Cell Biol.* **108:**2183–2192.

Haldar, K., N. Mohandas, B. U. Samuel, T. Harrison, N. L. Hiller, T. Akompong, and P. Cheresh. 2002. Protein and lipid trafficking induced in erythrocytes infected by malaria parasites. *Cell. Microbiol.* **4:**383–395.

Hanada, K., T. Mitamura, M. Fukasawa, P. A. Magistrado, T. Horii, and M. Nishijima. 2000. Neutral sphingomyelinase activity dependent on Mg2+ and anionic phospholipids in the intraerythrocytic malaria parasite Plasmodium falciparum. *Biochem. J.* **346:**671–677.

Hanada, K., N. M. Palacpac, P. A. Magistrado, K. Kurokawa, G. Rai, D. Sakata, T. Hara, T. Horii, M. Nishijima, and T. Mitamura. 2002. Plasmodium falciparum phospholipase C hydrolyzing sphingomyelin and lysocholinephospholipids is a possible target for malaria chemotherapy. *J. Exp. Med.* **195:**23–34.

Hannun, Y. A. 1994. The sphingomyelin cycle and second messenger function of ceramide. *J. Biol. Chem.* **269:**3125–3128.

Holz, G. G. 1977. Lipids and the malaria parasite. *Bull. W. H. O.* **55:**237–248.

Huitema, K., J. van den Dikkenberg, J. F. Brouwers, and J. C. Holthuis. 2004. Identification of a family of animal sphingomyelin synthases. *EMBO J.* **23:**33–44.

Jackson, K. E., N. Klonis, D. J. Ferguson, A. Adisa, C. Dogovski, and L. Tilley. 2004. Food vacuole-associated lipid bodies and heterogeneous lipid environments in the malaria parasite, Plasmodium falciparum. *Mol. Microbiol.* **54:**109–122.

Joshi, P., G. P. Dutta, and C. M. Crupta. 1987. An intracellular siman malarial parasite (*Plasmodium knowlesi*) induces stage-dependent alterations in membrane phospholipid organization of its host erythrocyte. *Biochem. J.* **146:**103–108.

Kent, C. 1995. Eukaryotic phospholipid biosynthesis. *Annu. Rev. Biochem.* **64:**315–343.

Krishnegowda, G., and D. C. Gowda. 2003. Intraerythrocytic Plasmodium falciparum incorporates extraneous fatty acids to its lipids without any structural modification. *Mol. Biochem. Parasitol.* **132:**55–58.

Krugliak, M., Z. Waldman, and H. Ginsburg. 1987. Gentamicin and amikacin repress the growth of *Plasmodium falciparum* in culture, probably by inhibiting a parasite acid phospholipase. *Life Sci.* **40:**1253–1257.

Lang, F., P. A. Lang, K. S. Lang, V. Brand, V. Tanneur, C. Duranton, T. Wieder, and S. M. Huber. 2004. Channel-induced apoptosis of infected host cells—the case of malaria. *Pflugers Arch.* **448:**319–324.

Larvor, M. P., R. Cerdan, C. Gumila, L. Maurin, P. Seta, C. Roustan, and H. Vial. 2003. Characterization of the lipid-binding domain of the Plasmodium falciparum CTP:phosphocholine cytidylyltransferase through synthetic-peptide studies. *Biochem. J.* **375:**653–661.

Lauer, S., J. Van Wye, T. Harrison, H. McManus, B. U. Samuel, N. L. Hiller, N. Mohandas, and K. Haldar. 2000. Vacuolar uptake of host components, and a role for cholesterol and sphingomyelin in malarial infection. *EMBO J.* **19:**3556–3564.

Lauer, S. A., N. Ghori, and K. Haldar. 1995. Sphingolipid synthesis as a target for chemotherapy against malaria parasites. *Proc. Natl. Acad. Sci. USA* **92:**9181–9185.

Lehane, A. M., K. J. Saliba, R. J. Allen, and K. Kirk. 2004. Choline uptake into the malaria parasite is energized by the membrane potential. *Biochem. Biophys. Res. Commun.* **320:**311–317.

Le Roch, K. G., Y. Zhou, P. L. Blair, M. Grainger, J. K. Moch, J. D. Haynes, P. De La Vega, A. A. Holder, S. Batalov, D. J. Carucci, and E. A. Winzeler. 2003. Discovery of gene function by expression profiling of the malaria parasite life cycle. *Science* **301:**1503–1508.

Maguire, P. A., J. Prudhomme, and I. W. Sherman. 1991. Alterations in erythrocyte membrane phospholipid organization due to the intracellular growth of the human malaria parasite, *Plasmodium falciparum. Parasitology* **102:**179–186.

Maguire, P. A., and I. W. Sherman. 1990. Phospholipid composition, cholesterol content and cholesterol exchange in *Plasmodium falciparum*-infected red cells. *Mol. Biochem. Parasitol.* **38:**105–112.

Marechal, E., N. Azzouz, C. S. de Macedo, M. A. Block, J. E. Feagin, R. T. Schwarz, and J. Joyard. 2002. Synthesis of chloroplast galactolipids in apicomplexan parasites. *Eukaryot. Cell* **1:**653–656.

Martin, D., L. Gannoun-Zaki, S. Bonnefoy, P. Eldin, K. Wengelnik, and H. Vial. 2000. Characterization of *Plasmodium falciparum* CDP-diacylglycerol synthase, a proteolytically cleaved enzyme. *Mol. Biochem. Parasitol.* **110:**93–105.

Matesanz, F., M. M. Tellez, and A. Alcina. 2003. The Plasmodium falciparum fatty acyl-CoA synthetase family (PfACS) and differential stage-specific expression in infected erythrocytes. *Mol. Biochem. Parasitol.* **126:**109–112.

Mitamura, T., K. Hanada, E. P. Ko-Mitamura, M. Nishijima, and T. Horii. 2000. Serum factors governing intraerythrocytic development and cell cycle progression of *Plasmodium falciparum. Parasitol. Int.* **49:**219–229.

Moll, G. N., H. J. Vial, M. L. Ancelin, J. A. Op den Kamp, B. Roelofsen, and L. L. van Deenen. 1988. Phospholipid uptake by Plasmodium knowlesi infected erythrocytes. *FEBS Lett.* **232:**341–346.

Moll, G. N., H. J. Vial, F. C. van der Wiele, M. L. Ancelin, B. Roelofsen, A. J. Slotboom, G. H. de Haas, L. L. van Deenen, and J. A. Op den Kamp. 1990. Selective elimination of malaria infected erythrocytes by a modified phospholipase A2 in vitro. *Biochim. Biophys. Acta* **1024:**189–192.

Murphy, S. C., B. U. Samuel, T. Harrison, K. D. Speicher, D. W. Speicher, M. E. Reid, R. Prohaska, P. S. Low, M. J. Tanner, N. Mohandas, and K. Haldar. 2004. Erythrocyte detergent-resistant membrane proteins: their characterization and selective uptake during malarial infection. *Blood* **103:**1920–1928.

Nagao, E., K. B. Seydel, and J. A. Dvorak. 2002. Detergent-resistant erythrocyte membrane rafts are modified by a Plasmodium falciparum infection. *Exp. Parasitol.* **102:**57–59.

Naik, R. S., O. H. Branch, A. S. Woods, M. Vijaykumar, D. J. Perkins, B. L. Nahlen, A. A. Lal, R. J. Cotter, C. E. Costello, C. F. Ockenhouse, E. A. Davidson, and D. C. Gowda. 2000. Glycosylphosphatidylinositol anchors of Plasmodium falciparum: molecular characterization and naturally elicited antibody response that may provide immunity to malaria pathogenesis. *J. Exp. Med.* **192:**1563–1576.

Nawabi, P., A. Lykidis, D. Ji, and K. Haldar. 2003. Neutral-lipid analysis reveals elevation of acylglycerols and lack of cholesterol esters in Plasmodium falciparum-infected erythrocytes. *Eukaryot. Cell* **2:**1128–1131.

Omodeo-Sale, F., A. Motti, N. Basilico, S. Parapini, P. Olliaro, and D. Taramelli. 2003. Accelerated senescence of human erythrocytes cultured with Plasmodium falciparum. *Blood* **102:**705–711.

Palacpac, N. M. Q., Y. Hiramine, F. Mi-Ichi, M. Torii, K. Kita, R. Hiramatsu, T. Horii, and T. Mitamura. 2004. Developmental-stage-specific triacylglycerol biosynthesis, degradation and trafficking as lipid bodies in Plasmodium falciparum-infected erythrocytes. *J. Cell Sci.* **117:**1469–1480.

Pandey, A. V., V. K. Babbarwal, J. N. Okoyeh, R. M. Joshi, S. K. Puri, R. L. Singh, and V. S. Chauhan. 2003. Hemozoin formation in malaria: a two-step process involving histidine-rich proteins and lipids. *Biochem. Biophys. Res. Commun.* **308:**736–743.

Pessi, G., G. Kociubinski, and C. B. Mamoun. 2004. A pathway for phosphatidylcholine biosynthesis in Plasmodium falciparum involving phosphoethanolamine methylation. *Proc. Natl. Acad. Sci. USA* **101:**6206–6211.

Ralph, S. A., G. G. Van Dooren, R. F. Waller, M. J. Crawford, M. J. Fraunholz, B. J. Foth, C. J. Tonkin, D. S. Roos, and G. I. McFadden. 2004. Tropical infectious diseases: metabolic maps and functions of the Plasmodium falciparum apicoplast. *Nat. Rev. Microbiol.* **2:**203–216.

Roggero, R., R. Zufferey, M. Minca, E. Richier, M. Calas, H. Vial, and C. Ben Mamoun. 2004. Unraveling the mode of action of the antimalarial choline analog G25 in *Plasmodium falciparum* and *Saccharomyces cerevisiae*. *Antimicrob. Agents Chemother.* **48:**2816–2824.

Rontein, D., I. Nishida, G. Tashiro, K. Yoshioka, W. I. Wu, D. R. Voelker, G. Basset, and A. D. Hanson. 2001. Plants synthesize ethanolamine by direct decarboxylation of serine using a pyridoxal phosphate enzyme. *J. Biol. Chem.* **276:**35523–35529.

Salom-Roig, X., H. Hamzé, M. Calas, and H. Vial. 2005. Dual molecules as new antimalarials. *Comb. Chem. High Throughput Screen.* **8:**49–62.

Samuel, B. U., N. Mohandas, T. Harrison, H. McManus, W. Rosse, M. Reid, and K. Haldar. 2001. The role of cholesterol and glycosylphosphatidylinositol-anchored proteins of erythrocyte rafts in regulating raft protein content and malarial infection. *J. Biol. Chem.* **276:**29319–29329.

Santiago, T. C., R. Zufferey, R. S. Mehra, R. A. Coleman, and C. Ben Mamoun. 2004. The Plasmodium falciparum PfGatp is an endoplasmic reticulum membrane protein important for the initial step of malarial glycerolipid synthesis. *J. Biol. Chem.* **279:**9222–9232.

Sherman, I. W., S. Eda, and E. Winograd. 2003. Cytoadherence and sequestration in Plasmodium falciparum: defining the ties that bind. *Microbes Infect.* **5:**897–909.

Sherman, I. W., J. Prudhomme, and J. F. Tait. 1997. Altered membrane phospholipid asymmetry in Plasmodium falciparum-infected erythrocytes. *Parasitol. Today* **13:**242–243.

Sherman, L. 1979. Biochemistry of *Plasmodium* (malarial parasites). *Microbiol. Rev.* **43:**453–495.

Shibuya, I. 1992. Metabolic regulations and biological functions of phospholipids in *Escherichia coli*. *Prog. Lipid Res.* **31:**245–299.

Shohet, S. B. 1971. The apparent transfer of fatty acid from phosphatidylcholine to phosphatidylethanolamine in human erythrocytes. *J. Lipid Res.* **12:**139–141.

Simões, A. P., S. Fiebig, F. Wunderlich, H. Vial, B. Roelofsen, and J. A. Op den Kamp. 1993. Plasmodium chabaudi-parasitized erythrocytes: phosphatidylcholine species of parasites and host cell membranes. *Mol. Biochem. Parasitol.* **57:**345–348.

Simões, A. P., G. N. Moll, B. Beaumelle, H. J. Vial, B. Roelofsen, and J. A. Op den Kamp. 1990. *Plasmodium knowlesi* induces alterations in phosphatidylcholine and phosphatidylethanolamine molecular species composition of parasitized monkey erythrocytes. *Biochim. Biophys. Acta* **1022:**135–145.

Smith, J. D. 1993. Phospholipid biosynthesis in protozoa. *Prog. Lipid Res.* **32:**47–60.

Sohlenkamp, C., I. M. Lopez-Lara, and O. Geiger. 2003. Biosynthesis of phosphatidylcholine in bacteria. *Prog. Lipid Res.* **42:**115–162.

Stahl, U., A. S. Carlsson, M. Lenman, A. Dahlqvist, B. Huang, W. Banas, A. Banas, and S. Stymne. 2004. Cloning and functional characterization of a phospholipid:diacylglycerol acyltransferase from Arabidopsis. *Plant Physiol.* **135:**1324–1335.

Staines, H. M., and K. Kirk. 1998. Increased choline transport in erythrocytes from mice infected with the malaria parasite Plasmodium vinckei vinckei. *Biochem. J.* **334:**525–530.

Surolia, A., T. Ramya, V. Ramya, and N. Surolia. 2004. 'FAS't inhibition of malaria. *Biochem. J.* **383:**401–412.

Surolia, N., and A. Surolia. 2001. Triclosan offers protection against blood stages of malaria by inhibiting enoyl-ACP reductase of *Plasmodium falciparum*. *Nat. Med.* **7:**167–173.

Tellez, M., F. Matesanz, and A. Alcina. 2003. The C-terminal domain of the Plasmodium falciparum acyl-CoA synthetases PfACS1 and PfACS3 functions as ligand for ankyrin. *Mol. Biochem. Parasitol.* **129:**191–198.

Van Deenen, L. L. M., and J. De Gier. 1975. Lipids of the red cell membrane, p. 147–211. *In* G. Surgenor (ed.), *The Red Blood Cell*. Academic Press, New York, N.Y.

Van der Schaft, P. H., B. Beaumelle, H. Vial, B. Roelofsen, J. A. Op den Kamp, and L. L. Van Deenen. 1987. Phospholipid organization in monkey erythrocytes upon Plasmodium knowlesi infection. *Biochim. Biophys. Acta* **901:**1–14.

Vial, H., S. Wein, C. Farenc, F. Bressolle, C. Kocken, A. Thomas, and M. Calas. 2004. Prodrugs of bisthiazolium salts are orally potent antimalarials. *Proc. Natl. Acad. Sci. USA* **101:**15458–15463.

Vial, H. J., and M. L. Ancelin. 1992. Malarial lipids. An overview. *Subcell. Biochem.* **18:**259–306.

Vial, H. J., M. L. Ancelin, J. R. Philippot, and M. J. Thuet. 1990. Biosynthesis and dynamics of lipids in *Plasmodium*-infected mature mammalian erythrocytes. *Blood Cells* **16:**531–555.

Vial, H. J., M. L. Ancelin, M. J. Thuet, and J. R. Philippot. 1989. Phospholipid metabolism in *Plasmodium*-infected erythrocytes: guidelines for further studies using radioactive precursor incorporation. *Parasitology* **98:**351–357.

Vial, H. J., and M. Calas. 2001. Inhibitors of phospholipid metabolism, p. 347–365. *In* P. Rosenthal (ed.), *Antimalarial Chemotherapy, Mechanisms of Action, Modes of Resistance, and New Directions in Drug Development.* Humana Press, Totowa, N.J.

Vial, H. J., P. Eldin, A. G. Tielens, and J. J. van Hellemond. 2003. Phospholipids in parasitic protozoa. *Mol. Biochem. Parasitol.* **126:**143–154.

Vial, H. J., M. J. Thuet, M. L. Ancelin, J. R. Philippot, and C. Chavis. 1984a. Phospholipid metabolism as a new target for malaria chemotherapy. Mechanism of action of D-2-amino-1-butanol. *Biochem. Pharmacol.* **33:**2761–2770.

Vial, H. J., M. J. Thuet, J. L. Broussal, and J. R. Philippot. 1982a. Phospholipid biosynthesis by Plasmodium knowlesi-infected erythrocytes: the incorporation of phospholipid precursors and the identification of previously undetected metabolic pathways. *J. Parasitol.* **68:**379–391.

Vial, H. J., M. J. Thuet, and J. R. Philippot. 1984b. Cholinephosphotransferase and ethanolaminephosphotransferase activities in Plasmodium knowlesi-infected erythrocytes. Their use as parasite-specific markers. *Biochim. Biophys. Acta* **795:**372–383.

Vial, H. J., M. J. Thuet, and J. R. Philippot. 1982b. Phospholipid biosynthesis in synchronous Plasmodium falciparum cultures. *J. Protozool.* **29:**258–263.

Vielemeyer, O., M. T. McIntosh, K. A. Joiner, and I. Coppens. 2004. Neutral lipid synthesis and storage in the intraerythrocytic stages of Plasmodium falciparum. *Mol. Biochem. Parasitol.* **135:**197–209.

Wang, L., N. Mohandas, A. Thomas, and R. L. Coppel. 2003. Detection of detergent-resistant membranes in asexual blood-stage parasites of Plasmodium falciparum. *Mol. Biochem. Parasitol.* **130:**149–153.

Wang, Q., S. Brown, D. S. Roos, V. Nussenzweig, and P. Bhanot. 2004. Transcriptome of axenic liver stages of Plasmodium yoelii. *Mol. Biochem. Parasitol.* **137:**161–168.

Wengelnik, K., V. Vidal, M. L. Ancelin, A. M. Cathiard, J. L. Morgat, C. H. Kocken, M. Calas, S. Herrera, A. W. Thomas, and H. J. Vial. 2002. A class of potent antimalarials and their specific accumulation in infected erythrocytes. *Science* **295:**1311–1314.

Yeo, H. J., M. P. Larvor, M. L. Ancelin, and H. J. Vial. 1997. *Plasmodium falciparum* CTP:phosphocholine cytidylyltransferase expressed in *Escherichia coli*: purification, characterization and lipid regulation. *Biochem. J.* **324:**903–910.

Yeo, H. J., J. Sri Widada, O. Mercereau-Puijalon, and H. J. Vial. 1995. Molecular cloning of CTP:phosphocholine cytidylyltransferase from *Plasmodium falciparum. Eur. J. Biochem.* **233:**62–72.

PLASMODIUM RIBOSOMES AND OPPORTUNITIES FOR DRUG INTERVENTION

Indu Sharma and Thomas F. McCutchan

18

The translation of mRNA, a fundamental property of all organisms, is carried out by the ribosome. Ribosomes consist of a complex of rRNA and protein. The ribosomes from eukaryotes can be sedimented at 80S and dissociated into two smaller subunits of 60S and 40S, whereas bacterial ribosomes sediment at 70S and can be dissociated into 50S and 30S subunits. Although the ribosomes of *Plasmodium* sp. are typically eukaryotic in their sedimentation properties, they do differ from the host by the rRNA having a low G+C base composition (Rogers et al., 1998).

"The crystal structures of the large ribosomal subunit complexed with substrate and product analogues show that only rRNA is involved in the positioning of the A-site and P-site substrates, and only RNA is in a position to chemically facilitate peptide-bond formation." In other words, "the ribosome is a ribozyme" (Steitz and Moore, 2003). Hence, ribosomal RNA can be a primary target of antibiotics. When the rRNA genes of *Plasmodium* were first characterized, stage-specific expression was unknown, so the genes were arbitrarily assigned letters. Later, it was found that the A-type pattern of expression was found in the asexual stages and the S-type was found in the sporozoite stage. The parasite's cytoplasmic ribosomes expressed in the vertebrate (A-type) seem less likely to provide a target for clinical development of drugs because they are structurally similar to the ribosomes of the host; the cytoplasmic ribosomes of the gametocyte and sporozoite stages (S-type), however, are so structurally different from both the A-type and host ribosomes that an investigation of the S-type ribosome might provide the insight necessary to develop drugs that could interfere with the transmission of the parasite. Indeed, antibiotics that selectively disable protein synthesis over defined periods of the developmental cycle or organellar function could play an important role in a better understanding of the molecular events during the developmental cycle of the malaria parasite.

PLASMODIUM RIBOSOME GENES

The arrangement of the cytoplasmic ribosomal gene (rRNA gene unit) of *Plasmodium* is similar to that seen in other eukaryotic organisms:

Indu Sharma and Thomas F. McCutchan, Molecular Biology Section, Laboratory of Malaria Vector Research, National Institute of Allergy and Infectious Diseases, National Institutes of Health, Bethesda, MD 20892.

Molecular Approaches to Malaria, Edited by Irwin W. Sherman
© 2005 ASM Press, Washington, D.C.

a copy of the small subunit rRNA (SSU rRNA), an internal transcribed spacer (ITS1), the 5.8S rRNA, another internal transcribed spacer (ITS2), and the large subunit rRNA (LSU rRNA) genes proceeding in a 5'-to-3' direction. Copies of *Plasmodium* rRNA gene units are dispersed as individual units on different chromosomes, unlike those of most other eukaryotes, which maintain tandem repeated copies. Normally, the rRNA gene units are dosage-response genes in that the deletion of a large number of copies has serious phenotypic consequences for the organism, presumably because cellular requirements for rRNA are not being met (Mitchell et al., 1992). In contrast, *Plasmodium* species have only four to eight rRNA genes (Dame and McCutchan, 1983; Langsley et al., 1983). It is unlikely that more than one or two of the genes are being transcribed at any one time (Rogers et al., 1995); mutations that occur in rRNA genes of *Plasmodium* thus have a leveraged effect not seen in most other organisms. A mutation in a single gene may affect 50% of the transcripts and would be expected to result in phenotypic changes.

CORRELATION OF rRNA EXPRESSION AND THE LIFE CYCLE

The life cycle of *Plasmodium falciparum* is shown in Fig. 1 along with corresponding changes in ribosomal type. The A-type gene, predominantly expressed in the asexual blood-stage parasite, has features typical of a eukaryotic ribosome. The S-type RNA, however, is structurally distinct in regions known to be the target for antibiotics. This is important in the development of gametocytes and the mosquito forms of the parasite and may therefore provide useful targets for the development of antimalarial drugs. This distinction is critical in regard to the treatment of malaria. The goal is to cure both the symptoms of the disease and to prevent subsequent trans-

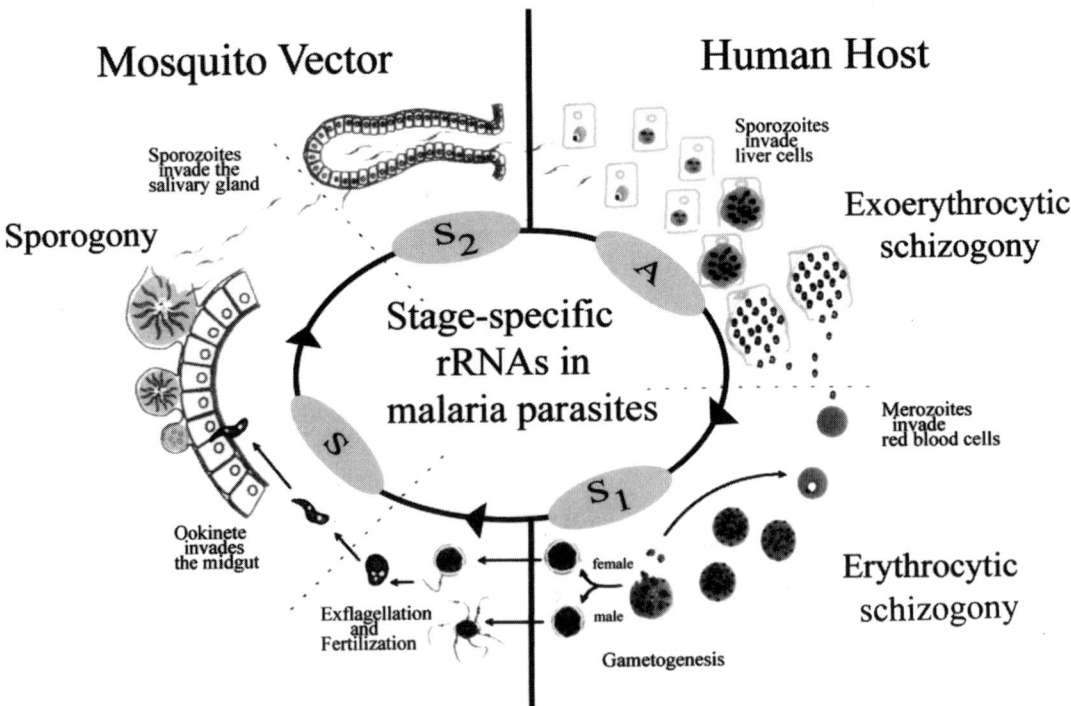

FIGURE 1 Life cycle of *P. falciparum*. Changes in rRNA transcription correlate with the development cycle. Active transcription periods for three different rRNA gene units are indicated.

mission; a priori, it would seem that curing the disease would also prevent transmission, but this is not the case. Drugs that bring symptomatic relief often have little effect on the mature sexual stages of the parasite. This is because the stages are distinctly different forms and respond differently to both drugs and the immune system. For example, mature gametocytes are arrested in the S phase of their cycle, long lived and presumably nearly quiescent. The result of this is that the parasite can be transmitted long after the patient recovers from the illness.

ANALYSIS OF THE TARGET

Our most detailed knowledge of ribosomal chemistry comes from the study of prokaryotic ribosomes. Given the degree of universal conservation seen among organisms with regard to peptide-bond formation, we include information about the prokaryotic systems. We will also review specific facts about the ribosomes of *Plasmodium* species as they relate to drug interaction.

RIBOSOMES AND REGULATION

Within the ribosomal complex lies the machinery that decodes information from the messenger RNA and catalyzes the ordered assembly of amino acids into proteins. The component parts of this complex are remarkably similar among organisms. In fact, the translational apparatus of one species often produces protein from the mRNA of a range of distantly related species. The structural similarity among eukaryotic ribosomes makes it difficult to design drugs that distinguish host from parasite, at least to an extent useful for clinical treatment. Malaria parasites may be the exception (see below).

Ribosomes are found in both free and membrane-bound forms. The type of protein being synthesized tends to segregate according to where it is produced, free or attached. Free ribosomes synthesize proteins that are retained in the cell, while ribosomes bound to the cytosolic side of the membranes constitute the rough endoplasmic reticulum (ER). After synthesis, ribosomes bound to the endoplasmic reticulum are mainly involved in synthesis of secretory and membrane proteins. Proteins synthesized on the rough ER are inserted through the cisternal membrane and released into the ER lumen. These proteins become part of membranes, are packaged into vesicles for storage in the cytoplasm, or are exported to the cell exterior.

Regulation of ribosome production is essential to all cells. The number of ribosomes present in a cell is directly related to the protein-synthesizing activity and to the size of the cell. In an actively growing bacterium, there may be roughly 20,000 ribosomes per cell (Stainer et al., 1988). An actively growing eukaryotic cell may require 10 million ribosomes to meet its requirements for protein synthesis, with criteria varying in relation to factors such as state of differentiation, available nutrients, etc. (Alberts et al., 1994). Feedback regulation that balances the needs of the cell with its situation is also essential. Cell growth and function require precise regulation of protein synthesis, including the means to speed up, slow down, or remain stationary in response to its environment, because any of these states can be advantageous in specific situations. A number of strategies for ribosome synthesis regulation will be briefly described.

Nutrients are primary regulators of signaling pathways that control transcription and translation in all organisms. One mechanism of control, the stringent response, occurs in both eukaryotic and prokaryotic organisms. External signals (e.g., glucose or amino acid concentrations) are sensed by receptors on the cell surface, leading to adaptive changes to the cell's interior; in this case, response is related directly to the available carbon source. In another regulator of rRNA synthesis, growth rate-dependent control, the actual growth rate of the cell is a key regulating factor. This may be achieved by the cell's capacity to sense nucleoside triphosphate (NTP) levels in the cytosol (Green and Noller, 1997). A number of eukaryotic organisms respond to their environment with a combination of stringent and growth-rate responses.

Mechanistically, the control of ribosomal RNA biosynthesis can sometimes be traced to the level of methylation in the upstream promoter region of the rRNA gene cassette—as is the case with human rRNA gene, where hypomethylation of the promoter region of tumor cells leads to increased transcription of rRNA genes (Ghoshal et al., 2004). Thus, there is a need to explore the relationship between methylation and regulation of gene expression, especially with regard to the switch from A- to S-type expression.

Synthesis of ribosomal proteins (r-proteins) is coordinated with synthesis of rRNA by balancing the transcription of ribosomal RNA and the mRNA of ribosomal protein; i.e., ribosomal protein production can be controlled by a feedback mechanism at the translational level. Excess ribosomal proteins act to self-regulate translation of their own mRNA. The translation is only repressed when the level of r-protein exceeds that of rRNA, as is the case with S4 and S15 ribosomal proteins (Baker and Draper, 1995; Ehresmann et al., 1995). In eukaryotic organisms, translational repression can also be achieved by establishing a balance between phosphorylation and dephosphorylation of initiation factors (IFs).

Further discrimination during protein production is achieved by localizing mRNA to particular sites in the cell. Both temporal and locational variables can affect the production of individual proteins. Selection of mRNA subset for producing protein is often achieved by localizing mRNA to particular sites in the cell. This can serve to concentrate the expressed protein to a particular area of the cell, prevent it from being expressed elsewhere, or simply store it until it is to be used.

INTERACTION BETWEEN SMALL AND LARGE SUBUNITS

Positioning of ribosomal subunits with respect to one another is key to forming functionally active ribosomes. The crystal structure of *Thermus thermophilus* and *Escherichia coli* 70S ribosome, along with biochemical data, suggests that ~80% of intersubunit bridges are contributed by RNA-RNA interactions. The intersubunit regions involving RNA-RNA bridges are static, maintaining the stability of subunit association and showing only a small ($<3Å$) local rearrangement (Yusupov et al., 2001; Gao et al., 2003). The RNA-RNA interactions are the actual moieties that perform the chemical reactions of peptide synthesis. These intersubunit bridges are directly involved in ribozyme action (Color Plate 12A). Peripheral interactions between protein and RNA or between different proteins participate in peptide synthesis as well and are mainly involved in conformational changes that are essential in the logistics of peptide formation. The precision of the bridge restricts conformational mobility to those motions that permit sequential translocation of tRNA from A- to P- to E-sites (Color Plate 12B and C). Orchestrated conformational change over this region is essential for progression through the various protein synthesis phases.

PEPTIDE-BOND FORMATION

When the two subunits are aligned appropriately, a channel is formed through which the messenger RNA is ratcheted, and the process of peptide elongation commences. Crystallographic studies of the 50S subunit have yielded sufficient information for visual representation of the process (Color Plate 13). The channel is depicted in the diagram of a split large subunit (Color Plate 13a) running through the center of the ribosome. The diagram in Color Plate 13a is based upon crystallographic studies of the ribosome interacting with its substrates under conditions where peptide bond formation occurs in the crystals. Also shown in Color Plate 13 is the peptidyl transferase center (PTC), which is located within the large subunit. It is composed entirely of 23S rRNA and is an important drug target. Drugs interacting with this site work either by inhibiting peptide bond formation or by blocking the passage of growing polypeptide.

RIBOSOME AND DRUG INTERACTION

The protein biosynthesis machinery is essential to all living cells and is one of the major targets for antibiotics, which may be divided in two groups. Group 1 includes drugs that interact with the small subunit of ribosome. The crystal structure of the 30S subunit of *T. thermophilus* has yielded functional insight (Wimberley et al., 2000; Carter et al., 2000). The 30S subunit consists of 16S rRNA and ~20 proteins. It plays a crucial role in decoding mRNA by monitoring the base pairing between the codon on mRNA and the anticodon on tRNA. The structure shows that r-proteins are concentrated in the top, sides, and back of the 30S subunit and that the central core of 30S is mainly composed of 16S rRNA. Most of the antibiotics that bind to the small subunit act either by affecting the selection of aminoacyl-tRNA by the ribosome or by inducing errors in the decoding of the mRNA (Table 1). Group 2 consists of drugs that interact with the large subunit RNA, specifically the PTC, and possibly interfering with the A- or P-site, peptide-bond formation, or blocking the peptide exit tunnel (Table 2). Color Plate 14 displays putative interactions between drugs and the crystal structure of both the PTC and the peptide exit tunnel in the ribosomal large subunit (Hansen et al., 2003). The polypeptide exit tunnel of the large subunit is a passage about 100 Å long that begins immediately below the PTC and ends at the back of the large subunit (Color Plate 13). The wall in the first third of its length is formed by portions of ribosomal proteins L4 and L22, which are highly conserved. The wall is a mosaic of hydrophobic and hydrophilic patches that allows the passage of nascent peptides of all sequence types. A number of antibiotics bind at the exit tunnel and block the passage of nascent peptide.

Also included in group 2 are the thiazole peptide antibiotics, which bind to the L11-binding domain of 23S rRNA. They act by destabilizing the binding of elongation factor G (EF-G) and initiation factor 2 (IF-2) to the GTPase center and prevent the conformational changes required for GTPase activity. Although eukaryotes are usually not sensitive to thiostrepton, *P. falciparum* growth is severely inhibited in culture (Rogers et al., 1997), presumably by interaction with the apicoplast. Thiostrepton has some limited topical applications and to date has not been very useful clinically. The problems with these drugs are their large size, solubility in aqueous solution, and cellular uptake. There is a need to explore the derivatives of thiostrepton or newer thiazole antibiotics that are soluble, can be readily taken up by the cells, and have good pharmacokinetic properties.

POTENTIALLY VULNERABLE SITES IN *PLASMODIUM* SPECIES

Structural analysis of *P. falciparum* A and S-type RNA reveals differences in the GTPase region of the molecule (Fig. 2) (Rogers et al., 1996). While the sequence of the A-type 28S RNA resembles that found in other eukaryotes, the S-type contains a compensatory base-pair change in the loop that joins two helices. This S-type variation is not seen in other eukaryotes (Gutell et al., 1993). It has been demonstrated that the *P. falciparum* S-type GTPase cannot complement the yeast rRNA gene unit, although the A-type gene can (Velichutina et al., 1998). Clearly, there are functional differences between the ribosomes carrying the different GTPase sites, but the significance of this finding is unknown. We have investigated eight different species of *Plasmodium* and found similar variations between the GTPase site of A and S genes in each. The exception is in the model system *Plasmodium berghei*, where both A and S genes have the same GTPase site. Whether this is reflective of *P. berghei* in nature or the result of continued laboratory passage is not known.

In experiments designed to determine whether each rRNA gene was essential to de-

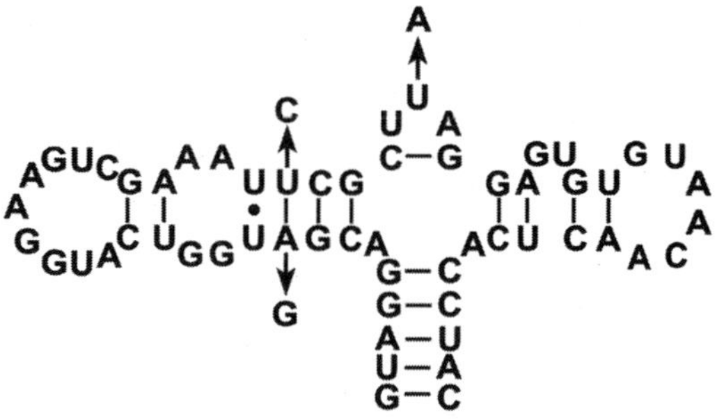

FIGURE 2 Two-dimensional structure of the *P. falciparum* rRNA GTPase site, showing the location of sequence differences between the A-type and S-type RNAs.

TABLE 1 Antibiotics interacting with 16S rRNA

Class	Antibiotic or drug	Binding site	Mode of action	Clinical use	Reference(s)
Aminoglycoside[a]	Neomycin	Selectively binds to A site internal loop within the deep groove of helix H44 of 16S rRNA	Interferes with two conformationally flexible adenine residues (A1492, A1493) involved in the selection of cognate aminoacyl-tRNA during translation	Used in combination for treatment of protozoan infections	Brodersen et al., 2000
	Hygromycin B	Binds to A site in 30S subunit	Sequesters A site-bound tRNA by restricting a conformational change in H44 required for ribosome translocation	Universal inhibitor	Carter et al., 2000
	Streptomycin	Binds tightly to 16S rRNA in small subunit forming contacts with phosphate backbone of rRNA helix: H27, H18, H44, and S12 protein	Interferes with initial tRNA selection and proofreading by locking the ram state	Used for bacterial infections; resistance reported for mycobacterium	
	Spectinomycin	Binds to shallow groove at one end of H34 in the 3′ major domain of 16S rRNA; makes single contact with 2′OH and makes hydrogen bonds to a number of bases	Interacts with RNA moieties and inhibits EF-G-catalyzed translocation of the peptidyl-tRNA from A to P site	Universal inhibitor	Carter et al., 2000
	Apramycin	Binds to 16S rRNA in the A site	Interferes with the flexible adenine A1492 and A1493; locks these bases within RNA helix		
Nonaminoglycoside	Pactomycin	Interacts with the hairpin loop H23b and H24b or 16S rRNA close to the tRNA-binding cleft	May act as a dinucleotide mimic; partially displaces mRNA from the ribosomal E site	Universal inhibitor	Brodersen et al., 2000
	Tetracycline	Interacts with 16S rRNA in the shallow groove of helix H31	Interferes with aminoacyl-tRNA binding to A site	Frequently used for bacterial infections; no resistance reported when used as antimalarial	Brodersen et al., 2000; Pioletti et al., 2001
	Viomycin (tuberactinomycins)	Binds to the ribosomal subunit interface in proximity to the A site contacting the central pseudoknot of 16S rRNA	Stabilizes subunit cohesion and inhibits translocation by sequestration of peptidyl-tRNA in the A site	Successfully used against tuberculosis	

[a] Aminoglycosides inhibit the protein synthesis by preventing conformational change in the small-subunit moieties to A site in 16S rRNA (small subunit).

TABLE 2 Antibiotics interacting with 23S rRNA

Antibiotic class; action	Antibiotics	Binding site	Mode of action	Clinical use	Reference
Structural analogs of the 3' terminus of aminoacyl-tRNA	Sparsomycin	Binds to ribosome in presence of peptidyl-tRNA in the P-site	Stabilizes peptidyl tRNA in P-site and prevents it from proceeding to the peptidyl transfer step	Universal inhibitor	Lazaro et al., 1991
	Puromycin	Structural and functional analog of aminoacyl-tRNA; binds to A-site	Terminates peptide elongation	Universal inhibitor	
	Anisomycin	Interacts with 23S rRNA in PTC	Inhibits the peptide bond formation by sterical interference with aminoacyl-tRNA 3' acceptor	Universal inhibitor	
	Chloramphenicol	Binds to peptidyl transferase loop in 23S rRNA and contacts exclusively with RNA residues; displaces Mg^{2+} ions from native RNA on binding	Blocks peptidyl transferase activity by sterical interference with the amino-acyl moiety in the A-site; prevents formation of transition state during peptide bond formation	Only inhibits bacterial translation; low level of toxicity in humans	Schlunzen et al., 2001
Macrolide; binds to 23S rRNA at the entrance of polypeptide exit tunnel immediately adjacent to PTC	Azithromycin, erythromycin, tylosin, spiramycin, carbomycin A	Binds in the peptide tunnel of the large subunit, immediately adjacent to PTC	Blocks passage of nascent polypeptide through exit tunnel	Widely used for bacterial infection	Hansen et al., 2002
Ketolides; semisynthetic derivatives of erythromycin, the first macrolide in clinical use	Telithromycin	Macrolactone ring binds to domain V of 23S rRNA through hydrogen bonds involving its 3-keto group and via hydrophobic interaction; interacts with 23S rRNA domain II	Blocks ribosomal exit tunnel	Elevated antimicrobial action against several macrolide-resistant strains	Berisio et al., 2003

(*Continued*)

TABLE 2 (Continued)

Antibiotic class; action	Antibiotics	Binding site	Mode of action	Clinical use	Reference
Lincosamide; binds to peptidyl transferase loop in 23S rRNA and contacts exclusively with RNA residues; displaces Mg^{2+} ions from native RNA on binding; interferes with positioning of amino acyl group at A-site-bound tRNA and peptidyl group at P-site tRNA	Clindamycin	Used as antimalarial in combination with quinine and fosmidomycin but not as monotherapy; clindamycin in combination safe in children and pregnant women	Not effective for treatment of vivax malaria; relapse was observed in all patients	An important candidate when drug resistance is increasing against quinine, sulfonamide-pyrimethamine combination and artesunate, especially in areas of endemicity	
Thiazole antibiotics; bind to L11-binding domain of 23S rRNA (GTPase center of LSU)	Thiostrepton	Cyclic peptide moiety binds at interface of domain II of 23S rRNA and L11 protein; linear extension of thiostrepton interacts with RNA in narrow deep pocket formed between two-hairpin loop of L11-binding domain	Acts by destabilizing the binding of EF-G and IF-2; prevents conformational change required for GTPase activity	Large in size, less soluble; need to explore its structural analogues and derivatives	
	Micrococcin	Binds to L11-binding domain of 23S rRNA (GTPase center of LSU)	Acts by destabilizing the binding of EF-G and IF-2		
	Nocathiacins	Binds to 23S rRNA of the 50S subunit at the same site as L11 protein	Acts by preventing the conformational transition that occurs at 23S-L11 interaction and results in stalling of translation affecting the elongation step of bacterial protein synthesis	More soluble, active against resistant bacteria, potent antimicrobial activity	Pucci et al., 2004

			Cocito et al., 1997
Streptogramins; bind to 50S subunit or 70S free ribosomes only	Streptogramins A	Blocks the substrate attachment to both A and P sites	Inhibitor of bacterial protein synthesis; types A and B have synergistic effect: bacteriostatic if given alone, but bactericidal in combination
	Streptogramins B	Blocks peptide bond formation	
Oligosaccharide antibiotics	Eveninomycin	Binds to hairpin loop H91 and H89 of 23S rRNA extending from multijunctional PTC	Inhibits the translation by preventing interaction of IF-2 with LSU

velopment, van Spaendonk et al. (2001) disrupted the (S-type) rRNA gene units of *P. berghei* and analyzed the resulting phenotype. They demonstrated that the deletion of both S-type rRNA gene units did not prevent full development of the parasite but that there was retardation of oocyst growth. The effect of these genetic changes on the parasite survival in nature is unknown.

PSEUDOKNOTS

The pseudoknot of the A-type small subunit RNA is typical of eukaryotic ribosomes, while no similar structure exists in the S-type rRNA. RNA pseudoknots are formed when nucleotides in a hairpin loop bind to complementary positions within a second stem loop (Fig. 3). The functional role of the pseudoknot varies considerably depending on its source. In 5′ untranslated regions (UTR) of mRNA near the ribosome-binding site, they are often involved in autoregulation (Baker and Draper, 1995; Ehresmann et al., 1995); pseudoknots are also involved in the folding of catalytic sites of ribozymes. In the case of 16S rRNA, a pseudoknot forms part of the core structure and is central to in vivo translational control.

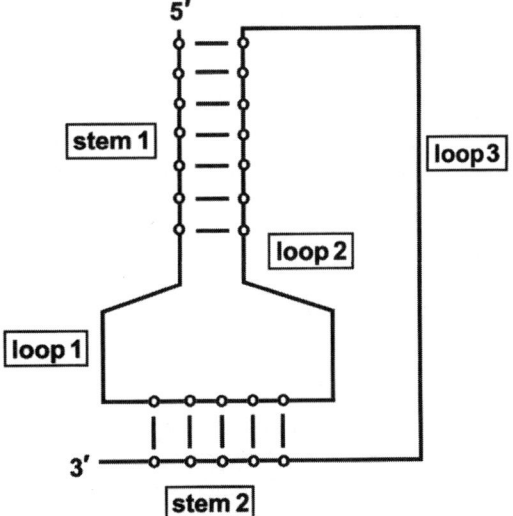

FIGURE 3 Schematic presentation of the classical pseudoknot configuration.

Some antibiotics interact with pseudoknots, impacting their inhibitory effect. One example is viomycin, a peptide antibiotic that has an affinity for some forms of pseudoknot. It is shown to bind to *E. coli* ribosome and to prevent ribosome subunit dissociation and translocation. Viomycin, also known as tuberactinomycin, is being successfully used against tuberculosis.

ORGANELLAR RIBOSOMES

Apicomplexa species contain both a mitochondrion and another plastid, the apicoplast, in which protein synthesis occurs (see Chapter 14). These can also be antimalarial drug targets. Protein synthesis in the mitochondrion is a fascinating but as-yet-unsolved mystery. No ribosomal RNA has been associated with this organelle, although the functional regions necessary for peptide-bond formation appear to be encoded in noncontiguous stretches of DNA. The apicoplast has its own genome that encodes two rRNA gene units and several proteins. A number of drugs that are effective against malaria are thought to interfere with the function of the organelles. This phenomenon has been investigated extensively in the laboratories of David Roos and Elmer Pfefferkorn. Using *Toxoplasma* as a model, much has been learned. Treatment of *Toxoplasma* infection with clindamycin has no effect on intracellular survival of the treated parasite, lysis of the host cell, extracellular survival of the parasite, or invasion of the next host cell. The treatment interferes with formation of the parasitophorous vacuole at the point of reinvasion, regardless of whether the clindamycin is present or not. One wonders whether this delayed death effect will affect the development of resistance to this class of drugs.

CONCLUSIONS

One primary factor in the resurgence of malaria worldwide has been the development of drug resistance to such first-line drugs as chloroquine. The development and spread of drug resistance are unquestionably tied to the population dynamics of parasite, host, and vector. One event in the development of drug resistance is a mutation(s) in the pathogen that allows its survival in the presence of the drug. We assume that this resistant phenotype carries with it some loss in fitness when the drug is removed. Next, and perhaps more frightening, genetic changes allow the pathogen to compete with the wild-type phenotype once drug pressure is removed. This compensatory mutation ensures that the resistant phenotype will remain in the population and spread to other areas. Once this happens, the drug is lost to our repertoire.

The search for drug replacements is an ongoing task in malariology as it is in bacteriology and virology. The use of ribosome-based drugs on eukaryotic pathogens is often impossible because of the similarity between the ribosomes of host and pathogen. It is difficult to find an appropriate drug that will kill the invader and leave the host free from side effects. *Plasmodium* species are likely to be an exception, and drugs may serve to treat the disease and/or its transmission. Further, targeting ribosomal RNA may provide drugs that permit broad scale and extended use, as has been the case in combating bacterial infections. There are several reasons for this. (i) The drugs affect the very heart of the organism, and biological redundancy will not be a factor in the evolution of resistance. (ii) The small copy number of *Plasmodium* genes encoding these RNAs would make mutations conferring resistance carry with them a very significant loss in fitness once the drug was removed. (iii) Drugs that target the gametocyte would interfere with a bottleneck in the cycle rather than a point where the parasite number is expanding exponentially.

REFERENCES

Alberts, B., D. Bray, J. Lewis, M. Raff, K. Roberts, and J. D. Watson. 1994. *Molecular Biology of the Cell*, 3rd ed., p. 335–399. Garland Publishing, Inc., New York, N.Y.

Baker, A. M., and D. E. Draper. 1995. Messenger RNA recognition by fragments of ribosomal protein S4. *J. Biol. Chem.* **270:**22939–22945.

Berisio, R., J. Harms, F. Schluenzen, R. Zarivach, H. A. Hansen, P. Fucini, and A. Yonath. 2003. Structural insight into the antibiotic action of telithromycin against resistant mutants. *J. Bacteriol.* **185:**4276–4279.

Brodersen, D. E., W. M. Clemons, Jr., A. P. Carter, R. J. Morgan-Warren, B. T. Wimberly, and V. Ramakrishnan. 2000. The structural basis for the action of the antibiotics tetracycline, pactamycin, and hygromycin B on the 30S ribosomal subunit. *Cell* **103:**1143–1154.

Carter, A. P., W. M. Clemons, D. E. Brodersen, R. J. Morgan-Warren, B. T. Wimberly, and V. Ramakrishnan. 2000. Functional insights from the structure of the 30S ribosomal subunit and its interactions with antibiotics. *Nature* **407:**340–348.

Cocito, C., M. Di Giambattista, E. Nyssen, and P. Vannuffel. 1997. Inhibition of protein synthesis by streptogramins and related antibiotics. *J. Antimicrob. Chemother.* **39**(Suppl A)**:**7–13.

Dame, J. B., and T. F. McCutchan. 1983. The four ribosomal DNA units of the malaria parasite *Plasmodium berghei*: identification, restriction map, and copy number analysis. *J. Biol. Chem.* **258:**6984–6990.

Ehresmann, C., C. Philippe, E. Westhof, L. Benard, C. Portier, and B. Ehresmann. 1995. A pseudoknot is required for efficient translational initiation and regulation of the *Escherichia coli* rpsO gene coding for ribosomal protein S15. *Biochem. Cell Biol.* **73:**1131–1140.

Gao, H., J. Sengupta, M. Valle, A. Korostelev, N. Eswar, S. M. Stagg, P. Van Roey, R. K. Agrawal, S. C. Harvey, A. Sali, M. S. Chapman, and J. Frank. 2003. Study of the structural dynamics of the *E. coli* 70S ribosome using real-space refinement. *Cell* **113:**789–801.

Ghoshal, K., S. Majumder, J. Datta, T. Motiwala, S. Bai, S. M. Sharma, W. Frankel, and S. T. Jacob. 2004. Role of human ribosomal RNA (rRNA) promoter methylation and of methyl-CpG-binding protein MBD2 in the suppression of rRNA gene expression. *J. Biol. Chem.* **279:**6783–6793.

Green, R., and H. F. Noller. 1997. Ribosome and translation. *Annu. Rev. Biochem.* **66:**679–716.

Gutell, R. R., M. W. Gray, and M. N. Schnare. 1993. A compilation of large subunit (23S and 23S-like) ribosomal RNA structures. *Nucleic Acids Res.* **21:**3055–3074.

Hansen, J. L., P. B. Moore, and T. A. Steitz. 2003. Structures of five antibiotics bound at the peptidyl transferase center of the large ribosomal subunit. *J. Mol. Biol.* **330:**1061–1075.

Langsley, G., J. E. Hyde, M. Goman, and J. G. Scaife. 1983. Cloning and characterisation of the rRNA genes from the human malaria parasite *Plasmodium falciparum*. *Nucleic Acids Res.* **11:**8703–8717.

Lazaro, E., L. A. van den Broek, A. San Felix, H. C. Ottenheijm, and J. P. Ballesta. 1991. Biochemical and kinetic characteristics of the interaction of the antitumor antibiotic sparsomycin with prokaryotic and eukaryotic ribosomes. *Biochemistry* **30:**9642–9648.

Li, J., W. E. Collins, R. A. Wirtz, D. Rathore, A. Lal, and T. F. McCutchan. 2000. Synergy between *Plasmodium vivax* and mosquito vectors can result in the genetic isolation of parasite populations. *Emerg. Infect. Dis.* **7:**35–42.

Mitchell, P., M. Osswald, and R. Brimacombe. 1992. Identification of intermolecular RNA crosslinks at the subunit interface of the *Escherichia coli* ribosome. *Biochemistry* **31:**3004–3011.

Pioletti, M., F. Schlunzen, J. Harms, R. Zarivach, M. Gluhmann, H. Avila, A. Bashan, H. Bartels, T. Auerbach, C. Jacobi, T. Hartsch, A. Yonath, and F. Franceschi. 2001. Crystal structures of complexes of the small ribosomal subunit with tetracycline, edeine and IF3. *EMBO J.* **20:**1829–1839.

Pucci, M. J., J. J. Bronson, J. F. Barrett, K. L. DenBleyker, L. F. Discotto, J. C. Fung-Tomc, and Y. Ueda. 2004. Antimicrobial evaluation of nocathiacins, a thiazole peptide class of antibiotics. *Antimicrob. Agents Chemother.* **48:**3697–3701.

Rogers, M. J., Y. V. Bukhman, T. F. McCutchan, and D. E. Draper. 1997. Interaction of thiostrepton with an RNA fragment derived from the plastid-encoded ribosomal RNA of the malaria parasite. *RNA* **3:**815–820.

Rogers, M. J., R. R. Gutell, S. H. Damberger, J. Li, G. A. McConkey, A. P. Waters, and T. F. McCutchan. 1996. Structural features of the large subunit rRNA expressed in *Plasmodium falciparum* sporozoites that distinguish it from the asexually expressed large subunit rRNA. *RNA* **2:**134–145.

Rogers, M. J., J. Li, and T. F. McCutchan. 1998. The Plasmodium rRNA genes: developmental regulation and drug target, p. 203–217. *In* I. W. Sherman (ed.), *Malaria: Parasite Biology, Pathogenesis, and Protection.* ASM Press, Washington, D.C.

Rogers, M. J., G. A. McConkey, J. Li, and T. F. McCutchan. 1995. The ribosomal DNA loci in *Plasmodium falciparum* accumulate mutations independently. *J. Mol. Biol.* **254:**881–891.

Schlunzen, F., R. Zarivach, J. Harms, A. Bashan, A. Tocilj, R. Albrecht, A. Yonath, and F. Franceschi. 2001. Structural basis for the interaction of antibiotics with the peptidyl transferase centre in eubacteria. *Nature* **413:**814–821.

Schmeing, T. M., A. C. Seila, J. L. Hansen, B. Freeborn, J. K. Soukup, S. A. Scaringe, S. A. Strobel, P. B. Moore, and T. A. Steitz. 2002. A pretranslocational intermediate in protein synthesis observed in crystals of enzymatically active 50S subunits. *Nat. Struct. Biol.* **9:**225–230.

Stanier, R. Y., J. L. Ingraham, M. L. Wheelis, and P. R. Painter. 1988. *General Microbiology*, p. 102–144. Macmillan Education Ltd., London, United Kingdom.

Steitz, T. A., and P. B. Moore. 2003. RNA, the first macromolecular catalyst: the ribosome is a ribozyme. *Trends Biochem. Sci.* **28:**411–418.

van Spaendonk, R. M. L., J. Ramesar, A. Wigcheren, W. Eling, A. Beetsma, G. J. Gemert, J. Hooghof, C. J. Janse, and A. P. Waters. 2001. Functional equivalence of structurally distinct ribosomes in the malaria parasite *Plasmodium berghei*. *J. Biol. Chem.* **276:**22638–22647.

Velichutina, I. V., M. J. Rogers, T. F. McCutchan, and S. W. Liebman. 1998. Chimeric rRNAs containing the GTPase centers of the developmentally regulated ribosomal rRNAs of *Plasmodium falciparum* are functionally distinct. *RNA* **4:**594–602.

Waters, A. P., W. White, and T. F. McCutchan. 1995. The structure of the large subunit rRNA expressed in blood stages of *Plasmodium falciparum*. *Mol. Biochem. Parasitol.* **72:**227–237.

Wimberly, B. T., D. E. Brodersen, W. M. Clemons, Jr., R. J. Morgan-Warren, A. P. Carter, C. Vonrhein, T. Hartsch, and V. Ramakrishnan. 2000. Structure of the 30S ribosomal subunit. *Nature* **407:**327–339.

Yusupov, M. M., G. Z. Yusupova, A. Baucom, K. Lieberman, T. N. Earnest, J. H. Cate, and H. F. Noller. 2001. Crystal structure of the ribosome at 5.5 A resolution. *Science* **292:**883–896.

OXIDATIVE STRESS AND ANTIOXIDANT DEFENSE IN MALARIAL PARASITES

Katja Becker, Sasa Koncarevic, and Nicholas H. Hunt

19

Malarial parasites are continuously exposed to oxidative and nitrosative stress. Such stress might be exogenously produced by the immune response of the host as well as endogenously by the high metabolic rate of *Plasmodium* and the hemoglobin degradation within the parasite. Oxidants derived from the diet are another potential source of exogenous oxidative stress. In recent years, *Plasmodium* has been shown to possess a whole range of antioxidant defense mechanisms, namely a complete glutathione (GSH) system comprising NADPH, highly active glutathione reductase (GR), glutathione, and different glutaredoxin-like proteins (Becker et al., 2003b; Becker et al., 2004) as well as a functional glutathione-dependent glyoxalase system and a glutathione S-transferase (GST) with peroxidase activity. In addition, a complete thioredoxin (Trx) system comprising NADPH, thioredoxin reductase (TrxR), different thioredoxin-like proteins, and thioredoxin-dependent peroxidases (TPx) has been characterized (Rahlfs and Becker, 2001; Muller, 2004), and two functional superoxide dismutases (SODs), as well as two lipoamide dehydrogenase-like proteins, are present.

As indicated by human glucose 6-phosphate dehydrogenase (G6PD) deficiency, limited availability of reducing equivalents in the form of NADPH confers protection from malaria, which indicates that malarial parasites are highly susceptible to oxidative stress. Furthermore, the antioxidant enzymes catalase and glutathione peroxidase seem to be absent in the parasite. This constellation offers a great potential for the development of urgently required novel chemotherapeutic agents that act by disturbing the redox equilibrium of the parasite (Becker et al., 2004; Turrens, 2004). In addition, the redox metabolism may be involved in the pathology of malaria and has been demonstrated to play a major role in the action of and resistance to clinically used antimalarial drugs.

In this chapter, we summarize the currently available knowledge on sources of oxidative and nitrosative stress in malarial parasites, the different available detoxification pathways, and the impact on mechanisms of drug action.

Katja Becker and Sasa Koncarevic, Interdisciplinary Research Center, Heinrich-Buff-Ring 26-32, Justus-Liebig-University, D-35392 Giessen, Germany. *Nicholas H. Hunt,* Institute for Biomedical Research, Department of Pathology (D06), University of Sydney, New South Wales 2006, Australia.

Molecular Approaches to Malaria, Edited by Irwin W. Sherman
© 2005 ASM Press, Washington, D.C.

SOURCES OF OXIDATIVE STRESS IN MALARIAL PARASITES

Plasmodium-infected red blood cells (*Plasmodium* IRBCs) appear to be under exogenous and endogenous oxidative stress. H_2O_2 has been shown to be produced in *Plasmodium berghei* IRBCs and $O_2^{\bullet-}$ in *Plasmodium vinckei* IRBCs, and lipid peroxidation has been demonstrated to be increased in *Plasmodium falciparum* (Pf). The major source of this stress appears to be the spontaneous oxidation and degradation of ingested hemoglobin in the acidic environment of the parasite's food vacuole (Famin and Ginsburg, 2003; Becker et al., 2004).

Phagocytosis

Malaria parasites induce oxidative stress in their host red blood cell. In the membrane of *P. falciparum*-parasitized cells, increasing amounts of hemichromes and band 3 aggregates have been demonstrated. These changes account for the observed deposition of band 3-specific autologous immunoglobulin G and for the consequent deposition of fragments of complement C3c. Such alterations of the surface of the IRBCs are sufficient for the recognition of the cells by macrophages, and this would be predicted to induce both their phagocytosis and the production of reactive oxygen species (ROS) on the basis of phagocyte NADPH oxidase activity. In both human and murine malaria infections, circulating phagocytes are "primed" for an augmented oxidative burst (production of ROS) when stimulated (Hunt and Stocker, 1990). This, together with the sensitivity of IRBC to oxidative damage, suggests that phagocyte-derived ROS might be an important component of host antimalarial defenses. However, the use of genetically manipulated mice whose phagocytes lack the ability to produce ROS strongly suggests that this may not be the case (Harada et al., 2001; M. S. Potter, A. Mitchell, W. B. Cowden, L. Sanni, M. C. Dinauer, A. de Haan and N. H. Hunt, manuscript submitted for publication). Studies using inducible nitric oxide (NO) synthase gene knockout mice also cast doubt on a role for phagocyte-derived nitric oxide in the host immune response against malaria (Favre et al., 1999).

Similar membrane alterations and subsequent phagocytosis occur in normally senescent RBCs. Interestingly, IRBCs bearing a malaria-protective mutation, such as those causing G6PD deficiency, thalassemia, or sickle cell disease, display all the membrane alterations described above and are recognized by macrophages already at the ring stage. Enhanced phagocytosis of ring forms developing in pathologic RBCs might thus explain the protective effects of these genetic alterations (Becker et al., 2004).

Hemoglobin Degradation

The mature human red blood cell contains about 20 mM hemoglobin monomers. Malaria parasites feed by degrading at least 75% of this hemoglobin in an acidic food vacuole (estimated pH 5.2). Hemoglobin degradation represents an essential source of amino acids for the parasite, creates sufficient space for parasite growth, and helps maintain the osmotic integrity of the infected cell (Lew et al., 2003). However, free heme (FP), which is released in large quantities, must be detoxified by the parasite (heme [ferroprotoporphyrin IX] contains Fe^{2+}, whereas in hemin [ferriprotoporphyrin IX], the iron is oxidized to Fe^{3+}) (Tilley et al., 2001). In the acidic environment, ferroprotoporphyrin IX is oxidized to ferriprotoporphyrin IX, a reaction that produces superoxide radicals, which are converted to H_2O_2 and O_2 by superoxide dismutase. Both FP and H_2O_2 are toxic molecules that the parasite needs to destroy or neutralize.

Most of the heme is sequestered into hemozoin, a crystalline form of FP, which leads to the formation of the malaria pigment. A smaller percentage of the FP escapes the crystallization process and diffuses into the parasite cytosol, reaching concentrations of up to 100 µM in IRBCs. The redistribution of even a small fraction of the 20 mM cellular load of FP is likely to cause damage to host proteins and membranes. Free FP inhibits parasite enzymes (Famin and Ginsburg, 2003; Campanale et al., 2003), lyses erythrocytes, and causes substantial redox damage (Tilley et al., 2001; Becker et al., 2004; Muller, 2004).

Thus, the parasite possesses additional mechanisms for dealing with free FP that escapes mineralization into hemozoin. One possible route is the reaction with GSH, which leads to the release of free iron that might, however, produce oxidative stress in the Fenton reaction. Interference with this GSH-dependent FP breakdown has been suggested as a mechanism of action of antimalarial drugs (Famin and Ginsburg, 2003). A second possibility is the binding of FP to proteins like the histidine-rich proteins, HRP2 and HRP3, which contain 60 to 70% histidine and alanine residues. HRP2 has been shown to bind FP with a stoichiometry of up to 1:50, and HRP2 enhances the conversion of FP to β-hematin in vitro. Also, glutathione S-transferase, which is present in the cytosol of *Plasmodium* in micromolar concentrations, has been shown to serve as a ligandin for free heme by binding it in an uncompetitive manner (Tilley et al., 2001; Becker et al., 2004; Muller, 2004).

Dietary Oxidants

J. B. S. Haldane first proposed the insightful concept that malaria might be a selection pressure that has ensured the survival of several genetic traits that would otherwise appear to be disadvantageous to humans, and this concept has been extended and debated by many other authors (Hunt and Stocker, 1990). One of the most thoroughly investigated genetic traits in this context is G6PD deficiency. As discussed below, the link may be the inability of erythrocytes from G6PD-deficient individuals to produce sufficient reducing equivalents to maintain the reduced intracellular environment that seems to favor growth of the malaria parasite.

An additional factor has been suggested to be the occurrence in some human diets of agents capable of inducing oxidative stress in G6PD-deficient RBC. Fava beans contain precursors to divicine and isouramil, molecules that redox cycle inside RBC and not only induce oxidative stress (Golenser et al., 1983) but also kill malaria parasites (Clark et al., 1984). Fava beans are a staple of Mediterranean diets, and G6PD deficiency is widespread in that area, as malaria used to be. Whether equivalent oxidative stressors of dietary origin exist in other regions of the world has not been established.

REDOX METABOLISM AND PATHOPHYSIOLOGY

In children with malaria, plasma lipid peroxides are increased, especially in those with concomitant riboflavin deficiency. Erythrocyte lipid peroxidation is also increased, and erythrocyte GSH, catalase, and tocopherol are significantly lower in malaria patients than in control subjects. Altered redox metabolism at the level of the host cell (especially endothelial cells) may play a part in disease manifestations, and enhanced oxidative stress on erythrocytes may contribute to hemolysis and the development of anemia (Dondorp et al., 2002; Becker et al., 2004, for review).

P. falciparum IRBCs sequester in the deep circulation, adhering to endothelial cell receptors, which contributes to the pathophysiology of (for example) cerebral malaria. Hypoxia-reoxygenation events, free radicals, cytokines, and excitatory amino acids may contribute to central nervous system dysfunction. Expression of some of these endothelial adhesion molecule receptors for IRBCs, including E selectin, intercellular cell adhesion molecule 1, P selectin, and vascular cell adhesion molecule 1, is upregulated by oxidant stress (Becker et al., 2004). Oxidative stress may also increase adhesion of IRBCs through increased expression of phosphatidylserine on their surface, which appears at least partially to mediate adhesion of IRBCs to CD36 (Eda and Sherman, 2002).

In acute lung injury—a common manifestation of severe malaria in Asian adults, and especially in pregnant women—ROS have been shown to be increased (Abraham, 2000). In Thai adults with severe malaria, the antioxidant *N*-acetylcysteine increased the rate of normalization of plasma lactate, possibly by improving red blood cell deformability or by replenishing glutathione. Reducing agents such as *N*-acetylcysteine may have a therapeutic role in malaria complications and could reduce the microvascular obstruction to blood flow (Becker et al., 2004).

The pathological consequences of malaria infection are considered to be, at least in part, due to immunopathological reactions consequent upon the host immune response against the circulating malaria parasite. Thus, the involvement of leukocytes and their products, including cytokines and ROS, in malaria illness has been intensively investigated. Studies in humans have shown increased circulating levels in malaria infection of cytokines including gamma interferon, tumor necrosis factor, interleukin-10, and lymphotoxin, although such correlations do not prove a causative link. In mice, at least, it appears that phagocyte-derived ROS are not involved in the pathogenesis of cerebral malaria (Sanni et al., 1999), but lymphotoxin is a key effector molecule (Engwerda et al., 2002).

ANTIOXIDANT DEFENSE IN MALARIAL PARASITES

IRBC are very sensitive to oxidative damage both in vivo and in vitro (Hunt and Stocker, 1990), much more so than uninfected erythrocytes. Peroxidized IRBCs generate 4-hydroxyalk-2-enals and alka-2,4-dienals (Buffinton et al., 1988), and these aldehydes are toxic to *P. falciparum* in vitro. Thus, the antioxidant capabilities of the parasite and RBC are of considerable significance. For an overview on redox metabolism in malarial parasites, see Fig. 1. One GPx-like TPx and four peroxiredoxins are likely to contribute to ROS/reactive nitrogen species detoxification. Data refer to PfTrx1; however, five additional thioredoxin-like proteins have been described in *P. falciparum*, one of them being plasmoredoxin (Table 1). The formation of GSH adducts is likely to be the main function of PfGST. However, the protein also exhibits peroxidase activity, which might become significant due to high intracellular concentrations of PfGST. Data refer to PfGrx1; however, three other glutaredoxin-like proteins have been detected in *P. falciparum* (Table 1). Relevant genes and proteins are summarized in Table 1.

NADPH Production in *Plasmodium*

The glutathione and thioredoxin system depend on reducing equivalents provided by NADPH, which is mainly produced by the hexose-monophosphate shunt (HMS). As shown by gene knockout experiments with mammalian cells, the absence of G6PD activity, the key enzyme of the HMS, leads to high sensitivity to oxidative stress. Approximately 80% of the total HMS activity of an intact IRBC is provided by the parasite, and HMS activity of the host cell compartment is increased some 24-fold when compared with nonparasitized RBCs. G6PD has been identified in a number of *Plasmodium* species, isolated, and characterized. The gene has been cloned, and its stage-dependent transcription was determined. In addition to the HMS, glutamate dehydrogenase and isocitrate dehydrogenase activities have been postulated as alternative mechanisms for $NADP^+$ reduction in malarial parasites. The three-dimensional structure of *P. falciparum* glutamate dehydrogenase was determined (Becker et al., 2004; Muller, 2004; R. L. Krauth-Siegel, personal communication).

Glutathione

Like in many other organisms, the cysteine-containing tripeptide glutathione is the most abundant low-molecular-weight antioxidant in malarial parasites (Becker et al., 2003b). At millimolar concentrations, GSH plays a pivotal role in the antioxidant defense of cells through the maintenance of the redox state of protein-SH moieties, the reduction of noxious hydrogen and lipid peroxides, and the extrusion of toxic compounds, including drugs. Relevant glutathione species occurring in trophozoite cytosol include reduced glutathione (GSH, $\sim$2 mM), the thiolate anion GS^-, glutathione disulfide (GSSG, present at 10 μM or less), and S-glutathionylated proteins. Furthermore, species like S-nitrosoglutathione (GSNO), GSO^-, the anion of the sulfenic acid GSOH, conjugated xenobiotics of the arylsulfatase G type, resulting from a conjugation of aromatic compounds with GSH catalyzed by the enzyme glutathione S-transferase, glutathionylated coenzyme A (coenzyme A-S-S G) (as substrates of the glyoxalase I and II), hemithioacetals of glutathione, and glutathione thioesters such as D-lactoyl glu-

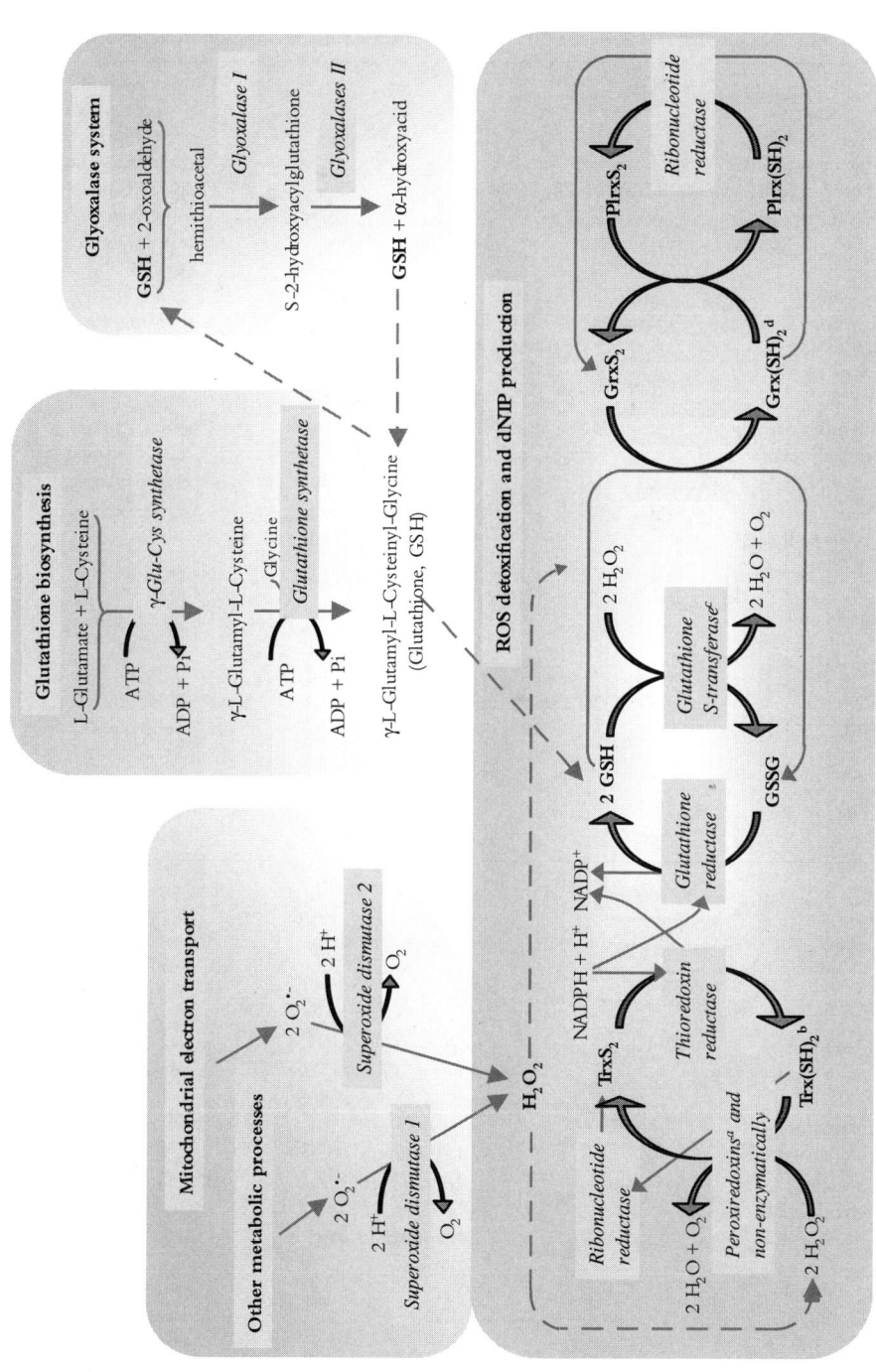

FIGURE 1 A schematic representation of the antioxidant defense in *Plasmodium*.

TABLE 1 Overview of enzymes involved in *Plasmodium falciparum* redox metabolism[a]

Enzyme class	Name (EC no.)	Abbreviation	Synonym(s)	Chromosomal/ subcellular localization
Disulfide reductases	Glutathione reductase (EC 1.8.1.7; formerly 1.6.4.2)	PfGR		Chr14/cytosolic, N-terminally extended version putatively leading to mito
	Thioredoxin reductase (EC 1.8.1.9; formerly 1.6.4.5)	PfTrxR		Chr9/cytosolic, putatively N-terminally extended version
	Dihydrolipoamide dehydrogenase (EC 1.8.1.4)	PfLipDH1	Lipoamide dehydrogenase	Chr12/mitochondrial
	Dihydrolipoamide dehydrogenase (EC 1.8.1.4)	PfLipDH2	Lipoamide dehydrogenase	Chr8/ts, apicoplast transit peptide predicted
Glutathione synthesis	Glutamate-cysteine ligase (EC 6.3.2.2)	PfGCS	Glu-Cys synthetase, γ-glutamyl-cysteine synthetase	Chr9/cytosolic
	Glutahione synthetase (EC 6.3.2.3)	PfGS		Chr5/cytosolic
Glutathione consumption	GST (EC 2.5.1.18)	PfGST		Chr14/cytosolic
	Glutaredoxin 1	PfGrxl		Chr3/cytosolic
	Glutaredoxin-like protein 1	PfGlp1	1-Cys-glutaredoxin-like protein 1	Chr3/targeting sequence, putatively mitochondrial
	Glutaredoxin-like protein 2	PfGlp2	1-Cys-glutaredoxin-like protein 2	Chr6/putatively cytosolic
	Glutaredoxin-like protein 3	PfGlp3	1-Cys-glutaredoxin-like protein 3; CG6	Chr7
Glyoxalase system	Cytosolic glyoxalase I (EC 4.4.1.5)	cPfGlo1	Lactoylglutathione lyase; methylglyoxalase	Chr11/cytosolic
	Glyoxalase I-like protein (EC 4.4.1.5)	PfGILP		Chr6/ts, apicoplast transit peptide predicted
	Cytosolic glyoxalase II (EC 3.1.2.6)	cPfGloII	Hydroxyacylglutathione hydrolase	Chr4/cytosolic

Accession no.[b]	mRNA peak[c] intensity[d]	MS-expression evidence[e]	Reference[f]	Biochemical characterization (reference)
PF14_1092 CAA63747 X93463	ET, LT, ES (2,563)	G, M, S, T, Sch, iRBCM	Faber et al., 1996	2 GSH + NADP$^+$ $\leftrightarrow$ GSSG + NADPH + H$^+$; FAD dependent, homodimeric; structure solved (Sarma et al., 2003)
PFI1170c CAA60574 X87095	ET, LT, ES, LS (91,947; Oligo obfi17632)	M	Muller et al., 1995; Nickel et al., unpublished	NADPH + H$^+$ + Trx(S$_2$) $\leftrightarrow$ Trx(SH)^2 + NADP$^+$; FAD dependent, homodimeric
PFL1550w CAF34426 AJ630268	ES, LS (2,862; Oligo f46067_3)	G, M, S, T	McMillan et al., 2005	Protein N^6-(dihydrolipoyl)lysine + NAD+ $\leftrightarrow$ protein N^6-(lipoyl)lysine + NADH; FAD-dependent, homodimeric
PF08_0066 CAD51214 AL844507	LT, ES (323)	None	McMillan et al., 2005	See above
PFI0925w CAA74990 AY14674	LR, ET (2,862; Oligo F46067_3)	M	Luersen et al., 1999	ATP + L-glutamate + L-cysteine $\leftrightarrow$ ADP + phosphate + γ-L-glutamyl-L-cysteine
PFE0605c CAC59841.1 AJ309927	LR, ET, LT (1,576)	G, M	Meierojohann et al., 2002	ADP + γ-L-glutamyl-L-cysteine + glycine $\leftrightarrow$ ADP + phosphate + GSH; homodimer
PF14_0187 AAK00582 AY014840	ET, LT (6,943)	M, T	Harwaldt et al., 2002	CDNB turnover, ligandin binding heme, peroxidase activity, homodimer, structure solved (Fritz-Wolf et al., 2003)
PFC0271c AAK91589 AF276083	ET, LT, ES (7,442)	G, M	Rahlfs et al., 2003; Becker et al., 2003a	Classic, dithiol glutaredoxin 1, active with PfAOP, reduces insulin, reduces plasmoredoxin, and transfers electrons to ribonucleotide reductase
PFC0205c AAK00581 AY014839	ET, LT, ES (3,544)	None	Rahlfs et al., 2003, Deponte et al., unpublished	Monothiol glutaredoxin 1, negative in HEDS assay, reacts with glutathione, reduces insulin
MAL6P1.72 CAG2539 CR382398	No data	M	Deponte et al., unpublished	Monothiol glutaredoxin 2, negative in HEDS assay, reacts with glutathione
Pf07_0036 CAD50844 AAC47843	ES (512)	None	Deponte et al., unpublished	Monothiol glutaredoxin 3
PF11_0145 AAQ05975 AF486284	Not clear (530)	M, S	Iozef et al., 2003	Glyoxalase I family glutathione + methylglyoxal $\leftrightarrow$ (R)-S-lactoylglutathione [Zn^{2+}], gene duplication
MAL6P1.50 NP_703709	LT, ES	None	Akoachere et al., 2005	Glyoxalase I family similarity
chr4.gen_37 AY494055	ER, LR	None	Akoachere et al., 2005	(S)-(2-Hydroxyacyl) glutathione + H$_2$O $\leftrightarrow$ α-2-hydroxy acid anion + glutathione [Fe^{2+}]?, [Zn^{2+}]?

(continued)

TABLE 1 (*Continued*)

Enzyme class	Name (EC no.)	Abbreviation	Synonym(s)	Chromosomal/ subcellular localization
	Targeted glyoxalase II (EC 3.1.2.6)	tPfGloII	Hydroxyacylglutathione hydrolase	Chr 12/ts
Thioredoxin-like proteins	Thioredoxin 1	PfTrx1		Chr14/cytosolic
	Thioredoxin 2	PfTrx2		Chr13/targeting sequence
	Thioredoxin-like protein 1	PfTlp1	Dynein light chain, thioredoxin-like protein 1	Chr14/cytosolic
	Thioredoxin-like protein 2	PfTlp2		Chr9/cytosolic
	Thioredoxin-like protein 3	PfTlp3		Chr9/cytosolic transit peptide predicted
	Plasmoredoxin	PfPlrx	FRED, thioredoxin-like redox-active protein	Chr3/cytosolic
Peroxiredoxins	GPx-like TPx (EC 1.11.1.-)	PfTPx$_{Gl}$ PfGPx	Glutathione peroxidase-like TPx	Chr12/apicoplast transit peptide predicted
	2-Cys peroxiredoxin 1 (EC 1.11.1.-)	PfTPx1 Pf2-CysPrx1 PfPrx1 PfPx1 PfTrx-Px1	Thioredoxin peroxidase 1	Chr14/cytosolic
	2-Cys peroxiredoxin 2 (EC 1.11.1.-)	Pf2-CysPrx2 PfTPx2 PfPrx2 PfPx2 PfTrx-Px2	Thioredoxin peroxidase 2	Chr12/mitochondrial
	1-Cys peroxiredoxin (EC 1.11.1.-)	Pf1-Cys Prx Pf1-CysTPx Pf1-CysTrx-Px Pf1-Cys Px PfPx1	1-Cys peroxiredoxin peroxidase 2	Chr8/cytosolic

Accession no.[b]	mRNA peak[c] intensity[d]	MS-expression evidence[e]	Reference[f]	Biochemical characterization (reference)
PFL0285w AAQ05976 AF486285	ET, LT, ES (1,646)	G, T	Akoachere et al., 2005	See cytosolic glyoxalase II, [Zn^{2+}]
PF14_0545 AAF34541 AF202664	ET, LT, ES	M, S	Kanzok et al., 2001	Active with PfTrxR, reduces insulin disulfides and transfers electrons to ribonucleotide reductase; structure solved according to MMDB (Robin and Hol)
MAL13P1.225 AAQ05974 AF484689	LR, ET, M (8,250)	T	Nickel et al., unpublished; Rahlfs et al., 2003	Atypical active site active with PfTrxR and PfTPx1
PF14_0590 AAN37203 AE014826	LT, E (311)	None	Nickel et al., unpublished; Rahlfs et al., 2003	Reduces insulin disulfides, dynein light chain
chr9.phat_260 AAQ07983 AF508130	No data	None	Nickel et al., unpublished; Rahlfs et al., 2003	
chr9.phat_172	No data	None	Nickel et al., unpublished; Rahlfs et al., 2003	
chr3.phat_63 AAF87222 AF234633	LR, ET, LT (722)	None	Becker et al., 2003	Reduces insulin disulfides and transfers electrons to ribonucleotide reductase, reacts with PfTPxl and AOP (Nickel et al., unpublished)
PPF00255 CAA92396 Z68200	LT, ES (2,053)	iRBCM G, T, M, Sch	Sztajer et al., 2001	Thioredoxin-dependent detoxification of peroxides
PF14_0368 AF225977 BAA97121	LR, ET, LT, ES (17,453; Oligo opfn0251)	G, M, S, T, Sch, iRBCM	Rahlfs and Becker, 2001	Detoxification of peroxides with Trx1 and Plrx as reductants; involved in $ONOO^-$ detoxification (Nickel et al., unpublished; R. Radi et al., unpublished)
PFL0725w AF225978 AAK20024	LT, ES, LS (857)	G, S, T, Sch, iRBCM	Rahlfs and Becker, 2001	Detoxification of peroxides; reducing equivalent needs to be established
PF08_0131 BAA78369 AB020595	ET, LT, ES (7,937)	M, S, T T, Sch, iRBCM	Kawazu et al., 2000; Krnajski et al., 2001	Detoxification of peroxides, slight activity with thioredoxin system measured, detoxification of peroxides with GSH, DTT

(continued)

TABLE 1 (*Continued*)

Enzyme class	Name (EC no.)	Abbreviation	Synonym(s)	Chromosomal/ subcellular localization
	PfAOP (EC 1.11.1.-)	Pf antioxidant protein	Grx/Trx-dependent peroxidase	Chr7/ER (Nickel et al., unpublished)
H_2O_2 generation	Superoxide dismutase 1 (EC 1.15.1.1)	PfSOD1 PfFeSOD1		Chr8/cytosolic
	Superoxide dismutase 2 (EC 1.15.1.1)	PfSOD2 PfFeSOD2		Chr6/mitochondrial

[a] Abbreviations: ER, early rings; LR, late rings; ET, early trophozoite; LT, late trophozoites; ES, early schizonts; LS, late schizonts; G, gametocytes; M, merozoites; S, sporozoites; R, rings; T, trophozoites; Sch, schizonts; iRBCM, infected red blood cell membrane; Oligo, oligonucleotides; mito, mitochondrion; FRED, falciparum redoxin; MMDB, molecular modeling database; HEDS, β-hydroxyethyl disulfide; DTT, dithiothreitol.

[b] PlasmoDB accession numbers are from http://www.plasmodb.org; protein and nucleotide accession numbers are from http://www.ncbi.nlm.nih.gov/entrez. The protein entries in the Entrez search and retrieval system were compiled from a variety of sources, including Swiss Prot, PIR, PRF, PDB, and translations from annotated coding regions in GenBank and RefSeq. The Entrez Nucleotides database is a collection of sequences from several sources, including GenBank, RefSeq, and PDB.

[c] mRNA peak, asexual blood stages. Data from http://www.plasmodb.org (Le Roch et al., 2003; Bozdech et al., 2003; Ben Mamoun et al., 2001). Average median intensity is taken from http://www.malaria.ucsf.edu/ (developmental series, glass slide oligonucleotide array; Bozdech et al., 2003).

[d] Data obtained from http://www.plasmodb.org.

[e] Due to limited space, further important references are not included here.

tathione, are most likely to occur. The thiyl radical GS$^\bullet$ and the glutathione disulfide radical GSSG$^{\bullet-}$ are intermediates in reaction sequences that terminate radical-based chain reactions.

In the host cell compartment of trophozoite-infected RBCs, the GSH/GSSG ratio is decreased to ~28, whereas in normal erythrocytes and in the parasite compartment, values of ~300 were determined. This indicates adequate antioxidant defense of the parasite in the absence of exogenous oxidative challenge but a redox burden on the host cell compartment (Atamna and Ginsburg, 1997). This burden is also reflected by a considerable GSSG loss (40 to 60 times the rate observed in normal erythrocytes) from infected cells. This efflux of GSSG is mediated by new permeability pathways that the parasite induces in the red blood cell plasma membrane (Atamna and Ginsburg, 1997; Muller, 2004). Both γ-glutamyl-cysteine and GSH can penetrate via the parasite-induced pathways from the extracellular space into the host cytosol but not into that of the parasite. This implies that the host cannot directly supply GSH to the parasite. Exogenous γ-Glu-Cys is not converted to GSH in the host cell, indicating that RBC GSH synthetase may not be functional. A possible explanation could be the low Mg^{2+} and K^+ levels in the host compartment. Taken together, the results demonstrate that the parasite has a high capacity for de novo synthesis of GSH, much of which is expelled in the form of GSSG. Part of this exported GSSG is reduced and used by the host cell compartment (Atamna and Ginsburg, 1997).

The redox potential E_0' of the GSSG/2GSH system at pH 7.0, physiologic ionic strength, and 25°C is −240 mV. For the conditions in the trophozoite cytosol (37°C and pH 7.2), a value of −270 mV has been estimated (Becker et al., 2003b). As the GSH concentration (1 to 5 mM) is much higher than that of NADPH (<100 μM) and Trx(SH)$_2$ (<50 μM), the overall redox environment in the trophozoite results largely from the contribution of the glutathione redox couple; this implies that glutathione is the major redox buffer in the parasites. GR establishes a high GSH/GSSG ratio; GR is abundant in

Accession no.[b]	mRNA peak[c] intensity[d]	MS-expression evidence[e]	Reference[f]	Biochemical characterization (reference)
MAL7P1.159 CAD51033 AL844506	ET, LT (400)	T, Sch, iRBCM	Nickel et al., unpublished; Sarma et al., unpublished	Detoxification of peroxides with PfGrx1, PfTrx1, and Plrx
PF08_0071 CAA89971 Z49819	ET, LT, ES (6,893)	G, M, S, T	Becuwe et al., 1996	$2\,O_2^{\cdot-} + 2\,H^+ \leftrightarrow O_2 + H_2O_2$; iron dependent, homodimeric
PF1130c AAT11554 AY586514	LS	G, S, iRBCM	Sienkewicz et al., 2004	$2\,O_2^{\cdot-} + 2\,H^+ \leftrightarrow O_2 + H_2O_2$; likely to be iron dependent, homodimeric

trophozoites, having an activity of up to 10 U of cytosol/ml. Under conditions when GR activity is insufficient, GSSG can be reduced also by other processes such as protein thiol glutathionylation, by reduction with Trx(SH)$_2$ (see below), and probably by dihydrolipoate-dependent reactions catalyzed by glutaredoxin. The system of GSSG export compensated by highly efficient synthesis of intraparasitic GSH from the constituent amino acids also has a net function of being an efficient GSSG reduction system (Becker et al., 2003b).

The role of other low-molecular-weight antioxidants like tocopherols, ascorbic acid, lipoic acid, or ubiquinone in maintaining the redox equilibrium of malarial parasites has not yet been studied in detail. However, preliminary results point to the notion that the peroxidase system of *P. falciparum* can at least partially be fueled by low-molecular-weight electron donors (C. Nickel et al., unpublished data). The metabolism of vitamin C in malaria-infected mice has been studied as well as plasma α-tocopherol, retinol, and carotinoides in children with malaria (Das et al., 1996). *P. vinckei* IRBCs have an increased content of vitamin C, and this may serve to reduce α-tocopherol, the major lipid phase antioxidant, if it becomes oxidized. It has been known for many years (Eaton et al., 1976) that malaria parasites grow poorly in mice deficient in α-tocopherol. Erythrocyte membrane α-tocopherol and the content of polyunsaturated fatty acids are decreased in malaria patients, which might contribute to erythrocyte loss in severe anemia (Griffiths et al., 2001). The ratio of α-tocopherol to polyunsaturated fatty acids in murine *P. vinckei* IRBCs is decreased relative to that in normal RBCs, perhaps in part explaining why IRBCs are so susceptible to spontaneous lipid peroxidation in vitro. Furthermore, much of the α-tocopherol in IRBCs may be associated with the RBC membrane rather than the malaria parasite (Hunt and Stocker, 1990). Although the function of ubiquinone as an antioxidant in malarial parasites has not yet been studied, the presence of an active isoprenoid pathway contributing to the biosynthesis of coenzyme Q (De Macedo et al., 2002) and its participation in the respiratory chain have been demonstrated (Kita et al., 2002).

Peroxidases

P. falciparum possesses at least five proteins with homologies to peroxidases and peroxiredoxins, respectively. Peroxiredoxins form a family of antioxidant enzymes that act as peroxidases by reducing hydrogen peroxide and other hydroperoxides to water or the corresponding alcohol at the expense of reduced thioredoxin [Trx(SH)$_2$]. The protein previously described as glutathione peroxidase (GPx) (Gamain et al., 1996) was shown to be thioredoxin rather than GSH dependent and is not a member of the peroxiredoxin family (Sztajer et al., 2001). As a nonselenocysteine GPx homolog, its reactions with hydroperoxides and GSH are 3 orders of magnitude lower than those of typical selenoperoxidases.

The majority of the peroxide-detoxifying capacity, however, seems to be provided by peroxiredoxins. Apart from the GPx-like TPx mentioned above, two peroxiredoxins of the 2-Cys family (Rahlfs et al., 2002; Becker et al., 2004; Muller, 2004), one 1-Cys Prx with homologies to atypical 2-Cys Prx, named *P. falciparum* antioxidant protein (PfAOP), and one typical 1-Cys peroxiredoxin have been identified. One of the 2-Cys Prx, PfTPx-1, has been shown to be thioredoxin dependent and seems to be the major cytosolic peroxidase in *P. falciparum*. As shown by gene disruption, the protein is not essential for *P. falciparum* but contributes to defense against ROS and reactive nitrogen species (Komaki-Yasuda et al., 2003). Recent kinetic studies with isolated recombinant PfTPx-1 have furthermore shown that the enzyme turns over peroxynitrite (R. Radi, personal communication) and can (apart from thioredoxin) be reduced by plasmoredoxin (Plrx) (Nickel et al., unpublished). PfAOP has recently been crystallized in the Giessen laboratory; the X-ray structure demonstrates that the protein acts on the basis of a 1-Cys mechanism and is present as a dimer (G. N. Sarma et al., unpublished data).

Glutathione *S*-transferase exhibits glutathione peroxidase activity, which might contribute to the total peroxide-reducing capacity of the parasite, since the enzyme is present at very high concentrations (Harwaldt et al., 2002; Fritz-Wolf et al., 2003). As discussed above, part of the hydrogen peroxide originating in the trophozoite is probably detoxified in the host cell compartment (Ginsburg and Atamna, 1994).

Superoxide Dismutase

Superoxide dismutase (SOD) is the major enzyme involved in catabolizing the superoxide anion, resulting in production of molecular oxygen and hydrogen peroxide. The superoxide anion promotes cellular oxidative stress by leading to the production of hydroxyl radicals, interacting with iron-sulfur clusters, and being involved in the production of peroxynitrite (Dive et al., 2003). Endogenously superoxide is produced by the enzyme dihydroorotate dehydrogenase, putatively by the endoplasmic reticulum-resident oxidoreductin (Bozdech and Ginsburg, 2004) and during hemoglobin digestion in the food vacuole (Dive et al., 2003). Exogenous superoxide anions result, e.g., from hemoglobin autoxidation and from dietary oxidants (see above).

As the parasite resides in the host cell, it was earlier proposed that RBC SOD might be adopted by *P. falciparum*. Other results indicate that the host Cu,Zn SOD resides in the food vacuole, where it should be digested and where a spontaneous dismutation of $O_2^{\bullet-}$ to H_2O_2 and O_2 may occur at the low pH of 5.2 (Ginsburg and Atamna, 1994); in this case, the host SOD would not play a role in the parasite's superoxide detoxification. So far, two plasmodial SODs have been identified, namely PfFeSOD1 and PfFeSOD2 (Sienkewicz et al., 2004). PfFeSOD1 has been shown to be a homodimeric, iron-dependent cytosolic enzyme. It is highly expressed in *P. falciparum* blood cycle stages, especially in trophozoites and early schizonts (Bozdech and Ginsburg, 2004). The second enzyme, PfFeSOD2, is a mitochondrial enzyme and is expressed more weakly than PfFeSOD1. The enzyme is predicted to be iron dependent and homodimeric like PfFeSOD1. The PfFeSOD2 has not yet been functionally characterized, and predicted features need to be proven, especially since the mitochondrial enzymes are often Mn dependent. No gene coding for catalase has been identified in the malaria parasite genome so far and it is doubtful that it exists. Considering the fast transport of $O_2^{\bullet-}$ and the high diffusion rates of H_2O_2 and $HO_2^{\bullet}$, a contribution of the host cell (catalase) to the detoxification of ROS originating in the parasite is most likely.

Glutathione Reductase

High levels of GSH in malarial parasites are NADPH-dependently maintained by the homodimeric flavin adenine dinucleotide (FAD)-containing enzyme GR (Schirmer et al., 2002).

P. falciparum GR (PfGR) has been characterized biochemically and kinetically (Becker et al., 2003b; Becker et al., 2004). A number of reports indicate that a lack of FAD, the prosthetic group of GR, or of the cosubstrate NADPH, as observed with glucose-6-phosphate dehydrogenase deficiency, leads to substantial protection from severe malaria. These clinical and epidemiological observations suggest that both human GR and PfGR are potential targets for the development of antimalarial drugs. The three-dimensional structure of PfGR was solved at a resolution of 2.6 Å (Sarma et al., 2003). Although the overall architecture of PfGR is very similar to human GR, specific features of PfGR have been identified that might serve as a starting point for the development of specific inhibitors. At therapeutically used concentrations, PfGR is the only verified enzymatic target of the antimalarial dye methylene blue. As a novel class of potential antimalarial agents, double-headed prodrugs (containing a PfGR inhibitor linked via a hydrolase-labile ester bond to a 4-anilinoquinoline) were synthesized and successfully tested in vitro and in vivo (Davioud-Charvet et al., 2001).

Thioredoxin Reductase

The *P. falciparum* thioredoxin reductase (Rahlfs et al., 2002; Becker et al., 2004; Muller et al., 2004) is a homodimeric, FAD-dependent oxidoreductase. As is the case for mammalian TrxRs, the role of the C-terminal redox center in catalysis has been proven, and its interaction with the active site disulfide-dithiol has been studied in detail. The fact that PfTrxR is nonselenium dependent and differs in its C-terminal active site motif (CGGGKC) from human TrxR (Cys-Sec) represents a good starting point for the development of antiparasitic drugs. As recently demonstrated by knockout studies, PfTrxR is indeed essential for erythrocytic stages of *P. falciparum* (Krnajski et al., 2002). For high-throughput screening of inhibitors directed against PfTrxR, 5,5'-dithiobis(2-nitrobenzamides) have been synthesized and established as alternative substrates of the enzymes.

In addition, two genes encoding lipoamide dehydrogenases, which belong to the same disulfide reductase family as GR and TrxR, have been identified in the genome of *P. falciparum*. Both genes have been amplified from a cDNA library. One of the deduced proteins carries an apicoplast targeting sequence, and the second one carried a mitochondrial targeting sequence (S. Rahlfs and K. Becker, unpublished data; Muller, 2004).

Thioredoxin-Like Proteins

Thioredoxins are a group of small (ca. 12-kDa) redox-active proteins with the typical active site motif Cys-Gly-Pro-Cys. They are reduced by TrxR and are involved in antioxidant defense, ribonucleotide reduction, reduction of peroxidases, and transcription factors (Rahlfs et al., 2002). Many organisms have more than one Trx; often, a mitochondrial system exists in parallel to the cytosolic one. *P. falciparum* possesses a typical thioredoxin, PfTrx1 (Rahlfs et al., 2002; Becker et al., 2004; Muller, 2004). PfTrx1 provides reducing equivalents to peroxidases and ribonucleotide reductase and is reduced by PfTrxR. Furthermore, a nonenzymatic reduction of GSSG by PfTrx1 has been discovered. On the basis of this reaction, the PfTrx system was shown to support GSSG fluxes up to 200 μM/min (25°C). The reaction is particularly efficient in malarial parasites, although it has also been shown to exist in other organisms, including *Anopheles gambiae* and humans (Kanzok et al., 2001).

Since thioredoxins are known to be involved in direct nonenzymatic buffering of ROS, the reduction of H_2O_2, *t*-butylhydroperoxide and cumene hydroperoxide by PfTrx1 has recently been characterized (Rahlfs et al., 2003). Although the peroxide buffering efficiency of PfTrx1 is rather low, it might contribute to the antioxidant defense of the parasite, which lacks a classical glutathione peroxidase and catalase. GSNO represents an important transport form of NO in biological systems, and NO has been proposed to be involved in the pathology of cerebral malaria. GSNO has been shown to be an inhibitor of GR. Furthermore, GSNO is not reduced by PfGR and it represents but a weak substrate of PfTrxR ($k_{cat} = 9.4$ min^{-1}). PfTrx1, however, does reduce GSNO nonenzymatically with a rate constant k_2 of 0.034 μM^{-1} min^{-1}. Apart from PfTrx1, five thioredoxin-like proteins have been detected in *P. falciparum* (Rahlfs et al., 2003). One of them is likely to represent a part of dynein, thus being not directly involved in redox reactions. A second Trx, located on chromosome 13 and possessing a targeting sequence, is also reduced by PfTrxR and provides electrons to PfTPx1 (Nickel et al., unpublished). Two additional Trx-like proteins located on chromosome 9 are presently under investigation. In addition, a novel 22-kDa active redox protein with homologies to Trx in *P. falciparum* has been described (Becker et al., 2003a). The protein, named plasmoredoxin (Plrx), is highly conserved and found exclusively in malaria parasites. Plrx is a member of the thioredoxin superfamily. However, it clusters separately from other members in a phylogenetic tree. It can be reduced by glutathione but is reduced much faster by dithiols like thioredoxin and glutaredoxin.

Glutaredoxins

Glutaredoxins (Grx) are ubiquitous thiol-disulfide oxidoreductases of ~13 kDa that protect cells from oxidative damage and are associated with the control of transcription and apoptosis. *P. falciparum* possesses a classical 2-Cys glutaredoxin (PfGrx1) and a redox-active 1-Cys glutaredoxin-like protein (PfGLP-1). As hypothesized for other 1-Cys Grx, the latter might be involved in the synthesis of Fe/S cluster proteins, in protein deglutathionylation, and in interaction with protein kinase C. In addition to PfGLP-1, two more GLPs have been detected in *P. falciparum* and have been characterized (M. Deponte et al., unpublished results).

GSH Biosynthesis

The rate of glutathione synthesis in trophozoites was determined to be 184 nmol per 10^{10} cells per h (corresponding to a rate of ≥3 μM per min) (Atamna and Ginsburg, 1997). In most organisms, the generation of GSH, which takes place constitutively and in a regulated fashion, involves two ATP-dependent enzymes (Fig. 1). The first reaction is catalyzed by glutamate-cysteine ligase, also known as γ-glutamylcysteine synthetase (GCS), and the second reaction is catalyzed by glutathione synthetase (GS). GCS provides the rate-limiting reaction. The importance of endogenous glutathione biosynthesis in *Plasmodium* is supported by the facts that chloroquine resistance is accompanied by an increase in GSH content and that partial loss of this resistance can be caused by specific inhibition of GSH synthesis. GCS has been characterized for *P. berghei* and *P. falciparum*. The *P. berghei* gene transcript is present throughout the schizogonic cycle. For the *P. falciparum* gene, strain-specific differences in length have been demonstrated: the chloroquine-resistant K1 strain contains a repetitive motif comprising 168 bp, which is not present in chloroquine-sensitive 3D7. The PfGCS gene is stage-specifically transcribed, which correlates well with the intensity of metabolism and hemoglobin digestion: the transcription starts in the young trophozoite (12 to 18 h), is highest in the late trophozoite stage, and is diminished in the schizont stage (Muller, 2004).

GS from *P. falciparum* has been characterized as a 655-amino-acid protein. Many residues contributing to substrate and Mg^{2+} binding are conserved between the *P. falciparum* and human enzymes. Differences include a number of insertions in the PfGS sequence and the multitude of 11 cysteine residues in the parasite enzyme. Interestingly, the apparent K_m for glycine is four times higher than in the human enzyme, whereas the values for ATP and γ-glutamyl-aminobutyrate are four- and sixfold lower (Meierjohann et al., 2002).

GSH as well as γ-Glu-Cys can leave and enter the IRBC via parasite-dependent permeability pathways. In contrast, GSH and γ-Glu-Cys can permeate neither the membrane of an uninfected RBC nor the parasite membrane. It is therefore assumed that the parasite does not use host-cell glutathione, as suggested previously, but rather supplies GSSG to the host compartment, where it is rapidly reduced by erythrocyte GR. Thus, GSH and ATP represent crucial metabolites that are produced by the parasite and supplied to the host cell (Atamna and Ginsburg, 1997).

Glyoxalases

The glyoxalase system consists of glyoxalase I (GloI), glyoxalase II (GloII), and the coenzyme glutathione. It is a cyclic metabolic pathway removing toxic 2-oxoaldehydes like methylglyoxal by converting them to the corresponding nontoxic 2-hydroxycarboxylic acids like D-lactate. Glyoxalase I catalyzes the formation of S-2-hydroxyacylglutathione (a thiol ester of GSH), and glyoxalase II then hydrolyzes the ester, thus producing GSH and a free 2-hydroxycarboxylic acid. With the absence of a typical citric acid cycle and the levels of electron transport activity in the respiratory chain appearing to be much lower than in mammals (Kita et al., 2002), *Plasmodium* is bound to maintain high levels of glycolytic activity. This is indicated by the 100-fold increase in glucose consumption in *P. falciparum*-infected erythrocytes in comparison with noninfected cells, which exposes the parasite to higher fluxes of methylglyoxal produced by the nonenzymatic fragmentation of triose phosphates. In IRBCs, the formation of D-lactate from methylglyoxal was found to be increased by a factor of 30, and there were higher levels of GloI and GloII activity than in uninfected erythrocytes. The antimalarial activity of Glo$^-$ inhibitors was demonstrated by the growth inhibition caused by S-p-bromobenzylglutathione diethyl ester with 50% inhibitory concentration values in the lower micromolar range (Thornalley et al., 1994).

A complete glyoxalase system in the cytosol (and putatively the apicoplast) of *P. falciparum* has recently been described (Iozef et al., 2003; Akoachere et al., 2005) which comprises one typical cytosolic GloI (cGloI) and two GloIIs (cGloII and with a targeting sequence GloII). The kinetic properties of targeting sequence GloII are similar to human GloII, whereas cGloII and cGloI exhibited higher substrate affinity and catalytic efficiency than the host enzymes. cGloI evolved on the basis of a gene duplication event (Iozef et al., 2003). Different S-(N-hydroxy-N-arylcarbamoyl)glutathiones were tested as PfGlo-inhibitors and found to be active in the lower nanomolar range. In addition to the typical cGloI, a second protein with high homologies to GloI in *P. falciparum* has been identified. This protein is unique for malarial parasites and has an-as-yet-undefined substrate specificity (Akoachere et al., 2005).

Glutathione S-Transferase

GSTs exist in many organisms in different isoforms and serve in the intracellular detoxification of noxious chemicals by linking them to glutathione. GSTs can furthermore detoxify lipid peroxidation products and serve as carrier proteins, so-called ligandins, of certain organic molecules, which leads to the inactivation and immobilization of these compounds. In chloroquine-resistant parasites, GST activity is directly and positively related to drug pressure. The *P. falciparum* GST gene is located on chromosome 14 and encodes the only GST isoenzyme in the parasite (Harwaldt et al., 2002; Liebau et al., 2002). PfGST is a homodimer of 26-kDa subunit size. A number of PfGST inhibitors, including ferriprotoporphyrin IX, have been identified. The enzyme represents ~1% of cellular protein and might therefore serve as an efficient in vivo buffer for parasitotoxic hemin, which binds uncompetitively to GST (K_i = 6.5 μM = [GST × GSH] [hemin]/[GST × GSH × hemin]).

PfGST was found to exhibit a slight peroxidase activity. The three-dimensional structure of the enzyme has been elucidated at 1.9-Å resolution (Fritz-Wolf et al., 2003). PfGST represents a new GST isoform that cannot be assigned to any of the known GST classes. The unique structure of the substrate-binding site predisposes PfGST as a drug target.

TRANSCRIPTOME ANALYSES

Recent transcriptome analyses (Le Roch et al., 2003; Bozdech et al., 2003) serve as a data mine for many malaria researchers. Bozdech and Ginsburg (2004) analyzed the transcription of genes coding for redox-active proteins and viewed them in the time frame of the intraerythrocytic cycle of *P. falciparum* (DeRisi transcriptome database at http://malaria.ucsf.edu/; further data are available at http://carrier.gnf.org/publications/CellCycle and http://www.PlasmoDB.

org). In many cases, there is a functionally explicable coordination of time-dependent transcription. But still, our current biochemical knowledge about participating transcripts and their products' interactions and functions is not sufficient to explain expression profiles. The following section deals with transcriptional profiles of some elements of redox metabolism.

Glutathione Metabolism

The genes coding for enzymes of GSH synthesis and GSSG reduction are transcribed sequentially in a time-dependent manner. The transcript level of GCS is highest at 17 h postinvasion (HPI) in late ring and early trophozoite stages. The next step in the metabolic sequence is catalyzed by the GS, which has a maximum at 20 HPI. GR is constitutively transcribed throughout the intraerythrocytic life cycle. Relative transcript levels of GR are low in the ring stage and rise in the early trophozoite stages. The highest level is reached in the trophozoite and early schizont stage with a maximum at 22 HPI. GST, Grx1, Glp1, Glp3, and Plrx have high transcription levels when glutathione biosynthesis should be maximal at 20 HPI. Interestingly, Grx1 and GST already have high transcription levels (10 to 15 HPI) when those for glutathione synthetase are still relatively low. For Grx1, thioredoxin and plasmoredoxin can serve as alternative substrates. Cytosolic glyoxalase II already peaks in the ring stage, suggesting an important detoxifying function.

Peroxidases

Also according to the transcriptome data, cytosolic TPx1 seems to represent the major defense against peroxides. The transcript level is peaking at 11 HPI but stays high until 30 HPI. This is supported by the very high average median intensity of 17,453 (see Table 1). AOP has a similar profile at lower intensity but seems to be localized in the ER. In contrast, 1-Cys Prx is transcribed in higher amounts from ~20 to 30 HPI, a maximum at about 26 HPI, and a very high average median intensity of 7,937. These facts imply that the 1-Cys Prx might partially be able to complement the detoxification functions of TPx1 in the cytosol. The putatively mitochondrial TPx2 (2-Cys Prx2) and glutathione peroxidase-like thioredoxin peroxidase (TPx$_{Gl}$), which has a predicted apicoplast transit peptide, have maxima at 36 and 34 HPI, respectively. The transcription levels of Trx1 and the genes of some Trx1-dependent enzymes (TrxR, AOP, TPx$_{Gl}$) seem to be coordinated.

Superoxide Dismutases

The transcript of cytosolic PfFeSOD1 peaks at 30 HPI, at the trophozoite stage; the average mean intensity is about 6,893 in the DeRisi array. The mitochondrial PfFeSOD2 is expressed in much lower amounts and peaks clearly later in the schizogony. Interestingly, the same holds true for the mitochondrial peroxidase 2-Cys Prx2, suggesting that the production of $O_2^{\bullet-}$ in the mitochondrion by leakage of the respiratory chain may be maximized in this phase of the intraerythrocytic cycle.

Further analyses of the stage-dependent protein, substrate, or product concentrations, as well as analyses concerning interaction partners, localization, and substrate specificities, are required for thoroughly interpreting transcriptional profiles.

REDOX EQUILIBRIUM AND DRUG ACTION

It has been known for some years that a range of oxidants of dissimilar structure are capable of effectively killing malaria parasites both in vivo and in vitro (Hunt and Stocker, 1990). Later, it was realized that a number of clinically effective antimalarial drugs might exert their effects through oxidative mechanisms. Furthermore, for the reasons discussed above, the redox metabolism represents an excellent target for future drug development (Becker et al., 2004). The three clinically used drugs chloroquine, primaquine, and artemisinin will be discussed here briefly.

Chloroquine (CQ), a weakly basic quinoline antimalarial, accumulates in the food vacuole of malarial parasites and interferes with hemoglobin breakdown by interacting with the μ-oxo dimer form of oxidized heme (FP-FeIII).

This interaction prevents the formation of the β-hematin dimer, which is presumably an intermediate in the formation of the β-hematin crystal. The consequent buildup of free FP has been proposed to damage membranes and enzymes, due to the detergent-like properties of FP and its ability to participate in damaging redox reactions (Tilley et al., 2001; Sullivan, 2002). A number of studies support the involvement of oxidative events in the effects of CQ and the involvement of GSH in CQ resistance. The expression of γ-GCS has been shown to be related to drug resistance, and GSH-dependent FP degradation is competitively inhibited by CQ. Modulation of GSH levels in *P. falciparum* alters the parasite's sensitivity to CQ, and correspondingly, the levels of GSH were found to increase with CQ resistance in *P. falciparum* and in *P. berghei*.

Primaquine, an 8-aminoquinoline, is still the only drug effective against tissue stages of *P. vivax*. Although the mechanism of action of 8-aminoquinolines is still not understood, the hydroxylated metabolites of primaquine are thought to be responsible for both the mechanism of action of primaquine and its hemolytic side effects. Redox cycling of the quinone metabolite exerts a substantial oxidative stress on the erythrocyte. The resulting depletion of GSH and NADPH impairs the cell's ability to detoxify ROS. Indeed, the clinical use of 8-aminoquinolines is limited by the production of methemoglobin, particularly in patients with G6PDH deficiency where hemolysis can occur (Becker et al., 2004).

In artemisinin and its active derivatives, the endoperoxide moiety is an essential feature. The endoperoxides are thought to accumulate in the parasite cytosol and membranes and to interact with FP-FeII, forming cytotoxic radical intermediates (Meshnick, 2002). The antimalarial activity of artemisinin against *P. falciparum* in culture is enhanced by increased oxygen tension and by prooxidant compounds. Artemisinin has been shown to react with GSH and to increase levels of lipid peroxidation. Furthermore, dihydroartemisinin, the active metabolite of artemisinin, reduced catalase and GPx activities, as well as GSH concentrations in parasitized red blood cells. Together, these results indicate that activated oxygen is an important mediator of the antimalarial activity of endoperoxide antimalarials (Becker et al., 2004).

ACKNOWLEDGMENTS

Our work is supported by the Deutsche Forschungsgemeinschaft (Be 1540/4-3 and SFB 535 to K.B.) and by the National Health and Medical Research Council of Australia (N.H.).

REFERENCES

Abraham, E. 2000. Coagulation abnormalities in acute lung injury and sepsis. *Am. J. Respir. Cell Mol. Biol.* **22:**401–404.

Akoachere, M., R. Iozef, S. Rahlfs, M. Deponte, B. Mannervik, D. J. Creighton, H. Schirmer, and K. Becker. 2005. Characterization of the glyoxalases of the malarial parasite *Plasmodium falciparum* and comparison with their human counterparts. *Biol. Chem.* **386:**41–52.

Atamna, H., and H. Ginsburg. 1997. The malaria parasite supplies glutathione to its host cell—investigation of glutathione transport and metabolism in human erythrocytes infected with *Plasmodium falciparum*. *Eur. J. Biochem.* **250:**670–679.

Becker, K., S. M. Kanzok, R. Iozef, M. Fischer, R. H. Schirmer, and S. Rahlfs. 2003a. Plasmoredoxin, a novel redox-active protein unique for malarial parasites. *Eur. J. Biochem.* **270:**1057–1064.

Becker, K., S. Rahlfs, C. Nickel, and R. H. Schirmer. 2003b. Glutathione—function and metabolism in the malarial parasite *Plasmodium falciparum*. *Biol. Chem.* **348:**551–566.

Becker, K., L. Tilley, J. L. Vennerstrom, D. Roberts, S. Rogerson, and H. Ginsburg. 2004. Oxidative stress in malaria parasite-infected erythrocytes: host-parasite interactions. *Int. J. Parasitol.* **34:**163–189.

Ben Mamoun, C., I. Y. Gluzman, C. Hott, S. K. MacMillan, A. S. Amarakone, D. L. Anderson, J. M. Carlton, J. B. Dame, D. Chakrabarti, R. K. Martin, B. H. Brownstein, and D. E. Goldberg. 2001. Co-ordinated programme of gene expression during asexual intraerythrocytic development of the human malaria parasite *Plasmodium falciparum* revealed by microarray analysis. *Mol. Microbiol.* **39:**26–36.

Bozdech, Z., and H. Ginsburg. 2004. Antioxidant defense in *Plasmodium falciparum*—data mining of the transcriptome. *Malar. J.* **3:**23.

Bozdech, Z., M. Llinás, B. L. Pulliam, E. D. Wong, J. Zhu, and J. L. DeRisi. 2003. The transcriptome of the intraerythrocytic developmental cycle of *Plasmodium falciparum*. *PLoS Biol.* **1:**E5.

Buffinton, G. D., N. H. Hunt, W. B. Cowden, and I. A. Clark. 1988. Detection of short-chain carbonyl products of lipid peroxidation from malaria-infected erythrocytes exposed to oxidative stress. *Biochem. J.* **249**:63–68.

Campanale, N., C. Nickel, C. Daubenberger, D. Wehlan, J. Gorman, M. Foley, N. Klonis, K. Becker, and L. Tilley. 2003. Identification and characterisation of a series of haem-interacting proteins of the malaria parasite, *Plasmodium falciparum*. *J. Biol. Chem.* **278**:27354–27361.

Clark, I. A., W. B. Cowden, N. H. Hunt, and E. J. Mackie. 1984. Activity of divicine in *Plasmodium vinckei*-infected mice has implications for treatment of favism and epidemiology of G-6-PD deficiency. *Br. J. Haematol.* **57**:479–487.

Das, B. S., D. I. Thurnham, and D. B. Das. 1996. Plasma alpha-tocopherol, retinol, and carotenoids in children with falciparum malaria. *Am. J. Clin. Nutr.* **64**:94–100.

Davioud-Charvet, E., S. Delarue, C. Biot, B. Schwobel, C. C. Boehme, A. Mussigbrodt, L. Maes, C. Sergheraert, P. Grellier, R. H. Schirmer, and K. Becker. 2001. A prodrug form of a *Plasmodium falciparum* glutathione reductase inhibitor conjugated with a 4-anilinoquinoline. *J. Med. Chem.* **44**:4268–4276.

De Macedo, C. S., M. L. Uhrig, E. A. Kimura, and A. M. Katzin. 2002. Characterization of the isoprenoid chain of coenzyme Q in *Plasmodium falciparum*. *FEMS Microbiol. Lett.* **207**:13–22.

Dive, D., S. Gratepanche, H. Yera, P. Becuwe, W. Daher, P. Delplace, C. Odberg-Ferragut, M. Capron, and J. Khalife. 2003. Superoxide dismutase in *Plasmodium*: a current survey. *Redox Rep.* **8**:265–267.

Dondorp, A. M., M. Nyanoti, P. A. Kager, S. Mithwani, J. Vreeken, and K. Marsh. 2002. The role of reduced red cell deformability in the pathogenesis of severe falciparum malaria and its restoration by blood transfusion. *Trans. R. Soc. Trop. Med. Hyg.* **96**:282–286.

Eaton, J. W., J. R. Eckman, E. Berger, and H. S. Jacob. 1976. Suppression of malaria infection by oxidant-sensitive host erythrocytes. *Nature* **264**:758–760.

Eda, S., and I. W. Sherman. 2002. Cytoadherence of malaria-infected red blood cells involves exposure of phosphatidylserine. *Cell. Physiol. Biochem.* **12**:373–384.

Engwerda, C. R., T. L. Mynott, S. Sawhney, J. B. De Souza, Q. D. Bickle, and P. M. Kaye. 2002. Locally up-regulated lymphotoxin alpha, not systemic tumor necrosis factor alpha, is the principle mediator of murine cerebral malaria. *J. Exp. Med.* **195**:1371–1377.

Famin, O., and H. Ginsburg. 2003. The treatment of *Plasmodium falciparum*-infected erythrocytes with chloroquine leads to accumulation of ferriprotoporphyrin IX bound to particular parasite proteins and to the inhibition of the parasite's 6-phosphogluconate dehydrogenase. *Parasite* **10**:39–50.

Favre, N., B. Ryffel, and W. Rudin. 1999. Parasite killing in murine malaria does not require nitric oxide production. *Parasitology* **118**:139–143.

Fritz-Wolf, K., A. Becker, S. Rahlfs, P. Harwaldt, R. H. Schirmer, W. Kabsch, and K. Becker. 2003. X-ray structure of glutathione S-transferase from the malarial parasite *Plasmodium falciparum*. *Proc. Natl. Acad. Sci. USA* **100**:13821–13826.

Gamain, B., G. Langsley, M. N. Fourmaux, J. P. Touzel, D. Camus, D. Dive, and C. Slomianny. 1996. Molecular characterization of the glutathione peroxidase gene of the human malaria parasite *Plasmodium falciparum*. *Mol. Biochem. Parasitol.* **78**:237–248.

Ginsburg, H., and H. Atamna. 1994. The redox status of malaria-infected erythrocytes: an overview with an emphasis on unresolved problems. *Parasite* **1**:5–13.

Golenser, J., J. Miller, D. T. Spira, T. Navok, and M. Chevion. 1983. Inhibitory effect of a fava bean component on the in vitro development of *Plasmodium falciparum* in normal and glucose-6-phosphate dehydrogenase deficient erythrocytes. *Blood* **61**:507–510.

Grellier, P., J. Sarlauskas, Z. Anusevicius, A. Maroziene, C. Houee-Levin, J. Schrevel, and N. Cenas. 2001. Antiplasmodial activity of nitroaromatic and quinoidal compounds: redox potential vs. inhibition of erythrocyte glutathione reductase. *Arch. Biochem. Biophys.* **393**:199–206.

Griffiths, M. J., F. Ndungu, K. L. Baird, D. P. Muller, K. Marsh, and C. H. Newton. 2001. Oxidative stress and erythrocyte damage in Kenyan children with severe *Plasmodium falciparum* malaria. *Br. J. Haematol.* **113**:486–491.

Harada, M., M. Owhashi, S. Suguri, A. Kumatori, M. Nakamura, H. Kanbara, H. Matsuoka, and A. Ishii. 2001. Superoxide-dependent and -independent pathways are involved in the transmission blocking of malaria. *Parasitol. Res.* **87**:605–608.

Harwaldt, P., S. Rahlfs, and K. Becker. 2002. Glutathione S-transferase of the malarial parasite *Plasmodium falciparum*: characterization of a potential drug target. *Biol. Chem.* **383**:821–830.

Iozef, R., S. Rahlfs, T. Chang, R. H. Schirmer, and K. Becker. 2003. Glyoxalase 1 of the malarial parasite *Plasmodium falciparum*. Evidence for subunit fusion in glyoxalase I. *FEBS Lett.* **554**:284–288.

Kanzok, S. M., A. Fechner, H. Bauer, J. K. Ulschmid, H. M. Muller, J. Botella-Munoz, S. Schneuwly, R. Schirmer, and K. Becker. 2001. Substitution of the thioredoxin system for glutathione reductase in *Drosophila melanogaster*. *Science* **291**:643–646.

Kita, K., K. Hirawake, H. Miyadera, H. Amino, and S. Takeo. 2002. Role of complex II in anaerobic respiration of the parasite mitochondria from *Ascaris suum* and *Plasmodium falciparum*. *Biochim. Biophys. Acta* **1553:**123–139.

Komaki-Yasuda, K., S. Kawazu, and S. Kano. 2003. Disruption of the *Plasmodium falciparum* 2-Cys peroxiredoxin gene renders parasites hypersensitive to reactive oxygen and nitrogen species. *FEBS Lett.* **547:**140–144.

Le Roch, K. G., Y. Zhou, P. L. Blair, M. Grainger, J. K. Moch, J. D. Haynes, P. De la Vega, A. A. Holder, S. Batalov, D. J. Carucci, and E. A. Winzeler. 2003. Discovery of gene function by expression profiling of the malaria parasite life cycle. *Science* **301:**1503–1508.

Lew, V. L., T. Tiffert, and H. Ginsburg. 2003. Excess hemoglobin digestion and the osmotic stability of *Plasmodium falciparum*-infected red blood cells. *Blood* **101:**4189–4194.

Liebau, E., B. Bergmann, A. M. Campbell, P. Teesdale-Spittle, P. M. Brophy, K. Luersen, and R. D. Walter. 2002. The glutathione S-transferase from *Plasmodium falciparum*. *Mol. Biochem. Parasitol.* **124:**85–90.

McMillan, P. J., L. M. Stimmler, B. J. Foth, G. I. McFadden, and S. Muller. 2005. The human malaria parasite Plasmodium falciparum possesses two distinct dihydrolipoamide dehydrogenases. *Mol. Microbiol.* **55:**27–38.

Meierjohann, S., R. D. Walter, and S. Muller. 2002. Regulation of intracellular glutathione levels in erythrocytes infected with chloroquine-sensitive and chloroquine-resistant *Plasmodium falciparum*. *Biochem. J.* **368:**761–768.

Meshnick, S. R. 2002. Artemisinin: mechanisms of action, resistance and toxicity. *Int. J. Parasitol.* **32:**1655-1660.

Muller, S. 2004. Redox and antioxidant systems of the malaria parasite *Plasmodium falciparum*. *Mol. Microbiol.* **53:**1291–1305.

Rahlfs, S., and K. Becker. 2001. Thioredoxin peroxidases of the malarial parasite *Plasmodium falciparum*. *Eur. J. Biochem.* **268:**1404–1409.

Rahlfs, S., C. Nickel, M. Deponte, R. H. Schirmer, and K. Becker. 2003. *Plasmodium falciparum* thioredoxins and glutaredoxins as central players in redox metabolism. *Redox Rep.* **8:**246–250.

Rahlfs, S., R. H. Schirmer, and K. Becker. 2002. The thioredoxin system of *Plasmodium falciparum* and other parasites. *Cell. Mol. Life Sci.* **59:**1024–1041.

Sanni, L. A., S. Fu, R. T. Dean, G. Bloomfield, R. Stocker, G. Chaudhri, M. C. Dinauer, and N. H. Hunt. 1999. Are reactive oxygen species involved in the pathogenesis of murine cerebral malaria? *J. Infect. Dis.* **179:**217–222.

Sarma, G. N., S. N. Savvides, K. Becker, M. Schirmer, R. H. Schirmer and P. A. Karplus. 2003. Glutathione reductase of the malarial parasite *Plasmodium falciparum*: crystal structure and inhibitor development. *J. Mol. Biol.* **328:**893–907.

Schirmer, R. H., H. Bauer, and K. Becker. 2002. Glutathione reductase, p. 1471–1476. *In* T. E. Creighton (ed.), *Wiley Encyclopedia of Molecular Medicine*. John Wiley and Sons, New York, N.Y.

Sienkiewicz, N., W. Daher, D. Dive, C. Wrenger, E. Viscogliosi, R. Wintjens, H. Jouin, M. Capron, S. Muller, and J. Khalife. 2004. Identification of a mitochondrial superoxide dismutase with an unusual targeting sequence in *Plasmodium falciparum*. *Mol. Biochem. Parasitol.* **137:**121–132.

Sullivan, D. J. 2002. Theories on malarial pigment formation and quinoline action. *Int. J. Parasitol.* **32:**1645–1653.

Sztajer, H., B. Gamain, K. D. Aumann, C. Slomianny, K. Becker, R. Brigelius-Flohé, and L. Flohé. 2001. The putative glutathione peroxidase gene of *Plasmodium falciparum* codes for a thioredoxin peroxidase. *J. Biol. Chem.* **276:**7397–7403.

Thornalley, P. J., M. Strath, and R. J. Wilson. 1994. Antimalarial activity in vitro of the glyoxalase I inhibitor diester, S-p-bromobenzylglutathione diethyl ester. *Biochem. Pharmacol.* **47:**418–420.

Tilley, L., P. Loria, and M. Foley. 2001. Chloroquine and other quinoline antimalarials, p. 87–122. *In* P. J. Rosenthal (ed.), *Antimalarial Chemotherapy*. Humana Press, Totowa, N.J.

Turrens, J. F. 2004. Oxidative stress and antioxidant defense: a target for the treatment of diseases caused by parasitic protozoa. *Mol. Aspects Med.* **25:**211–220.

NEW PERMEATION PATHWAYS

Serge L. Thomas and Stéphane Egée

20

The intraerythrocytic stage of the life cycle allows the malaria parasite to escape the host immune system for production of new parasites. However, in doing so, it enters a hostile environment where nutrients are not present in sufficient amounts, and accumulation of potentially hazardous metabolic end products may rapidly become critical. Production of new parasites generates metabolic processes that demand more energy than the host red blood cell (RBC) can provide. The RBC membrane is naturally endowed with a variety of membrane transporters, mainly geared to optimize the respiratory function and to maintain cell homeostasis at minimal metabolic cost. Therefore, to survive within a red blood cell, the malaria parasite must alter the permeability of the host's plasma membrane by up-regulation of existing carriers or by creation of new permeation pathways (NPP). These pathways, indispensable for parasite growth, could be possible antimalarial targets for selective inhibition, as well as routes for drug delivery. It is therefore important to characterize these in detail.

Discovering potent and specific inhibitors of the NPP is an important therapeutic challenge, but many questions remain unanswered. Do the NPP correspond to a single path or multiple pathways? Are they parasite-derived proteins? Are they up-regulated or modified endogenous quiescent red blood cell proteins? Flux and hemolysis measurements have established that the functional and pharmacological properties of NPP display many similarities to anion-selective channels present in other cell types. Electrophysiological techniques, such as the patch clamp, are ideal for the study of channels that are permeable to charged solutes, even though the NPP are also used for transport of electroneutral and organic osmolytes in malaria-infected RBCs. These techniques have demonstrated that anion channels are activated in the RBC's plasma membrane following infection, but the origin, number, and molecular identity of these channels remain the subject of continuing investigation and debate.

THE INTRAERYTHROCYTIC STAGE OF THE PARASITE LIFE CYCLE

The life cycle of a malaria parasite is complex and requires the parasite to migrate through and

---

Serge L. Thomas and Stéphane Egée, CNRS, FRE 2775, Station Biologique, Place G. Teissier, BP 74, 29682 Roscoff, France.

multiply within both a vertebrate host and a mosquito vector. *Plasmodium falciparum* invades the human red blood cell during 48 h for production of up to 30 new parasites, but in this environment the nutrient supply and waste removal routes are largely insufficient. In addition, they enter a milieu with high K^+ (140 mM) and low Na^+ (10 mM) and Ca^{2+} (0.1 μM) concentrations. Despite these hostile conditions, they gradually increase the glycolytic rate of the infected RBCs 100-fold; generate new metabolic processes; increase the traffic of nutrients, waste products, and cations; and dissipate the normal Na and K gradients across the host cell membrane. This strategy raises several fascinating scientific questions, including how malaria parasites are able to obtain nutrients, remove waste products, and grow without the red blood cell swelling and bursting prematurely.

The intracellular parasite obtains most of the nutrients it requires from the digestion of components of the host's cytosol (predominantly hemoglobin) (Rosenthal and Meshnick, 1998). Nevertheless, for a *P. falciparum* parasite to grow, several essential nutrients have to be supplied from the external milieu of the infected cell. For instance, human hemoglobin does not contain the amino acid isoleucine, and parasite survival is totally dependent on supply of exogenous isoleucine (Sherman, 1977). To reach the parasite, these substrates must first cross the host red blood cell plasma membrane. In some cases, e.g., the case of glucose (the primary energy source for the parasite), the endogenous (i.e., native) specific transporters in the human red blood cell plasma membrane are capable of maintaining an adequate supply. In other cases, the human red blood cell plasma membrane simply has no transport pathways for essential nutrients including, for example, the vitamin pantothenic acid or the amino acid glutamate (Divo et al., 1985).

It has long been recognized that as the intracellular parasites mature, the transport properties of the host cell membrane become markedly different from the normal (uninfected) red blood cell membrane (Ancelin et al., 1991; Homewood and Neame, 1974; Staines et al., 2002; Staines and Kirk, 1998). Mature infected cells possess malaria-induced transport pathways which increase the transport of many specific solutes (for comprehensive reviews, see Ginsburg, 1994; Ginsburg and Kirk, 1998; and Kirk, 2001). There is evidence for the up-regulation of endogenous red blood cell transport systems to increase substrate delivery (monosaccharides, small polyols, amino acids, peptides, nucleosides, monocarboxylates, and monovalent inorganic anions and cations). The parasite also produces large amounts of lactate from glycolysis and waste amino acids from the digestion of hemoglobin, which need to be removed (Poole and Halestrap, 1993). Additionally, these waste products exert significant osmotic effects, and since the parasite occupies a significant fraction of the intracellular compartment, induced pathways may play an important role in cell volume regulation. If left to accumulate, these products would build up to toxic levels and lead to cell swelling, followed by cell rupture, before the parasite finished multiplication (Krugliak et al., 2002; Lew et al., 2003).

These inward and outward movements are essential for parasite growth. The mechanisms underlying up-regulation of endogenous transporters are still unidentified, but it is known that in addition to changes in cytosolic concentrations of ions and protein, the altered biophysical characteristics of the parasitized host membrane (Vial and Ancelin, 1998) are likely to induce dramatic modifications in the activity of transport systems.

NEW PERMEATION PATHWAYS

In cases where endogenous systems do not exist or cannot maintain an adequate supply, the parasite activates NPP within 10 to 20 h postinvasion, which allows the transport of a diverse range of structurally unrelated, low-molecular-weight solutes. Most of what is currently known about the functional and pharmacological properties of NPP results from flux and hemolysis measurements performed by Hagai Ginsburg and colleagues at the Hebrew University of Jerusalem (Ginsburg, 1994; Ginsburg and Kirk, 1998; Ginsburg et al., 1983; Ginsburg et al., 1985) and by Kiaran Kirk and colleagues at the University of Oxford, Oxford, United Kingdom and the Australian National University (Kirk, 2001; Kirk et al., 1994; Kirk et al., 1999). Their research characterized the nature of the transported substrates, their rates of transport, the selectivity properties of the path, and the main inhibitors. This early functional and pharmacological characterization provided crucial clues about the possible nature of NPP, suggesting that it was a large, poorly selective anion channel which showed many of the characteristics of the volume-sensitive anion channels (volume-regulated anion channels, volume-sensitive organic osmolyte/anion channels, swelling-activated Cl^- current) found in most mammalian cells (Kirk and Strange, 1998; Nilius and Droogmans, 2003). It is generally accepted that these channels allow both organic and inorganic anions, electroneutral molecules, and organic and inorganic monovalent cations to pass, although the latter at much lower rates. Whether these NPP represent modification of host membrane proteins (e.g., aggregation and/or proteolysis via parasite-derived peptides or enzymes) or insertion of novel parasite-engendered channels is still not resolved. This question has been around for a long time and it is unlikely to be answered until the NPP are identified at a molecular level. From an immunological point of view, the parasite is less likely to be detected if it can use endogenous transport pathways, but it is also known that the parasite expresses several of its own proteins at the red blood cell surface (Su et al., 1995). The recent publication of the *P. falciparum* genome (Gardner et al., 2002) increased the wealth of data on parasite-encoded transporters but did not improve our understanding of the NPP.

The *P. falciparum*-induced NPP have been characterized as being a single type of pathway with the general properties of a channel (i.e., linear concentration dependence, low energy of activation, and the inability to distinguish between stereoisomers of permeant solutes). The NPP are selective for anionic solutes (i.e., negatively charged ions such as Cl^- and lactate) and a range of electroneutral (e.g., polyols, amino acids, and nucleosides) and cationic (i.e., positively charged ions such as Na^+ and K^+) solutes with a permeability sequence Cl^- > lactate > thymidine > adenosine > pantothenate > taurine > sorbitol > glutamate > carnitine > choline > Rb^+ > K^+ > Na^+. In addition, compounds known to block other anion-selective transport pathways also inhibit the NPP. The main NPP inhibitors, furosemide, 5-nitro-2-(3-phenylpropylamino)benzoic acid (NPPB), niflumic acid, and glibenclamide, are typical anion channel inhibitors. Thus, ion selectivity and pharmacological data strongly support the notion that NPP are part of the anion channel family. Although the Na and K permeabilities of the NPPs are 10,000-fold lower than that of Cl, they are sufficiently high to allow rapid dissipation of the normal Na and K gradients of RBCs, so that by ~35 to 40 h after invasion, the parasite is surrounded by a high-Na, low-K environment, similar to that in extracellular fluids (Staines et al., 2001). The parasite's internal milieu, on the other hand, is maintained with high K and low Na concentrations by the combined performance of active and passive membrane transporters in the parasite plasma membrane (Allen and Kirk, 2004). The result of a recent analysis (Ginsburg and Stein, 2004) of available data from tracer fluxes and isosmotic lysis suggests that the NPP consist of two types of channels. The first type is present in a small number of copies per cell (about four) and is charge and size selective. It discriminates against cations by a factor of 4. The second type is 100-fold more abundant (about 400 copies per cell) and allows the movement of anions and nucleosides.

ELECTROPHYSIOLOGICAL STUDIES

Since a wealth of transport studies have demonstrated the importance of parasite-induced transport pathways for anions and other solutes in infected RBCs and since the NPP show many similarities to anion-selective channels, the patch clamp electrophysiological technique represented the best method available to investigate channel-mediated transport of charged solutes (Hamill et al., 1981) through the infected red blood cell membrane. By measuring the electrical current across a membrane (produced by the charged solutes) in relation to a known applied potential difference, it is possible to obtain a detailed characterization of ion-transporting channels at both the tissue and single-cell level. However, until recently, electrophysiological studies of human RBCs have been difficult, due to the fragility and small size of the cells, and little is known of the anionic conductive pathways present in the RBC membrane in health and disease. The work on the human red blood cell is in marked contrast to early studies on nucleated, e.g., amphibian (Hamill, 1983), and more recently fish red blood cells (Egée et al., 1998; Lapaix et al., 2002) where patch clamping proved easy and productive; the difference was attributed to the size of these cells. To circumvent this problem, some researchers have therefore chosen nucleated avian red blood cells (which are infected by the malarial parasite *Plasmodium gallinaceum*) (Thomas et al., 2001) as a system more amenable to study the malaria-infected cell by patch clamp methods.

Under a microscope, a small glass tube, the patch electrode, thinned to a cone-shaped end (usually ~ 1 μm^2 in area with ~ 10 MΩ tip resistance) and filled with an appropriate saline solution is brought in contact with the surface of a cell with the purpose of generating a very high resistance electrical seal around the edge of the tip. Since the object is to detect and measure tiny electrical currents through channels in the cell membrane, the seal has to be highly resistant ($>10^9$ Ω) to prevent artefacts resulting from leaky seals. Once the seal is established, the membrane potential across the patch or whole cell is fully under experimental control and may be held (holding potential) or varied at will. Currents are measured by applying voltage steps across the membrane, usually following preset sequences (ramps).

The patch clamp technique can be used in three distinct configurations, as follows.

1. Cell attached. This configuration is used for measuring the current through the approximately 1 μm^2 of membrane within the tip of the pipette. This configuration maintains the integrity of the cell and is therefore the most physiological; it does not, however, provide access to the cytosolic side of the membrane.

2. Excised inside-out. This configuration allows full control of the ionic composition of the cytosolic side of the membrane for selectivity and substitution experiments and for pharmacological assays. The two configurations of the excised mode (inside and outside) enable measurements of unitary currents at the single channel level, thus allowing a biophysical characterization of all channel types present in the membrane under study. In these configurations, the success rate of experiments depends largely on the surface density of channels. Considering the small fraction of membrane under the patch, it takes hundreds of observations to build up a reliable and comprehensive picture of the density and variety of channels present in the membrane.

3. Whole-cell configuration, obtained from the cell-attached configuration. When the patch of membrane within the tip is ruptured by suction or by application of brief electric shocks, the currents measured are those flowing through all the ionic channels present and active in the cell. In this configuration, the cell content changes dramatically as the cell cytosol equilibrates with the fluid within the patch pipette.

In the cell-attached configuration, the recordings of flickering currents provide important information on how single-channel conductance and open-state probabilities vary with voltage. In whole-cell recordings, the conductance through the whole-membrane area measures the number of active channels per cell if it is assumed that the current is carried by one specific type of channel of known single-channel conductance and open-state probability distribution. The current voltage curves provide additional information on the collective kinetics of the channels. In general, current voltage curves that deviate from the usual linear ohmic behavior are described as rectified; conductance measured under patch clamp conditions is conventionally described as an outward rectifier if the current is reduced at negative-membrane voltages (relative to the extracellular medium), and as an inward rectifier if reduced at positive voltages. When applied to single channels, rectification properties reflect the kinetics of the molecular components controlling the aperture of the conductive path.

ENDOGENOUS CHANNELS AS CANDIDATES TO REPRESENT NPP

Red blood cells have proven to be extremely useful as a model system to study the different membrane transport pathways, and there is a plethora of publications aimed at a detailed description of pumps, cotransporters, and specific carriers in the red blood cell membrane (Bernhardt and Ellory, 2003). In contrast, until recently, there have been few reports detailing ionic channels in human RBCs. Previous electrophysiological studies of noninfected RBCs have identified two different cation channels: an intermediate conductance Ca^{2+}-activated K^+ channel, known as the Gàrdos channel (Grygorczyk and Schwarz, 1983; Hamill, 1981) and a voltage-dependent nonselective cation channel (Christophersen and Bennekou, 1991). The Ca^{2+}-activated K^+ Gàrdos channel does not seem to be involved in volume regulation or K^+ homeostasis in malaria-infected human red blood cells (Kirk et al., 1992), even at physiologically relevant external Ca^{2+} concentrations (1.0 to 1.3 mM) and can only be activated in the presence of ionophore A23187. An early study by Schwartz and coworkers (1989) reported small electrophysiological events, which are almost certainly due to anion transport; several studies reported biochemical evidence that anion channels are probably present in the red blood cell membrane (Abraham et al., 2001; Freedman et al., 1994; Schwartz et al., 1997; Sprague et al., 1998). So far, most studies of nucleated erythrocytes using electrophysiological techniques have demonstrated the existence of cationic and anionic channels spontaneously active in the membrane and involved in volume regulation (Egée et al., 1998; Lapaix et al., 2002; Thomas et al., 2001). The recent surge of electrophysiological activity resulting from recent improvement in techniques has revealed the existence of quiescent anionic channels in the red blood cell membrane (Egée et al., 2002). The absence of spontaneously active channels in the mature anucleated red blood cell is in keeping with the low metabolic demand and such dormant channel is probably linked to the physiology of premature nucleated stages of these cells.

Under steady-state conditions, H^+ and Cl^- distribution across the RBC membrane is purely passive and in accordance with a Donnan equilibrium; the intraerythrocytic proton concentration results from a physicochemical equilibrium where Na^+ and K^+ movements across the RBC membrane are extremely slow and the Cl^- and HCO_3^- movements are approximately 1 million times faster. Chloride movements occur through a highly efficient, electrically silent $Cl^-/HCO3^-$ exchange mechanism that is fundamental to the CO_2-carrying capacity of the blood. Using the values for extra- and intracellular concentrations for Cl^- (145 mM and 95 mM, respectively), it is possible to calculate the reversal potential for Cl^- $E_{cl^-} = -10$ mV, which is similar to the membrane potential estimated from techniques such as microelectrode measurements (Lassen and Sten-Knudsen, 1968) or fluorescent dye staining (Hoffman and Laris, 1974). This is an indication that a chloride conductance normally determines the resting potential. Under these conditions, the functional significance of a chloride conductance may be to set the RBC resting potential at E_{Cl} so that there is no electrochemical gradient for conductive anion movement (Hamill, 1983). This voltage clamping of E_{Cl} may optimize the role of the Cl^-/HCO_3^-.

Using the whole-cell configuration of the patch clamp technique, uninfected RBCs display negligible conductance. The membrane conductance (G_m) is in the range of 40 to 50 pS.

Using the excised inside-out configuration to identify the different channel types present in the plasma membranes of uninfected RBCs, two types of anion channels were found (Egée et al., 2002). The first type has linear conductance and is apparent in >80% of membrane patches under appropriate conditions. The second channel type displays outward rectification and is present in <5% of uninfected cell patches. The linear conductance anion channel is not spontaneously active, but it can be activated by exposure of the cytosolic side of the membrane to the catalytic subunit of protein kinase A (PKA) (100 nM) in the presence of ATP (1 mM). The unit conductance is ~15 pS. ATP alone cannot induce this channel activity. Similar channel activity can also be observed in the absence of PKA by the application of suction to the pipette of quiescent patches. Under this condition, the open probability (P_o) values measured at a given membrane potential are related to the degree of depression imposed on the membrane patch. The unit conductance is identical to the value calculated by PKA-induced activation. This linear conductance anion channel exhibits two characteristic substates corresponding to one-third and two-thirds of the full amplitude, has a halide selectivity of $I^- > Br^- > Cl^-$, and is inhibited by 100 μM NPPB, 100 μM niflumic acid, 1 mM diphenylamino-2-carboxylic acid, and 10 μM tamoxifen. 4,4′-Diiso-thiocyano-2,2′-disulfonic acid stilbene (DIDS), at a concentration of 100 μM, induces a variable partial block. The second, rare anion channel displays strong outward rectification. The channel slope conductance is ~80 pS at positive membrane potentials. Addition of the chloride channel blockers NPPB, DIDS, and 9-anthracenecarboxylic acid on the cytosolic face of excised patches (100 μM) leads to 90 to 100% inhibition of channel activity. The presence of PKA and ATP is never observed to induce activity of this channel type.

ANION CHANNELS IN INFECTED RBCs

The first unambiguous electrophysiological evidence that anion-selective channels are spontaneously active in human infected red blood cells was obtained by Desai and coworkers in Bethesda, Md. (2000), by the patch clamp technique. They demonstrated that the membrane conductance of *P. falciparum*-infected human RBCs is 150 times larger than that measured in uninfected RBCs and that this increased conductance results from the activation of small anion channels showing functional and pharmacological properties in keeping with those reported previously for the NPPs. They concluded that the small anion channels formed the NPP. In addition, they reported that the conductance rectifies inwardly (i.e., the conductance decreases at positive potentials and increases at negative potentials) and has an anion selectivity of $I^- > Br^- > Cl^- >$ lactate. However, whether these channels are up-regulated and/or modified endogenous RBC proteins or are parasite engendered was not addressed.

The Thomas group in France reported similar results (Egée et al., 2002). They found that *P. falciparum*-infected RBCs have a membrane conductance G_m, calculated between −100 and 0 mV of around 5,000 pS. NPPB, niflumic acid, and glibenclamide added in the bathing solution reduce the membrane conductance of infected cells immediately with 50% inhibitory concentration values, the concentration of inhibitor that decreases transport via the NPP by 50% of 0.8, 2.5, and 35 μM, respectively. Furosemide (200 μM), a potent blocker of malaria-induced solute transport, only reduces the membrane conductance by 45%; 100 μM DIDS produces no significant inhibition. With the exception of the weak effect of furosemide, these data are in good agreement with the work of Desai and coworkers. The Ellory group (Oxford, United Kingdom) and the De Jonge group (Rotterdam, The Netherlands) confirmed the presence of an inwardly rectifying anion conductance (Staines et al., 2003; Verloo et al., 2004) in the membranes of infected cells.

In the cell-attached and excised configurations, *P. falciparum*-infected RBCs show a very different pattern of channel activity from uninfected cells (Egée et al., 2002). Spontaneous channel activity is observed in infected RBCs as bursts of channel openings separated by short closures. These channels exhibit voltage-dependent gating, with low P_o values of between +50 and +100 mV, and increasing P_o values of between +50 mV and −100 mV. In addition, for a given membrane patch, the number of simultaneously active channels is directly proportional to the membrane potential (V_m), with the maximum number activated at negative potentials. The mean unit conductance is ~18 pS.

The unitary conductance, substrate selectivity, and pharmacology of the observed parasite-induced channel are identical to those of the endogenous linear conductance anion channel measured in uninfected RBCs. While the majority of comparable data are consistent with the observations of Desai and coworkers, the mean unit conductance of the anion channel measured in malaria-infected RBCs by the Thomas group is several times higher than that reported by Desai's group (~18 pS versus ~3 pS). Desai and coworkers used the cell-attached configuration with 1150 mM Cl^- in the bath and pipette solutions which approximated a channel conductance of only 3 pS under physiological conditions (155 mM Cl^-). Using the data presented above, if we assume that the whole-cell current can be completely accounted for by one channel type, it is possible to calculate that approximately 275 copies of an 18-pS channel (Egée et al., 2002) or 1,500 copies of a 3-pS channel (Desai et al., 2000) could carry the measured whole-cell currents in infected RBCs.

A study by the Lang group (Germany), using only the whole-cell configuration (Huber et al., 2002), confirmed an inwardly rectifying anion conductance in *P. falciparum*-infected RBCs but also reported an outwardly rectifying anion conductance. With additional inhibitor studies, they concluded that at least two anion-selective channel types must be induced by the parasite. However, this group could not confirm the existence of outwardly rectifying channel by single-channel recording.

Only inwardly rectifying channels were detected in cell-attached configurations in infected cells (not a single outward rectifier was detected by the Thomas group). Thus, there is no experimental justification for attributing whole-cell outwardly rectifying currents to outwardly rectifying channels. Therefore, inwardly rectifying channels are at present the only candidates to represent NPP. In addition, the fact that an outwardly rectifying channel was seen in uninfected but not in infected cells raises the interesting possibility that this channel could be inactivated or modified by the parasite. Indeed, we can ask the following question. Why does the outwardly rectifying channel, present in about 3% of patches in noninfected cells, suddenly disappear from infected cell membranes? A reasonable working hypothesis is that this channel in the cell-attached configuration in infected cells escapes electrophysiological recordings because it carries electrically silent solutes. Considering the 3% frequency of occurrence in noninfected cells, the size of the patches, and the total surface of a red blood cell, one could estimate the number of copies of outwardly rectifying channels to <10, which would fit quite well with calculations by Ginsburg and Stein (2004) for the nonelectrolyte channel. However, we cannot exclude the possibility that the channels that produce outward whole-cell currents are too small to be seen in single-channel recordings. Of course, another possibility is that the inward or outward rectifying behavior in whole-cell configurations represents alternative kinetic modalities of the same channels, dependent on conditions. For instance, it was shown recently (Staines et al., 2003) in whole-cell configuration that a combination of the use of a negative holding potential and the presence of a factor contained in serum (and used during in vitro parasite culture) alters the conductance phenotype from inwardly to outwardly rectifying. By contrast, in the cell-attached configuration (S. L. Thomas et al., unpublished data) these maneuvers failed to activate outwardly rectifying channels in infected cells, no matter the value of the holding potentials.

In view of the above data it is reasonable to conclude that the membrane conductance of *P. falciparum*-infected RBCs results from the upregulation of endogenous, normally quiescent, channels. The Lang group presented evidence that anionic conductances with characteristics essentially identical to those that they observed with *P. falciparum*-infected RBCs could be induced in uninfected RBCs by oxidation (Duranton et al., 2002; Huber et al., 2002). Although using different stimuli, they also concluded that the parasite induces a quiescent endogenous pathway(s). In addition, when the patch clamp method was used with avian RBCs infected by *P. gallinaceum*, it was impossible to identify novel parasite-induced pathways in infected chicken RBCs (Thomas et al., 2001); rather, the avian malaria parasite up-regulates existing pathways in the host RBC membrane. The channels induced by malaria in chicken RBCs were mostly nonselective cationic channels but they were clearly identified as up-regulated endogenous channels, a finding which is consistent with the hypothesis that *P. falciparum* up-regulates native channels in the host plasma membrane. NPP could thus have resulted from the opportunistic control of endogenous resources by the parasite. Of course, it remains possible that the parasite also exports its own channel to the host membrane, as suggested by Desai and coworkers, who used the name *Plasmodium* surface anion channel (Alkhalil et al., 2004; Desai, 2004) for this putative protein. However, since no suitable candidates have turned up in the parasite genome, the presence of an up-regulated endogenous channel offers an interesting alternative to parasite-encoded channel proteins. If this is the case, the observed differences in gating and P_o between the native anion channel and the malaria-induced channel may well be due to the changes occurring in the erythrocyte cytosol and the manner by which the parasite up-regulates the native channels. This hypothesis is reinforced by the differences in the gating of the anion channel measured, using two different modes of activation (PKA/ATP and deformation).

Thomas and coworkers observed that it is possible to induce a whole-cell current in uninfected RBCs by the addition of PKA and ATP, which mimics the membrane current observed in infected cells, albeit to a lower extent (Egée et al., 2002). In addition, they showed that the membrane conductance of malaria-infected RBCs is totally and immediately abolished in the presence of alkaline phosphatase (0.1 U/ml) (Decherf et al., 2004). This suggests that the mechanism of up-regulation used by the parasite may involve phosphorylation steps. It is known that the malaria parasite produces many kinases (Kappes et al., 1999), and parasite-dependent kinase activities have been detected at the membrane of the host RBC (Chishti et al., 1994; Droucheau et al., 2004). Furthermore, the host PKA present in the erythrocyte appears to be modified, following infection with *P. falciparum*, which produces its own PKA molecule (Syin et al., 2001). A better understanding of a possible role for these kinases in activating anion channels could be important in developing strategies for future malarial chemotherapies.

MOLECULAR NATURE OF THE ACTIVATED ANION CHANNELS

The molecular natures of the two anion channels observed in the plasma membrane of human RBCs are as yet unknown. However, they may well be related to known anion channels, which are ubiquitous in many cell types (Jentsch, 2002; Nilius and Droogmans, 2003): the ATP binding cassette family of chloride channels (including the cystic fibrosis [CF] transmembrane regulator protein, CFTR), the volume-regulated anion channels, and the chloride channel family (ClC). In addition, the anion exchanger band 3 is not a channel, but the malaria parasite modifies band 3 (Crandall and Sherman, 1991), which could acquire channel-like properties.

The Thomas group showed evidence that CFTR-like channels may be involved directly in the formation of the NPP (Egée et al., 2002). Indeed, the linear, small-conduction Cl^- channel identified in uninfected and malaria-infected human RBCs displays some of the functional characteristics (conductance, pharmacology, and mode of activation) of CFTR channels (Tabcharani et al., 1991). These features support their putative identity as CFTR-like channels, even though mechanosensitivity was never observed previously for CFTR channels. The existence of CFTR in the plasma membrane of human RBCs is supported by the biochemical observations of Abraham and coworkers (2001) showing the presence of different ATP binding cassette proteins (CFTR, multidrug resistance-associated protein 1, and multidrug resistance protein) in the RBC membrane. The mechanosensitivity of the linear conductance channel type is interesting and may relate to the ATP transport pathway reported by Sprague and coworkers (Sprague et al., 1998; Sprague et al., 2001) in human RBCs, which is activated by mechanical deformation and reduced in RBCs from patients with CF. This would support a model in which mechanical deformation of RBCs in the vasculature would result in the activation of anionic channels. These channels would then provide a pathway for ATP release and, thereby, play an important role in the regulation of vascular resistance in vivo (by the activation of puringeric receptors and synthesis of nitric oxide) (Busse et al., 1988). It may also be possible that CFTR acts as a regulator. The De Jonge group inferred from their whole-cell observations that two types of anionic channels are activated by the malaria parasite (Verloo et al., 2004). One is a CFTR channel, but according to these authors, it would only play a role of regulator of a second inwardly rectifying and shrinkage-activated channel. According to these authors, only the second type of channel, which displays characteristics identical to malaria-induced channels, could account for the membrane conductance found in infected cells. The Lang group concentrated predominantly on the involvement of anion channel ClC-2 (Huber et al., 2004). Western blots and fluorescence-activated cell sorter analysis showed protein and functional ClC-2 expression in human erythrocytes. Supporting evidence for ClC-2 in malaria-infected mice RBCs with and without the ClC-2 protein included pharmacology (inhibition by Zn^{2+}) and modulation (sensitivity to cell volume).

NPP AS THERAPEUTIC TARGET OR ROUTE FOR DRUG DELIVERY?

It has been suggested that NPP could be used as therapeutic targets (Ginsburg, 1994; Krishna et al., 2002). Several compounds corresponding to large hydrophobic organic anions (e.g., based on the loop diuretic furosemide) have been identified as potent inhibitors of NPP with 50% inhibitory concentration values in the submicromolar concentration range (Kirk and Horner, 1995; Staines et al., 2004). However, although these compounds inhibit parasite protein synthesis in culture conditions, their therapeutic effectiveness is reduced primarily because they are not specific and inhibit other transport pathways (e.g., the anion exchanger and band 3). In addition, their potency is reduced in the presence of plasma (since they bind strongly to albumin and other plasma proteins), and it is not known how much NPP activity is required for parasite growth (particularly under culture conditions that provide an optimal growing environment).

The De Jonge group made a recent, intriguing observation that in absence of a functional CFTR, as is the case in the red blood cell membrane of CF patients, the *P. falciparum* infection is not accompanied by increased membrane conductance, while the parasite remains able to grow normally (Verloo et al., 2004). This casts serious doubt on the possibility of using the anionic channels as therapeutic targets and moreover affirms that the parasite-induced anionic channels are the NPP. However, these conclusions were not confirmed by recent data from the Thomas group showing that the membrane conductance in infected cells from CF patients is only reduced by 50% compared to infected cells from non-CF patients (G. Decherf, 2004). Again, the discrepant results might well find an explanation in the different experimental protocols (e.g., stage of infection, bathing solutions).

It has also been proposed that the NPP act as a specific drug delivery route for malariacidal agents (Ginsburg and Stein, 1987). Saliba and Kirk (1998) reported that partial inhibition of the NPP reduced the antimalarial activity of the protease inhibitor, pepstatin A, which uses the NPP to gain access to the internal parasite. The Gero group in Australia has shown that conjugates built up by attaching known antimalarial agents to NPP-permeant solutes are extremely toxic to the parasite but are incapable of entering other mammalian cells (Gero et al., 2003). Recent work by the Ward group (Biagini et al., 2004) has demonstrated that bis-amidine and bis-quaternary ammonium compounds penetrate the host erythrocyte membrane via the NPP and that the parasite choline transporter mediates the delivery of these compounds to the intracellular parasite. This type of cooperative transport system shows great potential as a route for selective delivery of antimalarial drugs.

NPP, A CONTINUING CHALLENGE

The reasons for the differences between the results of the different groups may be related to the different experimental conditions used. The immediate question arising from the above analysis is this: which behavior best represents physiological conditions? Flux data are obtained under conditions closer to physiological conditions than patch clamp data, which makes it difficult to directly compare results from both techniques. Patch clamp experimental conditions proved to be very critical; improvement will have to come from future work focused on physiological conditions. This implies working under isotonic conditions in the presence of serum, bicarbonate ions, and a CO_2- and pH-controlled environment, as well as carefully evaluating the effects of changes in the composition of cell perfusates in whole-cell patch configurations.

In summary, the use of electrophysiological techniques has enhanced the investigation of the NPP by demonstrating a very high membrane conductance carried by endogenous (normally quiescent) inwardly rectifying anionic channels in infected cells and displaying many biophysical and pharmacological characteristics previously found for NPP by flux and lysis experiments. However, the exact nature of the NPP remains to be resolved; part of the present confusion, due to discrepant results, comes from the lack of background information on the channels present in noninfected red blood cell membranes.

REFERENCES

Abraham, E. H., K. M. Sterling, R. J. Kim, A. Y. Salikhova, H. B. Huffman, M. A. Crockett, N. Johnston, H. W. Parker, W. E. Boyle, Jr., A. Hartov, E. Demidenko, J. Efird, J. Kahn, S. A. Grubman, D. M. Jefferson, S. C. Robson, J. H. Thakar, A. Lorico, G. Rappa, A. C. Sartorelli, and P. Okunieff. 2001. Erythrocyte membrane ATP binding cassette (ABC) proteins: MRP1 and CFTR as well as CD39 (ecto-apyrase) involved in RBC ATP transport and elevated blood plasma ATP of cystic fibrosis. *Blood Cells Mol. Dis.* **27:**165–180.

Alkhalil, A., J. V. Cohn, M. A. Wagner, J. S. Cabrera, T. Rajapandi, and S. A. Desai. 2004. Plasmodium falciparum likely encodes the principal anion channel on infected human erythrocytes. *Blood* **104:**4279–4286.

Allen, R. J., and K. Kirk. 2004. The membrane potential of the intraerythrocytic malaria parasite Plasmodium falciparum. *J. Biol. Chem.* **279:**11264–11272.

Ancelin, M. L., M. Parant, M. J. Thuet, J. R. Philipot, and H. J. Vial. 1991. Increased permeability to choline in simian erythrocytes after *Plasmodium knowlesi* infection. *Biochem. J.* **273:**701–709.

Bernhardt, I., and J. C. Ellory. 2003. Red Cell Membrane Transport in Health and Disease. Springer Verlag, Berlin, Germany.

Biagini, G. A., E. M. Pasini, R. Hughes, H. P. De Koning, H. J. Vial, P. M. O'Neill, S. A. Ward, and P. G. Bray. 2004. Characterization of the choline carrier of Plasmodium falciparum: a route for the selective delivery of novel antimalarial drugs. *Blood* **104:**3372–3377.

Busse, R., A. Ogilvie, and U. Pohl. 1988. Vasomotor activity of diadenosine triphosphate and diadenosine tetraphosphate in isolated arteries. *Am. J. Physiol.* **254:**823–828.

Chishti, A. H., G. J. Maalouf, S. Marfatia, J. Palek, W. Wang, D., Fisher, and S. C. Liu. 1994. Phosphorylation of protein 4.1 in Plasmodium falciparum-infected human red blood cells. *Blood* **83:**3339–3345.

Christophersen, P., and P. Bennekou. 1991. Evidence for a voltage-gated, non-selective cation channel in the human red cell membrane. *Biochim. Biophys. Acta* **1065:**103–106.

Crandall, I., and I. W. Sherman. 1991. Plasmodium falciparum (human malaria)-induced modifications in human erythrocyte band 3 protein. *Parasitology* **3:**335–340.

Decherf, G. 2004. Identification et caractérisation de canaux CFTR-like dans la membrane des érythrocytes humains: rôle physiologique et implication dans la nouvelle voie de perméabilité induite au cours de l'infection par Plasmodium falciparum. Ph.D. thesis, University Pierre et Marie Curie, Paris, France.

Decherf, G., S. Egée, H. Staines, J. C. Ellory, and S. L. Thomas. 2004. Anionic channels in malaria-infected human red blood cells. *Blood Cells Mol. Dis.* **32:**366–371.

Desai, S. A. 2004. Targeting ion channels of Plasmodium falciparum-infected human erythrocytes for antimalarial development. *Curr. Drug Targets Infect. Disord.* **4:**79–86.

Desai, S. A., S. M. Bezrukov, and J. Zimmerberg. 2000. A voltage-dependent channel involved in nutrient uptake by red blood cells infected with the malaria parasite. *Nature* **406:**1001–1005.

Divo, A. A., T. G. Geary, N. L. Davis, and J. B. Jensen. 1985. Nutritional requirements of Plasmodium falciparum in culture. I. Exogenously supplied dialyzable components necessary for continuous growth. *J. Protozool.* **32:**59–64.

Droucheau, E., A. Primot, V. Thomas, D. Mattei, M. Knockaert, C. Richardson, P. Sallicandro, P. Alano, A. Jafarshad, B. Baratte, C. Kunick, D. Parzy, L. Pearl, C. Doerig, and L. Meijer. 2004. *Plasmodium falciparum* glycogen synthase kinase-3: molecular model, expression, intracellular localisation and selective inhibitors. *Biochim. Biophys. Acta* **1697:**181–196.

Duranton, C., S. M. Huber, and F. Lang. 2002. Oxidation induces a Cl$^-$-dependent cation conductance in human red blood cells. *J. Physiol.* **539:**847–855.

Egée, S., F. Lapaix, G. Decherf, H. M. Staines, J. C. Ellory, C. Doerig, and S. L. Thomas. 2002. A stretch-activated anion channel is up-regulated by the malaria parasite Plasmodium falciparum. *J. Physiol.* **542:**795–801.

Egée, S., O. Mignen, B. J. Harvey, and S. Thomas. 1998. Chloride and non-selective cation channels in unstimulated trout red blood cells. *J. Physiol.* **511:**213–224.

Freedman, J. C., T. S. Novak, J. D. Bisognano, and P. R. Pratap. 1994. Voltage dependence of DIDS-insensitive chloride conductance in human red blood cells treated with valinomycin or gramicidin. *J. Gen. Physiol.* **104:**961–983.

Gardner, M. J., N. Hall, E. Fung, O. White, M. Berriman, R. W. Hyman, J. M. Carlton, A. Pain, K. E. Nelson, S. Bowman, I. T. Paulsen, K. James, J. A. Eisen, K. Rutherford, S. L. Salzberg, A. Craig, S. Kyes, M. S. Chan, V. Nene, S. J. Shallom, B. Suh, J. Peterson, S. Angiuoli, M. Pertea, J. Allen, J. Selengut, D. Haft, M. W. Mather, A. B. Vaidya, D. M. Martin, A. H. Fairlamb, M. J. Fraunholz, D. S. Roos, S. A. Ralph, G. I. McFadden, L. M. Cummings, G. M. Subramanian, C. Mungall, J. C. Venter, D. J. Carucci, S. L. Hoffman, C. Newbold, R. W. Davis, C. M. Fraser, and B. Barrell. 2002. Genome sequence of the human malaria parasite Plasmodium falciparum. *Nature* **419:**498–511.

Gero, A. M., C. G. Dunn, D. M. Brown, K. Pulenthiran, E. L. Gorovits, T. Bakos, and A. L. Weis. 2003. New malaria chemotherapy developed by utilization of a unique parasite transport system. *Curr. Pharm. Des.* **9:**867–877.

Ginsburg, H. 1994. Transport pathways in the malaria-infected erythrocyte. Their characterization and their use as potential targets for chemotherapy. *Biochem. Pharmacol.* **48:**1847–1856.

Ginsburg, H., and K. Kirk. 1998. Membrane transport in the malaria-infected erythrocyte, p. 219–232. *In* I. W. Sherman (ed.), *Malaria: Parasite Biology, Pathogenesis, and Protection.* ASM Press, Washington, D.C.

Ginsburg, H., M. Krugliak, O. Eidelman, and Z. I. Cabantchick. 1983. New permeability pathways induced in membranes of *Plasmodium falciparum* infected erythrocytes. *Mol. Biochem. Parasitol.* **8:**177–190.

Ginsburg, H., S. Kutner, M. Krugliak, and Z. I. Cabantchick. 1985. Characterization of permeation pathways appearing in the host membrane of *Plasmodium falciparum* infected red blood cells. *Mol. Biochem. Parasitol.* **14:**185–199.

Ginsburg, H., and W. D. Stein. 1987. Biophysical analysis of novel transport pathways induced in red blood cell membranes. *J. Membr. Biol.* **96:**1–10.

Ginsburg, H., and W. D. Stein. 2004. The new permeability pathways induced by the malaria parasite in the membrane of the infected erythrocyte: comparison of results using different experimental techniques. *J. Membr. Biol.* **197:**113–122.

Grygorczyk, R., and W. Schwarz. 1983. Properties of the CA^{2+}-activated K^+ conductance of human red cells as revealed by the patch-clamp technique. *Cell Calcium* **4:**499–510.

Hamill, O. P. 1981. Potassium channel currents in human red blood cells. *J. Physiol.* **319:**97–98.

Hamill, O. P. 1983. Potassium and chloride channels in red blood cells, p. 501. *In* B. Sakmann and E. Neher (ed.), *Single-Channel Recording.* Plenum Press, New York, N.Y.

Hamill, O. P., A. Marty, E. Neher, B. Sakmann, and F. J. Sigworth. 1981. Improved patch-clamp techniques for high-resolution current recording from cells and cell-free membrane patches. *Pflugers Arch.* **391:**85–100.

Hoffman, J. F., and P. C. Laris. 1974. Determination of membrane potentials in human and *Amphiuma* red blood cells by means of a fluorescent probe. *J. Physiol.* **239:**519–552.

Homewood, C. A., and K. D. Neame. 1974. Malaria and the permeability of the host erythrocyte. *Nature* **252:**718–719.

Huber, S. M., C. Duranton, G. Henke, van de C. Sand, V. Heussler, E. Shumilina, C. D. Sandu, V. Tanneur, V. Brand, R. S. Kasinathan, K. S. Lang, P. G. Kremsner, C. A. Hübner, M. B. Rust, K. Dedek, T. Jentsch, and F. Lang. 2004. Plasmodium induces swelling-activated ClC-2 anion channels in the host erythrocyte. *J. Biol. Chem.* **279:**41444–41452.

Huber, S. M., A. C. Uhlemann, N. L. Gamper, C. Duranton, P. G. Kremsner, and F. Lang. 2002. Plasmodium falciparum activates endogenous Cl$^-$ channels of human erythrocytes by membrane oxidation. *EMBO J.* **21:**22–30.

Jentsch, T. J. 2002. Chloride channels are different. *Nature* **415:**276–277.

Kappes, B., C. D. Doerig, and R. Graeser. 1999. An overview of Plasmodium protein kinases. *Parasitol. Today* **15:**449–454.

Kirk, K. 2001. Membrane transport in the malaria-infected erythrocyte. *Physiol. Rev.* **81:**495–537.

Kirk, K., B. C. Elford, and J. C. Ellory. 1992. The increased K^+ leak of malaria-infected erythrocytes is not via a Ca^{2+}-activated K^+ channel. *Biochim. Biophys. Acta* **1135:**8–12.

Kirk, K., and H. A. Horner. 1995. In search of a selective inhibitor of the induced transport of small solutes in Plasmodium falciparum-infected erythrocytes: effects of arylaminobenzoates. *Biochem. J.* **311:**761–768.

Kirk, K., H. A. Horner, B. C. Elford, J. C. Ellory, and C. I. Newbold. 1994. Transport of diverse substrates into malaria-infected erythrocytes via a pathway showing functional characteristics of a chloride channel. *J. Biol. Chem.* **269:**3339–3347.

Kirk, K., and K. Strange. 1998. Functional properties and physiological roles of organic solute channels. *Annu. Rev. Physiol.* **60:**719–739.

Kirk, K., L. Tilley, and H. Ginsburg. 1999. Transport and trafficking in the malaria-infected erythrocyte. *Parasitol. Today* **15:**355–357.

Krishna, S., U. Eckstein-Ludwig, T. Joet, A. C. Uhlemann, C. Morin, R. Webb, C. Woodrow, J. F. Kun, and P. G. Kremsner. 2002. Transport processes in Plasmodium falciparum-infected erythrocytes: potential as new drug targets. *Int. J. Parasitol.* **32:**1567–1573.

Krugliak, M., J. Zhang, and H. Ginsburg. 2002. Intraerythrocytic Plasmodium falciparum utilizes only a fraction of the amino acids derived from the digestion of host cell cytosol for the biosynthesis of its proteins. *Mol. Biochem. Parasitol.* **119:**249–256.

Lapaix, F., S. Egée, L. Gibert, G. Decherf, and S. L. Thomas. 2002. ATP-sensitive K^+ and Ca^{2+}-activated K^+ channels in lamprey (Petromyzon marinus) red blood cell membrane. *Pflugers Arch.* **445:**152–160.

Lassen, U. V., and O. Sten-Knudsen. 1968. Direct measurement of membrane potential and membrane resistance of human red cells. *J. Physiol.* **195:**681–696.

Lew, V. L., T. Tiffert, and H. Ginsburg. 2003. Excess hemoglobin digestion and the osmotic stability of Plasmodium falciparum-infected red blood cells. *Blood* **101:**4189–4194.

Nilius, B., and G. Droogmans. 2003. Amazing chloride channels: an overview. *Acta Physiol. Scand.* **177:**119–147.

Poole, R. C., and A. P. Halestrap. 1993. Transport of lactate and other monocarboxylates across mammalian plasma membranes. *Am. J. Physiol.* **264:**C761–C782.

Rosenthal, P. J., and S. R. Meshnick. 1998. Hemoglobin processing and the metabolism of amino acids, heme and iron, p. 145–159. *In* I. W. Sherman (ed.), *Malaria: Parasite Biology, Pathogenesis, and Protection.* ASM Press, Washington, D.C.

Saliba, K. J., and K. Kirk. 1998. Uptake of an antiplasmodial protease inhibitor into Plasmodium falciparum-infected human erythrocytes via a parasite-induced pathway. *Mol. Biochem. Parasitol.* **94:**297–301.

Schwartz, R. S., A. C. Rybicki, and R. L. Nagel. 1997. Molecular cloning and expression of a chloride channel-associated protein pICln in human young red blood cells: association with actin. *Biochem. J.* **327:**609–616.

Schwartz, W., R. Gryorczyk, and D. Hof. 1989. Recording single-channel currents from human red cells. *Methods Enzymol.* **173:**112–121.

Sherman, I. W. 1977. Transport of amino acids and nucleic acid precursors in malarial parasites. *Bull. W. H. O.* **55:**211–225.

Sprague, R. S., M. L. Ellsworth, A. H. Stephenson, M. E. Kleinhenz, and A. J. Lonigro. 1998. Deformation-induced ATP release from red blood cells requires CFTR activity. *Am. J. Physiol.* **275:**H1726–H1732.

Sprague, R. S., M. L. Ellsworth, A. H. Stephenson, and A. J. Lonigro. 2001. Participation of cAMP in a signal-transduction pathway relating erythrocyte deformation to ATP release. *Am. J. Physiol. Cell Physiol.* **281:**C1158–C1164.

Staines, H. M., B. C. Dee, O'Brien, M., H. J. Lang, H. A. Horner, J. C. Ellory, and K. Kirk. 2004. Furosemide analogues as potent inhibitors of the new permeability pathways of Plasmodium falciparum-infected human erythrocytes. *Mol. Biochem. Parasitol.* **133:**315–318.

Staines, H. M., J. C. Ellory, and K. Kirk. 2001. Perturbation of the pump-leak balance for Na^+ and K^+ in malaria- infected erythrocytes. *Am. J. Physiol. Cell Physiol.* **280:**C1576–C1587.

Staines, H. M., E. M. Godfrey, F. Lapaix, S. Egée, S. Thomas, and J. C. Ellory. 2002. Two functionally distinct organic osmolyte pathways in Plasmodium gallinaceum-infected chicken red blood cells. *Biochim. Biophys. Acta* **1561:**98–108.

Staines, H. M., and K. Kirk. 1998. Increased choline transport in erythrocytes from mice infected with the malaria parasite Plasmodium vinckei vinckei. *Biochem. J.* **334:**525–530.

Staines, H. M., T. Powell, J. C. Ellory, S. Egée, F. Lapaix, G. Decherf, S. L. Y. Thomas, C. Duranton, F. Lang, and S. Huber. 2003. Modulation of whole-cell currents in Plasmodium falciparum-infected human red blood cells by holding potential and serum. *J. Physiol.* **552:**177–183.

Su, X. Z., V. M. Heatwole, S. P. Wertheimer, F. Guinet, J. A. Herrfeldt, D. S. Peterson, J. A. Ravetch, and T. E. Wellems. 1995. The large diverse gene family var encodes proteins involved in cytoadherence and antigenic variation of Plasmodium falciparum-infected erythrocytes. *Cell* **82:**89–100.

Syin, C., D. Parzy, F. Traincard, I. Boccaccio, M. B. Joshi, D. T. Lin, X. M. Yang, K. Assemat, C. Doerig, and G. Langsley. 2001. The H89 cAMP-dependent protein kinase inhibitor blocks Plasmodium falciparum development in infected erythrocytes. *Eur. J. Biochem.* **268:**4842-4849.

Tabcharani, J. A., X. B. Chang, J. R. Riordan, and J. W. Hanrahan. 1991. Phosphorylation-regulated Cl- channel in CHO cells stably expressing the cystic fibrosis gene. *Nature* **352:**628-631.

Thomas, S. L., S. Egée, F. Lapaix, L. Kaestner, H. M. Staines, and J. C. Ellory. 2001. Malaria parasite Plasmodium gallinaceum up-regulates host red blood cell channels. *FEBS Lett.* **500:**45-51.

Verloo, P., C. H. Kocken, A. Van Der Wel, B. C. Tilly, B. M. Hogema, M. Sinaasappel, A. W. Thomas, and H. R. De Jonge. 2004. *Plasmodium falciparum*-activated chloride channels are defective in erythrocytes from cystic fibrosis patients. *J. Biol. Chem.* **279:**10316-10322.

Vial, H. J., and M. L. Ancelin. 1998. Malaria lipids, p. 159-175. *In* I. W. Sherman (ed.), *Malaria: Parasite Biology, Pathogenesis, and Protection*. ASM Press, Washington, D.C.

IMMUNE EVASION

IV

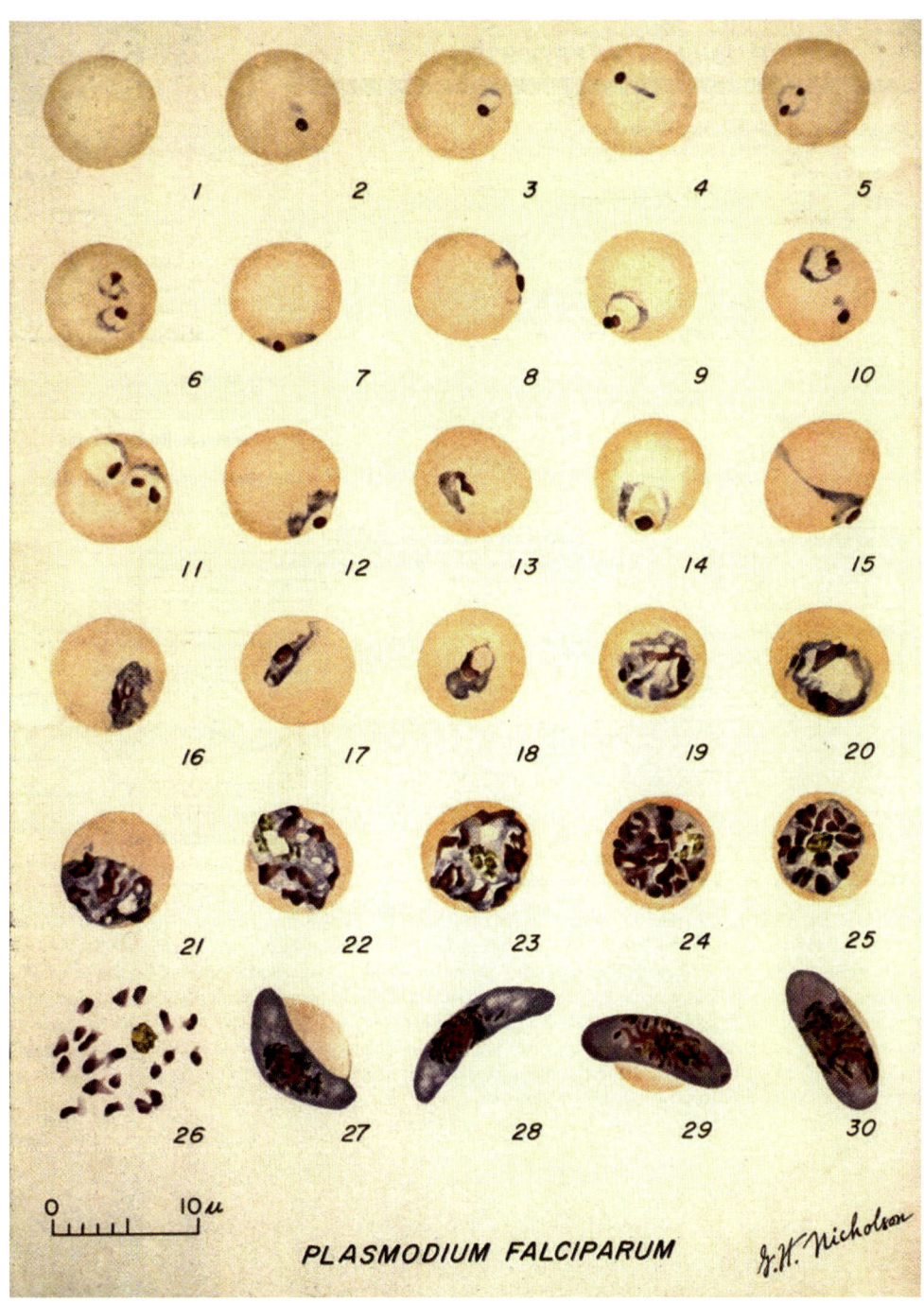

COLOR PLATE 1 (Chapter 1) Appearance of *P. falciparum* by light microscopy after staining.

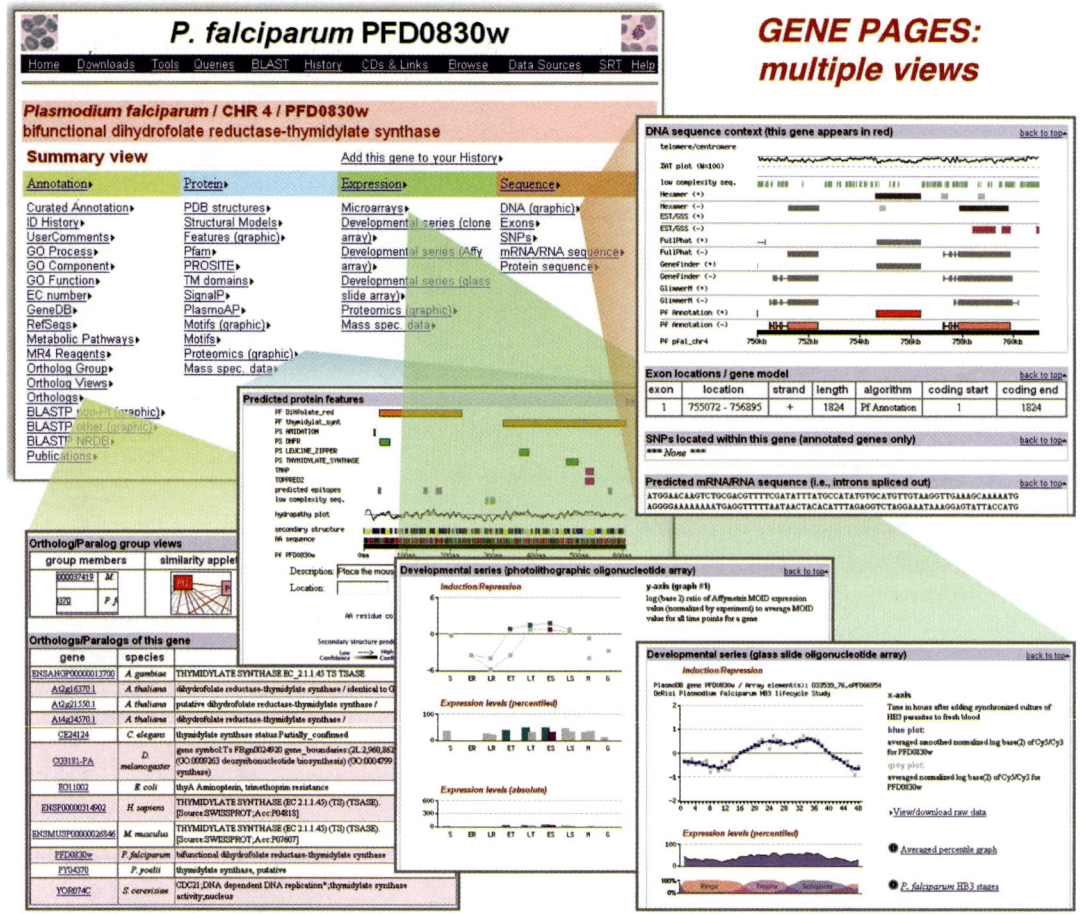

COLOR PLATE 2 (Chapter 2) Gene Pages on PlasmoDB. The summary view (upper left) provides a tabulation of the various forms of information available and scrolls to reveal highlights of general interest, such as curated annotations (and comments from the research community, which may also be entered by the user), DNA and protein features, transcript and protein expression data, information on protein structure, reagents, and publications. Insets illustrate for the *P. falciparum* DHFR-TS gene (PFD0830w) (i) a list of orthologs in other species, (ii) mapped protein motifs and features, (iii) transcript profiling data, including information on both abundance and induction-repression, based on both photolithographic and glass slide microarrays, and (iv) the gene model(s), sequence, and information in local chromosomal DNA contexts.

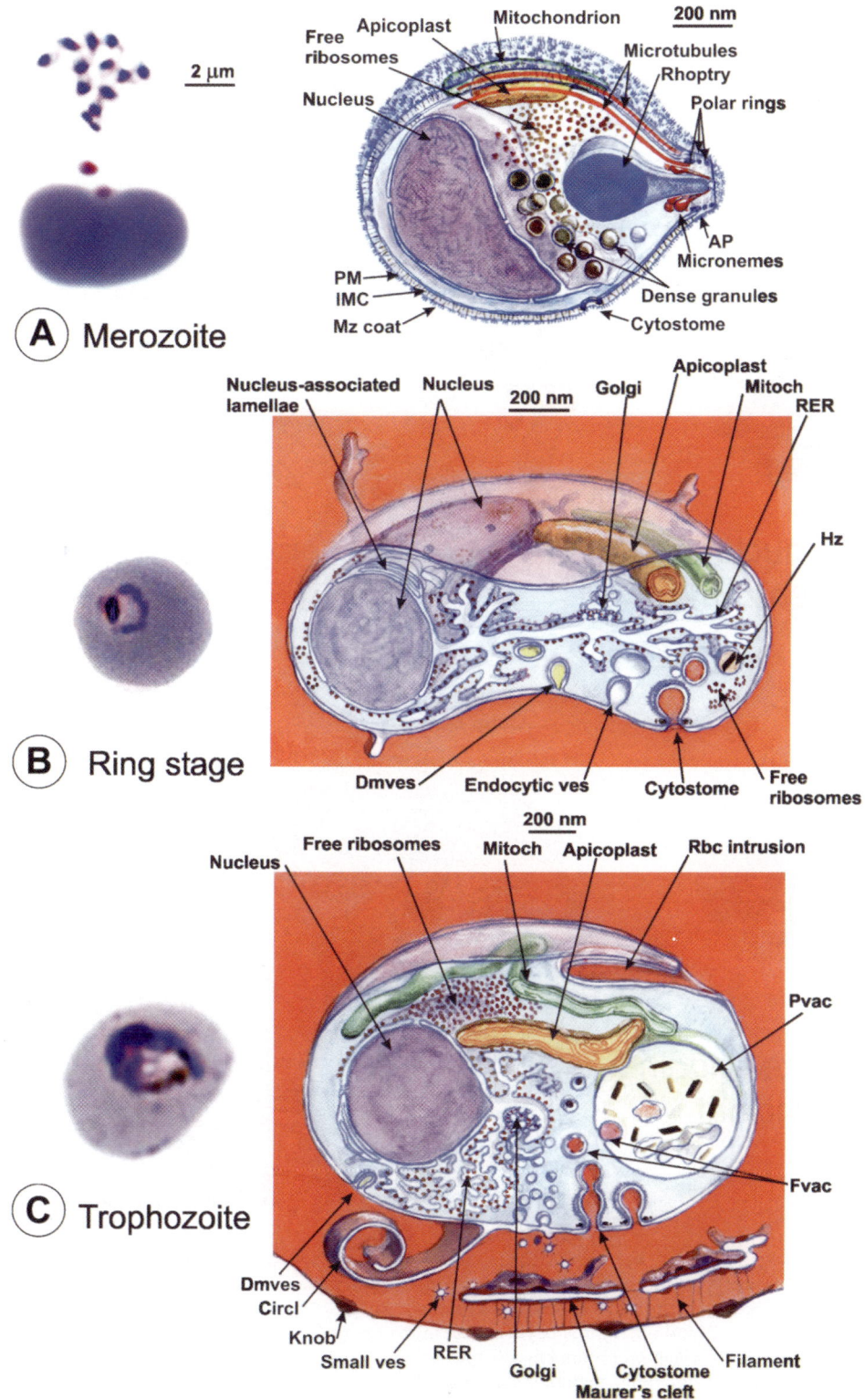

COLOR PLATE 3 (Chapter 3) Illustrations summarizing the structural organizations of the major erythrocytic stages of *P. falciparum*, including a group of free and two invading merozoites (A), ring (B), trophozoite (C), schizont

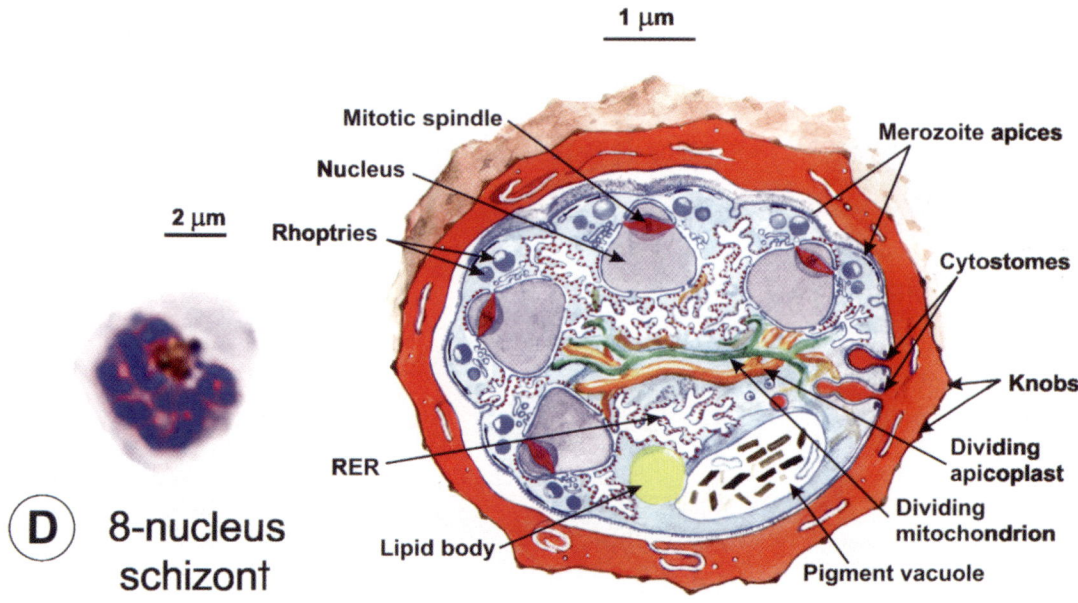

D 8-nucleus schizont

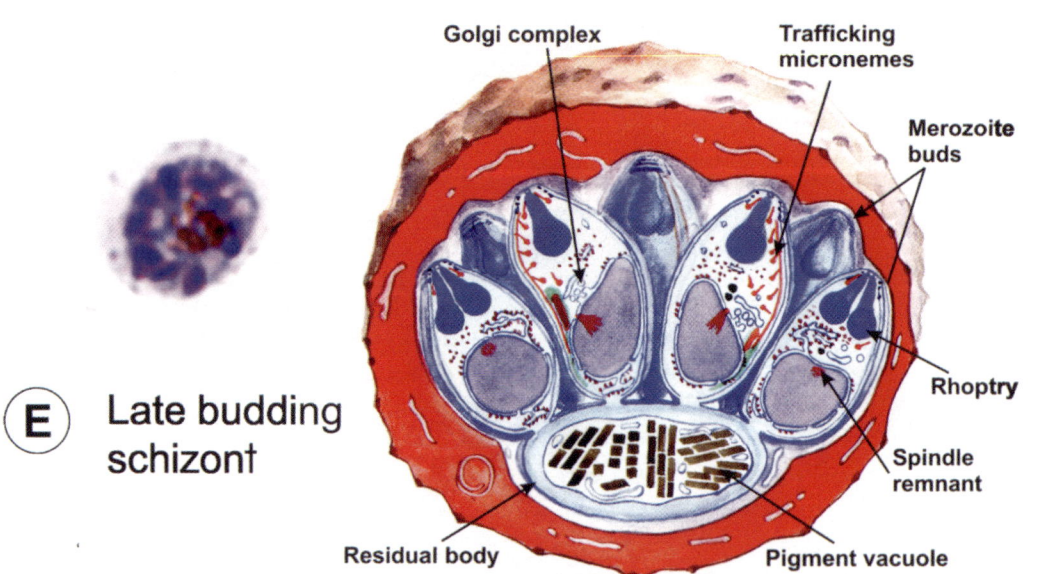

E Late budding schizont

at the eight dividing nucleus (D), and late (E) developmental stages, and a mature macrogametocyte (F), shown with part of the surrounding RBC. Accompanying each schema is an example of a Giemsa-stained parasite from a film preparation of approximately the same stage. In panel F, three gametocytes are shown; from left to right, a mature (stage V) macrogametocyte and microgametocyte and an immature (stage IV) macrogametocyte with typical sharp ends (the oat-grain form). AP, apical prominence; Circl, circular cleft; Dmves, double membrane vesicle; Fvac, food vacuole; Hz, hemozoin; IMC, inner membrane complex; Mitoch, mitochondrion; Mz, merozoite; PM, plasma membrane; Pvac, parasitophorous vacuole; RER, rough endoplasmic reticulum; STIC, sexual-stage intraerythrocytic tubular compartment; ves, vesicle.

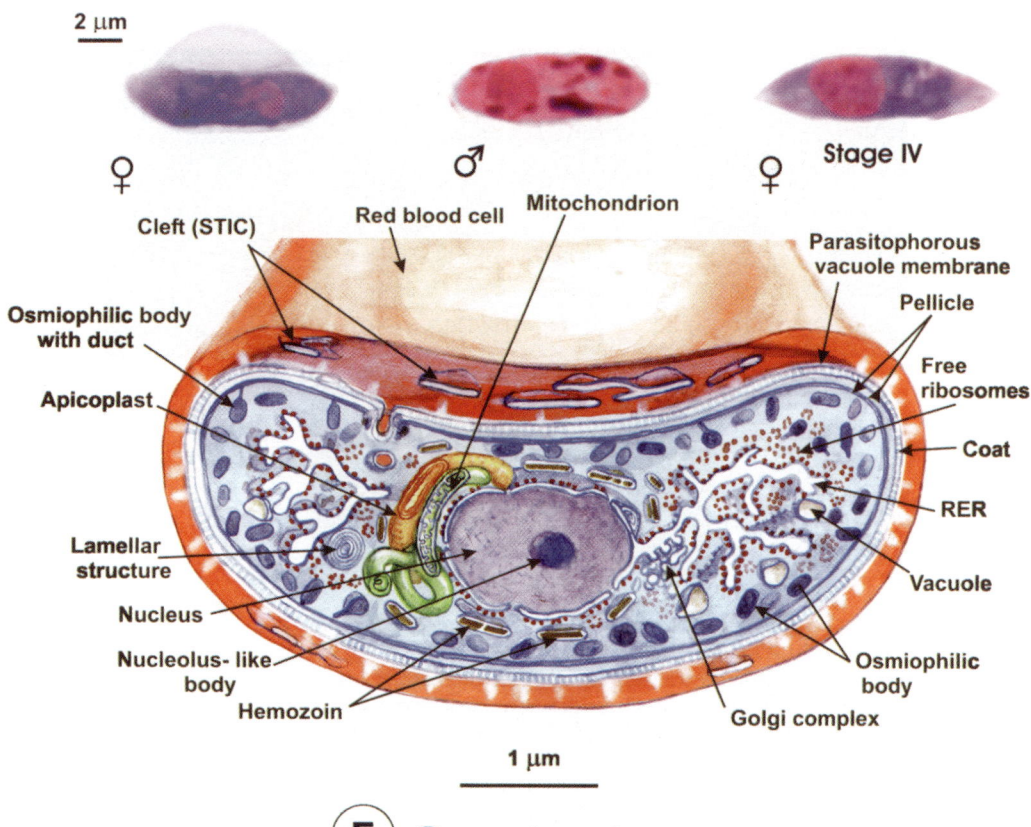

F Gametocytes

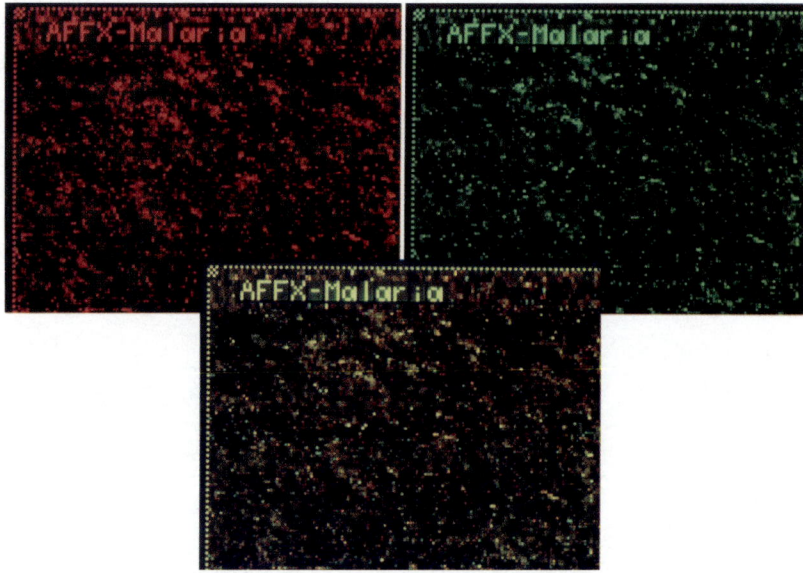

COLOR PLATE 4 (Chapter 5) Comparison of the cRNA hybridizing intensity level by the Affymetrix array for the trophozoite (red) and schizont (green) stages. Each colored square represents the hybridization signal from a single 25-bp probe to the *P. falciparum* genome sequence. The intensity of the color indicates hybridization level for each stage. Yellow indicates equal hybridization to both trophozoite and schizont stages, red indicates hybridization for the trophozoite stages but not for the schizont stage, and green indicates hybridization for the schizont stage but not for the trophozoite stage.

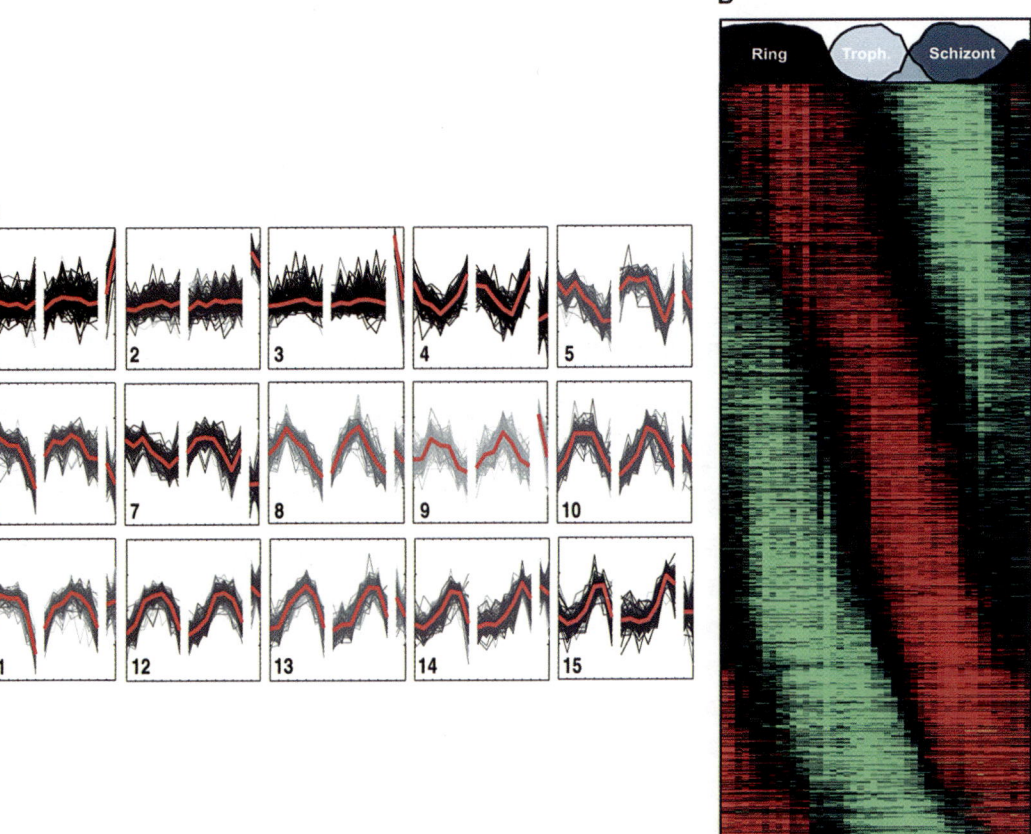

COLOR PLATE 5 (Chapter 5)   Comparison of the two-microarray analysis using the robust k-mean algorithm (A) with the short oligonucleotide array (reproduced from *Science* [Le Roch et al., 2003] with permission from the publisher) or FTT analysis (B) with the long 70-mer array (reproduced from *PLoS Biol.* [Bozdech et al., 2003a] with permission from the publisher).

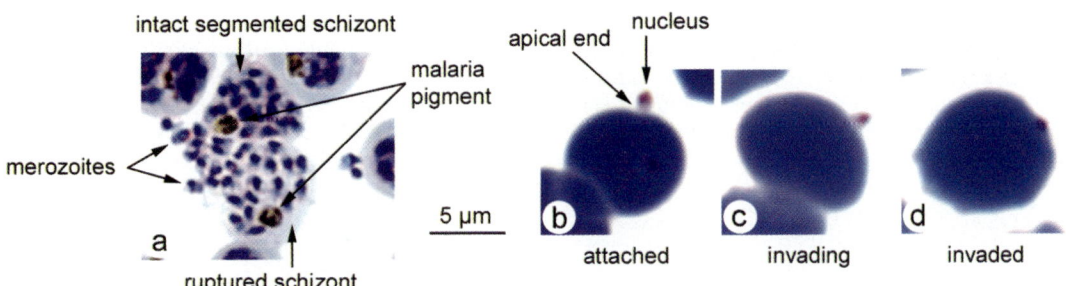

COLOR PLATE 6 (Chapter 8)   Light microscopy of Giemsa-stained *P. falciparum* mature schizonts (strain IT04) and invading merozoites (strain NF54). (a) Both intact (upper) and rupturing (lower) schizonts are shown. (b) Initial apically attached merozoite; (c) partially invaded merozoite; (d) almost or fully invaded merozoite. The *P. falciparum* (strain NF54) images were provided by Gabriele Margos (KCL). The images were generated with an Axiocam camera and a Zeiss Axiophot microscope.

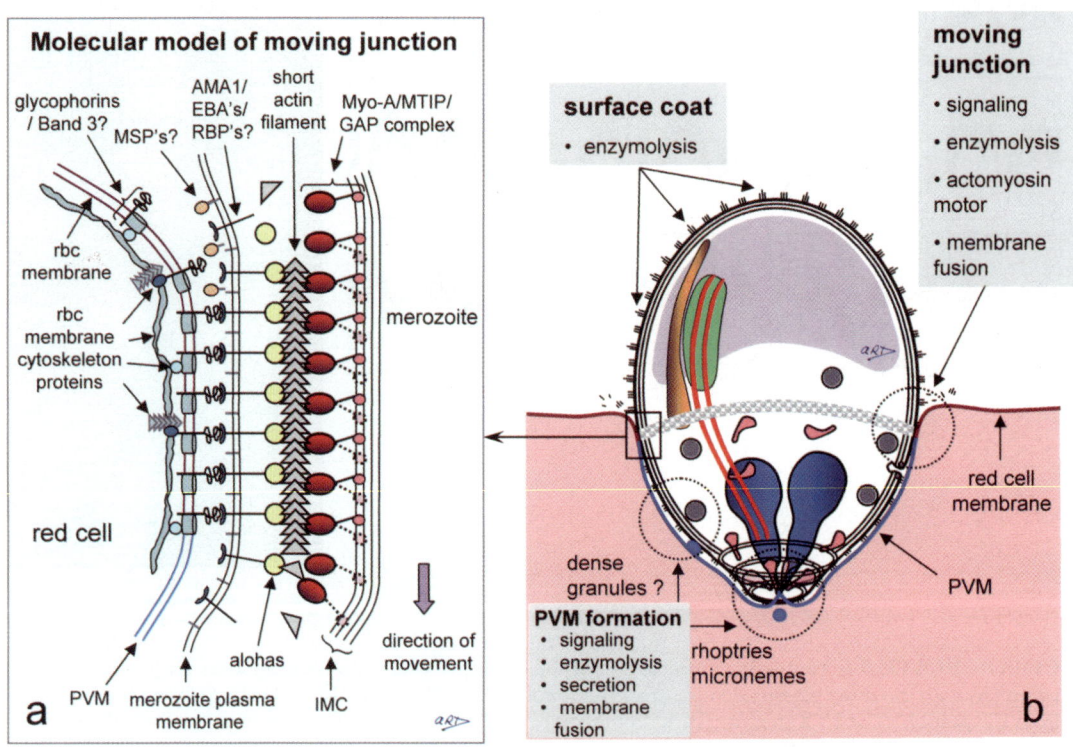

COLOR PLATE 7 (Chapter 8) Diagrammatic representation of an invading merozoite and moving junction. (a) A theoretical molecular view of the electron-dense tight moving junction, showing the intimately associated red blood cell membrane and merozoite pellicle, along with current predicted molecular players, as detailed in the text. Here, a novel single-headed merozoite myosin, which is linked to the IMC, pulls the merozoite into the red blood cell on a short filament of F-actin, connected via aldolase to a membrane-bound merozoite ligand, which in turn binds to a red cell receptor. (b) A partially invaded merozoite highlighting major structures, parasitophorous vacuole formation, and likely biochemical events as entry proceeds. Organellar secretion, membrane generation, and host cell membrane restructuring contribute to the formation of the PVM. Points at which enzymolysis may occur are indicated. The moving junction "embracing the merozoite" as it enters the erythrocyte is detailed as a stippled band.

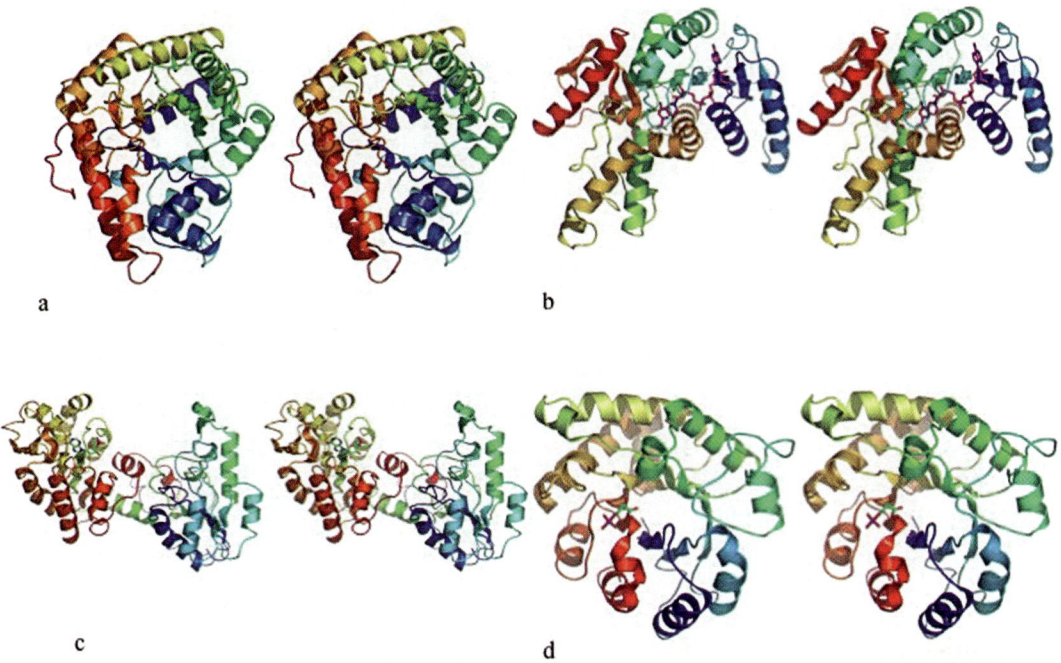

COLOR PLATE 8 (Chapter 11) *P. falciparum* fructose-1,6-bisphosphate aldolase (1A5C) (A), *P. falciparum* L-LDH complexed with NADH and oxamate (1LDG) (B), *P. falciparum* phosphoglycerate kinase adenosine monophosphate (1LTK) (C), and *P. falciparum* triose phosphate isomerase coupled to substrate analog 3-phosphoglycerate (1M70) (D).

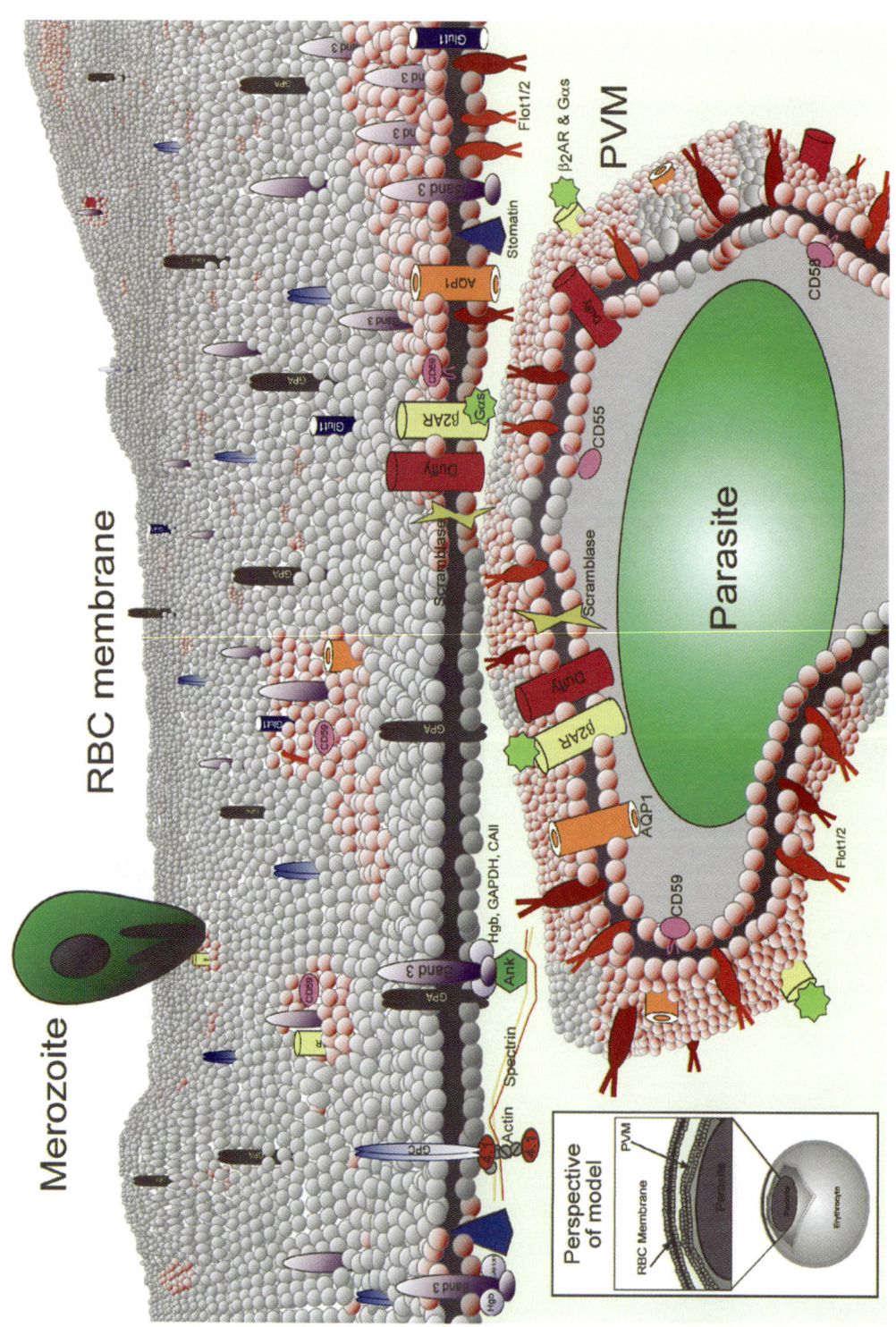

COLOR PLATE 9 (Chapter 13)   Model of erythrocyte DRM rafts and their enrichment in the malarial vacuolar membrane. The uninfected erythrocyte membrane contains a dynamic milieu of generalized lipid domains (gray spheres) and raft microdomains (pink spheres) containing various proteins. Some proteins partition heavily into raft domains (i.e., flotillins), while others are only minimally present in rafts (i.e., band 3 and Glut1). During malaria infection, merozoite-stage parasites invade erythrocytes to reside in a membrane-bound parasitophorous vacuole. The PVM becomes selectively cholesterol enriched, and 10 of the known raft proteins are internalized to the PVM (flotillin 1 and flotillin 2, Gs, 2-AR, AQP1, Duffy, CD55, CD58, CD59, and scramblase). Most of the abundant erythrocyte membrane proteins are not internalized to the PVM (i.e., glycophorins A and C and cytoskeleton-associated band 3). The lower-left inset shows the perspective of the model, which depicts a whole infected erythrocyte and a magnified view through the plasma membrane and PVM of a malaria-infected erythrocyte. Because the PVM is formed by invagination of the plasma membrane, proteins that are cytoplasmically oriented in uninfected cells remain so upon infection; protein structures exposed to the extracellular space face the vacuolar space upon infection. 4.1, band 4.1; Gs, G-protein αs; 2-AR, β2-adrenergic receptor; AQP1, aquaporin 1.

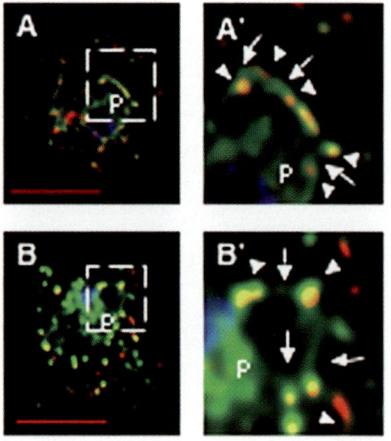

COLOR PLATE 10 (Chapter 13) Maurer's clefts are intermediates in PfHRPII transport and are targeted by its histidine-rich sequences. Single optical section of trophozoite-infected cells expressing PfHRPIIGFP (A) and SSGFPHis154 to -327 (B). Cells were fixed, permeabilized, subjected to indirect immunofluorescence assays, and probed with antibodies to GFP (green) and a Maurer's cleft resident protein, PfSBP (red). Magnifications of the indicated boxes in A and B are shown as A′ and B′, respectively. The extent of colocalization is shown in yellow. In A′ and B′, arrows indicate GFP-labeled tubules or vesicles connecting clefts (arrowheads) to each other or the vacuolar parasite (P). Nuclei were stained with Hoechst (blue). Bar, 5 μm.

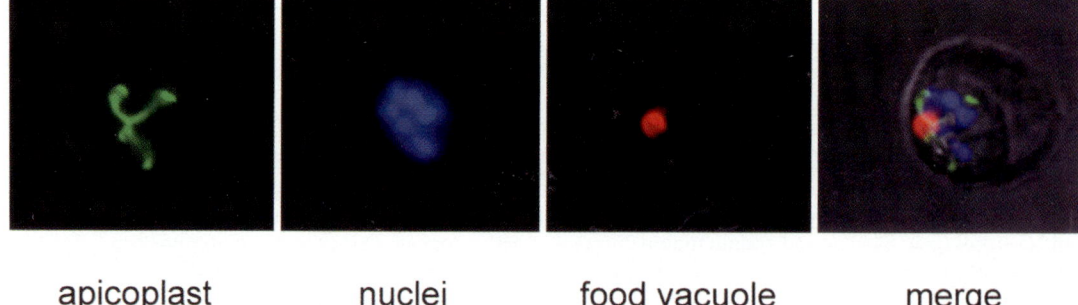

apicoplast   nuclei   food vacuole   merge

COLOR PLATE 11 (Chapter 14) Plastid division in *Plasmodium falciparum*. The dividing apicoplast is labeled with green fluorescent protein fused to the apicoplast protein ACP. The nuclei are stained with 4′,6′-diamidino-2-phenylindole (DAPI; blue), and the food vacuole is stained with the marker of acidic compartments, Lysotracker (red) (Molecular Probes). Although apicoplast division in *Toxoplasma gondii* is synchronous with nuclear division, the dividing apicoplast (green) in *P. falciparum* does not appear to be strictly associated with the six or more newly forming nuclei (blue) in the early schizont. By the end of schizogony, a new plastid must segregate with each new nucleus. The microscopy was performed by students of the Biology of Parasitism Course, Marine Biological Laboratory, Woods Hole, Mass., in 2002.

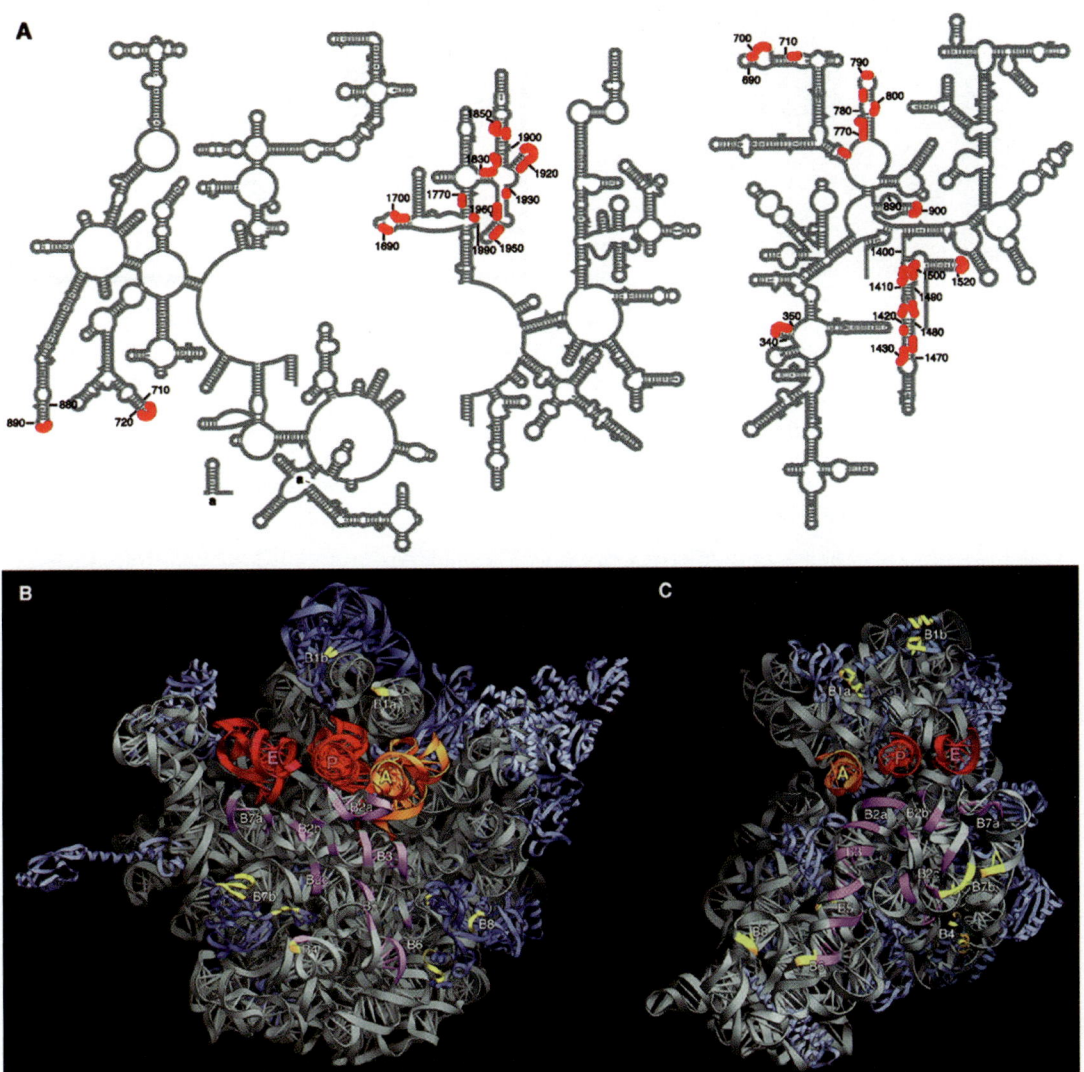

COLOR PLATE 12 (Chapter 18) Intersubunit bridges (A) of secondary structure of 16S and 23S rRNAs showing the features involved in intersubunit contacts (red). (B and C) Interface view of the 50S and 30S subunits with the bridges numbered. RNA-RNA contacts are shown in magenta; protein-RNA and protein-protein contacts are shown in yellow. A, P, and E indicate the three tRNAs (23S rRNA) or tRNA anticodon stem loops (16S rRNA) (Yusupov et al., 2001).

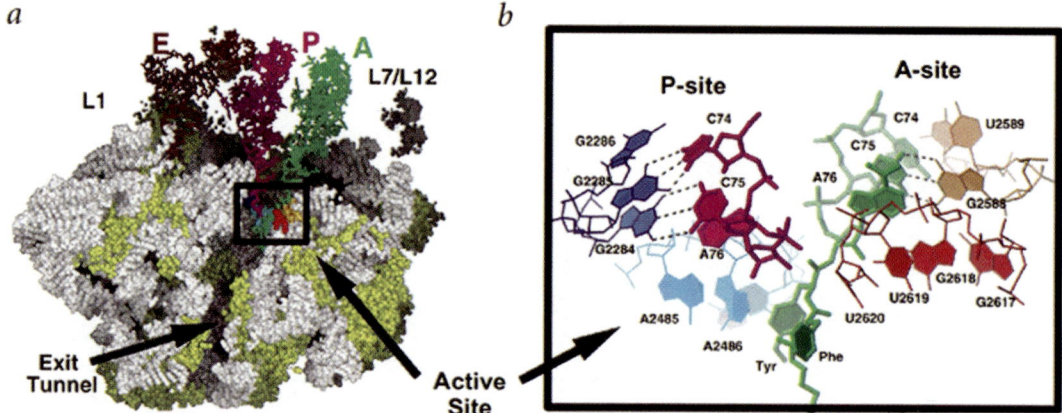

COLOR PLATE 13 (Chapter 18) Structure of the center of the large ribosomal subunit of *Haloarcula marismortui*. (a) Space-filling representation of the 50S particle (RNA in white and protein in yellow) in complex with three tRNAs (Lazaro et al., 1991). The subunit has been split to reveal the tunnel and the peptidyl transferase site (boxed). The orientation is the crown view, with the L1 protein to the left and the L7-L12 stalk to the right. (b) Close-up view of the active site shows the P-site, A-site, and the transfer RNA carrying the elongating peptide. The N3 of A2486 (A2451) (light blue) is in proximity to the 3′ OH of the deacylated product, and the base of U2620 (U2585) (red) has moved near to the newly formed peptidyl ester link and the 3′ OH of dimethyl A76. (Reproduced with permission from Schmeing et al. [2002].)

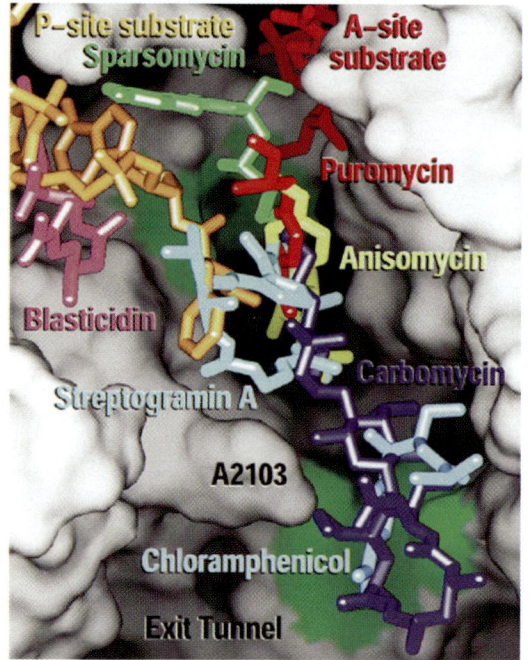

COLOR PLATE 14 (Chapter 18) Superposition of antibiotic-binding locations and hydrophobic crevices. Surface representation of the ribosome (grey contour) shows that many antibiotics (stick figures) interact in part with the A-site crevice (green contour, upper middle) or with the exit tunnel hydrophobic crevice (green contour, lower right). These antibiotics overlap the binding site of a P-site substrate (orange) or that of an A-site substrate (red). (Reprinted from the *Journal of Molecular Biology* with permission of the publisher [Hansen et al., 2003].)

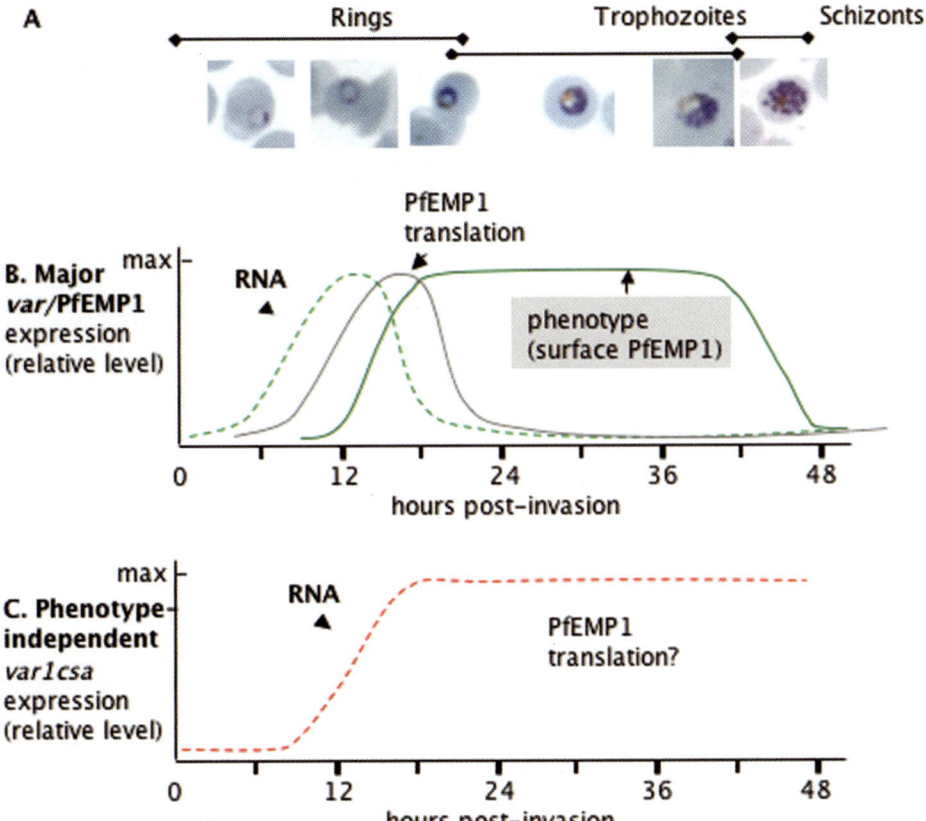

COLOR PLATE 15 (Chapter 21) Temporal regulation of *var* gene expression. (A) IE undergo several morphological and phenotypic alterations during the asexual life cycle, changes that correlate roughly with stages of *var* gene and protein expression. Examination of Giemsa-stained IE from synchronized cultures indicates that ring-stage parasites occur immediately after merozoite invasion of the erythrocyte up to roughly 15 to 20 hpi. The pigmented trophozoite stage follows from between approximately 20 to 40 hpi, with segmenting and schizont stages developing between 40 to 48 hpi. (B) *var* transcription is tightly synchronized within ring-stage parasites and is followed by translation of a single major *var* gene encoding the PfEMP1 variant exposed on the IE surface. PfEMP1 is first detected on the IE surface from 16 to 18 hpi and correlates with the first detection of a cytoadherence phenotype. At this stage, the level of full-length *var* RNA transcript decreases, and it is no longer detected in mature pigmented trophozoites. PfEMP1 protein synthesis presumably ceases at this point. Segmenting and/or schizont stages correspond with poorly detected phenotype in in vitro studies, possibly due to fragile nature of the cells. Both RNA and phenotype are difficult to assess during this stage. (C) Phenotype-independent *var* gene expression of *var1CSA*. Transcripts for this gene are present in all parasites regardless of their cytoadherent phenotype, at a very low level in ring stages and at a very high level in pigmented trophozoites (Kyes et al., 2003). It is unclear whether this protein is translated or what role it might have in cytoadherence; in 3D7, *var1CSA* is a pseudogene. It was originally thought that this *var* gene encoded a PfEMP1 mediating the CSA-binding phenotype, but current data show that another gene, *var2CSA*, encodes a more plausible candidate (Salanti et al., 2003).

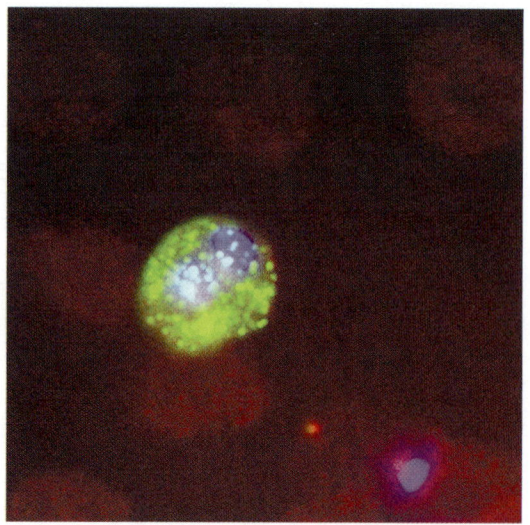

COLOR PLATE 16 (Chapter 22) Immunofluorescence assay with a mouse anti-human IgM monoclonal antibody and Alexa488 secondary antibody to show nonimmune IgM (green) on the surface of a rosetting infected erythrocyte. Parasite nuclei are stained blue with 4′,6′-diamidino-2-phenylindole, and red blood cells are stained red with concanavalin A-tetramethyl rhodamine isocyanate. (Reproduced from the *American Journal of Tropical Medicine and Hygiene* [Rowe et al., 2002b] with permission from the publisher.)

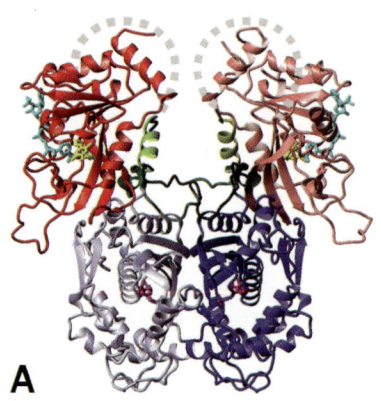

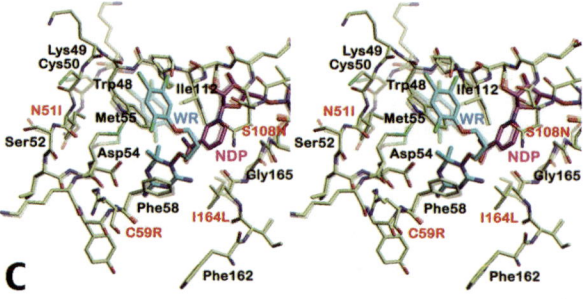

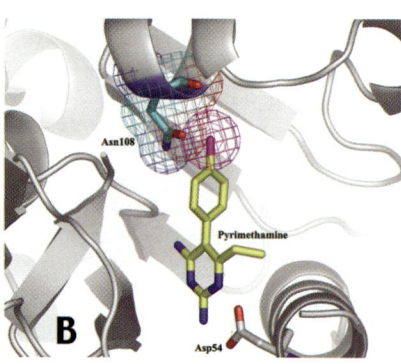

COLOR PLATE 17 (Chapter 23) (A) Ribbon diagram of the overall structure of *P. falciparum* dihydrofolate reductase-thymidylate synthase (PfDHFR-TS). The two DHFR monomers are shown in red and pink, while the two TS monomers are shown in grey and blue. (B) The structure of the complex of pyrimethamine in the active site of a mutant enzyme harboring the S108N mutation. The asparagine side chain clashes with the Cl group of pyrimethamine, resulting in decreased binding affinity. (C) Comparison of enzyme-inhibitor interactions at the active sites of wild-type (lighter model) and the quadruple mutant PfDHFR-TS, in stereo view. Both enzymes are shown as a complex with WR99210 (WR, in cyan) and NADPH (NDP, in magenta). The flexible tail of WR99210 allows binding in a conformation that is unaffected by the pyrimethamine-resistant mutations (N51I+C59R+S108N+I164L) labeled in red.

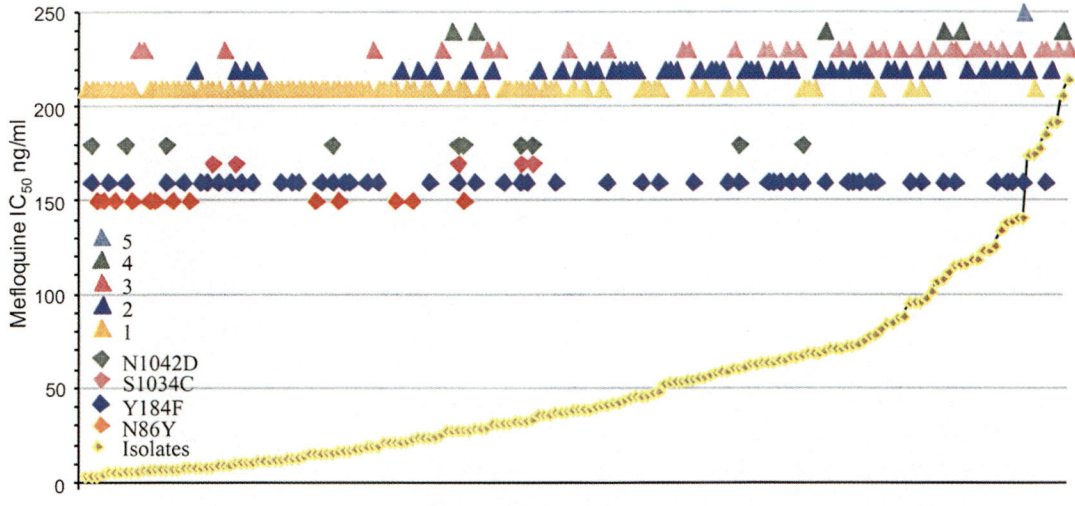

COLOR PLATE 18 (Chapter 23) Distribution of MFQ IC_{50} values by copy number (triangles) and codon mutations (diamonds). Isolates (diamonds) are arranged in ascending order of MFQ resistance. (Reprinted from *The Lancet* [Price et al., 2004] with permission from the publisher.)

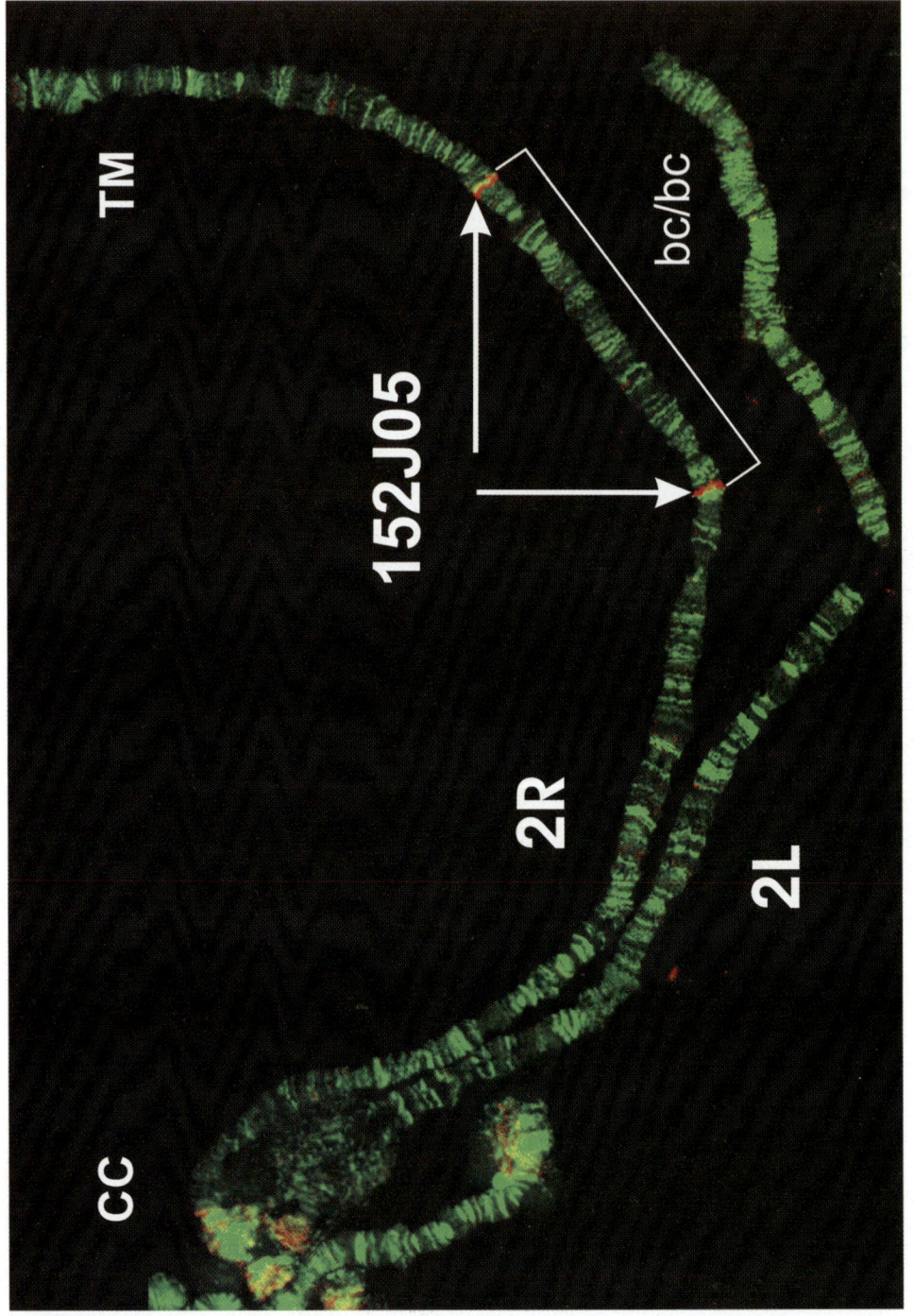

COLOR PLATE 19 (Chapter 26) In situ hybridization of *Anopheles gambiae* BAC clone 24D05 to the polytene chromosome complement of *A. gambiae*. BAC DNA was labeled with Cy3 from Amersham; polytene chromosomes were counterstained with Yoyo-1 from Molecular Probes.

MOLECULAR ASPECTS OF ANTIGENIC VARIATION IN *PLASMODIUM FALCIPARUM*

Paul Horrocks, Susan A. Kyes, Peter C. Bull, and Kirk W. Deitsch

21

Antigenic variation is a key survival strategy employed by a wide range of infectious organisms, allowing them to colonize and persist in a vertebrate host in the face of an evolving immune response. This subsequent chronic infection of the host maximizes the opportunity for these infectious organisms to be transmitted to a new host. Variant surface antigens (VSA) expressed on the surface of malaria-infected erythrocytes (IE) are targets of antibody-mediated immune responses (Marsh and Howard, 1986), resulting in clearance of the recognized parasite population (Bull et al., 1998). Clonal switches in the VSA expressed by a subpopulation of parasites allow them to survive the mounting immune response against the previous variant, thus ensuring the chronicity of infection (Borst et al., 1995).

VSA of the parasite responsible for the most severe form of human malaria, *Plasmodium falciparum*, also contribute significantly to the pathology of disease (Miller et al., 1994; Kyes et al., 2001). A strain-specific component of *P. falciparum* IE surface antigens was shown during the 1980s to be a family of large polymorphic proteins, termed *P. falciparum* erythrocyte membrane protein 1 (PfEMP1) (Leech et al., 1984). Presence of VSA on the IE surface was shown to correlate with the ability of IE to adhere to host cells (David et al., 1983), with many studies since showing that PfEMP1 specifically mediates IE adhesion to a variety of different host ligands. Adhesion to microvascular endothelial cells results in the sequestration of IE within major organs which, among other complications, can lead to reduced blood perfusion through occlusion of the microvasculature (Newbold, et al., 1997b; Kyes et al., 2001). Specific host adhesion properties of PfEMP1 have been associated with severe disease pathologies. For example, adhesion by PfEMP1 to intercellular adhesion molecule 1 (ICAM-1; CD54) expressed on endothelial cells in the brain may play a role in cerebral malaria (Turner et al., 1994). Similarly, adhesion of IE to complement receptor 1 (CR1; CD35) (Rowe et al., 1997) on uninfected red blood cells leads to the formation of rosettes, a phenotype associated with severe

---

Paul Horrocks, Susan A. Kyes, and Peter C. Bull, Weatherall Institute of Molecular Medicine, University of Oxford, John Radcliffe Hospital, Oxford OX3 9DS, United Kingdom. *Kirk W. Deitsch,* Department of Microbiology and Immunology, Weill Medical College of Cornell University, 1300 York Ave., W-704, Box 62, New York, NY 10021.

Molecular Approaches to Malaria, Edited by Irwin W. Sherman
© 2005 ASM Press, Washington, D.C.

disease outcome (Rowe et al., 1995; Rowe et al., 1997). Binding to chondroitin sulfate A (CSA) expressed on syncytiotrophoblasts in the placenta is a factor contributing to pregnancy-associated malaria (Fried and Duffy, 1996; Buffet et al., 1999; Reeder et al., 1999). Furthermore, PfEMP1-mediated adhesion of IE to dendritic cells modulates the activity of these important professional antigen-presenting cells (Urban et al., 1999). As PfEMP1 forms the target of protective immune responses, with acquired immunity developing following repeated infection with IE expressing different PfEMP1 variant types (Marsh and Howard, 1986; Bull et al., 1998), PfEMP1 is thus central to both the pathogenesis of falciparum malaria and the induction of protective immunity.

The discovery in the mid-1990s of the gene family that encodes PfEMP1, termed *var*, opened up the potential for the molecular investigation of this molecule (Baruch et al., 1995; Smith et al., 1995; Su et al., 1995). More recently, the completion of the *P. falciparum* genome project has revealed the organization and distribution of the entire complement of the *var* gene family (Gardner et al., 2002). Given the intense focus of the malaria research community on PfEMP1 and *var* and their roles in pathology and immune evasion, many reviews have been published on these broad themes. Here, we have concentrated on more contemporary aspects of the molecular basis of PfEMP1-*var* expression and switching. In doing so, we hope to highlight aspects that may not have been covered to date and attempt to show how a better understanding of the molecular biology of antigenic variation will help us understand this fascinating phenomenon at the organismal level.

PfEMP1 AND *var* IN THE POSTGENOMIC ERA

The Malaria Genome Sequencing Project was completed in 2002 with the entire sequence of the 3D7 parasite clone made available to the scientific community through several Internet-based sites (www.ncbi.nlm.nih.gov/projects/Malaria, www.sanger.ac.uk/Projects/P_falciparum, and www.PlasmoDB.org) (Bahl et al., 2003). Following the completion of this landmark project, several new insights into *var* gene organization, structure, sequence diversity, and predicted PfEMP1 cytoadhesive properties have emerged as a consequence of a whole-genome approach in their investigation. Although the *var* gene repertoires of different isolates are largely nonoverlapping and only the complete sequence of the 3D7 clone is currently available, the observations regarding the repertoire organization summarized here are likely to be applicable in all *P. falciparum* isolates.

Genomic Distribution of *var* Genes

PfEMP1 is a collective term for a family of variant surface antigens encoded by approximately 60 *var* genes per haploid genome (Baruch et al., 1995; Smith et al., 1995; Su et al., 1995). *var* genes are principally located near the subtelomeric repetitive sequence elements of *P. falciparum*'s 14 chromosomes, along with members of two further multigene families, *stevor* and *rif* (Fig. 1A). However, more than a third of the *var* gene repertoire is located within chromosome-internal regions. Both the subtelomeric and centrally located *var* gene variants contribute to antigenic variation (Scherf et al., 1998). As *var* genes can be activated in situ, what role does this pattern of chromosomal distribution play? High meiotic and/or mitotic recombination rates within the more plastic subtelomeric regions may serve as a mechanism to diversify the antigenic repertoire or perhaps assist gene regulation through physical clustering of chromosome ends at the nuclear periphery (Freitas-Junior et al., 2000; Taylor et al., 2000; Scherf et al., 2001). Additionally, it has been proposed that the *var* genes located within the relatively stable internal chromosomal regions may serve as a library of basic *var* sequence information (Fischer et al., 1997). However, the few highly conserved *var* genes identified to date lie within the subtelomeric domains (Rowe et al., 2002; Salanti et al., 2002; Kraemer and Smith, 2003; Salanti et al., 2003). Certainly, the

A.

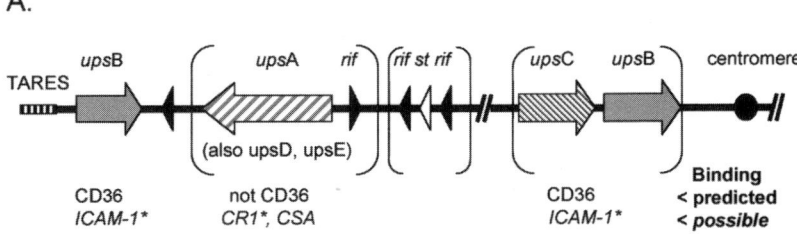

B.

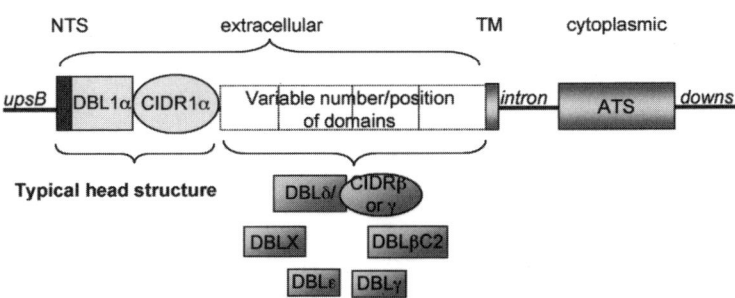

FIGURE 1 (A) Chromosomal distribution and orientation of the *var* multigene family. *var* genes are indicated as block arrows with the upstream sequence type (*ups*) indicated using letters A, B, and C. The most common subtelomeric organization of *var* genes is an *upsB var* gene in tail-to-tail orientation with a *rif* gene immediately adjacent to the telomere-associated repeat sequences (TARES). Additional features which may or may not be present are shown in brackets. Some subtelomeric regions contain an *upsA var* gene and/or other members of the *rifin* (*rif*) and *stevor* (*st*) multigene families. *upsD* and *upsE var* genes are always in the same relative location and/or orientation as *upsA*-type *var* genes. When present, chromosome-central *var* gene variants are either *upsB* or *upsC* type and are typically clustered in a head-to-tail orientation. The CD36-binding phenotype of the encoded PfEMP1 appears to be predictable from *var* gene location, as demonstrated in heterologous protein expression studies (Robinson et al., 2003). CSA binding is likely to be mediated by the *upsD*-type *var2CSA*-encoded PfEMP1 (Salanti et al., 2003). *var* genes encoding PfEMP1 for other severe disease phenotypes, such as ICAM-1 binding and CR1-mediated rosetting, have been confirmed with isolate IT/FCR3 but not with the genome project isolate 3D7. These *var* genes are type *upsB*, *upsC* (ICAM-1) (Smith et al., 1998; Smith et al., 2000a; Springer et al., 2004) and *upsA* (CR1) (Rowe et al., 1997). (B) *var* gene structure and sequence features. The basic coding sequence units of a *var* gene include an N-terminal sequence (NTS), at least two DBL domains, and the transmembrane (TM) region with exon 2 encoding the semiconserved acidic terminal sequence (ATS). Noncoding features include the relatively conserved 5' region (*ups*), the intron, and the 3' region (*downs*). A common polymorphic *var* gene is depicted, indicating the formation of the NTS, first DBLα and then CIDR, into what is termed the head structure. The presence of CIDR1α in the head structure predicts that the encoded PfEMP1 binds to CD36. Other domains that may be present in the PfEMP1 molecule are indicated below the basic PfEMP1 organization. Adhesion by these domains to certain host ligands, e.g., DBL1α1 to CR1, DBLβC2 to ICAM-1, and DBLγ to CSA (Buffet et al., 1999; Reeder et al., 1999; Gamain et al., 2004), has been reported, although the presence of these domains in the encoded PfEMP1 molecule should not always be assumed to indicate a particular adhesive property of the IE.

significance of distinct chromosomal compartments for *var* genes remains unresolved.

var Gene Structure within the Genomic Context

Characterization of *var* gene coding sequences is difficult due to their extreme diversity, but there are certain key conserved features in *var* gene structure. The first exon is highly variable in sequence and length (3.5 to 9.0 kb) (Gardner et al., 2002) and encodes the immunologically exposed portion of PfEMP1, as well as the transmembrane domain (Fig. 1B). The second exon is relatively short (1.0 to 1.5 kb) and highly conserved and encodes the cytoplasmic tail which anchors PfEMP1 to the IE cytoskeleton through noncovalent interactions with a number of parasite-encoded and host proteins (Oh et al., 2000; Waller et al., 2002). Exon 1 encodes between two and seven Duffy-binding-like (DBL) domains and, in most cases, at least one cysteine-rich interdomain region (CIDR). Nomenclature for position and sequence subgroups of DBL and CIDR domains is described by Smith et al. (2000b) (Fig. 1B; Tables 1 and 2).

In striking contrast to exon 1 sequences, the noncoding regions 5' of *var* genes fit into highly conserved sequence groups which are specific to particular genomic locations, as initially reported by Voss et al. (2000) in a comparison of several different isolates. Analysis of the entire 3D7 genome sequence confirmed that most *var* gene upstream sequences belong to one of three groups: *upsA*, *upsB*, or *upsC* (Table 1) (Gardner et al., 2002; Kraemer and Smith, 2003; Lavstsen et al., 2003). *upsA var* genes are always subtelomeric, *upsC var* genes are always chromosome internal, and *upsB var* genes can be in either location (Fig. 1A). Additionally, at chromosome ends, the transcriptional orientation of *upsA var* genes (if present) is always towards the telomere, whereas *upsB var* genes are transcribed away from it. Chromosome-internal *upsB* and *upsC var* genes are usually organized in head-to-tail clusters, with no other obvious bias in orientation. Single copies of two further 5' sequence types, *upsD* and *upsE*, are associated with two unusual conserved *var* genes which are subtelomeric and transcribed towards the telomere (Fig. 1A). These general rules in *var* promoter orientation seem to apply to other *P. falciparum* isolates, allowing prediction of genomic location on the basis of 5' noncoding sequence (Voss et al., 2000).

Introns and sequences downstream of *var* genes exhibit a less-pronounced degree of conservation. Only those *var* genes with *upsA* type 5' noncoding regions have similar 3' noncoding sequences (Gardner et al., 2002; Kraemer and Smith, 2003). Intron sequences vary greatly in length, ranging from ~150 bp up to 1.2 kbp. Although some introns can be divided into three regions of broad sequence conservation (Calderwood et al., 2003), any association of particular intron sequence types with chromosomal location remains to be tested.

The conservation in upstream sequences, in terms of both sequence and organization, may indicate some evolutionary pressure that restricts recombination between limited subsets of *var* genes (Voss et al., 2000). The identification of conserved *var* gene types, all located at chromosome ends and transcribed towards the telomere, supports this speculation (Fig. 1A; Table 2). Comparison of these genes in unrelated isolates shows that they share an identical organization of DBL domains (Table 2), each with a reasonably high degree of peptide sequence conservation (dependent on DBL type). These "conserved" *var*

TABLE 1 *ups* promoter-type and exon 1 comparison for polymorphic *var* genes in 3D7 genome

ups	Head structure	CD36 binding[a]	Remainder of exon 1	3D7 (n)[b]
B or C	DBL1α/CIDR1α	+	DBL2β/CIDRγ orCIDRβ	38
B orC	DBL1α/cIDR1α	+	Variable number of domains	10
A	DBL1α/CIDR1α or CIDR1γ	−	Variable number of domains	5

[a]Predicted CD36 binding based on presence or absence of CIDR1α (Robinson et al., 2003).
[b]Number of variants of this type present in the 3D7 genome (Gardener et al., 2002).

TABLE 2 *ups* promoter-type and exon 1 comparison for polymorphic *var* genes in 3D7 genome, conserved within and between genomes

ups	Head structure[a]	Remainder of exon 1	3D7 (n)	Name
D	DBLX	DBL2X, DBL3X, DBL4ε, DBL5ε, DBL6ε	1	PFL0030w[b]
E	DBL1α/CIDR1α1	DBL2βC2, DBL3γ, DBL4ε, DBL5γ, DBL6βC, DBL7ε	1Ψ	PFE1640w[c]
A	DBL1[a]	DBL2ε	3	PFA0015c[d]
				PFF0020c
				PFI1820w
A	DBL1α1 CIDR1α1	DBL2βC2, DBL3βC2, DBL4γ DBL5β-CIDRβ	2	PFD1235w[e] MAL7P1.1

[a] Absence of CIDR1α suggests that none of these PfEMP1 is likely to bind to CD36.
[b] *var2CSA* codes for likely CSA-binding PfEMP1 (Salanti et al., 2003). *upsD* contains a short open reading frame (Lavstsen et al., 2003). An unusual head structure was observed (no DBL1α or CIDR1) (Gardner et al., 2002; Kraemer and Smith, 2003).
[c] *var1CSA* is a pseudogene (Ψ) in 3D7 (Rowe et al., 2002); in other isolates, it has an unusual temporal regulation (Kyes et al., 2003). Its role in CSA binding is unclear (Rowe and Kyes, 2004).
[d] Type 3 *var* contains no CIDR; the shortest *var* genes, with shortest introns; and three copies in the 3D7 genome that are similar but not identical to one another (Gardner et al., 2002).
[e] *var4* is duplicated in the 3D7 genome (Gardner et al., 2002); upregulation of the gene is associated with recognition of IE by immune serum from children with severe malaria (Jensen et al., 2004).

genes are *var1CSA* (Rowe et al., 2002; Salanti et al., 2002), *var2CSA* (Kraemer and Smith, 2003; Salanti et al., 2003), type 3 *var* (Gardner et al., 2002; Kraemer and Smith, 2003), and *var4* (Jensen et al., 2004). Two of these genes have unique upstream sequences, *upsD* (*var2CSA*) and *upsE* (*var1CSA*), the other two being *upsA* types (Fig. 1B, Table 2). Given the instability of subtelomeric regions, there must be strong positive selection to maintain conservation within these *var* coding sequences, as well as their upstream sequences.

PfEMP1 Domain Structure and Association with Cytoadhesive Phenotypes

The adhesion of different DBL and CIDR domain types to host cell-surface molecules has been investigated in heterologous expression studies, establishing domains responsible for particular adhesive phenotypes and pinpointing regions critical for binding. Each of these phenotypes is mediated by distinct PfEMP1 domain types (Fig. 1B). It is worth mentioning here that the presence of a particular domain within a PfEMP1 variant is not sufficient to ascribe an adhesive phenotype to an IE expressing this variant. Heterologous adhesion studies consider each domain out of context of the whole PfEMP1 molecule in a completely different dynamic environment.

Analysis of adhesive phenotypes has highlighted another potential restriction in the genomic distribution of the *var* repertoire. A majority of field and laboratory parasite isolates adhere to CD36 (Newbold et al., 1997a; Newbold et al., 1997b). Heterologous expression studies indicate that CD36 binding is mediated by the CIDR1α domain (Baruch et al., 1997; Robinson et al., 2003). A majority of *var* genes contain the CIDR1α domain and are typically encoded by *upsB*- or *upsC*-type *var* genes (Table 1). In contrast, the non-CD36-binding CIDR1α1 or CIDR1γ domains are usually encoded by *upsA var* genes (Robinson et al., 2003). Further examples from field isolates are necessary to investigate the possibility of any correlation between *var* gene genomic position (and/or 5' sequence type) and other cytoadhesive phenotypes and, in particular, any association with disease severity (Jensen et al., 2004).

CONTROL OF *var* GENE EXPRESSION

Clonal switching of *var* gene expression during chronic infection reflects the sum of several molecular processes, including temporal regulation of *var* gene expression during intraerythrocytic development, mutually exclusive expression of only one *var* variant per IE, and the ability to switch *var* expression in progeny parasites. Of necessity, we will consider these regulatory features separately, although a central role for transcriptional control suggests that these regulatory processes may actually be integrated.

PfEMP1 expression is developmentally regulated, first appearing at the IE surface some 16 to 18 h postinvasion (HPI) in late-ring stages, when IE cytoadherent phenotypes are first detected (Color Plate 15) (Gardner et al., 1996; Kriek et al., 2003). Investigation of the control of *var* gene transcription has typically been carried out on long-term in vitro-adapted cultures that are highly selected in order to be homogeneous in their adhesion phenotype. Different experimental approaches taken in the molecular investigation of temporal *var* transcription patterns have led to two possible models of control.

Northern blot analysis indicates that the major *var* mRNA transcript encoding the PfEMP1 variant destined for the IE surface is present at its highest relative level in mid-ring-stage parasites (Color Plate 15), declining at steady-state level until it is barely detectable in pigmented trophozoites (Kyes et al., 2000). Other *var* transcript types are undetectable by Northern blot analysis of ring stages from these highly selected parasites. The difference in timing of *var* transcript accumulation and PfEMP1 expression on the IE surface presumably reflects the time necessary for PfEMP1 to be translated and exported from the intracellular parasite to the IE surface (Kriek et al., 2003). Together, these data suggest that temporal control of *var* gene expression is regulated at either the level of transcription initiation or a stage-specific targeted posttranscriptional degradation of mRNA (Color Plate 15). By contrast, studies using reverse transcription-PCR (RT-PCR) suggest that many *var* genes are transcribed in ring-stage parasites but that only one major *var* gene transcript is present in pigmented trophozoite stages (Chen et al., 1998; Scherf et al., 1998). These data would imply a posttranscriptional mechanism for the control of temporal gene expression. Abortive or incomplete transcription has been invoked as a possible molecular control mechanism, since numerous ring-stage transcripts are not detected by Northern blotting, although a report has described multiple spliced *var* transcript types in pigmented trophozoites (Noviyanti et al., 2001).

How can we reconcile these conflicting data? A partial explanation may lie in the recent discovery that one *var* gene has an unusual temporal regulation: it is transcribed in pigmented trophozoite stages of most parasites regardless of their adhesive phenotype (Color Plate 15) (Kyes et al., 2003). Similarly, but equally important, the level of transcript detected by either experimental approach has to be meaningful in the context of the overall population or individual cell phenotype. For example, RT-PCR detects not only multiple *var* transcripts but also sporozoite-specific transcripts in asexual stages (Fidock et al., 1994). This suggests that RT-PCR may readily pick up background noise, present in either slightly heterogeneous parasite populations or transcripts present at extremely low levels, which are then interpreted as the real signal (Duffy et al., 2003). By contrast, Northern blots, which detect only full-length RNA transcripts in large amounts, lack the sensitivity of detection necessary to investigate more subtle underlying mechanisms. The best investigative approach, described as yet only for the *var* gene exhibiting unusual temporal regulation and is therefore not representative (Scherf et al., 1998), relies on nuclear run-on transcription studies that describe where and when the RNA polymerases are active over the *var* repertoire.

Unlike antigenic variation in African trypanosomes where genes encoding variant surface glycoproteins can be activated through duplicative transposition into a limited number of specific expression sites, *var* genes are activated in situ (Scherf et al., 1998; Borst and Ulbert, 2001; Borst, 2002). While it would appear that most members of the *var* repertoire are equally capable of transcriptional activation at any time (see below for possible exceptions), only one is expressed in any IE (Smith et al., 1995). The expression of only one *var* gene variant while the rest of the repertoire remains silent is termed mutually exclusive expression (Fig. 2).

So, the key question is, how is only one *var* gene variant expressed while the rest of the repertoire is silent? Recent work characterizing a cooperative interaction between two regulatory regions within each *var* locus provides im-

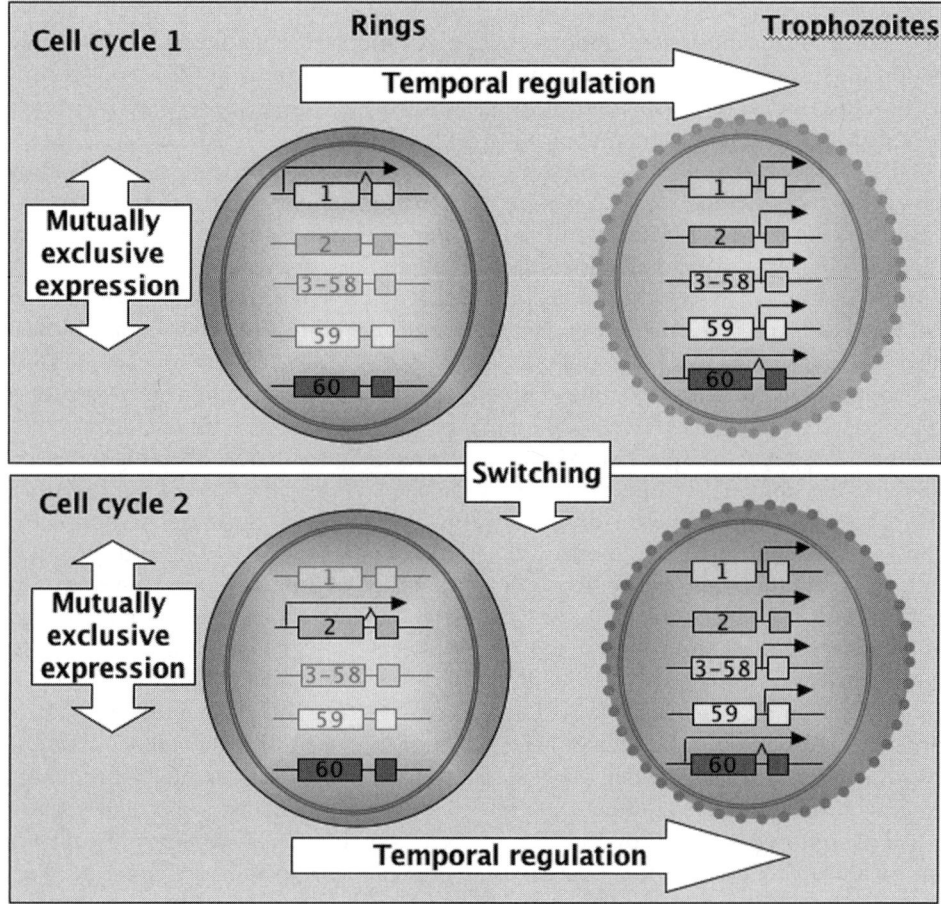

FIGURE 2 Control of *var* gene expression. The control of *var* gene expression is complex and includes temporal regulation during a cell cycle, mutually exclusive expression of a single variant in an IE, as well as switching of variants expressed in progeny clones. The status of expression for 5 of the 60 *var* genes present per haploid genome is represented in the ring and pigmented trophozoite stages during two cell cycles. Changes to the IE surface are indicated schematically, from smooth (rings) to patterned (pigmented trophozoites), emphasizing that PfEMP1 absent from the IE surface for most of the ring stage is present during trophozoite stages. In the first cell cycle, *var* variant 1 encodes a PfEMP1 variant (light grey dots) on the trophozoite surface, whereas in the second cycle a switch to a variant (dark grey dots) encoded by *var* variant 2 has occurred. During the ring stage of cell cycle 1, variant 1 is exclusively transcribed, giving rise to the corresponding PfEMP1 variant on the trophozoite IE surface while the remainder of the repertoire is transcriptionally silent (at least in terms of mRNA that encodes surface exposed PfEMP1). In trophozoite stage IE, all *var* gene variants are silenced (note transcription from the intronic promoter, although possibly not from that of variant 1) except for one variant (here variant 60, analogous to the conserved *var1CSA*). Thus *var* genes are subject to both temporal and mutually exclusive transcriptional regulation within a cell cycle. During an infection a small proportion of the parasite population may switch expression of PfEMP1. This is represented here by a switch to expression of the PfEMP1 variant 2 in cell cycle 2. *var* variant 2 is subject to the same temporal and mutually exclusive transcriptional regulation as *var* variant 1 during cell cycle 1.

portant clues to how *var* genes are silenced (Deitsch et al., 2001). Each *var* gene appears to have two promoters or, more specifically, two start sites of transcription (Deitsch et al., 2001; Calderwood et al., 2003). The first, approximately 1 kb from the start of exon I, presumably gives rise to the functional mRNA present in ring-stage parasites that is ultimately translated into PfEMP1. The second promoter is found within a relatively conserved sequence of the *var* intron and drives transcription of abundant, noncoding "sterile" RNA molecules that accumulate within pigmented trophozoite-stage IE (Su et al., 1995; Calderwood et al., 2003). While a single *var* gene variant is apparently transcribed from its upstream promoter and the remainder of the *var* repertoire is silent, or at least not producing full-length mRNA, most or all of the intron promoters seem to be active at the same time (Fig. 2) (Calderwood et al., 2003).

The interaction of these promoters in regulating mutually exclusive expression has been investigated using transfection technology. The cooperative nature of upstream and intronic *var* sequences was first demonstrated when the constitutive activity of a *var* upstream sequence on an episomal plasmid was silenced by the addition of a *var* intron to the plasmid construct (Deitsch et al., 2001). The specificity of this cooperative activity was demonstrated when *var* intronic sequences were unable to silence the transcriptional activity of an unrelated promoter. Interestingly, the ability of the intron to induce silencing was dependent on transition of the transfected parasites through S phase of the cell cycle. S-phase-dependent silencing is a characteristic of gene regulation based on both chromatin assembly and modification, which are thought to take place simultaneously with DNA replication during S phase (Kirchmaier and Rine, 2001; Li et al., 2001). Electrophoretic mobility shift assays have identified a conserved element upstream of *var* genes that is bound by either a protein or protein complex found in nuclear extracts (Voss et al., 2003). While this element is likely to be important for regulating *var* gene expression, its exact role remains unclear, although it is noteworthy that this element appears to be bound during S phase of the cell cycle. Together, these data imply that cooperative interaction between sequence elements within the introns and the *var* upstream regions control the transcriptional status of each individual *var* gene.

var introns thus appear to be powerful, S-phase-dependent transcriptional silencing elements that possess an independent promoter activity. Deletions within the intron both disrupt transcriptional activity from this site and eliminate its ability to silence a *var* upstream promoter (Calderwood et al., 2003). The intron appears to exert its silencing effect through its promoter activity, with the sterile transcripts produced presumably affecting the assembly of a silent epigenetic state over the *var* upstream promoter. Such a model is consistent with mechanisms of imprinting and allelic exclusion in other organisms, including autosomal imprinting and X-inactivation in mammals (Sleutels et al., 2002; Akhtar, 2003). In these examples, the transcriptional state of the gene is determined by alterations in chromatin structure and the action of noncoding RNAs. Exactly how these RNA molecules affect chromatin assembly is not yet understood in detail, and it is not yet known whether transcriptionally active *var* genes also produce sterile transcripts (Fig. 2). Aspects of epigenetic mechanisms that may contribute to regulating gene expression are considered below (Clemson et al., 1996; Francis and Kingston, 2001; Mechali, 2001).

While duplications and rearrangements within the *var* gene family do not appear to play a role in the control of *var* gene switching, some role for deletions has been indicated (Scherf et al., 1998; Deitsch et al., 1999; Kyes et al., 2001; Horrocks et al., 2002; Horrocks et al., 2004a). Deletions are unlikely, however, to represent a major switching mechanism, since this would result in loss of *var* repertoire. Moreover, the lack of any apparent alterations to DNA methylation or nucleosomal phasing over transcriptionally active *var* genes similarly eliminates them as potential regulatory mechanisms. Since the majority of parasites continue to express the same *var* gene through multiple cell cycles, this implies that there is some form of epigenetic imprinting or cellular memory that maintains a single *var* gene in the active state from one cell generation to the next, while the remaining *var* gene variants are silenced. Thus, molecular investigations are currently being directed toward understanding how silencing is lifted over the *var* gene variant that is being activated while simultaneously imposing it on the previously active copy.

IS THE CONTROL OF MUTUALLY EXCLUSIVE *var* GENE EXPRESSION EPIGENETIC?

Our understanding of mechanisms of epigenetic gene regulation and genome programming indicates that modifications to histones appear to play a central role in this process (Grant, 2001). Histone modifications such as acetylation, phosphorylation, methylation, and ubiquitination have led to the proposal of a histone code that determines the expression state of a gene surrounded by this modified chromatin (Grant, 2001; Jenuwein and Allis, 2001). The genes encoding proteins involved in modifying histones (i.e., histone acetyl transferases, deacetylases, and methyltransferases) have been identified in a number of organisms. Interestingly, while the completed sequence of the *P. falciparum* genome showed a paucity of transcription factors, a full complement of genes involved in chromatin modification and assembly has been identified (Aravind et al., 003). Thus, chromatin remodeling may be one of the few areas where *Plasmodium* transcriptional regulation processes bear some similarity to those described in other organisms.

Insights into the mechanism of *var* silencing have also been derived from our understanding of silencing mechanisms in yeast (Feuerbach et al., 2002; Hediger and Gasser, 2002). Genes in yeast appear to be subject to a telomeric silencing effect, mediated in part through the organization of chromosomes within the nuclear architecture. Yeast telomeres cluster together at the nuclear periphery, possibly associating with the nuclear matrix, a region known to contain high concentrations of proteins involved in transcriptional silencing. For example, in yeast the SIR and KU families of proteins, necessary for the assembly and maintenance of silent chromatin, are found to be more highly concentrated at the nuclear periphery. The structure of *Plasmodium* telomeres indicates that a similar nuclear organization exists in malaria parasites (Freitas-Junior et al., 2000; Scherf et al., 2001; Figueiredo et al., 2002). Therefore, by virtue of their location in the subtelomeric regions of the chromosomes, a large portion of the *var* repertoire is positioned within a region of the chromosomes and an area of the nucleus that is particularly prone to being transcriptionally silent. A model has been proposed for the spreading of the condensed chromatin structure from telomeres into the adjacent *var* genes, thus contributing to the silencing of these genes (Scherf et al., 2001). How such a model would apply to the clusters of *var* genes found in the internal regions of the chromosomes or whether these internal *var* genes also localize to the nuclear periphery is currently under investigation.

VARIABLE *var* SWITCHING RATES

Chronic infection with malaria is characterized by periodic peaks of parasitemia. By analogy to other systems, each peak of parasitemia consists of a population of parasites predominantly expressing the same variant surface antigen. *P. falciparum*, with only 60 *var* genes, has fewer members in its variant antigen repertoire than many other parasitic organisms infecting humans. Given the huge populations of *P. falciparum* that can develop during peaks of parasitemia (up to 10^{10} parasites), switching rates would be expected to be uniformly low; otherwise, all variants would be exposed to immune selection very early during the infection. Such uniformly low switching rates are unsuitable, however, for the later stages of a chronic infection where parasitemia levels are much smaller, and high switch rates would be required to seed antigenic diversity to maintain the infection. So how does *P. falciparum* effectively utilize its limited *var* repertoire to maintain a chronic infection?

Like many other parasites, *P. falciparum* does not appear to randomly switch expression between different *var* variants. Instead, a more efficient, statistically preferred order of expression is suggested, with the hierarchy of variant antigen expression during a chronic infection determined by differences in switching rates. Variable *var* switching rates had been predicted from computer models of chronic infections and from empirical observations relating to rare and commonly observed PfEMP1 variants in parasite cultures (Molineaux and Dietz, 1999; Kyes et al., 2001; Molineaux et al., 2001; Paget-McNicol et al., 2002). However, it is only more recently that the existence of variable *var* switching rates has been demonstrated both in vivo and in vitro (Peters et al., 2002; Horrocks et al., 2004b). Analysis of *var* transcripts present in parasites isolated from naïve volunteers inoculated via an infective mosquito bite indicated that initial rates of *var* switching were high but that subsequent switching rates were lower (Peters et al., 2002). Similarly, in vitro analysis of *var* switching in a panel of isogenic parasite clones indicated that not only were *var* switching rates variable, but that in parasite clones expressing the same variant, these rates were nearly identical (Horrocks et al., 2004b). The latter observation suggests that the switching rate for any given *var* gene is an intrinsic property of that variant.

In vitro experiments describe a range of switching rates between 2.0% per generation to an immeasurably low rate (Roberts et al., 1992; Horrocks et al., 2004b). However, modeling of the switch rates required to drive the observed *var* expression profile in experimentally infected humans indicates that transcriptional switch rates may be as high as 18% per generation (Gatton et al., 2003). Although calculating precise switching rates is technically challenging, the range in switching rates determined may actually reflect more upon the experimental strategy adopted in their measurement. For example, measuring switch rates by quantitative Northern blot analysis limited the variants investigated to those with more stable (or slower) switching rates. Similarly, although the models used to determine in vivo switching rates have correctly assumed that variant-specific and variant-transcending immunity would have a negligible selective effect on these early variants in a naïve host, the potential for host selection, mediated through the different cytoadhesive properties of PfEMP1, in altering the apparent variant switch rate was not considered. Although the range of PfEMP1 switching rates measured reflects those of VSA in murine malaria or recently fly-transmitted trypanosomes, more data are required to better understand their contribution to the order of variant appearance during infection (Brannan et al., 1994; Turner, 1997).

The molecular mechanisms that underpin variable *var* switching rates include either variant-specific growth rates, variant-specific transcription switch rates, or possibly a combination of both. Variant-specific growth rates in *P. falciparum* were first predicted from models of antigenic variation where the effective multiplication rate depended upon not only innate, variant-specific, and variant-transcending immune responses, but also a different intrinsic multiplication rate for each variant (Molineaux et al., 2001). Although essential for producing models that accurately reflect chronic parasitemia, no experimental evidence exists for such variant-specific growth rates in *P. falciparum*. Additionally, determination of *var* switching rates in vitro indicated that these rates were independent of parasite growth rates (Horrocks et al., 2004b). Variations in transcriptional switch rates are therefore more likely to be responsible for the observed variation in *var* switch rates. As switches in variant expression are almost certainly controlled at the level of transcriptional initiation with *var* silencing similarly regulated at the transcriptional level, it would be reasonable to suggest that the regulation of variable switch rates also acts at the same level. The intrinsic nature of these rates suggests that some aspect of local noncoding sequence determines these rates. Currently, however, there is no evidence to suggest a simple association between switch rate and *ups* promoter type (Horrocks et al., 2004b).

VARIANT EXPRESSION DURING CHRONIC INFECTION

A consensus model for variant expression during chronic infections can now be proposed. Following merozoite release from the liver, an initial PfEMP1 variant dominates the first cycle of intraerythrocytic infection. Evidence for the preferential expression of a single variant type at the start of a chronic infection was first shown in the murine malaria *Plasmodium chabaudi* (Brannan et al., 1993). That this also appears to be the case in *P. falciparum* is suggested from evidence of restricted variant expression in naïve humans following mosquito transmission of the parasite (Peters et al., 2002). Following expression of an initial variant type, the subsequent order of variant expression is predicted to follow a hierarchical order, determined principally by transcriptional switch rates. During the early stages of chronic infection, variants with high transcriptional switch rates would be predicted to dominate, with the later stages characterized by variants with low transcriptional switch rates.

Thus, the order in which variants appear during a chronic infection is a reflection of their intrinsic transcriptional switch rates. In vitro analysis of *var* switch rates also revealed the potential for a second factor that may affect the order of variant expression (Horrocks et al., 2004b). Analysis of a panel of clones derived from the IT clone indicated that transcriptional switch rates were an intrinsic, fixed property of a *var* gene and were not affected by the variant previously expressed. Whether a switch to a particular variant could occur, however, was dependent upon the previously expressed variant. This observation supports a model for three states of *var* gene expression: on, off (but capable of activation), and heavily silenced (Fig. 3). How control is exerted over this heavily silenced state has not been determined, although to date DNA methylation has been excluded (Horrocks et al., 2004b). Thus, the order of variant expression is determined by not only the intrinsic transcriptional switch rate, but also transition from the highly silenced state to one capable of activation (Fig. 3). Rather than a linear variant switching pathway, like that of *Plasmodium fragile* in toque monkeys (Handunnetti et al., 1987), *P. falciparum* may instead have a branched switching pathway. These branched pathways would predict that not all rapid switching variants are used during the initial stages of the infection, maintaining some of the *var* repertoire for the later stages of chronic infection. This not only is a potentially important adaptive mechanism, but could also be a key aspect in the parasite being able to establish a long-term chronic infection with only the available 60 *var* variants.

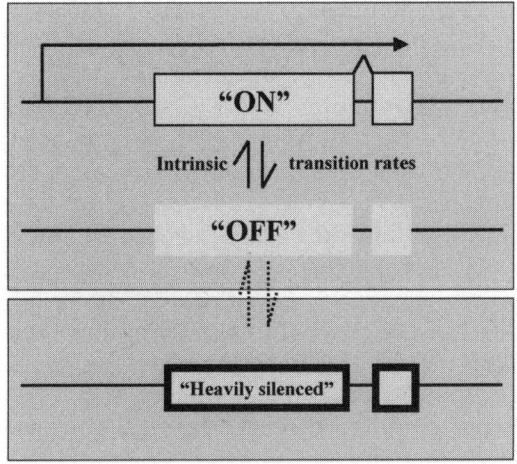

FIGURE 3 Do *var* genes exist in three transcriptional states? One interpretation of *var* switching rates measured in vitro was that each *var* gene is capable of existing in one of three transcriptional states. First, the on state is where a *var* gene is transcribed and encodes the PfEMP1 molecule on the IE surface. Second, the *var* gene is transcriptionally silent, off but capable of being activated. Transition rates between these two transcriptional states (solid arrows) have been measured and shown to be an intrinsic property of each *var* gene variant. A third transcriptional state was suggested from in vitro data, indicating that the variant expression history impacts on the ability of a *var* gene variant to be switched on. This is termed the heavily silenced transcriptional state. How *var* genes transition to and from this state, and the rates at which this occurs, is not known (broken arrows); their speculative nature is emphasized here by their separation in a second (lower) grey box. Thus, it is proposed that the overall transcriptional status of a *var* gene variant is dependent not only on the intrinsic switch rate between on and off states, but also on whether the gene exists in the heavily silenced state.

The generation of successive waves of parasitemia during a chronic infection, each expressing a predominant variant, is attributed to the expansion of subpopulations of different antigenic types in the presence of immune pressure.

Modeling of antigenic variation in *Trypanosoma brucei*, *Borrelia hermsii*, and *P. falciparum* supports this antibody-mediated selection of antigenic types (Agur et al., 1989; Borst et al., 1997; Deitsch et al., 1997; Turner, 1999; Nordstrand et al., 2000; Molineaux et al., 2001). A recent model proposed by Recker and colleagues (2004) suggests that this process may actually be more complex than initially thought. Essentially, this model suggests that each PfEMP1 variant has two types of epitope: a unique major epitope against which a long-term immune response is raised and a minor epitope which is shared between subsets of different variants against which short-lived immune responses are raised. This combination of major and minor epitopes can structure the appearance of pfEMP1 variants, ensuring that sequentially expressed PfEMP1 variants are immunologically distinct from one another. This model not only provides a mechanism for the sequential exposure of PfEMP1 variants over time, but also impacts on a longstanding challenge to our understanding of age distribution of parasite prevalence in epidemiological data (Recker et al., 2004). This model proposes that whereas the duration of chronic infection is longer in older children, although with a reduced parasite density, these children actually have the highest levels of cross-reactive immune responses. Thus, as older children become more capable of mounting cross-reactive responses, this paradoxically actually helps structure the appearance of PfEMP1 variants and thus extends the length of the chronic infection. The model, however, does not consider variable transcriptional switch rates or the effect of switching history, instead assigning all variants an equal transition rate. It will be interesting to reconcile the contribution that transient immune responses and variable transcriptional switch rates have on the hierarchical expression of variants.

PERSPECTIVES

With the advances in our understanding of the molecular basis of antigenic variation, a priority now is to attempt to bridge the gap between these molecular studies, often based on in vitro or computer models, with "real" epidemiological data from the field to better describe the host-parasite relationship in vivo. The mathematical model of Recker et al. (2004) was developed to reproduce the characteristics of a "normal" chronic infection incorporating both long-lived highly specific antibody responses and a network of short-lived responses that recognize epitopes shared between subsets of PfEMP1; both types of anti-VSA responses have been observed in field studies (Marsh and Howard, 1986; Forsyth et al., 1989; Giha et al., 1998; Bull et al., 2002; Chattopadhyay et al., 2003). These models, however, have their limitations. For example, a striking feature of epidemiological data is that these antibody responses show a high level of variation between individuals (Giha et al., 1999; Kinyanjui et al., 2003). Why this is the case and what impact they have on the susceptibility to severe malaria are currently being investigated (Bull et al., 2002; Tebo et al., 2002).

VSA expressed by parasites from either very young children or individuals with severe malaria tend to be better recognized by antibodies carried by other children (Bull et al., 2000; Nielsen et al., 2002; Lindenthal et al., 2003), an observation that extends to children even from diverse geographical locations (Nielsen et al., 2004). This observation has led to an idea of "common" VSA, where a hypothetical subset of these molecules termed VSA_{FoRH} (VSA with a high frequency of recognition) (Bull et al., in press) is preferentially expressed by parasites in more immunologically naïve hosts. VSA_{FoRH} have also been called VSA_{SM}, as these molecules are more likely to be expressed in individuals with severe malaria (Nielsen et al., 2004). VSA expressed on the IE are considered to be subject to two selective pressures in the host. The first is the ability of VSA to mediate optimal colonization of the host, presumably through its cytoadhesive properties, with the second a requirement of VSA to express novel epitopes not yet recognized by preexisting antibodies. Thus, the expression of VSA by parasites is shaped not only by the immune status of the host but also by the functional (adhesive) properties of the molecule, an activity which may constrain its subsequent epitope diversity. Selection of IE in vitro using sera isolated from postacute infections selects for expression of a restricted subset of *var* genes (Jensen et al., 2004). These *var* genes all contained CIDR1α1 domains and hence were predicted to lack the ability to bind CD36 (*upsA* type variants). Studies are now under way to characterize what role genomic distribution or switching rates may have in defining a possible VSA_{FoRH} subgroup and their potential role in severe disease.

Several key questions are best addressed by bringing molecular and epidemiological investigations together. What impact does the immune status of the infected host have in the subsequent immunological and/or epidemiological properties of the expressed PfEMP1 (VSA_{SM}/VSA_{FoRH})? What are the key epitopes of PfEMP1 and where are they located within the molecule? How do the kinetics of the antibody response to homologous VSA (those from the child in question) and heterologous VSA (those from other children) predict host susceptibility to severe malaria? Answers to these questions are important in assessing the feasibility of vaccines based on VSA_{FoRH} which would potentially accelerate naturally acquired immunity against severe *P. falciparum* malaria infections.

ACKNOWLEDGMENTS

We thank our colleagues Chris Newbold, Alister Craig, Gloria Rudenko, Christian Epp, Matthias Frank, and Andrew Serazin for their expert input during the preparation of the manuscript.

P.H., S.A.K., and P.C.B. are supported by Wellcome Trust. K.W.D. is a Stavros S. Niarchos Scholar and is supported by grants from the Ellison Medical Foundation and the National Institutes of Health. The Microbiology and Immunology Department at Weill Medical College of Cornell University acknowledges the support of the William Randolph Hearst Foundation.

REFERENCES

Agur, Z., D. Abiri, and L. H. Van der Ploeg. 1989. Ordered appearance of antigenic variants of African trypanosomes explained in a mathematical model based on a stochastic switch process and immune-selection against putative switch intermediates. *Proc. Natl. Acad. Sci. USA* **86:**9626–9630.

Akhtar, A. 2003. Dosage compensation: an intertwined world of RNA and chromatin remodelling. *Curr. Opin. Genet. Dev.* **13:**161–169.

Aravind, L., L. M. Iyer, T. E. Wellems, and L. H. Miller. 2003. *Plasmodium* biology: genomic gleanings. *Cell* **115:**771–785.

Bahl, A., B. Brunk, J. Crabtree, M. J. Fraunholz, B. Gajria, G. R. Grant, H. Ginsburg, D. Gupta, J. C. Kissinger, P. Labo, L. Li, M. D. Mailman, A. J. Milgram, D. S. Pearson, D. S. Roos, J. Schug, C. J. Stoeckert, Jr., and P. Whetzel. 2003. PlasmoDB: the *Plasmodium* genome resource. A database integrating experimental and computational data. *Nucleic Acids Res.* **31:**212–215.

Baruch, D. I., X. C. Ma, H. B. Singh, X. Bi, B. L. Pasloske, and R. J. Howard. 1997. Identification of a region of PfEMP1 that mediates adherence of *Plasmodium falciparum* infected erythrocytes to CD36: conserved function with variant sequence. *Blood* **90:**3766–3775.

Baruch, D. I., B. L. Pasloske, H. B. Singh, X. Bi, X. C. Ma, M. Feldman, T. F. Taraschi, and R. J. Howard. 1995. Cloning the *P. falciparum* gene encoding PfEMP1, a malarial variant antigen and adherence receptor on the surface of parasitized human erythrocytes. *Cell* **82:**77–87.

Borst, P. 2002. Antigenic variation and allelic exclusion. *Cell* **109:**5–8.

Borst, P., W. Bitter, R. McCulloch, F. Van Leeuwen, and G. Rudenko. 1995. Antigenic variation in malaria. *Cell* **82:**1–4.

Borst, P., G. Rudenko, P. A. Blundell, F. van Leeuwen, M. A. Cross, R. McCulloch, H. Gerrits, and I. M. Chaves. 1997. Mechanisms of antigenic variation in African trypanosomes. *Behring Inst. Mitt.* **March:**1–15.

Borst, P., and S. Ulbert. 2001. Control of VSG gene expression sites. *Mol. Biochem. Parasitol.* **114:**17–27.

Brannan, L. R., S. A. McLean, and R. S. Phillips. 1993. Antigenic variants of *Plasmodium chabaudi chabaudi* AS and the effects of mosquito transmission. *Parasite Immunol.* **15:**135–141.

Brannan, L. R., C. M. Turner, and R. S. Phillips. 1994. Malaria parasites undergo antigenic variation at high rates in vivo. *Proc. R. Soc. Lond. B Biol. Sci.* **256:**71–75.

Buffet, P. A., B. Gamain, C. Scheidig, D. Baruch, J. D. Smith, R. Hernandez-Rivas, B. Pouvelle, S. Oishi, N. Fujii, T. Fusai, D. Parzy, L. H. Miller, J. Gysin, and A. Scherf. 1999. *Plasmodium falciparum* domain mediating adhesion to chondroitin sulfate A: a receptor for human placental infection. *Proc. Natl. Acad. Sci. USA* **96:**12743–12748.

Bull, P. C., M. Kortok, O. Kai, F. Ndungu, A. Ross, B. S. Lowe, C. I. Newbold, and K. Marsh. 2000. *Plasmodium falciparum*-infected erythrocytes: agglutination by diverse Kenyan plasma is associated with severe disease and young host age. *J. Infect. Dis.* **182:**252–259.

Bull, P. C., B. S. Lowe, N. Kaleli, F. Njuga, M. Kortok, A. Ross, F. Ndungu, R. W. Snow, and K. Marsh. 2002. *Plasmodium falciparum* infections are associated with agglutinating antibodies to parasite-infected erythrocyte surface antigens among healthy Kenyan children. *J. Infect. Dis.* **185:**1688–1691.

Bull, P. C., B. S. Lowe, M. Kortok, C. S. Molyneux, C. I. Newbold, and K. Marsh. 1998. Parasite antigens on the infected red cell surface are targets for naturally acquired immunity to malaria. *Nat. Med.* **4:**358–360.

Bull, P. C., A. Pain, F. M. Ndungu, S. M. Kinyanjui, D. J. Roberts, C. I. Newbold, and K. Marsh. *Plasmodium falciparum* antigenic variation: relationships between *in-vivo* selection, the acquired antibody response and disease severity. *J. Infect. Dis.*, in press.

Calderwood, M. S., L. Gannoun-Zaki, T. E. Wellems, and K. W. Deitsch. 2003. *Plasmodium falciparum var* genes are regulated by two regions with separate promoters, one upstream of the coding region and a second within the intron. *J. Biol. Chem.* **278:**34125–34132.

Chattopadhyay, R., A. Sharma, V. K. Srivastava, S. S. Pati, S. K. Sharma, B. S. Das, and C. E. Chitnis. 2003. *Plasmodium falciparum* infection elicits both variant-specific and cross-reactive antibodies against variant surface antigens. *Infect Immun.* **71:**597–604.

Chen, Q., V. Fernandez, A. Sundstrom, M. Schlichtherle, S. Datta, P. Hagblom, and M. Wahlgren. 1998. Developmental selection of *var* gene expression in *Plasmodium falciparum*. *Nature* **394:**392–395.

Clemson, C. M., J. A. McNeil, H. F. Willard, and J. B. Lawrence. 1996. XIST RNA paints the inactive X chromosome at interphase: evidence for a novel RNA involved in nuclear/chromosome structure. *J. Cell Biol.* **132:**259–275.

David, P. H., M. Hommel, L. H. Miller, I. J. Udeinya, and L. D. Oligino. 1983. Parasite sequestration in *Plasmodium falciparum* malaria: spleen and antibody modulation of cytoadherence of infected erythrocytes. *Proc. Natl. Acad. Sci. USA* **80:**5075–5079.

Deitsch, K. W., M. S. Calderwood, and T. E. Wellems. 2001. Malaria. Cooperative silencing elements in *var* genes. *Nature* **412:**875–876.

Deitsch, K. W., A. del Pinal, and T. E. Wellems. 1999. Intra-cluster recombination and *var* transcription switches in the antigenic variation of *Plasmodium falciparum*. *Mol. Biochem. Parasitol.* **101:**107–116.

Deitsch, K. W., E. R. Moxon, and T. E. Wellems. 1997. Shared themes of antigenic variation and virulence in bacterial, protozoal, and fungal infections. *Microbiol. Mol. Biol. Rev.* **61:**281–293.

Duffy, M. F., J. C. Reeder, and G. V. Brown. 2003. Regulation of antigenic variation in *Plasmodium falciparum*: censoring freedom of expression? *Trends Parasitol.* **19:**121–124.

Feuerbach, F., V. Galy, E. Trelles-Sticken, M. Fromont-Racine, A. Jacquier, E. Gilson, J. C. Olivo-Marin, H. Scherthan, and U. Nehrbass. 2002. Nuclear architecture and spatial positioning help establish transcriptional states of telomeres in yeast. *Nat. Cell Biol.* **4:**214–221.

Fidock, D. A., E. Bottius, K. Brahimi, I. I. Moelans, M. Aikawa, R. N. Konings, U. Certa, P. Olafsson, T. Kaidoh, and A. Asavanich. 1994. Cloning and characterization of a novel *Plasmodium falciparum* sporozoite surface antigen, STARP. *Mol. Biochem. Parasitol.* **64:**219–232.

Figueiredo, L. M., L. H. Freitas-Junior, E. Bottius, J. C. Olivo-Marin, and A. Scherf. 2002. A central role for *Plasmodium falciparum* subtelomeric regions in spatial positioning and telomere length regulation. *EMBO J.* **21:**815–824.

Fischer, K., P. Horrocks, M. Preuss, J. Wiesner, S. Wunsch, A. A. Camargo, and M. Lanzer. 1997. Expression of *var* genes located within polymorphic subtelomeric domains of *Plasmodium falciparum* chromosomes. *Mol. Cell. Biol.* **17:**3679–3686.

Forsyth, K. P., G. Philip, T. Smith, E. Kum, B. Southwell, and G. V. Brown. 1989. Diversity of antigens expressed on the surface of erythrocytes infected with mature *Plasmodium falciparum* parasites in Papua New Guinea. *Am. J. Trop. Med. Hyg.* **41:**259–265.

Francis, N. J., and R. E. Kingston. 2001. Mechanisms of transcriptional memory. *Nat. Rev. Mol. Cell Biol.* **2:**409–421.

Freitas-Junior, L. H., E. Bottius, L. A. Pirrit, K. W. Deitsch, C. Scheidig, F. Guinet, U. Nehrbass, T. E. Wellems, and A. Scherf. 2000. Frequent ectopic recombination of virulence factor genes in telomeric chromosome clusters of *P. falciparum*. *Nature* **407:**1018–1022.

Fried, M., and P. E. Duffy. 1996. Adherence of *Plasmodium falciparum* to chondroitin sulfate A in the human placenta. *Science* **272:**1502–1504.

Gamain, B., J. D. Smith, M. Avril, D. I. Baruch, A. Scherf, J. Gysin, and L. H. Miller. 2004. Identification of a 67-amino-acid region of the *Plasmodium falciparum* variant surface antigen that binds chondroitin sulphate A and elicits antibodies reactive with the surface of placental isolates. *Mol. Microbiol.* **53:**445–455.

Gardner, J. P., R. A. Pinches, D. J. Roberts, and C. I. Newbold. 1996. Variant antigens and endothelial receptor adhesion in *Plasmodium falciparum*. *Proc. Natl. Acad. Sci. USA* **93:**3503–3508.

Gardner, M. J., N. Hall, E. Fung, O. White, M. Berriman, R. W. Hyman, J. M. Carlton, A. Pain,

K. E. Nelson, S. Bowman, I. T. Paulsen, K. James, J. A. Eisen, K. Rutherford, S. L. Salzberg, A. Craig, S. Kyes, M. S. Chan, V. Nene, S. J. Shallom, B. Suh, J. Peterson, S. Angiuoli, M. Pertea, J. Allen, J. Selengut, D. Haft, M. W. Mather, A. B. Vaidya, D. M. Martin, A. H. Fairlamb, M. J. Fraunholz, D. S. Roos, S. A. Ralph, G. I. McFadden, L. M. Cummings, G. M. Subramanian, C. Mungall, J. C. Venter, D. J. Carucci, S. L. Hoffman, C. Newbold, R. W. Davis, C. M. Fraser, and B. Barrell. 2002. Genome sequence of the human malaria parasite *Plasmodium falciparum*. *Nature* **419**:498–511.

Gatton, M. L., J. M. Peters, E. V. Fowler, and Q. Cheng. 2003. Switching rates of *Plasmodium falciparum var* genes: faster than we thought? *Trends Parasitol.* **19**:202–208.

Giha, H. A., T. Staalsoe, D. Dodoo, I. M. Elhassan, C. Roper, G. M. Satti, D. E. Arnot, T. G. Theander, and L. Hviid. 1999. Nine-year longitudinal study of antibodies to variant antigens on the surface of *Plasmodium falciparum*-infected erythrocytes. *Infect. Immun.* **67**:4092–4098.

Giha, H. A., T. G. Theander, T. Staalso, C. Roper, I. M. Elhassan, H. Babiker, G. M. Satti, D. E. Arnot, and L. Hviid. 1998. Seasonal variation in agglutination of *Plasmodium falciparum*-infected erythrocytes. *Am. J. Trop. Med. Hyg.* **58**:399–405.

Grant, P. A. 2001. A tale of histone modifications. *Genome Biol.* **2**:3.

Handunnetti, S. M., K. N. Mendis, and P. H. David. 1987. Antigenic variation of cloned *Plasmodium fragile* in its natural host *Macaca sinica*. Sequential appearance of successive variant antigenic types. *J. Exp. Med.* **165**:1269–1283.

Hediger, F., and S. M. Gasser. 2002. Nuclear organization and silencing: putting things in their place. *Nat. Cell Biol.* **4**:53–55.

Horrocks, P., R. Pinches, S. Kyes, N. Kriek, S. Lee, Z. Christodoulou, and C. I. Newbold. 2002. Effect of *var* gene disruption on switching in *Plasmodium falciparum*. *Mol. Microbiol.* **45**:1131–1141.

Horrocks, P., S. Kyes, R. Pinches, Z. Christodoulou, and C. Newbold. 2004a. Transcription of a subtelomerically located *var* gene variant in *Plasmodium falciparum* appears to require the truncation of an adjacent *var* gene. *Mol. Biochem. Parasitol.* **134**:193–199.

Horrocks, P., R. Pinches, Z. Christodoulou, S. A. Kyes, and C. I. Newbold. 2004b. Variable *var* transition rates underlie antigenic variation in malaria. *Proc. Natl. Acad. Sci. USA* **101**:11129–11134.

Jensen, A. T., P. Magistrado, S. Sharp, L. Joergensen, T. Lavstsen, A. Chiucchiuini, A. Salanti, L. S. Vestergaard, J. P. Lusingu, R. Hermsen, R. Sauerwein, J. Christensen, M. A. Nielsen, L. Hviid, C. Sutherland, T. Staalsoe, and T. G. Theander. 2004. *Plasmodium falciparum* associated with severe childhood malaria preferentially expresses PfEMP1 encoded by group A *var* genes. *J. Exp. Med.* **199**:1179–1190.

Jenuwein, T., and C. D. Allis. 2001. Translating the histone code. *Science* **293**:1074–1080.

Kinyanjui, S. M., P. Bull, C. I. Newbold, and K. Marsh. 2003. Kinetics of antibody responses to *Plasmodium falciparum*-infected erythrocyte variant surface antigens. *J. Infect. Dis.* **187**:667–674.

Kirchmaier, A. L., and J. Rine. 2001. DNA replication-independent silencing in *S. cerevisiae*. *Science* **291**:646–650.

Kraemer, S. M., and J. D. Smith. 2003. Evidence for the importance of genetic structuring to the structural and functional specialization of the *Plasmodium falciparum var* gene family. *Mol. Microbiol.* **50**:1527–1538.

Kriek, N., L. Tilley, P. Horrocks, R. Pinches, B. C. Elford, D. J. Ferguson, K. Lingelbach, and C. I. Newbold. 2003. Characterization of the pathway for transport of the cytoadherence-mediating protein, PfEMP1, to the host cell surface in malaria parasite-infected erythrocytes. *Mol. Microbiol.* **50**:1215–1227.

Kyes, S., P. Horrocks, and C. Newbold. 2001. Antigenic variation at the infected red cell surface in malaria. *Annu. Rev. Microbiol.* **55**:673–707.

Kyes, S., R. Pinches, and C. Newbold. 2000. A simple RNA analysis method shows *var* and *rif* multigene family expression patterns in *Plasmodium falciparum*. *Mol. Biochem. Parasitol.* **105**:311–315.

Kyes, S. A., Z. Christodoulou, A. Raza, P. Horrocks, R. Pinches, J. A. Rowe, and C. I. Newbold. 2003. A well-conserved *Plasmodium falciparum var* gene shows an unusual stage-specific transcript pattern. *Mol. Microbiol.* **48**:1339–1348.

Lavstsen, T., A. Salanti, A. T. Jensen, D. E. Arnot, and T. G. Theander. 2003. Sub-grouping of *Plasmodium falciparum* 3D7 *var* genes based on sequence analysis of coding and non-coding regions. *Malar. J.* **2**:27.

Leech, J. H., J. W. Barnwell, L. H. Miller, and R. J. Howard. 1984. Identification of a strain-specific malarial antigen exposed on the surface of *Plasmodium falciparum*-infected erythrocytes. *J. Exp. Med.* **159**:1567–1575.

Li, Y. C., T. H. Cheng, and M. R. Gartenberg. 2001. Establishment of transcriptional silencing in the absence of DNA replication. *Science* **291**:650–653.

Lindenthal, C., P. G. Kremsner, and M. Q. Klinkert. 2003. Commonly recognised *Plasmodium falciparum* parasites cause cerebral malaria. *Parasitol. Res.* **91**:363–368.

Marsh, K., and R. J. Howard. 1986. Antigens induced on erythrocytes by *P. falciparum*: expression of diverse and conserved determinants. *Science* **231**:150–153.

Mechali, M. 2001. DNA replication origins: from sequence specificity to epigenetics. *Nat. Rev. Genet.* **2:**640–645.

Miller, L. H., M. F. Good, and G. Milon. 1994. Malaria pathogenesis. *Science* **2640:**1878–1883.

Molineaux, L., H. H. Diebner, M. Eichner, W. E. Collins, G. M. Jeffery, and K. Dietz. 2001. *Plasmodium falciparum* parasitaemia described by a new mathematical model. *Parasitology* **122:**379–391.

Molineaux, L., and K. Dietz. 1999. Review of intra-host models of malaria. *Parassitologia* **41:**221–231.

Newbold, C., P. Warn, G. Black, A. Berendt, A. Craig, B. Snow, M. Msobo, N. Peshu, and K. Marsh. 1997a. Receptor-specific adhesion and clinical disease in *Plasmodium falciparum*. *Am. J. Trop. Med. Hyg.* **57:**389–398.

Newbold, C. I., A. G. Craig, S. Kyes, A. R. Berendt, R. W. Snow, N. Peshu, and K. Marsh. 1997b. PfEMP1, polymorphism and pathogenesis. *Ann. Trop. Med. Parasitol.* **91:**551–557.

Nielsen, M. A., T. Staalsoe, J. A. Kurtzhals, B. Q. Goka, D. Dodoo, M. Alifrangis, T. G. Theander, B. D. Akanmori, and L. Hviid. 2002. *Plasmodium falciparum* variant surface antigen expression varies between isolates causing severe and nonsevere malaria and is modified by acquired immunity. *J. Immunol.* **168:**3444–3450.

Nielsen, M. A., L. S. Vestergaard, J. Lusingu, J. A. Kurtzhals, H. A. Giha, B. Grevstad, B. Q. Goka, M. M. Lemnge, J. B. Jensen, B. D. Akanmori, T. G. Theander, T. Staalsoe, and L. Hviid. 2004. Geographical and temporal conservation of antibody recognition of *Plasmodium falciparum* variant surface antigens. *Infect. Immun.* **72:**3531–3535.

Nordstrand, A., A. G. Barbour, and S. Bergstrom. 2000. *Borrelia* pathogenesis research in the post-genomic and post-vaccine era. *Curr. Opin. Microbiol.* **3:**86–92.

Noviyanti, R., G. V. Brown, M. E. Wickham, M. F. Duffy, A. F. Cowman, and J. C. Reeder. 2001. Multiple *var* gene transcripts are expressed in *Plasmodium falciparum* infected erythrocytes selected for adhesion. *Mol. Biochem. Parasitol.* **114:**227–237.

Oh, S. S., S. Voigt, D. Fisher, S. J. Yi, P. J. LeRoy, L. H. Derick, S. Liu, and A. H. Chishti. 2000. *Plasmodium falciparum* erythrocyte membrane protein 1 is anchored to the actin-spectrin junction and knob-associated histidine-rich protein in the erythrocyte skeleton. *Mol. Biochem. Parasitol.* **108:**237–247.

Paget-McNicol, S., M. Gatton, I. Hastings, and A. Saul. 2002. The *Plasmodium falciparum var* gene switching rate, switching mechanism and patterns of parasite recrudescence described by mathematical modelling. *Parasitology* **124:**225–235.

Peters, J., E. Fowler, M. Gatton, N. Chen, A. Saul, and Q. Cheng. 2002. High diversity and rapid changeover of expressed *var* genes during the acute phase of *Plasmodium falciparum* infections in human volunteers. *Proc. Natl. Acad. Sci. USA* **99:**10689–10694.

Recker, M., S. Nee, P. C. Bull, S. Kinyanjui, K. Marsh, C. Newbold, and S. Gupta. 2004. Transient cross-reactive immune responses can orchestrate antigenic variation in malaria. *Nature* **429:**555–558.

Reeder, J. C., A. F. Cowman, K. M. Davern, J. G. Beeson, J. K. Thompson, S. J. Rogerson, and G. V. Brown. 1999. The adhesion of *Plasmodium falciparum*-infected erythrocytes to chondroitin sulfate A is mediated by *P. falciparum* erythrocyte membrane protein 1. *Proc. Natl. Acad. Sci. USA* **96:**5198–5202.

Roberts, D. J., A. G. Craig, A. R. Berendt, R. Pinches, G. Nash, K. Marsh, and C. I. Newbold. 1992. Rapid switching to multiple antigenic and adhesive phenotypes in malaria. *Nature* **357:**689–692.

Robinson, B. A., T. L. Welch, and J. D. Smith. 2003. Widespread functional specialization of *Plasmodium falciparum* erythrocyte membrane protein 1 family members to bind CD36 analysed across a parasite genome. *Mol. Microbiol.* **47:**1265–1278.

Rowe, A., J. Obeiro, C. I. Newbold, and K. Marsh. 1995. *Plasmodium falciparum* rosetting is associated with malaria severity in Kenya. *Infect. Immun.* **63:**2323–2326.

Rowe, J. A., and S. A. Kyes. 2004. The role of *Plasmodium falciparum var* genes in malaria in pregnancy. *Mol. Microbiol.* **53:**1011–1019.

Rowe, J. A., S. A. Kyes, S. J. Rogerson, H. A. Babiker, and A. Raza. 2002. Identification of a conserved *Plasmodium falciparum var* gene implicated in malaria in pregnancy. *J. Infect. Dis.* **185:**1207–1211.

Rowe, J. A., J. M. Moulds, C. I. Newbold, and L. H. Miller. 1997. *P. falciparum* rosetting mediated by a parasite-variant erythrocyte membrane protein and complement-receptor 1. *Nature* **388:**292–295.

Salanti, A., A. T. Jensen, H. D. Zornig, T. Staalsoe, L. Joergensen, M. A. Nielsen, A. Khattab, D. E. Arnot, M. Q. Klinkert, L. Hviid, and T. G. Theander. 2002. A sub-family of common and highly conserved *Plasmodium falciparum var* genes. *Mol. Biochem. Parasitol.* **122:**111–115.

Salanti, A., T. Staalsoe, T. Lavstsen, A. T. Jensen, M. P. Sowa, D. E. Arnot, L. Hviid, and T. G. Theander. 2003. Selective upregulation of a single distinctly structured *var* gene in chondroitin sulphate A-adhering *Plasmodium falciparum* involved in pregnancy-associated malaria. *Mol. Microbiol.* **49:**179–191.

Scherf, A., L. M. Figueiredo, and L. H. Freitas-Junior. 2001. *Plasmodium* telomeres: a pathogen's perspective. *Curr. Opin. Microbiol.* **4:**409–414.

Scherf, A., R. Hernandez-Rivas, P. Buffet, E. Bottius, C. Benatar, B. Pouvelle, J. Gysin, and M. Lanzer. 1998. Antigenic variation in malaria: in situ switching, relaxed and mutually exclusive transcription of *var* genes during intra-erythrocytic development in *Plasmodium falciparum*. *EMBO J.* **17:**5418–5426.

Sleutels, F., R. Zwart, and D. P. Barlow. 2002. The non-coding Air RNA is required for silencing autosomal imprinted genes *Nature* **415:**810–813.

Smith, J. D., C. E. Chitnis, A. G. Craig, D. J. Roberts, D. E. Hudson-Taylor, D. S. Peterson, R. Pinches, C. I. Newbold, and L. H. Miller. 1995. Switches in expression of *Plasmodium falciparum var* genes correlate with changes in antigenic and cytoadherent phenotypes of infected erythrocytes. *Cell* **82:**101–110.

Smith, J. D., A. G. Craig, N. Kriek, D. Hudson-Taylor, S. Kyes, T. Fagen, R. Pinches, D. I. Baruch, C. I. Newbold, and L. H. Miller. 2000a. Identification of a Plasmodium falciparum intercellular adhesion molecule-1 binding domain: a parasite adhesion trait implicated in cerebral malaria. *Proc. Natl. Acad. Sci. USA* **97:**1766–1771.

Smith, J. D., S. Kyes, A. G. Craig, T. Fagan, D. Hudson-Taylor, L. H. Miller, D. I. Baruch, and C. I. Newbold. 1998. Analysis of adhesive domains from the A4VAR *Plasmodium falciparum* erythrocyte membrane protein-1 identifies a CD36 binding domain. *Mol. Biochem. Parasitol.* **97:**133–148.

Smith, J. D., G. Subramanian, B. Gamain, D. I. Baruch, and L. H. Miller. 2000b. Classification of adhesive domains in the *Plasmodium falciparum* erythrocyte membrane protein 1 family. *Mol. Biochem. Parasitol.* **110:**293–310.

Springer, A. L., L. M. Smith, D. Q. Mackay, S. O. Nelson, and J. D. Smith. 2004. Functional interdependence of the DBLβ domain and C2 region for binding of the *Plasmodium falciparum* variant antigen to ICAM-1. *Mol. Biochem. Parasitol.* **137:**55–64.

Su, X. Z., V. M. Heatwole, S. P. Wertheimer, F. Guinet, J. A. Herrfeldt, D. S. Peterson, J. A. Ravetch, and T. E. Wellems. 1995. The large diverse gene family *var* encodes proteins involved in cytoadherence and antigenic variation of *Plasmodium falciparum*-infected erythrocytes. *Cell* **82:**89–100.

Taylor, H. M., S. A. Kyes, and C. I. Newbold. 2000. *Var* gene diversity in *Plasmodium falciparum* is generated by frequent recombination events. *Mol. Biochem. Parasitol.* **110:**391–397.

Tebo, A. E., P. G. Kremsner, K. P. Piper, and A. J. Luty. 2002. Low antibody responses to variant surface antigens of *Plasmodium falciparum* are associated with severe malaria and increased susceptibility to malaria attacks in Gabonese children. *Am. J. Trop. Med. Hyg.* **67:**597–603.

Turner, C. M. 1997. The rate of antigenic variation in fly-transmitted and syringe-passaged infections of *Trypanosoma brucei*. *FEMS Microbiol. Lett.* **153:**227–231.

Turner, C. M. 1999. Antigenic variation in *Trypanosoma brucei* infections: an holistic view. *J. Cell Sci.* **112:**3187–3192.

Turner, G. D., H. Morrison, M. Jones, T. M. Davis, S. Looareesuwan, I. D. Buley, K. C. Gatter, C. I. Newbold, S. Pukritayakamee, and B. Nagachinta. 1994. An immunohistochemical study of the pathology of fatal malaria. Evidence for widespread endothelial activation and a potential role for intercellular adhesion molecule-1 in cerebral sequestration. *Am. J. Pathol.* **145:**1057–1069.

Urban, B. C., D. J. Ferguson, A. Pain, N. Willcox, M. Plebanski, J. M. Austyn, and D. J. Roberts. 1999. *Plasmodium falciparum*-infected erythrocytes modulate the maturation of dendritic cells. *Nature* **400:**73–77.

Voss, T. S., M. Kaestli, D. Vogel, S. Bopp, and H. P. Beck. 2003. Identification of nuclear proteins that interact differentially with *Plasmodium falciparum var* gene promoters. *Mol. Microbiol.* **48:**1593–1607.

Voss, T. S., J. K. Thompson, J. Waterkeyn, I. Felger, N. Weiss, A. F. Cowman, and H. P. Beck. 2000. Genomic distribution and functional characterisation of two distinct and conserved *Plasmodium falciparum var* gene 5' flanking sequences. *Mol. Biochem. Parasitol.* **107:**103–115.

Waller, K. L., W. Nunomura, B. M. Cooke, N. Mohandas, and R. L. Coppel. 2002. Mapping the domains of the cytoadherence ligand *Plasmodium falciparum* erythrocyte membrane protein 1 (PfEMP1) that bind to the knob-associated histidine-rich protein (KAHRP). *Mol. Biochem. Parasitol.* **119:**125–129.

ROSETTING

J. Alexandra Rowe

22

The discovery that *Plasmodium falciparum*-infected erythrocytes can bind to uninfected erythrocytes to form rosette-like clumps of cells (Fig. 1) was first made in the late 1980s (Udomsangpetch et al., 1989; Wahlgren, 1986). Since then, this unusual host-parasite interaction has attracted attention because of its link with severe malaria in African children. Some of the parasite ligands and host uninfected erythrocyte receptors that mediate rosette formation have been identified, and work has begun to determine the potential for a rosette-inhibiting antidisease vaccine. Despite this progress, the function of rosetting remains unknown, and the exact role of rosetting in the pathogenesis of severe malaria remains controversial.

ROSETTING AND SEVERE MALARIA

It was realized early on that rosetting is a phenotype that varies between *P. falciparum* strains (Wahlgren et al., 1990). Many clinical isolates show no rosettes, whereas a minority show high levels of rosetting. In The Gambia in 1990 the exciting discovery was made that significantly higher levels of rosetting are seen in parasite isolates from cerebral malaria patients than in those from mild malaria patients (Carlson et al., 1990). Differences in parasite virulence have long been thought to play a role in the variable clinical outcome seen in malaria infections. However, rosetting was the first parasite property to be clearly linked to severe disease. A large number of studies of children with severe and uncomplicated malaria in various parts of Africa have confirmed the initial findings of Carlson et al. (Table 1). In the majority of these studies, rosetting is associated with cerebral malaria and other severe malaria syndromes such as severe malarial anemia, respiratory distress, and prostration (Table 1). High levels of rosetting are commonly found in isolates from severe malaria cases, whereas isolates from uncomplicated malaria cases predominantly show no or low rosetting (Fig. 2). From our recent studies in Mali, my colleagues and I have found that very high levels of rosetting (>30% of infected erythrocytes forming rosettes) are seen in approximately 25 to 30% of severe malaria cases (J. A. Rowe, unpublished data). If this result is indicative of the situation throughout sub-Saharan Africa (Breman, 2001), then rosetting could play

J. Alexandra Rowe, Institute of Immunology and Infection Research, University of Edinburgh, Edinburgh, United Kingdom.

Molecular Approaches to Malaria, Edited by Irwin W. Sherman
© 2005 ASM Press, Washington, D.C.

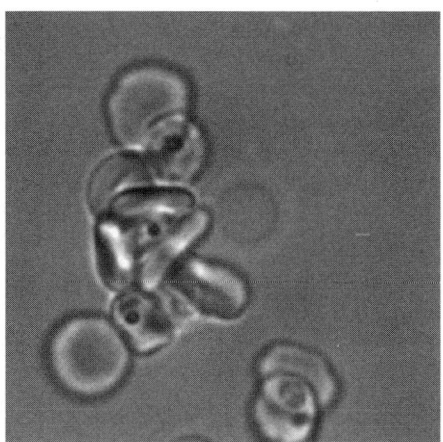

FIGURE 1 *P. falciparum* rosettes viewed by microscopy. The parasite-infected erythrocytes can be identified by the dark spot of pigment within the cell.

a substantial role in the deaths of up to about 500,000 children every year.

Although the association between rosetting and severe malaria in African children has consistently been shown, a different picture unexpectedly emerged from Southeast Asia and Papua New Guinea. In Southeast Asia, malaria transmission levels tend to be less intense than those in Africa, and there is less acquired immunity in the population, with severe disease affecting all age groups. The clinical features of severe malaria are also different, with multiorgan failure being common in Asia but rare in African children. Studies of rosetting and malaria severity in Thai adults show clearly that there is no significant association between rosetting and noncerebral forms of severe disease (mostly multiorgan failure) (Table 2). These studies do, however, suggest a link between cerebral malaria and rosetting, although in some cases the number of isolates studied were small and the results do not reach statistical significance (Table 2). Rosetting is positively correlated with parasitemia in some of the Thai studies, and it was suggested that the association between rosetting and cerebral malaria was merely due to the high parasitemias in the rosetting isolates (Chotivanich et al., 2004).

In Papua New Guinea, malaria transmission is high, and as in Africa, severe disease occurs mostly in young children. Unlike the data collected in Africa, however, a large study of 81 children with cerebral malaria found no evidence of an association between rosetting and cerebral disease (al-Yaman et al., 1995). It should be noted that the criteria for diagnosing cerebral malaria in this study were less stringent than in similar studies in Africa, and it may be that the contrasting results are merely due to different definitions of severe disease being used in the two areas.

The studies outlined above indicate that rosetting is linked to all forms of severe malaria in African children and may be linked to cerebral malaria but not multiorgan failure in Asian adults. It is likely that the pathogenic mechanisms leading to multiorgan failure in Southeast Asia are different from those responsible for severe malaria in African children, and this may be due to both parasite factors and human genetic polymorphisms that vary in different regions of the world.

IS ROSETTING A DIRECT CAUSE OF SEVERE MALARIA?

The association between rosetting and severe malaria in African children is intriguing; however, a statistically significant association does not prove a causal link between the two. Direct evidence addressing the role of rosetting in malaria pathogenesis is difficult to acquire because of the restraints on performing experiments with human subjects and the lack of an animal model that truly reflects severe malaria in humans. However, several lines of indirect evidence do support the idea that rosetting may be directly pathogenic. In an ex vivo experiment using the vasculature of a rat, it was shown that a *P. falciparum* rosetting clone caused greater obstruction to microvascular blood flow than an isogenic nonrosetting clone (Kaul et al., 1991). By videomicroscopy, it was observed that the rosettes were disrupted under the high shear forces in the arterial side of the circulation and then reformed and accumulated in the capillaries and postcapillary venules. The rosettes tended

TABLE 1 Rosetting and severe malaria in Africa

Site (reference)	Mean or median % rosette frequency (range); no. of isolates[h]			P value[h,i]
	Cerebral[a]	Severe[b]	Mild[c]	
The Gambia (Carlson et al., 1990)	35 (6–85); 24	ND	17 (0–71); 57	<0.001
The Gambia (Treutiger et al., 1992)	28.3 (0.5–70); 24	ND	8.5 (0–55); 106	<0.000001
Madagascar (Ringwald et al., 1993)	19.5 (5–28); 6	30.5 (20–43); 6	5 (0–19); 9	<0.05 (cerebral versus mild); <0.002 (severe versus mild)
Kenya (Rowe et al., 1995)	6 (0–94); 21	7 (0–97); 15	1 (0–82); 54	<0.003 (severe and cerebral versus mild)
Kenya (Newbold et al., 1997)	5.0 (0–37); 45	9.5[d] (0–46); 49	4.5 (0–26); 50	<0.05 (severe versus mild); (cerebral verus mild, NS)
Gabon (Kun et al., 1998)	ND	16; 47	8; 47	<0.05
Malawi (Rogerson et al., 1999)	13.9; 46	16.4; 18	15.0; 62	NS
Kenya (Pain et al., 2001)	ND	6.5[e] (2.0–34.9)[f]; 57	4.1 (0.9–10.8)[f]; 64	0.02/0.054[g]
Kenya (Heddini et al., 2001)	17.2; 11	29.7[d]; 21	12.9; 45	0.001 (severe versus mild); (cerebral versus mild, NS)
Kenya (Rowe et al., 2002b)	ND	14.0[e] (4–32)[f]; 25	3.5 (1–10)[f]; 49	<0.001

[a] Unrousable coma (Blantyre coma score of ≤2).
[b] Includes a variety of syndromes such as severe malarial anemia, respiratory distress, prostration, and hypoglycemia.
[c] Acute falciparum malaria with no complications of severe or cerebral disease.
[d] Severe malarial anemia only.
[e] Includes cerebral malaria cases.
[f] Interquartile range.
[g] Mann-Whitney U test/logistic regression.
[h] ND, not determined; NS, not significant.
[i] Mann-Whitney U test, Kruskal Wallis test, or Student's t test.

to accumulate at the intersection of vessels where rosetting infected cells bound to endothelial cells and uninfected erythrocytes simultaneously. If the same process occurs in human malaria infections, this could explain why rosetting is only associated with severe disease in falciparum malaria, even though rosetting also occurs in *Plasmodium vivax* (Udomsanpetch et al., 1995), *Plasmodium malariae* (Lowe et al., 1998), and *Plasmodium ovale* (Angus et al., 1996). In the latter three species, cytoadherence to the endothelium does not occur and infections rarely reach high levels of parasitemia. In falciparum malaria it may be the combination of rosetting and cytoadherence, together with high parasite burdens, that is particularly obstructive to microvascular blood flow (Fig. 3) and could lead to hypoxia, tissue damage, and severe malaria.

DO ROSETTES FORM IN VIVO?

Skeptics of rosetting claim that there is no evidence that rosettes form in vivo. Because mature infected erythrocytes are sequestered in falciparum malaria and only immature ring-stage parasites are seen in peripheral blood samples, it is necessary to culture parasites from patients for 12 to 36 h to allow the parasites to mature before any assessment of rosetting can be made. This leads to the possibility that rosetting could be an in vitro artifact. However, in rare cases patients with a very high parasite burden have some mature infected erythrocytes in the pe-

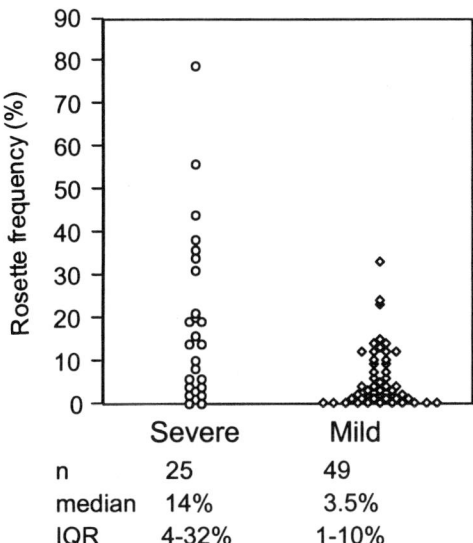

FIGURE 2 Rosetting and malaria severity in Kenya. Each point on the graph represents the rosette frequency (percentage of mature-infected erythrocytes in rosettes) of a single isolate. (Reproduced from the *American Journal of Tropical Medicine and Hygiene* [Rowe et al., 2002b] with permission from the publisher.)

et al., 1997; Nash et al., 1992). Indeed, the rosetting interaction was found to be five times stronger than the interaction between infected erythrocytes and endothelial cells that brings about sequestration (Nash et al., 1992). These data are suggestive that rosettes can and do form in vivo.

Further uncertainty regarding the pathological role of rosetting arises from the paucity of descriptions of rosettes seen in postmortem studies of fatal malaria. Indeed, rosette-like clumps of cells have either been described rarely (Pongponratn et al., 1991; Riganti et al., 1990) or not at all (MacPherson et al., 1985; Pongponratn et al., 2003). There are several possible explanations for these findings. Most postmortem studies have been carried out using material from adults in southeast Asia, a region where rosetting is not associated with all forms of severe malaria and where rosette-reducing human genetic polymorphisms are common (see below). Few postmortem studies have been carried out of African children dying from severe malaria because of cultural and infrastructural difficulties. Obviously, it would be extremely interesting and important to know whether rosetting is a common histopathological feature of fatal malaria in African children. It also remains possible that postmortem changes and/or procedures used for sample collection and preparation lead to the loss of rosettes from postmortem samples. For example, we have

ripheral blood; and in such cases, rosettes have been found in ex vivo blood samples examined for rosetting immediately after the blood was drawn (Chotivanich et al., 1998; Ho et al., 1991). Several studies indicate that rosettes are strong enough to withstand the shear forces experienced in vivo (Chotivanich et al., 2000; Chu

TABLE 2 Rosetting and severe malaria in Thailand

Reference	% Mean or median rosette frequency (range); no. of isolates			P value[d]
	Cerebral[a]	Severe[b]	Mild[c]	
Ho et al., 1991	18.3 (0–51.6); 9	9.0 (0–34); 16	10.7 (0–47.3); 10	>0.2
Udomsangpetch et al., 1996	20. (2–71); 10	8 (1–32); 21	5 (0–21); 29	Cerebral versus severe, 0.006; cerebral versus mild, 0.02
Angkasekwinai et al., 1998	11.1 (0–42); 19	5.1 (0–33); 82	4.1 (0–32); 162	>0.05
Chotivanich et al., 2004	3.7 (1.2–10); 15	3.3 (0.4–10.8); 104	1.6 (0–15.7); 63	0.002; NS[e]

[a] Unrousable coma.
[b] Evidence of multiorgan failure such as raised creatinine, bilirubin, or aspartate aminotransferase levels or pulmonary edema.
[c] Acute falciparum malaria with no complications of severe or cerebral disease.
[d] Kruskal-Wallis test.
[e] Logistic regression. NS, not significant.

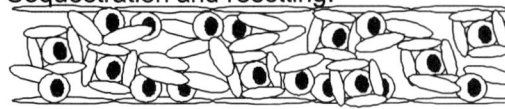

FIGURE 3 Rosetting and cytoadherence together cause greater obstruction to microvascular blood flow than cytoadherence alone.

found that the commonly used fixatives glutaraldehyde and paraformaldehyde make rosettes fall apart when added to in vitro cultures (Rowe, 1994).

HUMAN GENETICS AND ROSETTING

The complexity of the host-parasite interaction seen in malaria is well illustrated by the finding that human genetic polymorphisms of the red blood cell can influence the parasite's rosetting phenotype. α- and β-thalassemia and sickle cell trait have been shown to reduce rosetting, and it was suggested that this could play a role in the mechanism of protection against severe malaria provided by these disorders (Carlson et al., 1994; Udomsangpetch et al., 1993a). This remains controversial, however, as several other protective mechanisms for these genes have been described, such as increased susceptibility to phagocytosis (Ayi et al., 2004) or reduced erythrocyte invasion (Pattanapanyasat et al., 1999). My colleagues and I recently showed that a polymorphism in the human complement receptor 1 (CR1) gene that is associated with red blood cell CR1 deficiency and reduced rosetting (Rowe et al., 1997) occurs at extremely high frequencies in the malarious regions of Papua New Guinea and protects against severe malaria (Cockburn et al., 2004). Red blood cell CR1 deficiency only affects rosetting and has no other demonstrable effect on *P. falciparum* invasion or growth; therefore, it seems probable that in this case, the protection from severe malaria is due to reduced rosetting (Cockburn et al., 2004). The protective effect of a human rosette-reducing polymorphism provides strong evidence that rosetting plays a direct role in the pathogenesis of some cases of severe malaria. The red blood cell CR1 deficiency polymorphism also occurs at a high level in Thailand (Nagayasu et al., 2001) and Cambodia (Thomas et al., 2005) and may influence malaria severity throughout southeast Asia. In Thailand, it was found that heterozygotes for the CR1 deficiency gene showed a trend towards protection from severe malaria, but homozygotes had increased susceptibility to severe disease (Nagayasu et al., 2001). It may be that low levels of CR1 that reduce rosetting but still function to protect the red blood cell from complement damage are optimal, whereas very low levels of CR1 leave the red blood cell open to complement damage and immune complex disease. Studies in Africa have also suggested that red blood cell CR1 levels can influence disease outcome (Stoute et al., 2003; Waitumbi et al., 2004; Waitumbi et al., 2000). Additional polymorphisms have been described that cause amino acid substitutions in the CR1 molecule (Moulds et al., 2001). These CR1 polymorphisms occur at a high frequency in African populations but are rare in Caucasians (Moulds et al., 2000), and are associated with protection from severe malaria (Rowe, unpublished data).

MOLECULAR MECHANISMS OF ROSETTING

Uninfected Erythrocyte Receptors for Rosetting

A number of different red blood cell rosetting receptors have been described, including CR1 (Rowe et al., 1997), heparan sulfate-like molecules (Chen et al., 1998), ABO blood group sugars (Carlson and Wahlgren, 1992), and CD36 (Handunnetti et al., 1992) (Fig. 4). CR1 is an important rosetting receptor in the majority of rosetting field isolates and laboratory strains, and rosetting can be inhibited with a CR1 monoclonal antibody (Rowe et al., 1997; Rowe et al., 2000). Heparan sulfate-like molecules on the erythrocyte surface probably also play an important role but have not yet been fully characterized (Vogt et al., 2004). Heparan sulfates are

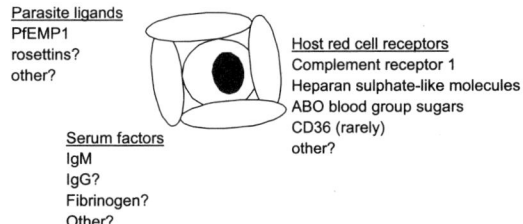

FIGURE 4 Summary of the molecular mechanisms of rosetting.

also found on endothelial cells; therefore, interaction with these molecules may allow some *P. falciparum* strains to simultaneously cytoadhere to the endothelium and form rosettes (Vogt et al., 2003). The ABO blood group sugars, which are small tri- or disaccharides attached to a variety of glycoproteins and glycolipids on the red blood cell surface, affect the size and strength of rosettes formed, with parasite isolates being divided into A-preferring or B-preferring types (Barragan et al., 2000; Carlson and Wahlgren, 1992; Udomsangpetch et al., 1993b). The CD36 glycoprotein, which is an important endothelial receptor for cytoadherence, is expressed at low levels on red blood cells (van Schravendijk et al., 1992) but only rarely acts as a rosetting receptor in field isolates (Rowe et al., 2000). It seems likely that rosetting is mediated by multiple receptor-ligand interactions (Chen et al., 2000; Heddini et al., 2001), and further work is needed to clarify the most clinically relevant rosetting receptors used by *P. falciparum* field isolates.

The *P. falciparum* Rosetting Ligand Is PfEMP1

The parasite ligands for rosetting have been shown to be particular members of the variant erythrocyte surface antigen family PfEMP1, encoded by *var* genes. In the R29 parasite clone, the Duffy-binding-like alpha (DBLα) domain of the *R29var1* variant expressed by the rosetting parasites was identified as the red blood cell-binding region (Rowe et al., 1997). The functionally important section was further localized to the central part of the domain between the 5th and the 12th cysteine residues (Mayor et al., 2004). The *R29var1* DBLα domain binds to CR1 on uninfected red blood cells (Rowe et al., 1997). Other work on the FCR3S1.2 parasite clone has confirmed the importance of DBLα in red blood cell binding, although in this case the binding occurs via heparan-sulfate-like molecules and blood group A sugars (Chen et al., 1998; Chen et al., 2000). Almost all *var* genes have a DBLα domain, yet only a minority mediate rosetting. The precise differences in sequence between DBLα domains that bind red blood cells compared to those that do not have not yet been described. However, preliminary data indicate that a small number of relatively well-conserved rosetting DBLα types may be found in diverse parasite isolates (Rowe, unpublished data). It is unknown whether any other parasite proteins apart from PfEMP1 play a role in rosetting. A family of low-molecular-weight proteins called rosettins were suggested as possible rosetting ligands (Helmby et al., 1993). The role of these proteins, which are probably members of the rifin multigene family (Kyes et al., 1999), remains to be clarified.

Serum Proteins—IgM

An important role in rosetting for immunoglobulins (Igs) derived from normal human serum was first shown by demonstrating the presence of human Igs in fibrillar strands connecting infected and uninfected red blood cells (Scholander et al., 1996). Initial reports indicated both IgG and IgM in the fibrillar strands, but subsequent work has shown predominantly IgM on the surface of rosetting infected erythrocytes (Color Plate 16) (Rowe et al., 2002b). In the rosetting parasite clone FCR3S1.2, it has been shown that the nonimmune IgM binds to the cysteine-rich interdomain region of PfEMP1 (Chen et al., 2000). For one parasite clone, PAR+ (called FCR3S1 in some papers), it has been shown that IgM is essential for the formation of rosettes, and it was suggested that the IgM might bridge between the infected and uninfected erythrocytes (Clough et al., 1998a; Scholander et al., 1996; Somner et al., 2000). PAR+ is an unusual parasite because it is knobless, and it is not yet clear if a dependence on IgM for rosetting is a feature of knob-positive clones, which are probably more typical of parasites seen in natural infections. The ability of anti-human immunoglobulin serum to inhibit rosetting in a wide range of field isolates is suggestive that Igs are important for rosetting in natural infections (Rowe et al., 2000). A role for other serum proteins such as fibrinogen and von Willebrand's factor has also been described (Treutiger et al., 1999), although it is unclear to what extent these proteins interact specifically with infected erythrocytes or whether their effect is due to their nonspecific interactions with all human erythrocytes, leading to the propensity to aggregate cells (Schechner et al., 2003). Somner et al. (2000) used serum fractionation techniques to demonstrate that, in addition to IgM, an acidic high-molecular-weight component and a smaller basic component are required for rosetting. Further research is needed to characterize these components in more detail.

WHY DO PARASITES ROSETTE?

All the plasmodial species infecting humans (Angus et al., 1996; Lowe et al., 1998; Udomsanpetch et al., 1995), and several that infect rodents (Mackinnon et al., 2002) and primates (David et al., 1988; Udomsangpetch et al., 1991) are known to form rosettes. This suggests that rosetting has an important biological function for plasmodia, but the precise nature of this function is unknown. A beneficial effect of rosetting on parasite growth or survival is supported by a positive correlation between rosetting and parasitemia in two large groups of *P. falciparum* field isolates (Rowe et al., 2002a). Rosetting might enhance invasion by targeting invasive merozoites into the uninfected cells in a rosette; however, careful studies of rosetting and nonrosetting isogenic clones showed no difference in invasion rates between the two (Clough et al., 1998b). An effect of rosetting on invasion might only be seen in the presence of host immunity, and we are currently investigating the invasion of rosetting and nonrosetting clones in the presence of invasion-blocking antibodies. Another possibility is that rosetting is an immune evasion mechanism, and that the covering of uninfected red blood cells acts as a "cloaking device" to prevent recognition of infected erythrocytes by phagocytic and antigen-presenting cells. There are no experimental data to support or refute this hypothesis at present. In an analogous system, it was shown that the rosetting of *Borrelia crocidurae* (a bacterium that causes African relapsing fever) inhibited the development of a specific immune response by several days compared to nonrosetting bacteria (Burman et al., 1998). Nonimmune IgM antibodies that bind to the surface of rosetting infected cells (Color Plate 16) could also contribute to immune evasion, perhaps by preventing the binding of specific IgG to PfEMP1.

ROSETTE-INHIBITING IMMUNE RESPONSES IN NATURAL INFECTIONS

The development of rosette-inhibiting immune responses in natural malaria infections has received relatively little attention. It is known that children develop rosette-inhibiting antibodies after infection with a rosetting parasite (Iqbal et al., 1993) and that acquisition of rosette-inhibiting antibodies is age-related and occurs at a younger age in areas where the rates of disease transmission are higher (Barragan et al., 1998). The extent to which antibodies raised in response to one particular isolate cross-react with the rosettes from another isolate is not clear and requires further study. A potentially important role for rosette-inhibiting antibodies in natural immunity to severe malaria was indicated by two studies in which children who developed severe malaria lacked anti-rosette antibodies, whereas those suffering mild malaria were able to inhibit rosettes (Carlson et al., 1990; Treutiger et al., 1992). However, these findings could not be replicated in two other studies (Rowe et al., 1995; Kun et al., 1998). Therefore, further research is needed to clarify the importance of rosette-inhibiting antibodies in naturally acquired immunity to severe malaria.

PROSPECTS FOR A ROSETTE-INHIBITING ANTIDISEASE VACCINE?

Because of the association between rosetting and severe malaria, there is considerable interest in the potential for a vaccine to elicit antibodies to inhibit rosetting and thereby attempt to prevent some deaths from severe malaria. Initial studies have shown that the DBLα domain of rosette-mediating PfEMP1 variants can be used to immunize mice, rats, and rabbits to raise antibodies that partially inhibit rosetting (Chen et al., 2004; Rowe, unpublished). It remains unclear whether these antibodies are entirely isolate specific or whether there is a cross-reactive component that will inhibit rosetting in a range of isolates. A cross-reactive response active against many isolates would clearly be beneficial for vaccine development. It is also unclear whether antibodies to DBLα will be sufficient to inhibit rosetting or whether receptor-ligand interactions between other PfEMP1 domains and red blood cell receptors also need to be inhibited to prevent rosetting.

CONCLUSIONS

Definitive proof that rosetting causes severe malaria is lacking. However, the body of evidence accumulated over the past 15 years does support a direct role for rosetting in the pathogenesis of some cases of life-threatening malaria. Therefore, rosette-inhibiting drugs or vaccines may have the potential to save many patients' lives. Researchers have begun to describe the molecular mechanisms of rosetting; however, further work on a range of natural *P. falciparum* isolates will be required to develop rosette-inhibiting interventions with a broad spectrum of activity.

ACKNOWLEDGMENTS

I am grateful to Helen Kyriacou and Anne-Marie Deans for comments on the manuscript and to Anne-Marie Deans for Fig. 1.

J.A.R. is funded by a Senior Research Fellowship from the Wellcome Trust (grant number 067431).

REFERENCES

al-Yaman, F., B. Genton, D. Mokela, A. Raiko, S. Kati, S. Rogerson, J. Reeder, and M. Alpers. 1995. Human cerebral malaria: lack of significant association between erythrocyte rosetting and disease severity. *Trans. R. Soc. Trop. Med. Hyg.* **89:**55–58.

Angkasekwinai, P., S. Looareesuwan, and S. C. Chaiyaroj. 1998 Lack of significant association between rosette formation and parasitized erythrocyte adherence to purified CD36. *Southeast Asian J. Trop. Med. Public Health* **29:**41–45.

Angus, B. J., K. Thanikkul, K. Silamut, N. J. White, and R. Udomsangpetch. 1996. Short report: rosette formation in Plasmodium ovale infection. *Am. J. Trop. Med. Hyg.* **55:**560–561.

Ayi, K., F. Turrini, A. Piga, and P. Arese. 2004. Enhanced phagocytosis of ring-parasitized mutant erythrocytes: a common mechanism that may explain protection against falciparum malaria in sickle trait and beta-thalassemia trait. *Blood* **104:**3364–3371.

Barragan, A., P. G. Kremsner, M. Wahlgren, and J. Carlson. 2000. Blood group A antigen is a coreceptor in Plasmodium falciparum rosetting. *Infect. Immun.* **68:**2971–2975.

Barragan, A., P. G. Kremsner, W. Weiss, M. Wahlgren, and J. Carlson. 1998. Age-related buildup of humoral immunity against epitopes for rosette formation and agglutination in African areas of malaria endemicity. *Infect. Immun.* **66:**4783–4787.

Breman, J. G. 2001. The ears of the hippopotamus: manifestations, determinants, and estimates of the malaria burden. *Am. J. Trop. Med. Hyg.* **64:**1–11.

Burman, N., A. Shamaei-Tousi, and S. Bergstrom. 1998. The spirochete *Borrelia crocidurae* causes erythrocyte rosetting during relapsing fever. *Infect. Immun.* **66:**815–819.

Carlson, J., H. Helmby, A. V. Hill, D. Brewster, B. M. Greenwood, and M. Wahlgren. 1990. Human cerebral malaria: association with erythrocyte rosetting and lack of anti-rosetting antibodies. *Lancet* **336:**1457–1460.

Carlson, J., G. B. Nash, V. Gabutti, F. al-Yaman, and M. Wahlgren. 1994. Natural protection against severe *Plasmodium falciparum* malaria due to impaired rosette formation. *Blood* **84:**3909–3914.

Carlson, J., and M. Wahlgren. 1992. *Plasmodium falciparum* erythrocyte rosetting is mediated by promiscuous lectin-like interactions. *J. Exp. Med.* **176:**1311–1317.

Chen, Q., A. Barragan, V. Fernandez, A. Sundstrom, M. Schlichtherle, A. Sahlen, J. Carlson, S. Datta, and M. Wahlgren. 1998. Identification of Plasmodium falciparum erythrocyte membrane protein 1 (PfEMP1) as the rosetting ligand of the malaria parasite *P. falciparum*. *J. Exp. Med.* **187:**15–23.

Chen, Q., A. Heddini, A. Barragan, V. Fernandez, S. F. Pearce, and M. Wahlgren. 2000. The semiconserved head structure of Plasmodium falciparum erythrocyte membrane protein 1 mediates binding to multiple independent host receptors. *J. Exp. Med.* **192:**1–10.

Chen, Q., F. Pettersson, A. M. Vogt, B. Schmidt, S. Ahuja, P. Liljestrom, and M. Wahlgren. 2004. Immunization with PfEMP1-DBL1α generates antibodies that disrupt rosettes and protect against the sequestration of *Plasmodium falciparum*-infected erythrocytes. *Vaccine* **22:**2701–2712.

Chotivanich, K., J. Sritabal, R. Udomsangpetch, P. Newton, K. A. Stepniewska, R. Ruangveerayuth, S. Looareesuwan, D. J. Roberts, and N. J. White. 2004. Platelet-induced autoagglutination of *Plasmodium falciparum*-infected red blood cells and disease severity in Thailand. *J. Infect. Dis.* **189:**1052–1055.

Chotivanich, K. T., A. M. Dondorp, N. J. White, K. Peters, J. Vreeken, P. A. Kager, and R. Udomsangpetch. 2000. The resistance to physiological shear stresses of the erythrocytic rosettes formed by cells infected with *Plasmodium falciparum*. *Ann. Trop. Med. Parasitol.* **94:**219–226.

Chotivanich, K. T., R. Udomsangpetch, B. Pipitaporn, B. Angus, Y. Suputtamongkol, S. Pukrittayakamee, and N. J. White. 1998. Rosetting characteristics of uninfected erythrocytes from healthy individuals and malaria patients. *Ann. Trop. Med. Parasitol.* **92:**45–56.

Chu, Y., Haigh, T. and G. B. Nash. 1997. Rheological analysis of the formation of rosettes by red blood cells parasitized by Plasmodium falciparum. *Br. J. Haematol.* **99:**777–783.

Clough, B., F. A. Atilola, J. Black, and G. Pasvol. 1998a. *Plasmodium falciparum*: the importance of IgM in the rosetting of parasite-infected erythrocytes. *Exp. Parasitol.* **89:**129–132.

Clough, B., F. A. Atilola, and G. Pasvol. 1998b. The role of rosetting in the multiplication of *Plasmodium falciparum*: rosette formation neither enhances nor targets parasite invasion into uninfected red cells. *Br. J. Haematol.* **100:**99–104.

Cockburn, I. A., M. J. Mackinnon, A., O'Donnell, S. J. Allen, J. M. Moulds, M. Baisor, M. Bockarie, J. C. Reeder, and J. A. Rowe. 2004. A human complement receptor 1 polymorphism that reduces Plasmodium falciparum rosetting confers protection against severe malaria. *Proc. Natl. Acad. Sci. USA* **101:**272–277.

David, P. H., S. M. Handunnetti, J. H. Leech, P. Gamage, and K. N. Mendi. 1988. Rosetting: a new cytoadherence property of malaria-infected erythrocytes. *Am. J. Trop. Med. Hyg.* **38:**289–297.

Handunnetti, S. M., M. R. van Schravendijk, T. Hasler, J. W. Barnwell, D. E. Greenwalt, and R. J. Howard. 1992. Involvement of CD36 on erythrocytes as a rosetting receptor for *Plasmodium falciparum*-infected erythrocytes. *Blood* **80:**2097–2104.

Heddini, A., F. Pettersson, O. Kai, J. Shafi, J. Obiero, Q. Chen, A. Barragan, M. Wahlgren, and K. Marsh. 2001. Fresh isolates from children with severe *Plasmodium falciparum* malaria bind to multiple receptors. *Infect. Immun.* **69:**5849–5856.

Helmby, H., L. Cavelier, U. Pettersson, and M. Wahlgren. 1993. Rosetting *Plasmodium falciparum*-infected erythrocytes express unique strain-specific antigens on their surface. *Infect. Immun.* **61:**284–288.

Ho, M., T. M. Davis, K. Silamut, D. Bunnag, and N. J. White. 1991. Rosette formation of *Plasmodium falciparum*-infected erythrocytes from patients with acute malaria. *Infect. Immun.* **59:**2135–2139.

Iqbal, J., P. Perlmann, and K. Berzins. 1993. Serological diversity of antigens expressed on the surface of erythrocytes infected with Plasmodium falciparum. *Trans. R. Soc. Trop. Med. Hyg.* **87:**583–588.

Kaul, D. K., E. F. J. Roth, R. L. Nagel, R. J. Howard, and S. M. Handunnetti. 1991. Rosetting of *Plasmodium falciparum*-infected red blood cells with uninfected red blood cells enhances microvascular obstruction under flow conditions. *Blood* **78:**812–819.

Kun, J. F., R. J. Schmidt-Ott, L. G. Lehman, B. Lell, D. Luckner, B. Greve, P. Matousek, and P. G. Kremsner. 1998. Merozoite surface antigen 1 and 2 genotypes and rosetting of *Plasmodium falciparum* in severe and mild malaria in Lambarene, Gabon. *Trans. R. Soc. Trop. Med. Hyg.* **92:**110–114.

Kyes, S. A., J. A. Rowe, N. Kriek, and C. I. Newbold. 1999. Rifins: a second family of clonally variant proteins expressed on the surface of red cells infected with *Plasmodium falciparum*. *Proc. Natl. Acad. Sci. USA* **96:**9333–9338.

Lowe, B. S., M. Mosobo, and P. C. Bull. 1998. All four species of human malaria parasites form rosettes. *Trans. R. Soc. Trop. Med. Hyg.* **92:**526.

Mackinnon, M. J., P. R. Walker, and J. A. Rowe. 2002. *Plasmodium chabaudi*: rosetting in a rodent malaria model. *Exp. Parasitol.* **101:**121–128.

MacPherson, G. G., M. J. Warrell, N. J. White, S. Looareesuwan, and D. A. Warrell. 1985. Human cerebral malaria. A quantitative ultrastructural analysis of parasitized erythrocyte sequestration. *Am. J. Pathol.* **119:**385–401.

Mayor, A., N. Bir, R. Sawhney, S. Singh, P. Pattnaik, S. K. Singh, A. Sharma, and C. E. Chitnis. 2004. Receptor-binding residues lie in central regions of Duffy-binding-like domains involved in red cell invasion and cytoadherence by malaria parasites. *Blood* **105:**2257–2263.

Moulds, J. M., L. Kassambara, J. J. Middleton, M. Baby, I. Sagara, A. Guindo, S. Coulibaly, D. Yalcouye, D. A. Diallo, L. Miller, and O. Doumbo. 2000. Identification of complement receptor one (CR1) polymorphisms in west Africa. *Genes Immun.* **1:**325–329.

Moulds, J. M., P. A. Zimmerman, O. K. Doumbo, L. Kassambara, I. Sagara, D. A. Diallo, J. P. Atkinson, M. Krych-Goldberg, R. E. Hauhart, D. E. Hourcade, D. T. McNamara, D. J. Birmingham, J. A. Rowe, J. J. Moulds, and L. H. Miller. 2001. Molecular identification of Knops blood group polymorphisms found in long homologous region D of complement receptor 1. *Blood* **97:**2879–2885.

Nagayasu, E., M. Ito, M. Akaki, Y. Nakano, M. Kimura, S. Looareesuwan, and M. Aikawa. 2001. CR1 density polymorphism on erythrocytes of falciparum malaria patients in Thailand. *Am. J. Trop. Med. Hyg.* **64:**1–5.

Nash, G. B., B. M. Cooke, J. Carlson, and M. Wahlgren. 1992. Rheological properties of rosettes formed by red blood cells parasitized by *Plasmodium falciparum*. *Br. J. Haematol.* **82:**757–763.

Newbold, C., P. Warn, G. Black, A. Berendt, A. Craig, B. Snow, M. Msobo, N. Peshu, and K. Marsh. 1997. Receptor-specific adhesion and clinical disease in *Plasmodium falciparum*. *Am. J. Trop. Med. Hyg.* **57:**389–398.

Pain, A., D. J. Ferguson, O. Kai, B. C. Urban, B. Lowe, K. Marsh, and D. J. Roberts. 2001. Platelet-mediated clumping of *Plasmodium falciparum*-infected erythrocytes is a common adhesive phenotype and is associated with severe malaria. *Proc. Natl. Acad. Sci. USA* **98:**1805–1810.

Pattanapanyasat, K., K. Yongvanitchit, P. Tongtawe, K. Tachavanich, W. Wanachiwanawin, S. Fucharoen, and D. S. Walsh. 1999. Impairment of *Plasmodium falciparum* growth in thalassemic red blood cells: further evidence by using biotin labeling and flow cytometry. *Blood* **93:**3116–3119.

Pongponratn, E., M. Riganti, B. Punpoowong, and M. Aikawa. 1991. Microvascular sequestration of parasitised erythrocytes in human falciparum malaria: a pathological study. *Am. J. Trop. Med. Hyg.* **44:**168–175.

Pongponratn, E., G. D. Turner, N. P. Day, N. H. Phu, J. A. Simpson, K. Stepniewska, N. T. Mai, P. Viriyavejakul, S. Looareesuwan, T. T. Hien, D. J. Ferguson, and N. J. White. 2003. An ultrastructural study of the brain in fatal Plasmodium falciparum malaria. *Am. J. Trop. Med. Hyg.* **69:**345–359.

Riganti, M., E. Pongponratn, T. Tegoshi, S. Looareesuwan, B. Punpoowong, and M. Aikawa. 1990. Human cerebral malaria in Thailand: a clinicopathological correlation. *Immunol. Lett.* **25:**199–206.

Ringwald, P., F. Peyron, J. P. Lepers, P. Rabarison, C. Rakotomalala, M. Razanamparany, M. Rabodonirina, J. Roux, and J. Le Bras. 1993. Parasite virulence factors during falciparum malaria: rosetting, cytoadherence, and modulation of cytoadherence by cytokines. *Infect. Immun.* **61:**5198–5204.

Rogerson, S. J., R. Tembenu, C. Dobano, S. Plitt, T. E. Taylor, and M. E. Molyneux. 1999. Cytoadherence characteristics of Plasmodium falciparum-infected erythrocytes from Malawian children with severe and uncomplicated malaria. *Am. J. Trop. Med. Hyg.* **61:**467–472.

Rowe, A., J. Obeiro, C. I. Newbold, and K. Marsh. 1995. *Plasmodium falciparum* rosetting is associated with malaria severity in Kenya. *Infect. Immun.* **63:**2323–2326.

Rowe, J. A. 1994. Rosetting of *Plasmodium falciparum* infected erythrocytes. Ph.D. thesis. University of Oxford, Oxford, United Kingdom.

Rowe, J. A., J. M. Moulds, C. I. Newbold, and L. H. Miller. 1997. *P. falciparum* rosetting mediated by a parasite-variant erythrocyte membrane protein and complement-receptor 1. *Nature* **388:**292–295.

Rowe, J. A., J. Obeiro, K. Marsh, and A. Raza. 2002a. Positive correlation between rosetting and parasitemia in *Plasmodium falciparum* clinical isolates. *Am. J. Trop. Med. Hyg.* **66:**458–460.

Rowe, J. A., S. J. Rogerson, A. Raza, J. M. Moulds, M. D. Kazatchkine, K. Marsh, C. I. Newbold, J. P. Atkinson, and L. H. Miller. 2000. Mapping of the region of complement receptor (CR) 1 required for *Plasmodium falciparum* rosetting and demonstration of the importance of CR1 in rosetting in field isolates. *J. Immunol.* **165:**6341–6346.

Rowe, J. A., J. Shafi, O. K. Kai, K. Marsh, and A. Raza. 2002b. Nonimmune IgM, but not IgG binds to the surface of *Plasmodium falciparum*-infected erythrocytes and correlates with rosetting and severe malaria. *Am. J. Trop. Med. Hyg.* **66:**692–699.

Schechner, V., I. Shapira, S. Berliner, D. Comaneshter, T. Hershcovici, J. Orlin, D. Zeltser, M. Rozenblat, K. Lachmi, M. Hirsch, and Y. Beigel. 2003. Significant dominance of fibrinogen over immunoglobulins, C-reactive protein, cholesterol and triglycerides in maintaining increased red blood cell adhesiveness/aggregation in the peripheral venous blood: a model in hypercholesterolaemic patients. *Eur. J. Clin. Investig.* **33:**955–961.

Scholander, C., C. J. Treutiger, K. Hultenby, and M. Wahlgren. 1996. Novel fibrillar structure confers adhesive property to malaria-infected erythrocytes. *Nat. Med.* **2:**204–208.

Somner, E. A., J. Black, and G. Pasvol. 2000. Multiple human serum components act as bridging molecules in rosette formation by *Plasmodium falciparum*-infected erythrocytes. *Blood* **95:**674–682.

Stoute, J. A., A. O. Odindo, B. O. Owuor, E. K. Mibei, M. O. Opollo, and J. N. Waitumbi. 2003. Loss of red blood cell-complement regulatory proteins and increased levels of circulating immune complexes are associated with severe malarial anemia. *J. Infect. Dis.* **187:**522–525.

Thomas, B. N., B. Donvito, I. Cockburn, T. Fandeur, J. A. Rowe, J. H. Cohen, and J. M. Moulds. 2005. A complement receptor-1 polymorphism with high frequency in malaria endemic regions of Asia but not Africa. *Genes Immun.* **6:**31–36.

Treutiger, C. J., I. Hedlund, H. Helmby, J. Carlson, A. Jepson, P. Twumasi, D. Kwiatkowski, B. M.

Greenwood, and M. Wahlgren. 1992. Rosette formation in *Plasmodium falciparum* isolates and antirosette activity of sera from Gambians with cerebral or uncomplicated malaria. *Am. J. Trop. Med. Hyg.* **46:**503–510.

Treutiger, C. J., C. Scholander, J. Carlson, K. P. McAdam, J. G. Raynes, L. Falksveden, and M. Wahlgren. 1999. Rouleaux-forming serum proteins are involved in the rosetting of *Plasmodium falciparum*-infected erythrocytes. *Exp. Parasitol.* **93:**215–224.

Udomsangpetch, R., A. E. Brown, C. D. Smith, and H. K. Webster. 1991. Rosette formation by *Plasmodium coatneyi*-infected red blood cells. *Am. J. Trop. Med. Hyg.* **44:**399–401.

Udomsangpetch, R., T. Sueblinvong, K. Pattanapanyasat, A. Dharmkrong-at, A. Kittikalayawong, and H. K. Webster. 1993a. Alteration in cytoadherence and rosetting of *Plasmodium falciparum*-infected thalassemic red blood cells. *Blood* **82:**3752–3759.

Udomsangpetch, R., B. J. Taylor, S. Looareesuwan, N. J. White, J. F. Elliott, and M. Ho. 1996. Receptor specificity of clinical Plasmodium falciparum isolates: nonadherence to cell-bound E-selectin and vascular cell adhesion molecule-1. *Blood* **88:**2754–2760.

Udomsangpetch, R., J. Todd, J. Carlson, and B. M. Greenwood. 1993b. The effects of hemoglobin genotype and ABO blood group on the formation of rosettes by Plasmodium falciparum-infected red blood cells. *Am. J. Trop. Med. Hyg.* **48:**149–153.

Udomsangpetch, R., B. Wahlin, J. Carlson, K. Berzins, M. Torii, M. Aikawa, P. Perlmann, and M. Wahlgren. 1989. *Plasmodium falciparum*-infected erythrocytes form spontaneous erythrocyte rosettes. *J. Exp. Med.* **169:**1835–1840.

Udomsanpetch, R., K. Thanikkul, S. Pukrittayakamee, and N. J. White. 1995. Rosette formation by *Plasmodium vivax. Trans. R. Soc. Trop. Med. Hyg.* **89:**635–637.

van Schravendijk, M. R., S. M. Handunnetti, J. W. Barnwell, and R. J. Howard. 1992. Normal human erythrocytes express CD36, an adhesion molecule of monocytes, platelets, and endothelial cells. *Blood* **80:**2105–2114.

Vogt, A. M., A. Barragan, Q. Chen, F. Kironde, D. Spillmann, and M. Wahlgren. 2003. Heparan sulfate on endothelial cells mediates the binding of Plasmodium falciparum-infected erythrocytes via the DBL1α domain of PfEMP1. *Blood* **101:**2405–2411.

Vogt, A. M., G. Winter, M. Wahlgren, and D. Spillmann. 2004. Heparan sulphate identified on human erythrocytes: a *Plasmodium falciparum* receptor. *Biochem. J.* **381:**593–597.

Wahlgren, M. 1986. Antigens and antibodies involved in humoral immunity to *Plasmodium falciparum.* Ph.D. thesis. Karolinska Institute, Stockholm, Sweden.

Wahlgren, M., J. Carlson, W. Ruangjirachuporn, D. Conway, H. Helmby, A. Martinez, M. E. Patarroyo, and E. Riley. 1990. Geographical distribution of *Plasmodium falciparum* erythrocyte rosetting and frequency of rosetting antibodies in human sera. *Am. J. Trop. Med. Hyg.* **43:**333–338.

Waitumbi, J. N., B. Donvito, A. Kisserli, J. H. Cohen, and J. A. Stoute. 2004. Age-related changes in red blood cell complement regulatory proteins and susceptibility to severe malaria. *J. Infect. Dis.* **190:**1183–1191.

Waitumbi, J. N., M. O. Opollo, R. O. Muga, A. O. Misore, and J. A. Stoute. 2000. Red cell surface changes and erythrophagocytosis in children with severe *Plasmodium falciparum* anemia. *Blood* **95:**1481–1486.

PROTECTION

V

MECHANISMS OF ANTIMALARIAL DRUG ACTION AND RESISTANCE

Anne-Catrin Uhlemann, Yongyuth Yuthavong, and David A. Fidock

23

Antimalarial chemotherapy remains the principal means of fighting malaria, yet current efforts are being thwarted by the ever-increasing prevalence of drug-resistant *Plasmodium falciparum* (White, 1996). Of particular concern is the appearance and spread of resistance to chloroquine (CQ), a 4-aminoquinoline, and sulfadoxine-pyrimethamine (SP, also known as Fansidar), an antifolate combination drug. Both of these drugs were previously characterized by their rapid efficacy, safety, and affordability, with costs as low as 10 to 20 cents per treatment regimen (Greenwood and Mutabingwa, 2002). In Africa, the earlier massive dependence on CQ is illustrated by the finding that malaria mortality rates increased dramatically following the appearance and dissemination of CQ resistance (CQR) (Trape, 2001). The shift to SP as a first-line drug has in turn led to a rapid expansion of SP-resistant parasites. Multidrug resistance (defined here as resistance to three or more drugs) has emerged in southeast Asia, and only the artemisinin family of agents (including artesunate) remains fully effective against all parasite strains. Compelling data have shown the excellent clinical benefit of using artesunate-based combination therapies to treat multidrug-resistant malaria in southeast Asia. These in turn have spurred intense efforts to finding ways of distributing artesunate-based combination therapies throughout malaria-endemic regions (Nosten et al., 2000; Sachs, 2005). Yet high costs, potential problems of global undersupply, and concerns about possible toxicity necessitate continuing efforts to also investigate other treatment options (Krishna et al., 2004). These include a broad range of discovery and development efforts focused on novel or established antimalarial targets (Fidock et al., 2004).

Antimalarial resistance can be defined as the "ability of a parasite strain to survive and/or multiply despite the administration and absorption of a drug given in doses equal to or higher than those usually recommended but within tolerance of the subject," assuming that

---

Anne-Catrin Uhlemann, Department of Infectious Diseases, St. George's Hospital Medical School, Cranmer Terrace, London SW17 0RE, United Kingdom. *Yongyuth Yuthavong*, National Center for Genetic Engineering and Biotechnology (BIOTEC), National Science and Technology Development Agency, Thailand Science Park, 111 Paholyothin Rd., Pathumthani 12120, Thailand. *David A. Fidock*, Department of Microbiology and Immunology, Albert Einstein College of Medicine, Forchheimer 403, 1300 Morris Park Ave., The Bronx, NY 10461.

Molecular Approaches to Malaria, Edited by Irwin W. Sherman
© 2005 ASM Press, Washington, D.C.

the drug gains access to the parasite or the infected red blood cells (iRBC) for the duration of time necessary for its normal action (Bruce-Chwatt, 1981). Treatment failures can manifest as the inability to clear parasites from the blood (parasitemia) or resolve clinical disease. Causes can include drug-resistant infections as well as other factors that include incorrect dosing, nonadherence to the dosing regimen, drug interactions, poor drug quality, poor absorption, or misdiagnosis. These factors can result in inadequate (i.e., selective) drug levels per parasite biomass in a patient, as can high parasitemia (Watkins and Mosobo, 1993; White, 2004). Drug resistance may also in some cases lead to increased gametocyte carriage, thereby facilitating the spread of resistance through the population (Price et al., 1999a). We note that the above definition specifies antimalarial resistance as relating to parasite factors that cause treatment failure in vivo. This must be clearly distinguished from in vitro resistance, which tends to be defined in terms of how much drug is required to inhibit in vitro parasite growth.

Genetically, resistance may develop when parasites undergo point mutations or gene amplification events, some of which can provide a fitness advantage, especially under conditions of repeated drug exposure. In endemic areas, resistance is difficult to reliably assess. In vivo tests aim to directly monitor clinical and parasitological parameters during follow-up for 7, 14, or 28 days or longer depending on the half-life of the drugs administered (World Health Organization, 2001). Results can be defined as parasite sensitivity or resistance at the RI, RII, or RIII levels. An alternative classification is therapeutic efficacy, which can be defined as adequate clinical response, early treatment failure, or late treatment failure (Dorsey et al., 2000). In vivo assessments are the gold standard, yet they require substantial logistical and financial support. In vitro tests are commonly used surveillance tools that frequently assist clinical studies. These assay the intrinsic drug susceptibility of in vitro-cultured parasites, typically stated as drug 50% inhibitory concentrations (IC_{50}s), independent of host factors such as immunity.

Several studies have attempted to define IC_{50} thresholds above which parasites would be expected to be resistant to a standard course of treatment, typically by comparing in vitro values with in vivo outcomes following drug treatment. As an example, Brasseur et al. (1992) used these comparisons to predict that *P. falciparum* strains with CQ IC_{50} values exceeding 80 nM in vitro would be CQ resistant in vivo. Other, more complex in vitro markers of resistance can also be useful; for example, testing whether ^{3}H-labeled CQ accumulation is reduced in iRBC or whether elevated CQ IC_{50} values in Asian and African strains are decreased in the presence of the reversal agent verapamil (VP) (Sidhu et al., 2002). In vitro tests, however, require a sophisticated infrastructure, and results can be difficult to compare between different laboratories. Further difficulties arise in how to interpret results and extrapolate data to the clinical response in vivo. Simpler tools are urgently needed to rapidly monitor drug-resistant parasites in an infected individual or in a population, which can lead to improved surveillance and treatment of malaria. This is made possible when resistance determinants are identified and molecular genotyping techniques are established. As described below, significant progress has recently been made in this area.

Here, we review the state of knowledge about the modes of action and mechanisms of *P. falciparum* resistance to the antifolate drugs SP, pyrimethamine, and cycloguanil, as well as the quinoline-based drugs, notably CQ, mefloquine (MFQ), and quinine (QN). Discussion of the resistance mechanisms will focus on the *P. falciparum* enzymes expressed by the following genes: *dhfr* (dihydrofolate reductase [DHFR]) and *dhps* (dihydropteroate synthase [DHPS]), which mediate antifolate resistance, and the transporters *pfcrt* (*P. falciparum* chloroquine resistance transporter gene) and *pfmdr1* (*P. falciparum* multidrug resistance gene 1), which are important determinants of resistance to the quinoline-based class of antimalarials. For more information on antimalarial modes of action and mechanisms of resistance, readers are referred to recent reviews by Ursos and Roepe (2002), Wongsrichanalai et al. (2002), Fitch (2004), Krishna et al. (2004), O'Neill and Posner (2004), Waller et al. (2004), White (2004), and Gregson and Plowe (2005). *Plasmodium vivax* drug resistance will not be addressed here and has been recently reviewed by Sina, 2002, and Baird, 2004.

ANTIFOLATES

Antifolates comprise a group of drugs that work through inhibition of folate metabolism of various organisms, including malaria parasites. Antifolate antimalarials such as pyrimethamine and cycloguanil (Fig. 1A) act by inhibiting DHFR, an enzyme responsible for reduction of dihydrofolate (DHF) to tetrahydrofolate. While pyrimethamine is the active form of the drug, cycloguanil is derived from proguanil, which is taken orally as a prodrug. These drugs were developed during World War II, when the supply of QN to the Allied Forces was interrupted. Sulfa drugs, including sulfonamides and sulfones, inhibit another enzyme, DHPS, responsible for forming a precursor of DHF two steps earlier in the folate metabolic pathway (Fig. 1B). Currently, antifolates are mainly used as combinations, most notably SP.

Antifolates: Antimalarials with a Well-Defined Target

Antifolates are among the very few antimalarials with well-defined targets and mechanisms of action. Together with other enzymes, DHFR and DHPS carry out catalytic actions required for de novo synthesis and salvage of folate cofactors (Fig. 2) (Wang et al., 2004) (see also http://sites.huji.ac.il/malaria/maps/folatebiopath.html). Although de novo synthesis appears to be the major route, some *P. falciparum* strains (e.g., 3D7 and Dd2) also appear to be capable of significant salvage of folate cofactors or their components from the external media (Fig. 2) (Wang et al., 1999).

In all *Plasmodium* species, as well as other protozoa and some plants, DHFR exists as a bifunctional enzyme that includes thymidylate synthase (TS), which forms deoxythymidylate (dTMP) from deoxyuridylate, while another substrate, methylenetetrahydrofolate, is converted to DHF. The DHF produced in this reaction, as part of the dTMP synthesis cycle (Fig. 2), as well as that formed from de novo synthesis and folate salvage, is reduced through DHFR catalysis to tetrahydrofolate. DHFR inhibition deprives parasites of folate cofactors essential for DNA synthesis and other metabolic conversions involving transfer of one-carbon fragments. Unlike its human counterpart, *Plasmodium* DHFR is sensitive to inhibition by pyrimethamine, cycloguanil, and other derivatives, which bind strongly with the active site of the malarial enzyme (Gregson and Plowe, 2005). Recent evidence suggests that *Plasmodium* sensitivity to these inhibitors also arises from the fact that the human host can respond to DHFR inhibitors by increasing DHFR expression, while the parasite cannot do so (Zhang and Rathod, 2002). In this study, *dhfr* mRNAs of both the parasite and the human host were found to bind to their respective DHFR polypeptides, but while drug treatment released host mRNA to produce new enzyme, no such regulation mechanism appeared to be present for the parasite.

DHPS, the target of sulfa drugs, catalyzes the condensation of *para*-aminobenzoic acid (PABA) with dihydropteridine pyrophosphate, yielding dihydropteroate, a DHFR precursor. DHPS is part of another bifunctional enzyme, which includes hydroxymethyldihydropteridine pyrophosphokinase that catalyzes the preceding step in the folate pathway. Sulfa drugs essentially mimic PABA, thereby inhibiting DHPS as well as acting as a false substrate and resulting in a product that further disrupts folate metabolism. Sulfa drugs are not used alone as antimalarials, owing to insufficient effectiveness, but are used in combination with DHFR inhibitors. Similar to their effective combinations in antibacterial chemotherapy, these form synergistic combinations against malaria parasites, due to the fact that they inhibit enzymes acting sequentially in the same metabolic pathway. Combinations of DHFR and DHPS inhibitors, such as SP and chlorproguanil-dapsone (LapDap), giving rise to chlorcycloguanil-dapsone in the body

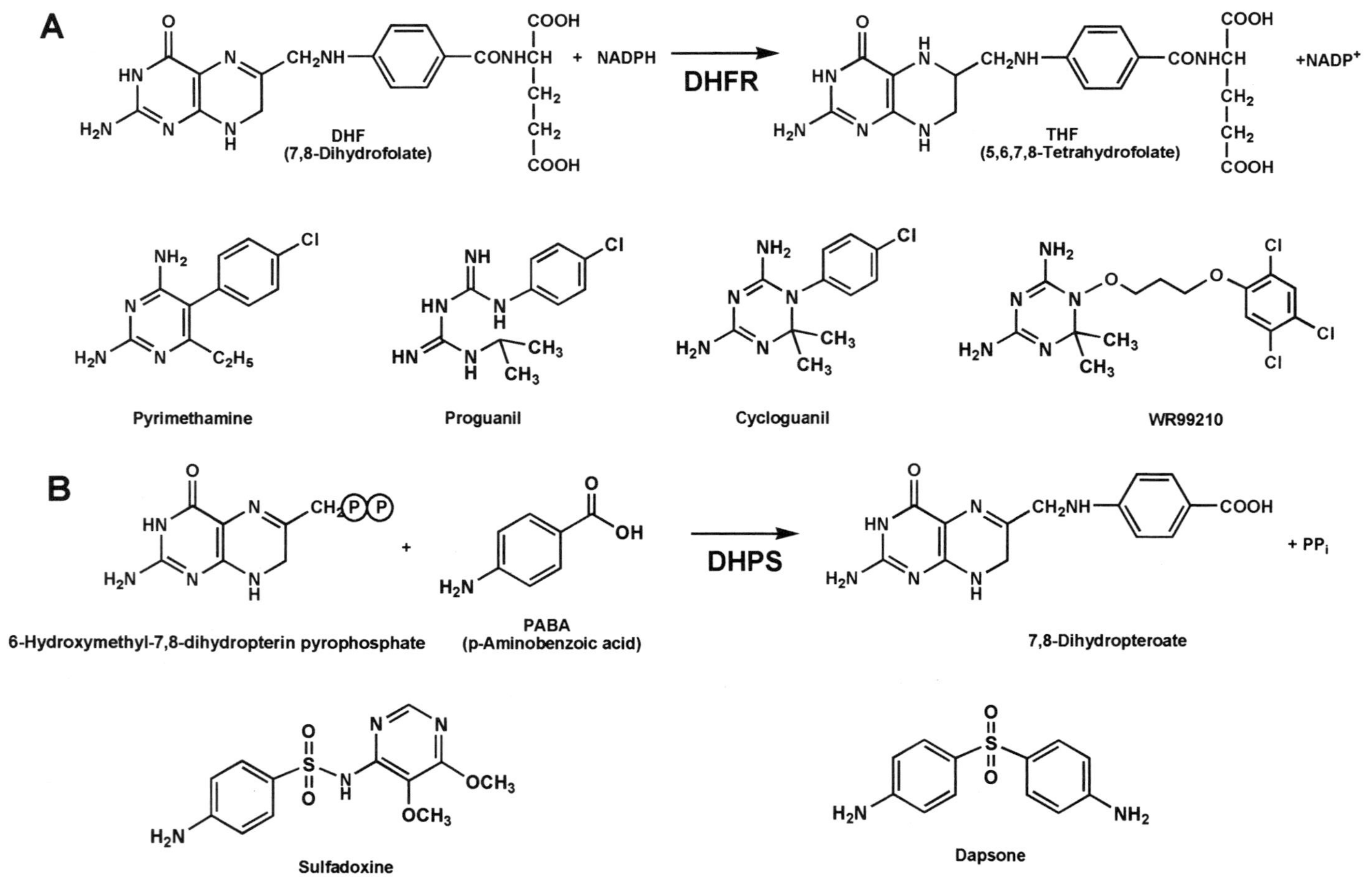

FIGURE 1 (A) Chemical reaction catalyzed by dihydrofolate reductase and structures of selected antimalarial DHFR inhibitors. The prodrug proguanil is converted into the DHFR inhibitor cycloguanil. (B) Chemical reaction catalyzed by dihydropteroate synthase, together with the structures of the DHPS inhibitors sulfadoxine and dapsone.

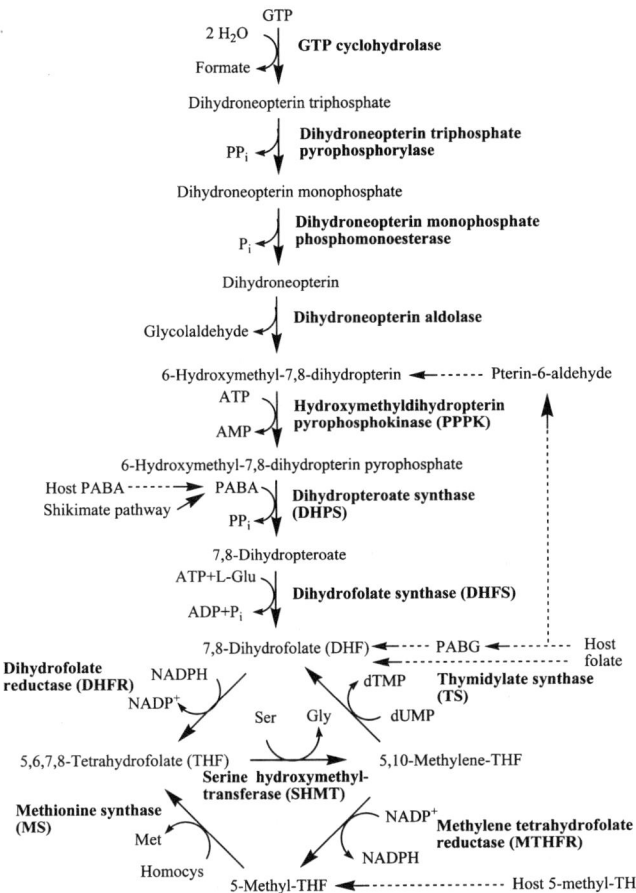

FIGURE 2 De novo synthesis, salvage, and utilization of folate cofactors in malaria parasites. Most enzymes in the de novo synthesis pathway in the *P. falciparum* genome have been identified, with the exception of dihydroneopterin aldolase. The salvage pathways have been investigated mainly through metabolic labeling analyses.

(Winstanley, 2001), are used as antimalarials, displaying efficacy even against parasites that have become moderately resistant to DHFR inhibitors. Unfortunately, resistance to some combinations such as SP has emerged (see below), prompting the need to search for new drugs and their combinations.

Although at present only inhibitors of DHFR and DHPS have been deployed as antimalarial drugs, the de novo pathway for synthesis of folate cofactors linked to the dTMP synthesis cycle and the methionine synthesis cycle (Fig. 2) together offer many as-yet-unexplored potential targets for antimalarials. In addition, salvage of folate cofactors and PABA from the host can also offer opportunities for drug interference.

Resistance to Antifolates Is Linked to Target Enzyme Mutations

Resistance to DHFR inhibitors, including pyrimethamine and cycloguanil, arose soon after their deployment as antimalarials. The addition of sulfa compounds created drug combinations that in many cases proved effective against the resistant parasites, although resistance quickly arose to these combinations, particularly SP. DHFR point mutations, resulting in decreased affinities of drug binding to the enzyme, have been shown to be the main cause of resistance (Foote et al., 1990a; Peterson et al., 1990). For *P. falciparum*, mutations have been found at various positions in the enzymes from resistant parasites, including those at codons specifying amino acid residues 16, 50, 51, 59,

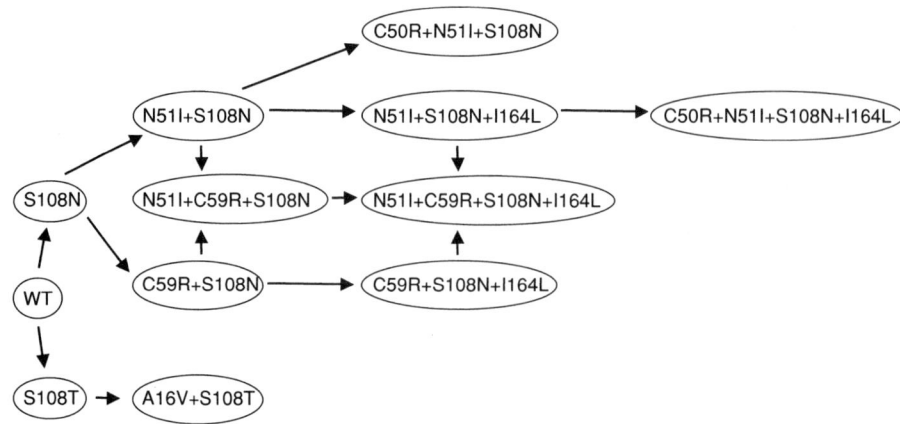

FIGURE 3 Possible evolution of antifolate resistance through cumulative mutations. The first mutation is considered to be S108N, followed by other mutations that confer increasing levels of resistance. The quadruple mutant N51I+C59R+S108N+I164L, resistant to both pyrimethamine and cycloguanil, is found in southeast Asia, while C50R+N51I+S108N+I164L might be present at very low levels in Africa. Another mutant, A16V+S108T, is resistant only to cycloguanil and not pyrimethamine.

108, and 164 (Cowman et al., 1988; Peterson et al., 1988; Plowe et al., 1997). Interestingly, these mutations appear to have arisen in Asia and spread to Africa (Roper et al., 2004). These mutations, which resulted in loss of inhibitor-binding affinities to the enzyme, generally correlated with increasing levels of resistance. Naturally, the enzyme catalytic activity, although also affected, must still be retained at a level sufficient for parasite viability. Activity in the presence of resistance-inducing mutations may be rescued by other genetic changes. For example, a five-amino-acid insertion after codon 30, which is a repeat of the preceding five and which is observed in areas of high resistance to SP, is believed to compensate for the loss of enzyme efficiency caused by resistance-inducing mutations (Gregson and Plowe, 2005).

Evidence that the enzyme mutations gave rise to parasite resistance was strengthened by the fact that recombinant mutant DHFRs showed decreased binding with pyrimethamine and cycloguanil (Sirawaraporn et al., 1997a; Kamchonwongpaisan et al., 2004), and transfection of wild-type parasites with constructs bearing mutant forms of *dhfr* gave rise to resistant phenotypes (Wu et al., 1996). One possible explanation for the evolution of resistance (Fig. 3) predicts stepwise mutations beginning with the mutation at codon 108 from serine to asparagine (S108N), which was shown to be the optimal mutation in that it caused a decrease in inhibitor-binding affinity while retaining normal enzyme activity (Sirawaraporn et al., 1997a; Sirawaraporn et al., 1997b). This produces a moderate level of resistance, while the subsequent mutations at other positions, namely, an asparagine to isoleucine at codon 51 (N51I), a cysteine to arginine at codons 59 or 50 (C59R; C50R), and an isoleucine to leucine at codon 164 (I164L) lead to increasingly higher levels of resistance (Gregson and Plowe, 2005). An alternative explanation, that the multiple mutants already existed in nature prior to antifolate exposure and were selected through drug deployment, has also been proposed (Thaithong et al., 2001).

Resistance to pyrimethamine conferred by these mutations was reported by Sirawaraporn et al. (1997a) to be mostly correlated with resistance to cycloguanil, reflecting in both cases a decrease in binding affinities with DHFR. However, parasites bearing some mutations, such as an alanine to valine at codon 16 (A16V) plus a serine to threonine at position 108 (S108T), are resistant only to cycloguanil, with corresponding decreases in the enzyme-binding affinity to this drug, while remaining sensitive to pyrimethamine, which still retains the original enzyme-binding affinity (Foote et al., 1990a). Furthermore, many derivatives of pyrimethamine and cycloguanil remain effective against parasites bearing multiple mutations, with corresponding retention of wild-type enzyme-binding affinities (Yuthavong et al., 2000; Kamchonwongpaisan et al., 2004; Yuthavong et al., 2005). The differences in cross-resistance patterns and point mutation positions have raised the possibility that new antifolates can be developed that would be effective against parasites resistant to conventional ones, if they can retain binding affinities against the mutant forms of the enzyme (Jensen et al., 2001; Hyde, 2002; Yuthavong, 2002; Yuthavong et al., 2005). The possibility of developing antifolates effective against resistant parasites is strengthened by the observation that mutations typically have a biochemical or fitness cost, as elegantly demonstrated by Reynolds and Roos (1998) using *Toxoplasma gondii* as a model. In addition, Sirawaraporn et al. (1997a) found that *P. falciparum* DHFR variants with multiple mutations and high levels of resistance to antifolates tended to have poor catalytic activities. Other factors, including inefficient expression and insufficient stability of the enzyme, can also contribute to limitations in mutation possibilities.

In addition to mutations and insertions causing changes in enzyme characteristics, there may be other mechanisms of antifolate resistance in *P. falciparum*, as suggested by the observation that the resistance level of parasites with mutant enzyme is not always correlated with the reduction in enzyme-binding affinity (Kamchonwongpaisan et al., 2004; Wang et al., 2004). Increased *dhfr-ts* copy number has been evoked as a mechanism of pyrimethamine resistance, although this has only been observed in drug-pressured laboratory lines (Thaithong et al., 2001). Increased DHFR-TS protein expression could conceivably also occur through increased gene transcription or translation. The capacity of many *P. falciparum* strains to salvage folates from RBC or from the blood plasma can also make them relatively resistant to antifolates (Wang et al., 2004). Decreased drug transport and enhanced drug efflux constitute possible alternative resistance mechanisms; however, to date these have not been demonstrated for *Plasmodium*.

Resistance to combinations of DHFR and DHPS inhibitors, such as SP, occurs when the parasite has multiple (three or more) mutations in *dhfr*, exacerbated by mutations in *dhps* (Plowe et al., 1997; Triglia et al., 1997). High levels of host plasma folate and enhanced salvage abilities tend to enhance the resistance levels, as would be expected if the drugs mainly inhibited de novo folate biosynthesis. As with *dhfr*, mutations in *dhps* may progress in a certain order, beginning primarily with an alanine to glycine change at codon 437 (A437G), followed by other mutations at codons 436, 540, 581, and 613 (S436A/F, K540E, A581G, and A613T/S) (Brooks et al., 1994; Triglia et al., 1997; Cowman, 1998), with the A437G and K540E mutations appearing to be the initial and most important ones for resistance.

As an alternative to SP, a combination of dapsone with chlorcycloguanil, administered as its prodrug chlorproguanil, has recently been developed (Winstanley, 2001; Lang and Greenwood, 2003). The latter combination, known as LapDap, has distinct pharmacokinetic advantages over pyrimethamine and sulfadoxine (Mutabingwa et al., 2001), although there is concern over dapsone toxicity for some patients. Pyrimethamine and sulfadoxine are slowly eliminated from the body, providing a powerful selection pressure for resistance generation with long exposure of the parasites to subtherapeutic drug concentrations (Watkins and Mosobo, 1993). In contrast, chlorproguanil and dapsone, which are more rapidly eliminated, offer the possibility of lower selection pressure for resistance. Importantly, this combination remains effective against parasites with low- or mid-level resistance to SP. Krudsood et al. (2005) reported that the antimalarial efficacy with chlorproguanil-dapsone was further enhanced by the addition of artesunate. However, some treatment failures were observed when the parasite carried DHFR mutations, including I164L. This mutation, which appears only to be viable in the presence of mutations at codons 108, 51, and 59, confers very high-level resistance to pyrimethamine and cycloguanil and appears to provide cross-resistance to chlorproguanil. *dhfr* quadruple mutant parasites (harboring I164L) are already quite widespread in Southeast Asia yet are rarely present if at all in Africa (Hastings et al., 2002). Should this fourth and most sinister *dhfr* mutation become widespread in Africa, it would most likely be the death knell of both SP and chlorproguanil-dapsone. Adding to concerns about whether chlorproguanil-dapsone presents a long-term solution for Africa, it should be noted that mutations inducing resistance to chlorcycloguanil (Hunt et al., 2005) or dapsone (Berglez et al., 2004) can be selected. Nevertheless, its improved pharmacokinetic and potency properties, compared to those of SP, may prolong its useful life, particularly if it is employed with a third antimalarial agent (Gregson and Plowe, 2005).

Structures of DHFR-TS and Modes of Drug Binding: Insights into Development of New, Effective Antifolates

Elucidation of the crystal structures of wild-type and mutant *P. falciparum* DHFR-TS complexed with antifolates (Color Plate 17A to C) (Yuvaniyama et al., 2003) has revealed how these antifolates bind to their target and has yielded important insights into how these mutations cause resistance. The overall structural features bear closest similarities to those of other protozoal DHFR-TS proteins. However, there are some unique features of the *P. falciparum* enzyme, mainly resulting from the fact that it has additional sequences, including the long junction region joining the two catalytic domains and two other inserts in the DHFR domain. The junction region of each *P. falciparum* DHFR-TS subunit contains a helical region that comes into close contact with the DHFR of the other subunit. Both inserts of the enzyme are also involved in interdomain contacts. These contacts are likely to be involved in communication between different subunits and regions during catalysis and may be susceptible to interference by external agents, offering additional possibilities for chemotherapy.

Co-occurrence of DHFR and TS as bifunctional enzymes in *Plasmodium* and other protozoa suggests functional linkage, a feature that is supported by inter- and intradomain contacts in the bifunctional enzymes from *Plasmodium* (Yuvaniyama et al., 2003), *Leishmania* (Knighton et al., 1994), and *Cryptosporidium* (O'Neil et al., 2003). Functional linkage between *P. falciparum* DHFR and TS was shown from the fact that while DHFR can function without TS, the latter cannot function without an almost fully intact DHFR-TS (Shallom et al., 1999; Wattanarangsan et al., 2003). It is unclear whether there is substrate channeling between the two *Plasmodium* enzymes, i.e., whether the product from TS (dihydrofolate) is immediately utilized as substrate by DHFR without leaving the bifunctional enzyme. Substrate channeling has been shown to occur in DHFR-TS from *Leishmania* (Meek et al., 1985; Knighton et al., 1994) and *Toxoplasma* (Trujillo et al., 1996), but not *Cryptosporidium* (Atreya and Anderson, 2004).

The mode of binding of inhibitors to *P. falciparum* DHFR was previously extensively studied by modeling (Lemcke et al., 1999; Rastelli et al., 2000; Warhurst, 2002), which predicted that the S108N mutation would give rise to steric conflict with pyrimethamine and cycloguanil binding, whereas the A16V mutation would give rise to steric conflict specifically with cycloguanil. These predictions have been mostly fulfilled by the results from crystal structure determinations of inhibitor-bound enzyme (Yuvaniyama et al., 2003).

The study of the binding modes of the flexible *P. falciparum* DHFR inhibitor WR99210 (Fig. 1A) (Canfield et al., 1993) has given particularly useful information for the design of novel inhibitors that would be capable of averting the effects of mutations that reduce the binding affinities of other, more rigid, inhibitors. WR99210 inhibits both wild-type and mutant enzymes, with excellent whole-cell potency even against *P. falciparum* strains harboring the quadruple *dhfr* mutant (Fidock and Wellems, 1997; Wooden et al., 1997). This compound is a triazine derivative similar to cycloguanil, but with a flexible side-chain (2,4,5-trichlorophenoxy) propyloxy group, instead of the rigid *p*-Cl-phenyl group of cycloguanil. Insight into the structural basis of resistance came from a comparison of the mode of binding of mutant DHFR-TS with pyrimethamine, which had lost its effectiveness, and WR99210, which still retained effectiveness (Rastelli et al., 2000; Warhurst, 2002; Yuvaniyama et al., 2003). The structure of the double-mutant (S108N+C59R) enzyme with bound pyrimethamine showed that the S108N mutation caused steric conflict for binding of the rigid *p*-chlorophenyl side chain, especially around the Cl atom (Color Plate 17B). In contrast, the structure of the quadruple-mutant enzyme (S108N+N51I+C59R+I164L) with bound WR99210 showed that its flexible side chain was oriented in such a way as to avoid this steric conflict (Color Plate 17C). The flexible side chain also interacted extensively with the enzyme, mainly through hydrophobic interactions. The quadruple mutant showed movement of residues 48 to 51 and residues 164 to 167, as the result of the N51I and I164L mutations, respectively, thereby widening the active site (Color Plate 17C). These changes likely contributed to the reduction in binding affinities of rigid inhibitors such as pyrimethamine, which could be overcome by the use of flexible inhibitors like WR99210. A number of other inhibitors with similarly flexible side chains have also been shown to retain high affinities for the mutant enzymes in the S108N series and maintain good antimalarial activities against parasites bearing the mutant enzymes (Jensen et al., 2001; Yuthavong et al., 2005). However, major drawbacks for WR99210 are its low oral bioavailability and gastrointestinal intolerance, which might be significantly averted by using its prodrug, PS-15 (Canfield et al., 1993). Many pyrimethamine derivatives, with the Cl atom removed from the *p*-Cl position, also show high binding affinities with the mutant enzymes, although their antimalarial activities are not as high as those of the flexible inhibitors (Sardarian et al., 2003; Kamchonwongpaisan et al., 2004).

In the case of cycloguanil resistance associated with the A16V mutation, modeling of the inhibitor-enzyme interaction showed that a methyl group of cycloguanil was in steric conflict with the bulky valine side chain, and cycloguanil derivatives without this methyl group could bind the mutant enzyme with high affinity and retain antimalarial activity (Rastelli et al., 2000; Yuthavong et al., 2000). WR99210 was able to adjust its orientation to avoid this conflict.

The important question of whether further DHFR mutations can occur so that resistance would arise against effective compounds such as WR99210 has been addressed through the use of bacterial (Chusacultanachai et al., 2002) or yeast (Ferlan et al., 2001) surrogate systems expressing the *P. falciparum dhfr-ts* gene. These studies reported that mutation-induced resistance could indeed occur. Nevertheless, the intrinsic limitation that mutant enzymes must retain significant catalytic activity and stability limits the scope of viable mutations and argues that the development of new effective antimalarial antifolates is still feasible. Furthermore, since a mutant enzyme may have different affinities for different inhibitors, it might be possible to identify combinations of DHFR inhibitors that work to neutralize mutant forms. DHFR inhibitors may also be synergistically combined with DHPS inhibitors as discussed above or with inhibitors of other enzymes in the folate pathway, such as TS. Thus, significant potential remains in pursuing further development of antimalarial antifolates that can be effective against *P. falciparum* strains harboring resistance to the current antifolate drugs.

QUINOLINES: MODES OF ACTION AND MECHANISMS OF RESISTANCE

The family of quinoline antimalarials is characterized by the presence of a quinoline ring (Fig. 4) and can be classified into two subclasses, the 4-aminoquinolines, including CQ, and the quinoline-4-methanol drugs, including QN and MFQ.

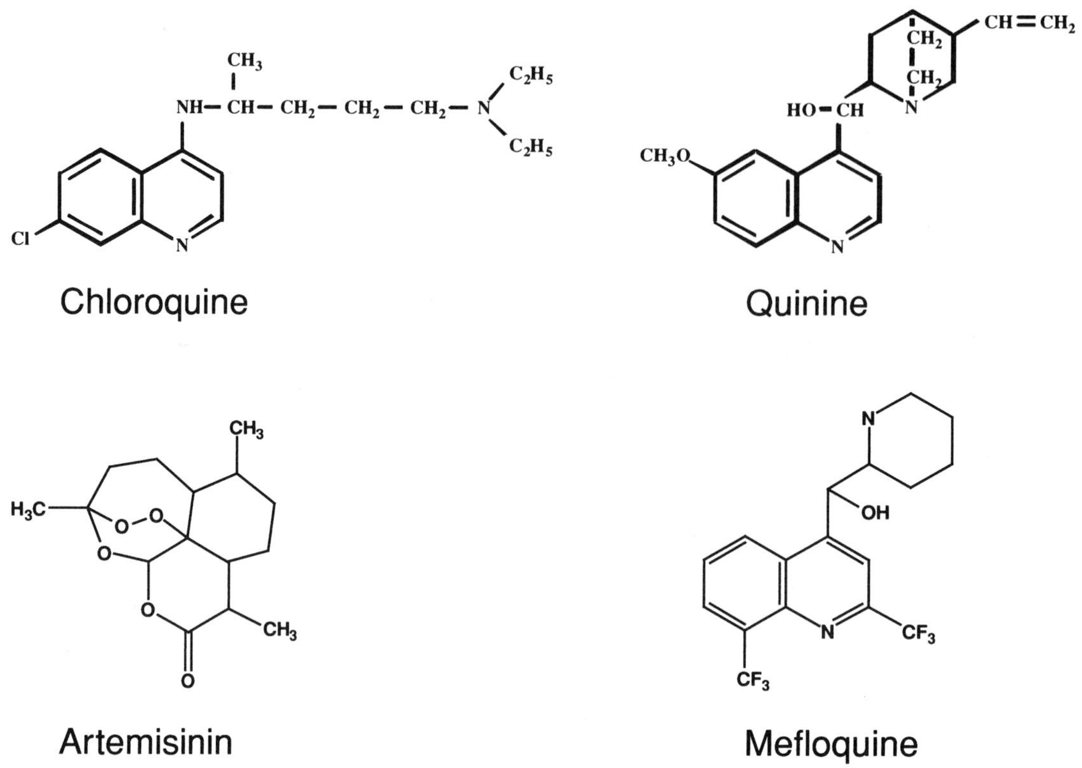

FIGURE 4 Chemical structures of major antimalarial quionolines, as well as the endoperoxide-containing drug artemisinin.

Chloroquine Action

Chloroquine use began worldwide in the late 1940s, and for several decades, this drug remained the gold standard in the prevention and treatment of uncomplicated malaria. Yearly CQ production in the mid-1980s was estimated at 1,300 tons, corresponding to around 900 million courses of a 3-day treatment. Defining characteristics of CQ included its rapid parasiticidal action, low cost, safety, and widespread availability.

Initial insights into the mechanistic basis of CQ sensitivity included the work of Wallace Peters (Peters, 1970), who observed CQ to be effective only on parasite intraerythrocytic stages that were actively degrading hemoglobin. This degradation occurs inside the digestive vacuole (DV), which was noted to swell upon CQ treatment (Aikawa, 1972; Jacobs et al., 1988). Accumulation in the DV of this highly soluble, diprotonable (dibasic) compound (pKa1 5 8.1, pKa2 5 10.2), demonstrated experimentally by Saliba et al. (1998), is believed to largely depend on the presence of a weak-base gradient between the acidic DV (pH 5.0 to 5.5) and the pH-neutral parasite cytosol (Yayon et al., 1984; Krogstad et al., 1985). Once inside this acidic compartment, CQ should become predominantly diprotonated and membrane impermeant. In support of this prediction, the degree of *P. falciparum* susceptibility to CQ has been found to be pH dependent (Yayon, 1985), and CQ action can be mitigated by the proton-pump inhibitors bafilomycin A1 and omeprazole (Bray et al., 1992; Skinner-Adams and Davis, 1999). However, CQ has been reported to accumulate in iRBC at much higher levels than those observed with similarly acidic compartments in eukaryotic cells, implicating more than just pH gradients (Macomber et al., 1966; Yayon, 1985; Krogstad and Schlesinger, 1987; Krogstad et al., 1992). Studies now point to interactions between CQ and hematin as the major factor leading to CQ retention inside the iRBC of CQ-sensitive (CQS) parasites, and a core element of CQ's mode of action (Chou et al., 1980; Bray et al., 1998; Dorn et al., 1998). Hematin, the Fe31 dimeric form of heme (Fe21-protoporphyrin IX), is produced upon hemoglobin degradation and is detoxified upon incorporation of hematin into inert hemozoin crystals (Pagola et al., 2000).

Debate continues on whether CQ binds primarily with membrane-associated (or soluble) hematin or instead with the terminal hematin structure present at the actively growing faces of the hemozoin crystal (Sullivan et al., 1996; Pagola et al., 2000). Presumably, either process would effectively allow the buildup of heme (or CQ-heme complexes) to levels that become irreversibly toxic to the parasite, possibly via heme-mediated DV membrane lipid peroxidation (Waller et al., 2004). Interestingly, studies of CQ action have found that drug accumulation follows a biphasic pattern, having a high-affinity, low-capacity component (K_d = 25 to 150 nM) plus a higher-capacity nonsaturable component that predominates at higher (micromolar) external CQ concentrations (Bray et al., 1998). The high-affinity, saturable component is believed to be responsible for CQ's antimalarial activity (Bray et al., 1998).

We note that whereas most of the current data implicate trapping of diprotonated CQ within the DV and CQ-heme binding as key to this drug's action, alternative models include CQ interference with glutathione-mediated detoxification of reactive oxygen radicals formed during hemoglobin degradation (Ginsburg et al., 1999).

Mefloquine and Quinine Action

While heme binding has been convincingly demonstrated as being central to the mode of action of CQ, this is less clear for MFQ and QN. Earlier studies identified weak interactions between MFQ and free heme (Chou et al., 1980; Chevli and Fitch, 1982), and MFQ was shown to inhibit heme polymerization in vitro at concentrations that were the same or lower than that of CQ (Slater and Cerami, 1992; Chou and Fitch, 1993; Raynes et al., 1996). Furthermore, ultrastructural studies revealed an effect of MFQ on the parasite DV and hemozoin (Peters et al., 1977). Yet while CQ treatment of *Plasmodium berghei*-infected mice was found to cause a decrease in hemozoin production, MFQ and QN had no such effect (Chou and Fitch, 1993). These data have led to the speculation that MFQ might interfere with a different step of the hemoglobin degradation pathway than does CQ (Geary et al., 1986). QN also has been reported to weakly interact with heme and can inhibit heme polymerization in vitro (Slater and Cerami, 1992; Chou and Fitch, 1993). Yet QN accumulation in the DV has not been documented. Further studies are needed to delineate the modes of action of these compounds and test whether heme is the sole or even the major target of these compounds.

Epidemiological Contexts of Quinoline Resistance

P. falciparum resistance to CQ was first documented in the late 1950s in Colombia and Thailand, extending over the next 20 years throughout South America and Southeast Asia. Arriving in the late 1970s in East Africa, CQR has since spread across all of sub-Saharan Africa. This drug is no longer used in Southeast Asia and South America, yet in many African countries it remains a common choice for home medication, despite widespread resistance (Djimdé et al., 2003). Interestingly, despite the tremendously higher burden of malaria and use of CQ in Africa, CQR does not appear to have independently evolved there. Instead, there is compelling evidence that *P. falciparum* CQR came from Asia (Wootton et al., 2002). As noted before ("Resistance to Antifolates Is Linked to Target Enzyme Mutations," above), this is also the case with pyrimethamine resistance (Roper et al., 2004). One plausible explanation is that many malarial infections in Africa go untreated and are held in check by premunition, resulting in a relatively low drug pressure in the local parasite population. This contrasts with the typical situation in Asia, where immunity is far more limited and infections typically progress to a disease state and are drug treated (White and Pongtavornpinyo, 2003; Yeung et al., 2004). Differences in population structure and gene flow may also play an important role (Ariey et al., 2003).

For QN, the first reports of decreased *P. falciparum* sensitivity date back to 1908 in Brazil (Gregson and Plowe, 2005). Yet QN still remains widely effective, perhaps in part because this drug has been far less used than CQ or SP and quite possibly also because QN resistance has a complex genetic basis. Reports of QN resistance have nevertheless slowly accumulated, mostly from Southeast Asia, where QN was widely used to treat CQ-resistant malaria (Watt et al., 1992; Pukrittayakamee et al., 2000). Efforts to increase QN efficacy and offset resistance have resulted in this drug often being used in combination with tetracycline or clindamycin (Kremsner et al., 1994). This has been particularly useful for treating uncomplicated CQ-resistant malaria in adults.

In contrast to QN, resistance to MFQ as monotherapy has arisen rapidly, as highlighted by the history of MFQ use in Thailand. There, MFQ was a highly potent drug when introduced as monotherapy in 1984. Yet substantial resistance occurred within 5 years, and cure rates in some areas dropped to <50%, leading to the development of multidrug resistance (Nosten et al., 1991; Mockenhaupt, 1995). Although MFQ has never been used extensively in Africa, some reports from West Africa have indicated intrinsic resistance to this drug (Brasseur et al., 1992). This was attributed in part to the widespread use of QN, contributing to the current thinking that resistance to these two drugs involves complex and partially overlapping mechanisms.

Initial Models of Chloroquine Resistance

Early investigations into the mechanism of CQR evoked widely differing models that nonetheless shared two principal tenets: (i) CQ-resistant parasites displayed reduced ^{3}[H]-CQ accumulation, and (ii) VP partially reversed CQR, particularly in the case of Asian and African strains, in effect "chemosensitizing" these strains to the effect of CQ. Initially, these models included (i) reduced CQ access to hematin; (ii) increased efflux of CQ from the parasite DV, possibly as a result of changes in the transport properties of a DV membrane protein (or proteins) that could transport drug; (iii) altered ionic flux or CQ uptake as a result of a physiological change at the parasite DV membrane that could include altered pH; or (iv) accelerated detoxification of CQ-heme complexes by a mechanism involving increased levels of glutathione.

REDUCED CQ ACCESS TO HEMATIN

In support of the model that CQR may result from reduced CQ access to hematin, Bray et al. (1998) reported that the antimalarial activity of CQ could be attributed to the saturable component of CQ accumulation and that CQ-resistant and CQS parasites could be distinguished by the K_d in saturable CQ binding. These data supported an earlier report (Bray et al., 1996b), indicating that the differences in CQ accumulation between CQ-resistant and CQS parasites were routinely smaller than the differences in their CQ IC_{50} values, suggesting that CQR involved both alterations in receptor accessibility and drug accumulation. Evidence in support of reduced CQ access to hematin was recently provided by biochemical investigations of *pfcrt*-modified recombinant lines (see "Confirmation by Allelic Exchange That Mutant *pfcrt* Confers CQR," below; Lakshmanan et al., 2005).

INCREASED CQ EFFLUX

Support for an increased CQ efflux model came initially from Krogstad et al. (1987), who showed that CQ-resistant parasites released preaccumulated CQ almost 50 times faster than CQS parasites. Subsequently, it was shown that addition of glucose to the medium markedly stimulated CQ accumulation in CQS parasites, yet reduced CQ steady-state accumulation in CQ-resistant parasites (Krogstad et al., 1992). These data would suggest that CQS parasites have an energy-dependent CQ uptake mechanism (with energy being required both to maintain the DV proton gradient and to traffic and digest hemoglobin, releasing heme) and that CQ-resistant parasites have an energy-dependent CQ efflux mechanism. These observations received support from a recent study showing that the reduction in CQ accumulation in CQ-resistant parasites, resulting from glucose addition, coincided with parasite ATP production (Sanchez et al., 2003). These data suggested that CQR might be caused by an energy-dependent effluxer, i.e., a transporter capable of moving the drug out of the DV against an electrochemical gradient (Sanchez et al., 2003). More recent work by Sanchez et al. (2004) provided further support for a model whereby CQ-resistant parasites had acquired a substrate-specific, inhibitable drug efflux system.

pH-DEPENDENT PHYSIOLOGICAL CHANGES AT THE DV MEMBRANE

Mechanistic models relating pH-dependent physiological changes in relation to CQR have recently seen intriguing yet contradictory developments. Initial modeling postulated that uptake of diprotonable CQ into the acidic DV resulted from weak-base ion trapping, as predicted by the Henderson-Hasselbach equation (Homewood et al., 1972; Yayon et al., 1984; Krogstad et al., 1985). According to the weak-base theory, increased DV pH in CQ-resistant parasites, causing a reduced pH gradient between the DV and the parasite cytosol, would therefore conceivably cause reduced CQ accumulation. Yet innovative developments in *P. falciparum* single cell imaging and photometry led Paul Roepe and colleagues to conclude that CQ-resistant parasites had a decreased DV pH relative to CQS parasites (Dzekunov et al., 2000; Bennett et al., 2004). One explanation was that the more acidic DV pH in CQ-resistant isolates might accelerate the rate of heme aggregation, thereby reducing the concentration of soluble hematin that retained bound CQ, and resulting in an overall reduction in cellular CQ uptake (Ursos and Roepe, 2002). Some kinetic studies, however, report that the quantity of heme receptor appears similar in CQ-resistant and CQS parasites (Sanchez et al., 1997; Bray et al., 1998; Sanchez et al., 2003); furthermore, the choice of dyes used as pH probes has been questioned (Bray et al., 2002; Wissing et al., 2002). Clearly, more work is required to define the DV pH of CQ-resistant parasites and its relationship to CQR.

GLUTATHIONE-MEDIATED DETOXIFICATION

Potentially, glutathione could participate in the CQR mechanism by its noted ability to bind to heme, resulting in decreased levels of accumulated CQ in the DV, protection of heme-mediated lipid peroxidation, and membrane lysis (Shviro and Shaklai, 1987). Earlier indications that glutathione-mediated detoxification of CQ-heme complexes might be involved in CQR included experiments using CQ-resistant rodent parasite models, which reported reduced hemozoin and higher levels of glutathione in CQ-resistant compared to CQS lines (Wood and Eaton, 1993; Dubois et al., 1995). Comparative studies of *P. falciparum* parasites demonstrated that increasing levels of glutathione could lead to increased CQR (Ginsburg et al., 1998). A separate study, however, showed no differences in the level of hemozoin accumulation between *P. falciparum* CQ-resistant and CQS parasites (Zhang et al., 1999), providing evidence that the mechanistic basis of CQR in *P. falciparum* and rodent *Plasmodium* species was likely to differ.

pfcrt

THE IDENTIFICATION OF *pfcrt*

Chloroquine's earlier tremendous benefit in treating malaria worldwide sparked significant interest in identifying the genetic basis of CQR. Ultimately, the successful approach to identifying the key determinant was adopted by Thomas Wellems and colleagues at the National Insitutes of Health, who implemented a genetic cross between a CQ-resistant clone (Dd2, from Indochina) and a CQS clone (HB3, from Honduras) (Wellems et al., 1990). Analysis of the ensuing haploid progeny revealed segregation of the VP-reversible CQR phenotype in a manner consistent with Mendelian inheritance of a single gene, which was subsequently localized to a 36-kb segment of chromosome 7 (Wellems et al., 1991; Su et al., 1997). After eliminating the initial candidate gene *cg2*, studies of this segment revealed the presence of *pfcrt*, a highly interrupted, previously cryptic gene (Fidock et al., 2000a; Fidock et al., 2000b). This encodes a 49-kDa transmembrane protein, termed PfCRT for *Plasmodium falciparum* chloroquine resistance transporter in recognition of the transporter-like nature of its predicted secondary structure (Fidock et al., 2000b). Immunofluorescence and immunoelectron microscopy studies localized PfCRT to the DV membrane (Fidock et al., 2000b; Cooper et al., 2002). Interestingly, sequence analysis of Dd2, HB3, and representative progeny identified eight PfCRT point mutations (M74I, N75E, K76T, A220S, Q271E, N326S, I356T, and R371I) that distinguished the resistant and sensitive parasites. These mutations typically resided within or near the boundaries of predicted transmembrane domains (Fig. 5).

THE EVOLUTION OF MUTANT PfCRT HAPLOTYPES

Linkage studies with culture-adapted isolates from around the world have since demonstrated a very close association between *pfcrt* mutations and the in vitro CQR phenotype (recently reviewed in Waller et al., 2004). This association is particularly strong with the K76T mutation, which is ubiquitous to CQ-resistant lines of *P. falciparum*. Mutant *pfcrt* alleles are now known to have arisen independently in Southeast Asia, South America, Papua New Guinea, and the Philippines (corresponding to allelic types E1a/E1b, W1a/W1b/W2, P1, and P2a/P2b, respectively) (Table 1). These alleles differ by four to eight point mutations from the canonical, wild-type (HB3) *pfcrt* allele. Mutant *pfcrt* alleles in Africa appear to have arisen in Asia, and this group (E1a and E1b) comprises two major alleles that differ only at codon 356 (Wootton et

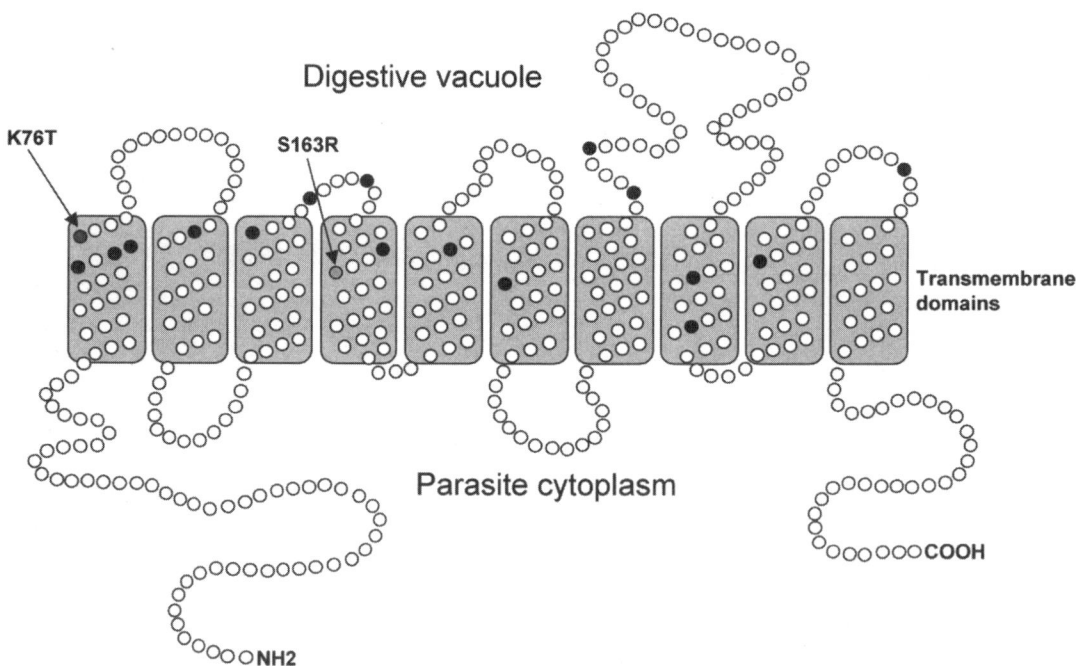

FIGURE 5 PfCRT predicted structure. PfCRT has been postulated to possess 10 transmembrane helices, with the N and C termini extending into the parasite cytoplasm. Filled circles indicate the positions of mutations published from full-length *pfcrt* cDNA sequences identified in CQ-resistant parasites from field samples, as well as additional mutations identified in amantadine- and halofantrine-resistant parasites selected in vitro (Table 1). (Reprinted from *Molecular Microbiology* [Bray et al., 2005] with permission from the publisher.)

TABLE 1 *pfcrt* allelic variants identified in *P. falciparum* field isolates and laboratory-adapted lines

Parasite origin, type (reference line)	Encoded amino acid and PfCRT position[a]														
	72	74	75	76	97	144	148	160	194	220	271	326	333	356	371
Chloroquine sensitive															
All regions, canonical wild type (HB3)[b]	C	M	N	K	H	A	L	L	I	A	Q	N	T	I	R
Africa (1061)[b]	C	I	E	K	H	A	L	L	I	S	E	S	T	I	I
Chloroquine resistant															
Africa and Southeast Asia, type E1a (Dd2)[b]	C	I	E	T	H	A	L	L	I	S	E	S	T	T	I
Africa and Southeast Asia, type E1b (FCB, K1)[b,c]	C	I	E	T	H	A	L	L	I	S	E	S	T	I	I
Papua New Guinea, type P1 (PNG 1935)[d]	S	M	N	T	H	A	L	L	I	S	Q	D	T	L	R
South America, type W1a (7G8)[b]	S	M	N	T	H	A	L	L	I	S	Q	D	T	L	R
South America, type W1b (Ecu 1110)[b]	C	M	E	T	H	A	L	L	I	S	Q	D	T	L	R
South America, type W2 (Jav)[b]	C	M	E	T	Q	A	L	L	I	S	Q	D	T	I	T
Philippines, type P2a (PH1)[e]	C	M	N	T	H	T	L	Y	I	A	Q	D	T	I	R
Philippines, type P2b (PH2)[e]	S	M	N	T	H	T	L	Y	I	A	Q	D	T	I	R
Indonesian Papua (2300)[f]	C	I	K	T	H	A	L	L	I	S	E	S	T	I	I
Cambodia (742)[g]	C	I	E	T	H	A	I	L	T	S	E	N	T	I	I
Cambodia (734)[g]	C	I	D	T	H	F	I	L	T	S	E	N	T	I	R
Cambodia (738)[g]	C	I	D	T	H	A	L	L	I	S	E	N	T	I	R
Cambodia (783)[g]	C	I	E	T	H	A	L	L	I	S	E	N	T	I	I

[a]Sequences are only listed in instances when the full-length *pfcrt* coding sequence was established. Thus, this table does not list several other polymorphic residues that were identified from field isolates for which *pfcrt* was only partially sequenced. (Reprinted from *Molecular Microbiology* [Bray et al., 2005] with permission from the publisher.)
[b]Sequence source, Fidock et al., 2000.
[c]Sequence source, Johnson et al., 2004.
[d]Sequence source, Mehlotra et al., 2001.
[e]Sequence source, Chen et al., 2003.
[f]Sequence source, Nagesha et al., 2003.
[g]Sequence source, Durrand et al., 2004.

al., 2002). A distinct *pfcrt* allele appears to have independently evolved with the identical PfCRT haplotype (W1a) in both South America and Papua New Guinea (Chen et al., 2001; Mehlotra et al., 2001). The particularly rapid dissemination of CQR throughout South America might be a result of the appearance and selection of four *pfcrt* alleles (two of which encode the same W1a PfCRT haplotype, though are distinct at the nucleotide level) (Wellems and Plowe, 2001). Despite the geographic diversity in *pfcrt* sequences, the K76T mutation is accompanied by an A220S mutation in all CQ-resistant lines, except for ones from the Philippines (the P2a/P2b types), where this mutation appears to be replaced by the A144T and L160Y paired mutations (Chen et al., 2003, 2005). Additional novel mutant alleles, possibly representing distinct origins, have also been identified in Indonesian Papua (formerly Irian Jaya) and Cambodia (Nagesha et al., 2003; Durrand et al., 2004). This brings the total number of variant residues in *pfcrt* to 15, making this an extraordinarily polymorphic gene. Interestingly, microsatellite typing of polymorphisms across all 14 chromosomes for 87 geographically dispersed isolates has revealed the *pfcrt* locus to be the most extensive site of recent sequence evolution, underscoring both the selective advantage to the parasite of mutations in this locus and also the tremendous selective pressure exerted by massive use of CQ in recent decades (Wootton et al., 2002).

ASSOCIATION OF *pfcrt* WITH IN VIVO AND IN VITRO CQR

Soon after its discovery in late 1998, the *pfcrt* sequence was forwarded to the World Health Organization and interested groups, resulting in a very rapid and thorough evaluation of the K76T mutation as a candidate marker of CQR. The first report, a clinical and molecular study published in 2001 by Djimdé et al. (2001a), found the *pfcrt* K76T mutation in 100% of the cases of CQ treatment failure, versus a baseline mutation frequency of 41% in samples obtained at the onset of treatment. This translated into a treatment failure odds ratio of 18.8. These authors subsequently extended this study to several sites in Mali with different rates of transmission. Data were standardized across sites using the genotype resistance index, defined as the prevalence of K76T divided by the prevalence of parasite resistance at a given site; and the genotype failure index, defined as the prevalence of K76T divided by the rates of CQ treatment failure (Djimdé et al., 2001b). Their results across areas of different endemicity revealed a remarkably uniform relationship between frequencies of *pfcrt* K76T mutations and rates of in vivo CQR (i.e., the higher the prevalence of K76T, the greater the rates of in vivo parasite resistance and treatment failure). This work had broad implications in inferring that the K76T prevalence could give a relatively accurate estimate of the percentage of CQ treatment failure across different malaria-endemic African settings.

Several subsequent studies reported absolute selection for *pfcrt* K76T in cases of CQ treatment failure, in settings where *pfcrt* mutations had not attained 100% prevalence. Other studies observed that *pfcrt* mutations were already present in all of the samples tested, suggestive of previous significant selective pressure (reviewed in (Wellems and Plowe, 2001). In vitro associations were typically greater than in vivo, implying that factors other than *pfcrt* were contributing to the different clinical outcomes. These factors may include differences in immunity or other host factors such as the presence of concomitant infections, frequency of exposure, nutritional status, and/or the requirement for additional parasite genetic determinants that may play an important role in CQ treatment failure. Important support for a key role of age-related acquisition of antiparasitic immunity in clearance of CQ-resistant infections came from the study of Djimdé et al. (2001a), who found that children over 10 years of age had a much greater rate of clearance of parasites harboring the *pfcrt* K76T mutation when compared to younger children ($P < 0.001$). The overall picture emerging from these studies is that *pfcrt* mutations appear to be necessary for CQR in vitro and for CQ treatment failure in vivo, but that other factors including immunity (and perhaps to a lesser extent a contribution from other parasite genes) are important in determining clinical outcome.

CONFIRMATION BY ALLELIC EXCHANGE THAT MUTANT *pfcrt* CONFERS CQR

Direct evidence in support of a determining role for *pfcrt* in CQR first came from transfection studies showing that coexpression of mutant *pfcrt* in CQS parasites produced low-level, VP-reversible CQR (Fidock et al., 2000b). More recently, a system was developed to replace the entire *pfcrt* allele in a CQS parasite with *pfcrt* alleles representative of mutant sequences prevalent in CQ-resistant parasites from Asia, Africa, South America, and Papua New Guinea (Sidhu et al., 2002). The resulting recombinant clones, while displaying reduced *pfcrt* expression as a result of the genetic modifications, demonstrated all the hallmarks of a bona fide CQR phenotype, i.e., increased CQ IC_{50} values (exceeding 100 nM), acquisition of VP reversibility, and decreased CQ accumulation. Further transfection studies also provided evidence that the degree of CQR could be affected in vitro by the level of PfCRT expression (Waller et al., 2003).

More recent allelic exchange studies provided conclusive evidence that the *pfcrt* K76T mutation is critical for CQR (Lakshmanan et al., 2005). In these studies, the mutant T76 codon was replaced with the wild-type K76 codon, in both Dd2 and 7G8 parasites. The resulting back-mutant lines reverted to a fully CQS phenotype and were no longer VP reversible. Taking advantage of the fact that these lines retained low-level QN resistance, these studies also found a total loss of VP reversibility of QN resistance. Introduction of the Dd2 mutations surrounding codon 76 into the 7G8 *pfcrt* allele demonstrated that these flanking mutations also contributed to determining the extent to which VP can reverse CQR, confirming an earlier hypothesis that *pfcrt* sequence differences in this region explained the different degrees of VP reversibility seen with South American lines compared to Asian and African lines (Mehlotra et al., 2001). These studies also identified a very tight specificity of the *pfcrt*-mediated CQR mechanism for the CQ structure. Addition or removal of a single ethylene group from the CQ side chain resulted in limited cross-resistance; however, resistance was totally ablated upon the removal or addition of two or more ethylene groups.

Biochemical analysis of these *pfcrt*-modified lines demonstrated CQ:heme binding and saturable CQ uptake properties that correlated tightly with changes in their CQ response profiles (Sidhu et al., 2002; Lakshmanan et al., 2005). As an example, CQ-resistant lines engineered to express a *pfcrt* allele encoding the wild-type K76 instead of the mutant T76, resulting in a total loss of VP-reversible CQR, displayed dramatically increased CQ accumulation at equilibrium and CQ-heme binding. Conversely, CQS parasites engineered to express mutant *pfcrt*, resulting in CQR, displayed reduced CQ equilibrium accumulation and CQ-heme binding. These data were consistent with PfCRT conferring CQR by controlling the access of CQ to its hematin receptor, a property that was critically influenced by the K76T mutation (Lakshmanan et al., 2005).

FUNCTIONAL PREDICTIONS

Possible functions for PfCRT were recently proposed following bioinformatic analyses that placed this protein in the drug-metabolite transporter superfamily (Martin and Kirk, 2004; Tran and Saier, 2004). Homology was greatest with the drug-metabolite effluxer family, especially in regions of transmembrane domains 4 and 9 that harbor putative substrate-binding motifs (Fig. 5). These studies also predicted that PfCRT is a homodimer, with the N and C termini oriented to the parasite cytosol. In this orientation, the key CQR-conferring K76T mutation was predicted to lie near the surface of the vacuolar side of the membrane, within a region of the transporter postulated to be involved in substrate specificity.

Substrates for drug-metabolite effluxer transporters include amino acids, weak bases, and organic cations, and transport is typically H^+ coupled (Jack et al., 2001). This led Martin and Kirk (2004) to predict that PfCRT functions to export metabolites, possibly amino acids or peptides, from the DV, potentially in symport with H^+. Other possible transport solutes could include fatty acids, carbohydrates, vitamins, sterols, or nucleotides, which are effluxed from lysosomes in other organisms (Eskelinen et al., 2003). Alternative proposed functions include chloride transport or activation or modulation of other transporters (Warhurst et al., 2002; Zhang et al., 2002; Nessler et al., 2004). The development of several heterologous expression assays, including with yeast, *Xenopus laevis* oocytes and *Dictyostelium discoideum* (Zhang et al., 2002; Nessler et al., 2004; Zhang et al., 2004, Naude et al., 2005), permits expanded efforts to identify the native function of PfCRT, and also explore whether mutant forms can physically interact with and facilitate CQ movement (see below).

We note that recent studies expressing mutant or wild-type *pfcrt* in *Dictyostelium discoideum* support the idea that mutant PfCRT can confer CQR by enabling outward-directed CQ efflux (Naude et al., 2005). These data are consistent with recombinant yeast studies that provided evidence that mutant PfCRT could bind CQ (Zhang et al., 2004).

THE PfCRT CHARGED DRUG LEAK MODEL

One model that has recently been proposed for how PfCRT might mediate CQR is the charged drug leak hypothesis, which proposed that this mutant protein mediates the leak of charged (protonated) forms of CQ from the DV, down its concentration gradient (Johnson et al., 2004; Bray et al., 2005). This model was based on the observation that the *pfcrt* mutations that cause CQR are associated with a loss of basic and hydrophobic residues (Warhurst et al., 2002). These mutations (including K76T) have been predicted to be in the PfCRT transporter barrel on the DV side of the membrane (Fig. 5) (Martin and Kirk, 2004). The presence of the positively charged lysine residue at position 76 in wild-type PfCRT was postulated to repulse the doubly protonated (cationic) form of CQ (CQ^{2+}, the predominant species present within the acidic vacuole) and prevent its interaction with the transporter. The CQR-conferring mutation of lysine 76 to threonine (or isoleucine or asparagine) (Fidock et al., 2000b; Cooper et al., 2002) removes the positive charge, possibly allowing CQ^{2+} to interact with PfCRT and exit the vacuole down the steep outward CQ^{2+} concentration gradient (perhaps in symport with H^+). This would result in a decrease in the overall concentration of CQ at its site of action within the DV and hence a decreased CQ sensitivity of the parasite.

Indirect support for this model came from recent work in which *P. falciparum* CQ-resistant K1 parasites were selected in vitro for resistance to halofantrine (a phenanthrene methanol) or amantadine (an antiinfluenza agent). Surprisingly, as the parasites became resistant to these drugs, they became sensitive to CQ (Nateghpour et al., 1993; Johnson et al., 2004). Detailed examination of these parasite lines revealed that CQ accumulation and access to hematin were restored and the resistance-reversing effect of VP was lost (Johnson et al., 2004). Both the amantadine-resistant and the halofantrine-resistant lines exhibited novel mutations of *pfcrt*, including one common mutation (S163R). Surprisingly, both lines retained the *pfcrt* K76T mutation, making these the first fully characterized examples of CQS parasite lines harboring K76T. One explanation, based on the charged drug leak model, is that the novel S163R mutation replaced a positive charge in the transporter barrel in both drug-selected lines, thereby compensating for the loss of positive charge associated with the K76T mutation (Johnson et al., 2004; Bray et al., 2005). This single amino acid change in PfCRT might block the leak of diprotonated CQ and thereby return the parasites to full CQS status, restoring access of CQ to heme and ablating VP synergism. This model predicts that the K76T mutation alters the substrate specificity of PfCRT to include CQ^{2+}. In the CQS strain harboring both the K76T and the S163R mutation, the transporter would also be predicted to recognize CQ^{2+} as a substrate; however, the presence of a positive charge in the binding motif would destabilize the CQ^{2+}-transporter complex and inhibit the translocation of CQ^{2+}. Thus, while PfCRT of CQS, wild-type parasites is predicted to neither recognize nor translocate CQ^{2+}, PfCRT containing the CQR-conferring K76T mutation could be predicted to do both, while PfCRT harboring both the K76T and S163R mutations could be predicted to recognize, but not transport, CQ^{2+}.

pfmdr1

IDENTIFICATION AND GENERAL FEATURES

The initial identification of *pfmdr1* (Foote et al., 1989; Wilson et al., 1989) was stimulated by the findings that *P. falciparum* CQR could be partially reversed by VP. This well-known calcium channel blocker displayed the important property of reversing multidrug resistance in mammalian drug-resistant cell lines, thought to be a result of VP inhibition of P-glycoproteins encoded by the multidrug resistance (*mdr*) gene family (Krogstad et al., 1987; Martin et al., 1987). *mdr* genes were thought to confer resistance as a result of gene amplification and overexpression of P-glycoprotein. In analogy to mammalian cells, *pfmdr1* was found to have been amplified in some CQ-resistant *P. falciparum* strains, as part of an extensive size polymorphism on chromosome 5 (Foote et al., 1989; Wilson et al., 1989; Wilson et al., 1993). Amplification was shown to result in elevated mRNA levels and increased expression of its protein product Pgh-1 (for P-glycoprotein homolog 1) (Cowman et al., 1991). This 162-kDa protein, localized to the DV, has 12 predicted transmembrane domains and shares sequence similarities to primary active ATP-binding cassette (ABC) transporters. These transporters, which include P-glycoproteins and the cystic fibrosis transport regulator, share the same basic structure of a tandem repeat of six transmembrane domains and a C-terminal two-nucleotide-binding domain, containing a glycine-rich Walker A consensus motif and a hydrophobic Walker B consensus motif. Intriguingly, *pfmdr1* is one of as many as 15 predicted ABC transporter genes in the *P. falciparum* genome (as listed in PlasmoDB, version 4.1).

POINT MUTATIONS IN *pfmdr1* AND DRUG SENSITIVITIES

Subsequent to the discovery of gene amplification in *pfmdr1* and its association with antimalarial drug resistance (Foote et al., 1989; Wilson et al., 1989), sequence analysis revealed a series of point mutations in drug-resistant strains (Foote et al., 1990b). The K1 (Thailand) strain encoded the *pfmdr1* N86Y mutation, whereas the 7G8 (Brazil) strain encoded Y184F, S1034C, N1042D, and D1246Y. These haplotypes were readily detected in CQ-resistant parasites from Southeast Asia and South America, suggesting a possible role for *pfmdr1* in CQR. Subsequent epidemiological studies, however, generated conflicting results. As an example, the N86Y mutation appeared to be associated with high-level CQR in some African parasites, although this mutation alone was not sufficient to predict CQ response (Babiker et al., 2001; Djimdé et al., 2001a). Studies from Thailand detected an association between N86Y and increased sensitivity to MFQ, pointing to an opposite relationship between N86Y and parasite response to these quinoline drugs (Price et al., 1999b; Price et al., 2004).

For the 3′ mutations S1034C, N1042D, and D1246Y, most studies found little or no association with CQR (Chaiyaroj et al., 1999; McCutcheon et al., 1999; Basco and Ringwald, 2002), although a low-level association was reported in some studies of Asian or African isolates (Flueck et al., 2000; Nagesha et al., 2001; Pickard et al., 2003). These mutations were found to be highly prevalent in South America, where all the strains are already CQ resistant (Povoa et al., 1998; Zalis et al., 1998; Pillai et al., 2003; Huaman et al., 2004). To date, only a few studies have addressed *pfmdr1* 3′ mutations and parasite responses to MFQ and QN. Work from Vietnam suggested that *P. falciparum* infections harboring the N1042D mutation may be more MFQ sensitive and more QN resistant (Ngo et al., 2003). A similar trend between N1042D and increased MFQ susceptibility was also observed in samples from different areas in southeast Asia (Pickard et al., 2003) and Brazil (Zalis et al., 1998), but here no association was found between these mutations and differences in QN response. A recent study from Southeast Asia reported some association between higher artesunate IC_{50} values and the presence of either the S1034C or N1042D mutations (Price et al., 2004). Here, all but 1 of more than 600 samples analyzed had a single-copy *pfmdr1*, pointing to the possibility that point mutations serve as negative markers of gene amplification (see below).

Genetically controlled investigations into the role of *pfmdr1* 3′ mutations in vitro were first performed by Reed et al. (2000). These authors observed that the introduction of the S1034C, N1042D, and D1246Y mutations into the *pfmdr1* gene of a CQS line (D10, Papua New Guinea) conferred increased resistance to QN, contrasting with increased susceptibility to MFQ and an unaltered response to CQ. When these point mutations were removed from a CQ-resistant strain (7G8), the resulting parasite clones showed decreased resistance to CQ and QN, contrasting with decreased susceptibility to MFQ. A subsequent transfection-based study also found that the *pfmdr1* 3′ mutations conferred increased MFQ and artemisinin susceptibility and identified a notable impact of the N1042D mutation in contributing to QN resistance (Sidhu et al., 2005). This study found no effect of these mutations on CQR, tested using either CQ or its metabolite monodesethyl-CQ.

pfmdr1 AMPLIFICATION

Amplification of *pfmdr1* was first detected in lines that were CQ or MFQ resistant, implicating this as a candidate drug resistance determinant (Foote et al., 1989; Wilson et al., 1989). Intriguingly, subsequent studies actually revealed that *pfmdr1* amplification tended to be found in lines that in fact were more sensitive to CQ (Cowman et al., 1994). Furthermore, CQ drug pressuring and in vitro selection of high-grade CQR resulted in lines that had undergone *pfmdr1* deamplification (Barnes et al., 1992). Separate studies selecting for MFQ resistance produced lines that had amplified their *pfmdr1* locus and that showed cross-resistance to halofantrine and decreased susceptibility to CQ (Wilson et al., 1989; Cowman et al., 1994). However, some MFQ-resistant lines did not reveal increases in *pfmdr1* copy number (Cowman et al., 1994; Price et al., 1999b), suggesting that MFQ resistance could be multifactorial. Molecular studies of the amplification of *pfmdr1* and surrounding regions of chromosome 5 found that this event could arise independently at different breakpoints, implicating this area of the genome as likely to be under intense selective pressure (Triglia et al., 1991). *pfmdr1*-containing amplicons were found to be typically arranged in a head-to-tail orientation, in contrast to mammalian cells where the most common arrangement of amplified fragments is in inverted repeating units. Taken together, these data suggested that the major mechanism of how *pfmdr1* contributes to resistance to multiple drugs is via gene amplification.

Technical difficulties in assessing *pfmdr1* copy number, however, limited its analysis in clinical studies to small sample sizes, precluding comprehensive conclusions (Chaiyaroj et al., 1999). For example, labor-intensive techniques such as Southern blotting required substantial amounts of parasite material, which are virtually impossible to obtain from field samples. The development of a tandem-competitive PCR method helped to establish an in vitro correlation between *pfmdr1* copy number and increased MFQ IC_{50} values (Price et al., 1997; Price et al., 1999b). Yet this methodology was also too time-consuming to be suitable for large field studies. The recent development of real-time PCR techniques to quantify *pfmdr1* copy number has now permitted accurate evaluation of the role of *pfmdr1* gene amplification in multidrug resistance in the field. These assays are reliable, sensitive, and rapid and can be applied to large numbers of samples. An initial, small study using real-time PCR observed that increased gene copy number (≥ 3) was associated with decreased susceptibility to MFQ and artemisinin compounds (Pickard et al., 2003). A more recent study using real-time PCR investigated a total of >600 parasite isolates for their in vitro and in vivo sensitivities to multiple antimalarials (Price et al., 2004). These samples derived from the western border of Thailand, where highly multidrug-resistant falciparum malaria first emerged and is now firmly established.

IN VITRO MOLECULAR CORRELATES

Genotyping of >180 samples collected over 10 years at the Thai-Burmese border revealed that *pfmdr1* amplification was associated with an up to 40-fold decrease in MFQ susceptibility in vitro (Fig. 6) (Price et al., 2004). Among those isolates with multiple copies of *pfmdr1*, increasing copy number was positively correlated with increasing MFQ IC_{50} values. Amplifications in *pfmdr1* could not account for about 20% of the cases of MFQ resistance. When compared to parasite isolates with a single copy of *pfmdr1*, parasites with an increased copy number had significantly higher median IC_{50} values for QN, halofantrine, artesunate, and dihydroartemisinin. For those antimalarials, differences in the number of multiple copies did not correlate with levels of parasite susceptibility. These findings suggested that the critical step in resistance is gene duplication, and that further amplifications contribute relatively less to high-grade MFQ resistance.

This study also enabled an analysis of the relationship between *pfmdr1* copy number and point mutations, in field-derived clinical samples. As illustrated in Fig. 6, almost 30% of isolates with a single *pfmdr1* copy number had mutations in codon 86, 1034, or 1042, compared to only one isolate out of 102 that had increased copy number (this isolate carried a single mutation at codon 1034). This would suggest that these point mutations serve as surrogate markers of single copy number (Price et al., 2004). The Y184F mutation presented a different scenario, being present in 43% of isolates with a single copy of *pfmdr1*, compared to 20% of cases with multiple *pfmdr1* copies (Fig. 6). The D1246Y mutation was not detected in any of the 147 samples studied and may not be relevant in Southeast Asia. Interestingly, among isolates with a single *pfmdr1* copy, the N86Y mutation was associated with lower MFQ IC_{50} values than parasites with the wild-type N86 codon. In this study, N86Y gave reasonable sensitivity but unacceptably low specificity for detecting MFQ resistance. Increased copy number in field isolates was not observed in the presence of N86Y and only occasionally occurred with mutations at positions 1034 and 1042. Of note, the observed frequency of N86Y was very low compared to that in regions in Africa. The earliest samples analyzed were collected in 1990, about 10 years after MFQ began to be widely used in Southeast Asia. Possibly, years of MFQ use may have significantly decreased the N86Y frequency. So far it remains to be determined why N86Y is restricted to single-copy *pfmdr1* in parasites collected in the field, and suggests a fitness disadvantage for N86Y under high drug pressure.

In this setting in Thailand, mutant *pfcrt* was close to 100% prevalence, perhaps as a result of the frequent use of CQ by the local population to treat *P. vivax* malaria, which sustains population CQ pressure on the local *P. falciparum* populations. The dissociation between *pfcrt* and *pfmdr1* mutations observed in Thailand contrasts with Africa, where the *pfcrt* K76T and *pfmdr1* N86Y polymorphisms are frequently associated (Hayton and Su, 2004).

Remarkably, sensitivity to artemisinins, which appear unrelated to the arylaminoalcohols in their mode of action (Eckstein-Ludwig et al., 2003), was also associated with *pfmdr1* copy number through a 25-fold range of IC_{50} values. The target of the artemisinin derivatives has recently been postulated to be pfATP6, a Ca^{2+}-transporting ATPase with homology to mammalian SERCA (Eckstein-Ludwig et al., 2003; Uhlemann et al., 2005), but a newly identified polymorphism in this gene did not reveal an association with higher IC_{50} values for artemisinins. These data suggested that *pfmdr1* copy number is the main apparent modulator of sensitivity to artesunate, although it should be emphasized that no clinically relevant resistance to this class of drug has been established. The fact that sensitivities to both MFQ and artesunate may be modulated by a common molecular mechanism may seem worrying. Paradoxically, if *pfmdr1* expression increases the export of antimalarials such as artemisinin, then this could reduce the chances of selection for artemisinin-resistant variants of *pfATP6* by maintaining a low selection pressure on this target. At first sight, MFQ and artesunate targets may be localized to different sides of membranous compartments: MFQ apparently interferes with hemoglobin digestion in the DV while artesunate has been

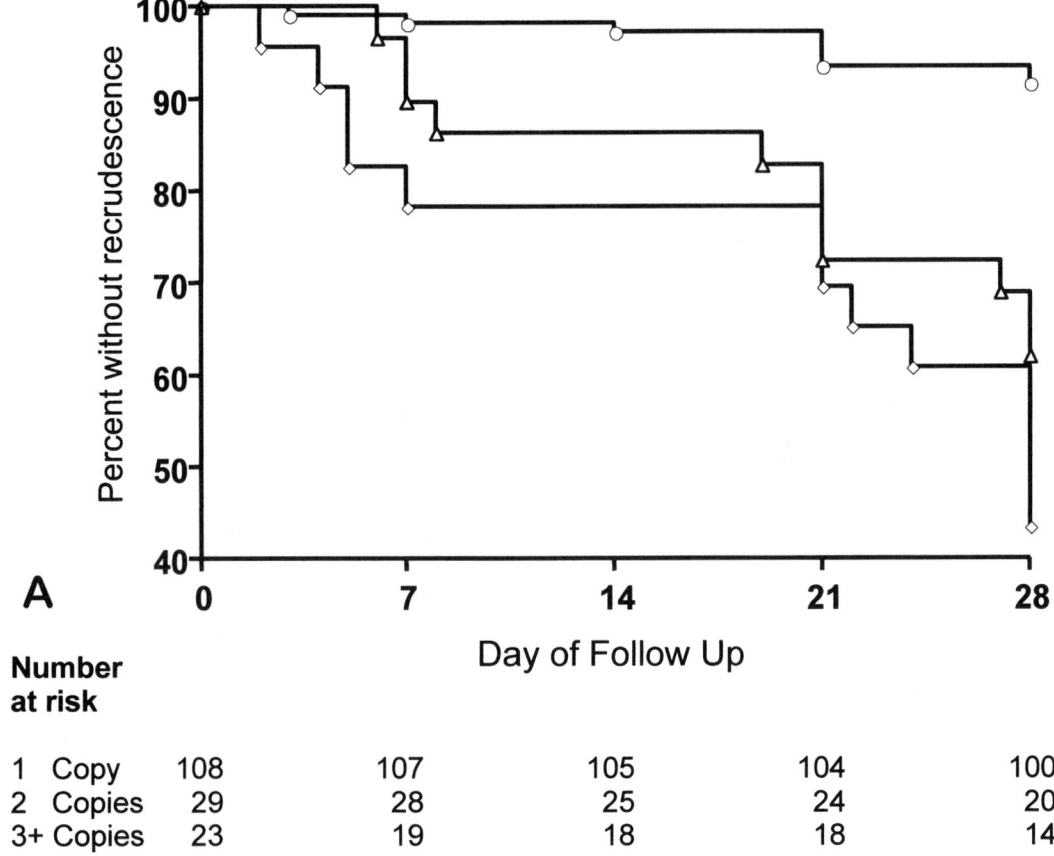

FIGURE 6 Cumulative percentage of patients free from malaria after treatment with MFQ monotherapy (A) or MFQ and 3 days of artesunate (B). Open circles, 1 copy; triangles, 2 copies; diamonds, 3+ copies. (Reprinted from *The Lancet* [Price et al., 2004] with permission from the publisher.)

proposed to act on PfATP6, which is in the parasite cytosolic membranes. Nevertheless, many details are still missing about the membrane connections (both functional and direct) between SERCA-type structures in the parasite cytosol and the membrane components of the DV.

IN VIVO MOLECULAR CORRELATES OF *pfmdr1* AND MEFLOQUINE RESISTANCE

The relationship between *pfmdr1* polymorphisms and in vivo resistance to MFQ and related antimalarials has not been extensively explored until recently. In a study from Peru, MFQ and MFQ-artesunate showed 100% clinical efficacy after 28 days of follow-up, although about half of these samples contained the *pfmdr1* D1246 codon previously linked to MFQ resistance in vitro (Pillai et al., 2003). In another study from Gabon (Mawili-Mboumba et al., 2002), no strong association was observed between point mutations and the results of treatment with ultra-low-dose MFQ. Treatment failures in this study, however, were probably due more to the treatment regimen than to any influence from *pfmdr1* mutations. Ngo et al. (2003) assessed the *pfmdr1* status and drug response profiles of parasites tested before and after 5 days of artemisinin treatment in Vietnamese patients. Interestingly, their data revealed an in-

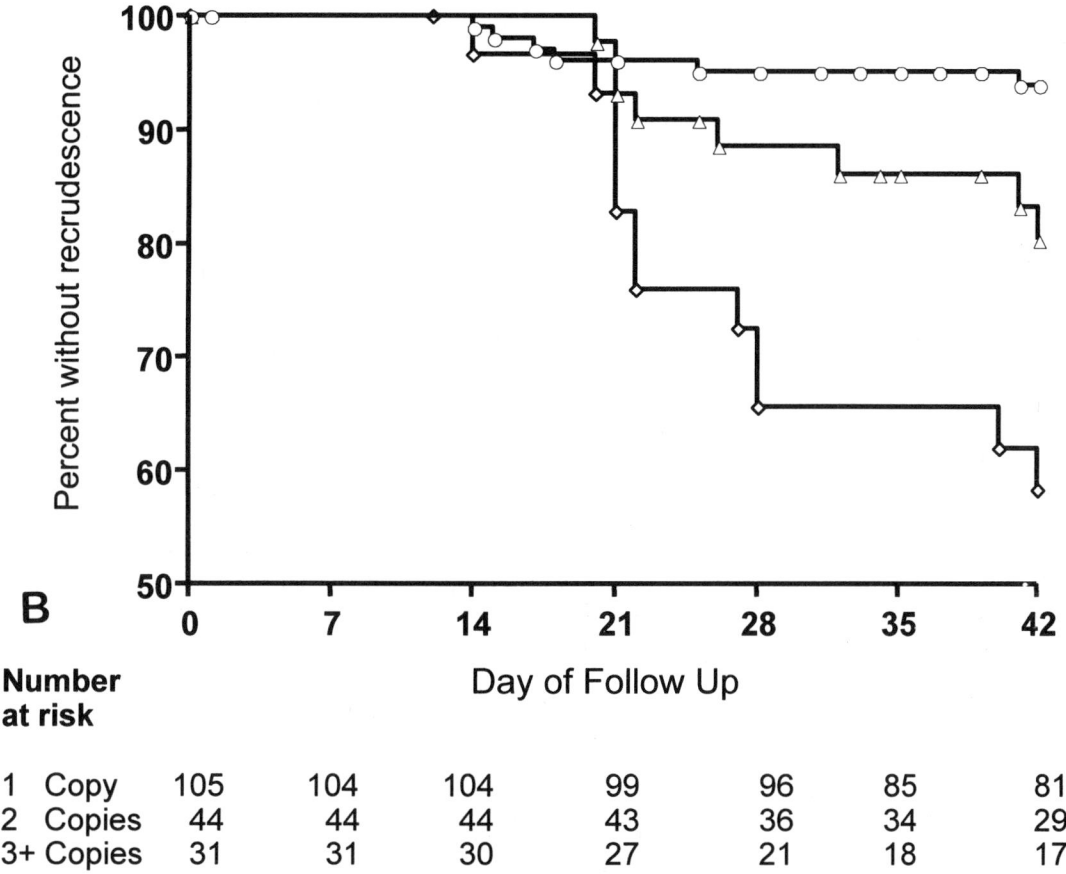

crease in MFQ IC_{50} values after artemisinin treatment and were suggestive of an association between higher MFQ IC_{50} values and the presence of the wild-type N1042 residue. Given the recent advances in understanding in vitro MFQ resistance, as discussed above, one might speculate that the increased MFQ IC_{50} values in some posttreatment samples resulted from the selection of parasites with increased *pfmdr1* copy number. In one patient, treatment also apparently selected for a different parasite population (as ascertained by PCR analysis) that was less MFQ susceptible, with parasites being N1042D pre- and N1042 posttreatment.

The recent development of real-time PCR methods made it feasible to comprehensively investigate the pharmacogenomic relationship between *pfmdr1* and in vivo therapeutic response (Pickard et al., 2003; Price et al., 2004). Analysis of blood samples analyzed from uncomplicated malaria patients living near the Thai-Burmese border found that patients harboring parasites with an increased copy number had about a 5 times higher risk of MFQ monotherapy (25 mg/kg) treatment failure compared to those infected with single-copy isolates (Color Plate 18) (Price et al., 2004). In this study, increased *pfmdr1* copy number predicted treatment failure by day 28 with a sensitivity of 71%, a specificity of 78%, and a positive predictive value of 44%. Amplification of *pfmdr1* was also correlated with decreased early therapeutic response. By day 3, one-third of the patients infected with parasites having two or more *pfmdr1* copies were still parasitemic, compared to about 12% of patients with parasites having single copies. Multiple *pfmdr1* copy number was also associated with a significantly increased risk of treatment failures by day 7. By this time, only 1.9% of patients infected with single *pfmdr1* copies had failed therapy, compared to 13.8% of those with two copies and 21.7% of those with three or more copies. Although the N86Y mutation was associated with increased MFQ susceptibility in vitro (as discussed in "Point Mutations in *pfmdr1* and Drug Sensitivities," above), it did not significantly impact in vivo responses, nor did any of the other point mutations contribute significantly to the risk of failure by day 28 after stratification by *pfmdr1* copy number. Again, the Y86 variant was restricted to single-copy parasites. Taken together, these data argue that N86 is a surrogate marker for *pfmdr1* amplification, at least in southeast Asia, and that the main genetic factor determining in vivo MFQ response is *pfmdr1* copy number. Even when taking into account the previously identified risk factors of baseline parasitemia, fever, or vomiting, increased *pfmdr1* copy number was the single most important predictor of failure at day 28 in multivariate analyses, and it alone accounted for 63% of failures in this study population. Interestingly, a threefold increase in the prevalence of parasites with amplified *pfmdr1* was observed between 1990 and 1994, when the efficacy of MFQ monotherapy dropped dramatically, providing further support for a key contribution of *pfmdr1* amplification to the spread of MFQ resistance in the field.

In response to the emergence of multidrug resistance, artemisinin-based combination treatment regimens have been systematically introduced in southeast Asia over the past decade (Nosten et al., 2000; Tran et al., 2004). These remain highly efficacious and are well tolerated and seem to promote protection against the emergence and spread of resistance. The early therapeutic response is predominantly attributable to rapid killing by the artesunate component. Residual parasites still present after 3 days are eliminated by the slower-acting, longer-lasting MFQ component. In Thailand, *pfmdr1* copy number was also found to be the best overall predictor of treatment failure after 3 days of MFQ given with artesunate (MAS3), with a sensitivity of 77% and specificity of 65% (Price et al., 2004). Again, in multivariable analysis, *pfmdr1* copy number was a better predictor of failure than several clinical predictors that had been identified in this population (such as age, parasitemia, emesis leading to poor drug absorption, and mixed infections). Genetic analysis also suggested that it was the MFQ component that was failing in the small proportion of patients who received MAS3 treatment and presented with late (>7 days) recrudescent infections. It may be some time before findings from studies such as these can be replicated in other geographical areas or that such assays can be applied prospectively to guide treatment of patients at risk of failure; nevertheless, their potential for increasing understanding of treatment failure is clear, and the data obtained suggest that *pfmdr1* copy number will become a useful tool in population-based surveillance of drug resistance.

Other Determinants of Antimalarial Drug Resistance

In the case of CQR, convincing evidence now supports the idea that *pfcrt* is the key determinant of in vitro and in vivo CQR. A more minor role may apply to *pfmdr1*, which can show a nonrandom association with mutant *pfcrt*. Presumably, mutant *pfmdr1* may serve to either augment the degree of CQR or compensate for altered function of mutant PfCRT (bearing in mind that both proteins are located on the DV membrane and may well have related functions). Genome-wide single nucleotide polymorphism (SNP) analyses recently identified some possible associations between decreased sensitivity to CQ and SNPs in several putative transporters, in addition to *pfcrt* and *pfmdr1* (Mu et al., 2003). Significant linkage disequilibria were also detected between some genes. These data suggested that some of these genes could account for the variance in CQ IC_{50} values observed between lines carrying the same *pfcrt* and *pfmdr1* alleles. These associations have not been confirmed, however, nor do they necessarily imply that these other genes actually contribute to decreased drug susceptibility levels. One tantalizing possibility is that some of these linkages reflect selection for alleles of other genes either that compensate for altered physiological function of mutant PfCRT inside the iRBC or that enhance the transmission of parasites carrying mutant *pfcrt*. The existence of transmission-enhancing genes is supported by recent studies from The Gambia indicating a transmission advantage for gametocytes that carry either *pfcrt* or *pfmdr1* (Sutherland et al., 2002).

For QN resistance, both *pfmdr1* and *pfcrt* have been found by allelic exchange strategies to contribute important roles (Reed et al., 2000; Sidhu et al., 2002; Lakshmanan et al., 2005). Quantitative trait loci analysis of the progeny of the HB3 × Dd2 cross identified *pfcrt* and (to a lesser extent) *pfmdr1* as components of low-level QN resistance and identified an additional locus mapping to chromosome 13 (Ferdig et al., 2004). This linkage group contained the *P. falciparum* sodium-proton exchanger (PfNHE), which also revealed intriguing associations with QN resistance in culture-adapted geographically diverse field isolates. Further work may well uncover yet more genes contributing to a multifactorial basis of QN resistance.

In the case of MFQ resistance, the study by Price et al. (2004) convincingly attributes a very significant role for *pfdmr1* amplification in causing treatment failure. Again, there are likely to be additional determinants that contribute to some instances of MFQ resistance, although the speed at which this resistance can arise in the field argues for a less complex basis of resistance than appears to be the case with QN. Identifying these genetic factors is a complex task for *Plasmodium* but will benefit from recent significant advances in *Plasmodium* genetics and genomics, including high-density microsatellite scanning, SNP analysis, real-time quantitative PCR for amplification events, microarray and proteomic capacities, and transfection-based transgene expression and allelic exchange studies.

CONCLUDING REMARKS

Tremendous progress has been made in the past few years in elucidating modes of action and mechanisms of resistance to the antifolate and quinoline-based classes of antimalarials. We now know the mutations in *dhps* and *dhfr* that mediate resistance to SP, and diagnostic screens have been developed to rapidly assess for resistance in field isolates. Biochemical studies have assessed properties of these enzyme inhibitors and also improved our understanding of the fitness cost associated with these mutations. Molecular studies have proven the role of these point mutations and have also defined where resistance arose and spread. Similarly, for the quinoline-based compounds including CQ, QN, and MFQ, in vitro and clinical studies have clearly defined a determining role for *pfcrt* in CQR and an important role for this gene in contributing to QN resistance. Whereas *pfmdr1* was thought earlier to be primarily be involved in CQR, extensive studies now reveal a relatively minor role in CQR, contrasting with a more significant role in QN and MFQ resistance, with amplification of this gene now clearly implicated in MFQ treatment failure. Biochemical studies, some involving genetically modified clones, have also enabled a clearer understanding of the biochemical basis of CQR.

Extensive work is still required to answer a multitude of pressing questions, including a more comprehensive understanding of the genetic basis of QN resistance, elucidation of PfCRT and Pgh-1 structures and functions, an explanation of the genetic basis of the observed increased transmissibility of parasites carrying mutant *pfcrt* or *pfmdr1* alleles, and insights into the immune effector mechanisms that contribute to in vivo clearance of drug-resistant infections. Besides providing a fascinating area of biological investigation, research in the area of antimalarial modes of action and mechanisms of resistance must continue to focus on applications that include improved diagnosis of drug-resistant infections and the discovery, development, and distribution of new antimalarials that benefit from existing knowledge and that circumvent current resistance mechanisms. Sadly, despite the tremendous gains in knowledge in this area, the reality is that malaria morbidity and mortality rates are not being reduced and may well dramatically worsen with the aggressive spread of CQ- and SP-resistant *P. falciparum* and the lack of affordable alternatives. Focused and applied efforts from the academic, industrial, funding, and health care sectors on a significantly greater scale are vital to achieving any success in reducing the devastating impact that malaria maintains on the poorest nations of this world.

REFERENCES

Aikawa, M. 1972. High-resolution autoradiography of malarial parasites treated with ^{3}H-chloroquine. *Am. J. Pathol.* **67:**277–284.

Ariey, F., J. B. Duchemin, and V. Robert. 2003. Metapopulation concepts applied to falciparum malaria and their impacts on the emergence and spread of chloroquine resistance. *Infect. Genet. Evol.* **2:**185–192.

Atreya, C. E., and K. S. Anderson. 2004. Kinetic characterization of bifunctional thymidylate synthase-dihydrofolate reductase (TS-DHFR) from Cryptosporidium hominis: a paradigm shift for ts activity and channeling behavior. *J. Biol. Chem.* **279:**18314–18322.

Babiker, H. A., S. J. Pringle, A. Abdel-Muhsin, M. Mackinnon, P. Hunt, and D. Walliker. 2001. High-level chloroquine resistance in Sudanese isolates of Plasmodium falciparum is associated with mutations in the chloroquine resistance transporter gene pfcrt and the multidrug resistance gene pfmdr1. *J. Infect. Dis.* **183:**1535–1538.

Baird, J. K. 2004. Chloroquine resistance in *Plasmodium vivax*. *Antimicrob. Agents Chemother.* **48:**4075–4083.

Barnes, D. A., S. J. Foote, D. Galatis, D. J. Kemp, and A. F. Cowman. 1992. Selection for high-level chloroquine resistance results in deamplification of the pfmdr 1 gene and increased sensitivity to mefloquine in Plasmodium falciparum. *EMBO J.* **11:**3067–3075.

Basco, L. K., and P. Ringwald. 2002. Molecular epidemiology of malaria in Cameroon. X. Evaluation of PFMDR1 mutations as genetic markers for resistance to amino alcohols and artemisinin derivatives. *Am. J. Trop. Med. Hyg.* **66:**667–671.

Bennett, T. N., A. D. Kosar, L. M. Ursos, S. Dzekunov, A. B. S. Sidhu, D. A. Fidock, and P. D. Roepe. 2004. Drug resistance-associated pfCRT mutations confer decreased *Plasmodium falciparum* digestive vacuolar pH. *Mol. Biochem. Parasitol.* **133:**99–114.

Berglez, J., P. Iliades, W. Sirawaraporn, P. Coloe, and I. Macreadie. 2004. Analysis in Escherichia coli of Plasmodium falciparum dihydropteroate synthase (DHPS) alleles implicated in resistance to sulfadoxine. *Int. J. Parasitol.* **34:**95–100.

Brasseur, P., J. Kouamouo, R. Moyou-Somo, and P. Druilhe. 1992. Multi-drug resistant falciparum malaria in Cameroon in 1987–1988. I. Stable figures of prevalence of chloroquine- and quinine-resistant isolates in the original foci. *Am. J. Trop. Med. Hyg.* **46:**1–7.

Bray, P. G., R. E. Howells, and S. A. Ward. 1992. Vacuolar acidification and chloroquine sensitivity in Plasmodium falciparum. *Biochem. Pharmacol.* **43:**1219–1227.

Bray, P. G., S. R. Hawley, M. Mungthin, and S. A. Ward. 1996a. Physicochemical properties correlated with drug resistance and the reversal of drug resistance in Plasmodium falciparum. *Mol. Pharmacol.* **50:**1559–1566.

Bray, P. G., S. R. Hawley, and S. A. Ward. 1996b. 4-Aminoquinoline resistance of Plasmodium falciparum: insights from the study of amodiaquine uptake. *Mol. Pharmacol.* **50:**1551–1558.

Bray, P. G., M. Mungthin, R. G. Ridley, and S. A. Ward. 1998. Access to hematin: the basis of chloroquine resistance. *Mol. Pharmacol.* **54:**170–179.

Bray, P. G., K. J. Saliba, J. D. Davies, D. G. Spiller, M. R. White, K. Kirk, and S. A. Ward. 2002. Distribution of acridine orange fluorescence in *Plasmodium falciparum*-infected erythrocytes and its implications for the evaluation of digestive vacuole pH. *Mol. Biochem. Parasitol.* **119:**301–304.

Bray, P. G., R. E. Martin, L. Tilley, S. A. Ward, K. Kirk, and D. A. Fidock. 2005. Defining the role of PfCRT in *P. falciparum* chloroquine resistance. *Mol. Microbiol.* **56:**323–333.

Brooks, D. R., P. Wang, M. Read, W. M. Watkins, P. F. Sims, and J. E. Hyde. 1994. Sequence variation of the hydroxymethyldihydropterin pyrophosphokinase: dihydropteroate synthase gene in lines of the human malaria parasite, Plasmodium falciparum, with differing resistance to sulfadoxine. *Eur. J. Biochem.* **224:**397–405.

Bruce-Chwatt, L. J. (ed.). 1981. *Chemotherapy of Malaria*, 2nd ed. World Health Organisation, Geneva, Switzerland.

Canfield, C. J., W. K. Milhous, A. L. Ager, R. N. Rossan, T. R. Sweeney, N. J. Lewis, and D. P. Jacobus. 1993. PS-15: a potent, orally active antimalarial from a new class of folic acid antagonists. *Am. J. Trop. Med. Hyg.* **49:**121–126.

Chaiyaroj, S. C., A. Buranakiti, P. Angkasekwinai, S. Looressuwan, and A. F. Cowman. 1999. Analysis of mefloquine resistance and amplification of *pfmdr1* in multidrug-resistant *Plasmodium falciparum* isolates from Thailand. *Am. J. Trop. Med. Hyg.* **61:**780–783.

Chen, N., B. Russell, J. Staley, B. Kotecka, P. Nasveld, and Q. Cheng. 2001. Sequence polymorphisms in pfcrt are strongly associated with chloroquine resistance in Plasmodium falciparum. *J. Infect. Dis.* **183:**1543–1545.

Chen, N., D. E. Kyle, C. Pasay, E. V. Fowler, J. Baker, J. M. Peters, and Q. Cheng. 2003. pfcrt allelic types with two novel amino acid mutations in chloroquine-resistant *Plasmodium falciparum* isolates from the Philippines. *Antimicrob. Agents Chemother.* **47:**3500–3505.

Chen, N., D. W. Wilson, C. Pasay, D. Bell, L. B. Martin, D. Kyle, and Q. Cheng. 2005. Origin and dissemination of chloroquine-resistant *Plasmodium falciparum* with mutant *pfcrt* alleles in the Philippines. *Antimicrob. Agents Chemother.* **49:**2102–2105.

Chevli, R., and C. D. Fitch. 1982. The antimalarial drug mefloquine binds to membrane phospholipids. *Antimicrob. Agents Chemother.* **21:**581–586.

Chou, A. C., R. Chevli, and C. D. Fitch. 1980. Ferriprotoporphyrin IX fulfills the criteria for identification as the chloroquine receptor of malaria parasites. *Biochemistry* **19:**1543–1549.

Chou, A. C., and C. D. Fitch. 1993. Control of heme polymerase by chloroquine and other quinoline derivatives. *Biochem. Biophys. Res. Commun.* **195:**422–427.

Chusacultanachai, S., P. Thiensathit, B. Tarnchompoo, W. Sirawaraporn, and Y. Yuthavong. 2002. Novel antifolate resistant mutations of Plasmodium falciparum dihydrofolate reductase selected in Escherichia coli. *Mol. Biochem. Parasitol.* **120:**61–72.

Cooper, R. A., M. T. Ferdig, X. Z. Su, L. M. Ursos, J. Mu, T. Nomura, H. Fujioka, D. A. Fidock, P. D. Roepe, and T. E. Wellems. 2002. Alternative mutations at position 76 of the vacuolar transmembrane protein PfCRT are associated with chloroquine resistance and unique stereospecific quinine and quinidine responses in Plasmodium falciparum. *Mol. Pharmacol.* **61:**35–42.

Cowman, A. F., M. J. Morry, B. A. Biggs, G. A. Cross, and S. J. Foote. 1988. Amino acid changes linked to pyrimethamine resistance in the dihydrofolate reductase-thymidylate synthase gene of Plasmodium falciparum. *Proc. Natl. Acad. Sci. USA* **85:**9109–9113.

Cowman, A. F., S. Karcz, D. Galatis, and J. G. Culvenor. 1991. A P-glycoprotein homologue of *Plasmodium falciparum* is localized on the digestive vacuole. *J. Cell Biol.* **113:**1033–1042.

Cowman, A. F., D. Galatis, and J. K. Thompson. 1994. Selection for mefloquine resistance in *Plasmodium falciparum* is linked to amplification of the *pfmdr1* gene and cross-resistance to halofantrine and quinine. *Proc. Natl. Acad. Sci. USA* **91:**1143–1147.

Cowman, A. F. 1998. The molecular basis of resistance to the sulfones, sulfonamides and dihydrofolate reductase inhibitors, p. 317–330. *In* I. W. Sherman (ed.), *Malaria: Parasite Biology, Pathogenesis and Protection*. ASM Press, Washington, D.C.

Djimdé, A., O. K. Doumbo, J. F. Cortese, K. Kayentao, S. Doumbo, Y. Diourte, A. Dicko, X. Z. Su, T. Nomura, D. A. Fidock, T. E. Wellems, and C. V. Plowe. 2001a. A molecular marker for chloroquine-resistant falciparum malaria. *N. Engl. J. Med.* **344:**257–263.

Djimdé, A., O. K. Doumbo, R. W. Steketee, and C. V. Plowe. 2001b. Application of a molecular marker for surveillance of chloroquine-resistant falciparum malaria. *Lancet* **358:**890–891.

Djimdé, A. A., O. K. Doumbo, O. Traore, A. B. Guindo, K. Kayentao, Y. Diourte, S. Niare-Doumbo, D. Coulibaly, A. K. Kone, Y. Cissoko, M. Tekete, B. Fofana, A. Dicko, D. A. Diallo, T. E. Wellems, D. Kwiatkowski, and C. V. Plowe. 2003. Clearance of drug-resistant parasites as a model for protective immunity in *Plasmodium falciparum* malaria. *Am. J. Trop. Med. Hyg.* **69:**558–563.

Dorn, A., S. R. Vippagunta, H. Matile, C. Jaquet, J. L. Vennerstrom, and R. G. Ridley. 1998. An assessment of drug-haematin binding as a mechanism for inhibition of haematin polymerisation by quinoline antimalarials. *Biochem. Pharmacol.* **55:**727–736.

Dorsey, G., M. R. Kamya, G. Ndeezi, J. N. Babirye, C. R. Phares, J. E. Olson, E. T. Katabira, and P. J. Rosenthal. 2000. Predictors of chloroquine treatment failure in children and adults with falciparum malaria in Kampala, Uganda. *Am. J. Trop. Med. Hyg.* **62:**686–692.

Dubois, V. L., D. F. Platel, G. Pauly, and J. Tribouley-Duret. 1995. Plasmodium berghei: implication of intracellular glutathione and its related enzyme in chloroquine resistance in vivo. *Exp. Parasitol.* **81:**117–124.

Durrand, V., A. Berry, R. Sem, P. Glaziou, J. Beaudou, and T. Fandeur. 2004. Variations in the sequence and expression of the Plasmodium falciparum chloroquine resistance transporter (Pfcrt) and their relationship to chloroquine resistance in vitro. *Mol. Biochem. Parasitol.* **136:**273–285.

Dzekunov, S. M., L. M. B. Ursos, and P. D. Roepe. 2000. Digestive vacuolar pH of intact intraerythrocytic *P. falciparum* either sensitive or resistant to chloroquine. *Mol. Biochem. Parasitol.* **110:**107–124.

Eckstein-Ludwig, U., R. J. Webb, I. D. Van Goethem, J. M. East, A. G. Lee, M. Kimura, P. M. O'Neill, P. G. Bray, S. A. Ward, and S. Krishna. 2003. Artemisinins target the SERCA of Plasmodium falciparum. *Nature* **424:**957–961.

Eskelinen, E. L., Y. Tanaka, and P. Saftig. 2003. At the acidic edge: emerging functions for lysosomal membrane proteins. *Trends Cell Biol.* **13:**137–145.

Ferdig, M. T., R. A. Cooper, J. Mu, B. Deng, D. A. Joy, X. Z. Su, and T. E. Wellems. 2004. Dissecting the loci of low-level quinine resistance in malaria parasites. *Mol. Microbiol.* **52:**985–997.

Ferlan, J. T., S. Mookherjee, I. N. Okezie, L. Fulgence, and C. H. Sibley. 2001. Mutagenesis of dihydrofolate reductase from Plasmodium falciparum: analysis in Saccharomyces cerevisiae of triple mutant alleles resistant to pyrimethamine or WR99210. *Mol. Biochem. Parasitol.* **113:**139–150.

Fidock, D. A., and T. E. Wellems. 1997. Transformation with human dihydrofolate reductase renders malaria parasites insensitive to WR99210 but does not affect the intrinsic activity of proguanil. *Proc. Natl. Acad. Sci. USA* **94:**10931–10936.

Fidock, D. A., T. V. Nguyen, H. Dodemont, W. M. Eling, and A. A. James. 1998. *Plasmodium falciparum*: identification of the ribosomal P2 protein gene whose expression is independent of the developmentally-regulated rRNAs. *Exp. Parasitol.* **89:**125–128.

Fidock, D. A., T. Nomura, R. A. Cooper, X.-Z. Su, A. K. Talley, and T. E. Wellems. 2000a. Allelic modifications of the cg2 and cg1 genes do not alter the chloroquine response of drug-resistant Plasmodium falciparum. *Mol. Biochem. Parasitol.* **110:**1–10.

Fidock, D. A., T. Nomura, A. K. Talley, R. A. Cooper, S. M. Dzekunov, M. T. Ferdig, L. M. Ursos, A. B. S. Sidhu, B. Naude, K. Deitsch, X.-Z. Su, J. C. Wootton, P. D. Roepe, and T. E. Wellems. 2000b. Mutations in the *P. falciparum* digestive vacuole transmembrane protein PfCRT and evidence for their role in chloroquine resistance. *Mol. Cell* **6:**861–871.

Fidock, D. A., P. J. Rosenthal, S. L. Croft, R. Brun, and S. Nwaka. 2004. Antimalarial drug discovery: efficacy models for compound screening. *Nat. Rev. Drug. Discov.* **3:**509–520.

Fitch, C. D. 2004. Ferriprotoporphyrin IX, phospholipids, and the antimalarial actions of quinoline drugs. *Life Sci.* **74:**1957–1972.

Flueck, T. P., T. Jelinek, A. H. Kilian, I. S. Adagu, G. Kabagambe, F. Sonnenburg, and D. C. Warhurst. 2000. Correlation of in vivo-resistance to chloroquine and allelic polymorphisms in Plasmodium falciparum isolates from Uganda. *Trop. Med. Int. Health* **5:**174–178.

Foote, S. J., J. K. Thompson, A. F. Cowman, and D. J. Kemp. 1989. Amplification of the multidrug resistance gene in some chloroquine-resistant isolates of P. falciparum. *Cell* **57:**921–930.

Foote, S. J., D. Galatis, and A. F. Cowman. 1990a. Amino acids in the dihydrofolate reductase-thymidylate synthase gene of Plasmodium falciparum involved in cycloguanil resistance differ from those involved in pyrimethamine resistance. *Proc. Natl. Acad. Sci. USA* **87:**3014–3017.

Foote, S. J., D. E. Kyle, R. K. Martin, A. M. Oduola, K. Forsyth, D. J. Kemp, and A. F. Cowman. 1990b. Several alleles of the multidrug-resistance gene are closely linked to chloroquine resistance in Plasmodium falciparum. *Nature* **345:**255–258.

Geary, T. G., L. C. Bonanni, J. B. Jensen, and H. Ginsburg. 1986. Effects of combinations of quinoline-containing antimalarials on Plasmodium falciparum in culture. *Ann. Trop. Med. Parasitol.* **80:**285–291.

Ginsburg, H., O. Famin, J. Zhang, and M. Krugliak. 1998. Inhibition of glutathione-dependent degradation of heme by chloroquine and amodiaquine as a possible basis for their antimalarial mode of action. *Biochem. Pharmacol.* **56:**1305–1313.

Ginsburg, H., S. A. Ward, and P. G. Bray. 1999. An integrated model of chloroquine action. *Parasitol. Today* **15:**357–360.

Greenwood, B., and T. Mutabingwa. 2002. Malaria in 2002. *Nature* **415:**670-672.

Gregson, A., and C. V. Plowe. 2005. Mechanisms of resistance of malaria parasites to antifolates. *Pharmacol. Rev.* **57:**117–145.

Hastings, M. D., S. J. Bates, E. A. Blackstone, S. M. Monks, T. K. Mutabingwa, and C. H. Sibley. 2002. Highly pyrimethamine-resistant alleles of dihydrofolate reductase in isolates of Plasmodium falciparum from Tanzania. *Trans. R. Soc. Trop. Med. Hyg.* **96:**674–676.

Hayton, K., and X. Z. Su. 2004. Genetic and biochemical aspects of drug resistance in malaria parasites. *Curr. Drug Targets Infect. Disord.* **4:**1–10.

Homewood, C. A., D. C. Warhurst, W. Peters, and V. C. Baggaley. 1972. Lysosomes, pH and the antimalarial action of chloroquine. *Nature* **235:**50–52.

Huaman, M. C., N. Roncal, S. Nakazawa, T. T. Long, L. Gerena, C. Garcia, L. Solari, A. J. Magill, and H. Kanbara. 2004. Polymorphism of the Plasmodium falciparum multidrug resistance and chloroquine resistance transporter genes and in vitro susceptibility to aminoquinolines in isolates from the Peruvian Amazon. *Am. J. Trop. Med. Hyg.* **70:**461–466.

Hunt, S. Y., B. B. Rezvani, and C. H. Sibley. 2005. Novel alleles of Plasmodium falciparum dhfr that confer resistance to chlorcycloguanil. *Mol. Biochem. Parasitol.* **139:**25–32.

Hyde, J. E. 2002. Mechanisms of resistance of Plasmodium falciparum to antimalarial drugs. *Mol. Biochem. Parasitol.* **4:**165–174.

Jack, D. L., N. M. Yang, and M. H. Saier, Jr. 2001. The drug/metabolite transporter superfamily. *Eur. J. Biochem.* **268:**3620–3639.

Jacobs, G. H., A. M. Oduola, D. E. Kyle, W. K. Milhous, S. K. Marin, and M. Aikawa. 1988. Ultrastructural study of the effects of chloroquine and verapmil on Plasmodium falciparum. *Am. J. Trop. Med. Hyg.* **39:**15–20.

Jensen, N. P., A. L. Ager, R. A. Bliss, C. J. Canfield, B. M. Kotecka, K. H. Rieckmann, J. Terpinski, and D. P. Jacobus. 2001. Phenoxypropoxybiguanides, prodrugs of DHFR-inhibiting diaminotriazine antimalarials. *J. Med. Chem.* **44:**3925–3931.

Johnson, D. J., D. A. Fidock, M. Mungthin, V. Lakshmanan, A. B. S. Sidhu, P. G. Bray, and S. A. Ward. 2004. Evidence for a central role for PfCRT in conferring *Plasmodium falciparum* resistance to diverse antimalarial agents. *Mol. Cell* **15:**867–877.

Kamchonwongpaisan, S., R. Quarrell, N. Charoensetakul, R. Ponsinet, T. Vilaivan, J. Vanichtanankul, B. Tarnchompoo, W. Sirawaraporn, G. Lowe, and Y. Yuthavong. 2004. Inhibitors of multiple mutants of Plasmodium falciparum dihydrofolate reductase and their antimalarial activities. *J. Med. Chem.* **47:**673–680.

Kirk, K., H. A. Horner, and J. Kirk. 1996. Glucose uptake in Plasmodium falciparum-infected erythrocytes is an equilibrative not an active process. *Mol. Biochem. Parasitol.* **82:**195–205.

Knighton, D. R., C. C. Kan, E. Howland, C. A. Janson, Z. Hostomska, K. M. Welsh, and D. A. Matthews. 1994. Structure of and kinetic channelling in bifunctional dihydrofolate reductase-thymidylate synthase. *Nat. Struct. Biol.* **1:**186–194.

Kremsner, P. G., S. Winkler, C. Brandts, S. Neifer, U. Bienzle, and W. Graninger. 1994. Clindamycin in combination with chloroquine or quinine is an effective therapy for uncomplicated Plasmodium falciparum malaria in children from Gabon. *J. Infect. Dis.* **169:**467–470.

Krishna, S., A. C. Uhlemann, and R. K. Haynes. 2004. Artemisinins: mechanisms of action and potential for resistance. *Drug Resist. Updat.* **7:**233–244.

Krogstad, D. J., P. H. Schlesinger, and I. Y. Gluzman. 1985. Antimalarials increase vesicle pH in Plasmodium falciparum. *J. Cell Biol.* **101:**2302–2309.

Krogstad, D. J., I. Y. Gluzman, D. E. Kyle, A. M. Oduola, S. K. Martin, W. K. Milhous, and P. H. Schlesinger. 1987. Efflux of chloroquine from *Plasmodium falciparum*: mechanism of chloroquine resistance. *Science* **238:**1283–1285.

Krogstad, D. J., and P. H. Schlesinger. 1987. The basis of antimalarial action: non-weak base effects of chloroquine on acid vesicle pH. *Am. J. Trop. Med. Hyg.* **36:**213–220.

Krogstad, D. J., I. Y. Gluzman, B. L. Herwaldt, P. H. Schlesinger, and T. E. Wellems. 1992. Energy dependence of chloroquine accumulation and chloroquine efflux in Plasmodium falciparum. *Biochem. Pharmacol.* **43:**57–62.

Krudsood, S., M. Imwong, P. Wilairatana, S. Pukrittayakamee, A. Nonprasert, G. Snounou, N. J. White, and S. Looareesuwan. 2005. Artesunate-dapsone-proguanil treatment of falciparum malaria: genotypic determinants of therapeutic response. *Trans. R. Soc. Trop. Med. Hyg.* **99:**142–149.

Lakshmanan, V., P. G. Bray, D. Verdier-Pinard, D. J. Johnson, P. Horrocks, R. A. Muhle, G. E. Alakpa, R. H. Hughes, D. J. Krogstad, A. B. S. Sidhu, and D. A. Fidock. A critical role for PfCRT K76T in *Plasmodium falciparum* verapamil-reversible chloroquine resistance. *EMBO J.*, in press.

Lang, T., and B. Greenwood. 2003. The development of Lapdap, an affordable new treatment for malaria. *Lancet Infect. Dis.* **3:**162–168.

Lemcke, T., I. T. Christensen, and F. S. Jorgensen. 1999. Towards an understanding of drug resistance in malaria: three-dimensional structure of Plasmodium falciparum dihydrofolate reductase by homology building. *Bioorg. Med. Chem.* **7:**1003–1011.

Macomber, P. B., R. L. O'Brien, and F. E. Hahn. 1966. Chloroquine: physiological basis of drug resistance in Plasmodium berghei. *Science* **152:**1374–1375.

Martin, R. E., and K. Kirk. 2004. The malaria parasite's chloroquine resistance transporter is a member of the drug/metabolite transporter superfamily. *Mol. Biol. Evol.* **21:**1938–1949.

Martin, S. K., A. M. Oduola, and W. K. Milhous. 1987. Reversal of chloroquine resistance in *Plasmodium falciparum* by verapamil. *Science* **235:**899–901.

Mawili-Mboumba, D. P., J. F. Kun, B. Lell, P. G. Kremsner, and F. Ntoumi. 2002. Pfmdr1 alleles and response to ultralow-dose mefloquine treatment in Gabonese patients. *Antimicrob. Agents Chemother.* **46:**166–170.

McCutcheon, K. R., J. A. Freese, J. A. Frean, B. L. Sharp, and M. B. Markus. 1999. Two mutations in the multidrug-resistance gene homologue of Plasmodium falciparum, pfmdr1, are not useful predictors of in-vivo or in-vitro chloroquine resistance in southern Africa. *Trans. R. Soc. Trop. Med. Hyg.* **93:**300–302.

Meek, T. D., E. P. Garvey, and D. V. Santi. 1985. Purification and characterization of the bifunctional thymidylate synthetase-dihydrofolate reductase from methotrexate-resistant Leishmania tropica. *Biochemistry* **24:**678–686.

Mehlotra, R. K., H. Fujioka, P. D. Roepe, O. Janneh, L. M. Ursos, V. Jacobs-Lorena, D. T. McNamara, M. J. Bockarie, J. W. Kazura, D. E. Kyle, D. A. Fidock, and P. A. Zimmerman. 2001. Evolution of a unique Plasmodium falciparum chloroquine-resistance phenotype in association with pfcrt polymorphism in Papua New Guinea and South America. *Proc. Natl. Acad. Sci. USA* **98:**12689–12694.

Mockenhaupt, F. P. 1995. Mefloquine resistance in Plasmodium falciparum. *Parasitol. Today* **11:**248–253.

Mu, J., M. T. Ferdig, X. Feng, D. A. Joy, J. Duan, T. Furuya, G. Subramanian, L. Aravind, R. A. Cooper, J. C. Wootton, M. Xiong, and X. Z. Su. 2003. Multiple transporters associated with malaria parasite responses to chloroquine and quinine. *Mol. Microbiol.* **49:**977–989.

Mutabingwa, T. K., C. A. Maxwell, I. G. Sia, F. H. Msuya, S. Mkongewa, S. Vannithone, J. Curtis, and C. F. Curtis. 2001. A trial of proguanil-dapsone in comparison with sulfadoxine-pyrimethamine for the clearance of Plasmodium falciparum infections in Tanzania. *Trans. R. Soc. Trop. Med. Hyg.* **95:**433–438.

Nagesha, H. S., S. Din, G. J. Casey, A. I. Susanti, D. J. Fryauff, J. C. Reeder, and A. F. Cowman. 2001. Mutations in the pfmdr1, dhfr and dhps genes of Plasmodium falciparum are associated with in-vivo drug resistance in West Papua, Indonesia. *Trans. R. Soc. Trop. Med. Hyg.* **95:**43–49.

Nagesha, H. S., G. J. Casey, K. H. Rieckmann, D. J. Fryauff, B. S. Laksana, J. C. Reeder, J. D. Maguire, and J. K. Baird. 2003. New haplotypes of the *Plasmodium falciparum* chloroquine resistance transporter (*pfcrt*) gene among chloroquine-resistant parasite isolates. *Am. J. Trop. Med. Hyg.* **68:**398–402.

Nateghpour, M., S. A. Ward, and R. E. Howells. 1993. Development of halofantrine resistance and determination of cross-resistance patterns in *Plasmodium falciparum*. *Antimicrob. Agents Chemother.* **37:**2337–2343.

Naude, B., J. A. Brzostowski, A. R. Kimmel, and T. E. Wellems. *Dictyostelium discoideum* expresses a malaria chloroquine resistance mechanism upon transfection with mutant, but not wild-type, *Plasmodium falciparum* transporter PfCRT. *J. Biol. Chem.*, in press.

Nessler, S., O. Friedrich, N. Bakouh, R. H. Fink, C. P. Sanchez, G. Planelles, and M. Lanzer. 2004. Evidence for activation of endogenous transporters in Xenopus laevis oocytes expressing the Plasmodium falciparum chloroquine resistance transporter, PfCRT. *J. Biol. Chem.* **279:**39438–39446.

Ngo, T., M. Duraisingh, M. Reed, D. Hipgrave, B. Biggs, and A. F. Cowman. 2003. Analysis of pfcrt, pfmdr1, dhfr, and dhps mutations and drug sensitivities in Plasmodium falciparum isolates from patients in Vietnam before and after treatment with artemisinin. *Am. J. Trop. Med. Hyg.* **68:**350–356.

Nosten, F., F. ter Kuile, T. Chongsuphajaisiddhi, C. Luxemburger, H. K. Webster, M. Edstein, L. Phaipun, K. L. Thew, and N. J. White. 1991. Mefloquine-resistant falciparum malaria on the Thai-Burmese border. *Lancet* **337:**1140–1143.

Nosten, F., M. van Vugt, R. Price, C. Luxemburger, K. L. Thway, A. Brockman, R. McGready, F. ter Kuile, S. Looareesuwan, and N. J. White. 2000. Effects of artesunate-mefloquine combination on incidence of Plasmodium falciparum malaria and mefloquine resistance in western Thailand: a prospective study. *Lancet* **356:**297–302.

O'Neil, R. H., R. H. Lilien, B. R. Donald, R. M. Stroud, and A. C. Anderson. 2003. The crystal structure of dihydrofolate reductase-thymidylate synthase from Cryptosporidium hominis reveals a novel architecture for the bifunctional enzyme. *J. Eukaryot. Microbiol.* **50**(Suppl):555–556.

O'Neill, P. M., and G. H. Posner. 2004. A medicinal chemistry perspective on artemisinin and related endoperoxides. *J. Med. Chem.* **47:**2945–2964.

Pagola, S., P. W. Stephens, D. S. Bohle, A. D. Kosar, and S. K. Madsen. 2000. The structure of malaria pigment beta-haematin. *Nature* **404:**307–310.

Peters, W. 1970. *Chemotherapy and Drug Resistance in Malaria.* Academic Press, London, United Kingdom.

Peters, W., R. E. Howells, J. Portus, B. L. Robinson, S. Thomas, and D. C. Warhurst. 1977. The chemotherapy of rodent malaria, XXVII. Studies on mefloquine (WR 142,490). *Ann. Trop. Med. Parasitol.* **71:**407–418.

Peterson, D. S., D. Walliker, and T. E. Wellems. 1988. Evidence that a point mutation in dihydrofolate reductase-thymidylate synthase confers resistance to pyrimethamine in falciparum malaria. *Proc. Natl. Acad. Sci. USA* **85:**9114–9118.

Peterson, D. S., W. K. Milhous, and T. E. Wellems. 1990. Molecular basis of differential resistance to cycloguanil and pyrimethamine in Plasmodium falciparum malaria. *Proc. Natl. Acad. Sci. USA* **87:**3018–3022.

Pickard, A. L., C. Wongsrichanalai, A. Purfield, D. Kamwendo, K. Emery, C. Zalewski, F. Kawamoto, R. S. Miller, and S. R. Meshnick. 2003. Resistance to antimalarials in southeast Asia and genetic polymorphisms in *pfmdr1*. *Antimicrob. Agents Chemother.* **47:**2418–2423.

Pillai, D. R., G. Hijar, Y. Montoya, W. Marouino, T. K. Ruebush II, C. Wongsrichanalai, and K. C. Kain. 2003. Lack of prediction of mefloquine and mefloquine-artesunate treatment outcome by mutations in the Plasmodium falciparum multidrug resistance 1 (pfmdr1) gene for P. falciparum malaria in Peru. *Am. J. Trop. Med. Hyg.* **68:**107–110.

Plowe, C. V., J. F. Cortese, A. Djimdé, O. C. Nwanyanwu, W. M. Watkins, P. A. Winstanley, J. G. Estrada-Franco, R. E. Mollinedo, J. C. Avila, J. L. Cespedes, D. Carter, and O. K. Doumbo. 1997. Mutations in Plasmodium falciparum dihydrofolate reductase and dihydropteroate synthase and epidemiologic patterns of pyrimethamine-sulfadoxine use and resistance. *J. Infect. Dis.* **176:**1590–1596.

Povoa, M. M., I. S. Adagu, S. G. Oliveira, R. L. Machado, M. A. Miles, and D. C. Warhurst. 1998. Pfmdr1 Asn1042Asp and Asp1246Tyr polymorphisms, thought to be associated with chloroquine resistance, are present in chloroquine-resistant and -sensitive Brazilian field isolates of Plasmodium falciparum. *Exp. Parasitol.* **88:**64–68.

Price, R., G. Robinson, A. Brockman, A. Cowman, and S. Krishna. 1997. Assessment of pfmdr1 gene copy number by tandem competitive polymerase chain reaction. *Mol. Biochem. Parasitol.* **85:**161–169.

Price, R., F. Nosten, J. A. Simpson, C. Luxemburger, L. Phaipun, F. ter Kuile, M. van Vugt,

T. Chongsuphajaisiddhi, and N. J. White. 1999a. Risk factors for gametocyte carriage in uncomplicated falciparum malaria. *Am. J. Trop. Med. Hyg.* **60:**1019–1023.

Price, R. N., C. Cassar, A. Brockman, M. Duraisingh, M. van Vugt, N. J. White, F. Nosten, and S. Krishna. 1999b. The *pfmdr1* gene is associated with a multidrug-resistant phenotype in *Plasmodium falciparum* from the western border of Thailand. *Antimicrob. Agents Chemother.* **43:**2943–2949.

Price, R. N., A.-C. Uhlemann, A. Brockman, R. McGready, E. Ashley, L. Phaipun, R. Patel, K. Laing, S. Looareesuwan, N. J. White, F. Nosten, and S. Krishna. 2004. Mefloquine resistance in *Plasmodium falciparum* and increased *pfmdr1* gene copy number. *Lancet* **364:**438–447.

Pukrittayakamee, S., A. Chantra, S. Vanijanonta, R. Clemens, S. Looareesuwan, and N. J. White. 2000. Therapeutic responses to quinine and clindamycin in multidrug-resistant falciparum malaria. *Antimicrob. Agents Chemother.* **44:**2395–2398.

Rastelli, G., W. Sirawaraporn, P. Sompornpisut, T. Vilaivan, S. Kamchonwongpaisan, R. Quarrell, G. Lowe, Y. Thebtaranonth, and Y. Yuthavong. 2000. Interaction of pyrimethamine, cycloguanil, WR99210 and their analogues with Plasmodium falciparum dihydrofolate reductase: structural basis of antifolate resistance. *Bioorg. Med. Chem.* **8:**1117–1128.

Raynes, K., M. Foley, L. Tilley, and L. W. Deady. 1996. Novel bisquinoline antimalarials. Synthesis, antimalarial activity, and inhibition of haem polymerisation. *Biochem. Pharmacol.* **52:**551–559.

Reed, M. B., K. J. Saliba, S. R. Caruana, K. Kirk, and A. F. Cowman. 2000. Pgh1 modulates sensitivity and resistance to multiple antimalarials in Plasmodium falciparum. *Nature* **403:**906–909.

Reynolds, M. G., and D. S. Roos. 1998. A biochemical and genetic model for parasite resistance to antifolates. Toxoplasma gondii provides insights into pyrimethamine and cycloguanil resistance in Plasmodium falciparum. *J. Biol. Chem.* **273:**3461–3469.

Roper, C., R. Pearce, S. Nair, B. Sharp, F. Nosten, and T. Anderson. 2004. Intercontinental spread of pyrimethamine-resistant malaria. *Science* **305:**1124.

Sachs, J. D. 2005. Achieving the Millennium Development Goals: the case of malaria. *N. Engl. J. Med.* **352:**115–117.

Saliba, K. J., P. I. Folb, and P. J. Smith. 1998. Role for the *Plasmodium falciparum* digestive vacuole in chloroquine resistance. *Biochem. Pharmacol.* **56:**313–320.

Sanchez, C. P., S. Wunsch, and M. Lanzer. 1997. Identification of a chloroquine importer in *Plasmodium falciparum*. Differences in import kinetics are genetically linked with the chloroquine-resistant phenotype. *J. Biol. Chem.* **272:**2652–2658.

Sanchez, C. P., W. Stein, and M. Lanzer. 2003. trans stimulation provides evidence for a drug efflux carrier as the mechanism of chloroquine resistance in Plasmodium falciparum. *Biochemistry* **42:**9383–9394.

Sanchez, C. P., J. E. McLean, W. Stein, and M. Lanzer. 2004. Evidence for a substrate specific and inhibitable drug efflux system in chloroquine resistant *Plasmodium falciparum* strains. *Biochemistry* **43:**16365–16373.

Sardarian, A., K. T. Douglas, M. Read, P. F. Sims, J. E. Hyde, P. Chitnumsub, R. Sirawaraporn, and W. Sirawaraporn. 2003. Pyrimethamine analogs as strong inhibitors of double and quadruple mutants of dihydrofolate reductase in human malaria parasites. *Org. Biomol. Chem.* **1:**960–964.

Shallom, S., K. Zhang, L. Jiang, and P. K. Rathod. 1999. Essential protein-protein interactions between Plasmodium falciparum thymidylate synthase and dihydrofolate reductase domains. *J. Biol. Chem.* **274:**37781–37786.

Shviro, Y., and N. Shaklai. 1987. Glutathione as a scavenger of free hemin. A mechanism of preventing red cell membrane damage. *Biochem. Pharmacol.* **36:**3801–3807.

Sidhu, A. B. S., D. Verdier-Pinard, and D. A. Fidock. 2002. Chloroquine resistance in *Plasmodium falciparum* malaria parasites conferred by *pfcrt* mutations. *Science* **298:**210–213.

Sidhu, A. B. S., S. G.-Valderramos, and D. A. Fidock. *pfmdr1* mutations contribute to quinine resistance and enhance mefloquine and artemisinin sensitivity in *Plasmodium falciparum*. *Mol. Microbiol.*, in press.

Sina, B. 2002. Focus on Plasmodium vivax. *Trends Parasitol.* **18:**287–289.

Sirawaraporn, W., T. Sathitkul, R. Sirawaraporn, Y. Yuthavong, and D. V. Santi. 1997a. Antifolate-resistant mutants of Plasmodium falciparum dihydrofolate reductase. *Proc. Natl. Acad. Sci. USA* **94:**1124–1129.

Sirawaraporn, W., S. Yongkiettrakul, R. Sirawaraporn, Y. Yuthavong, and D. V. Santi. 1997b. Plasmodium falciparum: asparagine mutant at residue 108 of dihydrofolate reductase is an optimal antifolate-resistant single mutant. *Exp. Parasitol.* **87:**245–252.

Skinner-Adams, T., and T. M. Davis. 1999. Synergistic in vitro antimalarial activity of omeprazole and quinine. *Antimicrob. Agents Chemother.* **43:**1304–1306.

Slater, A. F., and A. Cerami. 1992. Inhibition by chloroquine of a novel haem polymerase enzyme activity in malaria trophozoites. *Nature* **355:**167–169.

Su, X.-Z., L. S. Kirkman, and T. E. Wellems. 1997. Complex polymorphisms in a ~330 kDa protein are linked to chloroquine-resistant *P. falciparum* in southeast Asia and Africa. *Cell* **91:**593–603.

Sullivan, D. J., Jr., I. Y. Gluzman, D. G. Russell, and D. E. Goldberg. 1996. On the molecular mechanism of chloroquine's antimalarial action. *Proc. Natl. Acad. Sci. USA* **93:**11865–11870.

Sutherland, C. J., A. Alloueche, J. Curtis, C. J. Drakeley, R. Ord, M. Duraisingh, B. M. Greenwood, M. Pinder, D. C. Warhurst, and G. A. Targett. 2002. Gambian children successfully treated with chloroquine can harbour and transmit *Plasmodium falciparum* gametocytes carrying resistance genes. *Am. J. Trop. Med. Hyg.* **67:**578–585.

Thaithong, S., L. C. Ranford-Cartwright, N. Siripoon, P. Harnyuttanakorn, N. S. Kanchanakhan, A. Seugorn, K. Rungsihirunrat, P. V. Cravo, and G. H. Beale. 2001. Plasmodium falciparum: gene mutations and amplification of dihydrofolate reductase genes in parasites grown in vitro in presence of pyrimethamine. *Exp. Parasitol.* **98:**59–70.

Tran, C. V., and M. H. Saier, Jr. 2004. The principal chloroquine resistance protein of *Plasmodium falciparum* is a member of the drug/metabolite transporter superfamily. *Microbiology* **150:**1–3.

Tran, T. H., C. Dolecek, P. M. Pham, T. D. Nguyen, T. T. Nguyen, H. T. Le, T. H. Dong, T. T. Tran, K. Stepniewska, N. J. White, and J. Farrar. 2004. Dihydroartemisinin-piperaquine against multidrug-resistant Plasmodium falciparum malaria in Vietnam: randomised clinical trial. *Lancet* **363:**18–22.

Trape, J. F. 2001. The public health impact of chloroquine resistance in Africa. *Am. J. Trop. Med. Hyg.* **64:**12–17.

Triglia, T., S. J. Foote, D. J. Kemp, and A. F. Cowman. 1991. Amplification of the multidrug resistance gene *pfmdr1* in *Plasmodium falciparum* has arisen as multiple independent events. *Mol. Cell. Biol.* **11:**5244–5250.

Triglia, T., J. G. Menting, C. Wilson, and A. F. Cowman. 1997. Mutations in dihydropteroate synthase are responsible for sulfone and sulfonamide resistance in Plasmodium falciparum. *Proc. Natl. Acad. Sci. USA* **94:**13944–13949.

Trujillo, M., R. G. Donald, D. S. Roos, P. J. Greene, and D. V. Santi. 1996. Heterologous expression and characterization of the bifunctional dihydrofolate reductase-thymidylate synthase enzyme of Toxoplasma gondii. *Biochemistry* **35:**6366–6374.

Uhlemann, A. C., A. Cameron, U. Eckstein-Ludwig, W.-Y. Ho, W. C. Chan, J. Fischbarg, P. Iserovich, F. A. Zuniga, M. East, A. Lee, L. Brady, R. K. Haynes, and S. Krishna. A single amino acid residue can determine sensitivity of SERCAs to artemisinins. *Nature Struct. Mol. Biol.*, in press.

Ursos, L. M., and P. D. Roepe. 2002. Chloroquine resistance in the malarial parasite, Plasmodium falciparum. *Med. Res. Rev.* **22:**465–491.

Waller, K. L., R. A. Muhle, L. M. Ursos, P. Horrocks, D. Verdier-Pinard, A. B. Sidhu, H. Fujioka, P. D. Roepe, and D. A. Fidock. 2003. Chloroquine resistance modulated in vitro by expression levels of the *Plasmodium falciparum* chloroquine resistance transporter. *J. Biol. Chem.* **278:**33593-33601.

Waller, K. L., S. Lee, and D. A. Fidock. 2004. Molecular and cellular biology of chloroquine resistance in Plasmodium falciparum, p. 501–540. *In* A. P. Waters and C. J. Janse (ed.), *Malaria Parasites: Genomes and Molecular Biology*. Caister Academic Press, Wymondham, United Kingdom.

Wang, P., R. K. Brobey, T. Horii, P. F. Sims, and J. E. Hyde. 1999. Utilization of exogenous folate in the human malaria parasite Plasmodium falciparum and its critical role in antifolate drug synergy. *Mol. Microbiol.* **32:**1254–1262.

Wang, P., N. Nirmalan, Q. Wang, P. F. Sims, and J. E. Hyde. 2004. Genetic and metabolic analysis of folate salvage in the human malaria parasite Plasmodium falciparum. *Mol. Biochem. Parasitol.* **135:**77–87.

Warhurst, D. C. 2002. Resistance to antifolates in Plasmodium falciparum, the causative agent of tropical malaria. *Sci. Prog.* **85:**89–111.

Warhurst, D. C., J. C. Craig, and I. S. Adagu. 2002. Lysosomes and drug resistance in malaria. *Lancet* **360:**1527–1529.

Watkins, W. M., and M. Mosobo. 1993. Treatment of Plasmodium falciparum malaria with pyrimethamine-sulfadoxine: selective pressure for resistance is a function of long elimination half-life. *Trans. R. Soc. Trop. Med. Hyg.* **87:**75–78.

Watt, G., L. Loesuttivibool, G. D. Shanks, E. F. Boudreau, A. E. Brown, K. Pavanand, H. K. Webster, and S. Wechgritaya. 1992. Quinine with tetracycline for the treatment of drug-resistant falciparum malaria in Thailand. *Am. J. Trop. Med. Hyg.* **47:**108–111.

Wattanarangsan, J., S. Chusacultanachai, J. Yuvaniyama, S. Kamchonwongpaisan, and Y. Yuthavong. 2003. Effect of N-terminal truncation of Plasmodium falciparum dihydrofolate reductase on dihydrofolate reductase and thymidylate synthase activity. *Mol. Biochem. Parasitol.* **126:**97–102.

Wellems, T. E., L. J. Panton, I. Y. Gluzman, V. E. do Rosario, R. W. Gwadz, A. Walker-Jonah, and D. J. Krogstad. 1990. Chloroquine resistance not linked to *mdr*-like genes in a *Plasmodium falciparum* cross. *Nature* **345:**253–255.

Wellems, T. E., A. Walker-Jonah, and L. J. Panton. 1991. Genetic mapping of the chloroquine-resistance locus on *Plasmodium falciparum* chromosome 7. *Proc. Natl. Acad. Sci. USA* **88:**3382–3386.

Wellems, T. E., and C. V. Plowe. 2001. Chloroquine-resistant malaria. *J. Infect. Dis.* **184:**770–776.

White, N. J. 1996. The treatment of malaria. *N. Engl. J. Med.* **335:**800–806.

White, N. J., and W. Pongtavornpinyo. 2003. The de novo selection of drug-resistant malaria parasites. *Proc. R. Soc. Lond. B Biol. Sci.* **270:**545–554.

White, N. J. 2004. Antimalarial drug resistance. *J. Clin. Investig.* **113:**1084–1092.

Wilson, C. M., A. E. Serrano, A. Wasley, M. P. Bogenschutz, A. H. Shankar, and D. F. Wirth. 1989. Amplification of a gene related to mammalian mdr genes in drug-resistant Plasmodium falciparum. *Science* **244:**1184–1186.

Wilson, C. M., S. K. Volkman, S. Thaithong, R. K. Martin, D. E. Kyle, W. K. Milhous, and D. F. Wirth. 1993. Amplification of *pfmdr1* associated with mefloquine and halofantrine resistance in *Plasmodium falciparum* from Thailand. *Mol. Biochem. Parasitol.* **57:**151–160.

Winstanley, P. 2001. Chlorproguanil-dapsone (LAPDAP) for uncomplicated falciparum malaria. *Trop. Med. Int. Health* **6:**952–954.

Wissing, F., C. P. Sanchez, P. Rohrbach, S. Ricken, and M. Lanzer. 2002. Illumination of the malaria parasite Plasmodium falciparum alters intracellular pH. Implications for live cell imaging. *J. Biol. Chem.* **277:**37747–37755.

Wongsrichanalai, C., A. L. Pickard, W. H. Wernsdorfer, and S. R. Meshnick. 2002. Epidemiology of drug-resistant malaria. *Lancet Infect. Dis.* **2:**209–218.

Wood, P. A., and J. W. Eaton. 1993. Hemoglobin catabolism and host-parasite heme balance in chloroquine-sensitive and chloroquine-resistant *Plasmodium berghei* infections. *Am. J. Trop. Med. Hyg.* **48:**465–472.

Wooden, J. M., L. H. Hartwell, B. Vasquez, and C. H. Sibley. 1997. Analysis in yeast of antimalaria drugs that target the dihydrofolate reductase of *Plasmodium falciparum*. *Mol. Biochem. Parasitol.* **85:**25–40.

Wootton, J. C., X. Feng, M. T. Ferdig, R. A. Cooper, J. Mu, D. I. Baruch, A. J. Magill, and X. Z. Su. 2002. Genetic diversity and chloroquine selective sweeps in Plasmodium falciparum. *Nature* **418:**320–323.

World Health Organization. 2001. Drug resistance in malaria. WHO/CDC/CSR/DRS/2001.4.

Wu, Y., L. A. Kirkman, and T. E. Wellems. 1996. Transformation of *Plasmodium falciparum* malaria parasites by homologous integration of plasmids that confer resistance to pyrimethamine. *Proc. Natl. Acad. Sci. USA* **93:**1130–1134.

Yayon, A., Z. I. Cabantchik, and H. Ginsburg. 1984. Identification of the acidic compartment of *Plasmodium falciparum*-infected human erythrocytes as the target of the antimalarial drug chloroquine. *EMBO J.* **3:**2695–2700.

Yayon, A. 1985. The antimalarial mode of action of chloroquine. *Rev. Clin. Basic Pharm.* **5:**99–139.

Yeung, S., W. Pongtavornpinyo, I. M. Hastings, A. J. Mills, and N. J. White. 2004. Antimalarial drug resistance, artemisinin-based combination therapy, and the contribution of modeling to elucidating policy choices. *Am. J. Trop. Med. Hyg.* **71:**179–186.

Yuthavong, Y., T. Vilaivan, N. Chareonsethakul, S. Kamchonwongpaisan, W. Sirawaraporn, R. Quarrell, and G. Lowe. 2000. Development of a lead inhibitor for the A16V+S108T mutant of dihydrofolate reductase from the cycloguanil-resistant strain (T9/94) of Plasmodium falciparum. *J. Med. Chem.* **43:**2738–2744.

Yuthavong, Y. 2002. Basis for antifolate action and resistance in malaria. *Microbes Infect.* **4:**175–182.

Yuthavong, Y., J. Yuvaniyama, P. Chitnumsub, J. Vanichtanankul, S. Chusacultanachai, B. Tarnchompoo, T. Vilaivan, and S. Kamchonwongpaisan. 2005. Malaria (Plasmodium falciparum) dihydrofolate reductase-thymidylate synthase: structural basis for antifolate resistance and development of effective inhibitors. *Parasitology* **130:**249–259.

Yuvaniyama, J., P. Chitnumsub, S. Kamchonwongpaisan, J. Vanichtanankul, W. Sirawaraporn, P. Taylor, M. D. Walkinshaw, and Y. Yuthavong. 2003. Insights into antifolate resistance from malarial DHFR-TS structures. *Nat. Struct. Biol.* **10:**357–365.

Zalis, M. G., L. Pang, M. S. Silveira, W. K. Milhous, and D. F. Wirth. 1998. Characterization of Plasmodium falciparum isolated from the Amazon region of Brazil: evidence for quinine resistance. *Am. J. Trop. Med. Hyg.* **58:**630–637.

Zhang, H., E. M. Howard, and P. D. Roepe. 2002. Analysis of the antimalarial drug resistance protein Pfcrt expressed in yeast. *J. Biol. Chem.* **277:**49767–49775.

Zhang, H., M. Paguio, and P. D. Roepe. 2004. The antimalarial drug resistance protein Plasmodium falciparum chloroquine resistance transporter binds chloroquine. *Biochemistry* **43:**8290–8296.

Zhang, J., M. Krugliak, and H. Ginsburg. 1999. The fate of ferriprotorphyrin IX in malaria infected erythrocytes in conjunction with the mode of action of antimalarial drugs. *Mol. Biochem. Parasitol.* **99:**129–141.

Zhang, K., and P. K. Rathod. 2002. Divergent regulation of dihydrofolate reductase between malaria parasite and human host. *Science* **296:**545–547.

HOST GENETIC FACTORS IN RESISTANCE AND SUSCEPTIBILITY TO MALARIA

Dominic P. Kwiatkowski and Gaia Luoni[†]

24

People vary in their response to malaria infection. In regions where malaria is endemic, regions where everyone is repeatedly exposed to infection, some people die of malaria but the majority survive, despite having apparently similar levels of parasites in the blood. In the days when artificial infection with malaria was used as a form of treatment for neurosyphilis, some patients managed to control the infection much more effectively than others challenged with the same infectious dose (Kitchen, 1949).

To what extent are such differences determined by host genetic factors? They are clearly only part of the story. The genetic makeup of the parasite is almost certainly a determining factor, and there may be a wide variety of subtle immunological adaptations that occur after birth, which affect how an individual will respond when first exposed to infection. Curiously, age itself seems to be a determinant: it seems that when adults encounter malaria for the first time, they tend to suppress parasitemia more effectively than children, but they are also more likely to develop life-threatening complications of infection (Baird, 1995; Baird et al., 1998).

Well-controlled genetic studies are easier to conduct with mice than with humans, and it has long been known that certain strains of mice consistently show a greater degree of resistance to infection than other strains. It is clear that these differences are primarily genetic, rather than environmental or acquired, and later we will discuss specific genetic loci that have been identified as responsible for some of these differences (Greenberg and Kendrick, 1959; Greenberg et al., 1954; Nadel et al., 1955; Rest, 1982; Stevenson et al., 1982; Stevenson and Skamene, 1985). It has also been observed that mice which are relatively resistant to one strain of parasite may be relatively susceptible to a different strain: for example, some strains of mice are resistant to nonlethal strains but highly susceptible to lethal strains of *Plasmodium yoelii*; while other mouse strains show the opposite pattern (Sayles and Wassom, 1988). Thus, the message from experimental murine malaria is that both host and

---

Dominic P. Kwiatkowski and Gaia Luoni, Wellcome Trust Centre for Human Genetics and University Department of Paediatrics, University of Oxford, Oxford, United Kingdom.

[†]Dr. Gaia Luoni was a Marie Curie Fellow at the Wellcome Trust Centre for Human Genetics from 2002 to 2004 (http://www.gmap.net/oxford/gaia.htm). She died of cancer in September 2004.

Molecular Approaches to Malaria, Edited by Irwin W. Sherman
© 2005 ASM Press, Washington, D.C.

parasite genes and the specific way in which they are combined are important determinants of resistance or susceptibility to infection.

Humans, however, are not the same as inbred mice; and it is much more difficult to quantify the extent to which host genetic factors determine the variability in responses to malaria infection that is observed with natural human populations. Under certain circumstances, host genetic determinants undoubtedly play a crucial role. A classical example is the Duffy blood group antigen, which is essential for erythrocyte invasion by *Plasmodium vivax* (Miller et al., 1976). Most sub-Saharan African populations do not express the Duffy antigen on erythrocytes due to a specific genetic variant (discussed in more detail below) and are therefore completely resistant to infection by *P. vivax*. Thus, a single host genetic factor is believed to account for the near-absence of *P. vivax* infection throughout sub-Saharan Africa.

Genetic factors also appear to underlie some striking differences in resistance to malaria that have been observed between ethnic groups who live in the same area. A classical account of this phenomenon was made with the Fulani of Burkina Faso, who were found to have a lower prevalence of malaria parasitemia and fewer clinical attacks of malaria than other ethnic groups who lived in closely neighboring villages (Modiano et al., 1996). Although the Fulani have quite a different lifestyle from other groups, detailed epidemiological investigations indicate that their resistance to malaria arises primarily from genetic factors. Importantly, it has also been observed that the Fulani have high levels of antimalarial antibodies (Modiano et al., 1998; Modiano et al., 1999) and a low frequency of protective globin variants and other classical malaria resistance factors (Modiano et al., 2001a). There is, therefore, much interest in discovering the genetic factors that determine the high antibody responses seen in the Fulani; the possible role of the interleukin 4 gene (*IL4*) (Luoni et al., 2001) is discussed below.

Much more difficult to establish is the extent to which resistance to malaria varies within and between families. This may seem surprising, given how much we know about other genetic aspects of malaria, but it is largely due to practical obstacles. For example, if we focus on resistance to severe malaria, we need to ascertain whether malaria was responsible for previous episodes of severe illness and death within the family, and that is difficult since detailed medical records are rarely available in the communities that are most afflicted by malaria. An alternative strategy is to conduct longitudinal studies of quantitative traits such as malaria infection intensity or the frequency of malaria fever episodes. Such studies are epidemiologically demanding and require frequent monitoring, since infection intensity varies greatly over time within a single individual and is greatly affected by age and environment. However there have been some successes. A longitudinal study of Gambian twins showed that susceptibility to malaria fever episodes is partly determined by genetic factors (Jepson et al., 1995) with linkage to the major histocompatibility complex (MHC) region on chromosome 6 (Jepson et al., 1997). Longitudinal family studies of infection intensity in Cameroon and Burkina Faso indicate that complex genetic factors are involved (Garcia et al., 1998a; Rihet et al., 1998a) and that there is linkage to MHC and the 5q31-33 region (Flori et al., 2003; Garcia et al., 1998b; Rihet et al., 1998b). Familial segregation analysis of immunological responses to malaria antigens in Papua New Guinea has suggested that Mendelian effects might govern certain antigen responses, but the overall picture is complex (Stirnadel et al., 2000a; Stirnadel et al., 1999a, 2000b).

Much of the remainder of this chapter considers the growing list of genes that show some evidence of influencing resistance to malaria. A few, such as the Duffy blood group antigen and sickle hemoglobin genes, have a profound effect that is well documented. Quite a few, it must be admitted, appear to have weak effects that are supported by patchy epidemiological evidence. A third group is somewhere between these two extremes. With the explosion of information about human genomic diversity that is appearing on public databases such as dbSNP (http://www.ncbi.nlm.nih.gov/projects/SNP/) and the

International HapMap Project (http://www.hapmap.org), there is no question that the rate of discovery of new genetic associations with resistance or susceptibility to malaria is going to accelerate rapidly over the next few years.

The diverse genetic factors that determine resistance to malaria are biologically interesting, but are they of practical importance? The answer is almost certainly yes, because they may provide vital clues about molecular mechanisms of disease and immunity that are central to the development of an effective vaccine. For example, we have only a vague understanding of the molecular mechanisms by which malaria parasites invade erythrocytes, or the immunological mechanisms that are critical for clearing malaria parasites from the bloodstream. Genetic findings may offer radical new insights into those questions. The discovery that Duffy blood group-negative individuals are resistant to infection with *Plasmodium vivax* was the vital clue in understanding how this species of parasite invades erythrocytes and has led to the development of a *P. vivax* vaccine that is now undergoing testing. One of the many practical reasons for investing in large-scale epidemiological studies of malaria is to gain similar clues about the molecular mechanisms by which *Plasmodium falciparum* invades erythrocytes and other aspects of protective immunity that could be crucial for vaccine development.

There is a further reason why geneticists are particularly interested in malaria. It is the strongest known environmental force for evolutionary pressure on the human genome, responsible for the most common human Mendelian disorders: α- and β-thalassemia, sickle cell disease, glucose-6-phosphate dehydrogenase (G6PD) deficiency, and various other erythrocyte defects. If a genetic variant confers a significant degree of protection against death from malaria, then there is a good chance that it will undergo positive selection in regions where malaria is a major public health problem. That includes most of sub-Saharan Africa and large parts of the Indian subcontinent, Southeast Asia, and South America; in the relatively recent past, it also included parts of Europe. Thus, many human populations carry malaria resistance genes because their ancestors were exposed to malaria, including groups such as African Americans who have not been exposed to malaria for many generations, and this may be an important factor in susceptibility to contemporary diseases, in addition to the hemoglobinopathies and erythrocyte disorders. Many scientists suspect—though it has yet to be proven—that the historical influence of malaria on polymorphisms of immune and inflammatory genes may be a significant factor in the etiology of common diseases of affluent Western societies.

CURRENT UNDERSTANDING OF HUMAN GENES THAT DETERMINE RESISTANCE OR SUSCEPTIBILITY TO MALARIA

Erythrocyte Surface Molecules

As noted in the introduction, genetic variation in erythrocyte surface proteins is particularly interesting to vaccine developers, as it may provide important clues about how we might be able to prevent parasites from invading red blood cells.

FY ENCODING DUFFY ANTIGEN-CHEMOKINE RECEPTOR

The glycoprotein Duffy antigen, encoded by the *FY* gene, is expressed on the surface of erythrocytes in most people, excepting sub-Saharan Africa populations and their descendants. It is also expressed on various other cell types, and its functional role is as a chemokine receptor. There are three common alleles, designated FY*A, FY*B, and FY*O. The FY*O allele corresponds to the absence of the Duffy antigen on red blood cells and is defined by a single nucleotide polymorphism (SNP) in the *FY* promoter region which disrupts a binding site for the transcription factor GATA-1 (Tournamille et al., 1995). It was discovered almost 30 years ago that *P. vivax* requires the Duffy antigen to invade red blood cells (Miller et al., 1976), and this led to the biochemical and molecular biological characterization of a parasite molecule, *P. vivax* Duffy-binding protein, critical for erythrocyte invasion (Chitnis et al., 1994). This in turn has led to the development of a candidate vaccine against *P. vivax* (Yazdani et al., 2004) now undergoing clinical trials.

SLC4A ENCODING ERYTHROCYTE BAND 3 PROTEIN

SLC4A1 (also known as CD233 or erythrocyte band 3 protein) is an anion exchanger in the erythrocyte membrane, acting as a chloride-bicarbonate exchanger involved in transporting carbon dioxide from tissues to the lungs. Its correct conformation depends on its association with glycophorin A (see below). There are many different *SLC4A1* mutations, some of which cause hereditary spherocytosis. Southeast Asian ovalocytosis occurs in heterozygotes for a 27-bp deletion in *SLC4A1*. Studies in Papua New Guinea (Cattani et al., 1987) and Malaya (Foo et al., 1992) have found reduced levels of malaria infection in individuals with ovalocytosis. In an area of Papua New Guinea where 9% of healthy individuals have ovalocytosis, two independent studies have found it to be protective against cerebral malaria (Allen et al., 1999; Genton et al., 1995). The mechanism of this protective effect is not well understood; it may be due to inhibition of parasite invasion or growth, but it may also be due to an effect on parasite sequestration, as erythrocyte band 3 protein appears to be involved in the cytoadherence of parasitized erythrocytes to vascular endothelium (Sherman et al., 2003).

GYPA AND GYPB ENCODING GLYCOPHORINS A AND B

Glycophorins A and B, encoded by the homologous genes *GYPA* and *GYPB*, are major sialoglycoproteins of the erythrocyte membrane, which carries the antigenic determinants for various blood groups. Human erythrocytes that are genetically deficient in glycophorin A or B show resistance to invasion by *P. falciparum* (Facer, 1983). Specific sialic acid residues on the glycophorin A molecule are recognized by a Duffy-binding-like domain of *P. falciparum* erythrocyte-binding antigen 175 (Mayor et al., 2005; Orlandi et al., 1992). Analyses of *GYPA* and *GYPB* sequence variation among primates and humans show evidence of strong evolutionary selection (Baum et al., 2002; Wang et al., 2003). It has also been noted that the highest rate of nonsynonymous polymorphism observed in the genome of *P. falciparum* is in erythrocyte-binding antigen 175 (EBA175), suggesting that both the parasite ligand and the host receptor are engaged in an ongoing evolutionary battle (Wang et al., 2003).

GYPC ENCODING GLYCOPHORIN C

Glycophorin C (encoded by *GYPC*) is a minor component of the erythrocyte membrane that is important for the mechanical stability of red blood cells. *P. falciparum* EBA140 binds to glycophorin C. A deletion of *GYPC* exon 3 is responsible for Gerbich blood group negativity, and is found at frequencies approaching 50% in humans living in coastal areas of Papua New Guinea. It has been found that, in individuals who are Gerbich negative, EBA140 does not bind to glycophorin C and *P. falciparum* invasion is blocked (Maier et al., 2003). Curiously, epidemiological studies in Papua New Guinea have so far failed to find evidence that this *GYPC* deletion affects malaria infection, but possible effects on clinical severity have not yet been studied in detail (Patel et al., 2001; Patel et al., 2004).

Hemoglobin Genes

Adult hemoglobin, HbA, is a tetramer consisting of two α-globin chains (encoded by the identical *HBA1* and *HBA2* genes) and two β-globin chains (encoded by *HBB*). The selective pressure of malaria has been responsible for a remarkable variety of functional variants of these three genes.

HBB ENCODING β-GLOBIN

Hemoglobin S (HbS) results from a glutamic acid-to-valine substitution at codon 6 of the beta globin chain. HbS homozygotes have sickle cell disease, a debilitating and often fatal disorder caused by the red blood cell deformities that result from this structural defect, particularly at low oxygen concentrations. Heterozygotes, who do not generally have any clinical abnormality, have about 10-fold protection from life-threatening forms of malaria and somewhat lower levels of protection against milder forms of the disease (Allen et al., 1992; Allison, 1954; Gilles et al., 1967; Hill et al., 1991; Sokhna et al., 2000; Stirnadel et al., 1999b). The precise mechanism of this protective effect has not been proven but is thought to involve suppression of parasite growth in red blood cells (Pasvol et al., 1978) and enhanced splenic clearance of parasitized erythrocytes (Shear et al., 1993). The tradeoff between risk and benefit acts to maintain the HbS polymorphism at allele frequencies of around 10% in many parts of Africa, despite the grave consequences of the homozygous state. This is the most striking example of heterozygote advantage in human population genetics.

HbC is a different structural variant of exactly the same region of the *HBB* gene, where lysine replaces glutamic acid at codon 6. It is found in several parts of West Africa but is less common than HbS. HbC homozygotes have a relatively mild hemolytic anemia, i.e., it is a much less damaging phenotype than sickle cell disease. Heterozygotes do not experience a significant reduction in hemoglobin levels (Diallo et al., 2004). Several recent studies have shown that both HbC heterozygotes and homozygotes are protected against clinical episodes of malaria (Agarwal et al., 2000; Modiano et al., 2001b; Rihet et al., 2004).

HbE is yet another structural variant of the *HBB* gene, a glutamic acid-to-lysine substitution at codon 26. It is extremely common in southeast Asia, with carrier rates of 50% in some places, and homozygotes generally have symptomless anemia. In heterozygotes, it seems that *P. falciparum* can only invade a fraction of erythrocytes and it has been proposed that this acts to protect against high parasitemia and thus severe disease (Chotivanich et al., 2002). Analysis of haplotype structure suggests that the mutation is relatively recent and has risen rapidly in allele frequency (Ohashi et al., 2004).

HBA ENCODING α-GLOBIN

The thalassemias are a group of genetic disorders due to defective production of α- or β-globin chains, arising from a diverse set of deletions and other disruptions of the globin gene clusters on chromosomes 11 and 16. The clinical phenotypes are complex but, broadly speaking, the homozygous state results in severe disease or is fatal, while heterozygotes are healthy apart from mild anemia. An exception to this general rule is α+-thalassemia, which occurs when either the *HBA1* or the *HBA2* gene is disrupted but not both, so that some α-globin production is possible: α+-thalassemia homozygotes are only mildly anemic.

The thalassemias are the commonest Mendelian diseases of humans, and are believed to have resulted from evolutionary selection by malaria, because their global distribution corresponds so closely to the regions that have historically been exposed to malaria, including the Mediterranean and the Middle East as well as Africa, the Indian subcontinent, and Southeast Asia. In a detailed population genetic survey of Melanesia, it was found that the frequency of α+-thalassemia varied according to both altitude and latitude, in a manner that was highly correlated with malaria endemicity (Flint et al., 1986). A case-control study in Papua New Guinea found that the risk of severe malaria was reduced by 60% in α+-thalassemia homozygotes and by a smaller amount in heterozygotes, but curiously a similar effect was seen with other childhood infectious diseases (Allen et al., 1997). A study in Ghana found that α^+-thalassemia heterozygotes were protected against severe malaria (Mockenhaupt et al., 2004), while a study in Kenya found that both heterozygotes and homozygotes were protected (Williams et al., 2005). To thoroughly complicate matters, in Vanuatu it has been observed that young children with α+-thalassemia have a significantly higher incidence of malaria than nonthalassemic children (Williams et al., 1996). The authors rose to the challenge of explaining this totally unexpected result, by noting that most of the increased incidence was found in young children infected with *P. vivax* rather than *P. falciparum*; they proposed that if young children are highly exposed to the relatively benign parasite *P. vivax*, they may be protected against severe disease caused by *P. falciparum*.

Other Erythrocyte-Related Proteins

G6PD ENCODING GLUCOSE-6-PHOSPHATE DEHYDROGENASE

The *G6PD* gene on chromosome X encodes glucose-6-phosphate dehydrogenase, an enzyme that acts to produce NADPH, which is a key electron donor in defense against oxidizing agents. G6PD shows huge genetic diversity, causing wide variation in enzyme activity that can lead to hemolytic anemia. Common forms of G6PD deficiency are believed to have arisen as a result of evolutionary selection by malaria, because of their geographical distribution at both the global and the local level (Ganczakowski et al., 1995). Different methods of haplotype analysis agree that the G6PD locus has undergone recent evolutionary selection (Sabeti et al., 2002b; Tishkoff et al., 2001). Studies of G6PD enzyme phenotype (Gilles et al., 1967) and a common underlying mutation in African children (Ruwende et al., 1995) have confirmed that that G6PD deficiency confers protection against severe malaria in both hemizygous males and heterozygous females. This is thought to be because parasite replication is diminished in G6PD-deficient erythrocytes (Luzzatto et al., 1979), although it appears that the parasite is capable of evading this host-protective mechanism by synthesizing its own form of G6PD (Usanga and Luzzatto, 1985).

HP ENCODING HAPTOGLOBIN

Haptoglobin, encoded by the *HP* gene, is a protein found in plasma. Although it is not an erythrocyte protein, it is mentioned here because one of its major functions is to bind free hemoglobin, thus preventing hemoglobin-induced oxidative tissue damage. Studies in Sudan and Ghana have found an association between the haptoglobin 1-1 phenotype (determined by electropheresis of sera) and susceptibility to severe *P. falciparum* malaria (Elagib et al., 1998; Quaye et al., 2000), and this is consistent with experiments with mice which showed that HP gene knockout led to greater parasite burden (Hunt et al., 2001). However, a DNA-based study of haptoglobin polymorphisms in The Gambia failed to identify a significant association with disease susceptibility (Aucan et al., 2002).

Adhesion Molecules

Parasite sequestration in small blood vessels is central to the pathogenesis of severe malaria. It is caused by the ability of parasitized erythrocytes to bind to host adhesion molecules that are expressed on endothelium, platelets, and other erythrocytes. These binding properties are largely due to a parasite protein called *P. falciparum* erythrocyte membrane protein 1 (PfEMP-1) encoded by the *var* gene family (called *var* because each parasite contains multiple copies which are highly variable, and the parasite is able to switch expression between them). Here, we describe four well-characterized human adhesion molecules that serve as receptors for different forms of PfEMP1.

ICAM1 ENCODING INTERCELLULAR ADHESION MOLECULE 1

Intercellular adhesion molecule 1 (ICAM1), also known as CD54, is encoded by the *ICAM1* gene. It is expressed on endothelial cells and cells of the immune system and normally binds to integrins that are expressed by leukocytes. It is a strong binding receptor for some isolates of *P. falciparum* (Berendt et al., 1989). This binding is reduced by a polymorphism in the N-terminal domain, which is found at high frequency in African populations (Fernandez-Reyes et al., 1997). It was expected that this polymorphism would reduce susceptibility to severe malaria and that is what was found in a study in Gabon (Kun et al., 1999). However, a study in Kenya found the opposite (Fernandez-Reyes et al., 1997), and another in The Gambia found no significant association (Bellamy et al., 1998a).

CD36 ENCODING CD36 ANTIGEN

The CD36 antigen, encoded by the *CD36* gene, is a surface glycoprotein that acts as a receptor for various molecules including collagen, thrombospondin, anionic phospholipids, and long-chain fatty acids. It is highly expressed on the platelet surface and is also found on endothelial and dendritic cells. It serves as an endothelial binding receptor for many isolates of *P. falciparum* (Barnwell et al., 1989). It also allows parasites to bind to dendritic cells and to suppress their ability to present parasite antigens (Urban et al., 1999; Urban et al., 2001). A range of CD36 polymorphisms in West African and Southeast Asian populations where malaria is endemic have been described (Aitman et al., 2000; Omi et al., 2003), but the results of disease association studies are perplexing. A study that used both Gambian and Kenyan case-control samples found that homozygotes for the *CD36*+1264G allele (a nonsense mutation) were susceptible to cerebral malaria (Aitman et al., 2000) while a study that looked at the same allele in Kenya alone found that heterozygosity was associated with protection against severe malaria (Pain et al., 2001). In Thailand, the *CD36*+1264G allele was found to be absent or very rare, but protection against cerebral malaria was found to be associated with a dinucleotide repeat sequence in intron 3, which has been implicated in alternative splicing of *CD36* (Omi et al., 2003).

CR1 ENCODING COMPLEMENT RECEPTOR 1

CR1 encodes an erythrocyte membrane glycoprotein (also found on some dendritic cells) that acts as receptor for complement components C3b and C4b and regulates the activity of the complement cascade. It is a binding receptor for some *P. falciparum* isolates, leading to the phenomenon known as rosetting, where a parasitized erythrocyte binds to other erythrocytes (Rowe et al., 1997). Early studies of complement receptor 1 (CR1) polymorphisms in The Gambia, including the Knops blood group system, found no evidence of an association with disease severity (Bellamy et al., 1998a; Zimmerman et al., 2003), while a restriction fragment length polymorphism-based study in Thailand found evidence of increased susceptibility to severe malaria among homozygotes for an allele that was associated with reduced expression of CR1 on erythrocytes (Nagayasu et al., 2001). More recently, a study in malaria-endemic regions of Papua New Guinea found that up to 80% of the population had erythrocyte CR1 deficiency, which was associated both with polymorphisms in the *CR1* gene and, intriguingly, with α+-thalassemia which, as noted above, is common in this population. In a case-control analysis, it was found that *CR1* polymorphisms and α-thalassemia were independently associated with resistance to severe malaria (Cockburn et al., 2004).

PECAM1 ENCODING PLATELET-ENDOTHELIAL CELL ADHESION MOLECULE

The platelet-endothelial cell adhesion molecule (designated PECAM1 and also known as CD31) is a member of the immunoglobulin superfamily that is expressed on platelets, various leukocyte subsets, and endothelial cells. It is an endothelial binding receptor for *P. falciparum* (Treutiger et al., 1997). It has a common coding variant (Leu to Val at codon 125), which was analyzed in case-control studies of severe malaria in Papua New Guinea and Kenya, but no significant association was identified (Casals-Pascual et al., 2001). A study from Thailand has reported a PECAM1 haplotype that is more common in cerebral malaria than in other forms of severe malaria (Kikuchi et al., 2001).

MECHANISMS OF ACQUIRED IMMUNITY

HLA-B Encoding MHC Class IB

HLA-B encodes an MHC class I heavy chain which forms a heterodimer with β-2 microglobulin to make up the HLA-B antigen presentation complex. Hundreds of HLA-B alleles have been described. The HLA-B53 allele, which is much more common in West Africa than elsewhere in the world, is associated with a significant reduction in life-threatening complications of malaria in the Gambian population (Hill et al., 1991). This implies that MHC class I-dependent cytotoxic T-cell mechanisms afford a significant degree of immunological protection. Since MHC class I is expressed by liver cells but not by erythrocytes, this observation has bolstered efforts to develop a liver-stage malaria vaccine. Analysis of peptides that bind to HLA-B53 provides a way of screening for T-cell epitopes that may be effective targets for vaccine development (Hill et al., 1992). However, there is evidence that the parasite is rapidly evolving to evade the host antigen presentation system. A study of the *P. falciparum* circumsporozoite protein found that, within a single fragment of circumsporozoite protein, the Gambian parasite population expressed four common variants, of which only two (cp26 and cp29) bound to HLA-B53 and were therefore potential targets for cytotoxic T-lymphocyte (CTL) attack in HLA-B53$^+$ individuals. It was discovered in vitro that, although both cp26 and cp29 acted as effective targets for CTLs when expressed on their own, each variant acted to suppress the CTL response to the other variant (Gilbert et al., 1998). Moreover, when malaria parasites were recovered from the blood of infected individuals, it was found that the frequency of mixed infections with parasites expressing cp26 and cp29 was much greater than would be predicted from the individual frequencies of each variant. This seems to be an example of immunological antagonism, whereby the CTL response to one peptide fragment may be inhibited by a similar but nonidentical fragment, and it has been proposed that the parasite exploits this strategy of expressing antagonistic CTL epitopes as a mechanism of immune evasion.

HLA-DRB1 Encoding MHC Class II DRB1

HLA-DRB1 encodes an HLA class II beta chain which forms a heterodimer with an alpha chain to make up the HLA-DR antigen-presenting complex, expressed in B lymphocytes, dendritic cells, and macrophages, which presents peptides derived from extracellular proteins and is crucial for antibody production. HLA DRB1*1302-DQB1*0501 has been associated with protection from severe malaria in The Gambia (Hill et al., 1991). A longitudinal cohort study of children in The Gambia found that malaria attack rates were associated with the overall MHC class II distribution but individual DR-DQ haplotype associations were not conclusively identified (Bennett et al., 1993).

IL-4 Encoding Interleukin-4

Interleukin-4, produced by activated T cells, is important for the proliferation and differentiation of antibody-producing B cells. A study was carried out with the Fulani of Burkina Faso, who are more resistant to malaria attacks than neighboring ethnic groups and who also have higher levels of antimalarial antibodies. It was found that the *IL4*-524T allele is at high frequency in the Fulani and is associated with elevated antibody levels against malaria antigens, raising the possibility that this might be a factor in their increased resistance to malaria (Luoni et al., 2001).

TNFSF5 Encoding CD40 Ligand

TNFSF5, so called because it is a member of the tumor necrosis factor (TNF) superfamily, encodes a glycoprotein that is expressed on T cells, known as CD40 ligand. By engaging CD40 on the B-cell surface, it regulates B-cell function, particularly immunoglobulin class switching; rare coding mutations in CD40L can lead to life-threatening immunodeficiency. In a Gambian case-control study, a significant reduction in risk for severe malaria was associated with males hemizygous for the *TNFSF5*-726C allele, and this was confirmed by transmission disequilibrium test analysis of affected families. A similar but nonsignificant trend was found with female subjects (Sabeti et al., 2002a). Long-range haplotype analysis of this allele suggests that it has recently undergone positive evolutionary selection (Sabeti et al., 2002b).

FCGR2A Encoding Low-Affinity IIa Receptor for Fc Fragment of IgG

Receptors for the Fc portion of IgG are present on many leukocytes, and they are essential in removing antigen-antibody complexes from the circulation. In vitro studies indicate that IgG1 and IgG3, but not IgG2, are essential for inhibition of *P. falciparum* by antibody-dependent cellular mechanisms. An amino acid substitution of His to Arg at codon 131 of *FCGR2A* results in failure to bind to IgG2 and has been associated with protection against high levels of *P. falciparum* parasitemia in Kenya (Shi et al., 2001). Consistent with this observation, studies in Thailand and The Gambia found that homozygotes for the 131His genotype are susceptible to cerebral malaria (Cooke et al., 2003; Omi et al., 2002). In the Thai study, the *FCGR2A* association with severe malaria involved interaction with a polymorphism of the *FCGR3B* gene.

MEDIATORS OF INNATE IMMUNITY

TNF Encoding Tumor Necrosis Factor

Tumor necrosis factor (TNF) is a proinflammatory cytokine that is critical for innate immunity against malaria parasites but has also been implicated in the pathogenesis of severe malaria (reviewed in Kwiatkowski and Perlmann, 1999). The TNF locus is at the center of the linkage peaks identified in longitudinal family studies of malaria episodes in The Gambia and Burkina Faso (Flori et al., 2003; Jepson et al., 1997). Several TNF promoter polymorphisms have been independently associated with severe malaria. Gambian children who are homozygous for the TNF-308A allele have increased susceptibility to cerebral malaria (McGuire et al., 1994), and in Gabon it was found that those who carried this allele were more likely to encounter symptomatic reinfections with *P. falciparum* (Meyer et al., 2002). Among other investigations of the TNF-308A allele in malaria-endemic areas, a study in Sri Lanka found that the carriers of this allele had increased risk of severe infectious diseases in general (Wattavidanage et al., 1999), and a study in Kenya found an increase in infant mortality and malaria morbidity (Aidoo et al., 2001). The TNF-376A allele, which acts to recruit transcription factor OCT-1, has been associated with susceptibility to cerebral malaria (Knight et al., 1999). The TNF-238A allele has been associated with susceptibility to severe malarial anemia in The Gambia (McGuire et al., 1999). The functional role of TNF-308 and other TNF polymorphisms has generated a good deal of controversy, and the jury is still out (Abraham and Kroeger, 1999; Bayley et al., 2004; Knight et al., 2003). TNF resides in the MHC class III region, which is full of interesting immunological genes, and patterns of linkage disequilibrium are complex, with many extended haplotypes (Ackerman et al., 2003). Thus, it is conceivable that the reported TNF associations are simply markers for neighboring functional polymorphisms.

NOS2A Encoding Inducible Nitric Oxide Synthase

Nitric oxide is a reactive free radical which attracted much interest in malaria research because it kills parasites but also has the capacity to perturb immune function and neurotransmission (reviewed in Clark and Rockett, 1996). The latter property suggests a potential role in cerebral malaria. The *NOS2A* gene encodes an inducible nitric oxide synthase (NOS). The NOS2A-954C allele has been associated with high baseline NOS activity in cells from Gabonese individuals, and in this population it has been associated with protection from severe malaria and resistance to reinfection (Kun et al., 1998; Kun et al., 2001). Gambian and Tanzanian studies found no association with this SNP allele (Burgner et al., 2003; Levesque et al., 1999). The NOS2A-1173T allele has been associated with increased fasting urine and plasma NO metabolite concentrations in Tanzanian children; it is associated with protection from symptomatic malaria in Tanzania and with protection from severe malarial anemia in Kenya (Hobbs et al., 2002). This allele was not associated with severe malaria in The Gambia: in that population, an NOS2 microsatellite polymorphism was associated with susceptibility to fatal malaria, and a haplotype uniquely defined by the NOS2A-1659T allele was associated with cerebral malaria by both transmission disequilibrium test and case-control analysis (Burgner et al., 2003; Burgner et al., 1998).

IFNGR1 Encoding the Alpha Chain of the Gamma Interferon Receptor

IFNGR1 encodes the ligand-binding chain of the heterodimeric gamma interferon (IFN-γ) receptor. There is copious evidence that IFN-γ is a critical mediator of immunity to malaria, but it is also implicated in the pathogenesis of severe disease (reviewed in Stevenson and Riley, 2004). A Gambian case-control study found that in Mandinka, the major Gambian ethnic group, heterozygotes for the IFNGR1-56 polymorphism were protected against cerebral malaria (Koch et al., 2002). Reporter gene analysis suggests that the minor allele acts to reduce levels of *IFNGR1* gene expression (Juliger et al., 2003).

IFNAR1 Encoding Alpha Interferon Receptor 1

The protein encoded by the *IFNAR1* gene is a type I membrane protein that forms one of the two chains of a receptor for alpha and beta interferon. With a murine malaria model, it has been observed that alpha interferon inhibits parasite development within erythrocytes (Vigario et al., 2001). A Gambian case-control study found two *IFNAR1* SNPs that were associated with protection against severe malaria, and a resistance haplotype was identified (Aucan et al., 2003).

IL12B Encoding Interleukin-12 p40

This gene encodes a subunit of interleukin-12, a cytokine produced by activated macrophages that is essential for the development of Th1 cells. Homozygosity for an *IL12B* promoter polymorphism was associated with decreased cellular production of nitric oxide in blood samples and with fatality in Tanzanian children with cerebral malaria. However, the same investigators could not reproduce the latter association in Kenyan children with severe malaria (Morahan et al., 2002).

IL1A and *IL1B* Encoding Interleukin-1α and -β

The interleukin-1 family of cytokines, produced mainly by macrophages, are important mediators of the inflammatory response to infection and of fever. In a Gambian case-control study, one SNP in *IL1A* and another in *IL1B* showed a marginal association with susceptibility to malaria but the effect was small (Walley et al., 2004).

MBL2 Encoding Mannose-Binding Lectin

MBL2 encodes a serum mannose-binding protein, also known as protein C. It recognizes mannose and *N*-acetylglucosamine on bacterial pathogens and can activate the classical complement pathway. *MBL2* polymorphisms have been associated with susceptibility to various infectious diseases, and it is presently unclear whether the same applies to malaria. A study in Gabon found that children with severe malaria had low serum MBL levels compared to those with mild malaria and that mutations in codons 54 and 57 of *MBL2* (which lead to low protein levels) were present at higher frequency in those with severe malaria (Luty et al., 1998). However, a study in The Gambia found no evidence that these alleles were associated with susceptibility to severe malaria (Bellamy et al., 1998b).

MAPPING THE GENES RESPONSIBLE FOR DIFFERENCES IN RESISTANCE TO INFECTION AMONG INBRED MOUSE STRAINS

Plasmodium chabaudi Infection

VARIABILITY IN THE RESISTANCE OF INBRED MICE TO INFECTION WITH *P. CHABAUDI*

When the survival times of different inbred strains of mice are compared after infection with *P. chabaudi*, some strains are relatively resistant (e.g., C57BL/6J, C57L/J, DBA/2J, CBA/J, and B10.A/SgSn) while others are relatively susceptible (e.g., A/J, DBA/1J, BALB/c, C3H/HeJ, AKR/J, and SJL/J) (Stevenson et al., 1982). Crossbreeding experiments indicate that at least some of these interstrain differences show classical Mendelian patterns of segregation (indicating that a single gene is largely responsible, as opposed to many different genes) and that resistance is generally dominant over susceptibility. In crossbreeding experiments using susceptible A/J- and resistant C57BL-derived mice, it was found that the ability to suppress parasitemia was genetically linked to the magnitude of splenomegaly (Stevenson and Skamene, 1985).

MAPPING OF CHROMOSOMAL LOCI THAT DETERMINE RESISTANCE TO *P. CHABAUDI* INFECTION

Genome-wide linkage screens based on crossing susceptible C3H and SJL mice with resistant C57BL/6J mice revealed three interesting loci. The so-called *char1* locus on chromosome 9 determines death or survival (Foote et al., 1987). The *char2* locus on chromosome 8 determines the control of parasite density (Foote et al., 1987), a finding that has been confirmed in an independent study that crossed susceptible A/J mice with resistant C57BL/6J mice (Fortin et al., 1997), as well as a study of mice congenic for this locus (Burt et al., 2002). The *char3* locus in the MHC region of chromosome 17 influences parasite clearance rates at the time immediately after peak parasitemia (Burt et al., 1999).

The search for novel loci was extended by deriving recombinant congenic strains from susceptible A/J and resistant C57BL/6J mice. This led to the identification of a resistance locus designated *char4* that maps to a small congenic B6 fragment on chromosome 3 (Fortin et al., 2001). Sequencing of candidate genes across this region has identified a plausible functional mutation: a loss-of-function coding variant of the pyruvate kinase gene (*pklr*), whose protective action against malaria seems to be related to the fact that it causes hemolytic anemia, which is compensated for by constitutive reticulocytosis and splenomegaly (Min-Oo et al., 2004; Min-Oo et al., 2003).

Plasmodium berghei ANKA Infection in Inbred Mouse Strains

HISTORY OF DEVELOPMENT OF THE MODEL

Over 20 years ago, an attempt was made to develop a murine model of cerebral malaria by infecting several different mouse strains with *P. berghei* ANKA (PBA) (Rest, 1982). Most of the mouse strains tested did not have any significant neurological changes, but 100% of the A/J mice died after 5 to 8 days and developed brain hemorrhages as a terminal event. Cloned lines seem to differ in their tendency to cause these neurological changes, indicating that pathology is determined by a specific combination of host and parasite genotype (Amani et al., 1998). The lethal character of PBA, compared to other experimental murine parasites, may partly relate to the fact that it invades erythrocytes rather than reticulocytes (Jayawardena et al., 1983).

THE CEREBRAL PATHOLOGY OF *P. BERGHEI* ANKA INFECTION IN SUSCEPTIBLE MOUSE STRAINS

Parasite sequestration in cerebral capillaries, a hallmark of human cerebral malaria (Taylor et al., 2004), is notable by its absence in the PBA experimental model. Instead, there is mononuclear cell adhesion to endothelium (which is absent in human cerebral malaria) together with hemorrhage and cerebral endothelial cell damage, accompanied by breakdown of the blood-brain barrier and cerebral edema (Neill et al., 1993; Neill and Hunt, 1992; Thumwood et al., 1988). Thus, it is very difficult to sustain the argument that PBA is a good experimental model of human cerebral malaria—a claim that has been boldly put in many research publications. However, it is certainly an interesting experimental model of immunopathological processes arising from malaria infection and how these may be affected by host and parasite genotype.

VARIOUS INTERVENTIONS THAT HAVE BEEN SHOWN TO INHIBIT THE PATHOLOGY

PBA-induced cerebral pathology has proved to be a gold mine for experimental immunopathologists: it is a striking pathological phenotype which appears to be readily reversed by a vast range of different immunological interventions (Engwerda et al., 2002; Grau et al., 1987a; Grau et al., 1987b; Grau et al., 1989; Hunt et al., 1993; Kremsner et al., 1991; Schofield et al., 2002). These interventions include drugs (cyclosporin A and pentoxifylline), antibodies against various cytokines (TNF, granulocyte-macrophage colony-stimulating factor plus IL-3, IFN-γ, and CD11a), depletion of various cell types (neutrophils, gamma-delta T cells, CD4+ cells, CD8+ cells), disruption of various genes (endothelial P-selectin, *CCR5*, *LTA*, urokinase and urokinase receptor, *CD40* and CD40 ligand, *TNFR2*, *IFNGR1*, *ICAM1*, perforin, and *SOCS1*), various dietary changes (dietary restriction, antioxidants, dietary-induced oxidative stress, and fatty acid injections); and various other forms of immunomodulation (anti-glycosylphosphatidylinositol antibodies, immunization with killed blood-stage parasites, and recombinant IL-10 injection). Taken together, these findings suggest that many different immunological processes need to come together for this particular form of pathology to occur.

IDENTIFICATION OF CHROMOSOMAL LOCI THAT DETERMINE GENETIC SUSCEPTIBILITY TO PBA-INDUCED CEREBRAL PATHOLOGY

One approach to the discovery of genetic factors that determine PBA-induced cerebral pathology has been to enlarge the gene pool—which is very restricted in inbred strains—by deriving new mouse strains with a wide range of genetic polymorphisms from wild mouse populations (Bagot et al., 2002b). When a resistant wild mouse-derived inbred strain (WLA) was crossed with a susceptible laboratory strain (C57BL/6J), all of the F_1 progeny and 97% of the F_2 progeny displayed resistance. A genome-wide screen, performed after backcrossing the resistant wild strain onto the susceptible laboratory strain, identified significant linkage to two regions, dubbed *Berr1* and *Berr2*, on chromosome 1 and chromosome 11, respectively (Bagot et al., 2002a).

A study that crossed susceptible C57BL/6 mice with resistant DBA/2 mice identified a major locus on chromosome 18 (Nagayasu et al., 2002). Another study that crossed susceptible CBA mice with resistant DBA/2 mice has identified a different susceptibility locus at the MHC region on chromosome 17 (Ohno and Nishimura, 2004).

REFERENCES

Abraham, L. J., and K. M. Kroeger. 1999. Impact of the −308 TNF promoter polymorphism on the transcriptional regulation of the TNF gene: relevance to disease. *J. Leukoc. Biol.* **66:**562–566.

Ackerman, H. C., G. Ribas, M. Jallow, R. Mott, M. Neville, F. Sisay-Joof, M. Pinder, R. D. Campbell, and D. P. Kwiatkowski. 2003. Complex haplotypic structure of the central MHC region flanking TNF in a West African population. *Genes Immun.* **4:**476–486.

Agarwal, A., A. Guindo, Y. Cissoko, J. G. Taylor, D. Coulibaly, A. Kone, K. Kayentao, A. Djimde, C. V. Plowe, O. Doumbo, T. E. Wellems, and D. Diallo. 2000. Hemoglobin C associated with protection from severe malaria in the Dogon of Mali, a West African population with a low prevalence of hemoglobin S. *Blood* **96:**2358–2363.

Aidoo, M., P. D. McElroy, M. S. Kolczak, D. J. Terlouw, F. O. ter Kuile, B. Nahlen, A. A. Lal, and V. Udhayakumar. 2001. Tumor necrosis factor-alpha promoter variant 2 (TNF2) is associated with pre-term delivery, infant mortality, and malaria morbidity in western Kenya: Asembo Bay Cohort Project IX. *Genet. Epidemiol.* **21:**201–211.

Aitman, T. J., L. D. Cooper, P. J. Norsworthy, F. N. Wahid, J. K. Gray, B. R. Curtis, P. M. McKeigue, D. Kwiatkowski, B. M. Greenwood, R. W. Snow, A. V. Hill, and J. Scott. 2000. Malaria susceptibility and CD36 mutation. *Nature* **405:**1015–1016.

Allen, S. J., S. Bennett, E. M. Riley, P. A. Rowe, P. H. Jakobsen, A. O'Donnell, and B. M. Greenwood. 1992. Morbidity from malaria and immune responses to defined Plasmodium falciparum antigens in children with sickle cell trait in The Gambia. *Trans. R. Soc. Trop. Med. Hyg.* **86:**494–498.

Allen, S. J., A. O'Donnell, N. D. Alexander, M. P. Alpers, T. E. A. Peto, J. B. Clegg, and D. J. Weatherall. 1997. α+-Thalassemia protects children against disease caused by other infections as well as malaria. *Proc. Natl. Acad. Sci. USA* **94:**14736–14741.

Allen, S. J., A. O'Donnell, N. D. Alexander, C. S. Mgone, T. E. Peto, J. B. Clegg, M. P. Alpers, and D. J. Weatherall. 1999. Prevention of cerebral malaria in children in Papua New Guinea by southeast Asian ovalocytosis band 3. *Am. J. Trop. Med. Hyg.* **60:**1056–1060.

Allison, A. C. 1954. Protection afforded by sickle-cell trait against subtertian malareal infection. *Br. Med. J.* **4857:**290–294.

Amani, V., M. I. Boubou, S. Pied, M. Marussig, D. Walliker, D. Mazier, and L. Renia. 1998. Cloned lines of Plasmodium berghei ANKA differ in their abilities to induce experimental cerebral malaria. *Infect. Immun.* **66:**4093–4099.

Aucan, C., A. J. Walley, B. M. Greenwood, and A. V. Hill. 2002. Haptoglobin genotypes are not associated with resistance to severe malaria in The Gambia. *Trans. R. Soc. Trop. Med. Hyg.* **96:**327–328.

Aucan, C., A. J. Walley, B. J. Hennig, J. Fitness, A. Frodsham, L. Zhang, D. Kwiatkowski, and A. V. Hill. 2003. Interferon-alpha receptor-1 (IFNAR1) variants are associated with protection against cerebral malaria in The Gambia. *Genes Immun.* **4:**275–282.

Bagot, S., S. Campino, C. Penha-Goncalves, S. Pied, P. A. Cazenave, and D. Holmberg. 2002a. Identification of two cerebral malaria resistance loci using an inbred wild-derived mouse strain. *Proc. Natl. Acad. Sci. USA* **99:**9919–9923.

Bagot, S., M. Idrissa Boubou, S. Campino, C. Behrschmidt, O. Gorgette, J. L. Guenet, C. Penha-Goncalves, D. Mazier, S. Pied, and P. A. Cazenave. 2002b. Susceptibility to experimental cerebral malaria induced by *Plasmodium berghei* ANKA in inbred mouse strains recently derived from wild stock. *Infect. Immun.* **70:**2049–2056.

Baird, J. K. 1995. Host age as a determinant of naturally acquired immunity to Plasmodium falciparum. *Parasitol. Today* **11:**105–111.

Baird, J. K., S. Masbar, H. Basri, S. Tirtokusumo, B. Subianto, and S. L. Hoffman. 1998. Age-dependent susceptibility to severe disease with primary exposure to Plasmodium falciparum. *J. Infect. Dis.* **178:**592–595.

Barnwell, J. W., A. S. Asch, R. L. Nachman, M. Yamaya, M. Aikawa, and P. Ingravallo. 1989. A human 88-kD membrane glycoprotein (CD36) functions in vitro as a receptor for a cytoadherence ligand on Plasmodium falciparum-infected erythrocytes. *J. Clin. Investig.* **84:**765–772.

Baum, J., R. H. Ward, and D. J. Conway. 2002. Natural selection on the erythrocyte surface. *Mol. Biol. Evol.* **19:**223–229.

Bayley, J. P., T. H. Ottenhoff, and C. L. Verweij. 2004. Is there a future for TNF promoter polymorphisms? *Genes Immun.* **5:**315–329.

Bellamy, R., D. Kwiatkowski, and A. V. Hill. 1998a. Absence of an association between intercellular adhesion molecule 1, complement receptor 1 and interleukin 1 receptor antagonist gene polymorphisms and severe malaria in a West African population. *Trans. R. Soc. Trop. Med. Hyg.* **92:**312–316.

Bellamy, R., C. Ruwende, K. P. McAdam, M. Thursz, M. Sumiya, J. Summerfield, S. C. Gilbert, T. Corrah, D. Kwiatkowski, H. C. Whittle, and A. V. Hill. 1998b. Mannose binding protein deficiency is not associated with malaria, hepatitis B carriage nor tuberculosis in Africans. *QJM* **91:**13–18.

Bennett, S., S. J. Allen, O. Olerup, D. J. Jackson, J. G. Wheeler, P. A. Rowe, E. M. Riley, and B. M. Greenwood. 1993. Human leucocyte antigen (HLA) and malaria morbidity in a Gambian community. *Trans. R. Soc. Trop. Med. Hyg.* **87:**286–287.

Berendt, A. R., D. L. Simmons, J. Tansey, C. I. Newbold, and K. Marsh. 1989. Intercellular adhesion molecule-1 is an endothelial cell adhesion receptor for Plasmodium falciparum. *Nature* **341:**57–59.

Burgner, D., S. Usen, K. Rockett, M. Jallow, H. Ackerman, A. Cervino, M. Pinder, and D. P. Kwiatkowski. 2003. Nucleotide and haplotypic diversity of the NOS2A promoter region and its relationship to cerebral malaria. *Hum. Genet.* **112:**379–386.

Burgner, D., W. Xu, K. Rockett, M. Gravenor, I. G. Charles, A. V. Hill, and D. Kwiatkowski. 1998. Inducible nitric oxide synthase polymorphism and fatal cerebral malaria. *Lancet* **352:**1193–1194.

Burt, R. A., T. M. Baldwin, V. M. Marshall, and S. J. Foote. 1999. Temporal expression of an H2-linked locus in host response to mouse malaria. *Immunogenetics* **50:**278–285.

Burt, R. A., V. M. Marshall, J. Wagglen, F. R. Rodda, D. Senyschen, T. M. Baldwin, L. A. Buckingham, and S. J. Foote. 2002. Mice that are congenic for the *char2* locus are susceptible to malaria. *Infect. Immun.* **70:**4750–4753.

Casals-Pascual, C., S. Allen, A. Allen, O. Kai, B. Lowe, A. Pain, and D. J. Roberts. 2001. Short report: codon 125 polymorphism of CD31 and susceptibility to malaria. *Am. J. Trop. Med. Hyg.* **65:**736–737.

Cattani, J. A., F. D. Gibson, M. P. Alpers, and G. G. Crane. 1987. Hereditary ovalocytosis and reduced susceptibility to malaria in Papua New Guinea. *Trans. R. Soc. Trop. Med. Hyg.* **81:**705–709.

Chitnis, C. E., and L. H. Miller. 1994. Identification of the erythrocyte binding domains of Plasmodium vivax and Plasmodium knowlesi proteins involved in erythrocyte invasion. *J. Exp. Med.* **180:**497–506.

Chotivanich, K., R. Udomsangpetch, K. Pattanapanyasat, W. Chierakul, J. Simpson, S. Looareesuwan, and N. White. 2002. Hemoglobin E: a balanced polymorphism protective against high parasitemias and thus severe P falciparum malaria. *Blood* **100:**1172–1176.

Clark, I. A., and K. A. Rockett. 1996. Nitric oxide and parasitic disease. *Adv. Parasitol.* **37:**1–56.

Cockburn, I. A., M. J. Mackinnon, A. O'Donnell, S. J. Allen, J. M. Moulds, M. Baisor, M. Bockarie, J. C. Reeder, and J. A. Rowe. 2004. A human complement receptor 1 polymorphism that reduces Plasmodium falciparum rosetting confers protection against severe malaria. *Proc. Natl. Acad. Sci. USA* **101:**272–277.

Cooke, G. S., C. Aucan, A. J. Walley, S. Segal, B. M. Greenwood, D. P. Kwiatkowski, and A. V. Hill. 2003. Association of Fcγ receptor IIa (CD32) polymorphism with severe malaria in West Africa. *Am. J. Trop. Med. Hyg.* **69:**565–568.

Diallo, D. A., O. K. Doumbo, A. Dicko, A. Guindo, D. Coulibaly, K. Kayentao, A. A. Djimde, M. A. Thera, R. M. Fairhurst, C. V. Plowe, and T. E. Wellems. 2004. A comparison of anemia in hemoglobin C and normal hemoglobin A children with Plasmodium falciparum malaria. *Acta Trop.* **90:**295–299.

Elagib, A. A., A. O. Kider, B. Akerstrom, and M. I. Elbashir. 1998. Association of the haptoglobin phenotype (1-1) with falciparum malaria in Sudan. *Trans. R. Soc. Trop. Med. Hyg.* **92:**309–311.

Engwerda, C. R., T. L. Mynott, S. Sawhney, J. B. De Souza, Q. D. Bickle, and P. M. Kaye. 2002. Locally up-regulated lymphotoxin alpha, not systemic tumor necrosis factor alpha, is the principle mediator of murine cerebral malaria. *J. Exp. Med.* **195:**1371–1377.

Facer, C. A. 1983. Merozoites of P. falciparum require glycophorin for invasion into red cells. *Bull. Soc. Pathol. Exot. Filiales* **76:**463–469.

Fernandez-Reyes, D., A. G. Craig, S. A. Kyes, N. Peshu, R. W. Snow, A. R. Berendt, K. Marsh, and C. I. Newbold. 1997. A high frequency African coding polymorphism in the N-terminal domain of ICAM-1 predisposing to cerebral malaria in Kenya. *Hum. Mol. Genet.* **6:**1357–1360.

Flint, J., A. V. Hill, D. K. Bowden, S. J. Oppenheimer, P. R. Sill, S. W. Serjeantson, J. Bana-Koiri, K. Bhatia, M. P. Alpers, A. J. Boyce, et al. 1986. High frequencies of alpha-thalassaemia are the result of natural selection by malaria. *Nature* **321:**744–750.

Flori, L., S. Sawadogo, C. Esnault, N. F. Delahaye, F. Fumoux, and P. Rihet. 2003. Linkage of mild malaria to the major histocompatibility complex in families living in Burkina Faso. *Hum. Mol. Genet.* **12:**375–378.

Foo, L. C., V. Rekhraj, G. L. Chiang, and J. W. Mak. 1992. Ovalocytosis protects against severe malaria parasitemia in the Malayan aborigines. *Am. J. Trop. Med. Hyg.* **47:**271–275.

Foote, S. J., R. A. Burt, T. M. Baldwin, A. Presente, A. W. Roberts, Y. L. Laural, A. M. Lew, and V. M. Marshall. 1997. Mouse loci for malaria-induced mortality and the control of parasitaemia. *Nat. Genet.* **17:**380–381.

Fortin, A., A. Belouchi, M. F. Tam, L. Cardon, E. Skamene, M. M. Stevenson, and P. Gros. 1997. Genetic control of blood parasitaemia in mouse malaria maps to chromosome 8. *Nat. Genet.* **17:**382–383.

Fortin, A., L. R. Cardon, M. Tam, E. Skamene, M. M. Stevenson, and P. Gros. 2001. Identification of a new malaria susceptibility locus (Char4) in recombinant congenic strains of mice. *Proc. Natl. Acad. Sci. USA* **98:**10793–10798.

Ganczakowski, M., M. Town, D. K. Bowden, T. J. Vulliamy, A. Kaneko, J. B. Clegg, D. J. Weatherall, and L. Luzzatto. 1995. Multiple glucose 6-phosphate dehydrogenase-deficient variants correlate with malaria endemicity in the Vanuatu archipelago (southwestern Pacific). *Am. J. Hum. Genet.* **56:**294–301.

Garcia, A., M. Cot, J. P. Chippaux, S. Ranque, J. Feingold, F. Demenais, and L. Abel. 1998a. Genetic control of blood infection levels in human malaria: evidence for a complex genetic model. *Am. J. Trop. Med. Hyg.* **58:**480–488.

Garcia, A., S. Marquet, B. Bucheton, D. Hillaire, M. Cot, N. Fievet, A. J. Dessein, and L. Abel. 1998b. Linkage analysis of blood Plasmodium falciparum levels: interest of the 5q31-q33 chromosome region. *Am. J. Trop. Med. Hyg.* **58:**705–709.

Genton, B., F. al-Yaman, C. S. Mgone, N. Alexander, M. M. Paniu, M. P. Alpers, and D. Mokela. 1995. Ovalocytosis and cerebral malaria. *Nature* **378:**564–565.

Gilbert, S. C., M. Plebanski, S. Gupta, J. Morris, M. Cox, M. Aidoo, D. Kwiatkowski, B. M. Greenwood, H. C. Whittle, and A. V. Hill. 1998. Association of malaria parasite population structure, HLA, and immunological antagonism. *Science* **279:**1173–1177.

Gilles, H. M., K. A. Fletcher, R. G. Hendrickse, R. Lindner, S. Reddy, and N. Allan. 1967. Glucose-6-phosphate-dehydrogenase deficiency, sickling, and malaria in African children in south western Nigeria. *Lancet* **1:**138–140.

Grau, G. E., L. F. Fajardo, P. F. Piguet, B. Allet, P. H. Lambert, and P. Vassalli. 1987a. Tumor necrosis factor (cachectin) as an essential mediator in murine cerebral malaria. *Science* **237:**1210–1212.

Grau, G. E., D. Gretener, and P. H. Lambert. 1987b. Prevention of murine cerebral malaria by low-dose cyclosporin A. *Immunology* **61:**521–525.

Grau, G. E., H. Heremans, P. F. Piguet, P. Pointaire, P. H. Lambert, A. Billiau, and P. Vassalli. 1989. Monoclonal antibody against interferon gamma can prevent experimental cerebral malaria and its associated overproduction of tumor necrosis factor. *Proc. Natl. Acad. Sci. USA* **86:**5572–5574.

Greenberg, J., and L. P. Kendrick. 1959. Resistance to malaria in hybrids between Swiss and certain other strains of mice. *J. Parasitol.* **45:**263–267.

Greenberg, J., E. M. Nadel, and G. R. Coatney. 1954. Differences in survival of several inbred strains of mice and their hybrids infected with Plasmodium berghei. *J. Infect. Dis.* **95:**114–116.

Hill, A. V., C. E. Allsopp, D. Kwiatkowski, N. M. Anstey, P. Twumasi, P. A. Rowe, S. Bennett, D. Brewster, A. J. McMichael, and B. M. Greenwood. 1991. Common west African HLA antigens are associated with protection from severe malaria. *Nature* **352:**595–600.

Hill, A. V., J. Elvin, A. C. Willis, M. Aidoo, C. E. Allsopp, F. M. Gotch, X. M. Gao, M. Takiguchi,

B. M. Greenwood, A. R. Townsend, et al. 1992. Molecular analysis of the association of HLA-B53 and resistance to severe malaria. *Nature* **360**:434–439.

Hobbs, M. R., V. Udhayakumar, M. C. Levesque, J. Booth, J. M. Roberts, A. N. Tkachuk, A. Pole, H. Coon, S. Kariuki, B. L. Nahlen, E. D. Mwaikambo, A. L. Lal, D. L. Granger, N. M. Anstey, and J. B. Weinberg. 2002. A new NOS2 promoter polymorphism associated with increased nitric oxide production and protection from severe malaria in Tanzanian and Kenyan children. *Lancet* **360**:1468–1475.

Hunt, N. H., C. Driussi, and L. Sai-Kiang. 2001. Haptoglobin and malaria. *Redox Rep.* **6**:389–392.

Hunt, N. H., N. Manduci, and C. M. Thumwood. 1993. Amelioration of murine cerebral malaria by dietary restriction. *Parasitology* **107**:471–476.

Jayawardena, A. N., R. Mogil, D. B. Murphy, D. Burger, and R. K. Gershon. 1983. Enhanced expression of H-2K and H-2D antigens on reticulocytes infected with Plasmodium yoelii. *Nature* **302**:623–626.

Jepson, A., F. Sisay-Joof, W. Banya, M. Hassan-King, A. Frodsham, S. Bennett, A. V. Hill, and H. Whittle. 1997. Genetic linkage of mild malaria to the major histocompatibility complex in Gambian children: study of affected sibling pairs. *BMJ* **315**:96–97.

Jepson, A. P., W. A. Banya, F. Sisay-Joof, M. Hassan-King, S. Bennett, and H. C. Whittle. 1995. Genetic regulation of fever in Plasmodium falciparum malaria in Gambian twin children. *J. Infect. Dis.* **172**:316–319.

Juliger, S., M. Bongartz, A. J. Luty, P. G. Kremsner, and J. F. Kun. 2003. Functional analysis of a promoter variant of the gene encoding the interferon-gamma receptor chain I. *Immunogenetics* **54**:675–680.

Kikuchi, M., S. Looareesuwan, R. Ubalee, O. Tasanor, F. Suzuki, Y. Wattanagoon, K. Na-Bangchang, A. Kimura, M. Aikawa, and K. Hirayama. 2001. Association of adhesion molecule PECAM-1/CD31 polymorphism with susceptibility to cerebral malaria in Thais. *Parasitol. Int.* **50**:235–239.

Kitchen, S. F. 1949. Symptomatology: general considerations, p. 966–994. *In* M. F. Boyd (ed.), *Malariology*. W. B. Saunders, Philadelphia, Pa.

Knight, J. C., B. J. Keating, K. A. Rockett, and D. P. Kwiatkowski. 2003. In vivo characterization of regulatory polymorphisms by allele-specific quantification of RNA polymerase loading. *Nat. Genet.* **33**:469–475.

Knight, J. C., I. Udalova, A. V. Hill, B. M. Greenwood, N. Peshu, K. Marsh, and D. Kwiatkowski. 1999. A polymorphism that affects OCT-1 binding to the TNF promoter region is associated with severe malaria. *Nat. Genet.* **22**:145–150.

Koch, O., A. Awomoyi, S. Usen, M. Jallow, A. Richardson, J. Hull, M. Pinder, M. Newport, and D. Kwiatkowski. 2002. IFNGR1 gene promoter polymorphisms and susceptibility to cerebral malaria. *J. Infect. Dis.* **185**:1684–1687.

Kremsner, P. G., H. Grundmann, S. Neifer, K. Sliwa, G. Sahlmuller, B. Hegenscheid, and U. Bienzle. 1991. Pentoxifylline prevents murine cerebral malaria. *J. Infect. Dis.* **164**:605–608.

Kun, J. F., J. Klabunde, B. Lell, D. Luckner, M. Alpers, J. May, C. Meyer, and P. G. Kremsner. 1999. Association of the ICAM-1Kilifi mutation with protection against severe malaria in Lambarene, Gabon. *Am. J. Trop. Med. Hyg.* **61**:776–779.

Kun, J. F., B. Mordmuller, B. Lell, L. G. Lehman, D. Luckner, and P. G. Kremsner. 1998. Polymorphism in promoter region of inducible nitric oxide synthase gene and protection against malaria. *Lancet* **351**:265–266.

Kun, J. F., B. Mordmuller, D. J. Perkins, J. May, O. Mercereau-Puijalon, M. Alpers, J. B. Weinberg, and P. G. Kremsner. 2001. Nitric oxide synthase 2(Lambarene) (G-954C), increased nitric oxide production, and protection against malaria. *J. Infect. Dis.* **184**:330–336.

Kwiatkowski, D., and P. Perlmann. 1999. Inflammatory processes in the pathogenesis of malaria, p. 329–362. *In* M. Wahlgen and P. Perlmann (ed.), *Malaria: Molecular and Clinical Aspects*. Harwood Academic Publishers, The Netherlands.

Levesque, M. C., M. R. Hobbs, N. M. Anstey, T. N. Vaughn, J. A. Chancellor, A. Pole, D. J. Perkins, M. A. Misukonis, S. J. Chanock, D. L. Granger, and J. B. Weinberg. 1999. Nitric oxide synthase type 2 promoter polymorphisms, nitric oxide production, and disease severity in Tanzanian children with malaria. *J. Infect. Dis.* **180**:1994–2002.

Luoni, G., F. Verra, B. Arca, B. S. Sirima, M. Troye-Blomberg, M. Coluzzi, D. Kwiatkowski, and D. Modiano. 2001. Antimalarial antibody levels and IL4 polymorphism in the Fulani of West Africa. *Genes Immun.* **2**:411–414.

Luty, A. J., J. F. Kun, and P. G. Kremsner. 1998. Mannose-binding lectin plasma levels and gene polymorphisms in Plasmodium falciparum malaria. *J. Infect. Dis.* **178**:1221–1224.

Luzzatto, L., F. A. Usanga, and S. Reddy. 1969. Glucose-6-phosphate dehydrogenase deficient red cells: resistance to infection by malarial parasites. *Science* **164**:839–842.

Maier, A. G., M. T. Duraisingh, J. C. Reeder, S. S. Patel, J. W. Kazura, P. A. Zimmerman, and A. F. Cowman. 2003. Plasmodium falciparum erythrocyte invasion through glycophorin C and selection for Gerbich negativity in human populations. *Nat. Med.* **9:**87–92.

Mayor, A., N. Bir, R. Sawhney, S. Singh, P. Pattnaik, S. K. Singh, A. Sharma, and C. E. Chitnis. 2005. Receptor-binding residues lie in central regions of Duffy-binding-like domains involved in red cell invasion and cytoadherence by malaria parasites. *Blood* **105:**2557–2563.

McGuire, W., A. V. Hill, C. E. Allsopp, B. M. Greenwood, and D. Kwiatkowski. 1994. Variation in the TNF-alpha promoter region associated with susceptibility to cerebral malaria. *Nature* **371:**508–510.

McGuire, W., J. C. Knight, A. V. Hill, C. E. Allsopp, B. M. Greenwood, and D. Kwiatkowski. 1999. Severe malarial anemia and cerebral malaria are associated with different tumor necrosis factor promoter alleles. *J. Infect. Dis.* **179:**287–290.

Meyer, C. G., J. May, A. J. Luty, B. Lell, and P. G. Kremsner. 2002. TNFα-308A associated with shorter intervals of Plasmodium falciparum reinfections. *Tissue Antigens* **59:**287–292.

Miller, L. H., S. J. Mason, D. F. Clyde, and M. H. McGinniss. 1976. The resistance factor to Plasmodium vivax in blacks. The Duffy-blood-group genotype, FyFy. *N. Engl. J. Med.* **295:**302–304.

Min-Oo, G., A. Fortin, M. F. Tam, P. Gros, and M. M. Stevenson. 2004. Phenotypic expression of pyruvate kinase deficiency and protection against malaria in a mouse model. *Genes Immun.* **5:**168–175.

Min-Oo, G., A. Fortin, M. F. Tam, A. Nantel, M. M. Stevenson, and P. Gros. 2003. Pyruvate kinase deficiency in mice protects against malaria. *Nat. Genet.* **35:**357–362.

Mockenhaupt, F. P., S. Ehrhardt, S. Gellert, R. N. Otchwemah, E. Dietz, S. D. Anemana, and V. Bienzle. 2004. Alpha(+)-thalassemia protects African children from severe malaria. *Blood* **104:**2003–2006.

Modiano, D., A. Chiucchiuini, V. Petrarca, B. S. Sirima, G. Luoni, H. Perlmann, F. Esposito, and M. Coluzzi. 1998. Humoral response to Plasmodium falciparum Pf155/ring-infected erythrocyte surface antigen and Pf332 in three sympatric ethnic groups of Burkina Faso. *Am. J. Trop. Med. Hyg.* **58:**220–224.

Modiano, D., A. Chiucchiuini, V. Petrarca, B. S. Sirima, G. Luoni, M. A. Roggero, G. Corradin, M. Coluzzi, and F. Esposito. 1999. Interethnic differences in the humoral response to non-repetitive regions of the Plasmodium falciparum circumsporozoite protein. *Am. J. Trop. Med. Hyg.* **61:**663–667.

Modiano, D., G. Luoni, B. S. Sirima, A. Lanfrancotti, V. Petrarca, F. Cruciani, J. Simpore, B. M. Ciminelli, E. Foglietta, P. Grisanti, I. Bianco, G. Modiano, and M. Coluzzi. 2001a. The lower susceptibility to Plasmodium falciparum malaria of Fulani of Burkina Faso (west Africa) is associated with low frequencies of classic malaria-resistance genes. *Trans. R. Soc. Trop. Med. Hyg.* **95:**149–152.

Modiano, D., G. Luoni, B. S. Sirima, J. Simpore, F. Verra, A. Konate, E. Rastrelli, A. Olivieri, C. Calissano, G. M. Paganotti, L. D'Urbano, I. Sanou, A. Sawadogo, G. Modiano, and M. Coluzzi. 2001b. Haemoglobin C protects against clinical Plasmodium falciparum malaria. *Nature* **414:**305–308.

Modiano, D., V. Petrarca, B. S. Sirima, I. Nebie, D. Diallo, F. Esposito, and M. Coluzzi. 1996. Different response to Plasmodium falciparum malaria in West African sympatric ethnic groups. *Proc. Natl. Acad. Sci. USA* **93:**13206–13211.

Morahan, G., C. S. Boutlis, D. Huang, A. Pain, J. R. Saunders, M. R. Hobbs, D. L. Granger, J. B. Weinberg, N. Peshu, E. D. Mwaikambo, K. Marsh, D. J. Roberts, and N. M. Anstey. 2002. A promoter polymorphism in the gene encoding interleukin-12 p40 (IL12B) is associated with mortality from cerebral malaria and with reduced nitric oxide production. *Genes Immun.* **3:**414–418.

Nadel, E., J. Greenberg, G. E. Jay, and G. R. Coatney. 1955. Backcross studies on the genetics of resistance to malaria in mice. *Genetics* **40:**620–626.

Nagayasu, E., M. Ito, M. Akaki, Y. Nakano, M. Kimura, S. Looareesuwan, and M. Aikawa. 2001. CR1 density polymorphism on erythrocytes of falciparum malaria patients in Thailand. *Am. J. Trop. Med. Hyg.* **64:**1–5.

Nagayasu, E., K. Nagakura, M. Akaki, G. Tamiya, S. Makino, Y. Nakano, M. Kimura, and M. Aikawa. 2002. Association of a determinant on mouse chromosome 18 with experimental severe Plasmodium berghei malaria. *Infect. Immun.* **70:**512–516.

Neill, A. L., T. Chan-Ling, and N. H. Hunt. 1993. Comparisons between microvascular changes in cerebral and non-cerebral malaria in mice, using the retinal whole-mount technique. *Parasitology* **107:**477–487.

Neill, A. L., and N. H. Hunt. 1992. Pathology of fatal and resolving Plasmodium berghei cerebral malaria in mice. *Parasitology* **105:**165–175.

Ohashi, J., I. Naka, J. Patarapotikul, H. Hananantachai, G. Brittenham, S. Looareesuwan, A. G. Clark, and K. Tokunaga. 2004. Extended linkage disequilibrium surrounding the hemoglobin E variant due to malarial selection. *Am. J. Hum. Genet.* **74:**1198–1208.

Ohno, T., and M. Nishimura. 2004. Detection of a new cerebral malaria susceptibility locus, using CBA mice. *Immunogenetics* **56:**675–678.

Omi, K., J. Ohashi, J. Patarapotikul, H. Hananantachai, I. Naka, S. Looareesuwan, and K. Tokunaga. 2003. CD36 polymorphism is associated with protection from cerebral malaria. *Am. J. Hum. Genet.* **72:**364–374.

Omi, K., J. Ohashi, J. Patarapotikul, H. Hananantachai, I. Naka, S. Looareesuwan, and K. Tokunaga. 2002. Fcγ receptor IIA and IIIB polymorphisms are associated with susceptibility to cerebral malaria. *Parasitol. Int.* **51:**361–366.

Orlandi, P. A., F. W. Klotz, and J. D. Haynes. 1992. A malaria invasion receptor, the 175-kilodalton erythrocyte binding antigen of *Plasmodium falciparum* recognizes the terminal Neu5Ac((2-3)Gal- sequences of glycophorin A. *J. Cell Biol.* **116:**901–909.

Pain, A., B. C. Urban, O. Kai, C. Casals-Pascual, J. Shafi, K. Marsh, and D. J. Roberts. 2001. A nonsense mutation in Cd36 gene is associated with protection from severe malaria. *Lancet* **357:**1502–1503.

Pasvol, G., D. J. Weatherall, and R. J. Wilson. 1978. Cellular mechanism for the protective effect of haemoglobin S against P. falciparum malaria. *Nature* **274:**701–703.

Patel, S. S., C. L. King, C. S. Mgone, J. W. Kazura, and P. A. Zimmerman. 2004. Glycophorin C (Gerbich antigen blood group) and band 3 polymorphisms in two malaria holoendemic regions of Papua New Guinea. *Am. J. Hematol.* **75:**1–5.

Patel, S. S., R. K. Mehlotra, W. Kastens, C. S. Mgone, J. W. Kazura, and P. A. Zimmerman. 2001. The association of the glycophorin C exon 3 deletion with ovalocytosis and malaria susceptibility in the Wosera, Papua New Guinea. *Blood* **98:**3489-3491.

Quaye, I. K., F. A. Ekuban, B. Q. Goka, V. Adabayeri, J. A. Kurtzhals, B. Gyan, N. A. Ankrah, L. Hviid, and B. D. Akanmori. 2000. Haptoglobin 1-1 is associated with susceptibility to severe Plasmodium falciparum malaria. *Trans. R. Soc. Trop. Med. Hyg.* **94:**216–219.

Rest, J. R. 1982. Cerebral malaria in inbred mice. I. A new model and its pathology. *Trans. R. Soc. Trop. Med. Hyg.* **76:**410–415.

Rihet, P., L. Abel, Y. Traore, T. Traore-Leroux, C. Aucan, and F. Fumoux. 1998a. Human malaria: segregation analysis of blood infection levels in a suburban area and a rural area in Burkina Faso. *Genet. Epidemiol.* **15:**435-450.

Rihet, P., L. Flori, F. Tall, A. S. Traore, and F. Fumoux. 2004. Hemoglobin C is associated with reduced Plasmodium falciparum parasitemia and low risk of mild malaria attack. *Hum. Mol. Genet.* **13:**1–6.

Rihet, P., Y. Traore, L. Abel, C. Aucan, T. Traore-Leroux, and F. Fumoux. 1998b. Malaria in humans: Plasmodium falciparum blood infection levels are linked to chromosome 5q31-q33. *Am. J. Hum. Genet.* **63:**498–505.

Rowe, J. A., J. M. Moulds, C. I. Newbold, and L. H. Miller. 1997. P. falciparum rosetting mediated by a parasite-variant erythrocyte membrane protein and complement-receptor 1. *Nature* **388:**292–295.

Ruwende, C., S. C. Khoo, R. W. Snow, S. N. Yates, D. Kwiatkowski, S. Gupta, P. Warn, C. E. Allsopp, S. C. Gilbert, N. Peschu, et al. 1995. Natural selection of hemi- and heterozygotes for G6PD deficiency in Africa by resistance to severe malaria. *Nature* **376:**246-249.

Sabeti, P., S. Usen, S. Farhadian, M. Jallow, T. Doherty, M. Newport, M. Pinder, R. Ward, and D. Kwiatkowski. 2002a. CD40L association with protection from severe malaria. *Genes Immun.* **3:**286–291.

Sabeti, P. C., D. E. Reich, J. M. Higgins, H. Z. Levine, D. J. Richter, S. F. Schaffner, S. B. Gabriel, J. V. Platko, N. J. Patterson, G. J. McDonald, H. C. Ackerman, S. J. Campbell, D. Altshuler, R. Cooper, D. Kwiatkowski, R. Ward, and E. S. Lander. 2002b. Detecting recent positive selection in the human genome from haplotype structure. *Nature* **419:**832–837.

Sayles, P. C., and D. L. Wassom. 1988. Immunoregulation in murine malaria. Susceptibility of inbred mice to infection with Plasmodium yoelii depends on the dynamic interplay of host and parasite genes. *J. Immunol.* **141:**241–248.

Schofield, L., M. C. Hewitt, K. Evans, M. A. Siomos, and P. H. Seeberger. 2002. Synthetic GPI as a candidate anti-toxic vaccine in a model of malaria. *Nature* **418:**785–789.

Shear, H. L., E. F. Roth, Jr., M. E. Fabry, F. D. Costantini, A. Pachnis, A. Hood, and R. L. Nagel. 1993. Transgenic mice expressing human sickle hemoglobin are partially resistant to rodent malaria. *Blood* **81:**222–226.

Sherman, I. W., S. Eda, and E. Winograd. 2003. Cytoadherence and sequestration in *Plasmodium falciparum*: defining the ties that bind. *Microbes Infect.* **5:**897–909.

Shi, Y. P., B. L. Nahlen, S. Kariuki, K. B. Urdahl, P. D. McElroy, J. M. Roberts, and A. A. Lal. 2001. Fcγ receptor IIa (CD32) polymorphism is associated with protection of infants against high-density Plasmodium falciparum infection. VII. Asembo Bay Cohort Project. *J. Infect. Dis.* **184:**107–111.

Sokhna, C. S., C. Rogier, A. Dieye, and J. F. Trape. 2000. Host factors affecting the delay of reappearance of Plasmodium falciparum after radical treatment among a semi-immune population exposed to intense perennial transmission. *Am. J. Trop. Med. Hyg.* **62:**266–270.

Stevenson, M. M., J. J. Lyanga, and E. Skamene. 1982. Murine malaria: genetic control of resistance to *Plasmodium chabaudi*. *Infect. Immun.* **38:**80–88.

Stevenson, M. M., and E. M. Riley. 2004. Innate immunity to malaria. *Nat. Rev. Immunol.* **4:**169–180.

Stevenson, M. M., and E. Skamene. 1985. Murine malaria: resistance of AXB/BXA recombinant inbred mice to *Plasmodium chabaudi*. *Infect. Immun.* **47:**452–456.

Stirnadel, H. A., F. Al-Yaman, B. Genton, M. P. Alpers, and T. A. Smith. 2000a. Assessment of different sources of variation in the antibody responses to specific malaria antigens in children in Papua New Guinea. *Int. J. Epidemiol.* **29:**579–586.

Stirnadel, H. A., H. P. Beck, M. P. Alpers, and T. A. Smith. 2000b. Genetic analysis of IgG subclass responses against RESA and MSP2 of Plasmodium falciparum in adults in Papua New Guinea. *Epidemiol. Infect.* **124:**153-162.

Stirnadel, H. A., H. P. Beck, M. P. Alpers, and T. A. Smith. 1999a. Heritability and segregation analysis of immune responses to specific malaria antigens in Papua New Guinea. *Genet. Epidemiol.* **17:**16–34.

Stirnadel, H. A., M. Stockle, I. Felger, T. Smith, M. Tanner, and H. P. Beck. 1999b. Malaria infection and morbidity in infants in relation to genetic polymorphisms in Tanzania. *Trop. Med. Int. Health* **4:**187–193.

Taylor, T. E., W. J. Fu, R. A. Carr, R. O. Whitten, J. S. Mueller, N. G. Fosiko, S. Lewallen, N. G. Liomba, and M. E. Molyneux. 2004. Differentiating the pathologies of cerebral malaria by postmortem parasite counts. *Nat. Med.* **10:**143–145.

Thumwood, C. M., N. H. Hunt, I. A. Clark, and W. B. Cowden. 1988. Breakdown of the blood-brain barrier in murine cerebral malaria. *Parasitology* **96:**579–589.

Tishkoff, S. A., R. Varkonyi, N. Cahinhinan, S. Abbes, G. Argyropoulos, G. Destro-Bisol, A. Drousiotou, B. Dangerfield, G. Lefranc, J. Loiselet, A. Piro, M. Stoneking, A. Tagarelli, G. Tagarelli, E. H. Touma, S. M. Williams, and A. G. Clark. 2001. Haplotype diversity and linkage disequilibrium at human G6PD: recent origin of alleles that confer malarial resistance. *Science* **293:**455–462.

Tournamille, C., Y. Colin, J. P. Cartron, and C. Le Van Kim. 1995. Disruption of a GATA motif in the Duffy gene promoter abolishes erythroid gene expression in Duffy-negative individuals. *Nat. Genet.* **10:**224–228.

Treutiger, C. J., A. Heddini, V. Fernandez, W. A. Muller, and M. Wahlgren. 1997. PECAM-1/CD31, an endothelial receptor for binding Plasmodium falciparum-infected erythrocytes. *Nat. Med.* **3:**1405–1408.

Urban, B. C., D. J. Ferguson, A. Pain, N. Willcox, M. Plebanski, J. M. Austyn, and D. J. Roberts. 1999. Plasmodium falciparum-infected erythrocytes modulate the maturation of dendritic cells. *Nature* **400:**73–77.

Urban, B. C., N. Willcox, and D. J. Roberts. 2001. A role for CD36 in the regulation of dendritic cell function. *Proc. Natl. Acad. Sci. USA* **98:**8750–8755.

Usanga, E. A., and L. Luzzatto. 1985. Adaptation of Plasmodium falciparum to glucose 6-phosphate dehydrogenase-deficient host red cells by production of parasite-encoded enzyme. *Nature* **313:**793–795.

Vigario, A. M., E. Belnoue, A. Cumano, M. Marussig, F. Miltgen, I. Landau, D. Mazier, I. Gresser, and L. Renia. 2001. Inhibition of Plasmodium yoelii blood-stage malaria by interferon alpha through the inhibition of the production of its target cell, the reticulocyte. *Blood* **97:**3966–3971.

Walley, A. J., C. Aucan, D. Kwiatkowski, and A. V. Hill. 2004. Interleukin-1 gene cluster polymorphisms and susceptibility to clinical malaria in a Gambian case-control study. *Eur. J. Hum. Genet.* **12:**132–138.

Wang, H. Y., H. Tang, C. K. Shen, and C. I. Wu. 2003. Rapidly evolving genes in human. I. The glycophorins and their possible role in evading malaria parasites. *Mol. Biol. Evol.* **20:**1795–1804.

Wattavidanage, J., R. Carter, K. L. Perera, A. Munasingha, S. Bandara, D. McGuinness, A. R. Wickramasinghe, H. K. Alles, K. N. Mendis, and S. Premawansa. 1999. TNFα*2 marks high risk of severe disease during *Plasmodium falciparum* malaria and other infections in Sri Lankans. *Clin. Exp. Immunol.* **115:**350–355.

Williams, T. N., K. Maitland, S. Bennett, M. Ganczakowski, T. E. Peto, C. I. Newbold, D. K. Bowden, D. J. Weatherall, and J. B. Clegg. 1996. High incidence of malaria in alpha-thalassaemic children. *Nature* **383:**522–525.

Williams, T. N., S. Wambua, S. Uyoga, A. Macharia, J. K. Mwacharo, C. R. Newton, and K. Maitland. 2005. Both homozygous and heterozygous α^+-thalassemia protect against severe and fatal *Plasmodium falciparum* malaria on the coast of Kenya. *Blood* **106:**368–371.

Yazdani, S. S., A. R. Shakri, P. Mukherjee, S. K. Baniwal, and C. E. Chitnis. 2004. Evaluation of immune responses elicited in mice against a recombinant malaria vaccine based on Plasmodium vivax Duffy binding protein. *Vaccine* **22:**3727–3737.

Zimmerman, P. A., J. Fitness, J. M. Moulds, D. T. McNamara, L. J. Kasehagen, J. A. Rowe, and A. V. Hill. 2003. CR1 Knops blood group alleles are not associated with severe malaria in the Gambia. *Genes Immun.* **4:**368–373.

PROGRESS IN DEVELOPMENT OF A VACCINE TO AID MALARIA CONTROL

Vasee S. Moorthy and Filip Dubovsky

25

THE NEED FOR IMPROVED MALARIA CONTROL

The resurgence of malaria morbidity and mortality underlines the failure of hopes for the eradication of malaria of the middle part of the 20th century. By the 1950s, control measures based on chemotherapy with chloroquine and use of dichlorodiphenyltrichloroethane (DDT) had reduced the disease burden of malaria so substantially that many policymakers believed the task of malaria control was close to being completed. An example of this excessive optimism is a book published in 1955 entitled *Man's Mastery of Malaria*, edited by one of the leading scientists of the time. Central to the subsequent reversal of fortune of malaria control is the widespread dispersal of resistance to affordable drugs in *Plasmodium falciparum* and insecticide resistance in *Anopheles gambiae*, the most important vector for *P. falciparum* transmission in sub-Saharan Africa (Greenwood and Mutabingwa, 2002). There is also some evidence of a reduction in operational effectiveness of some national malaria control programs, contributions from increasing migration and refugee populations, and, for northern countries, increasing tourism (Moorthy et al., 2004). Estimates of numbers of cases and deaths have unquantified but probably wide confidence intervals. Recent estimates vary between 1 million and 2.7 million deaths globally due to malaria each year (Breman, 2001; Snow et al., 1999). Malaria-related mortality in sub-Saharan Africa for children under 5 years of age almost doubled from 1990 to 1998 compared to the period from 1982 to 1989 (World Health Organization, 2003). While this may be a result of improved disease surveillance, it occurred against a background of a decline of all-cause under-5-year mortality in sub-Saharan Africa (Snow et al., 2001). Indeed, it has been suggested that the leveling off of this decline in all-cause mortality may be due to the large rise in malaria mortality in recent decades. Recent estimates of the economic burden of malaria highlight the extent to which economic development may be hindered by malaria transmission in countries where malaria is endemic (Sachs and Malaney, 2002). These economic analyses do not take into account some less-

---

Vasee S. Moorthy and Filip Dubovsky, Malaria Vaccine Initiative, PATH, 7500 Old Georgetown Rd., 12th floor, Bethesda, MD 20814.

quantifiable indirect implications of malaria transmission, such as deterrences to inward travel of nonimmunes. Such indirect effects could be significant. Demographic projections suggest future increases in malaria among the population at risk (Hay et al., 2004). Malaria remains one of the big three killers among tropical diseases; without the introduction of new effective control measures, it is likely that the disease burden will continue to increase.

NONVACCINE CONTROL MEASURES

Existing control measures have much to offer. Some countries are implementing indoor residual spraying for vector control. Permethrin-treated bednets reduced all-cause mortality in children in western Kenya aged 0 to 4 years by 16% in one large trial (Phillips-Howard et al., 2003). A Cochrane review of randomized trials of insecticide-treated bednets found that efficacy against all-cause mortality in eligible trials was 17% (Lengeler, 2000). The Roll Back Malaria Partnership is coordinating implementation of existing control measures, improved case management and use of insecticide-treated bednets. This implementation impetus, first announced as the Abuja Declaration on HIV/AIDS, Tuberculosis, and Other Related Infectious Diseases at a summit in 2000, has yet to achieve the full impact of its laudable approach (Guerin et al., 2002). Even if more funds can be found to implement treated bednets and case management with more effective antimalarials, such as artemisinin derivatives, the path towards sustainable control of malaria is unclear.

Three intermittent doses of treatment with sulfadoxine-pyrimethamine or amodiaquine reduced mild malaria disease and anemia in Tanzanian infants without detectable rebound morbidity (Massaga et al., 2003; Schellenberg et al., 2001). Such a treatment schedule is called intermittent preventive treatment in infants (IPTi). IPTi may be feasible to implement, as these drugs promise to be available and affordable, and the schedule could involve administration during immunization visits in existing expanded program of immunization infrastructures. Many questions remain to be answered, including choice of drug and schedule, duration of efficacy, and efficacy in different transmission settings (Rosen and Breman, 2004). It is conceivable that IPTi may become an addition to the armory of malaria control measures.

In parallel to these crucial implementation efforts, development of an effective malaria vaccine is urgently needed, with the ultimate goal of integrating vaccination into national immunization and malaria control programs.

MALARIA VACCINES ARE FEASIBLE

In addition to protection experiments in ortholog models and other experimental systems, several lines of evidence lead us to believe that humans can be vaccinated against malaria. Furthermore, it is possible to deduce that *P. falciparum* can be impacted in more than one stage of its life cycle.

For the asexual blood stage of the life cycle it has been demonstrated that while those living in areas where malaria is endemic continue to develop parasitemia, they eventually become resistant to clinical disease. Transfer of gamma-globulin fractions from semi-immune patients to naïve humans mitigates malaria disease (Cohen et al., 1961). From this, one can infer that clinical protection from malaria is possible and that immune mechanisms, specifically immunoglobulin, can play a role in mitigating disease. Furthermore, there is evidence that a subunit vaccine is capable of impacting parasite density under conditions of natural exposure, lending support to the feasibility of this vaccine strategy (Genton et al., 2002).

For the preerythrocytic stage of the life cycle, inoculation of humans with irradiated sporozoites can prevent the emergence of blood-stage infection in subsequent experimental challenge (Cochrane et al., 1980). This demonstrates the possibility of inducing protection against infection under experimental conditions. This observation has been replicated using vaccines that prevent infection and delay parasitemia in a portion of vaccinated volunteers after experimental challenge (Stoute et al., 1997; Webster et al., in press). Importantly, short-lived protection from infection has also been demonstrated under conditions of natural exposure (Bojang et al., 2001).

SUBUNIT VACCINATION

The concept upon which vaccination is based, immunological memory, dates back to one of the earliest historians, Thucydides. He wrote about a plague that swept through Greece in 430 BC. "It was with those who had recovered from the disease that the sick and the dying found most compassion. These knew what it was from experience, and had now no fear for themselves; for the same man was never attacked twice; never at least fatally." The word vaccine is etymologically derived from vaccinia, which is the cowpox virus and the basis of the conventional smallpox vaccine. Most vaccines in widespread use are either attenuated or inactivated. One exception is hepatitis B virus vaccine (Crosnier et al., 1981), the first example of a licensed subunit vaccine against an infectious disease produced by recombinant DNA technology. Such subunit vaccines are necessary where inactivated or attenuated vaccines are technically unfeasible. This is probably the case for a malaria vaccine suitable for countries where it is endemic, although one group is pursuing an effort to produce a radiation-attenuated sporozoite vaccine (Luke and Hoffman, 2003). The concept of subunit vaccination is that partial or complete antigens or epitopes are selected from a pathogen's proteome and used to generate a protective immune response. In the case of hepatitis B virus vaccine, a single antigen (hepatitis B virus surface antigen) was produced in yeast expression systems as a recombinant protein. This antigen spontaneously self-assembles into virus-like particles, thereby increasing its inherent immunogenicity. When adjuvanted by binding to aluminum salts, potent and protective antibody induction is obtained.

Many subunit malaria vaccines are in early development (http://www.clinicaltrials.gov/ct/show/NCT00075049). Unfortunately, many of the critical epitopes are conformationally sensitive, and the proteins have a complex cysteine-bound tertiary structure that makes manufacture as recombinant proteins extremely difficult. Furthermore, many malaria proteins appear to exhibit low immunogenicity for antibody induction, both when unadjuvanted and when formulated with aluminum salts. New understanding of innate immunity, antigen processing, genetic engineering, and expression techniques, as well as novel delivery systems, are gradually increasing the field's ability to generate vaccines that can induce potent immune responses. Duration of the induced immune responses remains a considerable problem in malaria.

Recombinant protein vaccines are generally poor at induction of effector T-cell responses, such as $CD8^+$ cytotoxic T lymphocytes, which are thought by many as desirable for targeting infected hepatocytes. Other categories of subunit vaccines are DNA (Ulmer et al., 1993) and recombinant viral vaccines (Li et al., 1993). In this review, these DNA and recombinant viral vaccines will be referred to as DNA-based vaccines. DNA sequences from *P. falciparum* parasites can be delivered as DNA molecules (DNA vaccines) (Wang et al., 1998) or various recombinant attenuated DNA viruses (Schneider et al., 1998) to generate candidate DNA-based vaccines. In some cases, antigen transfer occurs from cells such as myocytes, transfected with DNA-based vaccines to antigen-presenting cells (APCs) (Corr et al., 1996). Some direct transfection of APCs by DNA-based vaccines may occur (Roy et al., 2000), depending on the carrier, delivery system, and route of administration. The transgene is processed and presented in T-cell epitope/HLA molecule complexes that prime naïve T cells to generate memory T cells (Gurunathan et al., 2000). Plasmid DNA vaccines can be passively taken up by cells; recombinant attenuated viral vaccines infect APCs and express the recombinant malaria proteins before aborting infection (Miyahira et al., 1998). These DNA-based subunit vaccines provide methods of inducing high levels of cell-mediated immunity (CMI) responses (Gurunathan et al., 2000; Paoletti, 1996). To date, antibody responses have been poor in clinical trials after DNA-based immunization.

CRITICAL STEPS IN MALARIA VACCINE DEVELOPMENT

Design of Constructs

There is an unimaginably large number of possible malaria constructs for inclusion in a malaria vaccine. For example, if there are 50 antigenic components that may warrant evaluation in a malaria vaccine and one decides to choose 5 for inclusion in a vaccine, there are over 250 million possible combinations. Given that there are about 5,300 antigens predicted from whole-genome sequencing (Gardner et al., 2002), many more constructs are possible. The field focuses on choices of whole or partial antigens that have some level of supporting rationale. Recent techniques for identification and characterization of novel antigens from genome-wide sequencing have yielded many new potential targets (Doolan et al., 2003; Haddad et al., 2004). Thus in the postgenomic era, the number of newly characterized antigens will continue to increase. To date, these new antigens have yet to progress into product development pathways, primarily because no method has been proven to predict an antigen's clinical utility. Current vaccines in development are largely based on 15 to 20 variably characterized antigens identified prior to the decoding of the malaria genome. Circumsporozoite protein (CS) and merozoite surface protein 1 (MSP-1) in particular may be overrepresented in the current global field of vaccine candidates and represent about half of all vaccine candidates currently in development.

Choice of Platform

In general, recombinant protein-based platforms are chosen where antibody induction is thought to be the most important effector mechanism. As there is previous global experience developing, licensing, and commercializing such vaccines, this approach remains the cornerstone of the malaria vaccine subunit field. The most straightforward recombinant protein platform is the monomeric protein. Production of these conformationally sensitive proteins has been challenging, due to the AT-rich genetic sequence, lack of posttranslational glycosylation, complex folded tertiary structure, and multiple internal disulfide bridges of the native proteins.

The hepatitis B virus vaccine is an example of a virus-like particle (VLP). These consist of multiple copies of a protein assembled into nanometer size particles (Boisgerault et al., 2002). GSK Biologicals and Apovia, Inc., possess two VLP platform technologies based on hepatitis B virus surface and core antigens, respectively, into which malaria epitopes have been engineered. For all diseases, DNA-based platforms such as plasmid DNA (Gurunathan et al., 2000), recombinant adenoviruses (Shiver and Emini, 2004), and recombinant poxviruses (Anderson et al., 2004) are at an earlier stage of development than recombinant protein platforms; there are currently no licensed products based on DNA-based platforms. However, they show promise in being superior in terms of Th1 and $CD8^+$ T-cell induction in animal models. Influenza virosomes, which are proteoliposome particles, have been licensed by Berna as carriers for a hepatitis A virus vaccine. This approach is being pursued in the malaria field by Pevion. The influenza hemagglutinin is believed to enhance entry into the HLA class II presentation pathway, facilitating generation of a humoral response. Sequential immunization with two different DNA-based platforms encoding the same antigen is known as prime-boost immunization. This approach has led to strong $CD4^+$ effector T-cell immunogenicity in clinical trials (McConkey et al., 2003).

Choice of Adjuvant

Adjuvants (from *adjuvare*, the Latin word meaning to help) are chemicals which, when administered in combination with an antigen, augment the immune response to the antigen. Only one class of adjuvant forms part of products licensed in the United States: aluminum salts including aluminum phosphate, hydroxide, and sulfate (Kenney and Edelman, 2003). In mice, however, alum adjuvants confer a Th2-biased immune response, and some clinical studies with malaria antigens adjuvanted with aluminum salts have yielded low immunogenicity and/or hypersensitivity reactions (reviewed in Edelman et al., 2002). MF59, a proprietary squalene oil-in-water emulsion, forms part of an influenza vaccine that has been licensed in much of Europe. RC-529, a synthetic monophosphoryl lipid A analog and Toll-like receptor 4 ligand (Evans et al., 2003), forms part of a hepatitis B virus vaccine licensed in Argentina. Recent advances in understanding of the role of adjuvants in modulation of antigen uptake, processing, and presentation by dendritic cells and T- and B-cell priming have led to a growth in the field of novel adjuvant development. Current thinking is that recombinant protein platforms are likely to require novel adjuvants to afford sufficient efficacy in malaria and that DNA-based vaccines may not require adjuvants due to intrinsic adjuvant properties. For example, some people believe that plasmid DNA vaccines contain CpG sequences, which are known to be ligands for Toll-like receptor 9, and thereby induce antigen-presenting cell activation (Klinman et al., 1999). Some DNA-based approaches use adjuvants such as delivery on polylactide coglycolide microparticles or inclusion of cytokine sequences to overcome limitations of the platform, such as poor induction of antibodies.

PRECLINICAL EVALUATION

All vaccines, including malaria vaccines, are evaluated in animal models to assess their safety and immunogenicity. While in the past preclinical general safety testing may have been sufficient for assessment of safety, many regulatory authorities now often require formal good laboratory practice toxicology studies. Immunogenicity assessment occurs, as for all vaccines, to evaluate the ability to induce an immune response. Immunoglobulin G (IgG) enzyme-linked immunosorbent assays (ELISAs) are the standard for antibody immunogenicity and are the most commonly used immunoassays in the malaria vaccine field; gamma interferon enzyme-linked immunospot assay (ELISPOT) and intracellular cytokine staining are most common for CMI. CMI is thought to be particularly important for preerythrocytic vaccines, although recent work raises the possibility that CMI may be desirable for blood-stage vaccine development (Hirunpetcharat et al., 2003; Pombo et al., 2002).

Assessment of function of the generated immune response is an important aspect of preclinical evaluation for malaria vaccines. Functional immune responses can be assessed in ortholog and experimental animal model systems, as well as in in vitro assays. While these systems provide supporting evidence that the antigen was correctly chosen and presented to the immune system to generate a signal, the predictive value of these tools cannot be determined until the results are correlated with clinical efficacy of vaccine candidates. Therefore, applying these results to make development decisions may result in discarding valuable vaccines or selecting candidates that may provide suboptimal efficacy. The current choice of vaccines for clinical development is necessarily empiric. Ongoing clinical trials should be designed to generate data to enable predictive preclinical and phase 1 clinical evaluation in the future; in time, the field may then be able to enter a less empirical phase. This process is known as iterative vaccine development—each cycle of clinical testing reduces the chances of technical failure in the next cohort of candidate vaccines.

In practice, development decisions for a given candidate are based on a combination of safety and immunological, functional, and challenge data from one or more animal models. For immunogenicity, magnitude, duration, and breadth are ideally all evaluated. Some commonly used models and assays are outlined in Table 1.

CLINICAL DEVELOPMENT

With malaria, it is possible to evaluate efficacy against three tiers of endpoints. In increasing relevance to public health these are infection following sporozoite challenge (phase 2a), parasitological endpoints in the field (phase 2b), and clinical endpoints in the field (phase 2b or phase 3). The two top tiers require relatively small clinical trials. Sporozoite challenge trials can be done with fewer than 20 volunteers, and trials looking at parasitological endpoints in the field

TABLE 1 Some models and assays used for preclinical assessment of malaria vaccine candidates

Immunologic assay	Functional assay	Preclinical challenge model
Total IgG ELISA	Growth inhibition assay	Mice; *P. berghei* or *P. yoelii*
IgG subtype ELISA	ADCI assay[a]	SCID mice; *P. falciparum*
γ-Interferon ELISPOT	Membrane-feeding assay[b]	*Aotus*; passaged *P. falciparum*
γ-Interferon intracellular cytokine staining	Inhibition of sporozoite invasion of hepatocytes	*Saimiri*; passaged *P. falciparum*
Immunofluorescence assay		

[a]ADCI, antibody-dependent cellular inhibition.
[b]For transmission-blocking vaccines.

require a few hundred volunteers. This ability to obtain surrogate efficacy results is a double-edged sword. The time taken to reach surrogate efficacy determination is much less than that needed for traditional clinical efficacy; however, relevance to clinical efficacy is yet to be determined. Most worryingly, product development could be stopped based on failure to detect efficacy against endpoints currently not known to be predictive of clinical efficacy. The terms phase 1 and phase 3 have meanings in the malaria vaccine field similar to those used in other drug and biologics areas. However, terminology for phase 2 trials is quite different. Phase 2a trials, also known as challenge trials, are the same size as phase 1 trials with addition of challenge of volunteers with highly characterized strains of *P. falciparum* parasites (3D7, NF54, or 7G8 strains are used) delivered by the bite of infected laboratory-reared *Anopheles stephensi* mosquitoes (Church et al., 1997). Volunteers are treated once they develop patent blood-stage parasitemia. Increasingly, molecular PCR monitoring is also conducted (Bejon et al., in press). Such trials are useful to evaluate the efficacy of preerythrocytic candidates but not blood-stage candidates as the volunteers are treated at first sign of infection, before one would expect to see an impact on blood-stage parasites. Several hundred volunteers have been challenged in different centers; this system is now well established and carries a low level of risk, providing volunteers are compliant with the study protocol. Phase 2b trials, by contrast, are field efficacy trials involving a few hundred to a few thousand volunteers with parasitological and sometimes a clinical primary endpoint. The epidemiological setting, the age group, and primary endpoint chosen determine the size of the phase 2b trial. For blood-stage candidates, phase 2b trials remain the generally accepted method of establishing proof of principle. Some early, promising work has been performed with blood-stage challenge by inoculation of ultra-low doses of infected erythrocytes. If this is further developed, it may form the basis of a challenge trial framework for phase 2a trials to evaluate blood-stage candidates. Once proof of principle is established, providing the product is suitable for licensure and commercialization, a phase 3 trial program leading to licensure may be conducted. There is some debate as to the priority and feasibility of determining efficacy against severe malaria or mortality measure prelicensure.

There has been partial harmonization of conduct of clinical trials with experimental products in recent years. To generate data suitable for submission for marketing authorization in the United States, the European Union, or Japan, it must satisfy International Council on Harmonization Good Clinical Practice Guidelines, as well as local guidelines. These guidelines impose a framework for conduct of clinical trials which both ensures data quality and greatly increases cost-time requirements for investigators. Not all clinical vaccine trials are yet conducted to International Council guidelines.

Regulatory Requirements

The regulatory environment affects vaccine development programs. Hence, an understanding of the relevant regulatory authority's requirements is a key element to product development. In general, manufacturing, quality control, toxicology testing, and documentation standards expected by regulatory authorities are increasing. Special scrutiny is placed on thorough product characterization (product integrity, purity, manufacturing consistency, stability, and potency). There is a trend toward a reduction in the previous regulatory distinction between academic-led translational research and industry-directed product development.

Some field trials are conducted in sub-Saharan African countries where the Ministry of Health serves in the role of the regulatory authority. In these cases, prior product approval by regulatory authorities in a developed country can give some comfort to local officials for product quality. Many countries that would most benefit from a *P. falciparum* vaccine do not have regulatory bodies with experience in licensing novel biologicals; therefore, it is anticipated that transnational agencies such as the World Health Organization and the European Medicines Agency will play a role.

The Pathway to Commercialization

For a vaccine to have any public health impact it must be licensed, manufactured, promoted, and distributed. History has taught us that this is most efficiently done through the private sector. However, except for an unproven traveler's market, malaria vaccines will be used by some of the world's poorest populations and are unlikely to be supported by significant sales in developed countries. If the vaccines have an efficacy less than the very high level provided by available prophylactic drugs, it is unlikely there will be demand in the traveler's market. Therefore, it is anticipated that the lion's share of the vaccine's cost will be borne by the donor community with little offset from markets in developed countries. To ensure maximal impact of a malaria vaccine, it would be beneficial to have the least expensive product manufactured in large enough quantities to vaccinate the population at risk. Therefore, technologies that are inherently low cost and can be manufactured at large scale are desirable. Complex intellectual property and licensing considerations can slow down research and development, and in some cases prevent novel technologies from being applied optimally. However, unless the public sector is willing to bear the entire cost of developing a malaria vaccine and is willing to forgo proprietary technologies, then intellectual property consideration must be taken into account so the private sector will be able to commercialize the product.

VACCINE INTRODUCTION

Once a vaccine that prevents clinical disease is licensed by a regulatory authority, it may yet be many years before the vaccine is added to local immunization programs in countries where malaria is endemic (Mahoney and Maynard, 1999). A discussion of introduction issues is outside the scope of this review; just a few key points are outlined here. Disease burden data will be needed from areas where malaria is endemic at different levels to inform introduction decisions. Cost-effectiveness analyses can be powerful ways of mobilizing governments and donors. A necessary intermediate step may be building an international consensus of public health experts for the introduction of the vaccine. It appears likely that the speed and breadth of introduction will occur in inverse relation to price, although cost-effectiveness analyses may also help inform the debate. It is anticipated that the international community—through organizations like GAVI (The Global Alliance for Vaccines and Immunization), The Global Fund, UNICEF (The United Nations Children's Fund), and bilateral donors—will largely drive the eventual introduction of malaria vaccines.

CURRENT CANDIDATE VACCINES

Preerythrocytic Vaccines

The irradiated sporozoite model strongly implies that humans are protected by CMI responses to liver-stage parasites. However, there is no consistent evidence that naturally acquired immune responses, whether antibody or CMI, contribute significantly to preerythrocytic immunity. In this sense, immunity conferred by preerythrocytic subunit vaccines would be "nonnatural." Candidate vaccine antigens from the preerythrocytic stages may be the targets of antibodies that prevent sporozoite invasion of hepatocytes or the targets of cellular immune responses that kill infected hepatocytes. A completely effective preerythrocytic vaccine would inactivate the parasite before it left the liver, leading to sterile immunity and prevention of disease. This goal appears to be beyond what is achievable with the vaccines that are currently being evaluated. Theoretically, a partially effective vaccine could decrease the number of merozoites exiting the liver, by decreasing the number of sporozoites entering the liver or killing parasites within hepatocytes, leading to clinical benefits analogous to insecticide-treated bednets. Partially effective preerythrocytic vaccines could lead to reductions in both the size and frequency of blood-stage inocula, which could result in reductions in mild disease, severe disease, and mortality. There is much indirect evidence for both utility and nonutility of partially effective preerythrocytic vaccines; the case remains to be proved. CS, the best characterized preerythrocytic antigen (Sinnis and Nardin, 2002), has been identified as the target of both protective antibody (Yoshida et al., 1980) and T-cell mediated (Weiss et al., 1990) immune responses. It is a major sporozoite surface protein (Aikawa et al., 1990; Fine et al., 1984) and has a role in motility of sporozoites (Stewart and Vanderberg, 1988). It has not been possible to generate viable CS knockout parasites (Kocken et al., 2002; Menard et al., 1997). One theoretical concern with CS as a vaccine antigen has been that T-cell-mediated immune responses are focused on variable regions (Good et al., 1988), possibly due to immune selection pressure, and therefore vaccines which depend on T-cell mediation of protection may confer strain-specific immunity. In addition, there is little evidence that naturally acquired immune responses to CS are protective (Hoffman et al., 1987).

Many CS-based candidates have been evaluated in clinical trials. The early CS candidates either resulted in only limited protection or were not deemed suitable to evaluate in phase 2a trials (Ballou et al., 1987; Brown et al., 1994; Gonzalez et al., 1994; Herrington et al., 1987; Herrington et al., 1992; Hoffman et al., 1994; Sherwood et al., 1991; Vreden et al., 1991). The lead malaria vaccine candidate of any life stage is RTS,S, a recombinant protein candidate based on the CS antigen and using GSK Biologicals' hepatitis B virus surface antigen VLP platform. The C terminus of CS was fused to part of the hepatitis B virus surface antigen to form a hybrid DNA sequence, RTS. When coexpressed in *Saccharomyces cerevisiae*, RTS binds hepatitis B virus surface antigen (S) to form RTS,S protein particles. Adjuvant down-selection was performed in preclinical and clinical studies and indicated optimal phase 2a protection on formulation with the AS02 adjuvant—a mixture of deacylated monophosphoryl lipid A, QS21, and a squalene oil-in-water emulsion. RTS,S/AS02 vaccination affords 30 to 60% short-term protection in phase 2a trials (Kester et al., 2001; Stoute et al., 1997). Because RTS,S vaccination induces high-titer antibodies to both CS and hepatitis B virus surface antigen, it may provide efficacy against both malaria and hepatitis B virus. A randomized controlled field trial of the efficacy of three 50-μg doses of RTS,S in Gambian adults reported vaccine efficacy of 34% ($P = 0.014$) over the 15-week surveillance period, but with 71% efficacy over the first 9 weeks and 0% over the next 6 weeks (Bojang et al., 2001). Protection was not strain specific (Alloueche et al., 2003). The primary endpoint of this trial was time to first infection in adults, and it is unclear whether this result will translate to protection against clinical disease in African children who bear the burden of malaria disease globally. Efficacy against clinical disease in this trial was 31% ($P = 0.096$), with few episodes of disease included in the analysis. RTS,S/AS02 is the first preerythrocytic vaccine to show clear protection against natural *P. falciparum* infection. Pediatric clinical development has proceeded in collaboration with the PATH Malaria Vaccine Initiative, a global vaccine development program. Phase 1 studies in Gambian children aged 1 to 11 showed that reactogenicity was acceptable in this population and that 25- and 50-μg doses were both highly immunogenic. A phase 1 study in Mozambique confirmed safety and immunogenicity of the 25-µg dose in children aged 1 to 4 years. Efficacy of RTS,S/AS02 against clinical disease is being evaluated in approximately 2,000 children aged 1 to 4 years in Mozambique at the 25-µg dose. A phase 2a trial is under way at the Walter Reed Army Institute of Research (WRAIR) comparing formulation of RTS,S in AS02A with AS01B, a different adjuvant (www.who.int/vaccine_research/documents/en/malaria_table.pdf). Studies with rhesus monkeys showed a longer duration of RTS,S-specific gamma interferon ELISPOT responses after immunization of RTS,S with AS01B compared to AS02A.

Several other preerythrocytic candidates have entered clinical trials in recent years. A preerythrocytic vaccine candidate known as ICC-1132, developed by Apovia, Inc., has been evaluated with alum and ISA720 formulations in the United States, Germany, and the United Kingdom. ICC-1132 is a VLP platform based on the hepatitis B virus core and genetically engineered to include CS B-cell epitopes on the tips of spikes of the protein particle, as well as two T-cell epitopes. Very high titers of biologically active CS antibody have been seen with preclinical studies (Birkett et al., 2002). A phase 2a challenge trial with a single dose of ICC-1132 formulated with ISA720 led to modest immunogenicity and no evidence of protection (Walther et al., in press). Clinical development continues with multidose formulations.

New York University (NYU) and the University of Lausanne have conducted clinical evaluation of other CS-based candidates. One NYU candidate vaccine is a multiple antigen peptide (Nardin et al., 2001; Nardin et al., 2000), a synthetic platform. It induced strong antibody responses. NYU has also evaluated a polyoxime construct, containing a universal T-cell epitope. The University of Lausanne candidate is a long synthetic peptide in ISA720 adjuvant, a water-in-oil emulsion. It induced $CD8^+$ T cells at low frequency, as well as antibody responses (Lopez et al., 2001). $CD8^+$ T-cell responses are thought to be important for elimination of infected hepatocytes. ISA720 is an adjuvant that is generally available and that often renders vaccines very immunogenic in humans; however, there have been candidate-specific local reactogenicity concerns with ISA720. A phase 2a trial of this University of Lausanne candidate is ongoing.

The U.S. Naval Medical Research Center pioneered plasmid DNA vaccination in the malaria field, demonstrating $CD8^+$ T-cell induction after repeated vaccination with up to 2.5 mg of DNA vaccine encoding CS (Wang et al., 1998, Wang et al., 2001). Robust $CD8^+$ T-cell induction published to date has been seen only after restimulation of cells in vitro. Perhaps more relevant, ex vivo $CD8^+$ T-cell induction has been documented in only a few isolated instances in clinical vaccine studies—an area that it is hoped new malaria candidates will improve upon. The Naval Medical Research Center immunized volunteers with mixtures of 5 plasmids encoding preerythrocytic antigens; no protection was seen after challenge (Richie et al., 2001). The University of Oxford has conducted a series of phase 1/2a trials with heterologous prime-boost combinations of two DNA-based platforms encoding the same multiple epitope–thrombospondin-related anonymous protein (ME-TRAP) construct in the United Kingdom and Gambia. The United Kingdom-based studies confirmed murine and rhesus data (Amara et al., 2001; Schneider et al., 1999) indicating that heterologous prime-boost immunization with plasmid DNA followed by recombinant modified vaccinia virus Ankara (MVA; an attenuated strain of the smallpox vaccine) induces very high effector $CD4^+$ T-cell frequencies, modest $CD8^+$ T-cell induction, and limited IgG responses to TRAP or CS (McConkey et al., 2003). Based on published data, DNA/MVA immunization represents the most immunogenic method currently available to induce effector T cells in humans. Replacing DNA with another attenuated poxvirus, fowlpox strain 9, recombinant for the same ME-TRAP insert, decreased induced frequencies slightly but in one trial induced protection of two of five volunteers against blood-stage infection. Volunteers who are not completely protected from blood-stage infection experience, on average, a delay in time to patent parasitemia estimated to be associated with killing of about 80% of infected hepatocytes (McConkey et al., 2003). Protection in terms of either prevention of blood-stage infection or delay in time to patent parasitemia is associated with the overall induced gamma interferon effector T-cell frequency as assayed by ELISPOT. A series of studies in The Gambia have confirmed that DNA-MVA and fowlpox strain 9-MVA immunization are safe and immunogenic in Gambian men (Moorthy et al., 2004; Moorthy et al., 2003). A randomized controlled trial of two doses of 2 mg of intramuscular DNA ME-TRAP followed by 1.5×10^8 PFU of intradermal MVA ME-TRAP compared to rabies vaccine as a control showed 10.3% efficacy (95% confidence interval, -22% to $+34\%$) against infection. Prime-boost studies using a CS construct are ongoing. These heterologous DNA-based prime-boost studies are synergistic, meaning induction of responses occurs that is not achievable through repeated immunization with one carrier alone. Protein-DNA-based immunization prime-boost studies (DNA vaccine encoding CS followed by RTS,S/AS02 [Wang et al., 2004] and RTS,S/AS02 followed or preceded by MVA encoding CS) have been conducted and proved to be additive but not synergistic. Therefore, to date, potent effector $CD8^+$ T-cell induction and antibody induction have not been achieved in clinical studies using a homologous vaccination regimen. Coadministration or combination vaccine studies may be able to achieve this. Alternatively, novel platforms not yet clinically evaluated may be able to deliver dual IgG and effector $CD8^+$ T-cell immunogenicity; one example of a promising platform in preclinical evaluation is a hybrid replicon particle (Perri et al., 2003).

Blood-Stage Vaccines

Naturally acquired immunity is thought to be conferred primarily by the gradual acquisition of antibodies to variant antigens expressed on the surface of infected erythrocytes (Marsh and Howard, 1986). Current understanding of these variant antigens has not highlighted promising candidate antigenic components for a vaccine. Naturally acquired antibodies which prevent merozoite invasion are associated with reduction of clinical disease (John et al., 2004). Most blood-stage candidate vaccines include proteins that are exposed on the merozoite surface, are involved in the red blood cell invasion process, or are expressed on the infected red blood cell. Antibodies with the correct specificity to these antigens may prevent merozoite invasion, thereby interrupting the parasite life cycle and preventing clinical disease. The tertiary structure of recombinant protein blood stage antigens is therefore likely to be critical for vaccine induction of antibody-based protection. In addition, there are early data that indicate that blood-stage-specific CMI responses in animal models (Makobongo et al., 2003) and humans (Pombo et al., 2002) can afford protection, although the mechanism of action of such an acquired CMI response is not well characterized.

The field is experiencing an expansion in the number of blood-stage candidates approaching clinical evaluation. This expansion has to date been mainly restricted to phase 1 trials, and as there is currently no widely used challenge trial system for blood-stage candidates, proof of principle rests upon phase 2b field efficacy trial evaluation. It is counterintuitive that blood-stage vaccine development in some senses lags behind preerythrocytic vaccines, given that naturally acquired immunity appears to be focused on the blood stages of the parasite. The only blood-stage vaccine phase 2b trial in recent years was that of the combination B vaccine conducted in Papua New Guinea. This vaccine, no longer in development, was a combination of three recombinant proteins based on MSP1 (Holder et al., 1999), MSP2, and ring-infected erythrocyte surface antigens (Saul et al., 1999). The vaccine was manufactured and tested by a Papua New Guinea-Australia-Europe academia-biotechnology collaboration. In the trial, pretreatment with sulfadoxine-pyrimethamine impacted parasite density for >4 weeks, and drug pretreatment adversely affected field efficacy data interpretation. Molecular monitoring greatly added value to this field trial: allele-specific

MSP2 genotyping showed that the vaccine reduced the infection rate with 3D7, the vaccine allele of MSP2, more than with the non-vaccine allelic family FC27 (Genton et al., 2002). Partly as a result, remanufacture of both 3D7 and FC27 MSP2 alleles is under way in Australia.

WRAIR has been a long-term partner with GSK Biologicals in development of RTS,S/AS02. In addition, they have manufactured and evaluated FMP1/AS02 in clinical trials. FMP1 is a recombinant protein based on the 3D7 allele of the 42-kDa component of MSP1. Phase 1 trials in adults in the United States have demonstrated safety and immunogenicity; pediatric clinical development is under way with field trials in Kisumu, Kenya. WRAIR has also collaborated with the University of Bamako, Bamako, Mali; Division of Microbiology and Infectious Diseases, National Institutes of Health; the University of Maryland, and GSK Biologicals in conducting an adult FMP1/AS02 trial in Mali.

The Malaria Vaccine Development Branch (MVDB), an intramural program of the National Institutes of Health, is manufacturing four blood-stage candidates, namely, 3D7 and FVO alleles of both AMA1 (Mitchell et al., 2004; Stowers et al., 2002) and the 42-kDa component of MSP1 (Singh et al., 2003). It is theorized that immunity to one allele of these antigens may be insufficient to afford durable clinical efficacy. Most blood-stage antigens contain highly variable regions and, in some cases, point mutations that may render protective IgG responses genotype specific. Improvement in efficacy with multiple alleles remains to be confirmed in field efficacy trials. Phase 1 clinical trials of MVDB's MSP1 and AMA1 vaccines are under way in adults in the United States.

Trials of two other blood-stage antigens have occurred in recent years: MSP3 and GLURP (for glutamate-rich protein). Naturally acquired cytophilic antibodies to MSP3 inhibit growth of parasites in vitro in conjunction with monocytes (Bouharoun-Tayoun et al., 1990; Oeuvray et al., 1994). This immune effector mechanism is known as antibody-dependent cellular inhibition. In animal models, protection induced by MSP3 vaccination is associated with antibody-dependent cellular inhibition activity (Badell et al., 2000). The predictive nature for clinical efficacy of such activity is unknown.

Phase 1 trials of an MSP3 long synthetic peptide have been conducted with adults in Switzerland and Burkina Faso. A phase 1 trial of a GLURP long synthetic peptide was conducted in The Netherlands. It is hoped that synthetic proteins are sufficiently similar to malaria proteins expressed in the parasite to induce protective immune responses.

A great deal of hope for effective vaccination centered on the SPf66 vaccine in the early 1990s. This synthetic vaccine was a 45-amino-acid construct including components from three blood-stage antigens and the CS antigen formulated on aluminum salt (Lopez et al., 1994). Early vaccine trials in South America showed some efficacy against mild *P. falciparum* malaria (Valero et al., 1993). The first African randomized controlled trial in Tanzania with 586 children aged 1 to 5 years reported 31% efficacy against mild malaria (Alonso et al., 1994). However, efficacy trials in The Gambia with 630 children aged 6 to 11 months (D'Alessandro et al., 1995) and 1,221 children aged 2 to 15 in Thailand (Nosten et al., 1996) showed no clinical efficacy. An efficacy trial in Tanzania with 1,207 infants again showed no efficacy (Acosta et al., 1999). Although experimentation continues with alternate formulations, the development of this vaccine has been suspended. One lesson from the experience with SPf66 was the importance of choosing vaccines with a robust, consistent manufacturing process for clinical development. Another is the importance of understanding the mechanism of action of partially effective vaccines.

Sexual-Stage Vaccines

With animal models, it has been shown that antibodies to antigens expressed in the mosquito midgut can impact oocyst formation. Thus, antibodies may be taken up by the mosquito during the blood meal and block parasite function at the gamete stage. This finding is the basis for the membrane-feeding assay, which allows functional assessment of sera raised against sexual-stage antigens for the ability to reduce oocyst formation in mosquitoes. This has provided the rationale for several antigens expressed during the sexual stages of the parasite. Currently, no evidence exists for protective CMI responses against gametocytes within humans. Inclusion of sexual-stage vaccine components in multistage vaccines is theoretically attractive. Some sexual-stage antigens, such as Pfs25, are not expressed in humans. Therefore, vaccination should not select for resistant parasites. Inclusion of such a component could impact the problems associated with immune selection to other nonsexual antigens by reducing the propagation of escape mutants. Efficacy testing of such vaccines will require special considerations, as efficacy cannot be measured directly in vaccinees, but rather in communities where those individuals live. Sexual-stage vaccines are also known as transmission-blocking vaccines because an effective sexual-stage vaccine would reduce the rate of transmission of malaria from human to human. In areas of low endemicity, depending on efficacy and vaccine coverage, transmission could be abolished through vaccination. Funding and development for sexual-stage vaccines have lagged behind other malaria vaccines, possibly because these vaccines definitely would have no application for travelers from developed countries. Manufacture of vaccine candidate antigens (for example, Pfs 25, Pfs 48/45, and Pfs 230) has been delayed by challenging technical difficulties (Kaslow, 2002). MVDB conducted a phase 1 trial at Johns Hopkins University with a recombinant protein Pvs25-aluminum hydroxide *Plasmodium vivax* transmission-blocking candidate.

Multistage Vaccines

There are two broad possibilities for development of vaccines targeting multiple stages of the parasite. Vaccines from individual stages can be first tested in coadministration studies and then ultimately coformulated into combination vaccines. The possibilities for such combination are dependent on the compatibility of different vaccines. The adjuvants used are a key component of compatibility. For example, combination of RTS,S/AS02 and FMP1/AS02 should be feasible, while combination of antigens formulated in ISA720 and ASO2 would be impossible. In a few cases, malaria antigens have exhibited immunological intereference, whereby presence of one antigen in a combination alters the immune response to another antigen. Development of combination vaccines may require the pursuit of two parallel approaches. A systematic incremental approach, demonstrating proof of principle in the field with components individually and lack of immunological interference would be followed by combination. In parallel, combination of multiple components simultaneously would have a higher chance of technical failure, but a much higher return in the event of success.

The latter approach can be followed most easily by using multivalent platforms which are amenable to inclusion of many antigenic components within one vaccine. High-capacity DNA-based platforms are one example. It is also possible to encode many epitopes and antigen fragments in large-capacity recombinant poxviruses. One example of this approach, NY-VAC-Pf7 (Ockenhouse et al., 1998), yielded impressive multicomponent antibody and T-cell immunogenicity, but development has ceased due to lack of demonstrated efficacy in sporozoite challenge trials.

The virosome platform (Moser et al., 2003) is now under clinical evaluation for malaria. Liposomal particles, including influenza proteins, form the 150-nm virosome particles, essentially empty reconstituted influenza virus envelopes. These have been manufactured with functionally relevant malaria antigen components engineered to reside on the surface of the particle. A phase 1 trial with malaria virosomes reported acceptable safety and reactogenicity and promising antibody induction to the malaria components in June 2004 (www.pevion.com).

MANAGING EXPECTATIONS WHILE ADVOCATING VACCINE DEVELOPMENT

With any effort to develop a new health intervention, the timeline to delivery is unknown. Any of the current malaria vaccine products could lead to a highly effective vaccine. The more likely scenario is that the first generation of vaccines will yield partial efficacy. Even a partially efficacious vaccine would have a tremendous public health impact. However, it is possible that cycles of iterative product development will be needed before a vaccine is sufficiently efficacious to be widely deployed. Therefore, managing expectation while ensuring appropriate advocacy for the development of an effective vaccine will continue to pose a challenge.

REFERENCES

Acosta, C. J., C. M. Galindo, D. Schellenberg, J. J. Aponte, E. Kahigwa, H. Urassa, J. R. Schellenberg, H. Masanja, R. Hayes, A. Y. Kitua, F. Lwilla, H. Mshinda, C. Menendez, M. Tanner, and P. L. Alonso. 1999. Evaluation of the SPf66 vaccine for malaria control when delivered through the EPI scheme in Tanzania. *Trop. Med. Int. Health* **4:**368–376.

Aikawa, M., C. T. Atkinson, L. M. Beaudoin, M. Sedegah, Y. Charoenvit, and R. Beaudoin. 1990. Localization of CS and non-CS antigens in the sporogonic stages of Plasmodium yoelii. *Bull. W. H. O.* **68**(Suppl.):165–171.

Alloueche, A., P. Milligan, D. J. Conway, M. Pinder, K. Bojang, T. Doherty, N. Tornieporth, J. Cohen, and B. M. Greenwood. 2003. Protective efficacy of the RTS,S/AS02 Plasmodium falciparum malaria vaccine is not strain specific. *Am. J. Trop. Med. Hyg.* **68:**97–101.

Alonso, P. L., T. Smith, J. R. Schellenberg, H. Masanja, S. Mwankusye, H. Urassa, I. Bastos de Azevedo, J. Chongela, S. Kobero, C. Menendez, et al. 1994. Randomised trial of efficacy of SPf66 vaccine against Plasmodium falciparum malaria in children in southern Tanzania. *Lancet* **344:**1175–1181.

Amara, R. R., F. Villinger, J. D. Altman, S. L. Lydy, S. P. O'Neil, S. I. Staprans, D. C. Montefiori, Y. Xu, J. G. Herndon, L. S. Wyatt, M. A. Candido, N. L. Kozyr, P. L. Earl, J. M. Smith, H. L. Ma, B. D. Grimm, M. L. Hulsey, J. Miller, H. M. McClure, J. M. McNicholl, B. Moss, and H. L. Robinson. 2001. Control of a mucosal challenge and prevention of AIDS by a multiprotein DNA/MVA vaccine. *Science* **292:**69–74.

Anderson, R. J., C. M. Hannan, S. C. Gilbert, S. M. Laidlaw, E. G. Sheu, S. Korten, R. Sinden, G. A. Butcher, M. A. Skinner, and A. V. Hill. 2004. Enhanced CD8+ T cell immune responses and protection elicited against Plasmodium berghei malaria by prime boost immunization regimens using a novel attenuated fowlpox virus. *J. Immunol.* **172:**3094–3100.

Badell, E., C. Oeuvray, A. Moreno, S. Soe, N. van Rooijen, A. Bouzidi, and P. Druilhe. 2000. Human malaria in immunocompromised mice: an in vivo model to study defense mechanisms against Plasmodium falciparum. *J. Exp. Med.* **192:**1653–1660.

Ballou, W. R., S. L. Hoffman, J. A. Sherwood, M. R. Hollingdale, F. A. Neva, W. T. Hockmeyer, D. M. Gordon, I. Schneider, R. A. Wirtz, and J. F. Young. 1987. Safety and efficacy of a recombinant DNA Plasmodium falciparum sporozoite vaccine. *Lancet* **i:**1277–1281.

Birkett, A., K. Lyons, A. Schmidt, D. Boyd, G. A. Oliveira, A. Siddique, R. Nussenzweig, J. M. Calvo-Calle, and E. Nardin. 2002. A modified hepatitis B virus core particle containing multiple epitopes of the *Plasmodium falciparum* circumsporozoite protein provides a highly immunogenic malaria vaccine in preclinical analyses in rodent and primate hosts. *Infect. Immun.* **70:**6860–6870.

Boisgerault, F., G. Moron, and C. Leclerc. 2002. Virus-like particles: a new family of delivery systems. *Expert Rev. Vaccines* **1:**101–109.

Bojang, K. A., P. J. Milligan, M. Pinder, L. Vigneron, A. Alloueche, K. E. Kester, W. R. Ballou, D. J. Conway, W. H. Reece, P. Gothard, L. Yamuah, M. Delchambre, G. Voss, B. M. Greenwood, A. Hill, K. P. McAdam, N. Tornieporth, J. D. Cohen, T. Doherty, H. Silveira, K. Bojang, and J. Cohen. 2001. Efficacy of RTS,S/AS02 malaria vaccine against Plasmodium falciparum infection in semi-immune adult men in The Gambia: a randomised trial. *Lancet* **358:**1927–1934.

Bouharoun-Tayoun, H., P. Attanath, A. Sabchareon, T. Chongsuphajaisiddhi, and P. Druilhe. 1990. Antibodies that protect humans against Plasmodium falciparum blood stages do not on their own inhibit parasite growth and invasion in vitro, but act in cooperation with monocytes. *J. Exp. Med.* **172:**1633–1641.

Breman, J. G. 2001. The ears of the hippopotamus: manifestations, determinants, and estimates of the malaria burden. *Am. J. Trop. Med. Hyg.* **64:**1–11.

Brown, A. E., P. Singharaj, H. K. Webster, J. Pipithkul, D. M. Gordon, J. W. Boslego, K. Krinchai, P. Su-archawaratana, C. Wongsrichanalai, and W. R. Ballou. 1994. Safety, immunogenicity and limited efficacy study of a recombinant Plasmodium falciparum circumsporozoite vaccine in Thai soldiers. *Vaccine* **12:**102–108.

Church, L. W., T. P. Le, J. P. Bryan, D. M. Gordon, R. Edelman, L. Fries, J. R. Davis, D. A. Herrington, D. F. Clyde, M. J. Shmuklarsky, I. Schneider, T. W. McGovern, J. D. Chulay, W. R. Ballou, and S. L. Hoffman. 1997. Clinical manifestations of Plasmodium falciparum malaria experimentally induced by mosquito challenge. *J. Infect. Dis.* **175:**915–920.

Cochrane, A. H., R. S. Nussenzweig, and E. H. Nardin. 1980. Immunization against sporozoites. In J. P. Kreier (ed.), *Malaria*, vol. 3. Academic Press, New York, N.Y.

Cohen, S., I. A. McGregor, and S. Carrington. 1961. Gamma globulin and acquired immunity to malaria. *Nature* **192:**733–737.

Corr, M., D. J. Lee, D. A. Carson, and H. Tighe. 1996. Gene vaccination with naked plasmid DNA: mechanism of CTL priming. *J. Exp. Med.* **184:**1555–1560.

Crosnier, J., P. Jungers, A. M. Courouce, A. Laplanche, E. Benhamou, F. Degos, B. Lacour, P. Prunet, Y. Cerisier, and P. Guesry. 1981. Randomised placebo-controlled trial of hepatitis B surface antigen vaccine in French haemodialysis units: I, medical staff. *Lancet* **i:**455–459.

D'Alessandro, U., A. Leach, C. J. Drakeley, S. Bennett, B. O. Olaleye, G. W. Fegan, M. Jawara, P. Langerock, M. O. George, and G. A. Targett. 1995. Efficacy trial of malaria vaccine SPf66 in Gambian infants. *Lancet* **346:**462–467.

Doolan, D. L., J. C. Aguiar, W. R. Weiss, A. Sette, P. L. Felgner, D. P. Regis, P. Quinones-Casas, J. R. Yates III, P. L. Blair, T. L. Richie, S. L. Hoffman, and D. J. Carucci. 2003. Utilization of genomic sequence information to develop malaria vaccines. *J. Exp. Biol.* **206:**3789–3802.

Edelman, R., S. S. Wasserman, J. G. Kublin, S. A. Bodison, E. H. Nardin, G. A. Oliveira, S. Ansari, C. L. Diggs, O. L. Kashala, B. J. Schmeckpeper, and R. G. Hamilton. 2002. Immediate-type hypersensitivity and other clinical reactions in volunteers immunized with a synthetic multi-antigen peptide vaccine (PfCS-MAP1NYU) against Plasmodium falciparum sporozoites. *Vaccine* **21:**269–280.

Evans, J. T., C. W. Cluff, D. A. Johnson, M. J. Lacy, D. H. Persing, and J. R. Baldridge. 2003. Enhancement of antigen-specific immunity via the TLR4 ligands MPL adjuvant and Ribi.529. *Expert Rev. Vaccines* **2:**219–229.

Fine, E., M. Aikawa, A. H. Cochrane, and R. S. Nussenzweig. 1984. Immuno-electron microscopic observations on Plasmodium knowlesi sporozoites: localization of protective antigen and its precursors. *Am. J. Trop. Med. Hyg.* **33:**220–226.

Gardner, M. J., N. Hall, E. Fung, O. White, M. Berriman, R. W. Hyman, J. M. Carlton, A. Pain, K. E. Nelson, S. Bowman, I. T. Paulsen, K. James, J. A. Eisen, K. Rutherford, S. L. Salzberg, A. Craig, S. Kyes, M. S. Chan, V. Nene, S. J. Shallom, B. Suh, J. Peterson, S. Angiuoli, M. Pertea, J. Allen, J. Selengut, D. Haft, M. W. Mather, A. B. Vaidya, D. M. Martin, A. H. Fairlamb, M. J. Fraunholz, D. S. Roos, S. A. Ralph, G. I. McFadden, L. M. Cummings, G. M. Subramanian, C. Mungall, J. C. Venter, D. J. Carucci, S. L. Hoffman, C. Newbold, R. W. Davis, C. M. Fraser, and B. Barrell. 2002. Genome sequence of the human malaria parasite Plasmodium falciparum. *Nature* **419:**498–511.

Genton, B., I. Betuela, I. Felger, F. Al-Yaman, R. F. Anders, A. Saul, L. Rare, M. Baisor, K. Lorry, G. V. Brown, D. Pye, D. O. Irving, T. A. Smith, H. P. Beck, and M. P. Alpers. 2002. A recombinant blood-stage malaria vaccine reduces Plasmodium falciparum density and exerts selective pressure on parasite populations in a phase 1-2b trial in Papua New Guinea. *J. Infect. Dis.* **185:**820–827.

Gonzalez, C., D. Hone, F. R. Noriega, C. O. Tacket, J. R. Davis, G. Losonsky, J. P. Nataro, S. Hoffman, A. Malik, E. Nardin, et al. 1994. Salmonella typhi vaccine strain CVD 908 expressing the circumsporozoite protein of Plasmodium falciparum: strain construction and safety and immunogenicity in humans. *J. Infect. Dis.* **169:**927–931.

Good, M. F., D. Pombo, I. A. Quakyi, E. M. Riley, R. A. Houghten, A. Menon, D. W. Alling, J. A. Berzofsky, and L. H. Miller. 1988. Human T-cell recognition of the circumsporozoite protein of Plasmodium falciparum: immunodominant T-cell domains map to the polymorphic regions of the molecule. *Proc. Natl. Acad. Sci. USA* **85:**1199–1203.

Greenwood, B., and T. Mutabingwa. 2002. Malaria in 2002. *Nature* **415:**670-672.

Guerin, P. J., P. Olliaro, F. Nosten, P. Druilhe, R. Laxminarayan, F. Binka, W. L. Kilama, N. Ford, and N. J. White. 2002. Malaria: current status of control, diagnosis, treatment, and a proposed agenda for research and development. *Lancet Infect. Dis.* **2:**564–573.

Gurunathan, S., D. M. Klinman, and R. A. Seder. 2000. DNA vaccines: immunology, application, and optimization. *Annu. Rev. Immunol.* **18:**927–974.

Haddad, D., E. Bilcikova, A. A. Witney, J. M. Carlton, C. E. White, P. L. Blair, R. Chattopadhyay, J. Russell, E. Abot, Y. Charoenvit, J. C. Aguiar, D. J. Carucci, and W. R. Weiss. 2004. Novel antigen identification method for discovery of protective malaria antigens by rapid testing of DNA vaccines encoding exons from the parasite genome. *Infect. Immun.* **72:**1594–1602.

Hay, S. I., C. A. Guerra, A. J. Tatem, A. M. Noor, and R. W. Snow. 2004. The global distribution and population at risk of malaria: past, present, and future. *Lancet Infect. Dis.* **4:**327–336.

Herrington, D. A., D. F. Clyde, G. Losonsky, M. Cortesia, J. R. Murphy, J. Davis, S. Baqar, A. M. Felix, E. P. Heimer, and D. Gillessen. 1987. Safety and immunogenicity in man of a synthetic peptide malaria vaccine against Plasmodium falciparum sporozoites. *Nature* **328:**257–259.

Herrington, D. A., G. A. Losonsky, G. Smith, F. Volvovitz, M. Cochran, K. Jackson, S. L. Hoffman, D. M. Gordon, M. M. Levine, and R. Edelman. 1992. Safety and immunogenicity in volunteers of a recombinant Plasmodium falciparum circumsporozoite protein malaria vaccine produced in Lepidopteran cells. *Vaccine* **10:**841–846.

Hirunpetcharat, C., J. Wipasa, S. Sakkhachornphop, T. Nitkumhan, Y. Z. Zheng, S. Pichyangkul, A. M. Krieg, D. S. Walsh, D. G. Heppner, and M. F. Good. 2003. CpG oligodeoxynucleotide enhances immunity against blood-stage malaria infection in mice parenterally immunized with a yeast-expressed 19 kDa carboxyl-terminal fragment of Plasmodium yoelii merozoite surface protein-1. *Vaccine* **21:**2923–2932.

Hoffman, S. L., R. Edelman, J. P. Bryan, I. Schneider, J. Davis, M. Sedegah, D. Gordon, P. Church, M. Gross, C. Silverman, et al. 1994. Safety, immunogenicity, and efficacy of a malaria sporozoite vaccine administered with monophosphoryl lipid A, cell wall skeleton of mycobacteria, and squalane as adjuvant. *Am. J. Trop. Med. Hyg.* **51:**603–612.

Hoffman, S. L., C. N. Oster, C. V. Plowe, G. R. Woollett, J. C. Beier, J. D. Chulay, R. A. Wirtz, M. R. Hollingdale, and M. Mugambi. 1987. Naturally acquired antibodies to sporozoites do not prevent malaria: vaccine development implications. *Science* **237:**639–642.

Holder, A. A., J. A. Guevara Patino, C. Uthaipibull, S. E. Syed, I. T. Ling, T. Scott-Finnigan, and M. J. Blackman. 1999. Merozoite surface protein 1, immune evasion, and vaccines against asexual blood stage malaria. *Parassitologia* **41:**409–414.

John, C. C., R. A. O'Donnell, P. O. Sumba, A. M. Moormann, T. F. de Koning-Ward, C. L. King, J. W. Kazura, and B. S. Crabb. 2004. Evidence that invasion-inhibitory antibodies specific for the 19-kDa fragment of merozoite surface protein-1 (MSP-1_{19}) can play a protective role against blood-stage *Plasmodium falciparum* infection in individuals in a malaria endemic area of Africa. *J. Immunol.* **173:**666–672.

Kaslow, D. C. 2002. Transmission-blocking vaccines. *Chem. Immunol.* **80:**287–307.

Kenney, R. T., and R. Edelman. 2003. Survey of human-use adjuvants. *Expert Rev. Vaccines* **2:**167–188.

Kester, K. E., D. A. McKinney, N. Tornieporth, C. F. Ockenhouse, D. G. Heppner, T. Hall, U. Krzych, M. Delchambre, G. Voss, M. G. Dowler, J. Palensky, J. Wittes, J. Cohen, and W. R. Ballou. 2001. Efficacy of recombinant circumsporozoite protein vaccine regimens against experimental Plasmodium falciparum malaria. *J. Infect. Dis.* **183:**640–647.

Klinman, D. M., K. M. Barnhart, and J. Conover. 1999. CpG motifs as immune adjuvants. *Vaccine* **17:**19–25.

Kocken, C. H., H. Ozwara, A. van der Wel, A. L. Beetsma, J. M. Mwenda, and A. W. Thomas. 2002. *Plasmodium knowlesi* provides a rapid in vitro and in vivo transfection system that enables double-crossover gene knockout studies. *Infect. Immun.* **70:**655–660.

Lengeler, C. 2000. Insecticide-treated bednets and curtains for preventing malaria. *Cochrane Database Syst. Rev.* **2000:**CD000363.

Li, S., M. Rodrigues, D. Rodriguez, J. R. Rodriguez, M. Esteban, P. Palese, R. S. Nussenzweig, and F. Zavala. 1993. Priming with recombinant influenza virus followed by administration of recombinant vaccinia virus induces CD8+ T-cell-mediated protective immunity against malaria. *Proc. Natl. Acad. Sci. USA* **90:**5214–5218.

Lopez, J. A., C. Weilenman, R. Audran, M. A. Roggero, A. Bonelo, J. M. Tiercy, F. Spertini, and G. Corradin. 2001. A synthetic malaria vaccine elicits a potent CD8(+) and CD4(+) T lymphocyte immune response in humans. Implications for vaccination strategies. *Eur. J. Immunol.* **31:**1989–1998.

Lopez, M. C., Y. Silva, M. C. Thomas, A. Garcia, M. J. Faus, P. Alonso, F. Martinez, G. Del Real, and C. Alonso. 1994. Characterization of SPf(66)n: a chimeric molecule used as a malaria vaccine. *Vaccine* **12:**585–591.

Luke, T. C., and S. L. Hoffman. 2003. Rationale and plans for developing a non-replicating, metabolically active, radiation-attenuated Plasmodium falciparum sporozoite vaccine. *J. Exp. Biol.* **206:**3803–3808.

Mahoney, R. T., and J. E. Maynard. 1999. The introduction of new vaccines into developing countries. *Vaccine* **17:**646–652.

Makobongo, M. O., G. Riding, H. Xu, C. Hirunpetcharat, D. Keough, J. de Jersey, P. Willadsen, and M. F. Good. 2003. The purine salvage enzyme hypoxanthine guanine xanthine phosphoribosyl transferase is a major target antigen for cell-mediated immunity to malaria. *Proc. Natl. Acad. Sci. USA* **100:**2628–2633.

Marsh, K., and R. J. Howard. 1986. Antigens induced on erythrocytes by P. falciparum: expression of diverse and conserved determinants. *Science* **231:**150–153.

Massaga, J. J., A. Y. Kitua, M. M. Lemnge, J. A. Akida, L. N. Malle, A. M. Ronn, T. G. Theander, and I. C. Bygbjerg. 2003. Effect of intermittent treatment with amodiaquine on anaemia and malarial fevers in infants in Tanzania: a randomised placebo-controlled trial. *Lancet* **361:**1853–1860.

McConkey, S. J., W. H. Reece, V. S. Moorthy, D. Webster, S. Dunachie, G. Butcher, J. M.

Vuola, T. J. Blanchard, P. Gothard, K. Watkins, C. M. Hannan, S. Everaere, K. Brown, K. E. Kester, J. Cummings, J. Williams, D. G. Heppner, A. Pathan, K. Flanagan, N. Arulanantham, M. T. Roberts, M. Roy, G. L. Smith, J. Schneider, T. Peto, R. E. Sinden, S. C. Gilbert, and A. V. Hill. 2003. Enhanced T-cell immunogenicity of plasmid DNA vaccines boosted by recombinant modified vaccinia virus Ankara in humans. *Nat. Med.* **9:**729–735.

Menard, R., A. A. Sultan, C. Cortes, R. Altzuler, M. R. van Dijk, C. J. Janse, A. P. Waters, R. S. Nussenzweig, and V. Nussenzweig. 1997. Circumsporozoite protein is required for development of malaria sporozoites in mosquitoes. *Nature* **385:**336–340.

Mitchell, G. H., A. W. Thomas, G. Margos, A. R. Dluzewski, and L. H. Bannister. 2004. Apical membrane antigen 1, a major malaria vaccine candidate, mediates the close attachment of invasive merozoites to host red blood cells. *Infect. Immun.* **72:**154–158.

Miyahira, Y., A. Garcia-Sastre, D. Rodriguez, J. R. Rodriguez, K. Murata, M. Tsuji, P. Palese, M. Esteban, F. Zavala, and R. S. Nussenzweig. 1998. Recombinant viruses expressing a human malaria antigen can elicit potentially protective immune CD8(+) responses in mice. *Proc. Natl. Acad. Sci. USA* **95:**3954–3959.

Moorthy, V. S., M. F. Good, and A. V. Hill. 2004a. Malaria vaccine developments. *Lancet* **363:**150–156.

Moorthy, V. S., E. B. Imoukhuede, S. Keating, M. Pinder, D. Webster, M. A. Skinner, S. C. Gilbert, G. Walraven, and A. V. Hill. 2004b. Phase 1 evaluation of 3 highly immunogenic prime-boost regimens, including a 12-month reboosting vaccination, for malaria vaccination in Gambian men. *J. Infect. Dis.* **189:**2213–2219.

Moorthy, V. S., M. Pinder, W. H. Reece, K. Watkins, S. Atabani, C. Hannan, K. Bojang, K. P. McAdam, J. Schneider, S. Gilbert, and A. V. Hill. 2003. Safety and immunogenicity of DNA/modified vaccinia virus ankara malaria vaccination in African adults. *J. Infect. Dis.* **188:**1239–1244.

Moser, C., I. C. Metcalfe, and J. F. Viret. 2003. Virosomal adjuvanted antigen delivery systems. *Expert Rev. Vaccines* **2:**189–196.

Nardin, E. H., J. M. Calvo-Calle, G. A. Oliveira, R. S. Nussenzweig, M. Schneider, J. M. Tiercy, L. Loutan, D. Hochstrasser, and K. Rose. 2001. A totally synthetic polyoxime malaria vaccine containing Plasmodium falciparum B cell and universal T cell epitopes elicits immune responses in volunteers of diverse HLA types. *J. Immunol.* **166:**481–489.

Nardin, E. H., G. A. Oliveira, J. M. Calvo-Calle, Z. R. Castro, R. S. Nussenzweig, B. Schmeckpeper, B. F. Hall, C. Diggs, S. Bodison, and R. Edelman. 2000. Synthetic malaria peptide vaccine elicits high levels of antibodies in vaccinees of defined HLA genotypes. *J. Infect. Dis.* **182:**1486–1496.

Nosten, F., C. Luxemburger, D. E. Kyle, W. R. Ballou, J. Wittes, E. Wah, T. Chongsuphajaisiddhi, D. M. Gordon, N. J. White, J. C. Sadoff, D. G. Heppner, et al. 1996. Randomised double-blind placebo-controlled trial of SPf66 malaria vaccine in children in northwestern Thailand. *Lancet* **348:**701–707.

Ockenhouse, C. F., P. F. Sun, D. E. Lanar, B. T. Wellde, B. T. Hall, K. Kester, J. A. Stoute, A. Magill, U. Krzych, L. Farley, R. A. Wirtz, J. C. Sadoff, D. C. Kaslow, S. Kumar, L. W. Church, J. M. Crutcher, B. Wizel, S. Hoffman, A. Lalvani, A. V. Hill, J. A. Tine, K. P. Guito, C. de Taisne, R. Anders, W. R. Ballou, et al. 1998. Phase I/IIa safety, immunogenicity, and efficacy trial of NYVAC-Pf 7, a pox-vectored, multiantigen, multistage vaccine candidate for Plasmodium falciparum malaria. *J. Infect. Dis.* **177:**1664–1673.

Oeuvray, C., H. Bouharoun-Tayoun, H. Gras-Masse, E. Bottius, T. Kaidoh, M. Aikawa, M. C. Filgueira, A. Tartar, and P. Druilhe. 1994. Merozoite surface protein-3: a malaria protein inducing antibodies that promote Plasmodium falciparum killing by cooperation with blood monocytes. *Blood* **84:**1594–1602.

Paoletti, E. 1996. Applications of pox virus vectors to vaccination: an update. *Proc. Natl. Acad. Sci. USA* **93:**11349–11353.

Perri, S., C. E. Greer, K. Thudium, B. Doe, H. Legg, H. Liu, R. E. Romero, Z. Tang, Q. Bin, T. W. Dubensky, Jr., M. Vajdy, G. R. Otten, and J. M. Polo. 2003. An alphavirus replicon particle chimera derived from Venezuelan equine encephalitis and Sindbis viruses is a potent gene-based vaccine delivery vector. *J. Virol.* **77:**10394–10403.

Phillips-Howard, P. A., B. L. Nahlen, M. S. Kolczak, A. W. Hightower, F. O. ter Kuile, J. A. Alaii, J. E. Gimnig, J. Arudo, J. M. Vulule, A. Odhacha, S. P. Kachur, E. Schoute, D. H. Rosen, J. D. Sexton, A. J. Oloo, and W. A. Hawley. 2003. Efficacy of permethrin-treated bed nets in the prevention of mortality in young children in an area of high perennial malaria transmission in western Kenya. *Am. J. Trop. Med. Hyg.* **68:**23–29.

Pombo, D. J., G. Lawrence, C. Hirunpetcharat, C. Rzepczyk, M. Bryden, N. Cloonan, K. Anderson, Y. Mahakunkijcharoen, L. B. Martin, D. Wilson, S. Elliott, D. P. Eisen, J. B. Weinberg, A. Saul, and M. F. Good. 2002. Immunity to malaria after administration of ultra-low doses of red cells infected with Plasmodium falciparum. *Lancet* **360:**610–617.

Richie, T. L., R. Wang, Y. Charoenvit, D. Freilich, J. E. Epstein, S. Kumar, J. Aguiar, G. Gray, S. E. Parker, P. Hobart, S. Kradjian, J. A. Norman, J. Sacci, T. C. Luke, S. L. Hoffman, et al. 2001. Safety, immunogenicity and efficacy of MuStDO5, a five gene sporozoite/hepatic stage *Plasmodium falciparum* DNA vaccine combined with human GM-CSF DNA. *Am. J. Trop. Med. Hyg.* **65**(Suppl.):230.

Rosen, J. B., and J. G. Breman. 2004. Malaria intermittent preventive treatment in infants, chemoprophylaxis, and childhood vaccinations. *Lancet* **363**:1386–1388.

Roy, M. J., M. S. Wu, L. J. Barr, J. T. Fuller, L. G. Tussey, S. Speller, J. Culp, J. K. Burkholder, W. F. Swain, R. M. Dixon, G. Widera, R. Vessey, A. King, G. Ogg, A. Gallimore, J. R. Haynes, and D. Heydenburg Fuller. 2000. Induction of antigen-specific CD8+ T cells, T helper cells, and protective levels of antibody in humans by particle-mediated administration of a hepatitis B virus DNA vaccine. *Vaccine* **19**:764–778.

Sachs, J., and P. Malaney. 2002. The economic and social burden of malaria. *Nature* **415**:680–685.

Saul, A., G. Lawrence, A. Smillie, C. M. Rzepczyk, C. Reed, D. Taylor, K. Anderson, A. Stowers, R. Kemp, A. Allworth, R. F. Anders, G. V. Brown, D. Pye, P. Schoofs, D. O. Irving, S. L. Dyer, G. C. Woodrow, W. R. Briggs, R. Reber, and D. Sturchler. 1999. Human phase I vaccine trials of 3 recombinant asexual stage malaria antigens with Montanide ISA720 adjuvant. *Vaccine* **17**:3145–3159.

Schellenberg, D., C. Menendez, E. Kahigwa, J. Aponte, J. Vidal, M. Tanner, H. Mshinda, and P. Alonso. 2001. Intermittent treatment for malaria and anaemia control at time of routine vaccinations in Tanzanian infants: a randomised, placebo-controlled trial. *Lancet* **357**:1471–1477.

Schneider, J., S. C. Gilbert, T. J. Blanchard, T. Hanke, K. J. Robson, C. M. Hannan, M. Becker, R. Sinden, G. L. Smith, and A. V. Hill. 1998. Enhanced immunogenicity for CD8+ T cell induction and complete protective efficacy of malaria DNA vaccination by boosting with modified vaccinia virus Ankara. *Nat. Med.* **4**:397–402.

Schneider, J., S. C. Gilbert, C. M. Hannan, P. Degano, E. Prieur, E. G. Sheu, M. Plebanski, and A. V. Hill. 1999. Induction of CD8+ T cells using heterologous prime-boost immunisation strategies. *Immunol. Rev.* **170**:29–38.

Sherwood, J. A., C. N. Oster, M. Adoyo-Adoyo, J. C. Beier, G. S. Gachihi, P. M. Nyakundi, W. R. Ballou, A. D. Brandling-Bennett, I. K. Schwartz, J. B. Were, et al. 1991. Safety and immunogenicity of a Plasmodium falciparum sporozoite vaccine: boosting of antibody response in a population with prior natural exposure to malaria. *Trans. R. Soc. Trop. Med. Hyg.* **85**:336–340.

Shiver, J. W., and E. A. Emini. 2004. Recent advances in the development of HIV-1 vaccines using replication-incompetent adenovirus vectors. *Annu. Rev. Med.* **55**:355–372.

Singh, S., M. C. Kennedy, C. A. Long, A. J. Saul, L. H. Miller, and A. W. Stowers. 2003. Biochemical and immunological characterization of bacterially expressed and refolded *Plasmodium falciparum* 42-kilodalton C-terminal merozoite surface protein 1. *Infect. Immun.* **71**:6766–6774.

Sinnis, P., and E. Nardin. 2002. Sporozoite antigens: biology and immunology of the circumsporozoite protein and thrombospondin-related anonymous protein. *Chem. Immunol.* **80**:70–96.

Snow, R. W., M. Craig, U. Deichmann, and K. Marsh. 1999. Estimating mortality, morbidity and disability due to malaria among Africa's non-pregnant population. *Bull. W. H. O.* **77**:624–640.

Snow, R. W., J. F. Trape, and K. Marsh. 2001. The past, present and future of childhood malaria mortality in Africa. *Trends Parasitol.* **17**:593–597.

Stewart, M. J., and J. P. Vanderberg. 1988. Malaria sporozoites leave behind trails of circumsporozoite protein during gliding motility. *J. Protozool.* **35**:389–393.

Stoute, J. A., M. Slaoui, D. G. Heppner, P. Momin, K. E. Kester, P. Desmons, B. T. Wellde, N. Garcon, U. Krzych, M. Marchand, et al. 1997. A preliminary evaluation of a recombinant circumsporozoite protein vaccine against Plasmodium falciparum malaria. *N. Engl. J. Med.* **336**:86–91.

Stowers, A. W., M. C. Kennedy, B. P. Keegan, A. Saul, C. A. Long, and L. H. Miller. 2002. Vaccination of monkeys with recombinant *Plasmodium falciparum* apical membrane antigen 1 confers protection against blood-stage malaria. *Infect. Immun.* **70**:6961-6967.

Ulmer, J. B., J. J. Donnelly, S. E. Parker, G. H. Rhodes, P. L. Felgner, V. J. Dwarki, S. H. Gromkowski, R. R. Deck, C. M. DeWitt, A. Friedman, et al. 1993. Heterologous protection against influenza by injection of DNA encoding a viral protein. *Science* **259**:1745–1749.

Valero, M. V., L. R. Amador, C. Galindo, J. Figueroa, M. S. Bello, L. A. Murillo, A. L. Mora, G. Patarroyo, C. L. Rocha, M. Rojas, et al. 1993. Vaccination with SPf66, a chemically synthesised vaccine, against Plasmodium falciparum malaria in Colombia. *Lancet* **341**:705–710.

Vreden, S. G., J. P. Verhave, T. Oettinger, R. W. Sauerwein, and J. H. Meuwissen. 1991. Phase I clinical trial of a recombinant malaria vaccine consisting of the circumsporozoite repeat region of Plasmodium falciparum coupled to hepatitis B surface antigen. *Am. J. Trop. Med. Hyg.* **45**:533–538.

Wang, R., D. L. Doolan, T. P. Le, R. C. Hedstrom, K. M. Coonan, Y. Charoenvit, T. R. Jones, P. Hobart, M. Margalith, J. Ng, W. R. Weiss, M. Sedegah, C. de Taisne, J. A. Norman, and S. L. Hoffman. 1998. Induction of antigen-specific cytotoxic T lymphocytes in humans by a malaria DNA vaccine. *Science* **282:**476–480.

Wang, R., J. Epstein, F. M. Baraceros, E. J. Gorak, Y. Charoenvit, D. J. Carucci, R. C. Hedstrom, N. Rahardjo, T. Gay, P. Hobart, R. Stout, T. R. Jones, T. L. Richie, S. E. Parker, D. L. Doolan, J. Norman, and S. L. Hoffman. 2001. Induction of CD4(+) T cell-dependent CD8(+) type 1 responses in humans by a malaria DNA vaccine. *Proc. Natl. Acad. Sci. USA* **98:**10817–10822.

Wang, R., J. Epstein, Y. Charoenvit, F. M. Baraceros, N. Rahardjo, T. Gay, J. G. Banania, R. Chattopadhyay, P. de la Vega, T. L. Richie, N. Tornieporth, D. L. Doolan, K. E. Kester, D. G. Heppner, J. Norman, D. J. Carucci, J. D. Cohen, and S. L. Hoffman. 2004. Induction in humans of CD8+ and CD4+ T cell and antibody responses by sequential immunization with malaria DNA and recombinant protein. *J. Immunol.* **172:**5561–5569.

Weiss, W. R., S. Mellouk, R. A. Houghten, M. Sedegah, S. Kumar, M. F. Good, J. A. Berzofsky, L. H. Miller, and S. L. Hoffman. 1990. Cytotoxic T cells recognize a peptide from the circumsporozoite protein on malaria-infected hepatocytes. *J. Exp. Med.* **171:**763–773.

World Health Organization. 2003. *Africa Malaria Report.*

Yoshida, N., R. S. Nussenzweig, P. Potocnjak, V. Nussenzweig, and M. Aikawa. 1980. Hybridoma produces protective antibodies directed against the sporozoite stage of malaria parasite. *Science* **207:**71–73.

VECTOR

VI

THE *ANOPHELES GAMBIAE* GENOME

Frank H. Collins and Catherine A. Hill

26

INTRODUCTION

A. gambiae: Malaria Vector

Anopheles gambiae is the most important vector of malaria in sub-Saharan Africa, where most of the world's human malaria cases and deaths occur each year (e.g., Coluzzi et al., 1979; White 1974). The distribution of this mosquito extends across all of sub-Saharan Africa north of approximately the Tropic of Capricorn (23.5° south latitude). *A. gambiae* is a particularly effective vector. This mosquito blood feeds almost exclusively on people; its larvae develop in transiently flooded bodies of water produced by human activity, such as borrow pits, hoofprints, tire tracks, and flooded agricultural fields; and the adults rest primarily inside buildings (Coluzzi et al., 1994). It is one member of a group of seven closely related and morphologically indistinguishable species called the *A. gambiae* complex (Coetzee et al., 2000). Although *A. gambiae* is the principal malaria vector in this complex, all of its sibling species, except for *Anopheles quadriannulatus* A and B, are also involved in malaria transmission. The most important among these is *Anopheles arabiensis*, which also has a continent-wide distribution and is also found only in association with humans. This species is somewhat less effective as a vector because it feeds on domestic animals as well as people, but it is more tolerant of drier and cooler environments, so it complements *A. gambiae* by extending malaria transmission into the drier seasons and geographically further north and south (Table 1).

Polymorphisms within *A. gambiae* Sensu Stricto

Genetically, *A. gambiae* sensu stricto is a very polymorphic taxon. Decades of study of the distribution of polymorphic inversions on the polytene chromosomes of this species have revealed significant population structure, particularly in populations from west and central Africa (e.g., Toure et al., 1998). In populations from Mali, for example, significant deficits in inversion heterozygote frequencies have been systematically observed among five polymorphic

---

Frank H. Collins, Center for Tropical Disease Research and Training, Galvin Life Sciences Building, University of Notre Dame, Notre Dame, IN 46556–0369. *Catherine A. Hill,* Department of Entomology, 901 West State St., Purdue University, West Lafayette, IN 47907–2089.

Molecular Approaches to Malaria, Edited by Irwin W. Sherman
© 2005 ASM Press, Washington, D.C.

TABLE 1 Species in the *A. gambiae* complex and their role in malaria transmission

Species	Distribution	Role in transmission
A. gambiae	Most of sub-Saharan Africa	Most important vector
A. arabiensis	Most of sub-Saharan Africa	Inportant vector
A. merus	Coastal east and southern Africa	Local vector
A. melas	Coastal west Africa	Local vector
A. bwambae	Ruwenzori Mountains, Uganda	Local vector
A. quadriannulatus A	Southern Africa	Not a vector
A. quiadriannulatus B	Ethiopia	Not a vector

inversions on the right arm of chromosome 2 or 2R (Coluzzi et al., 1985). The proposed explanation for these inversion frequency patterns is that three reproductively isolated, sympatric forms of *A. gambiae* sensu stricto exist in Mali. These are designated as the Bamako, Mopti, and Savanna chromosomal forms. Additional patterns of inversion heterozygote deficits among populations in other regions of west Africa have been interpreted as evidence for two additional chromosomal forms, referred to as the Bissau and Forest forms. Outside the geographic areas where these chromosomal forms were originally identified, these patterns of inversion-based population structure are less clear.

The discontinuities in population structure revealed by polymorphic chromosomal inversions have been supported to some extent by polymorphisms in molecular markers, including microsatellite loci, sequence polymorphisms, and in particular polymorphisms in rRNA gene sequence (Besansky et al., 1997; Lanzaro et al., 1998; Lehmann et al., 2003). Two distinct rRNA gene forms have been observed in *A. gambiae* sensu stricto (Favia et al., 1994; Favia et al., 1997). In Mali, where the population distribution of these molecular forms was first studied extensively, all individuals identified by karyotype as having Mopti chromosomal forms had what was designated as the M-form rRNA gene. All Savanna and Bamako chromosomal forms in Mali were of the S-form rRNA gene. In Burkina Faso, where only the Mopti and Savanna chromosomal forms have been identified, there is exact correspondence with the M and S molecular forms. In other parts of west and central Africa, however, populations identified chromosomally as having the Savanna form may have either the M or S rRNA gene form. The same is true for the Forest chromosomal form. Little data exist on the relationship between the Bissau chromosomal form and the rRNA gene M and S forms (della Torre et al., 2001; Gentile et al., 2001).

Heterozygous M/S rRNA gene forms are almost always absent or very rare, even in populations where both rRNA gene forms are sympatric and all individuals are of a single chromosomal form. Moreover, careful studies of the molecular rRNA gene form of sperm in the spermatheca of individuals collected from populations polymorphic for both M and S molecular forms show considerable premating isolation, with frequencies of females having mated with the other form being generally <1% (Tripet et al., 2001, 2003).

A. gambiae is clearly a species with a high degree of genetic population structure, particularly in west and central Africa. The distribution of the two rRNA gene forms and the chromosome 2R inversions suggests that *A. gambiae* is in the process of incipient speciation. The absence of concordance between these two types of genetic markers in Mali has been provisionally interpreted to reflect a kind of parallel chromosomal adaptation within each of the rRNA gene molecular forms. But regardless of the evolutionary status implied by these rRNA gene and chromosomal markers, *A. gambiae* is a taxon whose complex population structure probably reflects its adaptation to a highly varied set of environments in its evolutionary association with humans. Understanding the ecological and behavioral characteristics of these genetically differentiated populations could contribute to more efficient, vector-targeted malaria control strategies.

An Evolutionary Hypothesis

Coluzzi et al. (2002) have proposed that an ancestral form of *A. gambiae* with the standard chromosome 2 arrangement developed a close association with humans as people invaded the African equatorial forests between 2,800 and 2,500 years ago. Today, the typical forest-adapted *A. gambiae* is the Forest chromosomal form, which is fixed for the standard chromosome arrangement on chromosomes 2R and 2L. From this west African, equatorial forest environment, it is hypothesized that an S rRNA gene form of *A. gambiae* with the standard chromosome 2 arrangements expanded its range in association with human movement into the drier, savanna and sahel environments of Africa. Various chromosome 2 inversions were acquired in this expansion. Some, like inversions 2Rb and 2La, were probably acquired by interspecies introgression from the already Savanna-adapted sibling species *A. arabiensis*; others were acquired de novo. These inversions, by creating tight linkages between adaptive sets of alleles, could have been the genetic mechanism that permitted *A. gambiae* to expand its range ecologically into the many different types of habitats occupied today. Development of the M rRNA gene molecular form of *A. gambiae* and evolution of premating isolating mechanisms that restrict gene flow between the S and M molecular forms likely occurred in one of the west African Savanna environments, as the M form is not found in east African populations. But the actual role of these X chromosome rRNA gene types and the paracentric chromosome 2 inversions in genetic differentiation within the taxon *A. gambiae* remains unclear.

THE *A. GAMBIAE* GENOME PROJECT

Genome Size and Structure

The *A. gambiae* haploid genome has been estimated by C_0t analysis to be 260 Mb (Besansky and Powell, 1992). The mosquito has three pairs of chromosomes, an X and Y sex-determining pair and two autosomes, chromosomes 2 and 3. Both autosomes and the X chromosome undergo many rounds of replication without nuclear division in a number of different tissues to produce giant, polytene chromosomes. Coluzzi et al. (2002) produced a standard polytene chromosome map that partitions the chromosome complement into 46 numbered divisions, each of which is subdivided into between three and five lettered subdivisions (Color Plate 19, X chromosome). (Division 6 on the centromeric end of the telocentric X chromosome shown in Color Plate 19 is not subdivided.) The ability to physically map DNA clones to these chromosomes by in situ hybridization (Fig. 1) has been a valuable tool in assembling the *A. gambiae* genome (Kumar and Collins, 1993).

Strain Sequenced in Genome Project

The *A. gambiae* PEST (for pink eye standard) strain was sequenced in the *A. gambiae* genome project, primarily because it had been specifically selected to be fixed for both the standard chromosome arrangement and for an X-linked pink-eye mutation (Mukabayire and Besansky, 1996; Mason, 1967; Beard et al., 1995), and its DNA was cloned into two independent bacterial artificial chromosome (BAC) libraries that contributed to the genome project (Hong et al., 2003; Z. Ke and F. H. Collins, unpublished data). The PEST strain had been produced by crossing pink-eyed (PE) males from the LPE colony (established in 1951 from mosquitoes collected in Lagos, Nigeria) with virgin female offspring of wild caught *A. gambiae* mosquitoes from western Kenya. Pink-eyed males among the F2 progeny of this cross were again mated with female offspring of Kenyan mosquitoes. This outcrossing scheme was repeated a third time, and a colony homozygous for the pink-eye genotype was selected (Githeko et al., 1992). This pink-eye or PE colony was polymorphic for inversions 2La (32%) and 2Rbc (19%). PEST was a derivative of *A. gambiae* PE selected for the standard chromosome arrangement. The 2Rbc inversion is characteristic of the Mopti chromosomal form, and analysis of the rRNA gene of the PEST colony revealed it to be of the M rRNA gene form. Thus, the PEST colony

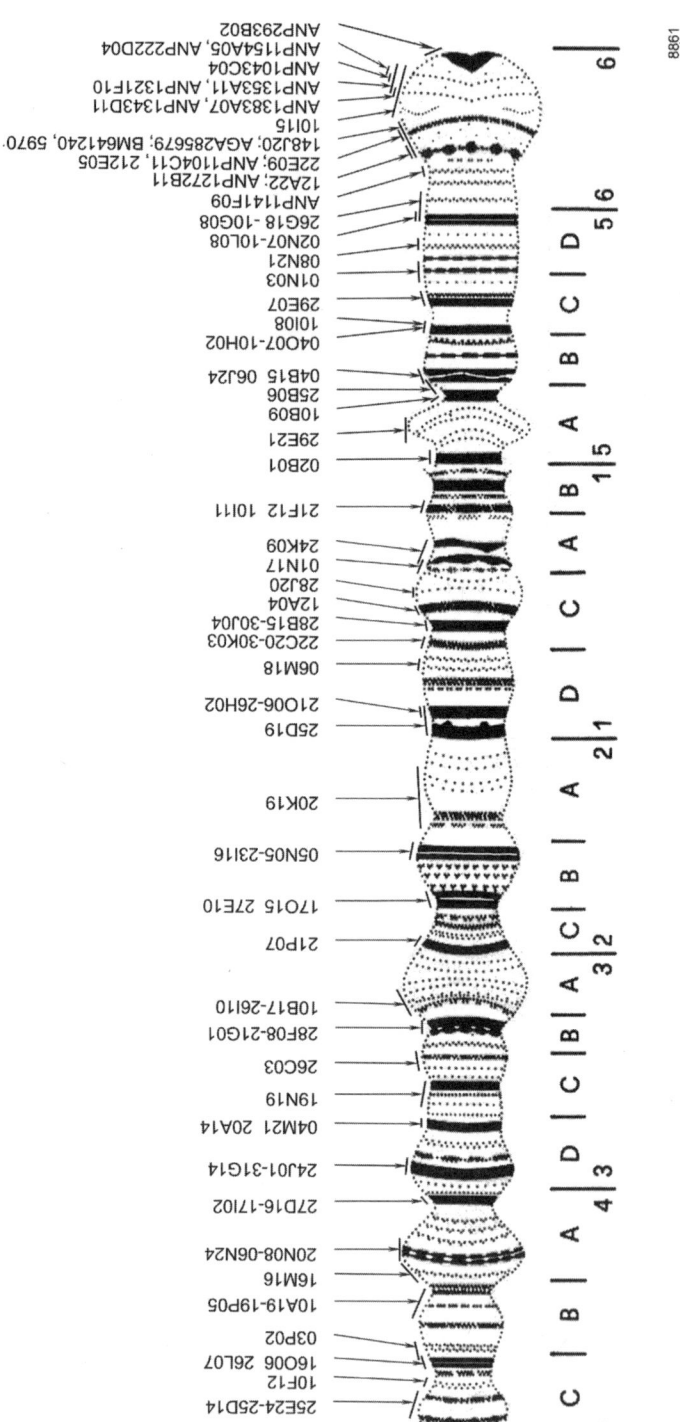

FIGURE 1 Schematic diagram by M. Coluzzi of the X chromosome of *A. gambiae* showing sites of hybridization of a sample of BAC clones and cDNA clones. The dark lines below the chromosome indicate the approximate location along the chromosome of individual scaffolds. Scaffolds are identified by the last four digits in the scaffold clone name (e.g., scaffold AAAB01008846 is marked 8846). Arrows indicate orientation of the scaffold from the first nucleotide pair to the last (at arrowhead). Some small scaffolds in division 6 have not been oriented.

appears to be a mix of Mopti and Savanna type DNA.

Genome Sequencing Project

The *A. gambiae* genome project involved contributions from a number of different investigators and consisted of essentially three coordinated projects. One project, supported by a Cooperative Agreement funding mechanism from the National Institute of Allergy and Infectious Diseases to a large group of U.S. and European laboratories, involved (i) the sequencing of approximately 60,000 clones from three normalized *A. gambiae* cDNA libraries, (ii) the production and sequencing of both ends of a ~14-fold coverage genomic library in a BAC vector (ND-TAM library; Young et al., 2003), (iii) the physical mapping of approximately 2,000 BAC clones by in situ hybridization to polytene chromosomes, and (iv) the sequencing of an approximately 500-kb contig (an overlapping cluster of contiguous sequence) of BAC clones (Thomasova et al., 2002). The second project, also funded by the National Institute of Allergy and Infectious Diseases, was directed by Celera Genomics. This project involved the shotgun sequencing of an ~10-fold coverage of the PEST strain genome and the additional single-pass sequencing of ~80,000 cDNA clones produced from adult female mosquitoes (half from blood-fed and half from non-blood-fed females). The shotgun sequencing was obtained from plasmid libraries containing 2.5-, 10-, and 50-kb sized inserts of PEST genomic DNA. The third major project was the contribution from the Genescope/Institut Pasteur in Paris. This group sequenced the ends of the second, ~5-fold coverage BAC library (ND-1 library; Ke and Collins, unpublished), contributed to the shotgun sequencing of clones produced at Celera, and sequenced an additional 60,000 cDNA sequences.

Genome Assembly and Annotation

The shotgun sequences from Celera and Genoscope were assembled with the Celera assembler into 8,987 scaffolds (ordered and oriented sets of contigs with gaps) that constitute the assembled genome. The process of assembly involved the comparison of every sequence read to every other read to produce contigs. Those contigs that appeared to be single copy, based on the number of overlapping reads, were then oriented with respect to other single-copy contigs by the orientation of mate-pair sequences from the opposite ends of the 10-kb, 50-kb, and BAC clones. The largest single scaffold in the assembly was 23.1 Mb, and the largest contig was 0.8 Mb.

Approximately 84% of the scaffolds constituting the initial assembly (Holt et al., 2002) were assigned by order and orientation to the chromosome arms (X, 2L, 2R, 3L, or 3R) (Color Plate 19). The remainder was assigned to an unknown chromosome. With a few exceptions, the scaffolds in this set are <100 kb; most were initially thought to represent the Y chromosome, which has still not been assembled, and transposon-rich regions of the genome, much of which seems to be associated with the centromeres. Most of these mapped scaffolds and their approximate chromosome location are illustrated in the *A. gambiae* foldout poster in the 4 October 2002 issue of *Science* which featured publications dealing with the genome of this vector (Coluzzi et al., 2002).

The initial annotation of the assembly was based on two independent, automatic annotations contributed by the Celera and Ensembl group at the European Bioinformatics Institute, European Molecular Biology Laboratory, both of which relied on combinations of ab initio gene-finding algorithms and sequence homology evidence. After comparing and combining these results, the Celera group identified 15,189 putative genes, 1,506 of which were probable transposable elements and 663 of which were similar to genes of possible bacterial origin. The scaffolds in this assembly and the associated annotation were submitted to GenBank as scaffolds AAAB01000001 to AAAB01008987. The *A. gambiae* genome and associated annotation are publicly displayed both at NCBI (http://www.ncbi.nlm.nih.gov/mapview/map_search.cgi?taxid=7165) and at the EBI/EMBL Ensembl Mosquito Genome Browser (http://www.ensembl.org/Anopheles_gambiae/).

The *A. gambiae* assembly described by Holt et al. (2002) was updated on 3 February 2003 (Ensembl Mosquito Genome Browser, version 10.2.1) to reflect some slight corrections to the assignment and orientation of scaffolds to chromosome location (Holt and Collins, in press). This assembly is reflected in the poster published on 4 October 2002. The annotation, display, and analysis schema have been updated or modified in a number of ways more than 2 dozen times since the initial version, but the principal changes are major new gene builds in 3 February 2003 (in association with the new assembly) and 1 October 2003. The current version, 27.2c.1, actually reflects the 1 October 2003 gene build. This current annotation predicts 14,707 genes within the 278-Mb assembly.

The *A. gambiae* genome is currently undergoing a major update of the assembly, which will then be followed by an entirely new annotation. This should be completed by the second or third month of 2005 in time for a careful comparison between the *A. gambiae* genome and the *Aedes aegypti* genome, an eightfold genome coverage of which is expected to be assembled and annotated by about that time (B. Loftus and D. W. Severson, personal communication).

The updated *A. gambiae* genome assembly will correct a number of errors and significantly increase the number of scaffolds whose genome location has been established by physical mapping of sequence-tagged clones. In the original automated assembly, the integrity of the assembly was evaluated by analyzing the orientation and distance between the paired sequence reads from the opposite ends of the 10- and 50-kb plasmid clones (Holt et al., 2002). Of 1,644,078 mate pairs analyzed, 27,703 showed distance violations and 10,166 showed orientation violations. These violations were clustered in 726 regions of the genome with average lengths of about 28 kb, for a total of 21.3 Mb, or about 7.7% of the total assembly. It is quite likely that many of these trouble spots reflect highly polymorphic regions of the PEST genome, and our current analysis suggests that many of these regions may reflect an artifact of the assembler incorporating two diverged haplotypes into adjacent locations in the assembly. One of the main goals of the current assembly update is a careful reexamination of these regions, and those that are not modified in the new assembly will be flagged in the Ensembl Mosquito Genome Browser.

The ongoing assembly update has made the following changes. A total of 261 scaffolds (~1.2 Mb) with strong sequence similarity to bacterial genomes are being removed from the assembly. Careful comparison of all scaffold sequences with other scaffolds has revealed a number of areas of possible concern. Areas where adjacent regions of the genome appear to be tandemly duplicated are simply being flagged. Many small scaffolds (>20 kb to ≥300 kb) that are highly similar in sequence to larger scaffolds have been removed from the assembly and will be displayed in an alternate haplotype track. We have identified 144 such scaffolds which add up to ~8.2 Mb. In addition, approximately 20 of the larger scaffolds (about 3 Mb) that have been physically assigned map locations have been found to overlap other mapped scaffolds. A further 18 small scaffolds (~6 Mb) have been newly mapped to specific chromosomal sites.

Incorporating these changes, the new haploid genome assembly will be ~269 Mb, and ~234 Mb (87%) of this will assigned to a genome location. This reduction in the estimated genome size estimate is consistent with the expected genome size as estimated by C_0t analysis, and it is quite likely that there remain a number of duplications within scaffolds that are a function of high levels of genome polymorphism in the PEST colony.

An especially important feature of the *A. gambiae* genome is the direct correspondence of the physically mapped scaffolds in the genome with the polytene chromosome complement. More than 2,000 sequence-tagged clones, including primarily BACs and cDNAs, have been physically mapped by in situ hybridization to specific bands on the Coluzzi polytene chromosome map (see Color Plate 19 for a sample of clones mapped to the X chromosome). The next major update of the *A. gambiae* genome display will include a field that physically associates landmark regions of the genome sequence with specific bands on the polytene chromosome map. This information will be especially valuable in the analysis of genome features such as polymorphic chromosome inversions that are associated with specific locations on the polytene chromosomes.

VectorBase: A GENOME ANALYSIS TOOL FOR VECTORS

VectorBase is a centralized relational database managed at the University of Notre Dame that will be the primary web interface for the research community interested in information dealing with arthropod vectors of human pathogens (http://www.vectorbase.org/). VectorBase will initially bring together the *A. gambiae* data currently managed and displayed in the Ensembl Mosquito Genome Browser and in AnoBase (http://skonops.imbb.forth.gr/AnoBase/), but it will be designed to rapidly expand to include new vector genomes as such projects are initiated and resources identified. The core architects of VectorBase include the Ensembl/EBI group, EMBL Heidelberg, the Harvard FlyBase team, the IMBB-Crete AnoBase group, and Notre Dame. VectorBase is also organized to receive key input from the teams of vector biologists working with the genome-sequencing centers to produce genome scale data for other vector species. In addition, VectorBase will be the repository for other types of data such as microarray and proteomic data and for tools that can be used by the community to analyze and query these data. A major milestone for the VectorBase developers will be the completion of the first draft of the *A. aegypti* genome and close comparison between the *Anopheles* and *Aedes* genomes.

VectorBase is based on a publicly available and supported database management system with a structured query language that can be searched using query by example. All information records in VectorBase will use XML or another explicitly defined, parsable format that includes a process for data representation (i.e., document type definition). The central feature of VectorBase will be based on the Ensembl Mosquito Genome display and analysis tool developed and managed by the European Bioinformatics Institute and the Sanger Institute (Mongin et al., 2004). Additional features of VectorBase have been modeled in many respects after FlyBase. VectorBase will take an active role in helping to establish formal international consortia of scientists who represent particular vector species or groups of species. Such consortia will serve several roles. They will represent a major conduit to the scientific community that will facilitate community use of VectorBase. They will also help develop the level of coordinated scientific interest in a particular vector that could lead to a genome project. In addition, VectorBase will recognize these consortia as primary sources of formal guidance. The present model for these consortia is the International *A. gambiae* Sequence Committee that was established to manage issues related to the *A. gambiae* genome sequencing project (http://www.ncbi.nlm.nih.gov/mapview/map_search.cgi?chr=agambiae.inf).

RESEARCH PROGRESS BASED ON THE *A. GAMBIAE* GENOME

In the relatively short time since the publication of the *A. gambiae* genome sequence, considerable progress has been made in a number of research areas, including *A. gambiae* behavior, population genetics, immunity, insecticide resistance, and comparative genomics. Many of these advances have a basis in some of the initial studies spawned by and published in association with the *A. gambiae* genome. Predictably, there has been an emphasis on the identification of new and improved ways to control *A. gambiae* and on understanding mosquito-parasite-human host relationships. In several fields, promising leads for the development of mosquito control products identified based on the genome sequence are being pursued. New avenues for vector biology research have been uncovered, scientists have been attracted to the field to utilize the genome data, and the volume of publications associated with this disease vector has increased exponentially. However, despite the research revolution engendered by the publication of the *A. gambiae* genome, several areas of vector biology research are yet to fully exploit the genome data.

Vector Genomics and Comparative Genomics

Postgenome studies using large-scale data sets involving expressed sequence tags (ESTs), microarray expression analysis, single nucleotide polymorphisms, and proteomics data, in addition to third-party annotations, are essential to inform on the annotation of the genome and to pinpoint unique and fundamental aspects of mosquito biology that could be exploited for control. The recent EST study of Ribeiro (2003) demonstrates the utility of genome data to illuminate new aspects of vector biology. The nonautogenous adult female *A. gambiae* mosquito requires a blood meal to develop eggs; ingestion of the blood meal triggers a series of complicated physiological processes in the mosquito that involve enzymatic digestion and conversion of the blood meal into proteins for egg development. cDNA libraries from blood-fed and non-blood-fed female *A. gambiae* mosquitoes were used to analyze the transcriptome of the female mosquito following a blood meal. A total of 435 transcripts were significantly increased or decreased following the blood meal. As anticipated, many genes associated with blood digestion, fat body metabolism, protein synthesis, and oogenesis were highly transcribed following the blood meal. Somewhat unexpected was the down-regulation post-blood meal of transcripts associated with mitochondrial metabolism, muscle proteins, and visual processes, indicating a decreased investment in activities associated with flight and host location while resources are redirected into the fat body and ovaries. This study provides a snapshot of the molecular processes that operate during mosquito hematophagy (blood feeding) and has generated many potential leads for the development of strategies that disrupt mosquito-specific processes such as blood feeding. Additional postgenome studies that employ ESTs, microarrays, gene silencing, and proteomics will have enormous utility in our understanding of many aspects of mosquito biology.

With the publication of the genome, *A. gambiae* has joined a relatively small but rapidly increasing group of sequenced eukaryotic genomes. The availability of enormous amounts of *A. gambiae* sequence data has facilitated whole-genome wide comparisons between species. Comparative genomics provides valuable insights into chromosomal and organismal evolution (Severson et al., 2004) and enables the identification of orthologous and paralogous genes and whole gene families. The genomes and proteomes of *A. gambiae* and the fruit fly *Drosophila melanogaster* reveal considerable similarities despite the fact that these two dipterans are estimated to have diverged approximately 250 million years ago (Zdobnov et al., 2002). While *A. gambiae* and *Drosophila* share similar body plans and other features, they are also substantially different in terms of their ecology, morphology, life history, and genome size; the development of hematophagy, the exploitation of human-made habitats for oviposition and parasite transmission, is specific to the mosquito. Differences between these two dipterans are clearly reflected in the genomes and proteomes of the mosquito and fly. The genes and gene products involved in these fundamental differences identified in studies such as that of Zdobnov et al. (2002) and Christophides et al. (2002) are obvious targets for the development of highly mosquito-specific control products.

As the first arthropod vector to be sequenced, *A. gambiae* has set the stage for comparative genomic analysis between arthropod vectors of disease. Several members of the family Culicidae (namely *A. aegypti* and *Culex pipiens*, the primary mosquito vectors for yellow fever and dengue fever and for lymphatic filariasis, respectively) as well as the tsetse fly vector of African trypanosomiasis (Butler, 2004) and the ixodid tick vector of Lyme disease (Hill and Wikel, 2005) are currently targeted for complete genome analysis. Comparisons between *A. aegypti*, *A. gambiae*, and *D. melanogaster* have been undertaken using initial *A. aegypti* genome data releases. The *A. aegypti* and *A. gambiae* genomes reflect interesting dichotomies in mosquito chromosome evolution between Culicinae and Anophelinae; the genomes are approximately 813 and 278 Mb, respectively, but basic chromosome number has remained constant within *A. aegypti*, reflecting an increase in repetitive DNA in the Culicinae. Despite extensive rearrangement reflecting numerous inversions, chromosome arm syntenies between the two species remain remarkably intact. However, the Anophelinae appear to have followed the *Drosophila* paradigm and exhibit genome evolution that is orders of magnitude higher than that for other eukaryotes (Severson et al., 2004). This result illustrates the utility of comparative genomics to analyze evolutionary processes between disease vectors and highlights the importance of genome projects for individual vector species in cases where the identification of genes and genome regions by comparative positioning is not possible (Severson et al., 2004). Tools such as anchor-tagged sequences for comparative genome analysis among members of the Culicidae are now being developed (Chambers et al., 2003). Resources such as VectorBase will provide an essential tool to expand the scope of comparative genomics and permit much needed analyses between closely related species, families, and even phyla of arthropod vectors.

Vector Behavior

A range of mosquito behaviors associated with host location, blood feeding, mating, and ovipositioning determine the life history strategies of mosquitoes. Visual, olfactory, gustatory, and audio processes play essential roles in these behaviors. However, despite its importance, vector behavior has been a largely neglected field of study. The *A. gambiae* genome is radically changing this situation, and this resource is being used to unravel the molecular basis of complex behavioral phenotypes. Recently, attention has refocused on mosquito behavior as a mechanism to identify new approaches to mosquito control. Furthermore, there is growing recognition that an understanding of mosquito behavior will be essential if scientists pursue the release of genetically modified mosquitoes with impaired ability to transmit diseases (Scott et al., 2002).

Some of the most promising recent studies of *A. gambiae* behavior focus largely on olfaction with respect to mosquito-host interactions. *A. gambiae* is highly anthropophilic, and its biting and host preference are largely influenced by its sense of smell (Zwiebel and Takken, 2004). Behavioral studies in wind tunnels and olfactometers have shown that human volatiles have a role as kairomones for many mosquito species, including *A. gambiae*. Both lactic acid, a component of human sweat, and carbon dioxide have been singled out as likely candidates involved in the attraction of mosquitoes to humans in laboratory and field studies. Many mosquitoes also use olfaction for oviposition, and it is not known whether mating is also mediated by smell. Now, the molecular mechanisms underlying these observations are finally coming to light.

Mosquito olfaction and other sensory processes such as taste and sight are presumably facilitated by G protein-coupled receptor (GPCR) signaling. GPCRs are a family of seven transmembrane-spanning receptor proteins that are coupled to heterotrimeric GTP binding (G-proteins) on the cytoplasmic surface of the plasma membrane (Zwiebel and Takken, 2004). GPCRs are widely expressed; for example, olfactory receptors are expressed on the sensilla of the antenna and maxillary palps, while photoreceptors are expressed in the photoreceptor membrane of the insect eye. GPCRs are capable of interacting with a wide range of chemical and nonchemical ligands, including peptides, neurohormones, small molecules, odorants, and light. When the GPCR binds a ligand such as an odorant molecule, it undergoes a conformational change that facilitates a signaling cascade and results in a neuronal or cellular response (Fig. 2).

Entire families of candidate GPCRs putatively involved in mosquito smell, taste, and sight have been identified from the assembled *A. gambiae* genome sequence (Fox et al., 2001; Fox et al., 2002; Hill et al., 2002). *A. gambiae* possesses 79 candidate odorant receptors, 76 putative

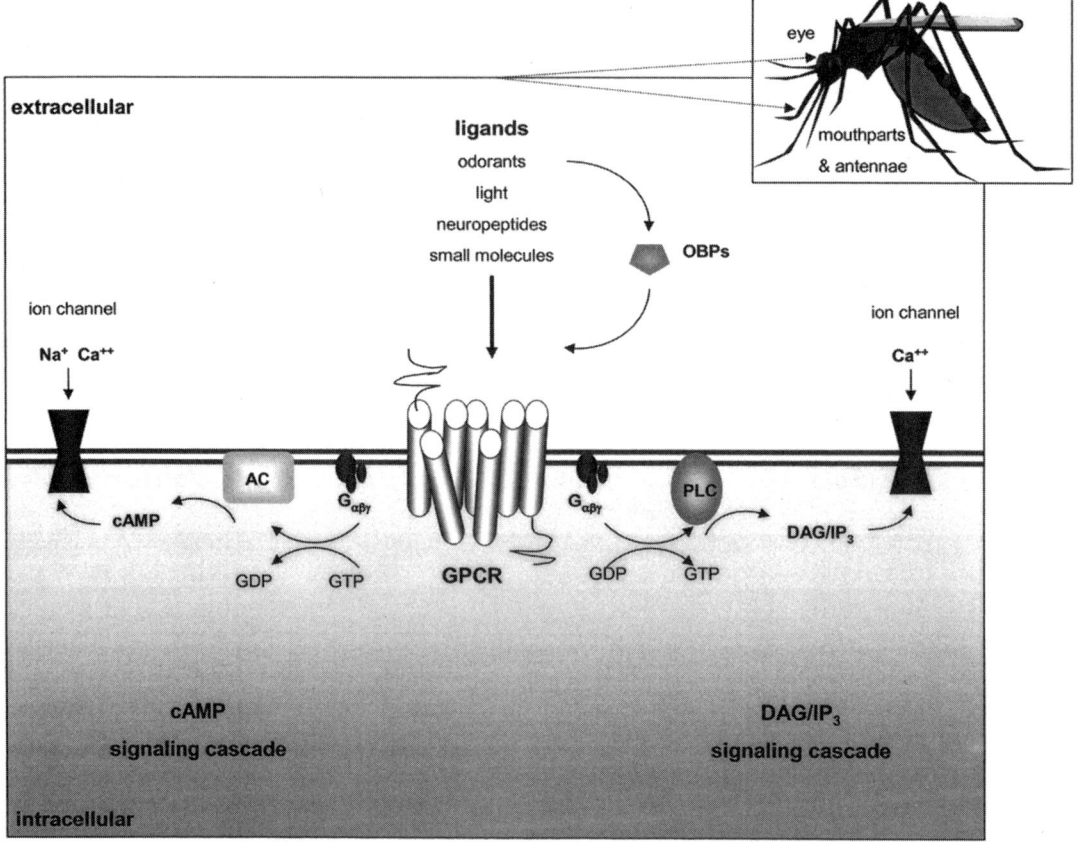

FIGURE 2 Schematic representation of the molecular processes involved in GPCR signal transduction pathways that may operate in a range of mosquito biological processes such as olfaction, taste, vision, and neurohormonal processes. GPCRs in sensilla of the mosquito antennae and mouthparts are thought to function in mosquito olfaction and taste, while GPCRs in the photoreceptor cells of the compound eye function in visual processes. In this diagram, a seven-transmembrane-spanning GPCR is shown interacting with a range of extracellular ligands, including odorant molecules in the case of olfaction and photons of light in the case of visual processes. Odorant receptors may interact with OBPs before interacting with the GPCR. Following interaction with a specific ligand, the GPCR undergoes a conformational change and interacts with heterotrimeric G protein complexes that activate downstream effector molecules, either adenyl cyclase (AC) or phospholipase C (PLC). This results in the synthesis of second messengers, namely cyclic AMP (cAMP), diacylglycerol (DAG), and inositol 1,4,5- triphosphate (IP_3), which regulate cation channels, resulting in a transduction current.

gustatory receptors, and 11 putative opsin photoreceptors (Hill et al., 2002) and displays a species-specific expansion of these receptor families relative to *Drosophila*. The contributions of many of these receptors to mosquito sensory processes are now being elucidated in follow-up studies. In addition, the identification of whole gene families based on the genome sequence has facilitated the discovery of orthologous receptors in other organisms.

Hallem et al. (2004) have subsequently shown that the *A. gambiae* odorant receptor 1 (GPRor1), a female-specific member of the odorant family of receptors, responds to a component of human sweat. The *GPRor1* gene was expressed in an engineered neuron of *Drosophila* with deleted endogenous receptors and the response of the neuron to individual odors was studied by single-unit electrophysiology. GPRor1 confers a strong response to the odorant 4-methylphenol, a component of human sweat, and this receptor may contribute to the anthropophilic host-seeking behavior of this mosquito. This idea is also supported by the fact that the *GPRor1* gene is expressed specifically in the olfactory tissue of female mosquitoes and its expression is down-regulated following a blood meal. In contrast, similar experiments with the *GPRor2* gene show that the receptor expressed by this gene has a different pharmacological profile and confers a strong response to 2-methylphenol but not 4-methylphenol.

GPRor7 is a member of a group of conserved, apparent single-gene orthologs from insects, including *D. melanogaster* (*DOr83b*), *Heliothis virescens* (*HvirR2*), *Apis mellifera* (*AmelR2*), and other insects. Initial characterization of this receptor by Pitts et al. (2004) showed that this gene is widely expressed in olfactory organs of both adults and immature stages of *A. gambiae*. Unlike other odorant receptors that show a high degree of divergence and a restricted expression pattern, the extreme conservation of *GPRor7* and its broad expression profile in olfactory and gustatory tissues suggest that the receptor expressed by this gene may play a significant and perhaps generalized role in chemosensory signal transduction in mosquitoes and possibly other insects. The putative ortholog of *GPRor7* has now been identified in another mosquito disease vector, *A. aegypti*, the yellow fever mosquito (*AaOr7*). This receptor is expressed specifically within most *A. aegypti* antennal and maxillary palp sensilla and a subset of proboscis sensilla, consistent with a general chemosensory role for AaOr7 in olfaction and gustation processes (Melo et al., 2004).

The *A. gambiae* genome sequence has also advanced the study of another class of molecules putatively involved in mosquito olfaction, the odorant-binding proteins (OBPs). These water-soluble molecules are highly expressed in the sensilla lymph and are hypothesized either to act as odorant carriers to facilitate the movement of the odor molecule from the porous cuticular surface of the antennal sensilla through the sensilla lymph to the odorant receptors on the dendritic membrane of the olfactory sensory neurons or to be integral in the catalytic removal of odorants from the lymph. Recently, several subfamilies of OBPs have been identified in *A. gambiae* (Biessmann et al., 2002; Vogt, 2002; Xu et al., 2003). Overall, there are a total of 57 candidate OBPs in the *A. gambiae* genome broadly divided into 29 classical OBPs based on amino acid sequence homology to OBPs in other organisms and a further 16 atypical OBPs. Justice et al. (2003) identified a range of cDNAs encoding OBPs and other novel antennal proteins from *A. gambiae*. Using filter array hybridizations and quantitative reverse transcription-PCR, they found significant differences in the steady-state levels of some of these genes between the sexes and after blood feeding in females. These carrier molecules are abundant and while their ligands remain unknown, they may include odor chemicals important in host-finding and nectar-feeding behavior and pheromones.

Additional studies are needed to further characterize the odorant and OBP receptors and to determine their chemical ligands. Different ligands could be screened for their activation or inhibition of receptors, potentially leading to new, more-effective insect traps or repellants. While mosquito olfaction is a critical behavioral response in the mosquito that significantly impacts host location and an obvious candidate for intervention strategies, other processes such as vision, taste, heat, and sound detection and their role in a wide range of mosquito behaviors should not be overlooked.

Vector Ecology and Vector Population Genetics

It is clear that the genome sequence will have a profound impact on all aspects of *A. gambiae* population biology and vector ecology research. Despite the widely recognized importance and interrelatedness of these two disciplines in terms of mosquito life strategies and vectorial capacity, studies that significantly exploit the *A. gambiae* genome data are currently lacking. Advances such as in vector behavior will benefit research in this area and help scientists to address any number of important ecoethological questions. Elucidating the genetic population structure within *A. gambiae* sensu stricto is necessary to determine those members that are vectors of malaria and to evaluate the ecological and ethological differences relevant to disease transmission (della Torre et al., 2001). The genome will facilitate the development of better tools to estimate phylogeny and population structure, allowing the identification of genes and regulatory networks underlying ecological differentiation, speciation, and vectorial capacity (Krzywinski and Besansky, 2003). Furthermore, understanding the population genetics of a disease vector such as *A. gambiae* has important implications for vector control because it may help us to better understand the evolution of insecticide resistance and the potential outcomes of transgenic mosquito release within a population framework (Donnelly et al., 2004).

Vector–Parasite Interactions and Vector Immunity

The *A. gambiae* innate immune system is activated during invasion by malaria parasites and may be implicated in the loss of ookinetes (Dimopoulos et al., 2002; Christophides et al., 2004). Vector immunity is an example of a field that was well poised to exploit *A. gambiae* genome data. Christophides et al. (2002) have identified 242 genes from 18 gene families from the genome sequence putatively implicated in innate immunity of *A. gambiae*. Many of these genes are markedly diversified from their *Drosophila* homologs, which may reflect adaptation to different pathogens. Approximately 100 of these genes are the subject of a large-scale functional screen using a gene-silencing technique established for *A. gambiae* (Blandin et al., 2002). Using RNA interference to silence gene expression, Osta et al. (2004a) have subsequently shown that the *A. gambiae* leucine-rich repeat protein (LRIM) acts as an antagonist and two C-type lectins (CTL4 and CTLMA2) act as protective agonists on the development of *Plasmodium* ookinetes to oocysts. Characterizing these and other molecules of similar function and identifying their ligands on the parasite surface will contribute to our understanding of the vector-parasite system and provide opportunities to disrupt the vectorial capacity of anopheline mosquitoes (Osta et al., 2004b).

Insecticide Resistance Mechanisms

Insecticides are an essential component to most malaria control programs. Synthetic pyrethroids are used as residual indoor house sprays, as aerosols, and to impregnate bednets, curtains, and screens. The emergence of insecticide resistance in the mosquito poses a serious threat to malaria control (Hemingway et al., 2004).

Three supergene families for enzymes thought to be primarily responsible for metabolic insecticide resistance (the carboxylesterases, glutathione transferases, and cytochrome P450s) in the *A. gambiae* genome were identified by Ranson et al. (2002). In results similar to that described by Hill et al. (2002), considerable expansion of these gene families was observed in the mosquito, suggesting rapid evolutionary adaptation to environmental changes. Postgenome transcriptome and proteome studies now promise to reveal the contributions of uncharacterized and hitherto unsuspected enzymes to insecticide resistance. In subsequent studies, Ranson et al. (2004) used microsatellite markers to scan the *A. gambiae* genome for quantitative trait loci associated with permethrin resistance. The authors identified two major and one minor quantitative trait loci, one of which colocalizes with the sodium channel gene located on chromosome 2L implicated in permethrin resistance and another that flanks a large cluster of cytochrome P450 genes. Genetic mapping studies of this nature will help to unravel the molecular mechanisms of metabolic insecticide resistance. Ortelli et al. (2003) have analyzed two allelic variants of glutathione transferases associated with a dichlorodiphenyl-trichloroethane (DDT) resistance locus in *A. gambiae*. Using heterologous expression and Western analysis, the researchers have shown that one variant, GSTE2-2, was able to metabolize DDT and that the expression of this protein was elevated in a DDT-resistant *A. gambiae* strain. Postgenome studies such as these enhance our understanding of resistance and its complexity (Hemingway et al., 2004).

IMPORTANT FUTURE PRIORITIES

The paradigm of whole-genome research is likely to dominate the future of vector biology. Our challenge will be to determine the function of a multitude of gene products and to establish precisely how they interact in time and space to affect vector biology and disease transmission (Cooke and Coppel, 2004). Complexity will be further increased by the availability of additional vector and parasite genome sequence data. Tools such as proteomic arrays, bioinformatics programs to analyze protein structure-function and transcription patterns, microarrays, mass gene knockout screens, interaction maps, and comparative genomics will yield important insights into protein function in the vector itself. Academia-industry partnerships and the continued deployment of large-scale resources (Hill et al., 2005) are essential to convert vector genome data into real disease control solutions.

REFERENCES

Beard, C. B., M. Q. Benedict, J. P. Primus, V. Finnerty, and F. H. Collins. 1995. Eye pigments in wild-type and eye-color mutant strains of the African malaria vector *Anopheles gambiae. J. Hered.* **86:**375–380.

Besansky, N. J., and J. R. Powell. 1992. Reassociation kinetics of *Anopheles gambiae* (Diptera: Culicidae) DNA. *J. Med. Entomol.* **29:**125–128.

Besansky, N. J., T. Lehmann, G. T. Fahey, D. Fontenille, L. E. Braack, W. A. Hawley, and F. H. Collins. 1997. Patterns of mitochondrial variation within and between African malaria vectors, *Anopheles gambiae* and *An. arabiensis*, suggest extensive gene flow. *Genetics* **147:**1817–1828.

Biessmann, H., M. F. Walter, S. Dimitratos, and D. Woods. 2002. Isolation of cDNA clones encoding putative odourant binding proteins from the antennae of the malaria-transmitting mosquito, *Anopheles gambiae. Insect Mol. Biol.* **11:**123–132.

Blandin, S., L. F. Moita, T. Kocher, M. Wilm, F. C. Kafatos, and E. A. Levashina. 2002. Reverse genetics in the mosquito *Anopheles gambiae*: targeted disruption of the Defensin gene. *EMBO Rep.* **3:**852–856.

Chambers, E. W., D. D. Lovin, and D. W. Severson. 2003. Utility of comparative anchor-tagged sequences as physical anchors for comparative genome analysis among the culicidae. *Am. J. Trop. Med. Hyg.* **69:**98–104.

Christophides, G. K., E. Zdobnov, C. Barillas-Mury, E. E. Birney, S. Blandin, C. Blass, P. T. Brey, F. H. Collins, A. Danielli, G. Dimopoulos, C. Hetru, N. T. Hoa, J. A. Hoffmann, S. M. Kanzok, I. Letunic, E. A. Levashina, T. G. Loukeris, G. Lycett, S. Meister, K. Michel, L. F. Moite, H. M. Muller, M. A. Osta, S. M. Paskewitz, J. M. Reichart, A. Rzhetsky, L. Troxler, K. D. Vernick, D. Vlachou, J. Volz, C. von Mering, J. Xu, L. Zheng, P. Bork, and F. C. Kafatos. 2002. Immunity-related genes and gene families in *Anopheles gambiae. Science* **298:**159–165.

Christophides, G. K., D. Vlachou, and F. C. Kafatos. 2004. Comparative and functional genomics of the innate immune system in the malaria vector *Anopheles gambiae. Immunol. Rev.* **198:**127–148.

Coetzee, M., M. Craig, and D. Le Sueur. 2000. Distribution of African malaria vector mosquitoes belonging to the *Anopheles gambiae* complex. 2000. *Parasitol. Today* **16:**74–77.

Coluzzi, M. 1994. Malaria and the Afrotropical ecosystems: impact of man-made environmental changes. *Parassitologia* **36:**223–227.

Coluzzi, M., V. Petrarca, and M. A. Di Deco. 1985. Chromosomal inversion integradation and incipient speciation in *Anopheles gambiae. Boll. Zool.* **52:**45–63.

Coluzzi, M., A. Sabatini, A. della Torre, M. A. Di Deco, and V. Petrarca. 2002. A polytene chromosome analysis of the *Anopheles gambiae* species complex. *Science* **298:**1415–1418.

Coluzzi, M., A. Sabatini, V. Petrarca, and M. A. Di Deco. 1979. Chromosomal differentiation and adaptation to human environments in the *Anopheles gambiae* complex. *Trans. R. Soc. Trop. Med. Hyg.* **73:**483–497.

Cooke, B. M., and R. L. Coppel. 2004. Blue skies or stormy weather: what lies ahead for malaria research? *Trends Parasitol.* **20:**611–614.

della Torre, A., C. Fanello, M. Akogbeto, J. Dossou-Yovo, and G. Favia. 2001. Molecular evidence of incipient speciation within *Anopheles gambaie* s.s. in West Africa. *Insect Mol. Biol.* **10:**9–18.

Dimopoulos, G., G. K. Christophides, S. Meister, J. Schultz, K. P. White, C. Barillas-Mury, and F. C. Kafatos. 2002. Genome expression analysis of *Anopheles gambiae*: responses to injury, bacterial challenge, and malaria infection. *Proc. Natl. Acad. Sci. USA* **99:**8814.

Donnelly, M. J., J. Pinto, R. Girod, N. J. Besansky, and T. Lehmann. 2004. Revisiting the role of introgression vs shared ancestral polymorphisms as key processes shaping genetic diversity in the recently separated sibling species of the *Anopheles gambiae* complex. *Heredity* **92:**61–68.

Favia, G., G. Dimopoulos, A. della Torre, Y. T. Toure, M. Coluzzi, and C. Louis. 1994. Polymorphisms detected by random PCR distinguish between different chromosomal forms of *Anopheles gambiae. Proc. Natl. Acad. Sci. USA* **91:**10315–10319.

Favia, G., A. della Torre, M. Bagayoko, A. Lanfrancotti, N. Sagnon, Y. T. Toure, and M. Coluzzi. 1997. Molecular identification of sympatric chromosomal forms of *Anopheles gambiae* and further evidence of their reproductive isolation. *Insect Mol. Biol.* **6:**377–383.

Fox, A., R. Pitts, H. Robertson, J. R. Carlson, and L. J. Zwiebel. 2001. Candidate odorant receptors from the malaria vector mosquito *Anopheles gambiae* and evidence of down-regulation in response to blood-feeding. *Proc. Natl. Acad. Sci. USA* **98:**14693–14697.

Fox, A. N., R. J. Pitts, and L. J. Zwiebel. 2002. A cluster of candidate odorant receptors from the malaria vector mosquito, *Anopheles gambiae*. *Chem. Senses* **27:**453.

Gentile, G., M. Slotman, V. Ketmaier, J. R. Powell, and A. Caccone. 2001. Attempts to molecularly distinguish cryptic taxa in *Anopheles gambiae* s.s. *Insect Mol. Biol.* **10:**25–32.

Githeko, A. K., A. D. Brandling-Bennett, M. Beier, F. Atieli, M. Owaga, and F. H. Collins. 1992. The reservoir of *Plasmodium falciparum* malaria in a holoendemic area of western Kenya. *Trans. R. Soc. Trop. Med. Hyg.* **86:**355–358.

Hallem, E. A., A. N. Fox, L. J. Zwiebel, and J. R. Carlson. 2004. Mosquito receptor for human-sweat odorant. *Nature* **427:**212–213.

Hemingway, J., N. J. Hawkes, L. McCarroll, and H. Ranson. 2004. The molecular basis of insecticide resistance in mosquitoes. *Insect Biochem. Mol. Biol.* **34:**653–665.

Hill, C. A., A. N. Fox, R. J. Pitts, L. B. Kent, P. L. Tan, M. A. Chrystal, A. Cravchik, F. H. Collins, H. M. Robertson, and L. J. Zwiebel. 2002. G protein-coupled receptors in *Anopheles gambiae*. *Science* **298:**176–178.

Hill, C. A., F. C. Kafatos, S. K. Stansfield, and F. H. Collins. 2005. Arthropod-borne diseases: vector control in the genomics era. *Nat. Rev. Microbiol.* **3:**262–268.

Hill, C. A., and S. K. Wikel. 2005. The Ixodes scapularis genome project: an opportunity for advancing tick research. *Trends Parasitol.* **21:**151–153.

Holt, R. A., and F. H. Collins. The malaria mosquito genome. *Encyclopedia of Molecular Cell Biology and Molecular Medicine*, in press.

Holt, R. A., G. M. Subramanian, A. Halpern, G. G. Sutton, R. Charlab, D. R. Nusskern, P. Wincker, A. G. Clark, J. M. Ribeiro, R. Wides, S. L. Salzberg, B. Loftus, M. Yandell, W. H. Majoros, D. B. Rusch, Z. Lai, C. L. Kraft, J. F. Abril, V. Anthouard, P. Arensburger, P. W. Atkinson, H. Baden, V. de Berardinis, D. Baldwin, V. Benes, J. Biedler, C. Blass, R. Bolanos, D. Boscus, M. Barnstead, S. Cai, A. Center, K. Chaturverdi, G. K. Christophides, M. A. Chrystal, M. Clamp, A. Cravchik, V. Curwen, A. Dana, A. Delcher, I. Dew, C. A. Evans, M. Flanigan, A. Grundschober-Freimoser, L. Friedli, Z. Gu, P. Guan, R. Guigo, M. E. Hillenmeyer, S. L. Hladun, J. R. Hogan, Y. S. Hong, J. Hoover, O. Jaillon, Z. Ke, C. Kodira, E. Kokoza, A. Koutsos, I. Letunic, A. Levitsky, Y. Liang, J. J. Lin, N. F. Lobo, J. R. Lopez, J. A. Malek, T. C. McIntosh, S. Meister, J. Miller, C. Mobarry, E. Mongin, S. D. Murphy, D. A. O'Brochta, C. Pfannkoch, R. Qi, M. A. Regier, K. Remington, H. Shao, M. V. Sharakhova, C. D. Sitter, J. Shetty, T. J. Smith, R. Strong, J. Sun, D. Thomasova, L. Q. Ton, P. Topalis, Z. Tu, M. F. Unger, B. Walenz, A. Wang, J. Wang, M. Wang, X. Wang, K. J. Woodford, J. R. Wortman, M. Wu, A. Yao, E. M. Zdobnov, H. Zhang, Q. Zhao, S. Zhao, S. C. Zhu, I. Zhimulev, M. Coluzzi, A. della Torre, C. W. Roth, C. Louis, F. Kalush, R. J. Mural, E. W. Myers, M. D. Adams, H. O. Smith, S. Broder, M. J. Gardner, C. M. Fraser, E. Birney, P. Bork, P. T. Brey, J. C. Venter, J. Weissenbach, F. C. Kafatos, F. H. Collins, and S. L. Hoffman. 2002. The genome sequence of the malaria mosquito *Anopheles gambiae*. *Science* **298:**129–149.

Hong, Y. S., J. R. Hogan, X. Wang, A. Sarkar, C. Sim, B. J. Loftus, C. Ren, E. R. Huff, J. L. Carlile, K. Black, H. B. Zhang, M. J. Gardner, and F. H. Collins. 2003. Construction of a BAC library and generation of BAC end sequence-tagged connectors for genome sequencing of the African malaria mosquito *Anopheles gambiae*. *Mol. Genet. Genomics* **268:**720–728.

Justice, R. W., S. Dimitratos, M. F. Walter, D. F. Woods, and H. Biessmann. 2003. Sexual dimorphic expression of putative antennal carrier protein genes in the malaria vector *Anopheles gambiae*. *Insect Mol. Biol.* **12:**581–594.

Krzywinski, J., and N. J. Besansky. 2003. Molecular systematics of *Anopheles*: from subgenera to subpopulations. *Annu. Rev. Entomol.* **48:**111–139.

Kumar, V., and F. H. Collins. 1994. A technique for nucleic acid in situ hybridization to polytene chromosomes of mosquitoes in the *Anopheles gambiae* complex. *Insect Mol. Biol.* **3:**41–47.

Lanzaro, G. C., Y. T. Toure, J. Carnahan, L. Zheng, G. Dolo, S. Traore, V. Petrarca, K. D. Vernick, and C. E. Taylor. 1998. Complexities in the genetic structure of *Anopheles gambiae* populations in west Africa as revealed by microsatellite DNA analysis. *Proc. Natl. Acad. Sci. USA* **95:**14260–14265.

Lehmann, T., M. Licht, N. Elissa, B. T. Maega, J. M. Chimumbwa, F. T. Watsenga, C. S. Wondji, F. Simard, and W. A. Hawley. 2003. Population structure of *Anopheles gambiae* in Africa. *J. Hered.* **94:**133–147.

Mason, G. F. 1967. Genetic studies on mutations in species A and B of the *Anopheles gambiae* complex. *Genet. Res.* **10**:205–217.

Melo, A. C., M. Rutzler, R. J. Pitts, and L. J. Zwiebel. 2004. Identification of a chemosensory receptor from the yellow fever mosquito, *Aedes aegypti*, that is highly conserved and expressed in olfactory and gustatory organs. *Chem. Senses* **29**:403–410.

Mongin, E., C. Louis, R. A. Holt, E. Birney, and F. H. Collins. 2004. The *Anopheles gambiae* genome: an update. *Trends Parasitol.* **20**:49–52.

Mukabayire, O., and N. J. Besansky. 1996. Distribution of T1, Q, Pegasus and mariner transposable elements on the polytene chromosomes of PEST, a standard strain of *Anopheles gambiae*. *Chromosoma* **104**:585–595.

Ortelli, F., L. C. Rossiter, J. Vontas, H. Ranson, and J. Hemingway. 2003. Heterologous expression of four glutathione transferase genes genetically linked to a major insecticide-resistance locus from the malaria vector *Anopheles gambiae*. *Biochem. J.* **373**:957–963.

Osta, M. A., G. K. Christophides, and F. C. Kafatos. 2004a. Effects of mosquito genes on *Plasmodium* development. *Science* **303**:2030–2032.

Osta, M. A., G. K. Christophides, D. Vlachou, and F. C. Kafatos. 2004b. Innate immunity in the malaria vector *Anopheles gambiae*: comparative and functional genomics. *J. Exp. Biol.* **207**:2551–2563.

Pitts, R. J., A. N. Fox, and L. J. Zwiebel. 2004. A highly conserved candidate chemoreceptor expressed in both olfactory and gustatory tissues in the malaria vector *Anopheles gambiae*. *Proc. Natl. Acad. Sci. USA* **103**:5058–5063.

Ranson, H., C. Claudianos, F. Ortelli, C. Abgrall, J. Hemingway, M. V. Sharakhova, M. F. Unger, F. H. Collins, and R. Feyereisen. 2002. Evolution of supergene families associated with insecticide resistance. *Science* **298**:179–181.

Ranson, H., M. G. Paton, B. Jensen, L. McCarroll, A. Vaughan, J. R. Hogan, J. Hemingway, and F. H. Collins. 2004. Genetic mapping of genes conferring permethrin resistance in the malaria vector, *Anopheles gambiae*. *Insect Mol. Biol.* **13**:379–386.

Ribeiro, J. M. C. 2003. A catalogue of *Anopheles gambiae* transcripts significantly more or less expressed following a blood meal. *Insect Biochem. Mol. Biol.* **33**:865–882.

Severson, D. W., B. deBruyn, D. D. Lovin, S. E. Brown, D. L. Knudson, and I. Morlas. 2004. Comparative genome analysis of the yellow fever mosquito *Aedes aegypti* with *Drosophila melanogaster* and the malaria vector mosquito *Anopheles gambiae*. *J. Hered.* **95**:103–113.

Scott, T. W., W. Takken, B. G. Knols, and C. Boete. 2002. The ecology of genetically modified mosquitos. *Science* **298**:117–119.

Thomasova, D., L. Q. Ton, R. R. Copley, E. M. Zdobnov, X. Wang, Y. S. Hong, C. Sim, P. Bork, F. C. Kafatos, and F. H. Collins. 2002. Comparative genomic analysis in the region of a major *Plasmodium*-refractoriness locus of *Anopheles gambiae*. *Proc. Natl. Acad. Sci. USA* **99**:8179–8184.

Toure, Y. T., V. Petrarca, S. F. Traore, A. Coulibaly, H. M. Maiga, O. Sankare, M. Sow, M. A. Di Deco, and M. Coluzzi. 1998. The distribution and inversion polymorphism of chromosomally recognized taxa of the *Anopheles gambiae* complex in Mali, West Africa. *Parassitologia* **40**:477–511.

Tripet, F., Y. T. Toure, G. Dolo, and G. C. Lanzaro. 2003. Frequency of multiple inseminations in field-collected *Anopheles gambiae* females revealed by DNA analysis of transferred sperm. *Am. J. Trop. Med. Hyg.* **68**:1–5.

Tripet, F., Y. T. Toure, C. E. Taylor, D. E. Norris, G. Dolo, and G. C. Lanzaro. 2001. DNA analysis of transferred sperm reveals significant levels of gene flow between molecular forms of *Anopheles gambiae*. *Mol. Ecol.* **10**:1725–1732.

Vogt, R. G. 2002. Odorant binding protein homologues of the malaria vector mosquito *Anopheles gambiae*, possible orthologues of the OS-E and OS-F OBPs of *Drosophila melanogaster*. *J. Chem. Ecol.* **28**:2371–2376.

White, G. B. 1974. *Anopheles gambiae* complex and disease transmission in Africa. *Trans. R. Soc. Trop. Med. Hyg.* **68**:278–301.

Xu, P. X., L. J. Zwiebel, and D. P. Smith. 2003. Identification of a distinct family of genes encoding atypical odorant-binding proteins from the malaria vector mosquito, *Anopheles gambiae*. *Insect Mol. Biol.* **12**:549-560.

Zdobnov, E. M., C. von Mering, I. Letunic, D. Torrents, M. Suyama, R. R. Copley, G. K. Christophides, D. Thomasova, R. A. Holt, G. M. Subramanian, H. M. Mueller, G. Dimopoulos, J. H. Law, M. A. Wells, E. Birney, R. Charlab, A. L. Halpern, E. Kokoza, C. L. Kraft, Z. Lai, S. Lewis, C. Louis, C. Barillas-Mury, D. Nusskern, G. M. Rubin, S. L. Salzberg, G. G. Sutton, P. Topalis, R. Wides, P. Wincker, M. Yandell, F. H. Collins, J. Ribeiro, W. M. Gelbart, F. C. Kafatos, and P. Bork. 2002. Comparative genome and proteome analysis of *Anopheles gambiae* and *Drosophila melanogaster*. *Science* **298**:149–159.

Zwiebel, L. J., and W. Takken. 2004. Olfactory regulation of mosquito-host interactions. *Insect Biochem. Mol. Biol.* **34**:645–652.

THE TRANSCRIPTOME OF HUMAN MALARIA VECTORS

Osvaldo Marinotti and Anthony A. James

27

Gene expression studies in vector mosquitoes are driven by the need to understand biological phenomena associated with the transmission of pathogenic organisms. It is anticipated that an understanding of these phenomena will result in novel technologies for reducing the morbidity and mortality associated with the pathogens (James et al., 1989; James et al., 1999). There is a general hypothesis that a detailed knowledge of the expression products of the mosquito genome, the transcriptome and proteome, will provide the tools needed for the development of these technologies (Holt et al., 2002). Thus, the focus of mosquito biology research has reflected a desire to investigate those tissues (for example, midgut and salivary glands) in which it is obvious that important interactions between vectors and pathogens take place (Meredith and James, 1990). Gene expression in these tissues could be exploited in the development of genetic antipathogen strategies. In addition, it was proposed that knowledge of the molecular basis of insecticide resistance could lead to innovative ways to prevent resistance and possibly lead to the development of new active compounds (Hemingway and Ranson, 2000). Thus, we see a progression in the study of gene expression in mosquitoes where initially individual genes from relevant tissues or those involved in insecticide resistance were subjected to in-depth analyses to define their products and controlling elements, to more recent mass screening techniques that allowed the isolation of large numbers of genes encoding many different products. This was followed by whole-genome analyses of gene expression involving hundreds of genes that may be coordinately regulated in response to pathogen or insecticide exposure. The challenge of these latter efforts was to find relevance in the large number of genes that modulate their expression. Concomitant with this ability to evaluate large numbers of genes was the tackling of complex phenotypes, those that result from the actions of more than one gene. Genetically determined behaviors such as oviposition site recognition and host preference may be described ultimately as the action of a number of coordinately reg-

Osvaldo Marinotti, Department of Molecular Biology & Biochemistry, University of California, Irvine, 3205 McGaugh Hall, Irvine, CA 92697–3900. *Anthony A. James,* Departments of Microbiology & Molecular Genetics and Molecular Biology & Biochemistry, University of California, Irvine, 3205 McGaugh Hall, Irvine, CA 92697–3900.

Molecular Approaches to Malaria, Edited by Irwin W. Sherman
© 2005 ASM Press, Washington, D.C.

ulated genes. Thus, the recent history of studying gene expression in vector mosquitoes moves from a starting point with a highly reductionist approach to a more general strategy of whole-genome analysis. The focus now is to reverse the trend and to be able to identify within the vast array of genes expressed and represented in the transcriptome those few key genes that define the outcome of vector-parasite interactions and determine the vector competence and vectorial capacity of mosquitoes.

This chapter traces the development of gene expression studies in mosquitoes from a historical perspective. We hope to illustrate how early efforts made possible the substantial research and resource commitments needed to mount whole-genome projects and that new insight into vector biology and mosquito-pathogen interactions have resulted from these efforts. In addition, we will review the relevant technologies that made the advances possible. While the principal focus is on the anopheline vectors of malaria, *Anopheles gambiae* in particular, we will cite relevant advances in other insects, most notably the yellow fever mosquito, *Aedes aegypti*.

THE PREGENOMICS ERA

The rapid development of molecular techniques and their applications in model organisms such as the vinegar fly, *Drosophila melanogaster*, fostered the reemergence of interest in whether genetics bolstered by molecular biology could provide useful tools in combating malaria transmission (World Health Organization, 1991). A number of strategies for population replacement and reduction were proposed, and the vast majority of them required knowledge of the expression profiles of specific genes. For example, replacement strategies in which antipathogen effector genes are introgressed into mosquito populations to make them incompetent for parasite transmission are predicated on driving the sex- and tissue-specific expression of the effector gene with endogenously derived promoter sequences (Meredith and James, 1990). Some population reduction strategies require the identification of sex-specific genes so that only male mosquitoes survive (Coleman and Alphey, 2004). These criteria focused early work on mosquito genes whose products were important in hematophagy, blood digestion, and reproduction.

The first protein-encoding gene isolated and characterized from any mosquito was *Maltase-like 1* (*Mal 1*) of *A. aegypti* (James et al., 1989). It was isolated in a molecular screen that was designed to identify a class of genes whose promoters were candidates to control transcription of effector genes in adult salivary glands. These early steps depended on the construction of highly representative cDNA libraries that reflected the accumulation of mRNAs and provided a record of the transcription state at the time the whole animal or tissue was harvested. Of particular interest were those mRNAs that corresponded to genes displaying sex-, stage-, and tissue-specific expression. Differential screening involving the hybridization of cDNA libraries with two radiolabeled probes (RNA or cDNA) made from mRNA extracted from specific tissues and/or developmental stages (St. John and Davis, 1979) was the first approach used to identify important classes of mosquito genes. This was the technique used to identify *Mal 1*, as well as a number of other genes (*D7, Apy I, Amy I, Andalys*) expressed in mosquito salivary glands (James et al., 1989, 1991; Champagne et al., 1995; Grossman et al., 1993; Moreira-Ferro et al., 1998) and other tissues such as the ovaries of *A. gambiae* (Zurita et al., 1997).

Screening of libraries with probes designed from the sequence of orthologous genes characterized previously in other organisms was also used to identify cDNA clones corresponding to specific mosquito genes. This technique is limited to those genes with nucleotide sequences sufficiently conserved to allow cross-hybridization between the probe and the target clones. However, this technique allowed the identification of a cyclodiene insecticide resistance gene, the γ-aminobutyric acid receptor in *A. aegypti*, with a probe derived from the *Rdl* gene of *D. melanogaster* (Thompson et al., 1993). In addition, α-glucosidase genes expressed in the midguts of *A. gambiae* were isolated with the *Mal 1* cDNA of *A. aegypti* as a probe (Zheng et al., 1995). Immunoscreening of expression cDNA libraries with antibodies reactive to mosquito proteins has also been effective for isolating specific cDNAs encoding midgut and immunity-related *A. gambiae* products, such as prophenoloxidases (Jiang et al., 1997), peritrophin (Shen and Jacobs-Lorena, 1998), and a midgut mucin (Shen et al., 1999).

Techniques of constructing and screening cDNA libraries are continually being optimized (Ying, 2004). Arca and collaborators employed a variation of the signal sequence trap technique (Tashiro et al., 1993) to identify cDNA clones corresponding to genes encoding *A. gambiae* salivary gland secreted products (Arca et al., 1999). Fifteen cDNAs were identified by this technique, of which 8 corresponded to genes that encode novel proteins with unknown functions, while the others encode products similar to previously characterized mosquito salivary gland proteins (encoded by *D7-like* and *apyrase* genes).

Methodological progress was made with the introduction of gene amplification techniques, most notably PCR, and this resulted in proven methods that mitigated the laborious and time-consuming construction and screening of cDNA libraries. Reverse transcription-PCR (RT-PCR) was widely applied for gene-specific amplification of mosquito cDNAs. As is the case for screening libraries with heterologous probes, success with this approach depends on the conservation of the nucleotide sequences. However, the PCR-based techniques enhance the possibilities of success because only the primer sequences need to be conserved to obtain the desired amplification product. If the exact sequence of the target gene or cDNA is unknown, degenerate oligonucleotides based on known nucleic acid sequences of orthologous or paralogous genes from other organisms can be used. PCR and RT-PCR techniques were used to amplify and characterize genes encoding serine proteases (Han et al., 1997; Siden-Kiamos et al., 1996), a cytoskeletal actin (Salazar et al., 1994), iron regulatory proteins (Zhang et al., 2002), laminins (Vlachou et al., 2001), and phenoloxidase (Cui et al., 2000) from *A. gambiae* and the *white* homologous genes in both *A. gambiae* and *A. aegypti* (Besansky et al., 1995; Coates et al., 1997). These techniques are also useful when mosquito proteins are purified and their amino acid sequences are determined. The design of oligonucleotides based on known amino acid sequences resulted in the amplification of genes and transcripts encoding mosquito hexamerins (Korochkina et al., 1997), a salivary antithrombin and gp65 of *Anopheles albimanus* (Valenzuela et al., 1999; Montero-Solis et al., 2004), and a *D7-like*-encoded protein of *Anopheles darlingi* (Calvo et al., 2002).

Differential display is a PCR-based technique that uses oligonucleotide primers of arbitrary sequence to amplify and identify mRNAs expressed differentially in two samples (Liang and Pardee, 1992, 1998). Using this method, Dimopoulos et al. (1996) identified *A. gambiae* genes expressed specifically in males, females, and midgut tissues of blood-fed or unfed females, as well as genes whose expression is induced by bacterial infection. Their work identified a serine protease related to the mosquito immune response to bacterial infection. In addition, several genes encoding products with unknown functions were discovered. In follow-up work, two genes encoding new immune responsive markers, a gram-negative bacteria-binding protein, and a serine protease-like molecule were identified (Dimopoulos et al., 1997).

Subtractive cDNA libraries, based mainly on suppression subtractive hybridization (Diatchenko et al., 1996) in which mRNAs accumulated differentially in two samples are cloned preferentially and are thus more easily identified, provided another powerful tool for the characterization of tissue- and stage-specific gene expression. This technique was applied to the identification of several genes involved in mosquito development and response to infection with bacteria and parasites. Subtraction libraries enriched in *A. gambiae* adult female midgut mRNAs led to the identification of an adult- and female-specific chitinase expressed only in the midgut (Shen et al., 1997). It was suggested that this enzyme modulates the physical properties of the peritrophic matrix, a chitin-containing extracellular structure secreted into the lumen of the midgut after a blood meal. The disruption of the formation of a normal peritrophic matrix could affect blood digestion and malaria parasite survival in the vector. The screening of the subtractive midgut-enriched cDNA library also resulted in the identification of a gene encoding a midgut-specific serine protease that is down-regulated after blood feeding; several genes encoding midgut-specific trypsins, chymotrypsins, aminopeptidases, and carboxypeptidase A; and several genes encoding proteins without significant similarity to anything in the National Center for Biotechnology Information (NCBI) database (Shen et al., 2000). The screening of subtractive cDNA libraries enriched in mRNAs up- or down-regulated by immune stimulation of adult *A. gambiae* with bacterial lipopolysaccharide revealed 23 immune-regulated genes, including genes encoding protease inhibitors, serine proteases, ribosomal proteins, a small heat shock protein, and others. Interestingly, two of these genes, *AgIMcr14* and *AgISPR5*, were up-regulated significantly in response to immune challenges of bacterial and malaria parasite infection (Oduol et al., 2000).

In addition to providing genes whose control sequences could be used to express antipathogen effector genes, the pregenomic era of gene characterization revealed a number of novel insights into mosquito biology. The analyses of genes encoding antihemostatic proteins in salivary glands revealed that while all mosquitoes have salivary anticoagulants, platelet antiaggregating factors, and vasodilatory substances, these arose from diverse gene families, emphasizing the convergent evolution of this important life-strategy (Stark and James, 1996). The serine protease gene family in *A. gambiae* and *A. aegypti* has different expression profiles in the two species, indicating divergences in the way these mosquitoes metabolize ingested blood (Noriega and Wells, 1999; Vizioli et al., 2001). Work with the immune system revealed that tissues such as the salivary gland and midgut express genes encoding antibacterial proteins that respond to bacteria and plasmodium infection (Dimopoulos et al., 1998). Thus, aspects of the mosquito immune response are organ based as well as systemic. Point mutations determining insecticide resistance were also identified in a number of species (Martinez-Torres, 1998). These findings validated the early approaches to studying the transcriptome.

GENOMICS AND EST ANALYSES

Early efforts in large-scale gene analysis were limited by the available techniques and relatively small amounts of starting materials that mosquitoes provide. Funding opportunities and recent technical advances have changed that landscape. High-throughput DNA sequencing is no longer prohibitively expensive, and technological innovations have facilitated and made short the time needed for the identification and annotation of a large number of mosquito cDNAs and their corresponding genes. This is reflected in the remarkable increase in the number of *Anopheles* nucleotide sequences deposited in central databases. The number of NCBI mRNA entries retrieved using *Anopheles* as a keyword increased nearly 30,000-fold in the last 10 years (Table 1). This significant increase results from expressed sequence tag (EST) projects for the rapid discovery of the full transcriptome of *A. gambiae*. Partial, and in most cases complete, sequences of large sets of genes and transcripts are now readily available. This has had the effect of refocusing research efforts from primary structural characterizations to more functional methods of analysis. Furthermore, the acquisition of data reflecting variations of expression in large sets of genes, even on a genome-wide scale, is now feasible, resulting in a wider, integrated picture of mosquito biology.

TABLE 1 Number of NCBI[a] entries for RNAs indexed by the keyword *Anopheles*

Yr	No. of entries
1993	7
2000	194
2002	80,000
2004	>200,000

[a]National Center for Biotechnology Information; http://www.ncbi.nlm.nih.gov/.

Pioneer large-scale EST sequencing projects generated data for thousands of cDNA clones. The sequences of 456 randomly selected clones of an *A. aegypti* adult female salivary gland cDNA library resulted in the identification of 238 clusters representing unique transcription products. The identified transcripts corresponded to genes encoding 118 general function (housekeeping) proteins, 38 secreted products, and 82 proteins with unknown function (Valenzuela et al., 2002). The same approach was applied to the characterization of the salivary gland transcriptomes of *A. gambiae* (691 cDNA clones) (Francischetti et al., 2002), *Anopheles stephensi* (1,127 clones) (Valenzuela et al., 2003), *A. darlingi* (593 clones) (Calvo et al., 2004), *Culex pipiens quinquefasciatus* (503 clones) (Ribeiro et al., 2004), and *Anopheles arabiensis* (J. M. Ribeiro, personal communication). These projects resulted in the identification of mosquito salivary components that are specific to anopheline and culicine mosquitoes. Many of the salivary products encoded by the genes identified in this work play active roles in hematophagy and may be involved directly with the pathogenesis of malaria. Previously mentioned and newly discovered molecules, including inhibitors of coagulation, inhibitors of platelet aggregation, and peptides with vasodilator activity, now have their molecular nature known, opening the opportunity for the development of new therapeutic products (Valenzuela, 2002; Ribeiro and Francischetti, 2003). Salivary components with putative T-cell immunomodulatory function, such as those found in *A. darlingi* (Calvo et al., 2004) and lectins characterized in the saliva of several mosquitoes, are likely candidates to test and determine if they interact and modulate the vertebrate immune system in a way that influences the outcome of a malaria infection. Interestingly, ~40% of all the sequenced mosquito salivary gland cDNAs correspond to genes encoding proteins with unknown function. These numbers indicate how much there is still to learn about mosquito biology. For example, the ubiquity of expression of *D7*, *D7-related*, and *D7-like* genes in mosquitoes and other hematophagous insect salivary glands is confounded by a general lack of knowledge of their function. Only one *D7-related* gene product, hamadarin, has a potential role, and it may be important for inhibiting activation of the plasma contact system (Isawa et al., 2002).

Extensive EST studies designed to dissect the molecular components of innate immunity in mosquitoes have generated large amounts of information on the mosquito transcriptome (Dimopoulos et al., 2000; Bartholomay et al., 2004; Osta et al., 2004a). Dimopoulos et al. (2000) characterized 3,242 cDNA clones, 2,380 of which represent individual genes, from normalized libraries made with RNA extracted from 4A-3A and 4A-3B *A. gambiae* hemocyte-like, immune-responsive cell lines. Genes whose products were putatively related to immunity were identified in 38 clusters, including genes encoding serine proteases with clip domains involved in prophenoloxidase activation, adhesive proteins that recognize and bind microorganisms leading to a series of antipathogen actions such as agglutination of bacteria and phagocytosis, effector molecules such as cecropin, and coagulation factors. The fraction of cDNAs corresponding to genes with known function was only 47%, leaving more than half of the generated sequences for future studies. A similar approach was used to generate cDNA libraries from hemocytes derived from immune-challenged (by bacteria) *A. aegypti* and *Armigeres subalbatus* adult females (Bartholomay et al., 2004). A total of 11,952 *A. aegypti* and 12,790 *A. subalbatus* ESTs were sequenced, resulting, respectively, in 2,686 and 2,107 unique sequence clusters. Sequences encoding proteins related to immunity were identified in the categories of effector molecules, pattern recognition, signal transduction, melanization, serine proteases, and serpins. Again, the data showed that >50% of the sequences encoded proteins of unknown function. As the knowledge of the mosquito immune system is growing rapidly, it is expected that immune factors and effector molecules capable of halting malaria parasite development within the vector will be identified. The identification of these molecules is a necessary foundation for one of the proposed control strategies of malaria transmission by genetic manipulation of anophelines.

An intriguing set of complex phenotypes is host location and host preference. Biessman et al. (2002) sequenced a large number of ESTs with the intent of identifying genes expressed in the antennae of *A. gambiae*. Two cDNA libraries were made from RNA extracted from male or female antennae, and randomly selected clones were sequenced. The analysis revealed 23 cDNAs corresponding to seven different genes with sequences similar to those that encode odorant-binding proteins (OBP) first identified in *D. melanogaster* and other insects (Vogt et al., 2002). Analyses revealed the differential expression of some of the genes in male or female mosquitoes. This analysis was followed up by a different technique, filter-array hybridization (Justice et al., 2003). Ten thousand clones from each library were arrayed on nylon filters and hybridized separately with antenna-specific male or female cDNA probes. Approximately 150 clones hybridized strongly and specifically or preferentially to antennal cDNAs, indicating their abundant and localized accumulation. All of the identified clones were sequenced and a catalog of genes expressed abundantly in *A. gambiae* antennae was generated. Among those, *OBP-1* and *OBP-8* were expressed preferentially in female antennae, while *Sap-1, AgamA5, ANP-1*, and *ANP-2* were expressed preferentially in males. Using an approach in silico with the *A. gambiae* genome sequence, Zwiebel and collaborators identified and analyzed the expression profiles of several odorant receptors and arrestins in the antenna of adult *A. gambiae* (Hallem et al., 2004). One of the *A. gambiae* odorant receptors identified (*AgOr1*) exhibits female-specific antennal expression and is down-regulated 12 h after blood feeding, a period during which mosquitoes display a reduced response to human odorants (Fox et al., 2001). The application of this information to the development of novel mosquito attractants for female traps and repellents remains to be seen.

Extensive cDNA sequencing also has been used to identify genes differentially expressed in insecticide-resistant and -susceptible mosquitoes (Wu et al., 2004). From 809 cDNA clones originally obtained from subtractive libraries of deltamethrin-resistant and -susceptible strains of *Culex pipiens pallens*, 16 were confirmed to be differentially expressed by hybridization to Northern blots and cDNA arrays. Moreover, while eight of them were expressed at higher levels in resistant mosquitoes, two others were expressed exclusively in the resistant strain, suggesting that multiple genes may be involved in the production of the resistant phenotype.

As part of the *A. gambiae* genome-sequencing project, approximately 40,000 ESTs were sequenced from each of two cDNA libraries made with RNA extracted from whole-blood-fed or non-blood-fed female mosquitoes. These EST sequences, along with the other EST projects described previously, contributed to the determination of the complement of the *A. gambiae* genome. Using this source of information, Ribeiro (2003) conducted an analysis of gene expression in adult mosquitoes and searched for genes expressed differentially in blood-fed and non-blood-fed females. The data showed that of the estimated ~15,000 genes encoded by the *A. gambiae* genome, at least 7,580 were expressed in adults, having produced at least one sequence among the 80,000 ESTs. The genes that were not represented in the analyzed pool were expressed at undetectable low levels, during other developmental stages (embryonic, larval, and pupal), in adult males or corresponded to gene and transcript prediction errors. About 6% (435 genes) of the genes expressed in adult females showed significant differential expression, with either increased or decreased numbers of mRNAs following a blood meal. Among these genes, 68 had no representation in one or the other cDNA library, indicating complete regulation by blood feeding of the expression of these genes. Up-regulation of genes encoding products associated with blood digestion (trypsins 1 and 2, chymotrypsins 1 and 2, and peritrophin I), protein synthesis (ribosomal proteins, translation initiation factors, and secretory pathway-associated proteins), and oogenesis (vitellogenin, vitellogenin receptor, vitelline membrane proteins, and maternal proteins Oskar and

Exuperantia) was confirmed by this analysis and in agreement with Holt et al. (2002). A general down-regulation of gene products related to striated-muscle activity (actin, myosin, troponin, and flightin), sensory organs (eye-lens cuticular protein, opsin, arrestin, and OBPs), and behavior (*takeout* circadian-controlled protein and glutamate decarboxylase) is noticeable, indicating that females tend to reduce flight activities and explore the environment less after taking a blood meal. These data support the general observation that female metabolism and behavior change after blood feeding from a host-seeking mode to reproductive mode. Unexpected down-regulation of genes whose products are associated with mitochondrial metabolism, such as those of oxidative phosphorylation and the tricarboxylic acid cycle, was explained as a decrease in expression of these genes in flight muscles, which comprise about 10% of the insect weight and in which mitochondria represent ~40% of the muscle volume (Ribeiro, 2003). However, a similar decrease in gene expression associated with the tricarboxylic acid cycle also was observed in the midguts of *A. aegypti* females after a blood meal (Sanders et al., 2003). Therefore, further analyses are necessary to reveal the biological meaning of this phenomenon. The sequencing of *A. gambiae* tissue-specific, normalized cDNA libraries is an ongoing project; ~150,000 EST sequences, including cDNAs from fat bodies and mosquito heads, are already available at the NCBI EST database (http://www.ncbi.nlm.nih.gov/dbEST/index.html).

As mentioned previously, ~15,000 predicted genes were identified for *A. gambiae* by gene-finding algorithms (Holt et al., 2002; Zdobnov et al., 2002). However, gene prediction in general is a complex problem in metazoans and is aggravated with *A. gambiae* by the fact that its genome or that of any closely related organism has yet to be annotated comprehensively. *D. melanogaster* is estimated to have last shared an ancestor with *A. gambiae* some 250 to 280 million years ago (Gaunt and Miles, 2002), and this must be kept in mind when using the vinegar fly genome to extrapolate to the mosquito. Alternative promoter usage, alternative splicing, and alternative polyadenylation are frequent in eukaryotes, which results in an increase in transcriptome diversity compared with gene number and contributes to protein diversity in metazoan organisms (Boue et al., 2003; Brett et al., 2002; Beaudoing and Gautheret, 2001). For example, estimates indicate that 60% of all human and at least 20% of the *D. melanogaster* genes have alternative splicing variants of their products (Misra et al., 2002). In both species, the percentage of alternatively spliced genes is most probably underestimated. Similarly, recent analyses revealed that the present estimate of the number of *A. gambiae* transcription products may underrepresent the total by ~20% (Mongin et al., 2004). Large-scale sequencing of clones from full-length, high-quality cDNA libraries and the products of RT-PCR and rapid amplification of cDNA ends (Schaefer, 1995) are the techniques of choice to identify alternative transcripts. Only extensive sequencing of *A. gambiae* ESTs derived from different organs and developmental stages will yield a better estimate of the frequency of posttranscriptional splicing events and provide information on their biological relevance. Full-length cDNA sequences are also valuable to accurately determine the 5'-end and 3'-end boundaries of a gene, assist in the identification of its immediate upstream basal promoter, provide the complete coding sequence, allow the deduction of the primary sequence of the encoded protein, and identify DNA sequences of untranslated segments possibly involved in posttranscriptional RNA localization and translation regulation (Das et al., 2001).

A recent example of how EST sequencing can improve the understanding of gene expression in mosquitoes was provided by the analysis of the glutathione transferase supergene family in *A. gambiae* (Ding et al., 2003). Twenty-eight genes putatively encoding cytosolic glutathione transferases were annotated manually, and their corresponding cDNAs were sequenced. Transcripts were found in adult mosquito samples for all but one of the predicted genes, and alternative splicing was found to occur during the transcription of at least two of them. Similar detailed studies with other genes and gene families need to be developed to support and or correct the computer-based predictions.

The amount of information made available through *A. gambiae* genome sequencing and assembly, coupled with the EST sequencing projects, provides the scientific community with a powerful resource for the further detailed analyses of gene expression. Two high-throughput techniques for gene expression analyses that benefit from the mosquito genome sequencing are the serial analysis of gene expression (SAGE) and microarray. Both of these approaches are powerful but have intrinsic limitations. It is imperative to validate the results obtained by each of these techniques by other independent analyses. Thus, RT-PCR and real-time PCR have been extensively used to validate microarray and SAGE results.

SAGE is based on the sequential analysis in large quantities of short cDNA sequence tags (Tuteja and Tuteja, 2004). Each tag is derived from a defined position within a transcript, and its size (14 nucleotides) is generally sufficient to identify the corresponding gene. The number of times each tag is observed is indexed against controls and provides an accurate measurement of its expression level. This technique is useful for the analysis of minute amounts of biological material, since tag populations can be amplified without altering their relative proportions. These advantages make SAGE ideal for the analysis of gene expression in small amounts of mosquito tissue that require extensive and specialized dissection skills, such as organs of the endocrine and nervous systems. Efforts are in progress to use SAGE tag expression technology to characterize how gene expression profiles are modulated in *A. gambiae* salivary glands in response to infection by malaria parasites (I. Rosinski-Chupin, personal communication).

The development of microarray technology has also provided the ability to monitor thousands of genes while measuring the abundance of their transcripts in cells or tissues of interest. Microarrays are made of large sets of DNA sequences (cDNAs or oligonucleotides) immobilized as probes on platforms consisting of materials such as nylon membranes or glass slides. The number and collection of transcripts represented in each microarray are selected by the user or are provided by a commercial source. Independent of the type and composition of microarray in use, the technical principle is the same and involves hybridization of the immobilized probe with a target made from RNA isolated from cells and tissues. The targets in most cases are made by the incorporation of fluorescently labeled nucleotides during the synthesis of cDNA or cRNA to enable quantification of the hybridization (Rogojina et al., 2003). Conditions for hybridization, washing, and data collection, which include image acquisition and processing, depend on the microarray platform used. Data normalization and analysis can be done by using a number of specific software packages that will not be discussed further here.

The first microarray-based application for high-throughput analysis of gene expression in mosquitoes was performed in a search for transcripts involved in insect defense reactions (Dimopoulos et al., 2002). The platform consisted of 3,840 PCR-amplified EST probes derived from mosquito immune-responsive cell lines and spotted onto aminosilane-coated glass slides. This large-scale analysis made it possible to confirm the regulation of specific genes observed in previous studies (Dimopoulos et al., 2000) and to identify numerous uncharacterized immunity-related genes. The data revealed the regulation of expression of sets of genes after gram-negative and gram-positive bacteria challenges, as well as *Plasmodium berghei* infections. Remarkably, the response to malaria infection partially overlapped the bacterial infection response. Genes activated by both bacterial and malaria infection include those encoding a peptidoglycan recognition protein LB receptor, the gram-negative bacteria-binding protein opsonin, an fibrinogen-like lectin, a thioester-containing putative opsonin, the 14-D serine protease, the CED-6-like phagocytic adaptor, and the leucine-rich repeat putative receptor. Genes identified as being activated by malaria but not by bacterial infection encode an isocitrate dehydrogenase, a double-stranded RNA (dsRNA)-binding RNase 3, and a mitochondrial phosphate carrier. Genes with reduced expression after malaria infection were also found and may represent immune suppression of the vector by the developing parasite.

The same microarray platform was further used for the study of the infection response of susceptible and refractory *A. gambiae* strains to malaria parasites (Kumar et al., 2003). Differences in gene expression between refractory and susceptible strains were observed in the absence of parasite infection, and steps were taken to normalize the data by comparing the expression patterns of uninfected, sugar-fed mosquitoes of each strain. These results emphasize the need for researchers to exercise care when comparing gene expression data originating from different laboratories and using different mosquito strains. After adequate controls, Kumar et al. (2003) were able to identify differentially expressed genes between strains and suggested that the refractory and susceptible mosquitoes had broad physiological differences related to the production and detoxification of reactive oxygen species. Higher expression of immunity-related genes, in particular of genes expressing immune-signaling pathways, and a gene expression profile that generates an abnormal chronic state of oxidative stress in mosquitoes of the refractory strain may be involved with the melanotic encapsulation of *P. berghei* by malaria-resistant *A. gambiae*.

In addition to mosquito-parasite interactions, microarrays are applicable for multiple purposes for genome-wide studies of gene expression. Hematophagy is essential for mosquito development and reproduction and plays a significant role in disease transmission. Despite the large amount of information available on morphological and physiological responses of the mosquito midgut to blood ingestion, the control of gene expression in this organ during digestion has to be explored further. Sanders et al. (2003) used a microarray platform containing 1,778 individual EST clone inserts printed onto CMT GAPSII-coated slides to study gene expression in *A. aegypti*. They identified 333 genes that are regulated by blood feeding in the midgut over a 72-h time course, including those involved in digestion, peritrophic matrix formation, nutrient uptake and metabolism, cellular stress responses, and ion balance. Seventy genes encoding proteins of unknown functions were identified as induced or repressed in response to a blood meal. Similar studies with other mosquito species will help to understand broadly this important process and may reveal conditions necessary for parasite survival within the midgut lumen.

The commercially available GeneChip Plasmodium/Anopheles Genome Array is an oligonucleotide-based platform that includes probe sets for 4,300 *P. falciparum* and 14,900 *A. gambiae* genes. With this tool, comparative gene expression analyses of whole mosquitoes between sugar-fed and blood-fed adult females 24 h after a meal were conducted (Marinotti et al., in press). A total of 5,230 probe sets showed significant variation in gene expression upon blood meal. Of those, 2,670 showed up-regulation while the remaining 2,556 showed down-regulation. The majority of the regulated genes displayed low levels of variation, with 52.2% of them up- or down-regulated by 1- to 2-fold, while only 2.9% of them varied by >10-fold and only 0.2% varied by >100-fold. Among the genes showing the strongest up-regulation were those related to oogenesis and blood digestion. Those results were expected, however; among the strongly up-regulated genes were a Gly-His-Tyr-rich protein, Anoga protein G12, and a possible phosphatidylinositol transfer protein with unknown function in the mosquito. These data again emphasize the necessity and importance of continued studies of the biological functions of mosquito proteins to contribute to the annotation of the existing *A. gambiae* gene databases.

TOOLS FOR FUNCTIONAL ANALYSES OF THE TRANSCRIPTOME

Functional studies are of primary importance to link the large amount of sequence and expression profile information with biochemical and physiological events. RNA interference (RNAi) or RNA silencing is a mechanism guided by dsRNA that triggers degradation of homologous transcripts. This mechanism was discovered first in *Caenorhabditis elegans* (Guo and Kemphues, 1995) and has subsequently been extensively used for sequence-specific silencing of genes from a variety of organisms including insects (Kennerdell and Carthrew, 2000; Shin et al., 2003; Ullu et al., 2004; Geley and Muller, 2004). The activity of an RNAi system was most likely responsible for the antisense mRNA silencing of a reporter gene in transformed *A. aegypti* (Johnson et al., 1999). Subsequent to this, Adelman et al. (2001) injected *A. aegypti* with double subgenomic Sindbis viruses with inserted nucleotide sequences from the genome of dengue fever viruses. The injected mosquitoes became highly resistant to challenge with homologous dengue fever viruses from which the nucleotide sequences were derived. By using the same Sindbis virus-based vector, Shiao et al. (2001) demonstrated that the RNAi technique was sufficient to reduce significantly hemolymph phenoloxidase activity in *A. subalbatus* expressing a phenoloxidase antisense RNA. When these mosquitoes were challenged with microfilaria of the dog heartworm, *Dirofilaria immitis*, the parasite melanization process was almost entirely inhibited.

RNAi methods were used in the functional characterization of a gene, *aTEP-I*, encoding a novel immune-related *A. gambiae* thioester protein, a hemocyte-specific acute-phase glycoprotein that can bind to *Escherichia coli* and *Staphylococcus aureus* (Levashina et al., 2001). dsRNA-guided *aTEP-I* knockout experiments demonstrated that the product of this gene is an important component of the immune system and strongly promotes phagocytosis of gram-negative bacteria. Remarkably, the *aTEP-I* product binds to the malaria parasite *Plasmodium berghei* and mediates its killing (Blandin et al., 2004). The dsRNA knockdown of *aTEP-I* in malaria-infected female mosquitoes completely abolished melanotic refractoriness in a genetically selected refractory strain while in susceptible mosquitoes the *aTEP-I* knockdown increased the number of developing parasites.

Blandin et al. (2002) conducted a functional analysis of *A. gambiae* defensin by injecting female mosquitoes with defensin-derived dsRNA. While the injection of *E. coli* or *S. aureus* induced the transcription of the defensin gene in control mosquitoes, only traces of defensin mRNA were detected in dsRNA-injected mosquitoes. The mosquitoes with reduced defensin expression were also challenged with *P. berghei*; however, parasite development was not altered, indicating that malaria-induced, endogenous *A. gambiae* defensin does not act as a significant antiparasitic factor in vivo. The same laboratory is currently conducting a large-scale functional screen of about 100 mosquito genes with the gene-silencing technique. A recent and important advance obtained from this large-scale analysis was to show that a leucine-rich repeat protein acts as an antagonist and that two C-type lectins act as protective agonists for the development of *P. berghei* ookinetes to oocysts (Osta et al., 2004b). RNAi can be linked to stably expressed transgenes, as demonstrated by the development of stable RNAi in *A. stephensi* (Brown et al., 2003) and *A. aegypti* (Travanty et al., 2004). The continued application of RNAi technique will be of great value for the systematic assignment of functions to *A. gambiae* genes.

FINAL REMARKS

The analysis of the transcriptome of *A. gambiae* is far from complete. Although good estimates put the number of genes to be ~15,000, the actual size of the transcriptome is certainly much larger. Of particular interest is the large fraction of genes, as much as 50%, that is regulated specifically but encodes products of as-of-yet unknown functions. The roles of some of these genes will be deciphered by analogy as the functions of orthologous genes are worked out in model organisms, but many will require direct analysis of the mosquito. Powerful tools such as RNAi exist to facilitate this work, and there is reason for optimism that these efforts will provide exciting new insights into basic biology and methods for disease control. Furthermore, mosquitoes exhibit behavior that is clearly genetic in its basis. The transcriptome provides a powerful resource to dissect these complex phenotypes. A principal focus will be to identify key genes in the hierarchy of control that result in these phenotypes.

A remarkable transition in the overall course of mosquito biology has been observed in recent years. New advances make possible moving gene expression studies from a perspective of a single molecule, tissue, process, or behavior to a more comprehensive view of the organism. Monitoring genome-wide changes in gene expression patterns in whole *A. gambiae* specimens is now feasible, and it is expected that this will be possible with other vectors in the near future. Access to genome-based data will enhance our ability to identify genes involved in each aspect of mosquito vectorial capacity. Nevertheless, the next challenge is to exploit this information and address specific questions that may lead to the development of effective malaria control tools.

REFERENCES

Adelman, Z. N., C. D. Blair, J. O. Carlson, B. J. Beaty, and K. E. Olson. 2001. Sindbis virus-induced silencing of dengue viruses in mosquitoes. *Insect Mol. Biol.* **10**:265–273.

Arca, B., F. Lombardo, M. de Lara Capurro, A. della Torre, G. Dimopoulos, A. A. James, and M. Coluzzi. 1999. Trapping cDNAs encoding secreted proteins from the salivary glands of the malaria vector *Anopheles gambiae*. *Proc. Natl. Acad. Sci. USA* **96**:1516–1521.

Bartholomay, L. C., W. L. Cho, T. A. Rocheleau, J. P. Boyle, E. T. Beck, J. F. Fuchs, P. Liss, M. Rusch, K. M. Butler, R. C. Wu, S. P. Lin, H. Y. Kuo, I. Y. Tsao, C. Y. Huang, T. T. Liu, K. J. Hsiao, S. F. Tsai, U. C. Yang, A. J. Nappi, N. T. Perna, C. C. Chen, and B. M. Christensen. 2004. Description of the transcriptomes of immune response-activated hemocytes from the mosquito vectors *Aedes aegypti* and *Armigeres subalbatus*. *Infect. Immun.* **72**:4114–4126.

Beaudoing, E., and D. Gautheret. 2001. Identification of alternate polyadenylation sites and analysis of their tissue distribution using EST data. *Genome Res.* **11**:1520–1526.

Besansky, N. J., J. A. Bedell, M. Q. Benedict, O. Mukabayire, D. Hilfiker, and F. H. Collins. 1995. Cloning and characterization of the white gene from *An. gambiae*. *Insect Mol. Biol.* **4**:217–231.

Biessmann, H., M. F. Walter, S. Dimitratos, and D. Woods. 2004. Isolation of cDNA clones encoding putative odourant binding proteins from the antennae of the malaria-transmitting mosquito, *Anopheles gambiae*. *Insect Mol. Biol.* **11**:123–132.

Blandin, S., L. F. Moita, T. Kocher, M. Wilm, F. C. Kafatos, and E. A. Levashina. 2002. Reverse genetics in the mosquito *Anopheles gambiae*: targeted disruption of the Defensin gene. *EMBO Rep.* **3**:852–856.

Blandin, S., S. H. Shiao, L. F. Moita, C. J. Janse, A. P. Waters, F. C. Kafatos, and E. A. Levashina. 2004. Complement-like protein TEP1 is a determinant of vectorial capacity in the malaria vector *Anopheles gambiae*. *Cell* **116**:661–670.

Boue, S., I. Letunic, and P. Bork. 2003. Alternative splicing and evolution. *Bioessays* **25**:1031–1034.

Brett, D., H. Pospisil, J. Valcartel, J. Reich, and P. Bork. 2002. Alternative splicing andgenome complexity. *Nat. Genet.* **30**:29–30.

Brown, A. E., L. Bugeon, A. Crisanti, and F. Catteruccia. 2003. Stable and heritable gene silencing in the malaria vector *Anopheles stephensi*. *Nucleic Acids Res.* **31**:e85.

Calvo, E., A. G. deBianchi, A. A. James, and O. Marinotti. 2002. The major acid soluble proteins of adult female *Anopheles darlingi* salivary glands include a member of the D7-related family of proteins. *Insect Biochem. Mol. Biol.* **32**:1419–1427.

Calvo, E., J. Andersen, I. M. Francischetti, M. de L. Capurro, A. G. de Bianchi, A. A. James, J. M. C. Ribeiro, and O. Marinotti. 2004. The transcriptome of adult female *Anopheles darlingi* salivary glands. *Insect Mol. Biol.* **13**:73–88.

Champagne, D. E., C. T. Smartt, J. M. Ribeiro, and A. A. James. 1995. The salivary gland-specific apyrase of the mosquito *Aedes aegypti* is a member of the 5'-nucleotidase family. *Proc. Natl. Acad. Sci. USA* **92**:694–698.

Coates, C. J., T. L. Schaub, N. J. Besansky, F. H. Collins, and A. A. James. 1997. The white gene from the yellow fever mosquito *Aedes aegypti*. *Insect Mol. Biol.* **6**:291–299.

Coleman, P. G., and L. Alphey. 2004. Genetic control of vector populations: an imminent prospect. *Trop. Med. Int. Health* **9**:433–437.

Cui, L., S. Luckhart, and R. Rosenberg. 2002. Molecular characterization of a prophenoloxidase cDNA from the malaria mosquito *Anopheles stephensi*. *Insect Mol. Biol.* **9**:127–137.

Das, M., I. Harvey, L. Chu, M. Sinha, and J. Pelletier. 2001. Full-length cDNAs: more than just reaching the ends. *Physiol. Genomics* **6**:57–80.

Diatchenko, L., Y. C., A. P. Lau, A. Campbell, F. Chenchik, G. Moqadam, B. Huang, S. Lukyanov, K. Lukyanov, N. Gurskaya, E. D. Sverdlov, and P. D. Siebert. 1996. Suppression subtractive hybridization: a method for generating differentially regulated or tissue-specific cDNA probes and libraries. *Proc. Natl. Acad. Sci. USA* **93:** 6025–6030.

Dimopoulos, G., G. K. Christophides, S. Meister, J. Schultz, K. P. White, C. Barillas-Mury, and F. C. Kafatos. 2002. Genome expression analysis of *Anopheles gambiae*: responses to injury, bacterial challenge, and malaria infection. *Proc. Natl. Acad. Sci. USA* **99:**8814–8819.

Dimopolous, G., A., Richman, A. della Torre, F. C. Kafatos, and C. Louis. 1996. Identification and characterization of differentially-expressed cDNAs of the vector mosquito, *An. gambiae*. *Proc. Natl. Acad. Sci. USA* **93:**13066–13071.

Dimopoulos, G., A. Richman, H. M. Muller, and F. C. Kafatos. 1997. Molecular immune responses of the mosquito *Anopheles gambiae* to bacteria and malaria parasites. *Proc. Natl. Acad. Sci. USA* **94:**11508–11513.

Dimopoulos, G., T. L. Casavant, S. Chang, T. Scheetz, C. Roberts, M. Donohue, J. Schultz, V. Benes, P. Bork, W. Ansorge, M. B. Soares, and F. C. Kafatos. 2000. *An. gambiae* pilot gene discovery project: identification of mosquito innate immunity genes from expressed sequence tags generated from immune-competent cell lines. *Proc. Natl. Acad. Sci. USA* **97:**6619–6624.

Dimopoulos, G., D. Seeley, A. Wolf, and F. C. Kafatos. 1998. Malaria infection of the mosquito *Anopheles gambiae* activates immune-responsive genes during critical transition stages of the parasite life cycle. *EMBO J.* **17:**6115–6123.

Ding, Y., F. Ortelli, L. C. Rossiter, J. Hemingway, and H. Ranson. 2003. The *Anopheles gambiae* glutathione transferase supergene family: annotation, phylogeny and expression profiles. *BMC Genomics* **4:**35.

Fox, A. N., R. J. Pitts, H. M. Robertson, J. R. Carlson, and L. J. Zwiebel. 2001. Candidate odorant receptors from the malaria vector mosquito *Anopheles gambiae* and evidence of down-regulation in response to blood feeding. *Proc. Natl. Acad. Sci. USA* **98:**14693–14697.

Francischetti, I. M., J. G. Valenzuela, V. M. Pham, M. K. Garfield, and J. M. C. Ribeiro. 2002. Toward a catalog for the transcripts and proteins (sialome) from the salivary gland of the malaria vector *An. gambiae*. *J. Exp. Biol.* **205:**2429–2451.

Gaunt, M. W., and M. A. Miles. 2002. An insect molecular clock dates the origin of the insects and accords with palaeontological and biogeographic landmarks. *Mol. Biol. Evol.* **19:**748–761.

Geley, S., and C. Muller. 2004. RNAi: ancient mechanism with a promising future. *Exp. Gerontol.* **39:** 985–998.

Grossman, G. L., and A. A. James. 1993. The salivary glands of the vector mosquito, *Aedes aegypti*, express a novel member of the amylase gene family. *Insect Mol. Biol.* **1:**223–232.

Guo, S., and K. J. Kemphues. 1995. Par-1, a gene required for establishing polarity in *C. elegans* embryos, encodes a putative Ser/Thr kinase that is asymmetrically distributed. *Cell* **81:**611–620.

Hallem, E. A., F. A. Nicole, L. J. Zwiebel, and J. R. Carlson. 2004. Olfaction: mosquito receptor for human-sweat odorant. *Nature* **427:**212–213.

Han, Y. S., C. E. Salazar, S. R. Reese-Stardy, A. Cornel, M. J. Gorman, F. H. Collins, and S. M. Paskewitz. 1997. Cloning and characterization of a serine protease from the human malaria vector, *An. gambiae*. *Insect Mol. Biol.* **6:**385–395.

Hemingway, J., and H. Ranson. 2000. Insecticide resistance in insect vectors of human disease. *Annu. Rev. Entomol.* **45:**371–391.

Holt, R. A., G. M. Subramanian, A. Halpern, G. G. Sutton, R. Charlab, D. R. Nusskern, P. Wincker, A. G. Clark, J. M. Ribeiro, R. Wides, S. L. Salzberg, B. Loftus, M. Yandell, W. H. Majoros, D. B. Rusch, Z. Lai, C. L. Kraft, J. F. Abril, V. Anthouard, P. Arensburger, P. W. Atkinson, H. de Baden, V. Berardinis, D. Baldwin, V. Benes, J. Biedler, C. Blass, R. Bolanos, D. Boscus, M. Barnstead, S. Cai, A. Center, K. Chaturverdi, G. K. Christophides, M. A. Chrysta, M. Clamp, A. Cravchik, V. Curwen, A. Dana, A. Delcher, I. Dew, C. A. Evans, M. Flanigan, A. Grundschober-Freimoser, L. Friedli, Z. Gu, P. Guan, R. Guigo, M. E. Hillenmeyer, S. L. Hladun, J. R. Hogan, Y. S. Hong, J. O. Jaillon, Z. Ke, C. Kodira, E. Kokoza, A. Koutsos, I. Letunic, A. Levitsky, Y. Liang, J. J. Lin, N. F. Lobo, J. R. Lopez, J. A. Malek, T. C. McIntosh, S. Meister, J. Miller, C. Mobarry, E. Mongin S. D., D. A. O' Brochta, C. Pfannkoch, R. Qi, M. A. Regier, K. Remington, H. Shao, M. V. Sharakhova, C. D. Sitter, J. Shetty, T. J. Smith, R. Strong, J. Sun, D. Thomasova, L. Q. Ton, P. Topalis, Z. Tu, M. F. Unger, B. Walenz, A. Wang, J. Wang, M. Wang, X. Wang, K. J. Woodford, J. R. Wortman, M. Wu, A. Yao, E. M. Zdobnov, H. Zhang, Q. Zhao, S. Zhao, S. C. Zhu, I. Zhimulev, M. Coluzzi, A. della Torre, C. W. Roth, C. Louis, F. Kalush, R. J. Mural, E. W. Myers, M. D. Adams, H. O. Smith, S. Broder, M. J. Gardner, C. M. Fraser, E. Birney, P. Bork, P. T. Brey, J. C. Venter, J. Weissenbach, F. C. Kafatos, F. H. Collins, and S. L. Hoffman. 2002. The genome sequence of the malaria mosquito *Anopheles gambiae*. *Science* **298:**129–149.

Isawa, H., M. Yuda, Y. Orito, and Y. Chinzei. 2002. A mosquito salivary protein inhibits activation of the plasma contact system by binding to factor XII and high molecular weight kininogen. *J. Biol. Chem.* **277:**27651–27658.

James, A. A., K. Blackmer, and J. V. Racioppi. 1989. A salivary gland-specific, maltase-like gene of the vector mosquito *Aedes aegypti. Gene* **75:**73–83.

James, A. A., K. Blackmer, O. Marinotti, C. R. Ghosn, and J. V. Racioppi. 1991. Isolation and characterization of the gene expressing the major salivary gland protein of the female mosquito *Aedes aegypti. Mol. Biochem. Parasitol.* **44:** 245–253.

James, A. A., B. T. Beerntsen, M. de L. Capurro, C. J. Coates, J. Coleman, N. Jasinskiene, and A. U. Krettli. 1999. Controlling malaria transmission with genetically-engineered, *Plasmodium*-resistant mosquitoes: milestones in a model system. *Parassitologia* **41:**461–471.

Jiang, H., Y. Wang, S. E. Korochkina, H. Benes, and M. R. Kanost. 1997. Molecular cloning of cDNAs for two pro-phenol oxidase subunits from the malaria vector, *An. gambiae. Insect Biochem. Mol. Biol.* **27:**693–699.

Johnson, B. W., K. E. Olson, T. Allen-Miura, J. O. Carlson, C. J. Coates, N. Jasinskiene, A. A. James, B. J. Beaty, and S. Higgs. 1999. Inhibition of luciferase expression in transgenic *Aedes aegypti* mosquitoes by Sindbis virus expression of antisense luciferase RNA. *Proc. Natl. Acad. Sci. USA* **96:**13399–13403.

Justice, R. W., S. Dimitratos, M. F. Walter, D. F. Woods, and H. Biessmann. 2003. Sexual dimorphic expression of putative antennal carrier protein genes in the malaria vector *An. gambiae. Insect Mol. Biol.* **12:**581–594.

Kennerdell, J. R., and R. W. Carthrew. 2000. Heritable gene silencing in Drosophila using double-stranded RNA. *Nat. Biotechnol.* **17:**896–898.

Korochkina, S. E., A. V. Gordadze, J. L. York, and H. Benes. 1997. Mosquito hexamerins: characterization during larval development. *Insect Mol. Biol.* **6:**11–21.

Kumar, S., G. K. Christophides, R. Cantera, B. Charles, Y. S. Han, S. Meister, G. Dimopoulos, F. C. Kafatos, and C. Barillas-Mury. 2003. The role of reactive oxygen species on Plasmodium melanotic encapsulation in *Anopheles gambiae. Proc. Natl. Acad. Sci. USA* **100:**14139–14144.

Levashina, E. A., L. F. Moita, S. Blandin, G. Vriend, M. Lagueux, and F. C. Kafatos. 2001. Conserved role of a complement-like protein in phagocytosis revealed by dsRNA knockout in cultured cells of the mosquito, *Anopheles gambiae. Cell* **104:**709–718.

Liang, P., and A. B. Pardee. 1992. Differential display of eukaryotic messenger RNA by means of the polymerase chain reaction. *Science* **257:**967–971.

Liang, P., and A. B. Pardee. 1998. Differential display. A general protocol. *Mol. Biotechnol.* **10:**261–267.

Marinotti, O., Q. K. Nguyen, E. Calvo, A. A. James, and J. M. C. Ribeiro. Microarray analysis of genes showing variable expression following a blood meal in *Anopheles gambiae. Insect. Mol. Biol.*, in press.

Martinez-Torres, D., F. Chandre, M. S. Williamson, F. Darriet, J. B. Berge, A. L. Devonshire, P. Guillet, N. Pasteur, and D. Pauron. 1998. Molecular characterization of pyrethroid knockdown resistance (kdr) in the major malaria vector *Anopheles gambiae* s. s. *Insect Mol. Biol.* **7:**179–184.

Meredith, S. E. O., and A. A. James. 1990. Biotechnology as applied to vectors and vector control. *Ann. Parasitol. Hum. Comp.* **65:**113–118.

Misra, S. M., A. Crosby, C. J. Mungall, B. B. Matthews, K. S. Campbell, P. Hradecky, Y. Huang, J. S. Kaminker, G. H. Millburn, S. E. Prochnik, C. D. Smith, J. L. Tupy, E. J. Whitfied, L. Bayraktaroglu, B. P. Berman, B. R. Bettencourt, S. E. Celniker, A. D. Grey, R. A. Drysdale, N. L. Harris, J. Richter, S. Russo, A. J. Schroeder, S. Q. Shu, M. Stapleton, C. Yamada, M. Ashburner, W. M. Gelbart, G. M. Rubin, and S. E. Lewis. 2002. Annotation of the Drosophila melanogaster euchromatic genome: a systematic review. *Genome Biol.* **3:** RESEARCH0083.

Mongin, E., C. Louis, R. A. Holt, E. Birney, and F. H. Collins. 2004. The *Anopheles gambiae* genome: an update. *Trends Parasitol.* **20:**49–52.

Montero-Solis, C., L. Gonzalez-Ceron, M. H. Rodriguez, B. E. Cirerol, F. Zamudio, L. C. Possanni, A. A. James, and F. de la Cruz Hernandez-Hernandez. 2004. Identification and characterization of gp65, a salivary-gland-specific molecule expressed in the malaria vector *Anopheles albimanus. Insect Mol. Biol.* **13:**155–164.

Moreira-Ferro, C. K., S. Daffre, A. A. James, and O. Marinotti. 1998. A lysozyme in the salivary glands of the malaria vector *Anopheles darlingi. Insect Mol. Biol.* **7:**257–264.

Noriega, F. G., and M. S. Wells. 1999. A molecular view of trypsin synthesis in the midgut of *Aedes aegypti. J. Insect Physiol.* **45:**613–620.

Oduol, F., J. Xu, O. Niare, R. Natarajan, and K. D. Vernick. 2000. Genes identified by an expression screen of the vector mosquito Anopheles gambiae display differential molecular immune response to malaria parasites and bacteria. *Proc. Natl. Acad. Sci. USA* **97:**11397–11402.

Osta, M. A., G. K. Christophides, D. Vlachou, and F. C. Kafatos. 2004a. Innate immunity in the malaria vector *Anopheles gambiae*: comparative and functional genomics *J. Exp. Biol.* **207:**2551–2563.

Osta, M. A., G. K. Christophides, and K. C. Kafatos. 2004b. Effects of mosquito genes on Plasmodium development. *Science* **303:**2030–2032.

Ribeiro, J. M., and I. M. Francischetti. 2003. Role of arthropod saliva in blood feeding: sialome and post-sialome perspectives. *Annu. Rev. Entomol.* **48:**73–88.

Ribeiro, J. M., R. Charlab, V. M. Pham, M. Garfield, and J. G. Valenzuela. 2004. An insight into the salivary transcriptome and proteome of the adult female mosquito *Culex pipiens quinquefasciatus*. *Insect Biochem. Mol. Biol.* **34:**543–563.

Ribeiro, J. M. 2003. A catalogue of *Anopheles gambiae* transcripts significantly more or less expressed following a blood meal. *Insect Biochem. Mol. Biol.* **33:**865–882.

Rogojina, A. T., W. E. Orr, B. K. Song, and E. E. Geisert, Jr. 2003. Comparing the use of Affymetrix to spotted oligonucleotide microarrays using two retinal pigment epithelium cell lines. *Mol. Vis.* **9:**482–496.

Salazar, C. E., D. M. Hamm, D. M. Wesson, C. B. Beard, V. Kumar, and F. H. Collins. 1994. A cytoskeletal actin gene in the mosquito *An. gambiae*. *Insect Mol. Biol.* **3:**1–13.

Sanders, H. R., A. M. Evans, L. S. Ross, and S. S. Gill. 2003. Blood meal induces global changes in midgut gene expression in the disease vector, *Aedes aegypti*. *Insect Biochem. Mol. Biol.* **33:**1105–1122.

Schaefer, B. C. 1995. Revolutions in rapid amplification of cDNA ends: new strategies for polymerase chain reaction cloning of full-length cDNA ends. *Anal. Biochem.* **227:**255–273.

Shen, Z., and M. Jacobs-Lorena. 1997. Characterization of a novel gut-specific chitinase gene from the human malaria vector *Anopheles gambiae*. *J. Biol. Chem.* **272:**28895–288909.

Shen, Z., and M. Jacobs-Lorena. 1998. A type I peritrophic matrix protein from the malaria vector *An. gambiae* binds to chitin. Cloning, expression, and characterization. *J. Biol. Chem.* **273:**17665–17670.

Shen, Z., G. Dimopoulos, and F. C. Kafatos. 1999. A cell surface mucin specifically expressed in the midgut of the malaria mosquito *Anopheles gambiae*. *Proc. Natl. Acad. Sci. USA.* **96:**5610–5615.

Shen, Z., M. J. Edwards, and M. Jacobs-Lorena. 2000. A gut-specific serine protease from the malaria vector *Anopheles gambiae* is down-regulated after blood ingestion. *Insect Mol. Biol.* **9:**223–229.

Shiao, H., S. Higgs, Z. Adelman, B. M. Christensen, S. H. Liu, and C. C. Chen. 2001. Effect of prophenoloxidase expression knockout on the melanization of microfilariae in the mosquito *Armigeres subalbatus*. *Insect Mol. Biol.* **10:**315–321.

Shin, S. W., V. A. Kokoza, and A. A. Raikhel. 2003. Transgenesis and reverse genetics of mosquito innate immunity. *J. Exp. Biol.* **206:**3835–3843.

Siden-Kiamos, I., G. Skavdis, J. Rubio, G. Papaginnakis, and C. Louis. 1996. Isolation and characterization of three serine protease genes in the mosquito *An. gambiae*. *Insect Mol. Biol.* **5:**61–71.

Stark, K. R., and A. A. James. 1996. The salivary glands of disease vectors, p. 333–348. *In* W. C. Marquardt and B. Beaty (ed.), *The Biology of Disease Vectors*. University Press of Colorado, Boulder, Colo.

St. John, T. P., and R. W. Davis. 1979. Differential screening of cDNA clones. *Cell* **16:**443–452.

Tashiro, K., H. Tada, R. Heilker, M. Shirozu, T. Nakanol, and T. Honjo. 1993. Signal sequence trap: a cloning strategy for secreted proteins and type I membrane proteins. *Science* **261:**600–603.

Thompson, M., R. Shotkoski, and R. ffrench-Constant. 1993. Cloning and sequencing of the cyclodiene insecticide resistance gene from the yellow fever mosquito *Aedes aegypti*. Conservation of the gene and resistance associated mutation with Drosophila. *FEBS Lett.* **325:**187–190.

Travanty, E. A., Z. N. Adelman, A. W. Franz, K. M. Keene, B. J. Beaty, C. D. Blair, A. A. James, and K. E. Olson. 2004. Using RNA interference to develop dengue virus resistance in genetically-modified *Aedes aegypti*. *Insect Biochem. Mol. Biol.* **34:**607–613.

Tuteja, R., and N. Tuteja. 2004. Serial analysis of gene expression (SAGE): unraveling the bioinformatics tools. *Bioessays* **26:**916–922.

Ullu, E., C. Tschudi, and T. Chakraborty. 2004. RNA interference in protozoan parasites. *Cell Microbiol.* **6:**509–519.

Valenzuela, J. G. 2002. High-throughput approaches to study salivary proteins and genes from vectors of disease. *Insect Biochem. Mol. Biol.* **32:**1199–1209.

Valenzuela, J. G., I. M. Francischetti, V. M. Pham, M. K. Garfield, and J. M. Ribeiro. 2003. Exploring the salivary gland transcriptome and proteome of the *Anopheles stephensi* mosquito. *Insect Biochem. Mol. Biol.* **33:**717–732.

Valenzuela, J. G., I. M. Francischetti, and J. M. Ribeiro. 1999. Purification, cloning, and synthesis of a novel salivary anti-thrombin from the mosquito *Anopheles albimanus*. *Biochemistry* **38:**11209–11215.

Valenzuela, J. G., V. M. Pham, M. K. Garfield, I. M. Francischetti, and J. M. Ribeiro. 2002. Toward a description of the sialome of the adult female mosquito *Aedes aegypti*. *Insect Biochem. Mol. Biol.* **32:**1101–1122.

Vizioli, J., F. Catteruccia, A. della Torre, I. Reckmann, and H. M. Muller. 2001. Blood digestion in the malaria mosquito *Anopheles gambiae*: molecular cloning and biochemical characterization of two inducible chymotrypsins. *Eur. J. Biochem.* **268:**4027–4035.

Vlachou, D., G. Lycett, I. Siden-Kiamos, C. Blass, R. E. Sinden, and C. Louis. 2001. *Anopheles gambiae* laminin interacts with the P25 surface protein of *Plasmodium berghei* ookinetes. *Mol. Biochem. Parasitol.* **112:**229–237.

Vogt, R. G., M. E. Rogers, M. D. Franco, and M. Sun. 2002. A comparative study of odorant binding protein genes: differential expression of the PBP1-GOBP2 gene cluster in *Manduca sexta* (Lepidoptera) and the organization of OBP genes in *Drosophila melanogaster* (Diptera). *J. Exp. Biol.* **205:**719–744.

World Health Organization. 1991. Prospects for malaria control by genetic manipulation of its vectors (TDR/BCV/MAL-ENT/91.3). World Health Organization, Geneva, Switzerland.

Wu, H.-W., H.-S. Tian, G.-L. Wu, G. Langdon, J. Kurtis, B. Shen, L. Ma, X.-K. Li, Y. Gu, X.-B. Hu, and C.-L. Zhu. 2004. *Culex pipiens pallens*: identification of genes differentially expressed in deltamethrin-resistant and -susceptible strains. *Pestic. Biochem. Physiol.* **79:**75–83.

Ying, S.-Y. 2004. Complementary DNA libraries: an overview. *Mol. Biotechnol.* **27:**245–252.

Zdobnov, E. M., C. von Mering, I. Letunic, D. Torrents, M. Suyama, R. R. Copley, G. K. Christophides, D. Thomasova, R. A. Holt, G. M. Subramanian, H. M. Mueller, G. Dimopoulos, J. H. Law, M. A. Wells, E. Birney, R. Charlab, A. L. Halpern, R. Kokoza, C. L. Kraft, Z. Lai, S. Lewis, C. Louis, C. Barillas-Mury, D. Nussker, G. M. Rubin, S. L. Salzberg, G. G. Sutton, P. Topalis, R. Wides, P. Wincker, M. Yandell, F. H. Collins, J. Ribeiro, W. M. Gelbart, F. C. Kafatos, and P. Bork. 2002. Comparative genome and proteome analysis of *Anopheles gambiae* and *Drosophila melanogaster*. *Science* **298:**149–159.

Zhang, D., G. Dimopoulos, A. Wolf, B. Minana, F. C. Kafatos, and J. J. Winzerling. 2002. Cloning and molecular characterization of two mosquito iron regulatory proteins. *Insect Biochem. Mol. Biol.* **32:**579–589.

Zheng, L., L. H. Whang, V. Kumarm, and F. C. Kafatos. 1995. Two genes encoding midgut-specific maltase-like polypeptides from *An. gambiae*. *Exp. Parasitol.* **81:**272–283.

Zurita, M., E. Reynaud, and F. C. Kafatos. 1997. Cloning and characterization of cDNAs preferentially expressed in the ovary of the mosquito, *Anopheles gambiae*. *Insect Mol. Biol.* **6:**55–62.

INDEX

Note: CP indicates a page on which a Color Plate concerning the subject is cited.

Actin microfilaments, in merozoite, 143
Adhesion molecules, as receptors for PfEMP1, 467
Aedes aegypti, 517
Alpha interferon receptor 1, *IFNAR1* encoding, 471
AMA-1, 133–134, 170
AMA1 vaccine, 490
American Type Culture Collection, 13–14
Amino acids, fate of, 320
δ-Aminolevulinate synthesis, 240
Aminopeptidases, 318–319
Aminoquinoline antimalarials, 321
Amodiaquine, 481
Anion channels, activated, molecular nature of, 391–392
    in infected red blood cells, 389–391
Anopheles, 6, 7, 175, 191
    NCBI mRNA entries retrieved for, 519
Anopheles gambiae, 9, 10
    antenna of, odorant receptors and arrestins in, 521
    as first arthropod vector sequenced, 508–509
    as malaria vector, 501
    data on, VectorBase and, 507
        whole-genome-wide comparisons between species and, 508–509
    evolution of, hypothesis of, 503
    G protein-coupled receptor signaling and, 509–510
    genetic population structure of, 502–503
    insecticide resistance in, 480
    population structures, discontinuities in, 502
    RNA interference (RNAi) and, 524
    species of, and malaria transmission, 501, 502
    vector behavior of, 509
    X chromosome of, 504
Anopheles gambiae genome, 501–515

    assembly and annotation in, CP19, 505–507
    metabolic insecticide resistance in, 512
    odorant-binding proteins in, 511
    research progress based on, 507–513
    size and structure of, CP19, 503, 504
    vector ecology and vector population genetics and, 511–512
    vector-parasite interactions in, and vector immunity, 512
Anopheles gambiae genome project, 503–507
    future priorities in, 513
    genome sequencing in, 504
    strain sequenced in, 503–505
Anopheles gambiae genome-sequencing project, expressed sequence tags in, 521
Anopheles gambiae odorant receptor 1, 510–511
Anopheles gambiae PEST, 503–505
Anopheles gambiae sensu stricto, polymorphisms within, 501–503
Anopheles gambiae tissue-specific normalized cDNA libraries, 522
Antibiotics, interacting with 16S rRNA, 358
    interacting with 23S rRNA, 359–361
    mitochondrial protein synthesis and, 246
Antidisease vaccine, rosette-inhibiting, prospects for, 422–423
Antifolates, as antimalarials with well-defined target, 431–433
    mechanism of action of, 431
    new, development of, CP17, 436–437
    P. falciparum resistance to, 430–431
    resistance to, evolution through mutations, 434
        target enzyme mutations and, 433–436
Antimalarial chemotherapy, lipid-based, 345–347
    of FASII pathway in apicoplast, 347
    phosphocholine analogs in, 346
    quaternary ammonium as choline analogs in, 346
    resistance to, 429

Antimalarial chemotherapy, lipid-based (*continued*)
   sphingolipid inhibitors in, 346–347
Antimalarial drugs, mitochondrial targets for, 243–247
Antimalarial resistance, definition of, 429–430
Antimalarial therapies, based on inhibition of hemoglobin hydrolysis, 321–322
Antioxidant defense, in malaria parasites, 368–374
Apical membrane antigen 1, 170
Apical membrane antigen-1 family, 130, 131, 132
Apical organelle secretion, by merozoite, 147–149
Apicomplexa, 276, 362
   and origin of parasitism in *Plasmodium*, 102
Apicomplexan Expressed Sequence Tag (EST) Database, 13
Apicoplast, 272–289
   anabolic pathways of, 284
   anabolic precursors of, 284–285
   aromatic amino acids and, 282–283
   delayed-death phenomenon, 277, 278
   division of, 274–277
   fatty acid synthesis and, 281–282
   function of, 280–285
   heme synthesis and, 283
   in *Plasmodium*, 78
   inhibitors of, 277–279
   iron-sulfur cluster assembly and, 283–284
   isopentenyl diphosphate synthesis, 282
   metabolites produced by, 280
   organelle division of, CP11, 274
   origin of, 275–277
Apicoplast genome, 272
   replication of, 274–275
Apicoplast-lacking mutants, 279–280
Apicoplast-lacking parasites, 280
Apicoplast-targeted proteins, 272–274
Aromatic amino acids, apicoplast and, 282–283
Artemisinin(s), 438, 449
   in malaria infection, 380
Artesunate-based combination therapy, 429
Aspartic protease inhibitors, 321
Atovaquone, 245–246
Atovaquone-proquanil combination, 243–246
ATP synthesis, mitochondrial, 239

Bacillus malariae, 6
Basic Local Alignment Tool (BLAST), 13
Bastianelli, Giovanni, 6
Bednets, insecticide-treated, 481
Bignami, Amico, 6
BLAST, PlasmoDB searched with, 225–227
BLASTP, 14
Blood cells, red. *See* Red blood cells
Blood-stage infection, malaria parasite-host interactions in, 267, 268
Blood-stage vaccines, 489–490
Burkina Faso, Fulani of, 463

Calcium, and calcium-dependent pathways, gametogenesis and, 209–211
Calcium signaling, and phosphoinositide pathways, 304
Caveolins, 257
CD31, 468
CD54, 467
CD233, 465
CD36 antigen, *CD36* gene encoding, 467–468
CD36 gene, encoding CD36 antigen, 467–468
CD40 ligand, *TNFSF5* encoding, 469
cDNA libraries, constructing and screening of, 518–519
CDP-choline pathway, 334–336
CDP-DAG, 330
   synthesis of, 331
CDP-diacylglycerol, synthesis of, 331
CDP-ethanolamine pathway, 334–335, 336
Cell cycle(s), eukaryotic, 290
   in yeast and mammalian cells, 305–306
Celli, Angelo, 6
Cerebral malaria, 367
Chemotherapy, antimalarial. *See* Antimalarial chemotherapy
Chlorcycloquanil, and dapsone, combination of, 435, 436
Chloroplasts, 280–281
Chloroquine, 321, 480
   action of, 438–439
   and pH-dependent physiological changes, 441
   efflux, increased, 440–441
   in malaria infection, 380
   reduced access to hematin, 440
   resistance to, 429
      models of, 440–441
Cholesterol, in *Plasmodium*-infected erythrocytes, 329
Choline, transport of, in acquisition of lipid precursors, 342–343
Choline analogs, quaternary ammonium as, in antimalarial chemotherapy, 346
Chromalveolata, 276
Ciprofloxacin, 277
Circular clefts (TVN), CP3C, 34, 35
Circumsporozoite protein, recombinant, 169
   as major surface protein, 176
Clindamycin, 279
Cluster analysis, of *P. falciparum* transcriptome, versus FFT analysis, CP5A, CP5B, 73–74
Complement receptor 1, *CR1* encoding, 468
CR1 gene, encoding complement receptor 1, 468
Cryptosporidium, 276–277
CS-based preerythrocytic vaccines, 487–488
CSP protein, vaccine development and, 97
Cyclic nucleotide-dependent pathways, 304
   gametogenesis and, 211
Cycloguanil, 431, 432, 433–434
   resistance to, 437
Cysteine protease inhibitors, 321–322

Cytochalasins, 142–143
Cytochrome bc_1 complex, 243–246
Cytochrome c, transfer of electrons from ubiquinol to, 243–244
Cytochrome c oxidase subunit, 238
Cytoplasmic phosphoprotein Pfg27, 193
    structure of, 194, 195

DAG, synthesis of, 331–333
Dapsone, chlorcycloquanil and, combination of, 435, 436
Data mining, and queries, gene pages and, 15–18
Databases, relational, storage in, 18
DBL/EBL adhesin family, 130, 131, 132
DBL-EBL adhesive invasion ligands, and erythrocyte receptors, 129–133
1-Deoxy-D-xylulose 5-phosphate reductoisomerase, 282
Diacylglycerol, synthesis of, 331–333
Dichlorodiphenyltrichloroethane (DDT), 480
    resistance in *Anopheles gambiae*, 512
Dietary oxidants, in malaria parasites, 367
Dihydrofolate reductase, 430, 431, 436
    and chemical reaction catalyzed by, 431, 432
    and dihydropteroate reductase combinations, resistance to, 435
    chemical reaction catalyzed by, 431, 432
Dihydrofolate reductase-TS, structures of, and modes of drug binding, CP17, 436–437
Dihydroorotate dehydrogenase, 236, 239
Dihydropteroate reductase, 430, 431
    and dihydrofolate reductase combinations, resistance to, 435
Dionisi, Antonio, 6
mtDNA, replication of, rolling circle mode of, 242–243
    transcription, translation, and replication of, 242–243
DNA/MVA preerythrocytic vaccine, 488, 489
DoTS namespace, 18
Doxycycline, 246
Drosophilia melanogaster, 517, 521, 522
Drugs, action of, redox equilibrium and, 375–381
    actions of, and resistance to, mechanisms of, 429–461
    and ribosomes, interaction of, CP13, CP14, 356–357
    resistance to, determinants of, 452–453
Duffy antigen-chemokine receptor, *FY* gene encoding of, 464–465
Duffy blood group antigen, 463, 464

Electron microscopy, 24, 25
Electron transport chain, generation of proton motive force, 236–238
Electrophysiological studies, to study new permeation pathways, 386–388
Entamoeba histolytica, 234

Environmental factors, malaria parasite multiplication and, 292–293
Enzyme mutations, parasite antifolate resistance and, 434
Enzymes, lipid biosynthetic, cellular localization of, 343–345
    phospholipid and neutral, of *P. falciparum*, 333
Erythrocyte(s). *See also* Red blood cells
    composition and structure of, *Plasmodium*-mediated changes in, 329
    cytoplasm of, parasite histidine-rich protein II and tubulovesicular structures in, CP10, 260
    export of parasite proteins into, 258–261
    infection of, 253–254
    initial adhesive contact between merozoite and, 123–128
    malaria parasite invasion of, 464
    mature, description of, 253
    membrane composition of, lipids and, 343–344
    merozoite entry into, and formation of parasitophorous vacuole, 140–151
    merozoite invasion of, alternate pathways for, 137–140
    parasite proteases and, 145–147
    merozoite release and attachment to, 121–128
    P. falciparum-infected, glycolytic reactions in, 223, 224
    Plasmodium-infected, cholesterol content of, 329
    lipid composition of, 328–329
    rupture of, and merozoite release, 121–123
    signaling in, upon infection with malaria, 256, 257
    surface molecules of, parasitic invasion and, 464–465
Erythrocyte band 3 protein, 465
Erythrocyte detergent-resistant membrane rafts, CP9, 254–257
    and G protein signaling, in malarial infection, 254–258
Erythrocyte invasion receptors, 128–129
Erythrocyte lipid raft proteins, 255, 256
Erythrocyte membrane, closure of, parasitophorous vacuole and, 150–151
Erythrocyte membrane glycophorins, 129
Erythrocyte raft proteins, 255, 258
    uptake of, tubulovesicular membrane network in, 258–261
Erythrocyte receptors, DBL-EBL adhesive invasion ligands and, 129–133
Erythrocyte-related proteins, susceptibility to malaria and, 467
Erythrocytic life, main phases of, 25
Erythrocytic schizogony, CDK- and cyclin-related protein appearance during, 294, 295, 296
Erythroreceptors, uninfected, for rosetting, 420–422
Eukaryotic cell cycle, 290
    phases of, 290
Exflagellation, mosquito species-specific inhibitors of, 209

Expressed sequence tag analyses (EST), genomics and, 519–524
   mosquito transcriptome information and, 520
Expression profiling, biological relevance of, 74

Falcilysin, 318
Falcipain cysteine proteases, 317–318
Fansidar. *See* Sulfadoxine-pyrimethamine
FASII pathway, antimalarial chemotherapy and, 347
   fatty acids and, 341
Fast Fourier-transform analysis, of *Plasmodium falciparum* parasite transcriptome, versus cluster analysis, CP5A, CP5B, 73–74
Fatty acids, biosynthesis of, 341
   phospholipids and, 340–342
   synthesis of, apicoplast and, 281–282
Fava beans, malaria parasites and, 367
FCGR2A, encoding low-affinity IIa receptor for Fc fragment of IgG, 470
Fluorescence-activated cell sorter, 280
FMP1/AS02 vaccine, 490
Folate cofactors, de novo synthesis and salvage of, 431, 433
Food vacuole, lipids of, and hemozoin formation, 345
Food vacuole proteases, cooperative activity of, 319
Fowlpox strain 9-MVA preerythrocytic vaccine, 488, 489
Fulani of Burkina Faso, 463
FY gene, encoding of Duffy antigen-chemokine receptor by, 464–465

G protein-coupled receptor signaling, *Anopheles gambiae* and, 509–510
G protein signaling, erythrocyte detergent-resistant membrane rafts and, in malarial infection, 254–258
Gametes, *Anopheles* mosquito and, 191
   fertilization of, 212
   gametocytes and, 191–219
      genes of *Plasmodium falciparum* and *Plasmodium berghei* in, 200–202
   male, development of, 205–206
Gametocyte(s), CP3E, CP3F, 41–44
   and gametes, 191–219
      genes of *Plasmodium falciparum* and *Plasmodium berghei* in, 200–202
   and structures exported into red blood cells, 40–41, 43
   distribution of, in peripheral circulation, 198–199
   emergence of, 205
   formation of, 193–197
   interactions with vertebrate host, 197–199
   origin of, 191
   osmiophilic bodies of, 40–41, 43
   pellicle of, CP1, 40–41, 42–43
   Plasmodium falciparum, cytoadhesive properties of, 197, 198
   production of, modulation of, 192–193
   sequestration of, 197–198
   sex ratio of, modulation of, 193
   specific organelles, 42
   stages of formation of, 40–41, 43–44
Gametocytogenesis, 193–199
   gene expression in, 199
Gametogenesis, 199–207
   and cell cycle, 207–212
   body temperature and, 208
   cell biology of, 203
   events in, and timing of, 203
   extracellular pH and, 208
   in *Plasmodium berghei*, 204–205
   male, malaria parasite multiplication and, 292–293
   regulation of, by extracellular triggers, 207–209
   signaling pathways in, 209–212
   XA and, 208–209
Gamma interferon receptor, alpha chain of, *IFNGR1* encoding, 471
Garnham, P. C. C., 7
Gene mapping, in inbred mouse strains' resistance to infection, 471–473
Gene pages, CP2, 13–14, 15
   data mining and queries and, 15–18
Gene queries, history function to integrate, 15, 16
Gene regulation, posttranslational, 206–207
GeneChip Plasmodium/Anopheles Genome Array, 524
GeneDB database, 20
Genes, and GO terms, 18
   human, determining resistance or susceptibility to malaria, 464–468
Genetic factors, of host, resistance and susceptibility to malaria and, 462–479
Genetic studies, in mice, 462–463
Genetics, human, rosetting and, 420
Genome project, novel sequences from, 227–228
Genome research, and expressed sequence tag analyses, 519–524
   and parasite evolution, 106
Giardia lamblia, 234
Gliding-associated proteins, 144
Glucose-6-phosphate dehydrogenase, *G6PD* encoding, 467
GLURP blood-stage antigen, 490
Glutaredoxins, in *P. falciparum*, 372
Glutathione, biosynthesis of, in *Plasmodium*, 372–374
   in malaria parasites, 368–369
   metabolism of, transcriptome analyses of, 375
Glutathione-mediated detoxification, 441
Glutathione peroxidase, 369–370
Glutathione reductase, in malaria parasites, 370–371
Glutathione S-transferase, 370
   in *Plasmodium*, 374
Glutathione system, 365
Glycerolipids, biosynthetic pathways of, 329–339
   metabolism of, steps of, 329–334
Glycolysis, functional genomic analysis of, 225–227

in asexual-stage parasites, 223–271
  inhibitor of, CP8, 228–230
Glycolytic enzymes, in *Plasmodium falciparum*, 225–227
Glycophorin C, *GYPC* encoding, 465
Glycophorins A and B, *GYPA* and *GYPB* encoding, 465
Glyoxalases, in *Plasmodium*, 374
GO terms, genes and, 18
Golgi, Camillo, 3–4
G6PD gene, encoding glucose-6-phosphate dehydrogenase, 467
GPROR1, 510–511
GPROR2, 511
GPROR7, 511
Grassi, Giovanni Battista, 6
GSK3-related proteins, 304
GUS, namespaces in, 18
GYPA and *GYPB*, encoding glycophorins A and B, 465
GYPC, encoding glycophorin C, 465

Haldane, J. B. S., 367
Haptoglobin, *HP* encoding, 467
HBA encoding α-globin, 465–466
HBB encoding β-globin, 465–466
HbC, 465
HbE, 465
HbS, 464–465
Heme, parasite-synthesized, iron and, fate of, 320
Heme biosynthesis, 239–240
Heme synthesis, apicoplast and, 283
Hemoglobin, degradation of, fate of products of, 319–321
  in malaria parasites, 366–367
  mechanism of, 313–314
  proteases and, 311–326
  hydrolysis of, 314
    antimalarial therapies inhibiting, 321–322
    explanations for, 312–313
    parasite amino acid supply and, 311–312
    proteases responsible for, 314–319
    sources of amino acids and, 312
  separation of components of, 313–314
  uptake of, 313
Hemoglobin C, 466
Hemoglobin E, 466
Hemoglobin genes, malaria susceptibility and, 465–467
Hemoglobin S, 465–466
Hemosporidia, placement with phylum Apicomplexa, 100–101
Hemozoin, formation of, 320–321
  lipids of food vacuole and, 345
Histoaspartic protease, 316
HLA-B, encoding MHC class IB, 468–469
HLA-DRB1, encoding MHC class II beta chain, 469
Host-targeting signal, 265–266, 267

HP gene, encoding haptoglobin, 467
Hydroxynaphthoquinones, 244

ICAM1 gene, encoding intercellular adhesion molecule 1, 467
ICC-1132 preerythrocytic vaccine, 487–488
IFNAR1 gene, encoding alpha interferon receptor 1, 471
IFNGR1 gene, encoding alpha chain of gamma interferon receptor, 471
IgG, low-affinity IIa receptor for Fc fragment of, *FCGR2A* encoding, 470
IL4, encoding interleukin-4, 469
IL1A and *IL1B*, encoding interleukin-1α and -β, 471
IL12β gene, encoding interleukin-12 p40, 471
Immune responses, rosette-inhibiting, in natural infections, 422
Immunity, acquired, mechanisms of, 468–470
  innate, mediators of, 470–471
Immunoglobulins, in rosetting, CP16, 421–422
Infection(s), drug-resistant, causes of, 430
  natural, rosette-inhibiting immune responses in, 422
Infrakingdom Alveolata, relationship to other infrakingdoms, 100–101
Intercellular adhesion molecule 1, *ICAM1* gene encoding, 467
Interleukin-4, *IL4* encoding, 469
Interleukin-12 p40, *IL12β* gene encoding, 471
Interleukin-1α and -β, *IL1A* and *IL1β* encoding, 471
Intermittent prevention treatment in infants (IPTi), 481
Intraerythrocytic loops, 261
Ionic exchange mechanisms, gametogenesis and, 211
Iron, and parasite-synthesized heme, fate of, 320
Iron-sulfur cluster assembly, apicoplast and, 283–284
Iron-sulfur cluster synthesis, 240–241
Isopentenyl diphosphate synthesis, apicoplast and, 282–283

Jensen, James B., 9

Knobs, CP3C, 34, 35–37
Koch, Robert, 6
Kupffer cell, sporozoite transmigration through, 182–183

Laveran, Alphonse, 3, 4, 6, 7, 199
Light microscopy, 24–25
Lipid biosynthetic enzymes, cellular localization of, and lipid trafficking, 343–345
Lipid precursors, mechanisms of acquisition of, 340–343
Lipids, biogenesis of, in *Plasmodium*, 327–328
  in *Plasmodium*-infected erythrocytes, 328–329
  metabolism of, in *Plasmodium falciparum* parasite, 78–79

Lipids, biogenesis of, in *Plasmodium* (*continued*)
   neutral, and phospholipids, pathways for synthesis of, 331, 332
   of food vacuole, and hemozoin formation, 345
   Plasmodium, metabolism of, analysis of, 328
      and function of, 327–352
   structural organization of, in host cell membrane, 343–344
   trafficking of, lipid biosynthetic enzymes and, 343–345
Liver, migration of sporozoite from skin to, 182–184
Lung injury, acute, in malaria infection, 367
Lysophosphatidic acid, synthesis of, 330

MacCallum, William, 4
Macrogametocytes, 206–207
Major histocompatibility complex class IB, *HLA-B* encoding, 468–469
Major histocompatibility complex class II beta chain, *HLA-DRB1* encoding, 469
Malaria, acute lung injury in, 367
   cerebral, 367
   chemotherapy in, lipid-based, 345–347
   control of, nonvaccine, 481
      vaccine for, development of, 480–497
   development of merozoites and, 113–114
   human, population dynamics of, 102–106
      relationship to model systems and other parasites, 100–102
   in New World, humans and, 100
   infection with, erythrocyte detergent-resistant membrane rafts and G protein signaling in, 254–258
      redox equilibrium and drug action in, 375–381
      redox metabolism and pathophysiology in, 367–368
   molecular approaches to, 223–271
   mortality related to, 480
   murine, studies in, 462–463
   research in, specimen preparation in, 25
   resistance to, and susceptibility to, host genetic factors in, 462–479
      within families, 463
   severe, rosetting and, 416–417, 418, 419
Malaria parasite(s), antioxidant defense in, 368–374
   oxidative stress and, 365–383
   asexual stage, glycolysis in, 223–271
   biology of, history of study of, 24
   cell cycle control proteins and, 293–295
   cell cycle organization in, cyclin-dependent kinases and, 292
   dietary oxidants in, 367
   evolution of, genome research and, 106
   evolutionary history and population genetics of, 95–109
   hemoglobin degradation in, 366–367
   –host interactions, blood-stage infections in, 267, 268
   human, in broad evolutionary context, 96–102
      relationships of, understanding of, 97–98
   in mammals, phylogenetic tree showing, 98
   in mosquito, 5
   in red blood cells, 384
   intracellular, nutrients for, 385
   invasion of erythrocytes by, 464
   life cycle of, intraerythrocytic stage of, 384–385
   membrane raft and, 344–345
   mitochondrial processes in, 237
   mitochondrion in, 235
   molecular evolution and, 95–96
   molecular phylogenetics of, history of, 96
   multiplication of, control of, 292–293
   of primates, 100
   oxidative stress in, sources of, 366–367
   phagocytosis in, 366
   protease repertoire of, 314–316
   rosetting by, reason for, 422
   TCA cycle in, 238–239
Malaria Parasite Metabolic Pathway database, 13
Malaria vectors, transcriptome of, 516–530
   tools for functional analyses of, 524–525
Malarone, 245
Malate-quinone oxidoreductase, 238
MAL6P1.100, 13
Maltaselike 1, of *A. aegypti*, 517
Mannose-binding lectin, *MBL2* encoding, 471
Manson, Patrick, 6
MAPK pathways, 303–304
Mass spectrometry, in high-throughput proteomics, 85–87
Maurer's clefts, CP3C, 34, 35, 90, 261–262
   in erythrocyte, CP10, 260–261
MBL2, encoding mannose-binding lectin, 471
Mefloquine, 430, 438
   and quinine action, 439
   resistance to, 453
      molecular correlates of *pfmdr1* and, CP18, 450–452
   treatment with, patients free from malaria after, 450–451
Membrane antigen/erthrocyte-binding-like protein, apical, sporozoites and, 177
Merozoite(s), alternate pathways for invasion of erythrocytes, 137–140
   and red blood cells, initial adhesive contact between, 123–128
   apical organelle secretion by, 147–149
   apicoplast in, CP3A, 26–27, 28–29, 34
   arrangement of merozoite surface proteins on, 126–127
   biogenesis of, 117–119
   cellular structure of, 115–117
   dense granules in, CP3A, 26–27, 27–28

entry into erythrocyte, erythrocyte membrane closure following, 150–151
erythrocyte entry by, and formation of parasitophorous vacuole, 140–151
formation of, 113
general morphology of, 25
invasion cascade, signaling of, 149–150
invasion of erythrocytes by, parasite proteases and, 145–147
micronemes in, 27
mitochondrion in, CP3A, 26–27, 28
moving junction of, 143–145
nucleus of, CP3A, 25, 26–27
phases of invasion by, CP7, 119–121
plasma membrane and organellar membranes of, 115–117
red blood cells and, 117–118
release of, and attachment to erythrocyte, 121–128
erythrocyte rupture and, 121–123
reorientation of, and irreversible attachment of, 128–136
rhoptries in, 25–27
secretory organelles of, 116
structural features of, 114
structures and protein expression of, development of, 119
subpellicular microtubules in, 28
surrounding *Plasmodium knowlesi* schizont, CP6A, 118
Merozoite actomyosin motor, and junctional movement, CP7, 142–143
Merozoite-capping protein 1, 144, 180
Merozoite surface proteins, 124–128
Mice, genetic studies in, 462–463
inbred strains of, resistance to infection among, gene mapping and, 471–473
Microarray technology, gene expression and, 523–524
mosquito-parasite interactions and, 524
Microgametes, 206
Microgametocytes, replication of, 206
Microneme proteins, 129–133
Microtubule organizing center, 170
Mitochondrial genome, 241–243
Mitochondrial membrane potential, 241
Mitochondrial protein synthesis, and antibiotics, 246
Mitochondrial targets, for antimalarial drugs, 243–247
Mitochondrion, 234–252
from *Plasmodium*, biochemical studies of, 235–236
in malaria parasites, 235
metabolic functions of, 235–241
processes of, in malaria parasites, 237
Mosquito, genetically determined behaviors of, 516–517
Mosquito biology research, focus of, 516
in pregenomics era, 517–519
Mosquito immunity, sporozoites and, 181

Mosquito species-specific inhibitors of exflagellation, 209
mRNA abundance profiles, *Plasmodium* proteome and, 91–93
MSP3 blood stage antigen, 490
MSP1 vaccine, 490
Multistage vaccines, 491
Myosin-A, 144

NADH dehydrogenases, 236
NADPH production, in *Plasmodium*, 368
Nitric oxide synthase, inducible, *NOS2A* encoding, 470–471
Nonmicronemal surface proteins, 144
Nonvaccine malaria control, 481
NOS2A, encoding inducible nitric oxide synthase, 470–471

Odorant-binding proteins, in *Anopheles gambiae* genome, 511
Oligonucleotide array, high-density, of *Plasmodium falciparum* parasite, CP4, 72–73
Oocyst, escape from, and traversal of hemococele, 174–175
malarial, cell cycle of, 172
maturation of, 172, 174
sporozoites in, 174
Opie, Eugene L., 4
Oscillaria malaria, 3
Oxidative stress, and antioxidant defense, in malaria parasites, 365–383

Parasite proteins, exported, secretome of virulence determinants and, 261–267
Parasites, malaria. *See* Malaria parasites
Parasitophorous vacuolar membrane, 254
composition of, 147–148
formation of, recruitment of parasite and DRM rafts during, 257, 258
host cell origin of, 148
Parasitophorous vacuole, and erythrocyte membrane closure, 150–151
composition and formation of, 147–149
formation of, erythrocyte entry by merozoite and, 140–151
Patch clamp, to study new permeation pathways, 384, 387
PECAM1, encoding platelet-endothelial cell adhesion molecule, 468
Peptide-bond formation, ribosomal subunits and, CP13, 356
Permeation pathways, new, 384–396
as continuing challenge, 393
as therapeutic target, 392–393
endogenous channels as candidates to represent, 388–389

Permeation pathways, new (*continued*)
    functional and pharmacological characterization of, 386
    inhibitors of, 384
Peroxidases, in *P. falciparum*, 369–370
    transcriptome analyses of, 375
Peroxiredoxins, 369
Pfcrk-1, 299–300
Pfcrk-3, 297
Pfcrk-4, 300–301
Pfcrk-5, 300
PfCRT, charged drug leak model, 446–447
    functions of, predictions of, 445–446
pfcrt, allelic variants, 443
    and drug resistance, 452–453
    association with chloroquine resistance, 444–445, 452
    haplotypes, mutant, evolution of, 442–444
    identification of, 441–442
    mutant, confers chloroquine resistance, 445
    predicted structure of, 442
Pfcyc-1, 297–298, 305
Pfcyc-2, 300–301, 305
Pfcyc-3, 300–301
Pfcyc-4, 300–301
PFD0830w (DHFR-TS), CP2, 19
PfEMP1, 264, 265, 399
    adhesion molecules as receptors for, 467
    and *var* in postgenomic era, 400–403
    domain structure, and associated cytoadhesive phenotypes, 403
    expression of, developmental regulation of, CP15, 404
    gene family encoding, 400
    variant, destined for IE surface in mid-ring-stage parasites, CP15, 404
        expression during chronic infection, 408–410
PFF0480w, 13
pfmdr1, amplification of, 448
    and drug resistance, 452–453
    identification and general features of, 447
    molecular correlates of, 449
        and mefloquine resistance, CP18, 450–452
    point mutations in, and drug sensitivities, 447–448
Pfmrk, 297, 306
PfPK5, 299, 305, 306
PfPK6, 298–299
PfPmt, 337
Pfstomatin, 257
Phosphatidate cytidylyltransferase, in CDP-DAG synthesis, 331
Phosphatidic acid, synthesis of, 330–331
Phosphatidylcholine, synthesis of, choline in, 342
Phosphatidylcholine biosynthesis, pathway for, 336–337
Phosphatidylethanolamine, synthesis of, 338

Phosphatidylserine, and lipid metabolism of serine, 337–339
    decarboxylation of, 338–339
    synthesis of, 337–338
Phosphocholine analogs, in antimalarial chemotherapy, 346
Phosphoethanolamine transmethylation, 336–337
Phosphoinositide pathways, calcium signaling and, 304
Phospholipids, and neutral lipids, pathways for synthesis of, 331, 332
Plasmepsin aspartic proteases, 316
Plasmid genes, wandering, 277
PlasmoDB database, 9, 12–23
    analysis tools, 14
    future plans for, 20–21
    gene queries using, 15, 16
    new data and data types in, 18–20
    searched with BLAST, 225–227
Plasmodium, apicoplast and, 284–285
    asexual multiplication of, in bloodstream, 191–192
    from biochemistry to genomics, 8–10
    life of, 3–11
    molecular evolution and, 96–97
    origin of parasitism in, phylum Apicomplexa and, 102
    orthologs of phosphate transporters of, 284
    placed with family Plasmodiidae, 100–101
    profiling studies in, 87–88
    proliferation and development of, protein kinases and, 290–310
    proteases of. *See* Proteases
    proteome of, 85–94
        mRNA abundance profiles and, 91–93
    ribosomes of, and opportunities for drug intervention, 353–364
    vulnerable sites in, 357–361
Plasmodium berghei, 8, 12, 27, 236
    gametocytes, maturation of, 196–197
    gametogenesis in, 204–205
    genes in gametocytes and gametes of, 200–202
    rhoptry-enriched proteins of, 89–90
    transfection in, 52
Plasmodium berghei ANKA, infection in inbred mouse strains, cerebral pathology of, 472
    chromosomal loci in, 473
    development of model of, 472
    interventions inhibiting, 473
Plasmodium chabaudi, 8, 12
    resistance of inbred mice to, chromosomal loci mapping and, 472
    mapping of, 471–472
Plasmodium cynomolgi, 7
Plasmodium falciparum, CP1, 3, 4, 7, 8, 9, 10, 24, 235, 236, 242, 253, 254
    allelic replacement in, gene targeting for, 59–61
    mutations in drug resistance and, 60–61

antigenic variation in, 399–415
blood stages of, 24–25
CDK-related kinases and cyclins of, 305–306
clonal reproduction of, 103
commitment to sexual differentiation, 192
drug resistance in, 480
emergence of, 98–99
fluorescent transgenes in, expression for trafficking analysis, 55–56
functional analysis of, transfection in, 53–61
gametocyte maturation, late phase of, 196
  middle phase of, 194–196
gametocytes, cytoadhesive properties of, 197, 198
gametocytogenesis, developmental steps in, 194
  early events in, 193–194
gene disruption events in, selection of, 58–59
gene sequences in, targeting for disruption and allelic replacement, 56–59
genes in gametocytes and gametes of, 200–202
genetic manipulation of, 50–67
  background and history of, 50–52
  efficiency of, 52–53
  episomal replication in, 53
  future outlook for, 62
  methodology and technical features of, 52–53
glycolytic enzymes in, 225–227
invasion mechanisms of, studies of, 139
invasion of red blood cells by, 121–122
life cycle of, 354
  rRNA expression and, 354–355
merozoite invasion and, 145
nine populations of, relationships between, 103, 104
parasite lines, lacking expression of specific merozoite proteins, 57–58
PlasmoDB and, 19, 20
population history of, 103–105
population structure of, and genome diversity, 102–103
promoter elements in, analysis by transient transfection, 53–54
promoter function in, and temporal expression of proteins, 54
redox metabolism, enzymes in, 376–380
resistance to antifolate drugs, 430–431
stable transgene expression in, 54–56
transfection in, biological processes investigated by, 51
  positive selectable markers for, 53
transgene expression in, polymorphisms in immune evasion and, 54–55
  protein function by, 54–55
two-dimensional structure of, 357
Plasmodium falciparum erythrocyte membrane protein 1. *See* PfEMP1
Plasmodium falciparum fructose 1,6-biphosphate aldolase, CP8
Plasmodium falciparum genome, properties of, 88
Plasmodium falciparum genome sequence, 12
Plasmodium falciparum parasite, biological relevance of expression profiling of, 74
  chromatin structure of, and gene regulation, 76
  complex life cycle of, 68
  efforts to increase understanding of, 69
  expression analysis of, before genome sequence completion, 72
  gene expression in, high-throughput techniques and, 70
  metabolic pathway of, analysis of, 74–75
  microarray technologies and full-genome coverage approach to, CP4, 72–73
  open reading frames and regulatory elements of, 69
  posttranscriptional control in, 76–77
  SAGE and antisense detection of, 71–72
  transcriptional change and stress, 77–78
  transcriptional control in, 75–76
  transcriptome of, 68–84
    and vaccine development, 79
    cluster analysis of, versus FFT analysis, CP5A, CP5B, 73–74
    to identify new drug targets, 78–79
Plasmodium falciparum rosettes, 416, 417
Plasmodium gallinaceum, 8, 12
Plasmodium genes, information about, PlasmoDB and, 15
Plasmodium genome, promoter regions of, 75
  transcription factors and, 75
Plasmodium Genome Database. *See* PlasmoDB
Plasmodium genome resource. *See* PlasmoDB
Plasmodium knowlesi, 12
  merozoite invasion sequence, CP7, 119–121
Plasmodium knowlesi schizont, merozoites surrounding, CP6A, 118
Plasmodium lophurae, 19, 122
Plasmodium malariae, 7, 102
Plasmodium ovale, 7, 102
Plasmodium parasites, conditional mutagenesis in, 61–62
Plasmodium post-genomics, making home for, 24–49
Plasmodium reichenowi, 12, 133, 135
Plasmodium vivax, 7, 12, 102
  Asian origin of, 99–100
  invasion of red blood cells by, 121
  life cycle of, 354
  populations of, studies at genome level, 105–106
  source of, 99
  triple membranes and surface coat of, 117
Plasmodium vivax merozoite proteins, 133–134
Plasmodium yoelii, 12, 19, 235, 236
Platelet-endothelial cell adhesion molecule, *PECAM1* encoding, 468
Preerythrocytic vaccines, 486–489
Primaquine, in malaria infection, 380
Profiling studies, in *Plasmodium*, 87–88
Proquanil, 245, 435
Proteases, and hemoglobin degradation, 311–326

Proteases, and hemoglobin degradation (*continued*)
  falcipain cysteine, 317–318
  food vacuole, cooperative activity of, 319
  in hemoglobin hydrolysis, 314–319
  in *Plasmodium falciparum* genome, 78
  parasite, merozoite invasion of erythrocytes and, 145–147
  plasmepsin aspartic, 316
Protein(s), cyclin-dependent kinases and, 290–295
  parasite, exported, secretome of virulence determinants and, 261–267
  pathways of export to erythrocyte, 265–267
  ribosomal, synthesis of, 356
  Thioredoxin-like, in *P. falciparum*, 371–372
Protein C, *MBL2* encoding, 471
Protein kinases, and proliferation and development of *Plasmodium*, 290–310
  cyclin-dependent, and associated proteins, 290–295
    and cell cycle organization in malaria parasites, 292
    appearance during erythrocytic schizogony, 294, 295, 296
    potential substrates of, 301–303
    upstream signaling pathways and, 291–292
    with mRNA expressed prior to DNA synthesis, 294, 295
    with mRNA not showing regulation during schizogony, 301
    with mRNA peaking during S phase and nuclear division, 298
    with mRNA peaking in late schizonts and segmenters, 301
    with potential upstream regulators and substrates, 302, 303
  plasmodial, and associated proteins, 294, 295, 296
Protein product *P. falciparum* hexose transporter, 228, 230
Protein synthesis, mitochondrial, and antibiotics, 246
Proteolysis, and merozoite invasion of erythrocytes, 145–147
Proteomics, fundamentals of, 85–87
  high-throughput, mass spectrometry in, 85–87
  subcellular, 88–91
Proton motive force, electron transport chain generation of, 236–238
Pseudoknots, of RNA, 361–362
PVM-enclosed merozoite structures, 121–122
Pyrimethamine, 431, 432, 433–434
  and sulfadoxine, 435
Pyrimidine biosynthesis, 239
Pyruvate kinase, 227–228

Quinine, 430, 431, 438
  resistance to, 453
Quinoline(s), modes of action of, and mechanisms of resistance, 437–453
  resistance to, epidemiological contexts of, 439–440
Quinolones, 247

RAD namespace, 18
RBL (RBP/NBP) adhesin family, 130, 131, 132
Red blood cells. *See also* Erythrocytes
  gametocyte structures exported into, 40–41, 43
  growth of *Plasmodium* parasite in, 327
  infected, anion channels in, 389–391
  malaria parasite in, 384
  merozoite biogenesis and, 117–118
  merozoite invasion of, 113–168
    phases of, CP7, 119–121
  parasite-infected, 235
  to study membrane transport pathways, 388
  trophozoite structures exported into, CP3C, 33–37
Repetitive interspersed family, 262
Reticulocyte-binding protein-like family, 131, 133–136
Rhoptry-associated membrane antigen, 142
Rhoptry-enriched proteins, subtractive analysis of, 89–90
Rhoptry protein complex, high-molecular-mass, 141–142
Rhoptry proteins, 141–142
  parasitophorous vacuolar membrane and, 148–149
Ribosomal genes, 353–354
Ribosomal proteins, synthesis of, 356
Ribosomes, and drugs, interaction of, CP13, CP14, 356–357
  and regulation of, 355–356
  composition of, 353
  organellar, 362
  peptide-bond formation and, CP13, 356
  Plasmodium, and opportunities for drug intervention, 353–364
  small and large subunits of, interaction between, CP12, 356
Ring-stage parasite, general morphology of, 29
  Golgi complex of, 30, 31
  mitochondrion and apicoplast of, 31
  nucleus, ribosomes, and RER of, 29
  PVM of, 31
  vacuoles and vesicles of, CP3B, 29–31
RNA, pseudoknots of, 361–362
RNA interference (RNAi), 525
Romanovsky, Dimitri, 4
Rosetting, 416–426
  and cytoadherence, 418, 420
  and formation of rosettes in vivo, 418–420
  and severe malaria, 416–417, 418, 419
  by parasites, reason for, 422
  human genetics and, 420
  immunoglobulins in, CP16, 421–422
  molecular mechanisms of, 420–421
  pathological role of, 419–420
  red blood cell rosetting receptors and, 420–421
  serum proteins in, CP16, 421–422
  significance of, in severe malaria, 417–418
Rosetting ligand, PfEMP1 family, 421
Ross, Ronald, 4–7

rRNA, expression of, correlation with *Plasmodium* life cycle, 354–355
in ribosomes, 353
16S rRNA, antibiotics interacting with, 358
23S rRNA, antibiotics interacting with, 359–361
RTS,S/AS02 preerythrocytic vaccine, 487–488, 490

SAGE, and antisense detection, of *Plasmodium falciparum* parasite, 71–72
Salivary glands, as barrier to sporozoite development, 175–177
   sporozoite invasion of, 176, 179
      mosquito physiology and, 180–181
      outcomes of, 180
   sporozoite maturation in, 179–181
Schaudinn, Fritz, 7
Schizont, general morphology of, CP3D, CP3E, 36–37, 38
   merozoite assembly in, CP3D, 36–37, 38, 39–41
   merozoite exit in, CP3E, 36–37, 41
   merozoite maturation in, 38, 39
   mitosis in, 38
   spindle pole body and merozoite assembly in, CP3D, 37–38, 39
   stages of development of, 37–38
   SzS 1 to 4, nuclear division of, 38–39
SDPM pathway, 336–337
Secretome, of virulence determinants, and exported parasite proteins, 261–267
   predicted, 262, 264
Serial analysis of gene expression (SAGE), 523
Serine, lipid metabolism of, phosphatidylserine and, 337–339
Serine decarboxylation, 336–337
Serine repeat antigen family of proteins, 122–123
Serum proteins, in rosetting, CP16, 421–422
Sexual-stage vaccines, 490–491
Sheddase enzyme, merozoite invasion and, 146
Shortt, H. E., 7
Sickle cell disease, 465
SLC4A1 anion exchanger, 465
SPf66 vaccine, 490
Sphingolipid inhibitors, antimalarial chemotherapy and, 346–347
Sphingolipid metabolism, 339–340
Sphingomyelin synthase, 259, 339–340
Sphingomyelinase, 340
Sporozoite(s), 169–190
   and mosquito immunity, 181
   biology of, history of research on, 169
      microarray methods for study of, 169
      proteomic analyses for study of, 169–170
      targeted gene disruptions for study of, 169
      transgenic technologies for study of, 169
   delivery to vertebrate host, 181–182
   development of, salivary glands as barrier to, 175–177
   formation of, 170–174

in oocyst, 174
invasion of salivary glands by, 176, 179, 180–181
invasive stages of, 170
irradiated, inoculation with, 481
maturation of, in salivary glands, 179–181
migration from skin to liver, 182–184
migration through Kupffer cell, 182–183
mobility of, and host cell invasion, 177–179
molecular motor, role in host cell invasion, 177–178
Staurosporine, merozoite invasion and, 149–150
Stress, malaria host-parasite interaction and, 77–78
   oxidative, and antioxidant defense in malaria parasites, 365–383
Subtelomeric variable open reading frame, 262
Subtilisin-like serine protease, 146
Succinate dehydrogenase, 236
Sulfadoxine, pyrimethamine and, 435
Sulfadoxine-pyrimethamine, 435, 481
   resistance to, 429
Superoxidase dismutase, in malaria parasites, 370
Superoxide dismutases, transcriptome analyses of, 375

TAG, biogenesis of, 334
   synthesis of, 333–334
TCA cycle, in malaria parasites, 238–239
Tetracyclines, as antimalarials, 246
Thalassemias, 466–467
Thioredoxin-like proteins, in *P. falciparum*, 371–372
Thioredoxin reductase, in malaria parasites, 371
Thiostrepton, 279
Thrombospondin-related anonymous protein, sporozoites and, 176
Thucydides, 482
Tight junction formation, irreversible attachment of merozoite through, 128–136
   merozoite entry of erythrocyte membrane at, 140–141
TNF alleles, encoding tumor necrosis factor, 470
TNFSF5, encoding CD40 ligand, 469
Topoisomerase, 247
Toxoplasma gondii, 272, 273, 274, 279
   host-targeted proteins of, 262
Trager, William, 9, 253
Trager-Jensen method, 9
Transcriptome analyses, 374–375
   of glutathione metabolism, 375
   of peroxidases, 375
   of superoxide dismutases, 375
TranslatedAASequence, 18
Triacylglycerol, biogenesis of, 334
   synthesis of, 333–334
Trophozoite, cytostomes of, CP3C, 32–33
   general morphology of, 32
   Golgi complex of, 32, 34
   mitochondrion and apicoplast of, 32, 34
   nucleus and ribosomes of, 32, 33

Trophozoite, cytostomes of (*continued*)
  structures exported by parasite into red blood cells and, CP3C, 33–37
  ultrastructure of, 33, 34
Trophozoite-to-schizont transitions, mRNA and protein fold changes during, 92–93
Tubulovesicular membrane network, in uptake of erythrocyte raft proteins, 258–261
Tubulovesicular membrane network, trafficking and, 253–271
Tumor necrosis factor, *TNF* alleles encoding, 470

Ubiquinol, transfer of electrons from to cytochrome *c*, 243–244
Ubiquinone, 247
  synthesis of, 241

Vaccine(s), antidisease, rosette-inhibiting, prospects for, 422–423
  blood-stage, 489–490
  commercialization of, 486
  development of, 480–497
    choice of adjuvant in, 483–484
    choice of platform for, 483
    clinical, regulatory requirements for, 485–486
    critical steps in, 483–484
    CSP protein and, 97
    design of constructs for, 483
    managing expectations and, 491–492
    transcriptome of *Plasmodium falciparum* parasite and, 79
  feasibility of, 481–482
  introduction of, 486
  preclinical evaluation of, 484, 485
  preerythrocytic, 486–489
  recombinant protein, 482
  subunit, 482
Vacuolar transport sequence, for protein export, 262–263
var gene(s), genomic distribution of, 400–402
  in transcription states, 409
  structure and sequence features of, 401, 402
  structure within genomic context, 402–403
var gene expression, control of, CP15, 403–407
  mutually exclusive, epigenetic control of, 407
var gene family, transcription within, understanding of, 70–71
var gene switching, deletions in, 406
var multigene family, chromosomal distribution and orientation of, 400–402
var switching rates, range of, 408
  variable, 407–408
Variant surface antigens, 399
  expressed by parasites, recognition of, 410–411
Vector genomics, and comparative genomics, 508–509
VectorBase, and *Anopheles gambiae* data, 507
Vectors, malaria, human, transcriptome of, 516–530
Vertebrate host, delivery of sporozoite to, 181–182
Vesicles, small, CP3C, 34, 35
Virulence determinants, secretome of, and exported parasite proteins, 261–267

WR99210, 436, 437
WU-BLAST, 14

Zanzarone, 6